Absolute Value Function

$$f(x) = |x|$$

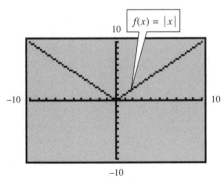

$$f(x) = |x|$$

$[-10, 10]$ by $[-10, 10]$

$Xscl = 1$ $\quad$ $Yscl = 1$

Rational Function

$$f(x) = \frac{1}{x}$$

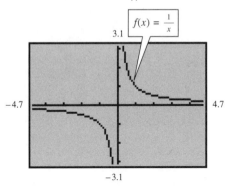

$$f(x) = \frac{1}{x}$$

$[-4.7, 4.7]$ by $[-3.1, 3.1]$

$Xscl = 1$ $\quad$ $Yscl = 1$

Exponential Function $a > 1$

$$f(x) = a^x,\ a > 1$$

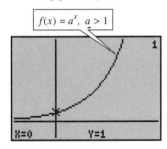

$$f(x) = a^x,\ a > 1$$

Exponential Function $a < 1$

$$f(x) = a^x,\ a < 1$$

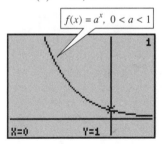

$$f(x) = a^x,\ 0 < a < 1$$

Logarithmic Function $a > 1$

$$f(x) = \log_a x,\ a > 1$$

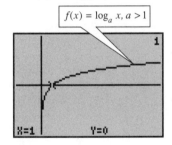

$$f(x) = \log_a x,\ a > 1$$

Logarithmic Function $a < 1$

$$f(x) = \log_a x,\ a < 1$$

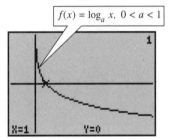

$$f(x) = \log_a x,\ 0 < a < 1$$

A GRAPHICAL APPROACH TO
Precalculus

E. John Hornsby, Jr.
University of New Orleans

Margaret L. Lial
American River College

HarperCollins*CollegePublishers*

Sponsoring Editor: Anne Kelly
Developmental Editor: Kathy Richmond
Project Editor: Ann-Marie Buesing
Design Administrator: Jess Schaal
Text Design: Lesiak/Crampton Design: Cindy Crampton
Cover Design: Lesiak/Crampton Design: Cindy Crampton
Cover Art: Tom James
Production Administrator: Randee Wire
Compositor: Interactive Composition Corporation
Printer and Binder: R. R. Donnelley & Sons Company
Cover Printer: The Lehigh Press, Inc.

A Graphical Approach to Precalculus

Library of Congress Cataloging-in-Publication Data
Hornsby, E. John.
    A graphical approach to precalculus / E. John Hornsby, Jr.,
Margaret L. Lial.
        p.   cm.
    Includes index.
    ISBN 0-673-99966-1
    1. Functions.   I. Lial, Margaret L.   II. Title.
QA331.H6453   1996
512′.1—dc20   95-15985
                                                         CIP

95 96 97 98 9 8 7 6 5 4 3 2 1

TO THE MEMORY OF JACK HORNSBY

CONTENTS

P R E F A C E

This book, intended for a precalculus course, is a culmination of four years of teaching experience with the graphics calculator and more years of teaching mathematics than we both like to admit. In it, we treat the standard topics of precalculus with complete integration of modern graphics calculators. We pursued this project with the firm commitment that it was not to be merely an adaptation of our traditional texts to graphics calculator analysis. In so doing, we realized that a completely new approach would be necessary, based on the premise that all students would have graphics calculators on the first day of class and would use this technology while studying the mathematical concepts throughout the book.

Development of the Project

We began teaching precalculus classes with a graphics calculator requirement during the 1991–1992 academic year. During the first year we attempted to adapt a traditional college precalculus text to this new approach and soon realized that a book designed to implement the graphics calculator at all stages was necessary. The approach used for so many years in our classes was just no longer applicable in light of the power of the technology available. For example, traditional college algebra texts usually do not cover graphing until the second or third chapter. The new technology demands that graphing be covered as early as possible. Furthermore, we now feel that *interpretation of graphs* plays a more important role than *graph-sketching techniques*. The modern approach to mathematics spawned by the statements of NCTM and AMATYC needed to be addressed. We felt it natural that, as mathematics authors for so many years in the traditional market, we could write a book to accomplish our goals. During the summer of 1993, we began writing the material and continued throughout the academic year 1993–1994, teaching the material after writing it. The class-testing and review processes continued during 1994–1995, while production commenced in late 1994.

The book you are reading is the culmination of literally thousands of hours of work by the authors, reviewers, answer-checkers, editors, class-testers, and students.

Philosophy of Our Approach

Throughout the first five chapters, we present the various classes of functions covered in a standard college algebra text. The first chapter introduces functions and relations, using the linear function as the basis for our presentation. In the chapter, we present the approach that follows throughout the succeeding chapters. After introducing a class of function and examining the nature of its graph, we discuss analytic methods of solution of equations based on the function and show how to provide graphical support using the graphics calculator. Once we establish methods of solving these equations, we move on to the analytic methods of solving the associated inequalities and how their solutions can be supported using graphs. We use two approaches to graphical analysis of equation and inequality solving: the x-intercept method and the intersection-of-graphs method. These two methods are constantly reviewed and reinforced. Finally, once the student has a feel for the particular class of function under consideration, we use our analytic and graphical methods to solve applications that lead to equations or inequalities involving that function. By consistently using this approach, students become aware that we are really doing the same thing, but just applying it to a new kind of function.

Overview of the Content

We have found that the function concept is a difficult one for many students to grasp. Rather than present the student with a variety of functions at the outset, we begin with the linear function in Chapter 1, analyzing its graph, solving equations and inequalities, and solving applications dealing exclusively with linear functions. Here we begin to explore the graphing capabilities of graphics calculators and continue this throughout the text. In Chapter 2 we examine the *basic* algebraic function graphs, how they exhibit symmetries, and how their graphs can be transformed. We use the absolute value function to extend the concepts presented in Chapter 1, once again using the graph/equation/inequality/application approach. In Chapter 3 we study polynomial functions, concentrating on quadratics at the outset and then expanding to higher degree functions. Chapter 4 covers rational and root functions using the same approach and concludes with a section on inverse functions that leads into Chapter 5 on exponential and logarithmic functions. Chapter 6 discusses conic sections and systems, and Chapter 7 shows the various applications of matrices and determinants. The appropriate capabilities of graphics calculators are addressed in these two chapters. Chapter 8 is the introductory trigonometry chapter, concentrating on basic concepts of angles and trigonometric applications. Chapter 9 introduces the circular functions, again using our approach to integrating graphs, equations, inequalities, and applications. Chapter 10 investigates vectors, polar coordinates, complex numbers, and parametric equations. The final chapter covers various other algebra topics. In this chapter, we address the sequence capabilities of some models of graphics calculators.

An appendix at the end of the book reviews some basic but important concepts of intermediate algebra for instructors who wish to cover them and students who wish to refer to them.

The Approach to Technology

Critics of technology often state that they oppose the idea of "pushing a button to get the answer." Anyone who has made an effort to teach with the graphics calculator knows that this is not the way it works. We constantly emphasize that it is essential to understand the mathematical concepts and apply them hand-in-hand with the calculator. The calculator helps us to understand the concepts, and we wrote the book with this idea. We agree with Peg Crider of Tomball College: "Your brain is the most powerful tool in the whole process."

Due to the ever-changing world of technology, we felt strongly that this text should not attempt to teach the student how to use a particular model of calculator. We do not provide specific keystrokes for various models, because experience has taught us that by the time a text is published, the technology is often out of date. However, we did decide to use actual graphics calculator-generated art in addition to the traditional art found in standard textbooks. We chose to use art generated by the TI-82 graphics calculator, manufactured by Texas Instruments. The TI-Graph Link software produced the graphics art, and we are grateful for the cooperation of Texas Instruments in this effort.

Features of the Text

Approach
The graphics calculator is introduced in Chapter 1 and is used consistently throughout the book. The functions are grouped by type, allowing a uniform presentation of graphs, equations, inequalities, and applications. This allows the student to progress from one type of function to the next, applying previously learned concepts to a new type of function. Students can see patterns, make connections, and more easily apply what they learn.

Pedagogy
Technological Notes supplement the exposition by discussing the power and the limitations of learning with technology. *For Group Discussion* activities allow instructors to use cooperative learning in their classes. *Chapter summaries* provide a quick review of the key ideas and terms in the chapter.

Exercises
There are many examples and exercises that draw students into the heart of mathematical concepts. True conceptual exercises are balanced by a suitable ratio of drill and applications. Many exercises involve real data and require brief essay answers. Most exercise sets include a group of exercises labeled *Relating Concepts* that make students aware of the connections between topics they studied earlier and the ones they are studying currently. *Further Explorations* are found in many sections, and offer further insight into the capabilities of the TABLE feature of the TI-82 graphics calculator.

Preview to the Text

The next four pages of the preface are a short guide through the text. These sample pages were chosen to show the features and pedagogical aids in the book that help reinforce the graphical approach to learning college algebra and trigonometry.

GRADED EXAMPLES

Well-explained examples prepare students for the exercises by illustrating the concepts. Many also focus on the graphical aspects as appropriate, and all of them are *titled* for ease of reference.

GRAPHICS CALCULATOR SCREENS

Actual graphics calculator-generated art is used (produced by TI-Graph Link software). Some other textbooks have such art rendered by an art house.

CAUTIONS/NOTES

Important comments are highlighted, as are common student difficulties, to help students avoid typical pitfalls.

TECHNOLOGICAL NOTES

Marginal notes supplement the exposition by discussing the power and limitations of learning with technology.

FOR GROUP DISCUSSION

This excellent feature is integrated throughout the text, and promotes discovery, examination of patterns, and cooperative learning by encouraging students to work with each other and their graphers. This feature provides meaningful extensions to the examples.

First page image (Chapter 1 excerpt):

14   CHAPTER 1   Rectangular Coordinates, Functions, and Analysis of Linear Functions

and 6, inclusive. Using interval notation,

the domain is $[-4, 4]$;
the range is $[-6, 6]$.

(b) In Figure 18(b), the arrowheads indicate that the line extends indefinitely left and right, as well as up and down. Therefore, both the domain and the range are the set of all real numbers, written $(-\infty, \infty)$.

(c) In Figure 18(c), the arrowheads indicate that the graph extends indefinitely left and right, as well as upward. The domain is $(-\infty, \infty)$. Because there is a least y-value, -3, the range includes all numbers greater than or equal to -3, written $[-3, \infty)$.

EXAMPLE 5
Finding Domain and Range from a Calculator Window

Later in this chapter, we will discuss graphing using a calculator. Figure 19 shows a graph on a screen with viewing window $[-5, 5]$ by $[-5, 5]$, with Xscl = 1 and Yscl = 1. By observation, give the domain and the range of this relation.

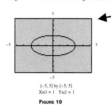

$[-5, 5]$ by $[-5, 5]$
Xscl = 1   Yscl = 1

FIGURE 19

SOLUTION   Since the scales on both axes are 1, we see that the graph *appears* to have a minimum x-value of -3, a maximum x-value of 3, a minimum y-value of -2, and a maximum y-value of 2. Therefore, observation leads us to conclude that the domain is $[-3, 3]$ and the range is $[-2, 2]$.

TECHNOLOGICAL NOTE
In Figure 19 we see a calculator-generated graph that is formed by a rather jagged curve. These are sometimes called *jaggies* and are typically found on low-resolution graphers, such as graphics calculators. In general we should remember that most curves we will study in this book are smooth, and the jaggies are just a part of the limitations of technology.

CAUTION   As we shall see many times while discussing calculator-generated graphs, simple observation is not enough to guarantee accuracy in determining domains and ranges. Calculators have capabilities which allow us to improve our accuracy, but even then, it is essential to understand the mathematical concepts behind graphing before we can be certain that our observations are correct. This is why in this book, we will study concepts and technological capabilities in an integrated fashion.

Functions

Look back at the relations F, G, and H introduced earlier in this section. Notice that in F and G, each x-value appears only once in the relation, while in H, the x-value -2 appears twice: it is paired with 1 in one ordered pair, while it is paired with 0 in another. Relations F and G are simple examples of a very important kind of relation, known as a function.

Second page image (Section 2.2 excerpt):

2.2   Vertical and Horizontal Shifts of Graphs of Functions   **127**

Vertical Shifts

TECHNOLOGICAL NOTE
You can save time when entering the functions in the first Group Discussion. If you let Y1 represent the basic function as shown, you can enter Y2 as Y1 + 3, Y3 as Y1 − 2, and Y4 as Y1 + 5. Then, you need only change Y1 in each function group, since the other functions are defined in terms of Y1.

FOR GROUP DISCUSSION

In each group of functions below, we give four related functions. Graph the four functions in the first group (Group A), and then answer the questions below regarding those functions. Then repeat the process for Group B, Group C, and Group D. Use the standard viewing window in each case.

A	B	C	D
$y_1 = x^2$	$y_1 = x^3$	$y_1 = \sqrt{x}$	$y_1 = \sqrt[3]{x}$
$y_2 = x^2 + 3$	$y_2 = x^3 + 3$	$y_2 = \sqrt{x} + 3$	$y_2 = \sqrt[3]{x} + 3$
$y_3 = x^2 - 2$	$y_3 = x^3 - 2$	$y_3 = \sqrt{x} - 2$	$y_3 = \sqrt[3]{x} - 2$
$y_4 = x^2 + 5$	$y_4 = x^3 + 5$	$y_4 = \sqrt{x} + 5$	$y_4 = \sqrt[3]{x} + 5$

1. How does the graph of y_2 compare to the graph of y_1?
2. How does the graph of y_3 compare to the graph of y_1?
3. How does the graph of y_4 compare to the graph of y_1?
4. If $c > 0$, how do you think the graph of $y_1 + c$ would compare to the graph of y_1?
5. If $c > 0$, how do you think the graph of $y_1 - c$ would compare to the graph of y_1?

Choosing your own value of c, support your answers to items 4 and 5 graphically. (Be sure that your choice is appropriate for the standard window.)

The objective of the preceding group discussion activity was to make conjectures about how the addition or subtraction of a constant c would affect the graph of a function $y = f(x)$. In each case, we obtained a vertical shift, or **translation,** of the graph of the basic function with which we started. Although our observations were based on the graphs of four different elementary functions, they can be generalized to any function.

VERTICAL SHIFTING OF THE GRAPH OF A FUNCTION
If $c > 0$, the graph of $y = f(x) + c$ is obtained by shifting the graph of $y = f(x)$ *upward* a distance of c units. The graph of $y = f(x) - c$ is obtained by shifting the graph of $y = f(x)$ *downward* a distance of c units.

In Figure 18 we give a graphical interpretation of the statement above.

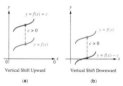

Vertical Shift Upward   Vertical Shift Downward
(a)   (b)

CALCULUS PREPARATORY DISCUSSIONS

As appropriate, the authors address the concepts students will encounter in calculus.

(Sample page 175)

2.6 Further Topics in the Study of Functions   **175**

EXAMPLE 2
Using the Operations on Functions

Let $f(x) = 8x - 9$ and $g(x) = \sqrt{2x - 1}$.

(a) $(f + g)(x) = f(x) + g(x) = 8x - 9 + \sqrt{2x - 1}$

(b) $(f - g)(x) = f(x) - g(x) = 8x - 9 - \sqrt{2x - 1}$

(c) $(fg)(x) = f(x) \cdot g(x) = (8x - 9)\sqrt{2x - 1}$

(d) $\left(\dfrac{f}{g}\right)(x) = \dfrac{f(x)}{g(x)} = \dfrac{8x - 9}{\sqrt{2x - 1}}$

(e) Find the domains of f, g, $f + g$, $f - g$, fg, and f/g.

SOLUTION The domain of f is the set of all real numbers, while the domain of $g(x) = \sqrt{2x - 1}$ includes just those real numbers that make $2x - 1 \geq 0$; the domain of g is the interval $[\frac{1}{2}, \infty)$. The domain of $f + g, f - g,$ and fg is thus $[\frac{1}{2}, \infty)$. With f/g, the denominator cannot be zero, so the value $\frac{1}{2}$ is excluded from the domain. The domain of f/g is $(\frac{1}{2}, \infty)$.

The Difference Quotient

Suppose that the point P lies on the graph of $y = f(x)$, and suppose that h is a positive number. If we let $(x, f(x))$ denote the coordinates of P and $(x + h, f(x + h))$ denote the coordinates of Q, then the line joining P and Q has slope

$$m = \frac{f(x + h) - f(x)}{(x + h) - x}$$

$$= \frac{f(x + h) - f(x)}{h}.$$

This expression, called the **difference quotient**, is important in the study of calculus.

Figure 58 shows the graph of the line PQ (called a secant line). If h is allowed to approach 0, the slope of this secant line approaches the slope of the line tangent to the curve at P. Important applications of this idea are developed in calculus, where the concepts of *limit* and *derivative* are investigated.

FIGURE 58

EXAMPLE 3
Finding the Difference Quotient

Let $f(x) = 2x^2 - 3x$. Find the difference quotient and simplify the expression.

SOLUTION To find $f(x + h)$, replace x in $f(x)$ with $x + h$, to get

$$f(x + h) = 2(x + h)^2 - 3(x + h).$$

(Sample page 372)

372   CHAPTER 4 Rational and Root Functions

Chapter 4   **SUMMARY**

Rational functions are defined by a quotient of polynomials. They often have graphs with one or more breaks. If the rule defining the function is in lowest terms, the graph may have one or more vertical asymptotes and one horizontal or oblique asymptote. Vertical asymptotes are found by setting the denominator equal to 0 and solving the equation. If the numerator is of lower degree than the denominator, the horizontal asymptote is $y = 0$. If the numerator and denominator have the same degree, the horizontal asymptote is $y = k$, where k is the quotient of the leading coefficients of the numerator and denominator. Rational functions, where the numerator is of degree one more than the denominator, may have an oblique asymptote. The graph may intersect a horizontal or oblique asymptote. If the rule of the rational function is not in lowest terms, the graph may have a" hole" instead of a vertical asymptote.

To solve a rational equation, multiply both sides of the equation by the least common denominator of all the terms in the equation. Because there may be values of the variable where the denominator is zero, all proposed solutions must be checked, or these values should be determined before beginning the solution process.

Inverse variation is an application of rational functions. We say y varies inversely as the nth power of x if a nonzero real number k exists, such that y is the quotient of k and x^n.

For positive integers n and $a > 0$, the notations $\sqrt[n]{a}$ and $a^{1/n}$ both represent the positive nth root of a. If n is even, the nth root of a is real only if a is nonnegative. With appropriate restrictions, $a^{m/n} = (\sqrt[n]{a})^m = \sqrt[n]{a^m}$. Important features and the graphs of root functions for n even and for n odd are introduced in Section 4.3. The terminology of Sections 2.2 and 2.3 is used to explain how variations of root functions are obtained from the basic graphs. Special functions defined by roots have graphs that are se micircles or semi-parabolas.

Radical equations are solved by raising both sides to a power that will eliminate the radical, and then solving the resulting equation. This method may produce extraneous solutions, so again it is important to check all proposed solutions in the *original* equation. A calculator may be used to graph the corresponding function and check the proposed solutions by observing the graph. Inequalities may be solved as they were with earlier types of functions, by observing where the graph of the corresponding function is positive or negative or by using a sign graph.

For a function to have an inverse function, it must be one-to-one. A function is one-to-one if different elements from the domain always lead to different elements in the range. The horizontal line test is a good way to verify that a function is one-to-one by observing its graph. A function that is increasing (or decreasing) over its domain is one-to-one. To find the inverse of a one-to-one function, exchange x and y in the rule for the function, then solve for y to get the rule of the inverse function. The graphs of a function and its inverse function are reflections across the line $y = x$.

Key Terms

SECTION 4.1

rational function
vertical asymptote
horizontal asymptote
oblique asymptote

SECTION 4.2

rational equation
rational inequality
extraneous solution

inverse variation
combined variation
joint variation

SECTION 4.3

radicand of a radical
root index of a radical
root function

SECTION 4.4

root equation
root inequality

SECTION 4.5

identity properties
inverse properties
identity function
inverse function
one-to-one function
horizontal line test

CHAPTER SUMMARY & KEY TERMS

At the end of each chapter, a review of concepts gives students an overview of the chapter (a built-in study guide). The friendly narrative (as opposed to listing only formulas) is focused on the connection of concepts.

In Massachusetts, speeding fines are determined by the linear function

$$y = 10(x - 65) + 50, \qquad x \geq 65,$$

where y is the cost in dollars of the fine if a person is caught driving x miles per hour. Use this information to work the problems in Exercises 49–52.

49. Jose had to make an 8:00 A.M. final examination, but overslept after a big weekend in Boston. Radar clocked his speed at 76 mph. How much was his fine?

50. While balancing his checkbook, Johnny ran across a cancelled check that his wife Gwen had written to the Department of Motor Vehicles for a speeding fine. The check was written for $100. How fast was Gwen driving?

51. Based on the formula above, at what speed do the troopers start giving tickets?

52. For what range of speeds is the fine greater than $200. Solve analytically and support graphically.

In Exercises 53–60, linear models are based on least squares regression techniques from statistics. Solve each problem analytically and support your solutions graphically.

53. The Measurements Standard Committee of the Pattern Fashion Industry provides a table of body measurements (in inches) corresponding to misses' sizes. For misses' sizes 6 through 20, bust measurement y corresponds to misses size x according to the model $y = .842x + 24.613$. If the size is 14, what is the corresponding bust size (rounded down to the nearest unit)? *

54. If alcohol "burned up" by the body since the time of the first drink is disregarded, the number of drinks (1 drink = 12 oz. beer, 4 oz. wine, 1 oz. hard liquor), x, and the blood alcohol level, y, of a 240-pound person are related by the function $y = .0156x$. What would be the blood alcohol level if this person had 4 drinks? If .100 blood alcohol level is considered legally drunk, after what whole number of drinks would this person be legally drunk?

55. According to information provided by Families USA Foundation, the national average family health care cost (in dollars) between 1980 and 2000 (projected) can be approximated by the linear function $y = 382.75x + 1742$, where $x = 0$ corresponds to 1980 and $x = 20$ corresponds to 2000. Based on this model, what would be the expected national average health care cost in 1996?

56. In what year will (or did) the national average family health care cost first exceed $7,000? (See Exercise 55.)

57. The percentage y of off-track-betting as a portion of all bets on horse racing between 1982 and 1992 can be estimated by the linear model $y = 1.8x + 15$, where $x = 0$ corresponds to 1982 and $x = 10$ corresponds to 1992. If this trend continues, what will the percentage of off-track-betting as a portion of all bets on horse racing be in 1998? (*Source:* Christiansen/Cummings Associates, Inc.)

58. For speeds between 10 and 60 mph, the stopping distance y in ft for a person going x mph is estimated by the linear model $y = 4.5x - 46.7$. What would the stopping distance be at 40 mph? *

59. From 1940 to 1960, the population y of California was approximated by the linear model $y = .4405x + 6.665$, where y is in millions and $x = 0$ corresponds to 1940 and $x = 20$ corresponds to 1960. If this trend had continued until 1965, what would have been the population then? *

60. It has been reported that the total length x and the tail length y of females of the snake species *Lampropeltis polyzona* are nearly linearly related by the model $y = .134x - 1.18$, where x is the length of the snake in millimeters. If a snake of this species measures 1000 millimeters, what is its tail length to the nearest millimeter? *

61. In 1787, Jacques Charles noticed that gases expand when heated and contract when cooled. Suppose that a particular gas follows the model $y = (\frac{5}{3})x + 455$ where x is the temperature in Celsius and y is the volume in cubic centimeters. *
(a) What is the volume when the temperature is 27° Celsius?
(b) What is the temperature when the volume is 605 cubic centimeters?
(c) Determine what temperature gives a volume of 0 cubic centimeters (in other words, absolute zero, or the coldest possible temperature).

* Problems 44, 53, and 58–62 are adapted from *A Sourcebook of Applications of School Mathematics* by Donald Bushaw et al. Copyright © 1980 by The Mathematical Association of America. Reprinted by permission.

APPLICATIONS WITH REAL DATA AND REFERENCES

Well thought-out, up-to-date applications help answer the students' question: "What is this good for?" Some also sharpen writing skills by requiring brief essay responses. Many also have multiple steps that lead students to a conclusion gradually.

EXERCISES

Match each equation with its calculator-generated graph in Exercises 1–10. Do this first without actually using your calculator. Then check your answer by generating a calculator graph of your own. (Every window has Xscl = Yscl = 1.)

1. $\dfrac{y^2}{16} + \dfrac{x^2}{4} = 1$

2. $\dfrac{x^2}{16} + \dfrac{y^2}{4} = 1$

3. $\dfrac{x^2}{4} - \dfrac{y^2}{16} = 1$

4. $\dfrac{y^2}{4} - \dfrac{x^2}{16} = 1$

5. $\dfrac{(y-4)^2}{25} + \dfrac{(x+2)^2}{9} = 1$

6. $\dfrac{(y+4)^2}{25} + \dfrac{(x-2)^2}{9} = 1$

7. $\dfrac{(x+2)^2}{9} - \dfrac{(y-4)^2}{25} = 1$

8. $\dfrac{(x-2)^2}{9} - \dfrac{(y+4)^2}{25} = 1$

9. $36x^2 + 4y^2 = 144$

10. $9x^2 - 4y^2 = 36$

11. Explain how a circle can be interpreted as a special case of an ellipse.

12. If an ellipse has vertices at $(-3, 0)$, $(3, 0)$, $(0, 5)$, and $(0, -5)$, what is its domain? What is its range?

BALANCED EXERCISES

There are many exercises that draw students into the heart of mathematical concepts. True conceptual exercises are balanced by a suitable ratio of drill, application, and writing exercises.

70. Show that the points $(1, -2)$, $(3, -18)$, and $(-2, 22)$ all lie on the same straight line. (*Hint:* Find the equation of the line joining two of the points and then show that the third point satisfies the equation.) Then use your calculator to support your result.

Further Explorations

The following TABLES show ordered pairs for equations in Y_1 and Y_2. For each TABLE, determine if Y_1 and Y_2 are parallel by finding the slopes of each line.

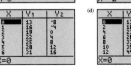

1.5 SOLUTION OF LINEAR EQUATIONS; ANALYTIC METHOD AND GRAPHICAL SUPPORT

Simplifying Like Terms ▪ Solving Linear Equations Analytically ▪ Graphical Support for Solutions of Linear Equations ▪ Identities and Contradictions

Simplifying Like Terms

Like terms are terms that have the exact same variable factors. Some pairs of like terms are

$$5x, \ -6x \qquad 3xy, \ -8xy \qquad \text{and} \qquad 4x^2y, \ -9x^2y.$$

Like terms may be added or subtracted by using the distributive property.

> **DISTRIBUTIVE PROPERTY**
> For all real numbers a, b, and c,
> $$ab + ac = a(b + c)$$
> and
> $$ab - ac = a(b - c).$$

---

356   CHAPTER 4   Rational and Root Functions

15. **(a)** $(x^2 + 2x)^{1/4} = \sqrt[4]{3}$
   (b) $(x^2 + 2x)^{1/4} > \sqrt[4]{3}$
   (c) $(x^2 + 2x)^{1/4} < \sqrt[4]{3}$

16. **(a)** $(x^2 + 6x)^{1/4} = 2$
   (b) $(x^2 + 6x)^{1/4} > 2$
   (c) $(x^2 + 6x)^{1/4} < 2$

17. **(a)** $(2x - 1)^{2/3} = x^{1/3}$
   (b) $(2x - 1)^{2/3} > x^{1/3}$
   (c) $(2x - 1)^{2/3} < x^{1/3}$

18. **(a)** $(x - 3)^{2/5} = (4x)^{1/5}$
   (b) $(x - 3)^{2/5} > (4x)^{1/5}$
   (c) $(x - 3)^{2/5} < (4x)^{1/5}$

19. **(a)** $\sqrt{3 - 3x} = 3 + \sqrt{3x + 2}$
   (b) $\sqrt{3 - 3x} > 3 + \sqrt{3x + 2}$
   (c) $\sqrt{3 - 3x} < 3 + \sqrt{3x + 2}$

20. **(a)** $\sqrt{2\sqrt{7x + 2}} = \sqrt{3x + 2}$
   (b) $\sqrt{2\sqrt{7x + 2}} > \sqrt{3x + 2}$
   (c) $\sqrt{2\sqrt{7x + 2}} < \sqrt{3x + 2}$

In Exercises 21–26, a pair of functions is given. The function defined by y_1 is a root function, while the function defined by y_2 is a polynomial function. Use your knowledge of the general appearance of each to draw by hand a sketch of both functions on the same set of axes. Then use the sketch to determine the number of points of intersection of the two graphs. Finally, use an analytic method to support your answer, and find the solution set of $y_1 = y_2$.

21. $y_1 = \sqrt{x}$
   $y_2 = -x + 3$

22. $y_1 = \sqrt{x}$
   $y_2 = x - 7$

23. $y_1 = \sqrt{x}$
   $y_2 = 2x + 5$

24. $y_1 = \sqrt{x}$
   $y_2 = 3x$

25. $y_1 = \sqrt[3]{x}$
   $y_2 = x^2$

26. $y_1 = \sqrt[3]{x}$
   $y_2 = 2x$

27. Use a hand-drawn graph to explain why $\sqrt{x} = -x - 5$ has no real solutions. (*Hint:* Sketch $y_1 = \sqrt{x}$ and $y_2 = -x - 5$ on the same axes. What do you notice about the number of points of intersection?)

28. Explain why the equation $\sqrt[3]{x} = ax + b$ must have at least one real solution for any values of a and b. (*Hint:* What is the range of $y_1 = \sqrt[3]{x}$? What kind of graph does $y_2 = ax + b$ have?)

Relating Concepts

Exercises 29–40 incorporate many concepts from Chapter 3 with the method of solving equations involving roots. They should be worked in order. Consider the equation
$$\sqrt[3]{4x - 4} = \sqrt{x + 1}.$$

29. Rewrite the equation using rational exponents.

30. What is the least common denominator of the rational exponents found in Exercise 29?

31. Raise both sides of the equation in Exercise 29 to the power indicated by your answer in Exercise 30.

32. Show that the equation in Exercise 31 is equivalent to $x^3 - 13x^2 + 35x - 15 = 0$.

33. Graph the cubic function defined by the polynomial on the left side of the equation in Exercise 32 in the window $[-5, 10]$ by $[-100, 100]$. How many real roots does the equation have?

34. Use synthetic division to show that 3 is a zero of $P(x) = x^3 - 13x^2 + 35x - 15$.

35. Use the result of Exercise 34 to factor $P(x)$ so that one factor is linear and the other is quadratic.

36. Set the quadratic factor of $P(x)$ from Exercise 35 equal to 0, and solve the equation using the quadratic formula.

37. What are the three proposed solutions of the original equation $\sqrt[3]{4x - 4} = \sqrt{x + 1}$?

38. Let $y_1 = \sqrt[3]{4x - 4}$ and $y_2 = \sqrt{x + 1}$. Graph $y_1 - y_2$ in the viewing window $[-2, 20]$ by $[-.5, .5]$ to determine the number of real solutions the original equation has.

The Supplement Package for *A Graphical Approach to Precalculus*

The text is supported with an extensive package of supplements for both the instructor and the student.

For the Instructor

Instructor's Annotated Exercises   With this volume, instructors have immediate access to the answers to every exercise in the text. Each answer is printed in bold type next to or below the pertinent exercise.

The *Instructor's Testing Manual*   offers four to six test items per text section for the instructor to use.

The *Instructor's Solution Manual*   provides worked solutions to all the even-numbered exercises in the text.

The HarperCollins Test Generator/Editor for Mathematics with QuizMaster
Available in IBM (both DOS and Windows applications) and Macintosh versions, the *Test Generator* is fully networkable. The *Test Generator* enables instructors to select questions by objective, section, or chapter, or to use a ready-made test for each chapter. The *Editor* enables instructors to edit any preexisting data or to create their own questions easily. The software is algorithm-driven, allowing the instructor to regenerate constants while maintaining problem type, providing a very large number of test or quiz items in multiple-choice and/or open-response formats for one or more test forms. The system features printed graphics and accurate mathematical symbols. *QuizMaster* enables instructors to create tests and quizzes using the *Test Generator/Editor* and save them to disk so students can take the test or quiz on a stand-alone computer or network. *QuizMaster* then grades the test or quiz and allows the instructor to create reports on individual students or entire classes. CLAST and TASP versions of this package are also available for IBM and Mac machines.

Graph Explorer   software, available for IBM and Macintosh hardware, allows students to learn through exploration. With this tool-oriented approach to algebra, students can graph rectangular, conic, polar, and parametric equations, zoom, transform functions, and experiment with families of equations. Students have the option to experiment with different solutions, display multiple representations, and print all work.

Videotapes   are available for key topics in algebra. A tape series is also available with lessons on appropriate operations using keystrokes of various calculators.

For the Student

The Student's Solution Manual   written by Norma James of New Mexico State University, contains worked solutions to all of the odd-numbered exercises in the text.

A Graphics Calculator Keystroke Guide, by Stuart Moskowitz of Butte College, provides keystroke operations for the following calculator models: TI-82®, TI-85®, and Casio CFX-9800G®. Examples are taken from the text.

Interactive Mathematics Tutorial Software with Management System   This innovative package is available in DOS, Windows, and Macintosh versions and is fully networkable. As with the *Test Generator/Editor,* this software is algorithm-driven, which automatically regenerates constants so that the numbers rarely repeat in a problem type when students revisit any particular section. The tutorial is objective-based, self-paced, and provides unlimited opportunities to review lessons and to practice problem solving. If students give a wrong answer, they can ask to see the problem worked out and get a textbook page reference.

Many problems include hints for first incorrect responses. Tools such as an online glossary and Quick Reviews provide definitions and examples, and an online calculator aids students in computation. The program is menu-driven for ease of use, and on-screen help can be obtained at any time with a single keystroke. Students' scores are calculated at the end of each lesson and can be printed for a permanent record. The optional *Management System* lets instructors record student scores on disk and print diagnostic reports for individual students or classes. CLAST and TASP versions of this tutorial are also available for both IBM and Mac machines. This software may also be purchased by students for home use. Student versions include record keeping and practice tests.

Acknowledgments

A project of this magnitude cannot be completed without the help of countless other individuals who offer support, suggestions, and criticisms. We would like to thank the following individuals who reviewed the text.

William A. Armstrong, Lakeland Community College ▐ John Baldwin, University of Illinois–Chicago ▐ Jim Birdsall, Santa Fe Community College ▐ Janis M. Cimperman, St. Cloud State University ▐ Dick J. Clark, Portland Community College ▐ John A. Dersch, Grand Rapids Community College ▐ William L. Grimes, Central Missouri State University ▐ Bruce Hoelter, Raritan Valley Community College ▐ Norma James, New Mexico State University ▐ Dick Little, Baldwin-Wallace University ▐ Dan Loprieno, Harper College ▐ Virginia E. Lund, Pensacola Junior College ▐ Karen Mitchell, Rowan-Cabarrus Community College ▐ Shelle A. Palaski, Northeast Missouri State University ▐ Richard Schori, Oregon State University ▐ Kathy Soderbom, Massoit Community College ▐ John P. Thomas, College of Lake County ▐ Mahbobeh Vezvaei, Kent State University ▐ Tom Williams, Rowan-Cabarrus Community College ▐ Karl M. Zilm, Lewis and Clark Community College ▐

To the hundreds of students who studied from the various stages of the manuscript we offer our thanks. Susan Danielson, an instructor at the University of New Orleans, offered great support and many excellent suggestions, and her support during the past few years has been most welcome. She is a true innovator, and our profession could use many more like her. Michael Schafferkötter, Bill Hebert, and Gerry Vidrine also helped to class-test the book. Norma James of New Mexico State University, Bill Armstrong of Lakeland Community College, and Janis M. Cimperman of St. Cloud State University assisted in checking the answers to the exercises. The *Further Explorations* exercises were written by Stuart Moskowitz of Butte College, who did an outstanding job. To Brent Simon, we extend our thanks just for being there for about twenty years now. Gwen Hornsby provided great assistance in the preparation of the graphics art. As always, our families gave up husband/father and wife/mother to many hours of work, and our thanks and love go out to them.

We feel that we work with the finest editorial, marketing, and production staff in college publishing today. The support of the Glenview Group at HarperCollins College Publishers has been unwavering from the first day we proposed this project. Anne Kelly signed the book. Ed Moura believed in it. Lisa Kamins and Emily Barman were fantastic in helping to provide manuscript copies for class testing. Laura Taber helped immensely with the art preparation. Kathy Richmond's wonderful laugh was always a pleasure to hear, especially when we were a few days late on our deadline. Ann-Marie Buesing did her usual excellent job in production. To these individuals and all the others up on the second floor, we offer our sincerest gratitude.

A Final Word

We hope that this book begins to make a difference in the manner in which precalculus is presented and learned as we move into the twenty-first century. We ask that both instructors and students pursue its contents with an open mind, ready to teach and to learn in a manner that only now, after so many thousands of years, is possible. We, like Newton, can do so only because we "have stood on the shoulders of giants."

We welcome your comments, suggestions, and criticisms. Please write to us at HarperCollins College Publishers, attention: George Duda, 1900 East Lake Avenue, Glenview, IL 60025, or email us at harperglenvw@delphi.com, and let us know what you think.

E. John Hornsby, Jr.
Margaret L. Lial

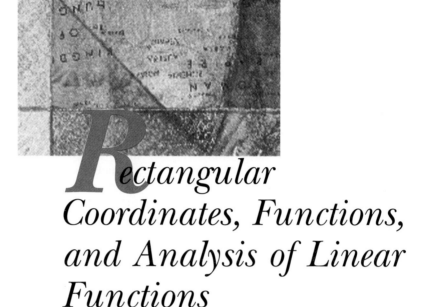

Rectangular Coordinates, Functions, and Analysis of Linear Functions

1.1 REAL NUMBERS AND COORDINATE SYSTEMS

Sets of Real Numbers ∎ Roots ∎ Coordinate Systems ∎ Viewing Windows

Sets of Real Numbers

The idea of counting goes back to the beginning of our civilization. When people first counted they used only the **natural numbers,** written in set notation as

$$\{1, 2, 3, 4, 5, \ldots\}.$$

Much more recent is the idea of counting *no* object—that is, the idea of the number 0. Including 0 with the set of natural numbers gives the set of **whole numbers.**

$$\{0, 1, 2, 3, 4, 5, \ldots\}$$

(These and other sets of numbers are summarized later in this section.)

About 500 years ago, people came up with the idea of counting backwards, from 4 to 3 to 2 to 1 to 0. There seemed no reason not to continue this process, calling the new numbers $-1, -2, -3$, and so on. Including these numbers with the set of whole numbers gives the very useful set of **integers,**

$$\{\ldots, -4, -3, -2, -1, 0, 1, 2, 3, \ldots\}.$$

Integers can be shown pictorially with a **number line.** (A number line is similar to a thermometer on its side.) As an example, the elements of the set $\{-3, -1, 0, 1, 3, 5\}$ are located on the number line in Figure 1.

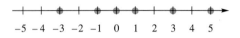

FIGURE 1

The result of dividing two integers, with a nonzero divisor, is called a *rational number.* By definition, the **rational numbers** are the elements of the set

$$\left\{ \frac{p}{q} \,\middle|\, p, q \text{ are integers and } q \neq 0 \right\}.$$

This definition, which is given in *set-builder notation,* is read "the set of all elements p/q such that p and q are integers and $q \neq 0$." Examples of rational numbers include $\frac{3}{4}$, $-\frac{5}{8}$, $\frac{7}{2}$, and $-\frac{14}{9}$. All integers are rational numbers, since any integer can be written as the quotient of itself and 1.

Rational numbers can be located on a number line by a process of subdivision. For example, $\frac{5}{8}$ can be located by dividing the interval from 0 to 1 into 8 equal parts, then labeling the fifth part $\frac{5}{8}$. Several rational numbers are located on the number line in Figure 2.

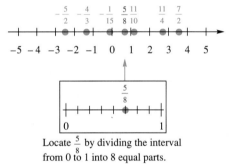

Locate $\dfrac{5}{8}$ by dividing the interval
from 0 to 1 into 8 equal parts.

FIGURE 2

The set of all numbers that correspond to points on a number line is called the set of **real numbers.** The set of real numbers is shown in Figure 3.

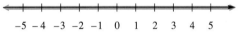

FIGURE 3

A real number that is not rational is called an **irrational number.** The set of irrational numbers includes $\sqrt{3}$ and $\sqrt{5}$ but not $\sqrt{1}$, $\sqrt{4}$, $\sqrt{9}$, . . . , which equal 1, 2, 3, . . . , and hence are rational numbers. Another irrational number is π, which is approximately equal to 3.14159. The numbers in the set $\{-\frac{2}{3}, 0, \sqrt{2}, \sqrt{5}, \pi, 4\}$ can be located on a number line as shown in Figure 4. (Only $\sqrt{2}$, $\sqrt{5}$, and π are irrational here. The others are rational.)

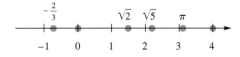

FIGURE 4

Real numbers can also be defined in another way, in terms of decimals. Using repeated subdivisions, any real number can be located (at least in theory) as a point on a number line. By this process, the set of real numbers can be defined as the set of all decimals. Every rational number has a decimal representation that either terminates (comes to an end) or repeats in a fixed "block" of digits. Here are some examples.

Rational Numbers Whose Decimals Terminate	**Rational Numbers Whose Decimals Repeat**
$\frac{1}{4} = .25$	$\frac{1}{3} = .3333\ldots$
$\frac{3}{8} = .375$	$\frac{5}{6} = .8333\ldots$
$\frac{7}{4} = 1.75$	$\frac{3}{7} = .428571428571\ldots$

The three dots at the end of the repeating decimals indicate that the pattern of digits established continues indefinitely. Another way of indicating the repeating digits is the use of a bar over the part that repeats. Thus, we would write the following.

$$\frac{1}{3} = .\overline{3} \qquad \frac{5}{6} = .8\overline{3} \qquad \text{and} \qquad \frac{3}{7} = .\overline{428571}$$

TECHNOLOGICAL NOTE
Graphics calculators will display only a finite number of decimal digits, and the overbar symbolism for repeating decimals is not available. Therefore you must be very careful in how you interpret the decimal digits you see on your screen, and you must be able to make the distinction between an exact value and an approximate value.

If at any time we use an approximation for a rational number, we use the $\approx$ symbol to indicate "is approximately equal to." Thus it would technically be incorrect to write $\frac{2}{3} = .67$; we should write $\frac{2}{3} \approx .67$ if an approximation is warranted. We call .67 an *approximation of $\frac{2}{3}$ to the nearest hundredth,* while $.\overline{6}$ is the *exact decimal representation for $\frac{2}{3}$.*

NOTE   In this text we will often make distinctions about whether an approximation or an exact value is required. As we progress in our work, more will be said about this.

The decimal representation of an irrational number will neither terminate nor repeat. The locations of $\sqrt{2}$, $\sqrt{5}$, and π on the number line in Figure 4 were determined by observing these calculator approximations:

$$\sqrt{2} \approx 1.414213562 \qquad \sqrt{5} \approx 2.236067977 \qquad \pi \approx 3.141592654.$$

EXAMPLE 1

Identifying Elements of Subsets of the Real Numbers

Let set $A = \{-8, -6, -.75, 0, .\overline{09}, \sqrt{2}, \sqrt{5}, 6, \frac{107}{4}\}$. List the elements from set A that belong to each of the following sets: (a) real numbers, (b) integers, (c) rational numbers, (d) irrational numbers, (e) whole numbers, and (f) natural numbers.

SOLUTION

(a)   Because every element of A can be represented by a point on a number line, all elements are real numbers.

(b)   The integers are -8, -6, 0, and 6.

(c)   The rational numbers are -8, -6, $-.75$, 0, $.\overline{09}$, 6, and $\frac{107}{4}$.

(d)   The irrational numbers are $\sqrt{2}$ and $\sqrt{5}$.

(e)   The whole numbers are 0 and 6.

(f)   The only natural number in the set is 6.   ∎

The relationships among the various subsets of the real numbers, along with examples in the sets, are shown in Figure 5.

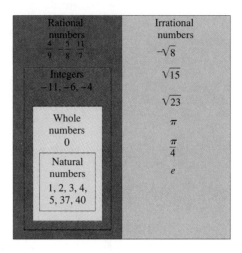

FIGURE 5

The Real Numbers

TECHNOLOGICAL NOTE Various models of calculators use different keystrokes for finding decimal values for roots. Be aware of how to find roots using your particular model.

Roots

The most common irrational numbers that we will encounter in this course are roots—**square roots, cube roots,** and so forth. A formal discussion of roots will follow in Chapter 4. For now, you should be able to use your calculator to find roots. Graphics calculators usually have dedicated keys for square and cube roots, and have functions that allow for other roots. You should consult your owner's manual to see how to find roots on your particular model.

EXAMPLE 2

Finding Roots on a Calculator

Use your calculator to verify the following decimal approximations for roots that are irrational numbers. (Depending on the model you have and how it is set for the number of digits displayed, there may be a slight discrepancy in the final digit.)

SOLUTION

(a) $\sqrt{23} \approx 4.795831523$

(b) $\sqrt[3]{87} \approx 4.431047622$

(c) $\sqrt[4]{12} \approx 1.861209718$ ◨

It is often convenient to use the fact that $\sqrt[n]{a} = a^{1/n}$ for appropriate values of n and a to find roots.

FOR **GROUP DISCUSSION**

Experiment with your calculator to show that the following pairs of expressions give the same approximations. You may need to refer to your owner's manual to see how to raise a number to a power.

1. $\sqrt{98.4}$, $98.4^{1/2}$ **2.** $\sqrt[3]{12.9}$, $12.9^{1/3}$ **3.** $\sqrt[4]{86}$, $86^{.25}$

Coordinate Systems

Figure 6 shows a number line with the points corresponding to several different numbers marked on the line. A number that corresponds to a particular point on a line is called the **coordinate** of the point. For example, the leftmost marked point in Figure 6 has coordinate -4. The correspondence between points on a line and the real numbers is called a **coordinate system** for the line. (The phrase "the point on a number line with coordinate a" will be abbreviated as "the point with coordinate a," or simply "the point a.")

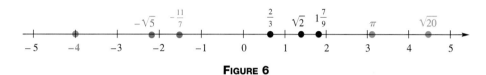

FIGURE 6

If the real number a is to the left of the real number b on a number line, then ***a* is less than *b*,** written $a < b$. If a is to the right of b, then ***a* is greater than *b*,** written $a > b$. For example, in Figure 6, $-\sqrt{5}$ is to the left of $-\frac{11}{7}$ on the number line, so $-\sqrt{5} < -\frac{11}{7}$, while $\sqrt{20}$ is to the right of π, indicating $\sqrt{20} > \pi$.

NOTE Remember that the "point" of the inequality symbol goes toward the smaller number.

As an alternative to this geometric definition of "is less than" or "is greater than," there is an algebraic definition: if a and b are two real numbers and if the difference $a - b$ is positive, then $a > b$. If $a - b$ is negative, then $a < b$. The geometric and algebraic statements of order are summarized as follows.

Statement	Geometric Form	Algebraic Form
$a > b$	a is to the right of b	$a - b$ is positive
$a < b$	a is to the left of b	$a - b$ is negative

The symbols $<$ and $>$ can be combined with the symbol for equality. The statement $a \leq b$ means "a is less than or equal to b" and is true if either $a < b$ or $a = b$. Similarly, $a \geq b$ means "a is greater than or equal to b" and is true if either $a > b$ or $a = b$. Statements involving $<$ or $>$ are called **strict** inequalities, while those involving $\leq$ or $\geq$ are called **nonstrict** inequalities. We can negate any of these symbols by using a slash bar ($/$).

EXAMPLE 3

Showing why Inequality Statements are True

The list below shows several statements and the reason why each is true.

Statement	Reason
$8 \leq 10$	$8 < 10$
$8 \leq 8$	$8 = 8$
$-9 \geq -14$	$-9 > -14$
$-8 \not> -2$	$-8 < -2$
$4 \not< 2$	$4 > 2$

The inequality $a < b < c$ says that b is *between* a and c, since

$$a < b < c$$

means $a < b$ and $b < c$.

In the same way, $a \leq b \leq c$

means $a \leq b$ and $b \leq c$.

CAUTION When writing these "between" statements, make sure that both inequality symbols point in the same direction, toward the smallest number. For example,

both $2 < 7 < 11$ and $5 > 4 > -1$

are true statements, but $3 < 5 > 2$ is meaningless. Generally, it is best to rewrite statements such as $5 > 4 > -1$ as $-1 < 4 < 5$, which is the order of these numbers on the number line.

A number line is an example of a one-dimensional coordinate system, and it is sufficient to graph real numbers. If we place two number lines at right angles, intersecting at their origins, we obtain a two-dimensional **rectangular coordinate system.** It is customary to have one of these lines vertical and the other horizontal. They intersect at the **origin** of the system, designated 0. The horizontal line is called the ***x*-axis,** and the vertical line is called the ***y*-axis.** On the x-axis, positive numbers are located to the right of the origin, while negative numbers are located to the left. On the y-axis, positive numbers are located above the origin, negative numbers below.

FIGURE 7

FIGURE 8

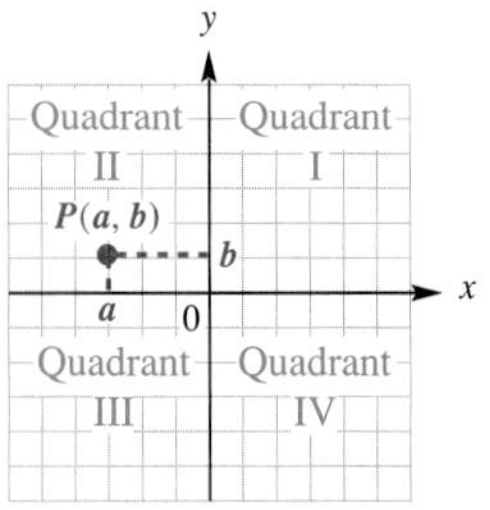

This rectangular coordinate system is also called the **cartesian coordinate system,** named after Rene Descartes (1596–1650). The plane into which the coordinate system is introduced is the **coordinate plane,** or ***xy*-plane.** The x-axis and y-axis divide the plane into four regions, or **quadrants,** labeled as shown in Figure 7. The points on the x-axis and y-axis belong to no quadrant.

Each point P in the xy-plane corresponds to a unique ordered pair (a, b) of real numbers. The numbers a and b are the **coordinates** of point P. We call a the x-coordinate and b the y-coordinate. To locate on the xy-plane the point corresponding to the ordered pair $(3, 4)$, for example, draw a vertical line through 3 on the x-axis and a horizontal line through 4 on the y-axis. These two lines intersect at point A in Figure 8. Point A corresponds to the ordered pair $(3, 4)$. Also in Figure 8, B corresponds to the ordered pair $(-5, 6)$, C to $(-2, -4)$, D to $(4, -3)$, and E to $(-3, 0)$. The point P corresponding to the ordered pair (a, b) often is written as $P(a, b)$ as in Figure 7 and referred to as "the point (a, b)."

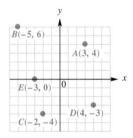

TECHNOLOGICAL NOTE
You should read your owner's manual to see how to alter the viewing window on your screen. Remember that different settings will result in different views of graphs. When you look out of your window at home, you may see something completely different than when your neighbor looks out of his or her window; the same goes for graphs generated by calculators.

Viewing Windows

The characteristic that distinguishes this text from traditional algebra and trigonometry texts is that it features full integration of modern-day graphics calculators. A graphics calculator differs from a typical scientific calculator in many ways, the most obvious of which is that it allows the user to plot points and a variety of graphs at the touch of keys.

NOTE In this text, all references to calculators are made with the understanding that the reader has access to a modern graphics calculator. Therefore, the term "calculator" is used to mean "graphics calculator."

The rectangular (cartesian) coordinate system theoretically extends indefinitely in all directions. We are limited to illustrating only a portion of such a system in a text figure. Similar limitations are found in portraying coordinate systems on calculator screens. For this reason, the student should become familiar with the key on the calculator that sets the limits for x- and y-coordinates. Some calculators use the word "range" for this function, while others use a more appropriate designation, "window." (There also may be other designations.) Figure 9 shows a calculator screen that has been set to have a minimum x-value of -10, a maximum x-value of 10, a minimum y-value of -10, and a maximum y-value of 10. Additionally, the tick marks on the axes have been set to be 1 unit apart. Throughout this book, this window will be called the *standard viewing window*.

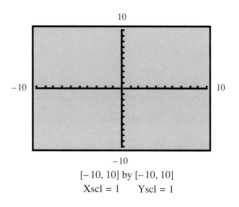

$$[-10, 10] \text{ by } [-10, 10]$$
$$\text{Xscl} = 1 \qquad \text{Yscl} = 1$$

FIGURE 9

In order to convey important information about a viewing window, in this text we will use the following abbreviations:

Xmin: minimum value of x Ymin: minimum value of y

Xmax: maximum value of x Ymax: maximum value of y

Xscl: scale (distance between   Yscl: scale (distance between
       tick marks) on the x-axis   tick marks) on the y-axis.

To further condense this information, we will often use the following symbolism:

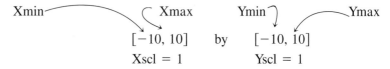

$$[-10, 10] \quad \text{by} \quad [-10, 10]$$
$$\text{Xscl} = 1 \qquad\qquad \text{Yscl} = 1$$

The symbols above indicate the viewing window information for the window in Figure 9.

All calculators have a standard viewing window. Viewing windows may be changed by manually entering the information, or by using the zoom feature of the calculator. The graphing screen is made up of pixels, which are small areas that, when illuminated, will represent points in the plane. The coordinates of the pixels may be found by using the trace feature of the calculator. When we begin our study of graphs later in this chapter, we will say more about the zoom and trace features.

Figure 10 on the following page shows several other viewing windows, with the important information. Notice that (b) and (c) look exactly alike, and unless we are told what the settings are, we have no way of distinguishing between them. Paying careful attention to window settings will be an important part of our work in this text.

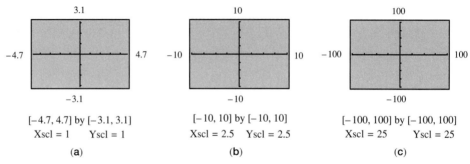

$[-4.7, 4.7]$ by $[-3.1, 3.1]$
Xscl = 1   Yscl = 1
(a)

$[-10, 10]$ by $[-10, 10]$
Xscl = 2.5   Yscl = 2.5
(b)

$[-100, 100]$ by $[-100, 100]$
Xscl = 25   Yscl = 25
(c)

FIGURE 10

1.1   EXERCISES

For each of the following sets, list all elements that belong to (a) natural numbers, (b) whole numbers, (c) integers, (d) rational numbers, (e) irrational numbers, and (f) real numbers.

1. $\left\{-6, -\dfrac{12}{4}, -\dfrac{5}{8}, -\sqrt{3}, 0, .31, .\overline{3}, 2\pi, 10, \sqrt{17}\right\}$ **2.** $\left\{-8, -\dfrac{14}{7}, -.245, 0, \dfrac{6}{2}, 8, \sqrt{81}, \sqrt{12}\right\}$

3. $\left\{-\sqrt{100}, -\dfrac{13}{6}, -1, 5.23, 9.\overline{14}, 3.14, \dfrac{22}{7}\right\}$ **4.** $\{-\sqrt{49}, -.405, -.\overline{3}, .1, 3, 18, 6\pi, 56\}$

Graph each set of numbers on a number line.

5. $\{2, 3, 4, 5\}$ **6.** $\{0, 2, 4, 6, 8\}$ **7.** $\{-4, -3, -2, -1, 0, 1\}$

8. $\{-6, -5, -4, -3, -2\}$ **9.** $\left\{-.5, .75, \dfrac{5}{3}, 3.5\right\}$ **10.** $\left\{-.6, \dfrac{9}{8}, 2.5, \dfrac{13}{4}\right\}$

Each rational number written in common fraction form in Exercises 11–20 has its decimal equivalent appearing in the column on the right. Without using a calculator, if possible, match the fraction with its decimal equivalent.

11. $\dfrac{1}{5}$ **12.** $\dfrac{2}{3}$ **A.** .5   **B.** .67

13. $\dfrac{67}{100}$ **14.** $\dfrac{12}{10}$ **C.** .75   **D.** .2

15. $\dfrac{3}{4}$ **16.** $\dfrac{10}{3}$ **E.** .$\overline{6}$ **F.** .125

17. $\dfrac{50}{100}$ **18.** $\dfrac{3}{11}$ **G.** .$\overline{27}$ **H.** 3.$\overline{3}$

19. $\dfrac{7}{5}$ **20.** $\dfrac{1}{8}$ **I.** 1.2   **J.** 1.4

Each rational number in Exercises 21–28 has a decimal equivalent that repeats. Use the bar symbolism to write the decimal. Use a calculator.

21. $\dfrac{5}{6}$ **22.** $\dfrac{1}{9}$ **23.** $-\dfrac{13}{3}$ **24.** $-\dfrac{9}{11}$

25. $\dfrac{6}{27}$ **26.** $\dfrac{5}{33}$ **27.** $\dfrac{9}{110}$ **28.** $\dfrac{77}{990}$

29. Explain the difference between the rational numbers $.87$ and $.\overline{87}$.

30. A student, using her powerful new calculator, found the decimal 1.414213562 when she evaluated $\sqrt{2}$. Is this decimal the exact value of $\sqrt{2}$, or just an approximation? Should she write $\sqrt{2} = 1.414213562$ or $\sqrt{2} \approx 1.414213562$?

Use a calculator to find a decimal approximation of each root or power. Give as many decimal places as your calculator shows.

31. $\sqrt{58}$ **32.** $\sqrt{97}$ **33.** $\sqrt[3]{33}$ **34.** $\sqrt[3]{91}$

35. $\sqrt[4]{86}$ **36.** $\sqrt[4]{123}$ **37.** $5^{1/2}$ **38.** $19^{1/2}$

39. $18^{1/3}$ **40.** $29^{1/3}$ **41.** $76^{.3}$ **42.** $98^{.275}$

Decide which of the following symbols may be placed in the blank to make a true statement: $<, \le, >, \ge$. There may be more than one correct answer.

43. -5 _____ -4 **44.** -1.3 _____ $-.6$ **45.** 8 _____ 4

46. 9 _____ 8.9 **47.** -6 _____ -6 **48.** 2 _____ 2

Locate the following points on a rectangular coordinate system. Identify the quadrant, if any, in which the point lies.

49. $(2, 3)$ **50.** $(-1, 2)$

51. $(-3, -2)$ **52.** $(1, -4)$

53. $(0, 5)$ **54.** $(-2, -4)$

55. $(-2, 4)$ **56.** $(3, 0)$

57. $(-2, 0)$ **58.** $(3, -3)$

Recall from elementary algebra that the product of two numbers with the same signs is positive and the product of two numbers with different signs is negative. A similar rule holds for quotients. Name the possible quadrants in which the point (x, y) can lie if the given condition is true.

59. $xy > 0$ **60.** $xy < 0$ **61.** $\dfrac{x}{y} < 0$ **62.** $\dfrac{x}{y} > 0$

63. If the x-coordinate of a point is 0, the point must lie on which axis?

64. If the y-coordinate of a point is 0, the point must lie on which axis?

It is important to become familiar with your graphics calculator so that you can operate it efficiently. Answer the following questions about your particular calculator, using a complete sentence or sentences.

65. How do you set the screen in order to obtain the standard viewing window?

66. What are the minimum and maximum values of x and y on your standard viewing window?

Using the notation described in the text, set the viewing window of your calculator to the following specifications.

67. $[-10, 10]$ by $[-10, 10]$
   Xscl $= 1$ Yscl $= 1$

68. $[-40, 40]$ by $[-30, 30]$
   Xscl $= 5$ Yscl $= 5$

69. $[-5, 10]$ by $[-5, 10]$
   Xscl $= 3$ Yscl $= 3$

70. $[-3.5, 3.5]$ by $[-4, 10]$
   Xscl $= 1$ Yscl $= 1$

71. $[-100, 100]$ by $[-50, 50]$
   Xscl $= 20$ Yscl $= 25$

72. $[-4.7, 4.7]$ by $[-3.1, 3.1]$
   Xscl $= .5$ Yscl $= .5$

73. Set your viewing window to $[-10, 10]$ by $[-10, 10]$ and then set Xscl to 0 and Yscl to 0. Do you notice any tick marks on the axes? Make a conjecture as to how to set a screen with no tick marks on the axes.

74. Set your viewing window to $[-50, 50]$ by $[-50, 50]$ and then set Xscl to 1 and Yscl to 1. Observe this screen and describe the appearance of the axes as compared to those seen in the standard window. Why do you think they appear this way? How can you change your scale settings so that this "problem" is alleviated?

1.2 INTRODUCTION TO RELATIONS AND FUNCTIONS

Set-builder and Interval Notation ▮ Relations, Domain, and Range ▮ Functions ▮ Function Notation

In this section we introduce some of the most important concepts in the study of mathematics: relation, function, domain, and range. In order to make our work simpler, various types of set notation are useful. We begin by discussing two types: set-builder and interval notation.

Set-Builder and Interval Notation

Inequalities and variables can be used to specify sets of real numbers. Suppose we wish to symbolize the set of real numbers greater than -2. One way to symbolize this is $\{x \mid x > -2\}$, read "the set of all x such that x is greater than -2." This is called set-builder notation, since the variable x is used to "build" the set. On a number line, we show the elements of this set (the set of all real numbers to the right of -2) by drawing a line from -2 to the right. We use a parenthesis at -2 since -2 is not an element of the given set. The result, shown in Figure 11, is called the **graph** of the set $\{x \mid x > -2\}$.

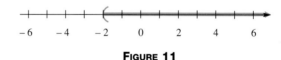

FIGURE 11

The set of numbers greater than -2 is an example of an **interval** on the number line. A simplified notation, called **interval notation,** is used for writing intervals. For example, using this notation, the interval of all numbers greater than -2 is written as $(-2, \infty)$. The **infinity symbol** ∞ does not indicate a number; it is used to show that the interval includes all real numbers greater than -2. The left parenthesis indicates that -2 is not included. A parenthesis is always used next to the infinity symbol in interval notation. The set of all real numbers is written in interval notation as $(-\infty, \infty)$.

EXAMPLE 1

Graphing an Inequality and Using Interval Notation

Write $\{x \mid x < 4\}$ in interval notation and graph the interval.

SOLUTION   The interval is written as $(-\infty, 4)$. The graph is shown in Figure 12. Since the elements of the set are all the real numbers *less* than 4, the graph extends to the left.   ▯

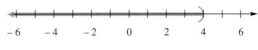

FIGURE 12

The set $\{x \mid x \le -6\}$ contains all the real numbers less than or equal to -6. To show that -6 itself is part of the set, a *square bracket* is used at -6, as shown in Figure 13. In interval notation, this set is written as $(-\infty, -6]$.

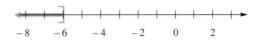

FIGURE 13

EXAMPLE **2**

Graphing an
Inequality and Using
Interval Notation

Write $\{x \mid x \ge -4\}$ in interval notation and graph the interval.

SOLUTION   This set is written in interval notation as $[-4, \infty)$. The graph is shown in Figure 14. A square bracket is used at -4 since -4 is part of the set.   ▯

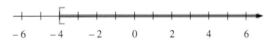

FIGURE 14

It is common to graph sets of numbers that are *between* two given numbers. For example, the set $\{x \mid -2 < x < 4\}$ is made up of all those real numbers between -2 and 4, but not the numbers -2 and 4 themselves. This set is written in interval notation as $(-2, 4)$. The graph has a heavy line between -2 and 4 with parentheses at -2 and 4. See Figure 15. The inequality $-2 < x < 4$ is read "x is greater than -2 and less than 4," or "x is between -2 and 4." It is an example of a *compound inequality*.

FIGURE 15

EXAMPLE **3**

Graphing a
Compound
Inequality and Using
Interval Notation

Write in interval notation and graph $\{x \mid 3 < x \le 10\}$.

SOLUTION   Use a parenthesis at 3 and a square bracket at 10 to get $(3, 10]$ in interval notation. The graph is shown in Figure 16. Read the inequality $3 < x \le 10$ as "3 is less than x and x is less than or equal to 10," or "x is between 3 and 10, excluding 3 and including 10."   ▯

FIGURE 16

A chart summarizing the names of various types of intervals follows. Whenever two real numbers a and b are used to write an interval in the chart, it is assumed that $a < b$.

Type of Interval	Set-Builder Notation	Interval Notation	Graph
Open interval	$\{x \mid a < x < b\}$	(a, b)	
Closed interval	$\{x \mid a \le x \le b\}$	$[a, b]$	
Half-open (or half-closed) interval	$\{x \mid a < x \le b\}$	$(a, b]$	
	$\{x \mid a \le x < b\}$	$[a, b)$	
Unbounded interval	$\{x \mid x > a\}$	(a, ∞)	
	$\{x \mid x \ge a\}$	$[a, \infty)$	
	$\{x \mid x < a\}$	$(-\infty, a)$	
	$\{x \mid x \le a\}$	$(-\infty, a]$	
All real numbers	$\{x \mid x \text{ is real}\}$	$(-\infty, \infty)$	

CAUTION   Notice how the interval notation for the open interval (a, b) looks exactly like the notation for the ordered pair (a, b). While this does not usually cause confusion, as the interpretation is determined by the context of the use, we will, when the need arises, distinguish between them by using "the interval (a, b)" or "the point (a, b)."

Relations, Domain, and Range

Suppose that you have made a study and found that in your new car, driving 55 miles per hour yields a mileage of 31 miles per gallon, while driving 65 miles per hour reduces your mileage to 28 miles per gallon. You have observed a relationship between speed and miles per gallon, and this information may be described using the ordered pairs (55, 31) and (65, 28). These are only two of infinitely many possible ordered pairs that you could have found. If we consider the set of all such ordered pairs, we have an example of a relation.

> **RELATION**
>
> A **relation** is a set of ordered pairs.

Here are three examples of relations.

$$F = \{(1, 2), (-2, 5), (3, -1)\}$$
$$G = \{(-2, 1), (-1, 0), (0, 1), (1, 2), (2, 2)\}$$
$$H = \{(-4, 1), (-2, 1), (-2, 0)\}$$

If we denote the ordered pairs of a relation by (x, y), the set of all x-values is called the **domain** of the relation and the set of all y-values is called the **range** of the relation. In the relations above,

Domain of $F = \{1, -2, 3\}$ Range of $F = \{2, 5, -1\}$

Domain of $G = \{-2, -1, 0, 1, 2\}$ Range of $G = \{1, 0, 2\}$

Domain of $H = \{-4, -2\}$ Range of $H = \{1, 0\}$.

Since a relation is a set of ordered pairs, it may be represented graphically in the rectangular coordinate plane. The graphs of F, G, and H are shown in Figure 17.

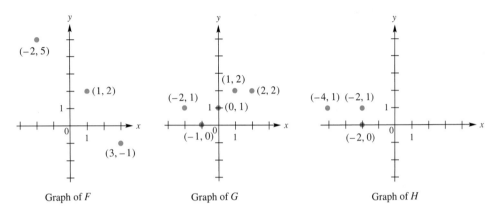

Graph of F Graph of G Graph of H

FIGURE 17

By observing the graph of a relation, we can determine its domain and range, as explained in Example 4.

EXAMPLE 4

Finding Domains and Ranges from Graphs

Three relations are graphed in Figure 18. Give the domain and range of each.

SOLUTION

(a) In Figure 18(a), the x-values of the points on the graph include all numbers between -4 and 4, inclusive. The y-values include all numbers between -6

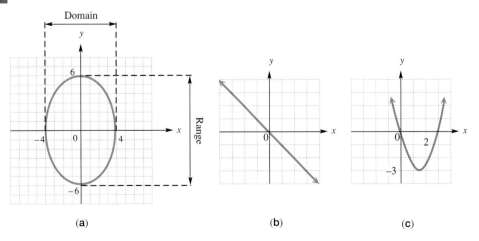

(a)   (b)   (c)

FIGURE 18

and 6, inclusive. Using interval notation,

the domain is $[-4, 4]$;
the range is $[-6, 6]$.

(b) In Figure 18(b), the arrowheads indicate that the line extends indefinitely left and right, as well as up and down. Therefore, both the domain and the range are the set of all real numbers, written $(-\infty, \infty)$.

(c) In Figure 18(c), the arrowheads indicate that the graph extends indefinitely left and right, as well as upward. The domain is $(-\infty, \infty)$. Because there is a least y-value, -3, the range includes all numbers greater than or equal to -3, written $[-3, \infty)$. ▯

Later in this chapter, we will discuss graphing using a calculator. Figure 19 shows a graph on a screen with viewing window $[-5, 5]$ by $[-5, 5]$, with Xscl $= 1$ and Yscl $= 1$. By observation, give the domain and the range of this relation.

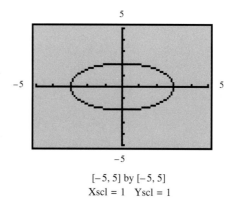

$[-5, 5]$ by $[-5, 5]$
Xscl $= 1$ Yscl $= 1$

Figure 19

SOLUTION   Since the scales on both axes are 1, we see that the graph *appears* to have a minimum x-value of -3, a maximum x-value of 3, a minimum y-value of -2, and a maximum y-value of 2. Therefore, observation leads us to conclude that the domain is $[-3, 3]$ and the range is $[-2, 2]$. ▯

TECHNOLOGICAL NOTE
In Figure 19 we see a calculator-generated graph that is formed by a rather jagged curve. These are sometimes called *jaggies* and are typically found on low-resolution graphers, such as graphics calculators. In general we should remember that most curves we will study in this book are smooth, and the jaggies are just a part of the limitations of technology.

CAUTION   As we shall see many times while discussing calculator-generated graphs, simple observation is not enough to guarantee accuracy in determining domains and ranges. Calculators have capabilities which allow us to improve our accuracy, but even then, it is essential to understand the mathematical concepts behind graphing before we can be certain that our observations are correct. This is why in this book, we will study concepts and technological capabilities in an integrated fashion.

Functions

Look back at the relations F, G, and H introduced earlier in this section. Notice that in F and G, each x-value appears only once in the relation, while in H, the x-value -2 appears twice: it is paired with 1 in one ordered pair, while it is paired with 0 in another. Relations F and G are simple examples of a very important kind of relation, known as a function.

> ### FUNCTION
>
> A **function** is a relation in which each element in the domain corresponds to exactly one element in the range. *

Suppose a group of students get together each Monday evening to study algebra (and perhaps watch football). A number giving the student's weight to the nearest kilogram can be associated with each member of this set of students. Since each student has only one weight at a given time, the relationship between the students and their weights is a function. The domain is the set of all students in the group, while the range is the set of all the weights of the students.

If x represents any element in the domain, the set of students, x is called the **independent variable.** If y represents any element in the range, the weights, then y is called the **dependent variable,** because the value of y *depends on* the value of x. That is, each weight depends on the student associated with it.

In most mathematical applications of functions, the correspondence between the domain and range elements is defined with an equation, like $y = 5x - 11$. The equation is usually solved for y, as it is here, because y is the dependent variable. As we choose values from the domain for x, we can easily determine the corresponding y values of the ordered pairs of the function. (These equations need not use only x and y as variables; any appropriate letters may be used. In physics, for example, t is often used to represent the independent variable *time*.)

EXAMPLE **6**
Deciding Whether a Relation is a Function

Decide whether the following sets are functions. Give the domain and range of each relation.

SOLUTION

(a) $\{(1, 2), (3, 4), (5, 6), (7, 8), (9, 10)\}$

The domain is the set $\{1, 3, 5, 7, 9\}$, and the range is $\{2, 4, 6, 8, 10\}$. Since each element in the domain corresponds to just one element in the range, this set is a function. The correspondence is shown below using D for the domain and R for the range.

$$D = \{1, 3, 5, 7, 9\}$$
$$\downarrow \ \downarrow \ \downarrow \ \downarrow \ \downarrow$$
$$R = \{2, 4, 6, 8, 10\}$$

(b) $\{(1, 1), (1, 2), (1, 3), (2, 4)\}$

The domain here is $\{1, 2\}$, and the range is $\{1, 2, 3, 4\}$. As shown in the correspondence below, one element in the domain, 1, has been assigned three different elements from the range, so this relation is not a function.

$$D = \{1, 2\}$$
$$R = \{1, 2, 3, 4\}$$

(c) $\{(-5, 2), (-4, 2), (-3, 2), (-2, 2), (-1, 2)\}$

Here, the domain is $\{-5, -4, -3, -2, -1\}$, and the range is $\{2\}$. Although every element in the domain corresponds to the same range element, this is a

---

*An alternate definition of function based on the idea of correspondence is given later in the section.

function because each element in the domain has exactly one range element assigned to it.

(d) $\{(x, y) \mid y = x - 2, x \text{ any real number}\}$

Since y is always found by subtracting 2 from x, each x corresponds to just one y, so this relation is a function. Any number can be used for x, and each x will give a number 2 smaller for y; thus, both the domain and the range are the set of real numbers, or in interval notation, $(-\infty, \infty)$.   ∎

There is a quick way to tell whether a given graph is the graph of a function. Figure 20 shows two graphs. In the graph for part (a), each value of x leads to only one value of y, so that this is the graph of a function. On the other hand, the graph in part (b) is not the graph of a function. For example, if $x = x_1$, the vertical line through x_1 intersects the graph at two points, showing that there are two values of y that correspond to this x-value. This idea is known as the *vertical line test* for a function.

VERTICAL LINE TEST

If each vertical line intersects a graph in no more than one point, the graph is the graph of a function.

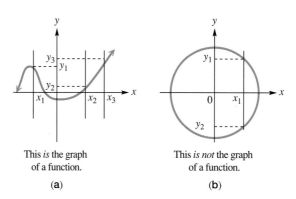

This *is* the graph of a function.

(a)

This *is not* the graph of a function.

(b)

FIGURE 20

(a)   Is the graph in Figure 21 the graph of a function? Specify the domain and the range using interval notation.

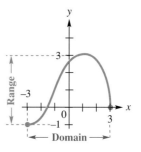

FIGURE 21

SOLUTION   The graph satisfies the vertical line test, and is therefore the graph of a function. As indicated by the annotations at the left and below the graph, the domain appears to be $[-3, 3]$ and the range appears to be $[-1, 3]$.

(b)   Assuming the graph in Figure 22 extends left and right indefinitely and upward indefinitely, does it appear to be the graph of a function? What are the domain and the range if Xscl = 1, Yscl = 1 (use observation)?

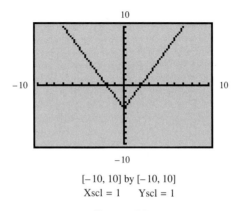

$[-10, 10]$ by $[-10, 10]$
Xscl = 1   Yscl = 1

FIGURE 22

SOLUTION   It appears that no vertical line will intersect the graph more than once, so we may conclude that it is the graph of a function. Since we are told that it extends left and right indefinitely, the domain is $(-\infty, \infty)$. It appears that the lowest point on the graph has the ordered pair $(0, -4)$, and since we know that the graph extends upward indefinitely, the range appears to be the interval $[-4, \infty)$.

(Graphs that are generated by graphics calculators will not exhibit arrowheads, and thus we will need to be aware of the type of function we are observing in order to determine the domain and the range.)   ❑

TECHNOLOGICAL NOTE
Some later model calculators have the capability of using function notation (as seen in *Further Explorations* in this section). Check to see if your model has it.

Function Notation

While the concept of function is crucial to the study of mathematics, the definition of function may vary in wording from text to text. We now give an alternate definition of function that will be helpful in understanding the function notation that follows.

> **ALTERNATE DEFINITION OF FUNCTION**
>
> A function is a correspondence in which each element x from a set called the domain is paired with one and only one element y from a set called the range.

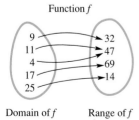

Function f

Domain of f Range of f

FIGURE 23

This idea of correspondence, or mapping, can be illustrated as shown in Figure 23, where the function $f = \{(9, 32), (11, 47), (4, 47), (17, 69), (25, 14)\}$ is depicted. In function f, the x-value 9 is paired with the y-value 32, the x-value 11 is paired with the y-value 47, and so on. Notice that each x-value is used only once in this correspondence. (There is a y-value, 47, that is used more than once, but this does not violate the definition of function.)

To say that y is a function of x means that for each value of x from the domain of the function, there is exactly one value of y. To emphasize that y *is a function of x*, or that y depends on x, it is common to write

$$y = f(x),$$

with $f(x)$ read "f of x." This notation is called **function notation.** For the function f illustrated in Figure 23, we have

$$f(9) = 32 \qquad \text{because } (9, 32) \text{ belongs to the correspondence,}$$
$$f(11) = 47 \qquad \text{because } (11, 47) \text{ belongs to the correspondence,}$$

and so on.

Function notation is used frequently when functions are defined by equations. For example, if a function is defined by the equation $y = 9x - 5$, we may name this function f and write

$$f(x) = 9x - 5.$$

Note that $f(x)$ is simply another name for y. In this function f, if $x = 2$, then we find y, or $f(2)$, by replacing x with 2.

$$f(2) = 9 \cdot 2 - 5$$
$$= 18 - 5$$
$$= 13.$$

The statement "if $x = 2$, then $y = 13$" is abbreviated with function notation as

$$f(2) = 13.$$

Also, $f(0) = 9 \cdot 0 - 5 = -5$, and $f(-3) = -32$.

These ideas and the symbols used to represent them can be explained as follows.

Name of the function   Defining expression

$$y = \overbrace{f(x)} = \overbrace{9x - 5}$$

Value of the function   Name of the independent variable

CAUTION   The symbol $f(x)$ *does not* indicate "f times x," but represents the y-value for the indicated x-value. As shown above, $f(2)$ is the y-value that corresponds to the x-value 2.

EXAMPLE 8

Using Function Notation

In each of the following, find $f(3)$.

(a) $f(x) = 3x - 7$

SOLUTION   Replace x with 3 to get

$$f(3) = 3(3) - 7$$
$$= 9 - 7$$
$$= 2.$$

This result means that the ordered pair (3, 2) belongs to the function, and this ordered pair lies on the graph of this function.

(b) function f depicted in Figure 24

SOLUTION In the correspondence shown, 3 in the domain is paired with 5 in the range, so $f(3) = 5$.

(c) the function f graphed in Figure 25

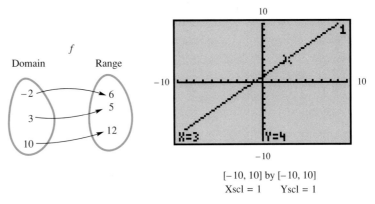

FIGURE 24

FIGURE 25

SOLUTION From the information displayed at the bottom of the screen, when $x = 3$, $y = 4$, so $f(3) = 4$. ◼

EXAMPLE 9

Using Function Notation

If $f(x) = x^2 + 2x - 3$, find $f(a + 1)$ and simplify.

SOLUTION We replace x with $a + 1$ in the expression $x^2 + 2x - 3$.

$$f(x) = x^2 + 2x - 3$$
$$f(a + 1) = (a + 1)^2 + 2(a + 1) - 3$$
$$= a^2 + 2a + 1 + 2a + 2 - 3$$
$$= a^2 + 4a$$

Notice that in our simplification, we squared the expression $a + 1$ and used the distributive property to write $2(a + 1)$ as $2a + 2$. These operations are covered in beginning and intermediate algebra courses. You should be able to perform them in this course. Refer to the appendix that appears at the end of this book if necessary. ◼

1.2 EXERCISES

Write each of the following using interval notation, and then graph each set on the real number line.

1. $\{x \mid -1 < x < 4\}$ **2.** $\{x \mid x \geq -3\}$ **3.** $\{x \mid x < 0\}$ **4.** $\{x \mid 8 > x > 3\}$

5. $\{x \mid 1 \leq x < 2\}$ **6.** $\{x \mid -5 < x \leq -4\}$ **7.** $\{x \mid -9 > x\}$ **8.** $\{x \mid 6 \leq x\}$

Using the variable x, write each of the following using set-builder notation.

9. $(-4, 3)$ **10.** $[2, 7)$ **11.** $(-\infty, -1]$ **12.** $(3, \infty)$

13.

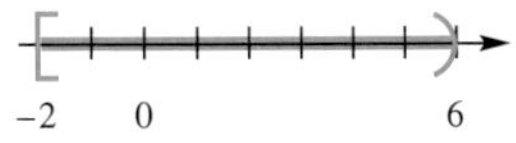

$-2 \quad 0 \qquad\qquad 6$

14.

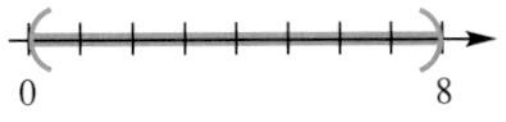

$0 \qquad\qquad 8$

15.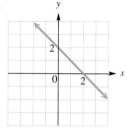

16.

17. Explain how to determine whether a parenthesis or a square bracket is used when graphing an inequality on a number line.

18. The three-part inequality $a < x < b$ means "a is less than x and x is less than b." Which one of the following inequalities is not satisfied by some real number x?
   (a) $-3 < x < 5$ **(b)** $0 < x < 4$ **(c)** $-3 < x < -2$ **(d)** $-7 < x < -10$

Determine the domain and the range of each relation, and tell whether the relation is a function. If it is calculator-generated, assume that the graph extends indefinitely.

19. $\{(5, 1), (3, 2), (4, 9), (7, 6)\}$

20. $\{(8, 0), (5, 4), (9, 3), (3, 8)\}$

21. $\{(2, 4), (0, 2), (2, 5)\}$

22. $\{(9, -2), (-3, 5), (9, 2)\}$

23. $\{(-3, 1), (4, 1), (-2, 7)\}$

24. $\{(-12, 5), (-10, 3), (8, 3)\}$

25. $\{(1, 3), (4, 7), (0, 6), (7, 2)\}$

26. $\{(8, 5), (3, 9), (-2, 11), (5, 3)\}$

27.

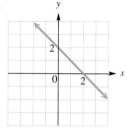

28.

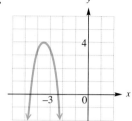

29.

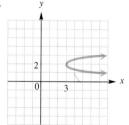

30.

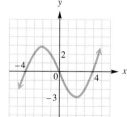

31.

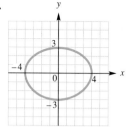

32.

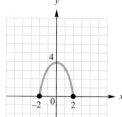

33.

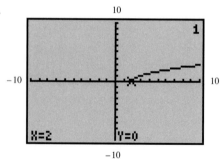

34.

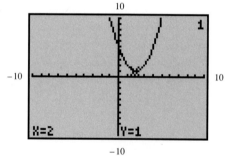

35.

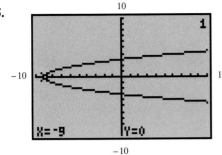

36.

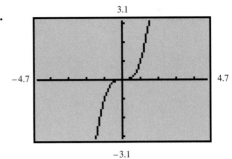

37.   *f*

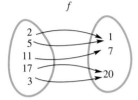

38.   *f*

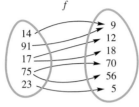

39.   *f*

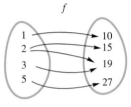

40. In your own words, explain what is meant by a function, using the terms *domain* and *range* in your explanation.

Find each of the following function values.

41. $f(3)$, if $f(x) = -2x + 9$

42. $f(6)$, if $f(x) = -2x + 8$

43. $f(0)$, if $f(x) = 4x + 8$

44. $f(0)$, if $f(x) = -8x + 1$

45. $f(11)$, for the function f in Exercise 37

46. $f(2)$, for the function f in Exercise 34

47. $f(2)$, for the function f in Exercise 33

48. $f(4)$

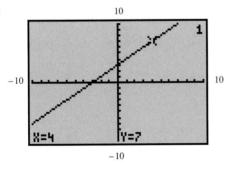

49. $f(8)$

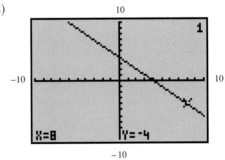

50. $f(3)$

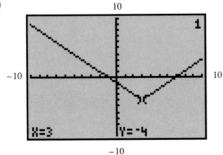

51. $f(-2)$

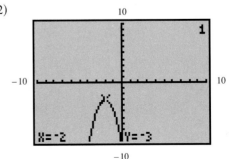

52. If f is the function graphed in Exercise 28 and $f(x) = 4$, what is x?

53. If f is the function graphed in Exercise 30, and $f(x) = 0$, what are the possible values of x?

54. Suppose that you were told the following: "For the function f graphed in Exercise 31, $f(0) = 3$ and $f(0) = -3$." What is wrong with this statement?

In Exercises 55–60, a function is given. Find the simplified form of the function value specified.

55. $f(x) = 3x - 7$; find $f(a + 2)$

56. $f(x) = -9x + 2$; find $f(t - 3)$

57. $g(x) = 2x^2 - 3x + 4$; find $g(r + 1)$

58. $g(x) = -3x^2 + 4x - 9$; find $g(s - 3)$

59. $F(x) = x$; find $F(7p + 2s - 1)$

60. $F(x) = -x$; find $F(\sqrt{2})$

*F*urther *Explorations*

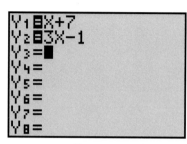

FIGURE 1

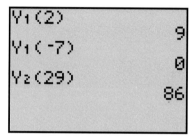

FIGURE 2

Some graphics calculators have the capability of evaluating functions using function notation. If your graphics calculator has such capabilities, enter the function in the $Y =$ menu (Figure 1), then use the home screen to evaluate the function for various values of X (Figure 2). Using the function notation feature of your graphics calculator, find each of the following function values.

1. $Y_1(4)$, if $Y_1 = -8x + 17$

2. $Y_1(-63)$, if $Y_1 = -8x + 17$

3. $Y_2(29)$, if $Y_2 = 27x - 82$

4. $Y_2(322)$, if $Y_2 = 27x - 82$

1.3 LINEAR FUNCTIONS

Functions Defined by Linear Equations ▌ Tracing a Graph with a Calculator ▌ Slope of a Line

Functions Defined by Linear Equations

A variable, a numeral, or a product of numerals and variables is called a **term,** or a **monomial.** Examples of terms, or monomials, are

$$x, \quad -yz, \quad 3, \quad 5x^2y, \quad \text{and} \quad -2x^3.$$

The numerical factor in a term is called the **numerical coefficient,** or simply **coefficient,** of the term. The coefficients of the terms above are 1 (understood), -1 (understood, indicated by the $-$ sign), 3, 5, and -2. The degree of a term in a single variable is simply the exponent of the variable. Thus, the degree of the term x is 1, since $x = x^1$, and the degree of $-2x^3$ is 3. A nonzero numeral, or **constant,** is defined to have degree 0. The number 0 has no degree.

Suppose that we add a term of degree 1 in x to a constant—for example,

$$3x + 6.$$

This is an example of a **binomial,** and its degree is the largest of the degrees of all of its terms. Therefore, $3x + 6$ is a binomial of degree 1, and is said to be **linear.** (The word linear refers to degree 1.) In this section, we will examine functions defined by linear binomials. For example, the function f defined by $3x + 6$ may be symbolized using function notation as

$$f(x) = 3x + 6.$$

This function f is called a linear function.

LINEAR FUNCTION

A function f defined by

$$f(x) = ax + b, \text{ where } a \text{ and } b \text{ are real numbers},$$

is called a linear function.

Similarly, an equation such as $y = 3x + 6$, is called a linear equation. A **solution** of such an equation is an ordered pair (x, y) that makes the equation true. Verify that $(0, 6)$, $(-1, 3)$, $(-2, 0)$, and $(1, 9)$ are all solutions of $y = 3x + 6$.

The traditional method of graphing linear equations involves plotting points whose coordinates are solutions of the equation, and then joining them with a straight line. Figure 26(a) shows the ordered pairs just mentioned for the linear equation $y = 3x + 6$. It is accompanied by a *table of values.* Some graphics calculators will generate such tables. Notice that the points appear to lie in a straight line; this is indeed the case. Since we may substitute *any* real number for x, we join these points with a line to obtain the graph of the function, as shown in Figure 26(b). This line is actually the graph of the relation $\{(x, y) \mid y = 3x + 6\}$.

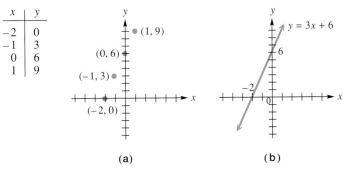

x	y
-2	0
-1	3
0	6
1	9

(a)   (b)

FIGURE 26

Using function notation, we can also say that the graph in Figure 26(b) is the graph of the linear function $f(x) = 3x + 6$.

The graph of the linear function $f(x) = ax + b$ is the same as the graph of the line whose equation is $y = ax + b$.

The graph of a linear function can be created on a calculator window. Most graphics calculators allow the user to graph several functions on the same window. These are sometimes entered by using subscripted y variables: y_1, y_2, y_3, and so on. To graph the function $f(x) = 3x + 6$ on a calculator, enter $3x + 6$ for one of the y-variables. If we choose $[-10, 10]$ by $[-10, 10]$ as our viewing window, with $\text{Xscl} = \text{Yscl} = 1$, we get the graph shown in Figure 27.

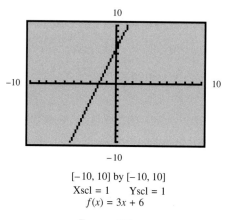

$[-10, 10]$ by $[-10, 10]$
$\text{Xscl} = 1 \qquad \text{Yscl} = 1$
$f(x) = 3x + 6$

FIGURE 27

Calculators generate graphs such as the one shown in Figure 27 by plotting a large number of points in a very short amount of time. The choice of the viewing window will give drastically different views of a graph, and we will often devote discussion to choosing windows appropriate to the problem at hand. Figure 28 shows three different views of the graph of $f(x) = 3x + 6$, with the viewing windows noted. Notice in particular the one in Figure 28(c), since it is graphed in a "square" viewing window. You should consult your calculator manual to see if your calculator has this capability, and how to obtain such a square window.

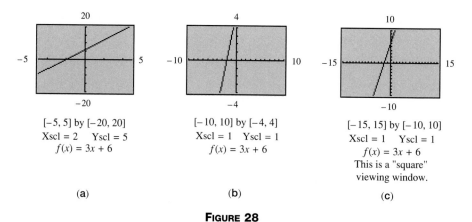

$[-5, 5]$ by $[-20, 20]$	$[-10, 10]$ by $[-4, 4]$	$[-15, 15]$ by $[-10, 10]$
$\text{Xscl} = 2 \quad \text{Yscl} = 5$	$\text{Xscl} = 1 \quad \text{Yscl} = 1$	$\text{Xscl} = 1 \quad \text{Yscl} = 1$
$f(x) = 3x + 6$	$f(x) = 3x + 6$	$f(x) = 3x + 6$
		This is a "square" viewing window.
(a)	**(b)**	**(c)**

FIGURE 28

TECHNOLOGICAL NOTE
One of the most important parts of learning mathematics with the aid of a graphics calculator involves choosing appropriate windows for graphs. Be sure that you know how to change the window on your model.

From geometry we know that two distinct points determine a line. Therefore, if we know the coordinates of a minimum of two points on a line, we can graph the line. For the equation $y = 3x + 6$, suppose we let $x = 0$ and find y: $y = 3x + 6 = 3(0) + 6 = 6$.

Now, let $y = 0$ and solve for x.

$$y = 3x + 6$$
$$0 = 3x + 6$$
$$-6 = 3x$$
$$-2 = x$$

We have found that the points $(0, 6)$ and $(-2, 0)$ lie on the graph of $y = 3x + 6$ and this is sufficient for obtaining the graph in Figure 26(b). The numbers 6 and -2 are called the **y- and x-intercepts** of the line.

---

x- AND y-INTERCEPTS

To find the y-intercept of the graph of $y = ax + b$, let $x = 0$ and solve for y.
To find the x-intercept, let $y = 0$ and solve for x (assuming $a \neq 0$).

EXAMPLE 1

Graphing a Line
Using Intercepts

Without using a calculator, graph the function $f(x) = -2x + 5$. Then support the answer with a calculator-generated graph.

SOLUTION   This is the same as the graph of $y = -2x + 5$. The table below shows the x- and y-intercepts.

x	y	
0	5	← y-intercept
x-intercept → 2.5	0	

The graph of $f(x) = -2x + 5$ is shown in Figure 29. A calculator-generated graph is shown in Figure 30. ∎

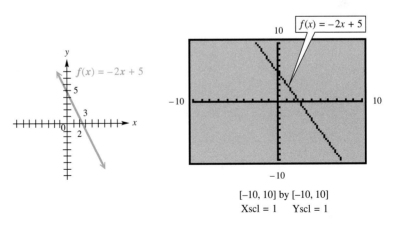

$[-10, 10]$ by $[-10, 10]$
Xscl = 1   Yscl = 1

FIGURE 29   **FIGURE 30**

Since it is possible to obtain infinitely many viewing windows for a graph, we will often be interested in choosing a window that shows the most important features of a particular graph. We will call such a graph a **complete graph.** (While the word "complete" may be actually a misnomer in that it is impossible to show the entire graph of most functions, it does convey the idea that the important features are shown.) Keep in mind that the choice of window for a complete graph is not unique—there will be many acceptable ones.

Each time we introduce a new kind of graph, we will state the requirements for a complete graph. For a line, we have the following.

> **COMPLETE GRAPH OF A LINE**
>
> A complete graph of a line will show all intercepts of the line.

EXAMPLE 2

Finding a Complete Graph of a Line

Find a complete graph of the function $g(x) = -.75x + 12.5$.

SOLUTION   There are many ways that we could choose a viewing window for this complete graph. The window $[-10, 10]$ by $[-10, 10]$ of Figure 31 does not show either intercept, so it will not do. We must increase Xmax and Ymax to show them, so if we use $[-10, 20]$ by $[-10, 20]$, for example, a complete graph is obtained. See Figure 32.   ∎

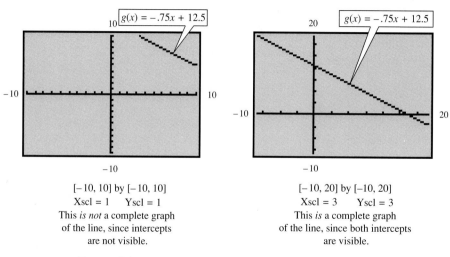

$g(x) = -.75x + 12.5$

$[-10, 10]$ by $[-10, 10]$
Xscl = 1   Yscl = 1
This *is not* a complete graph
of the line, since intercepts
are not visible.

FIGURE 31

$g(x) = -.75x + 12.5$

$[-10, 20]$ by $[-10, 20]$
Xscl = 3   Yscl = 3
This *is* a complete graph
of the line, since both intercepts
are visible.

FIGURE 32

Suppose that for a linear function $f(x) = ax + b$, we have $a = 0$. Then our function becomes

$$f(x) = b,$$

where b is some real number. Its graph is a special kind of straight line.

EXAMPLE 3

Sketching the Graph of $f(x) = b$

Consider the function $f(x) = -3$.

(a)   Sketch its graph on a rectangular coordinate system.

SOLUTION   Since y always equals -3, the value of y can never be 0. This means that the graph has no x-intercept. The only way a straight line can have no x-intercept is for it to be parallel to the x-axis, as shown in Figure 33.

(b) Plot a complete graph in an appropriate viewing window of a calculator.

SOLUTION Using the viewing window $[-5, 5]$ by $[-5, 2]$, we find the same horizontal line shown in part (a). See Figure 34 and compare it to Figure 33. ▮

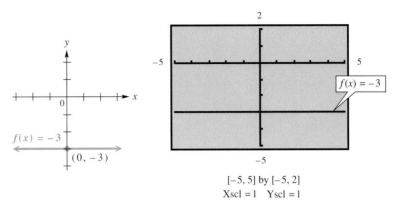

$[-5, 5]$ by $[-5, 2]$
Xscl $= 1$ Yscl $= 1$

FIGURE 33   **FIGURE 34**

The function considered in Example 3 is an example of a constant function.

CONSTANT FUNCTION

A function of the form

$$f(x) = b, \text{ where } b \text{ is a real number,}$$

is called a constant function. Its graph is a horizontal line with y-intercept b. It has no x-intercept.

We will agree that unless otherwise specified, the domain of a linear function will be the set of all real numbers. For $f(x) = ax + b$, where $a \neq 0$, any value of y may be obtained by letting $x = (y - b)/a$. (You are asked to verify this in Exercise 67.) Therefore, the range of a nonconstant linear function is also the set of all real numbers. For the constant function $f(x) = b$, the range is $\{b\}$.

EXAMPLE 4

Finding Domain and Range

(a) What are the domain and the range of $f(x) = 3x + 6$?

SOLUTION Since this is a nonconstant linear function, the domain is $(-\infty, \infty)$ and the range is $(-\infty, \infty)$. Verify this from Figure 26(b).

(b) What are the domain and the range of $f(x) = -3$?

SOLUTION Once again, the domain is $(-\infty, \infty)$. The range consists of only one number, -3, so the range is $\{-3\}$. See Figure 33. ▮

Tracing a Graph with a Calculator

Graphics calculators possess the capability of tracing a graph. A cursor can be activated to move along a graph while the coordinates of points on the graph are displayed on the screen. Later models allow us to actually input an x-value and obtain a y-value. In general, the x-coordinates are determined by the Xmin and

Xmax values used for the viewing window, and the y-coordinates are the function values of the x-coordinates generated. Figure 35 shows some typical screens with designated points identified for the graph of $y = 3x$. Notice that in each case, the y-value is three times the x-value, since this is how y is defined.

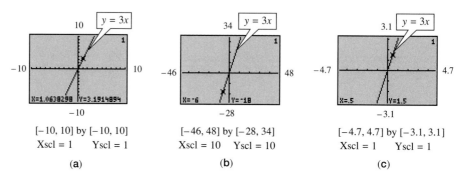

[−10, 10] by [−10, 10]
Xscl = 1   Yscl = 1
(a)

[−46, 48] by [−28, 34]
Xscl = 10   Yscl = 10
(b)

[−4.7, 4.7] by [−3.1, 3.1]
Xscl = 1   Yscl = 1
(c)

FIGURE 35

Later model calculators also provide the user with the options of using decimal or integer increments for x-values when tracing. Figure 36 shows screens for $y = 2x + 5$, using decimal increments and integer increments for x. Note that we have included the coordinates of a point on the graph at the bottom of the screen.

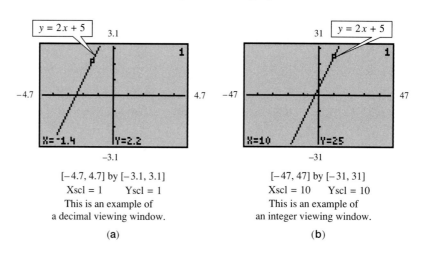

[−4.7, 4.7] by [−3.1, 3.1]
Xscl = 1   Yscl = 1
This is an example of
a decimal viewing window.
(a)

[−47, 47] by [−31, 31]
Xscl = 10   Yscl = 10
This is an example of
an integer viewing window.
(b)

Note: Different models of calculators
use different screens for decimal
and integer windows. Make a note
of what your particular model uses.

FIGURE 36

EXAMPLE 5

Using Tracing to
Support a Result

For the function $f(x) = 3x - 5$, find $f(2.1)$ analytically, and then support the answer by using an appropriate graphing feature of a calculator.

SOLUTION   Recall from Section 1.2 that to find $f(2.1)$, we substitute 2.1 for x.

$$f(2.1) = 3(2.1) - 5$$
$$= 6.3 - 5$$
$$= 1.3$$

Since $f(2.1) = 1.3$, the point $(2.1, 1.3)$ must lie on the graph of the line $y = 3x - 5$. This is supported in Figure 37, where the screen indicates that when $x = 2.1$, $y = 1.3$.   ⬛

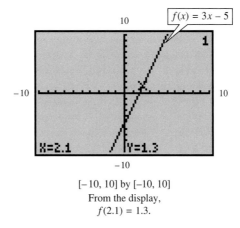

[−10, 10] by [−10, 10]
From the display,
$f(2.1) = 1.3$.

FIGURE 37

TECHNOLOGICAL NOTE
To duplicate Figure 37, you must learn how your particular model will plot a point with a designated x-value. This is an excellent example of the function concept: you input an x-value, and the calculator determines the y-value and shows you the point on the graph of the function.

Slope of a Line

The graph of the line $y = 3x + 1$ is shown in Figure 38. A table of selected points on the graph follows.

x	y
−2	−5
−1	−2
0	1
1	4
2	7

x-increase is 1. y-increase is 3.

FIGURE 38

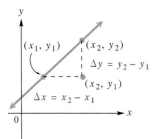

FIGURE 39

Notice that for each increase of 1 for the x-value, the y-value increases by 3. For example, when x increases from 1 to 2, y increases from 4 to 7. This idea is basic to the concept of slope of a line. Geometrically, the slope is a numerical measure of the steepness of the line. (This may be interpreted as the ratio of *rise* to *run*.) To find this measure, start with the line through the two distinct points (x_1, y_1) and (x_2, y_2), as shown in Figure 39, where $x_1 \neq x_2$. The difference

$$x_2 - x_1$$

is called the **change in x** and denoted by Δx (read "delta x"), where Δ is the Greek letter *delta*. In the same way, the **change in y** can be written

$$\Delta y = y_2 - y_1.$$

The *slope* of a nonvertical line is defined as the quotient of the change in y and the change in x, as follows.

SLOPE

The **slope** m of the line through the points (x_1, y_1) and (x_2, y_2) is

$$m = \frac{\Delta y}{\Delta x} = \frac{y_2 - y_1}{x_2 - x_1},$$

where $\Delta x \neq 0$.

CAUTION   When using the slope formula, be sure that it is applied correctly. It makes no difference which point is (x_1, y_1) or (x_2, y_2); however, it is important to be consistent. Start with the x- and y-value of *one* point (either one) and subtract the corresponding values of the *other* point.

EXAMPLE 6

Finding Slope Using the Slope Formula

Find the slope of the line through the points $(2, -1)$ and $(-5, 3)$.

SOLUTION   If $(2, -1) = (x_1, y_1)$ and $(-5, 3) = (x_2, y_2)$, then

$$m = \frac{y_2 - y_1}{x_2 - x_1}$$

$$= \frac{3 - (-1)}{-5 - (2)} = \frac{4}{-7} = -\frac{4}{7}.$$

See Figure 40. On the other hand, if $(2, -1) = (x_2, y_2)$ and $(-5, 3) = (x_1, y_1)$, the slope would be

$$m = \frac{-1 - 3}{2 - (-5)} = \frac{-4}{7} = -\frac{4}{7},$$

the same answer. This example suggests that the slope is the same no matter which point is considered first. Also, using similar triangles from geometry, it can be shown that the slope is the same no matter which two different points on the line are chosen.   ∎

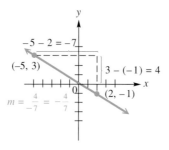

FIGURE 40

EXAMPLE 7

Finding Slope Using
the Slope Formula

The line $y = 4$ is graphed in Figure 41. What is the slope of this line?

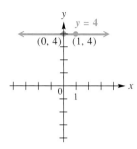

FIGURE 41

SOLUTION   We must choose any two points that lie on the line. For any choice of x, we have $y = 4$. Choose, for example, $(0, 4)$ and $(1, 4)$. Then by the slope formula,

$$m = \frac{4 - 4}{1 - 0} = \frac{0}{1} = 0.$$

The slope of this line is 0.   ▐

In Figure 38, a line with *positive* slope is graphed. In Figure 40, we show a line with *negative* slope, and in Figure 41, a line with slope 0 is given. Notice that the line with positive slope *rises* from left to right, the line with negative slope *falls* from left to right, and the line with slope 0 is *horizontal*. In general, we have the following.

GEOMETRIC ORIENTATION BASED ON SLOPE

For a line with slope m, if $m > 0$, the line rises from left to right. If $m < 0$, it falls from left to right. If $m = 0$, the line is horizontal.

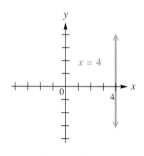

FIGURE 42

In the formula for slope, we have the condition $\Delta x = x_2 - x_1 \neq 0$. This means that $x_2 \neq x_1$. If we graph a line with two points having the same x-values, we get a vertical line. For example, the line with equation $x = 4$ is graphed in Figure 42. Notice that this is *not* the graph of a function, since 4 appears as the first number in more than one ordered pair. If we were to apply the slope formula, the denominator would be 0. As a result, the slope of such a line is *undefined*.

VERTICAL LINE

A vertical line with x-intercept a has an equation of the form

$$x = a.$$

Its slope is undefined.

TECHNOLOGICAL NOTE
It is essential that your calculator be in function mode (rather than parametric, polar, sequence, etc.) in order to study the material here. The other modes are covered in later chapters.

NOTE   *Unless we use parametric equations (covered later in the text), it is not possible to graph a vertical line on a graphics calculator, since it is not the graph of a function.*

If we go back to Figure 38, we see that the slope of the line, 3, is the same as the coefficient of x in its equation ($y = 3x + 1$). Also, the graph of $y = 4$, which may also be written $y = 0x + 4$, has slope 0 which is the coefficient of x. (See Figure 41.) Consider the linear function $f(x) = ax + b$. Suppose that we increase the value of x by 1, to get $x + 1$. Then by the slope formula,

$$
\begin{aligned}
m &= \frac{y_2 - y_1}{x_2 - x_1} = \frac{f(x + 1) - f(x)}{(x + 1) - x} \\
&= \frac{[a(x + 1) + b] - (ax + b)}{x + 1 - x} \\
&= \frac{ax + a + b - ax - b}{1} \\
&= \frac{a}{1} \\
&= a
\end{aligned}
$$

Therefore, for each increase of 1 for the x-value, the y-value increases by a. This is consistent with our earlier observations, and leads us to a very important result concerning linear functions.

> The slope of the graph of the linear function $f(x) = ax + b$ is a.
> The y-intercept of the graph is $f(0) = b$.

Because the slope of the graph of $f(x) = ax + b$ is a, it is often convenient to use m rather than a in the general form of the equation. Therefore, we will sometimes write

$$f(x) = mx + b \qquad \text{or} \qquad y = mx + b$$

to indicate a linear function. The slope is m and the y-intercept is b. This is generally called the *slope-intercept form* of the equation of a line.

EXAMPLE 8

Matching a Graph with an Equation

In Figure 43, we have four calculator-generated lines. Their equations are

$$y = 2x + 3, \qquad y = -2x + 3, \qquad y = 2x - 3, \qquad \text{and} \qquad y = -2x - 3,$$

but not necessarily in this order. Match each equation with its graph.

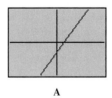 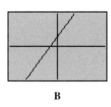 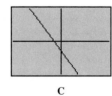 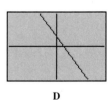

 A   B   C   D

FIGURE 43

SOLUTION   Keep in mind that the sign of m determines whether the graph rises or falls from left to right. Also, if $b > 0$, the y-intercept is *above* the x-axis, while if $b < 0$, the y-intercept is *below* the x-axis. Therefore,

$y = 2x + 3$ is shown in B, since the graph rises from left to right, and the y-intercept is positive;

$y = -2x + 3$ is shown in D, since the graph falls from left to right and the y-intercept is positive.

Verify that $y = 2x - 3$ is shown in A, and $y = -2x - 3$ is shown in C.   ∎

FOR **GROUP DISCUSSION**

With your calculator set for the standard viewing window, graph these four lines on the same screen.

$$y_1 = .5x + 1$$
$$y_2 = x + 1$$
$$y_3 = 2x + 1$$
$$y_4 = 4x + 1$$

Each line has a positive slope. As the slope m becomes larger, what do you notice about the steepness of the line?

Now repeat the experiment, but change each coefficient of x to its negative. As the absolute value of the slope becomes larger, what do you notice about the steepness of the line?

The result of the group discussion exercise should lead to the following.

STEEPNESS OF A LINE

For the line $y = mx + b$, as $|m|$ becomes larger, the line becomes steeper.

The final example shows how the geometric interpretation of slope can be used to graph a line by hand if we know the slope of the line and a point that lies on the line.

EXAMPLE 9

Using the Slope and a Point to Graph a Line

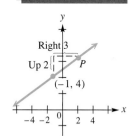

FIGURE 44

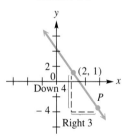

FIGURE 45

(a)   Graph the line that has slope $\frac{2}{3}$ and contains the point $(-1, 4)$.

SOLUTION   First locate the point $(-1, 4)$ on a graph as shown in Figure 44. Then, from the definition of slope,

$$m = \frac{\text{change in } y}{\text{change in } x} = \frac{2}{3}.$$

Move *up* 2 units in the y-direction and then 3 units to the *right* in the x-direction to locate another point on the graph (labeled P). The line through $(-1, 4)$ and P is the required graph.

(b)   Graph the line through $(2, 1)$ that has slope $-\frac{4}{3}$.

SOLUTION   Start by locating the point $(2, 1)$ on the graph. Find a second point on the line by using the definition of slope.

$$\text{slope} = \frac{\text{change in } y}{\text{change in } x} = \frac{-4}{3}$$

Move *down* 4 units from $(2, 1)$ and then 3 units to the *right*. Draw a line through this second point and $(2, 1)$, as shown in Figure 45. The slope also could be written as

$$\frac{\text{change in } y}{\text{change in } x} = \frac{4}{-3}.$$

In that case the second point is located *up* 4 units and 3 units to the *left*. Verify that this approach produces the same line.   ∎

The method of Example 9 can be adapted to graphing the linear function $f(x) = ax + b$. Since the y-intercept is b, we know that one point on the graph is $(0, b)$. Since a is the slope, we can determine another point by interpreting a as "rise over run." For example, to graph $f(x) = 1.5x + 3$, start at $(0, 3)$ and since $1.5 = 1\frac{5}{10} = \frac{15}{10} = \frac{3}{2}$, move 3 units up and 2 units to the right to find a second point on the line, $(2, 6)$. Join the points with a straight line to obtain the graph, as shown in Figure 46.

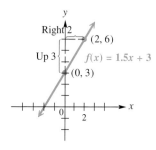

FIGURE 46

NOTE   Students often ask questions like the following:

> In an equation like the one for the line in Figure 46, should I use 1.5 or $\frac{3}{2}$ for the slope?

Remember that 1.5 and $\frac{3}{2}$ are just different forms of the same number, and either form is acceptable (unless otherwise specified).

1.3   EXERCISES

Determine whether the given function is linear.

1. $f(x) = 3x + 12$ **2.** $f(x) = 6x - 4$ **3.** $f(x) = .12 - .3x$

4. $f(x) = .53 + .25x$ **5.** $g(x) = x^2 - 4$ **6.** $g(x) = 1.5x^2 + 3.2$

7. $F(x) = \dfrac{1}{x}$ **8.** $F(x) = \dfrac{1}{x^2}$ **9.** $f(x) = \sqrt{x}$

10. Explain in your own words what is meant by a linear function, and describe the graph of such a function.

Graph each of the following linear functions by hand. You may wish to support your answer by graphing on a calculator. Also, give (a) the x-intercept, (b) the y-intercept, (c) the domain, (d) the range, and (e) the slope of the line.

11. $f(x) = x - 4$ **12.** $f(x) = -x + 4$ **13.** $f(x) = 3x - 6$ **14.** $f(x) = \dfrac{2}{3}x - 2$

15. $f(x) = -\dfrac{2}{5}x + 2$ **16.** $f(x) = \dfrac{4}{3}x - 3$ **17.** $f(x) = 3x$ **18.** $f(x) = -.5x$

19. Based on the graphs of the functions in Exercises 17 and 18, what conclusion can you make about one particular point that *must* lie on the graph of the line $y = ax$ (where $b = 0$)?

20. Using the geometric definition of slope and your answer to Exercise 19, give the equation of the line whose graph is shown here.

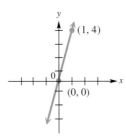

Graph each of the following by hand. For Exercises 21, 22, and 24, you may wish to support your answer by graphing on a calculator. Also, give (a) the x-intercept (if any), (b) the y-intercept (if any), (c) the domain, (d) the range, and (e) the slope of the line (if defined).

21. $f(x) = -3$ **22.** $f(x) = 5$ **23.** $x = -1.5$

24. $f(x) = \dfrac{5}{4}$ **25.** $x = 2$ **26.** $x = -3$

27. What special name is given to the functions found in Exercises 21, 22, and 24?

28. Give the equation of the line illustrated.

(a)
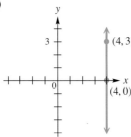

(b)
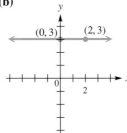

29. What is the equation of the *x*-axis?

30. What is the equation of the *y*-axis?

Graph each linear function on a calculator, using the two different windows given. One window gives a complete graph (as defined in this section), while the other does not. State which window gives the complete graph.

31. $f(x) = 4x + 20$
    Window A: $[-10, 10]$ by $[-10, 10]$
    Window B: $[-10, 10]$ by $[-5, 25]$

32. $f(x) = -5x + 30$
    Window A: $[-10, 10]$ by $[-10, 40]$
    Window B: $[-5, 5]$ by $[-5, 40]$

33. $f(x) = 3x + 10$
    Window A: $[-3, 3]$ by $[-5, 5]$
    Window B: $[-5, 5]$ by $[-10, 14]$

34. $f(x) = -6$
    Window A: $[-5, 5]$ by $[-5, 5]$
    Window B: $[-10, 10]$ by $[-10, 10]$

For each function in Exercises 35–38, find the function value indicated by analytic methods. Then, use the capabilities of your calculator to locate the point with the specified x-value, and show that the y-value corresponds to the one you found analytically.

35. $f(x) = -3x + .25$; $f(4.3)$ **36.** $f(x) = 4x - 1.3$; $f(-1.4)$

37. $f(x) = 2.9x + 10$; $f(-1.3)$ **38.** $f(x) = 5$; $f(6.7)$

Find the slope of the line that passes through the given points.

39. $(-2, 1)$ and $(3, 2)$ **40.** $(-2, 3)$ and $(-1, 2)$ **41.** $(8, 4)$ and $(-1, -3)$

42. $(-4, -3)$ and $(5, 0)$ **43.** $(-6, 5)$ and $(12, 5)$ **44.** $(3, 6)$ and $(3, 1)$

45. Based on your answer to Exercise 43, what special kind of straight line passes through these points? Sketch its graph. Is this the graph of a function?

46. Based on your answer to Exercise 44, what special kind of straight line passes through these points? Sketch its graph. Is this the graph of a function?

47. A linear function is graphed in a window, and tracing yields the *x*- and *y*-values shown. What is the slope of the line?

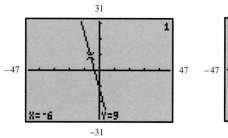

First View   Second View

48. Match each equation with the line that would most closely resemble its graph.

    (a) $y = 2$ **(b)** $y = -2$ **(c)** $x = 2$ **(d)** $x = -2$

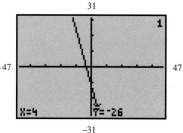

    A   **B**   **C**   **D**

49. Match each equation with its calculator-generated graph.

    (a) $y = 3x + 2$

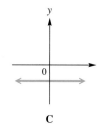 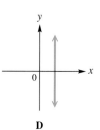

    (b) $y = -3x + 2$

         A   **B**

    (c) $y = 3x - 2$

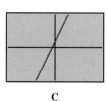 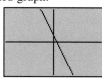

    (d) $y = -3x - 2$

         C   **D**

50. Without actually plotting points, sketch by hand a line on rectangular coordinate axes that would resemble the graph of $y = mx + b$ if

    (a) $m > 0, b > 0$ **(b)** $m > 0, b < 0$

    (c) $m < 0, b > 0$ **(d)** $m < 0, b < 0$.

In Exercises 51–56, two lines are graphed on the same screen. Decide whether y_1 or y_2 has a slope of larger absolute value. Also, determine whether the slope of each line is positive, negative, or zero.

51.

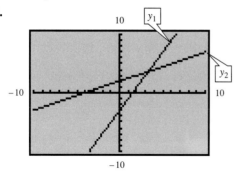

52.

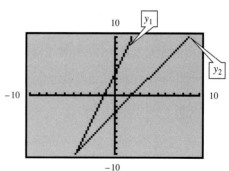

53.

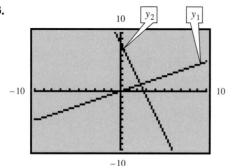

54.

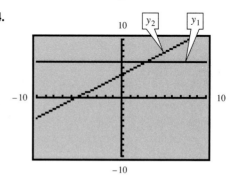

55.

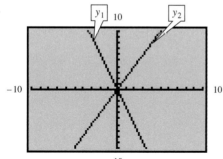

56.

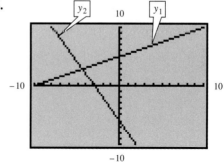

Sketch by hand the graph of the line passing through the given point and having the given slope. Indicate two points on the line.

57. Through $(-1, 3)$, $m = \dfrac{3}{2}$ **58.** Through $(-2, 8)$, $m = -1$ **59.** Through $(3, -4)$, $m = -\dfrac{1}{3}$

60. Through $(-2, -3)$, $m = -\dfrac{3}{4}$ **61.** Through $(-1, 4)$, $m = 0$ **62.** Through $\left(\dfrac{9}{4}, 2\right)$, undefined slope

63. Through $(0, -4)$, $m = \dfrac{3}{4}$ **64.** Through $(0, 5)$, $m = -2.5$

65. Give the equation of the line described in Exercise 63.

66. Give the equation of the line described in Exercise 64.

67. Show that if $f(x) = ax + b$, by letting $x = \frac{y-b}{a}$, we will get y as a result.

68. Use the result of Exercise 67 to show that for $f(x) = 7x + 12$, by letting $x = \frac{5 - 12}{7}$, we get $y = 5$.

69. Refer to Example 8 and accompanying Figure 43. Explain why the graph of $y = 2x - 3$ must be the line shown in A.

70. Repeat Exercise 69 for $y = -2x - 3$ and the line shown in C.

*F*urther *Explorations*

The following table was generated by a graphics calculator with a TABLE feature:

1. Find the slope of the line defined by the equation in Y_1.

2. Find the Y-intercept of the line in Y_1.

3. Find an equation for this line in $Y = mx + b$ form.

X	Y₁	
-2	-11	
-1	-8	
0	-5	
1	-2	
2	1	
3	4	
4	7	

X= -2

1.4 EQUATIONS OF LINES AND GEOMETRIC CONSIDERATIONS

Point-Slope Form of the Equation of a Line ▌ Other Forms of the Equation of a Line ▌ The Pythagorean Theorem, the Distance Formula, and the Midpoint Formula ▌ Parallel and Perpendicular Lines

Point-Slope Form of the Equation of a Line

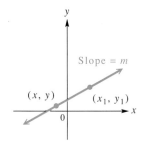

FIGURE 47

The equation of a line can be found if we know a point on the line and the slope of the line. Figure 47 shows a line passing through the fixed point (x_1, y_1) and having slope m. (Assuming that the line has a slope guarantees that it is not vertical.) Let (x, y) be any other point on the line. By the definition of slope, the slope of the line is

$$\frac{y - y_1}{x - x_1}.$$

Since the slope of the line is m,

$$\frac{y - y_1}{x - x_1} = m.$$

Multiplying both sides by $x - x_1$ gives

$$y - y_1 = m(x - x_1).$$

This result, called the *point-slope form* of the equation of a line, identifies points on a given line: a point (x, y) lies on the line through (x_1, y_1) with slope m if and only if

$$y - y_1 = m(x - x_1).$$

POINT-SLOPE FORM

The line with slope m passing through the point (x_1, y_1) has an equation

$$y - y_1 = m(x - x_1).$$

This is called the **point-slope form** of the equation of a line.

EXAMPLE 1

Using the Point-Slope Form (Given a Point and the Slope)

Write the equation of the line through $(-4, 1)$ with slope -3 in $y = mx + b$ form.

SOLUTION   Here $x_1 = -4$, $y_1 = 1$, and $m = -3$. Use the point-slope form of the equation of a line to get

$$y - 1 = -3[x - (-4)] \qquad x_1 = -4, y_1 = 1, m = -3$$
$$y - 1 = -3(x + 4)$$
$$y - 1 = -3x - 12 \qquad \text{Distributive property}$$
$$y = -3x - 11. \qquad y = mx + b \text{ form} \qquad \blacksquare$$

It is possible to find an equation of a line given the coordinates of two points on the line. The next example illustrates how this is done.

EXAMPLE 2

Using the Point-Slope Form (Given Two Points)

Find the equation of the line through $(-3, 2)$ and $(2, -4)$ in $y = mx + b$ form.

SOLUTION   Find the slope first. By the definition of slope,

$$m = \frac{-4 - 2}{2 - (-3)} = -\frac{6}{5}.$$

Either $(-3, 2)$ or $(2, -4)$ can be used for (x_1, y_1). Choosing $x_1 = -3$ and $y_1 = 2$ in the point-slope form gives

$$y - 2 = -\frac{6}{5}[x - (-3)]$$

$$5(y - 2) = -6(x + 3) \qquad \text{Multiply by 5.}$$
$$5y - 10 = -6x - 18 \qquad \text{Distributive property}$$
$$5y = -6x - 8 \qquad \text{Add 10.}$$
$$y = -\frac{6}{5}x - \frac{8}{5}. \qquad \text{Divide by 5.} \qquad \blacksquare$$

EXAMPLE 3

Finding an Equation of a Line (Given Its Intercepts)

Find the equation of the line having x-intercept -4 and y-intercept 5. Give it in $y = mx + b$ form.

SOLUTION   From the given information we know that two points on the line are $(-4, 0)$ and $(0, 5)$. Therefore, the slope is

$$m = \frac{5 - 0}{0 - (-4)} = \frac{5}{4}$$

and since the y-intercept is 5, using the discussion from Section 1.3 dealing with slope and y-intercept, we have the equation $y = \frac{5}{4}x + 5$, or $y = 1.25x + 5$. (Note: Once we have found the slope, we can also use the method of Example 2, employing the point-slope form.) $\blacksquare$

Other Forms of the Equation of a Line

Notice that the equations found in Examples 1–3 were all given in $y = mx + b$ form. As mentioned earlier, this very important form of the equation of a line is called the *slope-intercept form.*

SLOPE-INTERCEPT FORM

The **slope-intercept form** of the equation of a line with slope m and y-intercept b is

$$y = mx + b.$$

This form is probably the most useful form of the equation of a line, since at a single glance we can determine the slope and the y-intercept. In addition, for the purpose of graphing lines on a graphics calculator, this is the form required for input, as we saw in Section 1.3.

Other forms of the equation of a line include

$$Ax + By = C \qquad \text{and} \qquad Ax + By + C = 0,$$

such as $3x + 2y = 6$ and $3x + 2y - 6 = 0$. Texts often refer to these as **standard form.** One advantage of standard form is that it allows quick calculation for both intercepts. For example, if we begin with $3x + 2y = 6$, we can find the x-intercept by letting $y = 0$ and the y-intercept by letting $x = 0$.

$$x\text{-intercept: } 3x + 2(0) = 6$$
$$3x = 6$$
$$x = 2$$
$$y\text{-intercept: } 3(0) + 2y = 6$$
$$2y = 6$$
$$y = 3$$

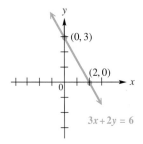

FIGURE 48

Having this information is particularly useful if we wish to sketch the graph of the line by hand. Simply plot the points $(2, 0)$ and $(0, 3)$ and join them with a straight line. See Figure 48.

EXAMPLE 4

Graphing $Ax + By = C$ with a Calculator

Graph $3x + 2y = 6$ in a viewing window of $[-10, 10]$ by $[-10, 10]$.

SOLUTION   To begin, we must first solve the equation for y.

$$3x + 2y = 6 \qquad \text{Given}$$
$$2y = -3x + 6 \qquad \text{Subtract } 3x.$$
$$y = -1.5x + 3 \qquad \text{Divide by 2.}$$

Using this last equation, we obtain the graph in the desired viewing window as shown in Figure 49.

This is the same graph as the one found in Figure 48, except that it is calculator-generated while the latter was drawn by hand.   ∎

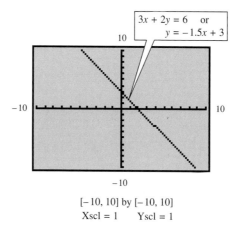

$$[-10, 10] \text{ by } [-10, 10]$$
$$\text{Xscl} = 1 \qquad \text{Yscl} = 1$$

FIGURE 49

NOTE Because of the usefulness of the slope-intercept form over the so-called standard form in the context of the graphics calculator approach of this text, we will emphasize the slope-intercept form in most of our work.

The Pythagorean Theorem, the Distance Formula, and the Midpoint Formula

The Pythagorean theorem from geometry, which gives a relationship concerning the lengths of the sides of a right triangle, is the basis for many results in mathematics. In a right triangle, the sides that form the right angle are called *legs* and the side opposite the right angle (the longest side) is called the *hypotenuse*.

PYTHAGOREAN THEOREM

In a right triangle, the sum of the squares of the lengths of the legs is equal to the square of the length of the hypotenuse.

$$a^2 + b^2 = c^2$$

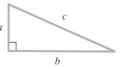

An outline of a proof of this theorem is found in Exercise 67.

EXAMPLE 5

Using the Pythagorean Theorem

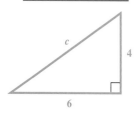

FIGURE 50

Use the Pythagorean theorem to find the length of the hypotenuse in the triangle in Figure 50.

SOLUTION By the theorem, the length of the hypotenuse is

$$c = \sqrt{a^2 + b^2} \qquad \text{Take square roots.}$$
$$= \sqrt{4^2 + 6^2} \qquad \text{Let } a = 4 \text{ and } b = 6.$$
$$= \sqrt{16 + 36}$$
$$= \sqrt{52} = \sqrt{4 \cdot 13}$$
$$= \sqrt{4} \cdot \sqrt{13}$$
$$= 2\sqrt{13}. \qquad ▯$$

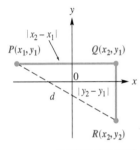

FIGURE 51

$$d(P,R) = \sqrt{(x_2 - x_1)^2 + (y_2 - y_1)^2}$$

FIGURE 52

CAUTION When using the equation $c^2 = a^2 + b^2$, be sure that the length of the hypotenuse is substituted for c, and that the lengths of the legs are substituted for a and b. Errors often occur because values are substituted incorrectly.

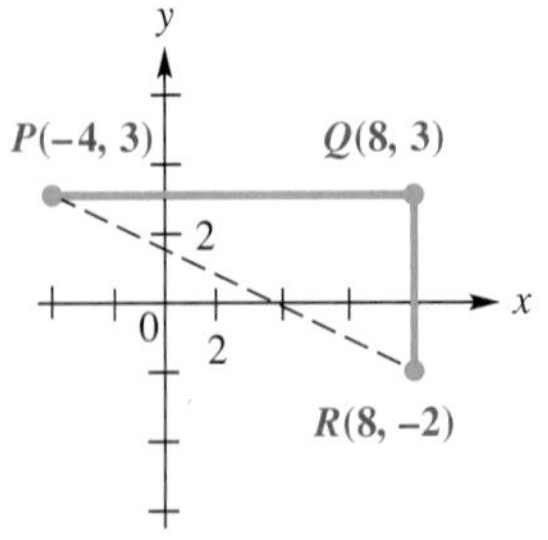

The *converse* of the Pythagorean theorem is also true. That is, if a, b, and c are lengths of the sides of a triangle and $a^2 + b^2 = c^2$, then the triangle is a right triangle with hypotenuse c.

An important formula that is derived using the Pythagorean theorem is the formula that allows us to find the distance between two points in a plane. Figure 51 shows the points $P(-4, 3)$ and $R(8, -2)$.

To find the distance between these two points, complete a right triangle as shown in the figure. This right triangle has its right angle at $(8, 3)$. The horizontal side of the triangle has length

$$\left|8 - (-4)\right| = 12,$$

where absolute value is used to make sure that the distance is not negative. The vertical side of the triangle has length

$$\left|3 - (-2)\right| = 5.$$

By the Pythagorean theorem, the length of the remaining side of the triangle is

$$\sqrt{12^2 + 5^2} = \sqrt{144 + 25} = \sqrt{169} = 13.$$

The distance between $(-4, 3)$ and $(8, -2)$ is 13.

To obtain a general formula for the distance between two points on a coordinate plane, let $P(x_1, y_1)$ and $R(x_2, y_2)$ be any two distinct points in a plane, as shown in Figure 52. Complete a triangle by locating point Q with coordinates (x_2, y_1). Using the Pythagorean theorem gives the distance between P and R, written $d(P, R)$, as

$$d(P, R) = \sqrt{(x_2 - x_1)^2 + (y_2 - y_1)^2}.$$

NOTE The use of absolute value bars is not necessary in this formula, since for all real numbers a and b, $\left|a - b\right|^2 = (a - b)^2$.

The distance formula can be summarized as follows.

DISTANCE FORMULA

Suppose that $P(x_1, y_1)$ and $R(x_2, y_2)$ are two points in a coordinate plane. Then the distance between P and R, written $d(P, R)$, is given by the **distance formula,**

$$d(P, R) = \sqrt{(x_2 - x_1)^2 + (y_2 - y_1)^2}$$

Although the figure used in the proof of the distance formula assumes that P and R are not on a horizontal or vertical line, the result is true for any two points.

EXAMPLE 6

Using the Distance Formula

Find the distance between $P(-8, 4)$ and $Q(3, -2)$.

SOLUTION According to the distance formula,

$$d(P, Q) = \sqrt{[3 - (-8)]^2 + (-2 - 4)^2} \qquad x_1 = -8, y_1 = 4,$$
$$= \sqrt{11^2 + (-6)^2} \qquad\qquad\qquad x_2 = 3, y_2 = -2$$
$$= \sqrt{121 + 36} = \sqrt{157}. \qquad ∎$$

NOTE   As shown in Example 6, it is customary to leave the distance between two points in radical form rather than approximating it with a calculator (unless, of course, it is otherwise specified).

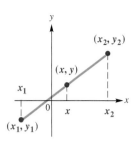

FIGURE 53

Given the coordinates of the endpoints of a line segment, it is possible to find the coordinates of the *midpoint* of the segment; that is, the point on the segment that lies the same distance from both endpoints. To develop the midpoint formula, let (x_1, y_1) and (x_2, y_2) be any two distinct points in a plane. (Although Figure 53 shows $x_1 < x_2$, no particular order is required.) Assume that the two points are not on a horizontal or vertical line. Let (x, y) be the midpoint of the segment connecting (x_1, y_1) and (x_2, y_2). Draw vertical lines from each of the three points to the x-axis, as shown in Figure 53.

Since (x, y) is the midpoint of the line segment connecting (x_1, y_1) and (x_2, y_2), the distance between x and x_1 equals the distance between x and x_2, so that

$$x_2 - x = x - x_1$$
$$x_2 + x_1 = 2x$$
$$x = \frac{x_1 + x_2}{2}.$$

By this result, the x-coordinate of the midpoint is the average of the x-coordinates of the endpoints of the segment. In a similar manner, the y-coordinate of the midpoint is $(y_1 + y_2)/2$, proving the following statement.

MIDPOINT FORMULA

The midpoint of the line segment with endpoints (x_1, y_1) and (x_2, y_2) is

$$\left(\frac{x_1 + x_2}{2}, \frac{y_1 + y_2}{2} \right).$$

In other words, the midpoint formula says that the coordinates of the midpoint of a segment are found by calculating the *average* of the x-coordinates and the average of the y-coordinates of the endpoints of the segment. In Exercise 68 you are asked to verify that the coordinates above satisfy the definition of midpoint.

EXAMPLE 7

Using the Midpoint Formula

Find the midpoint M of the segment with endpoints $(8, -4)$ and $(-9, 6)$.

SOLUTION   Use the midpoint formula to find that the coordinates of M are

$$\left(\frac{8 + (-9)}{2}, \frac{-4 + 6}{2} \right) = \left(-\frac{1}{2}, 1 \right). \quad ▮$$

EXAMPLE 8

Using the Midpoint Formula

A line segment has an endpoint at $(2, -8)$ and midpoint at $(-1, -3)$. Find the other endpoint of the segment.

SOLUTION   The formula for the x-coordinate of the midpoint is $(x_1 + x_2)/2$. Here the x-coordinate of the midpoint is -1. Letting $x_1 = 2$ gives

$$-1 = \frac{2 + x_2}{2}$$
$$-2 = 2 + x_2$$
$$-4 = x_2.$$

In the same way, $y_2 = 2$ and the endpoint is $(-4, 2)$.   ▮

Parallel and Perpendicular Lines

In the standard viewing window of your calculator, graph all four of the following lines.

$$y_1 = 2x - 6$$
$$y_2 = 2x - 2$$
$$y_3 = 2x$$
$$y_4 = 2x + 4$$

What is the slope of each line? What geometric term seems to describe the lines?

Two lines in a plane are *parallel* if they do not intersect. Although the exercise in the "For Group Discussion" box above does not actually prove the result that follows, it provides good visual support.

PARALLEL LINES

Two distinct nonvertical lines are parallel if and only if they have the same slope.

EXAMPLE 9

Using the Slope Relationship for Parallel Lines

Find the equation of the line that passes through the point $(3, 5)$ and is parallel to the line with the equation $2x + 5y = 4$. Then graph both lines in the window $[-10, 10]$ by $[-10, 10]$ to provide visual support for your answer.

SOLUTION Since it is given that the point $(3, 5)$ is on the line, we need only find the slope to use the point-slope form. Find the slope by writing the equation of the given line in slope-intercept form. (That is, solve for y.)

$$2x + 5y = 4$$
$$y = -\frac{2}{5}x + \frac{4}{5}$$

The slope is $-\frac{2}{5}$. Since the lines are parallel, $-\frac{2}{5}$ is also the slope of the line whose equation is to be found. Substituting $m = -\frac{2}{5}$, $x_1 = 3$, and $y_1 = 5$ into the point-slope form gives

$$y - y_1 = m(x - x_1)$$
$$y - 5 = -\frac{2}{5}(x - 3)$$
$$5(y - 5) = -2(x - 3)$$
$$5y - 25 = -2x + 6$$
$$5y = -2x + 31$$
$$y = -\frac{2}{5}x + \frac{31}{5}. \quad \text{Slope-intercept form of the desired line}$$

To provide visual support, we now graph the given line and the line just determined. They are $y = -\frac{2}{5}x + \frac{4}{5}$ (the slope-intercept form of the given equation) and $y = -\frac{2}{5}x + \frac{31}{5}$. It is often easier to enter decimal forms of fractions rather than $\frac{a}{b}$ (common fraction) forms since the latter often require use of parentheses to indicate grouping. Therefore, we will enter these two equations as

$$y_1 = -.4x + .8 \qquad \text{(the given line)}$$

and $\qquad y_2 = -.4x + 6.2 \qquad \text{(the desired line)}.$

As seen in Figure 54, the lines *seem* to be parallel, providing visual support for our result. ◧

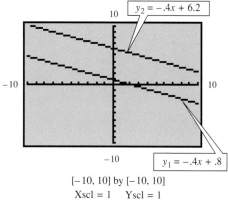

[−10, 10] by [−10, 10]
Xscl = 1   Yscl = 1

These lines appear to be parallel.
Because the slopes are equal, they
are indeed parallel.

FIGURE 54

When using graphics calculator technology, we must always be aware that visual support (as seen in Example 9) does not necessarily prove our result. For example, Figure 55 shows the graphs of $y_1 = -.5x + 4$ and $y_2 = -.5001x + 2$. Although they *seem* to be parallel by visual inspection, they are *not* parallel

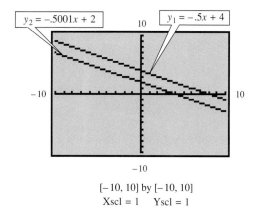

[−10, 10] by [−10, 10]
Xscl = 1   Yscl = 1

These lines appear to be parallel.
However, because the slopes are
not equal, they *are not* truly parallel.

FIGURE 55

because the slope of y_1 is $-.5$ and the slope of y_2 is $-.5001$, and these are unequal slopes. The fact that they are very close to each other causes the lines to *appear* to be parallel on the viewing screen. This discussion provides an excellent illustration as to why we can never completely rely on graphical methods; analytic methods *must* be studied hand-in-hand with graphical methods.

FOR GROUP DISCUSSION

With your calculator viewing window set for a "square" window, graph each pair of lines. Do each group separately, clearing the screen of one group before proceeding to the next.

I	II	III	IV
$y_1 = 4x + 1$	$y_1 = -\dfrac{2}{3}x + 3$	$y_1 = 6x - 3$	$y_1 = \dfrac{13}{7}x - 3$
$y_2 = -.25x + 3$	$y_2 = 1.5x - 4$	$y_2 = -\dfrac{1}{6}x + 4$	$y_2 = -\dfrac{7}{13}x + 4$

What geometric term applies to each pair of lines? What is the product of the slopes in each pair of lines?

Again, as in the earlier "For Group Discussion" activity, while we have not proved the result stated below, we have provided good visual support for it.

PERPENDICULAR LINES

Two lines, neither of which is vertical, are perpendicular if and only if their slopes have a product of -1.

For example, if the slope of a line is $-\frac{3}{4}$, the slope of any line perpendicular to it is $\frac{4}{3}$, since $\left(-\frac{3}{4}\right)\left(\frac{4}{3}\right) = -1$. We often refer to numbers like $-\frac{3}{4}$ and $\frac{4}{3}$ as "negative reciprocals." A proof of this result is outlined in Exercise 69.

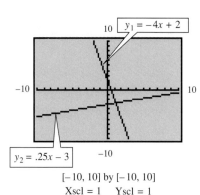

$y_1 = -4x + 2$

$y_2 = .25x - 3$

$[-10, 10]$ by $[-10, 10]$
Xscl = 1   Yscl = 1

Although the graphs of these lines are perpendicular, they do not seem to be when graphed in a standard viewing window.

(a)

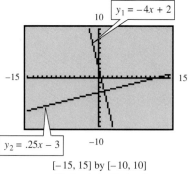

$y_1 = -4x + 2$

$y_2 = .25x - 3$

$[-15, 15]$ by $[-10, 10]$
Xscl = 1   Yscl = 1

Visual support for perpendicularity is more obvious using a "square" viewing window.

(b)

FIGURE 56

TECHNOLOGICAL NOTE
Read your owner's manual to determine the number of pixels your screen displays horizontally and vertically. This will determine the aspect ratio of the screen.

Because the standard viewing windows for graphics calculators are usually designed with an aspect ratio of approximately 3 to 2 (that is, the number of pixels across is about $\frac{3}{2}$ times the number of pixels up and down), in order to give visual support for perpendicularity, a "square" setting is necessary. Figure 56 illustrates this.

EXAMPLE **10**

Using the Slope Relationship for Perpendicular Lines

Find the equation of the line that passes through the point $(3, 5)$ and is perpendicular to the line with the equation $2x + 5y = 4$. Then graph both lines in a square viewing window to provide visual support for your answer.

SOLUTION   In Example 9 we found that the slope of the given line is $-\frac{2}{5}$, so the slope of any line perpendicular to it is $\frac{5}{2}$. Therefore, use $m = \frac{5}{2}$, $x_1 = 3$, and $y_1 = 5$ in the point-slope form.

$$y - 5 = \frac{5}{2}(x - 3)$$
$$2(y - 5) = 5(x - 3)$$
$$2y - 10 = 5x - 15$$
$$-5x + 2y = -5$$
$$2y = 5x - 5$$
$$y = 2.5x - 2.5$$

Graphing $y_1 = -.4x + .8$ (the point-slope form of the given equation) and $y_2 = 2.5x - 2.5$ (the point-slope form of the equation just determined) in a *square* viewing window provides support for our answer. See Figure 57.   ◨

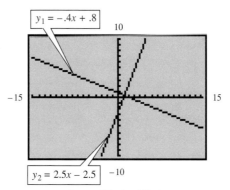

Square Viewing Window

FIGURE 57

1.4 EXERCISES

Write the equation of the line through the given point and having the indicated slope.
Give the equation in slope-intercept form, if possible.

1. Through $(1, 3)$, $m = -2$

2. Through $(2, 4)$, $m = -1$

3. Through $(-5, 4)$, $m = -\dfrac{3}{2}$

4. Through $(-4, 3)$, $m = \dfrac{3}{4}$

5. Through $(-8, 1)$, $m = -.5$

6. Through $(6, 1)$, $m = 0$

7. Through $(-1, 3)$ and $(3, 4)$

8. Through $(8, -1)$ and $(4, 3)$

9. x-intercept 3, y-intercept -2

10. x-intercept -2, y-intercept 4

11. Vertical, through $(-6, 5)$

12. Horizontal, through $(8, 7)$

13. Fill in each blank with the appropriate response: The line $x + 2 = 0$ has x-intercept
_____ . It __(does/does not)__ have a y-intercept. The slope of this line is
__(zero/undefined)__ . The line $4y = 2$ has y-intercept _____ . It
__(does/does not)__ have an x-intercept. The slope of this line is __(zero/undefined)__ .

14. What is true about y for every point on the x-axis?

15. What is true about x for every point on the y-axis?

16. Explain how you would go about graphing a line having an equation of the form $x = k$,
where k is a constant, using hand-drawn methods.

*In Exercises 17–20, you are given two different views of the same line in a viewing window,
with x- and y-coordinates of a point on the line in each view. Determine the slope-intercept
form of the equation of the line, and then verify your result by graphing it on your calculator.*

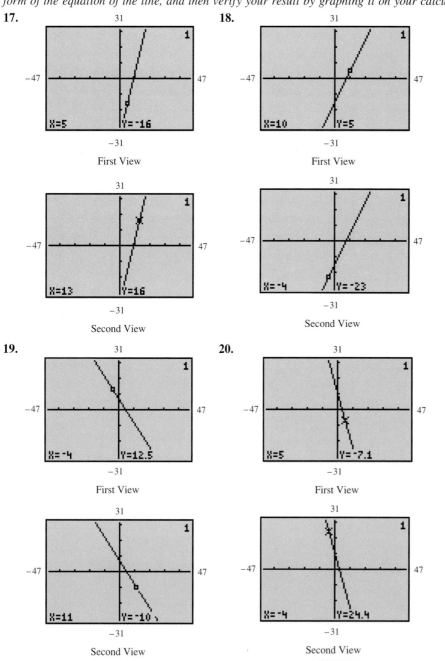

17.

First View

Second View

18.

First View

Second View

19.

First View

Second View

20.

First View

Second View

Graph each of the following lines by hand, finding intercepts to determine two points on the line, as explained in the discussion preceding Example 4.

21. $x - y = 4$ **22.** $x + y = 4$ **23.** $3x - y = 6$

24. $2x - 3y = 6$ **25.** $2x + 5y = 10$ **26.** $4x - 3y = 9$

A line having an equation of the form $y = kx$, where k is a real number, $k \neq 0$, will always pass through the origin. To graph such an equation by hand, we must determine a second point and then join the origin and that second point with a straight line. Use this method to graph each of the following.

27. $y = 3x$ **28.** $y = -2x$ **29.** $y = -.75x$ **30.** $y = 1.5x$

Write each equation in the form $y = mx + b$. Then using your calculator graph the line in the window indicated.

31. $5x + 3y = 15$
     $[-10, 10]$ by $[-10, 10]$

32. $6x + 5y = 9$
     $[-10, 10]$ by $[-10, 10]$

33. $-2x + 7y = 4$
     $[-5, 5]$ by $[-5, 5]$

34. $-.23x - .46y = .82$
     $[-5, 5]$ by $[-5, 5]$

35. $1.2x + 1.6y = 5.0$
     $[-6, 6]$ by $[-4, 4]$

36. $2y - 5x = 0$
     $[-10, 10]$ by $[-10, 10]$

Use the Pythagorean theorem to find the length of the unknown side of the right triangle. In each case, a and b represent the lengths of the legs and c represents the length of the hypotenuse.

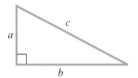

Typical Labeling

37. $a = 8, b = 15$; find c

38. $a = 7, b = 24$; find c

39. $a = 13, c = 85$; find b

40. $a = 14, c = 50$; find b

41. $a = 5, b = 8$; find c

42. $a = 9, b = 10$; find c

43. $a = \sqrt{13}, c = \sqrt{29}$; find b

44. $a = \sqrt{7}, c = \sqrt{11}$; find b

*Find **(a)** the distance between P and Q and **(b)** the coordinates of the midpoint of the segment joining P and Q.*

45. $P(5, 7), Q(13, -1)$ **46.** $P(-2, 5), Q(4, -3)$ **47.** $P(-8, -2), Q(-3, -5)$

48. $P(-6, -10), Q(6, 5)$ **49.** $P(3, -7), Q(-5, 19)$ **50.** $P(-4, 6), Q(8, -5)$

Find the other endpoint of each segment having endpoint and midpoint as given.

51. Endpoint $(-3, 6)$, midpoint $(5, 8)$ **52.** Endpoint $(-5, 3)$, midpoint $(-7, 6)$

53. Endpoint $(5, -4)$, midpoint $(12, 6)$ **54.** Endpoint $(-2, 7)$, midpoint $(-9, 8)$

Find the equation of the line satisfying the given conditions, giving it in slope-intercept form, if possible.

55. Through $(-1, 4)$, parallel to $x + 3y = 5$ **56.** Through $(3, -2)$, parallel to $2x - y = 5$

57. Through $(1, 6)$, perpendicular to $3x + 5y = 1$ **58.** Through $(-2, 0)$, perpendicular to $8x - 3y = 7$

59. Through $(-5, 7)$, perpendicular to $y = -2$ **60.** Through $(1, -4)$, perpendicular to $x = 4$

61. Through $(-5, 8)$, parallel to $y = -.2x + 6$ **62.** Through $(-4, -7)$, parallel to $x + y = 5$

63. Through the origin, perpendicular to $2x + y = 6$ **64.** Through the origin, parallel to $y = -3.5x + 7.4$

65. The figure shows the graphs of $y_1 = 2.3x + .57$ and $y_2 = 2.3001x - 4.8$ in a viewing window $[-10, 10]$ by $[-10, 10]$. The student unfamiliar with the concepts presented in this section may conclude that these two lines are parallel. Write a short paragraph explaining why they are or are not parallel.

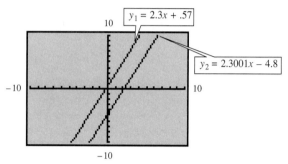

$[-10, 10]$ by $[-10, 10]$
Xscl = 1   Yscl = 1

66. The figure shows the graphs of $y_1 = -3x + 4$ and $y_2 = \frac{1}{3}x - 4$ in a typical standard viewing window. The student unfamiliar with the concepts presented in this section may conclude that these two lines are not perpendicular. Write a short paragraph explaining why they are or are not perpendicular. If they *are* perpendicular lines, explain how to set a graphics calculator so that this result may be more easily supported.

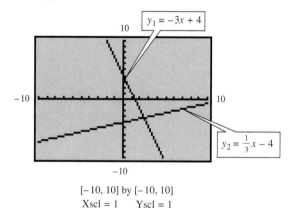

$[-10, 10]$ by $[-10, 10]$
Xscl = 1   Yscl = 1

67. The figure shown is a square made up of four right triangles and a smaller square. By use of the method of *equal areas*, the Pythagorean theorem may be proved. Fill in the blanks with the missing information.

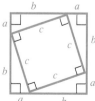

(a) The length of a side of the large square is _____ , so its area is (_____)^2 or _____ .

(b) The area of the large square may also be found by obtaining the sum of the areas of the four right triangles and the smaller square. The area of each right triangle is _____ , so the sum of the areas of the four right triangles is _____ . The area of the smaller square is _____ .

(c) The sum of the areas of the four right triangles and the smaller square is _____ .

(d) Since the areas in (a) and (c) represent the area of the same figure, the expressions there must be equal. Setting them equal to each other we obtain _____ = _____ .

(e) Subtract $2ab$ from each side of the equation in (d) to obtain the desired result _____ = _____ .

68. Suppose that the endpoints of a line segment have coordinates (x_1, y_1) and (x_2, y_2).

(a) Show that the distance between (x_1, y_1) and $(\frac{x_1 + x_2}{2}, \frac{y_1 + y_2}{2})$ is the same as the distance between (x_2, y_2) and $(\frac{x_1 + x_2}{2}, \frac{y_1 + y_2}{2})$.

(b) Show that the sum of the distances between (x_1, y_1) and $(\frac{x_1 + x_2}{2}, \frac{y_1 + y_2}{2})$, and (x_2, y_2) and $(\frac{x_1 + x_2}{2}, \frac{y_1 + y_2}{2})$ is equal to the distance between (x_1, y_1) and (x_2, y_2).

(c) From the results of parts (a) and (b), what conclusion can be made?

69. To prove that two perpendicular lines, neither of which is vertical, have slopes with a product of -1, go through the following steps. Let line L_1 have equation $y = m_1x + b_1$, and let line L_2 have equation $y = m_2x + b_2$. Assume that L_1 and L_2 are perpendicular, and complete right triangle MPN as shown in the figure. $\overline{PQ}$ is horizontal and $\overline{MN}$ is vertical.

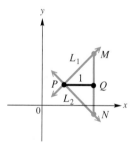

(a) Show that MQ has length m_1.
(b) Show that QN has length $-m_2$.
(c) Show that triangles MPQ and PNQ are similar.
(d) Show that $m_1/1 = 1/-m_2$ and that $m_1m_2 = -1$.

70. Show that the points $(1, -2)$, $(3, -18)$, and $(-2, 22)$ all lie on the same straight line. (*Hint:* Find the equation of the line joining two of the points and then show that the third point satisfies the equation.) Then use your calculator to support your result.

Further Explorations

The following TABLES show ordered pairs for equations in Y_1 and Y_2. For each TABLE, determine if Y_1 and Y_2 are parallel by finding the slopes of each line.

(a)

X	Y₁	Y₂
0	9	-5
1	11	-3
2	13	-1
3	15	1
4	17	3
5	19	5
6	21	7

X=0

(b)

X	Y₁	Y₂
-2	13	-20
-1	12.5	-20.5
0	12	-21
1	11.5	-21.5
2	11	-22
3	10.5	-22.5
4	10	-23

X=-2

(c)

X	Y₁	Y₂
0	13	-8
1	16	-4
2	19	0
3	22	4
4	25	8
5	28	12
6	31	16

X=0

(d)

X	Y₁	Y₂
0	13	42.5
2	18.2	47.7
4	23.4	52.9
6	28.6	58.1
8	33.8	63.3
10	39	68.5
12	44.2	73.7

X=0

1.5 SOLUTION OF LINEAR EQUATIONS; ANALYTIC METHOD AND GRAPHICAL SUPPORT

Simplifying Like Terms ∎ Solving Linear Equations Analytically ∎ Graphical Support for Solutions of Linear Equations ∎ Identities and Contradictions

Simplifying Like Terms

Like terms are terms that have the exact same variable factors. Some pairs of like terms are

$$5x, -6x \qquad 3xy, -8xy \qquad \text{and} \qquad 4x^2y, -9x^2y.$$

Like terms may be added or subtracted by using the distributive property.

DISTRIBUTIVE PROPERTY

For all real numbers a, b, and c,

$$ab + ac = a(b + c)$$

and

$$ab - ac = a(b - c).$$

Suppose that we want to find the sum of $3x$ and $2x$; that is, $3x + 2x$. By using the distributive property, we have

$$3x + 2x = (3 + 2)x = 5x.$$

Notice that to add these like terms, we add their coefficients and keep the same variable.

A graphics calculator can help us support visually our result $3x + 2x = 5x$. If we graph $y_1 = 3x + 2x$ and $y_2 = 5x$ in a standard viewing window, we obtain the display shown in Figure 58.

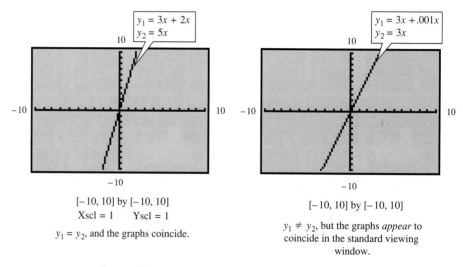

$y_1 = 3x + 2x$
$y_2 = 5x$
10

-10 10

-10

$[-10, 10]$ by $[-10, 10]$
Xscl = 1   Yscl = 1

$y_1 = y_2$, and the graphs coincide.

FIGURE 58

$y_1 = 3x + .001x$
$y_2 = 3x$
10

-10 10

-10

$[-10, 10]$ by $[-10, 10]$

$y_1 \neq y_2$, but the graphs *appear* to coincide in the standard viewing window.

FIGURE 59

Since the two lines appear to coincide, this is what we call **visual support** for our conclusion. However, it is dangerous to assume that visual support constitutes mathematical fact. Consider the graphs of $y_1 = 3x + .001x$ and $y_2 = 3x$ in Figure 59. Despite the fact that $y_1 = 3.001x$ and $y_2 = 3x$, there is no discernible difference between their graphs in a standard viewing window. Here, $y_1 \neq y_2$, and we know this by applying the analytic method to the sum $3x + .001x$.

CAUTION CONCERNING VISUAL SUPPORT

Visual support of an analytic result on a graph does *not* necessarily assure us of the validity of our result.

EXAMPLE 1

Using the Distributive Property to Combine Like Terms

Combine like terms in each part.

SOLUTION

(a) $3x - 4x + 12x - 8x = (3 - 4 + 12 - 8)x = 3x$

(b) $-7(2x + 3) - 5(x + 1) - (6x + 12)$

Apply the distributive property to remove parentheses. We get

$$-14x - 21 - 5x - 5 - 6x - 12.$$

Next, use the commutative and associative properties* to get

$$(-14x - 5x - 6x) + (-21 - 5 - 12).$$

Then, complete the work as follows.

$$(-14x - 5x - 6x) + (-21 - 5 - 12)$$
$$= (-14 - 5 - 6)x + (-38)$$
$$= -25x - 38$$

(c) $5x^2y^3 + 6xy^4 - 2x^2y^3 + 3xy^4 = (5 - 2)x^2y^3 + (6 + 3)xy^4$
$$= 3x^2y^3 + 9xy^4$$ ◘

EXAMPLE 2

Providing Visual
Support Graphically
that Two
Expressions are Not
Equal

Find a viewing window that will illustrate that $5x \neq 5x + .01$.

SOLUTION   The standard viewing window will not provide graphical support for this inequality, since there is no discernible difference between their graphs, as seen in Figure 60. However, if we use $[-.01, .01]$ by $[-.01, .01]$, we see that the graphs of $y_1 = 5x$ and $y_2 = 5x + .01$ are indeed different. (See Figure 61.) This window (among countless others) provides visual support for the inequality.   ◘

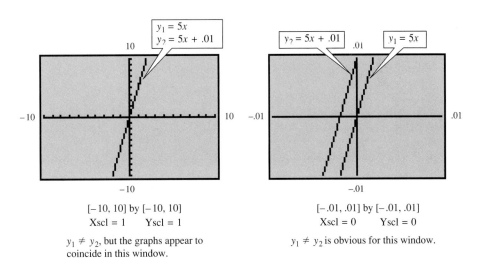

$[-10, 10]$ by $[-10, 10]$
Xscl = 1   Yscl = 1

$y_1 \neq y_2$, but the graphs appear to coincide in this window.

FIGURE 60

$[-.01, .01]$ by $[-.01, .01]$
Xscl = 0   Yscl = 0

$y_1 \neq y_2$ is obvious for this window.

FIGURE 61

---

* The commutative property for addition says that $a + b = b + a$ for all a and b. The associative property for addition says that $(a + b) + c = a + (b + c)$ for all a, b, and c.

Solving Linear Equations Analytically

In this text we will use two distinct approaches to equation solving. The **analytic approach** is the method you have probably seen in previous courses, where paper and pencil are used to transform complicated equations into simpler ones. In so doing, mathematical concepts are applied, and there is no reliance upon graphical representation. Most of the equations that we will encounter in this text are solvable by strictly analytic methods, and the student must realize that this approach *is not to be downplayed* just because the text is based on graphics calculator use.

The **graphical approach** is the method that distinguishes this text from most of the texts you have previously encountered. We will often *support* our analytic solutions by using graphical techniques. Occasionally we will encounter equations that are very difficult or even impossible to solve using analytic methods, and for those we may choose to present a solution that is strictly graphical in nature. It is important to realize that both analytic and graphical approaches will be used, and part of becoming a good mathematics student is learning when to use and when not to use the two approaches.

An **equation** is a statement that two expressions are equal. Examples of equations are

$$3(2x - 4) = 7 - (x + 5) \quad \text{and} \quad .06x + .09(15 - x) = .07(15).$$

To **solve** an equation means to find all numbers that make the equation a true statement. Such numbers are called **solutions** or **roots** of the equation. A number that is a solution of an equation is said to **satisfy** the equation, and the solutions of an equation make up its **solution set.**

In this section we will concentrate on solving equations that have the largest power of the variable equal to 1. These are called *linear equations.*

LINEAR EQUATION IN ONE VARIABLE

A **linear equation** in one variable is an equation that can be written in the form

$$ax + b = 0,$$

where $a \neq 0$.

One way to solve an equation is to rewrite it as a series of simpler equations, each of which has the same solution set as the original one. Such equations are said to be **equivalent equations.** These simpler equations are obtained by using the addition and multiplication properties of equality.

ADDITION AND MULTIPLICATION PROPERTIES OF EQUALITY

For real numbers a, b, and c:

$a = b$ and $a + c = b + c$ are equivalent. *(The same number may be added to both sides of an equation without changing the solution set.)*

If $c \neq 0$, then $a = b$ and $ac = bc$ are equivalent. *(Both sides of an equation may be multiplied by the same nonzero number without changing the solution set.)*

Extending the addition property of equality allows us to subtract the same number from both sides. Similarly, extending the multiplication property of equality allows us to divide both sides of an equation by the same nonzero number.

EXAMPLE 3
Solving a Linear Equation Analytically

Solve $10 + 3(2x - 4) = 17 - (x + 5)$.

SOLUTION Use the distributive property and then collect like terms to get the following series of simpler equivalent equations.

$$10 + 3(2x - 4) = 17 - (x + 5)$$

$10 + 6x - 12 = 17 - x - 5$ Distributive property

$-2 + 7x = 12$ Add x to each side; combine terms.

$7x = 14$ Add 2 to each side.

$x = 2$ Multiply both sides by $\frac{1}{7}$.

An *analytic* check to determine whether 2 is indeed the solution of this equation requires that we substitute 2 for x in the original equation to see if a true statement is obtained.

$10 + 3(2x - 4) = 17 - (x + 5)$ Original equation

$10 + 3(2 \cdot 2 - 4) = 17 - (2 + 5)$?   Let $x = 2$.

$10 + 3(4 - 4) = 17 - 7$?

$10 = 10$ True

Since replacing x with 2 results in a true statement, 2 is the solution of the given equation. The solution set is therefore {2}.   ◨

When fractions or decimals appear in an equation, our work can be made simpler by multiplying both sides by the least common denominator of all the fractions in the equation. Examples 4 and 5 illustrate these types of equations.

EXAMPLE 4
Solving a Linear Equation with Fractional Coefficients Analytically

Solve $\dfrac{x + 7}{6} + \dfrac{2x - 8}{2} = -4$.

SOLUTION Start by eliminating the fractions. Multiply both sides by 6.

$$6 \left[\frac{x + 7}{6} + \frac{2x - 8}{2} \right] = 6 \cdot (-4)$$

$6 \left(\dfrac{x + 7}{6} \right) + 6 \left(\dfrac{2x - 8}{2} \right) = 6(-4)$ Distributive property

$x + 7 + 3(2x - 8) = -24$

$x + 7 + 6x - 24 = -24$ Distributive property

$7x - 17 = -24$ Combine terms.

$7x = -7$ Add 17.

$x = -1$ Divide by 7.

Analytic check:

$$\frac{x+7}{6} + \frac{2x-8}{2} = -4 \qquad \text{Original equation}$$

$$\frac{(-1)+7}{6} + \frac{2(-1)-8}{2} = -4 \qquad ? \quad \text{Let } x = -1.$$

$$\frac{6}{6} + \frac{-10}{2} = -4 \qquad ?$$

$$1 + (-5) = -4 \qquad ?$$

$$-4 = -4 \qquad \text{True}$$

Our analytic check indicates that $\{-1\}$ is the solution set.   ∎

EXAMPLE 5

Solving a Linear Equation with Decimal Coefficients Analytically

Solve $.06x + .09(15 - x) = .07(15)$.

SOLUTION   Since each decimal number is given in hundredths, multiply both sides of the equation by 100. (This is done by moving the decimal points two places to the right.)

$$.06x + .09(15 - x) = .07(15)$$

$$6x + 9(15 - x) = 7(15) \qquad \text{Multiply by 100.}$$

$$6x + 9(15) - 9x = 105 \qquad \text{Distributive property}$$

$$-3x + 135 = 105 \qquad \text{Combine like terms.}$$

$$-3x = -30 \qquad \text{Subtract 135.}$$

$$x = 10 \qquad \text{Divide by } -3.$$

Analytic check:

$$.06x + .09(15 - x) = .07(15) \qquad \text{Original equation}$$

$$.06(10) + .09(15 - 10) = .07(15) \qquad ? \quad \text{Let } x = 10.$$

$$.6 + .09(5) = 1.05 \qquad ?$$

$$.6 + .45 = 1.05 \qquad ?$$

$$1.05 = 1.05 \qquad \text{True}$$

The solution set is $\{10\}$.   ∎

The equations solved in Examples 3, 4, and 5 each have a single solution. Such equations are called **conditional equations.** Later in this section, we will see that equations may have no solutions or infinitely many solutions.

Graphical Support for Solutions of Linear Equations

Let us go back to the linear equation solved in Example 3. We found that the number 2 makes the equation $10 + 3(2x - 4) = 17 - (x + 5)$ a true statement. We can also look at this situation from a standpoint of functions. The statement

"Solve $10 + 3(2x - 4) = 17 - (x + 5)$"

can be re-worded as follows:

"Find the value(s) in the domain of the functions
$$f(x) = 10 + 3(2x - 4)$$
and $\quad g(x) = 17 - (x + 5)$

that give the same function value(s) (i.e., range values)."

From a graphing perspective, this can be interpreted as follows:

"Find the x-value(s) of the point(s) of intersection of the graphs of $f(x) = 10 + 3(2x - 4)$ and $g(x) = 17 - (x + 5)$."

Since the graphs of both functions are straight lines (because they are linear functions), they will intersect in a single point, no point, or infinitely many points. See Figure 62.

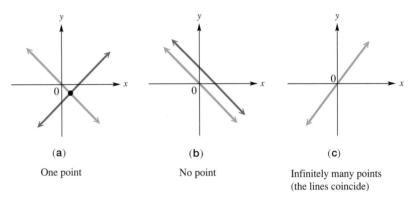

(a)	(b)	(c)
One point	No point	Infinitely many points (the lines coincide)

FIGURE 62

Since we are considering only conditional equations *for now,* we will need only consider the first of these situations, as seen in Figure 62(a). To support graphically our analytic work in Example 3, we graph the lines $y_1 = 10 + 3(2x - 4)$ and $y_2 = 17 - (x + 5)$ in the same viewing window, and use the capabilities of our calculator to find the point of intersection. We see in Figure 63 that the point of intersection is $(2, 10)$. The x-coordinate here, 2, is the solution of the equation, while the y-coordinate, 10, is the value that we get when we substitute 2 for x in both of the original expressions. (See the analytic check in Example 3.)

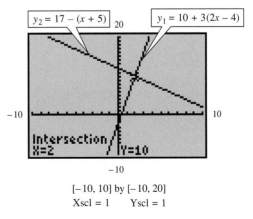

$[-10, 10]$ by $[-10, 20]$
Xscl = 1   Yscl = 1

FIGURE 63

Support the result of Example 4: that is,

$$\frac{x + 7}{6} + \frac{2x - 8}{2} = -4$$

has solution set $\{-1\}$, using the graphical method described above.

SOLUTION   Enter $y_1 = \frac{x+7}{6} + \frac{2x-8}{2}$ and $y_2 = -4$. See Figure 64. The two straight lines intersect at $(-1, -4)$, providing support that $\{-1\}$ is the solution set. Remember, if

$$(-1, -4)$$

is the point of intersection , -1 is the solution, and -4 is the value obtained when -1 is substituted into both expressions.   ❚

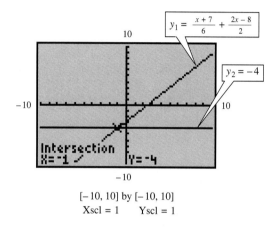

$$y_1 = \frac{x + 7}{6} + \frac{2x - 8}{2}$$

$$y_2 = -4$$

[−10, 10] by [−10, 10]
Xscl = 1   Yscl = 1

FIGURE 64

We now state the first of two methods of equation solving by graphical methods that we will use in this book.

INTERSECTION-OF-GRAPHS METHOD OF GRAPHICAL SOLUTION

To solve the equation

$$f(x) = g(x)$$

graphically, graph $y_1 = f(x)$ and $y_2 = g(x)$. The x-coordinate of any point of intersection of the two graphs is a solution of the equation.

At this point we must make two important observations. First, since we are studying *linear* functions in this chapter, our examples and exercises will consist of functions whose graphs are straight lines. The intersection-of-graphs method of solving equations will be applied to linear functions now, but we will, in later chapters, apply it to many other kinds of functions as well.

Second, we should realize that the most modern graphics calculators have the capability of determining the coordinates of the point of intersection of two lines to a great degree of accuracy, and will in many cases give the exact decimal values. Of course, if a coordinate is an irrational number, the decimal shown will be only an approximation. Figure 65(a) shows the former case, while Figure 65(b) shows the latter.

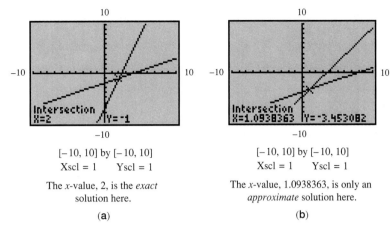

[−10, 10] by [−10, 10]
Xscl = 1   Yscl = 1

The *x*-value, 2, is the *exact* solution here.

(a)

[−10, 10] by [−10, 10]
Xscl = 1   Yscl = 1

The *x*-value, 1.0938363, is only an *approximate* solution here.

(b)

FIGURE 65

There is a second method of graphical support for solving equations. Suppose that we once again wish to solve

$$f(x) = g(x).$$

By subtracting $g(x)$ from both sides, we obtain

$$f(x) - g(x) = 0.$$

Notice that $f(x) - g(x)$ is simply a function itself. Let us call it $F(x)$. Then we only need to solve

$$F(x) = 0$$

to obtain the solution set of the original equation. In Section 1.3 we learned that any number that satisfies this equation is an *x*-intercept of the graph of $y = F(x)$. Using this idea, we now state another method of solving an equation graphically.

TECHNOLOGICAL NOTE
The words *root* and *solution* mean the same thing. If there is one thing that you should remember when studying methods of solving equations, it is this: *The real solutions (or roots) of an equation of the form f(x) = 0 correspond to the x-intercepts of the graph of y = f(x).* For this reason, modern graphics calculators are programmed so that *x*-intercepts (roots) can be located.

> **_x_-INTERCEPT METHOD OF GRAPHICAL SOLUTION**
>
> To solve the equation
>
> $$f(x) = g(x)$$
>
> graphically, graph $y = f(x) - g(x) = F(x)$. Any *x*-intercept of the graph of $y = F(x)$ is a solution of the equation.

The *x*-intercept method is used in the next example.

EXAMPLE 7

Providing Visual Support for an Analytic Solution

It can be shown analytically that the solution set of

$$6x - 4(3 - 2x) = 5(x - 4) - 10$$

is $\{-2\}$. Use the *x*-intercept method of graphical solution to support this result.

SOLUTION   Begin by letting

$$f(x) = 6x - 4(3 - 2x)$$

and $$g(x) = 5(x - 4) - 10.$$

Then, find $f(x) - g(x)$ and enter it as y_1 in a calculator.

$$y_1 = f(x) - g(x)$$

$$= 6x - 4(3 - 2x) - \left(5(x - 4) - 10\right)$$

Graph this function to get the straight line shown in Figure 66. By choosing a standard viewing window, we see that the x-intercept is -2, supporting the information given in the statement of the problem.   ◼

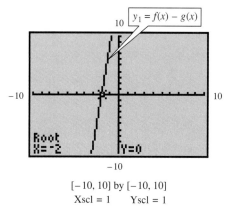

$[-10, 10]$ by $[-10, 10]$
Xscl = 1   Yscl = 1

The x-intercept, -2, is the **solution** (or **root**) of the equation $f(x) = g(x)$.

FIGURE 66

CAUTION   When using the x-intercept method of graphical solution, as in Example 7, it is a common error to forget to enter symbols of inclusion around the expression that is being subtracted. Notice how we used parentheses around $g(x) = 5(x - 4) - 10$ when we determined the expression for y_1. Graphics calculator technology provides us a new respect for the need to use symbols of inclusion correctly!

FOR GROUP DISCUSSION

Repeat Example 7, but this time, rather than graphing

$$y_1 = f(x) - g(x),$$

graph

$$y_2 = g(x) - f(x).$$

Observe the graph of y_2. Does it have the same x-intercept as y_1? Can you make a conjecture concerning the order in which the two functions are subtracted when using the x-intercept method of solution?

As stated earlier, we can use graphical methods to find approximate solutions of equations that, for one reason or another, are difficult to solve algebraically. In this book, we will occasionally require approximate solutions. We now state a method of rounding decimal numbers to a particular place value.

TECHNOLOGICAL NOTE
Consult your owner's manual to see how to set the number of decimal places your calculator will display.

RULES FOR ROUNDING

To round a number to a place value to the right of the decimal point:

Step 1   Locate the **place** to which the number is being rounded.

Step 2   Look at the next **digit to the right** of the place to which the number is being rounded.

Step 3A   If this digit is **less than 5,** drop all digits to the right of the place to which the number is being rounded. Do *not change* the digit in the place to which the number is being rounded.

Step 3B   If this digit is **5 or greater,** drop all digits to the right of the place to which the number is being rounded. *Add 1* to the digit in the place to which the number is being rounded.

To round a number to a place value to the left of the decimal point:

Step 1   Same as above.

Step 2   Same as above.

Step 3A   If this digit is **less than 5,** do *not change* the digit, and replace the digits to the right with zeros.

Step 3B   If this digit is **5 or greater,** *add 1* to it and replace all digits to the right with zeros.

If either situation requires that 1 be added to 9, replace the 9 with a 0 and add 1 to the digit to the left of 9.

EXAMPLE 8

Approximating a Solution of a Linear Equation Graphically

Use the intersection-of-graphs method to approximate the solution of

$$.51(\sqrt{2} + 3x) - .21(\pi x + 6.1) = 7$$

to the nearest hundredth.

SOLUTION   Let $y_1 = .51(\sqrt{2} + 3x) - .21(\pi x + 6.1)$ and let $y_2 = 7$. Since we do not have any idea of what the solution might be, graph them both in the standard viewing window. As seen in Figure 67, the point of intersection does indeed lie in the standard window. Using the capabilities of the calculator, we can find the coordinates of the point of intersection, as seen at the bottom of the screen. If we round the x-coordinate to the nearest hundredth, we have {8.69} as the solution set of the equation.   ◼

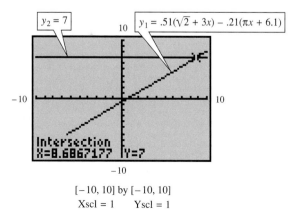

$y_2 = 7$ $y_1 = .51(\sqrt{2} + 3x) - .21(\pi x + 6.1)$

Intersection
X=8.6867177   Y=7

[−10, 10] by [−10, 10]
Xscl = 1   Yscl = 1

FIGURE 67

FOR **GROUP DISCUSSION**

How would you go about solving the equation in Example 8 using the *x*-intercept method of solution?

Identities and Contradictions

Every equation solved thus far in this section has been a conditional equation. However, since the graphs of two linear functions may not intersect or may overlap (as shown in Figures 62 (b) and (c)), there are two other situations that may occur when solving linear equations.

A **contradiction** is an equation that has no solution, as seen in the next example.

EXAMPLE 9

Identifying a Contradiction Analytically and Graphically

Consider the equation

$$-2x + 5x - 9 = 3(x - 4) - 5.$$

(a) Solve it analytically.

SOLUTION

$-2x + 5x - 9 = 3(x - 4) - 5$	Given equation
$3x - 9 = 3x - 12 - 5$	Combine like terms and use the distributive property.
$-9 = -17$	Subtract 3*x*.

Notice that this final equation is obviously false. When this happens, the equation is a contradiction, and the solution set is the *empty* or *null set,* symbolized ∅.

(b) Support the result of part (a) graphically.

SOLUTION Let

$$y_1 = -2x + 5x - 9$$

and $\qquad y_2 = 3(x - 4) - 5.$

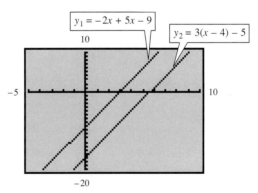

[−5, 10] by [−20, 10]
Xscl = 1   Yscl = 1
y_1 and y_2 appear to be parallel, and indeed are.

FIGURE 68

Graphing both in the viewing window $[-5, 10]$ by $[-20, 10]$ gives us the picture shown in Figure 68. Notice that the lines *appear* to be parallel, supporting our analytic work. Remember again that this graph is only support, and not an actual proof that there are no solutions. ∎

An **identity** is an equation that is true for all values in the domain of its variable. Example 10 illustrates an identity.

EXAMPLE 10

Identifying an Identity Analytically and Graphically

Consider the equation

$$9x + 4 - 3x = 2(3x + 4) - 4.$$

(a) Solve it analytically.

SOLUTION

$$\begin{array}{ll}
9x + 4 - 3x = 2(3x + 4) - 4 & \text{Given equation} \\
6x + 4 = 6x + 8 - 4 & \text{Combine like terms} \\
& \text{and use the distributive} \\
& \text{property.} \\
\\
4 = 4 & \text{Subtract } 6x \text{ and combine} \\
& 8 - 4 \text{ to get 4.}
\end{array}$$

This final equation is obviously true. When this happens, the equation is an identity, and for linear equations, the solution set is {all real numbers}, or $(-\infty, \infty)$. (In later chapters, we will see that some identities will have certain values excluded from their solution sets, but will still have infinitely many solutions.)

(b) Support the result of part (a) graphically.

SOLUTION Let

$$y_1 = 9x + 4 - 3x$$

and

$$y_2 = 2(3x + 4) - 4.$$

Proceeding as we did in Example 9, we graph both using a window that will give us a complete graph. As seen in Figure 69, a standard window appears to give us only one line, indicating an overlap of the graphs. This supports our conclusion in part (a) that all real numbers are solutions. Again, this is not a proof—only visual support. ∎

TECHNOLOGICAL NOTE
Some graphics calculators require that a multiplication symbol be used if a radical is used as a coefficient. For example, in Exercise 46, you may have to enter $\sqrt{2}x$ as $\sqrt{2} * x$. (This is easy to forget, so make a note of it.)

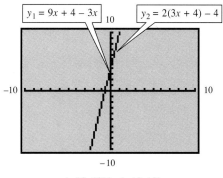

$y_1 = 9x + 4 - 3x$ $y_2 = 2(3x + 4) - 4$

$[-10, 10]$ by $[-10, 10]$
Xscl = 1   Yscl = 1

y_1 and y_2 appear to coincide, and indeed do.

FIGURE 69

1.5 EXERCISES

Combine similar terms, using the distributive property as necessary. In Exercises 1–6, support your answer graphically by graphing the given expression as y_1 and your answer as y_2 in an appropriate viewing window.

1. $23x + 2 - 7(4x - 8) + x - (3x + 2)$

2. $-9x - 6 - 4(-3x + 2) - 6x - (x + 9)$

3. $.54x - 8(.14x + 3) - 2.78$

4. $.98x - 3(.74x + 2) - 3.14$

5. $\dfrac{2}{3}\left(\dfrac{1}{2}x + 6\right) - \dfrac{1}{3}\left(\dfrac{9}{2}x - 1\right)$

6. $-\dfrac{3}{4}\left(\dfrac{1}{3}x + 20\right) - \dfrac{2}{5}\left(\dfrac{25}{4}x + 10\right)$

7. $-4x^4y^3 + 2(x^2y^2 - 4x^4y^3) + 7x^2y^2 - 19$

8. $12w^4xy^2 + 3wx^2y - 8(4w^4xy^2 + 5wx^2y) + xy^5$

9. Explain how the equality $4x + 2x = 6x$ is *supported* by the graphs of $y_1 = 4x + 2x$ and $y_2 = 6x$ in a standard viewing window.

10. Show that $4x + 2x \neq 6x + .01$ by choosing an appropriate viewing window. (There are many such possible viewing windows.)

In Exercises 11–16 two linear functions, y_1 and y_2, are graphed in a viewing window with the point of intersection of the graphs given in the display at the bottom. Using the intersection-of-graphs method of graphical solution, give the solution set of $y_1 = y_2$.

11.

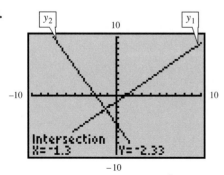

12.

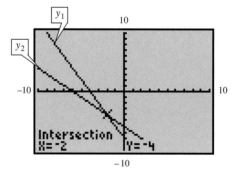

13.

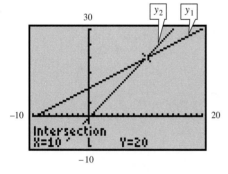

14.

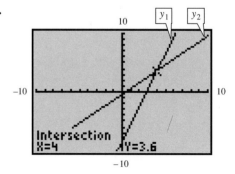

15.

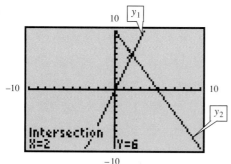

16.

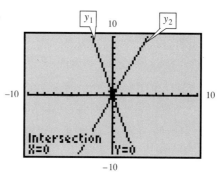

17. In Exercises 11–16 above, give the interpretation of the y-value shown at the bottom of the display.

18. If $y_1 = f(x)$ and $y_2 = g(x)$, and the solution set of $y_1 = y_2$ is $\{-4\}$, what is the value of $f(-4) - g(-4)$? How does your answer relate to the x-intercept method of graphical solution of equations?

In Exercises 19–22, linear functions y_1 and y_2 have been defined and then the graph of $y_1 - y_2$ has been graphed in an appropriate viewing window. Use the x-intercept method of graphical solution to solve the equation $y_1 = y_2$.

19.

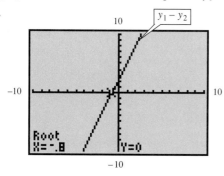

20.

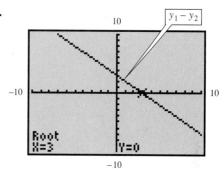

21.

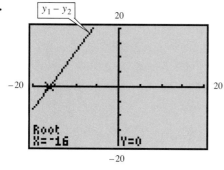

22.

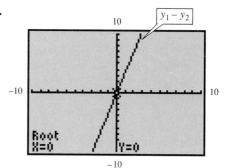

23. If the x-intercept method of graphical solution leads to the graph of a horizontal line above or below the x-axis, what is the solution set of the equation? What special name is this kind of equation given?

24. If the x-intercept method of graphical solution leads to a horizontal line that coincides with the x-axis, what is the solution set of the equation? What special name is this kind of equation given?

Solve each equation analytically. Check it analytically by direct substitution, and then support your solution graphically.

25. $2x - 5 = x + 7$

26. $9x - 17 = 2x + 4$

27. $.01x + 3.1 = 2.03x - 2.96$

28. $.04x + 2.1 = .02x + 1.92$

29. $-(x + 5) - (2 + 5x) + 8x = 3x - 5$

30. $-(8 + 3x) + 5 = 2x + 3$

31. $\dfrac{2x + 1}{3} + \dfrac{x - 1}{4} = \dfrac{13}{2}$

32. $\dfrac{x - 2}{4} + \dfrac{x + 1}{2} = 1$

33. $.40x + .60(100 - x) = .45(100)$

34. $1.30x + .90(.50 - x) = 1.00(50)$

35. $2[x - (4 + 2x) + 3] = 2x + 2$

36. $6[x - (2 - 3x) + 1] = 4x - 6$

37. $\dfrac{5}{6}x - 2x + \dfrac{1}{3} = \dfrac{1}{3}$

38. $\dfrac{3}{4} + \dfrac{1}{5}x - \dfrac{1}{2} = \dfrac{4}{5}x$

39. $5x - (8 - x) = 2[-4 - (3 + 5x - 13)]$

40. $-[x - (4x + 2)] = 2 + (2x + 7)$

Use the approach of Example 8 to find the approximate solution, to the nearest hundredth, of each equation in Exercises 41–46. Use the intersection-of-graphs method.

41. $4(.23x + \sqrt{5}) = \sqrt{2}\,x + 1$

42. $9(-.84x + \sqrt{17}) = \sqrt{6}\,x - 4$

43. $2\pi x + \sqrt[3]{4} = .5\pi x - \sqrt{28}$

44. $3\pi x - \sqrt[4]{3} = .75\pi x + \sqrt{19}$

45. $.23(\sqrt{3} + 4x) - .82(\pi x + 2.3) = 5$

46. $-.15(6 + \sqrt{2}\,x) + 1.4(2\pi x - 6.1) = 10$

Repeat Exercises 41–46, but use the x-intercept method of graphical solution.

47. Exercise 41

48. Exercise 42

49. Exercise 43

50. Exercise 44

51. Exercise 45

52. Exercise 46

Each equation in Exercises 53–58 is either an identity or a contradiction. Determine which of these is using the analytic method, and then support your answer graphically, as explained in Examples 9(b) and 10(b).

53. $6(2x + 1) = 4x + 8\left(x + \dfrac{3}{4}\right)$

54. $3(x + 2) - 5(x + 2) = -2x - 4$

55. $-4[6 - (-2 + 3x)] = 21 + 12x$

56. $4[6 - (1 + 2x)] + 10x = 2(10 - 3x) + 8x$

57. $7[2 - (3 + 4x)] - 2x = 9 + 2(1 - 15x)$

58. $-3[-5 - (-9 + 2x)] = 2(3x - 1)$

59. Verify the result in Example 5 graphically.

60. The figures at the top of the next page show two views of the equation
$$f(x) = 3,$$
where $y_1 = f(x)$ and $y_2 = 3$, as solved by the intersection-of-graphs method.

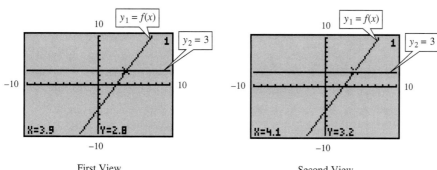

First View   Second View

Which one of these choices could be a possible solution for this equation? Why?
(a) 3.89   **(b)** 3.892   **(c)** 2.95   **(d)** 4.0

61. The figures show two views of a linear equation solved by the x-intercept method of graphical solution.

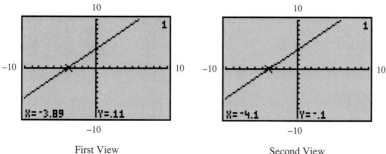

First View   Second View

Which one of these choices could be a possible solution for this equation? Why?
(a) -4.05 **(b)** .11   **(c)** $-.1$ **(d)** 0

62. If you had your choice to use the intersection-of-graphs method or the x-intercept method of graphical solution to approximate the solution of

$$\sqrt{1.7}x - 2[3x + \pi] - 4.79x = \pi[2.3x - (2x + 4)],$$

which method would you prefer? (There is no right or wrong answer here . . . it's just a matter of preference.)

63. If you were asked to solve

$$2x + 3 = 4x - 12$$

by the x-intercept method of graphical solution, why would you *not* get the correct answer by graphing $y_1 = 2x + 3 - 4x - 12$?

64. Try to solve each of these equations by the x-intercept method of graphical solution. What happens? What is the solution set in each case?
(a) $4x + 8 = 2(2x + 3)$ **(b)** $4x + 8 = 2(2x + 4)$

*F*urther Explorations

While the analytic solution of an equation can be supported graphically on your calculator, you can also get numerical support of the solution by using the TABLE feature of your graphics calculator. Figure 1 on the next page shows on the next page the intersection of Y_1 and Y_2 in SPLITSCREEN mode. SPLITSCREEN allows the user to view the graph and also the equations, window settings, or the home screen simultaneously. In Figure 2 the solution

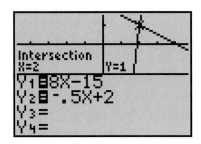

FIGURE 1 **FIGURE 2**

$x = 2$ is supported numerically in the TABLE because, for $x = 2$, Y_1 and Y_2 both have the same value. If your graphics calculator has a TABLE feature, use it to numerically support the following exercises.

(a) Exercise 25   **(b)** Exercise 26
(c) Exercise 27   **(d)** Exercise 28

1.6 SOLUTION OF LINEAR INEQUALITIES; ANALYTIC METHOD AND GRAPHICAL SUPPORT

Solving Linear Inequalities Analytically ∎ Graphical Support for Solutions of Linear Inequalities ∎ Three-part Linear Inequalities

Solving Linear Inequalities Analytically

An equation says that two expressions are equal, while an **inequality** says that one expression is greater than, greater than or equal to, less than, or less than or equal to, another. As with equations, a value of the variable for which the inequality is true is a solution of the inequality, and the set of all such solutions is the solution set of the inequality. Two inequalities with the same solution set are **equivalent inequalities.**

Inequalities are solved with the following properties of inequality.

PROPERTIES OF INEQUALITY

For real numbers a, b, and c:

a. $a < b$ and $a + c < b + c$ are equivalent.
(The same number may be added to both sides of an inequality without changing the solution set.)
b. If $c > 0$, then $a < b$ and $ac < bc$ are equivalent.
(Both sides of an inequality may be multiplied by the same positive number without changing the solution set.)
c. If $c < 0$, then $a < b$ and $ac > bc$ are equivalent.
(Both sides of an inequality may be multiplied by the same negative number without changing the solution set, as long as the direction of the inequality symbol is reversed.)

Replacing $<$ with $>$, $\leq$, or $\geq$ results in equivalent properties.

NOTE Because division is defined in terms of multiplication, the word "multiplied" may be replaced by "divided" in parts (b) and (c) of the properties of inequality. Similarly, in part (a) the words "added to" may be replaced with "subtracted from."

Pay careful attention to part (c): if both sides of an inequality are multiplied by a negative number, the direction of the inequality symbol must be reversed. For example, starting with the true statement $-3 < 5$ and multiplying both sides by the positive number 2 gives

$$-3 \cdot 2 < 5 \cdot 2$$
$$-6 < 10,$$

still a true statement. On the other hand, starting with $-3 < 5$ and multiplying both sides by the *negative* number -2 gives a true result only if the direction of the inequality symbol is reversed.

$$-3(-2) > 5(-2)$$
$$6 > -10$$

A similar situation exists when dividing both sides by a negative number. In summary, the following statement can be made.

> When multiplying or dividing both sides of an inequality by a negative number, we must reverse the direction of the inequality symbol to obtain an equivalent inequality.

A linear inequality in one variable is defined in a way similar to a linear equation in one variable.

> **LINEAR INEQUALITY IN ONE VARIABLE**
>
> A **linear inequality** in one variable is an inequality that can be written in one of the following forms, where $a \neq 0$:
>
> $$ax + b > 0 \qquad ax + b < 0$$
> $$ax + b \geq 0 \qquad ax + b \leq 0.$$

We solve a linear inequality analytically using the same steps as those used to solve a linear equation.

EXAMPLE 1

Solving a Linear Inequality Analytically

Solve the inequality $3x - 2(2x + 6) \leq 2(x + 3)$. Express the solution set using interval notation.

SOLUTION

$$3x - 2(2x + 6) \leq 2(x + 3)$$

$$3x - 4x - 12 \leq 2x + 6 \qquad \text{Distributive property}$$

$$-x - 12 \leq 2x + 6 \qquad \text{Combine like terms.}$$

$$-3x \leq 18 \qquad \text{Subtract } 2x \text{ and add 12;} \\ \text{combine like terms.}$$

$$x \geq -6 \qquad \text{Divide by } -3 \text{ and reverse the} \\ \text{direction of the inequality} \\ \text{symbol.}$$

The solution set is $[-6, \infty)$. ▯

If a linear inequality involves fractions or decimals as coefficients, we use the same procedure as described in Section 1.5 to clear them: Multiply both sides by the least common denominator (which, in the case of decimals, will be a power of 10).

FOR GROUP DISCUSSION

With the class divided into two groups, have each group solve one of the following inequalities analytically.

I

$$\frac{x + 7}{6} + \frac{2x - 8}{2} > -4$$

II

$$.06x + .09(15 - x) > .07(15)$$

After deciding on the correct solution for each, compare the solution set of inequality I with that in Example 4 of Section 1.5, and the solution set of inequality II with that in Example 5 of Section 1.5. What is the same in each case? How do they differ in each case?

From the result of this group discussion, we should conclude that solving a linear inequality will give us an interval whose endpoint (included if the symbol is $\leq$ or $\geq$, excluded if the symbol is $<$ or $>$) is the solution of the corresponding linear equation.

Graphical Support for Solutions of Linear Inequalities

In Section 1.5 we learned two methods of graphical support for solutions of equations. We will now extend these methods to solutions of inequalities. Suppose that two linear functions f and g are graphed, as shown in Figure 70, and the equation $f(x) = g(x)$ is conditional. Then, according to the figure, the solution set is $\{x_1\}$, since by applying the intersection-of-graphs method, it is the x-coordinate of the point of intersection of the two lines.

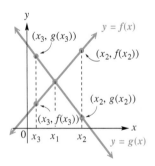

FIGURE 70

Notice that if we choose any x value *less than* x_1, such as x_3 in the figure, the point $(x_3, f(x_3))$ is *below* the point $(x_3, g(x_3))$ and on the same vertical line, indicating that $f(x_3) < g(x_3)$. Similarly, if we choose any x value *greater than* x_1, such as x_2 in the figure, the point $(x_2, f(x_2))$ is *above* the point $(x_2, g(x_2))$, indicating that $f(x_2) > g(x_2)$. This discussion leads to the following extension of the intersection-of-graphs method of solution of equations.

> ### INTERSECTION-OF-GRAPHS METHOD OF SOLUTION OF LINEAR INEQUALITIES
>
> Suppose that f and g are linear functions. The solution set of $f(x) > g(x)$ is the set of all real numbers x such that the graph of f is *above* the graph of g. The solution set of $f(x) < g(x)$ is the set of all real numbers x such that the graph of f is *below* the graph of g.

If an inequality involves one of the symbols $\geq$ or $\leq$, the same method is applied, with the solution of the corresponding equation included in the solution set. This is summarized as follows.

> ### SPECIFYING INTERVALS OF SOLUTION FOR LINEAR INEQUALITIES
>
> If f and g are linear functions, and $f(x) = g(x)$ has a single solution, k, the solution set of
> $$f(x) > g(x) \qquad \text{or} \qquad f(x) < g(x)$$
> will be of the form (k, ∞) or $(-\infty, k)$, with the endpoint of the interval not included. On the other hand, the solution set of
> $$f(x) \geq g(x) \qquad \text{or} \qquad f(x) \leq g(x)$$
> will be of the form $[k, \infty)$ or $(-\infty, k]$, with the endpoint of the interval included.

EXAMPLE **2**

Providing Graphical Support for an Analytic Solution

The inequality
$$3x - 2(2x + 6) \leq 2(x + 3),$$
solved in Example 1, has solution set $[-6, \infty)$. Support this result graphically.

SOLUTION   Start by entering the left side as y_1 and the right side as y_2.

$$y_1 = 3x - 2(2x + 6)$$
$$y_2 = 2(x + 3)$$

The graph, shown in Figure 71, indicates that the point of intersection of the two lines is $(-6, -6)$. The x-coordinate, -6, gives the included endpoint of the solution set of the inequality. Because the graph of y_1 is *below* the graph of y_2 when x is *greater than* -6, our solution set of $[-6, \infty)$ is supported.   ∎

TECHNOLOGICAL NOTE
When supporting the solution set in Example 2 using graphs, the calculator will not determine whether the endpoint of the interval is included or excluded. This must be done by looking at the inequality symbol in the given inequality.

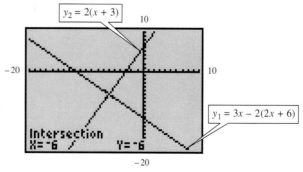

$[-20, 10]$ by $[-20, 10]$

Xscl = 1   Yscl = 1

Figure 71

CAUTION It is just coincidental that the y-value of the point of intersection of the graphs in Figure 71 is the same as the x-value (that is, -6). When determining the solution set of either an equation or an inequality using the intersection-of-graphs method, we are interested in finding the x-*values* (domain values of the functions) that give the same y-values (range values).

Since there is another method of providing graphical support for the solution of a linear equation (the x-intercept method), it too can be extended to linear inequalities. Suppose that we wish to solve

$$f(x) > g(x).$$

Subtracting $g(x)$ from both sides gives

$$f(x) - g(x) > 0.$$

If we call the expression on the left side $F(x)$, we are interested in solving

$$F(x) > 0.$$

We know that $F(x) = 0$ has as its solution the x-intercept of the graph of $y = F(x)$. Therefore, all solutions of $F(x) > 0$ will be the x-values of the points *above* the point at which the graph intersects the x-axis. Similarly, all solutions of $F(x) < 0$ will be the x-values of the points *below* the point at which the graph intersects the x-axis.

---

x-INTERCEPT METHOD OF SOLUTION OF LINEAR INEQUALITIES

The solution set of $F(x) > 0$ is the set of all real numbers x such that the graph of F is *above* the x-axis. The solution set of $F(x) < 0$ is the set of all real numbers x such that the graph of F is *below* the x-axis.

Figure 72 illustrates this discussion, and summarizes the solution sets for the appropriate inequalities.

TECHNOLOGICAL NOTE
If two functions defined by Y1 and Y2 are already entered into your calculator, you can enter Y3 as Y2 − Y1. Then if you direct the calculator to graph Y3 only, you can solve the equation Y1 = Y2 by finding the x-intercept of Y3. Consult your owner's manual to see how this is accomplished. It will save you a lot of time and effort.

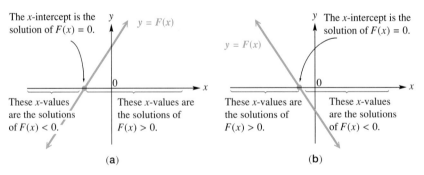

FIGURE 72

EXAMPLE 3

Providing Graphical Support for an Analytic Solution

Let $f(x) = -2(3x + 1)$ and $g(x) = 4(x + 2)$. If we let $y_1 = f(x)$ and $y_2 = g(x)$ and graph $y_1 - y_2$, we obtain the graph shown in Figure 73.

(a) Use the graph to solve the *equation* $f(x) = g(x)$, that is,

$$-2(3x + 1) = 4(x + 2).$$

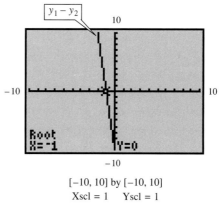

[−10, 10] by [−10, 10]
Xscl = 1   Yscl = 1

FIGURE 73

SOLUTION   The x intercept of $y_1 - y_2$ is -1, so the solution set of this *equation* is $\{-1\}$.

(b) Use the graph to solve the *inequality* $f(x) < g(x)$, that is,

$$-2(3x + 1) < 4(x + 2).$$

SOLUTION   Because the symbol is $<$, we want to find the x-values of the points *below* the x-axis. All such points are to the *right* of $x = -1$, leading to the solution set $(-1, \infty)$.

(c) Use the graph to solve the *inequality* $f(x) > g(x)$, that is,

$$-2(3x + 1) > 4(x + 2).$$

SOLUTION   Because the symbol is $>$, this time we choose the x-values of the points *above* the x-axis. All such points are to the *left* of $x = -1$, leading to the solution set $(-\infty, -1)$.

(d) Use the results of parts (b) and (c) to express the solution sets of

$$-2(3x + 1) \leq 4(x + 2) \qquad \text{and} \qquad -2(3x + 1) \geq 4(x + 2).$$

SOLUTION   Notice here that the only difference is that we now have the symbols that include equality. Therefore, the solution sets are

$$[-1, \infty) \qquad \text{and} \qquad (-\infty, -1], \quad \text{respectively.} \qquad ❚$$

FOR **GROUP DISCUSSION**

One of the authors of this text was once told the following fact by Steele Andrews, a wonderful man who taught him in graduate school:

"Equality is the boundary between less than and greater than."

Use Example 3 to explain this simple, but important, concept.

Either of the graphical methods of supporting the analytic solution of inequalities may be used to approximate the solutions of an inequality that involves complicated expressions or "messy" numbers. In Example 8 of Section 1.5, we solved such an equation using the intersection-of-graphs method. Now we will apply the x-intercept method to one of the related inequalities.

EXAMPLE **4**
Approximating a Solution of a Linear Inequality Graphically

Use the *x*-intercept method to find the solution set of the inequality below. Express the endpoint correct to the nearest hundredth.

$$.51(\sqrt{2} + 3x) - .21(\pi x + 6.1) > 7$$

SOLUTION If we let y_1 represent the left side of the inequality and y_2 respresent the right side, we have

$$y_1 > y_2$$
$$y_1 - y_2 > 0.$$

We now graph $y_1 - y_2$ and look for *x*-values of the points that lie *above* the *x*-axis. Figure 74 shows this graph after using a calculator to obtain the *x*-intercept. As we would expect, the *x*-intercept of the graph is, to the nearest hundredth, 8.69, confirming our result in Section 1.5. Since the graph lies *above* the *x*-axis for *x*-values *to the right* of 8.69, we express the solution set as the interval $(8.69, \infty)$. ∎

TECHNOLOGICAL NOTE
The display given for Y in Figure 74, 6E-13, means 6×10^{-13}, which is very close to 0. On some calculations, the calculator may not actually display the numeral 0 because of limitations in its computing routines. When you obtain a display such as the one for Y in Figure 74, simply think of it as 0.

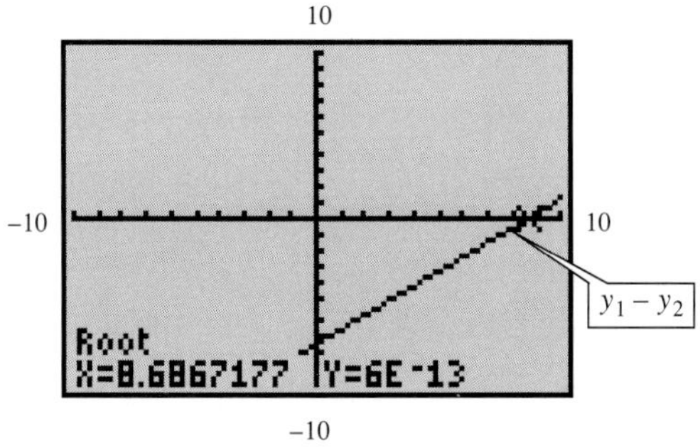

Root
X=8.6867177 Y=6E-13

$y_1 - y_2$

[−10, 10] by [−10, 10]
Xscl = 1   Yscl = 1

FIGURE 74

In Example 4 the endpoint given in the solution set was an approximation, to the nearest hundredth. In solving inequalities graphically, it may happen that due to the approximation process, the appropriate symbol for the endpoint, a parenthesis or a bracket, may not actually be valid. However, we will state the following agreement, to be used throughout this text.

> **AGREEMENT ON INCLUSION OR EXCLUSION OF ENDPOINTS FOR APPROXIMATIONS**
>
> When an approximation is used for an endpoint in specifying an interval, we will continue to use parentheses in specifying inequalities involving $<$ or $>$, and square brackets in specifying inequalities involving $\leq$ or $\geq$.

Three-Part Linear Inequalities

Let us now consider a linear inequality that is actually made up of two separate linear inequalities:

$$-2 < 5 + 3x < 20.$$

The solution set of this three-part inequality consists of all real numbers that make $5 + 3x$ lie in the interval $(-2, 20)$. Therefore, we are seeking the numbers that are solutions of the two inequalities

$$-2 < 5 + 3x \qquad \text{and} \qquad 5 + 3x < 20$$

at the same time. Rather than solve each one separately and then take the intersection of their solution sets, we may find the solution set more quickly by working with all three expressions at the same time, as shown in the following example.

EXAMPLE 5

Solving a Three-Part Inequality Analytically and Supporting the Solution Graphically

Solve the three-part inequality

$$-2 < 5 + 3x < 20$$

analytically. Then support the solution graphically.

SOLUTION   To begin, work with all three expressions at the same time.

$-2 < 5 + 3x < 20$	Given inequality
$-7 < 3x < 15$	Subtract 5 from each expression.
$\dfrac{-7}{3} < x < 5$	Divide each expression by 3.

The open interval $\left(-\frac{7}{3}, 5\right)$ is the solution set of the inequality, as determined by analytic methods.

To support this solution graphically, we graph

$$y_1 = -2$$
$$y_2 = 5 + 3x$$

and

$$y_3 = 20$$

in the viewing window $[-20, 6]$ by $[-20, 25]$. See the two views in Figure 75. The x-values of the points of intersection are $-\frac{7}{3} = -2.\overline{3}$, and 5, confirming that our analytic work is correct. Notice how the slanted line, y_2, lies *between* the graphs of $y_1 = -2$ and $y_3 = 20$ for x-values between $-\frac{7}{3}$ and 5. ∎

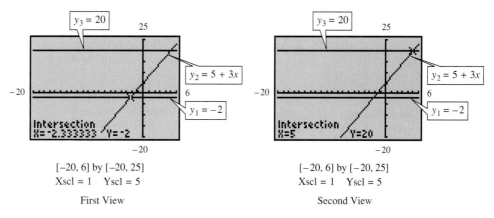

$[-20, 6]$ by $[-20, 25]$
Xscl = 1   Yscl = 5

First View

$[-20, 6]$ by $[-20, 25]$
Xscl = 1   Yscl = 5

Second View

FIGURE 75

EXAMPLE 6
Using Parentheses and Brackets as Appropriate

Use the analytic solution in Example 5 to express the solution set of each three-part inequality using interval notation.

(a) $-2 \leq 5 + 3x < 20$

SOLUTION Because the symbol at the left is $\leq$, we use a square bracket, and because the symbol at the right is $<$, we use a parenthesis. The solution set is the half-open interval $\left[-\frac{7}{3}, 5\right)$.

(b) $-2 \leq 5 + 3x \leq 20$

SOLUTION The solution set is the closed interval $\left[-\frac{7}{3}, 5\right]$. ∎

1.6 EXERCISES

Refer to the graphs of the linear functions $y_1 = f(x)$ and $y_2 = g(x)$ in the figure to find the solution set of each equation or inequality in Exercises 1–12. (Remember that the solution sets consist of x-values.)

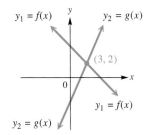

1. $f(x) = g(x)$ **2.** $y_1 = y_2$ **3.** $f(x) > g(x)$

4. $y_1 > y_2$ **5.** $y_1 < y_2$ **6.** $f(x) < g(x)$

7. $g(x) \geq f(x)$ **8.** $y_2 \geq y_1$ **9.** $f(x) - g(x) = 0$

10. $y_1 - y_2 = 0$ **11.** $g(x) - f(x) = 0$ **12.** $y_2 - y_1 = 0$

13. Explain why the solution sets in Exercises 1 and 9 are the same.

14. Explain why the solution sets in Exercises 10 and 12 are the same.

In Exercises 15–18, refer to the graph of the linear function $y = f(x)$ to solve the inequalities specified in parts (a)–(d). Express solution sets in interval notation.

15. (a) $f(x) > 0$ **16. (a)** $f(x) < 0$
     (b) $f(x) < 0$ **(b)** $f(x) \leq 0$
     (c) $f(x) \geq 0$ **(c)** $f(x) \geq 0$
     (d) $f(x) \leq 0$ **(d)** $f(x) > 0$

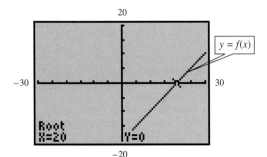

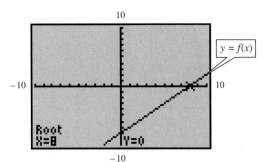

17. (a) $f(x) \le 0$
(b) $f(x) > 0$
(c) $f(x) < 0$
(d) $f(x) \ge 0$

18. (a) $f(x) < 0$
(b) $f(x) > 0$
(c) $f(x) \le 0$
(d) $f(x) \ge 0$

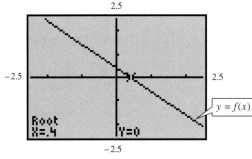

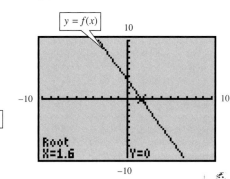

In Exercises 19 and 20, f and g are linear functions.

19. If the solution set of $f(x) \ge g(x)$ is $[4, \infty)$, what is the solution set of each?
(a) $f(x) = g(x)$ (b) $f(x) > g(x)$ (c) $f(x) < g(x)$

20. If the solution set of $f(x) < g(x)$ is $(-\infty, 3)$, what is the solution set of each?
(a) $f(x) = g(x)$ (b) $f(x) \ge g(x)$ (c) $f(x) \le g(x)$

Solve each inequality analytically, giving the solution set using interval notation. Support your answer graphically. (Hint: Once part (a) is done, part (b) follows from the answer to part (a).)

21. (a) $10x + 5 - 7x \ge 8(x + 2) + 4$
(b) $10x + 5 - 7x < 8(x + 2) + 4$

22. (a) $6x + 2 + 10x > -2(2x + 4) + 10$
(b) $6x + 2 + 10x \le -2(2x + 4) + 10$

23. (a) $x + 2(-x + 4) - 3(x + 5) < -4$
(b) $x + 2(-x + 4) - 3(x + 5) \ge -4$

24. (a) $-11x - (6x - 4) + 5 - 3x \le 1$
(b) $-11x - (6x - 4) + 5 - 3x > 1$

25. (a) $\frac{1}{3}x - \frac{1}{5}x \le 2$

(b) $\frac{1}{3}x - \frac{1}{5}x > 2$

26. (a) $\frac{3x}{2} + \frac{4x}{7} \ge -5$

(b) $\frac{3x}{2} + \frac{4x}{7} < -5$

27. (a) $\frac{x - 2}{2} - \frac{x + 6}{3} > -4$

(b) $\frac{x - 2}{2} - \frac{x + 6}{3} \le -4$

28. (a) $\frac{2x + 3}{5} - \frac{3x - 1}{2} < \frac{4x + 7}{2}$

(b) $\frac{2x + 3}{5} - \frac{3x - 1}{2} \ge \frac{4x + 7}{2}$

29. (a) $.6x - 2(.5x + .2) \le .4 - .3x$
(b) $.6x - 2(.5x + .2) > .4 - .3x$

30. (a) $-.9x - (.5 + .1x) \le -.3x - .5$
(b) $-.9x - (.5 + .1x) > -.3x - .5$

31. (a) $-\frac{1}{2}x + .7x > 5$

(b) $-\frac{1}{2}x + .7x \le 5$

32. (a) $\frac{3}{4}x - .2x > 6$

(b) $\frac{3}{4}x - .2x \le 6$

Use the approach of Example 4 to find the solution set of each inequality in Exercises 33–38. Give the solution set using interval notation, with endpoint rounded to the nearest hundredth. Use the x-intercept method.

33. $4(.28x + \sqrt{6}) - \sqrt{2}\,x > 1$

34. $9(-.78x + \sqrt{12}) - \sqrt{7}\,x > -4$

35. $2.3\pi + \sqrt[3]{7} \le .6\pi x - \sqrt{21}$

36. $\dfrac{3}{7}\pi - \sqrt[3]{9} \le \dfrac{2}{7}\pi x - \sqrt{23}$

37. $.29(\sqrt{6} + 5x) < .74(\pi x + 4.1) - 6$

38. $-.28(\sqrt{7} + 3x) < .89(2\pi x + 4.2) + 10$

Refer to the graphs of y_1, y_2, and y_3 in the two views of the figure shown here. Give the solution set for each equation or inequality. (Remember that solution sets consist of x-values.)

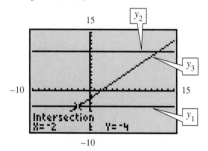

First View   Second View

39. $y_3 < y_2$ **40.** $y_1 < y_3$ **41.** $y_3 \ge y_2$

42. $y_1 \ge y_3$ **43.** $y_1 < y_3 < y_2$ **44.** $y_2 = y_3$

45. $y_2 = y_1$ (The graphs are   **46.** $y_2 > y_1$ **47.** $y_2 \le y_1$
      parallel lines.)

48. Draw a sketch indicating the following: "The functions $y_1 = f(x)$ and $y_2 = g(x)$ are both linear, and the solution set of $f(x) < g(x)$ is $(-2, \infty)$. Furthermore, $f(-2) = g(-2) = 6$."

Solve each of the following three-part inequalities analytically. Support your answer graphically, as explained in Example 5.

49. $4 \le 2x + 2 \le 10$ **50.** $-4 \le 2x - 1 \le 5$

51. $-10 > 3x + 2 > -16$ **52.** $4 > 6x + 5 > -1$

53. $-3 \le \dfrac{x - 4}{-5} < 4$ **54.** $1 < \dfrac{4x - 5}{-2} < 9$

If two linear functions y_1 and y_2 do not have a single point in common, their graphs will either be parallel (Figure A) or will coincide (Figure B). In these cases, the solution set of an inequality of one of the forms

$$y_1 < y_2 \qquad y_1 \le y_2 \qquad y_1 > y_2 \qquad y_1 \ge y_2$$

will either be $\emptyset$ or will consist of all real numbers.

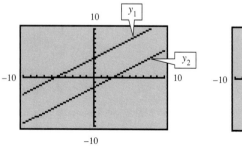

These lines are parallel.   These lines coincide.

        A   **B**

Use Figure A to find the solution set of each inequality in Exercises 55–58.

55. $y_1 > y_2$ **56.** $y_1 \geq y_2$ **57.** $y_1 \leq y_2$ **58.** $y_1 < y_2$

Use Figure B to find the solution set of each inequality in Exercises 59–62.

59. $y_1 > y_2$ **60.** $y_1 \geq y_2$ **61.** $y_1 \leq y_2$ **62.** $y_1 < y_2$

63. Graph the function $f(x) = 3x - 18$ so that a complete graph is shown, and then use the capabilities of your calculator to locate the *x*-intercept.
   (a) What is the *x*-intercept? What is the solution set of $f(x) = 0$?
   (b) Now trace to the right of the *x*-intercept. What do you notice about every *y*-value (in comparison to 0)? What is the solution set of $f(x) > 0$?
   (c) Return to the *x*-intercept, and then trace to the left of it. What do you notice about every *y*-value (in comparison to 0)? What is the solution set of $f(x) < 0$?

64. Sketch the graph of a linear function $y = f(x)$ satisfying these conditions: the *x*-intercept is 3, the graph passes through the point (4, 7). Now, use the graph to solve each of the following.
   (a) $f(x) = 0$ (b) $f(x) \geq 0$ (c) $f(x) \leq 0$
   (d) $f(x) = 12$ (*Hint:* You will have to use the methods of Section 1.4 to find the equation that defines *f*.)

*F*urther Explorations

1. The TABLE feature of a graphics calculator can be used to numerically support solutions for inequalities. For the equation $8x - 15 = -.5x + 2$, the solution set {2} is supported in the TABLE because at $x = 2$, the TABLE shows that $Y_1 = Y_2$ (see Figure 2). The same TABLE also shows that $Y_1 < Y_2$ for all values of *x* when $x < 2$ and $Y_1 > Y_2$ for all values of *x* when $x > 2$.

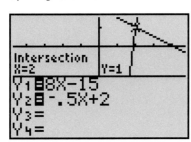

FIGURE 1   **FIGURE 2**

     Then for the inequality, $8x - 15 \geq -.5x + 2$, the solution is all values where $Y_1 > Y_2$ or $Y_1 = Y_2$. For the portion of the TABLE showing in Figure 2, this is true for $x = 2, x = 3, x = 4$, and $x = 5$. And if you scroll (with the down arrow) through the TABLE, it will continue to be true for all values of *x* where $x \geq 2$.
     Solve each of the following inequalities analytically. Support your answer numerically with the TABLE. (Note: If the endpoint of the solution interval is not an integer or a terminating decimal, that point will not be represented in the TABLE.)
   (a) Exercise 21   (b) Exercise 22
   (c) Exercise 23   (d) Exercise 24

2. Some graphics calculators have Boolean Algebra capabilities. This can provide strong visual support for solutions for inequalities. If the symbols $=, \neq, <, \leq, >$, or $\geq$ are used in an expression, the calculator will return a value of 1 if the expression is true and a 0 if the expression is false. Figure 3 demonstrates this feature on the home screen.

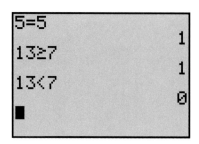

FIGURE 3

When these symbols are used in an expression in the Y= menu, the resulting graph will be a horizontal line at Y = 1 when the expression is true. When the expression is false, the points are plotted at Y = 0, which is hidden by the x-axis. The visual effect of this result is a solution set that is a graph that has the appearance of a number line. The MODE should be changed from CONNECTED to DOT. In Figures 4 and 5, note that Y_1 and Y_2 have been deselected so that only the graph of Y_3 is displayed.

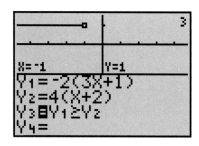

FIGURE 4

The TRACE feature shows that the endpoint is included in the solution set because at X = −1, Y = 1.

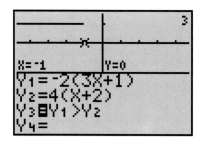

FIGURE 5

TRACE supports the fact that the endpoint of the interval is NOT part of the solution set because at X = −1, Y = 0.

Solve the following inequalities analytically. Support the solutions graphically using the Boolean Algebra feature of your graphics calculator.

(a) $5x + 7 \geq 12$ (b) $-2x + 4 > -3x - 1$

1.7 APPLICATIONS OF LINEAR FUNCTIONS AND MODELS

Formulas ▮ Problem-Solving Strategies ▮ Applications of Linear Equations ▮ Applications of Linear Inequalities ▮ Linear Modeling

Until now, our study of algebra has concentrated on analytic and graphical approaches to solving equations and inequalities derived from linear functions. We will now see how this mathematics can be applied to problem solving.

Formulas

In many applications we use formulas that give a general relationship among several quantities in a problem situation. For example, the formula

$$A = \frac{1}{2}h(b_1 + b_2)$$

gives the area A of a trapezoid in terms of its height (h) and its two parallel bases (b_1 and b_2). (See Figure 76.) Notice that A is alone on one side of the equation. Suppose, however, that we want to have the formula arranged in such a way that it is solved for b_1. The methods of solving linear equations analytically can be adapted so that this goal is accomplished.

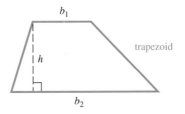

FIGURE 76

EXAMPLE 1

Solving a Formula for a Specified Linear Variable

Solve the formula $A = \frac{1}{2}h(b_1 + b_2)$ for b_1.

SOLUTION   We treat the equation as if b_1 were the only variable and all other variables are constants.

$$A = \frac{1}{2}h(b_1 + b_2) \qquad \text{Given formula}$$

$$2A = h(b_1 + b_2) \qquad \text{Multiply by 2.}$$

$$\frac{2A}{h} = b_1 + b_2 \qquad \text{Multiply by } \frac{1}{h}.$$

$$\frac{2A}{h} - b_2 = b_1 \qquad \text{Subtract } b_2.$$

$$b_1 = \frac{2A}{h} - b_2$$

An alternative method can also be applied after multiplying both sides by 2.

$$2A = h(b_1 + b_2)$$

$$2A = b_1 h + b_2 h \qquad \text{Distributive property; commutative property}$$

$$2A - b_2 h = b_1 h \qquad \text{Subtract } b_2 h.$$

$$\frac{2A - b_2 h}{h} = b_1 \qquad \text{Multiply by } \frac{1}{h}.$$

This is equivalent to the result found by the first method.   ∎

A list of some of the most useful formulas follows.

GEOMETRIC FORMULAS

Figure	Formulas	Examples
Square	Perimeter: $P = 4s$ Area: $A = s^2$	
Rectangle	Perimeter: $P = 2L + 2W$ Area: $A = LW$	
Triangle	Perimeter: $P = a + b + c$ Area: $A = \frac{1}{2}bh$	
Pythagorean Theorem (for Right Triangles)	$c^2 = a^2 + b^2$	
Sum of the Angles of a Triangle	$A + B + C = 180°$	
Circle	Diameter: $d = 2r$ Circumference: $C = 2\pi r = \pi d$ Area: $A = \pi r^2$	
Parallelogram	Area: $A = bh$ Perimeter: $P = 2a + 2b$	

GEOMETRIC FORMULAS

Figure	Formulas	Examples
Trapezoid	Area: $A = \frac{1}{2}h(b_1 + b_2)$ Perimeter: $\quad P = a + b_1 + c + b_2$	
Sphere	Volume: $V = \frac{4}{3}\pi r^3$ Surface area: $S = 4\pi r^2$	
Cone	Volume: $V = \frac{1}{3}\pi r^2 h$ Surface area: $\quad S = \pi r \sqrt{r^2 + h^2}$	
Cube	Volume: $V = e^3$ Surface area: $\quad S = 6e^2$	
Rectangular Solid	Volume: $V = LWH$ Surface area: $\quad S = 2HW + 2LW + 2LH$	
Right Circular Cylinder	Volume: $V = \pi r^2 h$ Surface area: $\quad S = 2\pi rh + 2\pi r^2$	
Right Pyramid	Volume: $V = \frac{1}{3}Bh$ $\quad B =$ area of the base	

> **SIMPLE INTEREST   MOTION   PERCENTAGE**
>
> **Simple Interest**   Interest = Principal × Rate × Time ($I = PRT$)
> **Motion**   Distance = Rate × Time ($D = RT$)
> **Percentage**   Percentage = Base × Rate ($P = BR$)

Problem-Solving Strategies

Probably the most famous study of problem-solving techniques was developed by George Polya (1888–1985), among whose many publications was the modern classic *How to Solve It*. In this book, Polya proposed a four-step process for problem solving.

> **POLYA'S FOUR-STEP PROCESS FOR PROBLEM SOLVING**
>
> 1. **Understand the problem.** You cannot solve a problem if you do not understand what you are asked to find. The problem must be read and analyzed carefully. You will probably need to read it several times. After you have done so, ask yourself "What must I find?"
> 2. **Devise a plan.** There are many ways to attack a problem and decide what plan is appropriate for the particular problem you are solving. (In this text, the plan will usually be to solve an equation or an inequality.)
> 3. **Carry out the plan.** Once you know how to approach the problem, carry out your plan. You may run into "dead ends" and unforeseen roadblocks, but be persistent. If you are able to solve a problem without a struggle, it isn't much of a problem, is it?
> 4. **Look back and check.** Check your answer to see that it is reasonable. Does it satisfy the conditions of the problem? Have you answered all the questions the problem asks? Can you solve the problem a different way and come up with the same answer?

A tool that we have to help us solve problems that George Polya did not is the technology of graphics calculators. We will use calculators whenever possible in our work, but we must remember that they will not *solve* problems for us—*we* must solve the problems by using our own ingenuity and skills, and let our calculators support our results. (Refer to the preface of this book, and read the quote from Peg Crider.)

Applications of Linear Equations

The next example illustrates an application of ratio.

EXAMPLE 2

Determining the Dimensions of a Television Screen

The Panasonic CinemaVision Projection television is one of a new generation of televisions that boasts a 16 : 9 aspect ratio technology. This means that the length of its rectangular screen is $\frac{16}{9}$ times its width. If the perimeter of the screen is 136 inches, find the length and the width of the screen.

SOLUTION   If we let x represent the width of the screen then $\frac{16}{9}x$ can represent the length. See Figure 77. The formula for the perimeter of a rectangle is $P = 2L + 2W$. Let $P = 136$, $L = \frac{16}{9}x$, and $W = x$ in the formula, and solve the equation analytically.

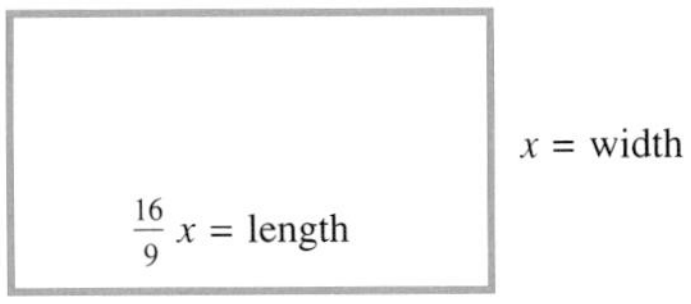

x = width

$\frac{16}{9}x$ = length

FIGURE 77

$$136 = 2\left(\frac{16}{9}x\right) + 2x$$

$$136 = \frac{32}{9}x + 2x$$

$$136 = \frac{50}{9}x \qquad \text{Add like terms.}$$

$$x = 24.48 \qquad \text{Multiply by } \tfrac{9}{50}.$$

Since x represents the width, the width of the screen is 24.48 inches. The length is $\frac{16}{9}(24.48) = 43.52$ inches.

To check our answer analytically, find the sum of the lengths of the sides (which is the meaning of *perimeter*):

$$24.48 + 43.52 + 24.48 + 43.52 = 136 \text{ inches.}$$

To check our answer graphically, we can use the intersection-of-graphs method, with $y_1 = 2(\frac{16}{9})x + 2x$ and $y_2 = 136$. As seen in Figure 78, the point of intersection of the graphs is (24.48, 136). The x-coordinate supports our answer of 24.48 inches for the width.   ▮

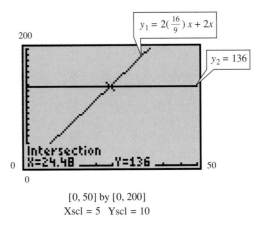

$y_1 = 2(\dfrac{16}{9})x + 2x$

$y_2 = 136$

[0, 50] by [0, 200]
Xscl = 5   Yscl = 10

FIGURE 78

NOTE Our checks provide support only for our graphical and analytic work. If the problem is not set up correctly, a check is worthless.

FOR GROUP DISCUSSION

Observe Figure 78, and remove the graph of $y_2 = 136$. What remains is the graph of a linear function that gives the perimeter y of all rectangles satisfying the 16 by 9 aspect ratio as a *function* of its width x. Trace along the graph, back and forth, and describe what the display at the bottom is actually giving us. Why would nonpositive values of x be meaningless here? If such a television has a perimeter of 117 inches, approximately what would be its width and length?

Television screens are usually advertised by their diagonal measure. The Panasonic model described in Example 2 is advertised as a "50-inch model." How can we further support our answer in Example 2? (*Hint:* Think Greek.)

EXAMPLE 3

Solving a Mixture-of-Concentrations Problem

How much pure alcohol should be added to 20 liters of a mixture that is 40% alcohol to increase the concentration to 50% alcohol?

SOLUTION   Since we are looking for the amount of pure alcohol, let x represent the number of liters of pure alcohol that must be added. The information can now be summarized in a "box diagram" as shown in Figure 79.

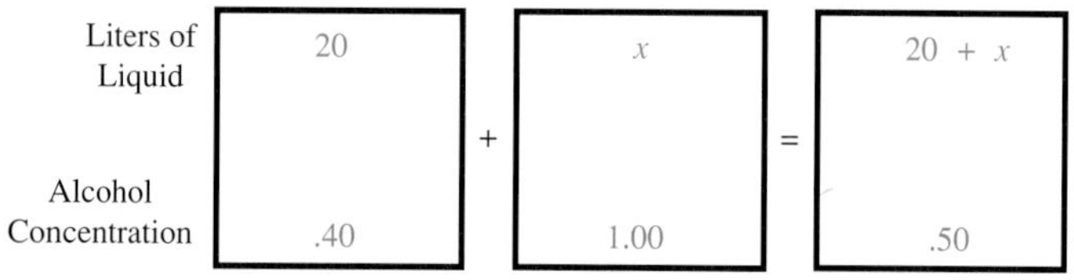

Liters of Liquid	20		x	20 + x
Alcohol Concentration	.40	+	1.00	=	.50

FIGURE 79

In each box we have the number of liters and the alcohol concentration. Using the formula $P = BR$ (Percentage = Base × Rate), we multiply the two items in each box to get the amount of pure alcohol in each case. The amount of pure alcohol on the left must equal the amount of pure alcohol on the right, so the equation to solve is

$$.40(20) \qquad + 1.00x \quad = \quad .50(20 + x).$$

liters of pure alcohol in starting mixture   liters of pure alcohol added   liters of pure alcohol in final mixture

Now solve this equation analytically.

$$40(20) + 100x = 50(20 + x) \qquad \text{Multiply by 100.}$$
$$800 + 100x = 1000 + 50x \qquad \text{Distributive property}$$
$$50x = 200 \qquad \text{Subtract 50x and subtract 800.}$$
$$x = 4 \qquad \text{Divide by 50.}$$

Therefore, 4 liters of pure alcohol must be added. We can check this solution analytically by direct substitution, and/or graphically using either method presented earlier. Figure 80 shows the graph of $y_1 - y_2$, where $y_1 = .40(20) + 1.00x$ and $y_2 = .50(20 + x)$. Using the x-intercept method of graphical solution, our answer of 4 liters is supported.   ◧

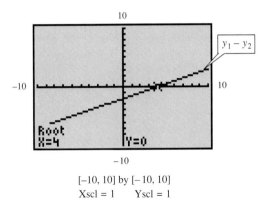

[−10, 10] by [−10, 10]
Xscl = 1   Yscl = 1

FIGURE 80

CAUTION   Just because we have correctly solved the *equation* that we have set up, it does not necessarily mean that we have solved the *problem.* What if our equation is set up incorrectly? This is why we should always check that our answer is reasonable in the context of the problem as stated.

A common application involving linear functions deals with quantities that are in direct proportion (or vary directly). A formal definition of direct proportion follows.

DIRECT PROPORTION

A number y is in direct proportion (or varies directly) to x if there exists a nonzero number k such that

$$y = kx.$$

The number k is called the **constant of variation.**

Notice that the graph of $y = kx$ is simply a straight line with slope k, passing through the origin. See Figure 81.

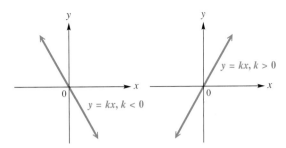

FIGURE 81

If we divide both sides of $y = kx$ by x, we get $\frac{y}{x} = k$, indicating that in a direct proportion, the quotient of the two quantities is constant.

EXAMPLE 4

Solving a Direct Proportion Problem

Hooke's Law for an elastic spring states that the distance (y) a spring stretches is directly proportional to the force (x) applied. If a force of 15 pounds stretches a spring 8 inches, how much will a force of 35 pounds stretch the spring? Give the answer correct to the nearest unit.

SOLUTION   Since there is direct proportion here, we know that $y = kx$ for some number k. Using the fact that when $x = 15$, $y = 8$, we can determine the value of k:

$$y = kx$$
$$8 = k \cdot 15 \qquad \text{Let } x = 15, y = 8.$$
$$k = \frac{8}{15} \qquad \text{Divide by 15.}$$

Therefore, for this spring we have the linear function

$$y = \frac{8}{15}x$$

describing the relationship between the force (x) and the distance stretched (y). To answer the question of the problem, we let $x = 35$ in the equation.

$$y = \frac{8}{15}(35) \qquad \text{Let } x = 35.$$
$$y = \frac{56}{3} = 18.\overline{6},$$

or approximately 19 inches (to the nearest unit).

Using a window of $[0, 95]$ by $[0, 63]$ and locating the point where $x = 35$, we see $y = 18.\overline{6}$, supporting our solution. See Figure 82.   ▯

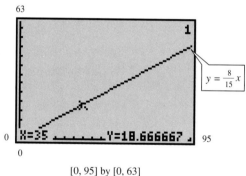

$$\frac{63}{0} \qquad y = \frac{8}{15}x$$

X=35   Y=18.66667   95

$[0, 95]$ by $[0, 63]$
Xscl = 5   Yscl = 5

FIGURE 82

FOR GROUP DISCUSSION

Look at Figure 82 and answer the following questions.

1. What does x represent in this situation?
2. What does y represent?
3. How is the constant of variation related to the graph?
4. If you locate the point where $x = 50$, what is the value of y? How can this result be put into *words* based on the problem in Example 4?
5. If you only had the graph of Figure 82 given to you, how could you find the equation of the line?

By expressing a company's cost of producing a product and the revenue from selling the product as linear functions, the company can determine at what point it will break even. In other words, we try to answer the question, "For what number of items sold will the revenue collected equal the cost of producing those items?"

EXAMPLE 5

Determining the
Break-Even Point

Peripheral Visions, Inc., produces studio-quality audiotapes of live concerts. The company places an ad in a trade newsletter. The cost of the ad is $100. Each tape costs $20 to produce, and the company charges $24 per tape.

(a) Express the cost C as a function of x, the number of tapes produced.

SOLUTION The *fixed cost* is $100, and for each tape produced, the *variable cost* is $20. Therefore, the cost C can be expressed as a function of x, the number of tapes produced:

$$C(x) = 20x + 100 \quad (C \text{ in dollars}).$$

(b) Express the revenue R as a function of x, the number of tapes sold.

SOLUTION Since each tape sells for $24, the revenue R is given by $R(x) = 24x$ (R in dollars).

(c) For what value of x does revenue equal cost?

SOLUTION The company will just break even (no profit and no loss) as long as revenue just equals cost, or $R(x) = C(x)$. This is true whenever

$$R(x) = C(x)$$
$$24x = 20x + 100 \quad \text{Substitute for } R(x) \text{ and } C(x).$$
$$4x = 100$$
$$x = 25.$$

When 25 tapes are sold, the company will break even.

(d) Graph $y_1 = 20x + 100$ and $y_2 = 24x$ in an appropriate window to support the answer in part (c).

SOLUTION Using a window of $[0, 95]$ by $[0, 1200]$, we find the graphs shown in Figure 83. Locating the point where the lines intersect, we find $x = 25$, confirming our solution. The y-value there, 600, indicates that when 25 tapes are sold both the cost and the revenue are $600. Verify analytically that $C(25) = R(25) = 600$. ∎

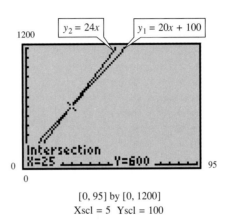

$[0, 95]$ by $[0, 1200]$
Xscl = 5 Yscl = 100

FIGURE 83

FOR GROUP DISCUSSION

Look at Figure 83. How would you describe the situation that exists for the company when $x < 25$? when $x > 25$? Use the fact that

$$\text{profit} = \text{revenue} - \text{cost}$$

to determine the company's profit when $x = 30$. What is its loss when $x = 20$? Support these results *analytically*, using the equations for $R(x)$ and $C(x)$ given in Example 5.

Applications of Linear Inequalities

If we generalize the problem in Example 5, the graphs of a linear function $y_1 = R(x)$ representing revenue taken in and a linear function $y_2 = C(x)$ representing cost might look like what is seen in Figure 84.

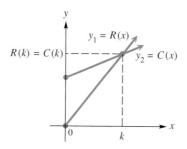

FIGURE 84

Notice that in each case, we have a function of x, where x represents the number of items in question. We choose a minimum domain value of 0, because, in practice, it makes no sense to deal with negative values of items. Now, recalling what we learned about inequalities and their graphical solutions in Section 1.6, we can make the following observations about the representation in the figure.

Observation 1: The break-even point is $(k, R(k))$, or equivalently, $(k, C(k))$. This means that when $C(x) = R(x)$, the company breaks even, selling k items and receiving $R(k)$ monetary units for these items.

Observation 2: The solution set of $R(x) < C(x)$ is $[0, k)$, indicating that when fewer than k items are produced and sold, revenue is less than cost and the company has not yet made a profit. Recall that $R(x)$ is less than $C(x)$ here, because the graph of $y_1 = R(x)$ is *below* that of $y_2 = C(x)$.

Observation 3: The solution set of $R(x) > C(x)$ is (k, ∞), indicating that when more than k items are produced and sold, revenue is greater than cost and the company is making a profit. Here, the graph of $y_1 = R(x)$ is *above* that of $y_2 = C(x)$.

EXAMPLE 6

Determining Intervals of Loss and Profit

Refer to Figure 83 and answer the following questions.

(a) For what numbers of tapes does the company not yet show a profit?

SOLUTION The company does not show a profit when the number of tapes produced and sold is less than 25.

(b) For what numbers of tapes does the company show a profit?

SOLUTION Since the graph of $y_2 = R(x)$ is above the graph of $y_1 = C(x)$ when x is greater than 25, the company is showing a profit if more than 25 tapes are produced and sold. ◼

EXAMPLE 7

Determining the Test Score Needed to Have a Certain Average

Angela Crist has scores of 84 and 92 on her first two tests in Mathematics with the Graphics Calculator. What score does she need on her third test in order to have an average of at least 90?

SOLUTION To solve this problem analytically, we first create a linear function $A(x)$ that will give us her average in terms of her third test grade, x:

$$A(x) = \frac{84 + 92 + x}{3}.$$

(Notice that this can also be written $A(x) = \frac{1}{3}x + \frac{176}{3}$, which is a linear function.)
Since she wants her average to be *at least* 90, we must solve $A(x) \geq 90$, or

$$\frac{84 + 92 + x}{3} \geq 90.$$

Using analytic methods described in Section 1.6, we find the solution set to be $[94, \infty)$. She must score 94 or greater to have an average of at least 90.
To support this result graphically, we graph $y = (84 + 92 + x)/3$ in a viewing window of $[5, 100]$ by $[0, 100]$, and see that if we trace past the point at which $x = 94$, the y-values will be greater than 90. See Figure 85. ◼

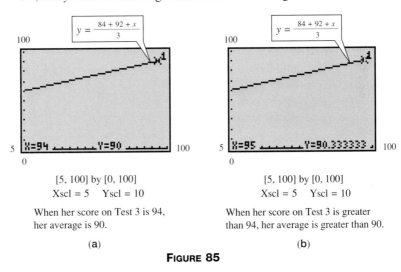

[5, 100] by [0, 100]
Xscl = 5   Yscl = 10

When her score on Test 3 is 94, her average is 90.

(a)

[5, 100] by [0, 100]
Xscl = 5   Yscl = 10

When her score on Test 3 is greater than 94, her average is greater than 90.

(b)

FIGURE 85

Linear Modeling

There are many cases in which two quantities are related by a linear function. If we know two ordered pairs that belong to the function, we can use the method of Example 2 in Section 1.4 to find an equation that describes the function. A classic example of this idea is found in the relationship between the Celsius and Fahrenheit temperature scales. In the following example, we derive the linear function, or linear model, that relates these two temperature scales.

EXAMPLE 8

Determining the Linear Function (Model) that Relates Celsius and Fahrenheit

(a) There is a linear relationship between the Celsius and Fahrenheit temperature scales. When C = 0°, F = 32°, and when C = 100°, F = 212°. Use this information to express F as a function of C.

SOLUTION Think of ordered pairs of temperatures (C, F), where C and F represent corresponding Celsius and Fahrenheit temperatures. The equation that relates the two scales has a straight-line graph that contains the points (0, 32) and (100, 212). The slope of this line can be found by using the slope formula.

$$m = \frac{212 - 32}{100 - 0} = \frac{180}{100} = \frac{9}{5}$$

Now, think of the point-slope form of the equation in terms of C and F, where C replaces x and F replaces y. Use $m = 9/5$, and $(C_1, F_1) = (0, 32)$.

$$F - F_1 = m(C - C_1)$$

$$F - 32 = \frac{9}{5}(C - 0) \qquad F_1 = 32, m = \frac{9}{5}, C_1 = 0$$

$$F - 32 = \frac{9}{5}C$$

$$F = \frac{9}{5}C + 32 \qquad \text{Solve for F.}$$

This final equation expresses F as a function of C.

(b) Graph this function on a graphics calculator using an integer viewing window, and trace along the graph. Interpret the displays obtained through tracing.

SOLUTION Enter the function as $y = \frac{9}{5}x + 32$ (or, equivalently, $y = 1.8x + 32$). Since y represents the Fahrenheit temperature that corresponds to the Celsius temperature x, the various displays shown in Figure 86 are interpreted as follows:

1. When the Celsius temperature is −5°, the Fahrenheit temperature is 23°.
2. When Celsius is 35°, Fahrenheit is 95°.
3. When Celsius is 0°, Fahrenheit is 32°.

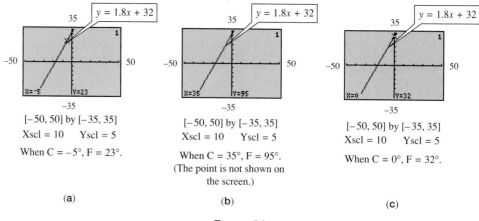

[−50, 50] by [−35, 35]
Xscl = 10   Yscl = 5
When C = −5°, F = 23°.

(a)

[−50, 50] by [−35, 35]
Xscl = 10   Yscl = 5
When C = 35°, F = 95°.
(The point is not shown on the screen.)

(b)

[−50, 50] by [−35, 35]
Xscl = 10   Yscl = 5
When C = 0°, F = 32°.

(c)

FIGURE 86

If you were to locate the point where x = 100, what would be the value of y?

NOTE Solving the equation $F = \frac{9}{5}C + 32$ for C in terms of F gives $C = \frac{5}{9}(F - 32)$, a form of this formula that is often seen.

The formula found in Example 8 is valid for all temperatures involving Celsius and Fahrenheit, because the relationship is an *exact* linear relationship. Very often we encounter relations whose ordered pairs lie in an *approximate* straight line. See Figure 87 for such a relation.

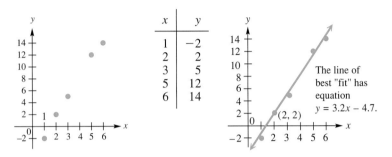

FIGURE 87

FIGURE 88

TECHNOLOGICAL NOTE The statistical capabilities of modern graphics calculators allow us to find regression equations. You may wish to read your owner's manual to see how this is done. Graphics calculators are currently changing the way that elementary statistics courses are taught.

A technique from statistics called **least squares regression** allows us to find the equation of the straight line that provides the best "fit" to the data points. Figure 88 shows this line for the points graphed in Figure 87. We will not go into how the equation of this line is determined, since it is a topic for a statistics course. However, we will in the next example and in several exercises give such an equation for analysis. Incidentally, graphics calculators have the capability of determining such equations.

EXAMPLE 9

Using a Linear Regression Equation to Forecast Future Data

The list below shows the winners of the Indianapolis 500 from 1980 through 1992 and their race speeds (in miles per hour).

1980	Johnny Rutherford	192.256
1981	Bobby Unser	200.546
1982	Gordon Johncock	207.004
1983	Tom Sneva	207.395
1984	Rick Mears	210.029
1985	Danny Sullivan	212.583
1986	Bobby Rahal	216.828
1987	Al Unser	215.390
1988	Rick Mears	219.198
1989	Emerson Fitipaldi	223.885
1990	Airie Luyendyk	225.301
1991	Rick Mears	224.113
1992	Al Unser, Jr.	232.482

A linear regression model for this data, where 1980 corresponds to $x = 0$ and 1992 corresponds to $x = 12$ is $y = 2.503x + 198.729$. Predict the winning race speed for the year 2000.

SOLUTION The predicted speed for the year 2000 is found by letting $x = 20$ in the equation.

$$y = 2.503x + 198.729 \qquad \text{Given equation}$$
$$y = 2.503(20) + 198.729 \qquad \text{Let } x = 20.$$
$$y = 248.789$$

By the year 2000, using this model we can predict the winning speed to be almost 250 miles per hour.

The graph of this equation (or, if you prefer, this prediction of speed as a function of time elapsed) is shown in Figure 89 in a viewing window of [0, 25] by [190, 250]. Notice that the display at the bottom verifies our analytic work. **◖◗**

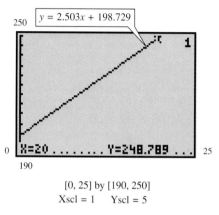

[0, 25] by [190, 250]
Xscl = 1   Yscl = 5

FIGURE 89

FOR **GROUP DISCUSSION**

1. How does the slope of the line in Figure 89, 2.503, relate to the data in the problem situation?
2. Is our prediction a "sure thing"?
3. Why does the function evaluated at $x = 0$ not equal *exactly* Johnny Rutherford's time in 1980?
4. How far off would this model be if we used it to try to determine Danny Sullivan's speed in 1985?
5. Discuss why we cannot predict far into the future with this model.

1.7 EXERCISES

Solve each formula for the specified variable.

1. $I = PRT$ for P (Simple interest)

2. $V = LWH$ for L (Volume of a box)

3. $P = 2L + 2W$ for W (Perimeter of a rectangle)

4. $P = a + b + c$ for c (Perimeter of a triangle)

5. $A = \frac{1}{2}h(b_1 + b_2)$ for h (Area of a trapezoid)

6. $S = 2LW + 2WH + 2HL$ for H (Surface area of a rectangular solid)

7. $S = 2\pi rh + 2\pi r^2$ for h (Surface area of a cylinder)

8. $V = \frac{1}{3}\pi r^2 h$ for h (Volume of a cone)

9. $F = \frac{9}{5}C + 32$ for C (Celsius to Fahrenheit)

10. $s = \frac{1}{2}gt^2$ for g (Distance traveled by a falling object)

Solve each of the following problems analytically, and support your solutions graphically.

11. The length of a rectangular mailing label is 3 centimeters less than twice the width. The perimeter is 54 centimeters. Find the dimensions of the label. (In the figure, x represents the width and so $2x - 3$ represents the length.)

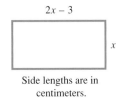

Side lengths are in
centimeters.

12. If the length of a side of a square is increased by 3 centimeters, the perimeter of the new square is 40 centimeters more than twice the length of the side of the original square. Find the dimensions of the original square.

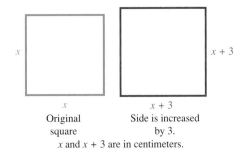

Original   Side is increased
square   by 3.
x and $x + 3$ are in centimeters.

13. The front face of a rectangular aquarium has a ratio of its length to its width of 5 to 3. The perimeter of the rectangle is 64 inches. Find the length and the width of the rectangle.

14. The aspect ratio of conventional television sets is 4:3. One such model Toshiba television has a rectangular viewing screen with a perimeter of 98 inches. What are the length and width of the screen? Since televisions are advertised by the diagonal measure of their screens, how would this set be advertised?

*Beginning in this exercise set, exercises called **Relating Concepts** will often appear. These exercises are designed to be worked so that you can see the connections between skills learned earlier and skills to be mastered in the section at hand. They will also at times "tie together" concepts that were presented earlier with those currently being studied. Section references will often be given if you wish to look back to review.*

Relating Concepts

Solve each equation for x analytically, and support your solution graphically. These are types of equations that are used in many applications. See Section 1.5.

15. $.15x + .30(3) = .20(3 + x)$ **16.** $.06x + .09(14{,}000 - x) = 1005$

Solve each of the following problems analytically, and support your solutions graphically.

17. A pharmacist wishes to strengthen a mixture that is 10% alcohol to one that is 30% alcohol. How much pure alcohol should be added to 7 liters of the 10% solution?

18. Janet Strehl, a chemistry student, needs 10% hydrochloric acid for an experiment. How much 5% acid should she mix with 60 milliliters of 20% acid to get a 10% solution?

Exercises 19 and 20 depend on the idea of the octane rating of gasoline, a measure of its antiknock qualities. In one measure of octane, a standard fuel is made with only two ingredients: heptane and isooctane. For this fuel, the octane rating is the percent of isooctane. An actual gasoline blend is then compared to a standard fuel. For example, a gasoline with an octane rating of 98 has the same antiknock properties as a standard fuel that is 98% isooctane.

19. How many gallons of 94-octane gasoline should be mixed with 400 gallons of 99-octane gasoline to obtain a mixture that is 97-octane?

20. How many gallons of 92-octane and 98-octane gasoline should be mixed together to provide 120 gallons of 96-octane gasoline?

21. How much water should be added to 20 liters of an 18% acid solution to reduce the concentration to 15% acid?

22. An automobile radiator holds 8 liters of fluid. There is currently a mixture in the radiator that is 80% antifreeze and 20% water. How much of this mixture should be drained and replaced by pure antifreeze so that the resulting mixture is 90% antifreeze? (The total amount at the end must be the same as what was there at the beginning.)

R*elating Concepts*

Solve each equation for x analytically, and support your solutions graphically. These will help you in working problems involving ratios and proportions. See Section 1.5.

23. $\dfrac{4}{3}x = 12$

24. $\dfrac{x}{12} = \dfrac{13}{18}$

Solve each of the following problems analytically, and support your solutions graphically.

25. The pressure exerted by a certain liquid at a given point is directly proportional to the depth of the point beneath the surface of the liquid. If the pressure at 30 feet is 15 pounds per square foot, what is the pressure exerted at 70 feet?

26. The rate at which impulses are transmitted along a nerve fiber is directly proportional to the diameter of the fiber. If the rate for a certain fiber is 40 meters per second when the diameter is 6 micrometers, what is the rate if the diameter is 8 micrometers?

27. The height of a vertical object is directly proportional to the length of its shadow assuming that the measure is made with the same angle of elevation of the sun. A certain tree casts a shadow 45 feet long. At the same time, the shadow cast by a vertical stick 2 feet high is 1.75 feet long. How tall is the tree?

28. (See Exercise 27.) A forest fire lookout tower casts a shadow 180 feet long at the same time that the shadow of a 15-foot tree is 9 feet long. What is the height of the lookout tower?

Biologists use direct proportions to estimate the number of individuals of a species in a particular area. They first capture a sample of individuals from the area and mark each specimen with a harmless tag. Then later they return and capture another sample from the same area. They base their estimate on the theory that the number of tagged specimens in the new sample is the same as the proportion of the tagged individuals in the entire area. Use this idea to work Exercises 29 and 30.

29. Biologists tagged 250 fish in City Park Lake on October 12. On a later date they found 7 tagged fish in a sample of 350. Estimate the total number of fish in the lake to the nearest hundred.

30. According to an actual survey in 1961, to estimate the number of seal pups in a certain breeding area in Alaska, 4963 pups were tagged in early August. In late August, a sample of 900 pups was examined and 218 of these were found to have been tagged earlier. Use this information to estimate, to the nearest hundred, the total number of seal pups in this breeding area. *

---

*From "Estimating the Size of Wildlife Populations" by S. Chatterjee in *Statistics by Example,* 1973, obtained from data in *Transactions of the American Fisheries Society,* July, 1968.

R*elating Concepts*

Graph each pair of functions in a window that shows complete graphs. Then,

    (a) solve graphically $y_1 = y_2$
    (b) solve graphically $y_1 > y_2$
    (c) solve graphically $y_1 < y_2$.

See Sections 1.5 and 1.6.

31. $y_1 = 4x$
    $y_2 = -2x + 30$

32. $y_1 = 1.5x$
    $y_2 = 4x - 50$

The graph of $y = f(x)$ is shown. Give the solution set of

    (a) $f(x) = 0$
    (b) $f(x) < 0$
    (c) $f(x) > 0$.

33.

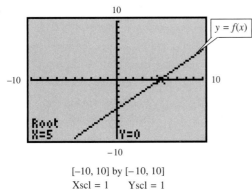

[−10, 10] by [−10, 10]
Xscl = 1   Yscl = 1

34.

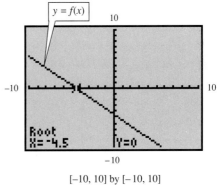

[−10, 10] by [−10, 10]
Xscl = 1   Yscl = 1

In each of Exercises 35–38,

    (a) Express the cost C as a function of x, where x represents the number of items as described.
    (b) Express the revenue R as a function of x.
    (c) Determine analytically the value of x for which revenue equals cost.
    (d) Graph $y_1 = C(x)$ and $y_2 = R(x)$ in the same viewing window and interpret the graphs.

35. Susan Danielson stuffs envelopes for extra income during her spare time. Her initial cost to obtain the necessary information for the job was $200.00. Each envelope costs $.02 and she gets paid $.04 per envelope stuffed. Let x represent the number of envelopes stuffed.

36. Walter Michaelis runs a copying service in his home. He paid $3,500 for the copier and a lifetime service contract. Each sheet of paper he uses costs $.01, and he gets paid $.05 per copy he makes. Let x represent the number of copies he makes.

37. Michael Shafferkötter operates a delivery service in a southern city. His start-up costs amounted to $2,300. He estimates that it costs him (in terms of gasoline, wear and tear on his car, etc.) $3.00 per delivery. He charges $5.50 per delivery. Let x represent the number of deliveries he makes.

38. Annie Boyle bakes cakes and sells them at county fairs. Her initial cost for the Washington Parish fair in 1992 was $40.00. She figures that each cake costs $2.50 to make, and she charges $6.50 per cake. Let x represent the number of cakes sold. (Assume that there were no cakes left over.)

39. Gerry Vidrine has grades of 88, 86, and 92 on her first three algebra tests. What score does she need on her fourth test to have an average of at least 90?

40. (See Exercise 39.) What range of scores on her fourth test would assure Gerry of having an average between and inclusive of 80 and 90?

R*elating Concepts*

Find the equation of the line described. Express it in the form $y = mx + b$. See Section 1.4.

41. the line joining the points $(-3, 6)$ and $(5, 12)$

42. the line with the two views pictured

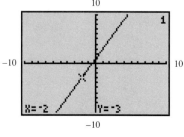

First View

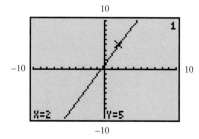

Second View

In Exercises 43–48, assume that a linear relationship exists between the two quantities.

43. A company finds that it can produce 10 solar heaters for $7,500, while producing 20 heaters costs $13,900.
  (a) Express the cost, y, as a function of the number of heaters, x.
  (b) Determine analytically the cost to produce 25 heaters.
  (c) Support the result of part (b) graphically.

44. At 68° Fahrenheit, a certain species of cricket chirps 124 times per minute. At 40°F, the same cricket chirps 86 times per minute. *
  (a) Express the number of chirps, y, as a function of the Fahrenheit temperature.
  (b) If the temperature is 60°F, how many times will the cricket chirp per minute? Determine your answer analytically, and support it graphically.
  (c) If you count the number of cricket chirps in one-half minute and hear 40 chirps, what is the temperature? Determine your answer analytically, and support it graphically.

45. In 1986, a house was purchased for $120,000. In 1996, it was appraised for $146,000.
  (a) If $x = 0$ represents 1986 and $x = 10$ represents 1996, express the appraised value of the house, y, as a function of the number of years, x, after 1986.
  (b) What will the house be worth in the year 2000? Determine your answer analytically and support it graphically.
  (c) What does the slope of the line represent (in your own words)?

46. A photocopying machine sold for $3000 in 1988 when it was purchased. Its value in 1996 had decreased to $600.
  (a) If $x = 0$ represents 1988 and $x = 8$ represents 1996, express the value of the machine, y, as a function of the number of years from 1988.
  (b) Graph the function from part (a) in a window $[0, 10]$ by $[0, 4000]$. How would you interpret the y-intercept in terms of this particular problem situation?
  (c) Use your calculator to determine the value of the machine in 1992, and verify this analytically.

47. Suppose a baseball is thrown at 85 mph. The ball will travel 320 ft when hit by a bat swung at 50 mph and will travel 440 ft when hit by a bat swung at 80 mph. Let y be the number of feet traveled by the ball when hit by a bat swung at x mph. Find the equation of the line. (*Note:* This function is valid for $50 \le x \le 90$, where the bat is 35 inches long, weighs 32 oz, and is swung slightly upward to drive the ball at an angle of 35°.**) How much farther will a ball travel for each one-mile-per-hour increase in the speed of the bat?

48. The amount of tropical rain forests in Central America decreased from 130,000 sq mi in 1969 to about 80,000 sq mi in 1985.† Let y be the amount (in ten-thousands of square miles) x years after 1965. Find the equation of the line. How large will the rain forests be in the year 1997?

** Robert K. Adair, *The Physics of Baseball.* New York; HarperCollins Publishers, 1990.

† From *Mathematics and Global Survival,* Second Edition, by Richard H. Schwartz. Copyright © 1991 by Ginn Press. Reprinted by permission.

In Massachusetts, speeding fines are determined by the linear function

$$y = 10(x - 65) + 50, \qquad x \geq 65,$$

where y is the cost in dollars of the fine if a person is caught driving x miles per hour. Use this information to work the problems in Exercises 49–52.

49. José had to make an 8:00 A.M. final examination, but overslept after a big weekend in Boston. Radar clocked his speed at 76 mph. How much was his fine?

50. While balancing his checkbook, Johnny ran across a cancelled check that his wife Gwen had written to the Department of Motor Vehicles for a speeding fine. The check was written for $100. How fast was Gwen driving?

51. Based on the formula above, at what speed do the troopers start giving tickets?

52. For what range of speeds is the fine greater than $200. Solve analytically and support graphically.

In Exercises 53–60, linear models are based on least squares regression techniques from statistics. Solve each problem analytically and support your solutions graphically.

53. The Measurements Standard Committee of the Pattern Fashion Industry provides a table of body measurements (in inches) corresponding to misses' sizes. For misses' sizes 6 through 20, bust measurement y corresponds to misses size x according to the model $y = .842x + 24.613$. If the size is 14, what is the corresponding bust size (rounded down to the nearest unit)? *

54. If alcohol "burned up" by the body since the time of the first drink is disregarded, the number of drinks (1 drink = 12 oz. beer, 4 oz. wine, 1 oz. hard liquor), $x,$ and the blood alcohol level, $y,$ of a 240-pound person are related by the function $y = .0156x$. What would be the blood alcohol level if this person had 4 drinks? If .100 blood alcohol level is considered legally drunk, after what whole number of drinks would this person be legally drunk?

55. According to information provided by Families USA Foundation, the national average family health care cost (in dollars) between 1980 and 2000 (projected) can be approximated by the linear function $y = 382.75x + 1742$, where $x = 0$ corresponds to 1980 and $x = 20$ corresponds to 2000. Based on this model, what would be the expected national average health care cost in 1996?

56. In what year will (or did) the national average family health care cost first exceed $7,000? (See Exercise 55.)

57. The percentage y of off-track-betting as a portion of all bets on horse racing between 1982 and 1992 can be estimated by the linear model $y = 1.8x + 15$, where $x = 0$ corresponds to 1982 and $x = 10$ corresponds to 1992. If this trend continues, what will the percentage of off-track-betting as a portion of all bets on horse racing be in 1998? (*Source:* Christiansen/Cummings Associates, Inc.)

58. For speeds between 10 and 60 mph, the stopping distance y in ft for a person going x mph is estimated by the linear model $y = 4.5x - 46.7$. What would the stopping distance be at 40 mph? *

59. From 1940 to 1960, the population y of California was approximated by the linear model $y = .4405x + 6.665$, where y is in millions and $x = 0$ corresponds to 1940 and $x = 20$ corresponds to 1960. If this trend had continued until 1965, what would have been the population then? *

60. It has been reported that the total length x and the tail length y of females of the snake species *Lampropeltis polyzona* are nearly linearly related by the model $y = .134x - 1.18$, where x is the length of the snake in millimeters. If a snake of this species measures 1000 millimeters, what is its tail length to the nearest millimeter? *

61. In 1787, Jacques Charles noticed that gases expand when heated and contract when cooled. Suppose that a particular gas follows the model $y = (\frac{5}{3})x + 455$ where x is the temperature in Celsius and y is the volume in cubic centimeters. *
  (a) What is the volume when the temperature is 27° Celsius?
  (b) What is the temperature when the volume is 605 cubic centimeters?
  (c) Determine what temperature gives a volume of 0 cubic centimeters (in other words, absolute zero, or the coldest possible temperature).

---

* Problems 44, 53, and 58–62 are adapted from *A Sourcebook of Applications of School Mathematics* by Donald Bushaw et al. Copyright © 1980 by The Mathematical Association of America. Reprinted by permission.

62. According to information in the Edmund Scientific Catalog #761, the price y in dollars for a group of 72 tubes of x milliliters is approximated by the linear function $y = .233x + 6.64$, for capacities between 4 milliliters and 36 milliliters. What would be the price for a group of 10-milliliter test tubes based on this model, rounded to the nearest 25 cents? *

Further Explorations

1. Employees in Massachusetts who determine fines for speeders most likely use a table (rather than the equation) with the fines listed for each mile above the speed limit. If your graphics calculator has a TABLE feature, solve Exercises 49–52 numerically using the TABLE.

2. Odds-n-Ends Clothing uses a 40% markup over the wholesale price to determine the retail price of their merchandise. Build a TABLE with your graphics calculator to find the markup and the retail price of products that wholesale for $25, $35, $45, and $55. Use $Y_1 = .4x$ for the markup and $Y_2 = 1.4x$ for the retail price.

Chapter 1   SUMMARY

Ordered pairs of real numbers are plotted in a rectangular coordinate system, and this can be done either by hand or with graphics calculator technology. The window used in graphing a set of ordered pairs with a graphics calculator can drastically alter the picture we view. Ordered pairs of real numbers are called relations, with the set of all first components forming the domain of the relation, and the set of all second components forming the range. A function is a special kind of relation that pairs with each element in its domain one and only one element in its range. One of the simplest, yet most important, kinds of functions is the linear function. Its graph is a straight line, and its equation can always be written in the form $f(x) = ax + b$. The coefficient of x, represented here by a but often represented by m, is the slope of the line, and b represents the y-intercept.

Linear equations are solved analytically by using the properties that allow us to transform equations into simpler equations that have the same solution set. The solution, or root, of a linear equation of the form $f(x) = 0$ is represented by the x-intercept of the associated linear function f. Alternatively, the solution of a linear equation of the form $f(x) = g(x)$ is represented by the x-coordinate of the point of intersection of the graphs of f and g. Linear inequalities are solved analytically using a method similar to solving linear equations. The solution set of the linear inequality $f(x) > 0$ consists of the domain values of the points on the graph of f that lie above the x-axis, while that of $f(x) < 0$ consists of the domain values of the points on the graph of f that lie below the x-axis. Alternatively, the solution set of $f(x) > g(x)$ consists of all domain values common to both f and g for which the graph of f lies above the graph of g. A similar statement can be made for $f(x) < g(x)$, with the word *above* replaced with *below*.

Many important applications lead to linear models, and by applying the concepts of this chapter, we can solve such applications analytically and support our solutions graphically.

---

*Adapted from *A Sourcebook of Applications of School Mathematics* by Donald Bushaw et al. Copyright © 1980 by The Mathematical Association of America. Reprinted by permission.

Key Terms

SECTION 1.1

natural numbers
whole numbers
number line
rational numbers
irrational numbers
real numbers
square root
cube root
coordinate of a point (on a number line)
coordinate system
a is less than (greater than) b
rectangular (cartesian) coordinate system
origin
x-axis
y-axis
coordinate (x-y) plane
quadrants
coordinates of a point (in the plane)
viewing window
standard viewing window
scale

SECTION 1.2

set-builder notation
interval notation
infinity symbol (∞)
compound inequality
relation
domain
range
function
vertical line test
function notation [$f(x)$ notation]

SECTION 1.3

term or monomial
numerical coefficient or coefficient
degree
constant
binomial
solution of an equation
linear function
square viewing window
y-intercept of a line
x-intercept of a line

complete graph of a line
constant function
tracing a graph
slope of a line
slope formula
vertical line
horizontal line

SECTION 1.4

point-slope form of the equation of a line
slope-intercept form of the equation of a line
standard form of the equation of a line
Pythagorean theorem
distance formula
midpoint formula
parallel lines
perpendicular lines

SECTION 1.5

analytic approach
graphical approach
solution set
visual support
linear equation
equivalent equations
addition and multiplication properties of equality
conditional equation
intersection-of-graphs method of graphical solution
x-intercept method of graphical solution
identity
contradiction

SECTION 1.6

inequality
equivalent inequalities
properties of inequality
linear inequality
formula

SECTION 1.7

Polya's problem-solving strategies
direct proportion
constant of variation
linear modeling
least squares regression

Chapter 1   **REVIEW EXERCISES**

Let A represent the point with coordinates $(-1, 16)$ and let B represent the point with coordinates $(5, -8)$.

1. Find the exact distance between points A and B.

2. Find the coordinates of the midpoint of the line segment joining points A and B.

3. Find the slope of the line AB.

4. Find the equation of the line passing through points A and B. Write it in $y = mx + b$ form. (*Hint:* To check your answer, enter the equation into your calculator and use the capabilities of your calculator to verify that the line does indeed contain the points $(-1, 16)$ and $(5, -8)$.)

Consider the line with equation $3x + 4y = 144$ in Exercises 5–8.

5. What is the slope of this line?

6. What is the x-intercept of this line?

7. What is the y-intercept of this line?

8. Give a viewing window that will show a complete graph. (There are many possible such windows.)

9. Suppose that f is a linear function such that $f(3) = 6$ and $f(-2) = 1$. Find $f(8)$.

10. Find the equation of the line perpendicular to the graph of $y = -4x + 3$, passing through the point $(-2, 4)$. Give it in $y = mx + b$ form. (*Hint:* To check your answer, use a square viewing window and graph the given equation, $y = -4x + 3$, and the equation you found. Then verify that the equation you found does indeed contain $(-2, 4)$. The lines should also appear perpendicular.)

Choose the letter of the graph that would most closely resemble the graph of $f(x) = mx + b$, given the conditions on m and b.

11. $m < 0, b < 0$ 14. $m > 0, b > 0$

12. $m > 0, b < 0$ 15. $m = 0$

13. $m < 0, b > 0$ 16. $b = 0$

A. 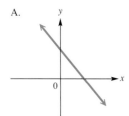   B. 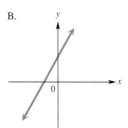   C.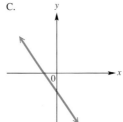

D. 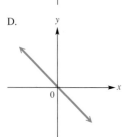   E. 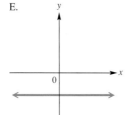   F.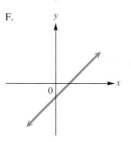

Refer to the graphs of the linear functions $y_1 = f(x)$ and $y_2 = g(x)$ in the figure below to match the solution set in the columns on the right with the equation or inequality on the left. Choices may be used once, more than once, or not at all.

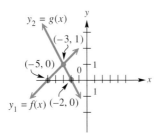

17. $f(x) = g(x)$

18. $f(x) > g(x)$

19. $f(x) < g(x)$

20. $g(x) \geq f(x)$

21. $y_2 - y_1 = 0$

22. $f(x) < 0$

23. $g(x) > 0$

24. $y_2 - y_1 < 0$

A. $(-\infty, -3]$

B. $(-\infty, -3)$

C. $\{3\}$

D. $\{2\}$

E. $\{(3, 2)\}$

F. $\{-5\}$

G. $\{-2\}$

H. $\{0\}$

I. $\{-3\}$

J. $[-3, \infty)$

K. $(-3, \infty)$

L. $[-3]$

M. $(-\infty, -5)$

N. $(-5, \infty)$

O. $(-\infty, -2)$

P. $(-2, \infty)$

Solve each equation using analytic methods.

25. $5[3 + 2(x - 6)] = 3x + 1$

26. $\dfrac{x}{4} - \dfrac{x + 4}{3} = -2$

27. Solve the inequality $-6 \leq \dfrac{4 - 3x}{7} < 2$ analytically.

Exercises 28–30 refer to the linear function
$$f(x) = 5\pi x + (\sqrt{3})x - 6.24(x - 8.1) + (\sqrt[3]{9})x.$$

28. Solve the equation $f(x) = 0$ using graphical methods. Give the solution to the nearest hundredth. Then give an explanation of how you went about solving this equation graphically.

29. Refer to the graph, and give the solution set of $f(x) < 0$.

30. Refer to the graph, and give the solution set of $f(x) \geq 0$.

31. What is the solution set of $f(x) > 0$, based on the screen shown?

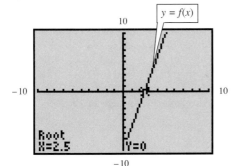

32. What is the solution set of $f(x) - g(x) = 0$, based on the screen shown?

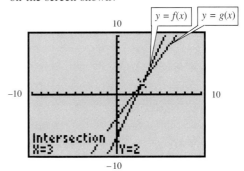

33. What is the domain of the function $f(x) = 3x - 6$?

34. What is the range of the function $g(x) = -6$?

35. If f is a linear function and the solution set of $f(x) \geq 0$ is $(-\infty, -3]$, what is the solution set of $f(x) < 0$?

36. What are the domain and the range of the relation graphed on the screen? (*Hint:* Pay attention to scale.)

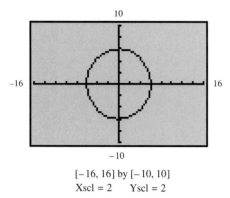

[−16, 16] by [−10, 10]
Xscl = 2   Yscl = 2

37. True or False? The graphs of $y_1 = 5.001x - 3$ and $y_2 = 5x + 6$ are shown on the accompanying screen. From this view, we may correctly conclude that these lines are parallel.

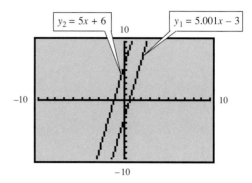

A company produces studio quality audiocassettes of live concerts. The company places an ad in a trade newsletter. The cost of the ad is $150. Each tape costs $30 to produce, and the company charges $37.50 per tape. Work Exercises 38–41.

38. Express the company's cost C as a function of x, where x is the number of tapes produced and sold.

39. Assuming that the company sells x tapes, express the revenue as a function of x.

40. Determine analytically the value of x for which revenue equals cost.

41. The graph shows $y = C(x)$ and $y = R(x)$. Use the graph to discuss how it illustrates when the company is losing money, when it is breaking even, and when it is making a profit.

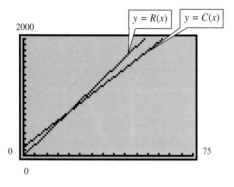

42. How much pure alcohol must be added to 40 liters of a mixture that is 20% alcohol to increase the concentration to 30% alcohol?

43. The equation $y = -3.52x + 58.6$ gives an approximation for temperature in degrees Fahrenheit above the surface of the earth, where x is in thousands of feet and y is the temperature.
  (a) When the height is 5000 feet, what is the temperature? Solve analytically.
  (b) When the temperature is $-15°F$, what is the height? Solve analytically.
  (c) Explain how the answers in parts (a) and (b) can be supported graphically.
        5.5 trillion?*

In Exercises 44–46, assume that a linear relationship exists.

44. The average size of a farm in the United States increased from 100 acres in 1920 to 700 acres in 1980. Let y be the average size x years after 1900. In what year was the average size 400 acres?*

45. The worldwide consumption of cigarettes increased from 2.5 trillion in 1960 to 4 trillion in 1980. Let y be the consumption of cigarettes (in trillions) x years

after 1940. In what year will the consumption reach 5.5 trillion?*

46. The number of farms in the United States declined from 6 million in 1920 to 2 million in 1980. Let y be the number of farms (in millions) x years after 1900. How many farms were there in 1960?*

* From *Mathematics and Global Survival*, Second Edition, by Richard H. Schwartz. Copyright © 1991 by Ginn Press. Reprinted with permission.

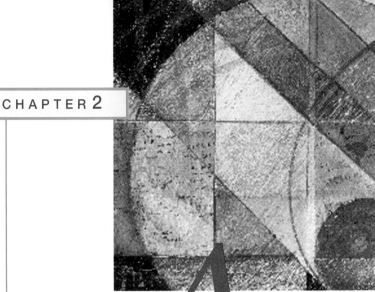

Analysis of Graphs of Functions

2.1 GRAPHS OF ELEMENTARY FUNCTIONS AND RELATIONS

Continuity; Increasing and Decreasing Functions ∎ The Identity
Function ∎ The Squaring Function and Symmetry with Respect to the
y-Axis ∎ The Cubing Function and Symmetry with Respect to the
Origin ∎ The Square Root and Cube Root Functions ∎ The Absolute Value
Function ∎ The Relation $x = y^2$ and Symmetry with Respect to the
x-Axis ∎ Even and Odd Functions

Continuity; Increasing and Decreasing Functions

In Chapter 1, our work dealt mainly with linear functions. The graph of a linear
function, a straight line, may be drawn by hand over any interval of its domain
without picking the pencil up from the paper. In mathematics we say that a
function with this property is **continuous** over any interval. The formal definition
of continuity requires concepts from calculus, but we can give an informal
definition at the college algebra level.

> **INFORMAL DEFINITION OF CONTINUITY**
> A function is continuous over an interval of its domain if its hand-drawn graph
> over that interval can be sketched without lifting the pencil from the paper.

If a function is not continuous at a point, then it may have a point of discontinuity (Figure 1(a)), or it may have a vertical *asymptote* (a vertical line which the graph does not intersect, as in Figure 1(b)). More will be said about asymptotes in Chapter 4.

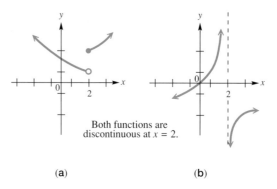

Both functions are
discontinuous at $x = 2$.

(a)   (b)

FIGURE 1

Notice that both graphs in Figure 1 are graphs of functions, since they pass the vertical line test.

> **EXAMPLE 1**
>
> Determining
> Intervals of
> Continuity

The figures show graphs of functions and the descriptions indicate the intervals of the domain over which they are continuous.

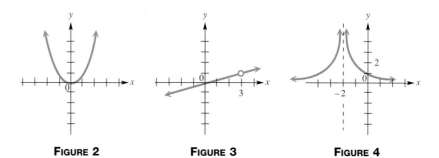

FIGURE 2   **FIGURE 3**   **FIGURE 4**

(a)   The function in Figure 2 is continuous over the entire domain of real numbers, $(-\infty, \infty)$.

(b)   The function in Figure 3 has a point of discontinuity at $x = 3$. It is continuous over the interval $(-\infty, 3)$ and the interval $(3, \infty)$.

(c)   The function in Figure 4 has a vertical asymptote at $x = -2$, as indicated by the dashed line. It is continuous over the interval $(-\infty, -2)$ and the interval $(-2, \infty)$.   ∎

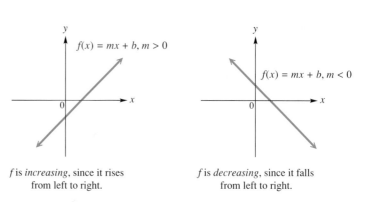

f is *increasing*, since it rises f is *decreasing*, since it falls
from left to right.   from left to right.

(a)   (b)

FIGURE 5

If a function is not constant over an interval, then its graph will either rise from left to right or will fall from left to right. We use the words *increasing* and *decreasing* to describe this behavior. For example, a linear function with a positive slope is increasing over its entire domain, while one with a negative slope is decreasing. See Figure 5.

Informally speaking, a function **increases** on an interval of its domain if its graph rises from left to right. It **decreases** on an interval if its graph falls from left to right. It is **constant** on an interval if its graph is horizontal on the interval.

The formal definitions of these concepts follow.

INCREASING, DECREASING, AND CONSTANT FUNCTIONS

Suppose that a function f is defined over an interval I.

a. f increases on I if, whenever $x_1 < x_2$, $f(x_1) < f(x_2)$;
b. f decreases on I if, whenever $x_1 < x_2$, $f(x_1) > f(x_2)$;
c. f is constant on I if, for every x_1 and x_2, $f(x_1) = f(x_2)$.

Figure 6 illustrates these ideas.

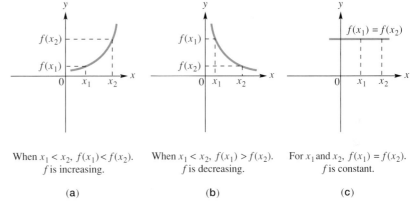

When $x_1 < x_2$, $f(x_1) < f(x_2)$.   When $x_1 < x_2$, $f(x_1) > f(x_2)$.   For x_1 and x_2, $f(x_1) = f(x_2)$.
　　　　f is increasing.　　　　　　　　　　f is decreasing.　　　　　　　　　　f is constant.

　　　　　　(a)　　　　　　　　　　　　　　(b)　　　　　　　　　　　　　　(c)

FIGURE 6

EXAMPLE 2

Determining Intervals Over Which a Function is Increasing, Decreasing, or Constant

Figure 7 shows the graph of a function. Determine the intervals over which the function is increasing, decreasing, or constant.

SOLUTION   In making our determination, we must always ask "What is the y-value doing as x is getting larger?" For this graph, we see that on the interval $(-\infty, 1)$, the y-values are *decreasing;* on the interval $(1, 3)$, the y-values are *increasing;* and on the interval $(3, \infty)$, the y-values are *constant* (all are 6). Therefore, the function is

　　　　decreasing on $(-\infty, 1)$,

　　　　increasing on $(1, 3)$,

　　　　constant on $(3, \infty)$.   ▯

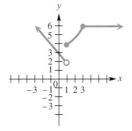

FIGURE 7

CAUTION   A common error involves writing range values when determining intervals like those in Example 2. Remember that we are determining intervals of the domain, and thus are interested in *x*-values for our interval designations.

FOR GROUP DISCUSSION

1. In the standard viewing window of your calculator, enter any linear function $y = mx + b$ with $m > 0$. Now trace the graph from *left to right*. Watch the *y*-values as *x* gets larger. What is happening to *y*? How does this reinforce the concepts presented so far in this section?
2. Repeat this exercise, but with $m < 0$.
3. Repeat this exercise, but with $m = 0$.

The next part of this section is devoted to the introduction of several important basic functions and relations that will be investigated in detail as we progress through the algebra chapters of this book.

The Identity Function

If we let $m = 1$ and $b = 0$ for the general form of the linear function $f(x) = mx + b$, we get the **identity function** $f(x) = x$. This function pairs every real number with itself.

IDENTITY FUNCTION

$f(x) = x$ (Figure 8)

Domain: $(-\infty, \infty)$

Range: $(-\infty, \infty)$

The identity function $f(x) = x$ increases on its entire domain $(-\infty, \infty)$, and is continuous on its entire domain.

x	x
0	0
1	1
−1	−1
2	2
−2	−2

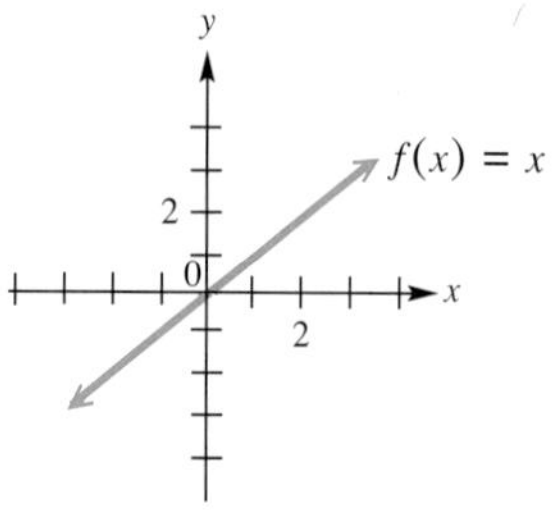

(a)

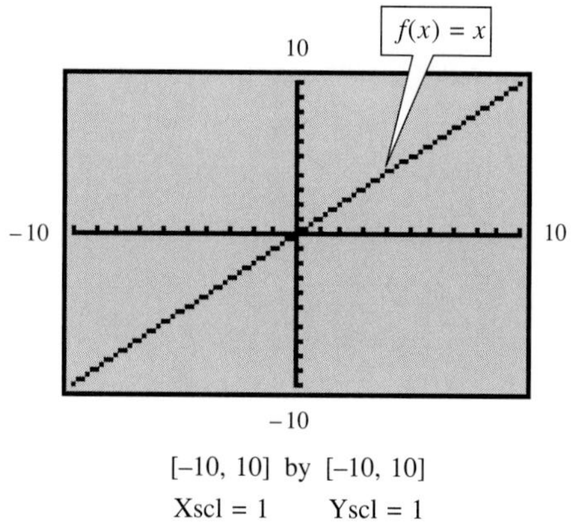

[−10, 10] by [−10, 10]
Xscl = 1   Yscl = 1

(b)

FIGURE 8

FOR **GROUP DISCUSSION**

Choose *any* viewing window on your graphics calculator and graph $y = x$. Trace, and compare the x and y values as you trace. What do you notice? Why do you think $f(x) = x$ is called the identity function?

The Squaring Function and Symmetry with Respect to the *y*-Axis

We now look at the graph of the simplest degree 2 function, the **squaring function** $f(x) = x^2$. (The word *quadratic* refers to degree 2; we will investigate the general quadratic function in Chapter 3.) This function pairs every real number with its square. Its graph is called a **parabola.**

SQUARING FUNCTION

$f(x) = x^2$ (Figure 9)

Domain: $(-\infty, \infty)$

Range: $[0, \infty)$

The squaring function $f(x) = x^2$ decreases on the interval $(-\infty, 0)$ and increases on the interval $(0, \infty)$. It is continuous on its entire domain.

x	x^2
0	0
1	1
-1	1
2	4
-2	4

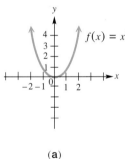

(a)

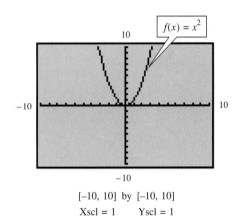

[–10, 10] by [–10, 10]
Xscl = 1   Yscl = 1

(b)

FIGURE 9

The point at which the graph changes from decreasing to increasing (the point $(0, 0)$) is called the **vertex** of the parabola.

Notice that if we were able to "fold" the graph of $f(x) = x^2$ along the y-axis, the two halves would coincide exactly. In mathematics we refer to this property as symmetry, and we say that the graph of $f(x) = x^2$ is **symmetric with respect to the *y*-axis.** This may be generalized as follows.

> **SYMMETRY WITH RESPECT TO THE *y*-AXIS**
>
> If a function f is defined so that
>
> $$f(x) = f(-x)$$
>
> for all x in its domain, then the graph of f is symmetric with respect to the *y*-axis.

Some *particular* cases illustrating that the graph of $f(x) = x^2$ is symmetric with respect to the *y*-axis are as follows:

$$f(-4) = f(4) = 16$$
$$f(-3) = f(3) = 9$$
$$f(-2) = f(2) = 4$$
$$f(-1) = f(1) = 1$$
$$f(-0) = f(0) = 0.$$

This pattern holds for any real number x, since $f(-x) = (-x)^2 = x^2 = f(x)$.

The Cubing Function and Symmetry with Respect to the Origin

The function $f(x) = x^3$ is the simplest degree 3 function, and it is an example of a *cubic* function. (Such functions will be examined more closely in Chapter 3.) It pairs with each real number the third power, or cube, of the number.

> **CUBING FUNCTION**
>
> $f(x) = x^3$ (Figure 10)
>
> Domain: $(-\infty, \infty)$
>
> Range: $(-\infty, \infty)$
>
> The cubing function $f(x) = x^3$ increases on its entire domain $(-\infty, \infty)$. It is also continuous on its entire domain $(-\infty, \infty)$.
>
> | x | x^3 |
> |-----|-------|
> | 0 | 0 |
> | 1 | 1 |
> | −1 | −1 |
> | 2 | 8 |
> | −2 | −8 |
>
>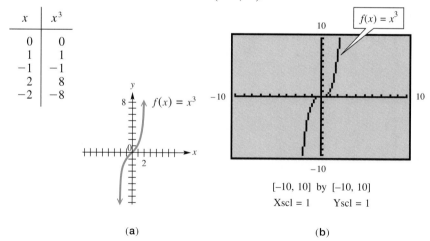
>
> (a)   (b)
>
> **FIGURE 10**
>
> The point at which the graph changes from "opening downward" to "opening upward" (the point $(0, 0)$) is called an **inflection point.**

Notice that if we were able to "fold" the graph of $f(x) = x^3$ along the y-axis and then along the x-axis, forming a "corner" at the origin, the two parts of the graph would coincide exactly. We say that the graph of $f(x) = x^3$ is **symmetric with respect to the origin.** This may be generalized as follows.

SYMMETRY WITH RESPECT TO THE ORIGIN

If a function f is defined so that
$$f(-x) = -f(x)$$
for all x in its domain, then the graph of f is symmetric with respect to the origin.

Some *particular* cases illustrating that the graph of $f(x) = x^3$ is symmetric with respect to the origin are as follows:

$$f(-2) = -f(2) = -8$$
$$f(-1) = -f(1) = -1$$
$$f(-0) = -f(0) = -0 = 0.$$

This pattern holds for any real number x, since

$$f(-x) = (-x)^3 = (-1)^3 x^3 = -x^3 = -f(x).$$

(a) Show analytically and support graphically the fact that $f(x) = x^4 - 3x^2 - 8$ has a graph that is symmetric with respect to the y-axis.

SOLUTION We must show that $f(-x) = f(x)$ for any x.

$$\begin{aligned} f(-x) &= (-x)^4 - 3(-x)^2 - 8 \\ &= (-1)^4 x^4 - 3(-1)^2 x^2 - 8 \\ &= x^4 - 3x^2 - 8 \\ &= f(x) \end{aligned}$$

This *proves* that there is symmetry with respect to the y-axis. The graph in Figure 11 supports this conclusion, since it appears to have this symmetry. (*Note:* Visual support is not a proof!) We will study graphs of this type in more detail in Chapter 3.

EXAMPLE 3

Determining Symmetry Analytically and Supporting it Graphically

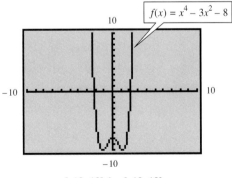

$f(x) = x^4 - 3x^2 - 8$

[-10, 10] by [-10, 10]
Xscl = 1   Yscl = 1

This graph is symmetric with respect to the y-axis.

FIGURE 11

(b) Show analytically and support graphically the fact that $f(x) = x^3 - 4x$ has a graph that is symmetric with respect to the origin.

SOLUTION In this case, we must show that $f(-x) = -f(x)$ for any x.

$$\begin{aligned}
f(-x) &= (-x)^3 - 4(-x) \\
&= (-1)^3 x^3 + 4x \\
&= -x^3 + 4x \qquad * \\
&= -(x^3 - 4x) \\
&= -f(x)
\end{aligned}$$

In the line denoted *, note that the signs of the coefficients are all *opposites* of those in $f(x)$. We completed the argument by factoring out -1, showing that the final result is $-f(x)$.

The graph in Figure 12 supports our conclusion that the graph is symmetric with respect to the origin, for folding it along the y-axis and then along the x-axis would lead to coinciding curves. ∎

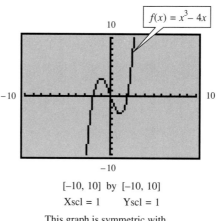

$f(x) = x^3 - 4x$

[−10, 10] by [−10, 10]
Xscl = 1   Yscl = 1

This graph is symmetric with respect to the origin.

FIGURE 12

The Square Root and Cube Root Functions

We now investigate functions that are defined by expressions involving radicals. The first of these is the **square root function,** $f(x) = \sqrt{x}$. Notice that for the function value to be a real number, we must have $x \geq 0$. Thus, the domain is restricted to nonnegative numbers.

SQUARE ROOT FUNCTION

$f(x) = \sqrt{x}$ (Figure 13)

Domain : $[0, \infty)$

Range : $[0, \infty)$

The square root function $f(x) = \sqrt{x}$ increases on $(0, \infty)$. It is also continuous on $[0, \infty)$.

x	$\sqrt{x}$
0	0
1	1
4	2

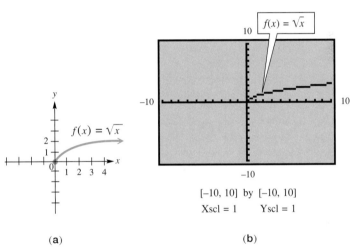

$f(x) = \sqrt{x}$

$f(x) = \sqrt{x}$

$[-10, 10]$ by $[-10, 10]$
Xscl = 1   Yscl = 1

(a)   (b)

FIGURE 13

(The definition of rational exponents allows us to also enter $\sqrt{x}$ as $x^{1/2}$ on a calculator.)

The **cube root function,** $f(x) = \sqrt[3]{x}$, differs from the square root function in that *any* real number, positive, zero, or *negative*, has a real cube root, and thus the domain is $(-\infty, \infty)$. Also, when $x > 0$, $\sqrt[3]{x} > 0$, when $x = 0$, $\sqrt[3]{x} = 0$, and when $x < 0$, $\sqrt[3]{x} < 0$. As a result, the range is also $(-\infty, \infty)$.

CUBE ROOT FUNCTION

$f(x) = \sqrt[3]{x}$ (Figure 14)

Domain: $(-\infty, \infty)$

Range: $(-\infty, \infty)$

The cube root function $f(x) = \sqrt[3]{x}$ increases on its entire domain $(-\infty, \infty)$. It is also continuous on $(-\infty, \infty)$.

x	$\sqrt[3]{x}$
0	0
-1	-1
1	1
-8	-2
8	2

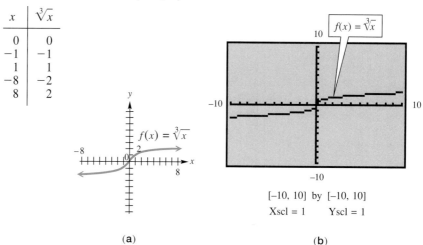

$[-10, 10]$ by $[-10, 10]$
Xscl = 1   Yscl = 1

(a)   (b)

FIGURE 14

(The definition of rational exponents allows us to also enter $\sqrt[3]{x}$ as $x^{1/3}$ on a calculator.)

FOR GROUP DISCUSSION

1. With a suitable viewing window, graph $y = \sqrt{x}$. Use the appropriate calculation feature to find the y-value when $x = 13$. Then find $\sqrt{13}$ using the square root key of the calculator. Compare the results. (These are *approximations*.)

2. With a suitable viewing window, graph $y = \sqrt[3]{x}$. Use the appropriate calculation feature to find the y-value when $x = 13$. Then find $\sqrt[3]{13}$ using the cube root key of the calculator. Compare the results. (These, too, are approximations.)

3. How could you use the graph of $y = \sqrt[n]{x}$ to find the nth root of a number x?

TECHNOLOGICAL NOTE
You should become familiar with the command on your particular calculator that allows you to graph the absolute value.

The Absolute Value Function

On a number line, the absolute value of a real number x, denoted $|x|$, represents its undirected distance from the origin, 0. The **absolute value function,** which pairs every real number with its absolute value, is defined as follows:

$$f(x) = |x| = \begin{cases} x & \text{if } x \geq 0 \\ -x & \text{if } x < 0. \end{cases}$$

Notice that this function is defined in two parts. We use $|x| = x$ if x is positive or zero, and we use $|x| = -x$ if x is negative. Since x can be any real number, the domain of the absolute value function is $(-\infty, \infty)$, but since $|x|$ cannot be negative, the range is $[0, \infty)$.

ABSOLUTE VALUE FUNCTION

$f(x) = |x|$ (Figure 15)

Domain: $(-\infty, \infty)$

Range: $[0, \infty)$

The absolute value function $f(x) = |x|$ decreases on the interval $(-\infty, 0)$ and increases on $(0, \infty)$. It is continuous on its entire domain.

x	$|x|$
0	0
1	1
−1	1
2	2
−2	2

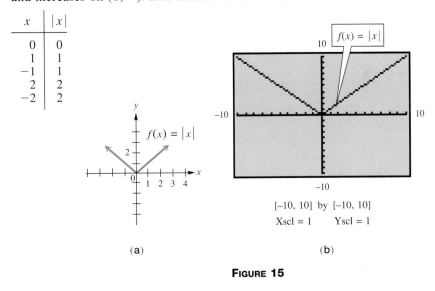

$[-10, 10]$ by $[-10, 10]$
Xscl = 1   Yscl = 1

(a)   (b)

FIGURE 15

FOR GROUP DISCUSSION

Based on the discussion so far in this section, answer the following questions.

1. Which functions have graphs that are symmetric with respect to the y-axis?
2. Which functions have graphs that are symmetric with respect to the origin?
3. Which functions have graphs that show neither of these symmetries?
4. Why is it not possible for the graph of a function to be symmetric with respect to the x-axis?

The Relation $x = y^2$ and Symmetry with Respect to the x-Axis

Our discussion in this text since Section 1.2 has dealt almost exclusively with functions. Recall from Chapter 1 that a function is a relation that satisfies the condition that every domain value is paired with one and only one range value. However, there are cases where we are interested in graphing relations that are not functions, and one of the simplest of these is the relation defined by the equation

$x = y^2$. Notice that the table of selected ordered pairs below indicates that this relation has two different y-values for positive values of x.

SELECTED ORDERED PAIRS FOR $x = y^2$

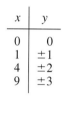

x	y	
0	0	
1	± 1	two different y-values
4	± 2	for the same x-value
9	± 3	

If we plot these points and join them with a smooth curve, we find that the graph of $x = y^2$ is a parabola opening to the right. See Figure 16.

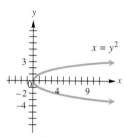

x	y
0	0
1	± 1
4	± 2
9	± 3

FIGURE 16

If a graphics calculator is set for the function mode, it is not possible to graph $x = y^2$ directly. (However, if it is set for the *parametric* mode, such a curve is possible with direct graphing.) To overcome this problem, we begin with $y^2 = x$ and take the square root on each side, remembering to choose both the positive and negative square roots of x:

$$x = y^2 \qquad \text{Given equation}$$
$$y^2 = x \qquad \text{Transform so that } y \text{ is on the left.}$$
$$y = \pm\sqrt{x}. \qquad \text{Take square roots.}$$

Now we have $x = y^2$ defined by two *functions*, $y_1 = \sqrt{x}$ and $y_2 = -\sqrt{x}$. Entering both of these into a calculator gives the graph shown in Figure 17.

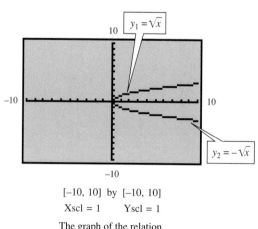

$$y_1 = \sqrt{x}$$
$$y_2 = -\sqrt{x}$$

[-10, 10] by [-10, 10]
Xscl = 1   Yscl = 1

The graph of the relation $x = y^2$ is symmetric with respect to the x-axis.

FIGURE 17

It appears that if we were to fold the graph of $x = y^2$ along the x-axis, the two halves of the parabola would coincide. This is indeed the case, and this graph exhibits symmetry with respect to the x-axis.

SYMMETRY OF A GRAPH WITH RESPECT TO THE x-AXIS

If replacing y with $-y$ in an equation results in the same equation, then the graph is symmetric with respect to the x-axis.

To illustrate this, if we begin with $x = y^2$ and replace y with $-y$, we get

$$x = (-y)^2$$
$$x = (-1)^2 y^2$$
$$x = y^2. \qquad \text{The same equation with which we started}$$

FOR GROUP DISCUSSION

With the two functions
$$y_1 = \sqrt{x} \qquad \text{and} \qquad y_2 = -\sqrt{x}$$
graphed on your calculator, trace to any point on y_1 and notice the x-value. Now switch back and forth from y_1 to y_2 at this x-value. What happens to the y-value. Why is this so?

A summary of the types of symmetry just discussed follows.

Type of Symmetry	Example	Basic Fact About Points on the Graph
y-axis symmetry	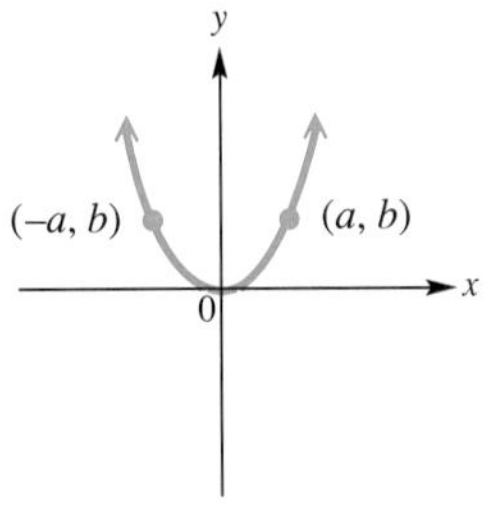	If (a, b) is on the graph, so is $(-a, b)$.
origin symmetry	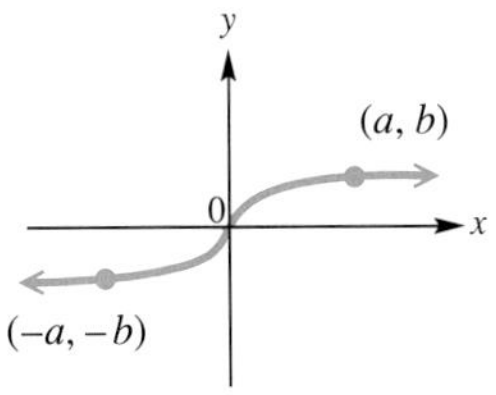	If (a, b) is on the graph, so is $(-a, -b)$.
x-axis symmetry (not possible for a function)	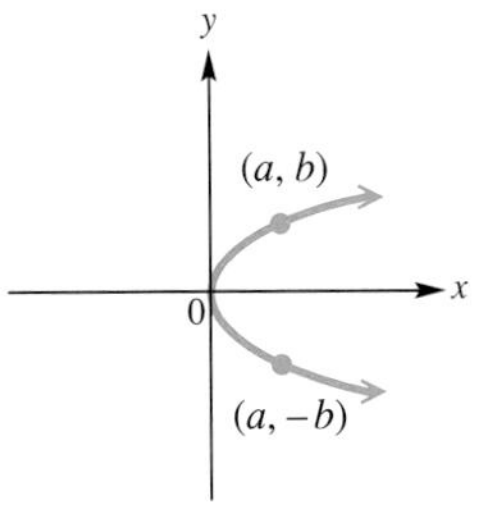	If (a, b) is on the graph, so is $(a, -b)$.

Even and Odd Functions

Closely associated with the concepts of symmetry with respect to the y-axis and symmetry with respect to the origin are the ideas of even and odd functions.

EVEN AND ODD FUNCTIONS

A function f is called an **even function** if $f(-x) = f(x)$ for all x in the domain of f. (Its graph is symmetric with respect to the y-axis.)

A function f is called an **odd function** if $f(-x) = -f(x)$ for all x in the domain of f. (Its graph is symmetric with respect to the origin.)

As an illustration, $f(x) = x^2$ is an even function because

$$f(-x) = (-x)^2 = x^2 = f(x).$$

The function $f(x) = x^3$ is an odd function because

$$f(-x) = (-x)^3 = -x^3 = -f(x).$$

A function may be neither even nor odd; for example, $f(x) = \sqrt{x}$ is neither even nor odd.

EXAMPLE 4

Determining Analytically Whether a Function is Even, Odd, or Neither

Decide if the functions defined as follows are even, odd, or neither.

(a) $f(x) = 8x^4 - 3x^2$

SOLUTION   Replacing x with $-x$ gives

$$f(-x) = 8(-x)^4 - 3(-x)^2 = 8x^4 - 3x^2 = f(x).$$

Since $f(x) = f(-x)$ for each x in the domain of the functon, f is an even function.

(b) $f(x) = 6x^3 - 9x$

SOLUTION   Here

$$f(-x) = 6(-x)^3 - 9(-x) = -6x^3 + 9x = -f(x).$$

This function is odd.

(c) $f(x) = 3x^2 + 2x$

SOLUTION

$$f(-x) = 3(-x)^2 + 2(-x)$$
$$= 3x^2 - 2x$$

Since $f(-x) \neq f(x)$ and $f(-x) \neq -f(x)$, f is neither even nor odd.   ∎

TECHNOLOGICAL NOTE
Some graphics calculators have the capability of drawing a tangent line to a curve at a specified point on the curve. (See Exercises 57–60.) You may wish to investigate whether your model can do this. It will be useful to know this if you go on to calculus.

2.1 EXERCISES

Determine the intervals of the domain over which the given function is continuous.

1.
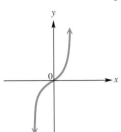

2.
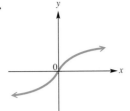

3.
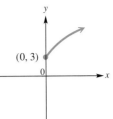
(0, 3)

4.
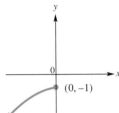
(0, −1)

5.
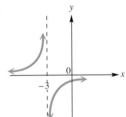
−3

6.
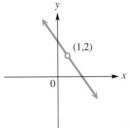
(1,2)

7.
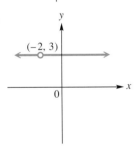
(−2, 3)

8.
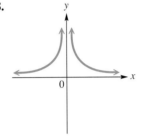

9. Graph the function $y = \frac{x^2 - 9}{x + 3}$ in the standard viewing window of your calculator. At first glance, does this graph seem to be continuous over the entire interval of the domain shown in the window? Now, try to locate the point for which $x = -3$. What happens? Why do you think this happens? (Functions of this kind, called *rational functions,* will be studied in detail in Chapter 4.)

10. Based on your work in Exercise 9, do you think that determination of continuity strictly by observation of a calculator-generated graph is foolproof?

*Determine the intervals of the domain over which the given function is **(a)** increasing, **(b)** decreasing, and **(c)** constant.*

11.
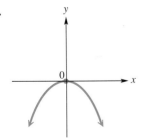

12.
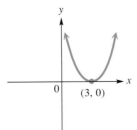
(3, 0)

13.
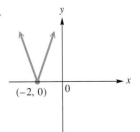
(−2, 0)

14.

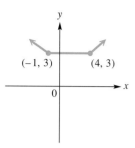

15.

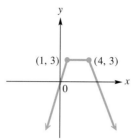

16.

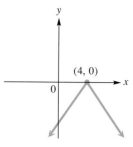

17.

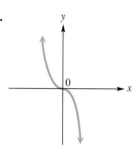

18.

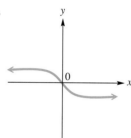

19.

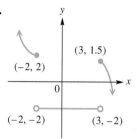

20.

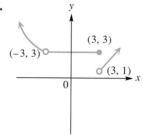

In Exercises 21–28, you are given a function and an interval. Graph the function in the standard viewing window of your calculator, and trace from left to right along a representative portion of the specified interval. Then fill in the blank of this sentence with either **increasing** or **decreasing**: *OVER THE INTERVAL SPECIFIED, THIS FUNCTION IS _____.*

21. $f(x) = x^5; (-\infty, \infty)$

22. $f(x) = -x^3; (-\infty, \infty)$

23. $f(x) = x^4; (-\infty, 0)$

24. $f(x) = x^4; (0, \infty)$

25. $f(x) = -|x|; (-\infty, 0)$

26. $f(x) = -|x|; (0, \infty)$

27. $f(x) = \pi x + 3; (-\infty, \infty)$

28. $f(x) = -\sqrt{2}x - 1; (-\infty, \infty)$

Relating Concepts

Use the results of Exercises 27 and 28, and the concepts of Section 1.3 to determine whether each function is increasing or decreasing without *actually graphing*. Then confirm your answer by graphing in the standard window.

29. $y = 2.36x - 1.56$

30. $y = \sqrt{5}\,x - \sqrt{3}$

31. $y = -\sqrt{6}\,x + .45$

32. $y = -.876\,x + \sqrt{5}$

*Based on a visual observation, determine whether each graph is symmetric with respect to the following: (**a**) x-axis, (**b**) y-axis, (**c**) origin.*

33.

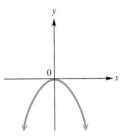

34.

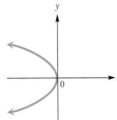

35.

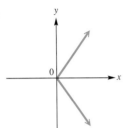

36.

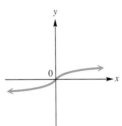

37.

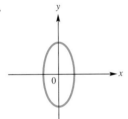

38.

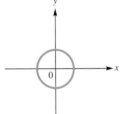

39.

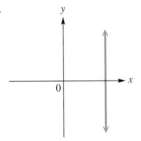

40.

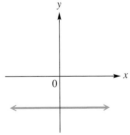

41.

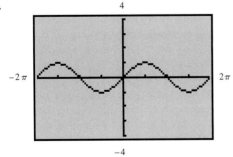

42.

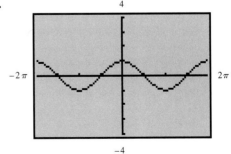

43.

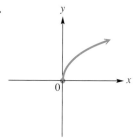

44.

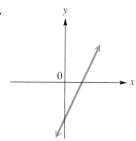

In Exercises 45–50, you are given a calculator-generated graph of a relation that exhibits a type of symmetry. Based on the given information at the bottom of each figure, determine the coordinates of another point that must also lie on the graph.

45. symmetric with respect to the *y*-axis

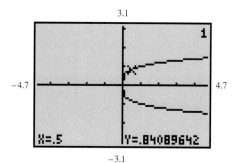

46. symmetric with respect to the *y*-axis

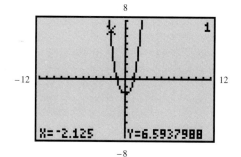

47. symmetric with respect to the *x*-axis

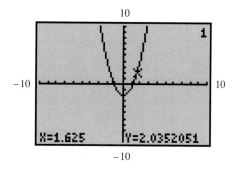

48. symmetric with respect to the *x*-axis

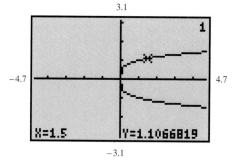

49. symmetric with respect to the origin

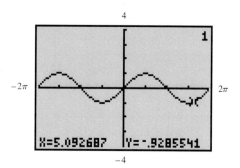

50. symmetric with respect to the origin

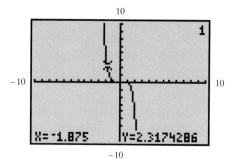

Use the method of Example 3 to determine whether the given function is symmetric with respect to the y-axis, symmetric with respect to the origin, or neither of these. Then graph the function on your calculator to support your conclusion, using the window specified.

51. $f(x) = -x^3 + 2x;$ $[-10, 10]$ by $[-10, 10]$

52. $f(x) = x^5 - 2x^3;$ $[-10, 10]$ by $[-10, 10]$

53. $f(x) = .5x^4 - 2x^2 + 1;$ $[-10, 10]$ by $[-10, 10]$

54. $f(x) = .75x^2 + |x| + 1;$ $[-10, 10]$ by $[-10, 10]$

55. $f(x) = x^3 - x + 3;$ $[-10, 10]$ by $[-10, 10]$

56. $f(x) = x^4 - 5x + 2;$ $[-10, 10]$ by $[-10, 10]$

*R*elating Concepts

The line in the sketch is said to be tangent to the curve at point P.

57. Sketch by hand the graph of $y = x^2$, and choose a point at which the function is *decreasing*. Draw a tangent line at that point. Is the slope of this line positive, negative, or zero? (See Section 1.4 if necessary.)

58. Repeat Exercise 57, but choose a point at which the function is *increasing*.

59. Repeat Exercise 57, but choose the point at which the function changes from *decreasing* to *increasing*.

60. Based on your answers to Exercises 57–59, what conclusions can be drawn? (These conclusions can be verified using methods discussed in calculus.)

Using the standard viewing windows below (or on a reproduction of them), graph by hand the basic functions described in this section, avoiding the temptation to look back at the figures of the section. Then check your work using your calculator. Finally, use the graphs to answer true *or* false *to the statements in Exercises 61–70.*

$f(x) = x$

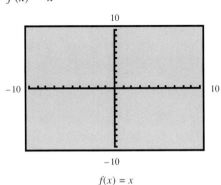

$f(x) = x$

$f(x) = x^2$

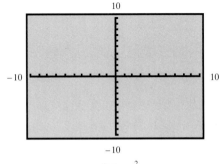

$f(x) = x^2$

$f(x) = x^3$

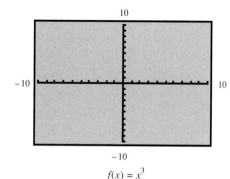

$f(x) = x^3$

$f(x) = \sqrt{x}$

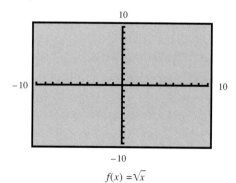

$f(x) = \sqrt{x}$

$f(x) = \sqrt[3]{x}$

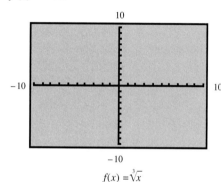

$f(x) = \sqrt[3]{x}$

$f(x) = |x|$

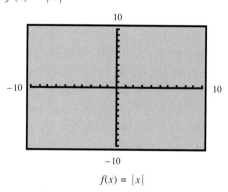

$f(x) = |x|$

61. The range of $f(x) = x^2$ is the same as the range of $f(x) = |x|$.

62. The functions $f(x) = x^2$ and $f(x) = |x|$ increase on the same interval.

63. The functions $f(x) = \sqrt{x}$ and $f(x) = \sqrt[3]{x}$ have the same domain.

64. The function $f(x) = \sqrt[3]{x}$ decreases on its entire domain.

65. The function $f(x) = x$ has its domain equal to its range.

66. The function $f(x) = \sqrt{x}$ is continuous on the interval $(-\infty, 0)$.

67. None of the functions shown decrease on the interval $(0, \infty)$.

68. Both $f(x) = x$ and $f(x) = x^3$ have graphs that are symmetric with respect to the origin.

69. Both $f(x) = x^2$ and $f(x) = |x|$ have graphs that are symmetric with respect to the y-axis.

70. None of the graphs shown are symmetric with respect to the x-axis.

*R*elating Concepts

Use the graphs of the elementary functions discussed in this section to solve each equation or inequality by the x-intercept method of graphical solution. See Sections 1.5 and 1.6 if necessary.

71. (a) $x^2 = 0$
   (b) $x^2 > 0$
   (c) $x^2 < 0$

72. (a) $x^3 = 0$
   (b) $x^3 > 0$
   (c) $x^3 < 0$

73. (a) $\sqrt{x} = 0$
   (b) $\sqrt{x} > 0$
   (c) $\sqrt{x} < 0$

74. (a) $\sqrt[3]{x} = 0$
    (b) $\sqrt[3]{x} > 0$
    (c) $\sqrt[3]{x} < 0$

75. (a) $|x| = 0$
    (b) $|x| > 0$
    (c) $|x| < 0$

76. Because the graphs of all of the elementary functions discussed in this section pass through the origin, the solution set of $f(x) = 0$ is ——————— for each function f.

Relating Concepts

Use the intersection-of-graphs method and the graphs shown to solve graphically the equations and inequalities in Exercises 77 and 78. See Sections 1.5 and 1.6 if necessary.

77. (a) $x^2 = \sqrt{x}$
    (b) $x^2 < \sqrt{x}$
    (c) $x^2 > \sqrt{x}$

78. (a) $x^3 = |x|$
    (b) $x^3 < |x|$
    (c) $x^3 > |x|$

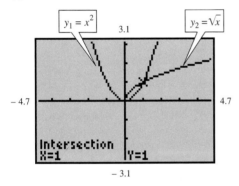

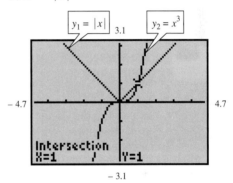

Refer to the function described and determine whether it is even, odd, or neither.

79. the function in Exercise 51

80. the function in Exercise 52

81. the function in Exercise 53

82. the function in Exercise 54

83. the function in Exercise 55

84. the function in Exercise 56

Further Explorations

1. Notice how Y_1 and Y_2 are defined. Observe the relationship between Y_1 and Y_2. Why does the TABLE return an error message when $X < 0$?

2. For each of the following pairs of TABLES, determine whether the functions in Y_1 are even functions, odd functions, or neither.

(a)

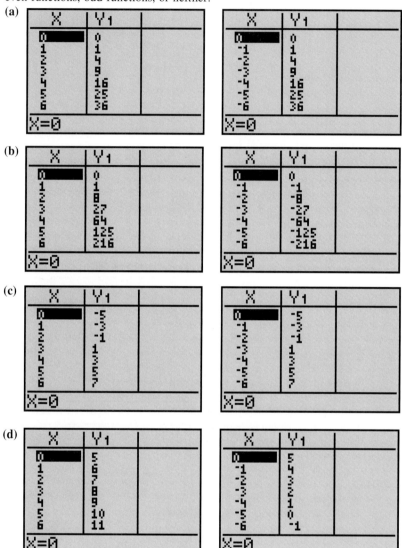

2.2 VERTICAL AND HORIZONTAL SHIFTS OF GRAPHS OF FUNCTIONS

Vertical Shifts ▮ Horizontal Shifts ▮ Combinations of Vertical and Horizontal Shifts ▮ Effects of Shifts on Domain and Range

In this section we will examine how the graphs of the elementary functions introduced in the previous section can be shifted vertically and horizontally in the plane. The basic ideas can then be generalized to apply to the graphs of any functions.

Vertical Shifts

TECHNOLOGICAL NOTE
You can save time when entering the functions in the first Group Discussion. If you let Y1 represent the basic function as shown, you can enter Y2 as Y1 + 3, Y3 as Y1 − 2, and Y4 as Y1 + 5. Then, you need only change Y1 in each function group, since the other functions are defined in terms of Y1.

FOR **GROUP DISCUSSION**

In each group of functions below, we give four related functions. Graph the four functions in the first group (Group A), and then answer the questions below regarding those functions. Then repeat the process for Group B, Group C, and Group D. Use the standard viewing window in each case.

A	B	C	D
$y_1 = x^2$	$y_1 = x^3$	$y_1 = \sqrt{x}$	$y_1 = \sqrt[3]{x}$
$y_2 = x^2 + 3$	$y_2 = x^3 + 3$	$y_2 = \sqrt{x} + 3$	$y_2 = \sqrt[3]{x} + 3$
$y_3 = x^2 - 2$	$y_3 = x^3 - 2$	$y_3 = \sqrt{x} - 2$	$y_3 = \sqrt[3]{x} - 2$
$y_4 = x^2 + 5$	$y_4 = x^3 + 5$	$y_4 = \sqrt{x} + 5$	$y_4 = \sqrt[3]{x} + 5$

1. How does the graph of y_2 compare to the graph of y_1?
2. How does the graph of y_3 compare to the graph of y_1?
3. How does the graph of y_4 compare to the graph of y_1?
4. If $c > 0$, how do you think the graph of $y_1 + c$ would compare to the graph of y_1?
5. If $c > 0$, how do you think the graph of $y_1 - c$ would compare to the graph of y_1?

Choosing your own value of c, support your answers to items 4 and 5 graphically. (Be sure that your choice is appropriate for the standard window.)

The objective of the preceding group discussion activity was to make conjectures about how the addition or subtraction of a constant c would affect the graph of a function $y = f(x)$. In each case, we obtained a vertical shift, or **translation,** of the graph of the basic function with which we started. Although our observations were based on the graphs of four different elementary functions, they can be generalized to any function.

VERTICAL SHIFTING OF THE GRAPH OF A FUNCTION

If $c > 0$, the graph of $y = f(x) + c$ is obtained by shifting the graph of $y = f(x)$ *upward* a distance of c units. The graph of $y = f(x) - c$ is obtained by shifting the graph of $y = f(x)$ *downward* a distance of c units.

In Figure 18 we give a graphical interpretation of the statement above.

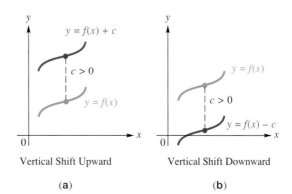

Vertical Shift Upward

(a)

Vertical Shift Downward

(b)

FIGURE 18

<table>
<tr><td>

EXAMPLE 1

Recognizing Vertical
Shifts on
Calculator-Generated
Graphs

</td></tr>
</table>

Figure 19 shows the graphs of four functions. The function labeled y_1 is the function $f(x) = |x|$. The viewing window is $[-10, 10]$ by $[-10, 10]$, with Xscl $= 1$ and Yscl $= 1$. Each of y_2, y_3, and y_4 are functions of the form $f(x) + c$ or $f(x) - c$, for $c > 0$. Give the rule for each of these functions.

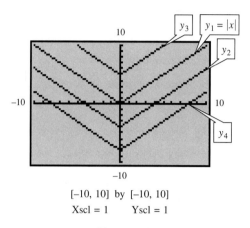

$[-10, 10]$ by $[-10, 10]$
Xscl $= 1$ Yscl $= 1$

FIGURE 19

SOLUTION   Because the graph of y_2 lies 4 units below the graph of y_1, we have $y_2 = |x| - 4$. The graph of y_3 is a shift of the graph of y_1 a distance of 5 units upward, so the equation for y_3 is $y_3 = |x| + 5$. Finally, the graph of y_4 is a vertical shift of the graph of y_1 8 units downward, so its equation is $y_4 = |x| - 8$.   ◘

When using a graphics calculator to investigate shifts of graphs, it is important to use an appropriate window; otherwise, the graph may not appear. For example, the graph of $y = |x| + 12$ does not appear in the viewing window $[-10, 10]$ by $[-10, 10]$. Why is this so? There are many windows that would show the graph. (Name one.)

Horizontal Shifts

<table>
<tr><td colspan="4">

FOR GROUP DISCUSSION

</td></tr>
<tr><td colspan="4">

This discussion parallels the one earlier in this section. Follow the same general directions.

</td></tr>
<tr>
<td>A</td>
<td>B</td>
<td>C</td>
<td>D</td>
</tr>
<tr>
<td>$y_1 = x^2$</td>
<td>$y_1 = x^3$</td>
<td>$y_1 = \sqrt{x}$</td>
<td>$y_1 = \sqrt[3]{x}$</td>
</tr>
<tr>
<td>$y_2 = (x - 3)^2$</td>
<td>$y_2 = (x - 3)^3$</td>
<td>$y_2 = \sqrt{x - 3}$</td>
<td>$y_2 = \sqrt[3]{x - 3}$</td>
</tr>
<tr>
<td>$y_3 = (x - 5)^2$</td>
<td>$y_3 = (x - 5)^3$</td>
<td>$y_3 = \sqrt{x - 5}$</td>
<td>$y_3 = \sqrt[3]{x - 5}$</td>
</tr>
<tr>
<td>$y_4 = (x + 4)^2$</td>
<td>$y_4 = (x + 4)^3$</td>
<td>$y_4 = \sqrt{x + 4}$</td>
<td>$y_4 = \sqrt[3]{x + 4}$</td>
</tr>
</table>

1. How does the graph of y_2 compare to the graph of y_1?
2. How does the graph of y_3 compare to the graph of y_1?
3. How does the graph of y_4 compare to the graph of y_1?
4. If $c > 0$, how do you think the graph of $y_5 = f(x - c)$ would compare to the graph of $y_1 = f(x)$?
5. If $c > 0$, how do you think the graph of $y_5 = f(x + c)$ would compare to the graph of $y_1 = f(x)$?

Choosing your own value of c, support your answers to items 4 and 5 graphically. Again, be sure that your choice is appropriate for the standard window.

The results of the above discussion should remind you of the results found earlier. There we saw how graphs of functions can be shifted vertically. Now, we see how they can be shifted *horizontally*. The observations can be generalized as follows.

> **HORIZONTAL SHIFTING OF THE GRAPH OF A FUNCTION**
>
> If $c > 0$, the graph of $y = f(x - c)$ is obtained by shifting the graph of $y = f(x)$ to the *right* a distance of c units. The graph of $y = f(x + c)$ is obtained by shifting the graph of $y = f(x)$ to the *left* a distance of c units.

CAUTION Errors of interpretation frequently occur when horizontal shifts are involved. In order to determine the direction and magnitude of horizontal shifts, find the value of x that would cause the expression within the parentheses to equal 0. For example, the graph of $f(x) = (x - 5)^2$ would be shifted 5 units to the *right,* because +5 would cause $x - 5$ to equal 0. On the other hand, the graph of $f(x) = (x + 4)^2$ would be shifted 4 units to the *left,* because −4 would cause $x + 4$ to equal 0.

Figure 20 illustrates the effect of horizontal shifts of the graph of a function $y = f(x)$.

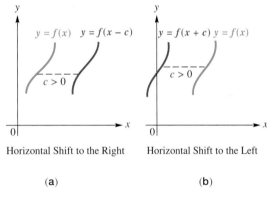

Horizontal Shift to the Right   Horizontal Shift to the Left

(a)   (b)

FIGURE 20

EXAMPLE **2**

Recognizing Horizontal Shifts on Calculator-Generated Graphs

Figure 21 shows the graphs of four functions. As in Example 1, the function labeled y_1 is the function $f(x) = |x|$. The viewing window is $[-10, 10]$ by $[-10, 10]$, with

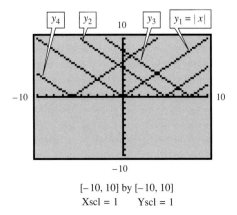

$[-10, 10]$ by $[-10, 10]$
Xscl = 1   Yscl = 1

FIGURE 21

Xscl = 1 and Yscl = 1. Each of y_2, y_3, and y_4 are functions of the form $f(x - c)$ or $f(x + c)$, where $c > 0$. Give the rule for each of these y's.

SOLUTION The graph of y_2 is the same as the graph of y_1, but it is shifted 5 units to the *right*. Therefore, we have $y_2 = |x - 5|$. Similarly, since y_3 is the graph of y_1 shifted 8 units to the *right*, $y_3 = |x - 8|$. The graph of y_4 is obtained by shifting that of y_1 6 units to the *left*, so its equation is $y_4 = |x + 6|$. ◻

TECHNOLOGICAL NOTE
Some later models of graphics calculators have the capability of showing dynamically how transformations of graphs can be made. Read your owner's manual to see if yours has this feature.

Combinations of Vertical and Horizontal Shifts

Now that we have seen how graphs of functions can be shifted vertically and shifted horizontally, it is not difficult to extend these ideas to graphs that are obtained by applying *both* types of translations.

EXAMPLE 3

Applying Both Vertical and Horizontal Shifts

Describe how the graph of $y = |x + 15| - 20$ would be obtained by translating the graph of $y = |x|$. Determine an appropriate viewing window, and support the results by plotting both functions with a graphics calculator.

SOLUTION The function defined by $y = |x + 15| - 20$ is translated 15 units to the *left* (because of the $|x + 15|$) and 20 units *downward* as compared to the graph of $y = |x|$. Because the point at which the graph changes from decreasing to increasing is now $(-15, -20)$, the standard viewing window is not appropriate. We must choose a window that contains the point $(-15, -20)$ in order to obtain a complete graph. While many such windows are possible, one such window is shown in Figure 22. The display at the bottom of the screen indicates that the point $(-15, -20)$ lies on the graph. ◻

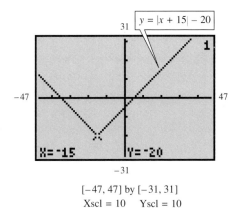

$[-47, 47]$ by $[-31, 31]$
Xscl = 10   Yscl = 10

FIGURE 22

FOR GROUP DISCUSSION

For each of the following pairs of functions, describe how the graph of y_2 can be obtained from the graph of y_1 by shifting, and then support your answer by graphing in an appropriate viewing window.

1. $y_1 = x^2$; $y_2 = (x - 12)^2 + 25$ 2. $y_1 = x^3$; $y_2 = (x + 10)^3 - 15$

3. $y_1 = \sqrt{x}$; $y_2 = \sqrt{x - 20} - 30$ 4. $y_1 = \sqrt[3]{x}$; $y_2 = \sqrt[3]{x + 20} + 40$

Effects of Shifts on Domain and Range

The domains and ranges of functions may or may not be affected by vertical and horizontal shifts. For example, if the domain of a function is $(-\infty, \infty)$, a horizontal shift will not affect the domain. Similarly, if the range is $(-\infty, \infty)$, a vertical shift will not affect the range. However, if the domain is not $(-\infty, \infty)$, a horizontal shift will affect the domain, and if the range is not $(-\infty, \infty)$, a vertical shift will affect the range. The final example illustrates this.

EXAMPLE 4
Determining Domains and Ranges of Shifted Graphs

The four functions graphed in Figures 23–26 are those discussed in "For Group Discussion" following Example 3. Give the domain and the range of each function.

SOLUTION   The graph of $y = (x - 12)^2 + 25$ is shown in Figure 23. It is a translation of the graph of $y = x^2$ 12 units to the right and 25 units upward. The original domain $(-\infty, \infty)$ is not affected. However, the range of this function is $[25, \infty)$, because of the vertical translation.

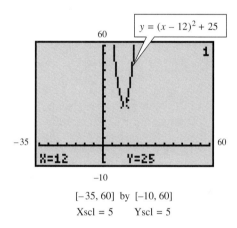

$y = (x - 12)^2 + 25$

$[-35, 60]$ by $[-10, 60]$
Xscl = 5   Yscl = 5

FIGURE 23

The graph in Figure 24, that of $y = (x + 10)^3 - 15$, was obtained by vertical and horizontal shifts of the graph of $y = x^3$, a function that has both domain and range equal to $(-\infty, \infty)$. Neither is affected here, and so the domain and range of $y = (x + 10)^3 - 15$ are also both $(-\infty, \infty)$.

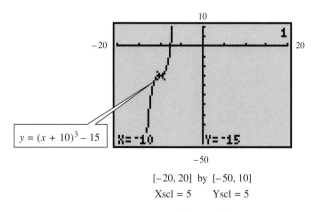

$y = (x + 10)^3 - 15$

$[-20, 20]$ by $[-50, 10]$
Xscl = 5   Yscl = 5

FIGURE 24

The function $y = \sqrt{x}$ has domain $[0, \infty)$. The function graphed in Figure 25, $y = \sqrt{x - 20} - 30$, was obtained by shifting the basic graph 20 units to the right, so the new domain is $[20, \infty)$. On the other hand, the original range, $[0, \infty)$ has been affected by the shift of the graph 30 units downward. The new range is $[-30, \infty)$.

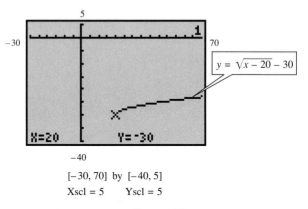

$[-30, 70]$ by $[-40, 5]$
Xscl = 5   Yscl = 5

FIGURE 25

The situation in Figure 26 is similar to that of Figure 24. A graph with domain and range both $(-\infty, \infty)$, that is, $y = \sqrt[3]{x}$, has been shifted 20 units to the left and 40 units upward. No matter what direction and magnitude these shifts might have been, the domain and the range are both unaffected. They both remain $(-\infty, \infty)$.   ◘

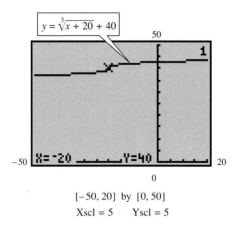

$[-50, 20]$ by $[0, 50]$
Xscl = 5   Yscl = 5

FIGURE 26

The feature of graphics calculators that allows us to locate a point on a graph helps us support the results obtained in Example 4. You might wish to experiment with yours to confirm those results.

2.2 EXERCISES

Exercises 1–25 are grouped in "fives." For each group of five functions, match the correct graph A, B, C, D, or E to the function without using your calculator. You should use the concepts developed in this section to work these exercises based on visual observation. Then, after you have answered each group of five, use your calculator to check your answers. Every graph in these groups is plotted in the standard viewing window.

1. $y = x^2 - 3$

2. $y = (x - 3)^2$

3. $y = (x + 3)^2$

4. $y = (x - 3)^2 + 2$

5. $y = (x + 3)^2 + 2$

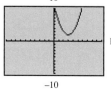

A

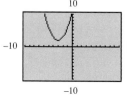

B

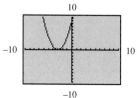

C

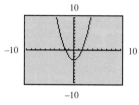

D

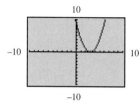

E

6. $y = |x| + 4$

7. $y = |x + 4|$

8. $y = |x - 4|$

9. $y = |x + 4| - 3$

10. $y = |x - 4| - 3$

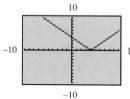

A

B

C

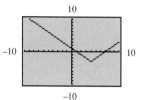

D

E

11. $y = \sqrt{x} + 6$
12. $y = \sqrt{x+6}$
13. $y = \sqrt{x-6}$
14. $y = \sqrt{x+2} - 4$
15. $y = \sqrt{x-2} - 4$

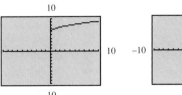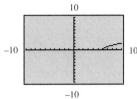

A   B   C

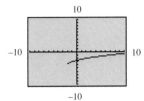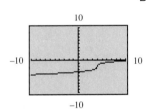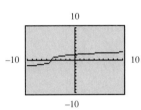

D   E

16. $y = \sqrt[3]{x} + 5$
17. $y = \sqrt[3]{x+5}$
18. $y = \sqrt[3]{x-4} + 2$
19. $y = \sqrt[3]{x+4} + 2$
20. $y = \sqrt[3]{x-4} - 2$

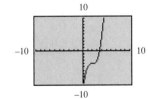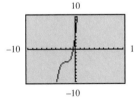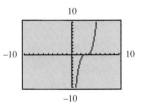

A   B   C

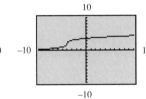

D   E

21. $y = x^3 + 3$
22. $y = (x-3)^3$
23. $y = (x+3)^3$
24. $y = (x+2)^3 - 4$
25. $y = (x-2)^3 - 4$

A   B   C

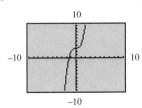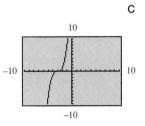

D   E

26. In which quadrant does the vertex of the graph of $y = (x - h)^2 + k$ lie, if $h < 0$ and $k < 0$?

*In Exercises 27–32, use the results of the specified earlier exercises and corresponding graphs to determine (**a**) the domain and (**b**) the range of the given function.*

27. $y = |x + 4| - 3$ (Exercise 9)

28. $y = |x - 4| - 3$ (Exercise 10)

29. $y = \sqrt{x - 2} - 4$ (Exercise 15)

30. $y = \sqrt{x + 2} - 4$ (Exercise 14)

31. $y = \sqrt[3]{x + 5}$ (Exercise 17)

32. $y = \sqrt[3]{x} + 5$ (Exercise 16)

The concepts introduced in this section can be applied to functions whose graphs may not be familiar to you. Given here is the graph of a function studied later in this text. We will call it $y = f(x)$. Each tick mark represents 1 unit.

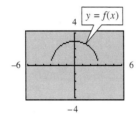

Now match the function specified with the appropriate graph from the choices A, B, C, or D.

33. $y = f(x) + 1$

34. $y = f(x + 1)$

35. $y = f(x - 1)$

36. $y = f(x) - 1$

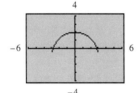

A

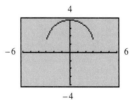

B

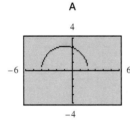

C

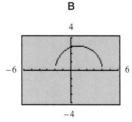

D

Given the graph shown below, sketch by hand the graph of the function described, indicating how the three points labeled on the original graph have been translated.

37. $y = f(x) + 2$

38. $y = f(x) - 2$

39. $y = f(x + 2)$

40. $y = f(x - 2)$

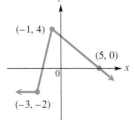

R*elating Concepts*

Recall from Chapter 1 that a unique line is determined by two different points on the line, and that the values of m and b can be then determined for the general form of the linear function $f(x) = mx + b$. Make the connections between the concepts introduced in Sections 1.3 and 1.4 to answer the following.

41. Sketch by hand the line that passes through the points $(1, -2)$ and $(3, 2)$.

42. Use the slope formula to find the slope of this line.

43. Find the equation of this line and write it in the form $y_1 = mx + b$.

44. Keeping the same two x-values as indicated in Exercise 41, add 6 to each y-value. What are the coordinates of the two new points?

45. Find the slope of the line through the points determined in Exercise 44.

46. Find the equation of this new line and write it in the form $y_2 = mx + b$.

47. Graph both y_1 and y_2 in the standard viewing window of your calculator, and describe how the graph of y_2 can be obtained by vertically translating the graph of y_1. What is the value of the constant by which this vertical translation occurs? Where do you think this comes from?

48. Fill in the blanks with the correct responses, based on your work in Exercises 41–47.

If the points (x_1, y_1) and (x_2, y_2) lie on a line, then when we add the positive constant c to each y-value, we obtain the points $(x_1, y_1 + $ _____$)$ and $(x_2, y_2 + $ _____$)$. The slope of the new line is _____ the slope of the original line.
                          (the same as/different from)
The graph of the new line can be obtained by shifting the graph of the original line _____ units in the _____ direction.

Suppose that h and k are both positive numbers. Match the equation with the correct graph in Exercises 49–52.

49. $y = (x - h)^2 - k$

50. $y = (x + h)^2 - k$

51. $y = (x + h)^2 + k$

52. $y = (x - h)^2 + k$

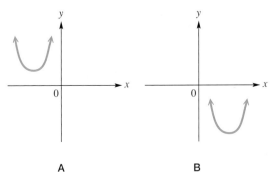

A   B

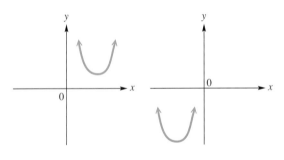

C   D

*R*elating Concepts

*Each function in Exercises 53–58 is a translation of one of the basic functions $y = x^2$, $y = x^3$, $y = \sqrt{x}$, $y = \sqrt[3]{x}$, or $y = |x|$. Using the concepts of increasing and decreasing functions discussed in Section 2.1, determine the interval of the domain over which the function is (**a**) increasing and (**b**) decreasing.*

53.

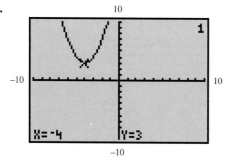

54.

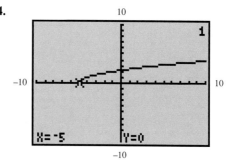

55.

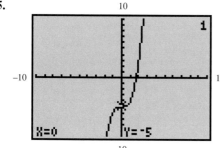

56.

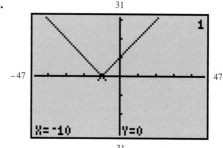

57. $y = \sqrt[3]{x - 12} + 18$

58. $y = (x - 10)^2 + 13$

59. Suppose that the graph of $y = x^2$ is translated in such a way that its domain is $(-\infty, \infty)$ and its range is $[38, \infty)$. What are the possible values of h and k if the new function is of the form $y = (x - h)^2 + k$?

60. The graph shown is a translation of $y = |x|$. What are the values of h and k if the equation is of the form $y = |x - h| + k$?

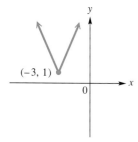

Relating Concepts

Use the x-intercept method of graphical solution of equations and inequalities to solve each equation or inequality, using the given graph of $y = f(x)$. See Sections 1.5 and 1.6.

61. (a) $f(x) = 0$
     (b) $f(x) > 0$
     (c) $f(x) < 0$

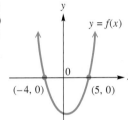

62. (a) $f(x) = 0$
     (b) $f(x) > 0$
     (c) $f(x) < 0$

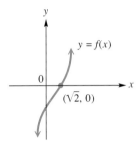

63. (a) $f(x) = 0$
     (b) $f(x) \geq 0$
     (c) $f(x) \leq 0$

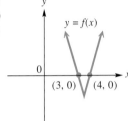

64. (a) $f(x) = 0$
     (b) $f(x) \geq 0$
     (c) $f(x) \leq 0$

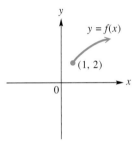

If c is a positive constant, tell whether or not the graph of the given function is symmetric with respect to the y-axis.

65. $f(x) = x^2 + c$

66. $f(x) = |x| + c$

67. $f(x) = \sqrt{x} + c$

68. $f(x) = \sqrt[3]{x} + c$

69. $f(x) = x^3 + c$

70. $f(x) = x^2 - c$

71. If $f(x) = \sqrt[3]{x - 7} + 4$,
     (a) find $f(15)$ analytically.
     (b) support your result of part (a) graphically.

72. If $f(x) = (x - 4)^3 + 7$,
     (a) find the value of x for which $f(x) = 15$ graphically, using the intersection-of-graphs method.
     (b) use the capabilities of your calculator to solve $f(x) = 0$ graphically, correct to the nearest hundredth.

2.3 STRETCHING, SHRINKING, AND REFLECTING GRAPHS OF FUNCTIONS

Vertical Stretching ∎ Vertical Shrinking ∎ Reflecting Across an Axis ∎ Combining Transformations of Graphs

We continue our discussion from the previous section on how the graphs of functions may be altered. We saw how adding or subtracting a constant can cause a vertical or horizontal shift. Now we will see how multiplication by a constant alters the graph of a function.

Vertical Stretching

In each group of functions below, we give four related functions. Graph the four functions in the first group (Group A), and then answer the questions below regarding those functions. Then repeat the process for Group B and Group C. Use the window specified for each group.

A	B	C
$[-5, 5]$ by $[-5, 20]$	$[-5, 15]$ by $[-5, 10]$	$[-20, 20]$ by $[-10, 10]$
$y_1 = x^2$	$y_1 = \sqrt{x}$	$y_1 = \sqrt[3]{x}$
$y_2 = 2x^2$	$y_2 = 2\sqrt{x}$	$y_2 = 2\sqrt[3]{x}$
$y_3 = 3x^2$	$y_3 = 3\sqrt{x}$	$y_3 = 3\sqrt[3]{x}$
$y_4 = 4x^2$	$y_4 = 4\sqrt{x}$	$y_4 = 4\sqrt[3]{x}$

1. How does the graph of y_2 compare to the graph of y_1?
2. How does the graph of y_3 compare to the graph of y_1?
3. How does the graph of y_4 compare to the graph of y_1?
4. If we choose $c > 4$, how do you think the graph of $y_5 = c \cdot y_1$ would compare to the graph of y_4?

Choosing your own value of c, support your answer to item 4 graphically.

In each group of functions in the preceding activity, we started with an elementary function y_1 and observed how the graphs of functions of the form $y = c \cdot y_1$ compared with y_1 for positive values of c that began at 2 and became progressively larger. In each case, we obtained a *vertical stretch* of the graph of the basic function with which we started. These observations can be generalized to any function.

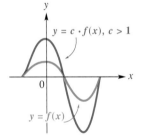

FIGURE 27

VERTICALLY STRETCHING THE GRAPH OF A FUNCTION

If $c > 1$, the graph of $y = c \cdot f(x)$ is obtained by vertically stretching the graph of $y = f(x)$ by a factor of c. In general, the larger the value of c, the greater the stretch.

In Figure 27 we give a graphical interpretation of the statement above.

Figure 28 (on the following page) shows the graphs of four functions. The function labeled y_1 is the function $f(x) = |x|$. The other three functions, y_2, y_3, and y_4 are defined as follows, but not necessarily in the given order: $2.4|x|, 3.2|x|, 4.3|x|$. Determine the correct rule for each graph.

SOLUTION   The values of c here are 2.4, 3.2, and 4.3. The vertical heights of the points with the same x-coordinates on the three graphs will correspond to the magnitudes of these c values. Thus, the graph just above $y_1 = |x|$ will be that of $y = 2.4|x|$, the "highest" graph will be that of $y = 4.3|x|$, and the graph of $y = 3.2|x|$ will lie "between" the others. Therefore, based on our observation of the graphs in the figure, we have $y_2 = 4.3|x|$, $y_3 = 2.4|x|$, and $y_4 = 3.2|x|$.

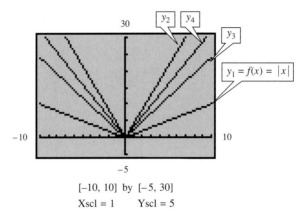

[−10, 10] by [−5, 30]
Xscl = 1   Yscl = 5

FIGURE 28

If we were to trace to any point on the graph of y_1 and then move our tracing cursor to the other graphs one by one, we would see that the y-values of the points would be multiplied by the appropriate values of c. You may wish to experiment with your calculator in this way. ▯

Vertical Shrinking

> ### FOR **GROUP DISCUSSION**
>
> This discussion parallels the one given earlier in this section. Follow the same general directions. (*Note:* The fractions $\frac{3}{4}$, $\frac{1}{2}$, and $\frac{1}{4}$ may be entered as their decimal equivalents when plotting the graphs.)
>
> | A | B | C |
> |---|---|---|
> | $[-5, 5]$ by $[-5, 20]$ | $[-5, 15]$ by $[-2, 5]$ | $[-10, 10]$ by $[-2, 10]$ |
> | $y_1 = x^2$ | $y_1 = \sqrt{x}$ | $y_1 = |x|$ |
> | $y_2 = \frac{3}{4}x^2$ | $y_2 = \frac{3}{4}\sqrt{x}$ | $y_2 = \frac{3}{4}|x|$ |
> | $y_3 = \frac{1}{2}x^2$ | $y_3 = \frac{1}{2}\sqrt{x}$ | $y_3 = \frac{1}{2}|x|$ |
> | $y_4 = \frac{1}{4}x^2$ | $y_4 = \frac{1}{4}\sqrt{x}$ | $y_4 = \frac{1}{4}|x|$ |
>
> **1.** How does the graph of y_2 compare to the graph of y_1?
> **2.** How does the graph of y_3 compare to the graph of y_1?
> **3.** How does the graph of y_4 compare to the graph of y_1?
> **4.** If we choose $0 < c < \frac{1}{4}$, how do you think the graph of $y_5 = c \cdot y_1$ would compare to the graph of y_4? Provide support by choosing such a value of c.

In this group discussion, we began with an elementary function y_1 and observed the graphs of $y = c \cdot y_1$, where we began with $c = \frac{3}{4}$ and chose progressively smaller positive values of c. In each case, the graph of y_1 was *vertically shrunk*. These observations, like the ones for vertical stretching, can be generalized to any function.

> ### VERTICALLY SHRINKING THE GRAPH OF A FUNCTION
> If $0 < c < 1$, the graph of $y = c \cdot f(x)$ is obtained by vertically shrinking the graph of $y = f(x)$ by a factor of c. In general, the smaller the value of c, the greater the shrink.

Figure 29 shows a graphical interpretation of vertical shrinking.

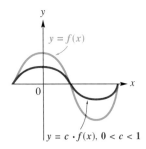

FIGURE 29

<table>
<tr><td>

EXAMPLE 2

Recognizing Vertical
Shrinks on
Calculator-Generated
Graphs

</td></tr>
</table>

Figure 30 shows the graphs of four functions. The function labeled y_1 is the function $f(x) = x^3$. The other three functions, y_2, y_3, and y_4 are defined as follows, but not necessarily in the given order: $.5x^3$, $.3x^3$, and $.1x^3$. Determine the correct rule for each graph.

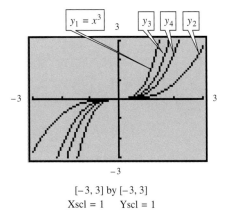

$[-3, 3]$ by $[-3, 3]$
Xscl = 1   Yscl = 1

FIGURE 30

SOLUTION   The smaller the positive value of c, where $0 < c < 1$, the more of a shrink toward the x-axis will there be. Since we have $c = .5, .3$, and $.1$, the function rules must be as follows: $y_2 = .1x^3$, $y_3 = .5x^3$, and $y_4 = .3x^3$.   ∎

Reflecting Across an Axis

In the previous section and so far in this one, we have seen how graphs can be transformed by shifting, stretching, and shrinking. We will now examine how graphs can be reflected across an axis.

FOR **GROUP DISCUSSION**

In each pair of functions, we give two related functions. Graph $y_1 = f(x)$ and $y_2 = -f(x)$ in the standard viewing window, and then answer the questions below for each pair.

A	B	C	D
$y_1 = x^2$	$y_1 = |x|$	$y_1 = \sqrt{x}$	$y_1 = x^3$
$y_2 = -x^2$	$y_2 = -|x|$	$y_2 = -\sqrt{x}$	$y_2 = -x^3$

With respect to the x-axis,

1. how does the graph of y_2 compare with the graph of y_1?
2. how would the graph of $y = -\sqrt[3]{x}$ compare with the graph of $y = \sqrt[3]{x}$, based on your answer to item 1? Confirm your answer by actual graphing.

Now, in each pair of functions, we give two related functions. Graph $y_1 = f(x)$ and $y_2 = f(-x)$ in the standard viewing window, and then answer the questions below for each pair.

A	B	C
$y_1 = \sqrt{x}$	$y_1 = \sqrt{x - 3}$	$y_1 = \sqrt[3]{x + 4}$
$y_2 = \sqrt{-x}$	$y_2 = \sqrt{-x - 3}$	$y_2 = \sqrt[3]{-x + 4}$

With respect to the y-axis,

3. how does the graph of y_2 compare with the graph of y_1?
4. how would the graph of $y = \sqrt[3]{-x}$ compare with the graph of $y = \sqrt[3]{x}$, based on your answer to item 3? Confirm your answer by actual graphing.

Based upon the preceding group discussion, we can see how the graph of a function can be reflected across an axis. The results of that discussion are now formally summarized.

REFLECTING THE GRAPH OF A FUNCTION ACROSS AN AXIS

For a function $y = f(x)$,

(a) the graph of $y = -f(x)$ is a reflection of the graph of f across the x-axis.

(b) the graph of $y = f(-x)$ is a reflection of the graph of f across the y-axis.

Figure 31 shows how the reflections described above affect the graph of a function in general.

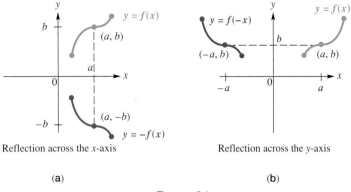

Reflection across the *x*-axis   Reflection across the *y*-axis

(a)   (b)

FIGURE 31

Figure 32 shows the graph of a function $y = f(x)$.

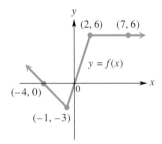

FIGURE 32

(a)   Sketch the graph of $y = -f(x)$.

SOLUTION   We must reflect the graph across the *x*-axis. This means that if a point (a, b) lies on the graph of $y = f(x)$, then the point $(a, -b)$ must lie on the graph of $y = -f(x)$. Using the labeled points to assist us, we find the graph of $y = -f(x)$ in Figure 33.

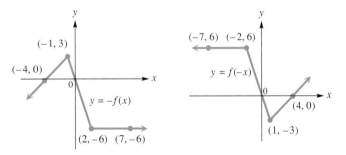

FIGURE 33   **FIGURE 34**

(b)   Sketch the graph of $y = f(-x)$.

SOLUTION   Here we must reflect the graph across the *y*-axis, meaning that if a point (a, b) lies on the graph of $y = f(x)$, then the point $(-a, b)$ must lie on the graph of $y = f(-x)$. Again using the labeled points to guide us, we obtain the graph of $y = f(-x)$ as shown in Figure 34.   ∎

To illustrate how reflections appear on calculator-generated graphs, observe the two graphs shown in Figures 35 and 36. We use a higher degree polynomial function (covered in detail in Chapter 3) here in Figure 35. The graph of y_2 is a reflection of the graph of y_1 across the x-axis. The display at the bottom of the screen shows that the point $(-2, 23)$ is on the graph of y_1. Therefore, if we were to use the capabilities of the calculator, we should locate the point $(-2, -23)$ on the graph of y_2. (*Note:* The function y_1 is defined as $y_1 = x^4 - 2x + 3$. You might wish to verify the above statement.)

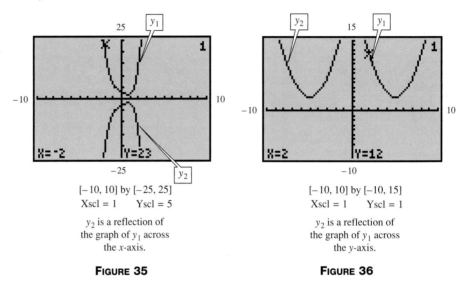

[−10, 10] by [−25, 25]
Xscl = 1   Yscl = 5

y_2 is a reflection of
the graph of y_1 across
the x-axis.

FIGURE 35

[−10, 10] by [−10, 15]
Xscl = 1   Yscl = 1

y_2 is a reflection of
the graph of y_1 across
the y-axis.

FIGURE 36

Figure 36 illustrates a reflection across the y-axis. The graph of y_2 is a reflection of the graph of y_1 across the y-axis. Notice that the point $(2, 12)$ lies on the graph of y_1. What point must lie on the graph of y_2? (To verify your answer, graph $y_1 = (x - 5)^2 + 3$. So we have $y_2 = (-x - 5)^2 + 3$. Now see if you were correct.)

Combining Transformations of Graphs

The graphs of $y_1 = x^2$ and $y_2 = -2x^2$ are shown in the same viewing window in Figure 37.

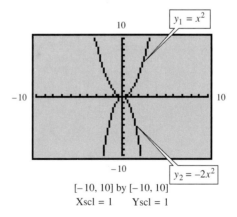

[−10, 10] by [−10, 10]
Xscl = 1   Yscl = 1

FIGURE 37

Notice that in terms of the types of transformations we have studied, the graph of y_2 is obtained by vertically stretching the graph of y_1 by a factor of 2 and then reflecting across the x-axis. Thus we have a case of a combination of transformations. As you might expect, we can create an infinite number of functions by vertically stretching or shrinking, shifting left or right, and reflecting across an axis. The next example investigates examples of this type of function.

EXAMPLE 4

Describing a Combination of Transformations of a Graph

(a) Describe how the graph of $y = -3(x - 4)^2 + 5$ can be obtained by transforming the graph of $y = x^2$.

SOLUTION   The fact that we have $(x - 4)^2$ in our function indicates that the graph of $y = x^2$ must be shifted 4 units to the *right*. Since the coefficient of $(x - 4)^2$ is -3 (a negative number with absolute value greater than 1), the graph is stretched vertically by a factor of 3 and then reflected across the x-axis. The constant $+5$ indicates that the graph is finally shifted up 5 units. Figure 38 shows the graph of both $y = x^2$ and $y = -3(x - 4)^2 + 5$.

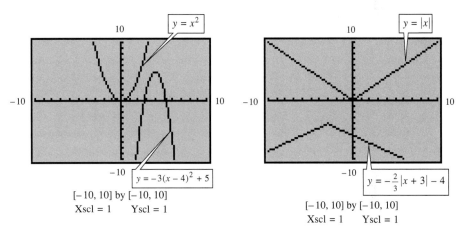

$y = x^2$

$y = -3(x - 4)^2 + 5$

$[-10, 10]$ by $[-10, 10]$
Xscl = 1   Yscl = 1

FIGURE 38

$y = |x|$

$y = -\dfrac{2}{3}|x + 3| - 4$

$[-10, 10]$ by $[-10, 10]$
Xscl = 1   Yscl = 1

FIGURE 39

(b) Give the equation of the function that would be obtained by starting with the graph of $y = |x|$, shifting 3 units to the left, vertically shrinking the graph by a factor of $\frac{2}{3}$, reflecting across the x-axis, and shifting the graph 4 units down, in this order.

SOLUTION   Shifting 3 units to the left means that $|x|$ is transformed to $|x + 3|$. Vertically shrinking by a factor of $\frac{2}{3}$ means multiplying $|x + 3|$ by $\frac{2}{3}$, and reflecting across the x-axis changes $\frac{2}{3}$ to $-\frac{2}{3}$. Finally, shifting 4 units down means subtracting 4. Putting this all together leads to the following equation:

$$y = -\frac{2}{3}|x + 3| - 4.$$

The graphs of both $y = |x|$ and our new function are shown in Figure 39.   ◼

NOTE   The order in which the transformations are made is important. If the same transformations are made in a different order, a different equation can result.

2.3 EXERCISES

Use the concepts of Section 2.2 and this section to draw a rough sketch by hand of the graphs of y_1, y_2, and y_3. Do not plot points. In each case, y_2 and y_3 can be graphed by one or more of these: a vertical and/or horizontal shift of the graph of y_1, a vertical stretch or shrink of the graph of y_1, or a reflection of the graph of y_1 across an axis. After you have made your sketches, check by graphing them in an appropriate viewing window of your calculator.

1. $y_1 = x$, $y_2 = x + 3$, $y_3 = x - 3$

2. $y_1 = x^2$, $y_2 = x^2 - 1$, $y_3 = x^2 + 1$

3. $y_1 = x^3$, $y_2 = x^3 + 4$, $y_3 = x^3 - 4$

4. $y_1 = \sqrt{x}$, $y_2 = \sqrt{x} + 6$, $y_3 = \sqrt{x} - 6$

5. $y_1 = |x|$, $y_2 = |x - 3|$, $y_3 = |x + 3|$

6. $y_1 = \sqrt[3]{x}$, $y_2 = \sqrt[3]{x} - 4$, $y_3 = \sqrt[3]{x} + 4$

7. $y_1 = |x|$, $y_2 = |x| - 3$, $y_3 = |x| + 3$

8. $y_1 = \sqrt[3]{x}$, $y_2 = \sqrt[3]{x} - 4$, $y_3 = \sqrt[3]{x} + 4$

9. $y_1 = \sqrt{x}$, $y_2 = \sqrt{x + 6}$, $y_3 = \sqrt{x - 6}$

10. $y_1 = x^3$, $y_2 = (x - 4)^3$, $y_3 = (x + 4)^3$

11. $y_1 = |x|$, $y_2 = 2|x|$, $y_3 = 2.5|x|$

12. $y_1 = |x|$, $y_2 = -2|x|$, $y_3 = -2.5|x|$

13. $y_1 = \sqrt[3]{x}$, $y_2 = -\sqrt[3]{x}$, $y_3 = -2\sqrt[3]{x}$

14. $y_1 = \sqrt[3]{x}$, $y_2 = 2\sqrt[3]{x}$, $y_3 = \dfrac{1}{2}\sqrt[3]{x}$

15. $y_1 = x^2$, $y_2 = (x - 2)^2 + 1$, $y_3 = -(x + 2)^2$

16. $y_1 = x^2$, $y_2 = -(x + 3)^2 - 2$, $y_3 = -(x - 4)^2$

17. $y_1 = |x|$, $y_2 = -2|x - 1| + 1$, $y_3 = -\dfrac{1}{2}|x| - 4$

18. $y_1 = |x|$, $y_2 = -|x + 1| - 4$, $y_3 = -|x - 1|$

In Exercises 19–24, fill in the blanks with the appropriate responses. (Remember that the vertical stretch or shrink factor is positive.)

19. The graph of $y = -4x^2$ can be obtained from the graph of $y = x^2$ by vertically stretching by a factor of _____ and reflecting across the _____ -axis.

20. The graph of $y = -6\sqrt{x}$ can be obtained from the graph of $y = \sqrt{x}$ by vertically stretching by a factor of _____ and reflecting across the _____ -axis.

21. The graph of $y = -\frac{1}{4}|x + 2| - 3$ can be obtained from the graph of $y = |x|$ by shifting horizontally _____ units to the _____, vertically shrinking by a factor of _____, reflecting across the _____ -axis, and shifting vertically _____ units in the _____ direction.

22. The graph of $y = -\frac{2}{5}|-x| + 6$ can be obtained from the graph of $y = |x|$ by reflecting across the _____ -axis, vertically shrinking by a factor of _____, reflecting a second time across the _____ -axis, and shifting vertically _____ units in the _____ direction.

23. The graph of $y = 6\sqrt[3]{(x - 3)}$ can be obtained from the graph of $y = \sqrt[3]{x}$ by shifting horizontally _____ units to the _____ and stretching vertically by a factor of _____.

24. The graph of $y = .5\sqrt[3]{(x + 2)}$ can be obtained from the graph of $y = \sqrt[3]{x}$ by shifting horizontally _____ units to the _____ and vertically shrinking by a factor of _____.

Give the equation of the function whose graph is described.

25. The graph of $y = x^2$ is vertically shrunk by a factor of $\frac{1}{2}$, and the resulting graph is shifted 7 units downward.

26. The graph of $y = x^3$ is vertically stretched by a factor of 3. This graph is then reflected across the x-axis. Finally, the graph is shifted 8 units upward.

27. The graph of $y = \sqrt{x}$ is shifted 3 units to the right. This graph is then vertically stretched by a factor of 4.5. Finally, the graph is shifted 6 units downward.

28. The graph of $y = \sqrt[3]{x}$ is shifted 2 units to the left. This graph is then vertically stretched by a factor of 1.5. Finally, the graph is shifted 8 units upward.

In Exercises 29 and 30, the graph of $y = f(x)$ has been transformed to the graph of $y = g(x)$. No shrinking or stretching is involved. Give the equation of $y = g(x)$.

29.

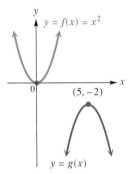

30.

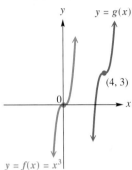

In each of Exercises 31–36, the figure shows the graph of a function $y = f(x)$. Sketch by hand the graphs of the functions in parts (a), (b) and (c), and answer the question of part (d).

31. (a) $y = -f(x)$ **(b)** $y = f(-x)$ **(c)** $y = 2f(x)$ **(d)** What is $f(0)$?

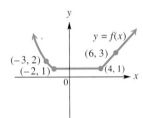

32. (a) $y = -f(x)$ **(b)** $y = f(-x)$ **(c)** $y = 3f(x)$ **(d)** What is $f(4)$?

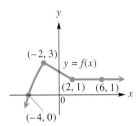

33. (a) $y = -f(x)$ **(b)** $y = f(-x)$ **(c)** $y = f(x + 1)$
(d) What are the x-intercepts of $y = f(x - 1)$?

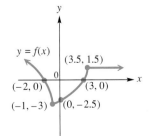

34. (a) $y = -f(x)$ **(b)** $y = f(-x)$ **(c)** $y = -2f(x)$
(d) On what interval of the domain is $f(x) < 0$?

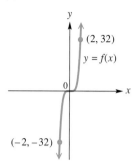

35. (a) $y = -f(x)$ **(b)** $y = f(-x)$ **(c)** $y = .5f(x)$
(d) What symmetry does the graph of $y = f(x)$ exhibit?

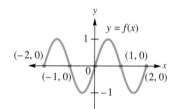

36. (a) $y = -f(x)$ **(b)** $y = f(-x)$ **(c)** $y = 3f(x)$
(d) What symmetry does the graph of $y = f(x)$ exhibit?

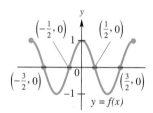

37. If r is an x-intercept of the graph of $y = f(x)$, what statement can be made about an x-intercept of the graph of each of the following? (*Hint:* Draw a picture.)
(a) $y = -f(x)$ **(b)** $y = f(-x)$ **(c)** $y = -f(-x)$

38. If b is the y-intercept of the graph of $y = f(x)$, what statement can be made about the y-intercept of the graph of each of the following? (*Hint:* Draw a picture.)
(a) $y = -f(x)$ **(b)** $y = f(-x)$ **(c)** $y = 5f(x)$ **(d)** $y = -3f(x)$

Relating Concepts

Each of the following functions will have a graph with an endpoint (a translation of the point $(0, 0)$). Enter each into your calculator in an appropriate viewing window, and with your knowledge of the graph of $y = \sqrt{x}$, determine the domain and the range of the function. (Hint: Locate the endpoint.) See Section 1.2.

39. $y = 10\sqrt{x - 20} + 5$ **40.** $y = -2\sqrt{x + 15} - 18$ **41.** $y = -.5\sqrt{x + 10} + 5$

42. Based on your observations in Exercise 39, what are the domain and the range of $f(x) = a\sqrt{x - h} + k$, if $a > 0$, $h > 0$, and $k > 0$?

*Determine the domain and the range of each function. State the intervals over which the function is (**a**) increasing, (**b**) decreasing, and (**c**) constant.*

43. the function graphed in Figure 32

44. the function graphed in Figure 33

45. the function graphed in Figure 34

46. $y = -\dfrac{2}{3}|x + 3| - 4$ (see Figure 39)

Shown here are the graphs of $y_1 = \sqrt[3]{x}$ and $y_2 = 5\sqrt[3]{x}$. The point whose coordinates are given at the bottom of the screen lies on the graph of y_1. Use this graph, and not your own calculator, to find the value of y_2 for the same value of x shown.

47.

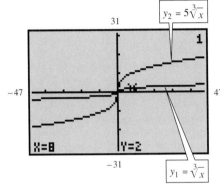

48.

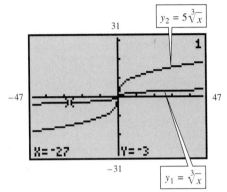

49.

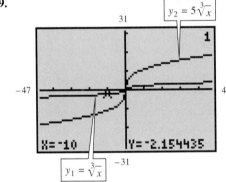

50.

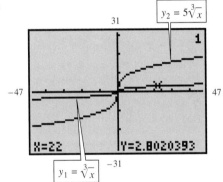

The following sketch shows an example of a function $y = f(x)$ that increases on the interval (a, b) of its domain.

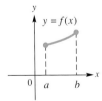

Use this graph as a visual aid, and apply the concepts of reflection introduced in this section to answer each of the following questions. (Make your own sketch if you wish.)

51. Does the function $y = -f(x)$ increase or decrease on the interval (a, b)?

52. Does the function $y = f(-x)$ increase or decrease on the interval $(-b, -a)$?

53. Does the function $y = -f(-x)$ increase or decrease on the interval $(-b, -a)$?

54. If $c > 0$, does the graph of $y = -c \cdot f(x)$ increase or decrease on the interval (a, b)?

55. If the graph of the function $y = f(x)$ is symmetric with respect to the y-axis, what can be said about the symmetry of **(a)** $y = f(-x)$ and **(b)** $y = -f(x)$?

56. If the graph of the function $y = f(x)$ is symmetric with respect to the origin, what can be said about the symmetry of **(a)** $y = f(-x)$ and **(b)** $y = -f(x)$?

R*elating Concepts*

Use either the x-intercept method of graphical solution or the intersection-of-graphs method of graphical solution to solve each equation or inequality. Express solutions or endpoints correct to the nearest hundredth. See Sections 1.5 and 1.6.

57. $\sqrt[3]{\pi x + 3.6} = |\sqrt{2}x - 4.8|$

58. $\sqrt[3]{\pi x + 3.6} > |\sqrt{2}x - 4.8|$

59. $\sqrt[3]{\pi x + 3.6} < |\sqrt{2}x - 4.8|$

60. $|2\pi x - 4| = 4(x - 1)^3 + 1.6$

61. $|2\pi x - 4| > 4(x - 1)^3 + 1.6$

62. $|2\pi x - 4| < 4(x - 1)^3 + 1.6$

2.4 THE ABSOLUTE VALUE FUNCTION: GRAPHS, EQUATIONS, AND INEQUALITIES

The Graph of $y = |f(x)|$ ▪ Properties of Absolute Value ▪ Equations and Inequalities Involving Absolute Value

Of the elementary functions introduced so far in this chapter, the identity, squaring, and cubing functions are all examples of polynomial functions, a class of functions that will be more closely examined in Chapter 3. The square root and cube root functions are examples of root functions, and this class will be studied in Chapter 4 (along with another important class, rational functions). This leaves only the absolute value function, and in this section we will investigate this function in detail.

The Graph of $y = |f(x)|$

Recall from Chapter 1 that the absolute value of a real number a is defined as follows:

$$|a| = \begin{cases} a & \text{if } a \geq 0 \\ -a & \text{if } a < 0. \end{cases}$$

Geometrically, the absolute value of a real number is its undirected distance from 0 on the number line. As a result of this, the absolute value of a real number is never negative; it is always greater than or equal to 0. The absolute value function was defined in Section 2.1 by simply replacing a with x. Thus, the function

$$f(x) = |x| = \begin{cases} x & \text{if } x \geq 0 \\ -x & \text{if } x < 0 \end{cases}$$

is just an extension of the definition of absolute value. The expression within the absolute value bars, x, is the defining expression for the identity function $y = x$.

Now let us extend the concept further, and consider the definition of a function defined by the *absolute value of a function f*:

$$|f(x)| = \begin{cases} f(x) & \text{if } f(x) \geq 0 \\ -f(x) & \text{if } f(x) < 0. \end{cases}$$

In order to graph a function of the form $y = |f(x)|$, the definition indicates that the graph is the same as that of $y = f(x)$ for values of $f(x)$ (that is, range values) that are nonnegative. The second part of the definition indicates that for range values that are negative, the graph of $y = f(x)$ is reflected across the x-axis.

The domain of $y = |f(x)|$ is the same as the domain of f, while the range of $y = |f(x)|$ will be a subset of $[0, \infty)$.

Figure 40 shows the graph of $y = (x - 4)^2 - 3$, which is the graph of $y = x^2$ shifted 4 units to the right and 3 units downward. Figure 41 shows the graph of $y = |(x - 4)^2 - 3|$. Notice that all points with negative y-values in the first graph have been reflected across the x-axis, while all points with nonnegative y-values are the same for both graphs.

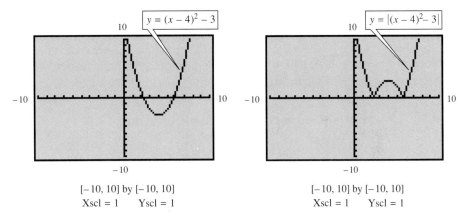

$$y = (x - 4)^2 - 3$$

$$y = |(x - 4)^2 - 3|$$

[−10, 10] by [−10, 10]   [−10, 10] by [−10, 10]
Xscl = 1   Yscl = 1   Xscl = 1   Yscl = 1

FIGURE 40   **FIGURE 41**

The domain of both functions is $(-\infty, \infty)$. The range of $y = (x - 4)^2 - 3$ is $[-3, \infty)$, while the range of $y = |(x - 4)^2 - 3|$ is $[0, \infty)$. ∎

Figure 42 shows the graph of a function $y = f(x)$. Use the figure to sketch the graph of $y = |f(x)|$. Give the domain and the range of each.

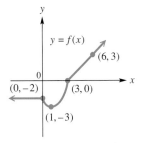

FIGURE 42

SOLUTION   As stated earlier, the graph will remain the same for points whose *y*-values are nonnegative, while it will be reflected across the *x*-axis for all other points. Figure 43 shows the graph of $y = |f(x)|$. The domain of both functions is $(-\infty, \infty)$. The range of $y = f(x)$ is $[-3, \infty)$, while the range of $y = |f(x)|$ is $[0, \infty)$.   █

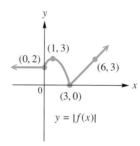

FIGURE 43

EXAMPLE **3**
Answering a Conceptual Question Concerning the Graph of $y = |f(x)|$

Why is the graph of $y = (x + 2)^2 + 1$ the same as the graph of $y = |(x + 2)^2 + 1|$?

SOLUTION   The function $y = (x + 2)^2 + 1$ has as its graph the graph of $y = x^2$ shifted 2 units to the left, and 1 unit upward, causing its range to be $[1, \infty)$. Since the range consists of only positive numbers, the graph of $y = |(x + 2)^2 + 1|$ will not be affected by reflection across the *x*-axis. Therefore, the two graphs are identical. (Verify this on your calculator.)   █

Properties of Absolute Value

There are several properties of absolute value that will be useful in our later work.

PROPERTIES OF ABSOLUTE VALUE

For all real numbers *a* and *b*:

1. $|ab| = |a| \cdot |b|$
The absolute value of a product is equal to the product of the absolute values.

2. $\left|\dfrac{a}{b}\right| = \dfrac{|a|}{|b|}$ $(b \neq 0)$
The absolute value of a quotient is equal to the quotient of the absolute values.

3. $|a| = |-a|$
The absolute value of a number is equal to the absolute value of its additive inverse.

4. $|a| + |b| \geq |a + b|$ (the triangle inequality)
The sum of the absolute values of two numbers is greater than or equal to the absolute value of their sum.

Among other applications, these properties can be used to explain the behavior of graphs of functions involving absolute value. For example, consider the function $y = |2x + 11|$ and observe the following sequence of transformations.

$$y = |2x + 11| \qquad \text{Given}$$

$$y = \left| 2\left(x + \frac{11}{2} \right) \right| \qquad \text{Factor out a 2.}$$

$$y = |2| \cdot \left| x + \frac{11}{2} \right| \qquad \text{Property 1}$$

$$y = 2\left| x + \frac{11}{2} \right| \qquad |2| = 2$$

Using the concepts of the previous two sections, we conclude that the graph of this function can be found by starting with the graph of $y = |x|$, shifting $\frac{11}{2}$ units to the left, and then vertically stretching by a factor of 2. The graphs of $y_1 = |x|$ and $y_2 = |2x + 11|$ in Figure 44 give support to this statement.

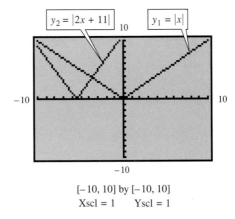

$y_2 = |2x + 11|$ $y_1 = |x|$

[−10, 10] by [−10, 10]
Xscl = 1   Yscl = 1

FIGURE 44

FOR **GROUP DISCUSSION**

Choose any elementary function studied so far, with any transformations you wish. Call it f. Find a complete graph of $y_1 = |f(x)|$ on your calculator. Then find a complete graph of $y_2 = |-f(x)|$ on your calculator. What do you notice? What property of absolute value justifies your conclusion?

We will often be interested in examining absolute value functions of the form

$$f(x) = |ax + b|,$$

where the expression inside the absolute value bars is linear. For purposes of discussion, we now give a definition of a complete graph of $f(x) = |ax + b|$.

COMPLETE GRAPH OF $f(x) = |ax + b|$

A complete graph of $f(x) = |ax + b|$ will include all intercepts and the lowest point on the "V-shaped" graph.

Equations and Inequalities Involving Absolute Value

There are numerous types of equations (and related inequalities) involving absolute value. We will investigate those types that involve absolute values of linear functions, and constants. As first explained in Chapter 1, we will solve them analytically, and then support our solutions graphically.

The summary that follows indicates the general method for solving some of the simpler absolute value equations and inequalities analytically.

SOLVING ABSOLUTE VALUE EQUATIONS AND INEQUALITIES

Let k be a positive number.

1. To solve $|ax + b| = k$, solve the compound equation

$$ax + b = k \quad \text{or} \quad ax + b = -k.$$

2. To solve $|ax + b| > k$, solve the compound inequality

$$ax + b > k \quad \text{or} \quad ax + b < -k.$$

3. To solve $|ax + b| < k$, solve the three-part inequality

$$-k < ax + b < k.$$

Inequalities involving $\leq$ or $\geq$ are solved similarly, using the equality part of the symbol as well.

<table>
<tr><td>

EXAMPLE 4

Solving
$|ax + b| = k,$
$k > 0$, Analytically

</td></tr>
</table>

Solve $|2x + 1| = 7$ analytically.

SOLUTION   For $|2x + 1|$ to equal 7, $2x + 1$ must be 7 units from 0 on the number line. This can happen only when $2x + 1 = 7$ or $2x + 1 = -7$. Solve this compound equation as follows.

$$
\begin{array}{rcl}
2x + 1 = 7 & \text{or} & 2x + 1 = -7 \\
2x = 6 & \text{or} & 2x = -8 \\
x = 3 & \text{or} & x = -4
\end{array}
$$

The solution set is $\{-4, 3\}$.   ∎

Figure 45 shows the graphs of $y_1 = |2x + 1|$ and $y_2 = 7$. Notice that the graphs appear to intersect at points where $x = -4$ and where $x = 3$, giving graphical support to our solution in Example 4. (Our calculator would confirm this observation.) Using the intersection-of-graphs method for solving *inequalities*, it would seem to suggest that the solution set of

$$|2x + 1| < 7$$

is the open interval $(-4, 3)$, since the graph of y_1 lies *below* that of y_2 there. On the other hand, the solution set of

$$|2x + 1| > 7$$

would be $(-\infty, -4) \cup (3, \infty)$, since for these intervals, the graph of y_1 lies *above* that of y_2. Notice that these observations agree with the way $|ax + b| < k$ and

$|ax + b| > k$ were defined. In the following example, we provide analytic justification for these observations.

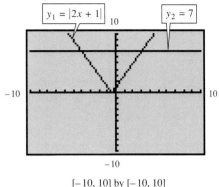

$[-10, 10]$ by $[-10, 10]$
Xscl = 1   Yscl = 1

The graphs appear to intersect
at points for which $x = -4$
and $x = 3$.

FIGURE 45

EXAMPLE 5

Solving
$|ax + b| < k$ and
$|ax + b| > k$,
$k > 0$, Analytically

(a) Solve $|2x + 1| < 7$ analytically.

SOLUTION Here the expression $2x + 1$ must represent a number that is less than 7 units from 0 on the number line. Another way of thinking of this is to realize that $2x + 1$ must be between -7 and 7. This is written as the three-part inequality

$$-7 < 2x + 1 < 7.$$

We solved such inequalities in Section 1.6 by working with all three parts at the same time.

$$-7 < 2x + 1 < 7$$
$$-8 < 2x < 6 \qquad \text{Subtract 1 from each part.}$$
$$-4 < x < 3 \qquad \text{Divide each part by 2.}$$

The solution set is the open interval $(-4, 3)$, as we observed earlier.

(b) Solve $|2x + 1| > 7$ analytically.

SOLUTION This absolute value inequality must be rewritten as

$$2x + 1 > 7 \qquad \text{or} \qquad 2x + 1 < -7,$$

because $2x + 1$ must represent a number that is *more* than 7 units from 0 on either side of the number line. Now solve the compound inequality.

$$2x + 1 > 7 \qquad \text{or} \qquad 2x + 1 < -7$$
$$2x > 6 \qquad \text{or} \qquad 2x < -8$$
$$x > 3 \qquad \text{or} \qquad x < -4$$

The solution set, $(-\infty, -4) \cup (3, \infty)$, again confirms our earlier observation. ▯

Provide support for the solutions in Examples 4 and 5 by letting $y_1 = |2x + 1|$, $y_2 = 7$, and graphing the function $y_1 - y_2$. Then use the x-intercept method of graphical solution, first explained for *linear* equations and inequalities in Sections 1.5 and 1.6.

An absolute value equation of the form $|ax + b| = k$, where $k < 0$, will have no solution, since the absolute value of a real number cannot be negative. The related inequalities, $|ax + b| > k$ and $|ax + b| < k$, will have solution sets $(-\infty, \infty)$ and $\emptyset$, respectively. (What is the solution of $|ax + b| = 0$?) The following example provides graphical support for a particular case.

EXAMPLE 6

Solving Absolute Value Equations and Inequalities With No Solutions or Infinitely Many Solutions Graphically

Solve graphically the following:

$$|3x + 10| = -5, \qquad |3x + 10| > -5, \qquad |3x + 10| < -5.$$

Use the x-intercept method of solution.

SOLUTION   Let $y_1 = |3x + 10|$ and $y_2 = -5$ and graph $y_1 - y_2$ in the standard viewing window. See Figure 46.

The equation $|3x + 10| = -5$ is equivalent to $|3x + 10| + 5 = 0$. The left side of this equation is our function $y_1 - y_2$. Since the graph has no x-intercepts, the solution of this equation is $\emptyset$.

The inequality $|3x + 10| > -5$ is equivalent to $|3x + 10| + 5 > 0$. Since the graph of $y_1 - y_2$ lies completely *above* the x-axis, the solution set of this inequality is $(-\infty, \infty)$. On the other hand, the solution set of the inequality $|3x + 10| < -5$ is $\emptyset$, since there are no x-values for which the graph lies *below* the x-axis.   ∎

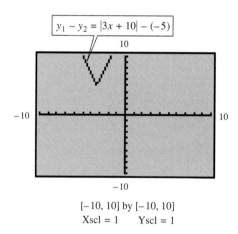

$y_1 - y_2 = |3x + 10| - (-5)$

$[-10, 10]$ by $[-10, 10]$
Xscl = 1   Yscl = 1

FIGURE 46

Confirm the results of Example 6 using the intersection-of-graphs methods.

If two quantities have the same absolute value, they must either be equal to each other or be negatives of each other. This fact allows us to solve absolute value equations of the form

$$|ax + b| = |cx + d|.$$

Once we have solved this type of equation analytically and supported our result graphically, we can then solve the related inequalities $|ax + b| < |cx + d|$ and $|ax + b| > |cx + d|$.

SOLVING $|ax + b| = |cx + d|$

To solve the equation $|ax + b| = |cx + d|$ analytically, solve the compound equation

$$ax + b = cx + d \qquad \text{or} \qquad ax + b = -(cx + d).$$

EXAMPLE 7

Solving an Equation and Related Inequalities Involving Two Absolute Values

(a) Solve the equation $|x + 6| = |2x - 3|$ analytically, and support the solution graphically.

SOLUTION   This equation is satisfied if

$$x + 6 = 2x - 3 \qquad \text{or} \qquad x + 6 = -(2x - 3).$$

Solve each equation.

$$x + 6 = 2x - 3 \qquad\qquad x + 6 = -2x + 3$$
$$9 = x \qquad\qquad\qquad 3x = -3$$
$$x = -1$$

The solution set of the equation is $\{-1, 9\}$. By graphing $y_1 = |x + 6|$ and $y_2 = |2x - 3|$ in the window $[-20, 20]$ by $[-20, 20]$, we see that the points of intersection of y_1 and y_2 appear to have x-coordinates -1 and 9. Use your calculator to verify this statement. (Why is the standard window not appropriate for this problem?) See Figure 47.

TECHNOLOGICAL NOTE
You might wish to support the results of Example 7 graphically by using the x-intercept method of solution. To do this, enter Y1 as $|x + 6|$, Y2 as $|2x - 3|$ and Y3 as Y1 − Y2. Then graph only Y3. You will notice that your graph differs from the one seen in Figure 47, but the support should be evident. (You may need to adjust the window to obtain a better view.)

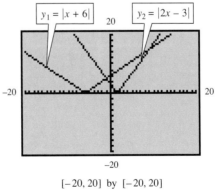

$y_1 = |x + 6|$ $y_2 = |2x - 3|$

$[-20, 20]$ by $[-20, 20]$
Xscl = 1   Yscl = 2

FIGURE 47

(b) Use Figure 47 and the results of part (a) to solve the inequalities

$$|x + 6| < |2x - 3| \qquad \text{and} \qquad |x + 6| > |2x - 3|.$$

SOLUTION   The graph of $y_1 = |x + 6|$ lies *below* the graph of $y_2 = |2x - 3|$ on the intervals $(-\infty, -1)$ and $(9, \infty)$. Therefore, the solution set of $|x + 6| < |2x - 3|$ is

$$(-\infty, -1) \cup (9, \infty).$$

The solution set of $|x + 6| > |2x - 3|$ consists of all x-values for which the graph of y_1 is *above* that of y_2. This happens between -1 and 9, so the solution set of $|x + 6| > |2x - 3|$ is

the interval $(-1, 9)$.   ◘

NOTE   In Examples 5, 6, and 7, all of the inequalities that were solved used the symbols $<$ or $>$. If such an inequality involves $\leq$ or $\geq$, the method of solution is the same, but the endpoints are included. The graphical support will *not* tell us this; we must pay attention to the type of inequality symbol in order to determine whether to use parentheses or brackets when writing the solution set.

EXAMPLE 8

Solving Equations and Inequalities Involving a Sum of Absolute Values

(a)   Solve graphically $|x + 5| + |x - 3| = 16$ by the intersection-of-graphs method and verify analytically.

SOLUTION   Let $y_1 = |x + 5| + |x - 3|$ and $y_2 = 16$. Graphing them in an appropriate window gives us the graphs shown in Figure 48. Using a calculator to locate the points of intersection of the graphs, we find the x-coordinates of the points are -9 and 7. To verify these solutions analytically, we substitute them into the equation.

$$\text{Let } x = -9: \quad |(-9) + 5| + |(-9) - 3| = |-4| + |-12|$$
$$= 4 + 12 = 16. \ \checkmark$$
$$\text{Let } x = 7: \quad |7 + 5| + |7 - 3| = |12| + |4|$$
$$= 12 + 4 = 16. \ \checkmark$$

Therefore, the solution set of the equation is $\{-9, 7\}$.

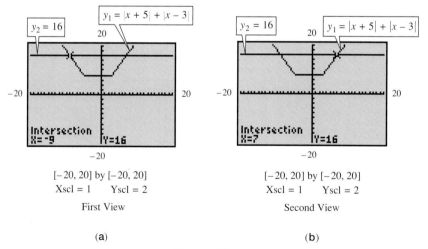

(a)   **(b)**

FIGURE 48

(b)   Use the results of part (a) to solve the inequalities

$$|x + 5| + |x - 3| \leq 16 \quad \text{and} \quad |x + 5| + |x - 3| \geq 16.$$

SOLUTION   To solve the first inequality, we look for the interval(s) on which the graph of the sum of the absolute values (y_1, the V-shaped graph with the "flat bottom" in Figure 48) is *below* or *coincides* with the graph of y_2 (the horizontal line). We see that this occurs on the closed interval $[-9, 7]$. The solution set of the second inequality consists of the intervals on which the graph of y_1 is *above* or *coincides* with the graph of y_2. This occurs on $(-\infty, -9] \cup [7, \infty)$.

Summarizing the two parts of this example, we have the following:

The solution set of $|x + 5| + |x - 3| = 16$ is $\{-9, 7\}$.

The solution set of $|x + 5| + |x - 3| \leq 16$ is the closed interval $[-9, 7]$.

The solution set of $|x + 5| + |x - 3| \geq 16$ is $(-\infty, -9] \cup [7, \infty)$.   ◼

FOR GROUP DISCUSSION

Figure 49 shows the graphs of two functions, $f(x) = |.5x + 6|$ and $g(x) = 3x - 14$. The coordinates of the point of intersection of the graphs are shown at the bottom of the screen. Decide which graph is that of f, which is that of g, and then solve each of these:

$$f(x) = g(x) \qquad f(x) > g(x) \qquad f(x) < g(x).$$

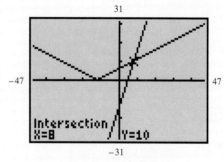

$[-47, 47]$ by $[-31, 31]$
Xscl = 10   Yscl = 10
Which is the graph of f?
Which is the graph of g?
$f(x) = |.5x + 6|$ $g(x) = 3x - 14$

FIGURE 49

2.4 EXERCISES

In Exercises 1–15, you are given graphs of functions $y = f(x)$. Sketch by hand the graph of $y = |f(x)|$.

1.
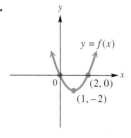

2.
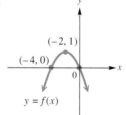

3.
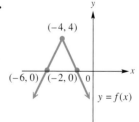

4.

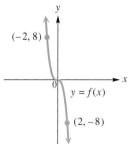

5.

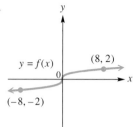

6.

7.

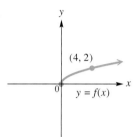

8.

9.

10.

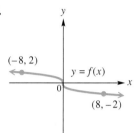

11.

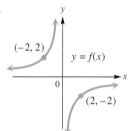

12.

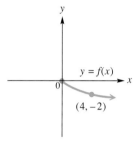

13.

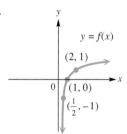

14.

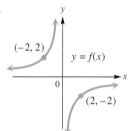

15.

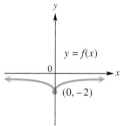

16. Explain in your own words the process you used to find the graphs of $y = |f(x)|$ in Exercises 1–15.

*R*elating Concepts

17. Shown here is the graph of a function $y = f(x)$. Sketch by hand, in order, the graph of each of the following. Use the concept of reflecting introduced in Section 2.3, and the concept of graphing $y = |f(x)|$ introduced in this section.
(a) $y = f(-x)$
(b) $y = -f(-x)$
(c) $y = |-f(-x)|$

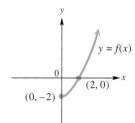

18. Repeat Exercise 17 for the graph of $y = f(x)$ shown here.

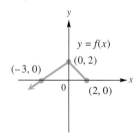

19. If the range of $y = f(x)$ is $[-2, \infty)$, what is the range of $y = |f(x)|$?

20. If the range of $y = f(x)$ is $(-\infty, -2]$, what is the range of $y = |f(x)|$?

In Exercises 21 and 22, one of the graphs is that of $y = f(x)$ and the other is that of $y = |f(x)|$. State which is which.

21. (a)   **(b)**

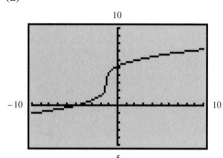
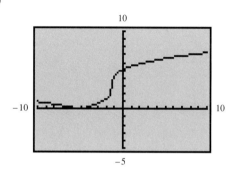

22. (a)   **(b)**

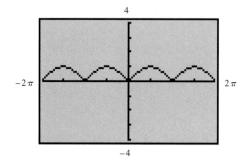
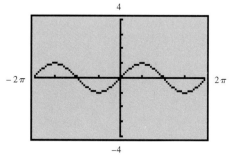

Verify properties 1, 2, 3, and 4 of absolute value for the given values of a and b.

23. $a = 7, b = 14$ **24.** $a = -10, b = -20$ **25.** $a = -26, b = 13$

26.
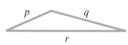

(a) In the figure above, how does the sum of the lengths p and q compare to length r?
   (*less than, greater than,* or *equal to*)

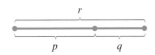

(b) In the figure above, how does the sum of the lengths p and q compare to length r? (*less than, greater than,* or *equal to*)

(c) Complete the following well-known fact: "The shortest distance between two points is _____."

(d) In the triangle inequality (absolute value property 4), when does the order operation $>$ hold? When does equality hold?

Solve each group of equations and inequalities analytically. Support your solutions graphically.

27. **(a)** $|x + 4| = 9$
   (b) $|x + 4| > 9$
   (c) $|x + 4| < 9$

28. **(a)** $|x - 3| = 5$
   (b) $|x - 3| > 5$
   (c) $|x - 3| < 5$

29. **(a)** $|2x + 7| = 3$
   (b) $|2x + 7| \geq 3$
   (c) $|2x + 7| \leq 3$

30. **(a)** $|3x - 9| = 6$
   (b) $|3x - 9| \geq 6$
   (c) $|3x - 9| \leq 6$

31. **(a)** $|6(x + 1)| = 1$
   (b) $|6(x + 1)| < 1$
   (c) $|6(x + 1)| > 1$

32. **(a)** $|-2(x + 3)| = 5$
   (b) $|-2(x + 3)| < 5$
   (c) $|-2(x + 3)| > 5$

33. **(a)** $|2x + 1| + 3 = 5$
   (b) $|2x + 1| + 3 \leq 5$
   (c) $|2x + 1| + 3 \geq 5$

34. **(a)** $|4x + 7| = 0$
   (b) $|4x + 7| > 0$
   (c) $|4x + 7| < 0$

35. **(a)** $|7 - 2x| = 0$
   (b) $|7 - 2x| > 0$
   (c) $|7 - 2x| < 0$

36. **(a)** $|7x - 5| = 0$
   (b) $|7x - 5| \geq 0$
   (c) $|7x - 5| \leq 0$

37. **(a)** $|\pi x + 8| = -4$
   (b) $|\pi x + 8| < -4$
   (c) $|\pi x + 8| > -4$

38. **(a)** $|\sqrt{2}x - 3.6| = -1$
   (b) $|\sqrt{2}x - 3.6| \leq -1$
   (c) $|\sqrt{2}x - 3.6| \geq -1$

Use the graph, along with the indicated points, to give the solution set for each equation or inequality.

39. **(a)** $y_1 = y_2$
   (b) $y_1 < y_2$
   (c) $y_1 > y_2$

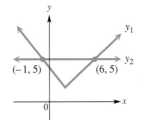

40. **(a)** $y_1 = y_2$
   (b) $y_1 < y_2$
   (c) $y_1 > y_2$

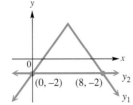

41. **(a)** $y_1 = y_2$
   (b) $y_1 \leq y_2$
   (c) $y_1 \geq y_2$

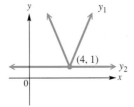

42. (a) $y_1 = y_2$
  (b) $y_1 \leq y_2$
  (c) $y_1 \geq y_2$

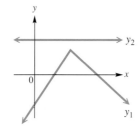

Solve the equation in part (a) by analytic methods. Then, using your graphics calculator, solve the related inequalities in parts (b) and (c).

43. (a) $|3x + 1| = |2x - 7|$
  (b) $|3x + 1| > |2x - 7|$
  (c) $|3x + 1| < |2x - 7|$

44. (a) $|7x + 12| = |x - 4|$
  (b) $|7x + 12| < |x - 4|$
  (c) $|7x + 12| > |x - 4|$

45. (a) $|.5x - 2| = |x - .5|$
  (b) $|.5x - 2| \leq |x - .5|$
  (c) $|.5x - 2| \geq |x - .5|$

46. (a) $|6x| = |9x + 5|$
  (b) $|6x| \geq |9x + 5|$
  (c) $|6x| \leq |9x + 5|$

47. (a) $|2x - 6| = |2x + 5|$
  (b) $|2x - 6| < |2x + 5|$
  (c) $|2x - 6| > |2x + 5|$

48. (a) $|3x - 1| = |3x + 4|$
  (b) $|3x - 1| > |3x + 4|$
  (c) $|3x - 1| < |3x + 4|$

The equations in Exercises 49–52 have integer solutions. Solve the equations and related inequalities graphically.

49. (a) $|x + 1| + |x - 6| = 11$
  (b) $|x + 1| + |x - 6| < 11$
  (c) $|x + 1| + |x - 6| > 11$

50. (a) $|2x + 2| + |x + 1| = 9$
  (b) $|2x + 2| + |x + 1| > 9$
  (c) $|2x + 2| + |x + 1| < 9$

51. (a) $|x| + |x - 4| = 8$
  (b) $|x| + |x - 4| \leq 8$
  (c) $|x| + |x - 4| \geq 8$

52. (a) $|.5x + 2| + |.25x + 4| = 9$
  (b) $|.5x + 2| + |.25x + 4| \geq 9$
  (c) $|.5x + 2| + |.25x + 4| \leq 9$

R*elating Concepts*

The equations in Exercises 53–56 are nonstandard types (as presented so far), but can be solved graphically using the intersection-of-graphs method or the x-intercept method. Find the solution sets. Express solutions rounded to the nearest hundredth.

53. $|3x - 13| = \sqrt{x + 5}$

54. $|(x - 4)^2 - 2| = \sqrt[3]{x} + 5$

55. $|\pi x - \sqrt{2}| = x^2$

56. $\sqrt[3]{x - 5} = -\sqrt{x - 1}$

Solve graphically the following equations and inequalities. Express solutions or endpoints of intervals rounded to the nearest hundredth if necessary.

57. $|2x + 7| = 6x - 1$

58. $-|3x - 12| \geq -x - 1$

59. $|x - 4| > .5x - 6$

60. $2x + 8 > -|3x + 4|$

61. $|3x + 4| < -3x - 14$

62. $|2\pi x - 5| = 5x + 3$

63. $|\pi x - \sqrt{7}| = -x + \sqrt{11}$

64. $|x - \sqrt{13}| + \sqrt{6} \leq -x - \sqrt{10}$

Solve each problem.

65. The temperatures on the surface of Mars in degrees Celsius approximately satisfy the inequality

$$|C + 84| \leq 56.$$

What range of temperatures corresponds to this inequality?

66. Dr. Tydings has found that, over the years, 95% of the babies that he has delivered have weighed y pounds, where $|y - 8.0| \leq 1.5$. What range of weights corresponds to this inequality?

67. The industrial process that is used to convert methanol to gasoline is carried out at a temperature range of 680°F to 780°F. Using F as the variable, write an absolute value inequality that corresponds to this range.

68. When a model kite was flown in crosswinds in tests to determine its limits of power extraction, it attained speeds of 98 to 148 feet per second in winds of 16 to 26 feet per second. Using x as the variable in each case, write absolute value inequalities that correspond to these ranges.

Further Explorations

1. Exercises 27–38 can be supported numerically with the TABLE feature of your graphics calculator.

    Example: Solve: $|x - 5| < 2$

    Solution: Let $Y_1 = |x - 5|$ and $Y_2 = 2$. Set TblMin $= 2$ and ΔTbl$=1$. Use the TABLE to see when $Y_1 < 2$.

    Note: Both boundaries of the solution set may not be visible at the same time in the TABLE. Scrolling may be necessary to view both boundaries.

2. Refer to Exercises 65 and 66. If your graphics calculator has a TABLE feature, use the TABLE to numerically support your solutions to these exercises. (For Exercise 66, set ΔTbl $= .5$.)

2.5 PIECEWISE-DEFINED FUNCTIONS

Graphing Functions Defined Piecewise ▮ The Greatest Integer Function ▮ An Application of a Step Function

TECHNOLOGICAL NOTE
As mentioned in the instructions for Exercises 37–40 in this section, some graphics calculators can graph piecewise-defined functions. This will often require the use of the logic (or Boolean) operations. You should read your owner's manual if you wish to investigate this topic.

Graphing Functions Defined Piecewise

The absolute value function, defined in Section 2.1, is a simple example of a function defined by different rules over different subsets of its domain. Recall that the domain of $f(x) = |x|$ is $(-\infty, \infty)$. For the interval $[0, \infty)$ of the domain, the rule that we use is $f(x) = x$. On the other hand, for the interval $(-\infty, 0)$, we use the rule $f(x) = -x$. Thus the graph of $f(x) = |x|$ is composed of two "pieces". One piece comes from the graph of $y = x$, and the other from $y = -x$.

EXAMPLE 1

Finding Function
Values for a
Piecewise-Defined
Function

Consider the function f defined piecewise:

$$f(x) = \begin{cases} x + 2 & \text{if } x \leq 0 \\ \dfrac{1}{2}x^2 & \text{if } x > 0. \end{cases}$$

Find the following function values.

(a) $f(-3)$

SOLUTION Since $-3 \leq 0$ (specifically, $-3 < 0$), we use the rule $f(x) = x + 2$. Thus, $f(-3) = -3 + 2 = -1$. Note that this means the graph of f will contain the point $(-3, -1)$.

(b) $f(0)$

SOLUTION Since $0 \leq 0$ (because $0 = 0$), we again use the rule $f(x) = x + 2$. So we have $f(0) = 0 + 2 = 2$. The point $(0, 2)$ will lie on the graph of f, meaning that the y-intercept of the graph will be 2.

(c) $f(3)$

SOLUTION The number 3 comes from the interval $(0, \infty)$, and the second part of the rule f indicates that we must now use the rule $f(x) = \frac{1}{2}x^2$. Therefore, $f(3) = \frac{1}{2}(3)^2 = \frac{1}{2}(9) = 4.5$. (What point on the graph of f does this result lead to?) ◖

If we look at the function f defined in Example 1, we can visualize that for negative values of x or 0, the graph of f will be a portion of a line because $y = x + 2$ defines a linear function. However, if x is positive, the graph of f will consist of a portion of a parabola. Specifically, it will be the "right half" of the graph of the squaring function $y = x^2$, with a shrink factor of $\frac{1}{2}$.

EXAMPLE 2

Graphing a Function
Defined Piecewise

Using your knowledge of the graphs of linear functions and the squaring function, sketch the graph of the function f in Example 1 by hand.

SOLUTION Graph the ray $y = x + 2$, choosing x so that $x \leq 0$, with a solid endpoint at $(0, 2)$. The ray has slope 1 and y-intercept 2. Then graph $y = \frac{1}{2}x^2$ for $x > 0$. This graph will be half of a parabola with an open endpoint at $(0, 0)$. See Figure 50. ◖

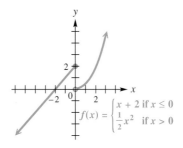

FIGURE 50

> **FOR GROUP DISCUSSION**
>
> **1.** Discuss the behavior of the graph in Figure 50 at the domain value 0.
> **2.** Why would it not be a function if the open endpoint at the origin were included on the graph.
> **3.** Why would it be *wrong* to join the points (0, 2) and (0, 0) with a "piece of graph"?
> **4.** What is the domain of the function? How can you determine the domain without looking at the graph? What is the range of the function?

Functions defined piecewise often exhibit discontinuities. The function graphed in Example 2 is an example of such a function. However, depending on the manner in which the function is defined, a piecewise-defined function may be continuous on its entire domain, as seen in Example 3.

EXAMPLE 3

Graphing a Function Defined Piecewise

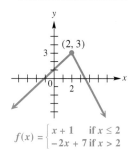

$$f(x) = \begin{cases} x + 1 & \text{if } x \leq 2 \\ -2x + 7 & \text{if } x > 2 \end{cases}$$

FIGURE 51

Graph the function by hand:

$$f(x) = \begin{cases} x + 1 & \text{if } x \leq 2 \\ -2x + 7 & \text{if } x > 2. \end{cases}$$

SOLUTION   We must graph each part of the domain separately. If $x \leq 2$, this portion of the graph has an endpoint at $x = 2$. Find the y-value by substituting 2 for x in $y = x + 1$ to get $y = 3$. Another point is needed to graph this part of the graph. Choose an x-value less than 2. Choosing $x = -1$ gives $y = -1 + 1 = 0$. Draw the graph through (2, 3) and (−1, 0) as a ray with an endpoint at (2, 3). Graph the ray for $x > 2$ similarly. This ray will have an open endpoint when $x = 2$ and $y = -2(2) + 7 = 3$. Choosing $x = 4$ gives $y = -2(4) + 7 = -1$. The ray through (2, 3) and (4, −1) completes the graph. In this example the two rays meet at (2, 3), although this is not always the case, as seen in Example 2. The graph is shown in Figure 51.   ▯

The Greatest Integer Function

An important example of a function defined piecewise is the **greatest integer function.** The notation $[\![x]\!]$ represents the greatest integer less than or equal to x. The definition of $[\![x]\!]$ follows.

$$f(x) = [\![x]\!] = \begin{cases} x \text{ if } x \text{ is an integer} \\ \text{the greatest integer less than } x \text{ if} \\ x \text{ is not an integer} \end{cases}$$

EXAMPLE 4

Evaluating $[\![x]\!]$ for Different Values of x

Evaluate $[\![x]\!]$ for each of the following values of x.

(a) 4

SOLUTION   Since 4 is an integer, $[\![4]\!] = 4$.

(b) −5

SOLUTION   Since −5 is an integer, $[\![-5]\!] = -5$.

(c) 2.46

SOLUTION Since 2.46 is not an integer, $[\![2.46]\!]$ is the greatest integer *less than* 2.46. Thus, $[\![2.46]\!] = 2$.

(d) $[\![\pi]\!]$

SOLUTION $[\![\pi]\!] = 3$, since $\pi \approx 3.14$.

(e) $[\![-6\frac{1}{2}]\!]$

SOLUTION $[\![-6\frac{1}{2}]\!] = -7$ ◻

The graph of $f(x) = [\![x]\!]$ is examined in the next example.

EXAMPLE 5

Graphing $f(x) = [\![x]\!]$

Graph $f(x) = [\![x]\!]$. Give the domain and the range.

SOLUTION For any value of x in the interval $[0, 1)$, $[\![x]\!] = 0$. Also, for x in $[1, 2)$, $[\![x]\!] = 1$. This process continues; for x in $[2, 3)$, the value of $[\![x]\!]$ is 2. The values of y are constant between integers, but they jump at integer values of x. This makes the graph, shown in Figure 52, a series of line segments. In each case, the left endpoint of the segment is included, and the right endpoint is excluded. The domain of the function is $(-\infty, \infty)$, while the range is the set of integers, $\{\ldots, -2, -1, 0, 1, 2, \ldots\}$. ◻

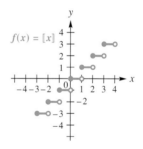

FIGURE 52

The greatest integer function is often called a **step function.** Do you see why?

Our discussion thus far in this section has concentrated on graphing piecewise-defined functions by hand. Although some graphics calculators do have the capability of graphing such functions, we will not discuss this here. You may wish to refer to your owner's manual to see how your particular model can be used this way.

Although graphics calculators provide excellent visual support in many situations involving functions that are continuous throughout their domains, we must be aware that graphing functions with discontinuities sometimes poses a problem for the calculator. For instance, we know from Example 5 what the graph of $f(x) = [\![x]\!]$ looks like. However, if we graph the greatest integer function on a calculator (designated "Int" on many popular models), and the calculator is in the *connected* graphing mode, the calculator attempts to literally connect the portions of the graph at integer values. This is why calculators can also be directed to graph in the *dot* graphing mode. In Figure 53(a), we show an accurate graph of $f(x) = [\![x]\!]$, with the calculator in the dot mode. Figure 53(b) shows a distorted graph of this function, since it was graphed in the connnected mode.

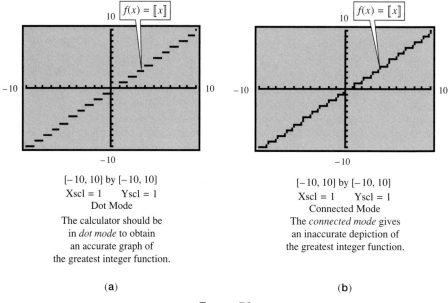

[−10, 10] by [−10, 10]
Xscl = 1   Yscl = 1
Dot Mode

The calculator should be
in *dot mode* to obtain
an accurate graph of
the greatest integer function.

(a)

[−10, 10] by [−10, 10]
Xscl = 1   Yscl = 1
Connected Mode

The *connected mode* gives
an inaccurate depiction of
the greatest integer function.

(b)

FIGURE 53

Notice from the graph in Figure 53(a) that the inclusion or exclusion of the endpoint for each segment is not readily determined from the calculator-generated graph. However, analysis of the hand-drawn graph in Figure 52 does show whether endpoints are included or excluded. Once again, we state the following important conclusion:

> Analysis of a calculator-generated graph is often not sufficient to draw correct conclusions. While this technology is incredibly powerful and pedagogically useful, we cannot rely on it alone in our study of the graphs of functions. We must understand the basic concepts of functional analysis as well.

EXAMPLE 6

Graphing a Step Function

Graph the function $y = [\![\frac{1}{2}x + 1]\!]$ **(a)** by hand and **(b)** with a calculator in dot mode. Give the domain and the range.

SOLUTION

(a)   Try some values of x in the equation to see how the values of y behave. Some sample ordered pairs are given here.

x	0	$\frac{1}{2}$	1	2	3	4	−1	−2	−3
y	1	1	1	2	2	3	0	0	−1

These ordered pairs suggest that if x is in the interval $[0, 2)$, then $y = 1$. For x in $[2, 4)$, $y = 2$, and so on. The graph is shown in Figure 54. Again, the domain is $(-\infty, \infty)$. The range is $\{\ldots, -1, 0, 1, 2, \ldots\}$.

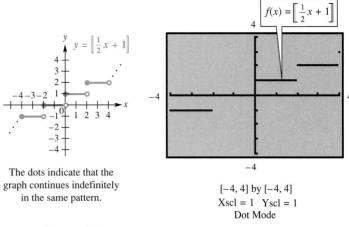

The dots indicate that the
graph continues indefinitely
in the same pattern.

FIGURE 54

[−4, 4] by [−4, 4]
Xscl = 1   Yscl = 1
Dot Mode

FIGURE 55

(b)   Figure 55 shows the graph of $y = [\![\frac{1}{2}x + 1]\!]$ in a window of $[-4, 4]$ by $[-4, 4]$. Use the trace feature to convince yourself that the range of the function is $\{ \ldots -2, -1, 0, 1, 2, \ldots \}$.   ▐

An Application of a Step Function

| FOR **GROUP DISCUSSION** |

1.   What does this mean to you? "Sir (or Madam), to ship this package will cost you $2.90 for the first ounce, and an additional $.50 per ounce for any ounce or fraction thereof after the first."
2.   Based on your answer to item 1, how much would it cost to send a 5.5-ounce package?

The type of pricing described in the group discussion above can be defined by a function that incorporates the greatest integer function concept.

EXAMPLE 7

Applying the
Greatest Integer
Function Concept

Downtown Parking charges a $5 base fee for parking through 1 hour, and $1 for each additional hour or fraction thereof. The maximum fee for 24 hours is $15. Sketch a graph of the function that describes this pricing scheme.

SOLUTION   For any amount of time during and up to the first hour, the rate is $5. Thus, some sample ordered pairs for the function in the interval $(0, 1]$ would be

$$(.25, 5), (.5, 5), (.75, 5), \text{ and } (1, 5).$$

After the first hour is completed, the price immediately jumps (or steps up) to $6, and remains $6 until the time equals 2 hours. It then jumps to $7 during the third hour, and so on. During the 11th hour, it will have jumped to $15, and will remain at $15 for the rest of the 24-hour period. Figure 56 on the next page shows the graph of this function, for the interval $(0, 24]$. The range of the function is $\{5, 6, 7, \ldots , 15\}$.   ▐

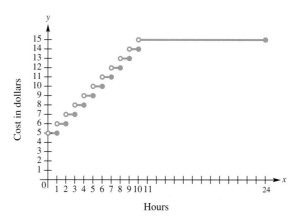

FIGURE 56

2.5   EXERCISES

For each of the following functions defined piecewise, find
(a) $f(-5)$ *(b)* $f(-1)$ *(c)* $f(0)$ *(d)* $f(3)$ *and*   *(e)* $f(5)$.

1. $f(x) = \begin{cases} 2x & \text{if } x \le -1 \\ x - 1 & \text{if } x > -1 \end{cases}$

2. $f(x) = \begin{cases} x - 2 & \text{if } x < 3 \\ 4 - x & \text{if } x \ge 3 \end{cases}$

3. $f(x) = \begin{cases} 3x + 5 & \text{if } x \le 0 \\ 4 - 2x & \text{if } 0 < x < 2 \\ x & \text{if } x \ge 2 \end{cases}$

4. $f(x) = \begin{cases} 4x + 1 & \text{if } x < 2 \\ 3x & \text{if } 2 \le x \le 5 \\ 3 - 2x & \text{if } x > 5 \end{cases}$

Sketch by hand the graph of each of the following piecewise defined functions.

5. $f(x) = \begin{cases} x - 1 & \text{if } x \le 3 \\ 2 & \text{if } x > 3 \end{cases}$

6. $f(x) = \begin{cases} 6 - x & \text{if } x \le 3 \\ 3x - 6 & \text{if } x > 3 \end{cases}$

7. $f(x) = \begin{cases} 4 - x & \text{if } x < 2 \\ 1 + 2x & \text{if } x \ge 2 \end{cases}$

8. $f(x) = \begin{cases} 2x + 1 & \text{if } x \ge 0 \\ x & \text{if } x < 0 \end{cases}$

9. $f(x) = \begin{cases} 2 + x & \text{if } x < -4 \\ -x & \text{if } -4 \le x \le 5 \\ 3x & \text{if } x > 5 \end{cases}$

10. $f(x) = \begin{cases} -2x & \text{if } x < -3 \\ 3x - 1 & \text{if } -3 \le x \le 2 \\ -4x & \text{if } x > 2 \end{cases}$

11. $f(x) = \begin{cases} |x| & \text{if } x > -2 \\ x & \text{if } x \le -2 \end{cases}$

12. $f(x) = \begin{cases} |x| - 1 & \text{if } x > -1 \\ x - 1 & \text{if } x \le -1 \end{cases}$

13. $f(x) = \begin{cases} -\dfrac{1}{2}x^2 + 2 & \text{if } x \le 2 \\ \dfrac{1}{2}x & \text{if } x > 2 \end{cases}$

14. $f(x) = \begin{cases} 5 & \text{if } x > 4 \\ 0 & \text{if } x = 4 \\ 3 & \text{if } x < 4 \end{cases}$

15. $f(x) = \begin{cases} x + 3 & \text{if } x \ne 2 \\ 7 & \text{if } x = 2 \end{cases}$

16. $f(x) = \begin{cases} \sqrt[3]{x} & \text{if } x < 0 \\ \sqrt{x} + 4 & \text{if } x \ge 0 \end{cases}$

17. $f(x) = \begin{cases} x^3 + 5 & \text{if } x \le 0 \\ -x^2 & \text{if } x > 0 \end{cases}$

18. $f(x) = \begin{cases} -|x| + 4 & \text{if } x \ne 6 \\ 3 & \text{if } x = 6 \end{cases}$

Give a rule of a function f defined piecewise for the graph shown. Give the domain and the range.

19.

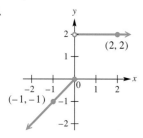

20.

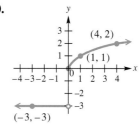

21.

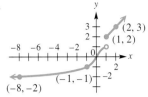

22.

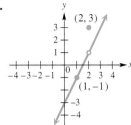

23. Graph the function $y = [\![x]\!]$ on your calculator, using the window $[-100, 100]$ by $[-100, 100]$, with $\text{Xscl} = 10$ and $\text{Yscl} = 10$. Use the dot mode of your calculator. Observe the graph, and discuss why an incorrect conclusion might be drawn concerning the behavior of this function if one does not know the mathematical definition of $[\![x]\!]$.

24. Graph the function $y = 3[\![x]\!]$ on your calculator, using the window $[0, 10]$ by $[0, 40]$, with $\text{Xscl} = 1$ and $\text{Yscl} = 3$. Use the dot mode of your calculator. Now trace from $x = 0$ to $x = 40$ by depressing the trace key *continuously*. Describe what is happening. What does this exercise reinforce about the special name given to functions of this type?

*R*elating Concepts

Describe, using the terminology of Sections 2.2 and 2.3, how the graph of the given function can be obtained from the graph of $y = [\![x]\!]$.

25. $y = [\![x]\!] - 1.5$

26. $y = [\![-x]\!]$

27. $y = -[\![x]\!]$

28. $y = [\![x + 2]\!]$

Work each of the following problems.

29. An express mail company charges $10 for a package weighing 1 pound or less. Each additional pound or part of a pound costs $3 more. Find the cost to send a package weighing 2 pounds; 2.5 pounds; 5.8 pounds. Graph the function on the interval $(0, 7]$. What is the range of the function for this domain?

30. Montreal taxi rates in a recent year were $1.80 for the first $\frac{1}{9}$ mi and $.20 for each additional $\frac{1}{9}$ mi or fraction of $\frac{1}{9}$. Let $C(x)$ be the cost for a taxi ride of $\frac{x}{9}$ mi. Find **(a)** $C(1)$, **(b)** $C(2.3)$, and **(c)** $C(8)$. **(d)** Graph $y = C(x)$ for $0 < x \le \frac{4}{9}$. **(e)** Give the domain and range of C, assuming a ride of unlimited length.

31. A mail order firm charges 30¢ to mail a package weighing one ounce or less, and then 27¢ for each additional ounce or fraction of an ounce. Let $M(x)$ be the cost of mailing a package weighing x oz. Find **(a)** $M(.75)$, **(b)** $M(1.6)$, and **(c)** $M(4)$. **(d)** Graph $y = M(x)$ for $0 < x \le 4$. **(e)** Give the domain and range of M, assuming an unlimited package weight.

32. A car rental costs $37 for one day, which includes 50 free miles. Each additional 25 mi or portion costs $10. Graph the function for $0 < x \le 150$, where x represents the number of miles driven.

33. For a lift truck rental of no more than three days, the charge is $300. An additional charge of $75 is made for each day or portion of a day after three. Graph the function for $0 < x \le 7$, where x represents the number of days the truck is rented.

34. Let f be a function that gives the cost to rent a floor polisher for x days. The cost is a flat $3 for cleaning the polisher plus $4 per day or fraction of a day for using the polisher. Graph the function for $0 < x \le 4$, where x represents the number of days rented.

35. When a diabetic takes long-acting insulin, the insulin reaches its peak effect on the blood sugar level in about 3 hr. This effect remains fairly constant for 5 hr, then declines, and is very low until the next injection. In a typical patient, the level of insulin might be given by the following function.

$$f(t) = \begin{cases} 40t + 100 & \text{if } 0 \le t \le 3 \\ 220 & \text{if } 3 < t \le 8 \\ -80t + 860 & \text{if } 8 < t \le 10 \\ 60 & \text{if } 10 < t \le 24 \end{cases}$$

Here $f(t)$ is the blood sugar level, in appropriate units, at time t measured in hours from the time of the injection. Chuck takes his insulin at 6 A.M. Find the blood sugar level at each of the following times.

(a) 7 A.M.   **(b)** 9 A.M.   **(c)** 10 A.M.
(d) noon   **(e)** 2 P.M.   **(f)** 5 P.M.
(g) midnight
(h) Graph $y = f(t)$ for $0 \leq t \leq 18$.

36. The snow depth in Michigan's Isle Royale National Park varies throughout the winter. In a typical winter, the snow depth in inches is approximated by the following function.

$$f(x) = \begin{cases} 6.5x & \text{if } 0 \leq x \leq 4 \\ -5.5x + 48 & \text{if } 4 < x \leq 6 \\ -30x + 195 & \text{if } 6 < x \leq 6.5 \end{cases}$$

Here, x represents the time in months with $x = 0$ representing the beginning of October, $x = 1$ representing the beginning of November, and so on.

(a) Graph the function for $0 \leq x \leq 6.5$.
(b) In what month is the snow deepest? What is the deepest snow depth?
(c) In what months does the snow begin and end?

Some models of graphics calculators are capable of graphing functions defined piecewise. Match the piecewise-defined function with its calculator-generated graph.

37. $f(x) = \begin{cases} x^2 - 4 & \text{if } x \geq 0 \\ -x + 5 & \text{if } x < 0 \end{cases}$

38. $g(x) = \begin{cases} |x - 4| & \text{if } x \geq -1 \\ -x^2 & \text{if } x < -1 \end{cases}$

39. $h(x) = \begin{cases} 6 & \text{if } x \geq 0 \\ -6 & \text{if } x < 0 \end{cases}$

40. $k(x) = \begin{cases} \sqrt{x} & \text{if } x \geq 0 \\ -x^2 & \text{if } x < 0 \end{cases}$

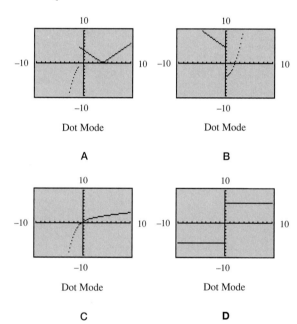

A   B

C   D

Further Explorations

Do the following exercises if your graphics calculator has the greatest integer function and a TABLE feature.

First class postage with the U.S. Postal Service in 1994 cost $.29 for the first ounce and $.23 for each ounce thereafter.

Let $Y_1 = .29 + .23$ int $(x - .001)$ and   let ΔTbl$=.1$

Use the TABLE to calculate the postage costs for packages with the following weights:

1. 1.7 ounces   **2.** 2.3 ounces

3. 24.8 ounces   **4.** 25.0 ounces

2.6 FURTHER TOPICS IN THE STUDY OF FUNCTIONS

Operations on Functions ▮ The Difference Quotient ▮ Composition of Functions ▮ Applications of Operations and Composition

Operations on Functions

Just as we add, subtract, multiply, and divide real numbers, we can also perform these operations on functions. Given two functions f and g, their *sum*, written $f + g$, is defined as

$$(f + g)(x) = f(x) + g(x),$$

for all x such that both $f(x)$ and $g(x)$ exist. Similar definitions can be given for the difference, $f - g$, product, fg, and quotient, f/g, of functions; however, the quotient,

$$\left(\frac{f}{g}\right)(x) = \frac{f(x)}{g(x)},$$

is defined only for those values of x where both $f(x)$ and $g(x)$ exist with the additional condition $g(x) \neq 0$. The various operations on functions are defined as follows.

OPERATIONS ON FUNCTIONS

If f and g are functions, then for all values of x for which both $f(x)$ and $g(x)$ exist, the **sum** of f and g is defined by

$$(f + g)(x) = f(x) + g(x),$$

the **difference** of f and g is defined by

$$(f - g)(x) = f(x) - g(x),$$

the **product** of f and g is defined by

$$(fg)(x) = f(x) \cdot g(x),$$

and the **quotient** of f and g is defined by

$$\left(\frac{f}{g}\right)(x) = \frac{f(x)}{g(x)}, \quad \text{where } g(x) \neq 0.$$

The domains of $f + g, f - g, fg$, and f/g are summarized below. (Recall that the intersection of two sets is the set of all elements belonging to *both* sets.)

DOMAINS OF $f + g, f - g, fg, f/g$

For functions f and g, the domains of $f + g, f - g$, and fg include all real numbers in the intersection of the domains of f and g, while the domain of f/g includes those real numbers in the intersection of the domains of f and g for which $g(x) \neq 0$.

As an example of a sum of functions, suppose that we let $f(x) = x^2 + 1$ and $g(x) = 3x + 5$. Then $f + g$ is found as follows:

$$(f + g)(x) = f(x) + g(x)$$
$$= (x^2 + 1) + (3x + 5)$$
$$= x^2 + 3x + 6.$$

Since the domains of both f and g are $(-\infty, \infty)$, the intersection of their domains is also $(-\infty, \infty)$, and this is the domain of $f + g$. The graph of $f + g$ can be found on a graphics calculator by entering it directly, or by letting $y_1 = x^2 + 1$, $y_2 = 3x + 5$, and $y_3 = y_1 + y_2$. Then instruct the calculator to graph y_3. Figure 57(a) shows the graphs of y_1 and y_2, and Figure 57(b) shows the graph of y_3. Notice that the graph of y_3 is a parabola. (In the next chapter, we will determine analytically the transformations on $y = x^2$ that will give us y_3.)

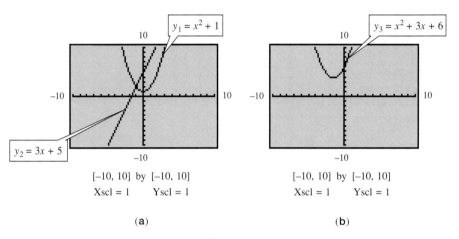

$[-10, 10]$ by $[-10, 10]$
Xscl = 1   Yscl = 1

(a)

$[-10, 10]$ by $[-10, 10]$
Xscl = 1   Yscl = 1

(b)

FIGURE 57

EXAMPLE 1

Using the Operations on Functions

Let $f(x) = x^2 + 1$, and $g(x) = 3x + 5$. Find each of the following.

(a) $(f + g)(1)$

SOLUTION Since $f(1) = 2$ and $g(1) = 8$, use the definition above to get

$$(f + g)(1) = f(1) + g(1) = 2 + 8 = 10.$$

This result indicates that the point $(1, 10)$ lies on the graph of $f + g$. This can be verified graphically by using the calculator to locate the point. It can also be verified by finding $(f + g)(1)$, using the rule for $(f + g)(x)$ found earlier: $(f + g)(x) = x^2 + 3x + 6$, so $(f + g)(1) = 1^2 + 3(1) + 6 = 10$.

(b) $(f - g)(-3) = f(-3) - g(-3) = 10 - (-4) = 14$

(c) $(fg)(5) = f(5) \cdot g(5) = 26 \cdot 20 = 520$

(d) $\left(\dfrac{f}{g}\right)(0) = \dfrac{f(0)}{g(0)} = \dfrac{1}{5}$ ∎

EXAMPLE 2

Using the
Operations on
Functions

Let $f(x) = 8x - 9$ and $g(x) = \sqrt{2x - 1}$.

(a) $(f + g)(x) = f(x) + g(x) = 8x - 9 + \sqrt{2x - 1}$

(b) $(f - g)(x) = f(x) - g(x) = 8x - 9 - \sqrt{2x - 1}$

(c) $(fg)(x) = f(x) \cdot g(x) = (8x - 9)\sqrt{2x - 1}$

(d) $\left(\dfrac{f}{g}\right)(x) = \dfrac{f(x)}{g(x)} = \dfrac{8x - 9}{\sqrt{2x - 1}}$

(e)   Find the domains of f, g, $f + g$, $f - g$, fg, and f/g.

SOLUTION   The domain of f is the set of all real numbers, while the domain of $g(x) = \sqrt{2x - 1}$ includes just those real numbers that make $2x - 1 \geq 0$; the domain of g is the interval $[\frac{1}{2}, \infty)$. The domain of $f + g$, $f - g$, and fg is thus $[\frac{1}{2}, \infty)$. With f/g, the denominator cannot be zero, so the value $\frac{1}{2}$ is excluded from the domain. The domain of f/g is $(\frac{1}{2}, \infty)$.   ∎

The Difference Quotient

Suppose that the point P lies on the graph of $y = f(x)$, and suppose that h is a positive number. If we let $(x, f(x))$ denote the coordinates of P and $(x + h, f(x + h))$ denote the coordinates of Q, then the line joining P and Q has slope

$$m = \frac{f(x + h) - f(x)}{(x + h) - x}$$
$$= \frac{f(x + h) - f(x)}{h}.$$

This expression, called the **difference quotient,** is important in the study of calculus.

Figure 58 shows the graph of the line PQ (called a secant line). If h is allowed to approach 0, the slope of this secant line approaches the slope of the line tangent to the curve at P. Important applications of this idea are developed in calculus, where the concepts of *limit* and *derivative* are investigated.

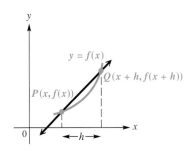

FIGURE 58

EXAMPLE 3

Finding the
Difference Quotient

Let $f(x) = 2x^2 - 3x$. Find the difference quotient and simplify the expression.

SOLUTION   To find $f(x + h)$, replace x in $f(x)$ with $x + h$, to get

$$f(x + h) = 2(x + h)^2 - 3(x + h).$$

Then

$$\frac{f(x + h) - f(x)}{h}$$

$$= \frac{2(x + h)^2 - 3(x + h) - (2x^2 - 3x)}{h}$$

$$= \frac{2(x^2 + 2xh + h^2) - 3x - 3h - 2x^2 + 3x}{h} \qquad \text{Square } x + h; \text{ use the distributive property.}$$

$$= \frac{2x^2 + 4xh + 2h^2 - 3x - 3h - 2x^2 + 3x}{h}$$

$$= \frac{4xh + 2h^2 - 3h}{h} \qquad \text{Combine terms.}$$

$$= \frac{h(4x + 2h - 3)}{h} \qquad \text{Factor out } h.$$

$$= 4x + 2h - 3. \qquad \text{Divide.} \qquad \blacksquare$$

CAUTION   Notice that $f(x + h)$ is not the same as $f(x) + f(h)$. For $f(x) = 2x^2 - 3x$, as shown in Example 3,

$$f(x + h) = 2(x + h)^2 - 3(x + h) = 2x^2 + 4xh + 2h^2 - 3x - 3h$$

but

$$f(x) + f(h) = (2x^2 - 3x) + (2h^2 - 3h) = 2x^2 - 3x + 2h^2 - 3h.$$

These expressions differ by $4xh$.

Composition of Functions

The diagram in Figure 59 shows a function f that assigns to each element x of set X some element y of set Y. Suppose also that a function g takes each element of set Y and assigns a value z of set Z. Using both f and g, then, an element x in X is assigned to an element z in Z. The result of this process is a new function h, that takes an element x in X and assigns an element z in Z. This function h is called the *composition* of functions g and f, written $g \circ f$, and is defined as follows.

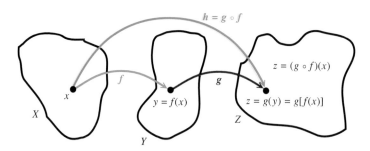

FIGURE 59

COMPOSITION OF FUNCTIONS

If f and g are functions, then the **composite function,** or **composition,** of g and f is

$$(g \circ f)(x) = g[f(x)]$$

for all x in the domain of f such that $f(x)$ is in the domain of g.

EXAMPLE 4

Finding and
Simplifying
Composite
Functions

Suppose that $f(x) = 2x^2 + 1$ and $g(x) = x - 1$.

(a) Find $f \circ g$ and simplify the expression.

SOLUTION

$$
\begin{aligned}
(f \circ g)(x) = f[g(x)] &= f(x - 1) \\
&= 2(x - 1)^2 + 1 \qquad * \\
&= 2(x^2 - 2x + 1) + 1 \\
&= 2x^2 - 4x + 2 + 1 \\
&= 2x^2 - 4x + 3
\end{aligned}
$$

By examining the form of $f \circ g$ in the line marked *, we can see that its graph is that of the function $y = x^2$, shifted 1 unit to the right, stretched vertically by a factor of 2, and shifted 1 unit upward.

(b) Find $g \circ f$ and simplify the expression.

SOLUTION

$$
\begin{aligned}
(g \circ f)(x) = g[f(x)] &= g(2x^2 + 1) \\
&= (2x^2 + 1) - 1 \\
&= 2x^2
\end{aligned}
$$

The simplified form shows us that the graph of $g \circ f$ is that of the function $y = x^2$, stretched vertically by a factor of 2. ∎

Comparing the results in the two parts in Example 4 shows that *in general, $f \circ g$ is not equal to $g \circ f$.*

TECHNOLOGICAL NOTE
You may wish to read the "Further Explorations" at the end of the exercise set in this section. It describes how composition of functions can be handled by one particular model of calculator.

FOR **GROUP DISCUSSION**

After reading through part (a) of Example 4, use the standard viewing window on your calculator and assign $y_1 = x - 1$ and $y_2 = 2y_1^2 + 1$. Then graph only y_2 on the screen. Describe in words the graph you see. Does this seem to correspond to the conclusion we reached in part (a)? Now graph $y_3 = 2x^2 - 4x + 3$ on the same screen as y_2. Do the graphs coincide?

Repeat this exercise for part (b), by letting $y_1 = 2x^2 + 1$ and $y_2 = y_1 - 1$. Does the graph of y_2 seem to be the same as that of the squaring function stretched vertically by a factor of 2?

In calculus it is sometimes useful to treat a function as a composition of two functions. The next example shows how this can be done.

EXAMPLE **5**
Finding Functions That Form a Given Composite

Suppose $h(x) = \sqrt{2x + 3}$. Find functions f and g, so that $(f \circ g)(x) = h(x)$.

SOLUTION Since there is a quantity, $2x + 3$, under a radical, one possibility is to choose $f(x) = \sqrt{x}$ and $g(x) = 2x + 3$. Then $(f \circ g)(x) = \sqrt{2x + 3}$, as required. Other combinations are possible. For example, we could choose $f(x) = \sqrt{x + 3}$ and $g(x) = 2x$. ◻

Applications of Operations and Composition

There are numerous applications in business, economics, physics, and other fields that can be viewed from the perspective of combining functions. For example, in manufacturing, the cost of producing a product usually consists of two parts. One part is a *fixed cost* for designing the product, setting up a factory, training workers, and so on. Usually the fixed cost is constant for a particular product and does not change as more items are made. The other part of the cost is a *variable cost* per item for labor, materials, packaging, shipping, and so on. The variable cost is often the same per item, so that the total amount of variable cost increases as more items are produced. A *linear cost function* has the form $C(x) = mx + b$, where m represents the variable cost per item and b represents the fixed cost. The revenue from selling a product depends on the price per item and the number of items sold, as given by the *revenue function, $R(x) = px$,* where p is the price per item and $R(x)$ is the revenue from the sale of x items. The profit is described by the *profit function* given by $P(x) = R(x) - C(x)$.

EXAMPLE **6**
Finding and Analyzing Cost, Revenue, and Profit

Suppose that a businessman invests $1500 as his fixed cost in a new venture that produces and sells a device that makes programming a VCR easier. Each such device costs $100 to manufacture.

(a) Write a cost function for the product, if x represents the number of devices produced. Assume that the function is linear.

SOLUTION Since the cost function is linear it will have the form $C(x) = mx + b$, with $m = 100$ and $b = 1500$. That is,

$$C(x) = 100x + 1500.$$

(b) Find the revenue function if each device in part (a) sells for $125.

SOLUTION The revenue function is

$$R(x) = px = 125x. \quad \text{Let } p = 125.$$

(c) Give the profit function for the item in part (a).

SOLUTION The profit function is given by

$$\begin{aligned}
P(x) &= R(x) - C(x) \\
&= 125x - (100x + 1500) \\
&= 125x - 100x - 1500 \\
&= 25x - 1500.
\end{aligned}$$

(d) How many items must be produced and sold before the company makes a profit?

SOLUTION To make a profit, $P(x)$ must be positive. Set $P(x) = 25x - 1500 > 0$ and solve for x.

$$25x - 1500 > 0$$
$$25x > 1500 \qquad \text{Add 1500 to each side.}$$
$$x > 60 \qquad \text{Divide by 25.}$$

At least 61 items must be sold for the company to make any profit.

(e) Support the result of part (d) graphically.

SOLUTION On a calculator, let $y_1 = 100x + 1500$ be the cost function $C(x)$ and let $y_2 = 125x$ be the revenue function $R(x)$. Then we can graph $y_3 = y_2 - y_1$, with y_3 representing the profit function (that is, $R(x) - C(x)$). We must decide on the smallest whole number value of x for which y_3 is greater than 0. Using an appropriate window, the calculator will allow us to find the point at which the graph intersects the x-axis; we see in Figure 60 that the x-intercept is 60. Thus, the company must sell at least 61 devices to earn a profit.

(Notice that in Figure 60, we have used the connected mode to emphasize that the profit function has a straight-line graph. However, we should understand that only whole number values of x are in the domain of this function, since it would not make sense to consider fractional parts of devices produced.) ◻

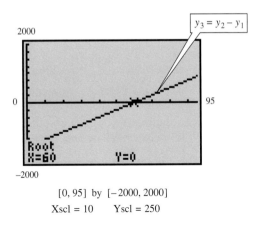

$[0, 95]$ by $[-2000, 2000]$
Xscl = 10   Yscl = 250

FIGURE 60

FOR **GROUP DISCUSSION**

1. Why would negative values of x not be realistic in the context of the problem examined in Example 6?
2. Review Example 5 in Section 1.7. What is the profit function in that example?

In Section 1.7 we saw how a quantity y is in direct proportion, or varies directly, with another quantity x if there exists a constant k such that $y = kx$. Sometimes y varies directly as a power of x.

DIRECT VARIATION AS A POWER

Let n be a positive real number. Then y **varies directly as the nth power** of x, or y is **directly proportional to the nth power** of x, if a nonzero real number k exists such that

$$y = kx^n.$$

While the definition above allows n to be any positive real number, we will now only consider positive integer powers. Furthermore, if $n = 2$, we have the case where one quantity varies as the square of another. The familiar formula $A = \pi r^2$ for the area of a circle is an example of this type of variation; that is, the area of a circle varies directly as the square of its radius. The constant of proportionality here is π.

The next example shows how composition of functions can be applied to a physical phenomenon.

EXAMPLE 7

Applying Composition of Functions

Suppose an oil well off the California coast is leaking, with the leak spreading oil in a circular layer over the surface. At any time t, in minutes, after the beginning of the leak, the radius of the circular oil slick is $r(t) = 5t$ feet. Express the area A of the leak as a function of the time that it has been spreading, and determine graphically how much of the surface is covered 20 minutes after the leak begins.

SOLUTION Since $A(r) = \pi r^2$ gives the area of a circle of radius r, the area can be expressed as a function of time by substituting $5t$ for r in $A(r) = \pi r^2$ to get

$$A(r) = \pi r^2$$
$$A[r(t)] = \pi(5t)^2 = 25\pi t^2.$$

The function $A[r(t)]$ is a composite function of the functions A and r.

To determine *graphically* how much of the surface is covered 20 minutes after the leak begins, we graph $y_1 = 25\pi x^2$. (Note that we use y rather than A, and x rather than t. The variables used are immaterial, and it is more appropriate for graphics calculator interpretation to use x and y.) Figure 61 indicates that if it is graphed in an appropriate window, when $x = 20$, $y \approx 31,415.927$. Thus, the leak will cover approximately 31,400 square feet 20 minutes after it begins. ∎

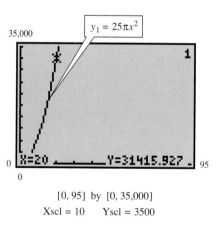

$[0, 95]$ by $[0, 35,000]$
Xscl = 10   Yscl = 3500

FIGURE 61

FOR **GROUP DISCUSSION**

1. How would you confirm *analytically* the result found graphically in Example 7.
2. Why would negative values of x be unrealistic in the situation described in Example 7?

EXAMPLE 8

Applying a Difference of Functions

The formula for the surface area S of a sphere is $S = 4\pi r^2$, where r is the radius of the sphere.

(a) Construct a model $S(r)$ that describes the amount of surface area gained if the radius r inches of a ball is increased by 2 inches.

In words, we have

Surface area gained = Larger surface area − smaller surface area.

Symbolically, this translates as a difference of functions:

$$S(r) = 4\pi(r + 2)^2 - 4\pi r^2.$$

(b) The function found in part (a) is graphed in Figure 62. What classification of function does this seem to be?

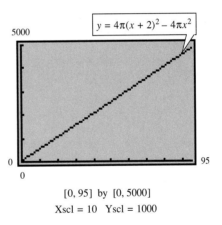

$y = 4\pi(x + 2)^2 - 4\pi x^2$

[0, 95] by [0, 5000]
Xscl = 10   Yscl = 1000

FIGURE 62

SOLUTION It appears to be a linear function. (You are asked to verify this in Exercise 63.)

(c) Use the graph to determine the amount of extra material needed to manufacture a ball of radius 22 inches as compared to a ball of radius 20 inches.

SOLUTION We observe that for the graph of S, where $x = 20$, $y \approx 1055.5751$. See Figure 63 on the next page. Thus it would take about 1056 extra square inches of material.

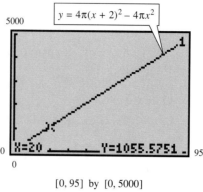

[0, 95] by [0, 5000]
Xscl = 10   Yscl = 1000

FIGURE 63

(d)   Confirm analytically the result obtained graphically in part (c).

SOLUTION   Let $r = 20$ in the model found in part (a).

$$S(r) = 4\pi(r + 2)^2 - 4\pi r^2$$
$$S(20) = 4\pi(20 + 2)^2 - 4\pi(20)^2$$
$$= 1936\pi - 1600\pi$$
$$= 336\pi$$
$$\approx 1056$$

This result confirms our earlier conclusion.   ◼

2.6   EXERCISES

For each pair of functions in Exercises 1–10, (a) find $f + g$, $f - g$, and fg, (b) give the domains of the functions in part (a), (c) find $\frac{f}{g}$ and give its domain, (d) find $f \circ g$ and give its domain, and (e) find $g \circ f$ and give its domain.

1. $f(x) = 4x - 1$, $g(x) = 6x + 3$

2. $f(x) = 9 - 2x$, $g(x) = -5x + 2$

3. $f(x) = |x + 3|$, $g(x) = 2x$

4. $f(x) = |2x - 4|$, $g(x) = x + 1$

5. $f(x) = \sqrt[3]{x + 4}$, $g(x) = x^3 + 5$

6. $f(x) = \sqrt[3]{6 - 3x}$, $g(x) = 2x^3 + 1$

7. $f(x) = \sqrt{x^2 + 3}$, $g(x) = x + 1$

8. $f(x) = \sqrt{2 + 4x^2}$, $g(x) = x$

9. $f(x) = (x - 2)^2$, $g(x) = 2x$

10. $f(x) = (x + 6)^2$, $g(x) = .5x$

Let $f(x) = 4x^2 - 2x$ and $g(x) = 8x + 1$. Find each of the following.

11. $(f \circ g)(x)$

12. $(f \circ g)(3)$

13. $(g \circ f)(x)$

14. $(g \circ f)(-2)$

15. $(f + g)(3)$

16. $(f + g)(-5)$

17. $(fg)(4)$

18. $(fg)(-3)$

19. $\left(\dfrac{f}{g}\right)(-1)$

20. $\left(\dfrac{f}{g}\right)(4)$

21. $(f - g)(m)$

22. $(f - g)(2k)$

23. $(f \circ g)(2)$

24. $(f \circ g)(-5)$

25. $(g \circ f)(2)$

26. $(g \circ f)(-5)$

For certain pairs of functions f and g, $(f \circ g)(x) = x$ and $(g \circ f)(x) = x$. Show that this is true for the pairs in Exercises 27–32.

27. $f(x) = 4x$, $g(x) = \dfrac{1}{4}x$

28. $f(x) = 3x - 5$, $g(x) = \dfrac{1}{3}(x + 5)$

29. $f(x) = 4x + 2, \ g(x) = \dfrac{1}{4}(x - 2)$

30. $f(x) = -3x, \ g(x) = -\dfrac{1}{3}x$

31. $f(x) = \sqrt[3]{5x + 4}, \ g(x) = \dfrac{1}{5}x^3 - \dfrac{4}{5}$

32. $f(x) = \sqrt[3]{x + 1}, \ g(x) = x^3 - 1$

R*elating Concepts*

33. Functions such as the pairs in Exercises 27–32 are called *inverse functions,* because upon composition in both directions, the result is the identity function. (Inverse functions will be discussed in detail in Section 4.5.) In a square viewing window, graph $y_1 = \sqrt[3]{x - 6}$ and $y_2 = x^3 + 6$, an example of a pair of inverse functions. Now graph $y_3 = x$. Describe how the graph of y_2 can be obtained from the graph of y_1, using the graph of $y_3 = x$ as a basis for your description. (*Hint:* Review the terminology of Section 2.3.)

34. Repeat Exercise 33 for $y_1 = 5x - 3$ and $y_2 = \frac{1}{5}(x + 3)$.

The graphs of two functions f and g are shown in the figure here.

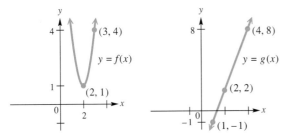

35. Find $(f \circ g)(2)$.

36. Find $(g \circ f)(3)$.

The tables below give some selected ordered pairs for functions f and g.

x	3	4	6
$f(x)$	1	3	9

x	2	7	1	9
$g(x)$	3	6	9	12

Find each of the following.

37. $(f \circ g)(2)$

38. $(f \circ g)(7)$

39. $(g \circ f)(3)$

40. $(g \circ f)(6)$

41. $(f \circ f)(4)$

42. $(g \circ g)(1)$

43. Why can you not determine $(f \circ g)(1)$ given the information in the tables for Exercises 37–42?

44. Extend the concept of composition of functions to evaluate $(g \circ (f \circ g))(7)$ using the tables for Exercises 37–42.

Determine the difference quotient

$$\frac{f(x + h) - f(x)}{h} \quad (h \neq 0)$$

for each function f in Exercises 45–50. Simplify completely.

45. $f(x) = 4x + 3$

46. $f(x) = 5x - 6$

47. $f(x) = -6x^2 - x + 4$

48. $f(x) = \dfrac{1}{2}x^2 + 4x$

49. $f(x) = x^3$

50. $f(x) = -2x^3$

In each of Exercises 51–56, a function h is defined. Find functions f and g such that
$(f \circ g)(x) = h(x)$. *(There are many possible ways to do this.)*

51. $h(x) = (6x - 2)^2$

52. $h(x) = (11x^2 + 12x)^2$

53. $h(x) = \sqrt{x^2 - 1}$

54. $h(x) = (2x - 3)^3$

55. $h(x) = \sqrt{6x} + 12$

56. $h(x) = \sqrt[3]{2x + 3} - 4$

For each of the following, if x represents the number of items produced, (a) write a cost function, (b) find a revenue function if each item sells for the price given, (c) give the profit function, (d) determine analytically how many items must be produced before a profit is shown (assume whole numbers of items), and (e) support the result of part (d) graphically.

57. The fixed cost is $500, the cost to produce an item is $10, and the selling price of the item is $35.

58. The fixed cost is $180, the cost to produce an item is $11, and the selling price of the item is $20.

59. The fixed cost is $2700, the cost to produce an item is $100, and the selling price of the item is $280.

60. The fixed cost is $1000, the cost to produce an item is $200, and the selling price of the item is $240.

61. An oil well off the Gulf Coast is leaking, with the leak spreading oil over the surface as a circle. At any time t, in minutes, after the beginning of the leak, the radius of the circular oil slick on the surface is $r(t) = 4t$ ft. Let $A(r) = \pi r^2$ represent the area of a circle of radius r.
  (a) Find $(A \circ r)(t)$.
  (b) Interpret $(A \circ r)(t)$.
  (c) What is the area of the oil slick after 3 minutes?
  (d) Support your result graphically.

62. When a thermal inversion layer is over a city, pollutants cannot rise vertically but are trapped below the layer and must disperse horizontally. Assume that a factory smokestack begins emitting a pollutant at 8 A.M. Assume that the pollutant disperses horizontally over a circular area. If t represents the time, in hours, since the factory began emitting pollutants ($t = 0$ represents 8 A.M.), assume that the radius of the circle of pollution is $r(t) = 2t$ mi. Let $A(r) = \pi r^2$ represent the area of a circle of radius r.
  (a) Find $(A \circ r)(t)$.
  (b) Interpret $(A \circ r)(t)$.
  (c) What is the area of the circular region covered by the layer at noon?
  (d) Support your result graphically.

Relating Concepts

63. Verify analytically that the function defined by $y = 4\pi(x + 2)^2 - 4\pi x^2$, derived in Example 8, is actually a linear function. What is the slope? If we allow 0 to be in the domain of the function, what is the y-intercept?

64. In part (a) of Exercises 57–60, you are asked to write a cost function. Describe how slope and y-intercept can be interpreted using the concepts of fixed cost and cost to produce an item.

65. (a) The formula for the volume of a sphere is $V = \frac{4}{3}\pi r^3$, where r represents the radius of the sphere. Construct a model representing the amount of volume gained when a sphere of radius r inches is increased by 3 inches.
  (b) Graph the model found in part (a), using y for V and x for r, in the window [0, 10] by [0, 1500]. What classification of function does this appear to be?
  (c) Use your calculator to find graphically the amount of volume gained when a sphere of 4-inch radius is increased to a 7-inch radius.
  (d) Verify your conjecture in part (b) analytically.

66. Rework Example 8, parts (a) and (b), but consider what happens when the radius is doubled (rather than increased by 2 inches). In part (b), change Ymax to 5000.

67. Suppose that the length of a rectangle is twice its width. Let x represent the width of the rectangle.

(a) Write a formula for the perimeter P of the rectangle in terms of x alone. Then use $P(x)$ notation to describe it as a function. What classification of function is this?

(b) Graph the function P found in part (a) in the window [0, 10] by [0, 100]. Trace along the graph and explain what x represents and what y represents.

(c) Locate on the graph of P a point with an integer x-value. Now sketch a rectangle satisfying the conditions described earlier, and evaluate its perimeter if its width is this x-value. Use the standard perimeter formula. How does the result compare with the y-value shown on your screen?

(d) Locate on the graph of P a point with an integer y-value. Describe in words the meaning of the x- and y-coordinates here.

68. The perimeter x of a square with side of length s is given by the formula

$$x = 4s.$$

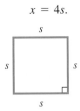

(a) Solve for s in terms of x.

(b) If y represents the area of this square, write y as a function of the perimeter x.

(c) Use the composite function of part (b) to analytically find the area of a square with perimeter 6.

(d) Support the result of part (c) graphically, and explain the result.

69. The area of an equilateral triangle with sides of length x is given by the function $A(x) = \frac{\sqrt{3}}{4}x^2$.

(a) Find $A(2x)$, the function representing the area of an equilateral triangle with sides of length twice the original length.

(b) Find analytically the area of an equilateral triangle with side length 16. Use the formula for $A(x)$ given above.

(c) Support graphically your result of part (b).

70. A textbook author invests his royalties in two accounts for 1 year.

(a) The first account pays 4% simple interest. If he invests x dollars in this account, write an expression for y_1 in terms of x, where y_1 represents the amount of interest earned.

(b) He invests in a second account $500 more than he invested in the first account. This second account pays 2.5% simple interest. Write an expression for y_2, where y_2 represents the amount of interest earned.

(c) What does $y_1 + y_2$ represent?

(d) Graph $y_1 + y_2$ in the window [0, 1000] by [0, 200]. Use the graph to find the amount of interest he will receive if he invests $250 in the first account.

(e) Support the result of part (d) analytically.

*F*urther Explorations

Some graphics calculators (such as the TI-82) can interpret function notation, similar to the notation used when writing by hand. For example: if $Y_1 = 3x$, then entering $Y_1(4)$ on the home screen returns a value of 12. This can be extended to the graphing screen. Let $Y_1 = 3x$ and $Y_2 = x - 5$, then enter $Y_3 = Y_1(Y_2)$. The TI-82 will interpret this as $Y_3 = 3(x - 5)$.

CAUTION   A calculator without the function notation capability will interpret $Y_3 = Y_1(Y_2)$ as multiplying Y_1 by Y_2 so that $Y_3 = (3x)(x - 5)$. If your graphics calculator will not interpret function notation, you can still see the graph of the composite function by first analytically calculating the composite, then entering this expression in Y_3.

1. Refer to Exercise 61.

   An oil well off the California coast is leaking, with the leak spreading oil over the surface as a circle. The radius of the circle is increasing at a constant rate of 2 feet per minute. The first figure shows a TABLE with the first seven minutes after the leak began where Y_1 is the radius of a circle, Y_2 is the radius of the circular leak as a function of time, and Y_3 is a composite function for the area of the leak as a function of time.

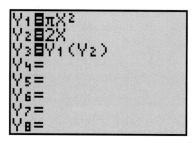

   (a) Change ΔTbl to an appropriate value and find the radius and area of the leak for each of the first two hours after the leak began.
   (b) Based on your observations in this problem, if the radius of the leak is spreading at a constant rate, what can you say about the volume of the oil leaking?

Chapter 2

SUMMARY

Several important elementary algebraic functions are introduced in this chapter. A student of college algebra should be familiar with the graphs of the following functions: identity, squaring, cubing, square root, cube root, and absolute value. The concepts of symmetry with respect to the y-axis, symmetry with respect to the origin, continuity, and increasing, decreasing, and constant functions are introduced, and these ideas are important in the study of functions. The relation defined by $x = y^2$ is also presented, and its graph exhibits symmetry with respect to the x-axis. A function may be classified as even, odd, or neither, according to the definitions found in Section 2.1.

The graphs of the basic functions can be altered by vertical and horizontal shifting, vertical stretching and shrinking, and reflections across an axis. These concepts, introduced in Sections 2.2 and 2.3, will be used many times in the sections that follow.

In Section 2.4 we use the absolute value function, $f(x) = |x|$, as an example to show how the methods of analytic and graphical solution, first introduced for linear functions in Chapter 1, can be applied to equations and inequalities involving another type of function. As we progress in succeeding chapters, these analytic and graphical methods will be applied repeatedly to other types of functions.

If a function is defined by different rules over different subsets of its domain, it is called a piecewise-defined function. Piecewise-defined functions often exhibit discontinuities. An important example of a function that exhibits discontinuities is the greatest integer function. Graphing this function with a graphics calculator requires the use of the dot (rather than connected) mode of the calculator in order to obtain a realistic picture.

The operations of addition, subtraction, multiplication, and division can be applied to functions in a manner similar to those for real numbers. Another operation, composition, is also defined for functions. The domains of the functions resulting from these operations may be different from those of the original functions. The operation of composition may be applied to various types of business and scientific applications.

Key Terms

SECTION 2.1

continuity
asymptote
increasing
decreasing
constant
identity function
squaring function
quadratic
parabola
symmetry with respect to the *y*-axis
cubing function
cubic
symmetry with respect to the origin
square root function
cube root function

absolute value function
symmetry with respect to the *x*-axis
even function
odd function

SECTION 2.2

vertical shift
translation
horizontal shift

SECTION 2.3

vertical stretch
vertical shrink
reflection across the *x*-axis
reflection across the *y*-axis

SECTION 2.4

properties of absolute value

SECTION 2.5

piecewise-defined function
greatest integer function
step function
connected mode
dot mode

SECTION 2.6

sum of *f* and *g*
difference of *f* and *g*
product of *f* and *g*
quotient of *f* and *g*
the difference quotient
composition of functions

Chapter 2   REVIEW EXERCISES

Match the equation with the graph that most closely resembles its graph.

1. $y = \sqrt{x} + 2$

2. $y = \sqrt{x + 2}$

3. $y = 2\sqrt{x}$

4. $y = -2\sqrt{x}$

5. $y = \sqrt[3]{x} - 2$

6. $y = \sqrt[3]{x - 2}$

7. $y = 2\sqrt[3]{x}$

8. $y = -2\sqrt[3]{x}$

A.
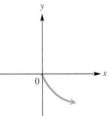

B.
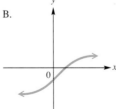

C.
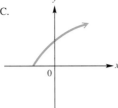

D.
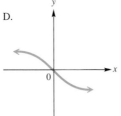

E.
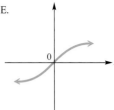

F.
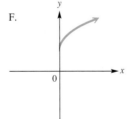

G.
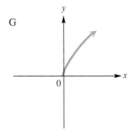

H.
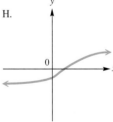

Give the interval that describes the following.

9. domain of $f(x) = \sqrt{x}$

10. range of $f(x) = |x|$

11. range of $f(x) = \sqrt[3]{x}$

12. domain of $f(x) = x^2$

13. the largest interval over which $f(x) = \sqrt[3]{x}$ is increasing

14. the largest interval over which $f(x) = |x|$ is increasing

15. domain of $x = y^2$

16. range of $x = y^2$

Consider the function whose graph is shown here.

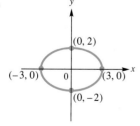

Give the interval(s) over which the function

17. is continuous.

18. increases.

19. decreases.

20. is constant.

21. What is the domain of the function graphed for Exercises 17–20?

22. What is the range of the function graphed for Exercises 17–20?

Consider the function $f(x) = -(x + 3)^2 - 5$. Give the interval(s) over which the function

23. is continuous.

24. increases.

25. decreases.

26. is constant.

27. What is the domain of the function described for Exercises 23–26?

28. What is the range of the function described for Exercises 23–26?

Determine whether the given relation has x-axis symmetry, y-axis symmetry, origin symmetry, or none of these symmetries. (More than one choice is possible.) Also, if the relation is a function, determine whether it is an even function, an odd function, or neither.

29.

30. $F(x) = x^3 - 6$

31. $y = |x| + 4$

32. $f(x) = \sqrt{x} - 5$

33. $y^2 = x - 5$

34. $f(x) = 3x^4 + 2x^2 + 1$

35. Use the terminology of Sections 2.2 and 2.3 to describe how the graph of $y = -3(x + 4)^2 - 8$ can be obtained from the graph of $y = x^2$.

36. Find the rule for the function whose graph is obtained by reflecting the graph of $y = \sqrt{x}$ across the y-axis, then reflecting across the x-axis, shrinking vertically by a factor of $\frac{2}{3}$, and finally translating 4 units upward.

The graph of y = f(x) is given here. Sketch by hand the graph of each function listed, and indicate three points on the graph.

37. $y = -f(x + 1) - 2$

38. $y = |f(x)| + 1$

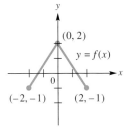

Solve the equation or inequality analytically.

39. $|2x + 5| = 7$ **40.** $|2x + 5| \leq 7$ **41.** $|2x + 5| \geq 7$

42. The graphs of $y_1 = |2x + 5|$ and $y_2 = 7$ are shown, along with the two points of intersection of the graphs. Write a paragraph explaining how this picture supports the answers in Exercises 39–41.

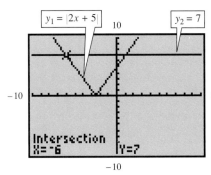

First View   Second View

43. Solve the equation $|x + 1| + |x - 3| = 8$ graphically. Then give an analytic check by substituting the values in the solution set directly into the left-hand side of the equation.

44. Use the graph found in Exercise 43, along with the solution set, to give the solution set for each inequality.
   (a) $|x + 1| + |x - 3| < 8$ **(b)** $|x + 1| + |x - 3| > 8$

Give the solution set of the equation or inequality, based on the graphs of y = f(x) and y = g(x).

45. $f(x) = g(x)$

46. $f(x) < g(x)$

47. $f(x) \geq g(x)$

48. $f(x) \geq 1$

49. $f(x) \leq 0$

50. $g(x) < 0$

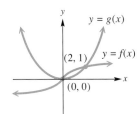

Consider the piecewise-defined function $f(x) = \begin{cases} -x^2 & \text{if } x < 0 \\ 3 & \text{if } 0 \leq x \leq 4 \\ x + 2 & \text{if } x > 4. \end{cases}$

51. Find $f(-1)$.   **52.** Find $f(0)$.

53. Find $f(4)$.   **54.** Sketch by hand the graph of f.

In Exercises 55–61, consider the functions $f(x) = 2x - 4$ *and* $g(x) = x^2 - 3$.

55. Find the rule for $f + g$.

56. Find the rule for $f - g$.

57. Find the rule for fg.

58. Find the domain for the functions in Exercises 55–57.

59. Find the rule for $\frac{g}{f}$ and give its domain.

60. Find the rule for $g \circ f$ and give its domain.

61. Find the rule for $f \circ g$ and give its domain.

62. If $f(x) = x^3 - 2x$ and $g(x) = 2x + 1$, find **(a)** $(f \circ g)(2)$ and **(b)** $(g \circ f)(2)$.

63. If $f(x) = 2x^2 + 1$, find the difference quotient

$$\frac{f(x + h) - f(x)}{h} \quad (h \neq 0)$$

and simplify the expression completely.

64. Given that $h(x) = \sqrt[3]{5x^2 - 3}$, find functions f and g such that $(f \circ g)(x) = h(x)$.

Jim Boyle operates a delivery service. His start-up cost was $2300. He estimates that it will cost him $3.00 per delivery and he charges $5.50 per delivery. Let x represent the number of deliveries he makes.

65. Find a cost function C.

66. Find a revenue function R.

67. Find the profit function P, using the fact that $P = R - C$.

68. Determine analytically how many deliveries must be made before a profit is shown.

69. Support your answer from Exercise 68 graphically.

70. Explain why the graph of a function cannot exhibit x-axis symmetry.

CHAPTER 3

Polynomial Functions

3.1 COMPLEX NUMBERS

Defining Complex Numbers ▌ Operations With Complex Numbers

Defining Complex Numbers

Our work in the two previous chapters has involved only real numbers—that is, numbers that are positive, negative, or zero. In order to fully develop some of the concepts in this chapter, we must now investigate some of the ideas of the complex number system.

Observe the graph of $y = x^2 + 1$ in Figure 1. Notice that the graph does not intersect the x-axis and therefore there are no real solutions to the equation $x^2 + 1 = 0$. This equation is equivalent to $x^2 = -1$, and we know from experience that no real number has a square of -1 (or *any* negative number, for that matter). To handle this situation, mathematicians have developed an expanded number system that includes the set of real numbers as a subset. This expanded system is called the complex number system, and its basic unit is i, which is defined to be a square root of -1. Thus, $i^2 = -1$.

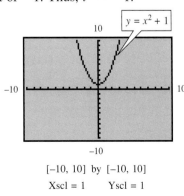

[−10, 10] by [−10, 10]
Xscl = 1   Yscl = 1

FIGURE 1

191

> **DEFINITION OF *i***
>
> $$i = \sqrt{-1} \quad \text{or} \quad i^2 = -1$$

Numbers of the form $a + bi$, where a and b are real numbers, are called **complex numbers.** Each real number is a complex number, since a real number a may be thought of as the complex number $a + 0i$. A complex number of the form $a + bi$, where b is nonzero, is called an **imaginary number.** Both the set of real numbers and the set of imaginary numbers are subsets of the set of complex numbers. (See Figure 2, which is an extension of Figure 5 in Section 1.1.) A complex number that is written in the form $a + bi$ or $a + ib$ is in **standard form.** (The form $a + ib$ is used to simplify certain symbols such as $i\sqrt{5}$, since $\sqrt{5}\,i$ could be too easily mistaken for $\sqrt{5i}$.) The real number a is called the **real part** of $a + bi$, and the real number b is called the **imaginary part.**

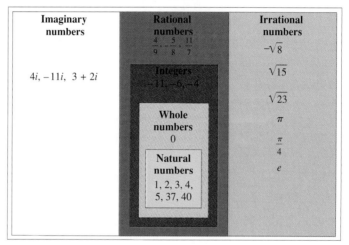

Complex numbers (Real numbers are shaded.)

FIGURE 2

EXAMPLE 1
Identifying Kinds of Complex Numbers

The following statements identify different kinds of complex numbers.

(a) -8, $\sqrt{7}$, and π are real numbers and complex numbers.

(b) $3i$, $-11i$, $i\sqrt{14}$, and $5 + i$ are imaginary numbers and complex numbers. ∎

EXAMPLE 2
Writing Complex Numbers in Standard Form

The list below shows several numbers, along with the standard form of each number.

Number	Standard Form
$6i$	$6i$
-9	-9
0	0
$-i + 2$	$2 - i$
$8 + i\sqrt{3}$	$8 + i\sqrt{3}$

∎

In later sections of this chapter, we will solve equations that will have solutions that lead to expressions involving terms of the form $\sqrt{-a}$, where $a > 0$. This kind of term may be rewritten as a product of a real number and i, using the following definition.

DEFINITION OF $\sqrt{-a}$

If $a > 0$, then

$$\sqrt{-a} = i\sqrt{a}.$$

EXAMPLE 3

Writing $\sqrt{-a}$ as $i\sqrt{a}$

Write each expression as the product of i and a real number.

(a) $\sqrt{-16} = i\sqrt{16} = 4i$

(b) $\sqrt{-70} = i\sqrt{70}$　❑

EXAMPLE 4

Simplifying the Form of a Complex Number

Use the rules of algebra and the definition of $\sqrt{-a}$ above to show that $\frac{4 + \sqrt{-24}}{6}$ is equivalent to $\frac{2}{3} + i\frac{\sqrt{6}}{3}$.

SOLUTION

$$\frac{4 + \sqrt{-24}}{6} = \frac{4 + i\sqrt{24}}{6} \qquad \sqrt{-a} = i\sqrt{a}$$

$$= \frac{4 + i\sqrt{4 \cdot 6}}{6} \qquad \text{Factor 24 as } 4 \cdot 6.$$

$$= \frac{4 + 2i\sqrt{6}}{6} \qquad \begin{array}{l}\sqrt{ab} = \sqrt{a} \cdot \sqrt{b}; \\ \sqrt{4} = 2\end{array}$$

$$= \frac{2(2 + i\sqrt{6})}{2 \cdot 3} \qquad \begin{array}{l}\text{Factor out a 2 in} \\ \text{both numerator and} \\ \text{denominator.}\end{array}$$

$$= \frac{2 + i\sqrt{6}}{3} \qquad \begin{array}{l}\text{Divide out the common} \\ \text{factor 2.}\end{array}$$

$$= \frac{2}{3} + i\frac{\sqrt{6}}{3} \qquad \begin{array}{l}\text{Write the complex number} \\ \text{in standard form.}\end{array}　❑$$

TECHNOLOGICAL NOTE
Some of the more powerful graphics calculators are capable of complex number arithmetic. The complex number $a + bi$ is represented as the ordered pair of real numbers (a, b) on the display of such calculators. You should consult your owner's manual to see if your particular model has this feature.

The procedure shown in Example 4 will be used extensively in simplifying solutions of quadratic equations, covered in the following sections.

Operations With Complex Numbers

Complex numbers may be added, subtracted, multiplied, and divided using the properties of real numbers, as shown by the following definitions and examples.

The *sum* of two complex numbers $a + bi$ and $c + di$ is defined as follows.

ADDITION OF COMPLEX NUMBERS

$$(a + bi) + (c + di) = (a + c) + (b + d)i$$

<table>
<tr><td>

EXAMPLE **5**

Adding Complex Numbers
</td></tr>
</table>

Find each sum.

(a) $(3 - 4i) + (-2 + 6i) = [3 + (-2)] + [-4 + 6]i$
$$= 1 + 2i$$

(b) $(-9 + 7i) + (3 - 15i) = -6 - 8i$ ◻

Since $(a + bi) + (0 + 0i) = a + bi$ for all complex numbers $a + bi$, the number $0 + 0i$ is called the *additive identity* for complex numbers. The sum of $a + bi$ and $-a - bi$ is $0 + 0i$ or 0, so the number $-a - bi$ is called the *negative* or *additive inverse* of $a + bi$.

Using this definition of additive inverse, *subtraction* of complex numbers $a + bi$ and $c + di$ is defined as

$$(a + bi) - (c + di) = (a + bi) + (-c - di)$$
$$= (a - c) + (b - d)i.$$

SUBTRACTION OF COMPLEX NUMBERS

$$(a + bi) - (c + di) = (a - c) + (b - d)i$$

EXAMPLE **6**

Subtracting Complex Numbers

Subtract as indicated.

(a) $(-4 + 3i) - (6 - 7i) = (-4 - 6) + [3 - (-7)]i$
$$= -10 + 10i$$

(b) $(12 - 5i) - (8 - 3i) = (12 - 8) + (-5 + 3)i$
$$= 4 - 2i$$ ◻

To summarize, we add complex numbers by adding their real parts and adding their imaginary parts. We subtract complex numbers by subtracting their real parts and subtracting their imaginary parts.

The *product* of two complex numbers can be found by multiplying as if the numbers were binomials and using the fact that $i^2 = -1$, as follows.

$$(a + bi)(c + di) = ac + adi + bic + bidi$$
$$= ac + adi + bci + bdi^2$$
$$= ac + (ad + bc)i + bd(-1)$$
$$(a + bi)(c + di) = (ac - bd) + (ad + bc)i$$

Based on this result, the product of the complex numbers $a + bi$ and $c + di$ is defined in the following way.

MULTIPLICATION OF COMPLEX NUMBERS

$$(a + bi)(c + di) = (ac - bd) + (ad + bc)i$$

In practice, we seldom use this definition to multiply complex numbers. Instead, we usually use the customary method of multiplying binomials (known as FOIL, standing for First, Outside, Inside, Last), then replace i^2 with -1, and combine the real and imaginary parts.

EXAMPLE 7

Multiplying Complex
Numbers

Find each of the following products.

(a) $(2 - 3i)(3 + 4i) = 2(3) + 2(4i) - 3i(3) - 3i(4i)$
$= 6 + 8i - 9i - 12i^2$
$= 6 - i - 12(-1)$ $i^2 = -1$
$= 18 - i$

(b) $(5 - 4i)(7 - 2i) = 5(7) + 5(-2i) - 4i(7) - 4i(-2i)$
$= 35 - 10i - 28i + 8i^2$
$= 35 - 38i + 8(-1)$
$= 27 - 38i$

(c) $(4 + 3i)^2 = 4^2 + 2(4)(3i) + (3i)^2$ Square of a binomial
$= 16 + 24i + (-9)$
$= 7 + 24i$

(d) $(6 + 5i)(6 - 5i) = 6^2 - 25i^2$ Product of the sum and
                                   difference of two terms
$= 36 - 25(-1)$ $i^2 = -1$
$= 36 + 25$
$= 61$ Standard form ∎

Example 7(d) illustrates an important property of complex numbers. Notice that the factors $6 + 5i$ and $6 - 5i$ have the same real parts but opposite imaginary parts. Pairs of complex numbers satisfying these conditions are called **complex conjugates,** or simply conjugates. An important property of conjugates is that their product is always a real number, determined by the sum of the squares of their real and imaginary parts.

PRODUCT OF COMPLEX CONJUGATES

$$(a + bi)(a - bi) = a^2 + b^2$$

EXAMPLE 8

Examining
Conjugates and
Their Products

The following list shows several pairs of conjugates, together with their products.

Number	Conjugate	Product
$3 - i$	$3 + i$	$(3 - i)(3 + i) = 9 + 1 = 10$
$2 + 7i$	$2 - 7i$	$(2 + 7i)(2 - 7i) = 53$
$-6i$	$6i$	$(-6i)(6i) = 36$

∎

The conjugate of the divisor is used to find the *quotient* of two complex numbers. The quotient is found by multiplying both the numerator and the denominator by the conjugate of the denominator. The result should be written in standard form.

EXAMPLE 9
Dividing Complex Numbers

Write the quotient $\frac{3+2i}{5-i}$ in standard form.

SOLUTION Multiply numerator and denominator by the conjugate of $5 - i$.

$$\frac{3+2i}{5-i} = \frac{(3+2i)(5+i)}{(5-i)(5+i)}$$

$$= \frac{15 + 3i + 10i + 2i^2}{25 - i^2} \qquad \text{Multiply.}$$

$$= \frac{13 + 13i}{26} \qquad i^2 = -1$$

$$= \frac{13}{26} + \frac{13i}{26} \qquad \frac{a+bi}{c} = \frac{a}{c} + \frac{bi}{c}$$

$$= \frac{1}{2} + \frac{1}{2}i \qquad \text{Lowest terms}$$

To check this answer, show that

$$(5 - i)\left(\frac{1}{2} + \frac{1}{2}i\right) = 3 + 2i. \qquad \blacksquare$$

By definition, $i^1 = i$ and $i^2 = -1$. Now observe the following pattern.

$$i^1 = i$$
$$i^2 = -1$$
$$i^3 = i^2 \cdot i = -1 \cdot i = -i$$
$$i^4 = i^3 \cdot i = -i \cdot i = -i^2 = -(-1) = 1$$

Because $i^4 = 1$, any larger power of i may be found by writing the power as a product of two powers of i, one exponent being a multiple of 4, and then simplifying. For example,

$$i^{13} = i^{12} \cdot i = (i^4)^3 \cdot i = 1^3 \cdot i = i$$

and

$$i^{56} = (i^4)^{14} = 1^{14} = 1.$$

3.1 EXERCISES

*For each complex number, (**a**) state the real part, (**b**) state the imaginary part, and (**c**) state whether the number is real or imaginary.*

1. $-9i$ **2.** 6 **3.** π **4.** $-\sqrt{7}$

5. $i\sqrt{6}$ **6.** $-3i$ **7.** $2 + 5i$ **8.** $-7 - 6i$

Write each of the following without negative radicands.

9. $\sqrt{-100}$ **10.** $\sqrt{-169}$ **11.** $-\sqrt{-400}$ **12.** $-\sqrt{-225}$

13. $-\sqrt{-39}$ **14.** $-\sqrt{-95}$ **15.** $5 + \sqrt{-4}$ **16.** $-7 + \sqrt{-100}$

17. $9 - \sqrt{-50}$ **18.** $-11 - \sqrt{-24}$

19. Explain why a real number must be a complex number, but a complex number need not be a real number.

20. If the complex number $a + bi$ is real, then what can be said about the value of b?

Use the rules of algebra and the definition of $\sqrt{-a}$ for $a > 0$ to show that the first expression is equivalent to the second expression.

21. $\dfrac{4 + \sqrt{-60}}{8}$; $\dfrac{1}{2} + \dfrac{1}{4}i\sqrt{15}$ **22.** $\dfrac{-2 - \sqrt{-88}}{6}$; $-\dfrac{1}{3} - \dfrac{1}{3}i\sqrt{22}$ **23.** $\dfrac{-12 - \sqrt{-18}}{6}$; $-2 - \dfrac{1}{2}i\sqrt{2}$

24. $\dfrac{-13 + \sqrt{-338}}{13}$; $-1 + i\sqrt{2}$ **25.** $\dfrac{-10 + \sqrt{-100}}{10}$; $-1 + i$ **26.** $\dfrac{5 - \sqrt{-25}}{5}$; $1 - i$

Add or subtract as indicated. Write each result in standard form.

27. $(3 + 2i) + (4 - 3i)$ **28.** $(4 - i) + (2 + 5i)$

29. $(-2 + 3i) - (-4 + 3i)$ **30.** $(-3 + 5i) - (-4 + 3i)$

31. $(2 - 5i) - (3 + 4i) - (-2 + i)$ **32.** $(-4 - i) - (2 + 3i) + (-4 + 5i)$

Multiply as indicated. Write each result in standard form.

33. $(2 + 4i)(-1 + 3i)$ **34.** $(1 + 3i)(2 - 5i)$ **35.** $(-3 + 2i)^2$ **36.** $(2 + i)^2$

37. $(2 + 3i)(2 - 3i)$ **38.** $(6 - 4i)(6 + 4i)$ **39.** $(\sqrt{6} + i)(\sqrt{6} - i)$ **40.** $(\sqrt{2} - 4i)(\sqrt{2} + 4i)$

41. $i(3 - 4i)(3 + 4i)$ **42.** $i(2 + 7i)(2 - 7i)$ **43.** $3i(2 - i)^2$ **44.** $-5i(4 - 3i)^2$

45. $(1 + i)^3$ **46.** $(-2 + i)^3$

*R*elating Concepts

Recall that a solution, or root, of an equation is a number that, when substituted for the variable, gives a true statement. In earlier chapters we have only considered real number solutions of equations. In this chapter we will see that equations may also have solutions that are not real numbers. For the equation

$$x^3 - x^2 - 7x + 15 = 0$$

show that each of the following is a solution by substituting it for x.

47. the real number -3

48. the complex number $2 - i$

49. the complex number $2 + i$

50. What relationship do the solutions in Exercises 48 and 49 have?

Divide as indicated. Write each result in standard form.

51. $\dfrac{1 + i}{1 - i}$ **52.** $\dfrac{2 - i}{2 + i}$ **53.** $\dfrac{4 - 3i}{4 + 3i}$ **54.** $\dfrac{5 - 2i}{6 - i}$

55. $\dfrac{3 - 4i}{2 - 5i}$ **56.** $\dfrac{1 - 3i}{1 + i}$ **57.** $\dfrac{-3 + 4i}{2 - i}$ **58.** $\dfrac{5 + 6i}{5 - 6i}$

59. $\dfrac{2}{i}$ **60.** $\dfrac{-7}{3i}$

61. Explain why the method of dividing complex numbers (that is, multiplying both the numerator and the denominator by the conjugate of the denominator) works. That is, what property justifies this process?

62. Suppose that your friend, Susan Katz, tells you that she has discovered a method of simplifying a posi-tive power of i. "Just divide the exponent by 4," she says, "and then look at the remainder. Then refer to the short table of powers of i in this section. The large power of i is equal to i to the power indicated by the remainder. And if the remainder is 0, the result is $i^0 = 1$." Explain why Susan's method works.

Simplify each of the following powers of i to i, 1, −i, or −1.

63. i^{15} **64.** i^{42} **65.** i^{61} **66.** i^{28}

67. i^{102} **68.** i^{19} **69.** i^{32} **70.** i^{69}

3.2 QUADRATIC FUNCTIONS AND THEIR GRAPHS

Basic Terminology ▮ Graphs of Quadratic Functions ▮ The Zero-Product Property ▮ Extreme Values, End Behavior, and Concavity

Basic Terminology

In Chapter 1 we introduced linear functions. Recall that a linear function is defined by an equation of the form $y = f(x) = ax + b$. It is the simplest example of a larger group of functions known as *polynomial functions*.

POLYNOMIAL FUNCTION

A **polynomial function of degree n in the variable x** is a function defined by

$$P(x) = a_n x^n + a_{n-1} x^{n-1} + \cdots + a_1 x + a_0,$$

where each a_i is a real number, $a_n \neq 0$, and n is a whole number.

While the letter used to name a function is immaterial, we will often use P (as above) to name a polynomial function. Polynomial functions of degree 1, 2 and 3 occur so often that we give them special names, as shown in the chart that follows.

Example	Degree	Special Name
$P(x) = 4x - 7$	1	linear
$P(x) = 2x^2 + 4x - 16$	2	quadratic
$P(x) = -3x^3 + 5x$	3	cubic

The coefficient a_n for a polynomial function of degree n is called the **leading coefficient,** and a_0 is called the **constant.** In the three examples shown above, the leading coefficients are, respectively, 4, 2, and −3. The constants are −7, −16, and 0. Notice (as in the example of the cubic polynomial function) that a power of the variable may not be present. There is no x^2 term in the cubic polynomial, but we may consider it to actually be there with coefficient 0.

In our study of polynomial functions, we will often be interested in finding the values of x that satisfy $P(x) = 0$. Such values are called zeros of the function.

ZERO OF A FUNCTION

For any function f, the number c is a **zero** of f if

$$f(c) = 0.$$

For example, the linear function $P(x) = 4x - 7$ has $\frac{7}{4}$ as its only zero, since $P(\frac{7}{4}) = 4(\frac{7}{4}) - 7 = 7 - 7 = 0$. We learned how to find zeros of linear functions in Chapter 1, and with respect to the graph, we can see that *a real zero of a function is an x-intercept of its graph.* We will devote a lot of effort to determining zeros of polynomial functions of higher degree in this chapter.

Graphs of Quadratic Functions

In Chapter 2 we saw that the graph of $y = x^2$ is a parabola. The function $P(x) = x^2$ is the simplest example of a quadratic function.

QUADRATIC FUNCTION

A function of the form

$$P(x) = ax^2 + bx + c, \; a \neq 0,$$

is called a **quadratic function.**

When discussing graphs of linear functions, we refer to a complete graph as a graph in a viewing window that shows all intercepts. For a quadratic function, whose graph is a parabola, we define a complete graph as follows.

COMPLETE GRAPH OF A QUADRATIC FUNCTION

A complete graph of a quadratic function will show all intercepts and the vertex of the parabola.

Let us consider $P(x) = 2x^2 + 4x - 16$. A complete graph is shown in a viewing window of $[-10, 10]$ by $[-20, 10]$ in Figure 3. When compared to the graph of $y = x^2$, also shown in the figure, it appears that the graph of P can be obtained by a vertical stretch with a factor greater than 1, a shift to the left, and a shift downward. It would be possible to determine the magnitudes if the function were written in the form

$$P(x) = a(x - h)^2 + k.$$

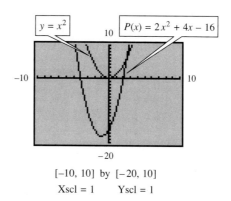

$[-10, 10]$ by $[-20, 10]$
Xscl = 1   Yscl = 1

FIGURE 3

It is possible to transform an equation of the form $P(x) = ax^2 + bx + c$ into this desired form by the method of *completing the square*. The steps in this procedure are summarized below.

COMPLETING THE SQUARE

To transform the equation $P(x) = ax^2 + bx + c$ into the form $P(x) = a(x - h)^2 + k$:

1. Divide both sides of the equation by a so that the coefficient of x^2 is 1.
2. Add to both sides the square of half the coefficient of x; that is, $\left(\frac{b}{2a}\right)^2$.
3. Factor the right-hand side as the square of a binomial and combine terms on the left.
4. Isolate the term involving $P(x)$ on the left.
5. Multiply both sides by a.

We will now apply this procedure to $P(x) = 2x^2 + 4x - 16$.

$$P(x) = 2x^2 + 4x - 16 \qquad \text{Given function}$$

$$\frac{P(x)}{2} = x^2 + 2x - 8 \qquad \text{Divide by 2 to make the coefficient of } x^2 \text{ equal to 1.}$$

$$\frac{P(x)}{2} + 8 = x^2 + 2x \qquad \text{Add 8 to both sides.}$$

$$\frac{P(x)}{2} + 8 + 1 = x^2 + 2x + 1 \qquad \text{Add } [\tfrac{1}{2}(2)]^2 = 1 \text{ to both sides to complete the square on the right.}$$

$$\frac{P(x)}{2} + 9 = (x + 1)^2 \qquad \text{Combine terms on the left and factor on the right.}$$

$$\frac{P(x)}{2} = (x + 1)^2 - 9 \qquad \text{Add } -9 \text{ to both sides.}$$

$$P(x) = 2(x + 1)^2 - 18 \qquad \text{Multiply both sides by 2.}$$

Using the concepts of Chapter 2, we can now make the following statements: The graph of $P(x) = 2x^2 + 4x - 16$ can be obtained from the graph of $y = x^2$ by a vertical stretch factor of 2, shifting 1 unit to the left, and 18 units down. The vertex of the parabola has coordinates $(-1, -18)$, its domain is $(-\infty, \infty)$, and its range is $[-18, \infty)$. It decreases on the interval $(-\infty, -1)$ and increases on the interval $(-1, \infty)$.

FOR **GROUP DISCUSSION**

Enter $y_1 = 2x^2 + 4x - 16$ and $y_2 = 2(x + 1)^2 - 18$ and graph in a viewing window of $[-10, 10]$ by $[-20, 10]$.

1. What do you notice about the graphs of y_1 and y_2?
2. Use the capabilities of your calculator to support the statement above that the point $(-1, -18)$ is the vertex of the graph.

<table>
<tr><td>

EXAMPLE 1

Analyzing the Graph of a Quadratic Function

</td></tr>
</table>

The graph of $P(x) = -x^2 - 6x - 8 = -(x + 3)^2 + 1$ is shown in Figure 4, using both a traditional graph and a calculator-generated graph in a viewing window of $[-8, 2]$ by $[-10, 2]$. Discuss the features of the graph.

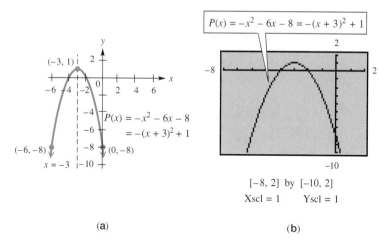

(a)   (b)

FIGURE 4

SOLUTION   The parabola is the graph of $y = x^2$, translated 3 units to the left and 1 unit upward. It opens downward because of the negative sign preceding $(x + 3)$. (This is a reflection across the x-axis.) The line $x = -3$ is its **axis of symmetry,** since if it were folded along this line, the two halves would coincide. The vertex, $(-3, 1)$, is the highest point on the graph. The domain is $(-\infty, \infty)$ and the range is $(-\infty, 1]$. The function increases on the interval $(-\infty, -3)$ and decreases on $(-3, \infty)$. Since $P(0) = -8$, the y-intercept is -8, and since $P(-4) = P(-2) = 0$, the x-intercepts are -4 and -2.   ◼

Our discussion up to this point leads to the following generalizations about the graph of a quadratic function in the form $P(x) = a(x - h)^2 + k$.

GRAPH OF $P(x) = a(x - h)^2 + k$

The graph of $P(x) = a(x - h)^2 + k$, $a \neq 0$,

a. is a parabola with vertex (h, k), and the vertical line $x = h$ as axis of symmetry;

b. opens upward if $a > 0$ and downward if $a < 0$;

c. is broader than $y = x^2$ if $0 < |a| < 1$ and narrower than $y = x^2$ if $|a| > 1$.

Determining the coordinates of the vertex of the graph of a quadratic function can be done by using the method described earlier for the function $P(x) = 2x^2 + 4x - 16$. Rather than go through the procedure for each individual function, we may generalize it for the standard form of the quadratic function, $P(x) = ax^2 + bx + c$.

$$P(x) = ax^2 + bx + c \quad (a \neq 0)$$ Standard form

$$y = ax^2 + bx + c \quad (a \neq 0)$$ Replace $P(x)$ with y to simplify notation.

$$\frac{y}{a} = x^2 + \frac{b}{a}x + \frac{c}{a}$$ Divide by a.

$$\frac{y}{a} - \frac{c}{a} = x^2 + \frac{b}{a}x$$ Subtract $\frac{c}{a}$.

$$\frac{y}{a} - \frac{c}{a} + \frac{b^2}{4a^2} = x^2 + \frac{b}{a}x + \frac{b^2}{4a^2}$$ Add $\frac{b^2}{4a^2}$.

$$\frac{y}{a} + \frac{b^2 - 4ac}{4a^2} = \left(x + \frac{b}{2a}\right)^2$$ Combine terms on left and factor on right.

$$\frac{y}{a} = \left(x + \frac{b}{2a}\right)^2 - \frac{b^2 - 4ac}{4a^2}$$ Get y term alone on the left.

$$y = a\left(x + \frac{b}{2a}\right)^2 + \frac{4ac - b^2}{4a}$$ Multiply by a.

$$P(x) = a\left[x - \left(-\frac{b}{2a}\right)\right]^2 + \underbrace{\frac{4ac - b^2}{4a}}_{k}$$ Write in the form $P(x) = a(x - h)^2 + k$.

$$\underbrace{\phantom{x - \left(-\frac{b}{2a}\right)}}_{h}$$

The final equation shows that the vertex (h, k) can be expressed in terms of a, b, and c. However, it is not necessary to memorize the expression for k, since it is equal to $P\left(-\frac{b}{2a}\right)$.

VERTEX FORMULA

The vertex of the graph of $P(x) = ax^2 + bx + c \ (a \neq 0)$ is the point

$$\left(-\frac{b}{2a}, P\left(-\frac{b}{2a}\right)\right).$$

TECHNOLOGICAL NOTE
Most current models of graphics calculators are capable of determining the coordinates of the "highest point" or "lowest point" in a designated interval of a graph. See if your calculator is capable of this; these are usually designated with commands like "maximum" and "minimum." With this capability, you can find the coordinates of the vertex of a parabola graphically, to support the analytic discussion in this section.

EXAMPLE 2

Using the Vertex Formula

Use the vertex formula to find the coordinates of the vertex of the graph of $P(x) = -.65x^2 + \sqrt{2}x + 4$. Give **(a)** the exact values of x and y and **(b)** the approximate values of x and y to the nearest hundredth. Finally, **(c)** support your answer in part **(b)** by graphing the function in an appropriate window and using the capabilities of your calculator to find the vertex.

SOLUTION

(a) For this function, $a = -.65$ and $b = \sqrt{2}$, so applying the vertex formula,

$$x = -\frac{b}{2a} = -\frac{\sqrt{2}}{2(-.65)} = \frac{\sqrt{2}}{2(.65)}$$

and

$$y = P\left(\frac{-b}{2a}\right) = -.65\left(\frac{\sqrt{2}}{2(.65)}\right)^2 + \sqrt{2}\left(\frac{\sqrt{2}}{2(.65)}\right) + 4$$

These are the *exact* (but not simplified) values of x and y.

(b) Using the arithmetic, squaring, and square root functions of a calculator, we find that to the nearest hundredth,

$$x \approx 1.09 \quad \text{and} \quad y \approx 4.77.$$

(c) Graphing the function in a window of $[-2, 4]$ by $[-2, 5]$ and locating the highest point on the graph, we see a display of $x = 1.0878556$ and $y = 4.7692308$ in Figure 5. These values support our answer in part (b). ∎

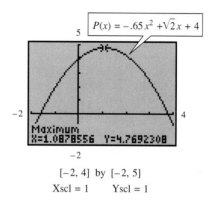

$$P(x) = -.65x^2 + \sqrt{2}x + 4$$

Maximum
X=1.0878556   Y=4.7692308

$[-2, 4]$ by $[-2, 5]$
Xscl = 1   Yscl = 1

FIGURE 5

The Zero-Product Property

An important property that will allow us to solve for the zeros of certain polynomial functions (and thus find the x-intercepts of their graphs) is the zero-product property.

ZERO-PRODUCT PROPERTY

If a and b are complex numbers and $ab = 0$, then $a = 0$ or $b = 0$ or both.

The zero-product property says that if the product of two complex numbers is 0, then at least one of the factors must equal 0.

Consider the following problems:

1. Find the zeros of the quadratic function $P(x) = 2x^2 + 4x - 16$.
2. Find the x-intercepts of the graph of $P(x) = 2x^2 + 4x - 16$.
3. Find the solution set of the equation $2x^2 + 4x - 16 = 0$.

All three problems may be solved the same way: we must find the numbers that make the expression $2x^2 + 4x - 16$ equal to 0. The following example shows how this can be done using the zero-product property.

EXAMPLE 3

Using the
Zero-Product
Property

Solve the equation $2x^2 + 4x - 16 = 0$.

SOLUTION To make our work easier, we may divide both sides by 2. Then factor, and set each factor equal to 0. Solve the resulting linear equations.

$$2x^2 + 4x - 16 = 0$$
$$x^2 + 2x - 8 = 0$$
$$(x + 4)(x - 2) = 0$$
$$x + 4 = 0 \quad \text{or} \quad x - 2 = 0$$
$$x = -4 \qquad\qquad x = 2$$

The solution set of the equation is $\{-4, 2\}$. ▮

Now, look back at Figure 3. It appears that the function graphed there, $P(x) = 2x^2 + 4x - 16$, has x-intercepts -4 and 2. Our work in Example 3 analytically proves that these are indeed the intercepts. And because $P(-4) = 0$ and $P(2) = 0$, -4 and 2 are zeros of the function.

EXAMPLE 4

Using the
Zero-Product
Property

Find all zeros of the quadratic function $P(x) = x^2 - 6x + 9$. Support your answer graphically.

SOLUTION We must solve $x^2 - 6x + 9 = 0$.

$$x^2 - 6x + 9 = 0$$
$$(x - 3)^2 = 0 \qquad \text{Factor.}$$
$$x - 3 = 0 \quad \text{or} \quad x - 3 = 0 \qquad \text{Use the zero-product}$$
$$x = 3 \qquad\qquad x = 3 \qquad \text{property.}$$

There is only one *distinct* zero, 3. It is sometimes called a double zero, or double solution (root) of the equation. Graphing the function in the standard viewing window allows us to support our result, as seen in Figure 6. The vertex has coordinates $(3, 0)$. ▮

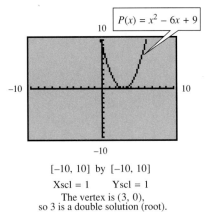

$[-10, 10]$ by $[-10, 10]$
Xscl = 1   Yscl = 1
The vertex is (3, 0),
so 3 is a double solution (root).

FIGURE 6

CAUTION A purely graphical approach may not prove to be as useful with a graph like the one in Figure 6, since the vertex may be slightly above or slightly below the x-axis. **This is why we need to understand the algebraic concepts presented in this section—only when we know the mathematics can we use the technology to its utmost.**

Figure 7 shows the possible numbers of x-intercepts of the graph of a quadratic function that opens upward.

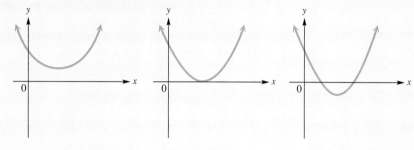

FIGURE 7

Figure 8 shows the possible numbers of x-intercepts of the graph of a quadratic function that opens downward.

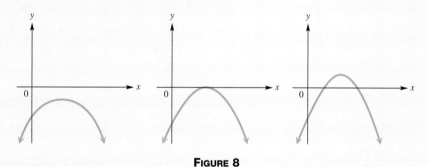

FIGURE 8

Use these figures to discuss the following.
1. What is the maximum number of real solutions of a quadratic equation?
2. What is the minimum number of real solutions of a quadratic equation?
3. If a quadratic function has only one real zero, what do we know about the vertex of its graph?

NOTE The zero-product property is quite limited in its practical applications. If a quadratic polynomial cannot easily be factored, then it is of little use. For this reason, we will develop a more powerful method of finding zeros of quadratic functions in the next section. It is called the quadratic formula.

Extreme Values, End Behavior, and Concavity

The vertex of the graph of $P(x) = ax^2 + bx + c$ is the lowest point on the graph of the function if $a > 0$, and is the highest point if $a < 0$. Such points are called **extreme points** (also **extrema**, singular: **extremum**). As we extend our study of polynomial functions, we will examine extrema on a more general basis.

> ### EXTREME POINT AND EXTREME VALUE OF A QUADRATIC FUNCTION
>
> For the quadratic function $P(x) = ax^2 + bx + c$,
>
> **a.** if $a > 0$, the vertex (h, k) is called the *minimum point* of the graph. The minimum value of the function is $P(h) = k$.
> **b.** if $a < 0$, the vertex (h, k) is called the *maximum point* of the graph. The *maximum value* of the function is $P(h) = k$.

Figure 9 illustrates the ideas above.

NOTE   Modern graphics calculators have the capabilities of locating extrema to great accuracy.

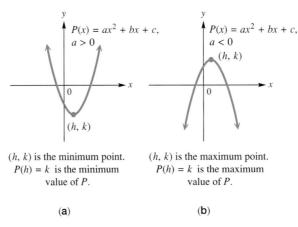

(h, k) is the minimum point.
$P(h) = k$ is the minimum
value of P.

(h, k) is the maximum point.
$P(h) = k$ is the maximum
value of P.

(a)

(b)

FIGURE 9

EXAMPLE **5**

Identifying Extreme Points and Extreme Values

Give the coordinates of the extreme point of the graph of each function, and the corresponding maximum or minimum value of the function.

(a) $P(x) = 2x^2 + 4x - 16$ (Figure 3)

SOLUTION   In the discussion preceding Example 1, we found that the vertex of the graph of this function is $(-1, -18)$. It opens upward since $a > 0$ (as seen in Figure 3), so the vertex $(-1, -18)$ is the minimum point and -18 is the minimum value of the function.

(b) $P(x) = -x^2 - 6x - 8$ (Example 1, Figure 4)

SOLUTION   The vertex $(-3, 1)$ is the maximum point and $P(-3) = 1$ is the maximum value of the function.

(c) $P(x) = -.65x^2 + \sqrt{2}x + 4$ (Example 2, Figure 5)

SOLUTION   Based on our work in Example 2, the vertex has approximate coordinates $(1.09, 4.77)$. It is the highest point on the graph, so it is a maximum point, and the maximum value of the function is approximately 4.77. (*Note:* The exact maximum value is the y-value indicated in part (a) of Example 2.)   ❚

FOR **GROUP DISCUSSION**

Use the capabilities of your calculator to approximate the maximum point and corresponding maximum value of the quadratic function $P(x) = -4.1x^2 - 28.2x + 13.8$. Give the answers to the nearest hundredth.

We know that if the value of a is positive for the quadratic function $P(x) = ax^2 + bx + c$, the graph opens upward, and if a is negative, the graph opens downward. The sign of a determines the *end behavior* of the graph. If $a > 0$, as x approaches $-\infty$ or ∞, (written $x \to -\infty$ or $x \to \infty$), the value of $P(x)$ approaches $+\infty$, (written $P(x) \to \infty$). The other situations similar to this are summarized in the following box.

END BEHAVIOR OF THE GRAPH OF A QUADRATIC FUNCTION

$P(x) = ax^2 + bx + c,$
$a > 0$

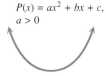

If $a > 0$,

as $x \to -\infty$, $P(x) \to \infty$.

as $x \to \infty$, $P(x) \to \infty$.

This end behavior will be symbolized in this text by

$P(x) = ax^2 + bx + c,$
$a < 0$

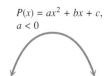

If $a < 0$,

as $x \to -\infty$, $P(x) \to -\infty$.

as $x \to \infty$, $P(x) \to -\infty$.

This end behavior will be symbolized in this text by

We conclude this section with a brief discussion of *concavity*. Using the quadratic function graph as an illustration, we see that if $a > 0$, the graph is at all times opening upward. If water were to be poured from above, the graph would, in a sense, "hold water." We say that this graph is *concave up* for all values in its domain. On the other hand, if $a < 0$, the graph opens downward at all times, and it would similarly "dispel water" if it were poured from above. In this case, the graph is *concave down* for all values in its domain. See Figure 10.

$P(x) = ax^2 + bx + c,$ $P(x) = ax^2 + bx + c,$
$a < 0$ $a > 0$

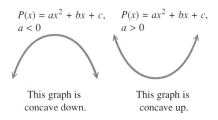

This graph is   This graph is
concave down.   concave up.

FIGURE 10

NOTE A *formal* discussion of concavity requires concepts beyond the scope of this text. It is studied more rigorously in calculus.

3.2 EXERCISES

*In Exercises 1–8, you are given an equation and the graph of a quadratic function. Without using your calculator, do each of the following: (**a**) Give the domain and the range. (**b**) Give the coordinates of the vertex. (**c**) Give the equation of the axis of symmetry. (**d**) Give the interval over which the function is increasing. (**e**) Give the interval over which the function is decreasing. (**f**) State whether the vertex is a maximum or minimum point, and give the corresponding maximum or minimum value of the function. (**g**) Tell whether the graph is concave up or concave down.*

1. $P(x) = (x - 2)^2$

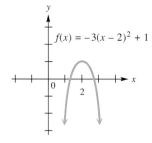

2. $P(x) = (x + 4)^2$

1. $P(x) = (x - 2)^2$

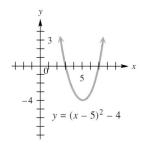

2. $P(x) = (x + 4)^2$

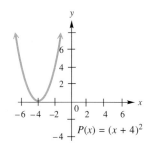

3. $y = (x + 3)^2 - 4$

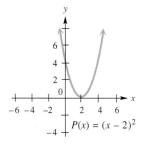

4. $y = (x - 5)^2 - 4$

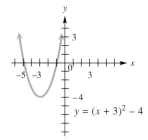

5. $f(x) = -2(x + 3)^2 + 2$

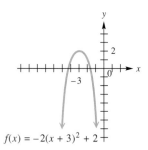

6. $f(x) = -3(x - 2)^2 + 1$

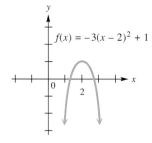

7. $P(x) = -.5(x + 1)^2 - 3$

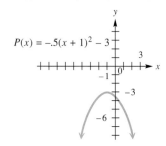

8. $P(x) = \frac{2}{3}(x - 2)^2 - 1$

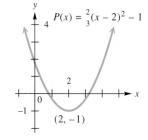

In Exercises 9–16, an equation of a quadratic function is given. Do each of the following.
(a) Find the coordinates of the vertex of the graph analytically, using the method of completing the square or the vertex formula given in this section.
(b) Find the x-intercepts of the graph using the zero-factor property.
(c) Find the y-intercept. (Hint: Evaluate the function for x = 0.)
(d) Support your answers to parts (a)–(c) using a calculator-generated graph in an appropriate viewing window that will show a complete graph of the function.

9. $P(x) = 2x^2 - 2x - 24$ **10.** $P(x) = 3x^2 + 3x - 6$ **11.** $y = x^2 - 2x - 15$

12. $y = -x^2 - 3x + 10$ **13.** $f(x) = -2x^2 + 6x$ **14.** $f(x) = 4x^2 - 4x$

15. $P(x) = 4x^2 - 22x - 12$ **16.** $P(x) = 6x^2 - 16x - 6$

The graphs of the functions in Exercises 17–20 are shown in Figures A–D. Match each function with its graph using the concepts of this section without actually entering it into your calculator. Then, after you have completed the exercises, check your answers with your calculator. Use the standard viewing window.

17. $y = (x - 4)^2 - 3$ **18.** $y = -(x - 4)^2 + 3$

19. $y = (x + 4)^2 - 3$ **20.** $y = -(x + 4)^2 + 3$

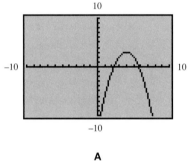

A

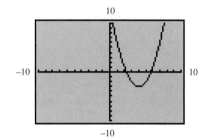

B

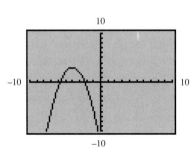

C

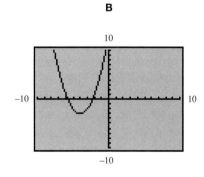

D

In Exercises 21–26, a quadratic function with decimal and/or irrational coefficients is given. Graph the function in a viewing window that will allow you to use your calculator to approximate (a) the coordinates of the vertex and (b) the x-intercepts. Give values to the nearest hundredth.

21. $P(x) = -.32x^2 + \sqrt{3}x + 2.86$ **22.** $P(x) = -\sqrt{2}x^2 + .45x + 1.39$

23. $y = 1.34x^2 - 3x + \sqrt{5}$ **24.** $y = 2.53x^2 - 2x + \sqrt{19}$

25. $f(x) = \sqrt{10}x^2 + 3.26x - 4.16$ **26.** $f(x) = \sqrt[3]{20}x^2 + 6.48x - \sqrt{2}$

*R*elating Concepts

Refer to the graphs in Figures A–F to answer the following. There may be one correct choice, more than one correct choice, or no correct choices.

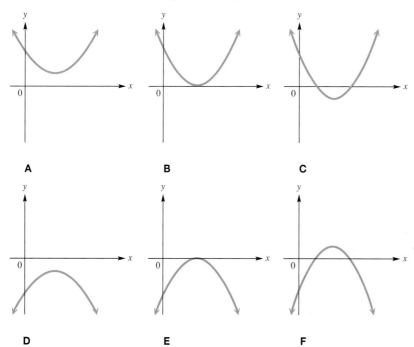

27. Which functions have two real zeros?

28. Which functions have exactly one real zero?

29. Which functions have no real zeros?

30. If each function is of the form $P(x) = ax^2 + bx + c$, for which functions is $a < 0$? For which is $a > 0$?

31. If each function is of the form $P(x) = ax^2 + bx + c$, for which functions is $c > 0$? For which is $c < 0$? For which is $c = 0$?

32. Which graphs are concave up? Which graphs are concave down?

33. If each function is of the form $P(x) = a(x - h)^2 + k$, which functions would satisfy the conditions that a, h, and k are all positive?

34. Which function could possibly have the equation $f(x) = -2(x - 4)^2$?

Draw an end behavior diagram (that is, $\nwarrow \nearrow$ or $\swarrow \searrow$) for each of the following quadratic functions. Do not rely on your calculator to do this.

35. $P(x) = 19(x + 12)^2 - 48$

36. $P(x) = 27(x - 3)^2 + 84$

37. $y = -200x^2 - 480x + 1993$

38. $y = -300x^2 + 1280x - 1936$

39. $f(x) = -\sqrt{483}(x + \sqrt{2})^2 - \sqrt{13}$

40. $f(x) = -\sqrt{276}(x - 1.7)^2 + .483$

41. $P(x) = 129\pi x^2 - \dfrac{\pi}{2}x + 12$

42. $P(x) = 486\pi x^2 - 13\pi x + \pi$

Relating Concepts

Recall that if we are given the graph of a function $y = f(x)$, the solutions of $f(x) = 0$ are the x-intercepts, the solutions of $f(x) < 0$ are those x-values for which the graph is below the x-axis, and the solutions of $f(x) > 0$ are those x-values for which the graph is above the x-axis. These concepts were first presented in Chapter 1. Use them to answer the following.

43. The figure shows the graph of $f(x) = 2x^2 + 5x - 3$. The x-intercepts, which may be determined by the zero-factor property, are -3 and $\frac{1}{2}$.
  (a) Give the solution set of $2x^2 + 5x - 3 = 0$.
  (b) Give the solution set of $2x^2 + 5x - 3 < 0$.
  (c) Give the solution set of $2x^2 + 5x - 3 > 0$.

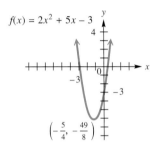

$f(x) = 2x^2 + 5x - 3$

$\left(-\frac{5}{4}, -\frac{49}{8}\right)$

44. For the function $f(x) = x^2 - x - 1$, use your calculator to determine the x-intercepts to the nearest hundredth, and then answer the following.
  (a) Give the solution set of $f(x) = 0$.
  (b) Give the solution set of $f(x) \geq 0$.
  (c) Give the solution set of $f(x) \leq 0$.

If we know the coordinates of the vertex of the graph of a quadratic function and the coordinates of another point on the graph, then it is possible to find the equation that defines the function. Suppose, for example, that a quadratic function P has vertex $(-7, -4)$ and the point $(-1, 104)$ lies on the graph. Then for some value a, the equation is $P(x) = a(x + 7)^2 - 4$. To find a, let $x = -1$ and $P(-1) = 104$:

$$104 = a(-1 + 7)^2 - 4$$
$$104 = 36a - 4$$
$$108 = 36a$$
$$a = 3.$$

Therefore, $P(x) = 3(x + 7)^2 - 4 = 3x^2 + 42x + 143$.

Use the method described above to find the equation of the quadratic function with the given vertex and given point on its graph. Then use the capabilities of your calculator to support your result. Express your answer in the form $P(x) = ax^2 + bx + c$.

45. vertex: $(-1, -4)$ point: $(5, 104)$ **46.** vertex: $(-2, -3)$ point: $(0, -19)$

47. vertex: $(8, 3)$ point: $(10, 5)$ **48.** vertex: $(-6, -12)$ point: $(6, 24)$

49. vertex: $(-4, -2)$ point: $(2, -26)$ **50.** vertex: $(5, 6)$ point: $(1, -6)$

51. Explain how you can determine, without actually graphing, the number of x-intercepts of the graph of a quadratic function if you know the quadrant in which the vertex lies and whether the graph is concave up or concave down.

52. Use your calculator to find an approximation of $\sqrt{2}$ in two ways:
  (a) Use the square root key.
  (b) Find the positive x-intercept of the graph of $P(x) = x^2 - 2$.

Further Explorations

Tracing can graphically support an analytical solution. If your graphics calculator has a TABLE feature, it can be used to provide further numerical support for the vertex and zeros of a quadratic function.

For $P(x) = 2x^2 + 4x - 16$ (Figure 3), the TABLE supports the analytical and graphical solutions: zeros at $(-4, 0)$ and $(2, 0)$ and the minimum point (vertex) at $(-1, -18)$.

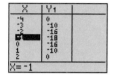

FIGURE A

The TABLE can also be used when vertices and zeros are not integers. Example 5c, $P(x) = -.65x^2 + \sqrt{2}x + 4$ generates the following TABLE:

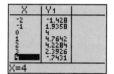

FIGURE B

From the TABLE, it can be seen that Y values are increasing when $x < 1$ and are decreasing when $x > 1$. This means there must be a "high point" between $x = 0$ and $x = 2$. The vertex cannot be at $x = 1$ since $x = 1$ would then be a line of symmetry and $P(0) \neq P(2)$.

It is possible to ZOOM IN on the TABLE to find the vertex and the zeros by decreasing the step increments for x. Use the TBLSET menu to change ΔTBL (Figure C). With ΔTBL $= .1$, the vertex is at $x = 1.1$, $y = 4.8$ (to the nearest tenth).

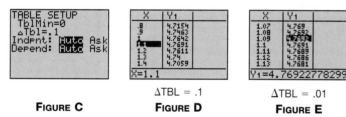

ΔTBL $= .1$	ΔTBL $= .01$
FIGURE C **FIGURE D**	**FIGURE E**

ZOOM IN again by letting ΔTBL $= .01$ to calculate the vertex to the nearest hundredth. Note that while the TABLE can only display up to four decimal places, the highlighted value is displayed with eleven decimal places at the bottom of the screen (Figure E). Therefore the vertex of the parabola is approximately $(1.09, 4.77)$.

The zeros can be found by finding the intervals in the TABLE where $P(x)$ changes sign. In Figure B, it can be seen that $P(x)$ changes sign between $x = -2$ and $x = -1$ and also between $x = 3$ and $x = 4$. Since the graph is a continuous graph (which means it can be drawn without lifting the pencil point off the paper), somewhere within each of these two intervals the graph must cross the x-axis. Change ΔTBL to .1 to generate the TABLE in Figure F. Now the zero of $P(x)$ is seen to be between -1.7 and -1.6 (or $P(x) = 0$ when $x = -1.65$ with an error of at most .05). With ΔTBL $= .01$ (Figure G), $P(x) = 0$ when $x = -1.625$ with an error of at most .005.

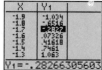
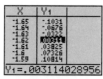

FIGURE F   **FIGURE G**

1. Find the other zero (to the nearest hundredth) of $P(x) = -.65x^2 + \sqrt{2}x + 4$.

2. Use the TABLE feature to find the vertex and zeros of $P(x) = x^2 - 3x - 3$.

3. Use the TABLE feature to find the vertex and zeros of $P(x) = .37x^2 + 15x - 3$.

3.3 SOLUTION OF QUADRATIC EQUATIONS AND INEQUALITIES

Solving $x^2 = k$ ▮ The Quadratic Formula and the Discriminant ▮ Solving Quadratic Equations ▮ Solving Quadratic Inequalities

We know that if $a \neq 0$, the solution of $ax + b = 0$ is $-b/a$. In this section we will develop methods of solving general quadratic equations, and extend them to solving quadratic inequalities.

> **QUADRATIC EQUATION IN ONE VARIABLE**
>
> An equation that can be written in the form
>
> $$ax^2 + bx + c = 0,$$
>
> where a, b, and c are real numbers with $a \neq 0$, is a **quadratic equation in standard form.**

Solving $x^2 = k$

We are often interested in solving quadratic equations of the form $x^2 = k$, where k is a real number. This type of equation can be solved by factoring using the following sequence of equivalent equations.

$$x^2 = k$$
$$x^2 - k = 0$$
$$(x - \sqrt{k})(x + \sqrt{k}) = 0$$
$$x - \sqrt{k} = 0 \qquad \text{or} \qquad x + \sqrt{k} = 0$$
$$x = \sqrt{k} \qquad \text{or} \qquad x = -\sqrt{k}$$

We have proved the following statement, which we will call the square root property for solving quadratic equations.

> **SQUARE ROOT PROPERTY FOR SOLVING QUADRATIC EQUATIONS**
>
> The solution set of $x^2 = k$ is
>
> **a.** $\{\pm\sqrt{k}\}$ if $k > 0$
> **b.** $\{0\}$ if $k = 0$
> **c.** $\{\pm i\sqrt{|k|}\}$ if $k < 0$.

As shown in Figure 11 on the next page, the graph of $y_1 = x^2$ intersects the graph of $y_2 = k$ twice if $k > 0$, once if $k = 0$, and not at all if $k < 0$.

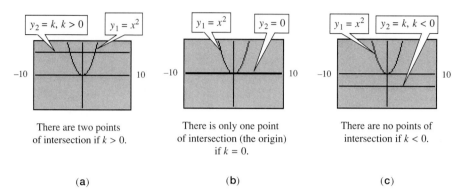

There are two points of intersection if $k > 0$.

There is only one point of intersection (the origin) if $k = 0$.

There are no points of intersection if $k < 0$.

(a)

(b)

(c)

FIGURE 11

EXAMPLE 1

Using the Square Root Property

Solve each of the following quadratic equations.

(a) $x^2 = 7$

SOLUTION Since $7 > 0$, there will be two real solutions.

$$x^2 = 7$$
$$x = \pm\sqrt{7}$$

This result may be supported graphically by using the intersection-of-graphs method, first introduced in Chapter 1. If we graph $y_1 = x^2$ and $y_2 = 7$ in a standard viewing window, and then locate the points of intersection, we will find that the x-coordinates are approximately -2.65 and 2.65, which are approximations for $\pm\sqrt{7}$. See Figure 12. The solution set is $\{\pm\sqrt{7}\}$.

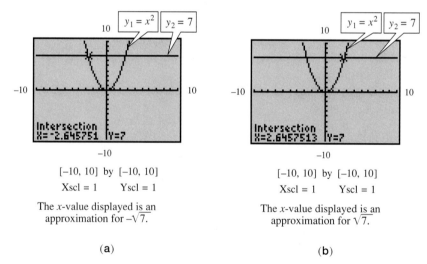

$[-10, 10]$ by $[-10, 10]$
Xscl = 1   Yscl = 1

The x-value displayed is an approximation for $-\sqrt{7}$.

(a)

$[-10, 10]$ by $[-10, 10]$
Xscl = 1   Yscl = 1

The x-value displayed is an approximation for $\sqrt{7}$.

(b)

FIGURE 12

(b) $x^2 = -5$

SOLUTION There is no real number whose square is -5. However, this equation has two complex imaginary solutions.

$$x^2 = -5$$
$$x = \pm\sqrt{-5}$$
$$x = \pm i\sqrt{5}$$

The solution set is $\{ \pm i\sqrt{5} \}$. Notice that the graphs of $y_1 = x^2$ and $y_2 = -5$ do not intersect. This indicates that there are no *real* solutions. See Figure 13. ∎

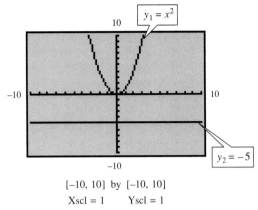

[−10, 10] by [−10, 10]
Xscl = 1   Yscl = 1

There are no points of intersection,
and thus no *real* solutions.

FIGURE 13

The Quadratic Formula and the Discriminant

We saw in Section 3.2 that certain quadratic equations can be solved by factoring and then using the zero-factor property. However, this method is quite limited—what if the polynomial cannot be factored? For this reason, we develop the quadratic formula.

We begin with the standard form of the general quadratic equation,

$$ax^2 + bx + c = 0 \quad (a \neq 0),$$

and will solve for x in terms of the constants a, b, and c. To do this, we employ the method of completing the square. For now, assume $a > 0$ and divide both sides by a to obtain

$$x^2 + \frac{b}{a}x + \frac{c}{a} = 0.$$

Add $-\frac{c}{a}$ to both sides.

$$x^2 + \frac{b}{a}x = -\frac{c}{a}$$

Now take half of $\frac{b}{a}$, and square the result:

$$\frac{1}{2} \cdot \frac{b}{a} = \frac{b}{2a} \quad \text{and} \quad \left(\frac{b}{2a}\right)^2 = \frac{b^2}{4a^2}.$$

Add the square to both sides, producing

$$x^2 + \frac{b}{a}x + \frac{b^2}{4a^2} = \frac{b^2}{4a^2} - \frac{c}{a}.$$

The expression on the left side of the equals sign can be written as the square of a binomial, while the expression on the right can be simplified.

$$\left(x + \frac{b}{2a}\right)^2 = \frac{b^2 - 4ac}{4a^2}$$

By the square root property, this last statement leads to

$$x + \frac{b}{2a} = \sqrt{\frac{b^2 - 4ac}{4a^2}} \qquad \text{or} \qquad x + \frac{b}{2a} = -\sqrt{\frac{b^2 - 4ac}{4a^2}}.$$

Since $4a^2 = (2a)^2$, or $4a^2 = (-2a)^2$,

$$x + \frac{b}{2a} = \frac{\sqrt{b^2 - 4ac}}{2a} \qquad \text{or} \qquad x + \frac{b}{2a} = \frac{-\sqrt{b^2 - 4ac}}{2a}.$$

Adding $-\frac{b}{(2a)}$ to both sides of each result gives

$$x = \frac{-b + \sqrt{b^2 - 4ac}}{2a} \qquad \text{or} \qquad x = \frac{-b - \sqrt{b^2 - 4ac}}{2a}.$$

It can be shown that these two results are also valid if $a < 0$. A compact form of these two equations, called the *quadratic formula*, follows.

QUADRATIC FORMULA

The solutions of the quadratic equation $ax^2 + bx + c = 0$, where $a \neq 0$, are

$$x = \frac{-b \pm \sqrt{b^2 - 4ac}}{2a}.$$

CAUTION Notice that the fraction bar in the quadratic formula extends under the $-b$ term in the numerator.

The expression under the radical in the quadratic formula, $b^2 - 4ac$, is called the **discriminant.** The value of the discriminant determines whether the quadratic equation has two real solutions, one real solution, or no real solutions. In the latter case, there will be two imaginary solutions. The following chart summarizes how the discriminant affects the number and nature of the solutions.

EFFECT OF THE DISCRIMINANT

If a, b, and c are real numbers, $a \neq 0$, then the complex solutions of $ax^2 + bx + c = 0$ are described as follows, based on the value of the discriminant, $b^2 - 4ac$.

Value of $b^2 - 4ac$	Number of Solutions	Nature of Solutions
Positive	Two	Complex, real
Zero	One (a double solution)	Complex, real
Negative	Two	Complex, imaginary

Furthermore, if a, b, and c are *integers*, $a \neq 0$, the real solutions are *rational* if $b^2 - 4ac$ is the square of an integer.

NOTE The final sentence in the box above suggests that the quadratic equation may be solved by factoring if $b^2 - 4ac$ is a "perfect square."

Solving Quadratic Equations

EXAMPLE 2

Using the Quadratic Formula

Solve the equation

$$x(x - 2) = 2x - 2$$

using the quadratic formula, and support your solutions graphically using the intersection-of-graphs method.

SOLUTION Before we can apply the quadratic formula, we must rewrite the equation in the form $ax^2 + bx + c = 0$.

$x(x - 2) = 2x - 2$	Given equation
$x^2 - 2x = 2x - 2$	Distributive property
$x^2 - 4x + 2 = 0$	Subtract 2x and add 2 on both sides.

Here $a = 1$, $b = -4$, and $c = 2$. Substitute these values into the quadratic formula to get

$$x = \frac{-b \pm \sqrt{b^2 - 4ac}}{2a}$$

$$= \frac{-(-4) \pm \sqrt{(-4)^2 - 4(1)2}}{2(1)} \qquad a = 1, b = -4, c = 2$$

$$= \frac{4 \pm \sqrt{16 - 8}}{2}$$

$$= \frac{4 \pm \sqrt{8}}{2} \qquad \text{The discriminant, 8, is positive, so there are two real solutions.}$$

$$= \frac{4 \pm 2\sqrt{2}}{2} \qquad \sqrt{16 - 8} = \sqrt{8} = 2\sqrt{2}$$

$$= \frac{2(2 \pm \sqrt{2})}{2} \qquad \text{Factor out a 2 in the numerator.}$$

$$= 2 \pm \sqrt{2} \qquad \text{Lowest terms}$$

The solution set is $\{2 + \sqrt{2}, 2 - \sqrt{2}\}$, abbreviated as $\{2 \pm \sqrt{2}\}$.

We can support our solution graphically by considering the graphs of $y_1 = x(x - 2)$ and $y_2 = 2x - 2$. (Note that the original form of our equation is $y_1 = y_2$.) By using the capabilities of the calculator, we can find that the x-coordinates of the points of intersection are approximately .59 and 3.41. (See Figure 14 on the next page.) These are also approximations of $2 - \sqrt{2}$ and $2 + \sqrt{2}$, supporting our results obtained by the quadratic formula. ∎

NOTE For the equation in Example 2,

$2 - \sqrt{2}$ and $2 + \sqrt{2}$	are the *exact* solutions;
.59 and 3.41	are *approximations* of the exact solutions.

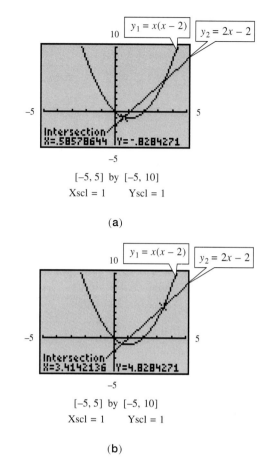

$[-5, 5]$ by $[-5, 10]$
Xscl = 1   Yscl = 1

(a)

$[-5, 5]$ by $[-5, 10]$
Xscl = 1   Yscl = 1

(b)

FIGURE 14

FOR **GROUP DISCUSSION**

If an equation is in the form $ax^2 = bx + c$, and it is considered as $y_1 = y_2$,

1. what kind of graph does y_1 have?
2. what kind of graph does y_2 have?
3. what are the possible numbers of points of intersection of these two kinds of graphs? How does this relate to the possible number of real solutions of a quadratic equation?

EXAMPLE 3

Using the Quadratic Formula

Solve

$$2x^2 - x + 4 = 0$$

by the quadratic formula and support your solutions graphically by the x-intercept method of solution.

SOLUTION Here we have $a = 2$, $b = -1$, and $c = 4$. By the quadratic formula,

$$x = \frac{-(-1) \pm \sqrt{(-1)^2 - 4(2)(4)}}{2(2)}$$

$$x = \frac{1 \pm \sqrt{1 - 32}}{4}$$

$$x = \frac{1 \pm \sqrt{-31}}{4}.$$

Because the discriminant is negative, we know that there are no real solutions. Writing the solutions in $a + bi$ form, we get

$$x = \frac{1 \pm i\sqrt{31}}{4}$$

$$x = \frac{1}{4} \pm i\frac{\sqrt{31}}{4}.$$

The solution set is $\left\{ \frac{1}{4} \pm i\frac{\sqrt{31}}{4} \right\}$.

If we graph $y = 2x^2 - x + 4$ we see that there are no x-intercepts, supporting our result that the only solutions are imaginary. See Figure 15. ∎

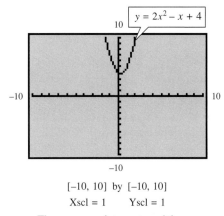

$y = 2x^2 - x + 4$

[-10, 10] by [-10, 10]
Xscl = 1   Yscl = 1

There are no x-intercepts and thus
no *real* solutions.

FIGURE 15

FOR **GROUP DISCUSSION**

Solve $x^2 = 4x - 4$ analytically as a class. Then graph $y_1 = x^2$ and $y_2 = 4x - 4$ and use the intersection-of-graphs method to support your result.

1. Do you encounter a problem if you use the standard viewing window?
2. Why do you think that analytic methods of solution are essential for understanding graphical methods?

Solving Quadratic Inequalities

Recall from Chapter 1 that the solution set of $P(x) = 0$ is determined by the x-intercepts of the graph of P. The solution set of $P(x) < 0$ consists of all values in the domain of P (that is, x-values) such that the graph of P lies *below* the x-axis.

And the solution set of $P(x) > 0$ consists of all values in the domain such that the graph of P lies *above* the x-axis. If P is a quadratic function, then these statements may be summarized in the following diagram.

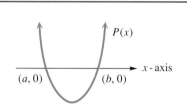

Solution Set of	is
$P(x) = 0$	$\{a, b\}$
$P(x) < 0$	the interval (a, b)
$P(x) > 0$	$(-\infty, a) \cup (b, \infty)$

Solution Set of	is
$P(x) = 0$	$\{a, b\}$
$P(x) < 0$	$(-\infty, a) \cup (b, \infty)$
$P(x) > 0$	the interval (a, b)

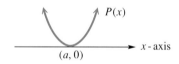

Solution Set of	is
$P(x) = 0$	$\{a\}$
$P(x) < 0$	$\emptyset$
$P(x) > 0$	$(-\infty, a) \cup (a, \infty)$

Solution Set of	is
$P(x) = 0$	$\{a\}$
$P(x) < 0$	$(-\infty, a) \cup (a, \infty)$
$P(x) > 0$	$\emptyset$

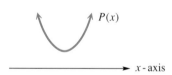

$P(x) = 0$ has no *real* solutions, but two complex imaginary solutions. Real solution set of $P(x) < 0$ is $\emptyset$. Real solution set of $P(x) > 0$ is $(-\infty, \infty)$.

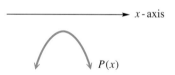

$P(x) = 0$ has no *real* solutions, but two complex imaginary solutions. Real solution set of $P(x) < 0$ is $(-\infty, \infty)$. Real solution set of $P(x) > 0$ is $\emptyset$.

Suppose that the graph of a quadratic polynomial intersects the x-axis in two points. Then the two solutions of the polynomial *equation* divide the real number line (x-axis) into three intervals. Within each interval, the polynomial is either always positive or always negative. This idea is used in explaining how a quadratic inequality may be solved using analytic methods, employing a *sign graph*.

EXAMPLE 4

Solving a Quadratic Inequality Analytically

Solve the quadratic inequality $x^2 - x - 12 < 0$ analytically using a sign graph, and then support the answer graphically using a calculator-generated graph.

SOLUTION Here we have $a = 1$, $b = -1$, and $c = -12$. The discriminant of the corresponding quadratic *equation* is $b^2 - 4ac = 49$. Since 49 is a perfect square, we can solve the quadratic equation $x^2 - x - 12 = 0$ by factoring.

$$x^2 - x - 12 = 0$$
$$(x + 3)(x - 4) = 0 \qquad \text{Factor.}$$
$$x = -3 \quad \text{or} \quad x = 4 \qquad \text{Use the zero-product property.}$$

These two points, -3 and 4, divide a number line into the three regions shown in Figure 16. If a point in region B, for example, makes the polynomial $x^2 - x - 12$ negative, then all points in region B will make that polynomial negative.

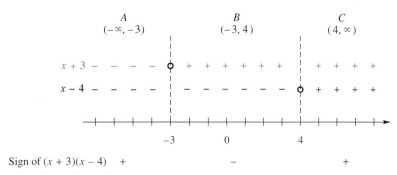

FIGURE 16

To find the regions that make $x^2 - x - 12$ negative (<0), draw a number line that shows where factors are positive or negative, as in Figure 16. First decide on the sign of the factor $x + 3$ in each of the three regions; then do the same thing for the factor $x - 4$. The results are shown in Figure 16.

Now consider the sign of the product of the two factors in each region. As Figure 16 shows, both factors are negative in the interval $(-\infty, -3)$; therefore their product is positive in that interval. For the interval $(-3, 4)$, one factor is positive and the other is negative, giving a negative product. In the last interval, $(4, \infty)$, both factors are positive, so their product is positive. The polynomial $x^2 - x - 12$ is negative (what the original inequality calls for) when the product of its factors is negative, that is, for the interval $(-3, 4)$. Therefore, the solution set is the open interval $(-3, 4)$.

The graph of $y = x^2 - x - 12$ is shown in Figure 17 on the next page. Notice that the graph lies *below* the x-axis between -3 and 4, supporting our analytic result. ∎

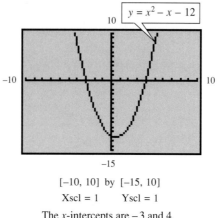

[−10, 10] by [−15, 10]
Xscl = 1   Yscl = 1

The x-intercepts are $−3$ and 4.
The graph lies *below* the x-axis in
the open interval $(−3, 4)$.

FIGURE 17

NOTE Graphical solution methods for solving inequalities will not be sufficient for determining whether endpoints should be included or excluded from the solution set. Therefore, we must make our decision based upon the symbol in the given inequality. The symbol < or the symbol > indicates that the endpoints are excluded, while ≤ or ≥ indicates that the endpoints are included.

The steps used in solving a quadratic inequality analytically are summarized below.

SOLVING A QUADRATIC INEQUALITY ANALYTICALLY

1. Solve the corresponding quadratic equation.
2. Identify the intervals determined by the solutions of the equation.
3. Use a sign graph to determine which intervals are in the solution set.

EXAMPLE 5

Solving a Quadratic
Inequality
Analytically

Solve the quadratic inequality

$$2x^2 \geq -5x + 12$$

analytically. Then support your answer graphically in two different ways.

SOLUTION We begin by writing the quadratic inequality in standard form: $2x^2 + 5x - 12 \geq 0$. The corresponding quadratic equation can be solved by factoring.

$$2x^2 + 5x - 12 = 0$$
$$(2x - 3)(x + 4) = 0$$
$$x = \frac{3}{2} = 1.5 \quad \text{or} \quad x = -4$$

These two points divide the number line into the three regions shown in the sign graph in Figure 18. Since both factors are negative in the first interval, their product, $2x^2 + 5x - 12$, is positive there. In the second interval, the factors have opposite signs, and therefore their product is negative. Both factors are positive in the third interval, and their product also is positive there. Thus, the polynomial $2x^2 + 5x - 12$ is positive or zero in the interval $(-\infty, -4]$ and also in the interval $[1.5, \infty)$. Since both of the intervals belong to the solution set, the result can be written as the *union** of the two intervals,

$$(-\infty, -4] \cup [1.5, \infty).$$

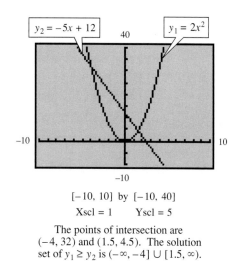

Sign of $(2x - 3)(x + 4)$ +   —   +

FIGURE 18

To support our answer in two different ways, we may begin by graphing $y_1 = 2x^2$ and $y_2 = -5x + 12$. See Figure 19. It appears that $y_1 \geq y_2$ for values of x less than or equal to -4 or for values of x greater than or equal to 1.5, since the graph of y_1 *lies above* or *intersects* the graph of y_2 in these intervals.

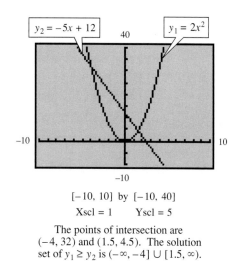

$[-10, 10]$ by $[-10, 40]$
Xscl = 1   Yscl = 5

The points of intersection are
$(-4, 32)$ and $(1.5, 4.5)$. The solution
set of $y_1 \geq y_2$ is $(-\infty, -4] \cup [1.5, \infty)$.

FIGURE 19

A second method of support involves writing the original inequality in the form $y_1 - y_2 \geq 0$ and then finding the domain values for which the graph of $y_1 - y_2$ *lies above* or *on* the x-axis. Once again, it appears that the intervals $(-\infty, -4]$ and $[1.5, \infty)$ give these values, supporting our earlier result. See Figure 20 on the next page. ∎

---

* The **union** of sets A and B, written $A \cup B$, is defined as $A \cup B = \{x \mid x$ is an element of A or x is an element of $B\}$.

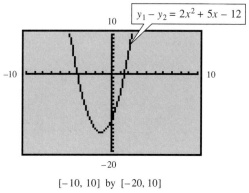

$$y_1 - y_2 = 2x^2 + 5x - 12$$

[−10, 10] by [−20, 10]
Xscl = 1   Yscl = 1

The x-intercepts are -4 and 1.5.
The solution set of $y_1 - y_2 \geq 0$ is
$(-\infty, -4] \cup [1.5, \infty)$.

FIGURE 20

NOTE   While we solved the quadratic equations in Examples 4 and 5 by factoring, we could also have used the quadratic formula. The formula will work for *any* quadratic equation, whether or not factoring is applicable.

▌ FOR GROUP DISCUSSION ▌

1.   The function $P(x) = x^2 - x - 20$ has integer zeros. Graph the function in a window that will show both x-intercepts, and without performing any analytic work, solve each of the following:

  a. $x^2 - x - 20 = 0$ **b.** $x^2 - x - 20 < 0$ **c.** $x^2 - x - 20 > 0$.

2.   The function $P(x) = x^2 + x + 20$ has no real zeros. Graph the function in a window that will show the vertex and the y-intercept. (Why won't the standard window work?) Then without performing any analytic work, solve each of the following:

  a. $x^2 + x + 20 = 0$ **b.** $x^2 + x + 20 < 0$ **c.** $x^2 + x + 20 > 0$.
  (Give only real solutions.)

From the above group discussion, we can see that solving a quadratic equation will lead quite easily to the solution sets of the corresponding inequalities.

EXAMPLE 6

Solving Quadratic Equations and Inequalities Graphically

Solve the quadratic equation

$$2.57x^2 - 1.56x - \sqrt{7.04} = 0$$

graphically, giving solutions to the nearest hundredth. Then use the graph to solve

$$2.57x^2 - 1.56x - \sqrt{7.04} > 0 \quad \text{and} \quad 2.57x^2 - 1.56x - \sqrt{7.04} < 0.$$

SOLUTION   A viewing window of $[-3, 3]$ by $[-3, 5]$ gives a complete graph of $P(x) = 2.57x^2 - 1.56x - \sqrt{7.04}$. See Figure 21. By using a root identification procedure, we find that the solutions, to the nearest hundredth, are $-.76$ and 1.36. The graph is below the x-axis between these x-intercepts, and above the x-axis to

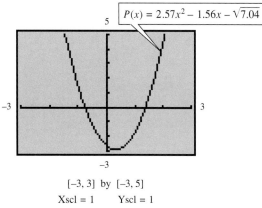

$[-3, 3]$ by $[-3, 5]$
Xscl = 1   Yscl = 1
The x-intercepts are approximately
$-.76$ and 1.36.

FIGURE 21

the left of $x = -.76$ and to the right of $x = 1.36$. Putting all this information together leads to the following conclusions:

The solution set of $2.57x^2 - 1.56x - \sqrt{7.04} = 0$ is $\{-.76, 1.36\}$.

The solution set of $2.57x^2 - 1.56x - \sqrt{7.04} < 0$ is the open interval $(-.76, 1.36)$.

The solution set of $2.57x^2 - 1.56x - \sqrt{7.04} > 0$ is $(-\infty, -.76) \cup (1.36, \infty)$.

Remember that the numbers given in the solutions above are approximations. Application of the quadratic formula would have given exact (but messy) answers. ∎

EXAMPLE 7

Solving Quadratic
Inequalities for
Exact Solutions

Use the graph in Figure 22 and the result of Example 2 to solve the following inequalities for intervals with *exact values* at endpoints.

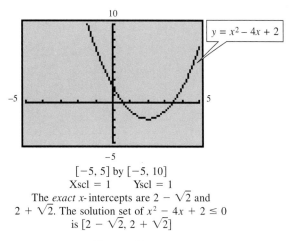

$[-5, 5]$ by $[-5, 10]$
Xscl = 1   Yscl = 1
The *exact* x-intercepts are $2 - \sqrt{2}$ and
$2 + \sqrt{2}$. The solution set of $x^2 - 4x + 2 \le 0$
is $[2 - \sqrt{2}, 2 + \sqrt{2}]$

FIGURE 22

(a) $x^2 - 4x + 2 \leq 0$

SOLUTION   We found by the quadratic formula that the exact solutions of $x^2 - 4x + 2 = 0$ are $2 - \sqrt{2}$ and $2 + \sqrt{2}$. Since the graph of $y = x^2 - 4x + 2$ lies below or intersects the x-axis between or at these values (as shown in Figure 22), the solution set for this inequality is the closed interval $[2 - \sqrt{2}, 2 + \sqrt{2}]$.

(b) $x^2 - 4x + 2 \geq 0$

SOLUTION   Using the same concept as in Example 6, we conclude that the solution set is $(-\infty, 2 - \sqrt{2}] \cup [2 + \sqrt{2}, \infty)$.   ∎

3.3 EXERCISES

Find all solutions, both real and imaginary, of the following quadratic equations. For the equations with real solutions, support your answers graphically.

1. $x^2 = 16$

2. $x^2 = 144$

3. $3x^2 = 27$

4. $2x^2 = 48$

5. $x^2 = -16$

6. $x^2 = -100$

7. $x^2 = -18$

8. $x^2 = -32$

Solve each of the following equations by the quadratic formula. Find all solutions, both real and imaginary. If the equation is not in the form $P(x) = 0$, you will need to write it this way in order to identify a, b, and c. For the equations with real solutions, support your answers graphically by using the x-intercept method, graphing $y = P(x)$, and then locating those intercepts.

9. $x^2 - 8x + 15 = 0$

10. $x^2 + 5x - 6 = 0$

11. $x^2 - 2x - 4 = 0$

12. $x^2 + 8x + 13 = 0$

13. $2x^2 + 2x = -1$

14. $9x^2 - 12x = -8$

15. $x(x - 1) = 1$

16. $x(x - 3) = 2$

17. $x^2 - 5x = x - 7$

18. $11x^2 - 3x + 2 = 4x + 1$

19. $4x^2 - 12x = -11$

20. $x^2 = 2x - 5$

21. $.5x^2 + .25x - 3 = 0$

22. $\frac{2}{3}x^2 + \frac{1}{4}x - 3 = 0$

Write each equation so that 0 is on the right side, and then evaluate the discriminant. Use the discriminant to determine the number of real solutions the equation has. If the equation has real solutions, tell whether they are rational or irrational. Do not actually solve the equation.

23. $x^2 + 8x + 16 = 0$

24. $x^2 - 5x + 4 = 0$

25. $3x^2 - 5x + 2 = 0$

26. $8x^2 = 14x - 3$

27. $4x^2 = 6x + 3$

28. $2x^2 - 4x + 1 = 0$

29. $9x^2 + 11x + 4 = 0$

30. $3x^2 = 4x - 5$

31. $8x^2 - 72 = 0$

32. Which one of the following equations has two real, distinct solutions? Do not actually solve.
  (a) $(3x - 4)^2 = -4$
  (b) $(4 + 7x)^2 = 0$
  (c) $(5x + 9)(5x + 9) = 0$
  (d) $(7x + 4)^2 = 11$

33. Which equations in Exercise 32 have only one distinct, real solution?

34. Which one of the equations in Exercise 32 has two imaginary solutions?

Exercises 35–50 refer to the graphs of the quadratic functions f, g, and h shown here.

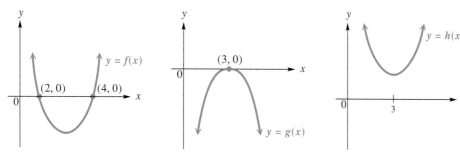

35. What is the solution set of $f(x) = 0$?

36. What is the solution set of $f(x) < 0$?

37. What is the solution set of $f(x) > 0$?

38. What is the solution set of $g(x) = 0$?

39. What is the solution set of $g(x) < 0$?

40. What is the solution set of $g(x) > 0$?

41. Solve $h(x) > 0$.

42. Solve $h(x) < 0$.

43. How many real solutions does $h(x) = 0$ have? How many complex solutions does it have?

44. What is the value of the discriminant of $g(x)$?

R*elating Concepts*

Recall the concepts presented in Section 3.2 involving vertex, axis of symmetry, and intercepts, and answer the following questions about the graphs of f, g, and h. (See the graphs accompanying Exercises 35–44.)

45. What is the x-coordinate of the vertex of the graph of $y = f(x)$?

46. What is the equation of the axis of symmetry of the graph of $y = g(x)$?

47. Does the graph of $y = g(x)$ have a y-intercept? If so, is it positive or is it negative?

48. Is the minimum value of h positive or negative?

49. Which function satisfies this description? It is concave up and the solution set of $y < 0$ is not empty.

50. Which function is decreasing on the interval $(3, \infty)$?

Solve each inequality analytically using a sign graph. Then support your answer graphically using a calculator-generated graph. Give exact values for endpoints.

51. (a) $x^2 + 4x + 3 \geq 0$
    (b) $x^2 + 4x + 3 < 0$

52. (a) $x^2 + 6x + 8 < 0$
    (b) $x^2 + 6x + 8 \geq 0$

53. (a) $2x^2 - 9x + 4 > 0$
    (b) $2x^2 - 9x + 4 \leq 0$

54. (a) $3x^2 + 13x + 10 \leq 0$
    (b) $3x^2 + 13x + 10 > 0$

55. (a) $-x^2 + 2x + 1 \geq 0$
    (b) $-x^2 + 2x + 1 < 0$

56. (a) $-x^2 - 5x + 2 > 0$
    (b) $-x^2 - 5x + 2 \leq 0$

57. (a) $4x^2 + 3x + 1 \leq 0$
    (b) $4x^2 + 3x + 1 > 0$

58. (a) $-3x^2 + x - 2 > 0$
    (b) $-3x^2 + x - 2 \leq 0$

Relating Concepts

We saw in Chapter 1 that in order to solve an inequality analytically, if we multiply both sides by a negative number we must reverse the direction of the inequality sign. Work through Exercises 59–64 in order. They illustrate this concept for quadratic inequalities.

59. Graph $y_1 = x^2 + 2x - 8$ in the standard viewing window. This function has two integer-valued x-intercepts. What are they?

60. Based on the graph, what is the solution set of $x^2 + 2x - 8 < 0$?

61. Now graph $y_2 = -y_1 = -x^2 - 2x + 8$ on the same screen. Using the terminology of Chapter 2, how is the graph of y_2 obtained by transforming the graph of y_1?

62. Based on the graph of y_2, what is the solution set of $-x^2 - 2x + 8 > 0$?

63. How do the two solution sets of the inequalities in Exercises 60 and 62 compare?

64. Write a short paragraph explaining how Exercises 59–63 illustrate the property involving multiplying an inequality by a negative number.

---

*In Exercises 65–70, **(a)** solve the equation $P(x) = 0$ graphically, giving solutions to the nearest hundredth. Then give the solution set of **(b)** $P(x) > 0$ and **(c)** $P(x) < 0$, based on your graph.*

65. $P(x) = 3.15x^2 + .65x - 3.24$

66. $P(x) = 2.78x^2 + .47x - 6.13$

67. $P(x) = -\pi x^2 + 9.8x - \sqrt{7}$

68. $P(x) = -\sqrt{5}x^2 + 5.4x - \sqrt{3}$

69. $P(x) = 2\pi x^2 - \sqrt{15}$

70. $P(x) = -4\pi x^2 + \sqrt[3]{9}$

71. Use symmetry of the graph to explain why the solutions of $P(x) = 0$ are negatives of each other in Exercise 69.

72. Without graphing, explain why the equation $2\pi x^2 + \sqrt{15} = 0$ has no real solutions.

Relating Concepts

The graphs of $y_1 = x^2$ and $y_2 = x + 2$ are shown in a viewing window. Two views are given, with the coordinates of a point of intersection given at the bottom of the screen. Recall from Chapter 1 that $y_1 = y_2$, $y_1 < y_2$, and $y_1 > y_2$ may be solved by the intersection-of-graphs method by determining the x-coordinate(s) of the point(s) of intersection, and observing the symbol $=$, $<$, or $>$ to determine whether we want the x-values for which the graphs intersect, or for which one graph lies above or below the other. Use these ideas in Exercises 73–75.

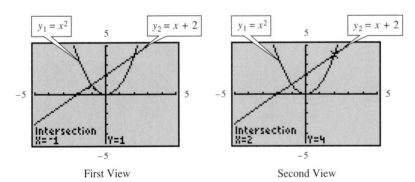

First View   Second View

73. Find the solution set of $x^2 = x + 2$.

74. Find the solution set of $x^2 < x + 2$.

75. Find the solution set of $x^2 > x + 2$.

76. Draw by hand the graph of a quadratic function y_1 and the graph of a linear function y_2 such that the solution set of the equation $y_1 = y_2$ is $\{-4, 5\}$, and the solution set of $y_1 < y_2$ is $(-\infty, -4) \cup (5, \infty)$. (There are many such pairs of functions.)

3.4 APPLICATIONS OF QUADRATIC FUNCTIONS AND MODELS

Applications of Quadratic Functions ∎ Quadratic Models

In Section 1.7 we saw how the concepts of linear functions, equations, and inequalities could be applied to solving certain types of problems. We will now see how certain problems can be solved using quadratic functions, and solving the corresponding equations and inequalities. Furthermore, determining the coordinates of the vertex of the graph of a quadratic function will often enable us to solve problems requiring the minimum or maximum value of the function. Such problems are called **optimization problems.**

Applications of Quadratic Functions

EXAMPLE 1

Solving a Problem
Involving Area of a
Rectangular Region

A farmer wishes to enclose a rectangular region. He has 120 feet of fencing, and plans to use one side of his barn as a part of the enclosure. See Figure 23. Let x represent the length of one of the parallel sides of the fencing, and respond to each of the following.

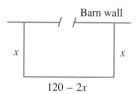

Barn wall

x x

$120 - 2x$

FIGURE 23

(a) Determine a function A that represents the area of the region in terms of x.

SOLUTION The lengths of the sides of the region bordered by the fencing are x, x, and $120 - 2x$, as shown in the figure. Since area = width × length, the function is

$$A(x) = x(120 - 2x) \quad \text{or} \quad A(x) = -2x^2 + 120x.$$

(b) For this particular problem, what are the restrictions on x?

SOLUTION Since x represents a length, we must have $x > 0$. Furthermore, the side of length $120 - 2x$ must also be positive. Therefore, $120 - 2x > 0$, or $x < 60$. Putting these two restrictions together, we have $0 < x < 60$.

(c) Find a viewing window that will show both x-intercepts and the vertex of the graph of this quadratic function.

SOLUTION   While many viewing windows will satisfy these requirements, one such window is shown in Figure 24.

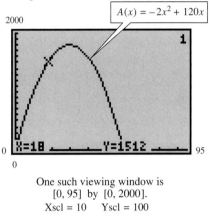

One such viewing window is
[0, 95] by [0, 2000].
Xscl = 10   Yscl = 100

FIGURE 24

TECHNOLOGICAL NOTE
In Figures 24 and 25, the display at the bottom obscures the view of the intercepts, even though the choice of window theoretically shows them. This problem is simple to overcome: just lower the minimum values of *x* and *y* enough so that when the display appears, the *x*- and *y*-axes are visible.

(d)   Figure 24 shows the cursor at (18, 1512). Interpret this information.

SOLUTION   If the parallel sides of fencing each measure 18 feet, the area of the enclosure is 1512 square feet. This can be written $A(18) = 1512$, and checked as follows: If the width is 18 feet, the length is $120 - 2(18) = 84$ feet, and $18 \times 84 = 1512$.

(e)   What is the maximum area the farmer can enclose? Determine the answer analytically, and support it graphically.

SOLUTION   We must find the maximum value of the function. This occurs at the vertex. For $A(x) = -2x^2 + 120x$, we have $a = -2$ and $b = 120$. Using the vertex formula from Section 3.2, $x = -\frac{b}{2a} = -\frac{120}{2(-2)} = 30$. Evaluating $A(30)$ gives $-2(30)^2 + 120(30) = 1800$. Therefore, the farmer can enclose a maximum of 1800 square feet when the parallel sides of fencing measure 30 feet.

   To support this answer, use a calculator to locate the vertex of the parabola and observe the *x* and *y* values there. Figure 25 gives support for our answer.   ◧

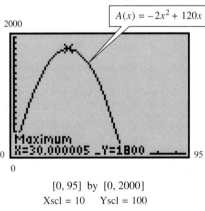

[0, 95] by [0, 2000]
Xscl = 10   Yscl = 100

The display supports our analytic result:
when $x = 30$, the area is maximized at
1800 square feet.

FIGURE 25

CAUTION As seen in Example 1(e), it is important to be careful when interpreting the meanings of the coordinates of the vertex in optimization problems. The first coordinate, x, gives the *domain* value for which the *function value* is a maximum or minimum. It is always necessary to read the problem carefully to determine whether you are asked to find the value of the independent variable x, the function value y, or both.

EXAMPLE 2

Solving a Problem Involving the Volume of a Box

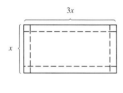

(a)

(b)

FIGURE 26

A piece of machinery is capable of producing rectangular sheets of metal satisfying the condition that the length is three times the width. Furthermore, equal size squares measuring 5 inches on a side can be cut from the corners so that the resulting piece of metal can be shaped into an open box by folding up the flaps. See Figure 26(a).

(a) Determine a function V that expresses the volume of the box in terms of the width x of the original sheet of metal.

SOLUTION If x represents the width, then $3x$ represents the length. Figure 26(b) indicates that the width of the bottom of the box is $x - 10$, the length of the bottom of the box is $3x - 10$, and the height is 5 inches (the length of the side of the cut out square). Since volume = length × width × height, the function is

$$V(x) = (3x - 10)(x - 10)(5)$$

or

$$V(x) = 15x^2 - 200x + 500.$$

(b) What restrictions must be placed on x in this particular problem?

SOLUTION Since the dimensions of the box must represent positive numbers, we must have $3x - 10 > 0$ and $x - 10 > 0$, or

$$x > \frac{10}{3} \quad \text{and} \quad x > 10.$$

Both conditions are satisfied when $x > 10$. Therefore, the theoretical domain of x in this problem is $(10, \infty)$.

(c) If specifications call for the volume of such a box to be 1435 cubic inches, what should the dimensions of the original piece of metal be? Solve analytically and support graphically.

SOLUTION We must find x such that $V(x) = 1435$.

$1435 = 15x^2 - 200x + 500$	Set $V(x) = 1435$.
$0 = 15x^2 - 200x - 935$	Subtract 1435.
$0 = (15x + 55)(x - 17)$	Factor.
$15x + 55 = 0 \quad \text{or} \quad x - 17 = 0$	Use the zero-product property.
$x = -\dfrac{11}{3} \quad \text{or} \quad x = 17$	Solve.

Of these two solutions, only 17 satisfies the condition that $x > 10$. Therefore, the dimensions of the original piece of metal should be 17 inches by $3(17) = 51$ inches.

Since $(51 - 10) \cdot (17 - 10) \cdot 5 = 1435$, our answer is correct. (Notice that we could have solved the quadratic equation by the quadratic formula as well.)

One way to support this result graphically is to graph $y_1 = 15x^2 - 200x + 500$ in a window with minimum x value 10 and $y_2 = 1435$ in the same window. The point at which they intersect should have an x-value of 17. See Figure 27.

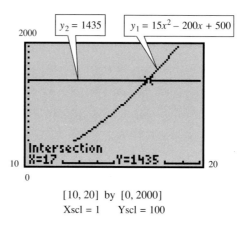

[10, 20] by [0, 2000]
Xscl = 1   Yscl = 100

FIGURE 27

NOTE If we were to graph y_1 and y_2 in a window containing the x-value $-\frac{11}{3}$, the parabola and the line would intersect again. (Where have we seen the analytic justification for this?) A window of $[-6, 20]$ by $[-200, 1600]$ will allow us to see this other point of intersection.

(d) What dimensions of the original piece of metal will assure us of a volume greater than 2000 but less than 3000? Solve graphically.

SOLUTION Using the graphs of the functions $y_1 = 15x^2 - 200x + 500$, $y_2 = 2000$, and $y_3 = 3000$, we find that the points of intersection of the graphs are *approximately* (18.7, 2000) and (21.2, 3000) for $x > 10$. See Figure 28. Therefore, the width of the rectangle should be between 18.7 and 21.2 inches, with the corresponding length three times these values (that is, between $3(18.7) \approx 56.1$ and $3(21.2) \approx 63.6$ inches). ∎

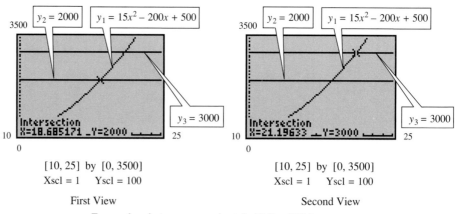

[10, 25] by [0, 3500]
Xscl = 1   Yscl = 100

First View

[10, 25] by [0, 3500]
Xscl = 1   Yscl = 100

Second View

For x-values between approximately 18.7 and 21.2, the y-values will be between 2000 and 3000.

FIGURE 28

An important application of quadratic functions deals with the height of a propelled object as a function of the time elapsed after it is propelled.

FORMULA FOR THE HEIGHT OF A PROPELLED OBJECT

If air resistance is neglected, the height s (in feet) of an object propelled directly upward from an initial height s_0 feet with initial velocity v_0 feet per second is described by the function

$$s(t) = -16t^2 + v_0 t + s_0,$$

where t is the number of seconds after the object is propelled.

In this formula, the coefficient of t^2 (that is, -16) is a constant based on the gravitational force of the earth. This constant varies on other surfaces, such as the moon and other planets. Here we have an example of a quadratic function in which height is a function of time. We use t to represent the independent variable; however, when graphing this type of function on our calculator, we will use x, since graphics calculators are so equipped. In reality, there is no difference between the functions

$$s(t) = -16t^2 + v_0 t + s_0 \qquad \text{and} \qquad s(x) = -16x^2 + v_0 x + s_0.$$

EXAMPLE 3

Solving a Problem Involving Projectile Motion

A ball is thrown directly upward from an initial height of 100 feet with an initial velocity of 80 feet per second.

(a) What is the function that describes the height of the ball in terms of the time t elapsed?

SOLUTION We use the projectile height formula with $v_0 = 80$ and $s_0 = 100$:

$$s(t) = -16t^2 + 80t + 100.$$

(b) Graph this function so that the y-intercept, the positive x-intercept, and the vertex are visible.

SOLUTION There are many suitable choices for such a window. One such choice is $[0, 10]$ by $[0, 300]$, as shown in Figure 29. It shows the graph of $y = -16x^2 + 80x + 100$. (Here, $x = t$.)

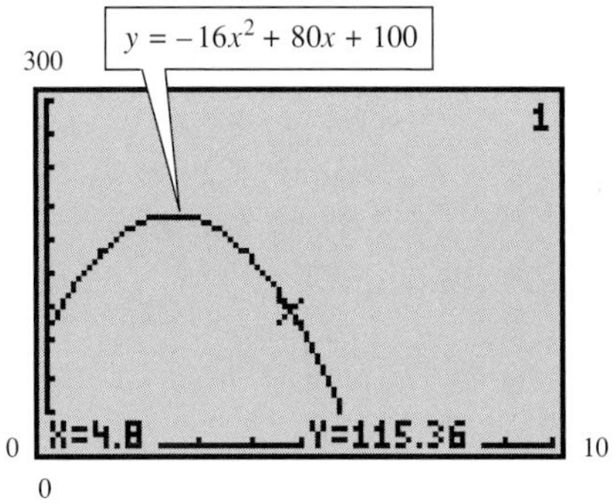

$[0, 10]$ by $[0, 300]$
Xscl = 1   Yscl = 30

FIGURE 29

(c)   The cursor in Figure 29 shows that the point (4.8, 115.36) lies on the graph of the function. What does this mean for this particular problem?

SOLUTION   When $x = 4.8$, $y = 115.36$. Therefore, when 4.8 seconds have elapsed, the projectile is at a height of 115.36 feet.

(d)   After how many seconds does the projectile reach its maximum height? What is this maximum height? Solve analytically and support graphically.

SOLUTION   To answer this question, we must find the coordinates of the vertex of the parabola. Using the vertex formula from Section 3.2, we find

$$x = -\frac{b}{2a} = -\frac{80}{2(-16)} = 2.5$$

and

$$y = -16(2.5)^2 + 80(2.5) + 100 = 200.$$

Therefore, after 2.5 seconds the ball reaches its maximum height of 200 feet.

To support this graphically, we use the capabilities of the calculator to find that the vertex coordinates are indeed (2.5, 200). See Figure 30.

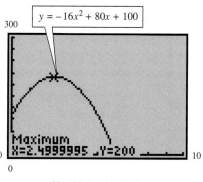

[0, 10] by [0, 300]
Xscl = 1   Yscl = 30

When the time is 2.5 seconds,
the height is a maximum, 200 feet.

FIGURE 30

CAUTION   It is easy to misinterpret the graph shown in Figure 30. This graph does not define the *path* followed by the ball; rather, it defines height as a function of time.

(e)   For what interval of time is the height of the ball greater than 160 feet? Determine the answer graphically.

SOLUTION   With $y_1 = -16x^2 + 80x + 100$ and $y_2 = 160$ graphed, we locate the two points of intersection. We find that the x-coordinates there are approximately .92 and 4.08. Therefore, between .92 and 4.08 seconds, the ball is more than 160 feet above the ground.

(f)   After how many seconds will the ball return to the ground? Determine the answer analytically and support graphically.

SOLUTION When the ball returns to the ground, its height will be 0 feet, so we must solve the quadratic equation

$$0 = -16x^2 + 80x + 100.$$

Using the quadratic formula, we find

$$x = \frac{-80 \pm \sqrt{80^2 - 4(-16)(100)}}{2(-16)}$$

$$x \approx -1.04, \ 6.04$$

We must choose the positive solution. Therefore, the ball will return to the ground about 6.04 seconds after it was projected. To support this result, we use the capabilities of the calculator to find the positive x-intercept of the graph; it is indeed about 6.04. See Figure 31. (Notice that if the negative x-intercept were shown, it would be about -1.04, the rejected solution above.)   ◧

TECHNOLOGICAL NOTE
The value of y displayed at the bottom of the screen in Figure 31 is as close as this particular model can get to 0 using the internal routine of the calculator. The display 1E-11 indicates that 1 is multiplied by 10^{-11}, yielding the decimal expression .00000000001, which is very close to 0.

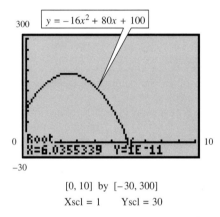

$$[0, 10] \ \text{by} \ [-30, 300]$$
$$\text{Xscl} = 1 \qquad \text{Yscl} = 30$$

FIGURE 31

Quadratic Models

In Section 1.7 we saw how a linear model can help us predict and interpret data in some cases. By extending the statistical concept of regression to polynomials, we can sometimes fit data to a quadratic function to provide a model for the data.

EXAMPLE 4

Using a Quadratic Model

In the United States, post-secondary degrees below bachelor's degrees earned during the years 1985 to 1990 can be approximated by the model

$$y = 5334.59x^2 - 23{,}024.00x + 617{,}519.11$$

where $x = 0$ corresponds to 1985. Use the graph of this function in the window $[0, 6]$ by $[500{,}000, 800{,}000]$ to determine the year during this interval that the number of these degrees earned reached a minimum. To the nearest 10,000, what was this minimum number?

SOLUTION Figure 32 on the following page shows the graph required. By using the capabilities of the calculator, we can determine the approximate coordinates of the vertex (the minimum point for the parabola) are (2.16, 592,676). Therefore, since $x = 2$ corresponds to 1987, the number of such degrees reached a minimum then, and to the nearest 10,000, the number of degrees was 590,000.   ◧

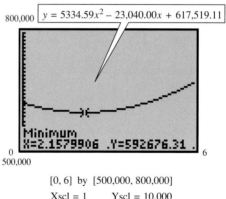

[0, 6] by [500,000, 800,000]

Xscl = 1   Yscl = 10,000

FIGURE 32

3.4 EXERCISES

1. Suppose that x represents one of two *positive* numbers whose sum is 30.
   (a) Represent the other of the two numbers in terms of x.
   (b) What are the restrictions on x?
   (c) Describe a function P that represents the product of these two numbers.
   (d) Determine analytically and support graphically the two such numbers whose product is a maximum. What is this maximum product?

2. Suppose that x represents one of two *positive* numbers whose sum is 45.
   (a) Represent the other of the two numbers in terms of x.
   (b) What are the restrictions on x?
   (c) Describe a function P that represents the product of these two numbers.
   (d) For what two such numbers is the product equal to 504? Determine analytically.
   (e) Determine analytically and support graphically the two such numbers whose product is a maximum.

3. American River College has plans to construct a rectangular parking lot on land bordered on one side by a highway. There are 640 feet of fencing available to fence the other three sides. Let x represent the length of each of the two parallel sides of fencing.

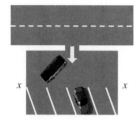

   (a) Express the length of the remaining side to be fenced in terms of x.
   (b) What are the restrictions on x?
   (c) Describe a function A that represents the area of the parking lot in terms of x.
   (d) Graph the function of part (c) in a window of [0, 320] by [0, 55,000]. Determine graphically

the values of x that will give an area of between 30,000 and 40,000 square feet.
   (e) What dimensions will give a maximum area and what will this area be? Determine analytically and support graphically.

4. A farmer wishes to enclose a rectangular region bordering a river with fencing as shown in the diagram. Suppose that x represents the length of each of the three parallel pieces of fencing. She has 600 feet of fencing available.

   (a) What is the length of the remaining piece of fencing in terms of x?
   (b) Describe a function A that represents the total

area of the enclosed region. Give any restrictions on x.

(c) What dimensions for the total enclosed region would give an area of 22,500 square feet? Determine the answer analytically.

(d) What is the maximum area that can be enclosed? Use a graph to answer this question.

5. If a rectangle has perimeter 120 feet, what is the largest area it can have?

6. Use a graph to show that there can be no two real numbers that have a difference of 3 and a product of -3.

7. A piece of cardboard is twice as long as it is wide. It is to be made into a box with an open top by cutting 2-inch squares from each corner and folding up the sides. Let x represent the width of the original piece of cardboard.

(a) Represent the length of the original piece of cardboard in terms of x.

(b) What will be the dimensions of the bottom rectangular base of the box? Give restrictions on x.

(c) Describe a function V that represents the volume of the box in terms of x.

(d) For what bottom dimensions will the volume be 320 cubic inches? Determine the answer analytically and support it graphically.

(e) Determine graphically (to the nearest tenth of an inch) the values of x if such a box is to have a volume between 400 and 500 cubic inches.

8. A piece of sheet metal is 2.5 times as long as it is wide. It is to be made into a box with an open top by cutting 3-inch squares from each corner and folding up the sides. Let x represent the width of the original piece of sheet metal.

(a) Represent the length of the original piece of sheet metal in terms of x.

(b) What are the restrictions on x?

(c) Describe a function V that represents the volume of the box in terms of x.

(d) For what values of x (that is, original widths) will the volume of the box be between 600 and 800 cubic inches? Determine the answer using graphical methods, and give values to the nearest tenth of an inch.

In Exercises 9–12, use the formula for the height of a propelled object, given in this section.

9. A rock is thrown from ground level with an initial velocity of 90 feet per second. Let t represent the amount of time elapsed after it is thrown.

(a) Explain why t cannot be a negative number in this problem situation.

(b) Explain why $s_0 = 0$ in this problem situation.

(c) Give the function s that describes the height of the rock as a function of t.

(d) How high will the rock be 1.5 seconds after it is thrown?

(e) What is the maximum height attained by the rock? After how many seconds will this happen? Determine the answer analytically and support graphically.

(f) After how many seconds will the rock hit the ground? Determine analytically and support graphically.

10. A toy rocket is launched from the top of a building 50 feet tall at an initial velocity of 200 feet per second. Let t represent the amount of time elapsed after the launch.

(a) Express the height s as a function of the time t.

(b) Determine both analytically and graphically the time at which it reaches its highest point. How high will it be at that time?

(c) For what time interval will the rocket be more than 300 feet above ground level? Determine the answer using graphical methods, and give times to the nearest tenth of a second.

(d) After how many seconds will it hit the ground? Determine the answer both analytically and graphically.

11. Determine graphically whether a ball thrown from ground level with an intial velocity of 150 feet per second will reach a height of 355 feet. If it will, determine the time(s) at which this happens. If it will not, explain why using a graphical interpretation.

12. Repeat Exercise 11 for a ball thrown from a height of 30 feet with an initial velocity of 250 feet per second.

In Exercises 13 and 14, assume that x and y represent positive numbers.

13. Find the minimum value of the expression $x^2 + y^2$, if the sum of x and y is 16.

14. Find the maximum value of the expression xy^2, if $x + y^2 = 6$.

In Exercises 15 and 16, use the fact that the total revenue obtained from selling x items at p dollars per item is p · x dollars.

15. Suppose that the price charged for a certain lottery ticket is $100 - .1x$ dollars and x lottery tickets are sold.
   (a) Find a function R that describes the revenue obtained from these sales.
   (b) Find the value of x that leads to a maximum revenue.
   (c) What is the maximum revenue?

16. The sale of cassette tapes of "lesser" performers is very sensitive to price. If a manufacturer charges $6 - .125x$ dollars per tape, then x tapes will be sold.
   (a) Find a function R that describes the revenue obtained from these sales.
   (b) Find the value of x that leads to a maximum revenue.
   (c) What is the maximum revenue?

17. The manager of an 80-unit apartment complex knows from experience that at a rent of $400 per month, all units will be rented. However, for each increase of $20 in rent, he can expect one unit to be vacated. Let x represent the number of $20 increases over $400.
   (a) Express, in terms of x, the number of apartments that will be rented if x increases of $20 are made. (For example, if 3 such increases are made, the number of apartments rented will be $80 - 3 = 77$.)
   (b) Express the rent per apartment if x increases of $20 are made. (For example, if he increases rent $60 = 3 \times \$20$, the rent per apartment is $400 + 3(20) = 460$ dollars.)
   (c) Describe a revenue function R in terms of x that will give the revenue generated as a function of the number of $20 increases.
   (d) For what number of increases will the revenue be $37,500?

(e) What rent should he charge in order to achieve a maximum revenue?

18. When *Money Means Power* charges $600 for a seminar on management techniques, it attracts 1000 people. For each decrease of $20 in the charge, an additional 100 people will attend the seminar. Let x represent the number of $20 decreases in the charge.
   (a) Describe a revenue function R that will give revenue generated as a function of the number of $20 decreases.
   (b) Find the value of x that maximizes the revenue. What should the company charge to maximize the revenue?
   (c) What is the maximum revenue the company can generate?

19. A local youth group is arranging a charter flight to Hawaii. The cost of the flight is $425 per person if there are 75 passengers. However, for every passenger in excess of 75, the airline will reduce the fare by $5. Let x represent the number of passengers in excess of 75.
   (a) Describe a revenue function R that expresses the revenue generated when there are x passengers in excess of 75.
   (b) What *total* number of passengers will maximize the revenue from the flight?
   (c) What is the maximum possible revenue?

20. The manager of a cherry orchard is trying to decide when to schedule the annual harvest. If the cherries are picked now, the average yield per tree will be 100 lb, and the cherries can be sold for 40 cents per pound. Past experience shows that the yield per tree will increase about 5 lb per week, while the price will decrease about 2 cents per pound per week. How many weeks should the manager wait to get an average revenue of $38.40 per tree?

In a certain situation, the function

$$s(x) = .1x^2 - 3x + 22$$

can be used to determine the appropriate landing speed of an airplane, where x is the initial landing speed in feet per second, and s(x) is the distance needed in feet to land.

21. Use the formula given to calculate the appropriate landing speed if 650 feet of runway are available.

22. If an airplane stalls at speeds less than 70 feet per second, find the minimum safe landing field length.

Exercises 23–26 refer to the following situation.

A frog leaps from a stump 3 feet high and lands 4 feet from the base of the stump. We can consider the initial position of the frog to be at (0, 3) and its landing position to be at (4, 0). See the figure.

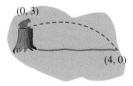

(0, 3)

(4, 0)

It is determined that the height of the frog as a function of its distance x from the base of the stump is given by the function

$$h(x) = -.5x^2 + 1.25x + 3,$$

where h is in feet.

23. How high was the frog when its horizontal distance from the base of the stump was 2 feet?

24. At what two times after it jumped from the base of the stump was the frog 3.25 feet above the ground?

25. At what distance from the base of the stump did the frog reach its highest point?

26. What was the maximum height reached by the frog?

Solve each problem.

27. The Gross State Product in current dollars from 1985 through 1989 is modeled by the quadratic function

$$f(x) = 18.14x^2 + 234.03x + 3954$$

where $x = 0$ corresponds to 1985 and $f(x)$ is in billions. If this model continues to apply, what would be the gross state product in 1996? (*Source:* U.S. Bureau of Economic Analysis)

28. The number of cases commenced by the U.S. Court of Appeals each year between 1984 and 1990 can be approximated by the quadratic model

$$f(x) = 68.90x^2 + 1165.29x + 31{,}676$$

where $x = 0$ corresponds to 1984. Based on this model, what would be the number of cases commenced in 1996?

29. Between 1985 and 1989, the number of female suicides by firearms in the United States each year can be modeled by

$$f(x) = -17x^2 + 44.6x + 2572$$

where $x = 0$ represents 1985. Based on this model, in what year did the number of such suicides reach its peak?

30. The number of infant deaths during the past decade has been decreasing. Between 1980 and 1989, the number of infant deaths per 1000 live births each year can be approximated by the function

$$f(x) = .0234x^2 - .5029x + 12.5$$

where $x = 0$ corresponds to 1980.

(a) Based on this model, how many deaths per 1000 live births were there during 1985?

(b) If the trend continued, how many deaths per 1000 live births would we have expected during the year 1990?

*F*urther Explorations

The TABLE feature of graphics calculators can numerically reinforce relationships between perimeter and area. It can show length, width, and area in each of the three columns. Use the TABLE to study rectangular areas given a constant perimeter. Suppose a rectangle has a perimeter of 40 inches. Build a TABLE that shows length and width in increments of 1 inch, and the corresponding areas. Let the length of a side be x. Then the corresponding width would be $(40 - 2x)/2$. Let $Y_1 = (40 - 2x)/2$. Since Area = Length $\times$ Width, let $Y_2 = x\,(Y_1)$. Figure A shows corresponding TABLES.

```
Y1B(40-2X)/2
Y2BX(Y1)
Y3=
Y4=
Y5=
Y6=
Y7=
Y8=
```

X	Y1	Y2
-3	23	-69
-2	22	-44
-1	21	-21
0	20	0
1	19	19
2	18	36
3	17	51

X=0

X	Y1	Y2
7	13	91
8	12	96
9	11	99
10	10	100
11	9	99
12	8	96
13	7	91

X=10

X	Y1	Y2
17	3	51
18	2	36
19	1	19
20	0	0
21	-1	-21
22	-2	-44
23	-3	-69

X=20

FIGURE A

1. What dimensions produce the greatest area?

2. If the above TABLE refers to fencing an enclosure, is it a more efficient use of materials to construct long skinny rectangles, or squares?

3. Why does the TABLE imply that lengths greater than 20 inches will produce enclosures with an area less than zero?

4. The TABLE in Figure B displays lengths (x), width (Y_1), and area (Y_2) of a rectangular enclosure for a fixed amount of fencing. What is the maximum area possible for this amount of fencing? How much fencing is being used in this situation?

X	Y1	Y2
0	30	0
5	25	125
10	20	200
15	15	225
20	10	200
25	5	125
30	0	0

X=0

FIGURE B

3.5 HIGHER DEGREE POLYNOMIAL FUNCTIONS AND THEIR GRAPHS

Introduction ▮ Extrema ▮ Cubic Functions and Other Functions of Odd Degree ▮ Number of x-Intercepts (Real Zeros) ▮ Quartic Functions and Other Functions of Even Degree ▮ Complete Graphs

Introduction

The linear functions discussed in Chapter 1 and the quadratic functions discussed earlier in this chapter are the simplest examples of a larger class of functions known as polynomial functions. The definition given in Section 3.2 is repeated here.

> ### POLYNOMIAL FUNCTION
> A **polynomial function of degree n in the variable x** is a function defined by
> $$P(x) = a_n x^n + a_{n-1} x^{n-1} + \cdots + a_1 x + a_0,$$
> where each a_i is a real number, $a_n \neq 0$, and n is a whole number.

We will see later in this section that the behavior of the graph of a polynomial function is due largely to the value of the coefficient a_n and the *parity* (that is, "evenness" or "oddness") of the exponent on the term of highest degree. For this reason, we will refer to a_n as the *leading coefficient,* and $a_n x^n$ as the *dominating term.* The term a_0 is the constant term of the polynomial function, and since $P(0) = a_0$, it is the y-intercept of the graph.

As we study the graphs of polynomial functions, we will use the following general properties (which are proved in higher courses):

1. A polynomial function (unless otherwise specified) has domain $(-\infty, \infty)$.

2. The graph of a polynomial function is a smooth, continuous curve with no sharp turns.

EXAMPLE 1

Observing Some Sample Graphs of Polynomial Functions

Figure 33 shows graphs of several typical polynomial functions generated on a graphics calculator. Each was graphed in the connected mode, in the specified window. ▯

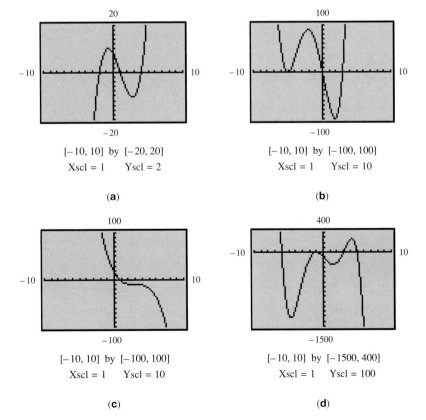

$[-10, 10]$ by $[-20, 20]$
Xscl $= 1$ Yscl $= 2$

(a)

$[-10, 10]$ by $[-100, 100]$
Xscl $= 1$ Yscl $= 10$

(b)

$[-10, 10]$ by $[-100, 100]$
Xscl $= 1$ Yscl $= 10$

(c)

$[-10, 10]$ by $[-1500, 400]$
Xscl $= 1$ Yscl $= 100$

(d)

FIGURE 33

Extrema

Notice in Figure 33 that several of the graphs have "peaks" and/or "valleys" where the function changes from increasing to decreasing or vice-versa. We first saw this when we studied quadratic functions, noticing that the vertex could be a maximum or minimum point on the graph. In general, the highest point at a "peak" is known as a **local maximum point,** and the lowest point at a "valley" is known as a **local minimum point.** Function values at such points are called **local maxima** (plural of maximum) and **local minima** (plural of minimum). Collectively they are called **extrema** (plural of **extremum**). Figure 34 and the accompanying chart illustrate these ideas for typical graphs.

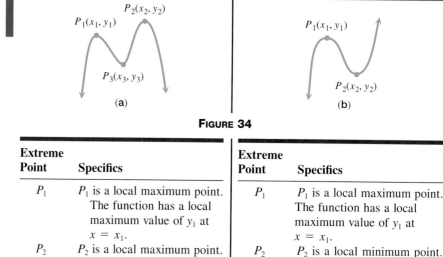

FIGURE 34

Extreme Point	Specifics	Extreme Point	Specifics
P_1	P_1 is a local maximum point. The function has a local maximum value of y_1 at $x = x_1$.	P_1	P_1 is a local maximum point. The function has a local maximum value of y_1 at $x = x_1$.
P_2	P_2 is a local maximum point. The function has a local maximum value of y_2 at $x = x_2$.	P_2	P_2 is a local minimum point. The function has a local minimum value of y_2 at $x = x_2$.
P_3	P_3 is a local minimum point. The function has a local minimum value of y_3 at $x = x_3$.		

Refer again to Figure 34(a). Notice that the point P_2 is the absolute highest point on the graph, and the range of the function is $(-\infty, y_2]$. We call P_2 the **absolute maximum point** on the graph, and y_2 the **absolute maximum value** of the function. Because the y-values approach $-\infty$, this function has no absolute minimum value. On the other hand, because the graph in Figure 34(b) is that of a function with range $(-\infty, \infty)$, it has neither an absolute maximum nor an absolute minimum.

Consider the graphs in Figure 35.

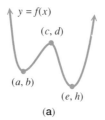 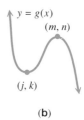

FIGURE 35

(a) Name and classify the local extrema of f.

SOLUTION The points (a, b) and (e, h) are local minimum points. The point (c, d) is a local maximum.

(b) Name and classify the local extrema of g.

SOLUTION The point (j, k) is a local minimum and the point (m, n) is a local maximum.

(c) Discuss absolute extrema for f and g.

SOLUTION The absolute minimum value of function f is the number h, since the range of f is $[h, \infty)$. It has no absolute maximum value. Function g has no absolute extrema, since its range is $(-\infty, \infty)$. ◫

We now state an important property of the graph of any polynomial function.

MAXIMUM NUMBER OF LOCAL EXTREMA

The maximum number of local extrema of the graph of a polynomial function of degree n is $n - 1$.

The property above can be applied to the polynomial functions we have previously studied. The degree of a linear function is 1 and since its graph is a straight line, it has no local extrema. A quadratic function is of degree 2, and since it has only 1 extreme point (its vertex), the property is satisfied. Notice that the property states that the *maximum* number of local extrema is $n - 1$ for a polynomial function of degree n. The graph may have fewer than $n - 1$ local extrema.

FOR **GROUP DISCUSSION**

1. Recall the graph of the function $f(x) = x^3$ studied in Chapter 2. How many local extrema does it have? How does your answer support the property stated above?
2. Consider the polynomial function graph in Figure 36.

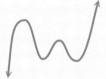

FIGURE 36

   a. What is the least possible degree of this function?
   b. Explain why this function cannot be of degree 4.
3. Repeat item 2 for the polynomial function graph in Figure 37.

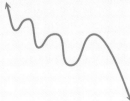

FIGURE 37

Cubic Functions and Other Functions of Odd Degree

A polynomial function of the form $P(x) = ax^3 + bx^2 + cx + d, a \neq 0$, is a third degree, or **cubic function.** We studied the simplest cubic function, $f(x) = x^3$, in Chapter 2. If we graph a cubic function in an appropriate window, the graph will resemble, in general, one of the shapes shown in Figure 38.

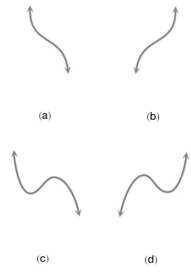

(a)   (b)

(c)   (d)

FIGURE 38

FOR GROUP DISCUSSION

1. Without graphing, tell which of the shapes in Figure 38 depicts the general form of the graph of $f(x) = x^3$.
2. Using the concepts of reflection of graphs from Chapter 2, tell which of the shapes depicted in Figure 38 most closely resembles the graph of $g(x) = -x^3$.
3. Graph each of the following functions in the window specified, and determine which one of the shapes in Figure 38 the graph most closely resembles.
   a. $P(x) = x^3 + 5x^2 + 5x - 2$, $[-10, 10]$ by $[-10, 10]$
   b. $f(x) = -x^3 + 5x - 1$, $[-10, 10]$ by $[-10, 10]$
   c. $g(x) = x^3 + 3x^2 + 3x + 1$, $[-10, 10]$ by $[-10, 10]$
   d. $h(x) = -2x^3 - 6x^2 - 6x - 1$, $[-2, 1]$ by $[-1, 2]$

In Section 3.2 we studied end behavior of quadratic function graphs. The end behavior of the graph of a polynomial function is determined by the sign of the leading coefficient and the parity of the degree. A cubic function is an example of an odd degree polynomial function, and we can make the following observations about the end behavior of an odd degree polynomial function.

END BEHAVIOR OF ODD DEGREE POLYNOMIAL FUNCTIONS

Suppose that ax^n is the dominating term of a polynomial function P of *odd degree*.

1. If $a > 0$, then as $x \to \infty$, $P(x) \to \infty$, and as $x \to -\infty$, $P(x) \to -\infty$. Therefore, the end behavior of the graph is of the type shown in Figure 39, and is symbolized ↗.

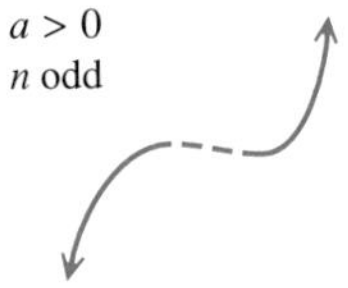

$a > 0$
n odd

FIGURE 39

2. If $a < 0$, then as $x \to \infty$, $P(x) \to -\infty$, and as $x \to -\infty$, $P(x) \to \infty$. Therefore, the end behavior of the graph is of the type shown in Figure 40, and is symbolized ↘.

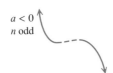

$a < 0$
n odd

FIGURE 40

EXAMPLE 3
Determining End Behavior Given the Defining Polynomial

One of the graphs shown in Figure 41 is that of

$$f(x) = -x^5 + 2x - 1,$$

and the other is that of

$$g(x) = 2x^3 - x^2 + x - 1.$$

Use end behavior to determine which is which.

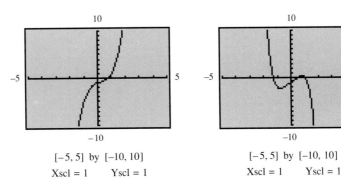

(a)   (b)

[-5, 5] by [-10, 10]   [-5, 5] by [-10, 10]
Xscl = 1   Yscl = 1   Xscl = 1   Yscl = 1

FIGURE 41

SOLUTION   The function $f(x) = -x^5 + 2x - 1$ is of odd degree and the dominating term, $-x^5$, has coefficient -1 (a negative number). Therefore, it has end behavior $\nwarrow\searrow$, and its graph is in part (b) of Figure 41. The function $g(x) = 2x^3 - x^2 + x - 1$ also is of odd degree, but the leading coefficient, 2, is positive. Therefore, it has end behavior $\swarrow\nearrow$ and its graph is in part (a) of Figure 41.   ■

FOR **GROUP DISCUSSION**

Graph $y = x$, $y = x^3$, $y = x^5$, and $y = x^7$ in the standard viewing window. Discuss the behavior of these graphs (all of which are examples of odd degree polynomial functions) as n gets larger. Then repeat for the window $[-1, 1]$ by $[-1, 1]$.

Number of x-Intercepts (Real Zeros)

Figure 42 shows how a cubic polynomial function may have 1, 2, or 3 x-intercepts.

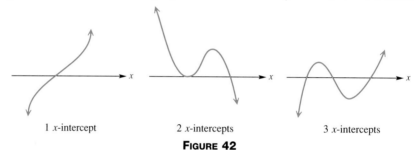

| 1 x-intercept | 2 x-intercepts | 3 x-intercepts |

FIGURE 42

We saw earlier how a linear function may have no more than one x-intercept, and a quadratic function may have no more than two x-intercepts. Now we see that a cubic polynomial function may have no more than three x-intercepts. These observations suggest an important property of polynomial functions.

NUMBER OF x-INTERCEPTS (REAL ZEROS) OF A POLYNOMIAL FUNCTION

A polynomial function of degree n will have a maximum of n x-intercepts. Also, a polynomial function of degree n will have at most n distinct zeros.

EXAMPLE 4

Determining
x-Intercepts
Graphically

Find the x-intercepts of the polynomial function

$$P(x) = x^3 + 5x^2 + 5x - 2$$

graphically.

SOLUTION   If we use the standard window, we get the graph shown in Figure 43. Notice that it has three x-intercepts. By using graphics calculator capabilities, we find that the x-intercepts are -2, approximately -3.30, and approximately $.30$.

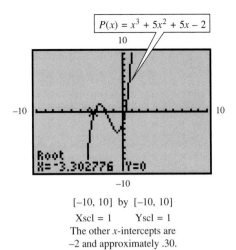

$[-10, 10]$ by $[-10, 10]$
Xscl = 1   Yscl = 1
The other x-intercepts are
-2 and approximately .30.

FIGURE 43

Equivalently, we may say that -2 is a real zero of P, and -3.30 and $.30$ are approximate real zeros of P. It can be shown, using the methods of Section 3.6, that the latter two real numbers are approximations of the exact real zeros

$$\frac{-3 - \sqrt{13}}{2} \quad \text{and} \quad \frac{-3 + \sqrt{13}}{2}.$$

EXAMPLE **5**

Analyzing a
Polynomial Function

For the fifth degree polynomial function

$$P(x) = x^5 + 2x^4 - x^3 + x^2 - x - 4,$$

do each of the following.

(a) Determine its domain.

SOLUTION Because it is a polynomial function, its domain is $(-\infty, \infty)$.

(b) Determine its range.

SOLUTION Because it is of odd degree, its range is $(-\infty, \infty)$.

(c) Use its graph to find approximations of its local extrema.

SOLUTION The graph of P is shown in Figure 44, using a window of $[-5, 5]$ by $[-20, 50]$. It appears that there are only two extreme points. (Methods of calculus

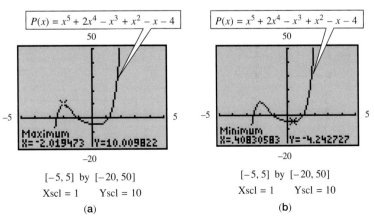

$[-5, 5]$ by $[-20, 50]$
Xscl = 1   Yscl = 10

(a)

$[-5, 5]$ by $[-20, 50]$
Xscl = 1   Yscl = 10

(b)

FIGURE 44

can verify that these are the only two.) Using the capabilities of a calculator, we find that the local maximum has approximate coordinates $(-2.02, 10.01)$ and the local minimum has approximate coordinates $(.41, -4.24)$.

NOTE By zooming in on the apparent extrema, we *may* encounter others. Sometimes functions have "hidden behavior." For example, compare $y = x^3 - 2x^2 + x - 2$ in the standard window and then in the window $[-2.5, 2.5]$ by $[-2.5, 2.5]$.

(d) Use its graph to find its approximate and/or exact x-intercepts.

SOLUTION Once again, using the capabilities of a calculator, we find that the x-intercepts are -1 (exact), 1.14 (approximate), and -2.52 (approximate). The first of these can be verified analytically quite easily by evaluating $P(-1)$:

$$P(-1) = (-1)^5 + 2(-1)^4 - (-1)^3 + (-1)^2 - (-1) - 4$$
$$= -1 + 2(1) - (-1) + 1 + 1 - 4$$
$$= -1 + 2 + 1 + 1 + 1 - 4$$
$$= 0.$$

This again points out the fact that an x-intercept of the graph of a function is a real zero of the function.

Notice that this function has only three x-intercepts, and thus three real zeros. This supports the earlier statement that a polynomial function of degree n will have *at most n x*-intercepts. It may have fewer, as in this case. ∎

FOR GROUP DISCUSSION

Based on the graph shown in Figure 44 and the coordinates of the local extrema determined in part (c) of Example 5, discuss the intervals over which the function decreases and over which the function increases.

Quartic Functions and Other Functions of Even Degree

A polynomial function of the form $P(x) = ax^4 + bx^3 + cx^2 + dx + e, a \neq 0$, is a fourth degree, or **quartic function.** The simplest quartic function, $P(x) = x^4$, is graphed in Figure 45. Notice that it resembles the graph of the squaring function; however, it is not actually a parabola (based on a formal definition of parabola, presented later in Chapter 6).

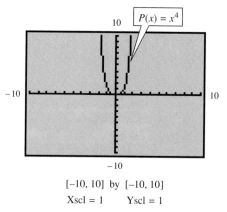

$[-10, 10]$ by $[-10, 10]$
Xscl = 1   Yscl = 1

FIGURE 45

If we graph a quartic function in an appropriate window, the graph will resemble in general one of the shapes shown in Figure 46. The dashed portions in C and D indicate that there may be irregular, but smooth, behavior.

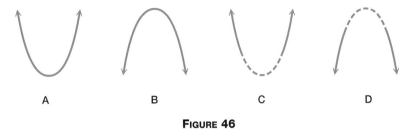

A   B   C   D

FIGURE 46

FOR GROUP DISCUSSION

1. Which one of the shapes in Figure 46 depicts the general form of the graphs of $f(x) = x^2$ and $g(x) = x^4$?
2. Using the concepts of reflection of graphs from Chapter 2, tell which of the shapes depicted in Figure 46 most closely resembles the graph of $h(x) = -x^4$.
3. Graph each of the following functions in the window specified, and determine which one of the shapes in Figure 46 the graph most closely resembles.
   a. $y = 3x^4 + x - 2$, $[-10, 10]$ by $[-10, 10]$
   b. $y = -2x^4 - x^3 + x - 3$, $[-10, 10]$ by $[-10, 10]$
   c. $y = x^4 - 5x^2 + 4$, $[-10, 10]$ by $[-10, 10]$
   d. $y = -x^4 + 12x^3$, $[-15, 15]$ by $[-5000, 2500]$

A quartic function is an example of an even degree polynomial function (as was a quadratic function). We can see from Figure 46 the end behavior of quartic functions corresponds to the type of end behavior exhibited by quadratic functions. This type of end behavior is typical of even degree polynomial functions, and we now make the following observations.

END BEHAVIOR OF EVEN DEGREE POLYNOMIAL FUNCTIONS

Suppose that ax^n is the dominating term of a polynomial function P of *even degree*.

1. If $a > 0$, then as $|x| \to \infty$, $P(x) \to \infty$. Therefore, the end behavior of the graph is of the type shown in Figure 47, and is symbolized ↖↗.
2. If $a < 0$, then as $|x| \to \infty$, $P(x) \to -\infty$. Therefore, the end behavior of the graph is of the type shown in Figure 48, and is symbolized ↙↘.

$a > 0$
n even

$a < 0$
n even

FIGURE 47   **FIGURE 48**

EXAMPLE 6

Determining End Behavior Given the Defining Polynomial

The graphs shown in Figure 49 are of the functions

$$f(x) = x^4 - x^2 + 5x - 4 \qquad g(x) = -x^6 + x^2 - 3x - 4$$
$$h(x) = 3x^3 - x^2 + 2x - 4 \quad \text{and} \quad k(x) = -x^7 + x - 4.$$

Based on dominating term analysis, determine which graph is which.

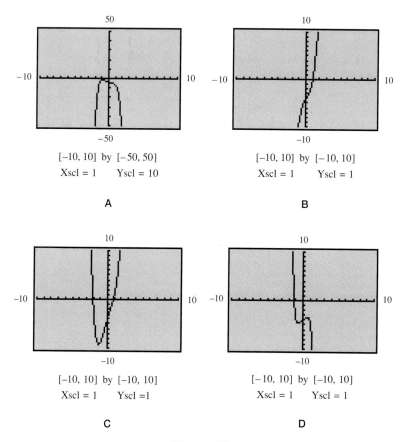

[−10, 10] by [−50, 50]
Xscl = 1   Yscl = 10

A

[−10, 10] by [−10, 10]
Xscl = 1   Yscl = 1

B

[−10, 10] by [−10, 10]
Xscl = 1   Yscl =1

C

[−10, 10] by [−10, 10]
Xscl = 1   Yscl = 1

D

FIGURE 49

SOLUTION   Because f is of even degree with positive leading coefficient, its graph is in C. Because g is of even degree with negative leading coefficient, its graph is in A. Using the ideas presented earlier (see Example 3), the graph of h is in B and the graph of k is in D. ∎

FOR **GROUP DISCUSSION**

Graph $y = x^2$, $y = x^4$, $y = x^6$, and $y = x^8$ in the standard viewing window. Discuss the behavior of these graphs (all of which are examples of even degree polynomial functions) as n gets larger. Then repeat for the window $[−1, 1]$ by $[−1, 1]$.

<table>
<tr><td>EXAMPLE 7</td></tr>
<tr><td>Analyzing a
Polynomial Function</td></tr>
</table>

For the fourth degree polynomial function

$$P(x) = x^4 + 2x^3 - 15x^2 - 12x + 36,$$

do each of the following.

(a) Determine its domain.

SOLUTION Because it is a polynomial function, its domain is $(-\infty, \infty)$.

(b) Use its graph to find its local extrema. Does it have an absolute minimum? What is the range of the function?

SOLUTION After experimenting with various windows, we see that a window of $[-6, 6]$ by $[-80, 50]$ provides a view of the extreme points, as well as all intercepts. See Figure 50. Using the capabilities of a calculator, we find that the two local minimum points have approximate coordinates $(-3.43, -41.61)$ and $(2.31, -18.63)$, and the local maximum has coordinates $(-.38, 38.31)$. Because the end behavior is ↖↗ and the point $(-3.43, -41.61)$ is the lowest point on the graph, the absolute minimum value of the function is approximately -41.61, and therefore the range is approximately $[-41.61, \infty)$.

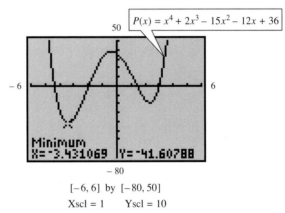

[−6, 6] by [−80, 50]
Xscl = 1   Yscl = 10
The two other extreme points are (−.38, 38.31) and (2.31, −18.63).

FIGURE 50

(c) Use its graph to find its x-intercepts.

SOLUTION This fourth degree function has the maximum number of x-intercepts possible (four). Using the calculator capabilities, we find that two seemingly exact values for the x-intercepts are -2 and 3, while to the nearest hundredth, the other two are -4.37 and 1.37. (Using concepts to follow in this chapter, we can show that these latter two are approximations for the *exact* values

$$\frac{-3 - \sqrt{33}}{2} \quad \text{and} \quad \frac{-3 + \sqrt{33}}{2}.)\quad \blacksquare$$

FOR **GROUP DISCUSSION**

Based on the graph shown in Figure 50 and the coordinates of the extreme points determined in part (b) of Example 7, discuss the intervals over which the function decreases and over which the function increases.

Complete Graphs

The most important features of the graph of a polynomial function are its intercepts, its extrema, and its end behavior. For this reason, we give the following criteria for the complete graph of a polynomial function.

> **COMPLETE GRAPH CRITERIA FOR A POLYNOMIAL FUNCTION**
>
> A complete graph of a polynomial function will exhibit the following features.
>
> **a.** all x-intercepts (if any)
> **b.** the y-intercept
> **c.** all extreme points (if any)
> **d.** enough of the graph to exhibit the correct end behavior

EXAMPLE 8

Determining an Appropriate Window for a Complete Graph

The window $[-1.25, 1.25]$ by $[-400, 50]$ is used in Figure 51 to give a view of the graph of

$$P(x) = x^6 - 36x^4 + 288x^2 - 256.$$

Is this a complete graph?

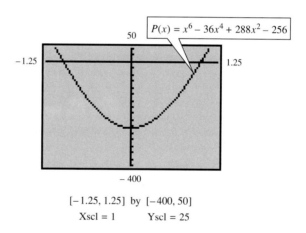

$P(x) = x^6 - 36x^4 + 288x^2 - 256$

$[-1.25, 1.25]$ by $[-400, 50]$
Xscl = 1   Yscl = 25

FIGURE 51

SOLUTION   Since the function is of even degree and the dominating term has positive coefficient, the end behavior seems to be correct. The y-intercept, -256, is shown, and two x-intercepts are shown. One local minimum is shown. This function may, however, have up to six x-intercepts, since it is of degree 6. By experimenting with other viewing windows, we see that a window of $[-8, 8]$ by $[-1000, 600]$ shows a total of *five* local extrema, and four x-intercepts that were not apparent in the earlier figure. See Figure 52. Since there can be no more than five local extrema, this second view (and *not* the first view, in Figure 51) gives us a complete graph.   ∎

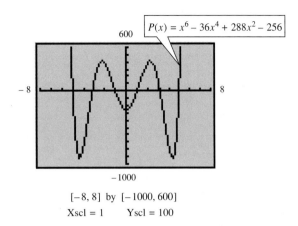

$$[-8, 8] \text{ by } [-1000, 600]$$
$$\text{Xscl} = 1 \qquad \text{Yscl} = 100$$

FIGURE 52

FOR **GROUP DISCUSSION**

How does Example 8 illustrate the warning "Don't always believe what you see."? How does it illustrate the shortcomings of learning technology without learning mathematical concepts?

3.5 EXERCISES

Figure 33 is repeated below. The graphs are of $y = x^3 - 3x^2 - 6x + 8$, $y = x^4 + 7x^3 - 5x^2 - 75x$, $y = -x^3 + 9x^2 - 27x + 17$, and $y = -x^5 + 36x^3 - 22x^2 - 147x - 90$, but not necessarily in that order. Assuming that each is a complete graph as specified in this section, answer each of the questions in Exercises 1–10.

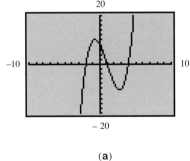

(a)

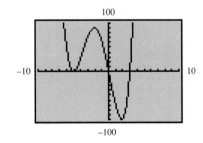

(b)

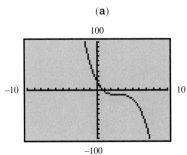

(c)

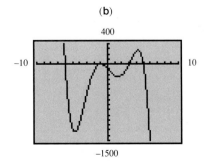

(d)

1. Which one of the graphs is that of $y = x^3 - 3x^2 - 6x + 8$?

2. Which one of the graphs is that of $y = x^4 + 7x^3 - 5x^2 - 75x$?

3. How many real zeros does the graph in (c) have?

4. Which one of (c) and (d) is the graph of $y = -x^3 + 9x^2 - 27x + 17$? (*Hint:* Look at the *y*-intercept.)

5. Which of the graphs cannot be that of a cubic polynomial function?

6. How many positive real zeros does the function graphed in (d) have?

7. How many negative real zeros does the function graphed in (a) have?

8. Is the absolute minimum value of the function graphed in (b) a positive number or a negative number?

9. Which one of the graphs is that of a function whose range is *not* $(-\infty, \infty)$?

10. One of the following is an approximation for the local maximum point of the graph in Figure (a). Which one is it?
    (a) $(.73, 10.39)$ (b) $(-.73, 10.39)$ (c) $(-.73, -10.39)$ (d) $(.73, -10.39)$

Use an end behavior diagram ($\nwarrow\nearrow$, $\swarrow\searrow$, $\nwarrow\searrow$, or $\swarrow\nearrow$) to describe the end behavior of the given function. Then verify your answer by graphing the function on your calculator.

11. $P(x) = \sqrt{5}x^3 + 2x^2 - 3x + 4$

12. $P(x) = -\sqrt{7}x^3 - 4x^2 + 2x - 1$

13. $P(x) = -\pi x^5 + 3x^2 - 1$

14. $P(x) = \pi x^7 - x^5 + x - 1$

15. $P(x) = 2.74x^4 - 3x^2 + x - 2$

16. $P(x) = \sqrt{6}x^6 - x^5 + 2x - 2$

17. $P(x) = -\pi x^6 + x^5 - x^4 - x + 3$

18. $P(x) = -2.84x^4 - 3.2x^3 + x^2 - x + 3$

The functions in Exercises 19–22 (at the top of the next page) are graphed in Figures A–D, but not necessarily in alphabetical order. Use end behavior and analysis of dominating term to match the equation with the correct graph.

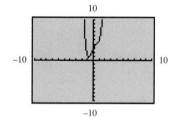

A

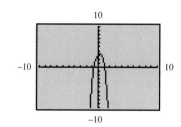

B

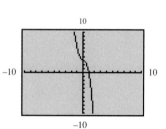

C

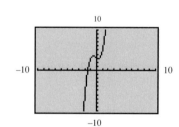

D

19. $f(x) = 2x^3 + x^2 - x + 3$

20. $g(x) = -2x^3 - x + 3$

21. $h(x) = -2x^4 + x^3 - 2x^2 + x + 3$

22. $k(x) = 2x^4 - x^3 - 2x^2 + 3x + 3$

23. Using a window of $[-1, 1]$ by $[-1, 1]$, graph the odd degree polynomial functions

$$y = x, \qquad y = x^3, \qquad \text{and} \qquad y = x^5.$$

Describe the behavior of these functions relative to each other. Predict the behavior of the graph of $y = x^7$ in the same window, and then graph it to support your prediction.

24. Repeat the activity of Exercise 23 for the even degree polynomial functions

$$y = x^2, \qquad y = x^4, \qquad \text{and} \qquad y = x^6.$$

Predict the behavior of the graph of $y = x^8$ in the same window, and then graph it to support your prediction.

25. The graphs of $f(x) = x^n$ for $n = 3, 5, 7, \ldots$ all resemble each other. Describe how they are the same. As n gets larger, what happens to the graph?

26. Repeat Exercise 25 for $f(x) = x^n$, where $n = 2, 4, 6, \ldots$.

For the functions given in Exercises 27–36, find a complete graph and do each of the following. Use your graphics calculator to its maximum capability.

   (a) Determine its domain.

   (b) Determine all local minimum points, and tell if any is an absolute minimum point. (Approximate coordinates to the nearest hundredth.)

   (c) Determine all local maximum points, and tell if any is an absolute maximum point. (Approximate coordinates to the nearest hundredth.)

   (d) Determine the range. (If an approximation is necessary, give it to the nearest hundredth.)

   (e) Determine all intercepts. For each function, there is at least one x-intercept that is an integer. For those that are not integers, give an approximation to the nearest hundredth. You should determine the y-intercept analytically.

27. $y = x^3 - 4x^2 + x + 6$

28. $y = x^3 + x^2 - 22x - 40$

29. $y = -2x^3 - 14x^2 + 2x + 84$

30. $y = -3x^3 + 6x^2 + 39x - 60$

31. $y = x^5 + 4x^4 - 3x^3 - 17x^2 + 6x + 9$

32. $y = -2x^5 + 7x^4 + x^3 - 20x^2 + 4x + 16$

33. $y = 2x^4 + 3x^3 - 17x^2 - 6x - 72$

34. $y = 3x^4 - 33x^2 + 54$

35. $y = -x^6 + 24x^4 - 144x^2 + 256$

36. $y = -3x^6 + 2x^5 + 9x^4 - 8x^3 + 11x^2 + 4$

Relating Concepts

In Section 3.2 we informally introduced the concept of concavity of a graph. (A formal discussion of concavity requires concepts from calculus.) A polynomial function graph may change concavity several times throughout its domain, or its concavity may remain the same. Use any method you wish to answer the following questions about concavity.

37. For a function of the form $P(x) = x^n$, where n is even, describe the concavity of the graph.

38. For a function of the form $P(x) = x^n$, where n is odd, discuss the concavity of the graph.

39. Graph the following function in the window $[-4, 4]$ by $[-20, 20]$, and discuss the concavity of the graph: $y = x^3 + x^2 + x$.

40. Explain why an even degree function with negative leading coefficient must be concave down for some interval of its domain.

Determine a window that will provide a complete graph (as defined in this section) of each of the following polynomial functions. (In each case, there are many possible such windows.)

41. $y = 4x^5 - x^3 + x^2 + 3x - 16$

42. $y = 3x^5 - x^4 + 12x^2 - 25$

43. $y = 2.9x^3 - 37x^2 + 28x - 143$

44. $y = -5.9x^3 + 16x^2 - 120$

45. $y = \pi x^4 - 13x^2 + 84$

46. $y = 2\pi x^4 - 12x^2 + 100$

47. $y = -\sqrt{6}x^6 + x^5 - x^3 + 12$

48. $y = -\sqrt{12}x^4 - x^3 + \pi x^2 - 18$

Relating Concepts

The concepts of stretching, translating, and reflecting graphs presented in Chapter 2 can be applied to polynomial functions of the form $P(x) = x^n$. For example, the graph of $y = -2(x + 4)^4 - 6$ can be obtained from the graph of $y = x^4$ by shifting 4 units to the left, vertically stretching by a factor of 2, reflecting across the x-axis, and shifting down 6 units. Thus, the graph should resemble the following:

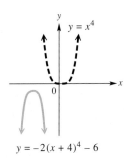

$$y = -2(x + 4)^4 - 6$$

If we simplify the expression $-2(x + 4)^4 - 6$ algebraically, we get

$$-2x^4 - 32x^3 - 192x^2 - 512x - 518.$$

Thus, the graph of $y = -2(x + 4)^4 - 6$ is the same as that of $y = -2x^4 - 32x^3 - 192x^2 - 512x - 518$.

In Exercises 49–52, two forms of the same polynomial function are given. Sketch by hand the general shape of the graph of the function using the concepts of Chapter 2 and describe the transformations. Then support your answer by graphing it on your calculator in a suitable window.

49. $y = 2(x + 3)^4 - 7$
$y = 2x^4 + 24x^3 + 108x^2 + 216x + 155$

50. $y = -3(x + 1)^4 + 12$
$y = -3x^4 - 12x^3 - 18x^2 - 12x + 9$

51. $y = -3(x - 1)^3 + 12$
$y = -3x^3 + 9x^2 - 9x + 15$

52. $y = .5(x - 1)^5 + 13$
$y = .5x^5 - 2.5x^4 + 5x^3 - 5x^2 + 2.5x + 12.5$

Without graphing, answer true or false to each of the following. Then support your answer, if you wish, by graphing.

53. The function $f(x) = x^3 + 2x^2 - 4x + 3$ has four real zeros.

54. The function $f(x) = x^3 + 3x^2 + 3x + 1$ must have at least one real zero.

55. If a polynomial function of even degree has a negative leading coefficient and a positive *y*-intercept, it must have at least two real zeros.

56. The function $f(x) = 3x^4 + 5$ has no real zeros.

57. The function $f(x) = -3x^4 + 5$ has two real zeros.

58. The graph of the function $f(x) = x^3 - 3x^2 + 3x - 1 = (x - 1)^3$ has exactly one *x*-intercept.

59. A fifth degree polynomial function cannot have a single real zero.

60. An even degree polynomial function must have at least one real zero.

There exists a useful theorem that helps us determine the number of positive and the number of negative real zeros of a polynomial function. It is known as Descartes' Rule of Signs.

DESCARTES' RULE OF SIGNS

Let $P(x)$ define a polynomial function with real coefficients and a nonzero constant term, with terms in descending powers of *x*.

a. The number of positive real zeros of *P* either equals the number of variations in sign occurring in the coefficients of $P(x)$, or is less than the number of variations by a positive even integer.

b. The number of negative real zeros of *P* either equals the number of variations in sign occurring in the coefficients of $P(-x)$, or is less than the number of variations by a positive even integer.

In the theorem, *variation in sign* is a change from positive to negative or negative to positive in successive terms of the polynomial. Missing terms (those with 0 coefficients) are counted as no change in sign and can be ignored. For example, consider the polynomial function $P(x) = x^4 - 6x^3 + 8x^2 + 2x - 1$.

$P(x)$ has 3 variations in sign:

$$+x^4 - 6x^3 + 8x^2 + 2x - 1.$$
$$\;\;\;1\quad\;\;2\qquad\qquad 3$$

Thus, by Descartes' Rule of Signs, *P* has either 3 or $3 - 2 = 1$ positive real zeros. Since

$$P(-x) = (-x)^4 - 6(-x)^3 + 8(-x)^2 + 2(-x) - 1$$
$$= x^4 + 6x^3 + 8x^2 - 2x - 1$$

has only one variation in sign, *P* has only one negative real zero. If you graph the function in the window $[-5, 5]$ by $[-10, 10]$, you can interpret the theorem in terms of *x*-intercepts. Verify that there are 3 positive *x*-intercepts, and 1 negative *x*-intercept.

Use Descartes' Rule of Signs to determine the possible number of positive real zeros and the possible number of negative real zeros for the following functions. Then use a graph to determine the actual numbers of positive and negative real zeros.

61. $P(x) = 2x^3 - 4x^2 + 2x + 7$

62. $P(x) = x^3 + 2x^2 + x - 10$

63. $P(x) = 5x^4 + 3x^2 + 2x - 9$

64. $P(x) = 3x^4 + 2x^3 - 8x^2 - 10x - 1$

65. $P(x) = x^5 + 3x^4 - x^3 + 2x + 3$

66. $P(x) = 2x^5 - x^4 + x^3 - x^2 + x + 5$

Relating Concepts

67. If the graph of a polynomial function P satisfies the property $P(x) = P(-x)$ for all real numbers x, then the graph of P is symmetric with respect to the y-axis, and P is called an even function. (These ideas were first presented in Sections 2.1 and 2.3.) Consider the polynomial function

$$P(x) = x^4 - 8x^2 - 9.$$

   (a) Show *analytically* that the graph of P is symmetric with respect to the y-axis.

   (b) Graph the function in the window $[-10, 0]$ by $[-30, 10]$ and use your calculator to find the single *negative* x-intercept of the graph.

   (c) Verify your result from part (b) analytically.

   (d) Now, using the results from parts (a)–(c), predict the value of the *positive* x-intercept of the graph, and then graph in the window $[-10, 10]$ by $[-30, 10]$ to support your result.

68. In Section 2.3 we learned that graphs of functions may be shifted horizontally. For example, the graph of $y = P(x + 4)$ may be obtained from the graph of $y = P(x)$ by shifting the graph of P four units to the left.

   (a) Graph the polynomial function $P(x) = x^3 - 3x^2 + x - 5$ in the standard viewing window.

   (b) Predict what the graph of $P(x - 4) = (x - 4)^3 - 3(x - 4)^2 + (x - 4) - 5$ will look like when compared to the graph of $y = P(x)$.

   (c) Support your prediction from part (b) by graphing $y_2 = P(x - 4)$ in the same window as $y_1 = P(x)$.

69. Use the capabilities of your calculator to find the intervals over which the function $y = 2x^3 - 5x^2 - 3x + 2$

   (a) is increasing.   **(b)** is decreasing.

   Express endpoints of intervals as approximations to the nearest hundredth.

70. Graph the functions

$$f(x) = \frac{2}{3}x^3 + \frac{1}{2}x^2 - 21x + 4 \quad \text{and} \quad g(x) = 2x^2 + x - 21$$

   in the window $[-8, 8]$ by $[-50, 75]$.

   (a) Use the capabilities of your calculator to show that the solutions of $g(x) = 0$ are the x-coordinates of the local extrema of f.

   (b) Confirm that the intervals over which f is increasing correspond to the intervals over which g is positive.

   (c) Confirm that the interval over which f is decreasing corresponds to the interval over which g is negative.

   (The ideas of this exercise relate to concepts from the study of calculus. The function g is called the *derivative* of f.)

3.6 TOPICS IN THE THEORY OF POLYNOMIAL FUNCTIONS

Root Location Theorem ▌ Division of Polynomials and Synthetic Division ▌ Remainder and Factor Theorems ▌ Complex Zeros and the Fundamental Theorem of Algebra ▌ Multiplicity of Zeros

In this section we will examine some important topics in the theory of polynomial functions. These topics will complement the graphical work done in Section 3.5, and will also help prepare us for the work to follow in Section 3.7 (equations, inequalities, and applications of polynomial functions).

Root Location Theorem

The graph of any polynomial function is a smooth, unbroken curve. This property is closely related to the following theorem.

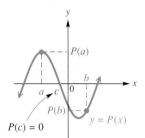

FIGURE 53

> **ROOT LOCATION THEOREM FOR POLYNOMIAL FUNCTIONS**
>
> Suppose that a polynomial function P is defined in such a way that for real numbers a and b, $P(a)$ and $P(b)$ differ in sign. Then there exists at least one real number c between a and b such that $P(c) = 0$.

This theorem helps to identify intervals where zeros of polynomials are located. For example, in Figure 53 $P(a)$ and $P(b)$ are opposite in sign, so 0 is between $P(a)$ and $P(b)$. Then, by the root location theorem, there must be a number c between a and b such that $P(c) = 0$.

EXAMPLE 1

Applying the Root Location Theorem

TECHNOLOGICAL NOTE
Another way that many models of graphics calculators can illustrate that the function in Example 1 has a zero between 2 and 3 is to exhibit a table of values, and show that there is a sign change in the y-values from $x = 2$ to $x = 3$. See the Further Explorations at the end of the exercise set for this section to learn more about the table capability.

(a) Show *analytically* that the polynomial function

$$P(x) = x^3 - 2x^2 - x + 1$$

has a real zero between 2 and 3.

SOLUTION Begin by evaluating $P(2)$ and $P(3)$.

$$P(2) = 2^3 - 2(2)^2 - 2 + 1 = -1$$
$$P(3) = 3^3 - 2(3)^2 - 3 + 1 = 7$$

Since $P(2) = -1$ and $P(3) = 7$ differ in sign, the root location theorem assures us that there is a real zero between 2 and 3.

(b) Support the result of part (a) *graphically*.

SOLUTION If we graph $P(x) = x^3 - 2x^2 - x + 1$ in the standard viewing window, we see that there is an x-intercept between 2 and 3, confirming our result that P has a zero between 2 and 3. See Figure 54. (P has two other real zeros, but this does not affect our discussion here.) Using the root-finding capabilities of the calculator, we can determine that, to the nearest hundredth, this zero is 2.25. ∎

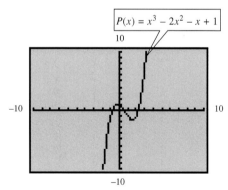

$$P(x) = x^3 - 2x^2 - x + 1$$

[−10, 10] by [−10, 10]
Xscl = 1   Yscl = 1
There is an x-intercept between 2 and 3.

FIGURE 54

CAUTION   Be careful how you interpret the root location theorem. If $P(a)$ and $P(b)$ are *not* opposite in sign, it does not necessarily mean that there is no zero between a and b. For example, in Figure 55, $P(a)$ and $P(b)$ are both negative, but -3 and -1, which are between a and b, are zeros of P.

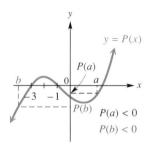

FIGURE 55

Division of Polynomials and Synthetic Division

As we shall see later in this section, it is important to be able to determine whether a binomial of the form $x - k$ is a factor of a polynomial. Just as we can use the old-fashioned "long division method" to determine whether one whole number is a factor of another, we can also use it to determine whether one polynomial is a factor of another.

Example 2 illustrates the method of long division of a polynomial by a binomial.

EXAMPLE 2

Dividing a Polynomial by a Binomial

Divide the polynomial $3x^3 - 2x + 5$ by $x - 3$. Determine the quotient and the remainder.

SOLUTION   In performing this division, we make sure that the powers of the variable in the dividend ($3x^3 - 2x + 5$) are descending, which they are. We also insert the term $0x^2$ to act as a placeholder. We then follow the same procedure as long division of whole numbers.

$$x - 3 \overline{)3x^3 + 0x^2 - 2x + 5} \qquad \text{— Missing term}$$

Start with $\frac{3x^3}{x} = 3x^2$.

$$\begin{array}{r} 3x^2 \\ x - 3 \overline{)3x^3 + 0x^2 - 2x + 5} \\ 3x^3 - 9x^2 \end{array}$$

$\leftarrow \frac{3x^3}{x} = 3x^2$

$\leftarrow 3x^2(x - 3)$

Subtract by changing the signs on $3x^3 - 9x^2$ and adding.

$$\begin{array}{r} 3x^2 \\ x - 3 \overline{)3x^3 + 0x^2 - 2x + 5} \\ \underline{3x^3 - 9x^2} \\ 9x^2 \end{array}$$

$\leftarrow$ Subtract.

Bring down the next term.

$$
\begin{array}{r}
3x^2 \\
x-3\overline{\smash{)}3x^3+0x^2-2x+5} \\
\underline{3x^3-9x^2} \\
9x^2-2x
\end{array}
$$
$\leftarrow$ Bring down $-2x$.

In the next step, $\frac{9x^2}{x}=9x$.

$$
\begin{array}{r}
3x^2+9x \\
x-3\overline{\smash{)}3x^3+0x^2-2x+5} \\
\underline{3x^3-9x^2} \\
9x^2-2x \\
\underline{9x^2-27x} \\
25x+5
\end{array}
$$
$\leftarrow \frac{9x^2}{x}=9x$

$\leftarrow 9x(x-3)$

$\leftarrow$ Subtract and bring down 5.

Finally, $\frac{25x}{x}=25$.

$$
\begin{array}{r}
3x^2+9x+25 \\
x-3\overline{\smash{)}3x^3+0x^2-2x+5} \\
\underline{3x^3-9x^2} \\
9x^2-2x \\
\underline{9x^2-27x} \\
25x+5 \\
\underline{25x-75} \\
80
\end{array}
$$
$\leftarrow \frac{25x}{x}=25$

$\leftarrow 25(x-3)$

$\leftarrow$ Subtract.

The quotient is $3x^2+9x+25$ with a remainder of 80.   ◼

TECHNOLOGICAL NOTE
If your calculator has a table feature, you can support the result of the discussion preceding Example 3 by showing that for $y_1=3x^3-2x+5$, when $x=3$, $y=80$.

The division shown in Example 2 allows us to make several observations about polynomial division in general. Notice that we divided a cubic polynomial (degree 3) by a linear polynomial (degree 1) and obtained a quadratic polynomial quotient (degree 2). Notice also that $3-1=2$, so we observe that the degree of the quotient polynomial is found by subtracting the degree of the divisor from the degree of the dividend. Also, since the remainder is a nonzero constant (80), we can write it as the numerator of a fraction with denominator $x-3$ to express the fractional part of the quotient:

Dividend $\rightarrow$
Divisor $\rightarrow$
$$
\frac{3x^3-2x+5}{x-3}=\underbrace{3x^2+9x+25}_{\substack{\text{quotient}\\\text{polynomial}}}+\underbrace{\frac{80}{x-3}}_{\substack{\text{fractional}\\\text{part of the}\\\text{quotient}}}
$$
$\leftarrow$ Remainder
$\leftarrow$ Divisor

In general, the following rules apply to the division of a polynomial by a binomial.

DIVISION OF A POLYNOMIAL BY $x-k$

1. If the degree n polynomial $P(x)$ is divided by $x-k$, the quotient polynomial, $Q(x)$, has degree $n-1$.
2. The remainder R is a constant (and may be 0). The complete quotient for $\frac{P(x)}{x-k}$ may be written as follows:

$$
\frac{P(x)}{x-k}=Q(x)+\frac{R}{x-k}.
$$

The procedure of long division of a polynomial by a binomial of the form $x - k$ can be condensed using **synthetic division.** To see how synthetic division works in the case of the division performed in Example 2, observe the following.

$$
\begin{array}{r}
3x^2 + 9x + 25 \\
x - 3\overline{)3x^3 + 0x^2 - 2x + 5} \\
\underline{3x^3 - 9x^2} \\
9x^2 - 2x \\
\underline{9x^2 - 27x} \\
25x + 5 \\
\underline{25x - 75} \\
80
\end{array}
$$

$$
\begin{array}{r}
3 \quad 9 \quad 25 \\
1 - 3\overline{)3 \quad 0 \quad -2 \quad 5} \\
(3) -9 \\
\underline{9 \quad -2} \\
(9) \quad -27 \\
\underline{25 \quad 5} \\
(25) -75 \\
80
\end{array}
$$

On the right, exactly the same division is shown written without the variables. All the numbers in parentheses on the right are repetitions of the numbers directly above them, so they may be omitted, as shown on the left below.

$$
\begin{array}{r}
3 \quad 9 \quad 25 \\
1 - 3\overline{)3 \quad 0 \quad -2 \quad 5} \\
-9 \\
\underline{9 \ (-2)} \\
-27 \\
\underline{25 \quad (5)} \\
-75 \\
80
\end{array}
$$

$$
\begin{array}{r}
3 \quad 9 \quad 25 \\
1 - 3\overline{)3 \quad 0 \quad -2 \quad 5} \\
-9 \\
\underline{9} \\
-27 \\
\underline{25} \\
-75 \\
80
\end{array}
$$

The numbers in parentheses on the left are again repetitions of the numbers directly above them; they too may be omitted, as shown on the right above.

Now the problem can be condensed. If the 3 in the dividend is brought down to the beginning of the bottom row, the top row can be omitted, since it duplicates the bottom row.

$$
\begin{array}{r}
1 - 3\overline{)3 \quad 0 \quad -2 \quad 5} \\
-9 \quad -27 \quad -75 \\
\underline{} \\
3 \quad 9 \quad 25 \quad 80
\end{array}
$$

Finally, the 1 at the upper left can be omitted. Also, to simplify the arithmetic, subtraction in the second row is replaced by addition. We compensate for this by changing the -3 at upper left to its additive inverse, 3. The result of doing all this is shown below.

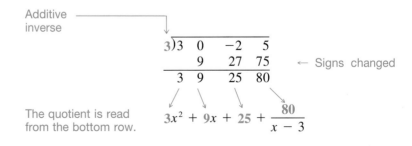

Additive inverse

← Signs changed

The quotient is read from the bottom row.

$3x^2 + 9x + 25 + \dfrac{80}{x - 3}$

EXAMPLE 3

Using Synthetic Division

Use synthetic division to divide $5x^3 - 6x^2 - 28x + 8$ by $x + 2$.

SOLUTION To use synthetic division, the divisor must be of the form $x - k$. Writing $x + 2$ as $x - (-2)$ shows that the value of k here is -2. We begin by writing

$$-2 \overline{)5 \quad -6 \quad -28 \quad 8.}$$

Next, bring down the 5.

$$-2 \overline{)5 \quad -6 \quad -28 \quad 8}$$
$$\overline{5}$$

Now, multiply -2 by 5 to get -10, and add it to -6 from the first row. The result is -16.

$$-2 \overline{)5 \quad -6 \quad -28 \quad 8}$$
$$ -10$$
$$\overline{5 \quad -16}$$

Next, multiply -2 by -16 to get 32. Add this to -28 from the first row.

$$-2 \overline{)5 \quad -6 \quad -28 \quad 8}$$
$$ -10 \quad 32$$
$$\overline{5 \quad -16 \quad 4}$$

Finally, $(-2)(4) = -8$. Add this result to 8 to obtain 0.

$$-2 \overline{)5 \quad -6 \quad -28 \quad 8}$$
$$ -10 \quad 32 \quad -8$$
$$\overline{5 \quad -16 \quad 4 \quad 0}$$

The coefficients of the quotient polynomial and the remainder are read directly from the bottom row. Since the degree of the quotient will always be one less than the degree of the polynomial to be divided, and here the remainder is 0,

$$\frac{5x^3 - 6x^2 - 28x + 8}{x + 2} = 5x^2 - 16x + 4.$$

Notice that $x + 2$ is a *factor* of $5x^3 - 6x^2 - 28x + 8$, and $5x^3 - 6x^2 - 28x + 8 = (x + 2)(5x^2 - 16x + 4)$. ∎

Remainder and Factor Theorems

In Example 2 we divided $3x^3 - 2x + 5$ by $x - 3$ and obtained a remainder of 80. If we evaluate the function $P(x) = 3x^3 - 2x + 5$ at $x = 3$, we get

$$P(3) = 3(3)^3 - 2(3) + 5 = 81 - 6 + 5 = 80.$$

Notice that the remainder is equal to $P(3)$. Also, in Example 3 we divided $5x^3 - 6x^2 - 28x + 8$ by $x - (-2)$ and obtained a remainder of 0. If we evaluate the function $P(x) = 5x^3 - 6x^2 - 28x + 8$ at $x = -2$, we get

$$P(-2) = 5(-2)^3 - 6(-2)^2 - 28(-2) + 8$$
$$= -40 - 24 + 56 + 8$$
$$= 0.$$

Notice that here also, the remainder is equal to $P(-2)$. These two examples illustrate an important theorem in the study of polynomial functions, the Remainder Theorem, and an important corollary to the theorem, the Factor Theorem. (A *corollary* is a theorem that follows directly from another theorem.)

THE REMAINDER THEOREM

If a polynomial $P(x)$ is divided by $x - k$, the remainder is equal to $P(k)$.

(A proof of the remainder theorem is outlined in Exercise 71).

EXAMPLE 4

Using the Remainder Theorem and Supporting the Result Graphically

(a) Use the remainder theorem and synthetic division to find $P(-2)$ if
$$P(x) = -x^4 + 3x^2 - 4x - 5.$$

SOLUTION We use synthetic division to find the remainder when $P(x)$ is divided by $x - (-2)$.

$$
\begin{array}{r|rrrrr}
-2) & -1 & 0 & 3 & -4 & -5 \\
    &   & 2 & -4 & 2 & 4 \\
\hline
    & -1 & 2 & -1 & -2 & -1 \leftarrow \text{Remainder}
\end{array}
$$

Since the remainder is -1, by the remainder theorem we have $P(-2) = -1$.

(b) Support the result of part (a) graphically.

SOLUTION If we graph $P(x) = -x^4 + 3x^2 - 4x - 5$, we should see that the point $(-2, -1)$ lies on the graph. Figure 56 supports this fact.

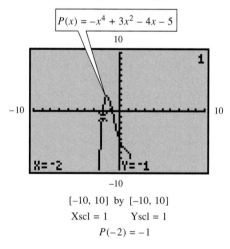

$$P(x) = -x^4 + 3x^2 - 4x - 5$$

X = -2   Y = -1

[−10, 10] by [−10, 10]
Xscl = 1   Yscl = 1
$$P(-2) = -1$$

FIGURE 56

EXAMPLE 5

Deciding Whether a Number is a Zero of a Polynomial Function

Decide whether the given number is a zero of the function defined by the given polynomial.

(a) 2; $P(x) = x^3 - 4x^2 + 9x - 10$

SOLUTION Use synthetic division.

$$
\begin{array}{r|rrrr}
2) & 1 & -4 & 9 & -10 \\
   &   & 2 & -4 & 10 \\
\hline
   & 1 & -2 & 5 & 0
\end{array}
$$

Since the remainder is 0, $P(2) = 0$, and 2 is a zero of the polynomial function $P(x) = x^3 - 4x^2 + 9x - 10$.

(b) -2; $P(x) = 3x^3 - 2x^2 + 4x$

SOLUTION   Remember to use a coefficient of 0 for the missing constant term in the synthetic division.

$$
\begin{array}{r}
-2\overline{)3 \quad -2 \quad\ \ 4 \quad\ \ \ 0} \\
-6 \quad\ 16 \quad -40 \\
\hline
3 \quad -8 \quad\ 20 \quad -40
\end{array}
$$

The remainder is not zero, so -2 is not a zero of P, where $P(x) = 3x^3 - 2x^2 + 4x$. In fact, $P(-2) = -40$. From this, we know that the point $(-2, -40)$ lies on the graph of P.   ∎

In Example 5(a), we showed that 2 is a zero of the polynomial function $P(x) = x^3 - 4x^2 + 9x - 10$. The first three numbers in the bottom row of the synthetic division process used there indicate the coefficients of the quotient polynomial, and thus

$$
\frac{P(x)}{x - 2} = x^2 - 2x + 5.
$$

Multiplying both sides of this equation by $x - 2$ gives

$$
P(x) = (x - 2)(x^2 - 2x + 5),
$$

indicating that $x - 2$ is a *factor* of $P(x)$.

By the remainder theorem, if $P(k) = 0$, then the remainder when $P(x)$ is divided by $x - k$ is zero. This means that $x - k$ is a factor of $P(x)$. Conversely, if $x - k$ is a factor of $P(x)$, then $P(k)$ must equal 0. This is summarized in the following theorem, a corollary of the remainder theorem.

FACTOR THEOREM

The polynomial $x - k$ is a factor of the polynomial $P(x)$ if and only if $P(k) = 0$.

EXAMPLE 6

Using the Factor Theorem

Determine whether the second polynomial listed is a factor of the first.

(a) $4x^3 + 24x^2 + 48x + 32$; $x + 2$

SOLUTION   We use synthetic division with $k = -2$, since $x + 2 = x - (-2)$.

$$
\begin{array}{r}
-2\overline{)4 \quad\ \ 24 \quad\ \ 48 \quad\ \ 32} \\
-8 \quad -32 \quad -32 \\
\hline
4 \quad\ 16 \quad\ \ 16 \quad\ \ \ 0 \qquad \leftarrow \text{Remainder is 0.}
\end{array}
$$

Since the remainder is 0, we know that $x + 2$ is a factor of $4x^3 + 24x^2 + 48x + 32$. A factored form (but not necessarily *completely* factored form) of the latter is $(x + 2)(4x^2 + 16x + 16)$.

(b) $2x^4 + 3x^2 - 5x + 7; \quad x - 1$

SOLUTION By the factor theorem, $x - 1$ will be a factor of $P(x)$ only if $P(1) = 0$. Use synthetic division and the remainder theorem to decide.

$$
\begin{array}{r|rrrr}
1 & 2 & 0 & 3 & -5 & 7 \\
  &   & 2 & 2 & 5 & 0 \\
\hline
  & 2 & 2 & 5 & 0 & 7
\end{array}
$$

Since the remainder is 7, $P(1) = 7$, not 0, so $x - 1$ is not a factor of $P(x)$. ◼

NOTE An easy way to determine $P(1)$ for a polynomial function P is simply to add the coefficients of $P(x)$. This method works since every power of 1 is equal to 1. For example, using $P(x) = 2x^4 + 3x^2 - 5x + 7$ as shown in Example 6(b), we have $P(1) = 2 + 3 - 5 + 7 = 7$, confirming our result found by synthetic division earlier.

The next example illustrates the close relationship among the ideas of x-intercepts of the graph of a polynomial function, real zeros of the function, and solutions of the corresponding polynomial equation.

EXAMPLE 7

Examining Relationships Among x-Intercepts, Zeros, and Solutions

---

Consider the polynomial function $P(x) = 2x^3 + 5x^2 - x - 6$.

(a) Show by synthetic division that -2, $-\frac{3}{2}$, and 1 are zeros of P, and write $P(x)$ in factored form with all factors linear.

SOLUTION

$$
\begin{array}{r|rrrr}
-2 & 2 & 5 & -1 & -6 \\
   &   & -4 & -2 & 6 \\
\hline
   & 2 & 1 & -3 & 0
\end{array} \quad \leftarrow P(-2) = 0
$$

Since $P(-2) = 0$, $x + 2$ is a factor, and thus $P(x) = (x + 2)(2x^2 + x - 3)$. Rather than show that $-\frac{3}{2}$ and 1 are zeros of $P(x)$, we need only show that they are zeros of $2x^2 + x - 3$. This can be shown by elementary factoring methods, or by synthetic division, as follows.

$$
\begin{array}{r|rrr}
-\dfrac{3}{2} & 2 & 1 & -3 \\
   &   & -3 & 3 \\
\hline
   & 2 & -2 & 0
\end{array} \quad \leftarrow P(-\tfrac{3}{2}) = 0. \quad -\tfrac{3}{2} \text{ is a zero of } 2x^2 + x - 3.
$$

$$
\begin{array}{r|rr}
1 & 2 & -2 \\
  &   & 2 \\
\hline
  & 2 & 0
\end{array} \quad \leftarrow P(1) = 0. \quad 1 \text{ is a zero of } 2x - 2.
$$

↑ This 2 is the constant factor.

The completely factored form of $P(x)$ is $2(x + 2)(x + \frac{3}{2})(x - 1)$, or

$$P(x) = (x + 2)(2x + 3)(x - 1).$$

(b) Graph P in a suitable viewing window and locate the x-intercepts.

SOLUTION Figure 57 shows the graph of this function. The calculator will determine the x-intercepts: -2, $-\frac{3}{2}$, and 1.

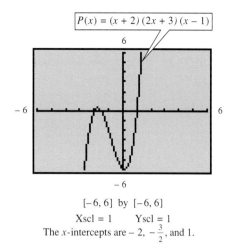

$[-6, 6]$ by $[-6, 6]$
Xscl = 1   Yscl = 1
The x-intercepts are -2, $-\frac{3}{2}$, and 1.

FIGURE 57

(c) Solve the polynomial equation $2x^3 + 5x^2 - x - 6 = 0$.

SOLUTION   The x-intercepts of the graph of P are the real solutions of the equation $P(x) = 0$. Therefore, the solution set of the equation is $\{-2, -\frac{3}{2}, 1\}$.

We can extend the zero-product property (Section 3.2) to more than two factors and solve this equation analytically. Since $P(x) = (x + 2)(2x + 3)(x - 1)$, we set each factor equal to 0 and solve.

$$x + 2 = 0 \quad\text{or}\quad 2x + 3 = 0 \quad\text{or}\quad x - 1 = 0$$

$$x = -2 \qquad\qquad x = -\frac{3}{2} \qquad\qquad x = 1$$

We see once again that the solution set is $\{-2, -\frac{3}{2}, 1\}$.   ◨

Complex Zeros and the Fundamental Theorem of Algebra

In Example 3 of Section 3.3 we found that the imaginary solutions of $2x^2 - x + 4 = 0$ are $\frac{1}{4} + i\frac{\sqrt{31}}{4}$ and $\frac{1}{4} - i\frac{\sqrt{31}}{4}$. Notice that these two solutions are complex conjugates. This is not a coincidence; it can be shown that if $a + bi$ is a zero of a polynomial function with *real* coefficients, then its complex conjugate $a - bi$ is also a zero. This is given as the next theorem. Its proof is left for the exercises.

> **CONJUGATE ZEROS THEOREM**
>
> If $P(x)$ is a polynomial having only real coefficients and if $a + bi$ is a zero of P then the conjugate $a - bi$ is also a zero of P.

EXAMPLE 8

Defining a
Polynomial Function
Satisfying Given
Conditions

(a) Find a cubic polynomial in standard form with real coefficients having zeros 3 and $2 + i$.

SOLUTION   By the conjugate zeros theorem, $2 - i$ must also be a zero of the function. Since the polynomial will be cubic, it will have three linear factors, and by the factor theorem they must be $x - 3$, $x - (2 + i)$, and $x - (2 - i)$.

Therefore, one such cubic polynomial $P(x)$ can be defined as follows:

$$P(x) = (x - 3)[x - (2 + i)][x - (2 - i)]$$
$$= (x - 3)(x - 2 - i)(x - 2 + i)$$
$$= x^3 - 7x^2 + 17x - 15.$$

Multiplying this polynomial by any real nonzero constant will also yield a function satisfying the given conditions, so a more general form of $P(x)$ is

$$a(x^3 - 7x^2 + 17x - 15).$$

(b) Find a polynomial function P satisfying the conditions of part (a), with the additional requirement $P(-2) = 4$. Support the result graphically.

SOLUTION   Let a represent a real nonzero constant. We must have

$$a(x^3 - 7x^2 + 17x - 15)$$

defined in such a way that $P(-2) = 4$. To find a, let $x = -2$ and set the result equal to 4. Then solve.

$$a[(-2)^3 - 7(-2)^2 + 17(-2) - 15] = 4$$
$$a(-8 - 28 - 34 - 15) = 4$$
$$-85a = 4$$
$$a = -\frac{4}{85}$$

Therefore, the desired function is

$$P(x) = -\frac{4}{85}(x^3 - 7x^2 + 17x - 15)$$

$$= -\frac{4}{85}x^3 + \frac{28}{85}x^2 - \frac{4}{5}x + \frac{12}{17}.$$

We can support this result graphically by graphing $P(x) = -\frac{4}{85}x^3 + \frac{28}{85}x^2 - \frac{4}{5}x + \frac{12}{17}$, and showing that the point $(-2, 4)$ lies on the graph. See Figure 58. ∎

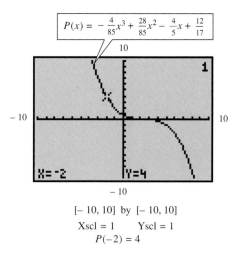

$$P(x) = -\frac{4}{85}x^3 + \frac{28}{85}x^2 - \frac{4}{5}x + \frac{12}{17}$$

$[-10, 10]$ by $[-10, 10]$
Xscl = 1   Yscl = 1
$P(-2) = 4$

FIGURE 58

The next theorem says that every polynomial of degree 1 or more has a zero, so that every such polynomial can be factored. This theorem was first proved by the mathematician Carl F. Gauss in his doctoral thesis in 1799 when he was 22 years old. Although many proofs of this result have been given, all of them involve mathematics beyond the algebra in this book, so no proof is included here.

FUNDAMENTAL THEOREM OF ALGEBRA

Every function defined by a polynomial of degree 1 or more has at least one complex zero.

From the fundamental theorem, if $P(x)$ is of degree 1 or more then there is some number k such that $P(k) = 0$. By the factor theorem, then

$$P(x) = (x - k) \cdot Q(x)$$

for some polynomial $Q(x)$. The fundamental theorem and the factor theorem can be used to factor $Q(x)$ in the same way. Assuming that $P(x)$ has degree n, repeating this process n times gives

$$P(x) = a(x - k_1)(x - k_2) \cdots (x - k_n),$$

where a is the leading coefficient of $P(x)$. Each of these factors leads to a zero of $P(x)$, so $P(x)$ has the n zeros $k_1, k_2, k_3, \ldots, k_n$. This result can be used to prove the next theorem. The proof is left for the exercises.

ZEROS OF A POLYNOMIAL FUNCTION (I)

A function defined by a polynomial of degree n has at most n distinct complex zeros.

Notice that the statement above says that a polynomial function has *at most n* complex (real or imaginary) zeros. Some zeros may be repeated, and this will be addressed later in this section.

EXAMPLE 9

Finding All Zeros of a Polynomial Function

Find all complex zeros of the function $P(x) = x^4 - 7x^3 + 18x^2 - 22x + 12$, given that $1 - i$ is a zero.

SOLUTION This quartic function will have at most four complex zeros. Since $1 - i$ is a zero and the coefficients are real numbers, by the conjugate zeros theorem $1 + i$ is also a zero. The remaining zeros are found by first dividing the original polynomial by $x - (1 - i)$.

$$
\begin{array}{r|rrrrr}
1 - i) & 1 & -7 & 18 & -22 & 12 \\
       &   & 1 - i & -7 + 5i & 16 - 6i & -12 \\
\hline
       & 1 & -6 - i & 11 + 5i & -6 - 6i & 0
\end{array}
$$

Rather than go back to the original polynomial, divide the quotient from the first division by $x - (1 + i)$ as follows.

$$
\begin{array}{r|rrrr}
1 + i) & 1 & -6 - i & 11 + 5i & -6 - 6i \\
       &   & 1 + i & -5 - 5i & 6 + 6i \\
\hline
       & 1 & -5 & 6 & 0
\end{array}
$$

Find the zeros of the function defined by the quadratic polynomial $x^2 - 5x + 6$ by solving the equation $x^2 - 5x + 6 = 0$. By the quadratic formula or by elementary factoring, we determine the other zeros to be 2 and 3. Thus, this function has exactly four complex zeros: $1 - i$, $1 + i$, 2, and 3. ▯

Multiplicity of Zeros

Consider the polynomial function

$$P(x) = x^6 + x^5 - 5x^4 - x^3 + 8x^2 - 4x$$
$$= x(x + 2)^2 (x - 1)^3.$$

Each factor will lead to a zero of the function. The factor x leads to a *single* zero, 0, the factor $(x + 2)^2$ leads to a zero of -2 appearing *twice,* and the factor $(x - 1)^3$ leads to a zero of 1 appearing *three* times. The number of times a zero appears is referred to as the **multiplicity of the zero.** We can now state the rule for the number of zeros of a polynomial function more precisely.

> ### ZEROS OF A POLYNOMIAL FUNCTION (II)
> A function defined by a polynomial of degree n has exactly n complex zeros if zeros of multiplicity m are counted m times.

EXAMPLE **10**

Defining a
Polynomial Function
Satisfying Given
Conditions

Find a polynomial function with real coefficients of lowest possible degree having a zero 2 of multiplicity 3, a zero 0 of multiplicity 2, and a zero i of single multiplicity.

SOLUTION This polynomial function must also have a zero $-i$ of single multiplicity. (Why?) Its lowest possible degree is 7. (Why?) By the factor theorem, one possible such polynomial function in factored form is

$$P(x) = x^2(x - 2)^3(x - i)(x + i).$$

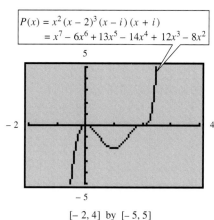

$P(x) = x^2 (x - 2)^3 (x - i) (x + i)$
$= x^7 - 6x^6 + 13x^5 - 14x^4 + 12x^3 - 8x^2$

$[-2, 4]$ by $[-5, 5]$
Xscl = 1   Yscl = 1
The graph is tangent to the x-axis at $x = 0$,
and crosses the x-axis at $x = 2$.

FIGURE 59

Multiplying the factors on the right leads to

$$P(x) = x^7 - 6x^6 + 13x^5 - 14x^4 + 12x^3 - 8x^2.$$

This is one of infinitely many such functions. Multiplying $P(x)$ by a nonzero constant will yield another function satisfying these conditions.

Graphing this function shows that it has only two distinct x-intercepts, corresponding to the real zeros 0 and 2. See Figure 59.   ◘

FOR **GROUP DISCUSSION**

Graph each of the following functions one at a time, and then respond to the items below.

$P(x) = (x + 3)(x - 2)^2$	$P(x) = (x + 3)^2(x - 2)^3$	$P(x) = x^2 (x - 1)(x + 2)^2$
Use the window	Use the window	Use the window
$[-10, 10]$ by $[-30, 30]$.	$[-4, 4]$ by $[-125, 50]$.	$[-4, 4]$ by $[-5, 5]$.

1. Describe the behavior of each graph at each x-intercept that corresponds to a zero of odd multiplicity.
2. Describe the behavior of each graph at each x-intercept that corresponds to a zero of even multiplicity.

The observations made in the preceding group discussion activity should indicate that the behavior of the graph of a polynomial function near an x-intercept depends upon the parity of multiplicity of the zero that leads to the x-intercept. If the zero is of odd multiplicity, the graph will cross the x-axis at the corresponding x-intercept. If the zero is of even multiplicity, the graph will be tangent to the x-axis at the corresponding x-intercept (that is, it will touch but not cross the x-axis). See Figure 60.

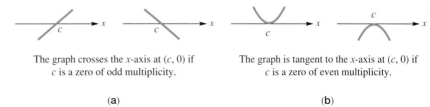

The graph crosses the x-axis at $(c, 0)$ if c is a zero of odd multiplicity.

The graph is tangent to the x-axis at $(c, 0)$ if c is a zero of even multiplicity.

(a)   (b)

FIGURE 60

By observing the dominating term and noting the parity of multiplicities of zeros of a polynomial function in factored form, we can sketch a rough graph of a polynomial function by hand. This is shown in the final example.

EXAMPLE 11

Sketching a Polynomial Function Graph by Hand

Consider the polynomial function

$$P(x) = -2x^5 - 18x^4 - 38x^3 + 42x^2 + 112x - 96,$$

with factored form

$$P(x) = -2(x + 4)^2(x + 3)(x - 1)^2.$$

Sketch the graph of P by hand, and then support the result with a calculator-generated graph.

SOLUTION   Because the dominating term is $-2x^5$, the end behavior of the graph will be ↖ ↘ . Because -4 and 1 are both x-intercepts determined by zeros of even multiplicity, the graph will be tangent to the x-axis at these x-intercepts. Because -3 is a zero of multiplicity one, the graph will cross the x-axis at $x = -3$. The y-intercept is easily determined to be -96. Combining all of this information leads to the following rough sketch of the graph. See Figure 61(a).

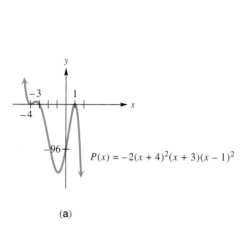

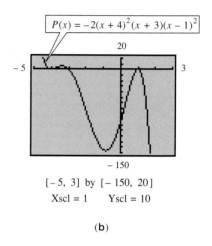

$$P(x) = -2(x + 4)^2(x + 3)(x - 1)^2$$

$[-5, 3]$ by $[-150, 20]$
Xscl = 1   Yscl = 10

(a)

(b)

FIGURE 61

Using a window of $[-5, 3]$ by $[-150, 20]$, we get the graph shown in Figure 61(b), confirming our rough sketch. (Notice that the hand-drawn method does not necessarily give a good indication of local extrema.)   ∎

3.6 EXERCISES

Use the root location theorem to show that each function has a real zero between the two numbers given. Then use your calculator to approximate the zero to the nearest hundredth.

1. $P(x) = 3x^2 - 2x - 6$;   1 and 2

2. $P(x) = x^3 + x^2 - 5x - 5$;   2 and 3

3. $P(x) = 2x^3 - 8x^2 + x + 16$;   2 and 2.5

4. $P(x) = 3x^3 + 7x^2 - 4$; $\frac{1}{2}$ and 1

5. $P(x) = 2x^4 - 4x^2 + 3x - 6$;   2 and 1.5

6. $P(x) = x^4 - 4x^3 - x + 1$;   1 and .3

7. Suppose that a polynomial function P is defined in such a way that $P(2) = -4$ and $P(2.5) = 2$. What conclusion does the root location theorem allow you to make?

8. Suppose that a polynomial function P is defined in such a way that $P(3) = -4$ and $P(4) = -10$. Can we be certain that there is no zero between 3 and 4? Explain, using a graph.

Use synthetic division to find the quotient polynomial Q(x) and the remainder R when P(x) is divided by the binomial following it.

9. $P(x) = x^3 + 2x^2 - 17x - 10$; $x + 5$

10. $P(x) = x^4 + 4x^3 + 2x^2 + 9x + 4$; $x + 4$

11. $P(x) = 3x^3 - 11x^2 - 20x + 3$; $x - 5$

12. $P(x) = x^4 - 3x^3 - 5x^2 + 2x - 16$; $x - 3$

13. $P(x) = x^4 - 3x^3 - 4x^2 + 12x$; $x - 2$

14. $P(x) = x^5 - 1$; $x - 1$

Use synthetic division to find P(k) for the given value of k and the given function P.

15. $k = 3$; $P(x) = x^2 - 4x + 5$

16. $k = -2$; $P(x) = x^2 + 5x + 6$

17. $k = -2$; $P(x) = 5x^3 + 2x^2 - x + 5$

18. $k = 2$; $P(x) = 2x^3 - 3x^2 - 5x + 4$

19. $k = 2 + i$; $P(x) = x^2 - 5x + 1$

20. $k = 3 - 2i$; $P(x) = x^2 - x + 3$

Use synthetic division to determine whether the given number is a zero of the polynomial.

21. 2; $P(x) = x^2 + 2x - 8$

22. -1; $P(x) = x^2 + 4x - 5$

23. 4; $P(x) = 2x^3 - 6x^2 - 9x + 6$

24. -4; $P(x) = 9x^3 + 39x^2 + 12x$

25. $2 + i$; $P(x) = x^2 + 3x + 4$

26. $1 - 2i$; $P(x) = x^2 - 3x + 5$

*R*elating Concepts

The close relationships among x-intercepts of a graph of a function, real zeros of the function, and real solutions of the associated equation should, by now, be apparent to you. Using the concepts presented so far in this text, consider the graph of the polynomial function $P(x) = x^3 - 2x^2 - 11x + 12$ shown below, and respond to Exercises 27–32.

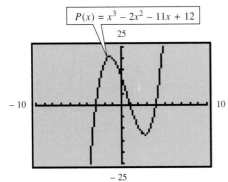

The x–intercepts are –3, 1, and 4.

27. What are the linear factors of $P(x)$?

28. What are the solutions of the equation $P(x) = 0$?

29. What are the zeros of the function P?

30. If $P(x)$ is divided by $x - 2$, what is the remainder? What is $P(2)$?

31. Give the solution set of $P(x) > 0$ using interval notation.

32. Give the solution set of $P(x) < 0$ using interval notation.

For each of the following, find a function P defined by a polynomial of degree 3 with real coefficients that satisfies the given conditions.

33. Zeros of -3, -1, and 4; $P(2) = 5$

34. Zeros of 1, -1, and 0; $P(2) = -3$

35. Zeros of -2, 1, and 0; $P(-1) = -1$

36. Zeros of 2, 5, and -3; $P(1) = -4$

For each of the following, find a polynomial function P of lowest possible degree, having real coefficients, with the given zeros.

37. 5 and -4

38. 6 and -2

39. -3, 2, and i

40. 5(multiplicity 2) and $-2i$

41. -3(multiplicity 2) and $2 + i$

42. 2(multiplicity 2) and $1 + 2i$

For each of the following, one zero is given. Find all others.

43. $P(x) = x^3 - x^2 - 4x - 6$; 3

44. $P(x) = x^3 - 5x^2 + 17x - 13$; 1

45. $P(x) = 2x^3 - 2x^2 - x - 6$; 2

46. $P(x) = 2x^3 - 5x^2 + 6x - 2$; $1 + i$

Factor P(x) into linear factors given that k is a zero of P.

47. $P(x) = 2x^3 - 3x^2 - 17x + 30$; $k = 2$

48. $P(x) = 2x^3 - 3x^2 - 5x + 6$; $k = 1$

49. $P(x) = 6x^3 + 25x^2 + 3x - 4$; $k = -4$

50. $P(x) = 8x^3 + 50x^2 + 47x - 15$; $k = -5$

51. Show that -2 is a zero of multiplicity 2 of P, where $P(x) = x^4 + 2x^3 - 7x^2 - 20x - 12$, and find all other complex zeros. Then write $P(x)$ in factored form.

52. Show that -1 is a zero of multiplicity 3 of P, where $P(x) = x^5 + 9x^4 + 33x^3 + 55x^2 + 42x + 12$, and find all other complex zeros. Then write $P(x)$ in factored form.

53. What are the possible numbers of real zeros (counting multiplicities) for a polynomial function with real coefficients of degree five?

54. Explain why a function defined by a polynomial of degree four with real coefficients has either 0, 2, or 4 real zeros (counting multiplicities).

55. Explain why it is not possible for a function defined by a polynomial of degree 3 with real coefficients to have zeros of 1, 2, and $1 + i$.

56. Suppose that k, a, b, and c are real numbers, $a \neq 0$, and a polynomial function $P(x)$ may be expressed in factored form as $(x - k)(ax^2 + bx + c)$.
 (a) What is the degree of P?
 (b) What are the possible numbers of *real* zeros of P?
 (c) What are the possible numbers of *imaginary* zeros of P?
 (d) Use the discriminant to explain how to determine the number and kind of zeros P has.

In each of the following, a polynomial function is given in both standard form and factored form. Use dominating term and multiplicity of zeros to draw by hand a rough sketch of the graph of the function. Then support your answer by using a calculator-generated graph.

57. $P(x) = 2x^3 - 5x^2 - x + 6$
$\quad = (x + 1)(2x - 3)(x - 2)$

58. $P(x) = x^3 + x^2 - 8x - 12$
$\quad = (x + 2)^2(x - 3)$

59. $P(x) = x^4 - 18x^2 + 81$
$\quad = (x - 3)^2(x + 3)^2$

60. $P(x) = x^4 - 8x^2 + 16$
$\quad = (x + 2)^2(x - 2)^2$

61. $P(x) = 2x^4 + x^3 - 6x^2 - 7x - 2$
$\quad = (2x + 1)(x - 2)(x + 1)^2$

62. $P(x) = 3x^4 - 7x^3 - 6x^2 + 12x + 8$
$\quad = (3x + 2)(x + 1)(x - 2)^2$

The following theorem, known as the *rational zeros theorem*, allows us to determine all *possible* rational zeros of polynomial functions with integer coefficients.

> ### RATIONAL ZEROS THEOREM
>
> Let $P(x) = a_n x^n + a_{n-1} x^{n-1} + \cdots + a_1 x + a_0$, where $a_n \neq 0$, define a polynomial function with integer coefficients. If p/q is a rational number written in lowest terms and if p/q is a zero of P, then p is a factor of the constant term a_0 and q is a factor of the leading coefficient a_n.

For example, suppose that we wish to find all possible rational zeros of

$$P(x) = 2x^4 - 11x^3 + 14x^2 - 11x + 12.$$

If p/q is to be a rational zero of P, by the rational zeros theorem p must be a factor of $a_0 = 12$ and q must be a factor of $a_4 = 2$. The possible values of p are ± 1, ± 2, ± 3, ± 4, ± 6, or ± 12, while q must be ± 1 or ± 2. The possible rational zeros are found by forming all possible quotients of the form p/q; any rational zero of P will come from the list

$$\pm 1, \quad \pm \frac{1}{2}, \quad \pm 2, \quad \pm 3, \quad \pm \frac{3}{2}, \quad \pm 4, \quad \pm 6, \quad \text{or} \quad \pm 12.$$

It can be shown that of the numbers in this list, only $\frac{3}{2}$ and 4 are rational zeros.

In the following exercises, use the rational zeros theorem to list all possible rational zeros of P. Then determine if any are indeed zeros of the function.

63. $P(x) = x^3 - 2x^2 - 13x - 10$

64. $P(x) = x^3 + 5x^2 + 2x - 8$

65. $P(x) = x^3 + 6x^2 - x - 30$

66. $P(x) = x^3 - x^2 - 10x - 8$

67. $P(x) = 6x^3 + 17x^2 - 31x - 12$

68. $P(x) = 15x^3 + 61x^2 + 2x - 8$

69. $P(x) = 12x^3 + 20x^2 - x - 6$

70. $P(x) = 12x^3 + 40x^2 + 41x + 12$

71. For any polynomial $P(x)$ and any complex number k, there exists a unique polynomial $Q(x)$ and number R such that

$$P(x) = (x - k) \cdot Q(x) + R.$$

This statement is known as the division algorithm. In order to prove the remainder theorem, let $x = k$ in this statement. Write out a proof of the remainder theorem.

72. Suppose that c and d represent complex numbers: $c = a + bi$ and $d = m + ni$. Let $\overline{c}$ and $\overline{d}$ represent the complex conjugates of c and d, respectively. Prove each of the following statements. (These properties will be used in Exercise 73 to prove the conjugate zeros theorem.)
  (a) $\overline{c + d} = \overline{c} + \overline{d}$ **(b)** $\overline{cd} = \overline{c} \cdot \overline{d}$
  (c) $\overline{x} = x$ for any real number x
  (d) $\overline{c^n} = (\overline{c})^n$, n is a positive integer

73. Complete the proof of the conjugate zeros theorem, outlined below. Assume that

$$P(x) = a_n x^n + a_{n-1} x^{n-1} + \cdots + a_1 x + a_0,$$

where all coefficients are real numbers.
  (a) Suppose the complex number z is a zero of P; find $P(z)$.
  (b) Take the conjugate of both sides of the result from part (a).
  (c) Use generalizations of the properties given in Exercise 72 on the result of part (b) to show that
  $$a_n (\overline{z})^n + a_{n-1} (\overline{z})^{n-1} + \cdots + a_1 (\overline{z}) + a_0 = 0.$$
  (d) Why does the result in part (c) mean that $\overline{z}$ is a zero of P?

74. The function $P(x) = x^2 - x + (i + 1)$ has i as a zero, but does not have its conjugate, $-i$, as a zero. Explain why this does not violate the conjugate zeros theorem.

*F*urther *Explorations*

The root location theorem states that for a polynomial function P, if $P(a)$ and $P(b)$ differ in sign then there exists at least one real number c between a and b such that $P(c) = 0$. We can use the TABLE and the Root Location Theorem to find the real zeros of polynomial functions. Consider $f(x) = Y_1 = x^3 - x^2 - 4x + 2$. The TABLE in the figure shows there is a sign change between $x = -2$ and $x = -1$, a second sign change between $x = 0$ and $x = 1$, and a third sign change between $x = 2$ and $x = 3$. "Zooming in" on the TABLE by decreasing ΔTbl shows that there is a zero between $x = -1.814$ and $x = -1.813$. Rounded to the nearest hundredth, there exists a zero at $x = -1.81$. Thus, $f(-1.81) \approx 0$.

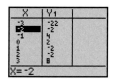 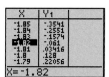 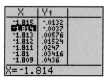

Use the TABLE to solve the following problems. Find solutions to the nearest hundredth.

1. Find the other two zeros of $f(x) = x^3 - x^2 - 4x + 2$.
2. Find the three real zeros of $f(x) = x^3 + 3x^2 - 5x - 1$.
3. Find the three real zeros of $f(x) = x^3 - 4x^2 - 5x$.

3.7 SOLUTION OF POLYNOMIAL EQUATIONS AND INEQUALITIES AND THEIR APPLICATIONS

Polynomial Equations and Inequalities ∎ Applications of Polynomial Functions ∎ Polynomial Models

While methods of solving quadratic equations were known to ancient civilizations, for hundreds of years mathematicians wrestled with finding methods of solving higher degree equations (by analytic methods, of course). It was not until the sixteenth century that progress in this area was made, and the European mathematicians Scipione del Ferro, Nicolo Fontana (a.k.a. Tartaglia), Girolamo Cardano, and François Viete were able to derive formulas allowing the solution of cubic equations. Work progressed and methods of solving quartics followed. While these methods were quite complicated, they showed that, in theory, third and fourth degree polynomial equations could be solved analytically. It was not until 1824 that the Norwegian mathematician Niels Henrik Abel proved that it is *impossible* to find a formula that will yield solutions to the general quintic (5th degree) equation. A similar result holds for polynomial functions of degree greater than five.

We can use elementary methods to solve *some* higher degree polynomial equations analytically, as we will show in this section. The technology of graphics calculators also allows us to support our analytic work, and allows us to find accurate approximations of solutions of such equations that cannot be solved easily by elementary methods or at all by analytic methods.

Polynomial Equations and Inequalities

EXAMPLE 1

Solving a Polynomial Equation and Associated Inequalities

(a) Solve the polynomial equation

$$x^3 + 3x^2 - 4x - 12 = 0$$

by using the zero-product property.

SOLUTION Since the right-hand side of the equation is 0, we begin by factoring the left side. Then set each factor equal to 0 and solve each equation.

$$x^3 + 3x^2 - 4x - 12 = 0 \qquad \text{Given equation}$$
$$(x^3 + 3x^2) + (-4x - 12) = 0 \qquad \text{Group terms with common factors.}$$
$$x^2(x + 3) - 4(x + 3) = 0 \qquad \text{Factor out common factors in each group.}$$
$$(x + 3)(x^2 - 4) = 0 \qquad \text{Factor out } x + 3.$$
$$(x + 3)(x - 2)(x + 2) = 0 \qquad \text{Factor the difference of two squares.}$$
$$x + 3 = 0 \quad \text{or} \quad x - 2 = 0 \quad \text{or} \quad x + 2 = 0 \qquad \text{Use the zero-product property. Solve.}$$
$$x = -3 \qquad\qquad x = 2 \qquad\qquad x = -2$$

The solution set is $\{-3, -2, 2\}$.

(b) Support the result of part (a) graphically.

SOLUTION   We graph $y = x^3 + 3x^2 - 4x - 12$ in the window $[-10, 10]$ by $[-15, 10]$ and see that the x-intercepts are -3, -2, and 2, supporting our analytic solution in part (a). See Figure 62.

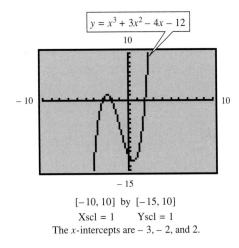

$[-10, 10]$ by $[-15, 10]$
Xscl = 1   Yscl = 1
The x-intercepts are -3, -2, and 2.

FIGURE 62

(c) Use the graph in Figure 62 to find the solution set of

$$x^3 + 3x^2 - 4x - 12 > 0.$$

SOLUTION   Recall that the solution set of $f(x) > 0$ consists of all numbers in the domain of f for which the graph lies *above* the x-axis. We see in Figure 62 that this occurs in the intervals $(-3, -2)$ and $(2, \infty)$. Therefore, the solution set of this inequality is $(-3, -2) \cup (2, \infty)$.

(d) Use the graph in Figure 62 to find the solution set of

$$x^3 + 3x^2 - 4x - 12 \le 0.$$

SOLUTION   Here we must locate the intervals where the graph lies *below or intersects* the x-axis. From the figure we find the solution set is

$$(-\infty, -3] \cup [-2, 2].$$

Notice that endpoints are included here due to the nonstrict inequality $\le$. ∎

FOR **GROUP DISCUSSION**

1. Consider the equation

$$x^3 + 3x^2 = 4x + 12.$$

   Why are the solutions of this equation the same as those in part (a) of Example 1?

2. Figure 63 shows the graph of

$$y_1 = x^3 + 3x^2 \qquad \text{and} \qquad y_2 = 4x + 12.$$

   How do we interpret the solutions of the equation in item 1 above using this graph?

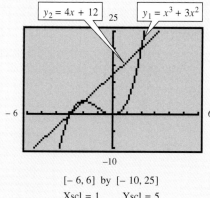

$[-6, 6]$ by $[-10, 25]$
Xscl = 1   Yscl = 5

FIGURE 63

3. Consider the inequality

$$x^3 + 3x^2 > 4x + 12.$$

   Why are the solutions of this inequality the same as those in part (c) of Example 1? How does the graph in Figure 63 support our result there?

4. Consider the inequality

$$x^3 + 3x^2 \leq 4x + 12.$$

   Why are the solutions of this inequality the same as those in part (d) of Example 1? How does the graph in Figure 63 support our result there?

EXAMPLE 2

Solving an Equation
Quadratic in Form

(a) Solve the polynomial equation

$$x^4 - 6x^2 - 40 = 0$$

analytically. Find all complex solutions.

SOLUTION This equation is said to be **quadratic in form.** Notice that if we let $t = x^2$, then the equation becomes quadratic in the variable t. Then we can solve for t using methods of solving quadratic equations, and finally go back and solve for x.

$$x^4 - 6x^2 - 40 = 0 \qquad \text{Given equation}$$
$$(x^2)^2 - 6x^2 - 40 = 0$$
$$t^2 - 6t - 40 = 0 \qquad \text{Let } t = x^2.$$
$$(t - 10)(t + 4) = 0 \qquad \text{Factor.}$$

$t = 10$	or	$t = -4$

Use the zero-product property.

$x^2 = 10$	or	$x^2 = -4$

Go back and solve for x.

$x = \pm\sqrt{10}$	or	$x = \pm 2i$

Use the square root property for solving quadratic equations.

The solution set is $\{-\sqrt{10}, \sqrt{10}, -2i, 2i\}$.

(b) Use a graph to support the real solutions of the equation in part (a).

SOLUTION We graph $y = x^4 - 6x^2 - 40$ in the window $[-4, 4]$ by $[-80, 50]$. See Figure 64. By using the root-finding feature of a calculator, we can find that the x-intercepts are approximately -3.16 and 3.16, which are approximations of $-\sqrt{10}$ and $\sqrt{10}$. Notice that the graph will not provide support for the imaginary solutions.

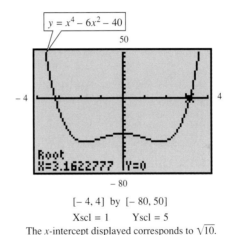

$[-4, 4]$ by $[-80, 50]$
Xscl = 1   Yscl = 5
The x-intercept displayed corresponds to $\sqrt{10}$.
The other x-intercept is $-\sqrt{10} \approx -3.1622777$.

FIGURE 64

(c) Use the graph in Figure 64 to solve the inequalities

$$x^4 - 6x^2 - 40 \geq 0 \qquad \text{and} \qquad x^4 - 6x^2 - 40 < 0.$$

Give endpoints of intervals in both exact and approximate form.

SOLUTION Since the graph lies above or intersects the x-axis for real numbers less than or equal to $-\sqrt{10}$ and for real numbers greater than or equal to $\sqrt{10}$, the solution set for $x^4 - 6x^2 - 40 \geq 0$ is

$$(-\infty, -\sqrt{10}] \cup [\sqrt{10}, \infty) \qquad \leftarrow \text{Exact form}$$

or $\qquad (-\infty, -3.16] \cup [3.16, \infty). \qquad \leftarrow \text{Approximate form}$

By similar reasoning, the solution set of $x^4 - 6x^2 - 40 < 0$ is

$$(-\sqrt{10}, \sqrt{10}) \qquad \leftarrow \text{Exact form}$$

or $\qquad (-3.16, 3.16). \qquad \leftarrow \text{Approximate form}$

(Note that the imaginary solutions do not affect the solution sets of the inequalities.)   ∎

EXAMPLE 3

Solving a
Polynomial Equation
and Associated
Inequalities

The graph of $P(x) = x^3 + 3x^2 - 11x + 2$ is shown in Figure 65.

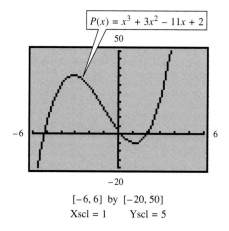

$P(x) = x^3 + 3x^2 - 11x + 2$

$[-6, 6]$ by $[-20, 50]$
Xscl = 1   Yscl = 5

FIGURE 65

(a) Explain why 2 is a real solution of the equation $x^3 + 3x^2 - 11x + 2 = 0$, and then find all solutions of this equation.

SOLUTION   We can use a calculator to confirm that 2 is an x-intercept of the graph of P, or we can show it by synthetic division:

$$
\begin{array}{r|rrrr}
2) & 1 & 3 & -11 & 2 \\
   &   & 2 & 10 & -2 \\
\hline
   & 1 & 5 & -1 & 0
\end{array}
$$

$\leftarrow P(2) = \mathbf{0}$ by the remainder theorem.

coefficients of the
quotient polynomial

By the factor theorem, $x - 2$ is a factor of $P(x)$, and

$$P(x) = (x - 2)(x^2 + 5x - 1).$$

To find the other zeros of P, we must solve

$$x^2 + 5x - 1 = 0.$$

We use the quadratic formula, with $a = 1$, $b = 5$, and $c = -1$.

$$x = \frac{-5 \pm \sqrt{5^2 - 4(1)(-1)}}{2(1)}$$

$$= \frac{-5 \pm \sqrt{29}}{2}$$

The complete solution set is $\left\{\frac{-5 - \sqrt{29}}{2}, \frac{-5 + \sqrt{29}}{2}, 2\right\}$.

(b) Support the result of part (a) graphically.

SOLUTION Using the capabilities of a calculator, we find that the x-intercepts are 2, approximately -5.19, and $.19$. Since these latter two approximations correspond to the approximations for $(-5 - \sqrt{29})/2$ and $(-5 + \sqrt{29})/2$, our analytic work is supported.

(c) Give the exact solution set of $x^3 + 3x^2 - 11x + 2 \le 0$.

SOLUTION By finding the intervals for which the graph of P lies below or intersects the x-axis, we conclude that the solution set is $\left(-\infty, \frac{-5 - \sqrt{29}}{2}\right] \cup \left[\frac{-5 + \sqrt{29}}{2}, 2\right]$. ∎

FOR **GROUP DISCUSSION**

1. Suppose that in Example 3(a), the quadratic factor had led to a negative discriminant. What kind of numbers would the other two solutions be? What special relationship would they have? How many x-intercepts would the graph of the function have?

2. Discuss how the solution in part (a) of Example 3 relates to the factor and remainder theorems, the theorem concerning the maximum number of zeros of a polynomial of degree n, and Descartes' Rule of Signs (see Exercises 61–66 in Section 3.5.)

EXAMPLE 4

Using Purely Graphical Methods to Solve a Polynomial Equation and Associated Inequalities

Let $P(x) = 2.45x^3 - 3.14x^2 - 6.99x + 2.58$. Use the graph of P to find the solution sets of

$$P(x) = 0, \qquad P(x) > 0, \qquad \text{and} \qquad P(x) < 0.$$

Express solutions of the equation and endpoints of the intervals for the inequalities to the nearest hundredth.

SOLUTION The graph of P is shown in Figure 66. Using the capabilities of a calculator, we find that the approximate x-intercepts are -1.37, $.33$, and 2.32.

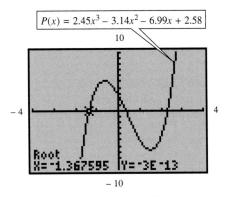

$P(x) = 2.45x^3 - 3.14x^2 - 6.99x + 2.58$

Root
X=-1.367595 Y=-3E-13

$[-4, 4]$ by $[-10, 10]$
Xscl = 1   Yscl = 1
The other two x-intercepts
are approximately .33 and 2.32.

FIGURE 66

Therefore, the solution set of the equation $P(x) = 0$ is

$$\{-1.37, .33, 2.32\}.$$

Based on the graph, the solution set of $P(x) > 0$ is

$$(-1.37, .33) \cup (2.32, \infty),$$

while that of $P(x) < 0$ is

$$(-\infty, -1.37) \cup (.33, 2.32). \quad \blacksquare$$

NOTE The graphical method of solving $P(x) = 0$ in Example 4 would not have yielded imaginary solutions had there been any. Only real solutions are obtained using this method.

If n is a positive integer and k is a nonzero complex number, then a solution of $x^n = k$ is called an **nth root of k.** For example, since -1 and 1 are solutions of $x^2 = 1$, they are called second or square roots of 1. Similarly, $-2i$ and $2i$ are called square roots of -4, since

$$(-2i)^2 = (-2)^2 i^2 = 4(-1) = -4$$

and

$$(2i)^2 = 2^2 i^2 = 4(-1) = -4.$$

The real number 2 is a sixth root of 64, since $2^6 = 64$. However, it can be shown that 64 has five more complex sixth roots. While a complete discussion of the following theorem requires concepts from trigonometry, we will state it and use it to solve particular problems involving nth roots.

COMPLEX nTH ROOTS THEOREM

If n is a positive integer and k is a nonzero complex number, then the equation $x^n = k$ has *exactly* n complex roots.

EXAMPLE 5

Finding the nth Roots of a Number

Find all six complex sixth roots of 64.

SOLUTION We must find all six complex roots of $x^6 = 64$.

$$
\begin{aligned}
x^6 &= 64 && \text{Equation to solve} \\
x^6 - 64 &= 0 && \text{Subtract 64.} \\
(x^3 - 8)(x^3 + 8) &= 0 && \text{Factor the difference of two squares.} \\
(x - 2)(x^2 + 2x + 4)(x + 2)(x^2 - 2x + 4) &= 0 && \text{Factor the difference of cubes and the sum of cubes.}
\end{aligned}
$$

Now we apply the zero-product theorem, to obtain the real roots 2 and -2. Setting the quadratic factors equal to zero and applying the quadratic formula twice gives us the remaining four complex roots (all imaginary).

$$x^2 + 2x + 4 = 0 \qquad \text{or} \qquad x^2 - 2x + 4 = 0$$

$$x = \frac{-2 \pm \sqrt{2^2 - 4(1)(4)}}{2(1)} \qquad\qquad x = \frac{2 \pm \sqrt{(-2)^2 - 4(1)(4)}}{2(1)}$$

$$= \frac{-2 \pm \sqrt{-12}}{2} \qquad\qquad = \frac{2 \pm \sqrt{-12}}{2}$$

$$= \frac{-2 \pm 2i\sqrt{3}}{2} \qquad\qquad = \frac{2 \pm 2i\sqrt{3}}{2}$$

$$= -1 \pm i\sqrt{3} \qquad\qquad = 1 \pm i\sqrt{3}$$

Therefore, the six complex sixth roots of 64 are

$$2, \quad -2, \quad -1 + i\sqrt{3}, \quad -1 - i\sqrt{3}, \quad 1 + i\sqrt{3}, \quad \text{and} \quad 1 - i\sqrt{3}.$$

The graph of $y = x^6 - 64$ confirms the two real sixth roots of 64. See Figure 67. The x-intercepts are -2 and 2. ∎

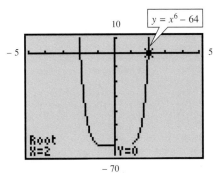

$[-5, 5]$ by $[-70, 10]$
Xscl = 1   Yscl = 10

The other x-intercept is -2. The two real
sixth roots of 64 are -2 and 2.

FIGURE 67

FOR GROUP DISCUSSION

Suppose that k is a positive number.

1. How many real nth roots will k have if n is even? (*Hint:* Think of the general shape of the graph of the function $y = x^n - k$ for n even, k positive.)
2. How many real nth roots will k have if n is odd?

Applications of Polynomial Functions

In Section 3.4 we saw how one class of polynomial functions, quadratics, can be used to solve certain types of applied problems. We will now see how higher degree polynomial functions can be used similarly.

EXAMPLE **6**

Using a Polynomial
Function to
Describe the
Volume of a Box

A box with an open top is to be constructed from a rectangular 12-inch by 20-inch piece of cardboard by cutting equal size squares from each corner and folding up the sides.

(a) If x represents the length of the side of each such square, determine a function V that describes the volume of the box in terms of x.

SOLUTION   As shown in Figure 68, the dimensions of the box to be formed will be

$$\text{length} = 20 - 2x$$
$$\text{width} = 12 - 2x$$
$$\text{height} = x. \qquad \text{All in inches}$$

Furthermore, x must be positive, and both $20 - 2x$ and $12 - 2x$ must be positive, implying that $0 < x < 6$. Since the volume of the box can be found by multiplying length times width times height, the desired function is

$$V(x) = (20 - 2x)(12 - 2x)x, \quad 0 < x < 6.$$
$$= 4x^3 - 64x^2 + 240x.$$

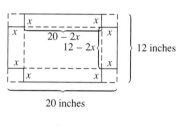

(a)

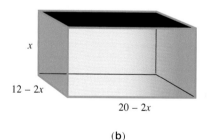

(b)

FIGURE 68

(b) Graph V in the window $[0, 6]$ by $[0, 300]$ and locate a point on the graph. Interpret the displayed values of x and y.

SOLUTION   Figure 69 shows the graph of V with the cursor at the arbitrarily chosen point $(3.6, 221.184)$. This means that when the side of each cut out square measures 3.6 inches, the volume of the resulting box is 221.184 cubic inches.

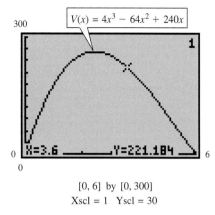

$$[0, 6] \text{ by } [0, 300]$$
$$\text{Xscl} = 1 \quad \text{Yscl} = 30$$

FIGURE 69

(c)   Determine the value of x for which the volume of the box is maximized. What is this volume?

SOLUTION   We use the capabilities of the calculator to find the local maximum point on the graph of V. To the nearest hundredth, the coordinates of this point are $(2.43, 262.68)$. Therefore, when $x \approx 2.43$ is the length of the side of each square, the volume of the box is at its maximum, approximately 262.68 cubic inches.

(d)   For what values of x is the volume equal to 200 cubic units? greater than 200 cubic units? less than 200 cubic units?

SOLUTION   Graphing $y_1 = V(x)$ and $y_2 = 200$ gives the graphs shown in Figure 70. The points of intersection of the line and the cubic curve are approximately $(1.17, 200)$ and $(3.90, 200)$. Using the intersection-of-graphs method of solving equations and inequalities, the volume is equal to 200 cubic units for $x \approx 1.17$ or 3.90, is greater than 200 cubic units for $1.17 < x < 3.90$, and is less than 200 cubic units for $0 < x < 1.17$ or $3.90 < x < 6$. (To solve the inequalities, here we use the idea of one graph above or below another.)   ∎

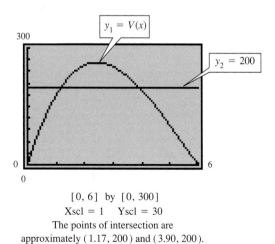

$$[0, 6] \text{ by } [0, 300]$$
$$\text{Xscl} = 1 \quad \text{Yscl} = 30$$

The points of intersection are
approximately $(1.17, 200)$ and $(3.90, 200)$.

FIGURE 70

FOR **GROUP DISCUSSION**

Refer to Example 6 to answer the following.

1. While we can enter $y_1 = (20 - 2x)(12 - 2x)x$ or $y_1 = 4x^3 - 64x^2 + 240x$ to obtain the desired graph, is it really necessary for us to multiply out the factors when using a graphing technique for solving? What would be a good reason for *not* actually performing the multiplication?

2. Figure 71 shows a complete graph of V in the window $[-5, 20]$ by $[-300, 300]$. Explain why a complete graph was not necessary in solving the problem.

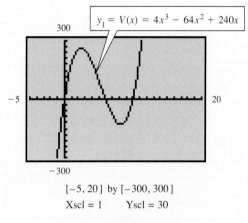

$[-5, 20]$ by $[-300, 300]$
Xscl = 1   Yscl = 30

FIGURE 71

3. Graph $y_3 = y_1 - 200$ in the window $[0, 6]$ by $[-50, 100]$ and explain how part (d) could be solved using the graph of y_3.

Polynomial Models

We saw in Chapter 1 that data can sometimes be fit to linear models, while earlier in this chapter we saw how quadratic functions can be used to model data. Higher degree polynomial functions may be used as well, and modern graphics calculators provide in their statistical functions the capability for determining many types of models. The final example shows how a cubic function can be used to model data.

EXAMPLE **7**

Analyzing a Cubic Polynomial Model

Based on information provided by the U.S. National Institute on Drug Abuse, the percent y of 18- to 25-year-olds who had used hallucinogens between 1974 and 1991 can be modeled by the function

$$y = .025x^3 - .70x^2 + 4.43x + 16.77$$

where $x = 0$ corresponds to the year 1974. In what year during this period did this type of drug use reach its maximum? Based on this model, what percent of 18- to 25-year-olds had used them?

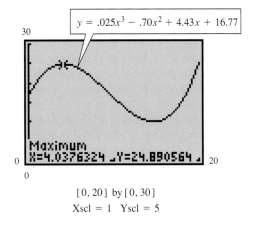

$$y = .025x^3 - .70x^2 + 4.43x + 16.77$$

[0, 20] by [0, 30]
Xscl = 1   Yscl = 5

FIGURE 72

SOLUTION The graph of this function in the window [0, 20] by [0, 30] is shown in Figure 72. We can use the capabilities of the calculator to find that the local maximum point is approximately (4.04, 24.89), meaning that during the fourth year after 1974 (that is, 1978), such drug use reached its maximum, with almost 25% of persons in this age group having reported use of them. ▯

FOR **GROUP DISCUSSION**

Use the graph in Figure 72 to describe the trends that occurred during the period. Relate the real-life trends to the concepts of increasing and decreasing functions.

3.7 EXERCISES

Solve each equation analytically for all complex solutions, giving exact *form in your solution set. Then graph the left-hand side of the equation as y_1 in the suggested viewing window, and support the real solutions using the capabilities of the calculator.*

1. $4x^4 - 25x^2 + 36 = 0$; [−5, 5] by [−5, 100]

2. $4x^4 - 29x^2 + 25 = 0$; [−5, 5] by [−50, 100]

3. $x^4 - 15x^2 - 16 = 0$; [−5, 5] by [−100, 100]

4. $9x^4 + 35x^2 - 4 = 0$; [−3, 3] by [−10, 100]

5. $x^3 - x^2 - 64x + 64 = 0$; [−10, 10] by [−300, 300]

6. $x^3 + 6x^2 - 100x - 600 = 0$; [−15, 15] by [−1000, 300]

7. $-2x^3 - x^2 + 3x = 0$; [−4, 4] by [−10, 10]

8. $-5x^3 + 13x^2 + 6x = 0$; [−4, 4] by [−2, 30]

9. $x^3 + x^2 - 7x - 7 = 0$; [−10, 10] by [−20, 20]

10. $x^3 + 3x^2 - 19x - 57 = 0$; [−10, 10] by [−100, 50]

11. $3x^3 + x^2 - 6x = 0$; $[-4, 4]$ by $[-10, 10]$

12. $-4x^3 - x^2 + 4x = 0$; $[-4, 4]$ by $[-10, 10]$

13. $3x^3 + 3x^2 + 3x = 0$; $[-5, 5]$ by $[-5, 5]$

14. $2x^3 + 2x^2 + 12x = 0$; $[-10, 10]$ by $[-20, 20]$

15. $x^4 + 17x^2 + 16 = 0$; $[-4, 4]$ by $[-10, 40]$

16. $36x^4 + 85x^2 + 9 = 0$; $[-4, 4]$ by $[-10, 40]$

17. $x^6 + 19x^3 - 216 = 0$; $[-4, 4]$ by $[-350, 200]$

18. $8x^6 + 7x^3 - 1 = 0$; $[-4, 4]$ by $[-5, 100]$

19. $3x^4 - 12x^2 + 1 = 0$; $[-10, 10]$ by $[-15, 10]$

20. $4x^4 - 13x^2 + 2 = 0$; $[-10, 10]$ by $[-10, 10]$

Relating Concepts

The graph of $y = x^4 - 28x^2 + 75$ is shown in the window $[-6, 0]$ by $[-150, 100]$ in the figure.

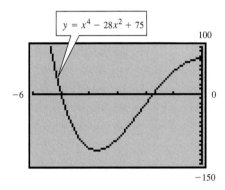

21. Based on the discussion of complete graphs of polynomial functions in Section 3.5, is this a complete graph of the function? Why or why not?

22. If this function is graphed over the domain of all real numbers, what symmetry is exhibited? (Recall the concepts of symmetry introduced in Section 2.1.)

23. The *x*-intercepts for the window shown are -5 and $-\sqrt{3}$. What is the *complete* solution set of $x^4 - 28x^2 + 75 = 0$?

24. Does the equation $x^4 - 28x^2 + 75 = 0$ have any imaginary solutions? Explain your answer.

In Exercises 25–30, a polynomial P(x) is given in both descending powers of the variable and factored form. Graph the polynomial by hand, as shown in Section 3.6, or in the window suggested, and solve the equation and the inequalities given.

25. $P(x) = x^3 - 3x^2 - 6x + 8$
$= (x - 4)(x - 1)(x + 2)$
Window: $[-10, 10]$ by $[-15, 15]$
(a) $P(x) = 0$
(b) $P(x) < 0$
(c) $P(x) > 0$

26. $P(x) = x^3 + 4x^2 - 11x - 30$
$= (x - 3)(x + 2)(x + 5)$
Window: $[-10, 10]$ by $[-40, 40]$
(a) $P(x) = 0$
(b) $P(x) < 0$
(c) $P(x) > 0$

27. $P(x) = 2x^4 - 9x^3 - 5x^2 + 57x - 45$
$= (x - 3)^2(2x + 5)(x - 1)$
Window: $[-5, 5]$ by $[-120, 50]$
(a) $P(x) = 0$
(b) $P(x) < 0$
(c) $P(x) > 0$

28. $P(x) = 4x^4 + 27x^3 - 42x^2 - 445x - 300$
$= (x + 5)^2(4x + 3)(x - 4)$
Window: $[-10, 10]$ by $[-1200, 400]$
(a) $P(x) = 0$
(b) $P(x) < 0$
(c) $P(x) > 0$

29. $P(x) = -x^4 - 4x^3 + 3x^2 + 18x$
$= x(2 - x)(x + 3)^2$
Window: $[-5, 5]$ by $[-30, 30]$
(a) $P(x) = 0$
(b) $P(x) \geq 0$
(c) $P(x) \leq 0$

30. $P(x) = -x^4 + 2x^3 + 8x^2$
$= x^2(4 - x)(x + 2)$
Window: $[-6, 6]$ by $[-10, 50]$
(a) $P(x) = 0$
(b) $P(x) \geq 0$
(c) $P(x) \leq 0$

Prior to the 1970s, courses in the theory of equations were taught at the undergraduate level and formed part of a typical mathematics major's curriculum. Such courses dealt with various algebraic techniques for solving cubic and quartic equations, as well as other topics dealing with analysis of polynomial functions. In Exercises 31–34, you are given equations with one of the real roots in its exact value, as determined in the classic text Theory of Equations *by J. V. Uspensky (New York: McGraw-Hill, 1948). Graph the function on the left-hand side of the equation as y_1 in the suggested window, and then use the capabilities of your calculator to locate the x-intercept whose approximation corresponds to that of the exact root given. Give the approximation of the root to as many places as your calculator will provide.*

31. $x^3 - 6x - 6 = 0$; Window: $[-6, 6]$ by $[-20, 10]$
Exact root: $\sqrt[3]{2} + \sqrt[3]{4}$

32. $x^3 - 12x - 34 = 0$; Window: $[-6, 6]$ by $[-100, 20]$
Exact root: $\sqrt[3]{2} + 2\sqrt[3]{4}$

33. $x^3 + 9x - 2 = 0$; Window: $[-5, 5]$ by $[-20, 20]$
Exact root: $\sqrt[3]{\sqrt{28} + 1} - \sqrt[3]{\sqrt{28} - 1}$

34. $x^4 + 5x^3 + x^2 - 13x + 6 = 0$; Window: $[-6, 6]$ by $[-5, 50]$
Exact root: $\dfrac{-3 + \sqrt{17}}{2}$

35. One of the most interesting stories in the history of mathematics involves the dispute between Nicolo Fontana (Tartaglia) and Girolamo Cardano, two sixteenth century Italian mathematicians. The source of the dispute was the origin of a formula for solving for one root of a cubic equation of the form $x^3 + mx = n$. The formula is

$$x = \sqrt[3]{\frac{n}{2} + \sqrt{\left(\frac{n}{2}\right)^2 + \left(\frac{m}{3}\right)^3}} - \sqrt[3]{-\left(\frac{n}{2}\right) + \sqrt{\left(\frac{n}{2}\right)^2 + \left(\frac{m}{3}\right)^3}}$$

(a) Solve the equation $x^3 + 9x = 26$ for its single real root using this formula.
(b) Support your answer in part (a) by finding the x-intercept of the graph of $y = x^3 + 9x - 26$.
(c) Find the two imaginary roots by synthetically dividing $x^3 + 9x - 26$ by $x - k$, where k is the real root, and then solving the equation $Q(x) = 0$, where $Q(x)$ is the quadratic quotient polynomial.

36. A method for solving fourth degree polynomial equations is described in Rees and Sparks, *College Algebra,* 5th Edition. (New York: McGraw-Hill, 1967.) As an example, the equation

$$x^4 + 2x^3 - x^2 + x + \frac{1}{4} = 0$$

is solved and shown to have the four solutions

$$\underbrace{\frac{-1 + \sqrt{3} \pm \sqrt{2}\sqrt{1 - \sqrt{3}}}{2}}_{\text{imaginary}}, \quad \underbrace{\frac{-1 - \sqrt{3} \pm \sqrt{2}\sqrt{1 + \sqrt{3}}}{2}}_{\text{real}}.$$

Use a calculator to support the two real solutions graphically.

Use strictly graphical methods (either intersection-of-graphs method or x-intercept method) to find all real solutions of the following equations. Express solutions rounded to the nearest hundredth.

37. $.86x^3 - 5.24x^2 + 3.55x + 7.84 = 0$ **38.** $-2.47x^3 - 6.58x^2 - 3.33x + .14 = 0$

39. $-\sqrt{7}x^3 + \sqrt{5}x^2 + \sqrt{17} = 0$ **40.** $\sqrt{10}x^3 - \sqrt{11}x - \sqrt{8} = 0$

41. $2.45x^4 - 3.22x^3 = -.47x^2 + 6.54x + 3$ **42.** $\sqrt{17}x^4 - \sqrt{22}x^2 = -1$

Find all n complex solutions of each of the following equations of the form $x^n = k$.

43. $x^2 = -1$ **44.** $x^2 = -4$ **45.** $x^3 = -1$

46. $x^3 = -8$ **47.** $x^3 = 27$ **48.** $x^3 = 64$

49. $x^4 = 16$ **50.** $x^4 = 81$ **51.** $x^6 = 1$

52. Consider the equation $x^8 = 1$. This equation has eight distinct solutions, each of which is an eighth root of 1. (Roots of 1 are often called *roots of unity.*)
    (a) Graph $y = x^8 - 1$ to determine the two real eighth roots of unity.
    (b) Show analytically that i is also an eighth root of unity.
    (c) Based on the conjugate zeros theorem and the result of part (b), what must be another eighth root of unity? Verify that this is a root analytically.
    (d) Using concepts from trigonometry, it can be shown that

$$\frac{\sqrt{2}}{2} + i\frac{\sqrt{2}}{2} \quad \text{and} \quad -\frac{\sqrt{2}}{2} + i\frac{\sqrt{2}}{2}$$

are imaginary eighth roots of unity. Based on the conjugate zeros theorem, what two other imaginary numbers must also be eighth roots of unity?
    (e) List all eight eighth roots of unity.

Solve each of the following problems. Use a graphical method to find numerical answers, and give approximations to the nearest hundredth.

53. A rectangular piece of cardboard measuring 12 inches by 18 inches is to be made into a box with an open top by cutting equal size squares from each corner and folding up the sides. Let x represent the length of a side of each such square.
    (a) Give the restrictions on x.
    (b) Describe a function V that gives the volume of the box as a function of x.
    (c) For what value of x will the volume be a maximum? What is this maximum volume?
    (d) For what values of x will the volume be greater than 80 cubic inches?

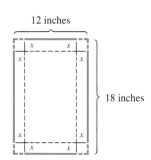

54. A piece of rectangular sheet metal is 20 inches wide. It is to be made into a rain gutter by turning up the edges to form parallel sides. Let x represent the length of each of the parallel sides.

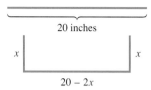

20 inches

x x

20 − 2x

(a) Give the restrictions on x.
(b) Describe a function A that gives the area of a cross section of the gutter.
(c) For what value of x will A be a maximum (and thus maximize the amount of water that the gutter will hold)? What is this maximum area?
(d) For what values of x will the area of a cross section be less than 40 square inches?

55. It has been determined that a spherical object of radius 4 inches with specific gravity .25 will sink in water to a depth of x inches, where x is a positive root of the equation $x^3 - 12x^2 + 64 = 0$. To what depth will this object sink given that $x < 10$?

56. Find the value of x in the figure that will maximize the area of rectangle $ABCD$ shown at the top of the next column.

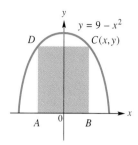

57. A certain right triangle has an area of 84 square inches. One leg of the triangle measures 1 inch less than the hypotenuse. Let x represent the length of the hypotenuse.

(a) Express the length of the leg mentioned above in terms of x.
(b) Express the length of the other leg in terms of x.
(c) Write an equation based on the information determined thus far. Square both sides and then write the equation with one side as a polynomial with integer coefficients, in descending powers, and the other side equal to 0.
(d) Solve the equation in part (c) graphically. Find the lengths of the three sides of the triangle.

58. A storage tank for butane gas is to be built in the shape of a right circular cylinder of altitude 12 feet, with a half sphere attached to each end. If x represents the radius of each half sphere, what radius should be used to cause the volume of the tank to be 144π cubic feet?

Solve each problem involving a polynomial model.

59. The number of military personnel on active duty in the United States during the period 1985 to 1990 can be determined by the cubic model $y = -7.66x^3 + 52.71x^2 - 93.43x + 2151$, where $x = 0$ corresponds to 1985, and y is in thousands. Based on this model, how many military personnel were on active duty in 1990? (*Source:* U.S. Department of Defense)

60. A technique for measuring cardiac output depends on the concentration of a dye in the bloodstream after a known amount is injected into a vein near the heart. For a normal heart, the concentration of dye in the bloodstream at time x (in seconds) is given by

the function defined as follows.

$$g(x) = -.006x^4 + .140x^3 - .053x^2 + 1.79x$$

What is the concentration after 20 seconds?

61. The polynomial function defined by

$$A(x) = -.015x^3 + 1.058x$$

gives the approximate alcohol concentration (in tenths of a percent) in an average person's bloodstream x hours after drinking about 8 oz of 100-proof whiskey. The function is approximately valid for x in the interval $[0, 8]$. Use a graph to estimate the time of maximum alcohol concentration.

F*urther Explorations*

1. A standard piece of notebook paper measuring 8.5 inches by 11 inches is to be made into a box with an open top by cutting equal size squares from each corner and folding up the sides. Let x represent the length of a side of each such square.
(a) Use the TABLE feature of your graphics calculator to find the maximum volume of the box.
(b) Use the TABLE feature to determine when the volume of the box will be greater than 40 cubic inches. (*Hint:* Let $Y_2 = 40$.)

C*hapter 3*   **SUMMARY**

The complex number system includes the real numbers and the imaginary numbers, numbers of the form $a + bi$, where $i = \sqrt{-1}$, so that $i^2 = -1$. Also, if a is positive, then $\sqrt{-a} = i\sqrt{a}$. With these definitions, complex numbers can be added, subtracted, and multiplied as though they were polynomials. The quotient of two complex numbers is found by multiplying numerator and denominator by the conjugate of the denominator.

Quadratic functions (second-degree polynomial functions) have graphs that can be obtained by stretching or shrinking, reflecting, or translating the graph of $f(x) = x^2$. To determine which of these changes apply to a particular function, equations in the form $P(x) = ax^2 + bx + c$ can be changed to the form $P(x) = a(x - h)^2 + k$ by completing the square. The vertex is (h, k); the graph opens upward if a is positive and downward if a is negative; the graph is broader than that of $y = x^2$ if $0 < |a| < 1$ and narrower if $|a| > 1$. The vertex can also be found using the formula $(h, k) = (-\frac{b}{2a}, f(-\frac{b}{2a}))$.

The zero-product property is used to find the zeros of polynomial functions $P(x)$ that can be factored by setting each factor equal to zero and solving for the variable. These values of the variable are also the x-intercepts and form the solution set of $P(x) = 0$.

Local extrema of a function are the maximum or minimum values of the function in some region. Some functions also have an absolute maximum and/or minimum. All quadratic functions have either a maximum or a minimum value at the vertex of the graph. The end behavior of the graph of a function is determined by the coefficient of the term with the highest power of the variable. In a quadratic function, if $a > 0$, both ends of the curve approach positive infinity; if $a < 0$, both ends approach negative infinity. The graph is concave upward in an interval if it "holds water"; it is concave downward if it "dispels water."

By the square root property the solution of an equation of the form $x^2 = k$ is $x = \sqrt{k}$ or $x = -\sqrt{k}$. If k is negative, the solutions are imaginary numbers. By the quadratic formula, the solution of any quadratic equation $ax^2 + bx + c = 0$ is

$$x = \frac{-b \pm \sqrt{b^2 - 4ac}}{2a}.$$

If the discriminant (the quantity $b^2 - 4ac$) is positive, there are two real solutions. If the discriminant is 0, there is one real solution, and there are two imaginary solutions if the discriminant is negative. Quadratic inequalities are solved analytically using a sign graph to determine the regions on the number line where the function is positive or negative. The endpoints of the regions are determined by the zeros of the function. These regions are found graphically as they were for linear functions in earlier work.

Optimization problems, which require finding the maximum or minimum value of functions are often modeled with quadratic functions. Other important applications of quadratic functions deal with the height of a propelled object as a function of time.

In general, polynomial functions have domain $(-\infty, \infty)$ and the graph is a smooth, continuous curve with no sharp turns. The graphs of polynomial functions may have both local extrema and absolute extrema. The maximum number of local extrema of a polynomial function of degree n is $n - 1$. A polynomial function of degree n will have at most n distinct zeros (or x-intercepts). A complete graph of a polynomial function should include all x-intercepts, the y-intercept, all extreme points, and the correct end behavior.

A cubic function (a third-degree polynomial function) is typical of polynomial functions of odd degree. The end behavior will be upward at one end and downward at the other, depending on the sign of the coefficient of the third degree term. The range is $(-\infty, \infty)$.

A quartic function (a fourth-degree polynomial function) is typical of polynomial functions of even degree. The end behavior will be upward at both ends or downward at both ends. The range will have the form $[k, \infty)$ or $(-\infty, k]$ and k will be an absolute maximum or minimum value.

The root location theorem is used to identify intervals where zeros of polynomials are located. If $P(a)$ and $P(b)$ differ in sign, there is at least one zero between a and b. Synthetic division is a short cut form of division of polynomials that is used when dividing by a binomial. This type of division is useful for applying the closely related remainder and factor theorems. If a polynomial is divided by $x - k$, the remainder is $P(k)$, and the polynomial $x - k$ is a factor of $P(x)$ if and only if $P(k) = 0$. The conjugate zeros theorem states that if $P(x)$ has only real coefficients and if $a + bi$ is a zero of P, then $a - bi$ is also a zero of P. These theorems are used to define a polynomial function that satisfies given conditions. By the fundamental theorem of algebra, every polynomial function of degree 1 or more has at least one complex zero. Furthermore, every polynomial function of degree n has at most n distinct complex zeros. The number of times a zero appears is the multiplicity of the zero, so a polynomial function of degree n has exactly n complex zeros if zeros of multiplicity m are counted m times.

Polynomial equations of the form $P(x) = 0$, where $P(x)$ can be factored, are solved with the zero-product property. A polynomial that is quadratic in form can be written as a quadratic equation with a suitable substitution for the variable. This type of equation can be solved by factoring or the quadratic formula. Most polynomials cannot easily be solved analytically and we rely on the graph to find solutions. Polynomial inequalities are also solved graphically. By the complex nth roots theorem, a generalization of the square root property, the equation $x^n = k$ ($k \neq 0$) has exactly n complex roots.

Polynomial functions are often appropriate models for many data sets. Other applications may involve polynomials to express volume.

Key Terms

SECTION 3.1

complex numbers
imaginary number
standard form of a complex number
additive identity for complex numbers
additive inverse of a complex number
complex conjugates

SECTION 3.2

polynomial function
leading coefficient
constant
zero of a function
quadratic function
axis of symmetry of a parabola
extreme point (extremum, extrema)
extreme value
minimum point
minimum value
maximum point
maximum value
end behavior
concavity
concave up
concave down

SECTION 3.3

quadratic equation in standard form
quadratic formula
discriminant
quadratic inequality
sign graph

SECTION 3.5

parity
dominating term
local minimum
local maximum
absolute maximum
absolute minimum
cubic function
quartic function

SECTION 3.6

synthetic division
multiplicity of a zero

SECTION 3.7

polynomial equation
polynomial inequality
quadratic in form
nth root

Chapter 3   REVIEW EXERCISES

Let $w = 17 - i$ and let $z = 1 - 3i$. Write each complex number in $a + bi$ form.

1. $w + z$ **2.** $w - z$ **3.** wz

4. w^2 **5.** $\dfrac{1}{z}$ **6.** $\dfrac{w}{z}$

Consider the function $P(x) = 2x^2 - 6x - 8$ for Exercises 7–17.

7. What is the domain of P?

8. Determine analytically the coordinates of the vertex of the graph.

9. Use an end behavior diagram to describe the end behavior of the graph of P.

10. Determine analytically the x-intercepts, if any, of the graph of P.

11. Determine analytically the y-intercept of the graph of P.

12. What is the range of P?

13. Over what interval is the function increasing? Over what interval is it decreasing?

14. Give the solution set of each of the following:
    (a) $2x^2 - 6x - 8 = 0$ **(b)** $2x^2 - 6x - 8 > 0$ **(c)** $2x^2 - 6x - 8 \leq 0$.

15. The graph of P is shown here. Explain how the graph supports your solution sets in Exercise 14.

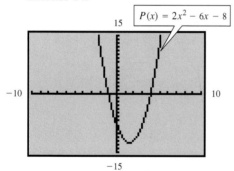

$P(x) = 2x^2 - 6x - 8$

16. What is the equation of the axis of symmetry of the graph in Exercise 15?

17. Discuss the concavity of the graph in Exercise 15.

Consider the function $P(x) = -2.64x^2 + 5.47x + 3.54$ for Exercises 18–22.

18. Use the discriminant to explain how you can determine the number of x-intercepts the graph of P will have even before graphing it on your calculator.

19. Graph the function in the standard window of your calculator, and use the root-finding capabilities to solve the equation $P(x) = 0$. Express solutions as approximations to the nearest hundredth.

20. Use your answer to Exercise 19 and the graph of P to solve **(a)** $P(x) > 0$ and **(b)** $P(x) < 0$.

21. Use the capabilities of your calculator to find the coordinates of the vertex of the graph. Express coordinates to the nearest hundredth.

22. Verify *analytically* that your answer in Exercise 21 is correct.

23. A piece of cardboard is 3 times as long as it is wide. Equal size squares measuring 4 inches on each side are to be cut from the corners of the piece of cardboard, and the flaps formed will be folded up to form a box with an open top.
    (a) Determine a function V that will describe the volume of the box as a function of x, where x is the original width in inches.
    (b) What should be the original dimensions of the piece of cardboard if the box is to have a volume of 2496 cubic inches? Solve this problem analytically.
    (c) Support the answer in part (b) graphically.

Consider the function $P(x) = x^3 - 2x^2 - 4x + 3$ for Exercises 24–28.

24. Graph the function in the window $[-10, 10]$ by $[-10, 10]$, giving a complete graph. Based on the graph, how many real solutions does the equation $x^3 - 2x^2 - 4x + 3 = 0$ have? Then use the root-finding capabilities of your calculator to find the real root that is an integer.

25. Use your answer in Exercise 24 along with synthetic division to factor $x^3 - 2x^2 - 4x + 3$ so that one factor is linear and the other factor is quadratic.

26. Find the exact values of any remaining zeros of P analytically.

27. Use the root-finding capabilities of your calculator to support your answer in Exercise 26.

28. Give the solution set of each inequality, using *exact* values:
(a) $x^3 - 2x^2 - 4x + 3 > 0$
(b) $x^3 - 2x^2 - 4x + 3 \leq 0$

29. Use an analytic method to find all solutions of the equation $x^3 + 2x^2 + 5x = 0$. Then without graphing, give the exact values of all *x*-intercepts of the graph. Using your knowledge of end behavior of the graph of $P(x) = x^3 + 2x^2 + 5x$, give the solution set of $P(x) > 0$ and of $P(x) < 0$.

30. The graph of $P(x) = x^4 - 5x^3 + x^2 + 21x - 18$ is shown here. Suppose that you know that all zeros of P are integers, and each linear factor is of degree 1 or 2. Give the factored form of $P(x)$. (*Hint:* Once you have determined your answer, graph the factored form to see if it matches the one shown here.)

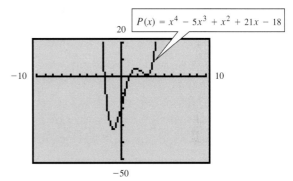

The *x*-intercepts are
−2, 1, and 3.

Complete graphs of polynomial functions f and g are shown here. They have only real coefficients. Answer Exercises 31–39 based on the graphs.

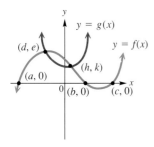

31. Is the degree of *g* even or odd?

32. Is the degree of *f* even or odd?

33. Is the leading coefficient of *f* positive or negative?

34. How many real solutions does $g(x) = 0$ have?

35. Express the solution set of $f(x) < 0$ in interval form.

36. What is the solution set of $f(x) > g(x)$?

37. What is the solution set of $f(x) - g(x) = 0$?

38. If $r + pi$ is an imaginary solution of $g(x) = 0$, what must be another imaginary solution?

39. Suppose that *f* is of degree 3. Explain why *f* cannot have imaginary zeros.

A projectile is fired vertically upward and its height s(t) in feet after t seconds is given by the function

$$s(t) = -16t^2 + 800t + 600.$$

Graph this function in the window [0, 60] *by* [0, 11,000] *and use either analytic or graphical methods to answer Exercises 40–44.*

40. From what height was the projectile fired?

41. After how many seconds will it reach its maximum height?

42. What is the maximum height it will reach?

43. Between what two times (in seconds, to the nearest tenth) will it be more than 5000 feet above the ground?

44. How long will the projectile be in the air? Give your answer to the nearest tenth of a second.

Answer true *or* false *to each statement in Exercises 45–50.*

45. The function $f(x) = 3x^7 - 8x^6 + 9x^5 + 12x^4 - 18x^3 + 26x^2 - x + 500$ has 8 x-intercepts.

46. The function f in Exercise 45 may have up to 6 local extrema.

47. The function f in Exercise 45 has a positive y-intercept.

48. Based on end behavior of the function f in Exercise 45 and your answer in Exercise 47, the graph must have at least one negative x-intercept.

49. If a polynomial function of even degree has a positive leading coefficient and a negative y-intercept, it must have at least two real zeros.

50. Because $-\dfrac{1}{2} + i\dfrac{\sqrt{3}}{2}$ is a complex zero of $f(x) = x^2 + x + 1$, another zero must be $\dfrac{1}{2} + i\dfrac{\sqrt{3}}{2}$.

Graph the function $P(x) = -2x^5 + 15x^4 - 21x^3 - 32x^2 + 60x$ *in the window* [-8, 8] *by* [-100, 200] *to obtain a complete graph. Then use your calculator and the concepts of this chapter to answer Exercises 51–55.*

51. How many local maxima does this function have?

52. One local minimum lies on the x-axis and has an integer as its x-value. What are the coordinates of this point?

53. The greatest x-intercept is 5. Therefore, $x - 5$ is a factor of $P(x)$. Use synthetic division to find the quotient polynomial $Q(x)$ obtained when $P(x)$ is divided by $x - 5$.

54. What is the range of P?

55. The graph of P has a local minimum with a negative x-value. Use your calculator to find its coordinates. Express them to the nearest hundredth.

56. Solve the equation $3x^3 + 2x^2 - 21x - 14 = 0$ analytically for all complex solutions, giving exact values in your solution set. Then graph the left-hand side of the equation as y_1 in the viewing window [-6, 6] by [-35, 35], and support the real solutions using the capabilities of the calculator.

57. Use the results of Exercise 56 to give the solution set of each inequality, expressing all values in exact form.
(a) $3x^3 + 2x^2 - 21x - 14 \geq 0$ (b) $3x^3 + 2x^2 - 21x - 14 < 0$

58. Consider the polynomial function $P(x) = -x^4 + 3x^3 + 3x^2 + 17x - 6$. The factored form of the polynomial is $(-x + 2)(x - 3)(x + 1)^2$. Graph the polynomial by hand as explained in Section 3.6, and solve the equation or inequality:
(a) $P(x) = 0$ (b) $P(x) > 0$ (c) $P(x) < 0$.

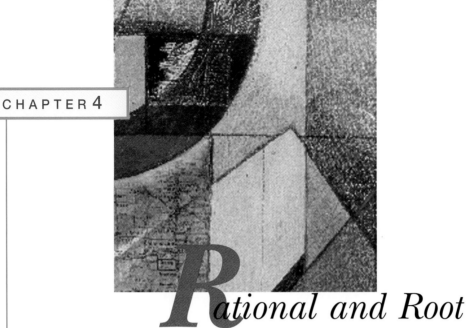

*R*ational and Root Functions

4.1 GRAPHS OF RATIONAL FUNCTIONS

Introduction to Rational Functions ▌ Graphics Calculator Considerations for Rational Functions ▌ Graphs of Simple Rational Functions ▌ Determination of Asymptotes ▌ Graphs of More Complicated Rational Functions ▌ Complete Graphs

Introduction to Rational Functions

Our study of the various types of functions now leads us to a new type: the rational function.

RATIONAL FUNCTION

A function f of the form $\frac{p}{q}$ defined by

$$f(x) = \frac{p(x)}{q(x)},$$

where $p(x)$ and $q(x)$ are polynomials, is called a **rational function.**

Since any values of x such that $q(x) = 0$ are excluded from the domain of a rational function, this type of function often has a graph that has one or more breaks in it.

Some examples of rational functions are listed below. Their graphs will be studied later in this section.

$$f(x) = \frac{1}{x} \qquad f(x) = \frac{x + 1}{2x^2 + 5x - 3} \qquad f(x) = \frac{3x^2 - 3x - 6}{x^2 + 8x + 16}$$

The simplest rational function with a variable denominator is defined by

$$f(x) = \frac{1}{x}.$$

The domain of this function is the set of all real numbers except 0. The number 0 cannot be used as a value of x, but for analysis it is helpful to find the values of $f(x)$ for some values of x close to 0. The following table shows what happens to $f(x)$ as x gets closer and closer to 0 from either side.

————— x approaches 0.

x	-1	$-.1$	$-.01$	$-.001$	.001	.01	.1	1
$f(x)$	-1	-10	-100	-1000	1000	100	10	1

————— $|f(x)|$ gets larger and larger.

The table suggests that $|f(x)|$ gets larger and larger as x gets closer and closer to 0, which is written in symbols as

$$|f(x)| \to \infty \text{ as } x \to 0.$$

(The symbol $x \to 0$ means that x approaches 0, without necessarily ever being equal to 0.) Since x cannot equal 0, the graph of $f(x) = \frac{1}{x}$ will never intersect the vertical line $x = 0$. This line is called a *vertical asymptote*.

On the other hand, as $|x|$ gets larger and larger, the values of $f(x) = \frac{1}{x}$ get closer and closer to 0, as shown in the following table.

x	$-10,000$	-1000	-100	-10	10	100	1000	10,000
$f(x)$	$-.0001$	$-.001$	$-.01$	$-.1$	.1	.01	.001	.0001

TECHNOLOGICAL NOTE
If your calculator has a TABLE feature, you can observe how the values of $1/x$ get closer and closer to 0 as $|x|$ gets larger and larger.

Letting $|x|$ get larger and larger without bound (written $|x| \to \infty$) causes the graph of $f(x) = \frac{1}{x}$ to move closer and closer to the horizontal line $y = 0$. This line is called a *horizontal asymptote*.

Notice that $f(-x) = \frac{1}{-x} = -\frac{1}{x} = -f(x)$, indicating that the graph of f is symmetric with respect to the origin. The graph of f is shown in Figure 1. (We show both traditional and graphics calculator-generated graphs.)

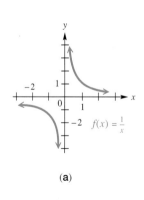

(a)

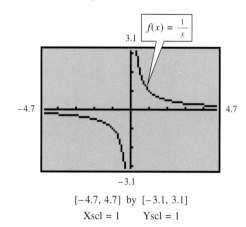

$[-4.7, 4.7]$ by $[-3.1, 3.1]$
Xscl = 1   Yscl = 1

(b)

Domain: $(-\infty, 0) \cup (0, \infty)$
Range: $(-\infty, 0) \cup (0, \infty)$

FIGURE 1

By definition, if $x \neq 0$, $x^{-1} = \frac{1}{x}$. Therefore, we may graph $f(x) = \frac{1}{x}$ by using the $\boxed{x^{-1}}$ key on our calculator. In general, the graph of $y = \frac{1}{x^n}$ for a positive integer n is the same as that of $y = x^{-n}$.

FOR **GROUP DISCUSSION**

1. Graph $y = x^{-1}$ in the window $[-3, 3]$ by $[-3, 3]$. Confirm that it is the same as that of $y = \frac{1}{x}$ graphed in the same window.
2. Graph $y = x^{-2}$ (or equivalently, $y = \frac{1}{x^2}$) in the window $[-3, 3]$ by $[-3, 3]$. What are the similarities between it and the graph of $y = x^{-1}$? What are the differences?

Graphics Calculator Considerations for Rational Functions

In Section 2.5 we studied the graph of the greatest integer function. Recall that because of the many discontinuities of the graph, in order to obtain a more realistic picture of the graph we used the dot mode of our calculator (rather than the connected mode). Because rational functions often have values for which the denominator is zero, there will be discontinuities in them as well. As a result, the dot mode of the calculator will often give a more realistic picture. If the calculator is in the connected mode, it may attempt to connect adjacent lighted pixels, giving the appearance of a vertical line on the screen. While this may be interpreted as a vertical asymptote, the student should be aware that this line is not a part of the graph.

TECHNOLOGICAL NOTE
Again, if your calculator has a TABLE feature, see what happens if the function y_1 is entered as $1/(x + 3)$ and x takes on the value -3. What kind of error message do you get?

To illustrate, consider the function $y = \frac{1}{x + 3}$. This graph is obtained by shifting the graph of $f(x) = \frac{1}{x}$ three units to the left, since it is the same as $f(x + 3)$. If this graph is generated by a calculator in connected mode in the window $[-6, 3]$ by $[-3, 3]$, we get the display shown in Figure 2. Notice the appearance of the vertical line at $x = -3$. This line cannot be part of the graph because the function is not defined for $x = -3$. (Why is this so?) Furthermore, we know that it cannot be part of the graph because the graph would then fail the vertical line test (Section 1.2).

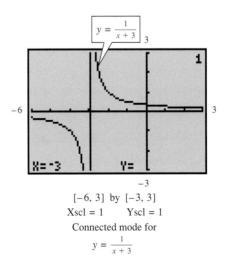

$[-6, 3]$ by $[-3, 3]$
Xscl = 1   Yscl = 1
Connected mode for
$$y = \frac{1}{x + 3}$$

FIGURE 2

On the other hand, if we graph this function in the dot mode, we get the more realistic picture found in Figure 3. However, we *must* realize that theoretically, the function is continuous on the intervals $(-\infty, -3)$ and $(-3, \infty)$.

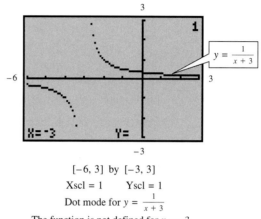

$[-6, 3]$ by $[-3, 3]$

Xscl = 1   Yscl = 1

Dot mode for $y = \dfrac{1}{x+3}$

The function is not defined for $x = -3$.

FIGURE 3

Graphs of Simple Rational Functions

The graph of $y = \frac{1}{x}$ can be shifted, translated, and reflected in the same way that we observed such transformations on other elementary functions in Chapter 2.

<table>
<tr><td>

EXAMPLE 1

Analyzing the Graph of a Simple Rational Function

</td></tr>
</table>

Discuss how the graph of $y = -\frac{2}{x}$ may be obtained from the graph of $f(x) = \frac{1}{x}$. Then graph the function both by hand and with a calculator.

SOLUTION   The expression $-\frac{2}{x}$ can be written as $-2\left(\frac{1}{x}\right)$, indicating that the graph may be obtained by stretching vertically by a factor of 2, and reflecting across the y-axis (or x-axis). The x- and y-axes remain the horizontal and vertical asymptotes. The domain and the range are still both $(-\infty, 0) \cup (0, \infty)$. Figure 4 shows both a traditional graph and a calculator-generated graph.   ◨

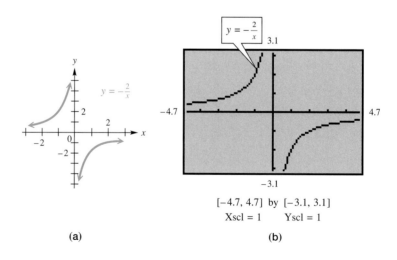

$[-4.7, 4.7]$ by $[-3.1, 3.1]$

Xscl = 1   Yscl = 1

(a)   (b)

FIGURE 4

EXAMPLE 2

Analyzing the Graph of a Simple Rational Function

Discuss how the graph of $y = \frac{2}{x+1}$ may be obtained from the graph of $f(x) = \frac{1}{x}$. Then graph the function both by hand and with a calculator.

SOLUTION   The expression $\frac{2}{x+1}$ can be written as $2(\frac{1}{x+1})$, indicating that the graph may be obtained by shifting the graph of $y = \frac{1}{x}$ one unit to the left, and stretching vertically by a factor of 2. The graphs are shown in Figure 5. Notice that the horizontal shift affects the domain; the domain of this new function is $(-\infty, -1) \cup (-1, \infty)$. The line $x = -1$ is the vertical asymptote. The range is still $(-\infty, 0) \cup (0, \infty)$.   ◼

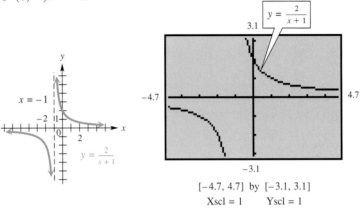

$[-4.7, 4.7]$ by $[-3.1, 3.1]$

Xscl = 1   Yscl = 1

(a)   (b)

FIGURE 5

EXAMPLE 3

Analyzing the Graph of a Simple Rational Function

Discuss how the graph of $g(x) = \frac{7}{x-3} + 2$ may be obtained from the graph of $f(x) = \frac{1}{x}$. Then graph the function both by hand and with a calculator.

SOLUTION   For the function g, we have

$$g(x) = 7 \cdot f(x - 3) + 2,$$

indicating a horizontal shift of 3 units to the right, a vertical stretch by a factor of 7, and a vertical shift of 2 units up. The graph is shown in Figure 6. Notice that the

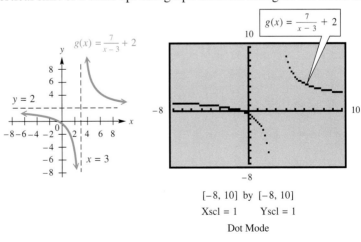

$[-8, 10]$ by $[-8, 10]$

Xscl = 1   Yscl = 1

Dot Mode

(a)   (b)

FIGURE 6

vertical asymptote has the equation $x = 3$, the horizontal asymptote has the equation $y = 2$, the domain is $(-\infty, 3) \cup (3, \infty)$, and the range is $(-\infty, 2) \cup (2, \infty)$. ◨

Determination of Asymptotes

In our discussion thus far we have seen how rational functions approach certain values as $|x| \to \infty$ and as $|x| \to a$ if the function is undefined at $x = a$. Determination of asymptotes is an important part of the analysis of rational functions. We now give formal definitions for vertical and horizontal asymptotes.

DEFINITIONS OF VERTICAL AND HORIZONTAL ASYMPTOTES

For the rational function f with $f(x) = \frac{p(x)}{q(x)}$, written in lowest terms, if $|f(x)| \to \infty$ as $x \to a$, then the line $x = a$ is a **vertical asymptote;** and if $f(x) \to a$ as $|x| \to \infty$, then the line $y = a$ is a **horizontal asymptote.**

Locating asymptotes is an important part of analysis of the graphs of rational functions. Vertical asymptotes are found by determining the values of x which make the denominator equal to 0 but do not make the numerator equal to 0. Horizontal asymptotes (and, in some cases, *oblique* asymptotes) are found by considering what happens to $f(x)$ as $|x| \to \infty$. The next example shows how to find asymptotes.

EXAMPLE 4

Finding Asymptotes of Graphs of Rational Functions

For each rational function f, find all asymptotes.

(a) $f(x) = \dfrac{x + 1}{2x^2 + 5x - 3}$

SOLUTION To find the vertical asymptotes, set the denominator equal to zero and solve.

$$2x^2 + 5x - 3 = 0$$
$$(2x - 1)(x + 3) = 0 \qquad \text{Factor.}$$
$$2x - 1 = 0 \quad \text{or} \quad x + 3 = 0 \qquad \text{Zero-product property}$$
$$x = \frac{1}{2} \quad \text{or} \quad x = -3$$

The equations of the vertical asymptotes are $x = \frac{1}{2}$ and $x = -3$.

To find the equation of the horizontal asymptote, we divide each term by the largest power of x in the expression for $f(x)$. Since the largest exponent on x in the expression here is 2, we divide each term by x^2.

$$f(x) = \frac{\dfrac{x}{x^2} + \dfrac{1}{x^2}}{\dfrac{2x^2}{x^2} + \dfrac{5x}{x^2} - \dfrac{3}{x^2}} = \frac{\dfrac{1}{x} + \dfrac{1}{x^2}}{2 + \dfrac{5}{x} - \dfrac{3}{x^2}}$$

As $|x|$ gets larger and larger, the quotients $\frac{1}{x}$, $\frac{1}{x^2}$, $\frac{5}{x}$, and $\frac{3}{x^2}$ all approach 0, and the value of $f(x)$ approaches

$$\frac{0 + 0}{2 + 0 - 0} = \frac{0}{2} = 0.$$

The line $y = 0$ (that is, the x-axis) is therefore the horizontal asymptote.

(b) $f(x) = \dfrac{2x + 1}{x - 3}$

SOLUTION Set the denominator equal to zero to find that the vertical asymptote has the equation $x = 3$. To find the horizontal asymptote, divide each term in the rational expression by x, since the greatest power of x in the expression is 1.

$$f(x) = \frac{2x + 1}{x - 3} = \frac{\dfrac{2x}{x} + \dfrac{1}{x}}{\dfrac{x}{x} - \dfrac{3}{x}} = \frac{2 + \dfrac{1}{x}}{1 - \dfrac{3}{x}}$$

As $|x|$ gets larger and larger, both $\frac{1}{x}$ and $\frac{3}{x}$ approach 0, and $f(x)$ approaches

$$\frac{2 + 0}{1 - 0} = \frac{2}{1} = 2,$$

so the line $y = 2$ is the horizontal asymptote.

(c) $f(x) = \dfrac{x^2 + 1}{x - 2}$

SOLUTION Setting the denominator equal to zero shows that the vertical asymptote has the equation $x = 2$. If we divide by the largest power of x as before (x^2 in this case), we see that there is no horizontal asymptote because

$$f(x) = \frac{\dfrac{x^2}{x^2} + \dfrac{1}{x^2}}{\dfrac{x}{x^2} - \dfrac{2}{x^2}} = \frac{1 + \dfrac{1}{x^2}}{\dfrac{1}{x} - \dfrac{2}{x^2}}$$

does not approach any real number as $|x| \rightarrow \infty$, since $\frac{1}{0}$ is undefined. This will happen whenever the degree of the numerator is greater than the degree of the denominator. In such cases, divide the denominator into the numerator to write the expression in another form.

$$
\begin{array}{r}
x + 2 \\
x - 2 \overline{)\, x^2 + 0x + 1} \\
\underline{x^2 - 2x } \\
2x + 1 \\
\underline{2x - 4} \\
5
\end{array}
$$

      Subtract.

      Subtract.

(Notice that we could have used synthetic division here.)
The function can now be written as

$$f(x) = \frac{x^2 + 1}{x - 2} = x + 2 + \frac{5}{x - 2}.$$

For very large values of $|x|$, $\frac{5}{x-2}$ is close to 0, and the graph approaches the line $y = x + 2$. This line is an **oblique asymptote** (neither vertical nor horizontal) for the graph of the function.

In general, if the degree of the numerator is exactly one more than the degree of the denominator, a rational function may have an oblique asymptote. The equation of this asymptote is found by dividing the numerator by the denominator and disregarding the remainder. ◧

The results of Example 4 can be summarized as follows.

DETERMINING ASYMPTOTES

In order to find asymptotes of a rational function defined by a rational expression *in lowest terms,* use the following procedures.

1. **Vertical Asymptotes**
   Find any vertical asymptotes by setting the denominator equal to 0 and solving for x. If a is a zero of the denominator but not the numerator, then the line $x = a$ is a vertical asymptote.
2. **Other Asymptotes**
   Determine any other asymptotes. We consider three possibilities:

   a. If the numerator has lower degree than the denominator, there is a horizontal asymptote, $y = 0$ (the x-axis).
   b. If the numerator and denominator have the same degree, and the function is of the form

   $$f(x) = \frac{a_n x^n + \cdots + a_0}{b_n x^n + \cdots + b_0}, \quad \text{where } b_n \neq 0,$$

   dividing by x^n in the numerator and denominator produces the horizontal asymptote

   $$y = \frac{a_n}{b_n}.$$

   c. If the numerator is of degree exactly one more than the denominator, there may be an oblique asymptote. To find it, divide the numerator by the denominator and disregard any remainder. Set the rest of the quotient equal to y to get the equation of the asymptote.*

NOTE The graph of a rational function may have more than one vertical asymptote, or it may have none at all. The graph cannot intersect any vertical asymptote. There can be only one other (non-vertical) asymptote, and the graph *may* intersect that asymptote. This will be seen in Example 7. The method of graphing a rational function that is not in lowest terms will be covered in Example 9.

Graphs of More Complicated Rational Functions

Because graphics calculators can lead easily to distorted views of rational function graphs, we will list a step-by-step method of graphing rational functions by hand. (We will continue to support our hand-drawn graphs using calculator-generated

---

* More involved rational functions, such as $f(x) = \frac{8x^3 - 1}{x}$, are not covered in this book.

graphs, being aware of some of the pitfalls associated with graphs of rational functions as seen on calculator viewscreens.)

> ## GRAPHING RATIONAL FUNCTIONS
>
> Let $f(x) = \frac{p(x)}{q(x)}$ define a function where the rational expression is written in lowest terms. To sketch its graph, follow the steps below.
>
> 1. Find any vertical asymptotes.
> 2. Find any horizontal or oblique asymptote.
> 3. Find the y-intercept by evaluating $f(0)$.
> 4. Find the x-intercepts, if any, by solving $f(x) = 0$. (These will be the zeros of the numerator, p.)
> 5. Determine whether the graph will intersect its nonvertical asymptote by solving $f(x) = k$, where k (or $mx + b$) is the y-value of the nonvertical asymptote.
> 6. Plot a few selected points, as necessary. Choose an x-value in each interval of the domain as determined by the vertical asymptotes and x-intercepts.
> 7. Complete the sketch.

The next example shows how the above guidelines can be used to graph a rational function.

EXAMPLE 5

Graphing a Rational Function Defined by an Expression with Degree of Numerator Less than Degree of Denominator

Graph $f(x) = \dfrac{x + 1}{2x^2 + 5x - 3}$.

SOLUTION

Step 1   As shown in Example 4(a), the vertical asymptotes have equations $x = \frac{1}{2}$ and $x = -3$.

Step 2   Again, as shown in Example 4(a), the horizontal asymptote is the x-axis.

Step 3   Since $f(0) = \frac{0 + 1}{2(0)^2 + 5(0) - 3} = -\frac{1}{3}$, the y-intercept is $-\frac{1}{3}$.

Step 4   The x-intercept is found by solving $f(x) = 0$.

$$\frac{x + 1}{2x^2 + 5x - 3} = 0$$

$$x + 1 = 0 \qquad \text{If a rational expression is equal to 0, then its numerator must equal 0.}$$

$$x = -1$$

The x-intercept is -1.

Step 5   To determine whether the graph intersects its horizontal asymptote, solve

$$f(x) = 0. \; \leftarrow y\text{-value of horizontal asymptote}$$

Since the horizontal asymptote is the x-axis, the solution of this equation was found in Step 4. The graph intersects its horizontal asymptote at $(-1, 0)$.

Step 6   Plot a point in each of the intervals determined by the *x*-intercepts and vertical asymptotes, $(-\infty, -3)$, $(-3, -1)$, $(-1, \frac{1}{2})$, and $(\frac{1}{2}, \infty)$ to get an idea of how the graph behaves in each region.

Step 7   Complete the sketch. Keep in mind that the graph approaches its asymptotes as the points on the graph become farther away from the origin. The traditional graph is shown in Figure 7(a), and the calculator-generated graph is shown in Figure 7(b).   ◼

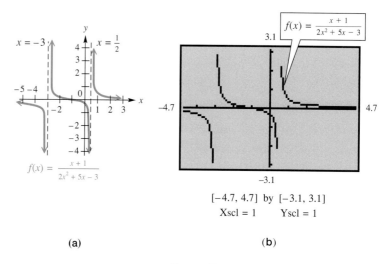

(a)   (b)

FIGURE 7

The function *g* discussed in Example 3 may be written in the form $\frac{p(x)}{q(x)}$ by using methods of adding rational expressions introduced in elementary algebra courses.

$$g(x) = \frac{7}{x - 3} + 2$$

$$= \frac{7}{x - 3} + \frac{2(x - 3)}{x - 3} \qquad \text{Find a common denominator.}$$

$$= \frac{7 + 2(x - 3)}{x - 3}$$

$$= \frac{7 + 2x - 6}{x - 3}$$

$$= \frac{2x + 1}{x - 3}$$

To be consistent with our step-by-step method of graphing rational functions presented earlier, let us rename this function *f*, so that $f(x) = \frac{2x + 1}{x - 3}$. Its graph is, of course, the same as that shown in Figure 6. We will use the step-by-step method to graph this function by hand in Example 6. (In the remaining examples, we will not specifically number the steps.)

EXAMPLE **6**

Graphing a Rational Function Defined by an Expression with Degree of Numerator Equal to Degree of Denominator

Graph $f(x) = \dfrac{2x + 1}{x - 3}$.

SOLUTION   As shown in Example 4(b), the equation of the vertical asymptote is $x = 3$ and the equation of the horizontal asymptote is $y = 2$. Since $f(0) = -\frac{1}{3}$, the y-intercept is $-\frac{1}{3}$. The solution of $f(x) = 0$ is $-\frac{1}{2}$, so the only x-intercept is $-\frac{1}{2}$. The graph does not intersect its horizontal asymptote, since $f(x) = 2$ has no solution. (Verify this.) The points $(-4, 1)$ and $(6, \frac{13}{3})$ are on the graph and can be used to complete the hand-drawn sketch, as shown in Figure 8(a). A calculator-generated graph is shown in Figure 8(b). Notice that these graphs are the same as those in Figure 6. ∎

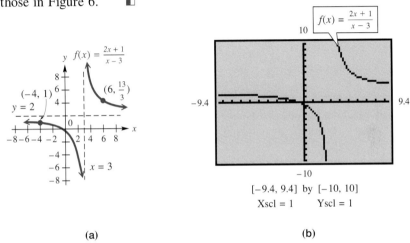

(a)

(b)

FIGURE 8

EXAMPLE **7**

Graphing a Rational Function Defined by an Expression with Degree of Numerator Equal to Degree of Denominator

Graph $f(x) = \dfrac{3x^2 - 3x - 6}{x^2 + 8x + 16}$.

SOLUTION   To find the vertical asymptote(s), solve $x^2 + 8x + 16 = 0$.

$$x^2 + 8x + 16 = 0$$
$$(x + 4)^2 = 0$$
$$x = -4$$

The only vertical asymptote has the equation $x = -4$. As explained in the preceding guidelines, by dividing all terms by x^2, the equation of the horizontal asymptote can be shown to be

$$y = \frac{3}{1} \qquad \begin{array}{l}\leftarrow \text{Leading coefficient of numerator} \\ \leftarrow \text{Leading coefficient of denominator}\end{array}$$

or $y = 3$. The y-intercept is $f(0) = -\frac{3}{8}$. To find the x-intercept(s), if any, we solve $f(x) = 0$.

$$f(x) = \frac{3x^2 - 3x - 6}{x^2 + 8x + 16} = 0$$
$$3x^2 - 3x - 6 = 0$$
$$x^2 - x - 2 = 0 \qquad \text{Divide by 3.}$$
$$(x - 2)(x + 1) = 0$$
$$x = 2 \quad \text{or} \quad x = -1$$

The x-intercepts are -1 and 2. By setting $f(x) = 3$ and solving, we can locate the point where the graph intersects the horizontal asymptote.

$$f(x) = \frac{3x^2 - 3x - 6}{x^2 + 8x + 16}$$

$$3 = \frac{3x^2 - 3x - 6}{x^2 + 8x + 16}$$

$$3x^2 - 3x - 6 = 3x^2 + 24x + 48 \qquad \text{Multiply by } x^2 + 8x + 16.$$

$$-3x - 6 = 24x + 48 \qquad \text{Subtract } 3x^2.$$

$$-27x = 54$$

$$x = -2$$

The graph intersects its horizontal asymptote at $(-2, 3)$.

Some other points that lie on the graph are $(-10, 9)$, $(-3, 30)$ and $(5, \frac{2}{3})$. These can be used to complete the hand-drawn graph, as shown in Figure 9(a). The calculator-generated graph in Figure 9(b) confirms our result. ∎

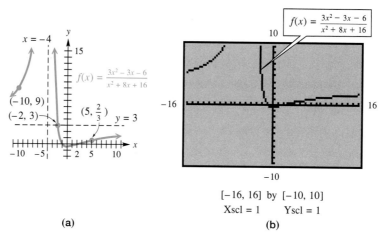

$$[-16, 16] \text{ by } [-10, 10]$$
$$\text{Xscl} = 1 \qquad \text{Yscl} = 1$$

(a)   (b)

FIGURE 9

Notice the behavior of the graph of the function in Example 7 near the line $x = -4$. As x approaches -4 from either side, we have $f(x)$ approaching ∞. On the other hand, if we examine the behavior of the graph in Figure 6 (or 8) near the line $x = 3$, we have $f(x) \to -\infty$ as x approaches 3 from the left, while $f(x) \to \infty$ as x approaches 3 from the right. The behavior of the graph of a rational function near a vertical asymptote $x = a$ will partially depend on the parity of the multiplicity of the factor $x - a$ in the denominator. The box that follows explains this.

BEHAVIOR OF GRAPHS OF RATIONAL FUNCTIONS
NEAR VERTICAL ASYMPTOTES

If n is the largest positive integer such that $(x - a)^n$ is a factor of the denominator of $f(x)$, the graph will behave in the manner illustrated if

n is even.

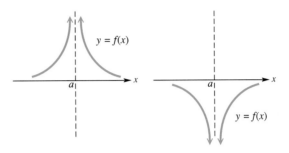

n is odd.

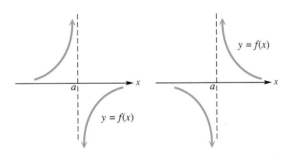

In both cases, we assume that $f(x)$ is in lowest terms.

The next example discusses a rational function defined by an expression having the degree of its numerator greater than the degree of its denominator.

EXAMPLE 8

Graphing a Rational Function Defined by an Expression with Degree of Numerator Greater than Degree of Denominator

Graph $f(x) = \dfrac{x^2 + 1}{x - 2}$.

SOLUTION   As shown in Example 4(c), the vertical asymptote has the equation $x = 2$, and the graph has an oblique asymptote with the equation $y = x + 2$. Refer to the box above to determine the behavior near the asymptote $x = 2$. The y-intercept is $-\frac{1}{2}$, and the graph has no x-intercepts, since the numerator, $x^2 + 1$, has no real zeros. It can be shown that the graph does not intersect its oblique asymptote. Using the intercepts, asymptotes, the points $(4, \frac{17}{2})$ and $(-1, -\frac{2}{3})$, and the general behavior of the graph near its asymptotes, we obtain the graph shown in Figure 10(a). Compare with the calculator-generated graph shown in Figure 10(b).   ∎

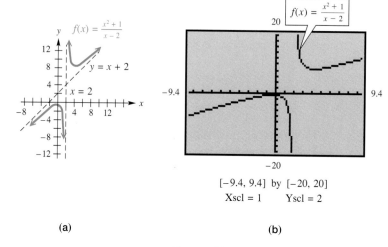

(a)

(b)

FIGURE 10

As mentioned earlier, a rational function must be defined by an expression in lowest terms before we can use the methods discussed thus far in this section to hand-sketch the graph. The final example shows a typical rational function defined by an expression that is not in lowest terms.

EXAMPLE 9

Graphing a Rational Function Defined by an Expression that Is not in Lowest Terms

Graph $f(x) = \dfrac{x^2 - 4}{x - 2}$.

SOLUTION Notice that the domain of this function cannot contain 2. The rational expression $\frac{x^2 - 4}{x - 2}$ can be reduced to lowest terms by factoring the numerator, and dividing both the numerator and denominator by $x - 2$.

$$f(x) = \frac{x^2 - 4}{x - 2} = \frac{(x + 2)(x - 2)}{x - 2} = x + 2 \quad (x \neq 2)$$

Therefore, the graph of this function will be the same as the graph of $y = x + 2$ (a straight line), with the exception of the point with x-value 2. A "hole" appears in the graph at $(2, 4)$. See Figure 11(a).

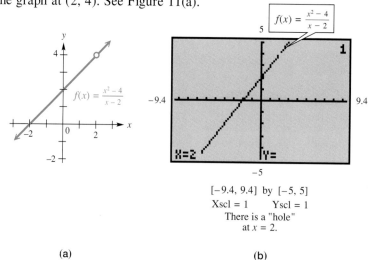

(a)

(b)

FIGURE 11

If the window of a graphics calculator is set in a manner so that an x-value for the location of the cursor is 2, then we can see from the display that the calculator cannot determine a value for y, as in Figure 11(b). A close examination of the screen will show an unlighted pixel at $x = 2$. However, such points of discontinuity will often *not* be evident from calculator-generated graphs—once again showing us a reason for studying the concepts hand-in-hand with the technology.   ◻

Complete Graphs

In order to sketch by hand the complete graph of a rational function (or observe the graph on a calculator viewscreen), we should be able to observe the features listed below.

> **COMPLETE GRAPH CRITERIA FOR A RATIONAL FUNCTION**
>
> A complete graph of a rational function will exhibit the following features.
>
> 1. all intercepts, both x- and y-
> 2. location of all asymptotes: vertical, horizontal, and/or oblique
> 3. the point at which the graph intersects its nonvertical asymptote (if there is any such point)
> 4. enough of the graph to exhibit the correct end behavior (i.e., behavior as the graph approaches its nonvertical asymptote)

4.1 EXERCISES

Use the terminology of Sections 2.2 and 2.3 involving shifting, stretching, shrinking, and reflecting to explain how the graph of f can be obtained from the graph of $y = \frac{1}{x}$ or $y = \frac{1}{x^2}$. Then draw a sketch of the graph of f by hand. You may wish to support your sketch with a calculator-generated graph.

1. $f(x) = \dfrac{2}{x}$

2. $f(x) = -\dfrac{3}{x}$

3. $f(x) = \dfrac{1}{x + 2}$

4. $f(x) = \dfrac{1}{x - 3}$

5. $f(x) = \dfrac{1}{x} + 1$

6. $f(x) = \dfrac{1}{x} - 2$

7. $f(x) = -\dfrac{2}{x^2}$

8. $f(x) = \dfrac{1}{x^2} - 3$

9. $f(x) = \dfrac{1}{(x - 3)^2}$

10. $f(x) = \dfrac{-2}{(x - 3)^2}$

The figures below show the four ways that the graph of a rational function can approach the vertical line $x = 2$ as an asymptote. Identify the graph of each of the rational functions in Exercises 11–14.

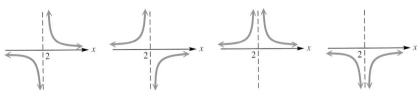

A   B   C   D

11. $f(x) = \dfrac{1}{(x - 2)^2}$

12. $f(x) = \dfrac{1}{x - 2}$

13. $f(x) = \dfrac{-1}{x - 2}$

14. $f(x) = \dfrac{-1}{(x - 2)^2}$

Give the equations of any vertical, horizontal, or oblique asymptotes for the graphs of the rational functions in Exercises 15–28. For those with horizontal asymptotes, graph the function on your calculator and support your answer by tracing and letting $x \to \infty$ and $x \to -\infty$. You should be able to see how the y-value approaches a constant.

15. $f(x) = \dfrac{2}{x - 5}$

16. $f(x) = \dfrac{-1}{x + 2}$

17. $f(x) = \dfrac{-8}{3x - 7}$

18. $f(x) = \dfrac{5}{4x - 9}$

19. $f(x) = \dfrac{2 - x}{x + 2}$

20. $f(x) = \dfrac{x - 4}{5 - x}$

21. $f(x) = \dfrac{3x - 5}{2x + 9}$

22. $f(x) = \dfrac{4x + 3}{3x - 7}$

23. $f(x) = \dfrac{2}{x^2 - 4x + 3}$

24. $f(x) = \dfrac{-5}{x^2 - 3x - 10}$

25. $f(x) = \dfrac{x^2 - 1}{x + 3}$

26. $f(x) = \dfrac{x^2 + 4}{x - 1}$

27. $f(x) = \dfrac{x^2 - 2x - 3}{2x^2 - x - 10}$

28. $f(x) = \dfrac{3x^2 - 6x - 24}{5x^2 - 26x + 5}$

29. Which one of the following has a graph that does not have a vertical asymptote?

(a) $f(x) = \dfrac{1}{x^2 + 2}$

(b) $f(x) = \dfrac{1}{x^2 - 2}$

(c) $f(x) = \dfrac{3}{x^2}$

(d) $f(x) = \dfrac{2x + 1}{x - 8}$

30. Which one of the following has a graph that does not have a horizontal asymptote?

(a) $f(x) = \dfrac{2x - 7}{x + 3}$

(b) $f(x) = \dfrac{3x}{x^2 - 9}$

(c) $f(x) = \dfrac{x^2 - 9}{x + 3}$

(d) $f(x) = \dfrac{x + 5}{(x + 2)(x - 3)}$

Relating Concepts

Consider the rational function

$$f(x) = \dfrac{x + 3}{x^2 + x + 4},$$

which is defined by a rational expression in lowest terms, whose denominator is a quadratic polynomial.

31. Explain the procedure you would use to find any vertical asymptotes of the graph of f.

32. Under what conditions does a quadratic equation have no real solutions? (See Section 3.3.)

33. Apply the procedure of Exercise 31 to this function f. What are the complex solutions of the equation you solved? What are the real solutions? Does f have any vertical asymptotes?

34. With your calculator in connected mode and using a window of $[-10, 10]$ by $[-1, 3]$ graph the function. Why is the connected mode acceptable here to get a realistic view of the graph?

Use the guidelines of this section to sketch by hand the graph of each of the following rational functions.

35. $f(x) = \dfrac{x + 1}{x - 4}$

36. $f(x) = \dfrac{x - 5}{x + 3}$

37. $f(x) = \dfrac{3x}{x - 1}$

38. $f(x) = \dfrac{4x}{-3x + 1}$

39. $f(x) = \dfrac{1}{x^2 + 2x - 8}$

40. $f(x) = \dfrac{-2}{x^2 - 5x - 6}$

41. $f(x) = \dfrac{x^2 + x - 6}{x^2 + 3x - 4}$

42. $f(x) = \dfrac{x^2 - x - 30}{x^2 - 2x - 3}$

43. $f(x) = \dfrac{x^2 - 2x - 3}{x^2 - 2x + 1}$

44. $f(x) = \dfrac{x^2 - 2x}{x^2 + 6x + 9}$

45. $f(x) = \dfrac{3x}{x^2 - 16}$

46. $f(x) = \dfrac{x}{x^2 - 9}$

47. $f(x) = \dfrac{1}{x^2 + 1}$

48. $f(x) = \dfrac{x^2 - 7x + 10}{x^2 + 9}$

49. $f(x) = \dfrac{5x}{x^2 - 1}$

50. $f(x) = \dfrac{2x + 1}{x^2 + 6x + 8}$

Relating Concepts

*Recall from Section 2.3 that if we are given the graph of $y = f(x)$, we can obtain the graph of $y = -f(x)$ by reflecting across the x-axis, and we can obtain the graph of $y = f(-x)$ by reflecting across the y-axis. In Exercises 51–54 you are given the graph of a rational function $y = f(x)$. Draw a sketch by hand of the graph of (**a**) $y = -f(x)$ and (**b**) $y = f(-x)$.*

51.

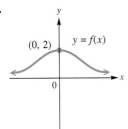

52.

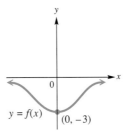

53.

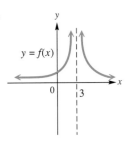

54.

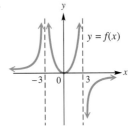

Each of the following rational functions has an oblique asymptote. Sketch the graph of the function by hand, and give the equation of the oblique asymptote.

55. $f(x) = \dfrac{2x^2 + 3}{x - 4}$

56. $f(x) = \dfrac{x^2 + 1}{x + 3}$

57. $f(x) = \dfrac{x^2 - x}{x + 2}$

58. $f(x) = \dfrac{x^2 + 2x}{2x - 1}$

59. Consider the rational function $f(x) = \frac{x^3 - 4x^2 + x + 6}{x^2 + x - 2}$. Divide the numerator by the denominator and use the method of Example 8 to determine the equation of the oblique asymptote. Then, determine the coordinates of the point where the graph of f intersects its oblique asymptote.

60. Use long division of polynomials to show that for the function

$$f(x) = \frac{x^4 - 5x^2 + 4}{x^2 + x - 12},$$

if we divide the numerator by the denominator, the quotient polynomial is $x^2 - x + 8$ and the remainder is $-20x + 100$. Then graph both f and $g(x) = x^2 - x + 8$ in the window $[-50, 50]$ by $[0, 1000]$. Comment on the appearance of the two graphs. Explain how the graph of f approaches that of g as $|x|$ gets very large.

Each of the following rational functions has a "hole" in its graph. Sketch the graph by hand, and show where the hole appears.

61. $f(x) = \dfrac{x^2 - 9}{x + 3}$

62. $f(x) = \dfrac{x^2 - 16}{x + 4}$

63. $f(x) = \dfrac{x^2 - 25}{5 - x}$

64. $f(x) = \dfrac{x^2 - 36}{6 - x}$

Relating Concepts

Consider the following "monster" rational function.

$$f(x) = \frac{x^4 - 3x^3 - 21x^2 + 43x + 60}{x^4 - 6x^3 + x^2 + 24x - 20}$$

Work Exercises 65–74 in order, referring to the indicated section if necessary.

65. Find the equation of the horizontal asymptote. (Section 4.1)

66. Given that -4 and -1 are zeros of the numerator, factor the numerator completely. (Section 3.6)

67. **(a)** Given that 1 and 2 are zeros of the denominator, factor the denominator completely. (Section 3.6)
    (b) Write the entire quotient for f so that the numerator and the denominator are in factored form.

68. **(a)** What is the common factor in the numerator and the denominator?
    (b) For what value of x will there be a point of discontinuity (i.e., a "hole")?

69. What are the x-intercepts of the graph of f? (Section 4.1)

70. What is the y-intercept of the graph of f? (Section 4.1)

71. Find the equations of the vertical asymptotes. (Section 4.1)

72. Determine the point or points of intersection of the graph of f with its horizontal asymptote. (Sections 4.1 and 3.3)

73. Sketch by hand the graph of f. (Section 4.1)

74. Use the graph of f to solve the inequalities
    (a) $f(x) < 0$ and   **(b)** $f(x) > 0$. (Section 1.6)

Further Explorations

1. Use the TABLE feature to compare the graphs of $\frac{1}{x}$ and $\frac{1}{x^2}$. As seen in the first figure, it is difficult to differentiate between the graphs in the standard viewing window. But the second figure shows a distinct difference in the decimal window.

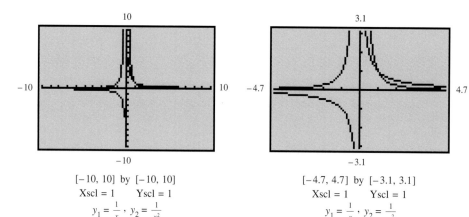

$[-10, 10]$ by $[-10, 10]$
Xscl = 1   Yscl = 1
$y_1 = \frac{1}{x}$, $y_2 = \frac{1}{x^2}$

$[-4.7, 4.7]$ by $[-3.1, 3.1]$
Xscl = 1   Yscl = 1
$y_1 = \frac{1}{x}$, $y_2 = \frac{1}{x^2}$

Use the TABLE to do the following.
(a) Begin with Tblmin $= -3$ and ΔTbl $= 1$. Let $Y_1 = \frac{1}{x}$ and $Y_2 = \frac{1}{x^2}$.
    i.   For what value(s) of x does $\frac{1}{x} = \frac{1}{x^2}$?
    ii. Explain the TABLE values when $x = 0$.
    iii. Scroll forward and backwards, and describe what happens to Y_1 and Y_2 as $x \to \infty$.

(b) Change TABLE SETUP so that Tblmin $= 1$ and ΔTbl $= -.01$ *Note:* Since ΔTbl < 0, the TABLE will show decreasing values for x.
   i. Scroll through the TABLE from $x = 1$ to $x = 0$. Describe the differences between $Y_1 = \frac{1}{x}$ and $Y_2 = \frac{1}{x^2}$.
   ii. Analytically confirm the values for Y_1 and Y_2 at $x = .5$, $x = .25$, $x = .2$, $x = .1$, $x = .05$, and $x = .01$.
   iii. Scroll through the TABLE from $x = 0$ to $x = -1$. Describe the differences between $Y_1 = \frac{1}{x}$ and $Y_2 = \frac{1}{x^2}$.
   iv. Analytically confirm the values for Y_1 and Y_2 at $x = -.5$, $x = -.25$, $x = -.2$, $x = -.1$, $x = -.05$, and $x = -.01$.

2. Use the TABLE feature to find the horizontal asymptote of

$$f(x) = \frac{6x^2 + 3}{2x^2}.$$

(*Hint:* Let Tblmin $= 0$, and ΔTbl $= 10$. Then scroll through the TABLE to see what happens to $f(x)$ as $x \to \infty$ and $x \to -\infty$).

3. Use the TABLE feature to find the horizontal asymptote of

$$f(x) = \frac{2x^3 + 3}{-5x^3 - x^2 + 8}.$$

4.2 RATIONAL EQUATIONS, INEQUALITIES, AND APPLICATIONS

Solving Rational Equations and Inequalities ∎ Inverse Variation ∎ Combined and Joint Variation ∎ Optimization Applications

Solving Rational Equations and Inequalities

In this section we will examine methods of solving equations, inequalities, and applications involving rational functions. A rational equation (or inequality) is an equation (or inequality) that involves at least one term having a variable expression in a denominator, or at least one term having a variable expression raised to a negative integer power. Some examples of such equations and inequalities are

$$\frac{2x - 1}{3x + 4} = 5, \quad \frac{2x - 1}{3x + 4} \le 5, \quad \frac{x}{x - 2} + \frac{1}{x + 2} = \frac{8}{x^2 - 4},$$

and

$$7x^{-4} - 8x^{-2} + 1 = 0.$$

In solving rational equations and inequalities, it is important to remember that an expression with a variable denominator or with a negative exponent may be undefined for certain values of the variable. Each time we begin solving such an equation or inequality, we will identify such values (which are actually values at which the associated rational function will have a vertical asymptote or "hole"). The general procedure for solving rational equations is to multiply both sides of the equation by the least common denominator of all the terms in the equation, as shown in the first example.

EXAMPLE **1**
Solving a Rational Equation and Supporting the Solution Graphically

Solve analytically the equation

$$\frac{2x - 1}{3x + 4} = 5$$

and support the solution graphically.

SOLUTION   We begin by noticing that the rational expression is undefined for $x = -\frac{4}{3}$. Multiplying both sides of the equation by $3x + 4$ gives

$$2x - 1 = 5(3x + 4).$$

Now solve this equation.

$$2x - 1 = 15x + 20$$
$$-13x = 21$$
$$x = -\frac{21}{13}$$

TECHNOLOGICAL NOTE
Some graphics calculators have the capability to convert decimals representing rational numbers to quotients of integers. For example, the solution of the equation in Example 1 has its decimal displayed in Figure 12. Consult your owner's manual to see if your model is capable of converting this decimal numeral to its fraction form, $-21/13$.

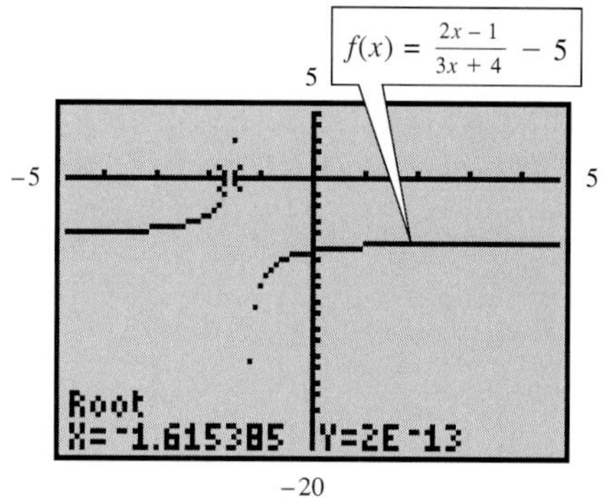

The solution set is $\{-\frac{21}{13}\}$. To support this solution, we notice that the original equation is equivalent to $\frac{2x-1}{3x+4} - 5 = 0$. If we let the left side of this equation be represented by $f(x)$, then the x-intercept of f is the solution of the equation $f(x) = 0$. The graph in Figure 12 suggests that $-\frac{21}{13} \approx -1.62$ is indeed the solution. By using the capabilities of our calculator, we can confirm that the decimal value of the x-intercept is approximately -1.62 which is the correct decimal approximation for $-\frac{21}{13}$ to the nearest hundredth.   ∎

$$f(x) = \frac{2x - 1}{3x + 4} - 5$$

```
Root
X=-1.615385   Y=2E-13
```

$[-5, 5]$ by $[-20, 5]$
Xscl = 1   Yscl = 1
Dot mode

FIGURE 12

In Example 2 we will solve the rational inequality $\frac{2x-1}{3x+4} \leq 5$ analytically. It would be incorrect to multiply both sides of the inequality by $3x + 4$, since it may represent a negative number and this would require reversing the inequality symbol. We will use the sign graph method, first introduced in Section 3.3 for solving quadratic inequalities analytically.

EXAMPLE 2

Solving a Rational
Inequality
Analytically and
Supporting
Graphically

Use a sign graph to solve

$$\frac{2x - 1}{3x + 4} \leq 5,$$

and use the graph in Figure 12 to support the answer.

SOLUTION   We begin by subtracting 5 on both sides and combining the terms on the left into a single fraction.

$$\frac{2x - 1}{3x + 4} \leq 5$$

$$\frac{2x - 1}{3x + 4} - 5 \leq 0 \qquad \text{Subtract 5.}$$

$$\frac{2x - 1 - 5(3x + 4)}{3x + 4} \leq 0 \qquad \text{Common denominator is } 3x + 4.$$

$$\frac{-13x - 21}{3x + 4} \leq 0 \qquad \text{Combine terms.}$$

To draw a sign graph, first solve the equations

$$-13x - 21 = 0 \qquad \text{and} \qquad 3x + 4 = 0,$$

getting the solutions

$$x = -\frac{21}{13} \qquad \text{and} \qquad x = -\frac{4}{3}.$$

Use the values $-\frac{21}{13}$ and $-\frac{4}{3}$ to divide the number line into three intervals. Now complete a sign graph and find the intervals where the quotient is negative. See Figure 13.

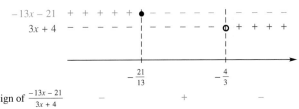

FIGURE 13

From the sign graph, values of x in the two intervals $(-\infty, -\frac{21}{13})$ and $(-\frac{4}{3}, \infty)$ make the quotient negative, as required. Because the inequality is not strict, the endpoint $-\frac{21}{13}$ satisfies it. However, the endpoint $-\frac{4}{3}$ causes the denominator to equal 0, so it is not included in the solution set. Therefore, the solution set should be written $(-\infty, -\frac{21}{13}] \cup (-\frac{4}{3}, \infty)$.

The original inequality is equivalent to $\frac{2x - 1}{3x + 4} - 5 \leq 0$. If we observe the graph in Figure 12, we see that it lies *below* the x-axis for x-values less than $-\frac{21}{13}$, and for x-values greater than the x-value of the vertical asymptote, which is $-\frac{4}{3}$. Therefore, the graph supports our analytic solution.   ◻

CAUTION   As suggested by Example 2, be very careful with the endpoints of the intervals in the solution of rational inequalities.

FOR GROUP DISCUSSION

Examples 1 and 2 show how to solve

$$\frac{2x-1}{3x+4}=5 \quad \text{and} \quad \frac{2x-1}{3x+4}\le 5.$$

1. Refer to the sign graph in Figure 13 and discuss how you would go about finding the solution set of $\frac{2x-1}{3x+4}\ge 5$.
2. The solution set of $\frac{2x-1}{3x+4}=5$ is $\{-\frac{21}{13}\}$, and the solution set of $\frac{2x-1}{3x+4}\le 5$ is $(-\infty, -\frac{21}{13}] \cup (-\frac{4}{3}, \infty)$. What must be the solution set of $\frac{2x-1}{3x+4}\ge 5$? How does the graph support this?

The next example shows the importance of determining the values for which the rational expressions in an equation are undefined before solving the equation analytically.

EXAMPLE 3

Solving a Rational Equation Analytically and Supporting Graphically

Solve the rational equation

$$\frac{x}{x-2}+\frac{1}{x+2}=\frac{8}{x^2-4}$$

analytically, and support the solution graphically.

SOLUTION   First, note that for this equation, $x \ne \pm 2$.

$$\frac{x}{x-2}+\frac{1}{x+2}=\frac{8}{x^2-4} \qquad \text{Given equation}$$
$$x(x+2)+1(x-2)=8 \qquad \text{Multiply by }(x-2)(x+2).$$
$$x^2+2x+x-2=8$$
$$x^2+3x-10=0 \qquad \text{Put in standard form.}$$
$$(x+5)(x-2)=0 \qquad \text{Factor.}$$
$$x+5=0 \quad \text{or} \quad x-2=0 \qquad \text{Use the zero-product property.}$$
$$x=-5 \qquad\qquad x=2$$

The numbers -5 and 2 are the *possible* solutions of the equation. Recall that 2 is not in the domain of the original equation, and therefore must be rejected. (Such a value is called *extraneous*.) The solution set is $\{-5\}$. (The number 2 is a solution of the equation obtained after the multiplication step. But multiplying by $x-2$ when $x=2$ is multiplying by 0, which does not lead to an equivalent equation.) To support this solution, we graph $f(x)=\frac{x}{x-2}+\frac{1}{x+2}-\frac{8}{x^2-4}$ and notice that the x-intercept is -5, as expected. The line $x=-2$ is a vertical asymptote, and for $x=2$, the graph has a hole. See Figure 14. ∎

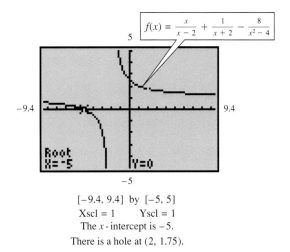

$$f(x) = \frac{x}{x-2} + \frac{1}{x+2} - \frac{8}{x^2-4}$$

[−9.4, 9.4] by [−5, 5]
Xscl = 1   Yscl = 1
The x-intercept is −5.
There is a hole at (2, 1.75).

FIGURE 14

FOR **GROUP DISCUSSION**

In solving the rational equation in Example 3, our first step was to multiply both sides by $(x - 2)(x + 2)$. This eventually led to the quadratic equation $x^2 + 3x - 10 = 0$. The graph of $y = x^2 + 3x - 10$ is shown in Figure 15.

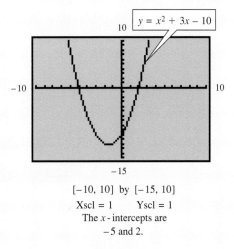

$$y = x^2 + 3x - 10$$

[−10, 10] by [−15, 10]
Xscl = 1   Yscl = 1
The x-intercepts are
−5 and 2.

FIGURE 15

1. Discuss how the solution of the equation in Example 3 and the extraneous solution found there relate to the solutions of the equation $x^2 + 3x - 10 = 0$. How does the extraneous solution show up on the graph of the parabola?
2. Why should you always pay attention to the domain for a rational equation?

EXAMPLE **4**
Solving a Rational Equation and Associated Inequalities

Consider the rational function

$$f(x) = 7x^{-4} - 8x^{-2} + 1.$$

(a) Solve the equation $f(x) = 0$ analytically and support the real solutions graphically.

SOLUTION The expression for $f(x)$ may be written equivalently as

$$\frac{7}{x^4} - \frac{8}{x^2} + 1.$$

We begin by noticing that for this expression, $x \neq 0$. Now we solve the equation.

$$\frac{7}{x^4} - \frac{8}{x^2} + 1 = 0$$

$$7 - 8x^2 + x^4 = 0 \qquad \text{Multiply by } x^4.$$

$$x^4 - 8x^2 + 7 = 0$$

$$(x^2 - 7)(x^2 - 1) = 0$$

$$(x + \sqrt{7})(x - \sqrt{7})(x + 1)(x - 1) = 0$$

$$x + \sqrt{7} = 0 \quad \text{or} \quad x - \sqrt{7} = 0 \quad \text{or} \quad x + 1 = 0 \quad \text{or} \quad x - 1 = 0$$

$$x = -\sqrt{7} \qquad\qquad x = \sqrt{7} \qquad\qquad x = -1 \qquad\qquad x = 1$$

The solution set is $\{\pm\sqrt{7}, \pm 1\}$. The graph in Figure 16 suggests the accuracy of our solutions, as the four x-intercepts correspond to the four solutions. This can be verified by using the calculator capabilities to find solutions, noting that $\sqrt{7} \approx 2.65$.

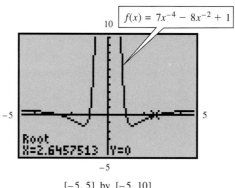

$[-5, 5]$ by $[-5, 10]$
Xscl = 1   Yscl = 1
A decimal approximation for $\sqrt{7}$
is displayed. The other solutions
are $-\sqrt{7}, -1$, and 1.

FIGURE 16

(b) Use the graph in Figure 16 to find the solution set of

$$f(x) \leq 0.$$

SOLUTION The graph of f lies *below* or *on* the x-axis for values for x between $-\sqrt{7}$ and -1 (inclusive of both values) or between 1 and $\sqrt{7}$ (inclusive of both values). Therefore, the solution set is

$$[-\sqrt{7}, -1] \cup [1, \sqrt{7}].$$

(c) Use the graph in Figure 16 to find the solution set of $f(x) \geq 0$.

SOLUTION The graph of f lies *above* or *on* the x-axis for the intervals $(-\infty, -\sqrt{7}], [-1, 0), (0, 1], [\sqrt{7}, \infty)$. Therefore, the solution set is the union of these intervals: $(-\infty, -\sqrt{7}] \cup [-1, 0) \cup (0, 1] \cup [\sqrt{7}, \infty)$. Notice that 0 is not included because it is not in the domain. ◘

Inverse Variation

In Section 1.7 we saw how direct variation is an example of an application of linear functions. Direct variation as a *power* was discussed in Section 2.6. Another type of variation, inverse variation, is an example of an application of rational functions.

INVERSE VARIATION

Let n be a positive real number. Then y **varies inversely** as the nth power of x, or y is **inversely proportional** to the nth power of x, if a nonzero real number k exists such that

$$y = \frac{k}{x^n}.$$

If $n = 1$, then $y = \frac{k}{x}$, and y **varies inversely** as x.

EXAMPLE 5

Solving an Inverse
Variation Problem

In a certain manufacturing process, the cost of producing a single item varies inversely as the square of the number of items produced. If 100 items are produced, each costs \$2. Find the cost per item if 400 items are produced.

SOLUTION Let x represent the number of items produced and y the cost per item, and write

$$y = \frac{k}{x^2}$$

for some nonzero constant k. Since $y = 2$ when $x = 100$,

$$2 = \frac{k}{100^2} \qquad \text{or} \qquad k = 20{,}000.$$

Thus, the relationship between x and y is given by

$$y = \frac{20{,}000}{x^2}.$$

When 400 items are produced, the cost per item is

$$y = \frac{20{,}000}{400^2} = .125, \text{ or } 12.5¢. ◘$$

Combined and Joint Variation

One variable may depend on more than one other variable. Such variation is called *combined variation*. More specifically, when a variable depends on the *product* of two or more other variables, it is referred to as *joint variation*.

> ## JOINT VARIATION
>
> Let m and n be real numbers. Then y **varies jointly** as the nth power of x and the mth power of z if a nonzero real number k exists such that
>
> $$y = kx^n z^m.$$

EXAMPLE 6

Solving a Joint Variation Problem

The area of a triangle varies jointly as the lengths of the base and the height. A triangle with a base of 10 feet and a height of 4 feet has an area of 20 square feet. Find the area of a triangle with a base of 3 centimeters and a height of 8 centimeters.

SOLUTION   Let A represent the area, b the base, the h the height of the triangle. Then

$$A = kbh$$

for some number k. Since A is 20 when b is 10 and h is 4,

$$20 = k(10)(4)$$

$$\frac{1}{2} = k.$$

Then $\qquad A = \frac{1}{2}bh,$

which is the familiar formula for the area of a triangle. When $b = 3$ centimeters and $h = 8$ centimeters,

$$A = \frac{1}{2}(3)(8) = 12 \text{ square centimeters.} \quad \blacksquare$$

EXAMPLE 7

Solving a Combined Variation Problem

Variation can be seen extensively in the field of photography. The formula $L = \frac{25F^2}{st}$ represents a combined variation. The luminance, L, varies directly as the square of the F-stop, F. It also varies inversely as the product of the film ASA number, s, and the shutter speed, t. The constant of variation is 25.

Suppose we want to use 200 ASA film and a shutter speed of $\frac{1}{250}$ when 500 footcandles of light are available. What would be an appropriate F-stop?

SOLUTION   We begin with the given formula $L = \frac{25F^2}{st}$ and substitute the given values for the variables: $L = 500$, $s = 200$, and $t = \frac{1}{250}$. Then solve for F.

$$L = \frac{25F^2}{st}$$

$$500 = \frac{25F^2}{200\left(\dfrac{1}{250}\right)}$$

$$400 = 25F^2$$

$$16 = F^2$$

$$4 = F \qquad (F > 0)$$

An F-stop of 4 would be appropriate. $\quad \blacksquare$

Optimization Applications

Sometimes the model for solving an optimization problem is a rational function. This is shown in the next example.

EXAMPLE 8

Solving an
Optimization
Problem

A manufacturer wants to construct cylindrical aluminum cans with a volume of 2000 cubic centimeters (2 liters). What radius and what height of the can will minimize the amount of aluminum used? What will this amount be?

SOLUTION   The two unknowns in this problem are the radius and the height of the can. We will label the radius x and the height h, as shown in Figure 17. Minimizing the amount of aluminum used requires minimizing the surface area of the can, which we will designate S. From the list of geometric formulas in Section 1.7, we find that the surface area S is given by the formula

$$S = 2\pi xh + 2\pi x^2. \qquad \text{(where } x \text{ is the radius and } h \text{ is the height)}$$

FIGURE 17

The formula involves two variables on the right, so we will solve for one in terms of the other in order to obtain a function of a single variable. Since the volume of the can is to be 2000 cubic centimeters, and the formula for volume is $V = \pi x^2 h$ (where x is the radius and h is the height), we have

$$V = \pi x^2 h$$
$$2000 = \pi x^2 h \qquad \text{Let } V = 2000.$$
$$h = \frac{2000}{\pi x^2} \qquad \text{Solve for } h.$$

Now we can write the surface area S as a function of x alone:

$$S(x) = 2\pi x\left(\frac{2000}{\pi x^2}\right) + 2\pi x^2$$
$$= \frac{4000}{x} + 2\pi x^2 \qquad\qquad\qquad *$$
$$= \frac{4000 + 2\pi x^3}{x}. \qquad \text{Combine terms.}*$$

Since x represents the radius, it must be a positive number. We graph the function using the window $[0, 20]$ by $[-500, 5000]$ and find that the local minimum is at approximately $(6.83, 878.76)$. See Figure 18 on the following page. Therefore, the radius should be 6.83 cm to the nearest hundredth of a centimeter, and the height should be approximately $\frac{2000}{\pi(6.83)^2}$, or 13.65 centimeters. These dimensions lead to a minimum amount of 878.76 cubic centimeters of aluminum used.   ∎

---

*These steps are not necessary to obtain the appropriate graph.

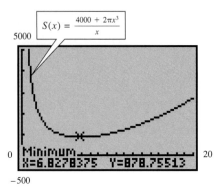

$$S(x) = \frac{4000 + 2\pi x^3}{x}$$

[0, 20] by [−500, 5000]
Xscl = 1   Yscl = 1000
The local minimum is
approximately (6.83, 878.76).

FIGURE 18

4.2 EXERCISES

In Exercises 1–6, the graph of a rational function y = f(x) is given. Use the graph to give the solution set of (a) f(x) = 0, (b) f(x) < 0, and (c) f(x) > 0. Use set braces for (a) and interval notation for (b) and (c).

1.

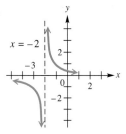

2.

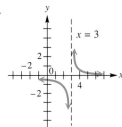

3.

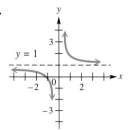

4.

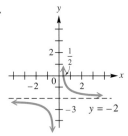

5.

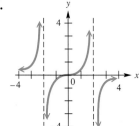

6.

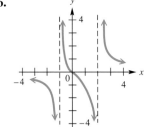

Use the methods of Examples 1 and 2 to solve the rational equation and rational inequalities given in each of Exercises 7–18. Then support your answer using the x-intercept method with a calculator-generated graph in the suggested window.

7. (a) $\dfrac{x-3}{x+5} = 0$

(b) $\dfrac{x-3}{x+5} \le 0$

(c) $\dfrac{x-3}{x+5} \ge 0$
    Window: $[-10, 10]$ by $[-5, 8]$

8. (a) $\dfrac{x+1}{x-4} = 0$

(b) $\dfrac{x+1}{x-4} \ge 0$

(c) $\dfrac{x+1}{x-4} \le 0$
    Window: $[-10, 10]$ by $[-5, 10]$

9. (a) $\dfrac{x-1}{x+2} = 1$

(b) $\dfrac{x-1}{x+2} > 1$

(c) $\dfrac{x-1}{x+2} < 1$
    Window: $[-10, 10]$ by $[-5, 10]$

10. (a) $\dfrac{x-6}{x+2} = -1$

(b) $\dfrac{x-6}{x+2} < -1$

(c) $\dfrac{x-6}{x+2} > -1$
    Window: $[-10, 10]$ by $[-10, 10]$

11. (a) $\dfrac{1}{x-1} = \dfrac{5}{4}$

(b) $\dfrac{1}{x-1} < \dfrac{5}{4}$

(c) $\dfrac{1}{x-1} > \dfrac{5}{4}$
    Window: $[-5, 5]$ by $[-5, 5]$

12. (a) $\dfrac{6}{5-3x} = 2$

(b) $\dfrac{6}{5-3x} \le 2$

(c) $\dfrac{6}{5-3x} \ge 2$
    Window: $[-5, 5]$ by $[-5, 5]$

13. (a) $\dfrac{4}{x-2} = \dfrac{3}{x-1}$

(b) $\dfrac{4}{x-2} \le \dfrac{3}{x-1}$

(c) $\dfrac{4}{x-2} \ge \dfrac{3}{x-1}$
    Window: $[-3, 3]$ by $[-20, 20]$

14. (a) $\dfrac{4}{x+1} = \dfrac{2}{x+3}$

(b) $\dfrac{4}{x+1} < \dfrac{2}{x+3}$

(c) $\dfrac{4}{x+1} > \dfrac{2}{x+3}$
    Window: $[-8, 5]$ by $[-10, 10]$

15. (a) $\dfrac{1}{(x-2)^2} = 0$

(b) $\dfrac{1}{(x-2)^2} < 0$

(c) $\dfrac{1}{(x-2)^2} > 0$
    Window: $[-5, 10]$ by $[-5, 10]$

16. (a) $\dfrac{-2}{(x+3)^2} = 0$

(b) $\dfrac{-2}{(x+3)^2} > 0$

(c) $\dfrac{-2}{(x+3)^2} < 0$
    Window: $[-10, 5]$ by $[-10, 5]$

17. (a) $\dfrac{5}{x+1} = \dfrac{12}{x+1}$

(b) $\dfrac{5}{x+1} > \dfrac{12}{x+1}$

(c) $\dfrac{5}{x+1} < \dfrac{12}{x+1}$
    Window: $[-10, 10]$ by $[-10, 10]$

18. (a) $\dfrac{7}{x+2} = \dfrac{1}{x+2}$

(b) $\dfrac{7}{x+2} \ge \dfrac{1}{x+2}$

(c) $\dfrac{7}{x+2} \le \dfrac{1}{x+2}$
    Window: $[-10, 10]$ by $[-10, 10]$

19. The graph of $y = f(x)$, where $f(x) = \frac{4}{x-2} - \frac{3}{x-1} = \frac{x+2}{x^2-3x+2}$ is shown in the window $[-2, 5]$ by $[-10, 10]$ in the accompanying figure. The solution set of $f(x) \le 0$ was required in Exercise 13(b) (in a different form, but equivalent to this one). From the graph, it appears that no part of the curve lies below the x-axis, which would lead to an empty solution set. Yet the solution set of $f(x) \le 0$ is $(-\infty, -2] \cup (1, 2)$. Use this observation to explain why relying on graphical analysis alone is not sufficient for solving equations and inequalities.

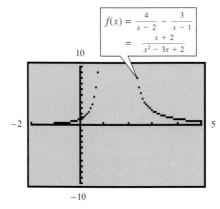

$$f(x) = \frac{4}{x-2} - \frac{3}{x-1}$$
$$= \frac{x+2}{x^2-3x+2}$$

Dot Mode

20. Consider the rational function $f(x) = \frac{1}{x^2+1}$. Without graphing, answer the following series of items in order.

   (a) For all real numbers x, $x^2 + 1$ is _____ 0.
                                         (equal to, greater than, less than)

   (b) Because $1 > 0$, for all real numbers x, $f(x)$ is _____ 0.
                                         (equal to, greater than, less than)

   (c) The solution set of $f(x) = 0$ is _____, of $f(x) < 0$ is _____, and of $f(x) > 0$ is _____.

   (d) Support your answers in part (c) with a graph and an explanation.

Use the methods explained in Examples 3 and 4 to find all complex solutions for each equation. Support your real solutions with a graph using an appropriate window.

21. $1 - \dfrac{13}{x} + \dfrac{36}{x^2} = 0$ **22.** $1 - \dfrac{3}{x} - \dfrac{10}{x^2} = 0$

23. $1 + \dfrac{3}{x} = \dfrac{5}{x^2}$ **24.** $4 + \dfrac{7}{x} = -\dfrac{1}{x^2}$

25. $\dfrac{x}{2-x} + \dfrac{2}{x} - 5 = 0$ **26.** $\dfrac{2x}{x-3} + \dfrac{4}{x} - 6 = 0$

27. $x^{-4} - 3x^{-2} - 4 = 0$ **28.** $x^{-4} - 5x^{-2} + 4 = 0$

29. $\dfrac{1}{x+2} + \dfrac{3}{x+7} = \dfrac{5}{x^2+9x+14}$ **30.** $\dfrac{1}{x+3} + \dfrac{4}{x+5} = \dfrac{2}{x^2+8x+15}$

31. $\dfrac{x}{x-3} + \dfrac{4}{x+3} = \dfrac{18}{x^2-9}$ **32.** $\dfrac{2x}{x-3} + \dfrac{4}{x+3} = \dfrac{24}{9-x^2}$

33. $9x^{-1} + 4(6x-3)^{-1} = 2(6x-3)^{-1}$ **34.** $x(x-2)^{-1} + x(x+2)^{-1} = 8(x^2-4)^{-1}$

Relating Concepts

In Example 2 we illustrated how a sign graph may be used to solve a rational inequality analytically. There is another method of solving rational inequalities analytically. For example, suppose that we wish to solve

$$\frac{x-2}{x+3} \le 2.$$

We first notice that -3 is not in the domain, so it cannot possibly be part of the solution set. We cannot multiply both sides by $x + 3$, since we do not know whether we are multiplying by a positive number or a negative number, and in the case of the latter, reversal of the inequality symbol would be necessary. However, we do know that for $x \ne -3$, the expression $(x + 3)^2$ *must* be positive. (Why?) Therefore, we may multiply both sides by $(x + 3)^2$ and keep the inequality sign in the same direction.

$$\frac{x-2}{x+3} \le 2 \qquad \text{Given inequality}$$

$$(x+3)^2 \left(\frac{x-2}{x+3} \right) \le 2(x+3)^2 \qquad \text{Multiply by } (x+3)^2 \text{, which is positive.}$$

$$(x+3)(x-2) \le 2(x^2+6x+9)$$

$$x^2 + x - 6 \le 2x^2 + 12x + 18$$

$$0 \le x^2 + 11x + 24$$

$$x^2 + 11x + 24 \ge 0$$

The polynomial on the left, $x^2 + 11x + 24$, defines a quadratic function whose graph opens upward. Its x-intercepts are the solutions of $x^2 + 11x + 24 = 0$:

$$x^2 + 11x + 24 = 0$$

$$(x+3)(x+8) = 0 \qquad \text{Factor.}$$

$$x = -3 \quad \text{or} \quad x = -8. \qquad \text{Use the zero-product property.}$$

The graph of $y = x^2 + 11x + 24$ indicates that it is *above* or *on* the x-axis for $(-\infty, -8] \cup [-3, \infty)$. See the accompanying figure. The inequality $x^2 + 11x + 24 \ge 0$ is equivalent to the original inequality except that -3 must be excluded from the solution set. Therefore, the solution set of the given inequality is $(-\infty, -8] \cup (-3, \infty)$.

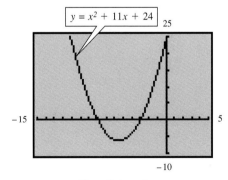

The x-intercepts are -8 and -3.

Use the method described above to solve the following rational inequalities.

35. $\dfrac{x + 5}{x - 3} \geq 1$

36. $\dfrac{2 - x}{x + 6} \geq -1$

37. $\dfrac{(x - 3)^2}{x + 1} \geq x + 1$

38. $\dfrac{(x - 6)^2}{x + 5} \geq x + 5$

39. $\dfrac{x + 3}{(x + 1)^2} < \dfrac{1}{x + 1}$

40. $\dfrac{x - 4}{(x + 2)^2} > \dfrac{1}{x + 2}$

Use a purely graphical method to solve the equation in part (a), expressing solutions to the nearest hundredth. Then use the graph to solve the associated inequalities in parts (b) and (c), expressing endpoints to the nearest hundredth.

41. (a) $\dfrac{\sqrt{2x + 5}}{x^3 - \sqrt{3}} = 0$

    (b) $\dfrac{\sqrt{2x + 5}}{x^3 - \sqrt{3}} > 0$

    (c) $\dfrac{\sqrt{2x + 5}}{x^3 - \sqrt{3}} < 0$

42. (a) $\dfrac{\sqrt[3]{7x^3} - 1}{x^2 + 2} = 0$

    (b) $\dfrac{\sqrt[3]{7x^3} - 1}{x^2 + 2} > 0$

    (c) $\dfrac{\sqrt[3]{7x^3} - 1}{x^2 + 2} < 0$

R*elating Concepts*

Consider the rational equation below, and work Exercises 43–46 in order.

$$1 - \frac{2}{x} - \frac{2}{x^2} + \frac{3}{x^3} = 0$$

43. Graph $f(x) = 1 - \frac{2}{x} - \frac{2}{x^2} + \frac{3}{x^3}$ and determine its x-intercepts graphically. (*Hint:* One is rational and two are irrational.) Give approximations of the irrational solutions correct to the nearest hundredth.

44. What is the domain of f?

45. Use the rational solution determined in Exercise 43 with synthetic division to obtain a quadratic equation that will give the other two solutions in their exact forms. (*Hint:* Multiply both sides by x^3 first.) (See Sections 3.3 and 3.6.)

46. Verify that the approximations found in Exercise 43 are indeed approximations of the two irrational solutions whose exact values were found in Exercise 45.

47. A student attempted to solve the inequality

$$\frac{2x - 1}{x + 2} \leq 0$$

by multiplying both sides by $x + 2$ to get

$$2x - 1 \leq 0$$

$$x \leq \frac{1}{2}.$$

He wrote the solution set as $(-\infty, \frac{1}{2}]$. Is his solution correct? Explain.

48. The inequality in Exercise 47 may be solved using this alternate analytic method:

Case 1: For $x > -2$, $x + 2$ is positive. Multiply both sides by $x + 2$, keeping the inequality symbol pointing in the same direction. Take the intersection of the interval where $x > -2$, $(-2, \infty)$, and the interval obtained in the solution process.

Case 2: For $x < -2$, $x + 2$ is negative. Multiply both sides by $x + 2$ and reverse the direction of the inequality symbol. Take the intersection of the interval where $x < -2$, $(-\infty, -2)$, and the interval obtained in the solution process.

The *union* of the intervals obtained in the two cases is the solution set of the original inequality. Solve the inequality in Exercise 47 using this method.

Solve the following problems involving various types of variation.

49. If m varies directly as x and y, and $m = 10$ when $x = 4$ and $y = 7$, find m when $x = 11$ and $y = 8$.

50. Suppose m varies directly as z and p. If $m = 10$ when $z = 3$ and $p = 5$, find m when $z = 5$ and $p = 7$.

51. Suppose r varies directly as the square of m, and inversely as s. If $r = 12$ when $m = 6$ and $s = 4$, find r when $m = 4$ and $s = 10$.

52. Suppose p varies directly as the square of z, and inversely as r. If $p = \frac{32}{5}$ when $z = 4$ and $r = 10$, find p when $z = 2$ and $r = 16$.

53. Let a be proportional to m and n^2, and inversely proportional to y^3. If $a = 9$ when $m = 4$, $n = 9$, and $y = 3$, find a if $m = 6$, $n = 2$, and $y = 5$.

54. If y varies directly as x, and inversely as m^2 and r^2, and $y = \frac{5}{3}$ when $x = 1$, $m = 2$, and $r = 3$, find y if $x = 3$, $m = 1$, and $r = 8$.

55. For $k > 0$, if y varies directly as x, when x increases, y _____ , and when x decreases, y _____ .

56. For $k > 0$, if y varies inversely as x, when x increases, y _____ , and when x decreases, y _____ .

Solve each problem.

57. In electric current flow, it is found that the resistance (measured in units called ohms) offered by a fixed length of wire of a given material varies inversely as the square of the diameter of the wire. If a wire .01 inch in diameter has a resistance of .4 ohm, what is the resistance of a wire of the same length and material with a diameter of .03 inch?

58. The illumination produced by a light source varies inversely as the square of the distance from the source. The illumination of a light source at 5 m is 70 candela. What is the illumination 12 m from the source?

59. The number of vibrations per second (the pitch) of a steel guitar string varies directly as the square root of the tension and inversely as the length of the string. If the number of vibrations per second is 5 when the tension is 225 kilograms and the length is .60 meter, find the number of vibrations per second when the tension is 196 kilograms and the length is .65 meter.

60. The period of a pendulum varies directly as the square root of the length of the pendulum and inversely as the square root of the acceleration due to gravity. Find the period when the length is 121 cm and the acceleration due to gravity is 980 cm per sec squared, if the period is 6π sec when the length is 289 cm and the acceleration due to gravity is 980 cm per sec squared.

61. Under certain conditions, the length of time that it takes for fruit to ripen during the growing season varies inversely as the average maximum temperature during the season. If it takes 25 days for friut to ripen with an average maximum temperature of 80°, find the number of days it would take at 75°.

62. The number of long-distance phone calls between two cities in a certain time period varies directly as the populations p_1 and p_2 of the cities, and inversely as the distance between them. If 10,000 calls are made between two cities 500 mi apart, having populations of 50,000 and 125,000, find the number of calls between two cities 800 mi apart having populations of 20,000 and 80,000.

63. A measure of malnutrition, called the *pelidisi,* varies directly as the cube root of a person's weight in grams and inversely as the person's sitting height in centimeters. A person with a pelidisi below 100 is considered to be undernourished, while a pelidisi greater than 100 indicates overfeeding. A person who weighs 48,820 g with a sitting height of 78.7 cm has a pelidisi of 100. Find the pelidisi (to the nearest whole number) of a person whose weight is 54,430 g and whose sitting height is 88.9 cm. Is this individual undernourished or overfed?

64. The force needed to keep a car from skidding on a curve varies inversely as the radius of the curve and jointly as the weight of the car and the square of the speed. It takes 3000 lb of force to keep a 2000-lb car from skidding on a curve of radius 500 ft at 30 mph. What force is needed to keep the same car from skidding on a curve of radius 800 ft at 60 mph?

65. The roof of a new sports arena rests on round concrete pillars. The maximum load a cylindrical column of circular cross section can hold varies directly as the fourth power of the diameter and inversely as the square of the height. The arena has 9 m tall columns that are 1 m in diameter and will support a load of 8 metric tons. How many metric tons will be supported by a column 12 m high and $\frac{2}{3}$ m in diameter?

66. The sports arena in Exercise 65 requires a beam 16 m long, 24 cm wide, and 8 cm high. The maxi-

Refer to Example 7 to work Exercises 67 and 68.

67. Determine the luminance needed when a photographer is using 400 ASA film, a shutter speed of $\frac{1}{60}$ of a second, and an F-stop of 5.6.

Solve the following problems.

69. Antique-car owners often enter their cars in a *concours d'elegance* in which a maximum of 100 points can be awarded to a particular car. Points are awarded for the general attractiveness of the car. The function defined by

$$C(x) = \frac{10x}{49(101 - x)}$$

expresses the cost, in thousands of dollars, of restoring a car so that it will win x points. Use the graph of this function to determine the number of points that would be awarded if $13,000 was spent in restoring the car.

70. In situations involving environmental pollution, a cost-benefit model expresses cost as a function of the percentage of pollutant removed from the environment. Suppose a cost-benefit model is expressed as

$$C(x) = \frac{6.7x}{100 - x},$$

where $C(x)$ is the cost in thousands of dollars of removing x percent of a certain pollutant. Use the graph of this model to determine the cost to remove 95 percent of the pollutant.

71. A truck burns fuel at the rate of

$$G(x) = \frac{1}{32}\left(\frac{64}{x} + \frac{x}{50}\right)$$

gallons per mile while traveling at x mph. Use the graph of the total cost function to find each of the following.
(a) the speed that will produce the minimum total cost for a 400-mile trip if fuel costs $1.60 per gallon
(b) the minimum total cost (*Hint:* The total cost is the product of the number of gallons per mile, the number of miles, and the cost per gallon.)

72. A rock-and-roll band travels from engagement to engagement in a large bus. This bus burns fuel at the

mum load of a horizontal beam that is supported at both ends varies directly as the width and square of the height and inversely as the length between supports. If a beam of the same material 8 m long, 12 cm wide, and 15 cm high can support a maximum of 400 kg, what is the maximum load the beam in the arena will support?

68. If 125 footcandles of light are available and an F-stop of 2 is used with 200 ASA film, what shutter speed should be used?

rate of

$$G(x) = \frac{1}{50}\left(\frac{200}{x} + \frac{x}{15}\right)$$

gallons per mile while traveling at x mph. Do each of the following.
(a) If fuel costs $2 per gallon, find the speed that will produce minimum total cost for a 250-mile trip.
(b) Find the minimum total cost. (See the hint for Exercise 71.)

73. If the cost of producing x items of a particular commodity is given by y, where $y = 200,000 + .50x$ (in dollars), then the *average cost per unit* for the commodity is given by the function

$$C(x) = \frac{y}{x} = \frac{200,000 + .50x}{x}, \quad x > 0.$$

Use the graph of $C(x)$ to find the average cost when 25,000 units are produced.

74. See Exercise 73. If

$$y = .2x^2 + 20x + 10, \quad x > 0,$$

approximate the number of units that should be produced to minimize the average cost per unit, by observing the graph of $C(x) = \frac{y}{x}$.

75. A company finds that if it places orders for stereo receivers x times per year, the expense of ordering and storage is given by the function

$$E(x) = 40x + \frac{1000}{x}, \quad x > 0,$$

dollars. Find the number of times that the company should order per year so that their ordering and storage expenses are minimized. (Use the graph of $y = E(x)$.)

76. A metal cylindrical can with an *open top* and *closed bottom* is to have a volume of 4 cubic feet. Find the dimensions that require the least amount of material. (Compare this problem to Example 8.)

Further Explorations

A packaging company is studying various shapes for cylindrical aluminum cans. The plan is to build a can with a volume of 1000 cubic centimeters. While minimizing the amount of material used ultimately will be a consideration, for now they are only interested in aesthetics and want to determine the height the can needs to be depending on a given radius. Use the TABLE feature of your graphics calculator to find the heights of cans when given a radius of 2, 4, 6, 8, 10, and 12 centimeters. (Let x be the radius, solve the volume formula for height. Then Y_1 will be the height of the can.)

4.3 GRAPHS OF ROOT FUNCTIONS

Radicals and Rational Exponents ▌ Graphs of $f(x) = \sqrt[n]{ax + b}$ ▌ Special Functions Defined by Roots

Radicals and Rational Exponents*

Up to this point, we have used radicals in a very informal manner. We observed the graphs of $y = \sqrt{x}$ and $y = \sqrt[3]{x}$ in Chapter 2; now we will investigate radical notation and rational numbers used as exponents in a more formal way.

DEFINITION OF $\sqrt[n]{a}$

Suppose that n is a positive integer greater than 1. Let a be any real number.

(i)   If $a > 0$, $\sqrt[n]{a}$ is the positive real number b such that $b^n = a$.
(ii) If $a < 0$ and n is odd, $\sqrt[n]{a}$ is the negative real number b such that $b^n = a$.
(iii) If $a < 0$ and n is even, then there is no real number $\sqrt[n]{a}$.
(iv) If $a = 0$, then $\sqrt[n]{a} = 0$.

In the radical expression $\sqrt[n]{a}$, a is called the **radicand** and n is called the **root index.**

EXAMPLE **1**
Using Radical Notation

The display in Figure 19 illustrates how a calculator will compute roots that are rational numbers.

$$\sqrt{25} = 5$$
$$\sqrt[3]{-27} = -3$$
$$\sqrt[5]{-32} = -2$$

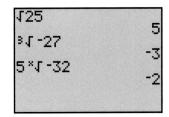

FIGURE 19

---

*At this point, you may wish to review the material on radicals and rational exponents in the appendix.

The display in Figure 20 illustrates how a calculator will give rational approximations of roots that are irrational numbers.

$$\sqrt[3]{-7} \approx -1.912931183$$
$$\sqrt[5]{36.7} \approx 2.055574459$$
$$\sqrt{21} \approx 4.582575695 \qquad \blacksquare$$

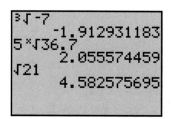

FIGURE 20

A familiar rule for exponents states that $(a^m)^n = a^{mn}$. If this rule is to be extended to rational numbers, then it makes sense to define $a^{1/n}$ as $\sqrt[n]{a}$, since $(a^{1/n})^n = a^1 = a$, while $(\sqrt[n]{a})^n = a^1 = a$ as well. We now define $a^{1/n}$.

DEFINITION OF $a^{1/n}$

(i) If n is an *even* positive integer and if $a > 0$, then $a^{1/n}$ is the *positive* real number whose nth power is a. That is, $a^{1/n} = \sqrt[n]{a}$.

(ii) If n is an *odd* positive integer and if a is any real number, then $a^{1/n}$ is the single real number whose nth power is a. That is, $a^{1/n} = \sqrt[n]{a}$.

EXAMPLE 2

Using $a^{1/n}$ Notation

The display in Figure 21 illustrates how a calculator will compute numbers raised to the power $\frac{1}{n}$, leading to rational number results. Compare with Figure 19.

$$25^{1/2} = 5$$
$$(-27)^{1/3} = -3$$
$$(-32)^{1/5} = -2$$

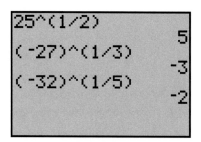

FIGURE 21

The display in Figure 22 illustrates how a calculator will compute numbers raised to the power $\frac{1}{n}$, leading to rational approximations of irrational results. Compare with Figure 20.

$$(-7)^{1/3} \approx -1.912931183$$
$$36.7^{1/5} \approx 2.055574459$$
$$21^{1/2} \approx 4.582575695 \quad \blacksquare$$

```
(-7)^(1/3)
           -1.912931183
(36.7)^(1/5)
            2.055574459
21^(1/2)
            4.582575695
```

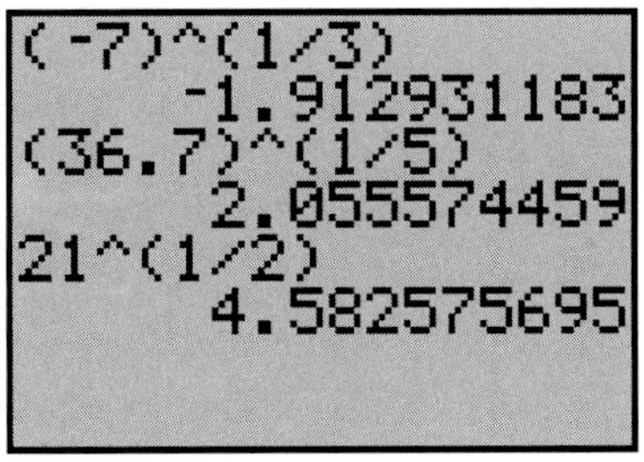

FIGURE 22

The definition of more general rational exponents is consistent with the familiar rules of exponents. For the power rule to hold, $(a^{1/n})^m$ must equal $a^{m/n}$.

DEFINITION OF $a^{m/n}$

If $\frac{m}{n}$ is a rational number where n is a positive integer greater than 1, and a is a real number such that $\sqrt[n]{a}$ is also real, then

$$a^{m/n} = (\sqrt[n]{a})^m = \sqrt[n]{a^m}.$$

EXAMPLE 3

Using $a^{m/n}$ Notation

(a) $125^{2/3} = (\sqrt[3]{125})^2 = 5^2 = 25$, or $\sqrt[3]{125^2} = \sqrt[3]{15{,}625} = 25$

(b) $32^{7/5} = (\sqrt[5]{32})^7 = 2^7 = 128$

(c) $(-27)^{2/3} = (\sqrt[3]{-27})^2 = (-3)^2 = 9$

(d) $16^{-3/4} = \dfrac{1}{16^{3/4}} = \dfrac{1}{(\sqrt[4]{16})^3} = \dfrac{1}{2^3} = \dfrac{1}{8}$ or .125

(e) $(-4)^{5/2}$ is not real because $\sqrt{-4}$ is not real.

(f) $-4^{5/2} = -(\sqrt{4})^5 = -(2^5) = -32$ $\quad \blacksquare$

CAUTION Notice the difference between the expressions in parts e and f of Example 3. In part e, the base is -4, while in part f, the base is 4. Be very careful when dealing with expressions of these types.

FOR GROUP DISCUSSION

Use your calculators to verify the results in Example 3. Some models of graphics calculators will not allow you to calculate the expression in part c, $(-27)^{2/3}$, by inputting the symbols as they appear. Use this fact to explain why it is essential for mathematical concepts to be studied hand-in-hand with technology.

Let us for the moment consider the function $f(x) = x^{3/2}$. Because $x^{3/2} = (\sqrt{x})^3$, the domain of this function must be $[0, \infty)$. It seems reasonable to expect that the graph of f will lie "between" the graphs of $y = x^1 = x$ and $y = x^2$, since $1 < \frac{3}{2} < 2$. Figure 23 supports this expectation.

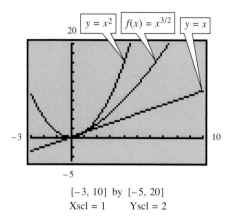

[−3, 10] by [−5, 20]
Xscl = 1   Yscl = 2

FIGURE 23

Now, let us examine several different ways to obtain a rational approximation of the number $14^{3/2}$. See Figure 24.

$$14^{3/2} \approx 52.38320341$$
$$(\sqrt{14})^3 \approx 52.38320341$$
$$\sqrt{14^3} \approx 52.38320341$$
$$14^{1.5} \approx 52.38320341 \qquad \text{(because } 1.5 = \tfrac{3}{2})$$
$$(14^{.5})^3 \approx 52.38320341$$
$$(14^3)^{.5} \approx 52.38320341$$

```
14^(3/2)
            52.38320341
(√14)³
            52.38320341
√(14³)
            52.38320341
```

(a)

TECHNOLOGICAL NOTE
The number of digits shown in the display can be set by the user of a graphics calculator. For example, the irrational number seen in Figure 24 has a total of 10 digits displayed, 8 of which follow the decimal point. Read your owner's manual to see how your model allows you to determine the number of digits that will be displayed.

```
14^1.5
            52.38320341
(14^.5)³
            52.38320341
(14³)^.5
            52.38320341
■
```

(b)

FIGURE 24

As seen in Figure 24, there are several options as to how this approximation can be found. Now, look at Figure 25, which shows the graph of $f(x) = x^{3/2}$. The display shows graphical support of what we examined numerically in Figure 24.

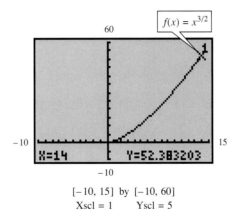

[−10, 15] by [−10, 60]
Xscl = 1   Yscl = 5

FIGURE 25

EXAMPLE **4**

Applying Rational
Exponents

Meteorologists can approximate the duration of a storm by using the formula $T = .07D^{3/2}$, where D is the diameter of the storm in miles and T is the time in hours. Suppose that radar shows a storm in the Gulf of Mexico to have a diameter of 3.8 miles. Approximately how long will the storm last?

SOLUTION   Let $D = 3.8$ and use a calculator to find T

$T = .07(3.8)^{3/2}$ Let $D = 3.8$.

$T = .52$ (to the nearest hundredth)

We see that the storm can be expected to last approximately .52 hour, or a little more than 30 minutes. This result can be supported by a graph of $y = .07x^{3/2}$. When $x = 3.8$, $y \approx .52$. See Figure 26.   ◼

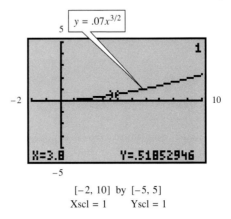

[−2, 10] by [−5, 5]
Xscl = 1   Yscl = 1

FIGURE 26

Graphs of $f(x) = \sqrt[n]{ax + b}$

Figure 13 in Section 2.2 shows the graph of $f(x) = \sqrt{x}$. When n is even, the graph of $f(x) = \sqrt[n]{x}$ resembles the graph of the square root function; as n gets larger, the graph lies closer to the x-axis as $x \to \infty$. Figure 27 shows the graphs of $y = \sqrt{x}$, $y = \sqrt[4]{x}$, and $y = \sqrt[6]{x}$, as examples of the graph of $f(x) = \sqrt[n]{x}$ where n is even, along with other pertinent information.

ROOT FUNCTION, n EVEN

$f(x) = \sqrt[n]{x}$ (Figure 27 for $n = 2, 4, 6$)

Domain: $[0, \infty)$

Range: $[0, \infty)$

For n even, the root function $f(x) = \sqrt[n]{x}$ increases on $(0, \infty)$. It is also continuous on $(0, \infty)$.

The definition of $x^{1/n}$ allows us to use a rational exponent as well as a radical when graphing a root function.

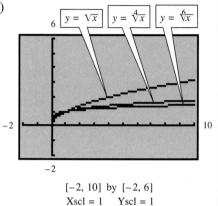

$[-2, 10]$ by $[-2, 6]$
Xscl = 1   Yscl = 1

FIGURE 27

A discussion similar to the one above pertains to the graph of $f(x) = \sqrt[n]{x}$, where n is odd. We saw the graph of $f(x) = \sqrt[3]{x}$, in Figure 14 of Section 2.2. When n is odd, the graph of $f(x) = \sqrt[n]{x}$ resembles the graph of the cube root function; as n gets larger, the graph lies closer to the x-axis as $|x| \to \infty$. Figure 28 shows the graphs of $y = \sqrt[3]{x}$, $y = \sqrt[5]{x}$, and $y = \sqrt[7]{x}$, as examples of the graph of $f(x) = \sqrt[n]{x}$ where n is odd, along with other pertinent information.

ROOT FUNCTION, n ODD

$f(x) = \sqrt[n]{x}$ (Figure 28 for $n = 3, 5, 7$)

Domain: $(-\infty, \infty)$

Range: $(-\infty, \infty)$

For n odd, the root function $f(x) = \sqrt[n]{x}$ increases on the domain $(-\infty, \infty)$. It is also continuous on $(-\infty, \infty)$.

The definition of $x^{1/n}$ allows us to use a rational exponent as well as a radical when graphing a root function.

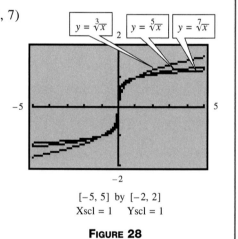

$[-5, 5]$ by $[-2, 2]$
Xscl = 1   Yscl = 1

FIGURE 28

In order to determine the domain of a function of the form $f(x) = \sqrt[n]{ax + b}$, we must note the parity of n. If n is even, $ax + b$ must be greater than or equal to 0; if n is odd, $ax + b$ can be any real number.

<table>
<tr><td>

EXAMPLE 5

Finding Domains of
Root Functions

</td></tr>
</table>

(a) Find the domain of the function $f(x) = \sqrt{4x + 12}$.

SOLUTION   For the function to be defined, $4x + 12$ must be greater than or equal to 0, since this is an even root ($n = 2$):

$$4x + 12 \geq 0$$
$$4x \geq -12$$
$$x \geq -3$$

The domain of f is $[-3, \infty)$.

(b) Find the domain of the function $g(x) = \sqrt[3]{-8x + 8}$.

SOLUTION   Because $n = 3$, an odd number, the domain of g is $(-\infty, \infty)$. ◼

<table>
<tr><td>

EXAMPLE 6

Transforming
Graphs of Root
Functions

</td></tr>
</table>

(a) Use the terminology of Sections 2.2 and 2.3 to explain how the graph of $y = \sqrt{4x + 12}$ can be obtained from the graph of $y = \sqrt{x}$. Then graph it in an appropriate window.

SOLUTION   We begin by writing the expression in an equivalent form.

$$y = \sqrt{4x + 12} \qquad \text{Given form}$$
$$y = \sqrt{4(x + 3)} \qquad \text{Factor.}$$
$$y = \sqrt{4}\sqrt{x + 3} \qquad \sqrt{ab} = \sqrt{a} \cdot \sqrt{b}$$
$$y = 2\sqrt{x + 3} \qquad \sqrt{4} = 2$$

The graph of this function can be obtained from the graph of $y = \sqrt{x}$ by shifting horizontally 3 units to the left and stretching vertically by a factor of 2. Figure 29 shows both graphs for comparison. Notice that the domain is $[-3, \infty)$, as determined in Example 5(a). The range is $[0, \infty)$.

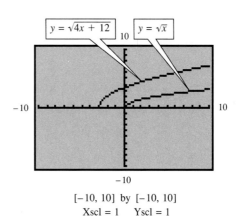

$[-10, 10]$ by $[-10, 10]$
Xscl = 1   Yscl = 1

FIGURE 29

(b) Repeat part (a) for the graph of $y = \sqrt[3]{-8x + 8}$, as compared to the graph of $y = \sqrt[3]{x}$.

SOLUTION $y = \sqrt[3]{-8x + 8} \qquad \text{Given form}$
$$y = \sqrt[3]{-8(x - 1)} \qquad \text{Factor.}$$
$$y = \sqrt[3]{-8} \cdot \sqrt[3]{x - 1} \qquad \sqrt[3]{ab} = \sqrt[3]{a} \cdot \sqrt[3]{b}$$
$$y = -2\sqrt[3]{x - 1} \qquad \sqrt[3]{-8} = -2$$

From this form we see that the graph can be obtained by shifting the graph of $y = \sqrt[3]{x}$ one unit to the right, stretching vertically by a factor of 2, and reflecting across the x-axis (because of the negative sign in -2). Figure 30 shows both graphs. The domain and range are both $(-\infty, \infty)$. ∎

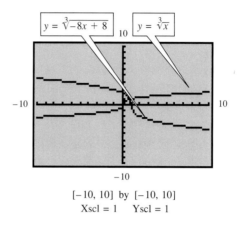

$$y = \sqrt[3]{-8x + 8}$$ $$y = \sqrt[3]{x}$$

[−10, 10] by [−10, 10]
Xscl = 1   Yscl = 1

FIGURE 30

(In Exercise 80, we will examine the function in Example 6(b) in a different manner.)

Special Functions Defined by Roots

Suppose that we consider the graph of all points (x, y) that lie a fixed distance r $(r > 0)$ from the origin. Then by the distance formula,

$$\sqrt{(x - 0)^2 + (y - 0)^2} = r.$$

Simplifying and squaring both sides of this equation leads to

$$x^2 + y^2 = r^2.$$

The graph of this set of points is a circle with center at $(0, 0)$ and radius r. See Figure 31. While we will study circles in more detail in Chapter 6, this brief introduction gives us an opportunity to investigate a special type of function defined by a root.

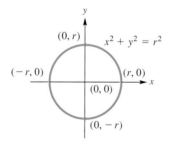

FIGURE 31

Because a circle is not the graph of a function, a graphics calculator in function mode is not appropriate for graphing a circle. However, if we imagine the circle as being the union of the graphs of two functions, one the top semicircle and the other the bottom semicircle, then we can graph both functions in the same square viewing window to obtain the desired graph. To accomplish this, we solve for y in the equation $x^2 + y^2 = r^2$.

$$x^2 + y^2 = r^2$$
$$y^2 = r^2 - x^2 \qquad \text{Subtract } x^2.$$
$$y = \pm\sqrt{r^2 - x^2} \qquad \text{Use the square root property of solving equations, from Section 3.3.}$$

The final line can be interpreted as two equations, both of which define functions:

$$y_1 = \sqrt{r^2 - x^2} \qquad \text{(the semicircle above the } x\text{-axis)}$$

and

$$y_2 = -\sqrt{r^2 - x^2}. \qquad \text{(the semicircle below the } x\text{-axis)}$$

Notice that $y_2 = -y_1$, indicating that the graph of the "bottom" semicircle is simply a reflection of the graph of the "top" semicircle across the x-axis.

Use a calculator in function mode to graph the circle $x^2 + y^2 = 36$.

SOLUTION Based on the argument above, this graph can be obtained by graphing both

$$y_1 = \sqrt{36 - x^2} \qquad \text{and} \qquad y_2 = -y_1 = -\sqrt{36 - x^2}$$

in the same window. In order to obtain a graph that is visually in correct proportions, we use a square window on the calculator. Figure 32 shows the graph of these two root functions, forming a circle. ❒

EXAMPLE **7**
Graphing a Circle Using a Calculator in the Function Mode

TECHNOLOGICAL NOTE
The circle shown in Figure 32 is formed by the union of two functions. Because of the manner in which the calculator plots points in the function mode, the two semicircles do not completely "connect". This will often happen, and you should realize that *mathematically, this is a complete circle.* This is an excellent example illustrating how we must understand the concepts in order to interpret what we see on the screen.

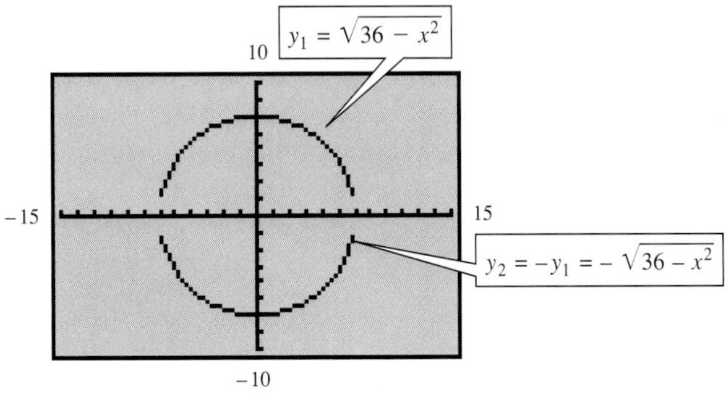

$y_1 = \sqrt{36 - x^2}$

$y_2 = -y_1 = -\sqrt{36 - x^2}$

$[-15, 15]$ by $[-10, 10]$
A square window is
necessary to obtain the
correct perspective.

FIGURE 32

FOR **GROUP DISCUSSION**

The two functions that form the circle in Example 7 both have the same domain.

1. Discuss how a sign graph can be used to solve the inequality $36 - x^2 \geq 0$. How does this inequality pertain to the graphs found in Example 7?
2. Figure 33 shows the graph of $y = 36 - x^2$. Use the graph to find the solution set of $36 - x^2 \geq 0$. Discuss how this solution set pertains to the graphs found in Example 7.

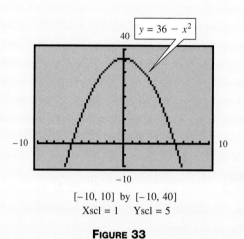

$[-10, 10]$ by $[-10, 40]$
Xscl = 1   Yscl = 5

FIGURE 33

From the discussion in Sections 3.2 and 3.3, we know that the graph of $y = ax^2 + bx + c$ $(a \neq 0)$ is a parabola with a vertical axis of symmetry. Such an equation defines a function. If we reverse the roles of x and y, however, we obtain an equation whose graph is also a parabola, but having a horizontal axis of symmetry. (These, too, will be examined more closely in Chapter 6.) Figure 34 shows the graph of $x = 2y^2 + 6y + 5$. Notice that this is not the graph of a function. However, if we consider it to be the union of the graphs of two functions, one the top half-parabola and the other the bottom, then it can be graphed by a calculator in the function mode. The two equations can be found by the method of completing the square. The final example shows how this can be done.

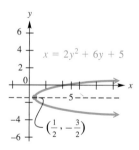

FIGURE 34

EXAMPLE 8

Graphing a
Horizontal Parabola
Using a Calculator
in Function Mode

Graph $x = 2y^2 + 6y + 5$ in the window $[-2, 8]$ by $[-8, 2]$.

SOLUTION We must solve for y by completing the square.

$$x = 2y^2 + 6y + 5$$

$$\frac{x}{2} = y^2 + 3y + \frac{5}{2} \qquad \text{Divide by 2.}$$

$$\frac{x}{2} - \frac{5}{2} = y^2 + 3y \qquad \text{Subtract } \tfrac{5}{2}.$$

$$\frac{x}{2} - \frac{5}{2} + \frac{9}{4} = y^2 + 3y + \frac{9}{4} \qquad \text{Add } [\tfrac{1}{2}(3)]^2 = \tfrac{9}{4}.$$

$$\frac{x}{2} - \frac{1}{4} = \left(y + \frac{3}{2}\right)^2 \qquad \begin{array}{l}\text{Combine terms on the left and}\\\text{factor on the right.}\end{array}$$

$$\left(y + \frac{3}{2}\right)^2 = \frac{x}{2} - \frac{1}{4}$$

$$y + \frac{3}{2} = \pm \sqrt{\frac{x}{2} - \frac{1}{4}} \qquad \text{Use the square root property.}$$

$$y = -\frac{3}{2} \pm \sqrt{\frac{x}{2} - \frac{1}{4}} \qquad \text{Subtract } \tfrac{3}{2}.$$

Two functions are now defined. It is easier to use decimal notation to input the equations into a calculator. Therefore, let

$$y_1 = -1.5 + \sqrt{.5x - .25} \quad \text{and} \quad y_2 = -1.5 - \sqrt{.5x - .25}.$$

The graphs of these two functions together form the parabola with horizontal axis of symmetry $y = -1.5$, both of which are shown in Figure 35. ∎

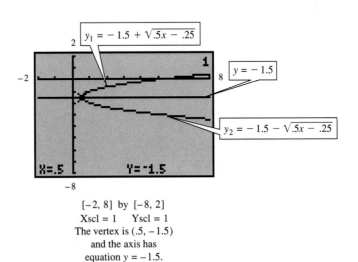

$[-2, 8]$ by $[-8, 2]$
Xscl $= 1$ Yscl $= 1$
The vertex is $(.5, -1.5)$
and the axis has
equation $y = -1.5$.

FIGURE 35

FOR **GROUP DISCUSSION**

The two functions that form the parabola in Example 8 both have the same domain.

1. Solve the inequality $.5x - .25 \geq 0$ using analytic methods. How does the solution set of this inequality pertain to the graphs found in Example 8?

2. Figure 36 shows the graph of $y = .5x - .25$. Use this graph to find the solution set of $.5x - .25 \geq 0$. Discuss how this solution set pertains to the graphs found in Example 8.

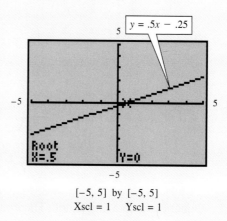

$[-5, 5]$ by $[-5, 5]$
Xscl = 1   Yscl = 1

FIGURE 36

4.3 EXERCISES

Evaluate each of the following without *the use of a calculator, using the definitions of* $\sqrt[n]{a}$ *and* $a^{m/n}$. *Then check your result with your calculator.*

1. $\sqrt{169}$ 2. $-\sqrt[3]{64}$ 3. $\sqrt[5]{-32}$ 4. $\sqrt[4]{16}$

5. $81^{3/2}$ 6. $27^{4/3}$ 7. $125^{-2/3}$ 8. $(\sqrt[3]{-27})^2$

9. $(-1000)^{2/3}$ 10. $(-125)^{-4/3}$

11. Some graphics calculators will not compute a value for expressions like $(-8)^{2/3}$, having a negative base and a rational exponent with an odd denominator and even numerator. Check to see if your particular model is one of these. If it is, use the fact that $(-8)^{2/3} = [(-8)^{1/3}]^2$ to calculate it. What rule for exponents applies here?

12. Graph $y = x^{2/3}$ in the standard window of your calculator. Refer to the concept explained in Exercise 11 if you get no graph for negative values of x.

Use your calculator to find each of the following roots. Tell whether the display represents the exact value or a rational approximation of an irrational number. Give as many digits as your display shows.

13. $\sqrt[6]{9}$ 14. $\sqrt[7]{12}$ 15. $\sqrt[3]{18.609625}$ 16. $\sqrt[5]{286.29151}$

17. $\sqrt[3]{-17}$ 18. $\sqrt[5]{-8}$ 19. $\sqrt[6]{\pi^2}$ 20. $\sqrt[6]{\pi^{-1}}$

Use your calculator to find each of the following powers. Tell whether the display represents the exact value or a rational approximation of an irrational number. Give as many digits as your display shows.

21. $5^{.1}$

22. $12^{.37}$

23. $\left(\dfrac{5}{6}\right)^{-1.3}$

24. $\left(\dfrac{4}{7}\right)^{-.6}$

25. π^{-3}

26. $(2\pi)^{4/3}$

27. $17^{1/17}$

28. $17^{-1/17}$

29. Consider the number $16^{-3/4}$.
  (a) Simplify this expression without the use of a calculator. Give the answer in both decimal and $\frac{a}{b}$ forms.
  (b) Write two different radical expressions that are equivalent to it, and use your calculator to evaluate them to show that the result is the same as the decimal form you found in part (a).
  (c) If your calculator has the capability to convert decimal numbers to fractions, use it to verify your results in part (a).

30. If a number is raised to a negative rational power, should we consider the negative sign to be in the numerator or in the denominator of the exponent to simplify the expression without the use of a calculator? Explain.

31. If we wish to compute a radical expression whose root index is a power of 2, we may simply use the $\boxed{\sqrt{}}$ key repeatedly to do so. For example,
$$\sqrt[4]{16} = \sqrt{\sqrt{16}} = 2 \quad \text{and} \quad \sqrt[8]{6561} = \sqrt{\sqrt{\sqrt{6561}}} = 3.$$
Explain why this is so, and calculate $\sqrt[16]{65{,}536}$ using this method. Then support your result by calculating $65{,}536^{1/16}$.

32. Consider the expression $5^{.47}$.
  (a) Use the exponentiation capability of your calculator to find an approximation. Give as many digits as your calculator displays.
  (b) Use the fact that $.47 = \frac{47}{100}$ to write the expression as a radical, and then use the root-finding capability of your calculator to find an approximation that agrees with the one found in part (a).

In Exercises 33–38, you are given the graph of a function defined by $y = x^n$ for some value of n, along with a display indicating a point on the graph. Use your calculator to support the display by using the exponentiation capability of the calculator and then by using the root-finding capability.

33. $y = x^{1/3}$

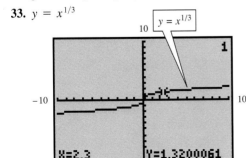

34. $y = x^{1/4}$

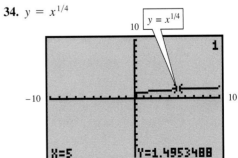

35. $y = x^{-1/2}$

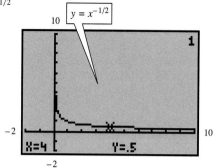

36. $y = x^{-1/3}$

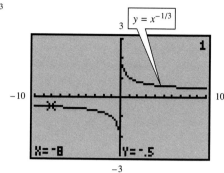

37. $y = x^{3/5}$

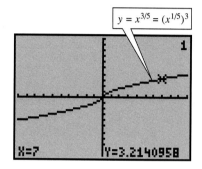

38. $y = x^{4/7}$

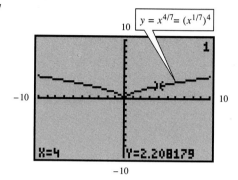

Solve each problem.

39. The illumination I in footcandles produced by a light source is related to the distance d in feet from the light source by the equation

$$d = \left(\frac{k}{I}\right)^{1/2},$$

where k is a constant. If $k = 400$, how far from the source will the illumination be 14 footcandles? Round to the nearest hundredth of a foot.

40. The velocity v of a meteorite approaching the earth is given by

$$v = kd^{-1/2}$$

measured in kilometers per second, where d is its distance from the center of the earth and k is a constant. If $k = 350$, what is the velocity of a meteorite that is 6000 kilometers away from the center of the earth? Round to the nearest tenth.

41. A formula for calculating the distance, d, one can see from an airplane to the horizon on a clear day is

$$d = 1.22x^{1/2},$$

where x is the altitude of the plane in feet and d is given in miles. How far can one see to the horizon in a plane flying at the following altitudes? Give answers to the nearest mile.
(a) 20,000 feet **(b)** 30,000 feet

42. The period of a pendulum in seconds depends on its length L in feet, and is given by

$$P = 2\pi\sqrt{\frac{L}{32}}.$$

If the length of a pendulum is 5 feet, what is its period? Round to the nearest tenth.

43. A biologist has shown that the number of different plant species S on a Galápagos Island is related to the area of the island, A, by

$$S = 28.6A^{1/3}.$$

How many plant species would exist on such an island with the following areas?
(a) 100 square miles **(b)** 1500 square miles

44. To estimate the speed s at which a car was traveling at the time of an accident, police sometimes use the following procedure. A police officer drives the car involved in the accident under conditions similar to those during which the accident took place, and then skids to a stop. If the car is driven at 30 miles per hour, the speed at the time of the accident is given by

$$s = 30\sqrt{\frac{a}{p}},$$

where a is the length of the skid marks and p is the length of the marks in the police test. Find s if $a = 900$ feet and $p = 97$ feet.

Determine analytically the domain of each of the following functions.

45. $f(x) = \sqrt{5 + 4x}$

46. $f(x) = \sqrt{9x + 18}$

47. $f(x) = -\sqrt{6 - x}$

48. $f(x) = -\sqrt{2 - .5x}$

49. $f(x) = \sqrt[3]{8x - 24}$

50. $f(x) = \sqrt[5]{x + 32}$

51. $f(x) = \sqrt{49 - x^2}$

52. $f(x) = \sqrt{81 - x^2}$

53. $f(x) = \sqrt{x^3 - x}$

54. Explain why determining the domain of functions of the form $f(x) = \sqrt[n]{ax + b}$ requires two different considerations, depending upon the parity of n.

Relating Concepts

55. Use a graph to determine the interval(s) over which the function $f(x) = x^2 - 6x - 7$ is nonnegative. (See Section 3.3.)

56. Use your calculator to graph $g(x) = \sqrt{x^2 - 6x - 7}$ in the window $[-6, 12]$ by $[-10, 10]$. What is the domain of g?

57. Graph f from Exercise 55 and g from Exercise 56 together in the window $[-6, 12]$ by $[-20, 10]$. Explain how the sign of $f(x)$ relates to the domain of g.

58. Without graphing, determine the domain of $h(x) = \sqrt{x^2 + 10}$. Why is this a relatively "easy" problem? Support your answer with a graph.

59. Discuss the symmetry of the graph of h from Exercise 58. (See Section 2.2.)

60. Use the graph of h in Exercise 58 to solve each of the following.
    (a) $h(x) = 0$ **(b)** $h(x) > 0$ **(c)** $h(x) < 0$

Graph each function in the window specified. The domain of each function was determined analytically in Exercises 45–53. Use the graph to (a) find the range, (b) give the interval over which the function is increasing, (c) give the interval over which the function is decreasing, and (d) solve graphically the equation $f(x) = 0$.

61. $f(x) = \sqrt{5 + 4x}$
Window: $[-2, 10]$ by $[-10, 10]$

62. $f(x) = \sqrt{9x + 18}$
Window: $[-10, 10]$ by $[-10, 10]$

63. $f(x) = -\sqrt{6 - x}$
Window: $[-10, 10]$ by $[-10, 10]$

64. $f(x) = -\sqrt{2 - .5x}$
Window: $[-10, 10]$ by $[-10, 10]$

65. $f(x) = \sqrt[3]{8x - 24}$
Window: $[-10, 10]$ by $[-10, 10]$

66. $f(x) = \sqrt[5]{x + 32}$
Window: $[-40, 5]$ by $[-10, 10]$

67. $f(x) = \sqrt{49 - x^2}$
Window: Standard square

68. $f(x) = \sqrt{81 - x^2}$
Window: Standard square

69. $f(x) = \sqrt{x^3 - x}$
Window: $[-5, 5]$ by $[-2, 10]$

Relating Concepts

Work Exercises 70–74 in order.

70. Use the standard viewing window to graph $y_1 = \sqrt{3x + 12} - 4$ and $y_2 = \sqrt[3]{3x + 12} - 6$. How many solutions does the equation $y_1 = y_2$ have?

71. Based on your answer to Exercise 70, how many x-intercepts does the graph of $f(x) = y_1 - y_2$ have? Use a graph to support your result.

72. Describe the appearance of the graph of $-f(x) = y_2 - y_1$. Use a graph to support your result. (See Section 2.3.)

73. Describe the appearance of the graph of $y = f(-x)$. Use a graph to support your result. (See Section 2.3.)

74. Use a calculator to solve graphically each of the following.
 (a) $f(x) - 5 = 0$ **(b)** $f(x) - 5 < 0$ **(c)** $f(x) - 5 > 0$

Follow the procedure of Example 6 to explain how the graph of the given function can be obtained from the graph of the appropriate root function ($y = \sqrt{x}$ *or* $y = \sqrt[3]{x}$).

75. $y = \sqrt{9x + 27}$ **76.** $y = \sqrt{16x + 16}$ **77.** $y = \sqrt{7x + 28} + 4$

78. $y = \sqrt{32 - 4x} - 3$ **79.** $y = \sqrt[3]{27x + 54} - 5$

80. In Example 6(b), we began by factoring -8 from the radicand. Repeat the problem, but start by factoring 8 from the radicand. Then explain how the graph of $y = \sqrt[3]{-8x + 8}$ can alternatively be obtained from the graph of $y = \sqrt[3]{x}$.

In Exercises 81–88, describe the graph of the equation as one of the following: circle or parabola with a horizontal axis of symmetry. Then analytically determine two functions, designated by y_1 and y_2, such that their union will give the graph of the given equation. Finally, graph the equation in the viewing window given.

81. $x^2 + y^2 = 100$;
 Square window

82. $x^2 + y^2 = 81$;
 Square window

83. $(x - 2)^2 + y^2 = 9$;
 Square window

84. $(x + 3)^2 + y^2 = 16$;
 Square window

85. $x = y^2 + 6y + 9$;
 Standard window

86. $x = y^2 - 8y + 16$;
 Standard window

87. $x = 2y^2 + 8y + 1$;
 Standard window

88. $x = -3y^2 - 6y + 2$;
 Standard window

89. Repeat the procedure of Exercises 81–88 for the equation $4x^2 + 9y^2 = 36$ in the viewing window $[-3, 3]$ by $[-2, 2]$. (This graph is called an *ellipse,* and ellipses will be studied in more detail in Chapter 6.)

90. Repeat the procedure of Exercises 81–88 for the equation $x^2 - y^2 = 25$ in the viewing window $[-15, 15]$ by $[-10, 10]$. (This graph is called a *hyperbola,* and hyperbolas will be studied in more detail in Chapter 6.)

4.4 ROOT EQUATIONS, INEQUALITIES, AND APPLICATIONS

Solving Equations Involving Roots ▮ Solving Inequalities Involving Roots ▮ Applications of Root Functions

Solving Equations Involving Roots

We will now look at the analytic procedure used to solve equations involving roots, such as

$$\sqrt{5 - 5x} + x = 1, \quad (11 - x)^{1/2} - x = 1 \quad \text{and} \quad \sqrt{2x + 3} - \sqrt{x + 1} = 1.$$

The procedure we will use is based on the following property.

> If P and Q are algebraic expressions, then every solution of the equation $P = Q$ is also a solution of the equation $(P)^n = (Q)^n$, for any positive integer n.

CAUTION Be very careful when using this result. It does *not* say that the equations $P = Q$ and $(P)^n = (Q)^n$ are equivalent; it says only that each solution of the original equation $P = Q$ is also a solution of the new equation $(P)^n = (Q)^n$.

When using this property to solve equations, we must be aware that the new equation may have *more* solutions than the original equation. For example, the solution set of the equation $x = -2$ is $\{-2\}$. If we square both sides of the equation $x = -2$, we get the new equation $x^2 = 4$, which has solution set $\{-2, 2\}$. Since the solution sets are not equal, the equations are not equivalent. Because of this, when an equation contains radicals or rational exponents, it is *essential* to check all proposed solutions in the original equation.

The analytic procedure for solving equations involving roots is outlined below.

SOLVING EQUATIONS INVOLVING ROOTS ANALYTICALLY

1. Isolate a term involving a root on one side of the equation.
2. Raise both sides of the equation to a power that will eliminate the radical or rational exponent.
3. Solve the resulting equation. (If a root is still present after Step 2, repeat Steps 1 and 2.)
4. Check all proposed solutions in the original equation.

EXAMPLE 1

Solving an Equation Involving Roots

Solve the equation

$$\sqrt{5 - 5x} + x = 1$$

analytically, and support the result graphically.

SOLUTION

$\sqrt{5 - 5x} + x = 1$	Given equation
$\sqrt{5 - 5x} = 1 - x$	Isolate the radical term.
$(\sqrt{5 - 5x})^2 = (1 - x)^2$	Square both sides.
$5 - 5x = 1 - 2x + x^2$	
$0 = x^2 + 3x - 4$	Put in standard form of a quadratic equation.
$0 = (x + 4)(x - 1)$	Factor.
$x + 4 = 0$ or $x - 1 = 0$	Use the zero-product property.
$x = -4$ $\qquad$ $x = 1$	

The proposed solutions are -4 and 1. They must be checked in the *original* equation.

Let $x = -4$.		Let $x = 1$.	
$\sqrt{5 - 5(-4)} + (-4) = 1$	?	$\sqrt{5 - 5(1)} + 1 = 1$	?
$\sqrt{25} + (-4) = 1$	?	$\sqrt{0} + 1 = 1$	?
$5 + (-4) = 1$	?	$0 + 1 = 1$	?
$1 = 1$	True	$1 = 1$	True

Both proposed solutions are indeed solutions, and the solution set is $\{-4, 1\}$. To support this result graphically, we may use the *x*-intercept method, letting

$y_1 = \sqrt{5 - 5x} + x$ and $y_2 = 1$. Graphing $y_1 - y_2$ produces the curve shown in Figures 37(a) and (b). Notice that the two x-intercepts are -4 and 1, as determined analytically.   ▯

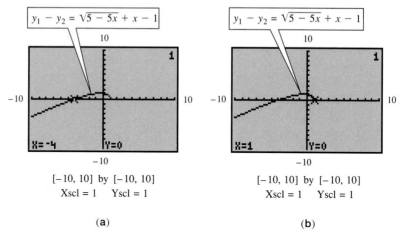

$$[-10, 10] \text{ by } [-10, 10]$$
$$\text{Xscl} = 1 \quad \text{Yscl} = 1$$

(a)   (b)

FIGURE 37

In Example 1, both proposed solutions proved to be actual solutions. However, this is not always the case, as seen in Example 2.

EXAMPLE 2

Solving an Equation
Involving Roots

Solve the equation

$$(11 - x)^{1/2} - x = 1$$

analytically, and support the result graphically.

SOLUTION   We use the same procedure as in Example 1.

$(11 - x)^{1/2} - x = 1$	Given equation
$(11 - x)^{1/2} = 1 + x$	Isolate the term involving the rational exponent.
$11 - x = 1 + 2x + x^2$	Square both sides.
$0 = x^2 + 3x - 10$	
$0 = (x + 5)(x - 2)$	
$x = -5 \quad \text{or} \quad x = 2$	

The proposed solutions are -5 and 2. These must be checked in the original equation.

Let $x = -5$.

$(11 - (-5))^{1/2} - (-5) = 1$?
$16^{1/2} + 5 = 1$?
$4 + 5 = 1$?
$9 = 1$ False

Let $x = 2$.

$(11 - 2)^{1/2} - 2 = 1$?
$9^{1/2} - 2 = 1$?
$3 - 2 = 1$?
$1 = 1$ True

The procedure of squaring both sides of the equation led to the *extraneous* root -5, as indicated by the false statement $9 = 1$. Therefore, the only solution of the equation is 2. The solution set is $\{2\}$.

While there are several ways to support our answer graphically, an interesting method involves using the second step in the solution above. The equation in the second step has the same solution set as the original equation, because the squaring of both sides has not yet been done. If we graph $y_1 = (11 - x)^{1/2}$ and $y_2 = 1 + x$, and observe the x-coordinate of the only point of intersection of the graphs, we see that it is 2, supporting our analytic solution. See Figure 38. ◘

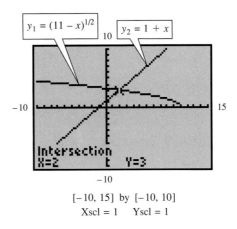

$[-10, 15]$ by $[-10, 10]$
Xscl = 1   Yscl = 1

FIGURE 38

FOR GROUP DISCUSSION

The third line in Example 2 is the "squaring" step. In the window $[-12, 12]$ by $[-5, 25]$, graph the left side, $11 - x$, as y_1 and the right side, $1 + 2x + x^2$ as y_2.

1. How many proposed solutions did the equation in Example 2 have? How many points of intersection does your window show?
2. What were the proposed solutions determined analytically in Example 2? What are the x-coordinates of the points of intersection in your window?
3. Discuss how the extraneous root, -5, appears in this graphical analysis.

EXAMPLE 3

Solving an Equation Involving Roots (Cube Root Radicals)

Solve the equation
$$\sqrt[3]{x^2 + 3x} = \sqrt[3]{5}$$
and support the result graphically.

SOLUTION Analytically, we have

$$\sqrt[3]{x^2 + 3x} = \sqrt[3]{5} \qquad \text{Given equation}$$
$$x^2 + 3x = 5 \qquad \text{Cube both sides.}$$
$$x^2 + 3x - 5 = 0. \qquad \text{Put in standard form.}$$

We must now use the quadratic formula, with $a = 1$, $b = 3$, and $c = -5$.

$$x = \frac{-3 \pm \sqrt{3^2 - 4(1)(-5)}}{2(1)} = \frac{-3 \pm \sqrt{29}}{2}$$

At this point, an analytic check would be rather messy, so we will use our calculator. The points of intersection of the curve $y_1 = \sqrt[3]{x^2 + 3x}$ and the line $y_2 = \sqrt[3]{5}$ should have x-values of

$$\frac{-3 + \sqrt{29}}{2} \approx 1.19 \qquad \text{and} \qquad \frac{-3 - \sqrt{29}}{2} \approx -4.19.$$

Figures 39(a) and (b) support this conclusion. Therefore, the solution set is $\left\{\frac{-3 \pm \sqrt{29}}{2}\right\}$. ∎

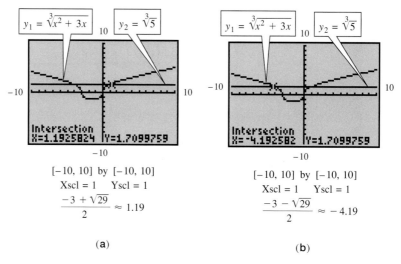

[-10, 10] by [-10, 10]
Xscl = 1   Yscl = 1
$\frac{-3 + \sqrt{29}}{2} \approx 1.19$

(a)

[-10, 10] by [-10, 10]
Xscl = 1   Yscl = 1
$\frac{-3 - \sqrt{29}}{2} \approx -4.19$

(b)

FIGURE 39

EXAMPLE 4

Solving an Equation Involving Roots (Squaring Twice)

Solve the equation

$$\sqrt{2x + 3} - \sqrt{x + 1} = 1$$

analytically, and support the result graphically.

SOLUTION   We begin by isolating a radical. When a choice must be made as to which radical to isolate, it is usually easier to isolate the more complicated radical first. We will isolate $\sqrt{2x + 3}$, and then proceed.

$$\sqrt{2x + 3} = 1 + \sqrt{x + 1}$$
$$(\sqrt{2x + 3})^2 = (1 + \sqrt{x + 1})^2 \qquad \text{Square both sides.}$$
$$2x + 3 = 1 + 2\sqrt{x + 1} + x + 1$$
$$x + 1 = 2\sqrt{x + 1} \qquad \text{Isolate the radical.}$$

One side of the equation still contains a radical; to eliminate it, square both sides again.

$$x^2 + 2x + 1 = 4(x + 1)$$
$$x^2 - 2x - 3 = 0$$
$$(x - 3)(x + 1) = 0$$
$$x = 3 \qquad \text{or} \qquad x = -1$$

Check these proposed solutions in the original equation.

Let $x = 3$.

$$\sqrt{2x + 3} - \sqrt{x + 1} = 1$$
$$\sqrt{2(3) + 3} - \sqrt{3 + 1} = 1 \qquad ?$$
$$\sqrt{9} - \sqrt{4} = 1 \qquad ?$$
$$3 - 2 = 1 \qquad ?$$
$$1 = 1 \qquad \text{True}$$

Let $x = -1$.

$$\sqrt{2x + 3} - \sqrt{x + 1} = 1$$
$$\sqrt{2(-1) + 3} - \sqrt{-1 + 1} = 1 \qquad ?$$
$$\sqrt{1} - \sqrt{0} = 1 \qquad ?$$
$$1 - 0 = 1 \qquad ?$$
$$1 = 1 \qquad \text{True}$$

Both proposed solutions 3 and -1 are solutions of the original equation, giving $\{3, -1\}$ as the solution set. The two solutions may be supported graphically by graphing $y_1 - y_2$, where $y_1 = \sqrt{2x + 3} - \sqrt{x + 1}$ and $y_2 = 1$. The x-intercepts indicated in Figures 40(a) and (b) are -1 and 3, as determined analytically.　▐

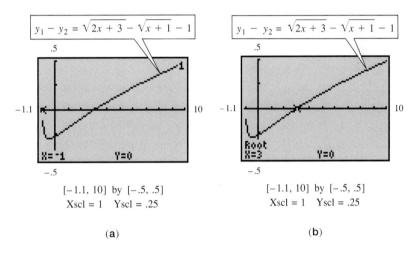

(a)　　　　　　　　　(b)

FIGURE 40

Solving Inequalities Involving Roots

We have seen in previous sections how the solution sets of inequalities can be determined once the solution set of the associated equations are found. The solution set of $f(x) > 0$ consists of all x-values for which the graph of f is above the x-axis, while that of $f(x) < 0$ consists of all x-values for which the graph is below the x-axis. We will use this approach in the following examples, referring to Examples 1–3 and their associated figures. We will always pay close attention to the domain of the function involved.

EXAMPLE 5

Solving an Inequality Involving Roots (Radical Form)

Find the solution set of

$$\sqrt{5 - 5x} + x \geq 1.$$

SOLUTION   The given inequality is equivalent to

$$\sqrt{5 - 5x} + x - 1 \geq 0.$$

Notice that the domain of $y = \sqrt{5 - 5x} + x - 1$ is $(-\infty, 1]$, because it is for only these real numbers that $5 - 5x$ is nonnegative. Figure 37 shows that the graph of this function lies above or on the x-axis in the interval $[-4, 1]$; therefore, this interval is the solution set of the inequality. ∎

FOR **GROUP DISCUSSION**

Refer to Examples 1 and 5, and Figure 37 to find the solution sets of the following inequalities. You should not have to do any pencil-and-paper work for these.

1. $\sqrt{5 - 5x} + x - 1 > 0$
2. $\sqrt{5 - 5x} + x - 1 < 0$
3. $\sqrt{5 - 5x} + x - 1 \leq 0$

EXAMPLE **6**
Solving an Inequality Involving Roots (Rational Exponent)

Solve each inequality.

(a) $(11 - x)^{1/2} - x > 1$ **(b)** $(11 - x)^{1/2} - x < 1$

SOLUTION   In Example 2 we found that the solution set of the equation $(11 - x)^{1/2} - x = 1$ is $\{2\}$. The domain of the function $y = (11 - x)^{1/2} - x - 1$ is $(-\infty, 11]$, and its x-intercept is 2, as seen in Figure 41.

The solution set of the inequality in (a) is the same as that of $y > 0$, which is $(-\infty, 2)$. The solution set of the inequality in (b) is the same as that of $y < 0$, which is $(2, 11]$. ∎

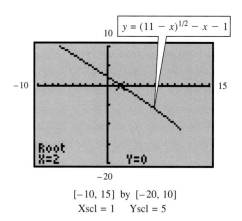

$y = (11 - x)^{1/2} - x - 1$

$[-10, 15]$ by $[-20, 10]$
Xscl = 1   Yscl = 5

FIGURE 41

EXAMPLE **7**
Solving an Inequality Involving Roots (Cube Root Radicals)

Solve the inequality

$$\sqrt[3]{x^2 + 3x} \leq \sqrt[3]{5}.$$

SOLUTION   The associated equation was solved in Example 3, and we found its solutions to be $\frac{-3 + \sqrt{29}}{2}$ and $\frac{-3 - \sqrt{29}}{2}$. We can use the x-intercept method to solve this inequality. As seen in Figure 42, the graph of $y = \sqrt[3]{x^2 + 3x} - \sqrt[3]{5}$ lies below or on the x-axis in the interval between the two x-intercepts, including the endpoints. Therefore, the solution set of this inequality is $\left[\frac{-3 - \sqrt{29}}{2}, \frac{-3 + \sqrt{29}}{2}\right]$. ∎

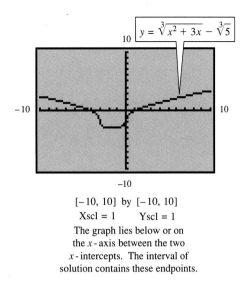

$y = \sqrt[3]{x^2 + 3x} - \sqrt[3]{5}$

$[-10, 10]$ by $[-10, 10]$
$\text{Xscl} = 1 \qquad \text{Yscl} = 1$
The graph lies below or on
the x-axis between the two
x-intercepts. The interval of
solution contains these endpoints.

FIGURE 42

FOR GROUP DISCUSSION

What is the domain of the function $y = \sqrt[3]{x^2 + 3x} - \sqrt[3]{5}$ graphed in Figure 42? What is the solution set of $y > 0$?

Applications of Root Functions

Certain types of problems can be solved using functions involving roots, as shown in the next example.

EXAMPLE 8

Solving a Problem Involving a Root Function

A company wishes to run a utility cable from point A on the shore (as shown in Figure 43) to an installation at point B on the island. The island is 6 miles from the shore. It costs \$400 per mile to run the cable on land and \$500 per mile underwater. Assume that the cable starts at A and runs along the shoreline, then angles and runs underwater to the island. Let x represent the distance from C at which the underwater portion of the cable run begins, and the distance between A and C be 9 miles.

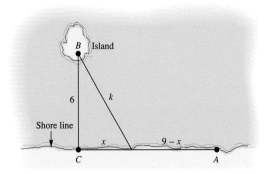

FIGURE 43

(a) What are the possible values of x in this problem?

SOLUTION The value of x must be a real number greater than 0 and less than 9, meaning that x must be in the interval $(0, 9)$. (If we assume that the cable may be laid so that it is all underwater, then 9 would be included, giving the interval $(0, 9]$.)

(b) Express the cost of laying the cable as a function of x.

SOLUTION The total cost is determined by adding the cost of the cable on land to the cost of the cable underwater. If we let k represent the length of the cable underwater, by the Pythagorean theorem,

$$k^2 = 6^2 + x^2$$
$$k^2 = 36 + x^2$$
$$k = \sqrt{36 + x^2} \qquad (k > 0).$$

The cost of the cable on land is $400(9 - x)$ dollars, while the cost of the cable underwater is $500k$, or $500\sqrt{36 + x^2}$ dollars. Therefore, the total cost is given by the function $C(x) = 400(9 - x) + 500\sqrt{36 + x^2}$, where $C(x)$ is in dollars.

(c) Find the total cost if 3 miles of cable are on land. Determine analytically, and support graphically.

SOLUTION According to Figure 43, if 3 miles of cable are on land, then $3 = 9 - x$, or $x = 6$. We must evaluate $C(6)$.

$$C(6) = 400(9 - 6) + 500\sqrt{36 + 6^2}$$
$$\approx 5442.64 \text{ dollars}$$

The graph of the function in Figure 44 supports this result, since the point $(6, 5442.64)$ lies on the graph.

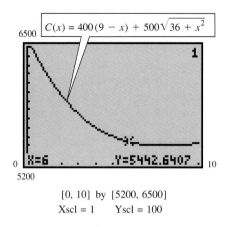

[0, 10] by [5200, 6500]
Xscl = 1   Yscl = 100

FIGURE 44

(d) Find the point at which the line should begin to angle in order to minimize the total cost. What is this total cost?

SOLUTION   The absolute minimum value of the function on the interval $(0, 9)$ is found when $x = 8$, meaning that $9 - 8 = 1$ mile should be along land and $\sqrt{36 + 1^2} \approx 6.08$ miles should be underwater. $C(8) = 5400$(dollars) is the minimum total cost. These results may be found using the capabilities of a graphics calculator. See Figure 45.   ▮

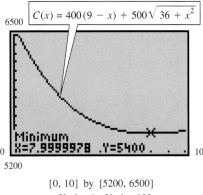

$$C(x) = 400(9 - x) + 500\sqrt{36 + x^2}$$

Minimum
X=7.9999978 .Y=5400.

[0, 10] by [5200, 6500]
Xscl = 1   Yscl = 100

FIGURE 45

4.4 EXERCISES

Use an analytic method to solve the equation in part (a). Then use a graph, along with your result from part (a) to solve the inequalities in parts (b) and (c). Use set notation to give the solution set for equations, and interval notation to give the solution set for inequalities.

1. (a) $\sqrt{3x + 7} - 3x = 5$
    (b) $\sqrt{3x + 7} - 3x > 5$
    (c) $\sqrt{3x + 7} - 3x < 5$

2. (a) $\sqrt{4x + 13} - 2x = -1$
    (b) $\sqrt{4x + 13} - 2x > -1$
    (c) $\sqrt{4x + 13} - 2x < -1$

3. (a) $\sqrt{3x + 4} = 8 - x$
    (b) $\sqrt{3x + 4} > 8 - x$
    (c) $\sqrt{3x + 4} < 8 - x$

4. (a) $\sqrt{1 + 5x} = 2x - 2$
    (b) $\sqrt{1 + 5x} > 2x - 2$
    (c) $\sqrt{1 + 5x} < 2x - 2$

5. (a) $(x + 5)^{1/2} - 2 = (x - 1)^{1/2}$
    (b) $(x + 5)^{1/2} - 2 \geq (x - 1)^{1/2}$
    (c) $(x + 5)^{1/2} - 2 \leq (x - 1)^{1/2}$

6. (a) $(2x - 5)^{1/2} - 2 = (x - 2)^{1/2}$
    (b) $(2x - 5)^{1/2} - 2 \geq (x - 2)^{1/2}$
    (c) $(2x - 5)^{1/2} - 2 \leq (x - 2)^{1/2}$

7. (a) $\sqrt[3]{4x + 3} = \sqrt[3]{2x - 1}$
    (b) $\sqrt[3]{4x + 3} > \sqrt[3]{2x - 1}$
    (c) $\sqrt[3]{4x + 3} < \sqrt[3]{2x - 1}$

8. (a) $\sqrt[3]{2x} = \sqrt[3]{5x + 2}$
    (b) $\sqrt[3]{2x} > \sqrt[3]{5x + 2}$
    (c) $\sqrt[3]{2x} < \sqrt[3]{5x + 2}$

9. (a) $\sqrt{3x + 4} - \sqrt{2x - 4} = 2$
    (b) $\sqrt{3x + 4} - \sqrt{2x - 4} > 2$
    (c) $\sqrt{3x + 4} - \sqrt{2x - 4} < 2$

10. (a) $\sqrt{5x - 6} = 2 + \sqrt{3x - 6}$
    (b) $\sqrt{5x - 6} > 2 + \sqrt{3x - 6}$
    (c) $\sqrt{5x - 6} < 2 + \sqrt{3x - 6}$

11. (a) $\sqrt[3]{(2x - 1)^2} - \sqrt[3]{x} = 0$
    (b) $\sqrt[3]{(2x - 1)^2} - \sqrt[3]{x} > 0$
    (c) $\sqrt[3]{(2x - 1)^2} - \sqrt[3]{x} < 0$

12. (a) $\sqrt[3]{x^2 - 2x} - \sqrt[3]{x} = 0$
    (b) $\sqrt[3]{x^2 - 2x} - \sqrt[3]{x} > 0$
    (c) $\sqrt[3]{x^2 - 2x} - \sqrt[3]{x} < 0$

13. (a) $\sqrt[4]{x - 15} = 2$
    (b) $\sqrt[4]{x - 15} > 2$
    (c) $\sqrt[4]{x - 15} < 2$

14. (a) $\sqrt[4]{3x + 1} = 1$
    (b) $\sqrt[4]{3x + 1} > 1$
    (c) $\sqrt[4]{3x + 1} < 1$

15. (a) $(x^2 + 2x)^{1/4} = \sqrt[4]{3}$
   (b) $(x^2 + 2x)^{1/4} > \sqrt[4]{3}$
   (c) $(x^2 + 2x)^{1/4} < \sqrt[4]{3}$

16. (a) $(x^2 + 6x)^{1/4} = 2$
   (b) $(x^2 + 6x)^{1/4} > 2$
   (c) $(x^2 + 6x)^{1/4} < 2$

17. (a) $(2x - 1)^{2/3} = x^{1/3}$
   (b) $(2x - 1)^{2/3} > x^{1/3}$
   (c) $(2x - 1)^{2/3} < x^{1/3}$

18. (a) $(x - 3)^{2/5} = (4x)^{1/5}$
   (b) $(x - 3)^{2/5} > (4x)^{1/5}$
   (c) $(x - 3)^{2/5} < (4x)^{1/5}$

19. (a) $\sqrt{3 - 3x} = 3 + \sqrt{3x + 2}$
   (b) $\sqrt{3 - 3x} > 3 + \sqrt{3x + 2}$
   (c) $\sqrt{3 - 3x} < 3 + \sqrt{3x + 2}$

20. (a) $\sqrt{2\sqrt{7x + 2}} = \sqrt{3x + 2}$
   (b) $\sqrt{2\sqrt{7x + 2}} > \sqrt{3x + 2}$
   (c) $\sqrt{2\sqrt{7x + 2}} < \sqrt{3x + 2}$

In Exercises 21–26, a pair of functions is given. The function defined by y_1 is a root function, while the function defined by y_2 is a polynomial function. Use your knowledge of the general appearance of each to draw by hand a sketch of both functions on the same set of axes. Then use the sketch to determine the number of points of intersection of the two graphs. Finally, use an analytic method to support your answer, and find the solution set of $y_1 = y_2$.

21. $y_1 = \sqrt{x}$
   $y_2 = -x + 3$

22. $y_1 = \sqrt{x}$
   $y_2 = x - 7$

23. $y_1 = \sqrt{x}$
   $y_2 = 2x + 5$

24. $y_1 = \sqrt{x}$
   $y_2 = 3x$

25. $y_1 = \sqrt[3]{x}$
   $y_2 = x^2$

26. $y_1 = \sqrt[3]{x}$
   $y_2 = 2x$

27. Use a hand-drawn graph to explain why $\sqrt{x} = -x - 5$ has no real solutions. (*Hint:* Sketch $y_1 = \sqrt{x}$ and $y_2 = -x - 5$ on the same axes. What do you notice about the number of points of intersection?)

28. Explain why the equation $\sqrt[3]{x} = ax + b$ must have at least one real solution for any values of a and b. (*Hint:* What is the range of $y_1 = \sqrt[3]{x}$? What kind of graph does $y_2 = ax + b$ have?)

R*elating Concepts*

Exercises 29–40 incorporate many concepts from Chapter 3 with the method of solving equations involving roots. They should be worked in order. Consider the equation

$$\sqrt[3]{4x - 4} = \sqrt{x + 1}.$$

29. Rewrite the equation using rational exponents.

30. What is the least common denominator of the rational exponents found in Exercise 29?

31. Raise both sides of the equation in Exercise 29 to the power indicated by your answer in Exercise 30.

32. Show that the equation in Exercise 31 is equivalent to $x^3 - 13x^2 + 35x - 15 = 0$.

33. Graph the cubic function defined by the polynomial on the left side of the equation in Exercise 32 in the window $[-5, 10]$ by $[-100, 100]$. How many real roots does the equation have?

34. Use synthetic division to show that 3 is a zero of $P(x) = x^3 - 13x^2 + 35x - 15$.

35. Use the result of Exercise 34 to factor $P(x)$ so that one factor is linear and the other is quadratic.

36. Set the quadratic factor of $P(x)$ from Exercise 35 equal to 0, and solve the equation using the quadratic formula.

37. What are the three proposed solutions of the original equation $\sqrt[3]{4x - 4} = \sqrt{x + 1}$?

38. Let $y_1 = \sqrt[3]{4x - 4}$ and $y_2 = \sqrt{x + 1}$. Graph $y_1 - y_2$ in the viewing window $[-2, 20]$ by $[-.5, .5]$ to determine the number of real solutions the original equation has.

39. Use both an analytic method and your calculator to find values of the real roots of the original equation.

40. Write a paragraph explaining how the solutions of the equation in Exercise 32 relate to the solutions of the original equation. Discuss any extraneous solutions that may be involved.

Solve each of the following applications of root functions.

41. Two vertical poles of lengths 12 feet and 16 feet are situated on level ground, 20 feet apart, as shown in the figure. A piece of wire is to be strung from the top of the 12-foot pole to the top of the 16-foot pole, attached to a stake in the ground at a point P on a line formed by the vertical poles. Let x represent the distance from P to D, the base of the 12-foot pole.

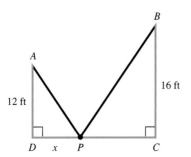

(a) Express the distance from P to C in terms of x.
(b) What are the restrictions on the value of x in this problem?
(c) Use the Pythagorean theorem to express the lengths AP and BP in terms of x.
(d) Form a function f that expresses the total length of the wire used.
(e) Graph f in the window $[0, 20]$ by $[0, 50]$. Use a function of your calculator to find $f(4)$, and interpret your result.
(f) Find the value of x that will minimize the amount of wire used.
(g) Write a short paragraph summarizing what this problem has examined, and the results you have obtained.

42. Repeat Exercise 41 if the heights of the poles are 9 feet and 12 feet, and the distance between the poles is 16 feet. Let P be x feet from the 9-foot pole. In part (e), use the window $[0, 16]$ by $[0, 50]$.

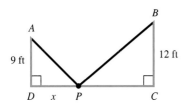

43. A hunter is at a point on a river bank. He wants to get to his cabin, located 3 miles north and 8 miles west. (See the figure.) He can travel 5 mph on the river but only 2 mph on this very rocky land. How far upriver should he go in order to reach the cabin in a minimum amount of time? (*Hint:* Distance = Rate × Time.)

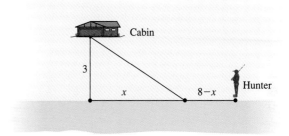

44. Homing pigeons avoid flying over large bodies of water, preferring to fly around them instead. (One possible explanation is the fact that extra energy is required to fly over water because air pressure drops over water in the daytime.) Assume that a pigeon released from a boat 1 mi from the shore of a lake (point B in the figure) flies first to point P on the shore and then along the straight edge of the lake to reach its home at L. If L is 2 mi from point A, the point on the shore closest to the boat, and if a pigeon needs $\frac{4}{3}$ as much energy to fly over water as over land, find the location of point P.

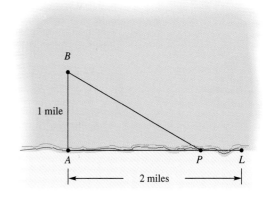

45. At noon, the cruise ship Queen Anne is 60 miles due south of the cruise ship King Bill and is sailing north at a rate of 30 mph. If the King Bill is sailing west at a rate of 20 mph, find the time at which the distance d between the ships is a minimum. What is this distance?

46. Brent Simon is in his bass boat, the Mido, 3 miles from the nearest point on the shore. He wishes to reach his camp at LaBranche, 6 miles farther down the shoreline. (See the figure.) If Brent's motor is disabled and he can row his boat at a rate of 4 mph, and he can walk at a rate of 5 mph, find the least amount of time that he will need to reach the camp.

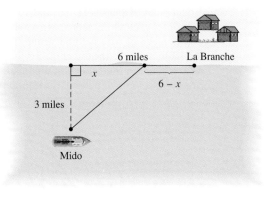

Further Explorations

Consider $f(x) = \sqrt{x - 15}$. Examine this function with the TABLE feature of your graphics calculator. Explain why the TABLE displays "ERROR" for all values of x less than 15.

4.5 INVERSE FUNCTIONS

Basic Concepts ▌ One-to-One Functions ▌ Inverse Functions and Their Graphs ▌ An Application of Inverse Functions

Basic Concepts

In ordinary arithmetic, the real numbers 0 and 1 serve as *identity* elements for addition and multiplication, respectively.

IDENTITY PROPERTIES

For all real numbers a, $a + 0 = 0 + a = a$.

For all real numbers a, $a \cdot 1 = 1 \cdot a = a$.

For the operation of composition of functions (Section 2.6), the function $f(x) = x$ serves as the identity function.

Every real number a has an *additive inverse,* a number which when added to a produces a sum of 0, the identity element for addition. Also, every nonzero real number a has a *multiplicative inverse,* or *reciprocal,* a number which when multiplied by a produces a product of 1, the identity element for multiplication.

INVERSE PROPERTIES

For every real number a, there exists a unique real number $-a$ such that $a + (-a) = (-a) + a = 0$.

For every nonzero real number a, there exists a unique real number a^{-1} such that $a \cdot a^{-1} = a^{-1} \cdot a = 1$.

It would seem that the operation of composition would lend itself to a natural extension of the concept of inverses. This is indeed the case, and in this section we will investigate inverse functions and the conditions under which they exist.

As a simple example, consider the functions

$$f(x) = 3x \qquad \text{and} \qquad g(x) = \frac{1}{3}x.$$

Let us choose a value for x, say $x = 6$. If we evaluate f at 6 we get $f(6) = 3(6) = 18$. Now, if we take this result and evaluate g, we get $g(18) = \frac{1}{3}(18) = 6$, which is the value with which we started. Recall from Section 2.6 that $(g \circ f)(x) = g(f(x))$, so

$$(g \circ f)(6) = g(f(6)) = g(18) = 6.$$

Similarly, we can show that $(f \circ g)(6) = 6$. This type of result is typical of *inverse functions.* For these functions, we can show that for all x,

$$(f \circ g)(x) = (g \circ f)(x) = x.$$

Notice that the final result is the identity function.

FOR **GROUP DISCUSSION**

With your graphics calculator in the standard window, graph $y_1 = 3x$ and $y_2 = \frac{1}{3}x$.

1. On y_1, locate the point where $x = 6$. What is the y-value?
2. On y_2, locate the point where $y = 18$. What is the x-value?
3. How do the results of items 1 and 2 above support the preceding explanation?
4. Repeat the process of items 1 and 2 for other values, letting x be an integer each time. Do you get the same sort of results?

One-to-One Functions

In order for a function to have an inverse, it must be *one-to-one*. For the function $y = 5x - 8$, any two different values of x produce two different values of y. On the other hand, for the function $y = x^2$, two different values of x can lead to the *same* value of y; for example, both $x = 4$ and $x = -4$ give $y = 4^2 = (-4)^2 = 16$. A function such as $y = 5x - 8$, where different elements from the domain always lead to different elements from the range, is called a *one-to-one function*.

ONE-TO-ONE FUNCTION

A function f is a **one-to-one function** if, for elements a and b from the domain of f,

$$a \neq b \quad \text{implies} \quad f(a) \neq f(b).$$

Deciding Whether a Function Is One-to-One

Decide whether each of the following functions is one-to-one.

(a) $f(x) = -4x + 12$

SOLUTION Suppose that $a \neq b$. Then $-4a \neq -4b$, and $-4a + 12 \neq -4b + 12$. Thus, the fact that $a \neq b$ implies that $f(a) \neq f(b)$, so f is one-to-one.

(b) $f(x) = \sqrt{25 - x^2}$

SOLUTION If $a = 3$ and $b = -3$, then $3 \neq -3$, but

$$f(3) = \sqrt{25 - 3^2} = \sqrt{25 - 9} = \sqrt{16} = 4$$

and

$$f(-3) = \sqrt{25 - (-3)^2} = \sqrt{25 - 9} = 4.$$

Here, even though $3 \neq -3, f(3) = f(-3)$. By definition, this is not a one-to-one function. ∎

As shown in Example 1(b), a way to show that a function is *not* one-to-one is to produce a pair of unequal numbers that lead to the same function value. There is also a useful graphical test that tells whether or not a function is one-to-one. This *horizontal line test* for one-to-one functions can be summarized as follows.

HORIZONTAL LINE TEST

If every horizontal line intersects the graph of a function in no more than one point, then the function is one-to-one.

NOTE In Example 1(b), the graph of the function is a semicircle. There are infinitely many horizontal lines that cut the graph of a semicircle in two points, so the horizontal line test shows that the function is not one-to-one.

Using the Horizontal Line Test

Use the horizontal line test to determine whether the graphs in Figures 46 and 47 are graphs of one-to-one functions.

(a)

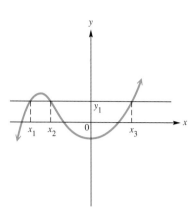

FIGURE 46

SOLUTION   Each point where the horizontal line intersects the graph has the same value of y but a different value of x. Since more than one (here three) different values of x lead to the same value of y, the function is not one-to-one.

(b)

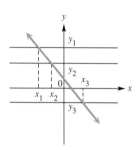

FIGURE 47

SOLUTION   Every horizontal line will intersect the graph in Figure 47 in exactly one point. This function is one-to-one.   ◻

FOR **GROUP DISCUSSION**

Based on your knowledge of the elementary functions studied so far in this text, answer the following questions. In each case, assume that the function has the largest possible domain.

1. Will a linear function (nonconstant) always be one-to-one?
2. Will an odd degree polynomial function always be one-to-one? If not, explain.
3. Will an even degree polynomial function ever be one-to-one? Why or why not?
4. Without graphing, tell whether each function is one-to-one. Then support your answer with a graph.

a. $f(x) = x$ **b.** $f(x) = x^n$, n odd   **c.** $f(x) = x^n$, n even

d. $f(x) = |x|$ **e.** $f(x) = \sqrt[n]{x}$, n odd   **f.** $f(x) = \sqrt[n]{x}$, n even

g. $f(x) = |2x + 4|$ **h.** $f(x) = \dfrac{1}{x}$ **i.** $f(x) = \dfrac{1}{x^2}$

Inverse Functions and Their Graphs

As mentioned earlier, certain pairs of one-to-one functions "undo" one another. For example, if

$$f(x) = 8x + 5 \quad \text{and} \quad g(x) = \frac{x-5}{8}$$

then

$$f(10) = 8 \cdot 10 + 5 = 85 \quad \text{and} \quad g(85) = \frac{85-5}{8} = 10.$$

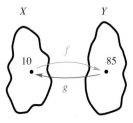

FIGURE 48

Starting with 10, we "applied" function f and then "applied" function g to the result, which gave back the number 10. See Figure 48. Similarly, for these same functions, check that

$$f(3) = 29 \quad \text{and} \quad g(29) = 3,$$
$$f(-5) = -35 \quad \text{and} \quad g(-35) = -5,$$
$$g(2) = -\frac{3}{8} \quad \text{and} \quad f\left(-\frac{3}{8}\right) = 2.$$

In particular, for these functions,

$$(f \circ g)(2) = 2 \quad \text{and} \quad (g \circ f)(2) = 2.$$

In general, it can be shown that for f and g,

$$(f \circ g)(x) = x \quad \text{and} \quad (g \circ f)(x) = x$$

for all real numbers x. Because of this property, g is called the inverse of f.

INVERSE FUNCTION

Let f be a one-to-one function. Then g is the **inverse function** of f if

$$(f \circ g)(x) = x \quad \text{for every } x \text{ in the domain of } g,$$

and

$$(g \circ f)(x) = x \quad \text{for every } x \text{ in the domain of } f.$$

EXAMPLE 3

Deciding Whether Two Functions Are Inverses

Let $f(x) = x^3 - 1$, and let $g(x) = \sqrt[3]{x+1}$. Is g the inverse function of f?

SOLUTION First, note that f is one-to-one. Then use the definition to find

$$(f \circ g)(x) = f[g(x)] = (\sqrt[3]{x+1})^3 - 1$$
$$= x + 1 - 1 = x$$
$$(g \circ f)(x) = g[f(x)] = \sqrt[3]{(x^3 - 1) + 1} = \sqrt[3]{x^3} = x.$$

Since $(f \circ g)(x) = x$ and $(g \circ f)(x) = x$, function g is the inverse of function f. Also, f is the inverse of function g. ∎

A special notation is often used for inverse functions: if g is the inverse function of f, then g can be written as f^{-1} (read "f-inverse"). In Example 3,

$$f^{-1}(x) = \sqrt[3]{x+1}.$$

CAUTION Do not confuse the -1 in f^{-1} with a negative exponent. The symbol $f^{-1}(x)$ does not represent $1/f(x)$; it represents the inverse function of f. Keep in mind that a function f can have an inverse function f^{-1} if and only if f is one-to-one.

The definition of inverse function can be used to show that the domain of f equals the range of f^{-1}, and the range of f equals the domain of f^{-1}. See Figure 49.

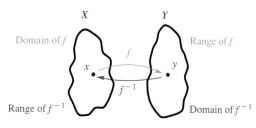

FIGURE 49

For the inverse functions f and g discussed at the beginning of this section, $f(6) = 18$ and $g(18) = 6$. Graphically, this means that the point $(6, 18)$ lies on the graph of f, and the point $(18, 6)$ lies on the graph of g. (This was examined in the group discussion that followed.)

The inverse of any one-to-one function f can be found by exchanging the components of the ordered pairs of f. The rule for the inverse of a function defined by $y = f(x)$ also is found by exchanging x and y. For example, if $f(x) = 7x - 2$, then $y = 7x - 2$. The function f is one-to-one, so that f^{-1} exists.

One method of finding the rule for f^{-1} is as follows:

$$y = 7x - 2 \qquad \text{Given one-to-one function}$$

$$x = 7y - 2 \qquad \text{Exchange } x \text{ and } y \text{ to find an equation that relates the variables in the inverse. (Step 1)}$$

$$x + 2 = 7y$$

$$\frac{x + 2}{7} = y \qquad \text{Solve for } y. \text{ (Step 2)}$$

$$f^{-1}(x) = \frac{x + 2}{7}. \qquad \text{Use inverse notation. (Step 3)}$$

We can now verify that $(f \circ f^{-1})(x) = (f^{-1} \circ f)(x) = x$ to complete our work.

EXAMPLE 4
Finding the Inverse of a Function

Find the inverse, if it exists, of $f(x) = \dfrac{4x + 6}{5}$.

SOLUTION   This function is linear and is therefore one-to-one. Thus it has an inverse. Let $f(x) = y$, and solve for x, getting

$$y = \frac{4x + 6}{5}$$

$$x = \frac{4y + 6}{5} \qquad \text{Exchange } x \text{ and } y.$$

$$5x = 4y + 6 \qquad \text{Multiply by 5.}$$

$$5x - 6 = 4y \qquad \text{Subtract 6.}$$

$$y = \frac{5x - 6}{4} \qquad \text{Divide by 4.}$$

or $\qquad f^{-1}(x) = \dfrac{5x - 6}{4}$.

The domain and range of both f and f^{-1} are the set of real numbers. In function f, the value of y is found by multiplying x by 4, adding 6 to the product, then dividing that sum by 5. In the equation for the inverse, x is *multiplied* by 5, then 6 is *subtracted,* and the result is *divided* by 4. This shows how an inverse function is used to "undo" what the function does to the variable x. ❑

Suppose f and f^{-1} are inverse functions, and $f(a) = b$ for real numbers a and b. Then, by the definition of inverse, $f^{-1}(b) = a$. This shows that if a point (a, b) is on the graph of f, then (b, a) will belong to the graph of f^{-1}. As shown in Figure 50, the points (a, b) and (b, a) are reflections of one another across the line $y = x$. Thus, the graph of f^{-1} can be obtained from the graph of f by reflecting the graph of f across the line $y = x$.

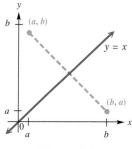

FIGURE 50

GEOMETRIC RELATIONSHIP BETWEEN THE GRAPHS OF f AND f^{-1}

If a function f is one-to-one, the graph of its inverse f^{-1} is a reflection of the graph of f across the line $y = x$.

The property discussed above is illustrated in the next example, using the pairs of inverse functions discussed thus far in this section.

EXAMPLE 5

Comparing the Graphs of f and f^{-1}

For each of the following pairs of inverse functions, a calculator-generated graph is shown. See Figures 51, 52, and 53. In each case a square window is used to give the correct perspective, and the line $y = x$ is also graphed to illustrate how the reflection appears.

(a) $f(x) = 3x$ and $f^{-1}(x) = \frac{1}{3}x$ (functions from the section introduction)

(b) $f(x) = x^3 - 1$ and $f^{-1}(x) = \sqrt[3]{x + 1}$ (functions from Example 3)

(c) $f(x) = \frac{4x + 6}{5}$ and $f^{-1}(x) = \frac{5x - 6}{4}$ (functions from Example 4) ❑

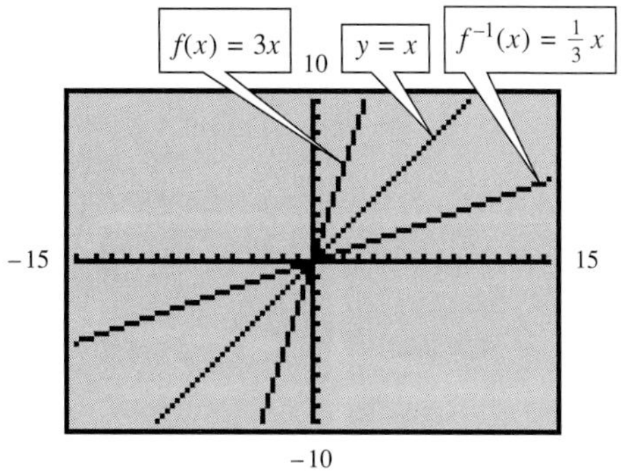

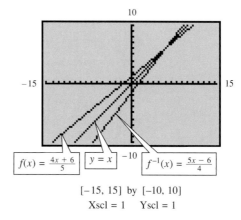

f(x) = 3x
y = x
f^{-1}(x) = 1/3 x
10
-15
15
-10

[−15, 15] by [−10, 10]
Xscl = 1 Yscl = 1

FIGURE 51

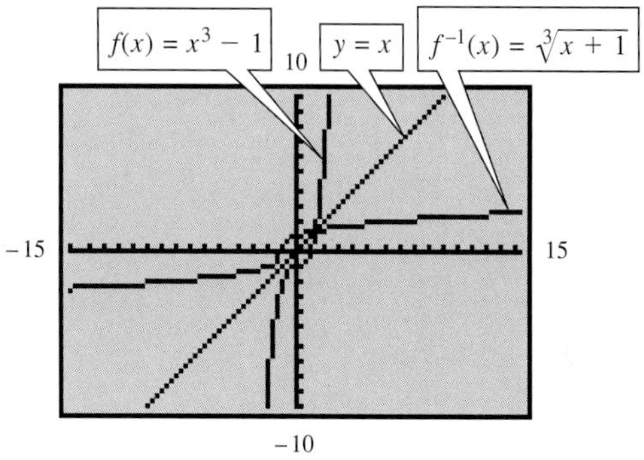

f(x) = x^3 − 1
y = x
f^{-1}(x) = ∛(x + 1)
10
-15
15
-10

[−15, 15] by [−10, 10]
Xscl = 1 Yscl = 1

FIGURE 52

10
-15
15
-10
f(x) = (4x + 6)/5
y = x
f^{-1}(x) = (5x − 6)/4

[−15, 15] by [−10, 10]
Xscl = 1 Yscl = 1

FIGURE 53

The next example considers a function with domain that is not $(-\infty, \infty)$.

EXAMPLE 6

Finding the Inverse of a Function Whose Domain is Not $(-\infty, \infty)$

Consider the function $f(x) = \sqrt{x + 5}$.

(a) What is the domain of f?

SOLUTION Using the methods described in Section 4.3. we know that for f to have real values, $x + 5$ must be nonnegative. Therefore,

$$x + 5 \geq 0$$
$$x \geq -5.$$

The domain is $[-5, \infty)$.

(b) Is f one-to-one?

SOLUTION We could use the horizontal line test, but another way to determine whether a function is one-to-one is to examine whether it is increasing over its entire domain, or whether it is decreasing over its entire domain. Figure 54 supports the first of these, and thus f is one-to-one.

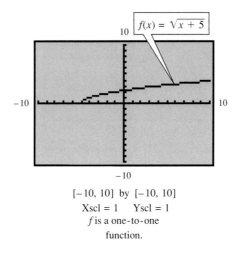

$$[-10, 10] \text{ by } [-10, 10]$$
$$\text{Xscl} = 1 \quad \text{Yscl} = 1$$
f is a one-to-one
function.

FIGURE 54

(c)   Find the rule for f^{-1}.

SOLUTION

$$y = \sqrt{x + 5}, \quad x \geq -5 \qquad \text{Replace } f(x) \text{ with } y.$$
$$x = \sqrt{y + 5}, \quad y \geq -5 \qquad \text{Exchange } x \text{ and } y.$$
$$x^2 = y + 5 \qquad\qquad\quad \text{Square both sides.}$$
$$y = x^2 - 5 \qquad\qquad\quad \text{Solve for } y.$$

The range of f^{-1} is indicated in the second line of the work above. The original function had a range of $[0, \infty)$, and thus the inverse must have this as its domain. Therefore,

$$f^{-1}(x) = x^2 - 5, \quad x \geq 0.$$

Note that the restriction on the domain of f^{-1} is necessary; otherwise, the function would not be one-to-one, as required. The graphs of f and f^{-1} are shown in Figure 55.   ◨

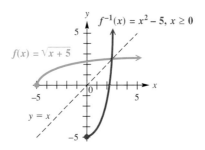

FIGURE 55

An Application of Inverse Functions

Inverse functions are used by government agencies and other businesses to send and receive coded information. The functions they use are usually very complicated. A simple example might use the function $f(x) = 2x + 5$. (Note that it is one-to-one.) Suppose that each letter of the alphabet is assigned a numerical value according to its position, as follows:

TECHNOLOGICAL NOTE
If your calculator has a TABLE feature, you may wish to experiment with it when studying how inverses can be applied to codes and code-breaking.

A	1	F	6	K	11	P	16	U	21
B	2	G	7	L	12	Q	17	V	22
C	3	H	8	M	13	R	18	W	23
D	4	I	9	N	14	S	19	X	24
E	5	J	10	O	15	T	20	Y	25
								Z	26

Using the function, the word ALGEBRA would be encoded as

$$7 \quad 29 \quad 19 \quad 15 \quad 9 \quad 41 \quad 7,$$

because $f(A) = f(1) = 2(1) + 5 = 7, f(L) = f(12) = 2(12) + 5 = 29$, and so on. The message would then be decoded by using the inverse of f, $f^{-1}(x) = \frac{x-5}{2}$. For example, $f^{-1}(7) = \frac{7-5}{2} = 1 = A$, $f^{-1}(29) = \frac{29-5}{2} = 12 = L$, and so on.

FOR **GROUP DISCUSSION**

1. You are an agent for a detective agency and know that today's function for your code is $f(x) = 4x - 5$. You receive the following message.

$$47 \quad 95 \quad 23 \quad 67 \quad -1 \quad 59 \quad 27 \quad 31 \quad 7 \quad 71 \quad 7 \quad -1 \quad 43 \quad 7 \quad 79 \quad 43 \quad -1 \quad 75$$
$$55 \quad 67 \quad 31 \quad 71 \quad 75 \quad 27 \quad 15 \quad 23 \quad 67 \quad 15 \quad -1 \quad 75 \quad 15 \quad 71 \quad 75 \quad 75 \quad 27 \quad 31$$
$$51 \quad 23 \quad 71 \quad 31 \quad 51 \quad 7 \quad 15 \quad 71 \quad 43 \quad 31 \quad 7 \quad 15 \quad 11 \quad 3 \quad 67 \quad 15 \quad -1 \quad 11$$

   Use the letter/number assignment described earlier to decode the message.

2. Why is a one-to-one function essential in this coding/decoding process?

To conclude this section, we list some important facts about inverses.

IMPORTANT FACTS ABOUT INVERSES

1. If f is one-to-one, then f^{-1} exists.
2. The domain of f is equal to the range of f^{-1} and the range of f is equal to the domain of f^{-1}.
3. If the point (a, b) lies on the graph of f, then (b, a) lies on the graph of f^{-1}.
4. To find the rule for f^{-1}, replace $f(x)$ with y, interchange x and y, and solve for y. This gives f^{-1}.
5. The graphs of f and f^{-1} are reflections of each other across the line $y = x$.

4.5 EXERCISES

Based on your reading of the examples and exposition in this section, answer each of the following.

1. In order for a function to have an inverse, it must be _____.

2. For a function f to be of the type mentioned in Exercise 1, if $a \neq b$, then _____.

3. If f and g are inverses, then $(f \circ g)(x) =$ _____ , and _____ $= x$.

4. The domain of f is equal to the _____ of f^{-1}, and the range of f is equal to the _____ of f^{-1}.

5. If the point (a, b) lies on the graph of f, and f has an inverse, then the point _____ lies on the graph of f^{-1}.

6. If the graphs of f and f^{-1} intersect, they do so at a point that satisfies what condition?

7. If a function f has an inverse, then the graph of f^{-1} may be obtained by reflecting the graph of f across the line with equation _____ .

8. If a function f has an inverse and $f(-3) = 6$, then $f^{-1}(6) =$ _____ .

9. If $f(-4) = 16$ and $f(4) = 16$, then f _____ have an inverse because _____ . (does/does not)

10. If f is a function that has an inverse, and the graph of f lies completely within the second quadrant, then the graph of f^{-1} lies completely within the _____ quadrant.

Decide whether the function graphed or defined is one-to-one. For graphs, assume that a complete graph is shown.

11.
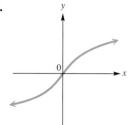

12.
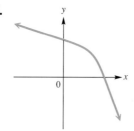

13.
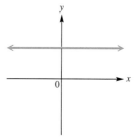

14.
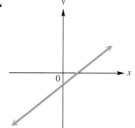

15.
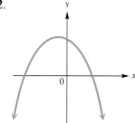

16.
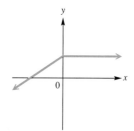

17.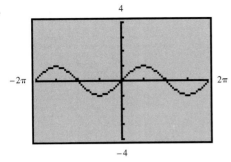

18.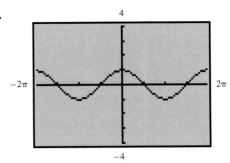

19.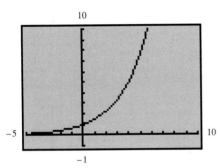

20.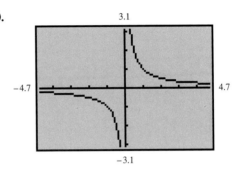

21. $y = 4x - 5$

22. $y = -x^2$

23. $y = (x - 2)^2$

24. $y = -(x + 3)^2 - 8$

25. $y = \sqrt{36 - x^2}$

26. $y = -\sqrt{100 - x^2}$

27. $y = 2x^3 + 1$

28. $y = -\sqrt[3]{x + 5}$

29. $y = \dfrac{1}{x + 2}$

30. $y = \dfrac{x + 2}{x - 3}$

31. Explain why a polynomial function of even degree cannot be one-to-one.

32. Explain why a polynomial function of odd degree *may* not be one-to-one.

In Exercises 33–38 an everyday activity is described. Keeping in mind that an inverse operation "undoes" what an operation does, describe the inverse activity.

33. Tying your shoelaces

34. Starting a car

35. Entering a room

36. Climbing the stairs

37. Taking off in an airplane

38. Filling a cup

39. Explain why the function $f(x) = x^4$ does not have an inverse. Give examples of ordered pairs to illustrate your explanation.

40. Explain why the function $f(x) = x^5$ is one-to-one.

Decide whether the functions in each pair are inverses of each other.

41.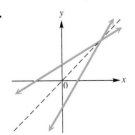

42.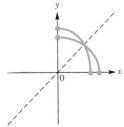

43.

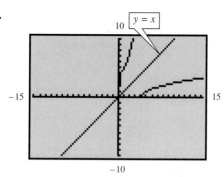

44.

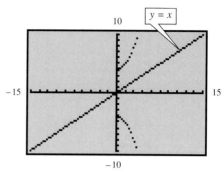

Dot Mode

45.

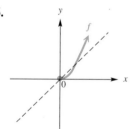

46.

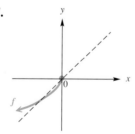

47. $f(x) = -\dfrac{3}{11}x$, $g(x) = -\dfrac{11}{3}x$ **48.** $f(x) = 2x + 4$, $g(x) = \dfrac{1}{2}x - 2$ **49.** $f(x) = 5x - 5$, $g(x) = \dfrac{1}{5}x + 1$

50. $f(x) = 8x - 7$, $g(x) = \dfrac{x + 8}{7}$ **51.** $f(x) = \dfrac{1}{x}$, $g(x) = \dfrac{1}{x}$ **52.** $f(x) = \dfrac{2x + 3}{x - 1}$, $g(x) = \dfrac{x + 3}{x - 2}$

Draw by hand the graph of the inverse of each function in Exercises 53–58.

53.

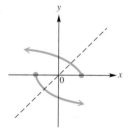

54.

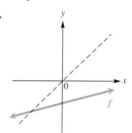

55.

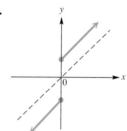

56.

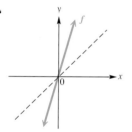

57.

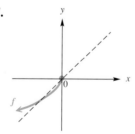

58.

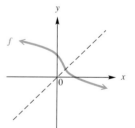

Each function f defined in Exercises 59–66 is one-to-one. Find f^{-1} analytically, and graph by hand both f and f^{-1} on the same axes.

59. $f(x) = 3x - 4$ **60.** $f(x) = 4x - 5$ **61.** $f(x) = \dfrac{1}{3}x$

62. $f(x) = -\dfrac{2}{5}x$ **63.** $f(x) = x^3 + 1$ **64.** $f(x) = \dfrac{1}{x}$

65. $f(x) = \sqrt{6 + x}$ **66.** $f(x) = \sqrt[3]{x - 5}$

While a function may not be one-to-one when defined over its "natural" domain, it may be possible to restrict the domain in such a way that it is one-to-one and the range of the function is unchanged. For example, if we restrict the domain of the function $f(x) = x^2$ (which is not one-to-one over $(-\infty, \infty)$) to $[0, \infty)$, we obtain a one-to-one function whose range is still the same. See the accompanying figure. Notice that we could also choose to restrict the domain of $f(x) = x^2$ to $(-\infty, 0]$ and obtain the graph of a one-to-one function, except that it would be the left half of the parabola.

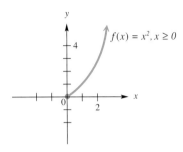

For each function in Exercises 67–72, decide on a suitable restriction on the domain so that the function is one-to-one and the range is not changed. You may wish to use a calculator-generated graph to help you decide.

67. $f(x) = -x^2 + 4$ **68.** $f(x) = (x - 1)^2$ **69.** $f(x) = |x - 6|$

70. $f(x) = x^4$ **71.** $f(x) = x^4 + x^2 - 6$ **72.** $f(x) = -\sqrt{x^2 - 16}$

Using the restrictions given here for the functions in Exercises 67–70, find a rule for f^{-1}.

73. $f(x) = -x^2 + 4$, $x \geq 0$ **74.** $f(x) = (x - 1)^2$, $x \geq 1$

75. $f(x) = |x - 6|$, $x \geq 6$ **76.** $f(x) = x^4$, $x \geq 0$

77. On a square sheet of paper, sketch the graph of any one-to-one function, using the entire sheet. Pattern your axes as shown in the figure. Now, hold the sheet with your left hand at the left end of the x-axis and your right hand at the right end of the x-axis. Rotate the sheet 180° across the x-axis, so that the back of the sheet is now face up. Then, rotate the sheet 90° in a counterclockwise direction. Hold the sheet up to the light, and observe the graph from the back side of the sheet. How does it compare to your original graph?

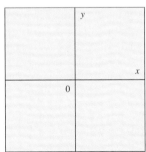

78. Some modern graphics calculators have the capability to "draw" the graph of the inverse of a function. Read your owner's manual to see if yours has this capability. If it does, use it to graph $f(x) = (x + 3)^3$ and f^{-1} on the same axes.

Use the alphabet coding assignment given in this section in Exercises 79 and 80.

79. The function $f(x) = 3x - 2$ was used to encode the following message:

37   25   19   61   13   34   22   1   55   1   52   52   25   64   13   10.

Find the inverse function and decode the message.

80. Encode the message SEND HELP using the one-to-one function $f(x) = x^3 - 1$. Give the inverse function that the decoder would need when the message is received.

Chapter 4   **SUMMARY**

Rational functions are defined by a quotient of polynomials. They often have graphs with one or more breaks. If the rule defining the function is in lowest terms, the graph may have one or more vertical asymptotes and one horizontal or oblique asymptote. Vertical asymptotes are found by setting the denominator equal to 0 and solving the equation. If the numerator is of lower degree than the denominator, the horizontal asymptote is $y = 0$. If the numerator and denominator have the same degree, the horizontal asymptote is $y = k$, where k is the quotient of the leading coefficients of the numerator and denominator. Rational functions, where the numerator is of degree one more than the denominator, may have an oblique asymptote. The graph may intersect a horizontal or oblique asymptote. If the rule of the rational function is not in lowest terms, the graph may have a "hole" instead of a vertical asymptote.

To solve a rational equation, multiply both sides of the equation by the least common denominator of all the terms in the equation. Because there may be values of the variable where the denominator is zero, all proposed solutions must be checked, or these values should be determined before beginning the solution process.

Inverse variation is an application of rational functions. We say y varies inversely as the nth power of x if a nonzero real number k exists, such that y is the quotient of k and x^n.

For positive integers n and $a > 0$, the notations $\sqrt[n]{a}$ and $a^{1/n}$ both represent the positive nth root of a. If n is even, the nth root of a is real only if a is nonnegative. With appropriate restrictions, $a^{m/n} = (\sqrt[n]{a})^m = \sqrt[n]{a^m}$. Important features and the graphs of root functions for n even and for n odd are introduced in Section 4.3. The terminology of Sections 2.2 and 2.3 is used to explain how variations of root functions are obtained from the basic graphs. Special functions defined by roots have graphs that are semicircles or semi-parabolas.

Radical equations are solved by raising both sides to a power that will eliminate the radical, and then solving the resulting equation. This method may produce extraneous solutions, so again it is important to check all proposed solutions in the *original* equation. A calculator may be used to graph the corresponding function and check the proposed solutions by observing the graph. Inequalities may be solved as they were with earlier types of functions, by observing where the graph of the corresponding function is positive or negative or by using a sign graph.

For a function to have an inverse function, it must be one-to-one. A function is one-to-one if different elements from the domain always lead to different elements in the range. The horizontal line test is a good way to verify that a function is one-to-one by observing its graph. A function that is increasing (or decreasing) over its domain is one-to-one. To find the inverse of a one-to-one function, exchange x and y in the rule for the function, then solve for y to get the rule of the inverse function. The graphs of a function and its inverse function are reflections across the line $y = x$.

Key Terms

SECTION 4.1

rational function
vertical asymptote
horizontal asymptote
oblique asymptote

SECTION 4.2

rational equation
rational inequality
extraneous solution

inverse variation
combined variation
joint variation

SECTION 4.3

radicand of a radical
root index of a radical
root function

SECTION 4.4

root equation
root inequality

SECTION 4.5

identity properties
inverse properties
identity function
inverse function
one-to-one function
horizontal line test

Chapter 4   REVIEW EXERCISES

Use the guidelines of Section 4.1 to sketch by hand the graph of each of the following rational functions. Use a graphics calculator in dot mode to support your work.

1. $f(x) = \dfrac{8}{x}$ **2.** $f(x) = \dfrac{2}{3x - 1}$ **3.** $f(x) = \dfrac{4x - 2}{3x + 1}$

4. $f(x) = \dfrac{6x}{(x - 1)(x + 2)}$ **5.** $f(x) = \dfrac{2x}{x^2 - 1}$ **6.** $f(x) = \dfrac{x^2 + 4}{x + 2}$

7. $f(x) = \dfrac{x^2 - 1}{x}$ **8.** $f(x) = \dfrac{-2}{x^2 + 1}$ **9.** $f(x) = \dfrac{4x^2 - 9}{2x + 3}$

10. Under what conditions will the graph of a rational function defined by an expression reduced to lowest terms have an oblique asymptote?

11. What is the equation of the oblique asymptote of the graph of $f(x) = \frac{2x^2 + x - 6}{x - 1}$?

Use the terminology of Chapter 2 to explain how the graph of the function can be obtained from the graph of $y = \frac{1}{x}$.

12. $y = -\dfrac{2}{x + 3}$ **13.** $y = -\dfrac{1}{x} + 6$ **14.** $y = \dfrac{4}{x} - 3$

Solve the rational equation in part (a) analytically. Then use a graph to determine the solution sets of the associated inequalities in parts (b) and (c).

15. (a) $\dfrac{3x - 2}{x + 1} = 0$ **16.** (a) $\dfrac{5}{2x + 5} = \dfrac{3}{x + 2}$

     (b) $\dfrac{3x - 2}{x + 1} < 0$ (b) $\dfrac{5}{2x + 5} < \dfrac{3}{x + 2}$

     (c) $\dfrac{3x - 2}{x + 1} > 0$ (c) $\dfrac{5}{2x + 5} > \dfrac{3}{x + 2}$

17. (a) $\dfrac{3}{x - 2} + \dfrac{1}{x + 1} = \dfrac{1}{x^2 - x - 2}$ **18.** (a) $1 - \dfrac{5}{x} + \dfrac{6}{x^2} = 0$

     (b) $\dfrac{3}{x - 2} + \dfrac{1}{x + 1} \leq \dfrac{1}{x^2 - x - 2}$ (b) $1 - \dfrac{5}{x} + \dfrac{6}{x^2} \leq 0$

     (c) $\dfrac{3}{x - 2} + \dfrac{1}{x + 1} \geq \dfrac{1}{x^2 - x - 2}$ (c) $1 - \dfrac{5}{x} + \dfrac{6}{x^2} \geq 0$

A complete graph of a rational function f is shown here. Use the figure to find the solution set of each of the following.

19. $f(x) = 0$

20. $f(x) > 0$

21. $f(x) < 0$

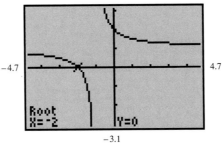

[-4.7, 4.7] by [-3.1, 3.1]
Xscl = 1   Yscl = 1
The graph has a vertical
asymptote at $x = -1$.

Solve each problem involving variation.

22. Suppose r varies directly as x and inversely as the square of y. If r is 10 when x is 5 and y is 3, find r when x is 12 and y is 4.

23. Suppose m varies jointly as n and the square of p, and inversely as q. If m is 20 when n is 5, p is 6, and q is 18, find m when n is 7, p is 11, and q is 2.

24. Suppose Z varies jointly as the square of J and the cube of M, and inversely as the fourth power of W. If Z is 125 when J is 3, M is 5 and W is 1, find Z if J is 2, M is 7, and W is 3.

25. The power a windmill obtains from the wind varies directly as the cube of the wind velocity. If a 10 kilometer per hour wind produces 10,000 units of power, how much power is produced by a wind of 15 kilometers per hour?

26. Hooke's law for an elastic spring states that the distance a spring stretches varies directly as the force applied. If a force of 32 pounds stretches a certain spring 48 inches, how much will a force of 24 pounds stretch the spring?

27. The weight w of an object varies inversely as the square of the distance d between the object and the center of the earth. If a man weighs 90 kg on the surface of the earth, how much would he weigh 800 km above the surface? (The radius of the earth is about 6400 km.)

Suppose that a and b are positive numbers. Draw by hand a sketch of the general shape of each of the following functions.

28. $y = -a\sqrt{x}$ **29.** $y = \sqrt[3]{x} + a$ **30.** $y = \sqrt[3]{x} - a$

31. $y = -a\sqrt[3]{x} - b$ **32.** $y = \sqrt{x + a} + b$

Evaluate each of the following without the use of a calculator. Then check your results with a calculator.

33. $-\sqrt[5]{-32}$ **34.** $36^{-3/2}$ **35.** $-1000^{2/3}$

Use your calculator to find each of the following roots. Tell whether the display represents the exact value or a rational approximation of an irrational number. Give as many digits as your display shows.

36. $\sqrt[5]{81.6}$ **37.** $\sqrt[4]{\dfrac{1}{16}}$ **38.** $12^{1/3}$ **39.** $\left(\dfrac{1}{8}\right)^{4/3}$

Consider the function $f(x) = -\sqrt{2x - 4}$.

40. Find the domain of f analytically.

41. Use a graph to determine the range of f.

42. Give the interval over which the function is increasing (if any).

43. Give the interval over which the function is decreasing (if any).

Consider the equation $x^2 + (y + 4)^2 = 25$.

44. Describe the graph of the equation.

45. Determine analytically two functions, y_1 and y_2, such that their union will give the graph of the equation.

46. Graph the equation in a square viewing window.

Suppose that $y_1 = f(x)$ is the root function graphed below, and $y_2 = g(x)$ is the absolute value function graphed on the same axes. Use the graphs to solve each equation or inequality.

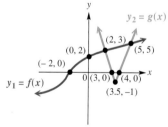

47. $f(x) = 0$ **48.** $g(x) = 0$ **49.** $f(x) = g(x)$ **50.** $f(x) \geq 0$

51. $f(x) < g(x)$ **52.** $f(x) > 3$ **53.** $f(x) - g(x) < 0$ **54.** $g(x) = -1$

Solve the radical equation in part (a) analytically. Then use a graph to determine the solution sets of the associated inequalities in parts (b) and (c).

55. (a) $\sqrt{5 + 2x} = x + 1$
  (b) $\sqrt{5 + 2x} > x + 1$
  (c) $\sqrt{5 + 2x} < x + 1$

57. (a) $\sqrt[3]{6x + 2} = \sqrt[3]{4x}$
  (b) $\sqrt[3]{6x + 2} \geq \sqrt[3]{4x}$
  (c) $\sqrt[3]{6x + 2} \leq \sqrt[3]{4x}$

56. (a) $\sqrt{2x + 1} - \sqrt{x} = 1$
  (b) $\sqrt{2x + 1} - \sqrt{x} > 1$
  (c) $\sqrt{2x + 1} - \sqrt{x} < 1$

58. (a) $(x - 2)^{2/3} - x^{1/3} = 0$
  (b) $(x - 2)^{2/3} - x^{1/3} \geq 0$
  (c) $(x - 2)^{2/3} - x^{1/3} \leq 0$

59. A company plans to package its product in a cylinder that is open at one end. The cylinder is to have a volume of 27π cubic inches. What radius should the circular bottom of the cylinder have to minimize the cost of the material? (*Hint:* The volume of a circular cylinder is $\pi r^2 h$, where r is the radius of the circular base and h is the height; the surface area of the cylinder open at one end is $2\pi rh + \pi r^2$.)

Determine whether the function is one-to-one. Assume that graphs shown are complete.

60.

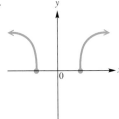

61. $f(x) = \sqrt{3x + 2}$

62.

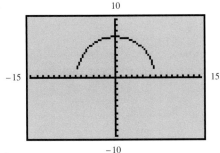

Consider the function $f(x) = \sqrt[3]{2x - 7}$ in Exercises 63–68.

63. What is the domain of f?

64. What is the range of f?

65. Explain why f^{-1} exists.

66. Find the rule for $f^{-1}(x)$.

67. Graph both f and f^{-1} in a square viewing window, along with the line $y = x$. Describe how the graphs of f and f^{-1} are related.

68. Verify analytically that $(f \circ f^{-1})(x) = x$ and $(f^{-1} \circ f)(x) = x$.

Use the letter/number coding assignment in Section 4.5 to decode the following, using the function $f(x) = 6x - 4$.

69. 56   122   110   116   20   86   50   116

70. 2   134   26   110   86   74   26   20   122   20   26

Exponential and Logarithmic Functions

5.1 INTRODUCTION TO EXPONENTIAL FUNCTIONS

Preliminary Considerations ▮ Graphs of Exponential Functions ▮
Exponential Equations (Type 1) ▮ The Number *e* ▮ Compound Interest

Preliminary Considerations

In Section 4.3 we studied how rational numbers are used as exponents. In particular, if r is a rational number and $r = \frac{m}{n}$, then for appropriate values of m and n,

$$a^{m/n} = (\sqrt[n]{a})^m.$$

For example,

$$16^{3/4} = (\sqrt[4]{16})^3 = 2^3 = 8,$$

$$27^{-1/3} = \frac{1}{27^{1/3}} = \frac{1}{\sqrt[3]{27}} = \frac{1}{3},$$

and

$$64^{-1/2} = \frac{1}{64^{1/2}} = \frac{1}{\sqrt{64}} = \frac{1}{8}.$$

In this section the definition of a^r is extended to include all real (not just rational) values of the exponent r. For example, the new symbol $2^{\sqrt{3}}$ might be evaluated by approximating the exponent $\sqrt{3}$ by the numbers 1.7, 1.73, 1.732, and so on. Since these decimals approach the value of $\sqrt{3}$ more and more closely, it seems reasonable that $2^{\sqrt{3}}$ should be approximated more and more closely by the numbers $2^{1.7}$, $2^{1.73}$, $2^{1.732}$, and so on. (Recall, for example, that $2^{1.7} = 2^{17/10} = \sqrt[10]{2^{17}}$.) In fact, this is exactly how $2^{\sqrt{3}}$ is defined (in a more advanced course).

With this interpretation of real exponents, all rules and theorems for exponents are valid for real-number exponents as well as rational ones. In addition to the usual rules for exponents (see the Appendix), several new properties are used in this chapter. For example, if $y = 2^x$, then each value of x leads to exactly one value of y, and therefore, $y = 2^x$ defines a function. Furthermore,

$$\text{if } 3^x = 3^4, \text{ then } x = 4,$$

and for $p > 0$,

$$\text{if } p^2 = 3^2, \text{ then } p = 3.$$

Also,

$$4^2 < 4^3 \qquad \text{but} \qquad \left(\frac{1}{2}\right)^2 > \left(\frac{1}{2}\right)^3,$$

so that when $a > 1$, increasing the exponent on a leads to a *larger* number, but if $0 < a < 1$, increasing the exponent on a leads to a *smaller* number.

These properties are generalized below. Proofs of the properties are not given here, as they require more advanced mathematics.

ADDITIONAL PROPERTIES OF EXPONENTS

a. If $a > 0$ and $a \neq 1$, then a^x is a unique real number for all real numbers x.
b. If $a > 0$ and $a \neq 1$, then $a^b = a^c$ if and only if $b = c$.
c. If $a > 1$ and $m < n$, then $a^m < a^n$.
d. If $0 < a < 1$ and $m < n$, then $a^m > a^n$.

Properties (a) and (b) require $a > 0$ so that a^x is always defined. For example, $(-6)^x$ is not a real number if $x = \frac{1}{2}$. This means that a^x will always be positive, since a is positive. In part (a), $a \neq 1$ because $1^x = 1$ for every real-number value of x, so that each value of x does not lead to a distinct real number. For Property (b) to hold, a must not equal 1 since, for example, $1^4 = 1^5$, even though $4 \neq 5$.

A graphics calculator can easily be used to find approximations of numbers raised to irrational powers. Using the exponentiation capabilities, we can find the following approximations:

$$2^{\sqrt{6}} \approx 5.462228786 \qquad\qquad 3^{\sqrt{2}} \approx 4.728804388$$

$$\left(\frac{1}{2}\right)^{\sqrt{3}} \approx .3010237439 \qquad\qquad .5^{-\sqrt{2}} \approx 2.665144143.$$

Later we will see how these approximations may also be found by using graphs of *exponential functions*.

Graphs of Exponential Functions

We now define a new kind of function, the **exponential function.**

EXPONENTIAL FUNCTION

If $a > 0$, $a \neq 1$, then

$$f(x) = a^x$$

is the exponential function with base a.

NOTE   We do not allow 1 as a base for the exponential function because $1^x = 1$ for all real x, and thus it leads to the constant function $f(x) = 1$.

As we shall see, the behavior of the graph of an exponential function depends, in general, on the magnitude of a. Figure 1 shows the graphs of $f(x) = a^x$ for $a = 2, 3,$ and 4.

Based on our earlier discussion, the domain of $f(x) = a^x$ is $(-\infty, \infty)$. From the graphs in Figure 1, we see that the range is $(0, \infty)$, and the function is *increasing* for $a = 2, 3,$ and 4. The x-axis is the horizontal asymptote as $x \to -\infty$, and the y-intercept is 1. As a becomes larger, the graph becomes "steeper."

TECHNOLOGICAL NOTE
Because of the limited resolution of the graphics calculator screen, it is difficult to interpret how the graph of the exponential function $f(x) = a^x$ behaves when the curve is close to the x-axis, as seen in Figures 1 and 2. Remember that there is no endpoint, and that the curve approaches but never touches the x-axis. Tracing along the curve and observing the y-coordinates as $x \to -\infty$ when $a > 1$ and as $x \to \infty$ when $0 < a < 1$ will help support this fact.

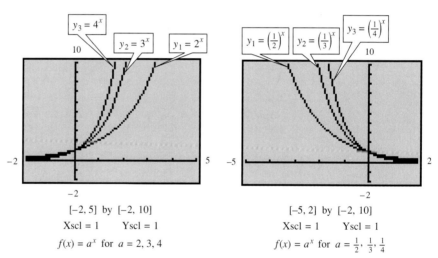

$[-2, 5]$ by $[-2, 10]$
Xscl = 1   Yscl = 1

$f(x) = a^x$ for $a = 2, 3, 4$

FIGURE 1

$[-5, 2]$ by $[-2, 10]$
Xscl = 1   Yscl = 1

$f(x) = a^x$ for $a = \frac{1}{2}, \frac{1}{3}, \frac{1}{4}$

FIGURE 2

If we choose the base a to be between 0 and 1, the graph of the exponential function $f(x) = a^x$ appears to be the same general type of curve, but with one important difference. Figure 2 shows the graphs of $f(x) = a^x$ for $a = \frac{1}{2}, \frac{1}{3},$ and $\frac{1}{4}$. As a gets closer to 0, the graph becomes "steeper."

Again, the domain is $(-\infty, \infty)$, the range is $(0, \infty)$, and the y-intercept is 1. However, the function is *decreasing* for $a = \frac{1}{2}, \frac{1}{3},$ and $\frac{1}{4}$, and the x-axis is the horizontal asymptote as $x \to \infty$. Our observations in Figures 1 and 2 lead to the following generalizations about the graphs of exponential functions.

EXPONENTIAL FUNCTION ($a > 1$)

$f(x) = a^x$, $a > 1$ (Figure 3)

Domain: $(-\infty, \infty)$

Range: $(0, \infty)$

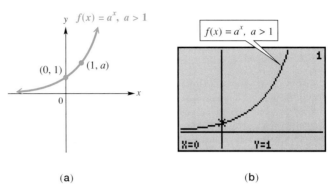

(a)

(b)

FIGURE 3

The exponential function $f(x) = a^x$, $a > 1$, increases on its entire domain, and is continuous as well. The x-axis is the horizontal asymptote as $x \to -\infty$. There is no x-intercept, and the y-intercept is 1. It is always concave up, and there are no extrema.

EXPONENTIAL FUNCTION ($0 < a < 1$)

$f(x) = a^x$, $0 < a < 1$ (Figure 4)

Domain: $(-\infty, \infty)$

Range: $(0, \infty)$

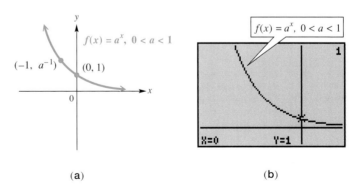

(a)

(b)

FIGURE 4

The exponential function $f(x) = a^x$, $0 < x < 1$, decreases on its entire domain, and is continuous as well. The x-axis is the horizontal asymptote as $x \to \infty$. There is no x-intercept, and the y-intercept is 1. It is always concave up, and there are no extrema.

FOR GROUP DISCUSSION

1. Using a standard viewing window, it is not possible to observe how the graph of the exponential function $f(x) = 2^x$ behaves for values of x less than about -3. Use the window $[-10, 0]$ by $[-.5, .5]$ and then trace to the left to see what happens.

2. Repeat item 1 for the exponential function $f(x) = (\frac{1}{2})^x$. Use a window of $[0, 10]$ by $[-.5, .5]$, and trace to the right.

3. Complete the following statement: For the graph of an exponential function $f(x) = a^x$, $a > 0$, $a \neq 1$, every range value appears exactly once, and thus f is a(n) _____ function. Because of this, f has a(n) _____. (*Hint:* Recall the concepts studied in Section 4.5.)

EXAMPLE 1

Using Exponential Function Graphs to Evaluate Powers

Evaluate each of the following powers in two ways. First, use the definition of an exponent or the exponentiation capability of a calculator. Then, use a graph to support the result.

(a) $4^{3/2}$

SOLUTION Using the definition of a rational number as an exponent, we have

$$4^{3/2} = (\sqrt{4})^3 = 2^3 = 8.$$

Figure 5 shows the graph of $y = 4^x$, with y evaluated for $x = \frac{3}{2} = 1.5$. Notice that when $x = 1.5$, $y = 8$.

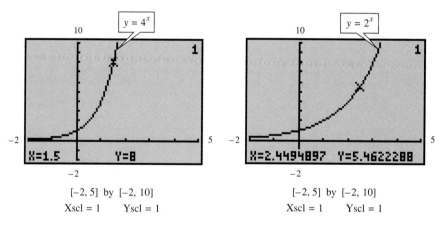

$[-2, 5]$ by $[-2, 10]$	$[-2, 5]$ by $[-2, 10]$
Xscl = 1 Yscl = 1	Xscl = 1 Yscl = 1

FIGURE 5   **FIGURE 6**

(b) $2^{\sqrt{6}}$

SOLUTION Earlier in this section we stated that the exponentiation capability of a graphics calculator will give the approximation of $2^{\sqrt{6}}$ as 5.462228786. By graphing $y = 2^x$ and letting $x = \sqrt{6} \approx 2.4494897$, we find $y \approx 5.4622288$, supporting our earlier result. See Figure 6.

(c) $.5^{-\sqrt{2}}$

SOLUTION Again, we saw earlier that this number is approximately 2.665144143 by using the exponentiation capability of a calculator. Using the graph of $y = .5^x$, with $x = -\sqrt{2} \approx -1.414214$, we find $y \approx 2.6651441$, supporting our earlier result. See Figure 7. ∎

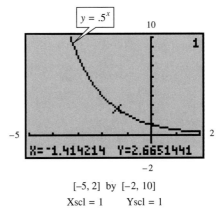

[−5, 2] by [−2, 10]

Xscl = 1   Yscl = 1

FIGURE 7

FOR **GROUP DISCUSSION**

Repeat the procedure of Example 1 for $3^{\sqrt{2}}$ and $(\frac{1}{2})^{\sqrt{3}}$.

EXAMPLE 2

Comparing the Graphs of $f(x) = 2^x$ and $g(x) = (\frac{1}{2})^x$

Use concepts of Section 2.3 to explain why the graph of $g(x) = (\frac{1}{2})^x$ is a reflection across the y-axis of the graph of $f(x) = 2^x$.

SOLUTION   We must show that $g(x) = f(-x)$ to prove analytically that the graph of g is a reflection of the graph of f across the y-axis.

$$g(x) = \left(\frac{1}{2}\right)^x \qquad \text{Given}$$

$$= (2^{-1})^x \qquad \text{Definition of } a^{-1}$$

$$= 2^{-x} \qquad (a^m)^n = a^{mn}$$

$$= f(-x) \qquad \text{Because } f(x) = 2^x, f(-x) = 2^{-x}.$$

Figure 8 shows traditional graphs of $f(x) = 2^x$, $g(x) = (\frac{1}{2})^x$, and a calculator-generated graph of both on the same set of axes.   ◨

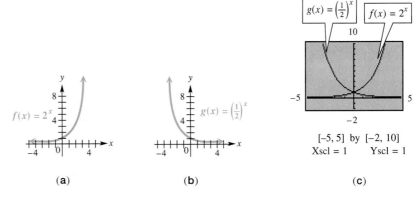

[−5, 5] by [−2, 10]

Xscl = 1   Yscl = 1

(a)   (b)   (c)

FIGURE 8

EXAMPLE **3**
Using Translations to Obtain the Graph of a Function

Use the terminology of Chapter 2 to explain how the graph of $y = -2^x + 3$ can be obtained from the graph of $y = 2^x$. Discuss other important features of the graph.

SOLUTION The graph of $y = -2^x$ is a reflection across the x-axis of the graph of $y = 2^x$. (Note that the base is not -2, but 2; negative numbers are not allowed as bases for exponential functions.) The $+3$ indicates that the graph is translated three units upward, making the horizontal asymptote the line $y = 3$ rather than the x-axis. The y-intercept is 2. The x-intercept can be approximated by using the root location capability of a calculator; it is approximately 1.58.

The graph of $y = -2^x + 3$ is shown in both traditional and calculator-generated forms in Figure 9. ∎

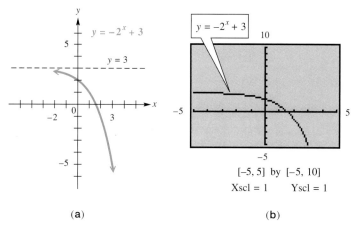

(a)   (b)

FIGURE 9

NOTE We will be able to find the x-intercept of the graph in Figure 9 analytically using concepts covered in a later section.

Exponential Equations (Type 1)

An equation such as $\left(\frac{1}{3}\right)^x = 81$ is different from any equation studied so far in this book because the variable appears in the exponent. Notice that the base on the left side $\left(\frac{1}{3}\right)$ and the constant on the right side (81) can both easily be written as powers of a common base, 3. We shall refer to such an equation as a Type 1 exponential equation. Using property (b) of the additional properties of exponents listed earlier in this section, we can solve Type 1 exponential equations.

EXAMPLE **4**
Solving a Type 1 Exponential Equation

Solve $\left(\frac{1}{3}\right)^x = 81$.

SOLUTION Because $\frac{1}{3} = 3^{-1}$ and $81 = 3^4$, we can solve the equation as follows.

$$\left(\frac{1}{3}\right)^x = 81$$
$$(3^{-1})^x = 3^4$$
$$3^{-x} = 3^4$$
$$-x = 4 \qquad \text{Property (b)}$$
$$x = -4$$

The solution set is $\{-4\}$. ∎

FOR **GROUP DISCUSSION**

Support the result of Example 4 using the graphs of $y = (\frac{1}{3})^x$ and $y = 81$, using the intersection-of-graphs method. Then support the result using the x-intercept method, by graphing $y = (\frac{1}{3})^x - 81$.

EXAMPLE 5

Solving a Type 1 Exponential Equation

Solve $1.5^{x+1} = (\frac{27}{8})^x$ analytically, and support the solution graphically using the x-intercept method.

SOLUTION   At first glance, this equation may not seem to be of Type 1. However, notice that $1.5 = \frac{3}{2}$ and $\frac{27}{8} = (\frac{3}{2})^3$. Thus, each base may easily be written as a power of a common base, $\frac{3}{2}$, and the method of solution used in Example 1 may be applied.

$$1.5^{x+1} = \left(\frac{27}{8}\right)^x$$

$$\left(\frac{3}{2}\right)^{x+1} = \left[\left(\frac{3}{2}\right)^3\right]^x \qquad \text{Write each base as a power of } \frac{3}{2}.$$

$$\left(\frac{3}{2}\right)^{x+1} = \left(\frac{3}{2}\right)^{3x} \qquad (a^m)^n = a^{mn}$$

$$x + 1 = 3x \qquad \text{Set exponents equal.}$$

$$1 = 2x$$

$$x = \frac{1}{2} \quad \text{or} \quad .5$$

The solution set is $\{.5\}$. Graphing $y = 1.5^{x+1} - (\frac{27}{8})^x$ and using the root location capability of a calculator confirms our analytic solution, as shown in Figure 10.

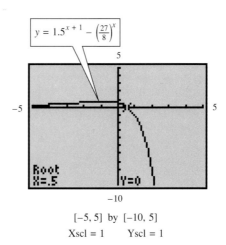

$[-5, 5]$ by $[-10, 5]$

Xscl = 1   Yscl = 1

FIGURE 10

EXAMPLE 6

Using a Graph to Solve Exponential Inequalities

Use the graph in Figure 10 to solve each inequality.

(a) $1.5^{x+1} - \left(\frac{27}{8}\right)^x > 0$

(b) $1.5^{x+1} - \left(\frac{27}{8}\right)^x < 0$

SOLUTION   It appears that the graph of $y = 1.5^{x+1} - (\frac{27}{8})^x$ in Figure 10 lies *above* the x-axis for values of x less than .5, and *below* the x-axis for values of x greater than .5. Tracing left and right will help to support this observation. The solution set for the inequality in part (a) is $(-\infty, .5)$ while the solution set for the inequality in part (b) is $(.5, \infty)$.   ▮

The Number e

Perhaps the most important exponential function has the irrational number e as its base. The number e is named after the Swiss mathematician Leonhard Euler (1707–1783). There are many mathematical expressions that can be used to approximate e, but one of the easiest to illustrate involves the expression $(1 + \frac{1}{x})^x$. If we let x take on larger and larger values (that is, approach ∞), the expression approaches e. The following table shows how this happens when we let x take on powers of 10.

TECHNOLOGICAL NOTE
Further Explorations at the end of the exercises in this section include an activity involving the table of values approaching e. The activity illustrates how the TABLE function of a graphics calculator supports what is seen in this table.

x	$(1 + \frac{1}{x})^x$
1	2
10	2.59374246
100	2.704813829
1000	2.716923932
10,000	2.718145927
100,000	2.718268237
1,000,000	2.718280469

You may wish to construct a similar table if your calculator has this feature.

It appears that as $x \to \infty$, $(1 + \frac{1}{x})^x$ is approaching some "limiting" number. (In fact, in calculus this number is called the *limit* of $(1 + \frac{1}{x})^x$ as x approaches ∞.) The number is e, and to nine decimal places, the value of e is as follows.

$$e \approx 2.718281828$$

Because e is such an important base for the exponential function, calculators have the capability to find powers of e. Using the exponentiation capability, with base e, we can find the following approximations.

$$e^{-1} \approx .3678794412 \qquad e^1 \approx 2.718281828$$
$$e^2 \approx 7.389056099 \qquad e^{2.5} \approx 12.18249396$$

Of course, by definition $e^0 = 1$. The graph of $f(x) = e^x$ is shown in Figure 11.

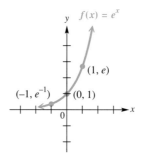

FIGURE 11

FOR **GROUP DISCUSSION**

Use a calculator-generated graph of $f(x) = e^x$ to support the approximations of e^{-1}, e^1, e^2, and $e^{2.5}$ given above, locating the points whose coordinates have x-values of -1, 1, 2, and 2.5.

We will see later in this section and in the final section of this chapter some important applications of the exponential function with base e.

Compound Interest

The formula for *compound interest* (interest paid on both principal and interest) is an important application of exponential functions. You may recall the formula for simple interest, $I = Prt$, where P is the amount left at interest, r is the rate of interest expressed as a decimal, and t is time in years that the principal earns interest. Suppose $t = 1$ year. Then at the end of the year the amount has grown to

$$P + Pr = P(1 + r),$$

the original principal plus the interest. If this amount is left at the same interest rate for another year, the total amount becomes

$$[P(1 + r)] + [P(1 + r)]r = [P(1 + r)](1 + r)$$
$$= P(1 + r)^2.$$

After the third year, this will grow to

$$[P(1 + r)^2] + [P(1 + r)^2]r = [P(1 + r)^2](1 + r)$$
$$= P(1 + r)^3.$$

Continuing in this way produces the following formula for compound interest.

COMPOUND INTEREST FORMULA

Suppose that a principal of P dollars is invested at an annual interest rate r (in percent), compounded n times per year. Then the amount A accumulated after t years is given by the formula

$$A = P\left(1 + \frac{r}{n}\right)^{nt}.$$

EXAMPLE 7

Using the Compound Interest Formula

Suppose that $1000 is invested at an annual rate of 8%, compounded quarterly (four times per year). Find the total amount in the account after 10 years if no withdrawals are made.

SOLUTION   Use the compound interest formula above, with $P = 1000$, $r = .08$, $n = 4$, and $t = 10$.

$$A = P\left(1 + \frac{r}{n}\right)^{nt}$$

$$A = 1000\left(1 + \frac{.08}{4}\right)^{4 \cdot 10}$$

$$A = 1000(1.02)^{40}$$

$$A \approx 2208.039664 \qquad \text{Use a calculator.}$$

To the nearest cent, there will be $2208.04 in the account after ten years. (Note that this means that $2208.04 − $1000 = $1208.04 interest was earned.)

Figure 12 supports this result graphically. The function $y = 1000(1.02)^x$ is graphed and evaluated for $x = 40$ (since $x = nt = 40$). The y-value is the amount in the account. (Tracing to the right on this graph gives new meaning to "watching your money grow.") ◼

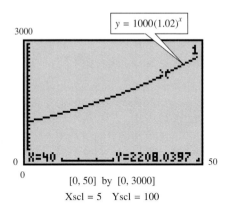

[0, 50] by [0, 3000]
Xscl = 5   Yscl = 100

FIGURE 12

The compounding formula given earlier applies if the financial institution compounds interest for a finite number of compounding periods annually. Theoretically, the number of compounding periods per year can get larger and larger (quarterly, monthly, daily, etc.), and if n is allowed to approach infinity, we say that interest is compounded *continuously*.

To derive the formula for continuous compounding, we begin with the earlier formula.

$$A = P\left(1 + \frac{r}{n}\right)^{nt}$$

Let $k = \frac{n}{r}$. Then $n = rk$, and with these substitutions, the formula becomes

$$A = P\left(1 + \frac{1}{k}\right)^{rkt}$$

$$A = P\left[\left(1 + \frac{1}{k}\right)^{k}\right]^{rt}$$

If $n \to \infty$, $k \to \infty$ as well, and the expression $(1 + \frac{1}{k})^k \to e$, as discussed earlier. This leads to the formula

$$A = Pe^{rt}.$$

CONTINUOUS COMPOUNDING FORMULA

If P dollars is deposited at a rate of interest r compounded continuously for t years, the final amount on deposit is

$$A = Pe^{rt}$$

dollars.

EXAMPLE **8**
Solving a Continuous Compounding Problem

Suppose $5000 is deposited in an account paying 8% compounded continuously for five years. Find the total amount on deposit at the end of five years.

SOLUTION   Let $P = 5000$, $t = 5$, and $r = .08$. Then

$$A = 5000e^{.08(5)} = 5000e^{.4}.$$

Using a calculator, we find that $e^{.4} \approx 1.491824698$, and, to the nearest cent,

$$A = 5000e^{.4} = 7459.12,$$

or $7459.12.

As Figure 13 shows, this result can be supported by the graph of $y = 5000e^{.08x}$. When $x = 5$, $y \approx 7459.12$.   ▯

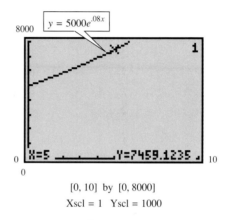

[0, 10] by [0, 8000]
Xscl = 1   Yscl = 1000

FIGURE 13

The continuous compounding formula is an example of an exponential growth function with base e. Other examples of exponential growth (and decay) will be given in Section 5.5.

5.1   EXERCISES

Use the exponentiation capabilities of a calculator to find an approximation for each of the following powers. Give the maximum number of decimal places that your calculator will display.

1. $2^{\sqrt{10}}$

2. $3^{\sqrt{11}}$

3. $\left(\dfrac{1}{2}\right)^{\sqrt{2}}$

4. $\left(\dfrac{1}{3}\right)^{\sqrt{6}}$

5. $4.1^{-\sqrt{3}}$

6. $6.4^{-\sqrt{3}}$

7. $\sqrt{7}^{\sqrt{7}}$

8. $\sqrt{13}^{-\sqrt{13}}$

Use the graph of the given exponential function and the capabilities of your calculator to graphically support the result found in the specified exercise.

9. $y = 2^x$, Exercise 1

10. $y = 3^x$, Exercise 2

11. $y = \left(\dfrac{1}{2}\right)^x$, Exercise 3

12. $y = \left(\dfrac{1}{3}\right)^x$, Exercise 4

In the given figure, the graphs of $y = a^x$ for $a = 1.8, 2.3, 3.2, .4, .75,$ and $.31$ are given. They are identified by letter, but not necessarily in the same order as the values of a just given. Use your knowledge of how the exponential function behaves for various powers of a to identify each lettered graph.

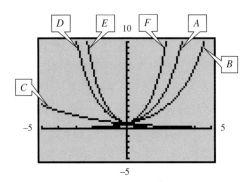

13. *A*   **14.** *B*   **15.** *C*   **16.** *D*   **17.** *E*   **18.** *F*

Evaluate each of the following powers in two ways. First, use the definition of an exponent. Then use a graph to support the result.

19. $8^{2/3}$ **20.** $27^{-4/3}$ **21.** $25^{-3/2}$ **22.** $16^{5/4}$

In Exercises 23 and 24, the graph of an exponential function with base a is given. Follow the directions in parts (a)–(f) in each exercise.

23.

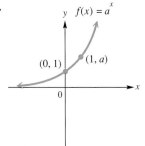

24.

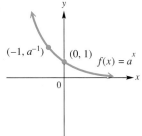

(a) Is $a > 1$ or is $0 < a < 1$?
(b) Give the domain and range of f.
(c) Sketch the graph of $g(x) = -a^x$.
(d) Give the domain and range of g.
(e) Sketch the graph of $h(x) = a^{-x}$.
(f) Give the domain and range of h.

(a) Is $a > 1$ or is $0 < a < 1$?
(b) Give the domain and range of f.
(c) Sketch the graph of $g(x) = a^x + 2$.
(d) Give the domain and range of g.
(e) Sketch the graph of $h(x) = a^{x+2}$.
(f) Give the domain and range of h.

*R*elating Concepts

Use the terminology of Chapter 2 to explain how the graph of the given function can be obtained from the graph of $y = 2^x$. (Recall that $\frac{1}{2} = 2^{-1}$, so $\left(\frac{1}{2}\right)^x = (2^{-1})^x = 2^{-x}$.)

25. $y = 2^{x+5} - 3$ **26.** $y = 2^{x-1} + 4$ **27.** $y = \left(\frac{1}{2}\right)^x + 1$

28. $y = \left(\frac{1}{2}\right)^x - 6$ **29.** $y = -3 \cdot 2^x$ **30.** $y = -4 \cdot \left(\frac{1}{2}\right)^x$

In part (a) of Exercises 31–42, an exponential equation of Type 1 is given. Solve the equation analytically. Use a calculator-generated graph to support your answer, applying either the intersection-of-graphs method or the x-intercept method. Then use the graph to solve the associated inequalities in parts (b) and (c).

31. (a) $125^x = 5$

(b) $125^x > 5$

(c) $125^x < 5$

32. (a) $\left(\frac{1}{2}\right)^x = 4$

(b) $\left(\frac{1}{2}\right)^x > 4$

(c) $\left(\frac{1}{2}\right)^x < 4$

33. (a) $\left(\frac{2}{3}\right)^x = \frac{9}{4}$

(b) $\left(\frac{2}{3}\right)^x \geq \frac{9}{4}$

(c) $\left(\frac{2}{3}\right)^x \leq \frac{9}{4}$

34. (a) $\left(\frac{3}{4}\right)^x = \frac{16}{9}$

(b) $\left(\frac{3}{4}\right)^x \geq \frac{16}{9}$

(c) $\left(\frac{3}{4}\right)^x \leq \frac{16}{9}$

35. (a) $2^{3-x} = 8$

(b) $2^{3-x} > 8$

(c) $2^{3-x} < 8$

36. (a) $5^{2x+1} = 25$

(b) $5^{2x+1} > 25$

(c) $5^{2x+1} < 25$

37. (a) $27^{4x} = 9^{x+1}$

(b) $27^{4x} < 9^{x+1}$

(c) $27^{4x} > 9^{x+1}$

38. (a) $32^x = 16^{1-x}$

(b) $32^x < 16^{1-x}$

(c) $32^x > 16^{1-x}$

39. (a) $\left(\frac{1}{8}\right)^{-2x} = 2^{x+3}$

(b) $\left(\frac{1}{8}\right)^{-2x} \geq 2^{x+3}$

(c) $\left(\frac{1}{8}\right)^{-2x} \leq 2^{x+3}$

40. (a) $3^{-x} = \left(\frac{1}{27}\right)^{1-2x}$

(b) $3^{-x} > \left(\frac{1}{27}\right)^{1-2x}$

(c) $3^{-x} < \left(\frac{1}{27}\right)^{1-2x}$

41. (a) $.5^{-x} = .25^{x+1}$

(b) $.5^{-x} < .25^{x+1}$

(c) $.5^{-x} > .25^{x+1}$

42. (a) $.4^x = 2.5^{1-x}$

(b) $.4^x < 2.5^{1-x}$

(c) $.4^x > 2.5^{1-x}$

Use the exponentiation capabilities of a calculator to find each of the following powers of e. Then use a calculator-generated graph of $y = e^x$ to support your result.

43. $e^{3.1}$

44. $e^{.15}$

45. $e^{-.25}$

46. $e^{-1.6}$

47. $e^{\sqrt{2}}$

48. $e^{-\sqrt{2}}$

Use the appropriate compound interest formula to find the amount that will be in an account, given the stated conditions.

49. $20,000 invested at 3% annual interest for 4 years compounded **(a)** annually; **(b)** semiannually

50. $35,000 invested at 4.2% annual interest for 3 years compounded **(a)** annually; **(b)** quarterly

51. $27,500 invested at 3.95% annual interest for 5 years compounded **(a)** daily ($n = 365$); **(b)** continuously

52. $15,800 invested at 4.6% annual interest for 6.5 years compounded **(a)** quarterly; **(b)** continuously

In Exercises 53 and 54, decide which of the two plans will provide a better yield.

53. Plan A: $40,000 invested for 3 years at 4.5%, compounded quarterly
Plan B: $40,000 invested for 3 years at 4.4%, compounded continuously

54. Plan A: $50,000 invested for 10 years at 4.75% compounded daily ($n = 365$)

Plan B: $50,000 invested for 10 years at 4.7%, compounded continuously

Relating Concepts

In this section we showed how Type 1 exponential equations can be solved using analytic techniques. In Section 5.4 we will show how to solve analytically exponential equations such as $7^x = 12$, where the bases cannot easily be written as powers of the same base. We will refer to such equations as Type 2 equations. However, we can use graphical methods to solve such equations for approximate solutions using either the x-intercept method or the intersection-of-graphs method. For example, to solve $7^x = 12$, we can graph $y_1 = 7^x$ and $y_2 = 12$ and find the x-coordinate of the point of intersection. Alternatively, we can graph $y = 7^x - 12$ and find the x-intercept. Use either of these methods to solve the equations in Exercises 55–62.

55. $7^x = 12$ **56.** $3^x = 5$ **57.** $2^{x-1} = 6$ **58.** $4^{x+3} = 9$

59. $e^x = 8$ **60.** $e^x = 5^{x+3}$ **61.** $10^x = 8$ **62.** $10^x = \left(\dfrac{1}{3}\right)^{2x+3}$

Further Explorations

As described for Figure 12, TRACE can be used to "watch your money grow." This can be done numerically with the TABLE, as well as graphically. An advantage of using the TABLE is that two functions can be "traced" simultaneously.

1. You have the choice of investing $1000.00 at an annual rate of 5%, compounded either annually or monthly. Let Y_1 represent the investment compounded annually, and let Y_2 represent the investment compounded monthly. Graph both Y_1 and Y_2 and observe the slight difference in each curve. Then use the TABLE to compare the graphs numerically. What is the difference between the returns on each investment after 1 year, 2 years, 5 years, 10 years, 20 years, 30 years, and 40 years?

2. You have the choice of investing $1000.00 at an annual rate of 7.5%, compounded daily, or at an annual rate of 7.75%, compounded annually. Let Y_1 represent the investment at 7.5% compounded daily, and let Y_2 represent the investment at 7.75% compounded annually. Graph both Y_1 and Y_2 and observe the slight difference in each curve. Then use the TABLE with $Y_3 = Y_1 - Y_2$ to compare the graphs numerically. What is the difference between the returns on each investment after 1 year, 2 years, 5 years, 10 years, 20 years, 30 years, and 40 years? Why does the lower interest rate yield the greater return?

3. With Y_1 assigned as $(1 + \frac{1}{x})^x$, TblMin $= 0$, and ΔTbl $= 100,000$, observe how the function values approach e. Scroll down the TABLE and see how the values of Y_1 change. (See the following figures.) Why does $x = 0$ lead to an error message?

```
TABLE SETUP
 TblMin=0
 ▵Tbl=100000
Indent: Auto Ask
Depend: Auto Ask
```

```
  X    │ Y₁
─────────────────
  0    │ ERROR
100000 │ 2.7183
200000 │ 2.7183
300000 │ 2.7183
400000 │ 2.7183
500000 │ 2.7183
600000 │ 2.7183
─────────────────
Y₁=2.7182682371?
```

5.2 LOGARITHMS AND THEIR PROPERTIES

Introduction to Logarithms ▮ Common Logarithms ▮ Natural Logarithms ▮
Properties of Logarithms ▮ The Change-of-Base Rule

Introduction to Logarithms

In order to introduce the concept of logarithm, let us consider the exponential
equation

$$2^3 = 8.$$

In this equation, 3 is the exponent to which 2 must be raised in order to obtain 8.
In this context, 3 is called the logarithm to the base 2 of 8, abbreviated

$$3 = \log_2 8.$$

It is important to remember that a logarithm is an exponent and as such will possess
the same properties as exponents.

LOGARITHM

For all positive numbers a, where $a \neq 1$,

$$a^k = x \qquad \text{is equivalent to} \qquad k = \log_a x.$$

A logarithm is an exponent, and $\log_a x$ is the exponent to which a must be raised
in order to obtain x. The number a is called the *base* of the logarithm, and x is
called the *argument* of the expression $\log_a x$. The value of x will always be
positive.

The first example shows conversions between exponential and logarithmic
statements.

EXAMPLE **1**
Converting Between Exponential and Logarithmic Statements

The chart below shows several pairs of equivalent statements. The same statement
is written in both exponential and logarithmic forms.

Exponential Form	Logarithmic Form
$2^3 = 8$	$\log_2 8 = 3$
$(\frac{1}{2})^{-4} = 16$	$\log_{1/2} 16 = -4$
$10^5 = 100{,}000$	$\log_{10} 100{,}000 = 5$
$3^{-4} = \frac{1}{81}$	$\log_3(\frac{1}{81}) = -4$
$5^1 = 5$	$\log_5 5 = 1$
$(\frac{3}{4})^0 = 1$	$\log_{3/4} 1 = 0$

▮

In some cases, the method for solving Type 1 exponential equations presented
in the previous section can be used to evaluate logarithms. The next example
illustrates this.

EXAMPLE **2**

Finding Logarithms
Analytically

Find each of the following logarithms by setting them equal to x and then solving a Type 1 exponential equation.

(a) $\log_8 4$

SOLUTION   Let $x = \log_8 4$ and write in exponential form.

$$x = \log_8 4$$
$$8^x = 4 \qquad \text{Exponential form}$$
$$(2^3)^x = 2^2 \qquad \text{Write as a power of the same base, 2.}$$
$$2^{3x} = 2^2 \qquad (a^m)^n = a^{mn}$$
$$3x = 2 \qquad \text{Set exponents equal.}$$
$$x = \frac{2}{3}$$

Therefore, $\log_8 4 = \frac{2}{3}$.

(b) $\log_{1/27} 81$

SOLUTION

$$x = \log_{1/27} 81$$
$$\left(\frac{1}{27}\right)^x = 81$$
$$(3^{-3})^x = 3^4$$
$$3^{-3x} = 3^4$$
$$-3x = 4$$
$$x = -\frac{4}{3}$$

Therefore, $\log_{1/27} 81 = -\frac{4}{3}$.   ∎

Common Logarithms

One of the two most important bases used for logarithms is 10. (The other is e.) Base 10 logarithms are called **common logarithms,** and the symbol that we use for the common logarithm of a positive number x is log x. Notice that when no base is indicated, the base is understood to be 10.

COMMON LOGARITHM

$\log x = \log_{10} x$ for all positive numbers x.

Remember that the argument of a common logarithm (and any base logarithm, for that matter) must be a positive number.

Graphics calculators have the capability of finding common logarithms (exact or approximate, depending upon the argument). Figure 14 shows a typical graphics calculator screen with several common logarithms evaluated.

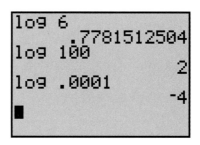

FIGURE 14

The first display indicates that .7781512504 is (approximately) the exponent to which 10 must be raised in order to obtain 6. The second says that 2 is the exponent to which 10 must be raised in order to obtain 100. This is reasonable, since $100 = 10^2$. The third display indicates that -4 is the exponent to which 10 must be raised in order to obtain .0001. Again, this is reasonable, since $.0001 = 10^{-4}$.

In Figure 15 we have graphed the functions $y_1 = 10^x$ and $y_2 = 6$. Notice that the point of intersection has an x-value of .77815125, supporting our observation in Figure 14 for the value of log 6.

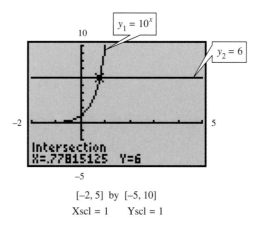

[-2, 5] by [-5, 10]
Xscl = 1   Yscl = 1

FIGURE 15

Common logarithms were originally used to aid mathematicians and scientists in paper-and-pencil calculations. With the advent of calculators and computers, this use is no longer necessary. However, there are still important applications of common logarithms. For example, in chemistry, the pH of a solution is defined as

$$pH = -\log[H_3O^+],$$

where $[H_3O^+]$ is the hydronium ion concentration in moles per liter.* The pH value is a measure of the acidity or alkalinity of solutions. Pure water has a pH of 7.0, substances with pH values greater than 7.0 are alkaline, and substances with pH values less than 7.0 are acidic.

---

*A *mole* is the amount of substance that contains the same number of molecules as the number of atoms in 12 grams of carbon 12.

> **EXAMPLE 3**
>
> Finding pH

(a) Find the pH of a solution with $[H_3O^+] = 2.5 \times 10^{-4}$.

SOLUTION

$$pH = -\log[H_3O^+]$$
$$pH = -\log(2.5 \times 10^{-4}) \qquad \text{Substitute.}$$
$$pH \approx 3.6 \qquad \begin{array}{l}\text{Use a calculator and round to} \\ \text{the nearest tenth.}\end{array}$$

(b) Find the hydronium ion concentration of a solution with pH = 7.1.

SOLUTION

$$pH = -\log[H_3O^+]$$
$$7.1 = -\log[H_3O^+] \qquad \text{Substitute.}$$
$$-7.1 = \log[H_3O^+] \qquad \text{Multiply by } -1.$$
$$[H_3O^+] = 10^{-7.1} \qquad \text{Write in exponential form.}$$

Graphics calculators have keys marked 10^x, usually in conjunction with the log x key, that allow you to raise 10 to a power. Use this key to find that

$$[H_3O^+] = 10^{-7.1} \approx 7.9 \times 10^{-8}. \qquad \blacksquare$$

Natural Logarithms

In many practical applications of logarithms, the number e (introduced in the previous section) is used as the base. Logarithms to base e are called **natural logarithms.** The symbol used for the natural logarithm of a positive number x is ln x (read "el en x").

> **NATURAL LOGARITHM**
>
> ln $x = \log_e x$ for all positive numbers x.

Natural logarithms of numbers can be found in much the same way as common logarithms using a calculator. The natural logarithm key is usually found in conjunction with the e^x key. Figure 16 shows a calculator display using three expressions involving natural logarithms.

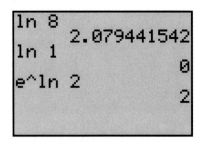

FIGURE 16

The first display indicates that 2.079441542 is (approximately) the exponent to which e must be raised in order to obtain 8. In Figure 17, this is supported by the fact that the graphs of $y_1 = e^x$ and $y_2 = 8$ intersect at a point with x-value 2.0794415.

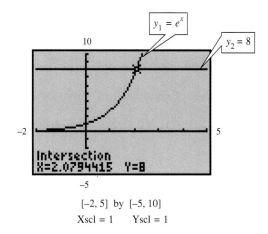

[-2, 5] by [-5, 10]
Xscl = 1   Yscl = 1

FIGURE 17

FOR GROUP DISCUSSION

1. The second display in Figure 16 indicates ln 1 = 0. Use your calculator to find log 1. Discuss your results. If 1 is the argument, does it matter what base is used? Why?
2. The third display in Figure 16 indicates $e^{\ln 2} = 2$. Use the same expression but replace 2 with the number of letters in your last name. What is the result? Can you make a generalization?
3. Based on your answer to item 2, predict the value of $10^{\log 14.6}$. Now verify with your calculator.

EXAMPLE 4

Using Natural Logarithms to Solve an Equation

In Section 5.1, we learned that the formula $A = Pe^{rt}$ is used to compute the amount of money in an account earning interest compounded continuously. Suppose that $1000 is invested at 3% annual interest compounded continuously. How long will it take for the amount to grow to $1500?

SOLUTION   We use the formula with $P = 1000$, $A = 1500$, and $r = .03$.

$$1500 = 1000e^{.03t}$$
$$1.5 = e^{.03t} \qquad \text{Divide by 1000.}$$

Because $.03t$ is the exponent to which e must be raised in order to obtain 1.5, it must equal the natural logarithm of 1.5, or ln 1.5.

$$.03t = \ln 1.5$$
$$t = \frac{\ln 1.5}{.03} \qquad \text{Divide by .03.}$$
$$t \approx 13.5 \qquad \text{Use a calculator.}$$

It will take about 13.5 years to grow to $1500.

This answer can be supported graphically by showing that the point $\left(\frac{\ln 1.5}{.03}, 1500\right)$ lies on the graph of $y = 1000e^{.03x}$. See Figure 18 on the next page. ∎

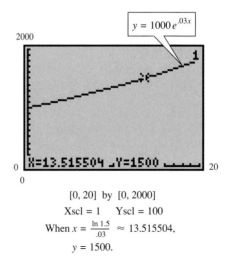

$[0, 20]$ by $[0, 2000]$
Xscl = 1   Yscl = 100
When $x = \frac{\ln 1.5}{.03} \approx 13.515504$,
$y = 1500$.

FIGURE 18

Properties of Logarithms

We will now formally state several properties of logarithms.

PROPERTIES OF LOGARITHMS

1. If $a > 0$, $a \neq 1$, then $\log_a 1 = 0$.
2. If $a > 0$, $a \neq 1$, and k is a real number, then $\log_a a^k = k$.
3. If $a > 0$, $a \neq 1$, and $k > 0$, then $a^{\log_a k} = k$.

Property 1 is true because $a^0 = 1$ for any nonzero value of a. Property 2 is verified by writing the equation in exponential form, giving the identity $a^k = a^k$. Property 3 is justified by the fact that $\log_a k$ is the exponent to which a must be raised in order to obtain k. Therefore, by the definition, $a^{\log_a k}$ must equal k.

Three additional rules of logarithms follow. These rules emphasize the close relationship between logarithms and exponents.

PROPERTIES OF LOGARITHMS

For $x > 0$, $y > 0$, $a > 0$, $a \neq 1$, and any real number r,

Product Rule   4. $\log_a xy = \log_a x + \log_a y$
(The logarithm of the product of two numbers is equal to the sum of the logarithms of the numbers.)

Quotient Rule   5. $\log_a \frac{x}{y} = \log_a x - \log_a y$
(The logarithm of the quotient of two numbers is equal to the difference between the logarithms of the numbers.)

Power Rule   6. $\log_a x^r = r \log_a x$
(The logarithm of a number raised to a power is equal to the exponent multiplied by the logarithm of the number.)

The proof of Property 4, the product rule, follows.

PROOF   Let

$$m = \log_a x \text{ and } n = \log_a y.$$

Then

$$a^m = x \quad \text{and} \quad a^n = y \qquad \text{Definition of logarithm}$$
$$a^m \cdot a^n = xy \qquad\qquad\qquad \text{Multiplication}$$
$$a^{m+n} = xy \qquad\qquad\qquad \text{Add exponents.}$$
$$\log_a xy = m + n. \qquad\qquad \text{Definition of logarithm}$$

Since $m = \log_a x$ and $n = \log_a y$,

$$\log_a xy = \log_a x + \log_a y \qquad \text{Substitution} \qquad ▯$$

Properties 5 and 6 are proved in a similar way. (See Exercises 95 and 96.)

FOR **GROUP DISCUSSION**

Let x represent the number of letters in your first name, and let y represent the number of letters in your last name. Use either the common logarithm or natural logarithm key of your calculator to do the following.

1. Find the logarithm of $x \cdot y$. Find the sum of the logarithms of x and y. Compare the results. What property of logarithms does this illustrate?
2. Find the logarithm of $\frac{x}{y}$. Subtract the logarithm of y from the logarithm of x. Compare the results. What property of logarithms does this illustrate?
3. Find the logarithm of x^y. Now multiply y by the logarithm of x. Compare the results. What property of logarithms does this illustrate?
4. Verify: $\log(x + y) \neq \log x + \log y$ and $\log \frac{x}{y} \neq \frac{\log x}{\log y}$.

EXAMPLE 5

Using the Properties of Logarithms

Assuming that all variables represent positive real numbers, use the properties of logarithms to rewrite each of the following expressions.

(a) $\log 8x$

$$\log 8x = \log 8 + \log x$$

(b) $\log_9 \dfrac{15}{7}$

$$\log_9 \frac{15}{7} = \log_9 15 - \log_9 7$$

(c) $\log_5 \sqrt{8}$

$$\log_5 \sqrt{8} = \log_5 8^{1/2} = \frac{1}{2} \log_5 8$$

(d) $\log_a \dfrac{x}{yz} = \log_a x - (\log_a y + \log_a z) = \log_a x - \log_a y - \log_a z$

(e) $\log_a \sqrt[3]{m^2} = \dfrac{2}{3} \log_a m$

(f) $\log_b \sqrt[n]{\dfrac{x^3 y^5}{z^m}} = \dfrac{1}{n} \log_b \dfrac{x^3 y^5}{z^m}$

$$= \dfrac{1}{n}(\log_b x^3 + \log_b y^5 - \log_b z^m)$$

$$= \dfrac{1}{n}(3 \log_b x + 5 \log_b y - m \log_b z)$$

$$= \dfrac{3}{n} \log_b x + \dfrac{5}{n} \log_b y - \dfrac{m}{n} \log_b z$$

Notice the use of parentheses in the second step. The factor $\frac{1}{n}$ applies to each term.   ∎

EXAMPLE **6**

Using the Properties of Logarithms

Use the properties of logarithms to write each of the following as a single logarithm with a coefficient of 1. Assume that all variables represent positive real numbers.

(a) $\log_3(x + 2) + \log_3 x - \log_3 2$

Using Properties 4 and 5,

$$\log_3(x + 2) + \log_3 x - \log_3 2 = \log_3 \dfrac{(x + 2)x}{2}.$$

(b) $2 \log_a m - 3 \log_a n = \log_a m^2 - \log_a n^3 = \log_a \dfrac{m^2}{n^3}$

Here we used Property 6, then Property 5.

(c) $\dfrac{1}{2} \log_b m + \dfrac{3}{2} \log_b 2n - \log_b m^2 n$

$$= \log_b m^{1/2} + \log_b (2n)^{3/2} - \log_b m^2 n \qquad \text{Property 6}$$

$$= \log_b \dfrac{m^{1/2}(2n)^{3/2}}{m^2 n} \qquad \text{Properties 4 and 5}$$

$$= \log_b \dfrac{2^{3/2} n^{1/2}}{m^{3/2}} \qquad \text{Rules for exponents}$$

$$= \log_b \left(\dfrac{2^3 n}{m^3}\right)^{1/2} \qquad \text{Rules for exponents}$$

$$= \log_b \sqrt{\dfrac{8n}{m^3}} \qquad \text{Definition of } a^{1/n} \qquad ∎$$

CAUTION   There is no property of logarithms to rewrite a logarithm of a *sum* or *difference*. That is why, in Example 6(a), $\log_3(x + 2)$ was not written as $\log_3 x + \log_3 2$. Remember, $\log_3 x + \log_3 2 = \log_3(x \cdot 2)$.

The Change-of-Base Rule

A natural question to ask at this point is "Can we use a calculator to find logarithms for bases other than 10 and e?" The answer is yes, but before we state the change-of-base rule, let us return to Figure 9(b). We have the calculator-generated graph of $y = -2^x + 3$. The root location capability of a calculator will show that the x-intercept of the graph is approximately 1.5849625. If we want to find the x-intercept *analytically*, we must let $y = 0$:

$$y = -2^x + 3$$
$$0 = -2^x + 3 \qquad \text{Let } y = 0.$$
$$2^x = 3 \qquad \text{Add } 2^x \text{ to both sides.}$$
$$x = \log_2 3. \qquad \text{Rewrite in logarithmic form.}$$

Based on the earlier analysis of the graph, $\log_2 3$ must be approximately equal to 1.5849625.

Now we develop the change-of-base rule that will allow us to verify this graphical result analytically.

CHANGE-OF-BASE RULE

For any positive real numbers x, a, and b, where $a \neq 1$ and $b \neq 1$:

$$\log_a x = \frac{\log_b x}{\log_b a}.$$

This rule is proved by using the definition of logarithm to write $y = \log_a x$ in exponential form.

PROOF Let

$$y = \log_a x.$$
$$a^y = x \qquad \text{Change to exponential form.}$$
$$\log_b a^y = \log_b x \qquad \text{Take logarithms on both sides.}$$
$$y \log_b a = \log_b x \qquad \text{Property 6 of logarithms}$$
$$y = \frac{\log_b x}{\log_b a} \qquad \text{Divide both sides by } \log_b a.$$
$$\log_a x = \frac{\log_b x}{\log_b a} \qquad \text{Substitute } \log_a x \text{ for } y. \qquad \blacksquare$$

Any positive number other than 1 can be used for base b in the change of base rule, but usually the only practical bases are e and 10, since calculators give logarithms only for these two bases.

EXAMPLE 7

Using the Change-of-Base Rule

Use the change-of-base rule to find an approximation for $\log_2 3$. Use a calculator, and apply the rule for both common and natural logarithms.

SOLUTION By the change-of-base rule,

$$\log_2 3 = \frac{\log 3}{\log 2} \quad \text{and} \quad \log_2 3 = \frac{\ln 3}{\ln 2}.$$

The display in Figure 19 shows that in each case, $\log_2 3 \approx 1.584962501$. Notice that this supports our earlier graphical analysis.   ∎

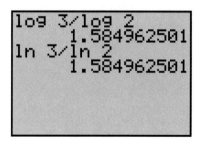

FIGURE 19

FOR **GROUP DISCUSSION**

1. Without using your calculator, determine the exact values of $\log_3 9$ and $\log_3 27$. Then, use the fact that 16 is between 9 and 27 to determine between what two consecutive integers $\log_3 16$ must lie. Finally, use the change-of-base rule to support your answer.
2. Without using your calculator, determine the exact values of $\log_5(\frac{1}{5})$ and $\log_5 1$. Then, use the fact that .68 is between $\frac{1}{5}$ and 1 to determine between what two consecutive integers $\log_5 .68$ must lie. Finally, use the change-of-base rule to support your answer.
3. Use your calculator and the change-of-base rule to support the results found analytically in Example 2: $\log_8 4 = \frac{2}{3}$ and $\log_{1/27} 81 = -\frac{4}{3}$.

5.2 EXERCISES

For each of the following statements, write an equivalent statement in logarithmic form.

1. $3^4 = 81$ **2.** $2^5 = 32$ **3.** $\left(\dfrac{1}{2}\right)^{-4} = 16$ **4.** $\left(\dfrac{2}{3}\right)^{-3} = \dfrac{27}{8}$

5. $10^{-4} = .0001$ **6.** $\left(\dfrac{1}{100}\right)^{-2} = 10{,}000$ **7.** $e^0 = 1$ **8.** $e^{1/2} = \sqrt{e}$

For each of the following statements, write an equivalent statement in exponential form.

9. $\log_6 36 = 2$ **10.** $\log_5 5 = 1$ **11.** $\log_{\sqrt{3}} 81 = 8$ **12.** $\log_4\left(\dfrac{1}{64}\right) = -3$

13. $\log_{10} .001 = -3$ **14.** $\log_3 \sqrt[3]{9} = \dfrac{2}{3}$ **15.** $\log \sqrt{10} = .5$ **16.** $\ln e^6 = 6$

17. Explain in your own words what $\log_a x$ means.

18. In the expression $\log_a x$, why can't x be 0? Why can't x be negative?

Find each of the following logarithms using the method of Example 2. (You may wish to verify your answer by also using the change-of-base rule.)

19. $\log_5 125$

20. $\log_3 81$

21. $\log_6 \dfrac{1}{216}$

22. $\log_{1/4} 16$

23. $\log_{\sqrt{3}} 3^{12}$

24. $\log_{\sqrt[3]{5}} 25$

25. $\log_4 \dfrac{\sqrt[3]{4}}{2}$

26. $\log_9 \dfrac{\sqrt[4]{27}}{3}$

27. $\log_{1/4} \dfrac{16^2}{2^{-3}}$

28. Simplify each of the following.
   (a) $3^{\log_3 7}$ **(b)** $4^{\log_4 9}$ **(c)** $12^{\log_{12} 4}$ **(d)** $a^{\log_a k}$ $(k > 0, a > 0, a \neq 1)$

29. Simplify each of the following.
   (a) $\log_3 3^{19}$ **(b)** $\log_4 4^{17}$ **(c)** $\log_{12} 12^{1/3}$ **(d)** $\log_a \sqrt{a}$ $(a > 0, a \neq 1)$

30. Simplify each of the following.
   (a) $\log_3 1$ **(b)** $\log_4 1$ **(c)** $\log_{12} 1$ **(d)** $\log_a 1$ $(a > 0, a \neq 1)$

Use a calculator to find a decimal approximation for each of the following common or natural logarithms.

31. $\log 43$

32. $\log 1247$

33. $\log .783$

34. $\log .014$

35. $\log 28^3$

36. $\log (47 \times 93)$

37. $\ln 43$

38. $\ln 1247$

39. $\ln .783$

40. $\ln .014$

41. $\ln 28^3$

42. $\ln (47 \times 93)$

43. **(a)** Use a calculator to find a decimal approximation of each of the following common logarithms: log 2.367, log 23.67, log 236.7, log 2367.
   (b) Write each of the following numbers in scientific notation: 2.367, 23.67, 236.7, 2367.
   (c) Compare the results in part (a) to the expressions in part (b). What similarities do you find? What differences do you find?

Relating Concepts

In Exercises 44–48, assume a > 1.

44. Is $f(x) = a^x$ a one-to-one function? If so, based on Section 4.5, what kind of related function exists for f?

45. If $f(x) = a^x$ has an inverse f^{-1}, sketch f and f^{-1} on the same set of axes.

46. If f^{-1} exists, find a rule for f^{-1} analytically using the method described in Section 4.5.

47. If $a = e$, what is the rule for $f^{-1}(x)$?

48. If $a = 10$, what is the rule for $f^{-1}(x)$?

For each of the following substances, find the pH from the given hydronium ion $[H_3O^+]$ concentration.

49. grapefruit, 6.3×10^{-4}

50. limes, 1.6×10^{-2}

51. crackers, 3.9×10^{-9}

52. sodium hydroxide (lye), 3.2×10^{-14}

Find the hydronium ion $[H_3O^+]$ concentration for each of the following substances for the given pH.

53. soda pop, 2.7

54. wine, 3.4

55. beer, 4.8

56. drinking water, 6.5

Suppose that $2500 is invested in an account that pays interest compounded continuously. Find the amount of time that it would take for the account to grow to the given amount at the given rate.

57. $3000, at 3.75%

58. $3500, at 4.25%

59. $5000, at 5%

60. $5000, at 6%

Use the product, quotient, and power rules of logarithms to rewrite each of the following logarithms. Assume that all variables represent positive real numbers.

61. $\log_3 \dfrac{2}{5}$

62. $\log_4 \dfrac{6}{7}$

63. $\log_2 \dfrac{6x}{y}$

64. $\log_3 \dfrac{4p}{q}$

65. $\log_5 \dfrac{5\sqrt{7}}{3}$

66. $\log_2 \dfrac{2\sqrt{3}}{5}$

67. $\log_4(2x + 5y)$

68. $\log_6(7m + 3q)$

69. $\log_k \dfrac{pq^2}{m}$

70. $\log_z \dfrac{x^5 y^3}{3}$

71. $\log_m \sqrt{\dfrac{5r^3}{z^5}}$

72. $\log_p \sqrt[3]{\dfrac{m^5 n^4}{t^2}}$

Use the product, quotient, and power rules of logarithms to rewrite each of the following as a single logarithm. Assume that all variables represent positive real numbers.

73. $\log_a x + \log_a y - \log_a m$

74. $(\log_b k - \log_b m) - \log_b a$

75. $2 \log_m a - 3 \log_m b^2$

76. $\dfrac{1}{2} \log_y p^3 q^4 - \dfrac{2}{3} \log_y p^4 q^3$

77. $2 \log_a(z - 1) + \log_a(3z + 2), z > 1$

78. $\log_b(2y + 5) - \dfrac{1}{2} \log_b(y + 3)$

79. $-\dfrac{2}{3} \log_5 5m^2 + \dfrac{1}{2} \log_5 25m^2$

80. $-\dfrac{3}{4} \log_3 16p^4 - \dfrac{2}{3} \log_3 8p^3$

Use the change-of-base rule to find an approximation for each of the following logarithms.

81. $\log_5 10$

82. $\log_9 12$

83. $\log_{15} 5$

84. $\log_{1/2} 3$

85. $\log_{100} 83$

86. $\log_{200} 175$

87. $\log_{2.9} 7.5$

88. $\log_{5.8} 12.7$

R*elating* *Concepts*

89. Use the terminology of Chapter 2 to explain how the graph of $y = -3^x + 7$ can be obtained from the graph of $y = 3^x$.

90. Graph $y_1 = 3^x$ and $y_2 = -3^x + 7$ in a window of $[-5, 5]$ by $[-10, 10]$ to support your answer in Exercise 89.

91. Use the root location capabilities of your calculator to find an approximation for the x-intercept of the graph of y_2 in Exercise 90.

92. Solve $0 = -3^x + 7$ for x, expressing x in terms of base 3 logarithms.

93. Use the change-of-base rule to find an approximation for the solution of the equation in Exercise 92.

94. Compare your results in Exercises 91 and 93.

95. Prove Property 5 of logarithms.

96. Prove Property 6 of logarithms.

5.3 INTRODUCTION TO LOGARITHMIC FUNCTIONS

Preliminary Considerations ▌ Graphs of Logarithmic Functions ▌ Connections: Earlier Results Supported by Logarithmic Function Graphs

TECHNOLOGICAL NOTE
If your calculator has the capability of drawing the inverse of a function, observe the graph of the inverse of $y = 2^x$. Repeat for $y = (\frac{1}{2})^x$. This will give you a preview of the types of graphs that will be discussed in this section.

Preliminary Considerations

The function $f(x) = a^x$, $a > 1$, is increasing on its entire domain. Likewise, if $0 < a < 1$, the function is decreasing on its entire domain. Therefore, for all allowable bases of the exponential function $f(x) = a^x$, the graph passes the horizontal line test and is one-to-one. As a result, f has an inverse.

FOR GROUP DISCUSSION

1. Refer to Exercise 77 in Section 4.5, and sketch the graph of $f(x) = a^x$ for $a > 1$. View your sketch according to the directions to observe the appearance of the graph of f^{-1}. Discuss the domain and the range of both f and f^{-1}.
2. Repeat item 1, for $0 < a < 1$.

We can find the rule for f^{-1} analytically, using the steps first described in Section 4.5.

$f(x) = a^x$	Exponential function
$y = a^x$	Replace $f(x)$ with y.
$x = a^y$	Exchange x and y to obtain an equation for the inverse.
$y = \log_a x$	Convert from exponential form to logarithmic form.
$f^{-1}(x) = \log_a x$	$y = f^{-1}(x)$

This final equation indicates that the logarithmic function with base a is the inverse of the exponential function with base a. To confirm this, we show that $(f \circ f^{-1})(x) = x$ and $(f^{-1} \circ f)(x) = x$:

$$(f \circ f^{-1})(x) = f[f^{-1}(x)] = a^{\log_a x} = x \quad \text{(from Section 5.2)}$$
$$(f^{-1} \circ f)(x) = f^{-1}[f(x)] = \log_a a^x = x \quad \text{(from Section 5.2)}.$$

RELATIONSHIP BETWEEN BASE a EXPONENTIAL AND LOGARITHMIC FUNCTIONS

The functions

$$f(x) = a^x \quad \text{and} \quad g(x) = \log_a x$$

are inverses.

Graphs of Logarithmic Functions

Recall from Section 4.5 that the graph of the inverse of a one-to-one function can be obtained by reflecting the graph of the function across the line $y = x$. (In the group discussion exercise earlier, you should have gotten a "preview" of what is to follow.) Figure 20 shows both traditional and calculator-generated graphs of a pair of inverse functions, $y = 2^x$ and $y = \log_2 x$. These are typical shapes for such graphs where $a > 1$.

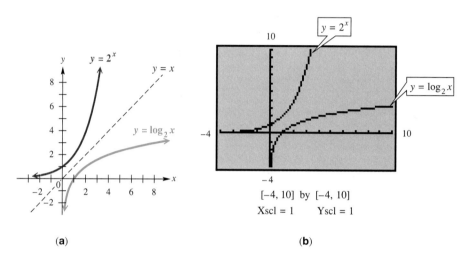

(a)   **(b)**

FIGURE 20

Figure 21 shows both traditional and calculator-generated graphs of $y = \left(\frac{1}{2}\right)^x$ and $y = \log_{1/2} x$. These are typical shapes for such graphs where $0 < a < 1$.

The most important logarithmic functions are $y = \ln x$ (base e) and $y = \log x$ (base 10). These are easily graphed on a graphics calculator, as seen in Figures 22 and 23.

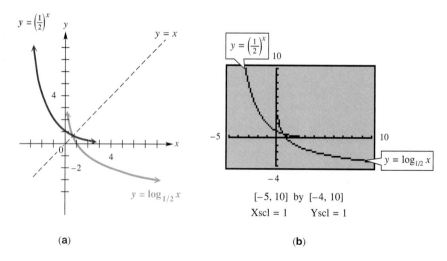

(a)   **(b)**

FIGURE 21

TECHNOLOGICAL NOTE
Just as we saw in Section 5.1 that it is difficult to interpret from a calculator screen the behavior of the graph of an exponential function as it approaches its horizontal asymptote, a similar problem occurs for graphs of logarithmic functions near their vertical asymptotes. Refer to Figures 22 and 23, and be aware that there are no endpoints and that as x approaches 0 from the right, y approaches $-\infty$.

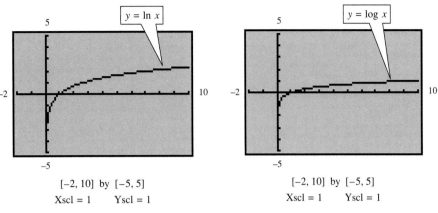

$[-2, 10]$ by $[-5, 5]$
Xscl = 1   Yscl = 1

FIGURE 22

$[-2, 10]$ by $[-5, 5]$
Xscl = 1   Yscl = 1

FIGURE 23

Because e and 10 are the only logarithmic bases that appear on a graphics calculator, if we wish to graph a logarithmic function for some other base, we must use the change-of-base rule. For example, to graph $y = \log_2 x$, we may graph either

$$y = \frac{\log x}{\log 2} \qquad \text{or} \qquad y = \frac{\ln x}{\ln 2}.$$

Similarly, if we wish to graph $y = \log_{1/2} x$, we may graph either

$$y = \frac{\log x}{\log \frac{1}{2}} \qquad \text{or} \qquad y = \frac{\ln x}{\ln \frac{1}{2}}.$$

These two graphs can be seen in Figures 20(b) and 21(b).

CAUTION   Calculator-generated graphs of logarithmic functions do not, in general, give an accurate picture of the behavior of the graphs near the vertical asymptotes. While it may seem as if the graph has an endpoint, this is not the case. The resolution of the calculator screen is not precise enough to indicate that the graph approaches the vertical asymptote as the value of x gets closer to it. Do not draw incorrect conclusions just because the calculator does not show this behavior.

We now summarize information about the graphs of logarithmic functions.

LOGARITHMIC FUNCTION $(a > 1)$

$f(x) = \log_a x$, $a > 1$ (Figure 24)

Domain: $(0, \infty)$

Range: $(-\infty, \infty)$

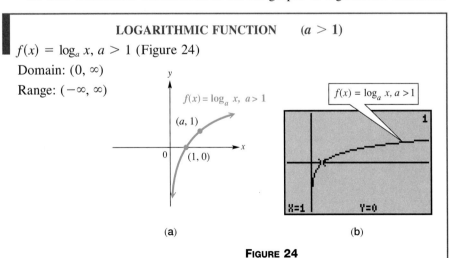

(a)

(b)

FIGURE 24

The logarithmic function $f(x) = \log_a x$, $a > 1$, increases on its entire domain, and is continuous as well. The y-axis is a vertical asymptote as x approaches 0 from the right. The x-intercept is 1, and there is no y-intercept. It is always concave down, and there are no extrema.

LOGARITHMIC FUNCTION   **$(0 < a < 1)$**

$f(x) = \log_a x, 0 < a < 1$ (Figure 25)

Domain: $(0, \infty)$

Range: $(-\infty, \infty)$

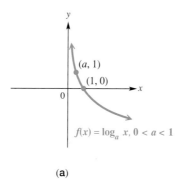

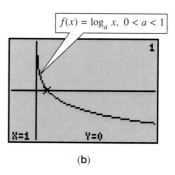

(a)

(b)

FIGURE 25

The logarithmic function $f(x) = \log_a x, 0 < a < 1$, decreases on its entire domain, and is continuous as well. The y-axis is a vertical asymptote as x approaches 0 from the right. The x-intercept is 1, and there is no y-intercept. It is always concave up, and there are no extrema.

In either case, to use a graphics calculator to graph $y = \log_a x$, we use the change-of-base rule,

$$\log_a x = \frac{\log x}{\log a} \qquad \text{or} \qquad \log_a x = \frac{\ln x}{\ln a}.$$

A function of the form $y = \log_a f(x)$ is defined only for values for which $f(x) > 0$. The first example shows how the domain of a function defined by a logarithm is determined.

EXAMPLE 1

Determining the Domain of a Function Defined by a Logarithm

Find the domain of each of the following functions analytically.

(a) $f(x) = \log_2(x - 1)$

SOLUTION   For this function to be defined, we must have $x - 1 > 0$, or equivalently, $x > 1$. The domain is $(1, \infty)$.

(b) $f(x) = \log_3 x - 1$

SOLUTION   Since we are interested in the logarithm to the base 3 of x (and not $x - 1$), we must have $x > 0$. The domain is $(0, \infty)$.

(c) $f(x) = \log_3 |x|$

SOLUTION $|x| > 0$ is true for all real numbers x except 0. Therefore, the domain is $(-\infty, 0) \cup (0, \infty)$, or equivalently, $\{x \mid x \neq 0\}$.

(d) $f(x) = \ln(x^2 - 4)$

SOLUTION   To solve $x^2 - 4 > 0$ analytically, we use a sign graph, first introduced in Section 3.3. Factor $x^2 - 4$ as $(x + 2)(x - 2)$, and then determine the signs of the factors in the intervals $(-\infty, -2)$, $(-2, 2)$, and $(2, \infty)$. See Figure 26. The product $(x + 2)(x - 2)$ is positive in the intervals $(-\infty, -2)$ and $(2, \infty)$. The domain is the union of these two intervals: $(-\infty, -2) \cup (2, \infty)$.   ∎

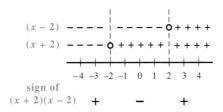

FIGURE 26

EXAMPLE **2**
Analyzing the Graph of a Function Defined by a Logarithm

Use the terminology of Chapter 2 to explain how the graph of $y = \log_2(x - 1)$ can be obtained from the graph of $y = \log_2 x$. Discuss other important features of the graph.

SOLUTION   Because the argument is $x - 1$, the graph of $y = \log_2(x - 1)$ is obtained from the graph of $y = \log_2 x$ by shifting one unit to the right. The vertical asymptote also moves one unit to the right, so its equation is $x = 1$. The x-intercept is 2. The domain is $(1, \infty)$, as found in Example 1(a), and the range is $(-\infty, \infty)$. It is always increasing. See Figure 27, which shows both traditional and calculator-generated graphs. (The graph in (b) was obtained by using $y = \frac{\log(x - 1)}{\log 2}$.)   ∎

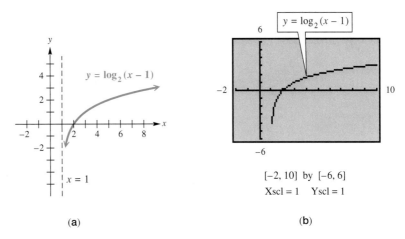

(a)   (b)

FIGURE 27

EXAMPLE **3**
Analyzing the Graph of a Function Defined by a Logarithm

Repeat the procedure of Example 2 for $y = \log_3 x - 1$.

SOLUTION   Here, the 1 is subtracted from $\log_3 x$, so the effect is that the graph of $y = \log_3 x$ is shifted down 1 unit. The vertical asymptote, the y-axis, is not affected. The x-intercept of $y = \log_3 x - 1$ is 3, and there is no y-intercept. The domain is $(0, \infty)$, as found in Example 1(b), and the range is $(-\infty, \infty)$. It is always increasing. See Figure 28 on the next page for both types of graphs. (The graph in (b) was obtained by using $y = \frac{\log x}{\log 3} - 1$.)   ∎

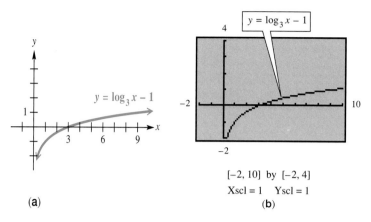

(a)

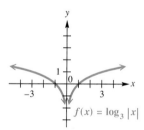

(b)

FIGURE 28

EXAMPLE 4

Analyzing the Graph of a Function Defined by a Logarithm

Discuss the symmetry exhibited by the graph of the function $f(x) = \log_3 |x|$.

SOLUTION   Recall that for all real numbers x, $|-x| = |x|$. Now, we find $f(-x)$.

$$f(-x) = \log_3 |-x| = \log_3 |x| = f(x).$$

Because $f(-x) = f(x)$, the graph is symmetric with respect to the y-axis and f is an even function. It consists of two parts, as shown in Figure 29. The domain is $(-\infty, 0) \cup (0, \infty)$, as determined in Example 1(c), and the y-axis serves both as a vertical asymptote and the axis of symmetry.   ◘

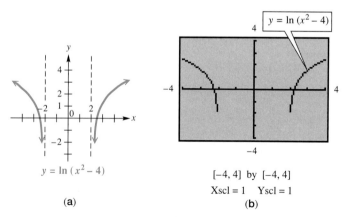

FIGURE 29

EXAMPLE 5

Supporting an Analytic Domain Determination Graphically

Use a graph to support the result obtained in Example 1(d): the domain of $f(x) = \ln(x^2 - 4)$ is $(-\infty, -2) \cup (2, \infty)$.

SOLUTION   Figure 30 shows both types of graphs of this function. Notice that for $[-2, 2]$, the graph does not exist, as expected.   ◘

TECHNOLOGICAL NOTE
If your calculator has a TABLE feature, support the result of Example 5 by showing that values of x between and including -2 and 2 lead to an error message.

(a)

(b)

FIGURE 30

CAUTION   It would be easy to misinterpret the calculator-generated graph in Figure 30(b), by thinking that there are "endpoints" on the two branches. The lines $x = -2$ and $x = 2$ are vertical asymptotes, and as x approaches -2 from the left, $f(x) \rightarrow -\infty$, while as x approaches 2 from the right, $f(x) \rightarrow -\infty$. Here we see another reason for learning the concepts. Calculator-generated graphs are limited in their resolution, and may not show important details in certain windows.

FOR **GROUP DISCUSSION**

1.   Use a calculator to evaluate $f(2.1)$, $f(2.01)$, $f(2.001)$, and $f(2.0001)$ for $f(x) = \ln(x^2 - 4)$. Describe what happens as x approaches 2 from the right. How does this tie in with the discussion in the CAUTION above?
2.   Repeat item 1 for $f(-2.1)$, $f(-2.01)$, $f(-2.001)$, and $f(-2.0001)$.

Connections: Earlier Results Supported by Logarithmic Function Graphs

Some of the results obtained in the previous two sections of this chapter can be supported through graphs of logarithmic functions. The remaining examples in this section illustrate this.

EXAMPLE 6

Supporting Graphically a Result Determined Analytically

In Example 2(a) of Section 5.2, we determined analytically $\log_8 4 = \frac{2}{3}$. Support this result graphically using the graph of $y = \log_8 x$.

SOLUTION   We graph $y = \log_8 x$ on a graphics calculator by using the change-of-base rule: $\log_8 x = \frac{\log x}{\log 8}$. Then, we evaluate y for $x = 4$. Figure 31 indicates that when $x = 4$, y is a decimal approximation for $\frac{2}{3}$, supporting our analytic result.

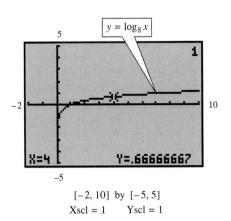

$[-2, 10]$ by $[-5, 5]$
Xscl = 1   Yscl = 1

FIGURE 31

EXAMPLE 7

Supporting Graphically a Result Determined by a Calculator Function

In Figure 14, we show a calculator approximation of log 6: log 6 ≈ .7781512504. Support this result using the graph of the common logarithm function $y = \log x$.

SOLUTION By graphing $y = \log x$ and letting $x = 6$, we find that $y \approx$ .77815125, supporting the result found earlier. See Figure 32. ◼

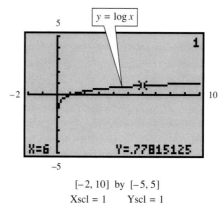

[−2, 10] by [−5, 5]
Xscl = 1   Yscl = 1

FIGURE 32

FOR **GROUP DISCUSSION**

In Figure 16, we show a calculator approximation of ln 8: ln 8 ≈ 2.079441542. Use a method similar to the one in Example 7 to support this result graphically.

EXAMPLE 8

Using a Property of Logarithms to Describe a Translation of a Graph

In Example 5(a) of Section 5.2, we used the product rule for logarithms to write log 8x as log 8 + log x. Use this result to explain how the graph of $y = \log 8x$ may be obtained from the graph of $y = \log x$ by a translation.

SOLUTION Because log 8 ≈ .90309, the graph of $y_2 = \log 8x = \log 8 + \log x$ is obtained by shifting the graph of $y_1 = \log x$ up log 8 ≈ .90309 units. Figure 33 shows the two graphs. To support our answer, we choose an arbitrary x-value, say 5, and determine the corresponding y-value for each graph. If we subtract log 5 from log 8(5) = log 40 using our common logarithm function, we get the same result, approximately .903089987, shown in Figure 34. ◼

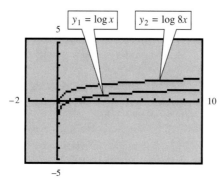

[−2, 10] by [−5, 5]
Xscl = 1   Yscl = 1
The vertical distance between y_2 and y_1 is log 8.

FIGURE 33

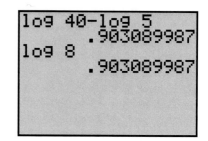

FIGURE 34

EXAMPLE **9**
Reinforcing the Inverse Relationship Between Exponential and Logarithmic Functions

The one-to-one function $f(x) = -2^x + 3$ is discussed and graphed in Example 3 of Section 5.1 and the corresponding figure (Figure 9). Find f^{-1} analytically, graph both f and f^{-1} in the same window, and discuss the inverse relationships exhibited by the functions and their graphs.

SOLUTION To find f^{-1}, we use the analytic method.

$$f(x) = -2^x + 3 \qquad \text{Given}$$
$$y = -2^x + 3 \qquad \text{Replace } f(x) \text{ with } y.$$
$$x = -2^y + 3 \qquad \text{Exchange } x \text{ and } y \text{ to obtain the inverse relationship.}$$
$$2^y = -x + 3 \qquad \text{Add } 2^y \text{ and subtract } x.$$
$$y = \log_2(-x + 3) \qquad \text{Write in logarithmic form.}$$
$$f^{-1}(x) = \log_2(-x + 3) \qquad \text{Replace } y \text{ with } f^{-1}(x).$$

Figure 35 shows both f and f^{-1} graphed. We now list some of the features of these inverses.

	Domain	Range	x-intercept	y-intercept	Asymptote
$f(x) = -2^x + 3$ $f^{-1}(x) = \log_2(-x + 3)$	$(-\infty, \infty)$ $(-\infty, 3)$	$(-\infty, 3)$ $(-\infty, \infty)$	$\log_2 3 \approx 1.58$ 2	2 $\log_2 3 \approx 1.58$	horizontal: $y = 3$ vertical: $x = 3$

Notice in the chart how the roles of x and y are reversed in f and f^{-1}.

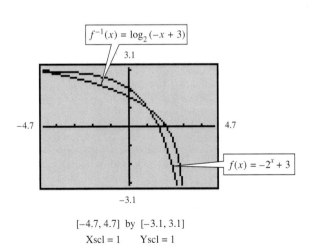

$[-4.7, 4.7]$ by $[-3.1, 3.1]$
Xscl = 1   Yscl = 1

FIGURE 35

5.3 EXERCISES

The graph of an exponential function f is given, with three points labeled. Sketch the graph of f^{-1} by hand, labeling three points on its graph. For f^{-1}, also state the domain, the range, whether it increases or decreases on its domain, and the equation of its vertical asymptote.

1.

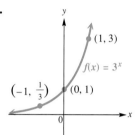

2.

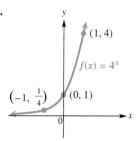

3.

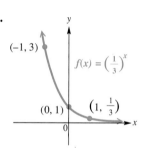

4.

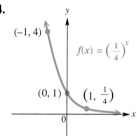

5.

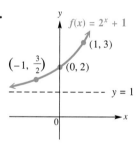

6.

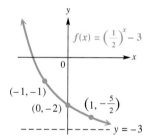

7. In Exercises 1–6, each function f is an exponential function. Therefore, each function f^{-1} is a(n) _____ function.

8. If $f(x) = a^x (a > 0, a \neq 1)$, what is the value of $f^{-1}(1)$?

Find the domain of each of the following logarithmic functions analytically. You may wish to support your answer graphically.

9. $y = \log 2x$

10. $y = \log \dfrac{x}{3}$

11. $y = \log(3x + 7)$

12. $y = \log(3 - 6x)$

13. $y = \ln(x^2 + 7)$

14. $y = \ln(-x^2 - 4)$

15. $y = \log_4(x^2 - 4x - 21)$

16. $y = \log_6(2x^2 - 7x - 4)$

17. $y = \log(x^2 + x - 1)$

18. $y = \ln(x^2 + x + 1)$

19. $y = \log(x^3 - x)$

20. $y = \log\left(\dfrac{x + 3}{x - 4}\right)$

*In Exercises 21–28, match the correct graph to the given equation. You should do this
by your knowledge of graphs, and not by generating your own graph on your calculator.*

21. $y = e^x + 3$

22. $y = e^x - 3$

23. $y = e^{x+3}$

24. $y = e^{x-3}$

25. $y = \ln x + 3$

26. $y = \ln x - 3$

27. $y = \ln(x - 3)$

28. $y = \ln(x + 3)$

A.

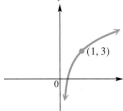

B.

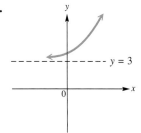

C.

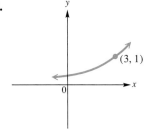

D.

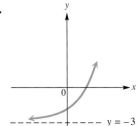

E.

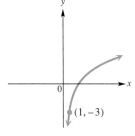

F.

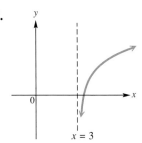

G.

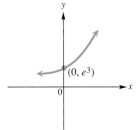

H.

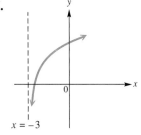

*Use the terminology of Chapter 2 to explain how the graph of the given function can be
obtained from the graph of $y = \log_2 x$.*

29. $y = \log_2(x + 4)$ **30.** $y = \log_2(x - 6)$ **31.** $y = 3 \log_2 x + 1$

32. $y = -4 \log_2 x - 8$ **33.** $y = \log_2(-x) + 1$ **34.** $y = -\log_2(-x)$

R*elating Concepts*

Work Exercises 35–40 in order. They apply to the function $f(x) = \log_4(2x^2 - x)$.

35. Determine analytically the domain of f.

36. Use the change-of-base rule to express, in terms of common logarithms, the equation
you would use to graph f on your calculator.

37. Determine analytically the x-intercepts of the graph of f.

38. Give the equations of the vertical asymptotes of the graph of f.

39. Explain why the graph of f has no y-intercept.

40. Graph f in the window $[-2.5, 2.5]$ by $[-5, 2.5]$. Based on this graph, give the solution sets of $f(x) = 0$, $f(x) < 0$, and $f(x) > 0$.

---

41. Graph $y = \log x^2$ and $y = 2 \log x$ on separate viewing screens. It would seem, at first glance, that by applying the power rule for logarithms, these graphs should be the same. Are they? If not, why not? (*Hint:* Consider the domain in each case.)

42. Graph $f(x) = \log_3 |x|$ in the window $[-4, 4]$ by $[-4, 4]$ and compare it to the traditional graph in Figure 29. How might one easily misinterpret the domain of the function simply by observing the calculator-generated graph? What is the domain of this function?

*In Exercises 43–46, evaluate the logarithm in three ways: (**a**) Use the method of Example 2 of Section 5.2 to find the exact value analytically. (**b**) Support the result of (a) by using the change-of-base rule and common logarithms on your calculator. (**c**) Support the result of (a) by locating the appropriate point on the graph of the function $y = \log_a x$.*

43. $\log_9 27$ **44.** $\log_4 \left(\dfrac{1}{8} \right)$ **45.** $\log_{16} \left(\dfrac{1}{8} \right)$ **46.** $\log_2 \sqrt{8}$

Find approximations for the following common and natural logarithms by using a graph. Then support your answer by using the common or natural logarithm key of your calculator.

47. $\log 7$ **48.** $\log 9$ **49.** $\ln 7$

50. $\ln 9$ **51.** $\log 14.7$ **52.** $\ln 14.7$

*R*elating Concepts

53. Complete the following statement of the quotient rule for logarithms: If x and y are positive numbers, $\log_a \frac{x}{y} = $ _____ .

54. Use the quotient rule to explain how the graph of $y = \log \frac{x}{4}$ can be obtained from the graph of $y = \log x$ by a vertical shift.

55. Graph $y_1 = \log x$ and $y_2 = \log \frac{x}{4}$ in the window $[-2, 10]$ by $[-3, 3]$, and explain how these graphs support your answer in Exercise 54.

56. Find approximations for $\log 8$ and $\log \frac{8}{4}$ (that is, $\log 2$) using the logarithm key of your calculator. Then subtract $\log 2$ from $\log 8$.

57. Use the graph of $y_1 = \log x$ to show that when $x = 4$, the value obtained in Exercise 56 corresponds to the y-value on the graph.

58. Use the graph of $y_2 = \log \frac{x}{4}$ and let $x = 16$ to find an approximation for $\log 4$. (You will need to adjust the window.) Does it agree with those found in Exercises 56 and 57?

In Exercises 59–62, a one-to-one exponential function f is given. Find f^{-1} analytically. Graph both f and f^{-1} in the same viewing window, and discuss the inverse relationship exhibited by the functions and their graphs. (See Example 9.)

59. $f(x) = 4^x - 3$ **60.** $f(x) = \left(\dfrac{1}{2} \right)^x - 5$

61. $f(x) = -10^x + 4$ **62.** $f(x) = -e^x + 6$

Relating Concepts

63. The graph of $y = 10^x$ is shown with the coordinates of a point displayed at the bottom of the screen. Write the *logarithmic* equation associated with the display.

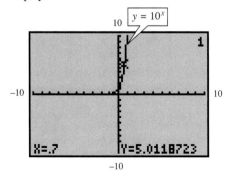

64. The graph of $y = e^x$ is shown with the coordinates of a point displayed at the bottom of the screen. Write the *logarithmic* equation associated with the display.

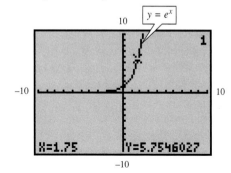

Further Explorations

The TABLE in the figure is for $Y_1 = \log(4 - x)$. Why do the values in Y_1 show ERROR for $x \geq 4$?

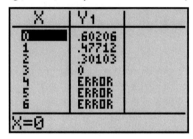

5.4 EXPONENTIAL AND LOGARITHMIC EQUATIONS AND INEQUALITIES

Preliminary Considerations ▮ Exponential Equations and Inequalities (Type 2) ▮ Logarithmic Equations and Inequalities ▮ Equations and Inequalities Involving Both Exponentials and Logarithms ▮ Formulas Involving Exponentials and Logarithms

Preliminary Considerations

In this section we will examine methods of solving equations involving exponential and logarithmic expressions. We have solved some types of these in the earlier sections. For example, in Section 5.1 we learned how to solve Type 1 exponential equations—those involving expressions that could easily be written as powers of the same base. We also saw how to solve exponential equations involving bases of e and 10.

General methods for solving exponential and logarithmic equations depend on the following properties. These properties follow from the fact that exponential and logarithmic functions are, in general, one-to-one. Property 1 was used in Section 5.1 to solve Type 1 exponential equations.

> **PROPERTIES OF LOGARITHMIC AND EXPONENTIAL FUNCTIONS**
> For $b > 0$ and $b \neq 1$:
> 1. $b^x = b^y$ if and only if $x = y$.
> 2. If $x > 0$ and $y > 0$,
>
> $$\log_b x = \log_b y \quad \text{if and only if} \quad x = y.$$

Exponential Equations and Inequalities (Type 2)

A Type 2 exponential equation or inequality is one in which the exponential expressions cannot easily be written as powers of the same base. Examples of Type 2 equations are

$$7^x = 12 \quad \text{and} \quad 2^{3x+1} = 3^{4-x}.$$

The general strategy in solving these equations is to use Property 2 by taking the same base logarithm of both sides (usually either common or natural) and then applying the power rule for logarithms to eliminate the variable exponents. Then the equation is solved using familiar algebraic techniques.

> ### FOR **GROUP DISCUSSION**
>
> Between what two consecutive integers must the solution of $7^x = 12$ lie? Why is this so?

EXAMPLE 1

Solving a Type 2 Exponential Equation

Solve the equation $7^x = 12$ analytically.

SOLUTION In Section 5.1, we saw that Property 1 cannot be used to solve this equation, so we apply Property 2. While any appropriate base b can be used to apply Property 2, the best practical base to use is base 10 or base e. Taking base e (natural) logarithms of both sides gives

$$7^x = 12$$
$$\ln 7^x = \ln 12$$
$$x \ln 7 = \ln 12 \qquad \text{Property (6) of logarithms}$$
$$x = \frac{\ln 12}{\ln 7}. \qquad \text{Divide by ln 7.}$$

The expression $\frac{\ln 12}{\ln 7}$ is the *exact* solution of $7^x = 12$. Had we used common logarithms instead, the solution would have the form $\frac{\log 12}{\log 7}$. In either case, we can use a calculator to find a decimal approximation for the exact solution. We find that to the nearest thousandth,

$$\frac{\ln 12}{\ln 7} = \frac{\log 12}{\log 7} \approx 1.277.$$

The solution set can be expressed with the exact solution as

$$\left\{ \frac{\ln 12}{\ln 7} \right\} \quad \text{or} \quad \left\{ \frac{\log 12}{\log 7} \right\},$$

while it is expressed as $\{1.277\}$ with an approximate solution.

The solution can be supported graphically by graphing $y_1 = 7^x$ and $y_2 = 12$, and then using the intersection-of-graphs method. The x-coordinate of the point of intersection is approximately 1.277, as seen in Figure 36(a). Figure 36(b) illustrates the x-intercept method of solution. The x-intercept of $y = 7^x - 12$ is also approximately 1.277.   ◼

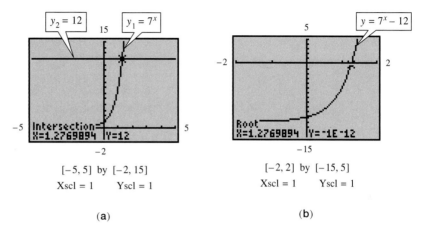

$y_2 = 12$ $y_1 = 7^x$

[−5, 5] by [−2, 15]
Xscl = 1   Yscl = 1

(a)

$y = 7^x - 12$

[−2, 2] by [−15, 5]
Xscl = 1   Yscl = 1

(b)

FIGURE 36

EXAMPLE 2

Solving a Type 2
Exponential
Inequality
Graphically

(a) Use Figure 36(a) to solve $7^x < 12$.

SOLUTION Because the graph of $y_1 = 7^x$ is below the graph of $y_2 = 12$ for all x-values less than $\frac{\ln 12}{\ln 7} \approx 1.277$, we can express the solution set as

$$\left(-\infty, \frac{\ln 12}{\ln 7}\right) \quad \text{or} \quad (-\infty, 1.277).$$

(b) Use Figure 36(b) to solve $7^x > 12$.

SOLUTION This inequality is equivalent to $7^x - 12 > 0$. Because the graph of $y = 7^x - 12$ is above the x-axis for all values of x greater than $\frac{\ln 12}{\ln 7} \approx 1.277$, we can express the solution set as

$$\left(\frac{\ln 12}{\ln 7}, \infty\right) \quad \text{or} \quad (1.277, \infty). \quad ◼$$

FOR GROUP DISCUSSION

These items refer to Examples 1 and 2 and the previous group discussion.

1. Does the result of Example 1 support your answer to the previous group discussion item?
2. How can you support the statement given in Example 1 that $\frac{\ln 12}{\ln 7} = \frac{\log 12}{\log 7}$?
3. Explain how Figure 36(a) can be used to solve $7^x > 12$, and how Figure 36(b) can be used to solve $7^x < 12$.

EXAMPLE **3**

Solving a Type 2
Exponential
Equation

Solve the equation $2^{3x+1} = 3^{4-x}$ analytically, and support the solution graphically.

SOLUTION   To solve this equation analytically, we take logarithms on both sides, use the power rule, and solve the resulting linear equation.

$$2^{3x+1} = 3^{4-x}$$ Given equation

$$\log 2^{3x+1} = \log 3^{4-x}$$ Take common logarithms on both sides.

$$(3x + 1) \log 2 = (4 - x) \log 3$$ Use the power rule of logarithms.

$$3x \log 2 + \log 2 = 4 \log 3 - x \log 3$$ Distributive property

$$3x \log 2 + x \log 3 = 4 \log 3 - \log 2$$ Transform so that all x terms are on one side.

$$x(3 \log 2 + \log 3) = 4 \log 3 - \log 2$$ Factor out x.

$$x = \frac{4 \log 3 - \log 2}{3 \log 2 + \log 3}$$ Divide by $3 \log 2 + \log 3$.

This expression is the exact solution in terms of common logarithms. A more compact form of this exact solution can be found by using properties of logarithms.

$$x = \frac{\log 3^4 - \log 2}{\log 2^3 + \log 3}$$ Power rule

$$x = \frac{\log 81 - \log 2}{\log 8 + \log 3}$$ $3^4 = 81; 2^3 = 8$

$$x = \frac{\log \frac{81}{2}}{\log 24}$$ Quotient and product rules

The solution set using the exact value is $\{\frac{\log(81/2)}{\log 24}\}$, while a calculator approximation of the solution is 1.165, giving the solution set $\{1.165\}$.

This solution can be supported graphically by locating the point of intersection of the graphs of $y_1 = 2^{3x+1}$ and $y_2 = 3^{4-x}$, and finding the x-coordinate of the point. As seen in Figure 37, the x-coordinate is approximately 1.165, supporting our analytic solution.   ∎

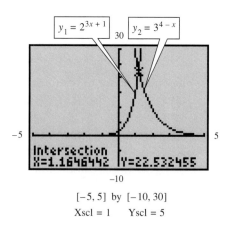

$[-5, 5]$ by $[-10, 30]$
Xscl = 1   Yscl = 5

FIGURE 37

FOR GROUP DISCUSSION

1. Use Figure 37 to determine the solution sets of

$$2^{3x+1} < 3^{4-x} \quad \text{and} \quad 2^{3x+1} > 3^{4-x}.$$

2. Discuss the difference between an exact solution and an approximate solution of an exponential equation. In general, can you find an exact solution from a graph?

Logarithmic Equations and Inequalities

The next examples show how to solve logarithmic equations and associated inequalities. The properties of logarithms given in Section 5.2 are useful here, as is Property 2 from this section. As we shall see, it is important to note the domain of the variable in the original form of a logarithmic equation, so that extraneous values can be rejected if they appear.

EXAMPLE 4

Solving a Logarithmic Equation

Solve

$$\log_3(x + 6) - \log_3(x + 2) = \log_3 x$$

analytically, and check. Support the result graphically.

SOLUTION   First, we note that x must be positive, since any nonpositive value of x would cause $\log_3 x$ to be undefined. Using the quotient property of logarithms, we can rewrite the equation as

$$\log_3 \frac{x + 6}{x + 2} = \log_3 x.$$

Now the equation is in the proper form to use Property 2.

$$\frac{x + 6}{x + 2} = x \qquad \text{Property 2}$$

$$x + 6 = x(x + 2) \qquad \text{Multiply by } x + 2.$$

$$x + 6 = x^2 + 2x \qquad \text{Distributive property}$$

$$x^2 + x - 6 = 0 \qquad \text{Get 0 on one side.}$$

$$(x + 3)(x - 2) = 0 \qquad \text{Use the zero-product property.}$$

$$x = -3 \quad \text{or} \quad x = 2$$

The negative solution ($x = -3$) cannot be used since it is not in the domain of $\log_3 x$ in the original equation. For this reason, the only valid solution is the positive number 2, giving the solution set $\{2\}$.

Figure 38 shows that the x-coordinate of the point of intersection of the graphs of $y_1 = \log_3(x + 6) - \log_3(x + 2)$ and $y_2 = \log_3 x$ is 2, supporting our analytic solution. Notice that the graphs do not intersect when $x = -3$, further supporting our conclusion that -3 is an extraneous value.   ◻

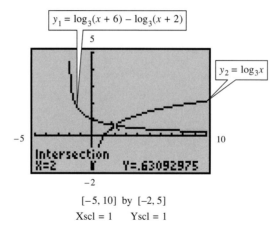

$[-5, 10]$ by $[-2, 5]$
Xscl = 1   Yscl = 1

FIGURE 38

NOTE The equation in Example 4 could also have been solved by transforming so that all logarithmic terms appeared on one side. Then by applying the properties of logarithms from Section 5.2 and rewriting in exponential form, the same equations would follow:

$$\log_3(x + 6) - \log_3(x + 2) - \log_3 x = 0$$

$$\log_3(x + 6) - [\log_3(x + 2) + \log_3 x] = 0$$

$$\log_3(x + 6) - \log_3(x + 2)x = 0 \qquad \text{Product rule}$$

$$\log_3 \frac{x + 6}{(x + 2)x} = 0 \qquad \text{Quotient rule}$$

$$\frac{x + 6}{(x + 2)x} = 3^0 \qquad \text{Exponential form}$$

$$\frac{x + 6}{(x + 2)x} = 1 \qquad 3^0 = 1$$

$$x + 6 = (x + 2)x \qquad \text{Multiply by } (x + 2)x.$$

$$x + 6 = x^2 + 2x \qquad \text{Distributive property}$$

From this point on, the work would be the same.

EXAMPLE 5

Solving a Logarithmic Equation

Solve
$$\log(3x + 2) + \log(x - 1) = 1$$

analytically, and support the solution graphically.

SOLUTION Recall from Section 5.2 that log x means $\log_{10} x$.

$$\log(3x + 2) + \log(x - 1) = 1 \qquad \text{Given equation}$$

$$\log(3x + 2)(x - 1) = 1 \qquad \text{Product rule}$$

$$(3x + 2)(x - 1) = 10^1 \qquad \text{Exponential form}$$

$$3x^2 - x - 2 = 10$$

$$3x^2 - x - 12 = 0 \qquad \text{Standard form}$$

Now use the quadratic formula with $a = 3$, $b = -1$, and $c = -12$ to get

$$x = \frac{1 \pm \sqrt{1 + 144}}{6}.$$

If $x = (1 - \sqrt{145})/6$, then $x - 1 < 0$; therefore, $\log(x - 1)$ is not defined and this proposed solution must be discarded, giving the solution set

$$\left\{ \frac{1 + \sqrt{145}}{6} \right\}.$$

To support this solution graphically, we first find that $(1 + \sqrt{145})/6 \approx 2.174$ using the arithmetic and square root functions of a calculator. Then we can choose to use the x-intercept method of solution, and graph $y = \log(3x + 2) + \log(x - 1) - 1$. As seen in Figure 39, the x-intercept agrees with the earlier approximation. ▯

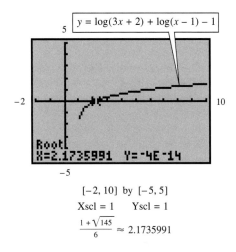

$[-2, 10]$ by $[-5, 5]$
Xscl = 1   Yscl = 1
$\frac{1 + \sqrt{145}}{6} \approx 2.1735991$

FIGURE 39

FOR GROUP DISCUSSION

Use Figure 39 to answer each of the following items.

1. Give the exact solution set of $\log(3x + 2) + \log(x - 1) - 1 \geq 0$.
2. Give the exact solution set of $\log(3x + 2) + \log(x - 1) - 1 \leq 0$.
    (*Hint:* Pay attention to the domain.)

Equations and Inequalities Involving Both Exponentials and Logarithms

We now look at equations that involve both exponentials and logarithms.

EXAMPLE 6

Solving a
Composite
Exponential
Equation

Solve $e^{-2 \ln x} = \frac{1}{16}$ analytically. Check analytically and support the solution graphically.

SOLUTION Use a property of logarithms to rewrite the exponent on the left side of the equation.

$$e^{-2 \ln x} = \frac{1}{16}$$

$$e^{\ln x^{-2}} = \frac{1}{16} \qquad \text{Property 6 of logarithms}$$

$$x^{-2} = \frac{1}{16} \qquad a^{\log_a X} = x$$

$$x^{-2} = 4^{-2} \qquad \begin{array}{l}\frac{1}{16} = \frac{1}{4^2} = 4^{-2}; \ -4 \text{ is not valid,} \\ \text{since } -4 < 0, \text{ and } x > 0.\end{array}$$

$$x = 4 \qquad \text{Property 1}$$

To check analytically, let $x = 4$ in the original equation.

$$e^{-2\ln x} = \frac{1}{16} \qquad \text{Given equation}$$

$$e^{-2\ln 4} = \frac{1}{16} \qquad ?$$

$$e^{\ln 4^{-2}} = \frac{1}{16} \qquad ?$$

$$e^{\ln(1/16)} = \frac{1}{16} \qquad ?$$

$$\frac{1}{16} = \frac{1}{16} \qquad \text{True}$$

To support graphically, we show that $y = e^{-2\ln x} - \frac{1}{16}$ has x-intercept 4. See Figure 40. The solution set is $\{4\}$. The associated inequalities may be solved graphically using Figure 40. Exercise 14 requires these solutions.

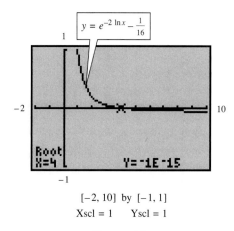

$[-2, 10]$ by $[-1, 1]$
Xscl = 1   Yscl = 1

FIGURE 40

EXAMPLE 7

Solving a Composite Logarithmic Equation

Solve

$$\ln e^{\ln x} - \ln(x - 3) = \ln 2$$

analytically. Give an analytic check.

SOLUTION On the left, $\ln e^{\ln x} = \ln x$. The equation becomes

$$\ln x - \ln(x - 3) = \ln 2$$

$$\ln \frac{x}{x - 3} = \ln 2 \qquad \text{Quotient Rule}$$

$$\frac{x}{x - 3} = 2 \qquad \text{Property 2}$$

$$x = 2x - 6 \qquad \text{Multiply by } x - 3.$$

$$6 = x.$$

To check, let $x = 6$ in the original equation.

$$\ln e^{\ln x} - \ln(x - 3) = \ln 2 \qquad \text{Given equation}$$

$$\ln e^{\ln 6} - \ln(6 - 3) = \ln 2 \qquad ? \quad \text{Let } x = 6.$$

$$\ln 6 - \ln 3 = \ln 2 \qquad ?$$

$$\ln\left(\frac{6}{3}\right) = \ln 2 \qquad ?$$

$$\ln 2 = \ln 2 \qquad \text{True}$$

The solution set is $\{6\}$. In Exercise 15, you are asked to support the solution graphically, and solve the associated inequalities.

A summary of the methods used for solving equations in this section follows.

SOLVING EXPONENTIAL AND LOGARITHMIC EQUATIONS

An exponential or logarithmic equation may be solved by changing the equation into one of the following forms, where a and b are real numbers, $a > 0$, and $a \neq 1$.

1. $a^{f(x)} = b$
   Solve by taking logarithms of each side. (Natural logarithms are the best choice if $a = e$.)

2. $\log_a f(x) = \log_a g(x)$
   From the given equation, $f(x) = g(x)$, which is solved analytically.

3. $\log_a f(x) = b$
   Solve by using the definition of logarithm to write the expression in exponential form as $f(x) = a^b$.

Formulas Involving Exponentials and Logarithms

In the next section we will examine many applications of exponential and logarithmic functions. We now present two examples that illustrate how formulas based on such functions may be solved for one variable in terms of the other.

EXAMPLE 8

Solving an Exponential Formula for a Particular Variable

The strength of a habit is a function of the number of times the habit is repeated. If N is the number of repetitions and H is the strength of the habit, then, according to psychologist C. L. Hull,

$$H = 1000(1 - e^{-kN}),$$

where k is a constant. Solve this equation for k.

SOLUTION   We must first solve the equation for e^{-kN}.

$$\frac{H}{1000} = 1 - e^{-kN} \qquad \text{Divide by 1000.}$$

$$\frac{H}{1000} - 1 = -e^{-kN} \qquad \text{Subtract 1.}$$

$$e^{-kN} = 1 - \frac{H}{1000} \qquad \text{Multiply by } -1.$$

Now solve for k. As shown earlier, we take logarithms on each side of the equation and use the fact that $\ln e^x = x$.

$$\ln e^{-kN} = \ln\left(1 - \frac{H}{1000}\right)$$

$$-kN = \ln\left(1 - \frac{H}{1000}\right) \qquad \text{In } e^x = x$$

$$k = -\frac{1}{N}\ln\left(1 - \frac{H}{1000}\right) \qquad \text{Multiply by } -\frac{1}{N}.$$

With the last equation, if one pair of values for H and N is known, k can be found, and the equation can then be used to find either H or N for given values of the other variable. ∎

EXAMPLE **9**
Solving a Logarithmic Formula for a Particular Variable

The formula

$$S = a \ln\left(1 + \frac{n}{a}\right)$$

gives the number of species in a sample, where n is the number of individuals in the sample, and a is a constant indicating the diversity of species in the community. Solve the equation for n.

SOLUTION We begin by solving for $\ln(1 + \frac{n}{a})$. Then we can change to exponential form and solve the resulting equation for n.

$$\frac{S}{a} = \ln\left(1 + \frac{n}{a}\right) \qquad \text{Divide by } a.$$

$$e^{S/a} = 1 + \frac{n}{a} \qquad \text{Write in exponential form.}$$

$$e^{S/a} - 1 = \frac{n}{a} \qquad \text{Subtract 1.}$$

$$n = a(e^{S/a} - 1) \qquad \text{Multiply by } a.$$

Using this equation and given values of S and a, the number of individuals in a sample can be found. ∎

5.4 EXERCISES

In Exercises 1–12, solve the equation in part (a) analytically, and express solutions in exact form. If logarithms are required to express solutions, use common logarithms. Then use a graph to support your solutions. Solve the associated inequalities in parts (b) and (c) by referring to the solutions in part (a) and the graph. Express endpoints of intervals using exact values, again using common logarithms if necessary.

1. (a) $3^x = 10$
 (b) $3^x > 10$
 (c) $3^x < 10$

2. (a) $.5^x = .3$
 (b) $.5^x > .3$
 (c) $.5^x < .3$

3. (a) $2^{x+3} = 5^x$
 (b) $2^{x+3} \geq 5^x$
 (c) $2^{x+3} \leq 5^x$

4. (a) $6^{x+3} = 4^x$
 (b) $6^{x+3} \geq 4^x$
 (c) $6^{x+3} \leq 4^x$

5. (a) $\log(x - 3) = 1 - \log x$
   (b) $\log(x - 3) < 1 - \log x$
   (c) $\log(x - 3) > 1 - \log x$

6. (a) $\log(x - 6) = 2 - \log(x + 15)$
   (b) $\log(x - 6) < 2 - \log(x + 15)$
   (c) $\log(x - 6) > 2 - \log(x + 15)$

7. (a) $\ln(4x - 2) = \ln 4 - \ln(x - 2)$
   (b) $\ln(4x - 2) \le \ln 4 - \ln(x - 2)$
   (c) $\ln(4x - 2) \ge \ln 4 - \ln(x - 2)$

8. (a) $\ln(5 + 4x) - \ln(3 + x) = \ln 3$
   (b) $\ln(5 + 4x) - \ln(3 + x) \le \ln 3$
   (c) $\ln(5 + 4x) - \ln(3 + x) \ge \ln 3$

9. (a) $\log_5(x + 2) + \log_5(x - 2) = 1$
   (b) $\log_5(x + 2) + \log_5(x - 2) \ge 1$
   (c) $\log_5(x + 2) + \log_5(x - 2) \le 1$

10. (a) $\log_2 x + \log_2(x - 7) = 3$
   (b) $\log_2 x + \log_2(x - 7) \ge 3$
   (c) $\log_2 x + \log_2(x - 7) \le 3$

11. (a) $e^{\ln x + \ln(x-2)} = 8$
   (b) $e^{\ln x + \ln(x-2)} > 8$
   (c) $e^{\ln x + \ln(x-2)} < 8$

12. (a) $3e^{2\ln x - \ln(x+2)} = 8$
   (b) $3e^{2\ln x - \ln(x+2)} > 8$
   (c) $3e^{2\ln x - \ln(x+2)} < 8$

13. Refer to Example 4 of Section 5.1. The exponential equation solved there is a Type 1 equation. Solve it using the method described in this section for a Type 2 equation, and show that the solution obtained is the same.

14. Use the result of Example 6 of this section and the corresponding figure (Figure 40) to solve the following inequalities.
   (a) $e^{-2\ln x} > \dfrac{1}{16}$ **(b)** $e^{-2\ln x} < \dfrac{1}{16}$

15. Use the result of Example 7 of this section to do each of the following.
   (a) Support the solution of the equation using a calculator-generated graph.
   (b) Solve the inequality $\ln e^{\ln x} - \ln(x - 3) > \ln 2$ using the graph from part (a).
   (c) Solve the inequality $\ln e^{\ln x} - \ln(x - 3) < \ln 2$ using the graph from part (a).

16. Use the fact that $\log 10 = 1$ to rewrite the equation in Example 5 with a logarithm on the right side. Then solve the equation using the method of Example 4.

In Exercises 17–28, follow the same directions as those for Exercises 1–12, but if necessary, write solutions and endpoints as decimals rounded to the nearest thousandth.

17. (a) $3^{x+2} = 5$
   (b) $3^{x+2} > 5$
   (c) $3^{x+2} < 5$

18. (a) $5^{2-x} = 12$
   (b) $5^{2-x} > 12$
   (c) $5^{2-x} < 12$

19. (a) $2e^{5x+2} = 8$
   (b) $2e^{5x+2} < 8$
   (c) $2e^{5x+2} > 8$

20. (a) $10e^{3x-7} = 5$
   (b) $10e^{3x-7} < 5$
   (c) $10e^{3x-7} > 5$

21. (a) $\ln x - \ln(x + 1) = \ln 5$
   (b) $\ln x - \ln(x + 1) \ge \ln 5$
   (c) $\ln x - \ln(x + 1) \le \ln 5$

22. (a) $\ln x + 1 = \ln(x - 4)$
   (b) $\ln x + 1 \ge \ln(x - 4)$
   (c) $\ln x + 1 \le \ln(x - 4)$

23. (a) $\ln e^x - \ln e^3 = \ln e^5$
   (b) $\ln e^x - \ln e^3 > \ln e^5$
   (c) $\ln e^x - \ln e^3 < \ln e^5$

24. (a) $\ln e^x - 2\ln e = \ln e^4$
   (b) $\ln e^x - 2\ln e > \ln e^4$
   (c) $\ln e^x - 2\ln e < \ln e^4$

25. (a) $\log_4 x + \log_4(x + 2) = 1$
   (b) $\log_4 x + \log_4(x + 2) \le 1$
   (c) $\log_4 x + \log_4(x + 2) \ge 1$

26. (a) $\log_5(x + 4) + \log_5(x - 1) = 0$
   (b) $\log_5(x + 4) + \log_5(x - 1) \le 0$
   (c) $\log_5(x + 4) + \log_5(x - 1) \ge 0$

27. (a) $\ln e^{\ln x} - \ln(x - 4) = \ln 3$
   (b) $\ln e^{\ln x} - \ln(x - 4) \ge \ln 3$
   (c) $\ln e^{\ln x} - \ln(x - 4) \le \ln 3$

28. (a) $\ln e^{\ln 2} - \ln(x - 1) = \ln 5$
   (b) $\ln e^{\ln 2} - \ln(x - 1) \ge \ln 5$
   (c) $\ln e^{\ln 2} - \ln(x - 1) \le \ln 5$

Each of the following formulas comes from an application of exponential and/or logarithmic functions. Solve the formula for the indicated variable.

29. $r = p - k \ln t$,   for t

30. $p = a + \dfrac{k}{\ln x}$,   for x

31. $T = T_0 + (T_1 - T_0)10^{-kt}$,   for t

32. $A = \dfrac{Pi}{1 - (1 + i)^{-n}}$,   for n

33. $A = T_0 + Ce^{-kt}$,   for k

34. $y = \dfrac{K}{1 + ae^{-bx}}$,   for b

35. $y = A + B(1 - e^{-Cx})$,   for x

36. $m = 6 - 2.5 \log\left(\dfrac{M}{M_0}\right)$,   for M

37. $\log A = \log B - C \log x$,   for A

38. $d = 10 \log\left(\dfrac{I}{I_0}\right)$,   for I

Relating Concepts

In Chapter 3 we introduced methods of solving quadratic equations. These methods can be applied to equations that may not necessarily be quadratic, but may be quadratic in form. Consider the equation

$$e^{2x} - 4e^x + 3 = 0$$

and work Exercises 39–44 in order.

39. The expression e^{2x} is equivalent to $(e^x)^2$. Explain why this is so.

40. The given equation is equivalent to $(e^x)^2 - 4e^x + 3 = 0$. Factor the left side of this equation.

41. Solve the equation in Exercise 40 by the zero-product property (from Section 3.2). Give exact values.

42. Support your solution(s) in Exercise 41 using a calculator-generated graph of $y = e^{2x} - 4e^x + 3$.

43. Use the graph from Exercise 42 to solve the inequality $e^{2x} - 4e^x + 3 > 0$.

44. Use the graph from Exercise 42 to solve the inequality $e^{2x} - 4e^x + 3 < 0$.

In general, it is not possible to find exact solutions analytically for equations that involve exponential or logarithmic functions together with polynomial, radical, and rational functions. However, it is possible to solve them graphically with a calculator using either the intersection-of-graphs method or the x-intercept method. Solve the following equations using a purely graphical method, and express solutions to the nearest thousandth if an approximation is appropriate.

45. $x^2 = 2^x$

46. $x^2 - 4 = e^{x-4} + 4$

47. $\log x = x^2 - 8x + 14$

48. $\ln x = -\sqrt[3]{x} + 3$

49. $e^x = \dfrac{1}{x + 2}$

50. $3^{-x} = \sqrt{x + 5}$

Use any method (analytic or graphical) to solve each of the following equations. If appropriate, round your solution to the nearest thousandth.

51. $100(1 + .02)^{3+x} = 150$

52. $500(1 + .05)^{x/4} = 200$

53. $\log_2 \sqrt{2x^2 - 1} = .5$

54. $\log_2(\log_2 x) = 1$

55. $\log x = \sqrt{\log x}$

56. $\log x^2 = (\log x)^2$

57. $\ln(\ln e^{-x}) = \ln 3$

58. $10^{5 \log x} = 32$

59. $e^x - 6 = -\dfrac{8}{e^x}$

60. $e^{x+\ln 3} = 4e^x$

5.5 APPLICATIONS OF EXPONENTIAL AND LOGARITHMIC FUNCTIONS

Physical Sciences ∎ Finance ∎ Biological Sciences and Medicine ∎ Economics ∎ Social Sciences

In the first two sections of this chapter, we saw how exponential functions can be used in computing interest and how logarithms are used to determine pH of substances in chemistry. These are two of many types of applications of exponential and logarithmic functions. In this section we will examine other applications in various fields of study.

Physical Sciences

The formula $A = Pe^{rt}$, introduced in Section 5.1 in conjunction with continuous compounding of interest, is an example of an exponential growth function. A function of the form $A(t) = A_0 e^{kt}$, where A_0 represents the initial quantity present, t represents time elapsed, $k > 0$ represents the growth constant associated with the quantity, and A represents the amount present at time t, is called an **exponential growth function.** As we would expect, this is an increasing function, because $e > 1$ and $k > 0$. On the other hand, a function of the form $A(t) = A_0 e^{-kt}$ is an **exponential decay function.** It is a decreasing function because

$$e^{-k} = (e^{-1})^k = \left(\frac{1}{e}\right)^k,$$

and $0 < \frac{1}{e} < 1$. In both cases, we usually restrict t to be nonnegative, giving a domain of $[0, \infty)$. (Why do you think this is so?) Figure 41 shows graphs of typical growth and decay functions.

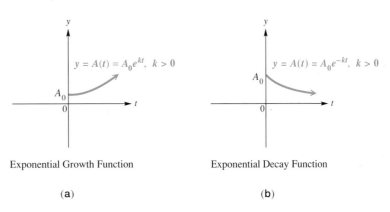

Exponential Growth Function   Exponential Decay Function

(a)   (b)

FIGURE 41

If a quantity decays exponentially, the amount of time that it takes to become one-half its original amount is called the **half-life.** The first example uses this idea.

EXAMPLE **1**

Analyzing an Exponential Decay Function from Chemistry

Nuclear energy derived from radioactive isotopes can be used to supply power to space vehicles. Suppose that the output of the radioactive power supply for a certain satellite is given by the function

$$y = 40e^{-.004t},$$

where y is measured in watts and t is the time in days.

(a)   What is the initial output of the power supply?

SOLUTION   Let $t = 0$ in the equation.

$$y = 40e^{-.004t} \qquad \text{Given function}$$

$$y = 40e^{-.004(0)} \qquad \text{Let } t = 0.$$

$$y = 40e^0$$

$$y = 40 \qquad e^0 = 1.$$

The initial output is 40 watts.

(b)   After how many days will the output be reduced to 35 watts?

SOLUTION   Let $y = 35$, and solve for t.

$$35 = 40e^{-.004t}$$

$$\frac{35}{40} = e^{-.004t} \qquad \text{Divide by 40.}$$

$$\ln\left(\frac{35}{40}\right) = \ln e^{-.004t} \qquad \text{Take the natural logarithm.}$$

$$\ln\left(\frac{35}{40}\right) = -.004t \qquad \ln e^k = k$$

$$t = \frac{\ln\left(\dfrac{35}{40}\right)}{-.004} \qquad \text{Divide by } -.004.$$

Using a calculator, we find that $t \approx 33.4$. It will take about 33.4 days for the output to be reduced to 35 watts.

(c)   After how many days will the output be half of its initial amount? (That is, what is its half-life?)

SOLUTION   Because the initial amount is 40, we must find the value of t for which $y = \frac{1}{2}(40) = 20$.

$$20 = 40e^{-.004t}$$

$$.5 = e^{-.004t} \qquad \text{Divide by 40.}$$

$$\ln .5 = \ln e^{-.004t} \qquad \text{Take the natural logarithm.}$$

$$\ln .5 = -.004t \qquad \ln e^k = k$$

$$t \approx 173 \qquad \text{Divide by } -.004.$$

The half-life is approximately 173 days. The graph in Figure 42 supports the result that when $x = t = 173$, $y \approx 20 = \frac{1}{2}(40)$.   ∎

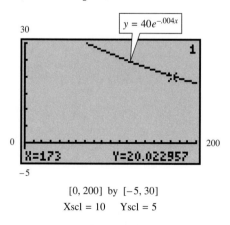

$$y = 40e^{-.004x}$$

X=173   Y=20.022957

[0, 200] by [−5, 30]
Xscl = 10   Yscl = 5

FIGURE 42

EXAMPLE 2

Using an
Exponential
Function to Find the
Age of a Fossil

Carbon-14 is a radioactive form of carbon that is found in all living plants and animals. After a plant or animal dies, the radiocarbon disintegrates. Scientists determine the age of the remains by comparing the amount of carbon-14 present with the amount found in living plants and animals. The amount of carbon-14 present after t years is given by the exponential equation

$$A(t) = A_0 e^{-kt}$$

with $k \approx (\ln 2)(\frac{1}{5700})$.

(a) Find the half-life.

SOLUTION   Let $A(t) = (\frac{1}{2})A_0$ and $k = (\ln 2)(\frac{1}{5700})$.

$$\frac{1}{2}A_0 = A_0 e^{-(\ln 2)(1/5700)t}$$

$$\frac{1}{2} = e^{-(\ln 2)(1/5700)t} \qquad \text{Divide by } A_0.$$

$$\ln \frac{1}{2} = \ln e^{-(\ln 2)(1/5700)t} \qquad \begin{array}{l}\text{Take logarithms on both}\\ \text{sides.}\end{array}$$

$$\ln \frac{1}{2} = -\frac{\ln 2}{5700}t \qquad \ln e^x = x$$

$$-\frac{5700}{\ln 2} \ln \frac{1}{2} = t \qquad \text{Multiply by } -\frac{5700}{\ln 2}.$$

$$-\frac{5700}{\ln 2}(\ln 1 - \ln 2) = t \qquad \begin{array}{l}\text{Quotient rule for}\\ \text{logarithms}\end{array}$$

$$-\frac{5700}{\ln 2}(-\ln 2) = t \qquad \ln 1 = 0$$

$$5700 = t$$

The half-life is 5700 years.

(b) Charcoal from an ancient fire pit on Java contained $\frac{1}{4}$ the carbon-14 of a living sample of the same size. Estimate the age of the charcoal.*

SOLUTION   Let $A(t) = \frac{1}{4}A_0$ and $k = (\ln 2)(\frac{1}{5700})$.

$$\frac{1}{4}A_0 = A_0 e^{-(\ln 2)(1/5700)t}$$

$$\frac{1}{4} = e^{-(\ln 2)(1/5700)t}$$

$$\ln \frac{1}{4} = \ln e^{-(\ln 2)(1/5700)t}$$

$$\ln \frac{1}{4} = -\frac{\ln 2}{5700}t$$

$$-\frac{5700}{\ln 2} \ln \frac{1}{4} = t$$

$$t = 11,400$$

The charcoal is about 11,400 years old.   ∎

---

*Adapted from *A Sourcebook of Applications of School Mathematics* by Donald Bushaw et al. Copyright © 1980 by The Mathematical Association of America. Reprinted by permission.

EXAMPLE 3

Measuring Sound Intensity

The loudness of sounds is measured in a unit called a *decibel*. To measure with this unit, we first assign an intensity of I_0 to a very faint sound, called the *threshold sound*. If a particular sound has intensity I, then the decibel rating of this louder sound is

$$d = 10 \log \frac{I}{I_0}.$$

Find the decibel rating of a sound with intensity $10,000I_0$.

SOLUTION   Let $I = 10,000I_0$ and find d.

$$d = 10 \log \frac{10,000I_0}{I_0}$$

$$= 10 \log 10,000$$

$$= 10(4) \qquad \log 10,000 = 4$$

$$= 40$$

The sound has a decibel rating of 40.   ◻

EXAMPLE 4

Measuring Age Using an "Atomic Clock"

Geologists sometimes measure the age of rocks by using "atomic clocks." By measuring the amounts of potassium-40 and argon-40 in a rock, the age t of the specimen in years is found with the formula

$$t = (1.26 \times 10^9) \frac{\ln[1 + 8.33(A/K)]}{\ln 2}.$$

A and K are respectively the numbers of atoms of argon-40 and potassium-40 in the specimen. The ratio $\frac{A}{K}$ for a sample of granite from New Hampshire is .212. How old is the sample?

SOLUTION   Since $\frac{A}{K}$ is .212, we have

$$t = (1.26 \times 10^9) \frac{\ln[1 + 8.33(.212)]}{\ln 2} \approx 1.85 \times 10^9.$$

The granite is about 1.85 billion years old.*   ◻

Finance

The formulas

$$A = P\left(1 + \frac{r}{n}\right)^{nt} \qquad \text{and} \qquad A = Pe^{rt}$$

were introduced in Section 5.1. They apply to compound interest: the first, to interest compounded a finite number of times annually, and the second to interest compounded continuously. We can use logarithms to determine how long it will take for a particular investment to grow to a desired amount.

---

*Adapted from *A Sourcebook of Applications of School Mathematics* by Donald Bushaw et al. Copyright © 1980 by the Mathematical Association of America. Reprinted by permission.

EXAMPLE 5
Using Compound Interest Formulas to Determine Time Required

(a) How long will it take $1000 invested at 6% compounded quarterly to grow to $2700?

SOLUTION   Here we must find t such that $A = 2700$, $P = 1000$, $r = .06$, and $n = 4$.

$$A = P\left(1 + \frac{r}{n}\right)^{nt} \qquad \text{Given formula}$$

$$2700 = 1000\left(1 + \frac{.06}{4}\right)^{4t} \qquad \text{Substitute.}$$

$$2700 = 1000(1.015)^{4t}$$

$$2.7 = 1.015^{4t} \qquad \text{Divide by 1000.}$$

To solve for t, we may use either common or natural logarithms. Choosing common logarithms, we obtain

$$\log 2.7 = \log 1.015^{4t}$$

$$\log 2.7 = 4t \log 1.015 \qquad \text{Power rule}$$

$$\frac{\log 2.7}{4 \log 1.015} = t \qquad \text{Divide by 4 log 1.015.}$$

$$t \approx 16.678. \qquad \text{Use a calculator.}$$

It will take about $16\frac{2}{3}$ years for the initial amount to grow to $2700.

(b) How long will it take for the money in an account that is compounded continuously at 8% interest to double?

SOLUTION   Use the formula for continuous compounding, $A = Pe^{rt}$, to find the time t that makes $A = 2P$. Substitute $2P$ for A and .08 for r; then solve for t.

$$A = Pe^{rt}$$

$$2P = Pe^{.08t} \qquad \text{Substitute.}$$

$$2 = e^{.08t} \qquad \text{Divide by } P.$$

Taking natural logarithms on both sides gives

$$\ln 2 = \ln e^{.08t}.$$

Use the property $\ln e^x = x$ to get $\ln e^{.08t} = .08t$.

$$\ln 2 = .08t$$

$$\frac{\ln 2}{.08} = t \qquad \text{Divide by .08.}$$

$$8.664 \approx t$$

It will take about $8\frac{2}{3}$ years for the amount to double.   ∎

If a quantity grows exponentially, the amount of time that it takes to become twice its original amount is called the **doubling time**. This is analogous to half-life for quantities that decay exponentially. In both cases, the actual amount present initially does not affect the doubling time or half-life.

If a loan is taken out from a lending institution, it is customary to pay the loan back in installments on a regular basis at a certain rate of interest for a length of time called the *term* of the loan. This process is called **amortization**.

AMORTIZATION OF A LOAN

If a principal of P dollars is amortized over a period of t years, and payments are made each $\frac{1}{n}$th of a year, with an annual interest rate of r (as a decimal), the payment p that must be made each period is given by the formula

$$p = \frac{Pr}{n\left[1 - \left(1 + \dfrac{r}{n}\right)^{-nt}\right]}.$$

The total interest I that will be paid during the term of the loan is given by the formula

$$I = npt - P.$$

EXAMPLE 6

Using the
Amortization
Formulas

After going through the tried-and-true American tradition of "haggling" with the used car manager at your local dealer's showroom, you decide on a price of $14,200 for a beautiful 1990 Cadillac Sedan DeVille. Your down payment is $4000, and you are able to find an institution that will lend you the balance of the price at 8.5%. The term of the loan is to be 3 years, and payments are to be made monthly. How much will your monthly payment be, and how much interest will you have paid when the loan is completely paid off?

SOLUTION To find the monthly payment p, use the first amortization formula with $P = 14,200 - 4000 = 10,200$, $r = .085$, $n = 12$, and $t = 3$.

$$p = \frac{10,200(.085)}{12\left[1 - \left(1 + \dfrac{.085}{12}\right)^{-12\cdot 3}\right]} \approx 321.99$$

The monthly payment will be $321.99.

The total amount of interest (I) paid over the life of the loan is found by using the second amortization formula, with $n = 12$, $p = 321.99$, $t = 3$, and $P = 10,200$.

$$I = 12(321.99)(3) - 10,200 \approx 1391.64.$$

The total amount of interest paid will be $1391.64. ∎

FOR GROUP DISCUSSION

Make a conjecture (without using the formula) as to what the monthly payment would be in Example 6 if all conditions were the same except

1. the rate is doubled.
2. the period of the loan is 5 years.
3. the down payment is $6000.

Then have class members divide into three groups and use the first amortization formula to answer the three items above. How good were the conjectures?

Biological Sciences and Medicine

Base two logarithms are used in a formula that measures the diversity of the species in a community, as seen in the next example.

EXAMPLE 7

Measuring the Diversity of Species

One measure of the diversity of the species in an ecological community is given by the formula

$$H = -[P_1 \log_2 P_1 + P_2 \log_2 P_2 + \cdots + P_n \log_2 P_n],$$

where $P_1, P_2, \ldots, P_n$ are the proportions of a sample belonging to each of n species found in the sample. Suppose that a community has two species, where there are 90 of one and 10 of the other.

(a) Find P_1 and P_2.

SOLUTION Since there are $90 + 10 = 100$ in total,

$$P_1 = \frac{90}{100} = .9 \quad \text{and} \quad P_2 = \frac{10}{100} = .1.$$

(b) Find H for the community.

SOLUTION

$$H = -[.9 \log_2 .9 + .1 \log_2 .1].$$

By the change-of-base rule,

$$\log_2 .9 = \frac{\log .9}{\log 2} \approx \frac{-.0457574906}{.3010299957} \approx -.152$$

and

$$\log_2 .1 = \frac{\log .1}{\log 2} \approx -3.322.$$

Using these in the formula for H above, we find

$$H \approx .469$$

If the number in each species is the same, the measure of diversity is 1, representing "perfect"diversity. In a community with little diversity, H is close to 0. In this example, since $H \approx .5$, there is neither great nor little diversity. ◧

The following example illustrates how the amount of medication left in the body after a certain period of time can be determined.

EXAMPLE 8

Determining the Amount of Medication Remaining

When physicians prescribe medication they must consider how the drug's effectiveness decreases over time. If, each hour, a drug is only 90% as effective as the previous hour, at some point the patient will not be receiving enough medication and must receive another dose. This situation can be modeled with a geometric sequence (see Section 3 in the final chapter). If the initial dose was 200 mg and the drug was administered 3 hours ago, the expression $200(.90)^2$ represents the amount of effective medication still available. Thus, $200(.90)^2 \approx 162$ mg are still in the system. (The exponent is equal to the number of hours since the drug was administered, less one.) How long will it take for this initial dose to reach the dangerously low level of 50 mg?

SOLUTION   We must solve the equation $200(.90)^x = 50$.

$$200(.90)^x = 50$$
$$(.90)^x = .25$$
$$\log (.90)^x = \log .25$$
$$x \log .90 = \log .25$$
$$x = \frac{\log .25}{\log .90} \approx 13.16$$

Since x represents one *less than* the number of hours since the drug was administered, the drug will reach a level of 50 mg in about 14 hours. The graph in Figure 43 supports the result that when $x \approx 13.16$, $y = 50$.   ∎

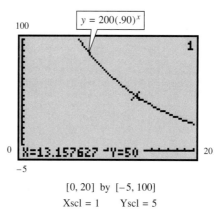

$$y = 200(.90)^x$$

[0, 20] by [−5, 100]
Xscl = 1   Yscl = 5

FIGURE 43

FOR GROUP DISCUSSION

Reproduce the graph shown in Figure 43 and trace from left to right. Discuss how the word "decreasing" applies to this activity in both mathematical and real-life contexts.

Economics

The U.S. Consumer Price Index can be modeled by an exponential function with base e, as shown in the next example.

EXAMPLE 9

Determining the U.S. Consumer Price Index Using a Model

The U.S. Consumer Price Index (CPI, or cost of living index) has risen exponentially over the years. From 1960 to 1990, the CPI is approximated by

$$A(t) = 34e^{.04t},$$

where t is time in years, with $t = 0$ corresponding to 1960. The index in 1960, at $t = 0$, was

$$A(0) = 34e^{(.04)(0)}$$
$$= 34e^0$$
$$= 34. \qquad e^0 = 1$$

To find the CPI for 1990, let $t = 1990 - 1960 = 30$, and find $A(30)$.

$$A(30) = 34e^{(.04)(30)}$$
$$= 34e^{1.2}$$
$$\approx 113 \qquad e^{1.2} \approx 3.3201$$

The graph of the function $y = 34e^{.04x}$ in Figure 44 supports the result that $A(30) \approx 113$.

The index measures the average change in prices relative to the base year 1983 (1983 corresponds to 100) of a common group of goods and services. Our result of 113 means that prices increased an average of $113 - 34 = 79$ percent over the 30-year period from 1960 to 1990. ◘

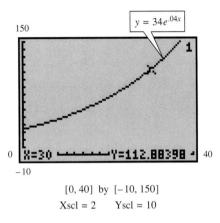

$$[0, 40] \text{ by } [-10, 150]$$
$$\text{Xscl} = 2 \qquad \text{Yscl} = 10$$

FIGURE 44

Social Sciences

The following example shows how logarithms can be used in the study of language development.

EXAMPLE 10

Determining Information About Language Development

The number of years, n, since two independently evolving languages split off from a common ancestral language is approximated by $n \approx -7600 \log r$, where r is the proportion of words from the ancestral language common to both languages.

(a) Find n if $r = .9$.

SOLUTION Using the formula,

$$n \approx -7600 \log .9 \approx 350.$$

Approximately 350 years have elapsed.

(b) Find r if $n = 4000$.

SOLUTION We again use the formula.

$$4000 = -7600 \log r \qquad \text{Let } n = 4000.$$
$$\frac{4000}{-7600} = \log r \qquad \text{Divide by } -7600.$$
$$r = 10^{(4000/-7600)} \qquad \text{Write in exponential form.}$$
$$r \approx .30 \qquad \text{Use a calculator.}$$

About 30% of the words from the ancestral language are common to both languages. ◘

5.5 EXERCISES

Solve the following applications from the physical sciences.

1. A sample of 500 grams of radioactive lead-210 decays to polonium-210 according to the function $A(t) = 500e^{-.032t}$, where t is time in years. Find the amount of the sample remaining after **(a)** 4 years **(b)** 8 years **(c)** 20 years. Then **(d)** find the half-life and **(e)** graph $y = A(t)$ in the window $[0, 300]$ by $[0, 750]$.

2. Repeat Exercise 1 for 500 grams of plutonium-241, which decays according to the function $A(t) = A_0e^{-.053t}$, where t is time in years.

3. Find the half-life of radium-226, which decays according to the function $A(t) = A_0e^{-.00043t}$, where t is in years.

4. How long will it take any quantity of iodine-131 to decay to 25% of its initial amount, knowing that it decays according to the function $A(t) = A_0e^{-.087t}$, where t is in years?

5. The air pressure in pounds per square inch h feet above sea level is given by the function $P(h) = 14.7e^{-.0000385h}$. At approximately what height is the pressure 10% of the pressure at sea level?

6. Find the decibel ratings of sounds having the following intensities. (See Example 3 for the interpretation of I_0.)
   (a) $100I_0$ **(b)** $1000I_0$ **(c)** $100,000I_0$
   (d) $1,000,000I_0$

7. Find the decibel ratings of the following sounds, having the intensities given.
   (a) whisper, $115I_0$
   (b) busy street, $9,500,000I_0$
   (c) rock music, $895,000,000,000I_0$

8. A jetliner at takeoff has a decibel rating of 140. Express the intensity of this sound as a multiple of I_0.

9. In a certain community there was a controversy about a proposed government limit on factory noise. One group wanted a maximum of 89 decibels, while another wanted a maximum of 86. Find the percent by which the 89-decibel intensity exceeds that of 86 decibels.

10. In the central Sierra Nevada mountains of California, the percent of moisture that falls as snow rather than rain is approximated reasonably well by the function $p(h) = 86.3 \ln h - 680$, where h is the altitude in feet, and $p(h)$ is the percent of snow. (This model is valid for $h \geq 3000$.) Find the percent of snow that falls at the following altitudes.
   (a) 3000 feet   **(b)** 4000 feet   **(c)** 7000 feet

The intensity of an earthquake, measured on the Richter scale, is given by $\log(\frac{I}{I_0})$, where I_0 is the intensity of an earthquake of a certain small size. Use this information in Exercises 11–13.

11. Find the Richter scale magnitude of an earthquake that has each of the following intensities.
    (a) $1000I_0$
    (b) $1,000,000I_0$
    (c) $100,000,000I_0$

12. On July 14, 1991, Peshawar, Pakistan, was shaken by an earthquake that measured 6.6 on the Richter scale.
    (a) Express this reading in terms of I_0.
    (b) In February of the same year a quake measuring 6.5 on the Richter scale killed about 900 people in the mountains of Pakistan and Afghanistan. Express the intensity of a 6.5 reading in terms of I_0.
    (c) How much greater was the force of the earthquake with a measure of 6.6?

13. The San Francisco Earthquake of 1906 had a Richter scale rating of 8.3.
    (a) Express the intensity of this earthquake as a multiple of I_0.
    (b) In 1989, the San Francisco region experienced an earthquake with a Richter scale rating of 7.1. Express the intensity of this earthquake as a multiple of I_0.
    (c) Compare the intensity of the two San Francisco earthquakes discussed above.

14. The magnitude M of a star is defined by the equation $M = 6 - \frac{5}{2} \log \frac{I}{I_0}$, where I_0 is the measure of a just-visible star and I is the actual intensity of the star being measured. The dimmest stars are of magnitude 6, and the brightest are of magnitude 1. Determine the ratio of light intensities between a star of magnitude 1 and a star of magnitude 3.

The information in Example 2 allows us to use the function $A(t) = A_0 e^{-.0001216t}$ to approximate the amount of carbon-14 remaining in a sample, where t is in years. Use this function in Exercises 15–18. $\left(Note: -.0001216 \approx \dfrac{\ln 2}{-5700}. \right)$

15. Suppose an Egyptian mummy is discovered in which the amount of carbon-14 present is only about one-third the amount found in the atmosphere. About how long ago did the Egyptian die?

16. A sample from a refuse deposit near the Strait of Magellan had 60% of the carbon-14 of a contemporary living sample. How old was the sample?

17. Paint from the Lascaux caves of France contains 15% of the normal amount of carbon-14. Estimate the age of the caves.

18. Estimate the age of a specimen that contains 20% of the carbon-14 of a comparable living specimen.

A large cloud of radioactive debris from a nuclear explosion has floated over the Pacific Northwest, contaminating much of the hay supply. Consequently, farmers in the area are concerned that the cows who eat this hay will give contaminated milk. (The tolerance level for radioactive iodine in milk is 0.) The percent of the initial amount of radioactive iodine still present in the hay after t days is approximated by $P(t) = 100e^{-.1t}$. Use this information in Exercises 19 and 20.

19. Some scientists feel that the hay is safe after the percent of radioactive iodine has declined to 10% of the original amount. Find the number of days before the hay can be used.

20. Other scientists believe that the hay is not safe until the level of radioactive iodine has declined to only 1% of the original level. Find the number of days this would take.

21. Use the function defined by
$$t = T \frac{\ln[1 + 8.33(A/K)]}{\ln 2}$$
to estimate the age of a rock sample, if tests show that A/K is .103 for the sample. Let $T = 1.26 \times 10^9$.

22. By Newton's law of cooling, the temperature of a body at time t after being introduced into an environment having constant temperature T_0 is
$$A(t) = T_0 + Ce^{-kt},$$
where C and k are constants. If $C = 100$, $k = .1$, and t is time measured in minutes, how long will it take a hot cup of coffee to cool to a temperature of 25°C in a room at 20°C?

Solve the following applications from finance. Use the formulas for compound interest, found in Section 5.1, in Exercises 23–26.

23. How long will it take for $1000 to grow to $5000 at an interest rate of 3.5% if interest is compounded **(a)** quarterly **(b)** continuously?

24. How long will it take for $5000 to grow to $8400 at an interest rate of 6% if interest is compounded **(a)** semiannually **(b)** continuously.

25. George Duda wants to buy a $30,000 car. He has saved $27,000. Find the number of years (to the nearest tenth) it will take for his $27,000 to grow to $30,000 at 6% interest compounded quarterly.

26. Find the doubling time of an investment earning 2.5% interest if interest is compounded **(a)** quarterly **(b)** continuously.

*The interest rate stated by a financial institution is sometimes called the **nominal rate.***
If interest is compounded, the actual rate is, in general, higher than the nominal rate,
*and is called the **effective rate.** If r is the nominal rate and n is the number of times*
interest is compounded annually, then

$$R = \left(1 + \frac{r}{n}\right)^n - 1$$

is the effective rate. Here, R represents the annual rate that the investment would earn
if simple interest were paid. Use this formula in Exercises 27 and 28.

27. Find the effective rate if the nominal rate is 6% and interest is compounded quarterly.

28. Find the effective rate if the nominal rate is 4.5% and interest is compounded daily ($n = 365$).

*In the formula $A = P(1 + \frac{r}{n})^{nt}$, we can interpret P as the **present value** of A dollars*
t years from now, earning annual interest r compounded n times per year. In this con-
*text, A is called the **future value.** If we solve the formula for P, we obtain*

$$P = A\left(1 + \frac{r}{n}\right)^{-nt}.$$

Use this formula in Exercises 29–32.

29. Find the present value of $10,000 five years from now, if interest is compounded semiannually at 12%.

30. Find the present value of $25,000 2.75 years from now, if interest is compounded quarterly at 6%.

31. Find the interest rate necessary for a present value of $25,000 to grow to a future value of $31,360, if interest is compounded annually for 2 years.

32. Find the interest rate necessary for a present value of $1200 to grow to a future value of $1780, if interest is compounded quarterly for 5 years.

33. Linda Youngman, who is self-employed, wants to invest $60,000 in a pension plan. One investment offers 7% compounded quarterly. Another offers 6.75% compounded continuously. Which investment will earn more interest in 5 years? How much more will the better plan earn?

34. If Ms. Youngman (see Exercise 33) chooses the plan with continuous compounding, how long will it take for her $60,000 to grow to $80,000?

*In Exercises 35–38, use the amortization formulas given in this section to find (**a**) the*
*monthly ($n = 12$) payment on a loan with the given conditions and (**b**) the total interest*
that will be paid during the term of the loan.

35. $8500 is amortized over 4 years with an interest rate of 7.5%

36. $9600 is amortized over 5 years with an interest rate of 9.2%

37. $55,000 is amortized over 15 years with an interest rate of 6.25%

38. $125,000 is amortized over 30 years with an interest rate of 7.25%

Solve the following applications from biological sciences and medicine.

39. *Escherichia coli* is a strain of bacteria that occurs naturally in many organisms. Under certain conditions, the number of bacteria present in a colony is approximated by the function $A(t) = A_0 e^{.023t}$, where t is in minutes. If $A_0 = 2,400,000$, find the number of bacteria at the following times.
(a) 5 minutes   (b) 10 minutes
(c) 60 minutes

40. The growth of bacteria in food products makes it necessary to time-date some products (such as milk) so that they will be sold and consumed before the bacteria count becomes too high. Suppose for a certain product that the number of bacteria present is given by

$$f(t) = 500e^{.1t},$$

under certain storage conditions, where t is time in days after packing of the product and the value of $f(t)$ is in millions.
(a) If the product cannot be safely eaten after the bacteria count reaches 3,000,000,000, how long will this take?
(b) If $t = 0$ corresponds to January 1, what date should be placed on the product?

41. The population of an animal species that is introduced into a certain area may grow rapidly at first but then grow more slowly as time goes on. A logarithmic function can provide an excellent description of such growth. Suppose that the population of foxes in an area t months after the foxes were first introduced there is

$$F = 500 \log (2t + 3).$$

Solve the equation for t. Then find t to the nearest tenth for the following values of F.
(a) 600   (b) 1000

For Exercises 43 and 44, refer to Example 8.

43. If 250 mg of a drug are administered, and the drug is only 75% as effective each subsequent hour, how much effective medicine will remain in the person's system after 6 hours?

44. A new drug has been introduced which is 80% as effective each hour as the previous hour. A minimum of 20 mg must remain in the patient's bloodstream during the course of treatment. If 100 mg are administered, how many hours may elapse before another dose is necessary?

Solve the following applications from economics.

47. The number of books, in millions, sold per year in the United States between 1985 and 1990, can be approximated by the function $A(t) = 1757e^{.0264t}$, where $t = 0$ corresponds to the year 1985. Based on this model, how many books would be sold in 1996? (*Source:* Book Industry Study Group)

48. Personal consumption expenditures for recreation in billions of dollars in the United States during the years 1984 through 1990 can be approximated by the function $A(t) = 185.4e^{.0587t}$, where $t = 0$ corresponds to the year 1984. Based on this model, how much would personal consumption expenditures be in 1996? (*Source:* U.S. Bureau of Economic Analysis)

49. In Example 9, the U.S. Consumer Price Index was approximated by $A(t) = 34e^{.04t}$, where t represents the number of years after 1960. Assuming the same equation continues to apply, find the year in which costs will be 50% higher than in 1983, that is, when the CPI equals 150.

50. Experiments have shown that the sales of a product, under relatively stable market conditions, but in the absence of promotional activities such as advertising, tend to decline at a constant yearly rate. This rate of sales decline varies considerably from product to product, but seems to remain the same for any particular product. The sales decline can be ex-

42. The number of Cesarean section deliveries in the United States has increased over the years. Between the years 1980 and 1989, the number of such births, in thousands, can be approximated by the function $f(t) = 625e^{.0516t}$, where $t = 1$ corresponds to the year 1980. Based on this function, what would be the approximate number of Cesarean section deliveries in 1996? (*Source:* U.S. National Center for Health Statistics)

45. Suppose a sample of a small community shows two species with 50 individuals each. Find the index of diversity H. (See Example 7.)

46. A virgin forest in northwestern Pennsylvania has 4 species of large trees with the following proportions of each: hemlock, .521; beech, .324; birch, .081; maple, .074. Find the index of diversity H. (See Example 7.)

pressed by a function of the form

$$S(t) = S_0 e^{-at},$$

where $S(t)$ is the rate of sales at time t measured in years, S_0 is the rate of sales at time $t = 0$, and a is the sales decay constant.
(a) Suppose the sales decay constant for a particular product is $a = .10$. Let $S_0 = 50,000$ and find $S(1)$ and $S(3)$.
(b) Find $S(2)$ and $S(10)$ if $S_0 = 80,000$ and $a = .05$.

51. Use the sales decline function given in Exercise 50. If $a = .1$, $S_0 = 50,000$, and t is time measured in years, find the number of years it will take for sales to fall to half the initial sales.

52. Assume the cost of a loaf of bread is $1. With continuous compounding, find the time it would take to triple at an annual inflation rate of 6%.

53. Historically, the consumption of electricity has increased at a continuous rate of 6% per year. If it continued to increase at this rate, find the number of years before twice as much electricity would be needed.

54. Suppose a conservation campaign together with higher rates caused demand for electricity to increase at only 2% per year. (See Exercise 53.) Find the number of years before twice as much electricity would be needed.

Solve the following applications from social sciences.

55. Since 1950, the growth in the world population in millions closely fits the exponential function defined by

$$A(t) = 2600e^{.018t},$$

where t is the number of years since 1950.
(a) The world population was about 3700 million in 1970. How closely does the function approximate this value?
(b) Use the function to approximate the population in 1990. (The actual 1990 population was about 5320 million.)

56. Vehicle theft in the United States has been rising exponentially since 1972. The number of stolen vehicles, in millions, is given by

$$f(x) = .88(1.03)^x,$$

where $x = 0$ represents the year 1972. Find the number of vehicles stolen in the following years.
(a) 1975   (b) 1980   (c) 1985   (d) 1990

57. Use the formula in Example 10 to find the number of years, n, since two independently evolving languages split off from a common ancestral language, given the following values of r. (Here r is the proportion of words from the ancestral language.)
(a) $r = 60\%$ (b) $r = 40\%$

58. (Refer to Example 10.) If $n = 2500$, what is the proportion of words remaining from the ancestral language?

59. One measure of living standards in the United States is given by $L = 9 + 2e^{.15t}$, where t is the number of years since 1982. Find L for the following years.

(a) 1982   (b) 1986   (c) 1992
(d) Graph L in the window $[0, 10]$ by $[0, 30]$.
(e) What can be said about the growth of living standards in the U.S. according to this equation?

60. A midwestern city finds its residents moving to the suburbs. Its population is declining according to the relationship $P = P_0 e^{-.04t}$, where t is time measured in years and P_0 is the population at time $t = 0$. Assume that $P_0 = 1,000,000$.
(a) Find the population at time $t = 1$.
(b) Estimate the time it will take for the population to be reduced to 750,000.
(c) How long will it take for the population to be half its initial amount?

61. A function of the form $f(x) = a \cdot b^x$ is called a **power regression** formula. Show that it can be written in the equivalent form $f(x) = a \cdot e^{(\ln b)x}$.

62. A **logistic function** is a function defined by the equation

$$y = \frac{k}{1 + be^{-ct}}$$

where b, c, and k are positive constants and t is a measure of time.
(a) Graph the logistic function with $b = 40$, $c = 1.15$, and $k = 100$. Use the window $[0, 10]$ by $[0, 110]$.
(b) Assuming that time is in days, what is the value of y after 4.25 days?
(c) If y represents a population, describe the behavior of the growth of this population based on the graph you found in part (a).

*F*urther Explorations

1. Use the TABLE feature of your graphics calculator to find how long it will take $1500 invested at 5.75% compounded daily to triple in value. Zoom in on the solution by systematically decreasing ΔTbl. Find the answer to the nearest day. (Find your answer to the nearest day by eventually letting ΔTbl $= 1/365$. The decimal part of the solution can be multiplied by 365 to determine the number of days greater than the nearest year. For example, if the solution is determined to be 16.2027 years, then multiply .2027 by 365 to get 73.9855. The solution is then, to the nearest day, 16 years and 74 days.) Confirm your answer analytically.

2. Use the TABLE feature of your graphics calculator to find how long it will take $2000 invested at 8% compounded daily to be worth $5000.

*C*hapter 5   SUMMARY

In Section 5.1 several properties of exponents are given that determine the character of exponential functions. The exponential function $f(x) = a^x$, $a > 1$, has domain $(-\infty, \infty)$ and range $(0, \infty)$. It is increasing on its entire domain and is continuous there. The x-axis is a horizontal asymptote as $x \rightarrow -\infty$. There is no x-intercept; the y-intercept is 1. If $0 < a < 1$, the graph of $f(x)$ has the same characteristics except that $f(x) = a^x$ decreases on its entire domain, and the x-axis is a horizontal asymptote as $x \to \infty$. Type 1 exponential equations are solved by expressing the bases on each side of the equation as powers of a common base. The exponents on the common base are then equal as a consequence of a property of exponents. The irrational number e, an important base for exponential functions, is shown to be the limit of an algebraic expression as the variable becomes infinitely large. The formulas for compound interest and continuous compounding are developed and applied in Section 5.1.

A logarithm is defined as the exponent to which the base a must be raised in order to obtain a given number k. That is, the two statements $a^k = x$ and $k = \log_a x$ are equivalent. (Note that k is the exponent in the first form.) Thus exponential statements can be written in logarithmic form and logarithmic statements can be written in exponential form. This provides a way to solve some logarithmic equations by writing them as Type 1 exponential equations. Base 10 logarithms are called common logarithms. One important application of common logarithms is in finding the pH of a solution. Natural logarithms are base e logarithms. These logarithms are important in growth and decay applications.

Several important properties of logarithms are given in Section 5.2, including the product, quotient, and power rules. These rules make it possible to rewrite the logarithm of a product, quotient, or power of a quantity in a different form. Calculator keys are available for common and natural logarithms. Logarithms to other bases are evaluated by using the change-of-base rule to write these logarithms as quotients of common or natural logarithms.

The functions $f(x) = a^x$ and $g(x) = \log_a x$ are inverses. The logarithmic function $g(x) = \log_a x$, $a > 1$, has domain $(0, \infty)$ and range $(-\infty, \infty)$. The function increases on its entire domain and is continuous there. The y-axis is a vertical asymptote as x approaches 0 from the right. The x-intercept is 1, and there is no y-intercept. If $0 < a < 1$, the graph of $g(x) = \log_a x$ has the same characteristics, except that $g(x)$ decreases on its entire domain.

Two properties are given in Section 5.4 that are used to solve exponential and logarithmic equations. The first of these was used to solve Type 1 exponential equations earlier by writing both sides as powers of the same base and then setting the exponents equal. The second is used to solve Type 2 exponential equations, which cannot easily be written as powers of the same base. The strategy is to take same base logarithms of both sides and then apply the power rule for logarithms to eliminate the variables from the exponents. Other rules for logarithms may be useful as well.

Many logarithmic equations may be solved by using the rules for logarithms to write the equation in a different form, or by writing a logarithmic equation as an exponential equation. The second property in Section 5.4 may also be used: if two logarithms are equal, their arguments are equal. Formulas involving logarithms can be solved for a particular variable by using these same techniques.

The exponential growth function $A(t) = A_0 e^{kt}$, $k > 0$, and the exponential decay function $A(t) = A_0 e^{-kt}$, $k > 0$, occur in a variey of useful applications. Several of these are discussed in Section 5.5.

Key Terms

SECTION 5.1
exponential function
exponential equation (Type 1)
e
compound interest
continuous compounding

SECTION 5.2
logarithm
common logarithm
natural logarithm
change-of-base rule

SECTION 5.3
logarithmic function

SECTION 5.4
exponential equation (Type 2)
logarithmic equation

SECTION 5.5
exponential growth function
exponential decay function
half-life
doubling time
amortization

Chapter 5   **REVIEW EXERCISES**

Match each equation with the graph that most closely resembles its graph. Assume that $a > 1$.

1. $y = a^{x+2}$

2. $y = a^x + 2$

3. $y = -a^x + 2$

4. $y = a^{-x} + 2$

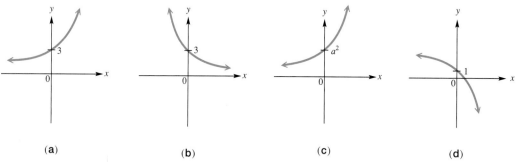

(a)   (b)   (c)   (d)

Consider the exponential function $y = f(x) = a^x$ graphed here. Answer the following based on the graph.

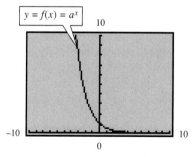

5. What is true about the value of a in comparison to 1?

6. What is the domain of f?

7. What is the range of f?

8. What is the value of $f(0)$?

9. Sketch the graph of $y = f^{-1}(x)$ by hand.

10. What is the expression that defines $f^{-1}(x)$?

Consider the function $f(x) = -2^{x-1} + 8$.

11. Graph it in the standard viewing window of your calculator.

12. Use the terminology of Chapter 2 to explain how the graph of f can be obtained from the graph of $y = 2^x$.

13. **(a)** What is the domain of f?
    (b) What is the range of f?

14. Does this graph have an asymptote? If so, is it vertical or horizontal, and what is its equation?

15. Find the x- and y-intercepts analytically, and use the graph from Exercise 11 to support your answers graphically.

16. The graphs of $y = x^2$ and $y = 2^x$ have the points $(2, 4)$ and $(4, 16)$ in common. There is a third point in common to the graphs whose coordinates can be approximated by using a graphics calculator. Find the coordinates, giving as many decimal places as your calculator will display.

Consider the equation $3^{x+4} = 27^{x+1}$.

17. Solve the equation analytically.

18. Let $y_1 = 3^{x+4}$ and $y_2 = 27^{x+1}$. Graph y_1 and y_2 on a graphics calculator, and find the coordinates of the point of intersection. (Use the window $[0, 10]$ by $[0, 300]$.) Explain how this supports your answer in Exercise 17.

19. Use the same functions for y_1 and y_2 as in Exercise 18. Graph $y_3 = y_1 - y_2$, and explain how the graph of y_3 supports your answer in Exercise 17.

20. (a) Solve the equation $(\frac{1}{8})^{2x-3} = 16^{x+1}$ analytically.
(b) Use a graph and the result of part (a) to solve the inequality $(\frac{1}{8})^{2x-3} > 16^{x+1}$.
(c) Use a graph and the result of part (a) to solve the inequality $(\frac{1}{8})^{2x-3} < 16^{x+1}$.

21. One of your friends is taking another mathematics course and tells you "I have no idea what an expression like $\log_5 27$ really means." Write a clear, coherent explanation of what it means, and how you can find an approximation for it using a calculator.

Use a calculator to find an approximation for each logarithm. Give the maximum number of digits possible on your calculator.

22. $\log 58.3$ **23.** $\log .00233$ **24.** $\ln 58.3$ **25.** $\ln .00233$

Evaluate each of the following, giving exact or approximate values as directed. In the case of approximations, give as many decimal places as your calculator shows.

26. $\log_{13} 1$ (exact) **27.** $\ln e^{\sqrt{6}}$ (exact) **28.** $\log_5 5^{12}$ (exact)

29. $7^{\log_7 13}$ (exact) **30.** $\log_4 9$ (approximate) **31.** x, if $3^x = 5$ (approximate)

32. Use the power, quotient, and product properties of logarithms to write the following expression as an equivalent expression.

$$\log \frac{m^3 \, n}{\sqrt{y}}$$

Solve for the indicated variable.

33. $3 = \dfrac{5^x - 5^{-x}}{2}$ for x

34. $\dfrac{2^x + 1}{2^x - 1} = -6$ for x

35. $\log_a(x - 1) = 1 + \log_a(x + 1)$ for x

36. $N = a + b \ln \dfrac{c}{d}$ for c

Solve for x. Give the solution set (a) with an exact value, using common logarithms, and (b) with an approximation to the nearest thousandth.

37. $6^{2-x} = 2^{3x+1}$

38. $10^{x+3} = 6$

Consider the equation $\log_2 x + \log_2(x + 2) = 3$.

39. Solve the equation analytically.

40. What is the extraneous solution from Exercise 39?

41. To support the solution in Exercise 39, we may graph $y_1 = \log_2 x + \log_2(x + 2) - 3$ and find the x-intercept. Write an expression for y_1 using the change-of-base rule, with base 10.

42. Graph y_1 from Exercise 41 to support the solution from Exercise 39.

43. Use the graph in Exercise 42 to solve the inequality $\log_2 x + \log_2(x + 2) > 3$.

44. Use the graph in Exercise 42 to solve the inequality $\log_2 x + \log_2(x + 2) < 3$.

45. A population is increasing according to the growth law $y = 2e^{.02t}$, where y is in millions and t is in years. Match each of the questions (a), (b), (c), and (d) with one of the solutions (A), (B), (C), or (D).

   (a) How long will it take for the population to triple?   (A) Evaluate $2e^{.02(1/3)}$.
   (b) When will the population reach 3 million?   (B) Solve $2e^{.02t} = 3 \cdot 2$ for t.
   (c) How large will the population be in 3 years?   (C) Evaluate $2e^{.02(3)}$.
   (d) How large will the population be in 4 months?   (D) Solve $2e^{.02t} = 3$ for t.

46. Suppose that $2000 is invested in an account that pays 3% annually and then is left untouched for 5 years.

   (a) How much will be in the account if interest is compounded quarterly (4 times per year)?
   (b) How much will be in the account if interest is compounded continuously?
   (c) To the nearest tenth of a year, how long will it take the $2000 to triple if interest is compounded continuously?

47. Suppose the gross national product (GNP) of a small country (in millions of dollars) is approximated by $G(t) = 15 + 2 \log t$ where t is time in years, for $1 \le t \le 6$. Find the GNP at the following times.

   (a) 1 year   (b) 2 years   (c) 5 years

48. The concentration of pollutants, in grams per liter, in the east fork of the Big Weasel River is approximated by $P(x) = .04e^{-4x}$, where x is the number of miles downstream from a paper mill that the measurement is taken.

   Find each of the following: (a) $P(.5)$ (b) $P(1)$
   (c) the concentration of pollutants 2 miles downstream
   (d) the number of miles downstream where the concentration of pollutants is .002 gram per liter.

49. A person learning certain skills involving repetition tends to learn quickly at first. Then learning tapers off and approaches some upper limit. Suppose the number of symbols per minute a textbook typesetter can produce is given by $p(t) = 250 - 120(2.8)^{-.5t}$, where t is the number of months the typesetter has been in training. Find each of the following: (a) $p(2)$ (b) $p(4)$ (c) $p(10)$.   (d) Graph $y = p(t)$ in the window $[0, 10]$ by $[0, 300]$ and support the answer of part (a).

Newton's law of cooling says that the rate at which a body cools is proportional to the difference in temperature between the body and the environment into which it is introduced. The temperature $f(t)$ of the body at time t in appropriate units after being introduced into an environment having constant temperature T_0 is $f(t) = T_0 + Ce^{-kt}$, where C and k are constants. Use this result in Exercises 50 and 51.

50. Boiling water, at 100°C, is placed in a freezer at 0°C. The temperature of the water is 50°C after 24 minutes. Find the temperature of the water after 96 minutes.

51. A piece of metal is heated to 300°C and then placed in a cooling liquid at 50°C. After 4 minutes the metal has cooled to 175°C. Find its temperature after 12 minutes.

52. A skydiver in free-fall travels at the speed of $f(t) = 176(1 - e^{-.18t})$ feet per second after t seconds. How long will it take for the skydiver to attain the speed of 147 feet per second?

The Conic Sections and Systems of Equations and Inequalities

6.1 CIRCLES AND PARABOLAS

Introduction to the Conic Sections ▮ Equations and Graphs of Circles ▮ An Application of Circles ▮ Equations and Graphs of Parabolas

In the first two sections of this chapter we will expand upon some ideas introduced earlier in the text. In Chapter 3 we saw that the graph of a quadratic function is a parabola, and in Section 4.4 we saw how horizontal parabolas can be graphed by using the union of root functions. In Section 4.4 we also saw that the graph of $x^2 + y^2 = r^2$ is a circle with center at the origin and radius r. By treating a circle as the union of two root functions, we can use a graphics calculator to graph the circle. We will now examine parabolas and circles in terms of their actual definitions, based on the distance formula (Section 1.4). Along with ellipses and hyperbolas, they form a group of curves known as the **conic sections**.

Introduction to the Conic Sections

By intersecting a cone with a plane, we may obtain curves called conic sections. Figure 1 on the next page illustrates these curves.

These curves can be defined mathematically by using the distance formula: the distance between the points $A(x_1, y_1)$ and $B(x_2, y_2)$, symbolized $d(A, B)$, is given by the expression $\sqrt{(x_2 - x_1)^2 + (y_2 - y_1)^2}$.

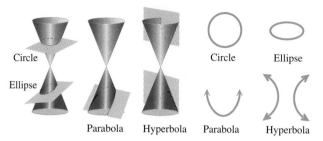

Circle

Ellipse

Parabola   Hyperbola

Circle   Ellipse

Parabola

Hyperbola

The Conic Sections
FIGURE 1

Equations and Graphs of Circles

We begin with the definition of a circle.

DEFINITION OF A CIRCLE

A **circle** is a set of points in a plane, each of which is equidistant from a fixed point. The distance is called the **radius** of the circle, and the fixed point is called the **center**.

Suppose that a circle has center (h, k) and radius $r > 0$, as shown in Figure 2. Then by the distance formula, if (x, y) is any point on the circle,

$$\sqrt{(x - h)^2 + (y - k)^2} = r.$$

Squaring both sides of this equation gives us the center-radius form of the equation of the circle.

CENTER-RADIUS FORM OF THE EQUATION OF A CIRCLE

The circle with center (h, k) and radius r has equation

$$(x - h)^2 + (y - k)^2 = r^2,$$

the **center-radius form** of the equation of a circle.

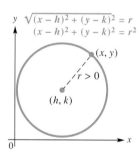

FIGURE 2

Notice that a circle is an example of a mathematical relation, but is not the graph of a function, as it does not pass the vertical line test (Section 1.2).

EXAMPLE **1**

Finding the Equation of a Circle

Find the center-radius form of the equation of a circle with radius 6 and center at $(-3, 4)$. Graph the circle by hand, and give the domain and the range of the relation.

SOLUTION Using the center-radius form with $h = -3$, $k = 4$, and $r = 6$, we find that the equation of the circle is

$$(x - (-3))^2 + (y - 4)^2 = 6^2$$
$$\text{or} \qquad (x + 3)^2 + (y - 4)^2 = 36.$$

Its graph is shown in Figure 3. As seen there, the domain is $[-9, 3]$ and the range is $[-2, 10]$. (We will see later how this circle can be graphed using a graphics calculator.) ◻

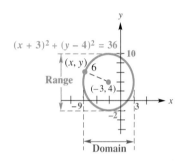

FIGURE 3

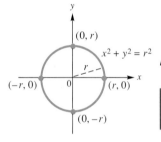

FIGURE 4

If a circle has center at the origin $(0, 0)$, then its equation is found by using $h = 0$ and $k = 0$ in the center-radius form.

EQUATION OF A CIRCLE WITH CENTER AT THE ORIGIN

A circle with center $(0, 0)$ and radius r has equation

$$x^2 + y^2 = r^2.$$

Figure 4 shows the graph of a circle with center at the origin and radius r.

EXAMPLE **2**

Finding the Equation of a Circle with Center at the Origin

Find the equation of a circle with center at the origin and radius 3. Give a traditional graph, and state the domain and the range of the relation.

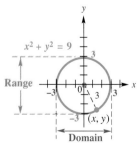

FIGURE 5

TECHNOLOGICAL NOTE When studying the graphs of circles in conjunction with graphics calculators, using a square screen is essential; otherwise, circles may appear to be ellipses. While some early models of graphics calculators did not allow for square screens, virtually all current models do.

SOLUTION Using the form $x^2 + y^2 = r^2$ with $r = 3$, we find that the equation of the circle is

$$x^2 + y^2 = 9. \qquad (r^2 = 3^2 = 9)$$

The graph is shown in Figure 5 on the previous page. Both the domain and the range are $[-3, 3]$. ❚

A graphics calculator in the function mode cannot directly graph a circle. In order to do so, we must first solve the equation of the circle for y, obtaining two functions y_1 and y_2. The union of these two graphs will be the graph of the entire circle.

NOTE In order to obtain an undistorted graph on a graphics calculator screen, a *square* screen must be used. See your instruction manual if necessary.

EXAMPLE 3

Graphing a Circle Using a Graphics Calculator

For each of the following circles, solve the equation for y and then use a graphics calculator to graph the circle in a square viewing window:

(a) $x^2 + y^2 = 9$

(b) $(x + 3)^2 + (y - 4)^2 = 36.$

SOLUTION In both cases, we must solve for y. Recall that if $k > 0$, $y^2 = k$ has two real solutions, $\sqrt{k}$ and $-\sqrt{k}$.

(a) $x^2 + y^2 = 9$

$$y^2 = 9 - x^2 \qquad \text{Subtract } x^2.$$
$$y = \pm\sqrt{9 - x^2} \qquad \text{Take square roots.}$$

TECHNOLOGICAL NOTE When entering expressions like those required for graphing functions like the two that produced the graphs in Figures 6(a) and 6(b), you must be careful to include the expressions under the radicals within parentheses. For example the function y_1 in Example 3(b) must be entered as $4 + \sqrt{(36 - (x + 3)^2)}$.

We graph two functions, $y_1 = \sqrt{9 - x^2}$ and $y_2 = -\sqrt{9 - x^2}$. See Figure 6(a), and compare to the traditional graph in Figure 5.

(b) $(x + 3)^2 + (y - 4)^2 = 36$

$$(y - 4)^2 = 36 - (x + 3)^2 \qquad \text{Subtract } (x + 3)^2.$$
$$y - 4 = \pm\sqrt{36 - (x + 3)^2} \qquad \text{Take square roots.}$$
$$y = 4 \pm \sqrt{36 - (x + 3)^2} \qquad \text{Add 4.}$$

Here, the two functions to be graphed are

$$y_1 = 4 + \sqrt{36 - (x + 3)^2} \quad \text{and} \quad y_2 = 4 - \sqrt{36 - (x + 3)^2}.$$

See Figure 6(b), and compare to Figure 3. ❚

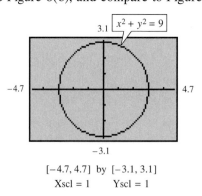

$[-4.7, 4.7]$ by $[-3.1, 3.1]$
Xscl = 1   Yscl = 1

(a)

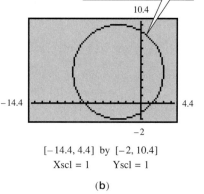

$[-14.4, 4.4]$ by $[-2, 10.4]$
Xscl = 1   Yscl = 1

(b)

FIGURE 6

FOR **GROUP DISCUSSION**

1. Notice in Example 3(b) the final expressions for y_1 and y_2 have radicands* $36 - (x + 3)^2$. If this were simplified analytically, the result would be $27 - x^2 - 6x$. Use your calculators to replace the given radicands with this simplified result. Do you get the same graphs? Does it matter to the calculator which form you use? What would be a possible drawback of attempting to use the simplified form. (*Hint*: Nobody's perfect.)

2. Suppose the functions y_1 and y_2 from Example 3(a) were graphed in a standard window. What might be a possible misinterpretation by a student who has not studied the mathematical theory along with the technological approach?

Starting with the center-radius form of the equation of a circle, $(x - h)^2 + (y - k)^2 = r^2$, and squaring $x - h$ and $y - k$ gives an equation of the form

$$x^2 + y^2 + cx + dy + e = 0, \qquad \text{[*]}$$

where c, d, and e are real numbers. This result is the **general form of the equation of a circle.** Also, starting with an equation in the form of (*), the process of completing the square can be used to get an equation of the form

$$(x - h)^2 + (y - k)^2 = m$$

for some number m. If $m > 0$, then $r^2 = m$, and the equation represents a circle with radius $\sqrt{m}$. If $m = 0$, then the equation represents the single point (h, k). If $m < 0$, no points satisfy the equation.

EXAMPLE 4

Finding the Center and Radius by Completing the Square

Find the center and the radius of the circle with equation

$$x^2 - 6x + y^2 + 4y - 3 = 0.$$

Then graph the circle using a graphics calculator.

SOLUTION Our goal is to obtain an equivalent equation of the form $(x - h)^2 + (y - k)^2 = r^2$. To do this, first write the equation with the constant on the right.

$$x^2 - 6x + y^2 + 4y = 3$$

Now we complete the square in both x and y. To complete the square in x, we add $[\frac{1}{2}(-6)]^2 = 9$ to both sides, and to complete the square in y, we add $[\frac{1}{2}(4)]^2 = 4$ to both sides. Insert parentheses as shown.

$$(x^2 - 6x + 9) + (y^2 + 4y + 4) = 3 + 9 + 4$$

Now factor on the left and add on the right.

$$(x - 3)^2 + (y + 2)^2 = 16 \qquad (*)$$
$$(x - 3)^2 + (y - (-2))^2 = 4^2 \qquad \text{Write } +2 \text{ as } -(-2) \text{ and } 16 \text{ as } 4^2.$$

---

* The **radicand** is the expression under the radical symbol.

The circle has its center at $(3, -2)$ and its radius is 4. A traditional graph is shown in Figure 7(a). To graph it using a graphics calculator, use the equation in the line marked (*) and solve for y to get $y = -2 \pm \sqrt{16 - (x - 3)^2}$. Let y_1 and y_2 define these two expressions (one with $+$ and the other with $-$) to obtain the calculator-generated graph shown in Figure 7(b).   ▯

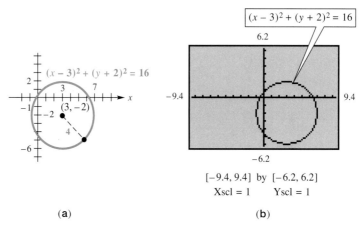

$[-9.4, 9.4]$ by $[-6.2, 6.2]$
Xscl = 1   Yscl = 1

(a)   (b)

FIGURE 7

An Application of Circles

Seismologists can locate the epicenter of an earthquake by determining the intersection of three circles. The radii of these circles represent the distances from the epicenter to each of three receiving stations. The centers of the circles represent the receiving stations.

EXAMPLE 5

Using Circles to
Locate the
Epicenter of an
Earthquake

Suppose that an earthquake is recorded by three receiving stations A, B, and C, located on a coordinate plane 2, 5, and 4 units respectively from the epicenter. Use Figure 8 to determine the location of the epicenter with respect to the coordinate plane.

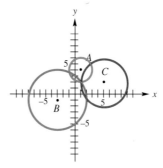

FIGURE 8

SOLUTION   Graphically, it appears that the epicenter is located at $(1, 2)$. To check this algebraically, determine the equation for each circle and substitute $x = 1$ and $y = 2$.

Station A: Station B:

$(x - 1)^2 + (y - 4)^2 = 4$? $(x + 3)^2 + (y + 1)^2 = 25$?

$(1 - 1)^2 + (2 - 4)^2 = 4$? $(1 + 3)^2 + (2 + 1)^2 = 25$?

$0 + 4 = 4$? $16 + 9 = 25$?

$4 = 4$ True $25 = 25$ True

Station C:

$(x - 5)^2 + (y - 2)^2 = 16$?

$(1 - 5)^2 + (2 - 2)^2 = 16$?

$16 + 0 = 16$?

$16 = 16$ True

Thus, we can be sure that the epicenter lies at $(1, 2)$.   ∎

FOR **GROUP DISCUSSION**

Have a volunteer from the class draw a circle on a blackboard using only a piece of string and a piece of chalk. Discuss how compasses allow us to also draw perfect circles. How do these relate to the definition of a circle given earlier?

Equations and Graphs of Parabolas

The definition of parabola is also based upon distance.

DEFINITION OF A PARABOLA

A **parabola** is the set of points in a plane equidistant from a fixed point and a fixed line. The fixed point is called the **focus,** and the fixed line, the **directrix,** of the parabola.

An equation of a parabola can be found from the definition as follows. Let the directrix be the line $y = -c$ and the focus be the point F with coordinates $(0, c)$, as shown in Figure 9. To get the equation of the set of points that are the same distance from the line $y = -c$ and the point $(0, c)$, choose one such point P and give it coordinates (x, y). Then, since $d(P, F)$ and $d(P, D)$ must have the same length, using the distance formula gives

$$d(P, F) = d(P, D)$$
$$\sqrt{(x - 0)^2 + (y - c)^2} = \sqrt{(x - x)^2 + (y - (-c))^2}$$
$$\sqrt{x^2 + (y - c)^2} = \sqrt{(y + c)^2}$$
$$x^2 + y^2 - 2yc + c^2 = y^2 + 2yc + c^2$$
$$x^2 = 4cy$$
$$y = \frac{1}{4c}x^2.$$

This discussion is summarized as follows.

PARABOLA WITH A VERTICAL AXIS

The parabola with focus at $(0, c)$ and directrix $y = -c$ has equation

$$y = \frac{1}{4c}x^2.$$

The parabola has a vertical axis, opens upward if $c > 0$, and opens downward if $c < 0$.

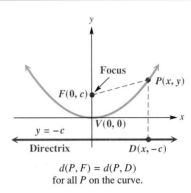

$d(P, F) = d(P, D)$
for all P on the curve.

FIGURE 9

If the directrix is the line $x = -c$, and the focus is at $(c, 0)$, using the definition of a parabola and the distance formula leads to the equation of a parabola with a horizontal axis. (See Exercise 99.)

PARABOLA WITH A HORIZONTAL AXIS

The parabola with focus at $(c, 0)$ and directrix $x = -c$ has equation

$$x = \frac{1}{4c}y^2.$$

The parabola opens to the right if $c > 0$, to the left if $c < 0$, and has a horizontal axis.

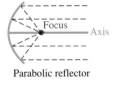

Parabolic reflector

FIGURE 10

The geometric properties of parabolas lead to many practical applications. For example, if a light source is placed at the focus of a parabolic reflector, as in Figure 10, light rays reflect parallel to the axis, making a spotlight or flashlight. The process also works in reverse. Light rays from a distant source come in parallel to the axis and are reflected to a point at the focus. (If such a reflector is aimed at the sun, a temperature of several thousand degrees may be obtained.) This use of parabolic reflection is seen in the satellite dishes used to pick up signals from communications satellites.

NOTE A parabola with a horizontal axis is not the graph of a function. However, since the equation $x = \frac{1}{4c}y^2$ is equivalent to

$$y_1 = 2\sqrt{cx} \quad \text{or} \quad y_2 = -2\sqrt{cx},$$

such a parabola can be graphed with a graphics calculator, using the same general procedure discussed for circles in Example 3(a). If $c > 0$, then $x \geq 0$, and if $c < 0$, then $x \leq 0$.

EXAMPLE 6

Determining Information About a Parabola From Its Equation

Find the focus, directrix, vertex, and axis of each of the following. Then use a graphics calculator to graph the parabola, comparing your result to the traditional graph shown.

(a) $y = \dfrac{1}{8}x^2$

SOLUTION The equation indicates that this is a vertical parabola. Because $4c = 8$, $c = 2$. Therefore, the focus is at $(0, 2)$, and the directrix has the equation $y = -2$. The vertex is at $(0, 0)$, and the axis is the y-axis. To graph this parabola (which defines a function), we simply enter y_1 as $\frac{1}{8}x^2$. Figure 11 shows both a traditional graph and a calculator-generated graph of this function.

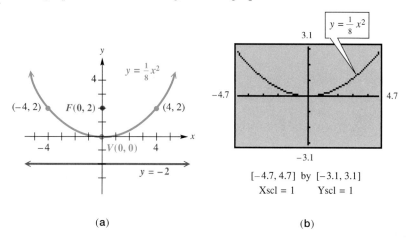

(a)   (b)

FIGURE 11

(b) $x = -\dfrac{1}{28}y^2$

SOLUTION Here we must think of the negative sign as being in the denominator to solve for c. Since $4c = -28$, $c = -7$. The parabola is horizontal (and is thus not the graph of a function), with focus $(-7, 0)$, directrix $x = 7$, vertex $(0, 0)$, and the x-axis is the axis of the parabola. Because c is negative, the graph opens to the left, as shown in Figure 12(a).

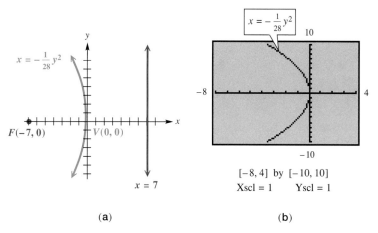

(a)   (b)

FIGURE 12

To graph this parabola with a graphics calculator, we must first transform it analytically into the union of two functions.

$$x = -\frac{1}{28}y^2$$

$$-28x = y^2$$

$$y = \pm\sqrt{-28x}$$

$$y_1 = \sqrt{-28x} \quad \text{or} \quad y_2 = -\sqrt{-28x}$$

Note from the equations for y_1 and y_2 that x must be nonpositive for y to represent a real number. See Figure 12(b) on the previous page for the calculator-generated graph. ◻

EXAMPLE 7

Determining the Equation of a Parabola from Given Information

Write an equation for each parabola described.

(a) focus $\left(\frac{2}{3}, 0\right)$ and vertex at the origin

SOLUTION Since the focus $\left(\frac{2}{3}, 0\right)$ is on the x-axis, the parabola is horizontal and opens to the right because $c = \frac{2}{3}$ is positive. The equation is of the form $x = \frac{1}{4c}y^2$, so

$$x = \frac{1}{4\left(\frac{2}{3}\right)}y^2$$

$$x = \frac{1}{\left(\frac{8}{3}\right)}y^2$$

$$x = \frac{3}{8}y^2$$

is the equation of the parabola. It is graphed in Figure 13.

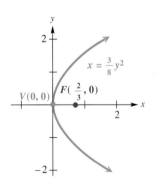

FIGURE 13

(b) vertical axis, vertex at the origin, through the point $(-2, 12)$.

SOLUTION The parabola will have an equation of the form $y = \frac{1}{4c}x^2$ because the axis is vertical. Since the point $(-2, 12)$ is on the graph, it must satisfy the equation. We substitute -2 for x and 12 for y to find c.

$$12 = \frac{1}{4c}(-2)^2$$

$$12 = \frac{1}{4c}(4)$$

$$\frac{1}{c} = 12$$

$$c = \frac{1}{12}$$

The equation of the parabola is $y = \frac{1}{4\left(\frac{1}{12}\right)}x^2$ or $y = 3x^2$. █

FOR **GROUP DISCUSSION**

The graph of the parabola described in Example 7(a) is shown in Figure 13 using traditional graphing methods. How would you go about graphing this parabola with a graphics calculator?

In Chapter 3 we showed that the graph of the quadratic function defined by the equation $y = a(x - h)^2 + k$ is a parabola with vertex at (h, k) and the line $x = h$ as its axis. The relation defined by $x = a(y - k)^2 + h$ also has a parabola as its graph, but since x and y are interchanged, the graph of this new relation is symmetric to the graph of $y = a(x - h)^2 + k$ with respect to the line $y = x$. This reflection changes the vertical axis to a horizontal axis; however, the vertex is still at (h, k).

TRANSLATION OF A HORIZONTAL PARABOLA

The parabola with vertex at (h, k) and the horizontal line $y = k$ as axis has an equation of the form

$$x = a(y - k)^2 + h.$$

The parabola opens to the right if $a > 0$ and to the left if $a < 0$.

Figure 14 shows the graph of $x = 2\left(y + \frac{3}{2}\right)^2 + \frac{1}{2}$. It is a horizontal parabola opening to the right, with vertex at $\left(\frac{1}{2}, -\frac{3}{2}\right)$. If you refer to Example 8 of Section 4.3, you will see that we obtained a calculator-generated graph of this same relation. We started with the equation $x = 2y^2 + 6y + 5$ in that example. Notice that this latter equation is equivalent to the one given above.

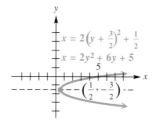

FIGURE 14

6.1 EXERCISES

Match each equation with its calculator-generated graph in Exercises 1–10. Do this first without actually using your calculator. Then check your answer by generating a calculator graph of your own. (Every window has Xscl = Yscl = 1.)

1. $y = x^2$

2. $x = y^2$

3. $x = 2(y + 3)^2 - 4$

4. $y = 2(x + 3)^2 - 4$

5. $y = -\dfrac{1}{3}x^2$

6. $x = -\dfrac{1}{3}y^2$

7. $x^2 + y^2 = 25$

8. $(x - 3)^2 + (y + 4)^2 = 25$

9. $(x + 3)^2 + (y - 4)^2 = 25$

10. $x^2 + y^2 = -4$

A.

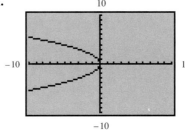

B.

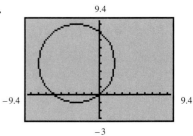

C.

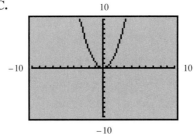

D.

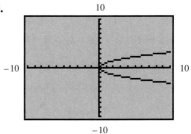

E.

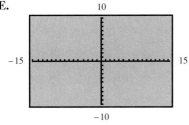

F.

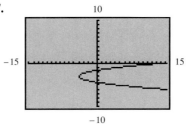

G.

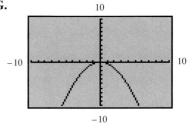

H.

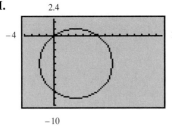

I.

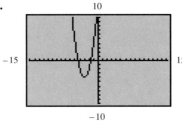

J.

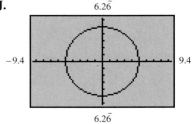

Find the center-radius form for each circle satisfying the given conditions.

11. Center $(1, 4)$, radius 3

12. Center $(-2, 5)$, radius 4

13. Center $(0, 0)$, radius 1

14. Center $(0, 0)$, radius 5

15. Center $(\frac{2}{3}, -\frac{4}{5})$, radius $\frac{3}{7}$

16. Center $(-\frac{1}{2}, -\frac{1}{4})$, radius $\frac{12}{5}$

17. Center $(-1, 2)$, passing through $(2, 6)$

18. Center $(2, -7)$, passing through $(-2, -4)$

19. Center $(-3, -2)$, tangent to the *x*-axis (*Hint: Tangent to* means touching at one point.)

20. Center $(5, -1)$, tangent to the *y*-axis

R*elating Concepts*

The figure shows a circle and a diameter of the circle. The endpoints of the diameter are $(-1, 3)$ and $(5, -9)$.

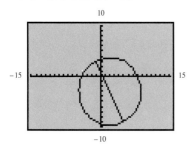

21. Find the coordinates of the midpoint of the circle. (See Section 1.4.)

22. Find the radius of the circle. (See Section 1.4.)

23. Find the center-radius form of the equation of the circle.

24. Read the owner's manual of your calculator to see if your model has a DRAW capability. If it does, use it to duplicate the figure shown for Exercises 21–23.

25. Find the center-radius form of the circle with endpoints of a diameter having coordinates $(3, -5)$ and $(-7, 3)$.

26. Suppose that a circle is tangent to both axes, is completely within the third quadrant, and has a radius of $\sqrt{2}$. Find the center-radius form of the equation of the circle.

Graph each of the following circles by hand (that is, without using a graphics calculator). Give the domain and the range.

27. $x^2 + y^2 = 36$

28. $(x - 2)^2 + y^2 = 36$

29. $(x + 2)^2 + (y - 5)^2 = 16$

30. $(x - 5)^2 + (y + 4)^2 = 49$

Graph each of the following circles using a graphics calculator. Follow the procedure explained in Example 3. Use a square viewing window.

31. $x^2 + y^2 = 81$

32. $x^2 + (y + 3)^2 = 49$

33. $(x - 4)^2 + (y - 3)^2 = 25$

34. $(x + 3)^2 + (y + 2)^2 = 36$

35. Describe the graph of the equation $(x - 3)^2 + (y - 3)^2 = 0$.

36. Describe the graph of the equation $(x - 3)^2 + (y - 3)^2 = -1$.

Find the center and the radius of each of the following circles.

37. $x^2 + 6x + y^2 + 8y = -9$

38. $x^2 - 4x + y^2 + 12y + 4 = 0$

39. $x^2 - 12x + y^2 + 10y + 25 = 0$

40. $x^2 + 8x + y^2 - 6y = -16$

41. $x^2 + 8x + y^2 - 14y + 64 = 0$

42. $x^2 - 8x + y^2 + 7 = 0$

43. $x^2 + y^2 = 2y + 48$

44. $x^2 + 4x + y^2 = 21$

Refer to Example 5 to solve the problems in Exercises 45 and 46.

45. Show analytically that if three receiving stations at $(1, 4)$, $(-6, 0)$, and $(5, -2)$ record distances to an earthquake epicenter of 4 units, 5 units, and 10 units, respectively, that the epicenter would lie at $(-3, 4)$.

46. Three receiving stations record the presence of an earthquake. The location of the receiving center and the distance to the epicenter are contained in the following three equations: $(x - 2)^2 + (y - 1)^2 = 25$, $(x + 2)^2 + (y - 2)^2 = 16$ and $(x - 1)^2 + (y + 2)^2 = 9$. Determine the location of the earthquake epicenter.

Each equation in Exercises 47–54 defines a parabola. Without actually graphing, match the description given in the column on the right with the equation.

47. $y = (x - 4)^2 - 2$ **A.** vertex $(2, -4)$, opens down

48. $y = (x - 2)^2 - 4$ **B.** vertex $(2, -4)$, opens up

49. $y = -(x - 4)^2 - 2$ **C.** vertex $(4, -2)$, opens down

50. $y = -(x - 2)^2 - 4$ **D.** vertex $(4, -2)$, opens up

51. $x = (y - 4)^2 - 2$ **E.** vertex $(-2, 4)$, opens left

52. $x = (y - 2)^2 - 4$ **F.** vertex $(-2, 4)$, opens right

53. $x = -(y - 4)^2 - 2$ **G.** vertex $(-4, 2)$, opens left

54. $x = -(y - 2)^2 - 4$ **H.** vertex $(-4, 2)$, opens right

55. For the graph of $y = a(x - h)^2 + k$, in what quadrant is the vertex if:
 (a) $h < 0, k < 0$;   **(b)** $h < 0, k > 0$;   **(c)** $h > 0, k < 0$;   **(d)** $h > 0, k > 0$?

56. Repeat parts (a)–(d) of Exercise 55 for the graph of $x = a(y - k)^2 + h$.

Give the coordinates of the focus, the equation of the directrix, and the axis of each of the following parabolas.

57. $y = \frac{1}{16}x^2$ **58.** $y = \frac{1}{4}x^2$ **59.** $y = -2x^2$ **60.** $y = 9x^2$

61. $x = 16y^2$ **62.** $x = -32y^2$ **63.** $x = -\frac{1}{16}y^2$ **64.** $x = -\frac{1}{4}y^2$

Write an equation for each of the following parabolas with vertex at the origin.

65. focus $(0, -2)$ **66.** focus $(5, 0)$ **67.** focus $\left(-\frac{1}{2}, 0\right)$ **68.** focus $\left(0, \frac{1}{4}\right)$

69. through $(2, -2\sqrt{2})$, opening to the right   **70.** through $(\sqrt{3}, 3)$, opening upward

71. through $(\sqrt{10}, -5)$, opening downward   **72.** through $(-3, 3)$, opening to the left

73. through $(2, -4)$, symmetric with respect to the y-axis

74. through $(3, 2)$, symmetric with respect to the x-axis

Graph each of the following parabolas either by hand or using a graphics calculator. Give the coordinates of the vertex, the axis, the domain, and the range. (Hint: Each of these is a parabola with a vertical axis and is thus a function. Therefore, each may be graphed directly with a graphics calculator in the function mode.)

75. $y = (x - 2)^2$ **76.** $y = (x + 4)^2$ **77.** $y = (x + 3)^2 - 4$

78. $y = (x - 5)^2 - 4$ **79.** $y = -2(x + 3)^2 + 2$ **80.** $y = -3(x - 2)^2 + 1$

81. $y = -\frac{1}{2}(x + 1)^2 - 3$ **82.** $y = \frac{2}{3}(x - 2)^2 - 1$ **83.** $y = x^2 - 2x + 3$

84. $y = x^2 + 6x + 5$ 　　　　　　**85.** $y = 2x^2 - 4x + 5$ 　　　　　　**86.** $y = -3x^2 + 24x - 46$

Graph each of the following parabolas either by hand or using a graphics calculator. Give the coordinates of the vertex, the axis, the domain, and the range. (Hint: Each of these is a parabola with a horizontal axis and is thus not a function. To graph with a graphics calculator in function mode, you must first rewrite the equation in terms of two functions, y_1 and y_2. See the explanation at the end of this section, and refer to Example 8 of Section 4.3.)

87. $x = y^2 + 2$ 　　　　　　**88.** $x = -y^2$ 　　　　　　**89.** $x = (y + 1)^2$ 　　　　　　**90.** $x = (y - 3)^2$

91. $x = (y + 2)^2 - 1$ 　　　　**92.** $x = (y - 4)^2 + 2$ 　　　　**93.** $x = -2(y + 3)^2$ 　　　　**94.** $x = \dfrac{2}{3}(y - 3)^2 + 2$

95. $x = y^2 + 2y - 8$ 　　　　**96.** $x = -4y^2 - 4y - 3$

97. The cable in the center portion of a bridge is supported as shown in the figure to form a parabola. The center support is 10 ft high, the tallest supports are 210 ft high, and the distance between the two tallest supports is 400 ft. Find the height of the remaining supports, if the supports are evenly spaced. (Ignore the width of the supports.)

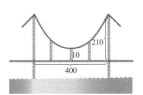

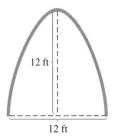

12 ft

12 ft

(Figure for Exercise 98)

99. Prove that the parabola with focus $(c, 0)$ and directrix $x = -c$ has the equation $x = \frac{1}{4c}y^2$.

100. Use the definition of a parabola to find an equation of the parabola with vertex at (h, k) and a vertical axis. Let the distance from the vertex to the focus and the distance from the vertex to the directrix be c, where $c > 0$.

98. An arch in the shape of a parabola has the dimensions shown in the figure. How wide is the arch 9 ft up?

Further Explorations

The domain of the graph of each half of a circle has restrictions. With our study of root functions, we have learned that the value under the radical sign cannot be negative. Now, with the circle, we have a squared variable under the radical. For example, to graph the circle

$$x^2 + y^2 = 9$$

we must solve for y to get

$$y_1 = \sqrt{9 - x^2} \quad \text{and} \quad y_2 = -\sqrt{9 - x^2}.$$

We can use the TABLE to help us determine the domain of these two functions. The first two figures show two views of the TABLE. You can scroll through the TABLE by using the

X	Y₁	Y₂
-6	ERROR	ERROR
-5	ERROR	ERROR
-4	ERROR	ERROR
-3	0	0
-2	2.2361	-2.236
-1	2.8284	-2.828
0	3	-3
X= -6		

X	Y₁	Y₂
0	3	-3
1	2.8284	-2.828
2	2.2361	-2.236
3	0	0
4	ERROR	ERROR
5	ERROR	ERROR
6	ERROR	ERROR
X=6		

(a) 　　　　　　　　　　(b)

up arrow and the down arrow. Notice that the TABLE shows ERROR for Y_1 and Y_2 when $x < -3$ and when $x > 3$. Y_1 and Y_2 are evaluated only when $-3 \le x \le 3$. We can now say the domain of both Y_1 and Y_2 is the closed interval $[-3, 3]$. This result can be supported graphically in a friendly window that evaluates the endpoints of each semicircle (a decimal window will usually work).

We can also use the TABLE to determine where the two semicircles intersect. The TABLES in the first figure show that $Y_1 = 0$ and $Y_2 = 0$ when $x = -3$ and when $x = 3$. We should expect the graph to show the semicircles intersecting at $(-3, 0)$ and $(3, 0)$. Again, using a decimal window is important here, or the graph may not be evaluated at $x = -3$ and $x = 3$. In this case, a gap will appear in the display window between the semicircles. (See figure (b) below.)

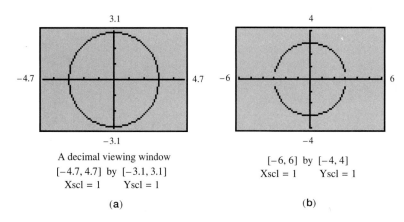

A decimal viewing window
$[-4.7, 4.7]$ by $[-3.1, 3.1]$
Xscl $= 1$ Yscl $= 1$

(a)

$[-6, 6]$ by $[-4, 4]$
Xscl $= 1$ Yscl $= 1$

(b)

CAUTION If the center of a circle has rational coordinates, and the radius is irrational, the graphics calculator will be unable to find the intersection of the semicircles in the TABLE and will also be unable to connect the semicircles in the graph. This is because the graphics calculator can only use decimal approximations for irrationals (very accurate approximations, but still approximations). As we have seen before, using a graphics calculator will not help if we do not understand the basic concepts.

Use the TABLE of your graphics calculator to find the point of intersection of the two semicircles, and the domain of each circle. Support your answers graphically in a DECIMAL window.

1. $x^2 + y^2 = 36$ **2.** $(x + 4)^2 + y^2 = 81$

3. $(x - 5)^2 + (y - 2)^2 = 16$ **4.** $(x + 1)^2 + (y + 3)^2 = 25$

6.2 ELLIPSES AND HYPERBOLAS

Equations and Graphs of Ellipses ❙ An Application of Ellipses ❙ Equations and Graphs of Hyperbolas

Equations and Graphs of Ellipses

We have studied two types of second-degree relations thus far: parabolas and circles. We now look at another type, the *ellipse*. The definition of an ellipse is also based on distance.

> ### ELLIPSE
> An **ellipse** is the set of all points in a plane such that the sum of their distances from two fixed points is always the same (constant). The two fixed points are called the **foci** (plural of *focus*) of the ellipse.

For example, the ellipse in Figure 15 has foci at points F and F'. By the definition, the ellipse is made up of all points P such that the sum $d(P, F) + d(P, F')$ is constant. This ellipse has its **center** at the origin. Points V and V' are the **vertices** of the ellipse, and the line segment connecting V and V' is the **major axis.** The foci always lie on the major axis. The line segment from B to B' is the **minor axis.** The major axis has length $2a$, and the minor axis has length $2b$.

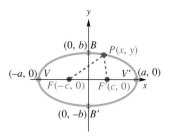

FIGURE 15

If the foci are chosen to be on the x-axis (or y-axis), with the center of the ellipse at the origin, then the distance formula and the definition of an ellipse can be used to obtain the following results. (See Exercise 64.)

> ### STANDARD FORMS OF EQUATIONS FOR ELLIPSES
> The ellipse with center at the origin and equation
>
> $$\frac{x^2}{a^2} + \frac{y^2}{b^2} = 1$$
>
> has vertices $(\pm a, 0)$, endpoints of the minor axis $(0, \pm b)$, and foci $(\pm c, 0)$, where $c^2 = a^2 - b^2$. The ellipse with center at the origin and equation
>
> $$\frac{y^2}{a^2} + \frac{x^2}{b^2} = 1$$
>
> has vertices $(0, \pm a)$, endpoints of the minor axis $(\pm b, 0)$, and foci $(0, \pm c)$, where $c^2 = a^2 - b^2$.

Do not be confused by the two standard forms—in one case a^2 is associated with x^2; in the other case a^2 is associated with y^2. However, when graphing an ellipse in a traditional manner, it is necessary only to find the intercepts of the graph—if the positive x-intercept is larger than the positive y-intercept, the major axis is horizontal, and otherwise it is vertical. When using the relationship $a^2 - b^2 = c^2$, choose a^2 and b^2 so that a^2 is larger than b^2.

In an equation of an ellipse, the coefficients of x^2 and y^2 must be different positive numbers. (What happens if the coefficients are equal?)

An ellipse is the graph of a relation. As suggested by the graph in Figure 15, if the ellipse has equation $\frac{x^2}{a^2} + \frac{y^2}{b^2} = 1$, the domain is $[-a, a]$ and the range is $[-b, b]$. Notice that the ellipse in Figure 15 is symmetric with respect to the x-axis, the y-axis, and the origin. More generally, every ellipse is symmetric with respect to its major axis, its minor axis, and its center.

<table>
<tr><td>**EXAMPLE 1**

Graphing an Ellipse Centered at the Origin</td></tr>
</table>

Transform the equation $4x^2 + 9y^2 = 36$ into the standard form for an ellipse. Graph the ellipse in a traditional manner by finding intercepts. Then solve for y and graph the ellipse using a graphics calculator.

SOLUTION   To get the form of the equation of an ellipse, divide both sides by 36.

$$\frac{x^2}{9} + \frac{y^2}{4} = 1$$

This ellipse is centered at the origin, wih x-intercepts 3 and -3, and y-intercepts 2 and -2. The domain of this relation is $[-3, 3]$, and the range is $[-2, 2]$. The graph is shown in Figure 16(a).

Solving the equation for y gives the two functions

$$y_1 = 2\sqrt{1 - \frac{x^2}{9}} \quad \text{and} \quad y_2 = -2\sqrt{1 - \frac{x^2}{9}}.$$

Graphing these in a square window gives the graph shown in Figure 16(b). Compare it to the traditional graph.   ∎

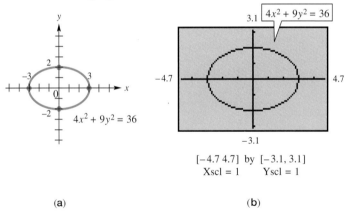

$$[-4.7\ 4.7] \ \text{by} \ [-3.1, 3.1]$$
$$\text{Xscl} = 1 \qquad \text{Yscl} = 1$$

(a)   (b)

FIGURE 16

<table>
<tr><td>**EXAMPLE 2**

Finding Foci of an Ellipse</td></tr>
</table>

Find the coordinates of the foci of the ellipse in Example 1.

SOLUTION   Since $9 > 4$, we can find the foci by letting $a^2 = 9$ and $b^2 = 4$ in the equation $c^2 = a^2 - b^2$. Then solve for c:

$$c^2 = a^2 - b^2$$
$$c^2 = 9 - 4$$
$$c^2 = 5$$
$$c = \sqrt{5}.$$

(By definition, $c > 0$. See Figure 15.) The major axis is along the x-axis, so the foci have coordinates $(-\sqrt{5}, 0)$ and $(\sqrt{5}, 0)$.   ∎

EXAMPLE **3**

Finding the Equation
of an Ellipse

Find the equation of the ellipse having center at the origin, foci at $(0, 3)$ and $(0, -3)$, and major axis of length 8 units.

SOLUTION   Since the major axis is 8 units long,

$$2a = 8$$
or $$a = 4.$$

Use the relationship $a^2 - b^2 = c^2$ to find b^2. Here $a = 4$ and $c = 3$. Substituting for a and c gives

$$a^2 - b^2 = c^2$$
$$4^2 - b^2 = 3^2$$
$$16 - b^2 = 9$$
$$b^2 = 7.$$

Since the foci are on the y-axis, the larger intercept, a, is used to find the denominator for y^2, giving the equation in standard form as

$$\frac{y^2}{16} + \frac{x^2}{7} = 1.$$

A traditional graph of this ellipse is shown in Figure 17.   ◘

Just as a circle need not have its center at the origin, an ellipse may also have its center translated away from the origin.

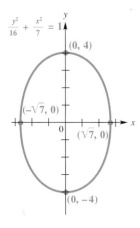

$\frac{y^2}{16} + \frac{x^2}{7} = 1$

$(0, 4)$

$(-\sqrt{7}, 0)$

$(\sqrt{7}, 0)$

$(0, -4)$

FIGURE 17

ELLIPSE CENTERED AT (h, k)

An ellipse centered at (h, k) with horizontal major axis of length $2a$ and vertical minor axis of length $2b$ has equation

$$\frac{(x - h)^2}{a^2} + \frac{(y - k)^2}{b^2} = 1.$$

There is a similar result for ellipses having a vertical major axis.

The definition of an ellipse can be used to prove the statement above.

EXAMPLE **4**

Graphing an Ellipse
Translated Away
From the Origin
(Traditional
Approach)

Use a traditional approach to graph $\frac{(y + 1)^2}{16} + \frac{(x - 2)^2}{9} = 1$.

SOLUTION   The graph of this equation is an ellipse centered at $(2, -1)$. As mentioned earlier, ellipses always have $a > b$. For this ellipse, then, $a = 4$ and $b = 3$. Since $a = 4$ is associated with y^2, the vertices of the ellipse are on the vertical line through $(2, -1)$. Find the vertices by locating two points on the vertical line through $(2, -1)$, one 4 units up from $(2, -1)$ and one 4 units down. The vertices are $(2, 3)$ and $(2, -5)$. Locate two other points on the ellipse by locating points on a horizontal line through $(2, -1)$, one 3 units to the right and one 3 units to the left. The graph is shown in Figure 18. As the graph suggests, the domain is $[-1, 5]$ and the range is $[-5, 3]$.   ◘

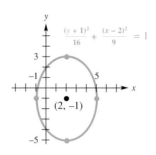

FIGURE 18

EXAMPLE **5**
Graphing an Ellipse Translated Away from the Origin (Calculator Approach)

Solve the equation in Example 4 for y, and give a calculator-generated graph of the ellipse.

SOLUTION

$$\frac{(y+1)^2}{16} + \frac{(x-2)^2}{9} = 1 \qquad \text{Given equation}$$

$$\frac{(y+1)^2}{16} = 1 - \frac{(x-2)^2}{9} \qquad \text{Subtract } \frac{(x-2)^2}{9}.$$

$$(y+1)^2 = 16 - \frac{16(x-2)^2}{9} \qquad \text{Multiply by 16.}$$

$$y + 1 = \pm\sqrt{16 - \frac{16(x-2)^2}{9}} \qquad \text{Take square roots.}$$

$$y = -1 \pm \sqrt{16 - \frac{16(x-2)^2}{9}} \qquad \text{Subtract 1.}$$

This final equation indicates that the graph may be obtained by graphing

$$y_1 = -1 + \sqrt{16 - \frac{16(x-2)^2}{9}} \quad \text{and} \quad y_2 = -1 - \sqrt{16 - \frac{16(x-2)^2}{9}}.$$

Using a square window gives the graph shown in Figure 19. ∎

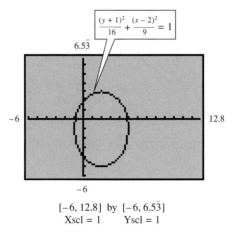

$$\frac{(y+1)^2}{16} + \frac{(x-2)^2}{9} = 1$$

$[-6, 12.8]$ by $[-6, 6.5\overline{3}]$
$\text{Xscl} = 1 \qquad \text{Yscl} = 1$

FIGURE 19

CAUTION Entering the expressions for y_1 and y_2 in Example 5 requires that parentheses be used with extreme care.

FOR **GROUP DISCUSSION**

The accompanying figure shows how an ellipse can be drawn using tacks and string. Have a class member volunteer to go to the board and using string and chalk, modify the method to draw a circle. Then have two class members work together to draw an ellipse. (*Hint*: Press hard!)

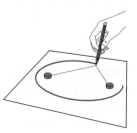

An Application of Ellipses

Ellipses have many useful applications. As the earth makes its year-long journey around the sun, it traces an ellipse. Spacecraft travel around the earth in elliptical orbits, and planets make elliptical orbits around the sun. An interesting recent application is the use of an elliptical tub in the nonsurgical removal of kidney stones. In this procedure, the reflective property of the ellipse is used. If a beam is projected from one focus onto the ellipse, it will reflect to the other focus. This feature has helped scientists develop the lithotripter, a machine that uses shock waves to crush kidney stones. The waves originate at one focus and are reflected to hit the kidney stone which is positioned at the second focus.

EXAMPLE 6

Applying Properties of an Ellipse to a Lithotripter

If a lithotripter is based on the ellipse $\frac{x^2}{36} + \frac{y^2}{27} = 1$, determine how many units the kidney stone and the wave source must be placed from the center of the ellipse.

SOLUTION Since 36 appears in the denominator of the term involving x^2, and $36 > 27$, the major axis will lie along the x-axis, with $a^2 = 36$. Since $b^2 = 27$, $c^2 = 36 - 27 = 9$. Therefore, $c = 3$, and the foci will be at $(-3, 0)$ and $(3, 0)$. Thus the kidney stone and the wave source must each be 3 units from the center of the ellipse on the longer axis. See Figure 20.   ▌

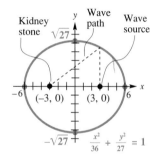

FIGURE 20

FOR **GROUP DISCUSSION**

What is a "whispering galley"? How does its special feature correspond to the foci of an ellipse?

Equations and Graphs of Hyperbolas

An ellipse was defined as the set of all points in a plane the sum of whose distances from two fixed points is a constant. A *hyperbola* is defined similarly.

HYPERBOLA

A **hyperbola** is the set of all points in a plane the *difference* of whose distances from two fixed points is constant. The two fixed points are called the **foci** of the hyperbola.

Suppose a hyperbola has center at the origin and foci at $F'(-c, 0)$ and $F(c, 0)$. The midpoint of the segment $F'F$ is the **center** of the hyperbola and the points $V'(-a, 0)$ and $V(a, 0)$ are the **vertices** of the hyperbola. The line segment $V'V$ is the **transverse axis** of the hyperbola. See Figure 21.

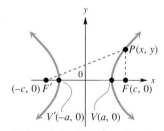

For this hyperbola, $d(P, F') - d(P, F) = 2a$.

FIGURE 21

Using the distance formula in conjunction with the definition of a hyperbola, the following standard forms of the equations for hyperbolas wih center at the origin can be verified. (See Exercise 65.)

STANDARD FORMS OF EQUATIONS FOR HYPERBOLAS

The hyperbola with center at the origin and equation

$$\frac{x^2}{a^2} - \frac{y^2}{b^2} = 1$$

has vertices $(\pm a, 0)$ and foci $(\pm c, 0)$, where $c^2 = a^2 + b^2$. The hyperbola with center at the origin and equation

$$\frac{y^2}{a^2} - \frac{x^2}{b^2} = 1$$

has vertices $(0, \pm a)$ and foci $(0, \pm c)$ where $c^2 = a^2 + b^2$.

FOR GROUP DISCUSSION

Graph the hyperbola with equation $x^2 - y^2 = 1$, using a graphics calculator and the window $[-7, 7]$ by $[-5, 5]$. Trace to $x = 5$ and zoom out. Trace to $x = 10$ and zoom out again. What is happening to the graph?

Starting with the equation for a hyperbola $\frac{x^2}{a^2} - \frac{y^2}{b^2} = 1$ and solving for y gives

$$\frac{x^2}{a^2} - 1 = \frac{y^2}{b^2}$$

$$\frac{x^2 - a^2}{a^2} = \frac{y^2}{b^2}$$

or $$y = \pm \frac{b}{a}\sqrt{x^2 - a^2}. \qquad \text{[*]}$$

If x^2 is very large in comparison to a^2, the difference $x^2 - a^2$ would be very close to x^2. If this happens, then the points satisfying equation (*) above would be very close to one of the lines

$$y = \pm \frac{b}{a}x.$$

Thus, as $|x|$ gets larger and larger, the points of the hyperbola $\frac{x^2}{a^2} - \frac{y^2}{b^2} = 1$ come closer to the lines $y = \pm \frac{b}{a}x$. These lines, called the **asymptotes** of the hyperbola, are very helpful when graphing the hyperbola using traditional graphing methods. The lines are the extended diagonals of the rectangle whose vertices are (a, b), $(-a, b)$, $(a, -b)$, and $(-a, -b)$. This rectangle is called the **fundamental rectangle** of the hyperbola.

Results similar to those above hold for a hyperbola of the form $\frac{y^2}{a^2} - \frac{x^2}{b^2} = 1$.

EXAMPLE 7

Graphing a Hyperbola Centered at the Origin (Traditional Approach)

Graph $\frac{x^2}{25} - \frac{y^2}{49} = 1$ using traditional graphing methods.

SOLUTION   For this hyperbola, $a = 5$ and $b = 7$. With these values, $y = \pm \frac{b}{a}x$ becomes $y = \pm \frac{7}{5}x$. The four points, $(5, 7)$, $(5, -7)$, $(-5, 7)$, and $(-5, -7)$, lead to the rectangle shown in Figure 22. The extended diagonals of this rectangle are the asymptotes of the hyperbola. The hyperbola has x-intercepts 5 and -5. The domain is $(-\infty, -5] \cup [5, \infty)$, and the range is $(-\infty, \infty)$. The final graph is shown in Figure 22.   ∎

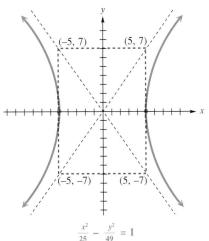

$$\frac{x^2}{25} - \frac{y^2}{49} = 1$$

FIGURE 22

EXAMPLE 8

Graphing a
Hyperbola Centered
at the Origin
(Calculator
Approach)

Solve the equation in Example 7 for y, and give a calculator-generated graph of the hyperbola.

SOLUTION   We solve for y as follows.

$$\frac{x^2}{25} - \frac{y^2}{49} = 1 \qquad \text{Given equation}$$

$$-\frac{y^2}{49} = 1 - \frac{x^2}{25} \qquad \text{Subtract } \tfrac{x^2}{25}.$$

$$\frac{y^2}{49} = \frac{x^2}{25} - 1 \qquad \text{Multiply by } -1.$$

$$\frac{y}{7} = \pm\sqrt{\frac{x^2}{25} - 1} \qquad \text{Take square roots.}$$

$$y = \pm 7\sqrt{\frac{x^2}{25} - 1} \qquad \text{Multiply by 7.}$$

The final equation indicates that the hyperbola is composed of the union of two functions, defined as

$$y_1 = 7\sqrt{\frac{x^2}{25} - 1} \qquad \text{and} \qquad y_2 = -7\sqrt{\frac{x^2}{25} - 1}.$$

Figure 23 shows the calculator-generated graph of this hyperbola. Again a square window gives the proper perspective. Compare it to the graph in Figure 22.   ∎

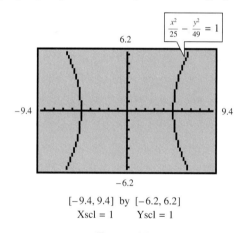

$$[-9.4, 9.4] \text{ by } [-6.2, 6.2]$$
$$\text{Xscl} = 1 \qquad \text{Yscl} = 1$$

FIGURE 23

FOR GROUP DISCUSSION

Determine the equation of the asymptote with positive slope for the hyperbola in Example 8. Then graph that asymptote, along with y_1 as defined in the example. Use various viewing windows to "watch" the hyperbola approach the asymptote.

EXAMPLE 9

Graphing a
Hyperbola Centered
at the Origin (Both
Approaches)

Graph $25y^2 - 4x^2 = 9$ using both traditional and graphics calculator approaches.

SOLUTION   First, let us use a traditional approach. We divide each side by 9 to get

$$\frac{25y^2}{9} - \frac{4x^2}{9} = 1.$$

To determine the values of a and b, write the equation as

$$\frac{y^2}{\frac{9}{25}} - \frac{x^2}{\frac{9}{4}} = 1.$$

This hyperbola is centered at the origin, has foci on the y-axis, and has y-intercepts $-\frac{3}{5}$ and $\frac{3}{5}$. Use the points $(\frac{3}{2}, \frac{3}{5})$, $(-\frac{3}{2}, \frac{3}{5})$, $(\frac{3}{2}, -\frac{3}{5})$, $(-\frac{3}{2}, -\frac{3}{5})$ to get the fundamental rectangle shown in Figure 24(a). Use the diagonals of this rectangle to determine the asymptotes for the graph. The domain is $(-\infty, \infty)$, and the range is $(-\infty, -\frac{3}{5}] \cup [\frac{3}{5}, \infty)$.

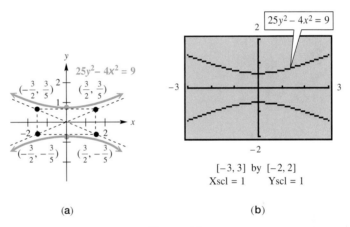

(a)   (b)

FIGURE 24

To graph this hyperbola using a graphics calculator, we go back to the original form of the equation and solve for y.

$$25y^2 - 4x^2 = 9 \qquad \text{Given equation}$$
$$25y^2 = 9 + 4x^2 \qquad \text{Add } 4x^2.$$
$$5y = \pm\sqrt{9 + 4x^2} \qquad \text{Take square roots.}$$
$$y = \pm\frac{1}{5}\sqrt{9 + 4x^2} \qquad \text{Multiply by } \tfrac{1}{5}.$$

Therefore, we graph

$$y_1 = \frac{1}{5}\sqrt{9 + 4x^2} \qquad \text{and} \qquad y_2 = -\frac{1}{5}\sqrt{9 + 4x^2}.$$

The union of the graphs of these two functions is the graph of the hyperbola. See Figure 24(b) for the calculator-generated graph.   ∎

Earlier we saw how ellipses can be translated away from the origin. This same translation can be made with hyperbolas, as shown in the next example.

EXAMPLE 10

Graphing a Hyperbola Translated Away From the Origin (Traditional Approach)

Graph $\frac{(y + 2)^2}{9} - \frac{(x + 3)^2}{4} = 1$.

SOLUTION   This hyperbola has the same graph as

$$\frac{y^2}{9} - \frac{x^2}{4} = 1,$$

except that it is centered at $(-3, -2)$. See Figure 25.   ▯

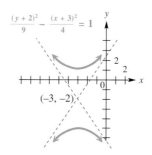

FIGURE 25

FOR GROUP DISCUSSION

How would you go about graphing the hyperbola in Example 10 on your graphics calculator? See if you can duplicate the calculator-generated graph of that hyperbola as seen in Figure 26.

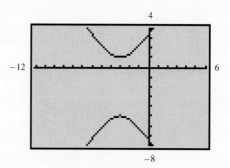

FIGURE 26

The two branches of a hyperbola are reflections about two different axes, and also about a point. What are the axes and the point for the hyperbola in Figure 26? (The reflecting property of the hyperbola was used on the Hubble Space Telescope.)

6.2 EXERCISES

Match each equation with its calculator-generated graph in Exercises 1–10. Do this first without actually using your calculator. Then check your answer by generating a calculator graph of your own. (Every window has Xscl = Yscl = 1.)

1. $\dfrac{y^2}{16} + \dfrac{x^2}{4} = 1$

2. $\dfrac{x^2}{16} + \dfrac{y^2}{4} = 1$

3. $\dfrac{x^2}{4} - \dfrac{y^2}{16} = 1$

4. $\dfrac{y^2}{4} - \dfrac{x^2}{16} = 1$

5. $\dfrac{(y-4)^2}{25} + \dfrac{(x+2)^2}{9} = 1$

6. $\dfrac{(y+4)^2}{25} + \dfrac{(x-2)^2}{9} = 1$

7. $\dfrac{(x+2)^2}{9} - \dfrac{(y-4)^2}{25} = 1$

8. $\dfrac{(x-2)^2}{9} - \dfrac{(y+4)^2}{25} = 1$

9. $36x^2 + 4y^2 = 144$

10. $9x^2 - 4y^2 = 36$

A.
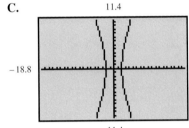

B.
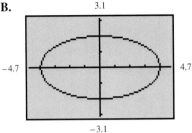

C.
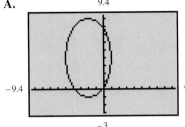

D.
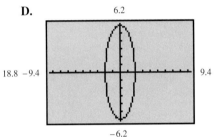

E.
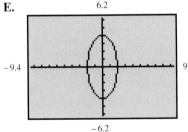

F.
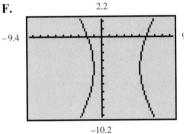

G.
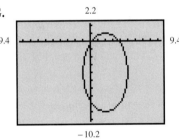

H.
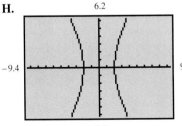

I.
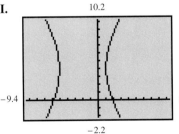

J.
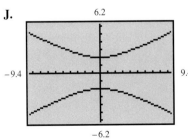

11. Explain how a circle can be interpreted as a special case of an ellipse.

12. If an ellipse has vertices at $(-3, 0)$, $(3, 0)$, $(0, 5)$, and $(0, -5)$, what is its domain? What is its range?

Graph each of the following ellipses by either using a traditional method or a graphics calculator. Give the domain and the range.

13. $\dfrac{x^2}{9} + \dfrac{y^2}{4} = 1$

14. $\dfrac{y^2}{36} + \dfrac{x^2}{16} = 1$

15. $9x^2 + 6y^2 = 54$

16. $12x^2 + 8y^2 = 96$

17. $\dfrac{x^2}{\frac{1}{9}} + \dfrac{y^2}{\frac{1}{16}} = 1$

18. $\dfrac{x^2}{\frac{4}{25}} + \dfrac{y^2}{\frac{9}{49}} = 1$

19. $\dfrac{25y^2}{36} + \dfrac{64x^2}{9} = 1$

20. $\dfrac{16y^2}{9} + \dfrac{121x^2}{25} = 1$

21. $\dfrac{(y + 3)^2}{25} + \dfrac{(x - 1)^2}{9} = 1$

22. $\dfrac{(y - 2)^2}{36} + \dfrac{(x + 3)^2}{16} = 1$

23. $\dfrac{(x - 2)^2}{16} + \dfrac{(y - 1)^2}{9} = 1$

24. $\dfrac{(y + 2)^2}{36} + \dfrac{(x + 3)^2}{25} = 1$

25. Discuss the symmetries exhibited by a hyperbola centered at the origin.

26. The ellipse $\frac{(y + 1)^2}{16} + \frac{(x - 2)^2}{9} = 1$ is graphed in Figure 18. What is the equation of its horizontal axis of symmetry? What is the equation of its vertical axis of symmetry?

Relating Concepts

In Example 5 we show how to graph the ellipse

$$\frac{(x - 2)^2}{9} + \frac{(y + 1)^2}{16} = 1$$

with a graphics calculator by solving for y. Its graph is the union of the graphs of the two functions

$$y_1 = -1 + \sqrt{16 - \frac{16(x - 2)^2}{9}} \quad \text{and} \quad y_2 = -1 - \sqrt{16 - \frac{16(x - 2)^2}{9}}.$$

The domain of this relation is $[-1, 5]$. Work Exercises 27–32 in order.

27. The relation is defined only when the radicand in y_1 and y_2 is greater than or equal to 0. Write the inequality that would need to be solved in order to find the domain analytically.

28. Let y represent the expression in x found in the radicand. What conic section is the graph of this function?

29. Graph the function defined by the radicand with a graphics calculator in the window $[-10, 10]$ by $[-10, 20]$.

30. Use the graph in Exercise 29 to solve the inequality of Exercise 27.

31. Explain how the solution set from Exercise 30 confirms what was found earlier using the graph of the original ellipse.

32. Solve the inequality of Exercise 27 analytically, using a sign graph (as explained in Section 3.3).

Graph each of the following hyperbolas by using either a traditional method or a graphics calculator. Give the domain and the range.

33. $\dfrac{x^2}{16} - \dfrac{y^2}{9} = 1$

34. $\dfrac{y^2}{9} - \dfrac{x^2}{9} = 1$

35. $49y^2 - 36x^2 = 1764$

36. $144x^2 - 49y^2 = 7056$

37. $\dfrac{4x^2}{9} - \dfrac{25y^2}{16} = 1$

38. $x^2 - y^2 = 1$

39. $\dfrac{(x-1)^2}{9} - \dfrac{(y+3)^2}{25} = 1$ **40.** $\dfrac{(x+3)^2}{16} - \dfrac{(y-2)^2}{36} = 1$ **41.** $\dfrac{(x-3)^2}{16} - \dfrac{(y+2)^2}{49} = 1$

42. $\dfrac{(y-5)^2}{4} - \dfrac{(x+1)^2}{9} = 1$ **43.** $\dfrac{(y+1)^2}{25} - \dfrac{(x-3)^2}{36} = 1$ **44.** $\dfrac{(x+2)^2}{16} - \dfrac{(y+2)^2}{25} = 1$

R*elating Concepts*

45. In Example 8 we show how to graph the hyperbola $\frac{x^2}{25} - \frac{y^2}{49} = 1$ on a graphics calculator by considering the union of the graphs of the two functions

$$y_1 = 7\sqrt{\dfrac{x^2}{25} - 1} \quad \text{and} \quad y_2 = -7\sqrt{\dfrac{x^2}{25} - 1}.$$

If we graph $y = \frac{x^2}{25} - 1$, the function defined by the radicand of y_1 and y_2, our graph is a parabola. Show how the solution set of $\frac{x^2}{25} - 1 \geq 0$ can be determined graphically, and explain how it relates to the domain of the given hyperbola. (*Hint:* Use the window $[-10, 10]$ by $[-2, 10]$.)

46. In Example 9 we show how to graph the hyperbola $25y^2 - 4x^2 = 9$ using a graphics calculator by considering the union of the graphs of the two functions

$$y_1 = \dfrac{1}{5}\sqrt{9 + 4x^2} \quad \text{and} \quad y_2 = -\dfrac{1}{5}\sqrt{9 + 4x^2}.$$

The function defined by the radicand, $y = 9 + 4x^2$, is a translation of the graph of the squaring function (Chapter 2).
  (a) Use the terminology of Chapter 2 to explain how the graph of $y = 9 + 4x^2$ can be obtained by a translation of the graph of $y = x^2$.
  (b) Which one of the following graphs most closely resembles the graph of $y = 9 + 4x^2$?

A. 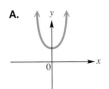   **B.**   **C.**   **D.**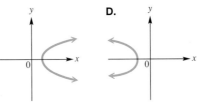

  (c) Explain how the domain of the graph of your choice in part (b) supports the conclusion concerning the domain of the original hyperbola found in Example 9.

Find an equation for each ellipse described.

47. x-intercepts ± 4; foci at $(-2, 0)$ and $(2, 0)$

48. y-intercepts ± 3; foci at $(0, \sqrt{3})$, $(0, -\sqrt{3})$

49. endpoints of major axis at $(6, 0)$, $(-6, 0)$; $c = 4$

50. vertices $(0, 5)$, $(0, -5)$; $b = 2$

51. center $(3, -2)$; $a = 5$, $c = 3$; major axis vertical

52. center $(2, 0)$; minor axis of length 6; major axis horizontal and of length 9

Find an equation for each hyperbola described.

53. x-intercepts ± 3; foci at $(-4, 0)$, $(4, 0)$

54. y-intercepts ± 5; foci at $(0, 3\sqrt{3})$, $(0, -3\sqrt{3})$

55. asymptotes $y = \pm\frac{3}{5}x$; y-intercepts 3 and -3

56. y-intercept -2; center at origin; passing through $(2, 3)$

Solve each application of ellipses or hyperbolas.

57. A patient is placed 12 units away from the source of the shock waves of a lithotripter. The lithotripter is based on an ellipse with a minor axis that measures 16 units. Find an equation of an ellipse that would satisfy this situation.

58. The orbit of Venus is an ellipse, with the sun at one focus. An approximate equation for the orbit is

$$\frac{x^2}{5013} + \frac{y^2}{4970} = 1,$$

where x and y are measured in millions of miles.
(a) Find the length of the major axis.
(b) Find the length of the minor axis.

59. The Roman Coliseum is an ellipse with major axis 620 ft and minor axis 513 ft. Find the distance between the foci of this ellipse.

60. A formula for the approximate circumference of an ellipse is

$$C \approx 2\pi\sqrt{\frac{a^2 + b^2}{2}},$$

where a and b are the lengths as shown in the figure. Use this formula to find the approximate circumference of the Roman Coliseum (see Exercise 59).

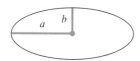

61. A one-way road passes under an overpass in the form of half of an ellipse, 15 ft high at the center and 20 ft wide. Assuming a truck is 12 ft wide, what is the height of the tallest truck that can pass under the overpass?

62. Ships and planes often use a location-finding system called LORAN. With this system, a radio transmitter at M on the figure sends out a series of pulses. When each pulse is received at transmitter S, it then sends out a pulse. A ship at P receives pulses from both M and S. A receiver on the ship measures the difference in the arrival times of the pulses. The navigator then consults a special map, showing certain curves according to the differences in arrival times. In this way, the ship can be located as lying on a portion of what type of curve?

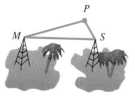

63. Microphones are placed at points $(-c, 0)$ and $(c, 0)$. An explosion occurs at point $P(x, y)$ having positive x-coordinate. (See the figure.) The sound is detected at the closer microphone t sec before being detected at the farther microphone. Assume that sound travels at a speed of 330 m per sec, and show that P must be on the hyperbola

$$\frac{x^2}{330^2 t^2} - \frac{y^2}{4c^2 - 330^2 t^2} = \frac{1}{4}.$$

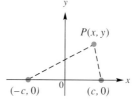

64. Suppose that $(c, 0)$ and $(-c, 0)$ are the foci of an ellipse. Suppose that the sum of the distances from any point (x, y) of the ellipse to the two foci is $2a$. See the accompanying figure.
(a) Use the distance formula to show that the equation of the resulting ellipse is

$$\frac{x^2}{a^2} + \frac{y^2}{a^2 - c^2} = 1.$$

(b) Show that a and $-a$ are the x-intercepts.
(c) Let $b^2 = a^2 - c^2$, and show that b and $-b$ are the y-intercepts.

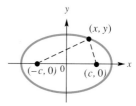

65. Suppose a hyperbola has center at the origin, foci at $F'(-c, 0)$ and $F(c, 0)$, and equation $d(P, F') - d(P, F) = 2a$. Let $b^2 = c^2 - a^2$, and show that an equation of the hyperbola is

$$\frac{x^2}{a^2} - \frac{y^2}{b^2} = 1.$$

66. (a) Use the result of Exercise 64(a) to find an equation of an ellipse with foci $(3, 0)$ and $(-3, 0)$, where the sum of the distances from any point of the ellipse to the two foci is 10.
(b) Use the result of Exercise 65 to find an equation of a hyperbola with center at the origin, foci at $(-2, 0)$ and $(2, 0)$, and the absolute value of the difference of the distances from any point of the hyperbola to the two foci equal to 2.

Further Explorations

Use the TABLE of your graphics calculator to find the points of intersection of the two halves of the following ellipses or hyperbolas, and the domains of each of the relations. Support your answers graphically in a DECIMAL window.

1. $\dfrac{x^2}{16} + \dfrac{y^2}{9} = 1$

2. $\dfrac{x^2}{4} - \dfrac{y^2}{25} = 1$

3. $\dfrac{x^2}{64} - \dfrac{(y-5)^2}{36} = 1$

4. $\dfrac{(x-3)^2}{16} + \dfrac{(y+2)^2}{81} = 1$

5. $\dfrac{x^2}{36} + \dfrac{(y-5)^2}{25} = 1$

6. $\dfrac{(y-6)^2}{4} - \dfrac{(x+1)^2}{16} = 1$

7. Use the TABLE to find and compare the domains of

$$\frac{x^2}{4} + \frac{y^2}{25} = 1 \quad \text{and} \quad \frac{x^2}{4} - \frac{y^2}{25} = 1.$$

In particular, note when each relation returns an ERROR. Support your answer graphically.

8. Use the TABLE to find and compare the domains of

$$\frac{x^2}{36} - \frac{y^2}{9} = 1 \quad \text{and} \quad \frac{y^2}{36} - \frac{x^2}{9} = 1.$$

Support your answer graphically.

6.3 SUMMARY OF THE CONIC SECTIONS

Characteristics of the Various Conic Sections ∎ Identification of Equations of Conic Sections ∎ Eccentricity

Characteristics of the Various Conic Sections

The conic sections presented in this chapter all have equations that can be written in the form

$$Ax^2 + Bx + Cy^2 + Dy + E = 0,$$

where either A or C must be nonzero. The special characteristics of each of the conic sections are summarized below.

EQUATIONS OF CONIC SECTIONS

Conic Section	Characteristic	Example
Parabola	Either $A = 0$ or $C = 0$, but not both.	$y = x^2$ $x = 3y^2 + 2y - 4$
Circle	$A = C \neq 0$	$x^2 + y^2 = 16$
Ellipse	$A \neq C, AC > 0$	$\frac{x^2}{16} + \frac{y^2}{25} = 1$
Hyperbola	$AC < 0$	$x^2 - y^2 = 1$

The following chart summarizes our work with conic sections.

SUMMARY OF THE CONIC SECTIONS

Equation and Graph	Description	Identification
$(x - h)^2 + (y - k)^2 = r^2$ Circle	Center is at (h, k), and radius is r.	x^2 and y^2 terms have the same positive coefficient.
$y = a(x - h)^2 + k$ Parabola	Opens upward if $a > 0$, downward if $a < 0$. Vertex is at (h, k)	x^2 term y is not squared.
$x = a(y - k)^2 + h$ Parabola	Opens to right if $a > 0$, to left if $a < 0$. Vertex is at (h, k).	y^2 term x is not squared.
$\dfrac{x^2}{a^2} + \dfrac{y^2}{b^2} = 1$ Ellipse	x-intercepts are a and $-a$. y-intercepts are b and $-b$.	x^2 and y^2 terms have different positive coefficients.
$\dfrac{x^2}{a^2} - \dfrac{y^2}{b^2} = 1$ Hyperbola	x-intercepts are a and $-a$. Asymptotes found from (a, b), $(a, -b)$, $(-a, -b)$, and $(-a, b)$.	x^2 has a positive coefficient. y^2 has a negative coefficient.
$\dfrac{y^2}{a^2} - \dfrac{x^2}{b^2} = 1$ Hyperbola	y-intercepts are a and $-a$. Asymptotes found from (b, a), $(b, -a)$, $(-b, -a)$, and $(-b, a)$.	y^2 has a positive coefficient. x^2 has a negative coefficient.

Identification of Equations of Conic Sections

In order to recognize the type of graph that a given conic section has, it is sometimes necessary to transform the equation into a more familiar form, as shown in the next examples.

EXAMPLE 1

Determining the Type of a Conic Section From Its Equation

Decide on the type of conic section represented by each of the following equations, and give each graph.

(a) $25y^2 - 4x^2 = 100$

SOLUTION Divide each side by 100 to get

$$\frac{y^2}{4} - \frac{x^2}{25} = 1.$$

This is a hyperbola centered at the origin, with foci on the y-axis, and y-intercepts 2 and -2. The points $(5, 2), (5, -2), (-5, 2), (-5, -2)$ determine the fundamental rectangle. The diagonals of the rectangle are the asymptotes, and their equations are

$$y = \pm \frac{2}{5}x.$$

The graph is shown in Figure 27. Both traditional and calculator-generated graphs are shown.

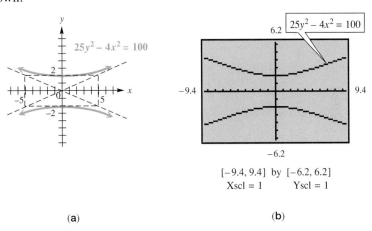

$[-9.4, 9.4]$ by $[-6.2, 6.2]$
Xscl = 1   Yscl = 1

(a)   (b)

FIGURE 27

(b) $x^2 = 25 + 5y^2$

SOLUTION Rewriting the equation as

$$x^2 - 5y^2 = 25$$

or

$$\frac{x^2}{25} - \frac{y^2}{5} = 1$$

shows that the equation represents a hyperbola centered at the origin, with asymptotes

$$y = \frac{\pm\sqrt{5}}{5}x.$$

The *x*-intercepts are ± 5; both types of graphs are shown in Figure 28.

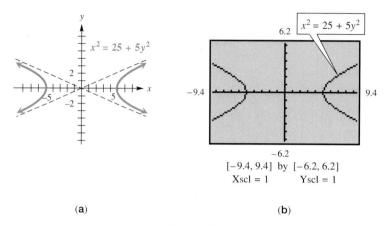

(a)   (b)

FIGURE 28

(c) $4x^2 - 16x + 9y^2 + 54y = -61$

SOLUTION Since the coefficients of the x^2 and y^2 terms are unequal and both positive, this equation might represent an ellipse. (It might also represent a single point or no points at all.) To find out, complete the square on *x* and *y*.

$$4(x^2 - 4x \quad\;\;) + 9(y^2 + 6y \quad\;\;) = -61 \qquad \text{Factor out a 4; factor out a 9.}$$

$$4(x^2 - 4x + 4 - 4) + 9(y^2 + 6y + 9 - 9) = -61 \qquad \text{Add and subtract the same quantity.}$$

$$4(x^2 - 4x + 4) - 16 + 9(y^2 + 6y + 9) - 81 = -61 \qquad \text{Regroup and distribute.}$$

$$4(x - 2)^2 + 9(y + 3)^2 = 36 \qquad \text{Add 97 and factor.}$$

$$\frac{(x - 2)^2}{9} + \frac{(y + 3)^2}{4} = 1 \qquad \text{Divide by 36.}$$

This equation represents an ellipse having center at $(2, -3)$ and graph as shown in Figure 29. (See if you can duplicate the calculator-generated graph.)

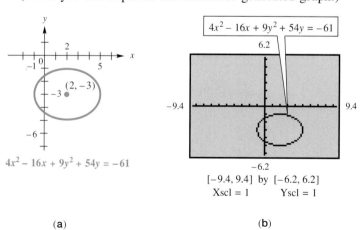

(a)   (b)

FIGURE 29

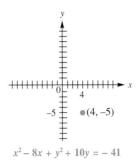

$x^2 - 8x + y^2 + 10y = -41$

FIGURE 30

(d) $x^2 - 8x + y^2 + 10y = -41$

SOLUTION Complete the square on both x and y, as follows:

$$(x^2 - 8x + 16) + (y^2 + 10y + 25) = -41 + 16 + 25$$
$$(x - 4)^2 + (y + 5)^2 = 0.$$

This result shows that the equation is that of a circle of radius 0; that is, the point $(4, -5)$. See Figure 30. Had a negative number been obtained on the right (instead of 0), the equation would have represented no points at all, and there would be no graph.

(e) $x^2 - 6x + 8y - 7 = 0$

SOLUTION Since only one variable is squared (x, and not y), the equation represents a parabola. Rearrange the terms with y (the variable that is not squared) alone on one side. Then complete the square on the other side of the equation.

$$8y = -x^2 + 6x + 7$$
$$8y = -(x^2 - 6x \quad) + 7 \qquad \text{Regroup and factor out } -1.$$
$$8y = -(x^2 - 6x + 9) + 7 + 9 \qquad \text{Add 0 in the form } -9 + 9.$$
$$8y = -(x - 3)^2 + 16 \qquad \text{Factor.}$$
$$y = -\frac{1}{8}(x - 3)^2 + 2 \qquad \text{Multiply both sides by } \tfrac{1}{8}.$$

The parabola has a vertex at $(3, 2)$, and opens downward, as shown in both graphs in Figure 31. ∎

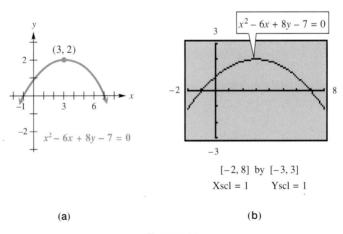

(a)   (b)

FIGURE 31

FOR GROUP DISCUSSION

Use an integer window and the equation in Example 1(d) to graph the single point $(4, -5)$ on your calculator screen.

Eccentricity

In Sections 6.1 and 6.2, we introduced the various definitions of the conic sections. The conic sections (or conics) can all be characterized by one general definition.

CONIC

A **conic** is the set of all points $P(x, y)$ in a plane such that the ratio of its distance from P to a fixed line and its distance from P to a fixed point is constant.

As with parabolas, the fixed line is the **directrix** and the fixed point is the **focus.** In Figure 32 the focus is $F(c, 0)$ and the directrix is the line $x = -c$. The constant ratio is called the **eccentricity** of the conic, written e. (This is not the same e as the base of natural logarithms.)

By definition, the distances $d(P, F)$ and $d(P, D)$ in Figure 32 are equal if the conic is a parabola. Thus, a parabola always has eccentricity $e = 1$.

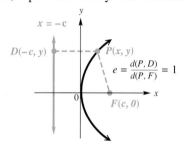

FIGURE 32

For the ellipse and hyperbola, it can be shown that the constant ratio of the definition is

$$e = \frac{c}{a},$$

where c is the distance from the center of the figure to a focus and a is the distance from the center to a vertex. By the definition of an ellipse, $a^2 > b^2$ and $c = \sqrt{a^2 - b^2}$. Thus, for the ellipse,

$$0 < c < a,$$

$$0 < \frac{c}{a} < 1,$$

$$0 < e < 1.$$

If a is constant, letting c approach 0 would force the ratio $\frac{c}{a}$ to approach 0, which also forces b to approach a (so that $\sqrt{a^2 - b^2} = c$ would approach 0). Since b leads to the y-intercepts, this means that the x- and y-intercepts are almost the same, producing an ellipse very close in shape to a circle when e is very close to 0.

In a similar manner, if e approaches 1, then b will approach 0. The path of the earth around the sun is an ellipse that is very nearly circular. If fact, for this ellipse, $e \approx .017$. On the other hand, the path of Halley's comet is a very flat ellipse, with $e \approx .98$.

EXAMPLE 2

Finding Eccentricity From the Equation of an Ellipse

Find the eccentricity of each ellipse.

(a) $\dfrac{x^2}{9} + \dfrac{y^2}{16} = 1$

SOLUTION Since $16 > 9$, let $a^2 = 16$, which gives $a = 4$. Also,

$$c = \sqrt{a^2 - b^2}$$
$$= \sqrt{16 - 9} = \sqrt{7}.$$

Finally, $e = \dfrac{\sqrt{7}}{4} \approx .66$.

(b) $5x^2 + 10y^2 = 50$

SOLUTION Divide by 50 to get

$$\frac{x^2}{10} + \frac{y^2}{5} = 1.$$

Here $a^2 = 10$, with $a = \sqrt{10}$. Now find c.

$$c = \sqrt{10 - 5} = \sqrt{5},$$

and $\quad e = \dfrac{\sqrt{5}}{\sqrt{10}} = \dfrac{1}{\sqrt{2}} = \dfrac{\sqrt{2}}{2} \approx .71$ ❚

As mentioned above, the hyperbola

$$\frac{x^2}{a^2} - \frac{y^2}{b^2} = 1 \quad \text{or} \quad \frac{y^2}{a^2} - \frac{x^2}{b^2} = 1,$$

where $c = \sqrt{a^2 + b^2}$, also has eccentricity

$$e = \frac{c}{a}.$$

By definition, $c = \sqrt{a^2 + b^2} > a$, so that $\frac{c}{a} > 1$, and so for a hyperbola,

$$e > 1.$$

EXAMPLE 3

Finding Eccentricity From the Equation of a Hyperbola

Find the eccentricity of the hyperbola

$$\frac{x^2}{9} - \frac{y^2}{4} = 1.$$

SOLUTION Here $a^2 = 9$; thus $a = 3$, $c = \sqrt{9 + 4} = \sqrt{13}$, and

$$e = \frac{c}{a} = \frac{\sqrt{13}}{3} \approx 1.2.$$ ❚

The following chart summarizes this discussion of eccentricity.

ECCENTRICITY OF CONICS

Conic	Eccentricity		
Parabola	$e = 1$		
Ellipse	$e = \dfrac{c}{a}$	and	$0 < e < 1$
Hyperbola	$e = \dfrac{c}{a}$	and	$e > 1$

FOR GROUP DISCUSSION

Answer the following questions in order.

1. As the two foci of an ellipse move closer and closer together, what familiar shape does the ellipse begin to resemble?
2. For a circle with center at the origin, how do the values of a and b compare?
3. What is the value of c as defined above for a circle?
4. What is the eccentricity of a circle?

EXAMPLE 4

Finding the Equation of a Conic Using Eccentricity

Find an equation for the following conics with centers at the origin.

(a) focus at $(3, 0)$ and eccentricity 2

SOLUTION Since $e = 2$, which is greater than 1, the conic is a hyperbola with $c = 3$. From $e = \dfrac{c}{a}$, find a by substituting $e = 2$ and $c = 3$.

$$e = \frac{c}{a}$$

$$2 = \frac{3}{a}$$

$$a = \frac{3}{2}$$

Now find b.

$$b^2 = c^2 - a^2 = 9 - \frac{9}{4} = \frac{27}{4}$$

The given focus is on the x-axis, so the x^2 term is positive and the equation is

$$\frac{x^2}{\dfrac{9}{4}} - \frac{y^2}{\dfrac{27}{4}} = 1,$$

or

$$\frac{4x^2}{9} - \frac{4y^2}{27} = 1.$$

(b) vertex at $(0, -8)$ and $e = \frac{1}{2}$

SOLUTION   The graph is an ellipse since $e = \frac{1}{2} < 1$. From the given vertex, we know that the vertices are on the y-axis and $a = 8$. Use $e = \frac{c}{a}$ to find c.

$$e = \frac{c}{a}$$

$$\frac{1}{2} = \frac{c}{8}$$

$$c = 4$$

Since $b^2 = a^2 - c^2$,

$$b^2 = 64 - 16 = 48,$$

and the equation is

$$\frac{y^2}{64} + \frac{x^2}{48} = 1. \quad \blacksquare$$

6.3 EXERCISES

In Exercises 1–12, the equation of a conic section is given in a familiar form. Identify the type of graph that the equation has, without actually graphing.

1. $x^2 + y^2 = 144$

2. $(x - 2)^2 + (y + 3)^2 = 25$

3. $y = 2x^2 + 3x - 4$

4. $x = 3y^2 + 5y - 6$

5. $x = -3(y - 4)^2 + 1$

6. $\dfrac{x^2}{25} + \dfrac{y^2}{36} = 1$

7. $\dfrac{x^2}{49} + \dfrac{y^2}{100} = 1$

8. $x^2 - y^2 = 1$

9. $\dfrac{x^2}{4} - \dfrac{y^2}{16} = 1$

10. $\dfrac{(x + 2)^2}{9} + \dfrac{(y - 4)^2}{16} = 1$

11. $\dfrac{x^2}{25} - \dfrac{y^2}{25} = 1$

12. $y = 4(x + 3)^2 - 7$

For each of the following equations that has a graph, identify the corresponding graph. It may be necessary to transform the equation.

13. $\dfrac{x^2}{4} = 1 - \dfrac{y^2}{9}$

14. $\dfrac{x^2}{4} = 1 + \dfrac{y^2}{9}$

15. $\dfrac{x^2}{4} + \dfrac{y^2}{4} = 1$

16. $\dfrac{x^2}{4} + \dfrac{y^2}{4} = -1$

17. $x^2 + 2x = x^2 + y - 6$

18. $y^2 - 4y = y^2 + 3 - x$

19. $x^2 = 25 + y^2$

20. $x^2 = 25 - y^2$

21. $9x^2 + 36y^2 = 36$

22. $x^2 = 4y - 8$

23. $\dfrac{(x + 3)^2}{16} + \dfrac{(y - 2)^2}{16} = 1$

24. $\dfrac{(x - 4)^2}{8} + \dfrac{(y + 1)^2}{2} = 0$

25. $y^2 - 4y = x + 4$

26. $11 - 3x = 2y^2 - 8y$

27. $(x + 7)^2 + (y - 5)^2 + 4 = 0$

28. $4(x - 3)^2 + 3(y + 4)^2 = 0$

29. $3x^2 + 6x + 3y^2 - 12y = 12$

30. $2x^2 - 8x + 2y^2 + 20y = 12$

31. $x^2 - 6x + y = 0$

32. $x - 4y^2 - 8y = 0$

33. $4x^2 - 8x - y^2 - 6y = 6$

34. $x^2 + 2x = x^2 - 4y - 2$

35. $4x^2 - 8x + 9y^2 + 54y = -84$

36. $3x^2 + 12x + 3y^2 = -11$

37. $6x^2 - 12x + 6y^2 - 18y + 25 = 0$

38. $4x^2 - 24x + 5y^2 + 10y + 41 = 0$

R*elating Concepts*

39. Suppose that both A and C are zero in the equation $Ax^2 + Bx + Cy^2 + Dy + E = 0$. What kind of graph does this equation have? (See Section 1.3.)

40. How can the graph of the equation discussed in Exercise 39 be obtained by a plane intersecting a cone?

Find the eccentricity of each of the following ellipses and hyperbolas.

41. $12x^2 + 9y^2 = 36$ **42.** $8x^2 - y^2 = 16$ **43.** $x^2 - y^2 = 4$ **44.** $x^2 + 2y^2 = 8$ **45.** $4x^2 + 7y^2 = 28$

46. $9x^2 - y^2 = 1$ **47.** $x^2 - 9y^2 = 18$ **48.** $x^2 + 10y^2 = 10$ **49.** $2x^2 + y^2 = 32$ **50.** $5x^2 - 4y^2 = 20$

Write an equation for each of the following conics. Each parabola has vertex at the origin, and each ellipse or hyperbola is centered at the origin.

51. focus at $(0, 8)$ and $e = 1$ **52.** focus at $(-2, 0)$ and $e = 1$ **53.** focus at $(3, 0)$ and $e = \frac{1}{2}$

54. focus at $(0, -2)$ and $e = \frac{2}{3}$ **55.** vertex at $(-6, 0)$ and $e = 2$ **56.** vertex at $(0, 4)$ and $e = \frac{5}{3}$

57. focus at $(0, -1)$ and $e = 1$

58. focus at $(2, 0)$ and $e = \frac{6}{5}$

59. vertical major axis of length 6 and $e = \frac{4}{5}$

60. vertical transverse axis of length 8 and $e = \frac{7}{3}$

61. Calculator-generated graphs are shown in Figures A–D. Arrange the figures in order so that the first in the list has the smallest eccentricity and the rest have eccentricities in increasing order.

A.

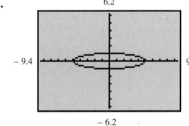

B.

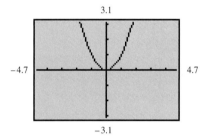

C.

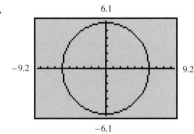

D.

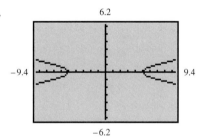

62. The orbit of Venus around the sun is an ellipse with equation

$$\frac{x^2}{5013} + \frac{y^2}{4970} = 1,$$

where x and y are measured in millions of miles. Find the eccentricity of this ellipse.

6.4 SYSTEMS OF LINEAR EQUATIONS IN TWO VARIABLES

Introduction ▮ The Substitution Method ▮ The Addition (or Elimination)
Method ▮ Connections: Relating Systems of Equations and the
Intersection-of-Graphs Method ▮ Applications of Systems

Introduction

When an applied problem requires that more than one unknown quantity must be
found, it is often helpful (or, in some cases, absolutely necessary) to write several
equations in several variables, and then solve this resulting *system of equations.* At
the college algebra level, we usually restrict the number of unknowns to two or
three. In this section, we limit the discussion to systems of equations in two
variables, because we cannot use graphical methods to solve systems with more
than two variables. We will introduce systems in three variables later in this chapter
and discuss graphics calculator methods of solving those systems in Chapter 7.
Graphics calculators eliminate the need for time-consuming computations. The
methods of this chapter and the next can help students appreciate the capabilities
of today's electronic marvels.

A group of equations that place restrictions on the same variables is called a
system of equations. The solution set of a system of equations is the intersection
of the solution sets of the individual equations. It is customary to write a system by
listing its equations. For example, the system of equations $2x + y = 4$ and
$x - y = 6$ is written as

$$2x + y = 4$$
$$x - y = 6.$$

In general, a **first-degree equation in n unknowns** is any equation of the form

$$a_1x_1 + a_2x_2 + \cdots + a_nx_n = k,$$

where $a_1, a_2, \ldots, a_n,$ and k are constants and $x_1, x_2, \ldots, x_n$ are variables. Such
equations are also called *linear equations.* We will concentrate on systems of linear
equations with two or three variables in this chapter, although the methods used
can be extended to systems with more variables.

The solution set of a linear equation in two variables is an infinite set of ordered
pairs. Since the graph of such an equation is a straight line, there are three possibilities for the solution set of a system of two linear equations in two variables. An
example of each possibility is shown in Figure 33.

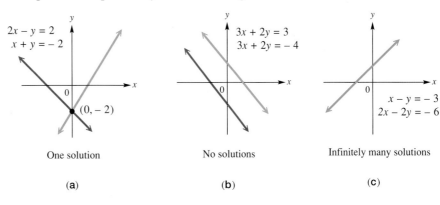

FIGURE 33

> **POSSIBLE GRAPHS OF A LINEAR SYSTEM WITH TWO EQUATIONS AND TWO VARIABLES**
>
> 1. The graphs of the two equations intersect in a single point. The coordinates of this point give the solution of the system. This is the most common case. See Figure 33(a).
> 2. The graphs are distinct parallel lines. In this case, the system is said to be **inconsistent.** That is, there is no solution common to both equations. The solution set of the linear system is empty. See Figure 33(b).
> 3. The graphs are the same line. In this case, the equations are said to be **dependent,** and any solution of one equation is also a solution of the other. Thus, there are infinitely many solutions. See Figure 33(c).

The Substitution Method

Although the *number* of solutions of a linear system can often be seen from the graph of the equations of the system, it is usually difficult to determine an exact solution from the graph. A general analytic method of finding the solution of a system of two linear equations, called the **substitution method,** is illustrated in the following example and used again in Section 6.5.

The substitution method involves substituting an expression for one variable in terms of the other in another equation of a system, as seen in the first example.

EXAMPLE 1

Solving a System by Substitution

Solve the system

$$3x + 2y = 11 \qquad \text{[1]}$$
$$-x + y = 3 \qquad \text{[2]}$$

by substitution. Support the solution graphically.

SOLUTION While there are several ways to approach this solution, one way is to solve equation (2) for y to get $y = x + 3$. We substitute $x + 3$ for y in equation (1) to get

$$3x + 2(x + 3) = 11. \qquad \text{[3]}$$

Notice the careful use of parentheses. Now we solve equation (3) for x.

$$3x + 2x + 6 = 11$$
$$5x + 6 = 11$$
$$5x = 5$$
$$x = 1$$

Replace x with 1 in $y = x + 3$ to find that $y = 1 + 3 = 4$. The solution set for this system is $\{(1, 4)\}$.

To support this solution graphically, we must graph equations (1) and (2) in an appropriate viewing window, and determine the coordinates of the point of intersection. First, we solve equation (1) for y to get $y_1 = -1.5x + 5.5$ and equation

(2) for y to get $y_2 = x + 3$. Graphing these in the standard viewing window (which is appropriate for this system) and using the capability of the calculator for finding the coordinates of a point of intersection of two graphs, we find that the point is indeed $(1, 4)$, as seen in Figure 34.

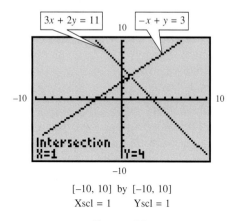

$$3x + 2y = 11 \qquad -x + y = 3$$

Intersection
X=1   Y=4

[−10, 10] by [−10, 10]
Xscl = 1   Yscl = 1

FIGURE 34

The Addition (or Elimination) Method

Another method of solving systems of two equations is the **addition method.** With this method, we first multiply the equations on both sides by suitable numbers, so that when they are added, one variable is eliminated. (Thus, this method is also called the *elimination method.*) The result is an equation in one variable that can be solved by methods from Chapter 1. The solution is then substituted into one of the original equations, making it possible to solve for the other variable. In this process the given system is replaced by new systems that have the same solution set as the original system. Systems that have the same solution set are called **equivalent systems.**

EXAMPLE 2

Solving a System by Addition

Solve the system

$$3x - 4y = 1 \qquad \textbf{[4]}$$
$$2x + 3y = 12 \qquad \textbf{[5]}$$

by addition, and support the solution graphically.

SOLUTION   To eliminate x, multiply both sides of equation (4) by -2 and both sides of equation (5) by 3 to get equations (6) and (7).

$$-6x + 8y = -2 \qquad \textbf{[6]}$$
$$6x + 9y = 36 \qquad \textbf{[7]}$$

Although this new system is not the same as the given system, it will have the same solution set.

Now add the two equations to eliminate x, and then solve the result for y.

$$\begin{array}{r} -6x + 8y = -2 \\ 6x + 9y = 36 \\ \hline 17y = 34 \\ y = 2 \end{array}$$

Substitute 2 for y in equation (4) or (5). Choosing equation (4) gives

$$3x - 4(2) = 1$$
$$3x = 9$$
$$x = 3$$

The solution set of the given system is $\{(3, 2)\}$, which can be checked analytically by substituting 3 for x and 2 for y in equation (5).

To support this solution graphically, we must solve equations (4) and (5) for y, and graph both in an appropriate viewing window.

$$\text{Equation (4): } 3x - 4y = 1$$
$$-4y = -3x + 1$$
$$y = \frac{3}{4}x - \frac{1}{4}$$

$$\text{Equation (5): } 2x + 3y = 12$$
$$3y = -2x + 12$$
$$y = -\frac{2}{3}x + 4$$

As seen in Figure 35, the graphs of $y_1 = \frac{3}{4}x - \frac{1}{4}$ and $y_2 = -\frac{2}{3}x + 4$ intersect at the point (3, 2), supporting our earlier analytic result.   ▌

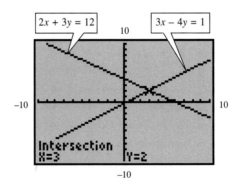

$[-10, 10]$ by $[-10, 10]$
$\text{Xscl} = 1 \qquad \text{Yscl} = 1$

The line with positive slope is $y_1 = \frac{3}{4}x - \frac{1}{4}$,

and the line with negative slope is $y_2 = -\frac{2}{3}x + 4$.

FIGURE 35

Connections: Relating Systems of Equations and the Intersection-of-Graphs Method

In Chapter 1 we learned how to support an analytic solution of a linear equation by two methods: the intersection-of-graphs method and the x-intercept method. The former is closely related to the concept of solving systems of linear equations. Suppose, for example, we wish to solve the linear equation

$$\frac{3}{4}x - \frac{1}{4} = -\frac{2}{3}x + 4$$

analytically. We can begin by multiplying both sides of the equation by 12, to get

$$12\left(\frac{3}{4}x - \frac{1}{4}\right) = 12\left(-\frac{2}{3}x + 4\right)$$

or $\qquad 9x - 3 = -8x + 48.$

Then, we complete the solution in the usual manner.

$$9x + 8x = 3 + 48$$
$$17x = 51$$
$$x = 3$$

To check this solution analytically, we substitute 3 for x in the original equation.

$$\frac{3}{4}(3) - \frac{1}{4} = -\frac{2}{3}(3) + 4 \qquad ?$$
$$\frac{9}{4} - \frac{1}{4} = -2 + 4 \qquad ?$$
$$\frac{8}{4} = 2 \qquad ?$$
$$2 = 2 \qquad \text{True}$$

Since both sides of the original equation yield 2 when $x = 3$, our analytic work is correct, and the solution set of the original equation is $\{3\}$.

Now compare this work with the system in Example 2 and the graphs in Figure 35. The linear equation we just solved had $y_1 = \frac{3}{4}x - \frac{1}{4}$ as its original left side and $y_2 = -\frac{2}{3}x + 4$ as its original right side. Its solution set, $\{3\}$, consists of the x-coordinate of the point of intersection of the two lines, while the y-coordinate of the point of intersection, 2, is the result obtained when 3 is substituted for x in the equation.

We can see from the preceding discussion that there is a close connection between the solution of a linear equation in one variable and the solution of a linear system in two variables. The former has a solution set consisting of only domain values, while the latter has a solution set consisting of both domain and range values, in the form of ordered pairs. Yet the connection is apparent.

FOR **GROUP DISCUSSION**

Suppose that we are given the following.

$$y_1 = f(x) = -2.5x - .5$$
$$\text{and} \qquad y_2 = g(x) = 2x - 5.$$

Figure 36 shows the graphs of y_1 and y_2 along with the coordinates of the point of intersection of the lines.

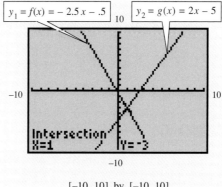

$$[-10, 10] \text{ by } [-10, 10]$$
$$Xscl = 1 \qquad Yscl = 1$$

FIGURE 36

1. What is the solution set of the linear equation $f(x) = g(x)$; that is, $-2.5x - .5 = 2x - 5$?
2. What is the solution set of the system of linear equations below?

$$y = -2.5x - .5$$
$$y = 2x - 5$$

3. Discuss the connections between your answers to items 1 and 2.

The examples presented thus far in this section have all been systems with a single solution. The next example illustrates the analytic procedure used to determine whether a system is inconsistent or has dependent equations.

EXAMPLE **3**

Solving an Inconsistent System and a System With Dependent Equations

Solve each system using the addition method, and support the solution graphically.

(a) $3x - 2y = 4$
 $-6x + 4y = 7$

SOLUTION   The variable x can be eliminated by multiplying both sides of the first equation by 2 and then adding.

$$\begin{array}{r} 6x - 4y = 8 \\ -6x + 4y = 7 \\ \hline 0 = 15 \quad \text{False} \end{array}$$

Both variables were eliminated here, leaving the false statement $0 = 15$, a signal that these two equations have no solutions in common. The system is inconsistent, and the solution set is $\emptyset$. In slope-intercept form, the first equation is written $y = 1.5x - 2$ and the second is written $y = 1.5x + 1.75$. Since the lines have the same slope but different y-intercepts, they are parallel, and have no point of intersection. This supports our analytic conclusion. See Figure 37(a).

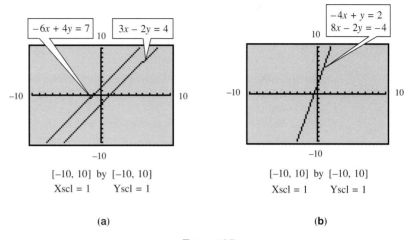

$[-10, 10]$ by $[-10, 10]$
Xscl $= 1$ Yscl $= 1$

$[-10, 10]$ by $[-10, 10]$
Xscl $= 1$ Yscl $= 1$

(a)   (b)

FIGURE 37

(b) $\quad -4x + y = 2$
$\qquad\quad 8x - 2y = -4$

SOLUTION   We can eliminate x by multiplying both sides of the first equation by 2 and then adding it to the second equation.

$$-8x + 2y = 4$$
$$\underline{8x - 2y = -4}$$
$$0 = 0 \qquad \text{True}$$

This true statement, $0 = 0$, indicates that a solution of one equation is also a solution of the other, so the solution set is an infinite set of ordered pairs. The graphs of the equations are the same line, since both equations are equivalent to $y = 4x + 2$. See Figure 37(b). The two equations are dependent.

We will write the solution of a system of dependent equations as an ordered pair by expressing x in terms of y as follows. Choose either equation and solve for x. Choosing the first equation gives

$$-4x + y = 2$$
$$x = \frac{2 - y}{-4} = \frac{y - 2}{4}.$$

We write the solution set as

$$\left\{ \left(\frac{y - 2}{4}, y \right) \right\}.$$

By selecting values for y and calculating the corresponding values for x, individual ordered pairs of the solution set can be found. For example, if $y = -2$, $x = \frac{(-2 - 2)}{4} = -1$ and the ordered pair $(-1, -2)$ is a solution.   ∎

NOTE In Example 3 we wrote the solution set in a form with the variable y arbitrary. However, it would be acceptable to write the ordered pair with x arbitrary. In this case, the solution set would be written

$$\{(x, 4x + 2)\}.$$

By selecting values for x and solving for y in the ordered pair above, individual solutions can be found. Verify that $(-1, -2)$ is a solution.

Applications of Systems

Many applied problems involve more than one unknown quantity. Although some problems with two unknowns can be solved using just one variable, many times it is easier to use two variables. To solve a problem using a system, determine the unknown quantities you are asked to find, and let different variables represent each of these quantities. Then write a system of equations, and solve it using one of the methods of this section. Be sure that you answer the question(s) posed in the problem, and check to see that your answer is reasonable.

EXAMPLE 4

Solving an Application Using a System

An animal feed is made from two ingredients: corn and soybeans. One unit of each ingredient provides units of protein and fat as shown in the table below. How many units of each ingredient should be used to make a feed that contains 7 units of protein and 9 units of fat?

	Protein	Fat
Corn	.25	.4
Soybeans	.4	.2

SOLUTION Let x represent the number of units of corn and y the number of units of soybeans required. Since the total amount of protein is to be 7 units,

$$.25x + .4y = 7.$$

For the 9 units of fat, $.4x + .2y = 9.$

Multiply the first equation on both sides by 100 and the second equation by 10 to get the system

$$25x + 40y = 700$$
$$4x + 2y = 90.$$

Now use the elimination method to solve this system. Multiply the bottom equation by -20 and add to get

$$\begin{aligned}
25x + 40y &= 700 \\
-80x - 40y &= -1800 \\
\hline
-55x &= -1100.
\end{aligned}$$

Multiplying both sides of $-55x = -1100$ by $-\frac{1}{55}$ gives

$$x = 20.$$

Now substitute 20 for x in the equation $4x + 2y = 90$.

$$4(20) + 2y = 90$$
$$2y = 10$$
$$y = 5$$

TECHNOLOGICAL NOTE
The statistics capabilities of modern graphics calculators allow for curve-fitting to be accomplished simply by entering the data points. We will not, however, suggest that the algebra student studying from this text rely on this capability at this time, as it would defeat the purpose of what is being introduced in this section.

The solution of the original system is $(20, 5)$. The feed should contain 20 units of corn and 5 units of soybeans to meet the given requirements. ▯

If we are given two distinct points, it is possible to find a linear function ($y = ax + b$) such that the points lie on its graph. This idea of finding the equation of a graph that contains certain points (often called *data points*) is called **curve-fitting,** and is studied extensively in statistics. It can be extended to finding a quadratic polynomial function containing three points, a cubic containing four points, and so on. The following example illustrates how curve-fitting is accomplished through solving a system of equations.

EXAMPLE 5

Using a System to Fit a Line to Two Data Points

Find the equation of the line $y = ax + b$ that passes through the data points $(2, 13)$ and $(4.5, 21.75)$, using a system of equations.

SOLUTION If $y = ax + b$ represents the equation of the line, we must find a and b. Since $(2, 13)$ lies on the line, we substitute 2 for x and 13 for y to get an equation in a and b.

$$y = ax + b$$
$$13 = a \cdot 2 + b \qquad x = 2, y = 13$$
$$2a + b = 13$$

A second equation in a and b can be found using the point $(4.5, 21.75)$.

$$y = ax + b$$
$$21.75 = a \cdot 4.5 + b \qquad x = 4.5, y = 21.75$$
$$4.5a + b = 21.75$$

We now have a system of two linear equations in a and b.

$$2a + b = 13$$
$$4.5a + b = 21.75$$

The variable b can be eliminated by multiplying the second equation by -1 and adding.

$$
\begin{array}{r}
2a + b = 13 \\
-4.5a - b = -21.75 \\
\hline
-2.5a = -8.75 \\
a = 3.5 \qquad \text{Divide by } -2.5.
\end{array}
$$

To find the value of b, we let 3.5 replace a in either of the original equations. Using $2a + b = 13$, we have

$$2(3.5) + b = 13$$
$$7 + b = 13$$
$$b = 6.$$

Therefore, the equation of the line is $y = 3.5x + 6$. Figure 38 shows the graph of this equation, and parts (a) and (b) illustrate that the given points do indeed lie on the line. ▮

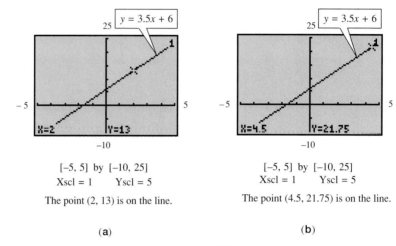

[−5, 5] by [−10, 25]
Xscl = 1   Yscl = 5

The point (2, 13) is on the line.

[−5, 5] by [−10, 25]
Xscl = 1   Yscl = 5

The point (4.5, 21.75) is on the line.

(a)

(b)

FIGURE 38

FOR **GROUP DISCUSSION**

Suppose that the data points of Example 5 represent ordered pairs of the form (miles traveled, cost of the tour) for a particular tour guide service. Use the results of Example 5 to predict the cost of a tour that covers 5 miles.

Usually, as the price of an item goes up, demand for the item goes down and the supply of the item goes up. Changes in gasoline prices illustrate this situation. The price where supply and demand are equal is called the *equilibrium price,* and the resulting supply or demand is called the *equilibrium supply* or *equilibrium demand.*

EXAMPLE 6

Solving a Supply and Demand Problem

(a) Suppose that the supply of a product is related to its price by the equation

$$p = \frac{2}{3}q,$$

where p is price in dollars and q is supply in appropriate units. (Here, q stands for quantity.) Find the price for the supply levels $q = 9$ and $q = 18$.

SOLUTION   When $q = 9$,

$$p = \frac{2}{3}q = \frac{2}{3}(9) = 6.$$

When $q = 18$,

$$p = \frac{2}{3}q = \frac{2}{3}(18) = 12.$$

(b) Suppose demand and price for the same product are related by

$$p = -\frac{1}{3}q + 18,$$

where p is price and q is demand. Find the price for the demand levels $q = 6$ and $q = 18$.

SOLUTION When $q = 6$,

$$p = -\frac{1}{3}q + 18 = -\frac{1}{3}(6) + 18 = 16,$$

and when $q = 18$,

$$p = -\frac{1}{3}q + 18 = -\frac{1}{3}(18) + 18 = 12.$$

(c) Using $x = q$ and $y = p$, graph both functions on the same axes.

SOLUTION Enter $y_1 = \frac{2}{3}x$ and $y_2 = -\frac{1}{3}x + 18$, and graph in an appropriate window. See Figure 39.

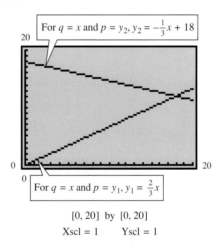

$$[0, 20] \text{ by } [0, 20]$$
$$\text{Xscl} = 1 \qquad \text{Yscl} = 1$$

FIGURE 39

(d) Use a system to find the equilibrium price, supply, and demand.

SOLUTION Solve the system

$$p = \frac{2}{3}q$$

$$p = -\frac{1}{3}q + 18$$

using substitution.

$$\frac{2}{3}q = -\frac{1}{3}q + 18$$

$$q = 18$$

This gives 18 units as the equilibrium supply or demand. Find the equilibrium price by substituting 18 for q in either equation. Using $p = \frac{2}{3}q$ gives

$$p = \frac{2}{3}(18) = 12$$

or \$12, the equilibrium price. The point $(18, 12)$ that gives the equilibrium values is shown in Figure 40. ∎

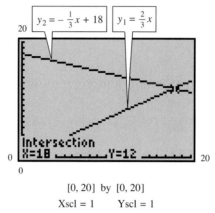

$$[0, 20] \text{ by } [0, 20]$$
$$\text{Xscl} = 1 \qquad \text{Yscl} = 1$$

The point $(18, 12)$ is the equillibrium point.

FIGURE 40

6.4 EXERCISES

In Exercises 1–6, a system of equations is given, along with the graphs of the two lines and the point of intersection of the lines (indicated at the bottom of the screen). Substitute the indicated values of x and y into both equations of the system to verify that the ordered pair (x, y) is indeed the solution of the system.

1. $\quad 4x - y = 3$
$\quad\quad -2x + 3y = 1$

2. $\quad 5x + 3y = 1$
$\quad\quad -3x - 4y = 6$

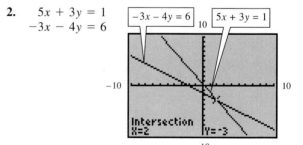

3. $2x - 5y - 10 = 0$
$\quad 3x + y - 15 = 0$

4. $\quad -3x + 5y - 2 = 0$
$\quad\quad 2x - 3y - 1 = 0$

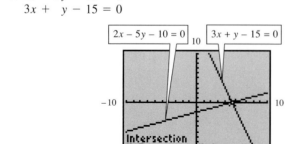

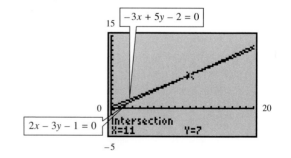

5. $4x + 2y = 3$
 $-3x - 3y = 0$

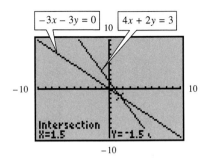

6. $.5x + y = -.2$
 $10x + 5y = .5$

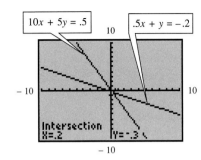

Without graphing, decide whether the system has a single solution, no solution, or infinitely many solutions.

7. One line has positive slope and one line has negative slope.

8. One line has slope 0 and one line has undefined slope.

9. Both lines have slope 0 and have the same y-intercept.

10. Both lines have undefined slope and have the same x-intercept.

11. $y = 3x + 6$
 $y = 3x + 9$

12. $y = 4x + 8$
 $y = 4x + 12$

13. $x + y = 4$
 $kx + ky = 4k \ (k \neq 0)$

14. $x + y = 10$
 $kx + ky = 9k \ (k \neq 0)$

*Use the substitution method to solve each system. Check your answers. (*Hint: *In Exercises 19 and 20, clear fractions first by multiplying through by the least common denominator.) Support your answers graphically if you wish.*

15. $y = 2x + 3$
 $3x + 4y = 78$

16. $y = 4x - 6$
 $2x + 5y = -8$

17. $3x - 2y = 12$
 $5x = 4 - 2y$

18. $8x + 3y = 2$
 $5x = 17 + 6y$

19. $\dfrac{x}{2} = \dfrac{7}{6} - \dfrac{y}{3}$

 $\dfrac{2x}{3} = \dfrac{3y}{2} - \dfrac{7}{3}$

20. $\dfrac{3x}{4} = 4 + \dfrac{y}{2}$

 $\dfrac{x}{3} + \dfrac{5y}{4} = -\dfrac{7}{6}$

21. Refer to Example 2 in this section. If we began solving the system by eliminating y, by what numbers might we have multiplied equations (4) and (5)?

22. Explain how one can determine whether a system is inconsistent or has dependent equations when using the substitution or elimination method.

Use the addition method to solve each system. Support your answers graphically if you wish.

23. $5x + 3y = 7$
 $7x - 3y = -19$

24. $2x + 7y = -8$
 $-2x + 3y = -12$

25. $3x + 2y = 5$
 $6x + 4y = 8$

26. $9x - 5y = 1$
 $-18x + 10y = 1$

27. $2x - 3y = -7$
 $5x + 4y = 17$

28. $4x + 3y = -1$
 $2x + 5y = 3$

29. $5x + 7y = 6$
 $10x - 3y = 46$

30. $12x - 5y = 9$
 $3x - 8y = -18$

31. $4x - y = 9$
 $-8x + 2y = -18$

32. $3x + 5y + 2 = 0$
 $9x + 15y + 6 = 0$

33. $\dfrac{x}{2} + \dfrac{y}{3} = 8$

 $\dfrac{2x}{3} + \dfrac{3y}{2} = 17$

34. $\dfrac{x}{5} + 3y = 31$

 $2x - \dfrac{y}{5} = 8$

Relating Concepts

The system

$$\frac{5}{x} + \frac{15}{y} = 16$$

$$\frac{5}{x} + \frac{4}{y} = 5$$

is not a linear system, because the variables appear in the denominator. However, it can be solved in a manner similar to the method for solving a linear system by using a substitution-of-variable technique. Let $t = \frac{1}{x}$ and let $u = \frac{1}{y}$.

35. Write a system of equations in t and u by making the appropriate substitutions.

36. Solve the system in Exercise 35 for t and u.

37. Solve the given system for x and y by using the equations relating t and x, and u and y.

38. Refer to the first equation in the given system, and solve for y in terms of x to obtain a rational function.

39. Repeat Exercise 38 for the second equation in the given system.

40. Using a viewing window of $[0, 10]$ by $[0, 2]$, show that the point of intersection of the graphs of the functions in Exercises 38 and 39 has the same x and y values as found in Exercise 37.

Use the substitution-of-variable technique to solve the systems in Exercises 41–43 analytically.

41. $\dfrac{2}{x} + \dfrac{1}{y} = \dfrac{3}{2}$

$\dfrac{3}{x} - \dfrac{1}{y} = 1$

42. $\dfrac{2}{x} + \dfrac{1}{y} = 11$

$\dfrac{3}{x} - \dfrac{5}{y} = 10$

43. $\dfrac{2}{x} + \dfrac{3}{y} = 18$

$\dfrac{4}{x} - \dfrac{5}{y} = -8$

Use only a graphical approach to solve the systems in Exercises 44–46. Express solutions to the nearest hundredth. In Exercise 46, use the e and π keys on your calculator.

44. $\sqrt{5}x + \sqrt[3]{6}y = 9$

$\sqrt{2}x + \sqrt[5]{9}y = 12$

45. $2.67x - .38y = 6$

$4.19x + .23y = 4$

46. $\pi x + ey = 3$

$ex + \pi y = 4$

Use the idea of curve-fitting as illustrated in Example 5 to find an equation of the form $y = ax + b$ whose graph contains the given data points.

47. $(-2, 1)$ and $(-1, -2)$

48. $(4, 6)$ and $(-5, -3)$

49. $(2, 5)$ and $(-1, 4)$

50. $(-3, 2)$ and $(-4, -5)$

51. $(1, 2.2)$ and $(5, 19.4)$

52. $(2, -6.1)$ and $(10, -42.9)$

Use a system of equations to solve each problem.

53. To start a new business, Shannon Mulkey borrowed money from two financial institutions. One loan was at 7% interest and the other was for one-third as much money at 8% interest. How much did she borrow at each rate if the total amount of annual interest was $1160?

54. Wally Smart has invested a total of $20,000 in two ways. Part of the money is in certificates of deposit paying 3.75% interest, while the rest is in municipal bonds that pay 8.2% interest. How much is there in each account if the total annual interest is $1417.50?

55. Alexis is a botanist who has patented a successful type of plant food that contains two chemicals, X and Y. Eight hundred kilograms of these chemicals will be used to make a batch of the food, and the ratio of X to Y must be 3 to 2. How much of each chemical should be used?

56. A manufacturer of portable compact disc players shipped 200 of the players to its two Quebec warehouses. It costs $3 per unit to ship to Warehouse A, and $2.50 per unit to ship to Warehouse B. If the total shipping cost was $537.50, how many were shipped to each warehouse?

57. Octane ratings show the percent of isooctane in gasoline. An octane rating of 98, for example, indicates a gasoline that is 98% isooctane. How many gallons of 98-octane gasoline should be mixed with 92-octane gasoline to produce 40 gallons of 94-octane gasoline?

58. A chemist needs 10 liters of a 24% alcohol solution. She has on hand a 30% alcohol solution and an 18% alcohol solution. How many liters of each should be mixed to get the required solution?

59. A supplier of poultry sells 20 turkeys and 8 chickens for $74. He also sells 15 turkeys and 24 chickens for $87. Find the cost of a turkey and the cost of a chicken.

60. Jose Ortega is a building contractor. If he hires 8 bricklayers and 2 roofers, his daily payroll is $960, while 10 bricklayers and 5 roofers require a daily payroll of $1500. What is the daily wage of a bricklayer and the daily wage of a roofer?

61. During summer vacation Hector and Ann earned a total of $6496. Hector worked 8 days less than Ann and earned $4 per day less. Find the number of days he worked and the daily wage made, if the total number of days worked by both was 72.

62. A bank teller has ten-dollar bills and twenty-dollar bills. He has 25 more twenties than tens. The value of the bills is $2900. How many of each kind does he have?

63. Mike Karelius plans to invest $30,000 he won in a lottery. With part of the money he buys a mutual fund, paying 4.5% a year. The rest he invests in utility bonds paying 5% per year. The first year his investments bring a return of $1410. How much is invested at each rate?

64. How much milk that is 3% butterfat should be mixed with milk that is 18% butterfat to get 25 gal of milk that is 4.8% butterfat?

The break-even point for a company is the point where its costs equal its revenues. If both cost and revenue are expressed as linear equations, the break-even point is the solution of a linear system. In each of the following exercises, C represents the cost to produce x items, and R represents the revenue from the sale of x items. Use the substitution method to find the break-even point in each case, that is, the point where C = R. Then find the value of C and R at that point.

65. $C = 1.5x + 252$
$R = 5.5x$

66. $C = 2.5x + 90$
$R = 3x$

67. $C = 20x + 10,000$
$R = 30x - 11,000$

68. $C = 4x + 125$
$R = 9x - 200$

In each of the following exercises, p is the price of an item, while q represents the supply in one equation and the demand in the other. Find the equilibrium price and the equilibrium supply/demand.

69. $p = 80 - \dfrac{3}{5}q$

$p = \dfrac{2}{5}q$

70. $p = 630 - \dfrac{3}{4}q$

$p = \dfrac{3}{4}q$

71. $3p = 84 - 2q$
$3p - q = 0$

72. $4p + q = 80$
$3p - 2q = 5$

73. Let the supply and demand equations for banana smoothies be

$$\text{supply: } p = \frac{3}{2}q \quad \text{and} \quad \text{demand: } p = 81 - \frac{3}{4}q.$$

(a) Graph these on the same axes. Let $x = q$, $y = p$, and use the window [0, 120] by [0, 100]. Find the point of intersection.
(b) Find the equilibrium demand analytically.
(c) Find the equilibrium price analytically.

74. Let the supply and demand equations for chocolate frozen yogurt be given by

$$\text{supply: } p = \frac{2}{5}q \quad \text{and} \quad \text{demand: } p = 100 - \frac{2}{5}q.$$

(a) Graph these on the same axes. Let $x = q$, $y = p$, and use the window [0, 250] by [0, 120]. Find the point of intersection.
(b) Find the equilibrium demand analytically.
(c) Find the equilibrium price analytically.

75. The figure shows a graph that accompanied a news article in *The Sacramento Bee* newspaper, July 20, 1994. The graph indicates that the percent of children living with a never-married parent has increased and appears about to overtake the percent of children living with a single divorced parent.*

(a) From the graph estimate the percents in 1983 and 1993 for each curve. Let 1983 be represented by 0, and find two data points for each curve.

(b) Use these data points to write a linear equation that approximates each curve.

(c) Use a calculator to graph the lines and find the point of intersection. Interpret your answer.

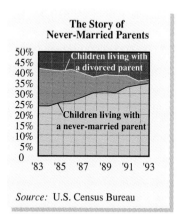

Source: U.S. Census Bureau

Relating Concepts

Consider the linear equation $-3(x + 2) - x = 2x + 3(2 - x)$.

76. Solve the equation analytically.

77. Check your solution in Exercise 76 by direct substitution. What is the value obtained on each side of the equation?

Consider the system of linear equations

$$y = -3(x + 2) - x$$
$$y = 2x + 3(2 - x).$$

78. Solve the system analytically using the substitution method.

79. Check your solution graphically using the viewing window $[-10, 10]$ by $[-10, 20]$.

80. Write a paragraph explaining the connections between the equation solved in Exercise 76 and the system solved in Exercise 78.

---

*From *The Sacramento Bee*, July 20, 1994. Copyright, The Sacramento Bee, 1995. Reprinted by permission.

6.5 NONLINEAR SYSTEMS OF EQUATIONS

Introduction ▮ Nonlinear Systems Involving Conic Sections ▮ Other
Nonlinear Systems ▮ An Application of a Nonlinear System: Polynomial
Connections

Introduction

A nonlinear system of equations is one in which at least one of the equations is not
a first degree equation. Some examples of nonlinear systems are

$$x^2 + y^2 = 9 \qquad 3x^2 - 2y = 5 \qquad\qquad x^2 + y^2 = 4$$
$$2x - y = 3, \qquad x + 3y = -4, \qquad \text{and} \qquad 2x^2 - y^2 = 8.$$

Since nonlinear systems vary, depending upon the type of equations in the system,
different solution methods are required for different systems. Although the substitution method discussed in Section 6.4 can almost always be used, it may not be the
most efficient approach to the solution. This section illustrates some of the methods
best suited to solving certain types of nonlinear systems.

Nonlinear Systems Involving Conic Sections

One of the most common types of nonlinear systems is one involving at least one
conic section. It is often helpful to visualize the types of graphs involved in a
nonlinear system to get an idea of the possible numbers of ordered pairs of real
numbers that may be in the solution set of the system. For example, a line and a
parabola may have 0, 1, or 2 points of intersection, as shown in Figure 41. A
parabola and an ellipse may have 0, 1, 2, 3, or 4 points of intersection, as shown
in Figure 42 on the following page.

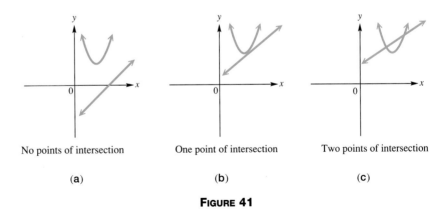

| No points of intersection | One point of intersection | Two points of intersection |
| (a) | (b) | (c) |

FIGURE 41

Nonlinear systems can be solved by the addition method, the substitution
method, or a combination of the two. The substitution method is usually the most
efficient when one of the equations is linear, as shown in the first example.

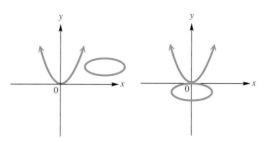

No points of intersection   One point of intersection

(a)   (b)

Two points of intersection   Three points of intersection   Four points of intersection

(c)   (d)   (e)

FIGURE 42

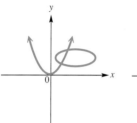

Solve the system

$$x^2 + y^2 = 9 \qquad [1]$$
$$2x - y = 3 \qquad [2]$$

by substitution and support the solution graphically.

SOLUTION   The graph of equation (1) is a circle and the graph of equation (2) is a line. Visualizing the possibilities indicates that there may be 0, 1, or 2 points of intersection. When solving a system of this type, it is best to solve the linear equation for one of the two variables; then substitute the resulting expression into the nonlinear equation to obtain an equation in one variable. Solving equation (2) for y gives

$$2x - y = 3 \qquad [2]$$
$$y = 2x - 3.$$

Substitute $2x - 3$ for y in equation (1) to get

$$x^2 + (2x - 3)^2 = 9$$

$$x^2 + 4x^2 - 12x + 9 = 9 \qquad \text{Square } 2x - 3.$$
$$5x^2 - 12x = 0. \qquad \text{Combine terms.}$$

Solve this quadratic equation by factoring.

$$x(5x - 12) = 0 \qquad \text{Common factor is } x.$$

$$x = 0 \qquad \text{or} \qquad x = \frac{12}{5}$$

Let $x = 0$ in the equation $y = 2x - 3$ to get $y = -3$. If $x = \frac{12}{5}$, then $y = \frac{9}{5}$. The solution set of the system is $\{(0, -3), (\frac{12}{5}, \frac{9}{5})\}$.

To support our solution graphically, we graph equation (1) as the union of the two graphs $y_1 = \sqrt{9 - x^2}$ and $y_2 = -\sqrt{9 - x^2}$, and we graph equation (2) as $y_3 = 2x - 3$. As seen in Figure 43, the points of intersection of the line and the circle are $(0, -3)$ and $(2.4, 1.8)$, confirming our analytic solution. ∎

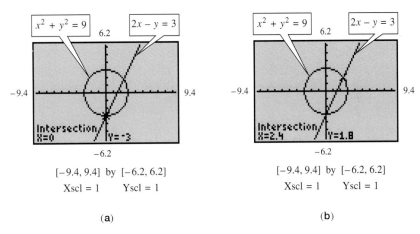

$[-9.4, 9.4]$ by $[-6.2, 6.2]$

Xscl = 1   Yscl = 1

(a)

$[-9.4, 9.4]$ by $[-6.2, 6.2]$

Xscl = 1   Yscl = 1

(b)

FIGURE 43

EXAMPLE **2**

Solving A Nonlinear System By Substitution

Use the substitution method to solve the system

$$3x^2 - 2y = 5 \tag{3}$$
$$x + 3y = -4, \tag{4}$$

and support graphically.

SOLUTION The graph of equation (3) is a parabola, and that of equation (4) is a line. There may be 0, 1, or 2 points of intersection of these graphs.

Although either equation could be solved for either variable, it is best to use the linear equation. Here, it is simpler to solve equation (4) for y, since x is squared in equation (3).

$$x + 3y = -4$$
$$3y = -4 - x$$
$$y = \frac{-4 - x}{3} \tag{5}$$

Substituting this value of y into equation (3) gives

$$3x^2 - 2\left(\frac{-4 - x}{3}\right) = 5.$$

Multiply both sides by 3 to eliminate the denominator.

$$9x^2 - 2(-4 - x) = 15$$
$$9x^2 + 8 + 2x = 15$$
$$9x^2 + 2x - 7 = 0$$
$$(9x - 7)(x + 1) = 0$$
$$x = \frac{7}{9} \quad \text{or} \quad x = -1$$

Substitute both values of x into equation (5) to find the corresponding y-values.

$$y = \frac{-4 - \left(\dfrac{7}{9}\right)}{3} \quad \text{or} \quad y = \frac{-4 - (-1)}{3}$$

$$y = -\frac{43}{27} \qquad\qquad y = -1$$

Check in the original system that the solution set is $\{(\frac{7}{9}, -\frac{43}{27}), (-1, -1)\}$.

TECHNOLOGICAL NOTE
The decimal values for x and y displayed at the bottom of Figure 44(b) represent the repeating decimal forms for the fractions 7/9 and -43/27, respectively.

To support our solution graphically, we graph equation (3) as $y_1 = 1.5x^2 - 2.5$ and equation (4) as $y_2 = -\frac{1}{3}x - \frac{4}{3}$. As Figure 44 indicates, the points of intersection are $(-1, -1)$ and $(.\overline{7}, -1.\overline{592})$, confirming our analytic solution.   ∎

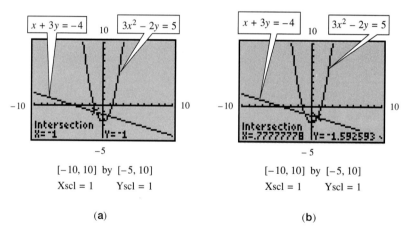

(a)   (b)

FIGURE 44

FOR **GROUP DISCUSSION**

The equation $9x^2 + 2x - 7 = 0$ arose in the solution of the system in Example 2. While it was solved by factoring, we could just have easily solved it by the quadratic formula (Section 3.3).

1. What is the discriminant of this equation?
2. How many solutions does the system in Example 2 have?
3. If the discriminant of the equation were zero, how many points of intersection would the line and the parabola have?
4. If the discriminant were negative, how many points of intersection would there be?

Some nonlinear systems can be solved by the addition method. This method works well if we can eliminate completely one variable from the system, as shown in the following example.

Solve the system

$$x^2 + y^2 = 4 \qquad \text{[6]}$$
$$2x^2 - y^2 = 8 \qquad \text{[7]}$$

by addition and support graphically.

SOLUTION   The graph of equation (6) is a circle and that of equation (7) is a hyperbola. Visualizing these suggests that there may be 0, 1, 2, 3, or 4 points of intersection.

Add the two equations to eliminate y^2.

$$
\begin{aligned}
x^2 + y^2 &= 4 \\
2x^2 - y^2 &= 8 \\
\hline
3x^2 \phantom{{}- y^2} &= 12 \\
x^2 &= 4
\end{aligned}
$$

$$x = 2 \qquad \text{or} \qquad x = -2$$

Substituting into equation (6) gives the corresponding values of y.

$$
\begin{array}{ccc}
2^2 + y^2 = 4 & \text{or} & (-2)^2 + y^2 = 4 \\
y^2 = 0 & & y^2 = 0 \\
y = 0 & & y = 0
\end{array}
$$

The solution set of the system is $\{(2, 0), (-2, 0)\}$.

To support the solution graphically, we graph equation (6) as the union of $y_1 = \sqrt{4 - x^2}$ and $y_2 = -\sqrt{4 - x^2}$, and equation (7) as the union of $y_3 = \sqrt{2x^2 - 8}$ and $y_4 = -\sqrt{2x^2 - 8}$. As suggested in Figure 45, the two points of intersection are indeed $(2, 0)$ and $(-2, 0)$.   ∎

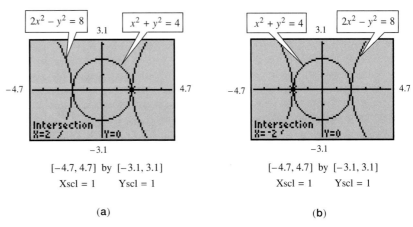

$[-4.7, 4.7]$ by $[-3.1, 3.1]$
Xscl = 1   Yscl = 1

$[-4.7, 4.7]$ by $[-3.1, 3.1]$
Xscl = 1   Yscl = 1

(a)   (b)

FIGURE 45

Nonlinear systems sometimes lead to solutions whose values include imaginary numbers, as shown in the next example.

Solve the system

$$x^2 + y^2 = 5 \qquad\qquad\qquad [8]$$

$$4x^2 + 3y^2 = 11. \qquad\qquad\qquad [9]$$

SOLUTION   A circle and an ellipse (equations (8) and (9), respectively) may intersect in 0, 1, 2, 3, or 4 points.

Multiplying equation (8) on both sides by -3 and adding the result to equation (9) gives

$$
\begin{aligned}
-3x^2 - 3y^2 &= -15 \\
4x^2 + 3y^2 &= 11 \\
\hline
x^2 &= -4.
\end{aligned}
$$

By the square root property,

$$x = \pm\sqrt{-4}$$

$$x = 2i \qquad \text{or} \qquad x = -2i.$$

Find y by substitution. Using equation (8) gives

$$-4 + y^2 = 5$$

$$y^2 = 9$$

$$y = 3 \qquad \text{or} \qquad y = -3,$$

for either $\qquad\qquad x = 2i \qquad \text{or} \qquad x = -2i.$

Checking the solutions in the given system shows that the solution set is $\{(2i, 3),$ $(2i, -3), (-2i, 3), (-2i, -3)\}$.

The graphs of $x^2 + y^2 = 5$ and $4x^2 + 3y^2 = 11$ are shown in Figure 46. Because each ordered pair in the solution set contains an imaginary number, there are no points of intersection of the two graphs. Only when solutions are ordered pairs composed of real numbers will points of intersection appear in the real number plane.   ◻

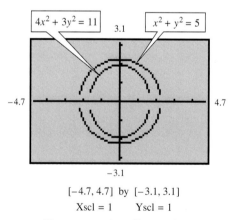

$[-4.7, 4.7]$ by $[-3.1, 3.1]$

Xscl = 1   Yscl = 1

There are no points of intersection.

FIGURE 46

Sometimes conic sections do not appear in the standard forms described in Sections 6.1 and 6.2, as shown in the next example. Analytic solution of the system requires a combination of methods.

EXAMPLE 5

Solving a Nonlinear
System by a
Combination of
Methods

Solve the system

$$x^2 + 3xy + y^2 = 22 \qquad \text{[10]}$$
$$x^2 - xy + y^2 = 6, \qquad \text{[11]}$$

and support graphically.

SOLUTION   Multiply both sides of equation (11) by -1, and then add the result to equation (10), as follows, to eliminate the x^2 and y^2 terms.

$$\begin{aligned} x^2 + 3xy + y^2 &= 22 \\ -x^2 + xy - y^2 &= -6 \\ \hline 4xy &= 16 \end{aligned} \qquad \text{[12]}$$

Now solve equation (12) for either x or y and substitute the result into one of the given equations. Solving for y gives

$$y = \frac{4}{x}, \quad \text{if } x \neq 0. \qquad \text{[13]}$$

(The restriction $x \neq 0$ is included since if $x = 0$ there is no value of y that satisfies the system.) Substituting for y in equation (11) (equation (10) could have been used) and simplifying gives

$$x^2 - x\left(\frac{4}{x}\right) + \left(\frac{4}{x}\right)^2 = 6.$$

Now solve for x.

$$x^2 - 4 + \frac{16}{x^2} = 6$$
$$x^4 - 4x^2 + 16 = 6x^2$$
$$x^4 - 10x^2 + 16 = 0$$
$$(x^2 - 2)(x^2 - 8) = 0$$
$$x^2 = 2 \quad \text{or} \quad x^2 = 8$$
$$x = \sqrt{2} \quad \text{or} \quad x = -\sqrt{2} \quad \text{or} \quad x = 2\sqrt{2} \quad \text{or} \quad x = -2\sqrt{2}$$

Substitute these x values into equation (13) to find corresponding values of y.

$$\text{If } x = \sqrt{2}, \, y = \frac{4}{\sqrt{2}} = 2\sqrt{2}.$$

$$\text{If } x = -\sqrt{2}, \, y = \frac{4}{-\sqrt{2}} = -2\sqrt{2}.$$

$$\text{If } x = 2\sqrt{2}, \, y = \frac{4}{2\sqrt{2}} = \sqrt{2}.$$

$$\text{If } x = -2\sqrt{2}, \, y = \frac{4}{-2\sqrt{2}} = -\sqrt{2}.$$

The solution set of the system is

$$\{(\sqrt{2}, 2\sqrt{2}), (-\sqrt{2}, -2\sqrt{2}), (2\sqrt{2}, \sqrt{2}), (-2\sqrt{2}, -\sqrt{2})\}.$$

To support these solutions graphically, we must first solve equations (10) and (11) for y. This requires the use of the quadratic formula. To solve equation (10) for y, first write in the form

$$y^2 + 3xy + x^2 - 22 = 0.$$

Then with $a = 1$, $b = 3x$, and $c = x^2 - 22$, use the quadratic formula to find y_1 and y_2:

$$y_1 = \frac{-3x + \sqrt{(3x)^2 - 4(1)(x^2 - 22)}}{2} = \frac{-3x + \sqrt{5x^2 + 88}}{2}$$

$$y_2 = \frac{-3x - \sqrt{(3x)^2 - 4(1)(x^2 - 22)}}{2} = \frac{-3x - \sqrt{5x^2 + 88}}{2}.$$

Similarly, we can solve equation (11) for y. Rewriting the equation as

$$y^2 - xy + x^2 - 6 = 0$$

TECHNOLOGICAL NOTE
If you are able to dupli-
cate the graph shown in
Figure 47 on your own
calculator, consider yourself an accomplished cal-
culator-grapher! It is probably the most difficult
screen to duplicate in this
entire text.

and using $a = 1$, $b = -x$, and $c = x^2 - 6$ gives us y_3 and y_4:

$$y_3 = \frac{x + \sqrt{(-x)^2 - 4(1)(x^2 - 6)}}{2} = \frac{x + \sqrt{-3x^2 + 24}}{2}$$

$$y_4 = \frac{x - \sqrt{(-x)^2 - 4(1)(x^2 - 6)}}{2} = \frac{x - \sqrt{-3x^2 + 24}}{2}.$$

Graphs of these four functions (composing the two graphs) are seen in Figure 47, suggesting that we have a hyperbola and an ellipse. The point of intersection indicated supports the solution $(\sqrt{2}, 2\sqrt{2})$ determined analytically. The other three solutions can be supported similarly.   ▢

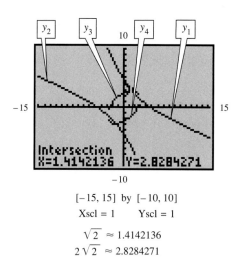

[−15, 15] by [−10, 10]
Xscl = 1   Yscl = 1

$$\sqrt{2} \approx 1.4142136$$
$$2\sqrt{2} \approx 2.8284271$$

FIGURE 47

Other Nonlinear Systems

A nonlinear system may involve an equation whose graph is neither a line nor a conic section. The following example shows one such system.

EXAMPLE 6

Solving a Nonlinear System with an Absolute Value Equation

Solve the system

$$x^2 + y^2 = 16 \tag{14}$$
$$|x| + y = 4. \tag{15}$$

SOLUTION The substitution method is required here. Equation (15) can be rewritten as $|x| = 4 - y$, and the definition of absolute value can be used to get

$$x = 4 - y \quad \text{or} \quad x = -(4 - y) = y - 4. \tag{16}$$

(Since $|x| \geq 0$ for all real x, $4 - y \geq 0$, or $4 \geq y$.) Substituting from either part of equation (16) into equation (14) gives the same result.

$$(4 - y)^2 + y^2 = 16 \quad \text{or} \quad (y - 4)^2 + y^2 = 16$$

Since $(4 - y)^2 = (y - 4)^2 = 16 - 8y + y^2$, either equation becomes

$$(16 - 8y + y^2) + y^2 = 16$$
$$2y^2 - 8y = 0$$
$$2y(y - 4) = 0$$
$$y = 0 \quad \text{or} \quad y = 4.$$

From equation (16),

$$\text{If } y = 0, \text{ then } x = 4 - 0 \quad \text{or} \quad x = 0 - 4.$$
$$x = 4 \qquad\qquad x = -4$$
$$\text{If } y = 4, \text{ then } x = 4 - 4 = 0.$$

The solution set is $\{(4, 0), (-4, 0), (0, 4)\}$. Figure 48 indicates that $(0, 4)$ is a solution, and suggests that $(4, 0)$ and $(-4, 0)$ are the two other solutions. ❑

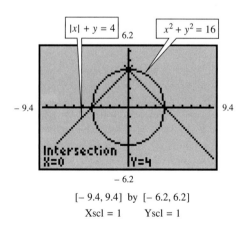

$$[-9.4, 9.4] \text{ by } [-6.2, 6.2]$$
$$\text{Xscl} = 1 \qquad \text{Yscl} = 1$$

FIGURE 48

Graphics calculators will often allow us to solve systems using purely graphical methods. Many systems are difficult or even impossible to solve using strictly analytic methods.

EXAMPLE **7**

Solving a Nonlinear
System Graphically

Solve the system

$$y = 2^x \qquad \qquad \text{[17]}$$
$$|x + 2| - y = 0 \qquad \qquad \text{[18]}$$

using graphical methods.

SOLUTION   We enter equation (17) as $y_1 = 2^x$ and equation (18) as $y_2 = |x + 2|$. As seen in Figure 49, the various windows indicate that there are three points of intersection of the graphs, and thus three solutions. Using the capabilities of the calculator we find that $(2, 4)$ is an exact solution, and $(-2.22, .22)$ and $(-1.69, .31)$ are approximate solutions. Therefore, the solution set is

$$\{(2, 4), (-2.22, .22), (-1.69, .31)\}. \qquad ■$$

TECHNOLOGICAL NOTE
Example 7 provides an
excellent illustration as to
why proficiency in setting
various windows is impor-
tant in studying algebra
with a graphics calculator.

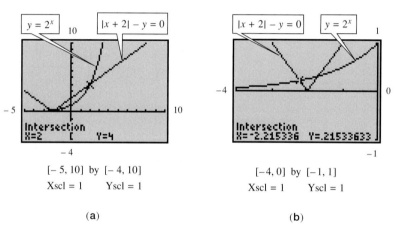

[−5, 10] by [−4, 10]
Xscl = 1   Yscl = 1

(a)

[−4, 0] by [−1, 1]
Xscl = 1   Yscl = 1

(b)

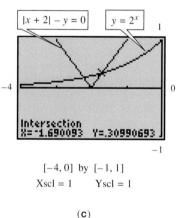

[−4, 0] by [−1, 1]
Xscl = 1   Yscl = 1

(c)

FIGURE 49

An Application of a Nonlinear System: Polynomial Connections

Sometimes applications lead to nonlinear systems, as shown in the next example. A knowledge of the theory of polynomial functions, introduced in Chapter 3, is often useful in solving such problems.

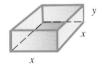

EXAMPLE 8

Solving an
Application Using a
Nonlinear System

A box with an open top has a square base and four sides of equal height. The volume of the box is 75 cubic inches, and the surface area is 85 square inches. What are the dimensions of the box?

SOLUTION In Figure 50 we show a sketch of such a box. If each side of the square base measures x inches and the height measures y inches, then the volume is

$$x^2 y = 75 \qquad \text{(Volume formula)}$$

and the surface area is

$$x^2 + 4xy = 85. \qquad \text{(Sum of areas of base and sides)}$$

FIGURE 50

Solve the first equation for y to get $y = \frac{75}{x^2}$, and substitute this into the second equation.

$$x^2 + 4x\left(\frac{75}{x^2}\right) = 85$$

$$x^2 + \frac{300}{x} = 85$$

$$x^3 + 300 = 85x \qquad \text{Multiply by } x.$$

$$x^3 - 85x + 300 = 0$$

The solutions of this equation are x-intercepts of the graph of $y = x^3 - 85x + 300$. A *complete* graph of this function indicates that there are three real solutions. However, one of them is negative and must be rejected. As Figure 51 indicates, one positive solution is 5 and the other positive solution is approximately 5.64. (In Exercise 87 you are asked to show that the exact value of this latter solution is $\frac{-5 + \sqrt{265}}{2}$.) By substituting back into the first equation, we find that when $x = 5$, $y = 3$, and when $x \approx 5.64$, $y \approx 2.36$. Therefore, this problem has two solutions: the box may have a base 5 inches by 5 inches and a height 3 inches, or it may have a base 5.64 inches by 5.64 inches and a height 2.36 inches.   ▮

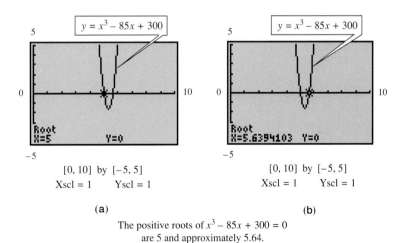

(a)   (b)

The positive roots of $x^3 - 85x + 300 = 0$
are 5 and approximately 5.64.

FIGURE 51

6.5   EXERCISES

In Exercises 1–8, a nonlinear system is given, along with a graph indicating the coordinates of one point of intersection of the graphs of the equations in the system. Show by direct substitution using the displayed values of x and y that the indicated point is indeed a solution of the system.

1. $y - x = 2$
   $x^2 + y^2 = 34$

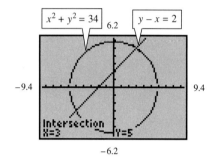

2. $y = 2x - 3$
   $y^2 - x^2 = 9$

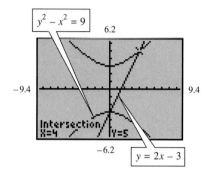

3. $y = x^2 - 13$
   $x^2 + y^2 = 25$

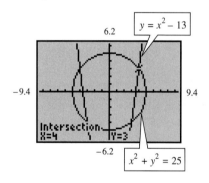

4. $x^2 - y^2 = 12$
   $x^2 + y^2 = 20$

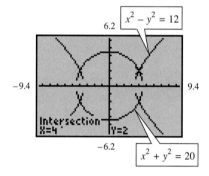

5. $y = x^2 + 3$
   $x^2 + y^2 = 9$

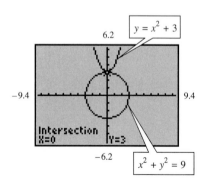

6. $y = \dfrac{4}{x}$
   $x + y = 5$

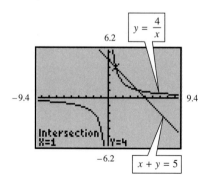

7. $y = \log_3 x$
$\quad x + y = 11$

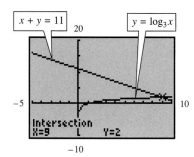

8. $y = |x + 3|$
$\quad y = 4x^2$

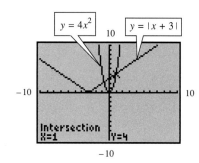

9. Determine visually the other solution of the system in Exercise 1. Then support your answer graphically.

10. Use symmetry to determine the other three solutions of the system in Exercise 4.

In Exercises 11–26, draw a sketch of the two graphs described, with the indicated number of points of intersection. (There may be more than one way to do this in some cases.)

11. a line and a circle;   no points

12. a line and a circle;   one point

13. a line and a circle;   two points

14. a line and an ellipse;   no points

15. a line and an ellipse;   one point

16. a line and an ellipse;   two points

17. a line and a hyperbola;   no points

18. a line and a hyperbola;   one point

19. a line and a hyperbola;   two points

20. a circle and an ellipse;   one point

21. a circle and an ellipse;   four points

22. a parabola and an ellipse;   one point

23. a parabola and an ellipse;   four points

24. a parabola and a hyperbola;   two points

25. a parabola and a hyperbola;   four points

26. a circle and a hyperbola;   two points

27. Write an explanation of the steps you would use to solve the system

$$x^2 + y^2 = 25$$
$$y = x - 1$$

by the substitution method.

28. Write an explanation of the steps you would use to solve the system

$$x^2 + y^2 = 12$$
$$x^2 - y^2 = 13$$

by the addition method.

Give all solutions of the following nonlinear systems of equations, including those with imaginary values. Use analytic methods, and support your solutions graphically.

29. $y = x^2$
$\quad x + y = 2$

30. $y = -x^2 + 2$
$\quad x - y = 0$

31. $y = (x - 1)^2$
$\quad x - 3y = -1$

32. $y = (x + 3)^2$
$\quad x + 2y = -2$

33. $y = x^2 + 4x$
$\quad 2x - y = -8$

34. $y = 6x + x^2$
$\quad 3x - 2y = 10$

35. $3x^2 + 2y^2 = 5$
$\quad x - y = -2$

36. $\quad x^2 + y^2 = 5$
$\quad -3x + 4y = 2$

37. $x^2 + y^2 = 8$
$\quad x^2 - y^2 = 0$

38. $\quad x^2 + y^2 = 10$
$\quad 2x^2 - y^2 = 17$

39. $5x^2 - y^2 = 0$
$\quad 3x^2 + 4y^2 = 0$

40. $\quad x^2 + y^2 = 4$
$\quad 2x^2 - 3y^2 = -12$

41. $3x^2 + y^2 = 3$
$\quad 4x^2 + 5y^2 = 26$

42. $\quad x^2 + 2y^2 = 9$
$\quad 3x^2 - 4y^2 = 27$

43. $2x^2 + 3y^2 = 5$
$\quad 3x^2 - 4y^2 = -1$

44. $3x^2 + 5y^2 = 17$
  $2x^2 - 3y^2 = 5$

45. $2x^2 + 2y^2 = 20$
  $3x^2 + 3y^2 = 30$

46. $x^2 + y^2 = 4$
  $5x^2 + 5y^2 = 28$

47. $9x^2 + 4y^2 = 1$
  $x^2 + y^2 = 1$

48. $2x^2 - 3y^2 = 8$
  $6x^2 + 5y^2 = 24$

49. $xy = -15$
  $4x + 3y = 3$

50. $xy = 8$
  $3x + 2y = -16$

51. $x^2 + 2xy - y^2 = 14$
  $x^2 - y^2 = -16$

52. $3x^2 + xy + 3y^2 = 7$
  $x^2 + y^2 = 2$

53. $x^2 - xy + y^2 = 5$
  $2x^2 + xy - y^2 = 10$

54. $3x^2 + 2xy - y^2 = 9$
  $x^2 - xy + y^2 = 9$

55. $x = |y|$
  $x^2 + y^2 = 18$

56. $2x + |y| = 4$
  $x^2 + y^2 = 5$

57. $y = |x - 1|$
  $y = x^2 - 4$

58. $2x^2 - y^2 = 4$
  $|x| = |y|$

Use a purely graphical method to solve each system in Exercises 59–64. Give x- and y-coordinates correct to the nearest hundredth.

59. $y = \log(x + 5)$
  $y = x^2$

60. $y = 5^x$
  $xy = 1$

61. $y = e^{x+1}$
  $2x + y = 3$

62. $x^2 - y^2 = 4.97$
  $x = 3y - 1$

63. $y = \sqrt[3]{x - 4}$
  $x^2 + y^2 = 6$

64. $y = \ln(2x + 3)$
  $x^2 + 2y^2 = 4$

Relating Concepts

Consider the nonlinear system

$$y - x^2 = -4$$
$$y = -x - 2.$$

65. Solve the first equation for y, and call it y_1. What kind of graph does y_1 have?

66. Let y in the second equation be called y_2. What kind of graph does y_2 have?

67. What are the possible numbers of solutions with real coordinates of this system?

68. Show by a graph that this system has two solutions.

69. Solve the system analytically, and support your solutions graphically.

70. Write $y_1 - y_2$ as a function of x, and call it y_3.

71. Show analytically that the equation $y_1 - y_2 = 0$ has as its solutions the x-coordinates of the points of intersection of the graphs in the system.

72. Support your result of Exercise 71 graphically.

Solve each problem using a system of equations in two variables.

73. Find two numbers whose ratio is 9 to 2 and whose product is 162.

74. Find two numbers whose ratio is 4 to 3 such that the sum of their squares is 100.

75. Does the straight line $3x - 2y = 9$ intersect the circle $x^2 + y^2 = 25$?

76. Do the parabola $y = x^2 + 4$ and the ellipse $2x^2 + y^2 - 4x - 4y = 0$ have any points in common?

77. Find the equation of the straight line through $(2, 4)$ that touches the parabola $y = x^2$ at only one point. (*Note:* Recall that a quadratic equation has a unique solution when the discriminant is 0.)

78. For what value of b will the line $x + 2y = b$ touch the circle $x^2 + y^2 = 9$ in only one point?

79. For what nonzero values of a do the graphs of $x^2 + y^2 = 25$ and $\frac{x^2}{a^2} + \frac{y^2}{25} = 1$ have exactly two points in common?

80. Find the equation of the line passing through the points of intersection of the graphs of $y = x^2$ and $x^2 + y^2 = 90$.

81. Suppose that you are given the equations of two circles that are known to intersect in exactly two points. Explain how you would find the equation of the only chord common to these circles.

82. In electronics, circuit gain is given by

$$G = \frac{Bt}{R + R_t}$$

where R is the value of a resistor, t is temperature, and B is a constant. The sensitivity of the circuit to temperature is given by

$$S = \frac{BR}{(R + R_t)^2}.$$

If $B = 3.7$ and t is 90K (Kelvin), find the values of R and R_t that will make $G = .4$ and $S = .001$.

83. The base of a box with a rectangular base and open top has length twice its width. Find the dimensions of the box if the perimeter of the base is 48 inches and the volume is 640 cubic inches.

84. The area of a certain right triangle is 546 square inches. The longer leg of the triangle measures 71 inches more than the shorter leg. Find the lengths of the sides of the triangle.

*R*elating Concepts

In Example 8 we solved a nonlinear system of equations by solving the related polynomial equation $x^3 - 85x + 300 = 0$.

85. Use synthetic division to show that a real solution of this equation is 5.

86. Factor $x^3 - 85x + 300$, using the results of Exercise 85.

87. From Exercise 86, show that the other positive solution has an exact value of $\frac{-5 + \sqrt{265}}{2}$.

88. Use the graph of $y = x^3 - 85x + 300$ to show that the negative solution of the polynomial equation is approximately -10.64.

89. What is the exact value of the negative solution?

90. Why was the negative solution rejected in the problem solved in Example 8?

*F*urther Explorations

Consider the following problem.

Los Angeles and Chico, CA, are 480 miles apart. Stuart leaves Chico at noon and heads for Los Angeles at 65 mph. John leaves Los Angeles, also at noon, and heads for Chico at 55 mph. If x represents the time traveled, how far from Los Angeles will they be when they pass, and what time will it be?

The first figure on the following page shows how Stuart's distance from Chico can be described as Y_1, and John's distance from Los Angeles can be described as Y_2. When they pass, $Y_1 = Y_2$. Solving analytically, we have

$$480 - 65x = 55x$$
$$480 = 120x$$
$$x = 4.$$

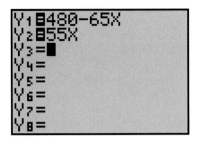

Stuart and John will meet at 4 P.M., 220 miles from Los Angeles. The next two figures support this analytic solution using both a TABLE and a graph.

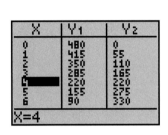

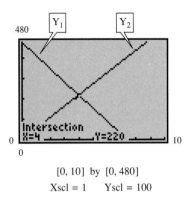

[0, 10] by [0, 480]
Xscl = 1   Yscl = 100

Now solve the following problem.

John leaves Slidell, LA, at 12 noon driving north. He maintains an average speed of 60 mph. Stuart leaves Slidell at 1:30 P.M. also heading north. Stuart averages 75 mph. Assuming Stuart does not get stopped for speeding, at what time will he overtake John, and how far will they be from Slidell at that time? Solve numerically using the TABLE. Support graphically, and confirm analytically.

6.6 SYSTEMS OF INEQUALITIES AND LINEAR PROGRAMMING

Graphs of Inequalities in Two Variables ▮ Systems of Inequalities ▮ Graphics Calculator Approach to Inequalities and Systems of Inequalities ▮ Linear Programming

Graphs of Inequalities in Two Variables

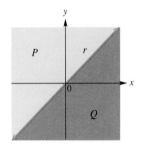

FIGURE 52

We begin our study of inequalities in two variables by considering the simplest type: linear inequalities. A line divides a plane into three sets of points: the points of the line itself and the points belonging to the two regions determined by the line. Each of these two regions is called a **half-plane.** In Figure 52 line r divides the plane into three different sets of points: line r, half-plane P, and half-plane Q. The points on r belong neither to P nor to Q. Line r is the **boundary** of each half-plane.

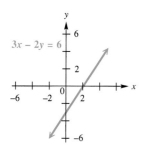

FIGURE 53

A **linear equality in two variables** is an inequality of the form

$$Ax + By > C, \qquad\qquad Ax + By \geq C,$$
$$Ax + By < C, \qquad or \qquad Ax + By \leq C,$$

where A, B, and C are real numbers, with A and B not both equal to 0.

Let us now see how a linear inequality in two variables is graphed using a traditional approach. Such a graph is a half-plane, perhaps with its boundary. For example, to graph the linear inequality $3x - 2y \leq 6$, first graph the boundary, $3x - 2y = 6$, as shown in Figure 53.

Since the points of the line $3x - 2y = 6$ satisfy $3x - 2y \leq 6$, this line is part of the solution. To decide which half-plane (the one above the line $3x - 2y = 6$ or the one below the line) is part of the solution, solve the original inequality for y.

$$3x - 2y \leq 6$$
$$-2y \leq -3x + 6$$
$$y \geq \frac{3}{2}x - 3 \qquad \text{Multiply by } -\tfrac{1}{2}; \text{ change } \leq \text{ to } \geq.$$

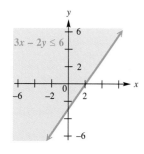

FIGURE 54

For a particular value of x, the inequality will be satisfied by all values of y that are *greater than* or equal to $\frac{3}{2}x - 3$. This means that the solution includes the half-plane *above* the line, as well as the line itself. The domain and range are both $(-\infty, \infty)$. See Figure 54.

There is an alternative method for deciding which side of the boundary line to shade. Choose as a test point any point not on the graph of the equation. The origin, $(0, 0)$, is often a good test point (as long as it does not lie on the boundary line). Substituting 0 for x and 0 for y in the inequality $3x - 2y \leq 6$ gives

$$3(0) - 2(0) \leq 6$$
$$0 \leq 6,$$

a true statement. Since $(0, 0)$ leads to a true result, it is part of the solution of the inequality. Shade the side of the graph containing $(0, 0)$, as shown in Figure 54.

Graph $x + 4y > 4$ using traditional graphing methods.

EXAMPLE 1

Graphing a Linear Inequality Using a Traditional Approach

SOLUTION The boundary here is the straight line $x + 4y = 4$. Since the points on this line do not satisfy $x + 4y > 4$, it is customary to graph the boundary as a dashed line, as in Figure 55. To decide which half-plane satisfies the inequality, use a test point. Choosing $(0, 0)$ as a test point gives $0 + 4 \cdot 0 > 4$, or $0 > 4$, a false statement. Since $(0, 0)$ leads to a false statement, shade the side of the graph *not* containing $(0, 0)$, as in Figure 55.

Alternatively, we can determine which half-plane should be shaded by solving the original inequality for y:

$$x + 4y > 4$$
$$4y > -x + 4$$
$$y > -\frac{1}{4}x + 1.$$

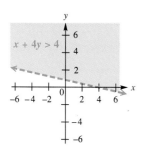

FIGURE 55

Since the inequality is true for any y value *greater than* $-\frac{1}{4}x + 1$, we shade the half-plane *above* the boundary line. This procedure will be helpful when we study the graphics calculator approach later in this section. ∎

> **EXAMPLE 2**
>
> Graphing a
> Nonlinear Inequality
> Using a Traditional
> Approach

Graph $y \leq 2x^2 - 3$ using traditional graphing methods.

SOLUTION   First graph the boundary, $y = 2x^2 - 3$. It is graphed as a solid curve, since the symbol in the given inequality includes the $=$ sign. The inequality is given in a convenient form; we want values of y that are *less than* $2x^2 - 3$, so we shade *below* the boundary. See Figure 56.   ◨

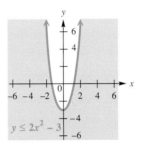

FIGURE 56

FOR GROUP DISCUSSION

Discuss how the method of using a test point could be applied in Example 2 to determine which side of the boundary line would be shaded.

> **EXAMPLE 3**
>
> Graphing
> Second-Degree
> Inequalities Using a
> Traditional
> Approach

Graph each inequality.

(a) $y^2 \leq 1 + 4x^2$

SOLUTION   Write the boundary $y^2 = 1 + 4x^2$ as $y^2 - 4x^2 = 1$, a hyperbola with y-intercepts 1 and -1, as shown in Figure 57. Select any point not on the hyperbola and test it in the original inequality. Since $(0, 0)$ satisfies this inequality, shade the area between the two branches of the hyperbola, as shown in Figure 57. The points on the hyperbola are part of the solution.

(b) $y^2 \geq 1 + 4x^2$

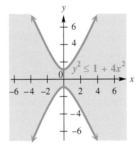   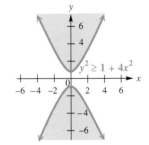

FIGURE 57   **FIGURE 58**

SOLUTION   The boundary is the same as in part (a). Since $(0, 0)$ makes this inequality false, the *two* regions above and below the branches of the hyperbola are shaded, as shown in Figure 58. Again, the points on the hyperbola itself are included in the solution.   ◨

Systems of Inequalities

Just as the solution set of a system of equations is the set of all solutions common to all the equations in the system, the solution set of a system of inequalities is the set of all solutions common to all the inequalities in the system. The solution set of a system of inequalities, such as

$$x + y < 3$$
$$y > \frac{1}{2}x^2,$$

is the *intersection* of the solution sets of the individual inequalities. This is best visualized by its graph. Using a traditional graphing approach, we graph each inequality separately and then identify the region of solution by shading heavily the region common to all the graphs.

EXAMPLE 4

Graphing a System of Inequalities Using a Traditional Approach

---

Use traditional graphing methods to graph the solution set of the system

$$x + y < 3$$
$$y > \frac{1}{2}x^2.$$

SOLUTION The first inequality is equivalent to $y < -x + 3$, so we shade *below* the dashed line $y = -x + 3$, as seen in Figure 59. To graph $y > \frac{1}{2}x^2$, we shade *above* the dashed parabola $y = \frac{1}{2}x^2$, as seen in Figure 60. The solution set of the system is shown in Figure 61, and consists of the intersection of the two regions. **▯**

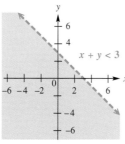

FIGURE 59

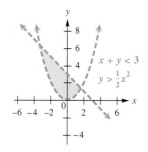

FIGURE 60   **FIGURE 61**

NOTE While we illustrated three graphs in the solution of Example 4, in practice it is customary to give only the final graph (Figure 61). The two individual inequalities were shown simply to illustrate the procedure.

EXAMPLE 5

Graphing a System of Inequalities Using Traditional Methods

Use traditional methods to graph the solution set of the system

$$y \geq 2^x$$
$$9x^2 + 4y^2 \leq 36$$
$$2x + y < 1.$$

SOLUTION Graph the three inequalities on the same axes and shade the region common to all three, as shown in Figure 62. Two boundary lines are solid and one is dashed. ∎

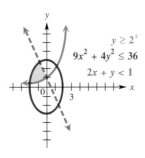

$$y \geq 2^x$$
$$9x^2 + 4y^2 \leq 36$$
$$2x + y < 1$$

FIGURE 62

Graphics Calculator Approach to Inequalities and Systems of Inequalities

Modern graphics calculators have the capability of shading above or below graphs of functions. Because the various makes and models differ in their shading commands, we suggest that you read your owner's manual carefully to determine how yours accomplishes shading.

<table>
<tr><td>

EXAMPLE 6

Graphing a Linear Inequality Using A Graphics Calculator

</td></tr>
</table>

Graph $x + 4y > 4$ using a graphics calculator.

SOLUTION As seen in Example 1, this inequality is equivalent to $y > -\frac{1}{4}x + 1$. We direct the calculator to graph $y = -\frac{1}{4}x + 1$, and shade above the line. See Figure 63. Notice that we cannot tell from the graph that the boundary line is not included in the solution set. Again, this illustrates how the mathematical concepts must be understood to correctly interpret what we see on the screen. ∎

TECHNOLOGICAL NOTE
If the shading feature of your calculator requires that you enter two functions, one of which to shade above and one of which to shade below, you can duplicate the graph in Figure 63 by directing it to shade above $y = (-1/4)x + 1$ and below $y = k$, where k is some number greater than or equal to the maximum y-value, 10. On the other hand, you can duplicate the graph in Figure 64 by directing the calculator to shade below $y = 2x^2 - 3$ and above $y = k$, where k is some number less than or equal to the minimum y-value, -5.

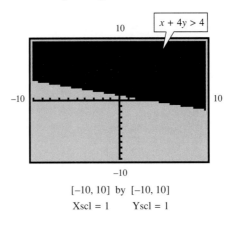

$x + 4y > 4$

$[-10, 10]$ by $[-10, 10]$
$X\text{scl} = 1 \qquad Y\text{scl} = 1$

FIGURE 63

EXAMPLE 7

Graphing a
Nonlinear Inequality
Using a Graphics
Calculator

Graph $y \leq 2x^2 - 3$ using a graphics calculator.

SOLUTION   The graph is that of the parabola $y = 2x^2 - 3$ and the region below the graph. Here, the parabola itself is understood to be included in the solution set. See Figure 64, and compare it to the traditional graph in Figure 56.   ▯

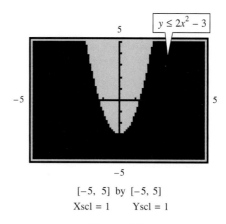

$[-5, 5]$ by $[-5, 5]$
Xscl = 1   Yscl = 1

FIGURE 64

EXAMPLE 8

Graphing a System
of Inequalities Using
A Graphics
Calculator

Graph the solution set of the system

$$x + y < 3$$
$$y > \frac{1}{2}x^2$$

using a graphics calculator.

SOLUTION   This system was first seen in Example 4. We must direct the calculator to shade below the line $y = -x + 3$ *and* above the parabola $y = \frac{1}{2}x^2$. This is done in Figure 65. Compare to the traditional graph in Figure 61.   ▯

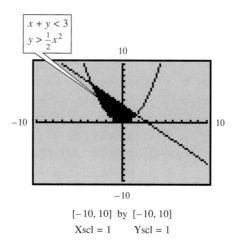

$[-10, 10]$ by $[-10, 10]$
Xscl = 1   Yscl = 1

FIGURE 65

Linear Programming

An important application of systems of inequalities is known as *linear programming*. This procedure is used to find minimum cost, maximum profit, the maximum amount of earning that can take place under given conditions, and so on. The procedure was developed by George Dantzig in 1947 when he was working on a problem of allocating supplies for the Air Force in a way that minimized total cost. The procedure involves a system of inequalities like the one seen in the next example.

Graph the solution set of the system

$$2x + 3y \geq 12$$
$$7x + 4y \geq 28$$
$$y \leq 6$$
$$x \leq 5$$

using traditional methods.

SOLUTION   The graph is obtained by graphing the four inequalities on the same axes and shading the region common to all four as shown in Figure 66. As the graph shows, the boundary lines are all solid.   ▯

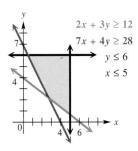

FIGURE 66

The system considered in Example 9 is typical of linear programming problems. To illustrate the method of linear programming, consider the following problem.

The Smith Company makes two products, tape decks and amplifiers. Each tape deck gives a profit of $3, while each amplifier produces $7 profit. The company must manufacture at least one tape deck per day to satisfy one of its customers, but no more than five because of production problems. Also, the number of amplifiers produced cannot exceed six per day. As a further requirement, the number of tape decks cannot exceed the number of amplifiers. How many of each should the company manufacture in order to obtain the maximum profit?

To begin, translate the statement of the problem into symbols by assuming

$$x = \text{number of tape decks to be produced daily}$$
$$y = \text{number of amplifiers to be produced daily.}$$

According to the statement of the problem given above, the company must produce at least one tape deck (one or more), so

$$x \geq 1.$$

Since no more than 5 tape decks may be produced,

$$x \leq 5.$$

Since no more than 6 amplifiers may be made in one day,

$$y \leq 6.$$

The requirement that the number of tape decks may not exceed the number of amplifiers translates as

$$x \leq y.$$

The number of tape decks and of amplifiers cannot be negative, so

$$x \geq 0 \quad \text{and} \quad y \geq 0.$$

These restrictions, or **constraints,** that are placed on production form the system of inequalities

$$x \geq 1, \quad x \leq 5, \quad y \leq 6, \quad x \leq y, \quad x \geq 0, \quad y \geq 0.$$

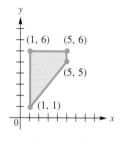

FIGURE 67

The maximum possible profit that the company can make, subject to these constraints, is found by sketching the graph of the solution of the system. See Figure 67. The only feasible values of x and y are those that satisfy all constraints. These values correspond to points that lie on the boundary or in the shaded region, called the **region of feasible solutions.**

Since each tape deck gives a profit of $3, the daily profit from the production of x tape decks is $3x$ dollars. Also, the profit from the production of y amplifiers will be $7y$ dollars per day. The total daily profit is thus given by the following **objective function:**

$$\text{Profit} = 3x + 7y.$$

The problem of the Smith Company may now be stated as follows: find values of x and y in the region of feasible solutions as shown in Figure 67 that will produce the maximum possible value of $3x + 7y$.

It can be shown that any optimum value (maximum or minimum) will always occur at a **vertex** (or **corner point**) of the region of feasible solutions. Locate the point (x, y) that gives the maximum profit by checking the coordinates of the vertex points, shown in Figure 67 and listed below. Find the profit that corresponds to each coordinate pair and choose the one that gives the maximum profit.

Point	Profit $= 3x + 7y$	
(1, 1)	$3(1) + 7(1) = 10$	
(1, 6)	$3(1) + 7(6) = 45$	
(5, 6)	$3(5) + 7(6) = 57$	← Maximum
(5, 5)	$3(5) + 7(5) = 50$	

The maximum profit of $57 is obtained when 5 tape decks and 6 amplifiers are produced each day.

EXAMPLE 10

Solving A Problem Using Linear Programming

Margaret Boyle, who is very health-conscious, takes vitamin pills. Each day, she must have at least 16 units of vitamin A, at least 5 units of vitamin B_1, and at least 20 units of vitamin C. She can choose between red pills, costing 10¢ each, which contain 8 units of A, 1 of B_1, and 2 of C, and blue pills, costing 20¢ each, which contain 2 units of A, 1 of B_1, and 7 of C. How many of each pill should she buy in order to minimize her cost yet fulfill her daily requirements?

SOLUTION Let x represent the number of red pills to buy, and let y represent the number of blue pills to buy. Then the cost in pennies per day is given by

$$\text{Cost} = 10x + 20y.$$

Since she buys x of the 10¢ pills and y of the 20¢ pills, she gets vitamin A as follows: 8 units from each red pill and 2 units from each blue pill. Altogether, she gets $8x + 2y$ units of A per day. Since she needs at least 16 units,

$$8x + 2y \geq 16.$$

Each red pill or each blue pill supplies 1 unit of vitamin B_1. Margaret needs at least 5 units per day, so

$$x + y \geq 5.$$

For vitamin C, the inequality is

$$2x + 7y \geq 20.$$

Also, $x \geq 0$ and $y \geq 0$, since Margaret cannot buy negative numbers of the pills.
The total cost of the pills can be minimized by finding the solution of the system of inequalities formed by the constraints. (See Figure 68.) The solution to this minimizing problem will also occur at a vertex point. Check the coordinates of the vertex points in the cost function to find the lowest cost.

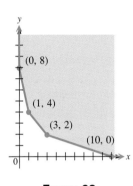

FIGURE 68

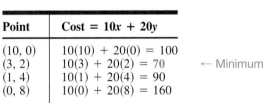

Point	Cost $= 10x + 20y$	
(10, 0)	$10(10) + 20(0) = 100$	
(3, 2)	$10(3) + 20(2) = 70$	← Minimum
(1, 4)	$10(1) + 20(4) = 90$	
(0, 8)	$10(0) + 20(8) = 160$	

Margaret's solution is to buy 3 red pills and 2 blue ones, for a total cost of 70¢ per day. She receives minimum amounts of vitamins B_1 and C but an excess of vitamin A. Even with an excess of A, this is still the best buy. ∎

To solve a linear programming problem in general, use the following steps.

SOLVING A LINEAR PROGRAMMING PROBLEM

1. Write the objective function and all necessary constraints.
2. Graph the feasible region.
3. Identify all vertices or corner points.
4. Find the value of the objective function at each vertex.
5. The solution is given by the vertex producing the optimum value of the objective function.

6.6 EXERCISES

Use traditional graphing methods to graph each of the following inequalities.

1. $x \le 3$

2. $y \le -2$

3. $x + 2y \le 6$

4. $x - y \ge 2$

5. $2x + 3y \ge 4$

6. $4y - 3x < 5$

7. $3x - 5y > 6$

8. $x < 3 + 2y$

9. $5x \le 4y - 2$

10. $2x > 3 - 4y$

11. $y < 3x^2 + 2$

12. $y \le x^2 - 4$

13. $y > (x - 1)^2 + 2$

14. $y > 2(x + 3)^2 - 1$

15. $x^2 + (y + 3)^2 \le 16$

16. $(x - 4)^2 + (y + 3)^2 \le 9$

17. $4x^2 \le 4 - y^2$

18. $x^2 + 9y^2 > 9$

19. $9x^2 - 16y^2 > 144$

20. $4x^2 \le 36 + 9y^2$

21. Which one of the following is a description of the graph of the inequality
$$(x - 5)^2 + (y - 2)^2 < 4?$$
    (a) the region inside a circle with center $(-5, -2)$ and radius 2
    (b) the region inside a circle with center $(5, 2)$ and radius 2
    (c) the region inside a circle with center $(-5, -2)$ and radius 4
    (d) the region outside a circle with center $(5, 2)$ and radius 4

22. Without graphing, write a description of the graph of the nonlinear inequality $y > 2(x - 3)^2 + 2$.

23. Which one of the following inequalities satisfies the following description: the region outside an ellipse centered at the origin, with x-intercepts 4 and -4, and y-intercepts 9 and -9?
    (a) $\dfrac{x^2}{4} + \dfrac{y^2}{9} > 1$ **(b)** $\dfrac{x^2}{16} - \dfrac{y^2}{81} > 1$ **(c)** $\dfrac{x^2}{16} + \dfrac{y^2}{81} > 0$ **(d)** $\dfrac{x^2}{16} + \dfrac{y^2}{81} > 1$

24. Explain how it is determined whether the boundary of an inequality is a solid line or a dashed line.

In Exercises 25–28, match the inequality with the appropriate calculator-generated graph. You should not use your calculator, but rather use your knowledge of the concepts involved in graphing inequalities.

25. $y \le 3x - 6$

26. $y \ge 3x - 6$

27. $y \le -3x - 6$

28. $y \ge -3x - 6$

A.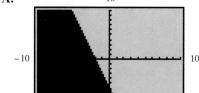

B.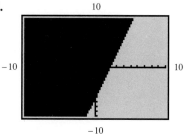

C.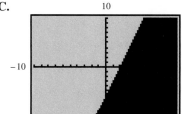

D.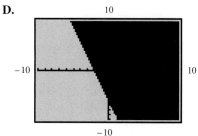

Use traditional graphing methods to graph each of the following systems of inequalities.

29. $x + y \leq 4$
$\quad\; x - 2y \geq 6$

30. $2x + y > 2$
$\quad\; x - 3y < 6$

31. $4x + 3y < 12$
$\quad\quad\; y + 4x > -4$

32. $3x + 5y \leq 15$
$\quad\;\; x - 3y \geq 9$

33. $\;\; x + y \leq 6$
$\quad\; 2x + 2y \geq 12$

34. $3x + 4y < 15$
$\quad\; 6x + 8y > 30$

35. $x + 2y \leq 4$
$\quad\;\; y \geq x^2 - 1$

36. $4x - 3y \leq 12$
$\quad\;\; y \leq x^2$

37. $y \leq -x^2$
$\quad\; y \geq x^2 - 6$

38. $x^2 + y^2 \leq 9$
$\quad\;\; x \leq -y^2$

39. $x^2 - y^2 < 1$
$\quad\; -1 < y < 1$

40. $x^2 + y^2 \leq 36$
$\quad\; -4 \leq x \leq 4$

41. $2x^2 - y^2 > 4$
$\quad\;\; 2y^2 - x^2 > 4$

42. $y \geq x^2 + 4x + 4$
$\quad\; y < -x^2$

43. $\dfrac{x^2}{16} + \dfrac{y^2}{9} \leq 1$

$\quad\; \dfrac{x^2}{4} - \dfrac{y^2}{16} \geq 1$

44. $\dfrac{x^2}{36} - \dfrac{y^2}{9} \geq 1$

$\quad\; \dfrac{x^2}{81} + y^2 \leq 1$

45. $\dfrac{x^2}{4} + \dfrac{y^2}{9} > 1$

$\quad\; x^2 - y^2 \geq 1$
$\quad\; -4 \leq x \leq 4$

46. $2x - 3y < 6$
$\quad\; 4x^2 + 9y^2 < 36$
$\quad\; x \geq -1$

47. $y \geq 3^x$
$\quad\; y \geq 2$

48. $y \leq \left(\dfrac{1}{2}\right)^x$

$\quad\; y \geq 4$

49. $|x| \geq 2$
$\quad\; |y| \geq 4$
$\quad\; y < x^2$

50. $|x| + 2 \geq 4$
$\quad\; |y| \leq 1$
$\quad\; \dfrac{x^2}{9} + \dfrac{y^2}{16} \leq 1$

51. $y \leq |x + 2|$
$\quad\; \dfrac{x^2}{16} - \dfrac{y^2}{9} \leq 1$

52. $y \leq \log x$
$\quad\; y \geq |x - 2|$

53. Which one of the following is a description of the solution set of the system below?

$$x^2 + 4y^2 < 36$$
$$y < x$$

(a) all points outside the ellipse $x^2 + 4y^2 = 36$ and above the line $y = x$
(b) all points outside the ellipse $x^2 + 4y^2 = 36$ and below the line $y = x$
(c) all points inside the ellipse $x^2 + 4y^2 = 36$ and above the line $y = x$
(d) all points inside the ellipse $x^2 + 4y^2 = 36$ and below the line $y = x$

54. Fill in the blanks with the appropriate responses. The graph of the system

$$y > x^2 + 2$$
$$x^2 + y^2 < 16$$
$$y < 7$$

consists of all points ＿＿＿＿＿＿＿ the parabola $y = x^2 + 2$, ＿＿＿＿＿＿＿ the
$\qquad\qquad\qquad\quad$ (above/below) $\qquad\qquad\qquad\qquad\qquad$ (inside/outside)

circle $x^2 + y^2 = 16$, and ＿＿＿＿＿＿＿ the line $y = 7$.
$\qquad\qquad\qquad$ (above/below)

In Exercises 55–58, match the system of inequalities with the appropriate calculator-generated graph. You should not use your calculator, but rather use your knowledge of the concepts involved in graphing systems of inequalities.

55. $y \geq x$
      $y \leq 2x - 3$

56. $y \geq x^2$
      $y < 5$

57. $x^2 + y^2 \leq 16$
      $y \geq 0$

58. $y \leq x$
      $y \geq 2x - 3$

A.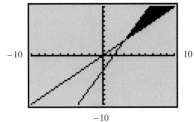

B.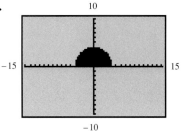

C.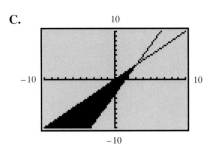

D.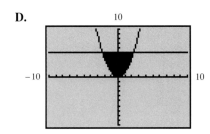

Use the shading capabilities of your graphics calculator to graph each inequality or system of inequalities.

59. $3x + 2y \geq 6$

60. $y \leq x^2 + 5$

61. $x + y \geq 2$
      $x + y \leq 6$

62. $y \geq |x + 2|$
      $y \leq 6$

63. $y \geq 2^x$
      $y \leq 8$

64. $y \leq x^3 + x^2 - 4x - 4$

The following systems of linear inequalities involve more than two inequalities. Use traditional graphing methods to graph the solution set of each system.

65. $3x - 2y \geq 6$
      $x + y \leq -5$
      $y \leq 4$

66. $2x + 3y \leq 12$
      $2x + 3y > -6$
      $3x + y < 4$
      $x \geq 0$
      $y \geq 0$

The graphs below represent regions of feasible solutions. Determine the maximum and minimum values of the given objective function.

67. Objective function: $3x + 5y$

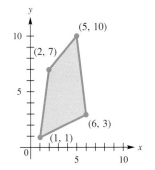

68. Objective function: $6x + y$

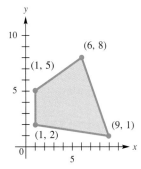

Solve each of the following linear programming problems.

69. Farmer Jones raises only pigs and geese. She wants to raise no more than 16 animals with no more than 12 geese. She spends $50 to raise a pig and $20 to raise a goose. She has $500 available for this purpose. Find the maximum profit she can make if she makes a profit of $80 per goose and $40 per pig.

70. A wholesaler of party goods wishes to display her products at a convention of social secretaries in such a way that she gets the maximum number of inquiries about her whistles and hats. Her booth at the convention has 12 sq m of floor space to be used for display purposes. A display unit for hats requires 2 sq m, and for whistles, 4 sq m. Experience tells the wholesaler that she should never have more than a total of 5 units of whistles and hats on display at one time. If she receives three inquiries for each unit of hats and two inquiries for each unit of whistles on display, how many of each should she display in order to get the maximum number of inquiries?

71. Gwen, who is on the Weight Watchers program, uses two food supplements, I and II. She can get these supplements from two different products, A and B. Product A provides 3 g per serving of supplement I and 2 g per serving of supplement II. Product B provides 2 g per serving of supplement I and 4 g per serving of supplement II. Her program director, Paulette, has recommended that she include at least 15 g of each supplement in her daily diet. If product A costs 25¢ per serving and product B costs 40¢ per serving, how can she satisfy her requirements most economically?

72. A manufacturer of refrigerators must ship at least 100 refrigerators to its two West coast warehouses. Each warehouse holds a maximum of 100 refrigerators. Warehouse A holds 25 refrigerators already, while warehouse B has 20 on hand. It costs $12 to ship a refrigerator to warehouse A and $10 to ship one to warehouse B. How many refrigerators should be shipped to each warehouse to minimize cost? What is the minimum cost?

73. A machine shop manufactures two types of bolts. Each can be made on any of three groups of machines, but the time required on each group differs, as shown in the table below.

		Machine Groups		
		I	**II**	**III**
Bolts	**Type 1**	.1 min	.1 min	.1 min
	Type 2	.1 min	.4 min	.5 min

Production schedules are made up one day at a time. In a day there are 240, 720, and 160 minutes available, respectively, on these machines. Type 1 bolts sell for 10¢ and type 2 bolts for 12¢. How many of each type of bolt should be manufactured per day to maximize revenue? What is the maximum revenue?

74. The manufacturing process requires that oil refineries manufacture at least 2 gal of gasoline for each gallon of fuel oil. To meet the winter demand for fuel oil, at least 3 million gal a day must be produced. The demand for gasoline is no more than 6.4 million gal per day. If the price of gasoline is $1.90 and the price of fuel oil is $1.50 per gal, how much of each should be produced to maximize revenue?

75. Earthquake victims in China need medical supplies and bottled water. Each medical kit measures 1 cubic foot and weighs 10 pounds. Each container of water is also 1 cubic foot but weighs 20 pounds. The plane can only carry 80,000 pounds with a total volume of 6,000 cubic feet. Each medical kit will aid 4 people, while each container of water will serve 10 people. How many of each should be sent in order to maximize the number of people aided?

76. If each medical kit could aid 6 people instead of 4, how would the results in Exercise 75 change?

6.7 SYSTEMS OF LINEAR EQUATIONS IN THREE VARIABLES

Geometric Considerations ∎ Analytic Solution of Systems in Three Variables ∎ Applications of Systems

Geometric Considerations

Our work with systems of equations and inequalities so far has dealt strictly with systems in two variables. We can extend the ideas of systems of equations to linear equations of the form $Ax + By + Cz = D$. A solution of such an equation is called an ordered triple, and is denoted (x, y, z). For example, $(1, 2, -4)$ is a solution of $2x + 5y - 3z = 24$. The solution set of such an equation is an infinite set of ordered triples. In geometry the graph of a linear equation in three variables is a plane in three-dimensional space. Considering the possible intersections of the planes representing three equations in three unknowns shows that the solution set of such a system may be either a single ordered triple (x, y, z), an infinite set of ordered triples (dependent equations), or the empty set (an inconsistent system). See Figure 69.

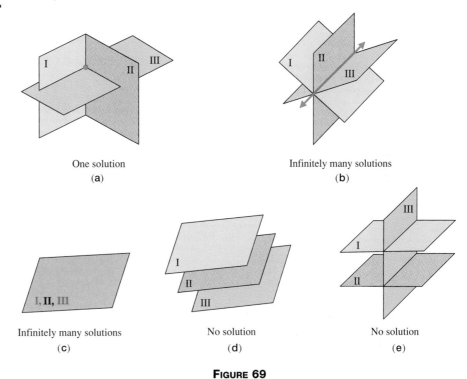

| One solution (a) | Infinitely many solutions (b) |

| Infinitely many solutions (c) | No solution (d) | No solution (e) |

FIGURE 69

Analytic Solution of Systems in Three Variables

A system of equations with three or more variables can be solved with repeated use of the addition method, as shown in the following example.

Solve the system

$$2x + y - z = 2 \qquad \text{[1]}$$
$$x + 3y + 2z = 1 \qquad \text{[2]}$$
$$x + y + z = 2. \qquad \text{[3]}$$

SOLUTION Several steps are needed to obtain an equation in one variable. First, eliminate the same variable from each of two pairs of equations. Then eliminate another variable from the two resulting equations. For example, to eliminate x, multiply both sides of equation (2) by -2 and add to equation (1).

$$
\begin{array}{ll}
2x + y - z = 2 & \quad (1) \\
-2x - 6y - 4z = -2 & \quad -2 \text{ times equation (2)} \\
\hline
-5y - 5z = 0 & \qquad\qquad\quad \text{[4]}
\end{array}
$$

The variable x must be eliminated again from a different pair of equations, say (2) and (3). Multiply both sides of (2) by -1, then add the result to equation (3).

$$
\begin{array}{ll}
-x - 3y - 2z = -1 & \text{-1 times equation (2)} \\
\underline{x + y + z = 2} & \text{(3)} \\
-2y - z = 1 & \qquad\qquad\qquad \textbf{[5]}
\end{array}
$$

Now solve the system formed by the two equations (4) and (5). One way to do this is to eliminate y by multiplying equation (4) by 2 and equation (5) by -5.

$$
\begin{array}{ll}
-10y - 10z = 0 & \text{2 times equation (4)} \\
\underline{10y + 5z = -5} & \text{-5 times equation (5)} \\
-5z = -5 & \\
z = 1 &
\end{array}
$$

Substitute 1 for z in equation (5) to find y. (Equation (4) could also have been used.)

$$
\begin{aligned}
-2y - z &= 1 \\
-2y - 1 &= 1 \\
-2y &= 2 \\
y &= -1
\end{aligned}
$$

Now use equation (3) and the values for y and z to find x. (Either equation (1) or (2) could also have been used.)

$$
\begin{aligned}
x + y + z &= 2 \\
x - 1 + 1 &= 2 \\
x &= 2
\end{aligned}
$$

The solution set of the system is $\{(2, -1, 1)\}$. ∎

NOTE In Example 1 we used a process that yielded a system of equations that is equivalent to the original system.

$$
\begin{array}{ll}
x + y + z = 2 & [3] \\
-2y - z = 1 & [5] \\
z = 1 &
\end{array}
$$

Notice the "triangular" form here. While it is not absolutely necessary to eliminate x first and then y to find z, it is a good preparation for the matrix method of solving systems, introduced in the next chapter.

EXAMPLE **2**
Solving a System Having Two Equations With Three Variables

Solve the system

$$
\begin{array}{ll}
x + 2y + z = 4 & \textbf{[6]} \\
3x - y - 4z = -9. & \textbf{[7]}
\end{array}
$$

SOLUTION Geometrically, the solution is the intersection of the two planes given by equations (6) and (7). The intersection of two different nonparallel planes is a line. Thus there will be an infinite number of ordered triples in the solution set, representing the points on the line of intersection. To describe these ordered triples, proceed as follows.

To eliminate x, multiply both sides of equation (6) by -3 and add to equation (7). (Either y or z could have been eliminated instead.)

$$\begin{array}{rcl} -3x - 6y - 3z &=& -12 \\ \underline{3x - y - 4z} &=& \underline{-9} \\ -7y - 7z &=& -21 \end{array} \qquad \text{[8]}$$

Now solve equation (8) for z.

$$\begin{aligned} -7y - 7z &= -21 \\ -7z &= 7y - 21 \\ z &= -y + 3 \end{aligned}$$

This gives z in terms of y. Now, express x also in terms of y by solving equation (6) for x and substituting $-y + 3$ for z in the result.

$$\begin{aligned} x + 2y + z &= 4 \\ x &= -2y - z + 4 \\ x &= -2y - (-y + 3) + 4 \\ x &= -y + 1 \end{aligned}$$

The system has an infinite number of solutions. For any arbitrary value of y, the value of z is given by $-y + 3$ and x equals $-y + 1$. For example, if $y = 1$, then $x = -1 + 1 = 0$ and $z = -1 + 3 = 2$, giving the solution $(0, 1, 2)$. Verify that another solution is $(-1, 2, 1)$.

With y arbitrary, the solution set is of the form $\{(-y + 1, y, -y + 3)\}$. Had equation (8) been solved for y instead of z, the solution would have had a different form but would have led to the same set of solutions. In that case we would have z arbitrary, and the solution set would be of the form $\{(-2 + z, 3 - z, z)\}$. By choosing $z = 2$, one solution would be $(0, 1, 2)$, which was verified above. ∎

A system like the one in Example 2 occurs when two of the equations in a system of three equations with three variables are dependent. In such a case, there are really only two equations in three variables, and Example 2 illustrates the method of solution. On the other hand, an inconsistent system is indicated by a false statement at some point in the solution, as in Example 3(a) of Section 6.4.

Applications of Systems

In Section 6.4 we discussed solving problems with systems of two equations in two unknowns. Those ideas can be extended to problems requiring three unknowns. With three unknowns, we must write a system of three equations, as shown in the next example.

EXAMPLE 3

Solving an Application Using a System of Three Equations

An animal feed is made from three ingredients: corn, soybeans, and cottonseed. One unit of each ingredient provides units of protein, fat, and fiber as shown in the table below. How many units of each ingredient should be used to make a feed that contains 22 units of protein, 28 units of fat, and 18 units of fiber?

	Corn	Soybeans	Cottonseed	Total
Protein	.25	.4	.2	22
Fat	.4	.2	.3	28
Fiber	.3	.2	.1	18

SOLUTION   Let x represent the number of units of corn, y, the number of units of soybeans, and z, the number of units of cottonseed that are required. Since the total amount of protein is to be 22 units,

$$.25x + .4y + .2z = 22.$$

Also, for the 28 units of fat,

$$.4x + .2y + .3z = 28,$$

and, for the 18 units of fiber,

$$.3x + .2y + .1z = 18.$$

Multiply the first equation on both sides by 100, and the second and third equations by 10 to get the system

$$25x + 40y + 20z = 2200$$
$$4x + 2y + 3z = 280$$
$$3x + 2y + z = 180.$$

Using the methods described earlier in this section, we can show that $x = 40$, $y = 15$, and $z = 30$. The feed should contain 40 units of corn, 15 units of soybeans, and 30 units of cottonseed to fulfill the given requirements.   ◻

NOTE   The table shown in Example 3 is useful in setting up the equations of the system, since the coefficients in each equation can be read from left to right. This idea is extended in the next chapter, where we introduce solution of systems by matrices.

In Section 6.4 we showed how to determine the equation of the line joining two points by using a system of equations (Example 5). Three noncollinear points lie on the graph of a parabola of the form $y = ax^2 + bx + c$. The procedure for finding a line to fit two data points can be extended to finding the equation of a parabola that fits three data points, as shown in the next example.

EXAMPLE 4

Using a System to Fit a Parabola to Three Data Points

Find the equation of the parabola $y = ax^2 + bx + c$ that passes through $(2, 4)$, $(-1, 1)$, and $(-2, 5)$.

SOLUTION   Since the three points lie on the graph of the equation $y = ax^2 + bx + c$, they must satisfy the equation. Substituting each ordered pair into the equation gives three equations with three variables.

$4 = a(2)^2 + b(2) + c$	or $4 = 4a + 2b + c$	**[9]**
$1 = a(-1)^2 + b(-1) + c$	or $1 = a - b + c$	**[10]**
$5 = a(-2)^2 + b(-2) + c$	or $5 = 4a - 2b + c$	**[11]**

This system can be solved by the addition method. First eliminate c using equations (9) and (10).

$$
\begin{array}{ll}
4 = 4a + 2b + c & \\
\underline{-1 = -a + b - c} & \text{-1 times equation (10)} \\
3 = 3a + 3b & \textbf{[12]}
\end{array}
$$

Now, use equations (10) and (11) to also eliminate c.

$$
\begin{array}{ll}
1 = a - b + c & \\
\underline{-5 = -4a + 2b - c} & \text{-1 times equation (11)} \\
-4 = -3a + b & \textbf{[13]}
\end{array}
$$

Solve the system of equations (12) and (13) in two variables by eliminating a.

$$\begin{array}{rl} 3 = & 3a + 3b \\ -4 = & -3a + b \\ \hline -1 = & 4b \end{array}$$

$$-\frac{1}{4} = b$$

Find a by substituting $-\frac{1}{4}$ for b in equation (12), which is equivalent to $1 = a + b$.

$$1 = a + b \qquad \text{Equation (12) divided by 3}$$

$$1 = a - \frac{1}{4} \qquad \text{Let } b = -\tfrac{1}{4}.$$

$$\frac{5}{4} = a$$

Finally, find c by substituting $a = \frac{5}{4}$ and $b = -\frac{1}{4}$ in equation (10).

$$1 = a - b + c$$

$$1 = \frac{5}{4} - \left(-\frac{1}{4}\right) + c \qquad a = \tfrac{5}{4}, b = -\tfrac{1}{4}$$

$$1 = \frac{6}{4} + c$$

$$-\frac{1}{2} = c$$

An equation of the parabola is $y = \frac{5}{4}x^2 - \frac{1}{4}x - \frac{1}{2}$, or equivalently, $y = 1.25x^2 - .25x - .5$.

This result may be supported graphically by graphing $y = 1.25x^2 - .25x - .5$ and showing that the points $(2, 4)$, $(-1, 1)$, and $(-2, 5)$ do indeed lie on the parabola. See Figure 70.   ◼

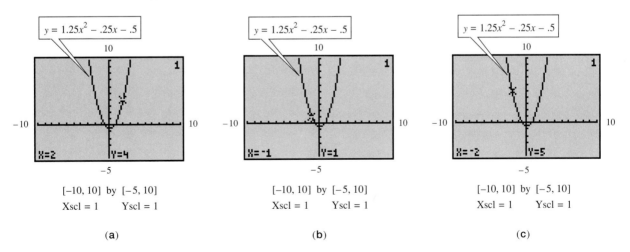

[−10, 10] by [−5, 10]	[−10, 10] by [−5, 10]	[−10, 10] by [−5, 10]
Xscl = 1 Yscl = 1	Xscl = 1 Yscl = 1	Xscl = 1 Yscl = 1
(a)	**(b)**	**(c)**

The points $(2, 4)$, $(-1, 1)$, and $(-2, 5)$ all
lie on the graph of $y = 1.25x^2 - .25x - .5$.

FIGURE 70

6.7 EXERCISES

Verify that the given ordered triple is a solution of the system by substituting the given values of x, y, and z into each equation.

1. $(-3, 6, 1)$

$$2x + y - z = -1$$
$$x - y + 3z = -6$$
$$-4x + y + z = 19$$

2. $\left(\dfrac{1}{2}, -\dfrac{3}{4}, \dfrac{1}{6}\right)$

$$2x + 8y - 6z = -6$$
$$x + y + z = -\dfrac{1}{12}$$
$$x + 3z = 1$$

3. $(-.2, .4, .5)$

$$5x - y + 2z = -.4$$
$$x + 4z = 1.8$$
$$-3y + z = -.7$$

4. $(0, 0, 0)$

$$\sqrt{5}x - \sqrt{2}y + \sqrt[3]{6}z = 0$$
$$8x - 8y + z = 0$$
$$\pi x + ey - (\ln 3)z = 0$$

Solve each of the following systems analytically. (Hint: In Exercises 19–22, let $t = \frac{1}{x}$, $u = \frac{1}{y}$, and $v = \frac{1}{z}$. Solve for t, u, and v, and then solve for x, y, and z.)

5.
$$x + y + z = 2$$
$$2x + y - z = 5$$
$$x - y + z = -2$$

6.
$$2x + y + z = 9$$
$$-x - y + z = 1$$
$$3x - y + z = 9$$

7.
$$x + 3y + 4z = 14$$
$$2x - 3y + 2z = 10$$
$$3x - y + z = 9$$

8.
$$4x - y + 3z = -2$$
$$3x + 5y - z = 15$$
$$-2x + y + 4z = 14$$

9.
$$x + 2y + 3z = 8$$
$$3x - y + 2z = 5$$
$$-2x - 4y - 6z = 5$$

10.
$$3x - 2y - 8z = 1$$
$$9x - 6y - 24z = -2$$
$$x - y + z = 1$$

11.
$$x + 4y - z = 6$$
$$2x - y + z = 3$$
$$3x + 2y + 3z = 16$$

12.
$$4x - 3y + z = 9$$
$$3x + 2y - 2z = 4$$
$$x - y + 3z = 5$$

13.
$$5x + y - 3z = -6$$
$$2x + 3y + z = 5$$
$$-3x - 2y + 4z = 3$$

14.
$$2x - 5y + 4z = -35$$
$$5x + 3y - z = 1$$
$$x + y + z = 1$$

15.
$$x - 3y - 2z = -3$$
$$3x + 2y - z = 12$$
$$-x - y + 4z = 3$$

16.
$$x + y + z = 3$$
$$3x - 3y - 4z = -1$$
$$x + y + 3z = 11$$

17.
$$2x + 6y - z = 6$$
$$4x - 3y + 5z = -5$$
$$6x + 9y - 2z = 11$$

18.
$$8x - 3y + 6z = -2$$
$$4x + 9y + 4z = 18$$
$$12x - 3y + 8z = -2$$

19.
$$\dfrac{1}{x} + \dfrac{1}{y} - \dfrac{1}{z} = \dfrac{1}{4}$$
$$\dfrac{2}{x} - \dfrac{1}{y} + \dfrac{3}{z} = \dfrac{9}{4}$$
$$-\dfrac{1}{x} - \dfrac{2}{y} + \dfrac{4}{z} = 1$$

20.
$$\dfrac{3}{x} + \dfrac{2}{y} - \dfrac{1}{z} = \dfrac{11}{6}$$
$$\dfrac{1}{x} - \dfrac{1}{y} + \dfrac{3}{z} = -\dfrac{11}{12}$$
$$\dfrac{2}{x} + \dfrac{1}{y} + \dfrac{1}{z} = \dfrac{7}{12}$$

21.
$$\dfrac{2}{x} - \dfrac{2}{y} + \dfrac{1}{z} = -1$$
$$\dfrac{4}{x} + \dfrac{1}{y} - \dfrac{2}{z} = -9$$
$$\dfrac{1}{x} + \dfrac{1}{y} - \dfrac{3}{z} = -9$$

22.
$$\dfrac{5}{x} - \dfrac{1}{y} - \dfrac{2}{z} = -6$$
$$-\dfrac{1}{x} + \dfrac{3}{y} - \dfrac{3}{z} = -12$$
$$\dfrac{2}{x} - \dfrac{1}{y} - \dfrac{1}{z} = 6$$

23. Consider the linear equation in three variables $x + y + z = 4$. Find a pair of linear equations that, when considered together with the given equation, will form a system having the following.

    (a) Exactly one solution   **(b)** No solution   **(c)** Infinitely many solutions

24. Refer to Example 2 in this section. Write the solution set with x arbitrary.

25. Give an example using your immediate surroundings of three planes that intersect in a single point.

26. Give an example using your immediate surroundings of three planes that intersect in a line.

Solve each of the following systems in terms of the arbitrary variable x. (Hint: Begin by eliminating either y or z.)

27. $x - 2y + 3z = 6$
$2x - y + 2z = 5$

28. $3x + 4y - z = 13$
$x + y + 2z = 15$

29. $5x - 4y + z = 9$
$x + y = 15$

30. $x - y + z = -6$
$4x + y + z = 7$

31. $3x - 5y - 4z = -7$
$y - z = -13$

32. $3x - 2y + z = 15$
$x + 4y - z = 11$

Use a system of three equations in three variables to solve the following problems.

33. A coin collection contains a total of 29 coins, made up of cents, nickels, and quarters. The number of quarters is 8 less than the number of cents. The total face value of the coins is $1.77. How many of each denomination are there?

34. A sparkling water distributor wants to make up 300 gallons of sparkling water to sell for $6.00 per gallon. She wishes to mix three grades of water selling for $9.00, $3.00, and $4.50 per gallon, respectively. She must use twice as much of the $4.50 water as the $3.00 water. How many gallons of each should she use?

35. A glue company needs to make some glue that it can sell for $120 per barrel. It wants to use 150 barrels of glue worth $100 per barrel, along with some glue worth $150 per barrel, and glue worth $190 per barrel. It must use the same number of barrels of $150 and $190 glue. How much of the $150 and $190 glue will be needed? How many barrels of $120 glue will be produced?

36. Billy Dixon and the Topics sell three kinds of concert tickets, "up close," "middle" and "farther back." "Up close" tickets cost $6 more than "middle" tickets, while "middle" tickets cost $3 more than "farther back" tickets. Twice the cost of an "up close" ticket is $3 more than 3 times the cost of a "farther back" seat. Find the price of each kind of ticket.

37. The perimeter of a triangle is 59 inches. The longest side is 11 inches longer than the medium side, and the medium side is 3 inches more than the shortest side. Find the length of each side of the triangle.

38. The sum of the measures of the angles of any triangle is 180°. In a certain triangle, the largest angle measures 55° less than twice the medium angle, and the smallest measures 25° less than the medium angle. Find the measures of each of the three angles.

39. Sam Abo-zahrah wins $100,000 in the Louisiana state lottery. He invests part of the money in real estate with an annual return of 5% and another part in a money market account at 4.5% interest. He invests the rest, which amounts to $20,000 less than the sum of the other two parts, in certificates of deposit that pay 3.75%. If the total annual interest on the money is $4450, how much was invested at each rate?

40. Laura Taber invests $10,000 received in an inheritance in three parts. With one part she buys mutual funds which offer a return of 4% per year. The second part, which amounts to twice the first, is used to buy government bonds paying 4.5% per year. She puts the rest into a savings account that pays 2.5% annual interest. During the first year, the total interest is $415. How much did she invest at each rate?

Use the method of Example 4 in Exercises 41–46.

41. Find a, b, and c so that the graph of the equation $y = ax^2 + bx + c$ passes through the points $(2, 3)$, $(-1, 0)$, and $(-2, 2)$.

42. Find a, b, and c so that $(2, 14)$, $(0, 0)$, and $(-1, -1)$ lie on the graph of $y = ax^2 + bx + c$.

Find the equation of the parabola shown. In each exercise, three views of the same curve are given.

43.

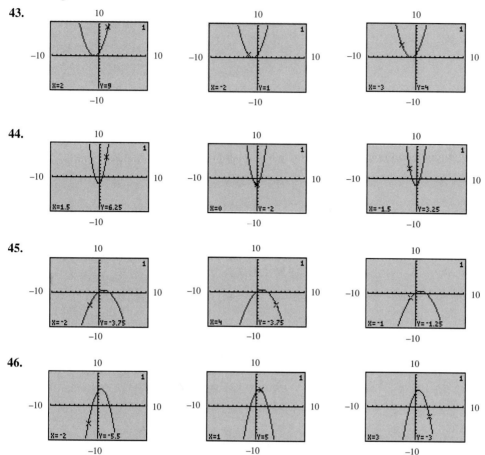

44.

45.

46.

Given three noncollinear points, there is one and only one circle that passes through them. Knowing that the equation of a circle may be written in the form

$$x^2 + y^2 + ax + by + c = 0,$$

find the equation of the circle described or graphed in Exercises 47 and 48.

47. passing through the points $(2, 1)$, $(-1, 0)$, and $(3, 3)$

48.

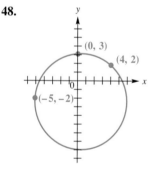

Suppose that the position of a particle moving along a straight line is given by the function $s(t) = at^2 + bt + c$, where t is time in seconds and a, b, and c are real numbers.

49. If $s(0) = 5$, $s(1) = 23$, and $s(2) = 37$, find a, b, and c. Then find $s(8)$.

50. If $s(0) = -10$, $s(1) = 6$, and $s(2) = 30$, find a, b, and c. Then find $s(10)$.

C*hapter 6*

SUMMARY

The circle, ellipse, parabola, and hyperbola are the conic sections. They are all defined using the distance formula. A circle is the set of points in a plane equidistant from a fixed point. The center-radius form of the equation of a circle is derived from the definition and gives the coordinates of the center of the circle and the length of the radius. The equation of a circle has both x and y squared with the same coefficients, so to graph a circle with a graphics calculator, the equation must be rewritten as the union of two functions. Multiplying out the center-radius form of the equation and collecting like terms on one side of the equation gives the general form.

A parabola is the set of points in a plane equidistant from a fixed point (the focus) and a fixed line (the directrix). The equation of a parabola has one variable squared and one variable to the first power. If x is squared, the parabola has a vertical axis; if y is squared, the axis is horizontal, and the graph is not that of a function. To graph a parabola with a horizontal axis, the equation must be rewritten as the union of two functions.

An ellipse is the set of all points in a plane such that the sum of their distances from two fixed points (foci) is constant. The equation of an ellipse has both variables squared with different coefficients having the same sign. The coefficients are used to find the x- and y-intercepts of the graph and the coordinates of the foci.

A hyperbola is the set of all points in a plane such that the difference of their distances from two fixed points (foci) is constant. The equation of a hyperbola has both variables squared with coefficients that have opposite signs. The positive coefficient is used to find either the x- or y-intercepts. The branches of a hyperbola approach the extended diagonals of the fundamental rectangle as $|x|$ and $|y|$ get very large.

A general definition of a conic is the set of all points $P(x, y)$ in a plane such that the ratio of the distance from P to a fixed line and the distance from P to a fixed point is constant. This ratio is called the eccentricity (e) of the conic. Parabolas have an eccentricity of 1. The eccentricity of an ellipse is between 0 and 1, and the eccentricity of a hyperbola is greater than 1.

The solution of a system of linear equations is either one ordered pair, an infinite set of ordered pairs that lie on a line, or no solution. To solve a linear system of two equations by substitution, solve one equation for one variable. Substitute the result in the other equation and solve for the other variable. Substitute back to get the value of the remaining variable. To use the addition (or elimination) method, multiply one or both equations by suitable numbers, so that when they are added, one variable is eliminated. Solve the resulting equation for the other variable, then substitute into any of the equations to find the value of the remaining variable. If a false statement, with no variables, occurs at some point in the process, the system has no solution. If a true statement, with no variables, occurs, there are an infinite number of solutions. Curve-fitting is a method of finding an equation that is satisfied by a set of given points. The method can be accomplished by solving a system of linear equations.

A nonlinear system of equations includes at least one nonlinear equation. The addition method, the substitution method, or a combination of the two are used to solve these systems analytically.

The graph of the solution of an inequality in two variables is the half-plane on one side of a boundary line and sometimes includes the points on the boundary. The solution of a system of inequalities is the intersection of the solutions of the inequalities in the system. Linear programming is an important application of a system of inequalities. It is used to find the optimum (maximum or minimum) value of some quantity such as profit or cost by solving a system of inequalities.

The solution of a system of three equations in three variables is an ordered triple. Systems of equations in three variables are solved with an extension of the addition method described for systems of equations in two variables. Three data points are needed to find an equation of a parabola or a circle, rather than the two points required to find the equation of a line. Thus, the solution of a system of three equations will be needed.

Key Terms

SECTION 6.1

conic section (conic)
circle
radius
center
center-radius form of the equation of a circle
general form of the equation of a circle
parabola
focus (foci)
directrix

SECTION 6.2

ellipse
vertices
major axis
minor axis
hyperbola
transverse axis
asymptotes of a hyperbola
fundamental rectangle

SECTION 6.3

eccentricity

SECTION 6.4

system of equations
first-degree equation in n unknowns
substitution method
addition (elimination) method
data points
curve-fitting

SECTION 6.5

nonlinear system

SECTION 6.6

linear inequality in two variables
half-plane
boundary
linear programming
region of feasible solutions
objective function
vertex (corner point)

SECTION 6.7

ordered triple
noncollinear points

Chapter 6   REVIEW EXERCISES

Write an equation for the circle satisfying the given conditions.

1. center $(-2, 3)$, radius 5

2. center $(\sqrt{5}, -\sqrt{7})$, radius $\sqrt{3}$

3. center $(-8, 1)$, passing through $(0, 16)$

4. center $(3, -6)$, tangent to the x-axis

Find the center and radius of the following circles.

5. $x^2 - 4x + y^2 + 6y + 12 = 0$

6. $x^2 - 6x + y^2 - 10y + 30 = 0$

7. $2x^2 + 14x + 2y^2 + 6y = -2$

8. $3x^2 + 3y^2 + 33x - 15y = 0$

9. Describe the graph of $(x - 4)^2 + (y - 5)^2 = 0$.

Give the focus, directrix, and axis for the parabola, and graph it by hand.

10. $x = -\dfrac{3}{2}y^2$

11. $x = \dfrac{1}{2}y^2$

12. $3x^2 = y$

13. $x^2 + 2y = 0$

Write an equation for the parabola with vertex at the origin that satisfies the given conditions.

14. focus $(4, 0)$

15. through $(2, 5)$, opening to the right

16. through $(3, -4)$, opening downward

Write an equation for each parabola.

17. vertex $(-5, 6)$, focus $(2, 6)$

18. vertex $(4, 3)$, focus $(4, 5)$

Graph the following ellipses and hyperbolas by hand, and give the coordinates of the vertices.

19. $\dfrac{y^2}{9} + \dfrac{x^2}{5} = 1$ **20.** $\dfrac{x^2}{16} + \dfrac{y^2}{4} = 1$ **21.** $\dfrac{x^2}{64} - \dfrac{y^2}{36} = 1$ **22.** $\dfrac{y^2}{25} - \dfrac{x^2}{9} = 1$

23. $\dfrac{(x-3)^2}{4} + (y+1)^2 = 1$ **24.** $\dfrac{(x-2)^2}{9} + \dfrac{(y+3)^2}{4} = 1$

25. $\dfrac{(y+2)^2}{4} - \dfrac{(x+3)^2}{9} = 1$ **26.** $\dfrac{(x+1)^2}{16} - \dfrac{(y-2)^2}{4} = 1$

Write equations for the following conic sections with centers at the origin.

27. ellipse; vertex at $(0, 4)$, focus at $(0, 2)$

28. ellipse; x-intercept 6, focus at $(-2, 0)$

29. hyperbola; focus at $(0, -5)$, transverse axis of length 8

30. hyperbola; y-intercept -2, passing through $(2, 3)$

31. focus at $(0, -3)$ and $e = \frac{2}{3}$

32. focus at $(5, 0)$ and $e = \frac{5}{2}$

33. Consider the circle with equation $x^2 + y^2 + 2x + 6y - 15 = 0$.
   (a) What are the coordinates of the center?
   (b) What is the radius?
   (c) What two functions must be graphed to graph this circle with your calculator?

Match each equation with its graph.

34. $4x^2 + y^2 = 36$

35. $x = 2y^2 + 3$

36. $(x-1)^2 + (y+2)^2 = 36$

37. $\dfrac{x^2}{36} + \dfrac{y^2}{9} = 1$

38. $(y-1)^2 - (x-2)^2 = 36$

39. $y^2 = 36 + 4x^2$

A.
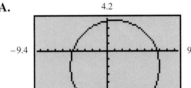

B.
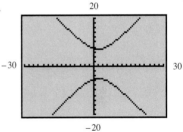

C.
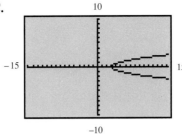

D.
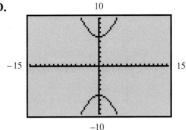

E.
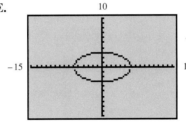

F.
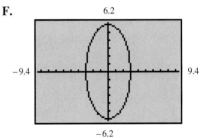

Find the eccentricity of each ellipse or hyperbola.

40. $9x^2 + 25y^2 = 225$ **41.** $4x^2 + 9y^2 = 36$ **42.** $9x^2 - y^2 = 9$

Use the substitution method to solve each of the following linear systems. Identify any systems with dependent equations or any inconsistent systems.

43. $4x - 3y = -1$
$3x + 5y = 50$

44. $10x + 3y = 8$
$5x - 4y = 26$

45. $7x - 10y = 11$
$\dfrac{3x}{2} - 5y = 8$

46. $\dfrac{x}{2} + \dfrac{2y}{3} = -8$
$\dfrac{3x}{4} + \dfrac{y}{3} = 0$

47. $\dfrac{x}{2} - \dfrac{y}{5} = \dfrac{11}{10}$
$2x - \dfrac{4y}{5} = \dfrac{22}{5}$

48. $4x + 5y = 5$
$3x + 7y = -6$

49. A student solves the system

$$x + y = 3$$
$$2x + 2y = 6$$

and gets the result $0 = 0$. The student gives the solution set as $\{(0, 0)\}$. Explain why this is or is not correct.

50. Use your calculator with a window of $[-18, 18]$ by $[-12, 12]$ to answer the following.
  (a) Do the circle $x^2 + y^2 = 144$ and the line $x + 2y = 8$ have any points in common?
  (b) Approximate any intersection points to the nearest tenth.
  (c) Find the exact values of the coordinates of the points of intersection analytically.

Solve the nonlinear systems of equations.

51. $y = x^2 - 1$
$x + y = 1$

52. $x^2 + y^2 = 2$
$3x + y = 4$

53. $x^2 + 2y^2 = 22$
$2x^2 - y^2 = -1$

54. $x^2 - 4y^2 = 19$
$x^2 + y^2 = 29$

55. $xy = 4$
$x - 6y = 2$

56. $x^2 + 2xy + y^2 = 4$
$x = 3y - 2$

57. Can a system of two linear equations in two variables have exactly two solutions? Explain.

58. Consider the system in Exercise 52.
  (a) To graph the first equation, what two functions must you enter into your calculator?
  (b) To graph the second equation, what function must you enter?
  (c) What would be an appropriate window to graph this system?

Graph the solution of each of the following systems of inequalities.

59. $x + y \le 6$
$2x - y \ge 3$

60. $x - 3y \ge 6$
$y^2 \le 16 - x^2$

61. $9x^2 + 16y^2 \ge 144$
$x^2 - y^2 \ge 16$

62. Find $x \ge 0$ and $y \ge 0$ such that
$$3x + 2y \le 12$$
$$5x + y \ge 5$$
and $2x + 4y$ is maximized.

63. Find $x \ge 0$ and $y \ge 0$ such that
$$x + y \le 50$$
$$2x + y \ge 20$$
$$x + 2y \ge 30$$
and $4x + 2y$ is minimized.

64. Sketch a line and a hyperbola on the same axes so that their intersection contains each of the following numbers of points.
  (a) 0   **(b)** 1   **(c)** 2

Use the addition method to solve each of the following linear systems.

65. $2x - 3y + z = -5$
$x + 4y + 2z = 13$
$5x + 5y + 3z = 14$

66. $x - 3y = 12$
$2y + 5z = 1$
$4x + z = 25$

67. $x + y - z = 5$
$2x + y + 3z = 2$
$4x - y + 2z = -1$

68. $5x - 3y + 2z = -5$
$2x + y - z = 4$
$-4x - 2y + 2z = -1$

Write a system of linear equations for each of the following, and then use the system to solve the problem.

69. Three hundred people attending a club's banquet paid a total of $4060. Each member paid $13 and each nonmember paid $15. How many members and how many nonmembers attended the banquet?

70. A cup of uncooked rice contains 15 g of protein and 810 cal. A cup of uncooked soybeans contains 22.5 g of protein and 270 cal. How many cups of each should be used for a meal containing 9.5 g of protein and 324 cal?

71. A candy company has 100 kg of chocolate-covered nuts and 125 kg of chocolate-covered raisins to be sold as two different mixtures. One mix will contain $\frac{1}{2}$ nuts and $\frac{1}{2}$ raisins, while the other mix will contain $\frac{1}{3}$ nuts and $\frac{2}{3}$ raisins. How much of each mixture should be made to maximize revenue if the first mix sells for $6.00 per kilogram and the second mix sells for $4.80 per kilogram?

72. Find an equation of the form $y = ax^2 + bx + c$ that passes through the points $(2, 3)$, $(-1, 0)$, and $(-2, 2)$.

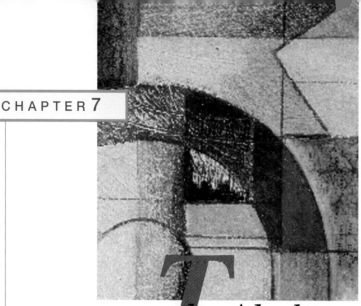

The Algebra and Applications of Matrices

7.1 SOLUTION OF LINEAR SYSTEMS USING MATRIX METHODS

Matrices and Technology ▌ Solving Systems by Matrix Row
Transformations ▌ Special Cases ▌ An Application

TECHNOLOGICAL NOTE
The earliest models of graphics calculators did not have the capability of working with matrices. Now all of them do. However, the manner in which matrices are entered and displayed varies greatly among different manufacturers, and sometimes even differs among the various models manufactured by the same corporation. As always, we suggest that you refer to your owner's manual when necessary.

Matrices and Technology

While modern graphics calculators are probably best known for their graphing capabilities, they possess other powerful features as well. One of these is the capability to handle work with *matrices* in an efficient, timesaving manner. As we shall see in this chapter, elementary matrix algebra requires much computational work, and graphics calculators are programmed to do those computations by simply entering the appropriate commands.

This chapter presents the theory of matrices in a traditional fashion. We feel that a knowledge of the concepts is essential in understanding how matrices are used; simply viewing results on a screen without the knowledge of how they were computed is not an acceptable method of using the matrix capabilities of graphics calculators. Examples and exercises are presented in a manner similar to that of a traditional algebra text. However, we urge the student to learn how to use his or her particular model by having it available at all times, working with it while reading the text. In many cases intermediate steps shown in print will not be necessary when using a calculator, but they are included for completeness.

| FOR **GROUP DISCUSSION** |

Refer to the owner's manual of your particular make and model calculator to find the pages that describe how to use the matrix capabilities of your calculator. Mark these pages with a bookmark or some other method so that you can readily refer to them while studying this chapter. (At the beginning of each exercise set in this chapter, we will address some calculator issues as notes titled *To the Student: Calculator Considerations.*)

Solving Systems by Matrix Row Transformations

The elimination method used to solve linear systems of equations in the previous chapter can be streamlined to a systematic method by using *matrices* (singular: *matrix*). In this section we describe one way matrices are used to solve linear systems. Matrix methods are particularly suitable for computer solutions of large systems of equations having many unknowns.

To begin, consider a system of three equations and three unknowns such as

$$a_1x + b_1y + c_1z = d_1$$
$$a_2x + b_2y + c_2z = d_2$$
$$a_3x + b_3y + c_3z = d_3.$$

This system can be written in an abbreviated form as

$$\begin{bmatrix} a_1 & b_1 & c_1 & d_1 \\ a_2 & b_2 & c_2 & d_2 \\ a_3 & b_3 & c_3 & d_3 \end{bmatrix}.$$

TECHNOLOGICAL NOTE
It is likely that the vertical bar separating the coefficients of the system from the constants will not appear on the graphics calculator screen.

Such a rectangular array of numbers enclosed by brackets is called a **matrix.** Each number in the array is an **element** or **entry.** The constants in the last column of the matrix can be set apart from the coefficients of the variables by using a vertical line, as shown in the following **augmented matrix.** (Because the matrix of coefficients has an extra column determined by the constants of the system, the coefficient matrix is *augmented.*)

$$\text{Rows} \rightarrow \begin{bmatrix} a_1 & b_1 & c_1 & \bigm| & d_1 \\ a_2 & b_2 & c_2 & \bigm| & d_2 \\ a_3 & b_3 & c_3 & \bigm| & d_3 \end{bmatrix}$$
$$\text{Columns}$$

As an example, the system

$$x + 3y + 2z = 1$$
$$2x + y - z = 2$$
$$x + y + z = 2$$

has the augmented matrix

$$\begin{bmatrix} 1 & 3 & 2 & \bigm| & 1 \\ 2 & 1 & -1 & \bigm| & 2 \\ 1 & 1 & 1 & \bigm| & 2 \end{bmatrix}.$$

This matrix has 3 rows (horizontal) and 4 columns (vertical). To refer to a number in the matrix, use its row and column numbers. For example, the number 3 is in the first row, second column position.

The rows of this augmented matrix can be treated just like the equations of the system of linear equations. Since the augmented matrix is nothing more than a short form of the system, any transformation of the matrix that results in an equivalent system of equations can be performed. Operations that produce such transformations are given below.

TECHNOLOGICAL NOTE
Refer to your owner's manual to see how these transformations are accomplished with your model.

MATRIX ROW TRANSFORMATIONS

For any augmented matrix of a system of linear equations, the following row transformations will result in the matrix of an equivalent system.

1. Any two rows may be interchanged.
2. The elements of any row may be multiplied by a nonzero real number.
3. Any row may be changed by adding to its elements a multiple of the corresponding elements of another row.

EXAMPLE 1

Using the Row Transformations

(a) The first row transformation is used to change the matrix

$$\begin{bmatrix} 1 & 3 & 5 \\ 0 & 1 & 2 \\ 1 & -1 & -2 \end{bmatrix} \text{ to } \begin{bmatrix} 0 & 1 & 2 \\ 1 & 3 & 5 \\ 1 & -1 & -2 \end{bmatrix}$$

by interchanging the first two rows.

(b) Using the second row transformation with $k = -2$ changes

$$\begin{bmatrix} 1 & 3 & 5 \\ 0 & 1 & 2 \\ 1 & -1 & -2 \end{bmatrix} \text{ to } \begin{bmatrix} -2 & -6 & -10 \\ 0 & 1 & 2 \\ 1 & -1 & -2 \end{bmatrix},$$

where the elements of the first row of the original matrix were multiplied by -2.

(c) The third row transformation is used to change

$$\begin{bmatrix} 1 & 3 & 5 \\ 0 & 1 & 2 \\ 1 & -1 & -2 \end{bmatrix} \text{ to } \begin{bmatrix} 0 & 4 & 7 \\ 0 & 1 & 2 \\ 1 & -1 & -2 \end{bmatrix},$$

by multiplying each element in the third row of the original matrix by -1 and adding the results to the corresponding elements in the first row of that matrix. That is, the elements in the new first row were found as follows.

$$\begin{bmatrix} 1 + 1(-1) & 3 + (-1)(-1) & 5 + (-2)(-1) \\ 0 & 1 & 2 \\ 1 & -1 & -2 \end{bmatrix} = \begin{bmatrix} 0 & 4 & 7 \\ 0 & 1 & 2 \\ 1 & -1 & -2 \end{bmatrix}$$

Rows two and three were left unchanged. ■

We will illustrate how matrix row transformations can be used to transform the augmented matrix of a system into one that is in **triangular** (or **echelon**) form. When in triangular form, the matrix will have 1s down the diagonal from upper left to lower right and 0s below each 1 as shown here.

$$\begin{bmatrix} 1 & 5 & 3 & | & 7 \\ 0 & 1 & 2 & | & 9 \\ 0 & 0 & 1 & | & 4 \end{bmatrix}$$

The 1s lie on the **main diagonal** of the matrix.

Before using matrices to solve a linear system, the system must be arranged in the proper form, with variable terms on the left side of the equation and constant terms on the right. The variable terms must be in the same order in each of the equations.

The method of using matrices to solve a system of linear equations, to be developed in this section, is called the **Gaussian reduction method,** after the mathematician Karl Friedrich Gauss (1777–1855). The following example illustrates this matrix method and compares it with the elimination method of the previous chapter.

EXAMPLE 2

Comparing the Elimination and Gaussian Reduction Methods

Solve the system

$$3x - 4y = 1$$
$$5x + 2y = 19.$$

SOLUTION The procedure for matrix solution is parallel to the elimination method used in the previous section, except for the last step. First, write the augmented matrix for the system. Here the system is already in the proper form.

Elimination Method

$3x - 4y = 1$ **[1]**
$5x + 2y = 19$ **[2]**

Divide both sides of equation (1) by 3 so that x has a coefficient of 1.

$x - \frac{4}{3}y = \frac{1}{3}$ **[3]**
$5x + 2y = 19$

Eliminate x from equation (2) by adding -5 times equation (3) to equation (2).

$x - \frac{4}{3}y = \frac{1}{3}$
$\frac{26}{3}y = \frac{52}{3}$ **[4]**

Multiply both sides of equation (4) by $\frac{3}{26}$ to get $y = 2$.

$x - \frac{4}{3}y = \frac{1}{3}$
$y = 2$

The system is in triangular form.

Gaussian Reduction Method

$$\begin{bmatrix} 3 & -4 & | & 1 \\ 5 & 2 & | & 19 \end{bmatrix}$$

Using row transformation (2), multiply each element of row 1 by $\frac{1}{3}$.

$$\begin{bmatrix} 1 & -\frac{4}{3} & | & \frac{1}{3} \\ 5 & 2 & | & 19 \end{bmatrix}$$

Using row transformation (3), add -5 times the elements of row 1 to the elements of row 2.

$$\begin{bmatrix} 1 & -\frac{4}{3} & | & \frac{1}{3} \\ 0 & \frac{26}{3} & | & \frac{52}{3} \end{bmatrix}$$

Multiply the elements of row 2 by $\frac{3}{26}$, using row transformation (2).

$$\begin{bmatrix} 1 & -\frac{4}{3} & | & \frac{1}{3} \\ 0 & 1 & | & 2 \end{bmatrix}$$

Write the corresponding equations.

$$x - \frac{4}{3}y = \frac{1}{3}$$
$$y = 2$$

Finish the solution in either method by substituting 2 for y in the first equation to get $x = 3$. The solution of the system is $(3, 2)$ with solution set $\{(3, 2)\}$. ◼

Use the Gaussian reduction method to solve the linear system

$$2x + 6y = 28$$
$$4x - 3y = -19.$$

SOLUTION We begin with the augmented matrix

$$\begin{bmatrix} 2 & 6 & | & 28 \\ 4 & -3 & | & -19 \end{bmatrix}.$$

It is best to work vertically, in columns, beginning in each column with the element which is to become 1. This is the same order used in Section 6.7 to arrange a system of equations in triangular form. The augmented matrix has a 2 in the row 1, column 1 position. To get 1 in this position, use the second transformation and multiply each entry in row 1 by $\frac{1}{2}$. This is indicated below with the notation $\frac{1}{2}R_1$ next to the new row 1.

$$\begin{bmatrix} 1 & 3 & | & 14 \\ 4 & -3 & | & -19 \end{bmatrix} \qquad \frac{1}{2}R_1$$

To get 0 in row 2, column 1, add -4 times row 1 to row 2.

$$\begin{bmatrix} 1 & 3 & | & 14 \\ 0 & -15 & | & -75 \end{bmatrix} \qquad -4R_1 + R_2$$

To get 1 in row 2, column 2, multiply each element of row 2 by $-\frac{1}{15}$, which gives

$$\begin{bmatrix} 1 & 3 & | & 14 \\ 0 & 1 & | & 5 \end{bmatrix}. \qquad -\frac{1}{15}R_2$$

This matrix corresponds to the system

$$x + 3y = 14 \qquad\qquad \textbf{[1]}$$
$$y = 5. \qquad\qquad \textbf{[2]}$$

Substitute 5 for y in equation (1) to get $x = -1$. The solution set of the system is thus $\{(-1, 5)\}$. ◼

The Gaussian reduction method can be extended to larger systems. The final matrix always will have zeros below the diagonal of ones on the left of the vertical bar. To transform the matrix, it is best to work column by column from upper left to lower right. For each column, first perform the step that gives the ones, then get the zeros below the 1 in that column.

Use the Gaussian reduction method to solve the system

$$x - y + 5z = -6$$
$$3x + 3y - z = 10$$
$$x + 3y + 2z = 5.$$

SOLUTION   We begin by writing the augmented matrix of the linear system.

$$\left[\begin{array}{ccc|c} 1 & -1 & 5 & -6 \\ 3 & 3 & -1 & 10 \\ 1 & 3 & 2 & 5 \end{array}\right]$$

There is already a 1 in row 1, column 1. The next thing to do is get 0s in the rest of column 1. First, add to row 2 the results of multiplying row 1 by -3.

$$\left[\begin{array}{ccc|c} 1 & -1 & 5 & -6 \\ 0 & 6 & -16 & 28 \\ 1 & 3 & 2 & 5 \end{array}\right] \qquad -3R_1 + R_2$$

Now add to row 3 the results of multiplying row 1 by -1.

$$\left[\begin{array}{ccc|c} 1 & -1 & 5 & -6 \\ 0 & 6 & -16 & 28 \\ 0 & 4 & -3 & 11 \end{array}\right] \qquad -1R_1 + R_3$$

To get 1 in row 2, column 2, multiply row 2 by $\frac{1}{6}$.

$$\left[\begin{array}{ccc|c} 1 & -1 & 5 & -6 \\ 0 & 1 & -\frac{8}{3} & \frac{14}{3} \\ 0 & 4 & -3 & 11 \end{array}\right] \qquad \frac{1}{6}R_2$$

Next, to get 0 in row 3, column 2, add to row 3 the results of multiplying row 2 by -4.

$$\left[\begin{array}{ccc|c} 1 & -1 & 5 & -6 \\ 0 & 1 & -\frac{8}{3} & \frac{14}{3} \\ 0 & 0 & \frac{23}{3} & -\frac{23}{3} \end{array}\right] \qquad -4R_2 + R_3$$

Now multiply the last row by $\frac{3}{23}$ to get 1 in row 3, column 3.

$$\left[\begin{array}{ccc|c} 1 & -1 & 5 & -6 \\ 0 & 1 & -\frac{8}{3} & \frac{14}{3} \\ 0 & 0 & 1 & -1 \end{array}\right] \qquad \frac{3}{23}R_3$$

While not absolutely necessary, we may obtain all integer elements by multiplying row 2 by 3.

$$\left[\begin{array}{ccc|c} 1 & -1 & 5 & -6 \\ 0 & 3 & -8 & 14 \\ 0 & 0 & 1 & -1 \end{array}\right] \qquad 3R_2$$

The final matrix is now written in triangular form and corresponds to the system of equations

$$x - y + 5z = -6 \qquad \text{[1]}$$
$$3y - 8z = 14 \qquad \text{[2]}$$
$$z = -1. \qquad \text{[3]}$$

We know that $z = -1$ from equation (3). Using back-substitution into equation (2), we find that $y = 2$, and then into equation (1) we find that $x = 1$. The solution set of the system is $\{(1, 2, -1)\}$.   ∎

NOTE   There is usually more than one way to approach the solution of a system using Gaussian reduction.

Special Cases

The next two examples show how to recognize inconsistent systems or systems with dependent equations when solving such systems using Gaussian reduction.

EXAMPLE 5

Using the Gaussian Reduction Method (No Solutions)

Solve the system

$$x + y = 2$$
$$2x + 2y = 5.$$

SOLUTION   We start with the augmented matrix.

$$\begin{bmatrix} 1 & 1 & | & 2 \\ 2 & 2 & | & 5 \end{bmatrix}$$

Next, add to row 2 the results of multiplying row 1 by -2.

$$\begin{bmatrix} 1 & 1 & | & 2 \\ 0 & 0 & | & 1 \end{bmatrix} \quad -2R_1 + R_2$$

This matrix gives the system of equations

$$x + y = 2$$
$$0 = 1,$$

an inconsistent system with no solution. The solution set is $\emptyset$.   ∎

Whenever a row of the augmented matrix is of the form

$$0\ 0\ 0 \ldots \,|\, a, \quad \text{where } a \neq 0,$$

the system is inconsistent and there will be no solution, since this row corresponds to the equation $0 = a$. A row of the matrix of a linear system in the form

$$0\ 0\ 0 \ldots \,|\, 0,$$

indicates that the equations of the system are *dependent*.

EXAMPLE 6

Using the Gaussian Reduction Method (Infinitely Many Solutions)

Use the Gaussian reduction method to solve the system

$$2x - 5y + 3z = 1$$
$$x + 4y - 2z = 8.$$

SOLUTION   Recall from Section 6.7 (Example 2) that a system with two equations and three variables has an infinite number of solutions. The Gaussian reduction method can be used to indicate the solution with one arbitrary variable. Start with the augmented matrix

$$\begin{bmatrix} 2 & -5 & 3 & | & 1 \\ 1 & 4 & -2 & | & 8 \end{bmatrix}.$$

Exchange rows to get a 1 in the row 1, column 1 position.

$$\begin{bmatrix} 1 & 4 & -2 & | & 8 \\ 2 & -5 & 3 & | & 1 \end{bmatrix}$$

Now multiply each element in row 1 by -2 and add to the corresponding element in row 2.

$$\begin{bmatrix} 1 & 4 & -2 & | & 8 \\ 0 & -13 & 7 & | & -15 \end{bmatrix} \qquad -2R_1 + R_2$$

Multiply each element in row 2 by $-\frac{1}{13}$.

$$\begin{bmatrix} 1 & 4 & -2 & | & 8 \\ 0 & 1 & -\frac{7}{13} & | & \frac{15}{13} \end{bmatrix} \qquad -\frac{1}{13}R_2$$

This is as far as we can go with the Gaussian reduction method. The equations which correspond to the final matrix are

$$x + 4y - 2z = 8 \qquad \text{and} \qquad y - \frac{7}{13}z = \frac{15}{13}.$$

Solve the second equation for y.

$$y = \frac{15}{13} + \frac{7}{13}z$$

Now substitute this result for y in the first equation and solve for x.

$$x + 4y - 2z = 8$$
$$x + 4\left(\frac{15}{13} + \frac{7}{13}z\right) - 2z = 8$$
$$x + \frac{60}{13} + \frac{28}{13}z - 2z = 8$$
$$x + \frac{60}{13} + \frac{2}{13}z = 8$$
$$x = 8 - \frac{60}{13} - \frac{2}{13}z$$
$$x = \frac{44}{13} - \frac{2}{13}z$$

TECHNOLOGICAL NOTE
The solution set of a system with dependent equations, such as the one shown in Example 6, cannot be obtained using the matrix capabilities of a graphics calculator. It must be found analytically, as explained in the solution of the system shown.

The solution set can now be written with z arbitrary as

$$\left\{ \left(\frac{44 - 2z}{13}, \frac{15 + 7z}{13}, z \right) \right\}. \qquad ▮$$

An Application

If an applied problem leads to a system of equations, the system may often be solved using Gaussian reduction.

EXAMPLE 7
Solving an Applied Problem

A manufacturer produces chairs in 3 styles: #604, #610, and #618. Each chair requires time on 3 machines as shown in the following chart. The total available time per week for each machine is also shown in the chart. All available time must be used each week. How many of each style chair can be made under these conditions each week?

Style	Number Made Each Week	Hours on Machine A	Hours on Machine B	Hours on Machine C
604	x	1	3	2
610	y	2	2	3
618	z	1	2	1
Total hours per week		30	56	44

SOLUTION The total number of hours per week on machine A for the three styles is $x + 2y + z$ which must equal 30, giving the equation

$$x + 2y + z = 30.$$

Similarly, the total hours on machine B leads to the equation

$$3x + 2y + 2z = 56,$$

and the total hours on machine C produces the equation

$$2x + 3y + z = 44.$$

These three equations form a system with the augmented matrix

$$\begin{bmatrix} 1 & 2 & 1 & | & 30 \\ 3 & 2 & 2 & | & 56 \\ 2 & 3 & 1 & | & 44 \end{bmatrix}.$$

Use the Gaussian reduction method on this matrix to get a matrix that gives the solution of the system, $(8, 6, 10)$. The manufacturer can produce 8 style 604 chairs, 6 style 610 chairs, and 10 style 618 chairs each week. ◼

FOR GROUP DISCUSSION

1. Some models of graphics calculators have the capability of performing matrix row transformations. Take an informal survey of the members of the class, and see what models are represented. Then read the owners' manuals to determine which models do have this capability.
2. Figure 1 shows three views of the graph of the same quadratic function $f(x) = ax^2 + bx + c$. Discuss how the values of a, b, and c can be found. (See Exercises 55–62 for a follow-up to this discussion.)

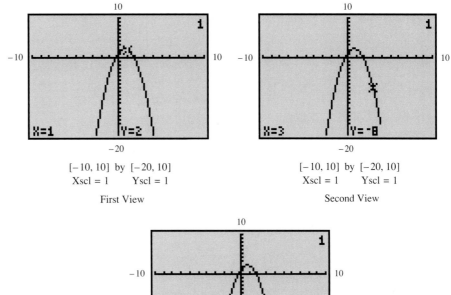

[−10, 10] by [−20, 10]
Xscl = 1   Yscl = 1

First View

[−10, 10] by [−20, 10]
Xscl = 1   Yscl = 1

Second View

[−10, 10] by [−20, 10]
Xscl = 1   Yscl = 1

Third View

FIGURE 1

7.1 EXERCISES

To the Student: Calculator Considerations

Some modern graphics calculators have the capability to perform the matrix row transformations described in this section. You should consult your owner's manual to learn whether your model has this capability. If it does, read it to learn how your calculator accomplishes them. Then use your calculator to help you in working the exercises of this section.

Use the row transformation described to transform each of the following matrices as indicated.

1. $\begin{bmatrix} 2 & 4 \\ 4 & 7 \end{bmatrix}$;   −2 times row 1

2. $\begin{bmatrix} -1 & 4 \\ 7 & 0 \end{bmatrix}$;   7 times row 1

3. $\begin{bmatrix} 1 & 5 & 6 \\ -2 & 3 & -1 \\ 4 & 7 & 0 \end{bmatrix}$;   row 1 added to row 2

4. $\begin{bmatrix} 2 & 5 & 6 \\ 4 & -1 & 2 \\ 3 & 7 & 1 \end{bmatrix}$;   row 3 added to row 1

5. $\begin{bmatrix} -3 & 1 & -4 \\ 2 & 1 & 3 \\ -7 & 5 & 2 \end{bmatrix}$; -5 times row 2 added to row 3

6. $\begin{bmatrix} 4 & 10 & -8 \\ 7 & 4 & 3 \\ -1 & 1 & 0 \end{bmatrix}$; -4 times row 3 added to row 2

Write the augmented matrix for each of the following systems. Do not solve the system.

7. $\begin{aligned} 2x + 3y &= 11 \\ x + 2y &= 8 \end{aligned}$

8. $\begin{aligned} 3x + 5y &= -13 \\ 2x + 3y &= -9 \end{aligned}$

9. $\begin{aligned} x + 5y &= 6 \\ x + 2y &= 8 \end{aligned}$

10. $\begin{aligned} 2x + 7y &= 1 \\ 5x &= -15 \end{aligned}$

11. $\begin{aligned} 2x + y + z &= 3 \\ 3x - 4y + 2z &= -7 \\ x + y + z &= 2 \end{aligned}$

12. $\begin{aligned} 4x - 2y + 3z &= 4 \\ 3x + 5y + z &= 7 \\ 5x - y + 4z &= 7 \end{aligned}$

13. $\begin{aligned} x + y &= 2 \\ 2y + z &= -4 \\ z &= 2 \end{aligned}$

14. $\begin{aligned} x &= 6 \\ y + 2z &= 2 \\ x - 3z &= 6 \end{aligned}$

Write the system of equations associated with each of the following augmented matrices. Do not try to solve.

15. $\left[\begin{array}{cc|c} 2 & 1 & 1 \\ 3 & -2 & -9 \end{array}\right]$

16. $\left[\begin{array}{cc|c} 1 & -5 & -18 \\ 6 & 2 & 20 \end{array}\right]$

17. $\left[\begin{array}{ccc|c} 1 & 0 & 0 & 2 \\ 0 & 1 & 0 & 3 \\ 0 & 0 & 1 & -2 \end{array}\right]$

18. $\left[\begin{array}{ccc|c} 1 & 0 & 1 & 4 \\ 0 & 1 & 0 & 2 \\ 0 & 0 & 1 & 3 \end{array}\right]$

19. $\left[\begin{array}{ccc|c} 3 & 2 & 1 & 1 \\ 0 & 2 & 4 & 22 \\ -1 & -2 & 3 & 15 \end{array}\right]$

20. $\left[\begin{array}{ccc|c} 2 & 1 & 3 & 12 \\ 4 & -3 & 0 & 10 \\ 5 & 0 & -4 & -11 \end{array}\right]$

Use the Gaussian reduction method to solve each of the following systems of equations.

21. $\begin{aligned} x + y &= 5 \\ x - y &= -1 \end{aligned}$

22. $\begin{aligned} x + 2y &= 5 \\ 2x + y &= -2 \end{aligned}$

23. $\begin{aligned} x + y &= -3 \\ 2x - 5y &= -6 \end{aligned}$

24. $\begin{aligned} 3x - 2y &= 4 \\ 3x + y &= -2 \end{aligned}$

25. $\begin{aligned} 2x - 3y &= 10 \\ 2x + 2y &= 5 \end{aligned}$

26. $\begin{aligned} 4x + y &= 5 \\ 2x + y &= 3 \end{aligned}$

27. $\begin{aligned} 2x - 3y &= 2 \\ 4x - 6y &= 1 \end{aligned}$

28. $\begin{aligned} x + 2y &= 1 \\ 2x + 4y &= 3 \end{aligned}$

29. $\begin{aligned} 6x - 3y &= 1 \\ -12x + 6y &= -2 \end{aligned}$

30. $\begin{aligned} x - y &= 1 \\ -x + y &= -1 \end{aligned}$

31. $\begin{aligned} x + y &= -1 \\ y + z &= 4 \\ x + z &= 1 \end{aligned}$

32. $\begin{aligned} x - z &= -3 \\ y + z &= 9 \\ x + z &= 7 \end{aligned}$

33. $\begin{aligned} x + y - z &= 6 \\ 2x - y + z &= -9 \\ x - 2y + 3z &= 1 \end{aligned}$

34. $\begin{aligned} x + 3y - 6z &= 7 \\ 2x - y + 2z &= 0 \\ x + y + 2z &= -1 \end{aligned}$

35. $\begin{aligned} -x + y &= -1 \\ y - z &= 6 \\ x + z &= -1 \end{aligned}$

36. $\begin{aligned} x + y &= 1 \\ 2x - z &= 0 \\ y + 2z &= -2 \end{aligned}$

37. $\begin{aligned} 2x - y + 3z &= 0 \\ x + 2y - z &= 5 \\ 2y + z &= 1 \end{aligned}$

38. $\begin{aligned} 4x + 2y - 3z &= 6 \\ x - 4y + z &= -4 \\ -x + 2z &= 2 \end{aligned}$

39. Compare the use of an augmented matrix as a shorthand way of writing a system of linear equations and the use of synthetic division as a shorthand way to divide polynomials.

40. Compare the use of the third row transformation on a matrix and the elimination method of solving a system of linear equations.

Solve each of the systems in Exercises 41–46 by Gaussian reduction. Let z be the arbitrary variable if necessary.

41. $\begin{aligned} x - 3y + 2z &= 10 \\ 2x - y - z &= 8 \end{aligned}$

42. $\begin{aligned} 3x + y - z &= 12 \\ x + 2y + z &= 10 \end{aligned}$

43. $\begin{aligned} x + 2y - z &= 0 \\ 3x - y + z &= 6 \\ -2x - 4y + 2z &= 0 \end{aligned}$

44. $\begin{aligned} 3x + 5y - z &= 0 \\ 4x - y + 2z &= 1 \\ -6x - 10y + 2z &= 0 \end{aligned}$

45. $\begin{aligned} x - 2y + z &= 5 \\ -2x + 4y - 2z &= 2 \\ 2x + y - z &= 2 \end{aligned}$

46. $\begin{aligned} 3x + 6y - 3z &= 12 \\ -x - 2y + z &= 16 \\ x + y - 2z &= 20 \end{aligned}$

Solve each applied problem by writing a system of equations and then solving the system using Gaussian reduction.

47. A working couple earned a total of $4352. The wife earned $64 per day; the husband earned $8 per day less. Find the number of days each worked if the total number of days worked by both was 72.

48. Midtown Manufacturing Company makes two products; plastic plates and plastic cups. Both require time on two machines—plates: 1 hr on machine A and 2 hr on machine B; cups: 3 hr on machine A and 1 hr on machine B. Both machines operate 15 hr per day. How many of each product can be produced in a day under these conditions?

49. A company produces two models of bicycles, model 201 and model 301. Model 201 requires 2 hr of assembly time, and model 301 requires 3 hr of assembly time. The parts for model 201 cost $25 per bike; those for model 301 cost $30 per bike. If the company has a total of 34 hr of assembly time and $365 available per day for these two models, how many of each can be made in a day?

50. George Esquibel deposits some money in a bank account paying 3% per year. He uses some addi-tional money, amounting to $\frac{1}{3}$ the amount placed in the bank, to buy bonds paying 4% per year. With the balance of his funds he buys a 4.5% certificate of deposit. The first year his investments bring a return of $400. If the total of the investments is $10,000, how much is invested at each rate?

51. To get necessary funds for a planned expansion, a small company took out three loans totaling $25,000. The company was able to borrow some of the money at 8%. It borrowed $2000 more than $\frac{1}{2}$ the amount of the 8% loan at 10%, and the rest at 9%. The total annual interest was $2220. How much did the company borrow at each rate?

52. A biologist has three salt solutions: some 5% solution, some 15% solution, and some 25% solution. She needs to mix some of each to get 50 liters of 20% solution. She wants to use twice as much of the 5% solution as the 15% solution. How much of each solution should she use?

We have seen how Gaussian reduction can be used to solve systems with up to three equations in three unknowns. It can be extended to larger systems as well. Use Gaussian reduction to solve each system in Exercises 53 and 54, and express your solutions in the form (x, y, z, w).

53.
$$\begin{aligned} x + 3y - 2z - w &= 9 \\ 4x + y + z + 2w &= 2 \\ -3x - y + z - w &= -5 \\ x - y - 3z - 2w &= 2 \end{aligned}$$

54.
$$\begin{aligned} 3x + 2y - w &= 0 \\ 2x + z + 2w &= 5 \\ x + 2y - z &= -2 \\ 2x - y + z + w &= 2 \end{aligned}$$

R*elating Concepts*

In item 2 of the group discussion exercise that ends this section, you should have reached the conclusion that the values a, b, and c can be found by substituting the corresponding x- and y-values in each case into the equation $y = ax^2 + bx + c$ to obtain equations in a, b, and c. Work Exercises 55–62 in order.

55. Give the equation obtained when $x = 1$ and $y = 2$.

56. Give the equation obtained when $x = -2$ and $y = -13$.

57. Give the equation obtained when $x = 3$ and $y = -8$.

58. Before solving the system formed by the equations in Exercises 55–57, what do you know from the graph about the *sign* of a? Why is this so?

59. What do you know about the *sign* of c?

60. Use Gaussian reduction to solve the system formed by the equations in Exercises 55–57. Give the values of a, b, and c.

61. If $f(x) = ax^2 + bx + c$, determine $f(-1.5)$ analytically for the function found in Exercise 60.

62. Support your result for Exercise 61 graphically.

---

63. Suppose that $g(x) = ax^3 + bx^2 + cx + d$, with $g(-2) = -38$, $g(0) = -2$, $g(1) = 1$, and $g(3) = 37$. Find the values of $a, b, c,$ and d, and then find $g(4)$ both analytically and graphically.

64. Solve the system

$$
\begin{aligned}
3x^2 + y^2 &= 4 \\
-5x^2 + 2y^2 &= -3
\end{aligned}
$$

for x^2 and y^2 using Gaussian reduction. Then solve for x and y using methods described in Section 6.5. (*Hint:* Start by letting $x^2 = t$ and $y^2 = s$.)

7.2 PROPERTIES OF MATRICES

Introduction ▌ Terminology of Matrices ▌ Operations on Matrices ▌ An Application of Matrix Algebra

Introduction

The use of matrix notation in solving a system of linear equations was illustrated in the previous section. In this section we will examine matrices as mathematical entities with properties analogous to those of real numbers. To motivate the concept of a matrix, consider the following.

Suppose you are the manager of a health food store and you receive the following shipments of vitamins from two suppliers: from Dexter, 2 cartons of vitamin A pills, 7 cartons of vitamin E pills, and 5 cartons of vitamin K pills; from Sullivan, 4 cartons of vitamin A pills, 6 cartons of vitamin E pills, and 9 cartons of vitamin K pills.

It might be helpful to rewrite the information in a chart to make it more understandable.

Manufacturer	Cartons of Vitamins		
	A	**E**	**K**
Dexter	2	7	5
Sullivan	4	6	9

The information is clearer when presented this way. In fact, as long as you remember what each number represents, you can remove all the labels and write the numbers as a matrix.

$$
\begin{bmatrix} 2 & 7 & 5 \\ 4 & 6 & 9 \end{bmatrix}
$$

This array of numbers gives all the information needed.

Terminology of Matrices

It is customary to use capital letters to name matrices. Also, subscript notation is often used to name the elements of a matrix, as in the following matrix A.

$$A = \begin{bmatrix} a_{11} & a_{12} & a_{13} & \cdots & a_{1n} \\ a_{21} & a_{22} & a_{23} & \cdots & a_{2n} \\ a_{31} & a_{32} & a_{33} & \cdots & a_{3n} \\ \vdots & \vdots & \vdots & & \vdots \\ a_{m1} & a_{m2} & a_{m3} & \cdots & a_{mn} \end{bmatrix}$$

With this notation, the row 1, column 1 element is a_{11}; the row 2, column 3 element is a_{23}; and in general, the row i, column j element is a_{ij}.

Matrices are classified by their **dimension,** that is, by the number of rows and columns that they contain. For example, the matrix

$$\begin{bmatrix} 2 & 7 & -5 \\ 3 & -6 & 0 \end{bmatrix}$$

has two rows and three columns, with dimension 2×3; a matrix with m rows and n columns has dimension $\boldsymbol{m \times n}$. The number of rows is always given first.

Certain matrices have special names: an $n \times n$ matrix is a **square matrix** of dimension $n \times n$. Also, a matrix with just one row is a **row matrix,** and a matrix with just one column is a **column matrix.**

Two matrices are **equal** if they have the same dimension and if each pair of corresponding elements, position by position, is equal. Using this definition, the matrices

$$\begin{bmatrix} 2 & 1 \\ 3 & -5 \end{bmatrix} \quad \text{and} \quad \begin{bmatrix} 1 & 2 \\ -5 & 3 \end{bmatrix}$$

are *not* equal (even though they contain the same elements and have the same dimension), since the corresponding elements differ.

TECHNOLOGICAL NOTE
The TEST function of some graphics calculators will allow you to determine whether two matrices are equal or unequal, returning a 1 if they are or a 0 if they are not. Refer to your owner's manual to see if your model is capable of this.

EXAMPLE 1

Classifying Matrices by Dimension

The following matrices are classified by dimension.

(a) The matrix $\begin{bmatrix} 6 & 5 \\ 3 & 4 \\ 5 & -1 \end{bmatrix}$ is a 3×2 matrix, because it has 3 rows and 2 columns.

(b) $\begin{bmatrix} 5 & 8 & 9 \\ 0 & 5 & -3 \\ -4 & 0 & 5 \end{bmatrix}$ is a 3×3 matrix. It is also a square matrix.

(c) $[1 \quad 6 \quad 5 \quad -2 \quad 5]$ is a 1×5 matrix. It is an example of a row matrix.

(d) $\begin{bmatrix} 3 \\ -5 \\ 0 \\ 2 \end{bmatrix}$ is a 4×1 column matrix. ∎

EXAMPLE **2**

Determining Equality
of Matrices

(a)   If $A = \begin{bmatrix} 2 & 1 \\ p & q \end{bmatrix}$ and $B = \begin{bmatrix} x & y \\ -1 & 0 \end{bmatrix}$, find the values of x, y, p, and q such that $A = B$.

SOLUTION   From the definition of equality given above, the only way that the statement

$$\begin{bmatrix} 2 & 1 \\ p & q \end{bmatrix} = \begin{bmatrix} x & y \\ -1 & 0 \end{bmatrix}$$

can be true is if $2 = x$, $1 = y$, $p = -1$, and $q = 0$.

(b)   Find the values of x and y, if possible, such that

$$\begin{bmatrix} x \\ y \end{bmatrix} = \begin{bmatrix} 1 \\ 4 \\ 0 \end{bmatrix}.$$

SOLUTION   The statement

$$\begin{bmatrix} x \\ y \end{bmatrix} = \begin{bmatrix} 1 \\ 4 \\ 0 \end{bmatrix}$$

can never be true, since the two matrices have different dimensions. (One is 2×1 and the other is 3×1.)   ◘

Operations on Matrices

At the beginning of this section we used the matrix

$$\begin{bmatrix} 2 & 7 & 5 \\ 4 & 6 & 9 \end{bmatrix},$$

TECHNOLOGICAL NOTE
The remaining material in
this section discusses the
addition, subtraction, and
multiplication of matrices.
As explained in the text,
these operations are
defined only when there
are appropriate restric-
tions on the dimensions
of the matrices being
added, subtracted, or
multiplied. Graphics cal-
culators will perform these
operations provided the
dimensions are compat-
ible for the operation. A
dimension error message
will occur if the operation
cannot be performed.

where the columns represent the numbers of cartons of three different types of vitamins (A, E, and K, respectively), and the rows represent two different manufac-turers (Dexter and Sullivan, respectively). For example, the element 7 represents 7 cartons of vitamin E pills from Dexter, and so on. Suppose another shipment from these two suppliers is described by the following matrix.

$$\begin{bmatrix} 3 & 12 & 10 \\ 15 & 11 & 8 \end{bmatrix}$$

Here, for example, 8 cartons of vitamin K pills arrived from Sullivan. The number of cartons of each kind of pill that were received from these two shipments can be found from these two matrices.

In the first shipment, 2 cartons of vitamin A pills were received from Dexter, and in the second shipment, 3 cartons of vitamin A pills were received. Altogether, $2 + 3$, or 5, cartons of these pills were received. Corresponding elements can be added to find the total number of cartons of each type of pill received.

$$\begin{bmatrix} 2 & 7 & 5 \\ 4 & 6 & 9 \end{bmatrix} + \begin{bmatrix} 3 & 12 & 10 \\ 15 & 11 & 8 \end{bmatrix} = \begin{bmatrix} 2+3 & 7+12 & 5+10 \\ 4+15 & 6+11 & 9+8 \end{bmatrix}$$

$$= \begin{bmatrix} 5 & 19 & 15 \\ 19 & 17 & 17 \end{bmatrix}$$

The last matrix gives the total number of cartons of each type of pill that were received. For example, 15 cartons of vitamin K pills were received from Dexter. Generalizing from this example leads to the following definition.

MATRIX ADDITION

The sum of two $m \times n$ matrices A and B is the $m \times n$ matrix $A + B$ in which each element is the sum of the corresponding elements of A and B.

CAUTION Only matrices with the same dimension can be added.

EXAMPLE 3

Adding Matrices

The following statements show how matrices are added.

(a) $\begin{bmatrix} 5 & -6 \\ 8 & 9 \end{bmatrix} + \begin{bmatrix} -4 & 6 \\ 8 & -3 \end{bmatrix} = \begin{bmatrix} 5+(-4) & -6+6 \\ 8+8 & 9+(-3) \end{bmatrix} = \begin{bmatrix} 1 & 0 \\ 16 & 6 \end{bmatrix}$

(b) $\begin{bmatrix} 2 \\ 5 \\ 8 \end{bmatrix} + \begin{bmatrix} -6 \\ 3 \\ 12 \end{bmatrix} = \begin{bmatrix} -4 \\ 8 \\ 20 \end{bmatrix}$

(c) The matrices

$$A = \begin{bmatrix} 5 & 8 \\ 6 & 2 \end{bmatrix} \quad \text{and} \quad B = \begin{bmatrix} 3 & 9 & 1 \\ 4 & 2 & 5 \end{bmatrix}$$

have different dimensions. Therefore, the sum $A + B$ does not exist. ◾

A matrix containing only zero elements is called a **zero matrix.** For example, $\begin{bmatrix} 0 & 0 & 0 \end{bmatrix}$ is the 1×3 zero matrix, while

$$\begin{bmatrix} 0 & 0 & 0 \\ 0 & 0 & 0 \end{bmatrix}$$

is the 2×3 zero matrix. A zero matrix can be written with any dimension.

The additive inverse of a real number a is the unique real number $-a$ such that $a + (-a) = 0$ and $-a + a = 0$. Given a matrix A, a matrix $-A$ can be found such that $A + (-A) = O$, where O is the appropriate zero matrix, and $-A + A = O$. For example, if

$$A = \begin{bmatrix} -5 & 2 & -1 \\ 3 & 4 & -6 \end{bmatrix},$$

then the elements of matrix $-A$ are the additive inverses of the corresponding elements of A. (Remember that each element of A is a real number and thus has an additive inverse.)

$$-A = \begin{bmatrix} 5 & -2 & 1 \\ -3 & -4 & 6 \end{bmatrix}$$

To check, first test that $A + (-A)$ equals O, the appropriate zero matrix.

$$A + (-A) = \begin{bmatrix} -5 & 2 & -1 \\ 3 & 4 & -6 \end{bmatrix} + \begin{bmatrix} 5 & -2 & 1 \\ -3 & -4 & 6 \end{bmatrix}$$

$$= \begin{bmatrix} 0 & 0 & 0 \\ 0 & 0 & 0 \end{bmatrix} = O$$

Then test that $-A + A$ is also O. Matrix $-A$ is the **additive inverse,** or **negative,** of matrix A. Every matrix has a unique additive inverse.

Just as subtraction of real numbers is defined in terms of the additive inverse, subtraction of matrices is defined in the same way.

MATRIX SUBTRACTION

If A and B are matrices with the same dimension, then

$$A - B = A + (-B).$$

EXAMPLE 4

Subtracting Matrices

The following statements show how matrices are subtracted.

(a) $\begin{bmatrix} -5 & 6 \\ 2 & 4 \end{bmatrix} - \begin{bmatrix} -3 & 2 \\ 5 & -8 \end{bmatrix} = \begin{bmatrix} -5 & 6 \\ 2 & 4 \end{bmatrix} + \begin{bmatrix} 3 & -2 \\ -5 & 8 \end{bmatrix} = \begin{bmatrix} -2 & 4 \\ -3 & 12 \end{bmatrix}$

(b) $\begin{bmatrix} 8 & 6 & -4 \end{bmatrix} - \begin{bmatrix} 3 & 5 & -8 \end{bmatrix} = \begin{bmatrix} 5 & 1 & 4 \end{bmatrix}$

(c) The matrices

$$\begin{bmatrix} -2 & 5 \\ 0 & 1 \end{bmatrix} \quad \text{and} \quad \begin{bmatrix} 3 \\ 5 \end{bmatrix}$$

have different dimensions and cannot be subtracted. ∎

If a matrix A is added to itself, each element in the sum is twice as large as the corresponding element of A. For example,

$$\begin{bmatrix} 2 & 5 \\ 1 & 3 \\ 4 & 6 \end{bmatrix} + \begin{bmatrix} 2 & 5 \\ 1 & 3 \\ 4 & 6 \end{bmatrix} = \begin{bmatrix} 4 & 10 \\ 2 & 6 \\ 8 & 12 \end{bmatrix} = 2\begin{bmatrix} 2 & 5 \\ 1 & 3 \\ 4 & 6 \end{bmatrix}.$$

In the last expression, the number 2 in front of the matrix is called a **scalar** to distinguish it from a matrix. (A scalar is just a special name for a real number.) The example above suggests the following definition of multiplication of a matrix by a scalar.

MULTIPLICATION OF A MATRIX BY A SCALAR

The product of a scalar k and a matrix X is the matrix kX, each of whose elements is k times the corresponding element of X.

EXAMPLE 5

Multiplying a Matrix by a Scalar

The following statements show how a matrix is multiplied by a scalar.

(a) $5\begin{bmatrix} 2 & -3 \\ 0 & 4 \end{bmatrix} = \begin{bmatrix} 10 & -15 \\ 0 & 20 \end{bmatrix}$ **(b)** $\frac{3}{4}\begin{bmatrix} 20 & 36 \\ 12 & -16 \end{bmatrix} = \begin{bmatrix} 15 & 27 \\ 9 & -12 \end{bmatrix}$ ◻

We now know how to multiply a matrix by a scalar. In order to motivate the definition of multiplying a matrix by another matrix, let us return to the example about vitamin pills earlier in this section. The matrix below shows the number of cartons of each type of vitamin received from Dexter and Sullivan, respectively.

$$\begin{bmatrix} 2 & 7 & 5 \\ 4 & 6 & 9 \end{bmatrix}$$

Now suppose that each carton of vitamin A pills costs the store \$12, each carton of vitamin E pills costs \$18, and each carton of vitamin K pills costs \$9.

To find the total cost of the pills from Dexter, we multiply as follows.

Vitamin	Number of Cartons	Cost Per Carton	Total Cost	
A	2	\$12	\$ 24	
E	7	\$18	\$ 126	
K	5	\$ 9	\$ 45	
			\$ 195	Total from Dexter

The Dexter pills cost a total of \$195.

This result is the sum of three products:

$$2(\$12) + 7(\$18) + 5(\$9) = \$195.$$

In the same way, using the second row of the matrix and the three costs, the total cost of the Sullivan pills is

$$4(\$12) + 6(\$18) + 9(\$9) = \$237.$$

The costs, \$12, \$18, and \$9, can be written as a column matrix.

$$\begin{bmatrix} 12 \\ 18 \\ 9 \end{bmatrix}$$

The total costs for each supplier, \$195 and \$237, also can be written as a column matrix.

$$\begin{bmatrix} 195 \\ 237 \end{bmatrix}$$

The product of the matrices

$$\begin{bmatrix} 2 & 7 & 5 \\ 4 & 6 & 9 \end{bmatrix} \quad \text{and} \quad \begin{bmatrix} 12 \\ 18 \\ 9 \end{bmatrix}$$

can be written as follows.

$$\begin{bmatrix} 2 & 7 & 5 \\ 4 & 6 & 9 \end{bmatrix} \begin{bmatrix} 12 \\ 18 \\ 9 \end{bmatrix} = \begin{bmatrix} 2 \cdot 12 + 7 \cdot 18 + 5 \cdot 9 \\ 4 \cdot 12 + 6 \cdot 18 + 9 \cdot 9 \end{bmatrix} = \begin{bmatrix} 195 \\ 237 \end{bmatrix}$$

Each element of the product was found by multiplying the elements of the *rows* of the matrix on the left and the corresponding elements of the *columns* of the matrix on the right, and then finding the sum of these products. Notice that the product of a 2 × 3 matrix and a 3 × 1 matrix is a 2 × 1 matrix.

Generalizing from this example gives the following definition of matrix multiplication.

MATRIX MULTIPLICATION

The product AB of an $m \times n$ matrix A and an $n \times k$ matrix B is found as follows. To get the ith row, jth column element of AB, multiply each element in the ith row of A by the corresponding element in the jth column of B. The sum of these products will give the element of row i, column j of AB. The dimension of AB is $m \times k$.

This definition requires the following restriction on matrix multiplication.

RESTRICTION ON MATRIX MULTIPLICATION

The product AB of two matrices A and B can be found only if the number of *columns* of A is the same as the number of *rows* of B. The final product will have as many rows as A and as many columns as B.

The next two examples illustrate how matrices are multiplied.

EXAMPLE 6

Finding the Product of Two Matrices

Find the product AB, where

$$A = \begin{bmatrix} 2 & 4 \\ 5 & 6 \end{bmatrix} \quad \text{and} \quad B = \begin{bmatrix} -3 & 5 \\ 4 & -2 \end{bmatrix}.$$

SOLUTION

Step 1 Multiply the elements of the first row of A and the corresponding elements of the first column of B, and add these products.

$$AB = \begin{bmatrix} 2 & 4 \\ 5 & 6 \end{bmatrix} \begin{bmatrix} -3 & 5 \\ 4 & -2 \end{bmatrix} \qquad 2(-3) + 4(4) = -6 + 16 = 10$$

The first-row, first-column entry of the product matrix AB is 10.

Step 2 Multiply the elements of the first row of A and the second column of B and then add the products to get the first-row, second-column entry of the product matrix.

$$AB = \begin{bmatrix} 2 & 4 \\ 5 & 6 \end{bmatrix} \begin{bmatrix} -3 & 5 \\ 4 & -2 \end{bmatrix} \qquad 2(5) + 4(-2) = 10 + (-8) = 2$$

Step 3

$$AB = \begin{bmatrix} 2 & 4 \\ 5 & 6 \end{bmatrix}\begin{bmatrix} -3 & 5 \\ 4 & -2 \end{bmatrix} \qquad 5(-3) + 6(4) = -15 + 24 = 9$$

The second-row, first-column entry of the product matrix is 9.

Step 4

$$AB = \begin{bmatrix} 2 & 4 \\ 5 & 6 \end{bmatrix}\begin{bmatrix} -3 & 5 \\ 4 & -2 \end{bmatrix} \qquad 5(5) + 6(-2) = 25 + (-12)$$
$$= 13$$

Finally, 13 is the second-row, second-column entry.

Step 5 Write the product. The four entries in the product matrix come from the four steps above.

$$AB = \begin{bmatrix} 2 & 4 \\ 5 & 6 \end{bmatrix}\begin{bmatrix} -3 & 5 \\ 4 & -2 \end{bmatrix} = \begin{bmatrix} 10 & 2 \\ 9 & 13 \end{bmatrix} \qquad \blacksquare$$

EXAMPLE 7

Finding the Product of Two Matrices

Find the product.

$$\begin{bmatrix} -3 & 4 & 2 \\ 5 & 0 & 4 \end{bmatrix}\begin{bmatrix} -6 & 4 \\ 2 & 3 \\ 3 & -2 \end{bmatrix}$$

SOLUTION

Step 1 $\begin{bmatrix} -3 & 4 & 2 \\ 5 & 0 & 4 \end{bmatrix}\begin{bmatrix} -6 & 4 \\ 2 & 3 \\ 3 & -2 \end{bmatrix} \qquad (-3)(-6) + 4(2) + 2(3) = 32$

Step 2 $\begin{bmatrix} -3 & 4 & 2 \\ 5 & 0 & 4 \end{bmatrix}\begin{bmatrix} -6 & 4 \\ 2 & 3 \\ 3 & -2 \end{bmatrix} \qquad (-3)(4) + 4(3) + 2(-2) = -4$

Step 3 $\begin{bmatrix} -3 & 4 & 2 \\ 5 & 0 & 4 \end{bmatrix}\begin{bmatrix} -6 & 4 \\ 2 & 3 \\ 3 & -2 \end{bmatrix} \qquad 5(-6) + 0(2) + 4(3) = -18$

Step 4 $\begin{bmatrix} -3 & 4 & 2 \\ 5 & 0 & 4 \end{bmatrix}\begin{bmatrix} -6 & 4 \\ 2 & 3 \\ 3 & -2 \end{bmatrix} \qquad 5(4) + 0(3) + 4(-2) = 12$

Step 5 Write the product.

$$\begin{bmatrix} -3 & 4 & 2 \\ 5 & 0 & 4 \end{bmatrix}\begin{bmatrix} -6 & 4 \\ 2 & 3 \\ 3 & -2 \end{bmatrix} = \begin{bmatrix} 32 & -4 \\ -18 & 12 \end{bmatrix}$$

As this example shows, the product of a 2 × 3 matrix and a 3 × 2 matrix is a 2 × 2 matrix. ∎

EXAMPLE **8**

Deciding Whether
Two Matrices can
be Multiplied

Suppose matrix A is 2×2, while matrix B is 2×4. Can the product AB be calculated? What is the dimension of the product?

SOLUTION   The following diagram helps answer these questions.

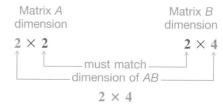

The product AB can be calculated because A has two columns and B has two rows. The dimension of the product is 2×4. (However, the product BA could not be found.)   ◻

EXAMPLE **9**

Multiplying Matrices
in Different Orders

If possible, find AB and BA, where

$$A = \begin{bmatrix} 1 & -3 \\ 7 & 2 \end{bmatrix} \quad \text{and} \quad B = \begin{bmatrix} 1 & 0 & -1 & 2 \\ 3 & 1 & 4 & -1 \end{bmatrix}.$$

SOLUTION   Use the definition of matrix multiplication to find AB.

$$AB = \begin{bmatrix} 1 & -3 \\ 7 & 2 \end{bmatrix}\begin{bmatrix} 1 & 0 & -1 & 2 \\ 3 & 1 & 4 & -1 \end{bmatrix}$$

$$= \begin{bmatrix} 1(1) + (-3)(3) & 1(0) + (-3)1 & 1(-1) + (-3)4 & 1(2) + (-3)(-1) \\ 7(1) + 2(3) & 7(0) + 2(1) & 7(-1) + 2(4) & 7(2) + 2(-1) \end{bmatrix}$$

$$= \begin{bmatrix} -8 & -3 & -13 & 5 \\ 13 & 2 & 1 & 12 \end{bmatrix}$$

Since B is a 2×4 matrix, and A is a 2×2 matrix, the product BA cannot be found.   ◻

EXAMPLE **10**

Multiplying Square
Matrices in Different
Orders

Find MN and NM, given

$$M = \begin{bmatrix} 1 & 3 \\ -2 & 4 \end{bmatrix} \quad \text{and} \quad N = \begin{bmatrix} 2 & 5 \\ 10 & -3 \end{bmatrix}.$$

SOLUTION   By the definition of matrix multiplication,

$$MN = \begin{bmatrix} 1 & 3 \\ -2 & 4 \end{bmatrix}\begin{bmatrix} 2 & 5 \\ 10 & -3 \end{bmatrix}$$

$$= \begin{bmatrix} 2 + 30 & 5 - 9 \\ -4 + 40 & -10 - 12 \end{bmatrix}$$

$$= \begin{bmatrix} 32 & -4 \\ 36 & -22 \end{bmatrix}.$$

TECHNOLOGICAL NOTE
After reading Examples 4–10, you may wish to experiment with your calculator to see if you can duplicate the results obtained.

Similarly,

$$NM = \begin{bmatrix} 2 & 5 \\ 10 & -3 \end{bmatrix}\begin{bmatrix} 1 & 3 \\ -2 & 4 \end{bmatrix}$$
$$= \begin{bmatrix} 2 - 10 & 6 + 20 \\ 10 + 6 & 30 - 12 \end{bmatrix}$$
$$= \begin{bmatrix} -8 & 26 \\ 16 & 18 \end{bmatrix}.$$ ◧

In Example 9 the product AB could be found, but not BA. In Example 10, although both MN and NM could be found, they were not equal, showing that multiplication of matrices is not commutative. This fact distinguishes matrix arithmetic from the arithmetic of real numbers.

CAUTION Since multiplication of square matrices is in general not commutative, be careful about the ordering when you multiply matrices.

An Application of Matrix Algebra

EXAMPLE 11

Applying Matrix Multiplication

A contractor builds three kinds of houses, models A, B, and C, with a choice of two styles, colonial or ranch. Matrix P below shows the number of each kind of house the contractor is planning to build for a new 100-home subdivision. The amounts for each of the main materials used depend on the style of the house. These amounts are shown in matrix Q below, while matrix R gives the cost in dollars for each kind of material. Concrete is measured here in cubic yards, lumber in 1000 board feet, brick in 1000s, and shingles in 100 square feet.

$$\begin{array}{c} \\ \text{Model A} \\ \text{Model B} \\ \text{Model C} \end{array}\begin{array}{c} \begin{array}{cc} \text{Colonial} & \text{Ranch} \end{array} \\ \begin{bmatrix} 0 & 30 \\ 10 & 20 \\ 20 & 20 \end{bmatrix} \end{array} = P$$

$$\begin{array}{c} \\ \text{Colonial} \\ \text{Ranch} \end{array}\begin{array}{c} \begin{array}{cccc} \text{Concrete} & \text{Lumber} & \text{Brick} & \text{Shingles} \end{array} \\ \begin{bmatrix} 10 & 2 & 0 & 2 \\ 50 & 1 & 20 & 2 \end{bmatrix} \end{array} = Q \qquad \begin{array}{c} \\ \text{Concrete} \\ \text{Lumber} \\ \text{Brick} \\ \text{Shingles} \end{array}\begin{array}{c} \text{Cost per unit} \\ \begin{bmatrix} 20 \\ 180 \\ 60 \\ 25 \end{bmatrix} \end{array} = R$$

(a) What is the total cost of materials for all houses of each model?

SOLUTION To find the materials cost for each model, first find matrix PQ, which will show the total amount of each material needed for all houses of each model.

$$PQ = \begin{bmatrix} 0 & 30 \\ 10 & 20 \\ 20 & 20 \end{bmatrix}\begin{bmatrix} 10 & 2 & 0 & 2 \\ 50 & 1 & 20 & 2 \end{bmatrix}$$

$$= \begin{array}{c} \begin{array}{cccc} \text{Concrete} & \text{Lumber} & \text{Brick} & \text{Shingles} \end{array} \\ \begin{bmatrix} 1500 & 30 & 600 & 60 \\ 1100 & 40 & 400 & 60 \\ 1200 & 60 & 400 & 80 \end{bmatrix} \end{array}\begin{array}{c} \\ \text{Model A} \\ \text{Model B} \\ \text{Model C} \end{array}$$

Multiplying PQ and the cost matrix R gives the total cost of materials for each model.

$$(PQ)R = \begin{bmatrix} 1500 & 30 & 600 & 60 \\ 1100 & 40 & 400 & 60 \\ 1200 & 60 & 400 & 80 \end{bmatrix} \begin{bmatrix} 20 \\ 180 \\ 60 \\ 25 \end{bmatrix} = \begin{bmatrix} 72{,}900 \\ 54{,}700 \\ 60{,}800 \end{bmatrix} \begin{matrix} \text{Model A} \\ \text{Model B} \\ \text{Model C} \end{matrix}$$

Cost

(b) How much of each of the four kinds of material must be ordered?

SOLUTION   The totals of the columns of matrix PQ will give a matrix whose elements represent the total amounts of each material needed for the subdivision. Call this matrix T and write it as a row matrix.

$$T = \begin{bmatrix} 3800 & 130 & 1400 & 200 \end{bmatrix}$$

(c)   What is the total cost of the materials?

SOLUTION   The total cost of all the materials is given by the product of matrix R, the cost matrix, and matrix T, the total amounts matrix. To multiply these and get a 1×1 matrix, representing the total cost, requires multiplying a 1×4 matrix and a 4×1 matrix. This is why in (b) a row matrix was written rather than a column matrix. The total materials cost is given by TR, so

$$TR = \begin{bmatrix} 3800 & 130 & 1400 & 200 \end{bmatrix} \begin{bmatrix} 20 \\ 180 \\ 60 \\ 25 \end{bmatrix} = [188{,}400].$$

The total cost of the materials is \$188,400.   ∎

7.2 EXERCISES

To the Student: Calculator Considerations
   If your graphics calculator has matrix capabilities, you should refer to your owner's manual to determine the keystrokes necessary for entering the dimension of a matrix, for entering the elements of a matrix, and for performing the arithmetic operations with matrices (as described in this section). Then use your calculator to help you in working the exercises of this section.

1. What is the maximum dimension for a matrix allowed by your particular calculator?

2. What condition is necessary for a matrix to be a square matrix? a zero matrix?

Find the dimension of each of the following matrices. Identify any square, column, or row matrices.

3. $\begin{bmatrix} -4 & 8 \\ 2 & 3 \end{bmatrix}$

4. $\begin{bmatrix} -9 & 6 & 2 \\ 4 & 1 & 8 \end{bmatrix}$

5. $\begin{bmatrix} -6 & 8 & 0 & 0 \\ 4 & 1 & 9 & 2 \\ 3 & -5 & 7 & 1 \end{bmatrix}$

6. $\begin{bmatrix} 8 & -2 & 4 & 6 & 3 \end{bmatrix}$

7. $\begin{bmatrix} 2 \\ 4 \end{bmatrix}$

8. $\begin{bmatrix} -9 \end{bmatrix}$

9. $\begin{bmatrix} -4 & 2 & 3 \\ -8 & 2 & 1 \\ 4 & 6 & 8 \end{bmatrix}$

10. $\begin{bmatrix} -4 & 2 \\ 3 & 5 \end{bmatrix}$

Find the values of the variables in each of the following matrices.

11. $\begin{bmatrix} 2 & 1 \\ 4 & 8 \end{bmatrix} = \begin{bmatrix} x & 1 \\ y & z \end{bmatrix}$

12. $\begin{bmatrix} -5 \\ y \end{bmatrix} = \begin{bmatrix} -5 \\ 8 \end{bmatrix}$

13. $\begin{bmatrix} x + 6 & y + 2 \\ 8 & 3 \end{bmatrix} = \begin{bmatrix} -9 & 7 \\ 8 & k \end{bmatrix}$

14. $\begin{bmatrix} 9 & 7 \\ r & 0 \end{bmatrix} = \begin{bmatrix} m - 3 & n + 5 \\ 8 & 0 \end{bmatrix}$

15. $\begin{bmatrix} 3 & 5 \\ 8 & 9 \end{bmatrix} + \begin{bmatrix} m & 3 \\ 5 & n \end{bmatrix} = \begin{bmatrix} 9 & 8 \\ 13 & 0 \end{bmatrix}$

16. $[8 \quad p + 9 \quad q + 5] + [9 \quad -3 \quad 12] = [k - 2 \quad 12 \quad 2q]$

17. $\begin{bmatrix} -7 + z & 4r & 8s \\ 6p & 2 & 5 \end{bmatrix} + \begin{bmatrix} -9 & 8r & 3 \\ 2 & 5 & 4 \end{bmatrix} = \begin{bmatrix} 2 & 36 & 27 \\ 20 & 7 & 12a \end{bmatrix}$

18. $\begin{bmatrix} a + 2 & 3z + 1 & 5m \\ 4k & 0 & 3 \end{bmatrix} + \begin{bmatrix} 3a & 2z & 5m \\ 2k & 5 & 6 \end{bmatrix} = \begin{bmatrix} 10 & -14 & 80 \\ 10 & 5 & 9 \end{bmatrix}$

19. Your friend missed the lecture on adding matrices. In your own words, explain to him how to add two matrices.

20. Explain to a friend in your own words how to subtract two matrices.

Perform each of the following operations, whenever possible.

21. $3\begin{bmatrix} 6 & -1 & 4 \\ 2 & 8 & -3 \\ -4 & 5 & 6 \end{bmatrix} + 5\begin{bmatrix} -2 & -8 & -6 \\ 4 & 1 & 3 \\ 2 & -1 & 5 \end{bmatrix}$

22. $4\begin{bmatrix} 1 & -4 \\ 2 & -3 \\ -8 & 4 \end{bmatrix} - 3\begin{bmatrix} -6 & 9 \\ -2 & 5 \\ -7 & -12 \end{bmatrix}$

23. $\begin{bmatrix} -8 & 4 & 0 \\ 2 & 5 & 0 \end{bmatrix} + \begin{bmatrix} 6 & 3 \\ 8 & 9 \end{bmatrix}$

24. $\begin{bmatrix} 2 \\ 3 \end{bmatrix} - \begin{bmatrix} 8 & 1 \\ 9 & 4 \end{bmatrix}$

25. $\begin{bmatrix} 9 & 4 & 1 & -2 \\ 5 & -6 & 3 & 4 \\ 2 & -5 & 1 & 2 \end{bmatrix} - \begin{bmatrix} -2 & 5 & 1 & 3 \\ 0 & 1 & 0 & 2 \\ -8 & 3 & 2 & 1 \end{bmatrix} + \begin{bmatrix} 2 & 4 & 0 & 3 \\ 4 & -5 & 1 & 6 \\ 2 & -3 & 0 & 8 \end{bmatrix}$

26. $\begin{bmatrix} 6 & -2 & 4 \\ -2 & 5 & 8 \\ 1 & 0 & 2 \end{bmatrix} + \begin{bmatrix} 3 & 0 & 8 \\ 1 & -2 & 4 \\ 6 & 9 & -2 \end{bmatrix} - \begin{bmatrix} -4 & 2 & 1 \\ 0 & 3 & -2 \\ 4 & 2 & 0 \end{bmatrix}$

27. $\begin{bmatrix} -4x + 2y & -3x + y \\ 6x - 3y & 2x - 5y \end{bmatrix} + \begin{bmatrix} -8x + 6y & 2x \\ 3y - 5x & 6x + 4y \end{bmatrix}$

28. $\begin{bmatrix} 4k - 8y \\ 6z - 3x \\ 2k + 5a \\ -4m + 2n \end{bmatrix} - \begin{bmatrix} 5k + 6y \\ 2z + 5x \\ 4k + 6a \\ 4m - 2n \end{bmatrix}$

Solve each problem.

29. Daniel Phillips bought 7 shares of Sears stock, 9 shares of IBM stock, and 8 shares of Chrysler stock. The following month, he bought 2 shares of Sears stock, no IBM, and 6 shares of Chrysler. Write this information first as a 3 × 2 matrix and then as a 2 × 3 matrix.

30. Margie Bezzone works in a computer store. The first week she sold 5 computers, 3 printers, 4 disk drives, and 6 monitors. The next week she sold 4 computers, 2 printers, 6 disk drives, and 5 monitors. Write this information first as a 2 × 4 matrix and then as a 4 × 2 matrix.

31. A recent study revealed that the average number of miles driven in the United States has been increasing steadily, with miles driven by women increasing much faster than miles driven by men. In 1969, 5411 thousand miles were driven by women and 11,352 thousand miles were driven by men. In 1990, 9371 thousand miles were driven by women and 15,956 thousand miles by men. Write this information as a 2 × 2 matrix in two ways.

32. The proportion of the population of China living in cities was slowed for a time by government-imposed birth-control policies. The urban population proportion has increased again in recent years. In 1952, the proportion was 12.5%; in 1960, 19.7%; in 1975, 12.1%; and in 1985, 19.7%. Write this information as a row matrix and as a column matrix.

Let $A = \begin{bmatrix} -2 & 4 \\ 0 & 3 \end{bmatrix}$ and $B = \begin{bmatrix} -6 & 2 \\ 4 & 0 \end{bmatrix}$. *Find each of the following matrices.*

33. $2A$ **34.** $-3B$ **35.** $-4A$ **36.** $5B$

37. $2A - B$ **38.** $-4A + 5B$ **39.** $3A - 11B$ **40.** $-2A + 4B$

The dimensions of matrices A and B are given. Find the dimensions of the product AB and of the product BA if the products are defined. If they are not defined, say so.

41. A is 4×2, B is 2×4 **42.** A is 3×1, B is 1×3

43. A is 3×5, B is 5×2 **44.** A is 4×3, B is 3×6

45. A is 4×2, B is 3×4 **46.** A is 7×3, B is 2×7

47. A is 4×3, B is 2×5 **48.** A is 1×6, B is 2×4

49. The product MN of two matrices can be found only if the number of _____ of M equals the number of _____ of N.

50. True or false: For matrices A and B, if AB can be found, then BA can always be found, too.

51. To find the product AB of matrices A and B, the first row, second column entry is found by multiplying the _____ elements in A and the _____ elements in B and then _____ these products.

52. If a matrix is multiplied by a zero matrix of appropriate dimension, what must the product be?

Find each of the following matrix products, whenever possible.

53. $\begin{bmatrix} p & q \\ r & s \end{bmatrix}\begin{bmatrix} a & c \\ b & d \end{bmatrix}$

54. $\begin{bmatrix} a & b & c \\ d & e & f \\ g & h & i \end{bmatrix}\begin{bmatrix} x \\ y \\ z \end{bmatrix}$

55. $\begin{bmatrix} 3 & -4 & 1 \\ 5 & 0 & 2 \end{bmatrix}\begin{bmatrix} -1 \\ 4 \\ 2 \end{bmatrix}$

56. $\begin{bmatrix} -6 & 3 & 5 \\ 2 & 9 & 1 \end{bmatrix}\begin{bmatrix} -2 \\ 0 \\ 3 \end{bmatrix}$

57. $\begin{bmatrix} 5 & 2 \\ -1 & 4 \end{bmatrix}\begin{bmatrix} 3 & -2 \\ 1 & 0 \end{bmatrix}$

58. $\begin{bmatrix} -4 & 0 \\ 1 & 3 \end{bmatrix}\begin{bmatrix} -2 & 4 \\ 0 & 1 \end{bmatrix}$

59. $\begin{bmatrix} 2 & 2 & -1 \\ 3 & 0 & 1 \end{bmatrix}\begin{bmatrix} 0 & 2 \\ -1 & 4 \\ 0 & 2 \end{bmatrix}$

60. $\begin{bmatrix} -9 & 2 & 1 \\ 3 & 0 & 0 \end{bmatrix}\begin{bmatrix} 2 \\ -1 \\ 4 \end{bmatrix}$

61. $\begin{bmatrix} -1 & 2 & 0 \\ 0 & 3 & 2 \\ 0 & 1 & 4 \end{bmatrix}\begin{bmatrix} 2 & -1 & 2 \\ 0 & 2 & 1 \\ 3 & 0 & -1 \end{bmatrix}$

62. $\begin{bmatrix} -2 & -3 & -4 \\ 2 & -1 & 0 \\ 4 & -2 & 3 \end{bmatrix}\begin{bmatrix} 0 & 1 & 4 \\ 1 & 2 & -1 \\ 3 & 2 & -2 \end{bmatrix}$

63. $\begin{bmatrix} -2 & 4 & 1 \end{bmatrix}\begin{bmatrix} 3 & -2 & 4 \\ 2 & 1 & 0 \\ 0 & -1 & 4 \end{bmatrix}$

64. $\begin{bmatrix} 0 & 3 & -4 \end{bmatrix}\begin{bmatrix} -2 & 6 & 3 \\ 0 & 4 & 2 \\ -1 & 1 & 4 \end{bmatrix}$

65. $\begin{bmatrix} -2 & 1 & 4 \\ 0 & 1 & 2 \end{bmatrix}\begin{bmatrix} -2 & 1 & 0 \\ 0 & -2 & 0 \\ 4 & 1 & 2 \end{bmatrix}$

66. $\begin{bmatrix} -1 & 0 & 0 \\ 2 & 1 & 4 \end{bmatrix}\begin{bmatrix} 4 & -2 & 5 \\ 0 & 1 & 4 \\ 2 & -9 & 0 \end{bmatrix}$

67. $\begin{bmatrix} -3 & 0 & 2 & 1 \\ 4 & 0 & 2 & 6 \end{bmatrix}\begin{bmatrix} -4 & 2 \\ 0 & 1 \end{bmatrix}$

68. $\begin{bmatrix} -1 & 2 & 4 & 1 \\ 0 & 2 & -3 & 5 \end{bmatrix}\begin{bmatrix} 1 & 2 & 4 \\ -2 & 5 & 1 \end{bmatrix}$

69. $\begin{bmatrix} -2 & 4 & 6 \end{bmatrix}\begin{bmatrix} 3 \\ -2 \\ 1 \end{bmatrix}$

70. $\begin{bmatrix} 4 & 0 & 2 \end{bmatrix}\begin{bmatrix} -5 \\ 1 \\ 6 \end{bmatrix}$

71. $\begin{bmatrix} 3 \\ -2 \\ 1 \end{bmatrix}\begin{bmatrix} -2 & 4 & 6 \end{bmatrix}$

72. $\begin{bmatrix} -5 \\ 1 \\ 6 \end{bmatrix}\begin{bmatrix} 4 & 0 & 2 \end{bmatrix}$

Let $A = \begin{bmatrix} -2 & 4 \\ 1 & 3 \end{bmatrix}$, $B = \begin{bmatrix} -2 & 1 \\ 3 & 6 \end{bmatrix}$, and $C = \begin{bmatrix} 5 & -2 & 1 \\ 0 & 3 & 7 \end{bmatrix}$. Find each of the following products.

73. AB
74. BA
75. AC
76. CA

77. Did you get the same answer in Exercises 73 and 74? What about Exercises 75 and 76? Do you think that matrix multiplication is commutative?

78. For any matrices P and Q, what must be true for both PQ and QP to exist?

Solve each problem.

79. Yummy Yogurt sells three types of yogurt: nonfat, regular, and super creamy at three locations. Location I sells 50 gallons of nonfat, 100 gallons of regular, and 30 gallons of super creamy each day. Location II sells 10 gallons of nonfat and Location III sells 60 gallons of nonfat each day. Daily sales of regular yogurt are 90 gallons at Location II and 120 gallons at Location III. At Location II, 50 gallons of super creamy are sold each day, and 40 gallons of super creamy are sold each day at Location III.

(a) Write a 3×3 matrix that shows the sales figures for the three locations.

(b) The income per gallon for nonfat, regular, and super creamy is $12, $10, and $15, respectively. Write a 1×3 or 3×1 matrix displaying the income.

(c) Find a matrix product that gives the daily income at each of the three locations.

(d) What is Yummy Yogurt's total daily income from the three locations?

80. The Bread Box, a small neighborhood bakery, sells four main items: sweet rolls, bread, cakes, and pies. The amount of each ingredient (in cups, except for eggs) required for these items is given by matrix A.

	Eggs	Flour	Sugar	Shortening	Milk	
Rolls (doz)	1	4	$\frac{1}{4}$	$\frac{1}{4}$	1	
Bread (loaves)	0	3	0	$\frac{1}{4}$	0	$= A$
Cakes	4	3	2	1	1	
Pies (crust)	0	1	0	$\frac{1}{3}$	0	

The cost (in cents) for each ingredient when purchased in large lots or small lots is given in matrix B.

	Cost Large lot	Small lot	
Eggs	5	5	
Flour	8	10	
Sugar	10	12	$= B$
Shortening	12	15	
Milk	5	6	

(a) Use matrix multiplication to find a matrix giving the comparative cost per item for the two purchase options.

Suppose a day's orders consist of 20 dozen sweet rolls, 200 loaves of bread, 50 cakes, and 60 pies.

(b) Write the orders as a 1×4 matrix and, using matrix multiplication, write as a matrix the amount of each ingredient needed to fill the day's orders.

(c) Use matrix multiplication to find a matrix giving the costs under the two purchase options to fill the day's orders.

For the following exercises, let $A = \begin{bmatrix} a & b \\ c & d \end{bmatrix}$, $B = \begin{bmatrix} e & f \\ g & h \end{bmatrix}$, and $C = \begin{bmatrix} j & m \\ k & n \end{bmatrix}$.

Decide which of the following statements are true for these three matrices. Then make a conjecture as to whether a similar property holds for any square matrices of the same order.

81. $(AB)C = A(BC)$ (associative property)

82. $A(B + C) = AB + AC$ (distributive property)

83. $k(A + B) = kA + kB$ for any real number k

84. $(k + p)A = kA + pA$ for any real numbers k and p

*The **transpose**, A^T, of a matrix A is found by exchanging the rows and columns of A. That is, if*

$$A = \begin{bmatrix} a & b \\ c & d \end{bmatrix}, \quad then \quad A^T = \begin{bmatrix} a & c \\ b & d \end{bmatrix}.$$

Show that each of the following equations is true for matrices A and B, where

$$B = \begin{bmatrix} m & n \\ p & q \end{bmatrix}.$$

85. $(A^T)^T = A$ **86.** $(A + B)^T = A^T + B^T$ **87.** $(AB)^T = B^T A^T$

88. Consider the product

$$\begin{bmatrix} 2 & -6 \\ 4 & -12 \end{bmatrix}\begin{bmatrix} 3 \\ 1 \end{bmatrix}.$$

    (a) What special matrix is this product?

    (b) If the product of two matrices is the zero matrix, must it follow that one of the factors is the zero matrix? Why or why not?

7.3 DETERMINANTS, INVERSES, AND SOLUTION OF SYSTEMS BY INVERSE MATRICES

Identity Matrices ∎ Determinants of Square Matrices of Dimension 2 × 2 ∎ Multiplicative Inverses of Square Matrices of Dimension 2 × 2 ∎ Solution of Linear Systems Using Inverse Matrices

Identity Matrices

The identity property for real numbers says that $a \cdot 1 = a$ and $1 \cdot a = a$ for any real number a. If there is to be a multiplicative identity matrix I, such that

$$AI = A \quad and \quad IA = A,$$

for any matrix A, then A and I must be square matrices of the same dimension. Otherwise it would not be possible to find both products. For example, let A be the 2 × 2 matrix

$$A = \begin{bmatrix} a_{11} & a_{12} \\ a_{21} & a_{22} \end{bmatrix},$$

and let

$$I_2 = \begin{bmatrix} x_{11} & x_{12} \\ x_{21} & x_{22} \end{bmatrix}$$

represent the 2 × 2 identity matrix. (The subscript 2 is used to denote the fact that it is a square matrix of dimension 2 × 2.) To find I_2, use the fact that $I_2 A = A$, so

$$\begin{bmatrix} x_{11} & x_{12} \\ x_{21} & x_{22} \end{bmatrix}\begin{bmatrix} a_{11} & a_{12} \\ a_{21} & a_{22} \end{bmatrix} = \begin{bmatrix} a_{11} & a_{12} \\ a_{21} & a_{22} \end{bmatrix}$$

    Multiplying the two matrices on the left side of this equation and setting the elements of the product matrix equal to the corresponding elements of A gives the following system of equations with variables x_{11}, x_{12}, x_{21}, and x_{22}.

$$x_{11}a_{11} + x_{12}a_{21} = a_{11}$$
$$x_{11}a_{12} + x_{12}a_{22} = a_{12}$$
$$x_{21}a_{11} + x_{22}a_{21} = a_{21}$$
$$x_{21}a_{12} + x_{22}a_{22} = a_{22}$$

Notice that this is really two systems of equations in two variables. Use one of the methods introduced earlier to find the solution of this system: $x_{11} = 1, x_{12} = x_{21} = 0$, and $x_{22} = 1$. From the solution of the system, the 2×2 identity matrix is

$$I_2 = \begin{bmatrix} 1 & 0 \\ 0 & 1 \end{bmatrix}.$$

Check that with this definition of I_2, both $AI_2 = A$ and $I_2A = A$.

EXAMPLE 1

Verifying the Identity Property

Let $M = \begin{bmatrix} -2 & 6 \\ 3 & 5 \end{bmatrix}$. Verify that $MI_2 = M$ and $I_2M = M$.

SOLUTION

$$MI_2 = \begin{bmatrix} -2 & 6 \\ 3 & 5 \end{bmatrix}\begin{bmatrix} 1 & 0 \\ 0 & 1 \end{bmatrix} = \begin{bmatrix} -2 & 6 \\ 3 & 5 \end{bmatrix} = M$$

$$I_2M = \begin{bmatrix} 1 & 0 \\ 0 & 1 \end{bmatrix}\begin{bmatrix} -2 & 6 \\ 3 & 5 \end{bmatrix} = \begin{bmatrix} -2 & 6 \\ 3 & 5 \end{bmatrix} = M \qquad ◼$$

The 2×2 identity matrix found above suggests the following generalization.

$n \times n$ IDENTITY MATRIX

For any value of n there is an $n \times n$ identity matrix having 1s down the diagonal and 0s elsewhere. The **$n \times n$ identity matrix** is given by I_n, where

$$I_n = \begin{bmatrix} 1 & 0 & \cdots & 0 \\ 0 & 1 & \cdots & 0 \\ \vdots & \vdots & a_{ij} & \vdots \\ 0 & 0 & \cdots & 1 \end{bmatrix}$$

Here $a_{ij} = 1$ when $i = j$ (the diagonal elements) and $a_{ij} = 0$ otherwise.

EXAMPLE 2

Stating and Verifying the 3×3 Identity Matrix

Let $K = \begin{bmatrix} -2 & 4 & 0 \\ 3 & 5 & 9 \\ 0 & 8 & -6 \end{bmatrix}$. Give the 3×3 identity matrix I_3 and show that $KI_3 = K$.

SOLUTION The 3×3 identity matrix is

$$I_3 = \begin{bmatrix} 1 & 0 & 0 \\ 0 & 1 & 0 \\ 0 & 0 & 1 \end{bmatrix}.$$

By the definition of matrix multiplication,

$$KI_3 = \begin{bmatrix} -2 & 4 & 0 \\ 3 & 5 & 9 \\ 0 & 8 & -6 \end{bmatrix} \begin{bmatrix} 1 & 0 & 0 \\ 0 & 1 & 0 \\ 0 & 0 & 1 \end{bmatrix} = \begin{bmatrix} -2 & 4 & 0 \\ 3 & 5 & 9 \\ 0 & 8 & -6 \end{bmatrix} = K. \qquad ∎$$

Determinants of Square Matrices of Dimension 2 × 2

Associated with every square matrix A is a real number called the **determinant** of A. There are several symbols used to represent the determinant of A, including $|A|$, $\delta(A)$, and det A. In this text we will use det A.

For now, we will only consider determinants of square matrices of dimension 2 × 2. In the next section, we will extend our discussion to include determinants of larger matrices.

DETERMINANT OF A 2 × 2 MATRIX

The **determinant of a 2 × 2 matrix** A,

$$A = \begin{bmatrix} a_{11} & a_{12} \\ a_{21} & a_{22} \end{bmatrix},$$

is defined as

$$\det A = a_{11}a_{22} - a_{21}a_{12}.$$

NOTE Be able to distinguish between a matrix and its determinant. A matrix is an array of numbers, while a determinant is a real number associated with a square matrix.

EXAMPLE 3

Evaluating the Determinant of a Dimension 2 × 2 Matrix

If $A = \begin{bmatrix} -3 & 4 \\ 6 & 8 \end{bmatrix}$, find det A.

SOLUTION $\det A = \det \begin{bmatrix} -3 & 4 \\ 6 & 8 \end{bmatrix} = -3(8) - 6(4) = -48.$ ∎

EXAMPLE 4

Solving an Equation Involving a Determinant

If $A = \begin{bmatrix} x & 3 \\ -1 & 5 \end{bmatrix}$ and det $A = 33$, find the value of x.

SOLUTION Since det $A = 5x - (-3)$, or $5x + 3$, we have

$$5x + 3 = 33$$
$$5x = 30$$
$$x = 6.$$

Check to see that $\det \begin{bmatrix} 6 & 3 \\ -1 & 5 \end{bmatrix} = 33.$ ∎

Multiplicative Inverses of Square Matrices of Dimension 2 × 2

Suppose that

$$A = \begin{bmatrix} a & b \\ c & d \end{bmatrix}.$$

Is it possible to find a matrix B such that $AB = I_2$ and $BA = I_2$? If so, then B is the inverse of A.

Let $B = \begin{bmatrix} x & y \\ z & w \end{bmatrix}$. We must find x, y, z, and w so that

$$\begin{bmatrix} a & b \\ c & d \end{bmatrix}\begin{bmatrix} x & y \\ z & w \end{bmatrix} = \begin{bmatrix} 1 & 0 \\ 0 & 1 \end{bmatrix}.$$

Multiplying the two matrices and setting the elements of the product equal to the corresponding elements in I_2 gives the following two systems of equations.

$$ax + bz = 1 \qquad ay + bw = 0$$
$$cx + dz = 0 \qquad cy + dw = 1$$

By solving these two systems we get the values of x, y, z, and w:

$$x = \frac{d}{ad - cb}, \quad y = \frac{-b}{ad - cb}, \quad z = \frac{-c}{ad - cb}, \quad \text{and} \quad w = \frac{a}{ad - cb}.$$

Verify that $AB = I_2$ and that $BA = I_2$. (See Exercise 61.) As a result, we can conclude that B is the inverse of A, written A^{-1}, provided that det $A = ad - cb \neq 0$.

> **MULTIPLICATIVE INVERSE OF A SQUARE MATRIX OF DIMENSION 2 × 2**
>
> If $A = \begin{bmatrix} a & b \\ c & d \end{bmatrix}$ and det $A \neq 0$, then
>
> $$A^{-1} = \frac{1}{\det A}\begin{bmatrix} d & -b \\ -c & a \end{bmatrix}$$
>
> or
>
> $$A^{-1} = \begin{bmatrix} \dfrac{d}{ad-cb} & \dfrac{-b}{ad-cb} \\ \dfrac{-c}{ad-cb} & \dfrac{a}{ad-cb} \end{bmatrix}.$$

EXAMPLE 5

Finding the Multiplicative Inverse of a Square Matrix of Dimension 2 × 2

Find A^{-1}, if it exists, for each of the following matrices.

(a) $A = \begin{bmatrix} 2 & 3 \\ 1 & -1 \end{bmatrix}$ **(b)** $A = \begin{bmatrix} 3 & -6 \\ 2 & -4 \end{bmatrix}.$

SOLUTION

(a) To find A^{-1} for $\begin{bmatrix} 2 & 3 \\ 1 & -1 \end{bmatrix}$, first we find det A and determine whether A^{-1} actually exists.

$$\det A = 2(-1) - 1(3) = -5.$$

Since $-5 \neq 0$, A^{-1} exists. By the rule given earlier,

$$A^{-1} = \frac{1}{-5} \begin{bmatrix} -1 & -3 \\ -1 & 2 \end{bmatrix}$$

$$= \begin{bmatrix} \frac{1}{5} & \frac{3}{5} \\ \frac{1}{5} & -\frac{2}{5} \end{bmatrix}.$$

(b) Here, A^{-1} does not exist because

$$\det A = 3(-4) - 2(-6) = 0. \quad \blacksquare$$

As seen in Example 5, it is always necessary for the determinant to be nonzero for the inverse to exist.

TECHNOLOGICAL NOTE
After reading Examples 1–5, you may wish to experiment with your calculator to see if you can duplicate the results obtained.

FOR GROUP DISCUSSION

1. Support the results of Example 5 using your graphics calculator.
2. We have seen that under certain conditions, the multiplicative inverse of a matrix does not exist. How does this compare to the multiplicative inverse property for real numbers? Does the multiplicative inverse of $\frac{x-4}{5}$ exist if $x = 4$?

Solution of Linear Systems Using Inverse Matrices

We used matrices to solve systems of linear equations by the Gaussian reduction method in Section 7.1. Another way to use matrices to solve linear systems is to write the system as a matrix equation $AX = B$, where A is the matrix of the coefficients of the variables of the system, X is the matrix of the variables, and B is the matrix of the constants. Matrix A is called the **coefficient matrix.**

To solve the matrix equation $AX = B$, first see if A^{-1} exists. Assuming A^{-1} exists and using the facts that $A^{-1}A = I$ and $IX = X$ gives

$$AX = B$$
$$A^{-1}(AX) = A^{-1}B \quad \text{Multiply both sides by } A^{-1}.$$
$$(A^{-1}A)X = A^{-1}B \quad \text{Associative property}$$
$$IX = A^{-1}B \quad \text{Multiplicative inverse property}$$
$$X = A^{-1}B. \quad \text{Identity property}$$

CAUTION When multiplying by matrices on both sides of a matrix equation, be careful to multiply in the same order on both sides of the equation, since multiplication of matrices is not commutative (unlike multiplication of real numbers).

EXAMPLE 6	Given

Solving a Matrix Equation

$$A = \begin{bmatrix} 2 & 2 \\ -1 & -2 \end{bmatrix} \quad \text{and} \quad B = \begin{bmatrix} 2 \\ 3 \end{bmatrix},$$

find a matrix X so that $AX = B$.

SOLUTION Since A is 2×2 and B is 2×1, matrix X will be a 2×1 matrix, like B.

To find X, first find A^{-1}. Verify that

$$A^{-1} = \begin{bmatrix} 1 & 1 \\ -\frac{1}{2} & -1 \end{bmatrix}.$$

As shown above,

$$X = A^{-1}B$$

$$\begin{bmatrix} x \\ y \end{bmatrix} = \begin{bmatrix} 1 & 1 \\ -\frac{1}{2} & -1 \end{bmatrix} \begin{bmatrix} 2 \\ 3 \end{bmatrix}$$

$$\begin{bmatrix} x \\ y \end{bmatrix} = \begin{bmatrix} 5 \\ -4 \end{bmatrix}.$$

Check:

$$AX = \begin{bmatrix} 2 & 2 \\ -1 & -2 \end{bmatrix} \begin{bmatrix} 5 \\ -4 \end{bmatrix} = \begin{bmatrix} 2 \\ 3 \end{bmatrix} = B. \quad \blacksquare$$

The result from Example 6 can be used to solve linear systems by first writing the system as a matrix equation $AX = B$, where X is a matrix of the variables of the system. Then the solution is $X = A^{-1}B$.

EXAMPLE 7	Use the inverse of the coefficient matrix to solve the system

Solving a System of Equations Using a Matrix Inverse

$$2x - 3y = 4$$
$$x + 5y = 2.$$

SOLUTION To represent the system as a matrix equation, use one matrix for the coefficients, one for the variables, and one for the constants, as follows.

$$A = \begin{bmatrix} 2 & -3 \\ 1 & 5 \end{bmatrix}, \quad X = \begin{bmatrix} x \\ y \end{bmatrix}, \quad \text{and} \quad B = \begin{bmatrix} 4 \\ 2 \end{bmatrix}$$

The system can then be written in matrix form as the equation $AX = B$, since

$$AX = \begin{bmatrix} 2 & -3 \\ 1 & 5 \end{bmatrix} \begin{bmatrix} x \\ y \end{bmatrix} = \begin{bmatrix} 2x - 3y \\ x + 5y \end{bmatrix} = \begin{bmatrix} 4 \\ 2 \end{bmatrix} = B.$$

A^{-1} exists, because det $A = 13 \neq 0$.

$$A^{-1} = \begin{bmatrix} \frac{5}{13} & \frac{3}{13} \\ -\frac{1}{13} & \frac{2}{13} \end{bmatrix}$$

TECHNOLOGICAL NOTE
If you perform the operation described in Example 7, it is possible that your calculator will return an expression like -1 E-14 for the y-value in the solution. We know that 0 is the actual y-value. The expression -1 E-14 represents $-.00000000000001$, which is as close to 0 as the calculator can get, based on the method which it uses to do the computation.

Next, find the product $A^{-1}B$.

$$A^{-1}B = \begin{bmatrix} \frac{5}{13} & \frac{3}{13} \\ -\frac{1}{13} & \frac{2}{13} \end{bmatrix}\begin{bmatrix} 4 \\ 2 \end{bmatrix} = \begin{bmatrix} 2 \\ 0 \end{bmatrix}$$

Since $X = A^{-1}B$,

$$X = \begin{bmatrix} x \\ y \end{bmatrix} = \begin{bmatrix} 2 \\ 0 \end{bmatrix},$$

and the solution set of the system is $\{(2, 0)\}$.   ∎

CAUTION   When solving a system using matrix inverses, you should always begin by finding the determinant of the coefficient matrix. If it is 0, then the system has either no solutions or infinitely many solutions, and should be solved by one of the methods described earlier in the text.

7.3 EXERCISES

To the Student: Calculator Considerations

Graphics calculators with matrix capabilities are capable of finding determinants of square matrices, and inverses of matrices (when they exist). While we have only discussed determinants and inverses for square matrices of dimension 2 × 2, these ideas will be extended to larger matrices in the next section. Read your owner's manual to find out how to evaluate determinants and how to find inverses on your particular model, using square matrices of dimension 2 × 2 to begin. Then it will be easy to extend these ideas in the next section.

Multiply the given pair of matrices in both directions to determine whether they are inverses of each other. Remember that the product must be the identity matrix for them to be inverses.

1. $\begin{bmatrix} 2 & 3 \\ 1 & 1 \end{bmatrix}, \begin{bmatrix} -1 & 3 \\ 1 & -2 \end{bmatrix}$

2. $\begin{bmatrix} 5 & 7 \\ 2 & 3 \end{bmatrix}, \begin{bmatrix} 3 & -7 \\ -2 & 5 \end{bmatrix}$

3. $\begin{bmatrix} 2 & 1 \\ 3 & 2 \end{bmatrix}, \begin{bmatrix} 2 & 1 \\ -3 & 2 \end{bmatrix}$

4. $\begin{bmatrix} -1 & 2 \\ 3 & -5 \end{bmatrix}, \begin{bmatrix} -5 & -2 \\ -3 & -1 \end{bmatrix}$

5. $\begin{bmatrix} 1 & -2 & -3 \\ 2 & -2 & -5 \\ -1 & 1 & 4 \end{bmatrix}, \begin{bmatrix} -1 & \frac{5}{3} & \frac{4}{3} \\ -1 & \frac{1}{3} & -\frac{1}{3} \\ 0 & \frac{1}{3} & \frac{2}{3} \end{bmatrix}$

6. $\begin{bmatrix} 1 & 2 & -1 \\ 2 & -1 & 3 \\ 3 & -2 & 3 \end{bmatrix}, \begin{bmatrix} \frac{3}{10} & -\frac{2}{5} & \frac{1}{2} \\ \frac{3}{10} & \frac{3}{5} & -\frac{1}{2} \\ -\frac{1}{10} & \frac{4}{5} & -\frac{1}{2} \end{bmatrix}$

7. $\begin{bmatrix} 1 & 2 & -1 \\ 0 & 1 & 3 \\ 2 & 1 & -2 \end{bmatrix}, \begin{bmatrix} 1 & 1 & 2 \\ 1 & 1 & 1 \\ 2 & 3 & 4 \end{bmatrix}$

8. $\begin{bmatrix} 2 & -1 & 4 \\ 0 & 5 & 0 \\ 3 & 2 & -1 \end{bmatrix}, \begin{bmatrix} 1 & 0 & 1 \\ 6 & 4 & 2 \\ 1 & 1 & 0 \end{bmatrix}$

9. Under what condition will the inverse of a square matrix not exist?

10. Explain why a square matrix of dimension 2 × 2 will not have an inverse if either a column or a row contains all zeros.

Find the determinant of each of the following matrices.

11. $\begin{bmatrix} 5 & 8 \\ 2 & -4 \end{bmatrix}$

12. $\begin{bmatrix} -3 & 0 \\ 0 & 9 \end{bmatrix}$

13. $\begin{bmatrix} -1 & -2 \\ 5 & 3 \end{bmatrix}$

14. $\begin{bmatrix} 6 & -4 \\ 0 & -1 \end{bmatrix}$

15. $\begin{bmatrix} 9 & 3 \\ -3 & -1 \end{bmatrix}$ **16.** $\begin{bmatrix} 0 & 2 \\ 1 & 5 \end{bmatrix}$ **17.** $\begin{bmatrix} 3 & 4 \\ 5 & -2 \end{bmatrix}$ **18.** $\begin{bmatrix} -9 & 7 \\ 2 & 6 \end{bmatrix}$

19. $\begin{bmatrix} 0 & 4 \\ 4 & 0 \end{bmatrix}$ **20.** $\begin{bmatrix} 1 & 0 \\ 0 & 2 \end{bmatrix}$ **21.** $\begin{bmatrix} 8 & 3 \\ 8 & 3 \end{bmatrix}$ **22.** $\begin{bmatrix} 9 & -4 \\ -4 & 9 \end{bmatrix}$

23. $\begin{bmatrix} x & 4 \\ 8 & 2 \end{bmatrix}$ **24.** $\begin{bmatrix} k & 3 \\ 0 & 4 \end{bmatrix}$ **25.** $\begin{bmatrix} y & 2 \\ 8 & y \end{bmatrix}$ **26.** $\begin{bmatrix} 3 & 8 \\ m & n \end{bmatrix}$

In Exercises 27–30, solve for x.

27. $A = \begin{bmatrix} 5 & x \\ -3 & 2 \end{bmatrix}$ and det $A = 6$ **28.** $A = \begin{bmatrix} -.5 & 2 \\ x & x \end{bmatrix}$ and det $A = 0$

29. $\det \begin{bmatrix} x & 3 \\ x & x \end{bmatrix} = 4$ **30.** $\det \begin{bmatrix} 2x & x \\ 11 & x \end{bmatrix} = 6$

Find the inverse, if it exists, of each matrix.

31. $\begin{bmatrix} 1 & -1 \\ 2 & 0 \end{bmatrix}$ **32.** $\begin{bmatrix} 3 & -1 \\ -5 & 2 \end{bmatrix}$ **33.** $\begin{bmatrix} -6 & 4 \\ -3 & 2 \end{bmatrix}$ **34.** $\begin{bmatrix} -1 & 2 \\ -2 & -1 \end{bmatrix}$

35. $\begin{bmatrix} -1 & -2 \\ 3 & 4 \end{bmatrix}$ **36.** $\begin{bmatrix} 5 & 10 \\ -3 & -6 \end{bmatrix}$ **37.** $\begin{bmatrix} .6 & .2 \\ .5 & .1 \end{bmatrix}$ **38.** $\begin{bmatrix} .8 & -.3 \\ .5 & -.2 \end{bmatrix}$

Write the matrix of coefficients, A, the matrix of variables, X, and the matrix of constants, B, for each of the following systems. Do not solve.

39. $\begin{aligned} x + y &= 8 \\ 2x - y &= 4 \end{aligned}$ **40.** $\begin{aligned} 2x + y &= 9 \\ x + 3y &= 17 \end{aligned}$ **41.** $\begin{aligned} 4x + 5y &= 7 \\ 2x + 3y &= 5 \end{aligned}$ **42.** $\begin{aligned} 2x + 3y &= -2 \\ 5x + 4y &= -12 \end{aligned}$

Solve each of the following systems by using the inverse of the coefficient matrix.

43. $\begin{aligned} -x + y &= 1 \\ 2x - y &= 1 \end{aligned}$ **44.** $\begin{aligned} x + y &= 5 \\ x - y &= -1 \end{aligned}$ **45.** $\begin{aligned} 2x - y &= -8 \\ 3x + y &= -2 \end{aligned}$ **46.** $\begin{aligned} x + 3y &= -12 \\ 2x - y &= 11 \end{aligned}$

47. $\begin{aligned} 2x + 3y &= -10 \\ 3x + 4y &= -12 \end{aligned}$ **48.** $\begin{aligned} 2x - 3y &= 10 \\ 2x + 2y &= 5 \end{aligned}$ **49.** $\begin{aligned} 2x - 5y &= 10 \\ 4x - 5y &= 15 \end{aligned}$ **50.** $\begin{aligned} 2x - 3y &= 2 \\ 4x - 6y &= 1 \end{aligned}$

Let $A = \begin{bmatrix} a & b \\ c & d \end{bmatrix}$. Show that the following are true.

51. $AA^{-1} = I_2$ and $A^{-1}A = I_2$ **52.** $I_2A = A$ **53.** $AI_2 = A$ **54.** $A \cdot O = O$

*R*elating Concepts

Consider the equation

$$\det \begin{bmatrix} x & x \\ 3 & x \end{bmatrix} = 7.$$

55. Write it as a polynomial equation by expressing the determinant in terms of x.

56. What is the degree of the polynomial if the equation is written in the form $P(x) = 0$?

57. Solve the equation using the quadratic formula. Give solutions as exact values.

58. Use your calculator to find approximations of the solutions to as many decimal places as it will give.

59. Graph the function $y = P(x)$ in the viewing window $[-10, 10]$ by $[-10, 10]$.

60. Use your graph to support your answers in Exercise 58.

---

61. Refer to the introductory discussion in **Multiplicative Inverses of Square Matrices of Dimension 2 × 2** in this section and show that $BA = I_2$.

7.4 MORE ON DETERMINANTS, MULTIPLICATIVE INVERSES, AND SYSTEMS

Determinants of Larger Matrices ∎ Multiplicative Inverses of Larger Matrices ∎ Solution of Larger Systems Using Inverse Matrices

Determinants of Larger Matrices

In the previous section we learned how to evaluate determinants of square matrices of dimension 2×2. We now investigate determinants of larger matrices, beginning with dimension 3×3. (Recall that a matrix is an *array of numbers,* while a determinant is a *real number* associated with a square matrix.)

DETERMINANT OF A 3 × 3 MATRIX

The **determinant of a square matrix** A of dimension 3×3,

$$A = \begin{bmatrix} a_{11} & a_{12} & a_{13} \\ a_{21} & a_{22} & a_{23} \\ a_{31} & a_{32} & a_{33} \end{bmatrix},$$

is defined as

$$\det A = \det \begin{bmatrix} a_{11} & a_{12} & a_{13} \\ a_{21} & a_{22} & a_{23} \\ a_{31} & a_{32} & a_{33} \end{bmatrix} = (a_{11}a_{22}a_{33} + a_{12}a_{23}a_{31} + a_{13}a_{21}a_{32}) - (a_{31}a_{22}a_{13} + a_{32}a_{23}a_{11} + a_{33}a_{21}a_{12}).$$

An easy method for calculating dimension 3×3 determinants is found by rearranging and factoring the terms given above to get

$$\det \begin{bmatrix} a_{11} & a_{12} & a_{13} \\ a_{21} & a_{22} & a_{23} \\ a_{31} & a_{32} & a_{33} \end{bmatrix} = a_{11}(a_{22}a_{33} - a_{32}a_{23}) - a_{21}(a_{12}a_{33} - a_{32}a_{13}) + a_{31}(a_{12}a_{23} - a_{22}a_{13}).$$

Each of the quantities in parentheses represents the determinant of a dimension 2×2 matrix that is the part of the dimension 3×3 matrix remaining when the row and column of the multiplier are eliminated, as shown below.

$$a_{11}(a_{22}a_{33} - a_{32}a_{23}) \qquad \begin{bmatrix} a_{11} & a_{12} & a_{13} \\ a_{21} & a_{22} & a_{23} \\ a_{31} & a_{32} & a_{33} \end{bmatrix}$$

$$a_{21}(a_{12}a_{33} - a_{32}a_{13}) \qquad \begin{bmatrix} a_{11} & a_{12} & a_{13} \\ a_{21} & a_{22} & a_{23} \\ a_{31} & a_{32} & a_{33} \end{bmatrix}$$

$$a_{31}(a_{12}a_{23} - a_{22}a_{13}) \qquad \begin{bmatrix} a_{11} & a_{12} & a_{13} \\ a_{21} & a_{22} & a_{23} \\ a_{31} & a_{32} & a_{33} \end{bmatrix}$$

These determinants of the dimension 3×3 matrices are called **minors** of an element in the dimension 3×3 matrix. The symbol M_{ij} represents the minor that results when row i and column j are eliminated. The following list gives some of the minors from the matrix above.

Element	Minor	Element	Minor
a_{11}	$M_{11} = \det\begin{bmatrix} a_{22} & a_{23} \\ a_{32} & a_{33} \end{bmatrix}$	a_{22}	$M_{22} = \det\begin{bmatrix} a_{11} & a_{13} \\ a_{31} & a_{33} \end{bmatrix}$
a_{21}	$M_{21} = \det\begin{bmatrix} a_{12} & a_{13} \\ a_{32} & a_{33} \end{bmatrix}$	a_{23}	$M_{23} = \det\begin{bmatrix} a_{11} & a_{12} \\ a_{31} & a_{32} \end{bmatrix}$
a_{31}	$M_{31} = \det\begin{bmatrix} a_{12} & a_{13} \\ a_{22} & a_{23} \end{bmatrix}$	a_{33}	$M_{33} = \det\begin{bmatrix} a_{11} & a_{12} \\ a_{21} & a_{22} \end{bmatrix}$

In a matrix of dimension 4×4, the minors are determinants of dimension 3×3 matrices. Similarly, a dimension $n \times n$ matrix has minors that are determinants of dimension $(n - 1) \times (n - 1)$ matrices.

To find the determinant of a dimension 3×3 or larger square matrix, first choose any row or column. Then the minor of each element in that row or column must be multiplied by $+1$ or -1, depending on whether the sum of the row numbers and column numbers is even or odd. The product of a minor and the number $+1$ or -1 is called a *cofactor*.

COFACTOR

Let M_{ij} be the minor for element a_{ij} in a dimension $n \times n$ matrix. The **cofactor** of a_{ij}, written A_{ij}, is

$$A_{ij} = (-1)^{i+j} \cdot M_{ij}.$$

Finally, the determinant of a dimension $n \times n$ matrix is found as follows.

FINDING THE DETERMINANT OF A MATRIX

Multiply each element in any row or column of the matrix by its cofactor. The sum of these products gives the value of the determinant.

The process of forming this sum of products is called **expansion by a given row or column.**

EXAMPLE 1
Finding the Cofactor of an Element

For the matrix

$$\begin{bmatrix} 6 & 2 & 4 \\ 8 & 9 & 3 \\ 1 & 2 & 0 \end{bmatrix},$$

find the cofactor of each of the following elements.

(a) 6

SOLUTION Since 6 is in the first row and first column of the matrix, $i = 1$ and $j = 1$.

$$M_{11} = \det\begin{bmatrix} 9 & 3 \\ 2 & 0 \end{bmatrix} = -6$$

The cofactor is $(-1)^{1+1} \cdot -6 = 1 \cdot -6 = -6$.

(b) 3

SOLUTION Here $i = 2$ and $j = 3$.

$$M_{23} = \det\begin{bmatrix} 6 & 2 \\ 1 & 2 \end{bmatrix} = 10$$

The cofactor is $(-1)^{2+3} \cdot 10 = -1 \cdot 10 = -10$.

(c) 8

SOLUTION We have $i = 2$ and $j = 1$.

$$M_{21} = \det\begin{bmatrix} 2 & 4 \\ 2 & 0 \end{bmatrix} = -8$$

The cofactor is $(-1)^{2+1} \cdot -8 = -1 \cdot -8 = 8$. ∎

EXAMPLE 2

Evaluating the Determinant of a Dimension 3 × 3 Matrix

Evaluate $\det\begin{bmatrix} 2 & -3 & -2 \\ -1 & -4 & -3 \\ -1 & 0 & 2 \end{bmatrix}$, expanding by the second column.

SOLUTION To find this determinant, first get the minors of each element in the second column.

$$M_{12} = \det\begin{bmatrix} -1 & -3 \\ -1 & 2 \end{bmatrix} = -1(2) - (-1)(-3) = -5$$

$$M_{22} = \det\begin{bmatrix} 2 & -2 \\ -1 & 2 \end{bmatrix} = 2(2) - (-1)(-2) = 2$$

$$M_{32} = \det\begin{bmatrix} 2 & -2 \\ -1 & -3 \end{bmatrix} = 2(-3) - (-1)(-2) = -8$$

Now find the cofactor of each of these minors.

$$A_{12} = (-1)^{1+2} \cdot M_{12} = (-1)^3 \cdot (-5) = (-1)(-5) = 5$$
$$A_{22} = (-1)^{2+2} \cdot M_{22} = (-1)^4 \cdot (2) = 1 \cdot 2 = 2$$
$$A_{32} = (-1)^{3+2} \cdot M_{32} = (-1)^5 \cdot (-8) = (-1)(-8) = 8$$

The determinant is found by multiplying each cofactor by its corresponding element in the matrix and finding the sum of these products.

$$\det\begin{bmatrix} 2 & -3 & -2 \\ -1 & -4 & -3 \\ -1 & 0 & 2 \end{bmatrix} = a_{12} \cdot A_{12} + a_{22} \cdot A_{22} + a_{32} \cdot A_{32}$$
$$= -3(5) + (-4)(2) + (0)(8)$$
$$= -15 + (-8) + 0 = -23 \qquad \blacksquare$$

CAUTION   Be very careful to keep track of all negative signs when evaluating determinants by hand. Work carefully, writing down each step as in the examples. Skipping steps frequently leads to errors in these computations.

Exactly the same answer would be found using any row or column of the matrix. One reason that column 2 was used in Example 2 is that it contains a 0 element, so that it was not really necessary to calculate M_{32} and A_{32} above. One learns quickly that zeros can be very useful in working with determinants.

Instead of calculating $(-1)^{i+j}$ for a given element, the following sign checkerboards can be used.

ARRAY OF SIGNS

For Dimension 3 × 3 Matrices	For Dimension 4 × 4 Matrices
+ − + − + − + − +	+ − + − − + − + + − + − − + − +

The signs alternate for each row and column, beginning with + in the first row, first column position. Thus, these arrays of signs can be reproduced as needed. If we expand a square matrix of dimension 3 × 3 about row 3, for example, the first minor would have a + sign associated with it, the second minor a − sign, and the third minor a + sign. These arrays of signs can be extended in this way for determinants of square matrices of dimension 5 × 5 and greater.

EXAMPLE 3

Evaluating the Determinant of a Dimension 4 × 4 Square Matrix

Evaluate

$$\det\begin{bmatrix} -1 & -2 & 3 & 2 \\ 0 & 1 & 4 & -2 \\ 3 & -1 & 4 & 0 \\ 2 & 1 & 0 & 3 \end{bmatrix}.$$

SOLUTION Expanding by minors about the fourth row gives

$$-2 \det \begin{bmatrix} -2 & 3 & 2 \\ 1 & 4 & -2 \\ -1 & 4 & 0 \end{bmatrix} + 1 \det \begin{bmatrix} -1 & 3 & 2 \\ 0 & 4 & -2 \\ 3 & 4 & 0 \end{bmatrix}$$

$$-0 \det \begin{bmatrix} -1 & -2 & 2 \\ 0 & 1 & -2 \\ 3 & -1 & 0 \end{bmatrix} + 3 \det \begin{bmatrix} -1 & -2 & 3 \\ 0 & 1 & 4 \\ 3 & -1 & 4 \end{bmatrix}$$

$$= -2(6) + 1(-50) - 0 + 3(-41)$$

$$= -185. \quad \blacksquare$$

NOTE As seen in these examples, the computation involved in finding determinants of dimension 3×3 and larger square matrices is quite tedious. For this reason, it is customary to use graphics calculators with matrix capabilities (or in some cases, computers) to evaluate such determinants. We have presented the theory behind these processes so that the student can see how such determinants are defined mathematically.

Multiplicative Inverses of Larger Matrices

If A is a matrix of dimension 2×2 and det $A \neq 0$, then A^{-1} exists. This was established in the previous section. Now we can extend the concept of inverse matrices to dimensions greater than 2×2.

EXAMPLE 4
Verifying Matrix Inverses

Consider the matrix $A = \begin{bmatrix} 1 & 0 & 1 \\ 2 & -2 & -1 \\ 3 & 0 & 0 \end{bmatrix}$.

(a) Show that det $A \neq 0$. **(b)** Verify that $A^{-1} = \begin{bmatrix} 0 & 0 & \frac{1}{3} \\ -\frac{1}{2} & -\frac{1}{2} & \frac{1}{2} \\ 1 & 0 & -\frac{1}{3} \end{bmatrix}$.

SOLUTION

(a) Expanding about the third row, we find

$$\det A = 3 \cdot \det \begin{bmatrix} 0 & 1 \\ -2 & -1 \end{bmatrix} = 3(0 + 2) = 6.$$

(Because $6 \neq 0$, A^{-1} exists.)

(b) First, show that $A \cdot A^{-1} = I_3$.

$$A \cdot A^{-1} = \begin{bmatrix} 1 & 0 & 1 \\ 2 & -2 & -1 \\ 3 & 0 & 0 \end{bmatrix} \cdot \begin{bmatrix} 0 & 0 & \frac{1}{3} \\ -\frac{1}{2} & -\frac{1}{2} & \frac{1}{2} \\ 1 & 0 & -\frac{1}{3} \end{bmatrix}$$

$$= \begin{bmatrix} 0+0+1 & 0+0+0 & \frac{1}{3}+0-\frac{1}{3} \\ 0+1-1 & 0+1+0 & \frac{2}{3}-1+\frac{1}{3} \\ 0+0+0 & 0+0+0 & 1+0+0 \end{bmatrix}$$

$$= \begin{bmatrix} 1 & 0 & 0 \\ 0 & 1 & 0 \\ 0 & 0 & 1 \end{bmatrix} = I_3$$

Next, show that $A^{-1} \cdot A = I_3$. This is left as Exercise 29. Because $A \cdot A^{-1} = A^{-1} \cdot A = I_3$, we know that the inverse of A is

$$\begin{bmatrix} 0 & 0 & \frac{1}{3} \\ -\frac{1}{2} & -\frac{1}{2} & \frac{1}{2} \\ 1 & 0 & -\frac{1}{3} \end{bmatrix}. \quad \blacksquare$$

While there are methods of finding inverses of square matrices of dimension 3×3 or greater by hand computation, we seldom do this because it is so tedious. Calculators and computers perform this function quickly and efficiently.

Solution of Larger Systems Using Inverse Matrices

The method introduced in the previous section for solving linear systems using inverse matrices can be extended to systems with n equations in n unknowns. The following example illustrates this for $n = 3$.

EXAMPLE 5

Solving a System of Equations Using a Matrix Inverse

Solve the system

$$\begin{aligned} 2x + y + 3z &= 1 \\ x - 2y + z &= -3 \\ -3x + y - 2z &= -4 \end{aligned}$$

using the inverse of the coefficient matrix.

SOLUTION   First, find the determinant of the coefficient matrix to be sure that it is not 0. Using the method described earlier or a calculator we find that for

$$A = \begin{bmatrix} 2 & 1 & 3 \\ 1 & -2 & 1 \\ -3 & 1 & -2 \end{bmatrix},$$

$\det A = -10 \neq 0$. Therefore, A^{-1} exists.

With

$$X = \begin{bmatrix} x \\ y \\ z \end{bmatrix} \quad \text{and} \quad B = \begin{bmatrix} 1 \\ -3 \\ -4 \end{bmatrix},$$

we use the calculator to evaluate $A^{-1}B$:

TECHNOLOGICAL NOTE
After reading Examples 1–5, you may wish to experiment with your calculator to see if you can duplicate the results obtained.

$$A^{-1}B = \begin{bmatrix} 2 & 1 & 3 \\ 1 & -2 & 1 \\ -3 & 1 & -2 \end{bmatrix}^{-1} \begin{bmatrix} 1 \\ -3 \\ -4 \end{bmatrix} = \begin{bmatrix} 4 \\ 2 \\ -3 \end{bmatrix}.$$

Since $X = A^{-1}B$, we have $x = 4$, $y = 2$, and $z = -3$. The solution set of the system is $\{(4, 2, -3)\}$. $\blacksquare$

CAUTION   Always evaluate the determinant of the coefficient matrix *before* using the inverse matrix method. If the determinant is 0, the system is either inconsistent or has dependent equations.

EXAMPLE **6**

Using a System to
Find the Equation
for a Cubic
Polynomial

Figure 2 shows four views of the graph of a polynomial function of the form $P(x) = ax^3 + bx^2 + cx + d$. Use the points indicated to write a system of four equations in the variables a, b, c, and d, and then use the inverse matrix method to solve the system. What is the equation that defines this graph?

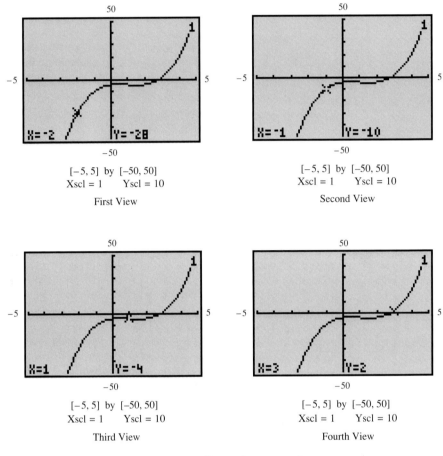

$$P(x) = ax^3 + bx^2 + cx + d$$

FIGURE 2

SOLUTION   From the graph we see that $P(-2) = -28$, $P(-1) = -10$, $P(1) = -4$, and $P(3) = 2$. From the first of these, we get

$$P(-2) = a(-2)^3 + b(-2)^2 + c(-2) + d = -28$$

or, equivalently,

$$-8a + 4b - 2c + d = -28.$$

Similarly, from the others, we find the following equations.

From $P(-1) = -10$: $-a + b - c + d = -10$

From $P(1) = -4$: $a + b + c + d = -4$

From $P(3) = 2$: $27a + 9b + 3c + d = 2$

Now we must solve the system formed by these four equations:

$$-8a + 4b - 2c + d = -28$$
$$-a + b - c + d = -10$$
$$a + b + c + d = -4$$
$$27a + 9b + 3c + d = 2.$$

We will use the inverse matrix method to solve the system. Let

$$A = \begin{bmatrix} -8 & 4 & -2 & 1 \\ -1 & 1 & -1 & 1 \\ 1 & 1 & 1 & 1 \\ 27 & 9 & 3 & 1 \end{bmatrix}, \quad X = \begin{bmatrix} a \\ b \\ c \\ d \end{bmatrix}, \quad \text{and} \quad B = \begin{bmatrix} -28 \\ -10 \\ -4 \\ 2 \end{bmatrix}.$$

Because $\det A = 240 \neq 0$, a unique solution exists. Based on our discussion earlier, $X = A^{-1}B$:

$$X = \begin{bmatrix} -8 & 4 & -2 & 1 \\ -1 & 1 & -1 & 1 \\ 1 & 1 & 1 & 1 \\ 27 & 9 & 3 & 1 \end{bmatrix}^{-1} \begin{bmatrix} -28 \\ -10 \\ -4 \\ 2 \end{bmatrix}.$$

Using a calculator, we find

$$X = \begin{bmatrix} 1 \\ -3 \\ 2 \\ -4 \end{bmatrix}.$$

Therefore, $a = 1$, $b = -3$, $c = 2$, and $d = -4$. The polynomial is $P(x) = x^3 - 3x^2 + 2x - 4$. ◼

FOR GROUP DISCUSSION

Use a window with dimensions $[-5, 5]$ by $[-50, 50]$ to reproduce the graph in Figure 2, based on the result of Example 6.

7.4 EXERCISES

To the Student: Calculator Considerations

From your work in Section 7.2, you should now be familiar with the keystrokes required to find determinants of square matrices of dimension 2×2. It is easy to extend this process to larger matrices, so that your work in this exercise set can be simplified.

Use any method to find the determinant of the given matrix.

1. $\begin{bmatrix} -2 & 0 & 1 \\ 3 & 2 & -1 \\ 1 & 0 & 2 \end{bmatrix}$

2. $\begin{bmatrix} 0 & -1 & 2 \\ 1 & 0 & 2 \\ 0 & -3 & 1 \end{bmatrix}$

3. $\begin{bmatrix} 1 & 2 & -1 \\ 2 & 3 & -2 \\ -1 & 4 & 1 \end{bmatrix}$

4. $\begin{bmatrix} 2 & -1 & 4 \\ 3 & 0 & 1 \\ -2 & 1 & 4 \end{bmatrix}$

5. $\begin{bmatrix} -2 & 0 & 0 \\ 4 & 0 & 1 \\ 3 & 4 & 2 \end{bmatrix}$

6. $\begin{bmatrix} 3 & -2 & 0 \\ 0 & -1 & 1 \\ 4 & 0 & 2 \end{bmatrix}$

7. $\begin{bmatrix} 1 & 2 & 0 \\ -1 & 2 & -1 \\ 0 & 1 & 4 \end{bmatrix}$

8. $\begin{bmatrix} 2 & 1 & -1 \\ 4 & 7 & -2 \\ 2 & 4 & 0 \end{bmatrix}$

9. $\begin{bmatrix} 10 & 2 & 1 \\ -1 & 4 & 3 \\ -3 & 8 & 10 \end{bmatrix}$

10. $\begin{bmatrix} 7 & -1 & 1 \\ 1 & -7 & 2 \\ -2 & 1 & 1 \end{bmatrix}$

11. $\begin{bmatrix} 1 & -2 & 3 \\ 0 & 0 & 0 \\ 1 & 10 & -12 \end{bmatrix}$

12. $\begin{bmatrix} 2 & 3 & 0 \\ 1 & 9 & 0 \\ -1 & -2 & 0 \end{bmatrix}$

13. $\begin{bmatrix} 3 & 3 & -1 \\ 2 & 6 & 0 \\ -6 & -6 & 2 \end{bmatrix}$

14. $\begin{bmatrix} 5 & -3 & 2 \\ -5 & 3 & -2 \\ 1 & 0 & 1 \end{bmatrix}$

15. $\begin{bmatrix} 4 & 0 & 0 & 2 \\ -1 & 0 & 3 & 0 \\ 2 & 4 & 0 & 1 \\ 0 & 0 & 1 & 2 \end{bmatrix}$

16. $\begin{bmatrix} -2 & 0 & 4 & 2 \\ 3 & 6 & 0 & 4 \\ 0 & 0 & 0 & 3 \\ 9 & 0 & 2 & -1 \end{bmatrix}$

17. $\begin{bmatrix} 1 & 1 & 0 & 1 \\ 2 & 1 & 0 & 2 \\ 0 & 1 & -1 & 1 \\ 1 & -1 & 1 & 1 \end{bmatrix}$

18. $\begin{bmatrix} 2 & 7 & 0 & -1 \\ 1 & 0 & 1 & 3 \\ 2 & 4 & -1 & -1 \\ -1 & 1 & 0 & 8 \end{bmatrix}$

Find the multiplicative inverse, if it exists, of each matrix.

19. $\begin{bmatrix} 1 & 0 & 0 \\ 0 & -1 & 0 \\ 1 & 0 & 1 \end{bmatrix}$

20. $\begin{bmatrix} 1 & 3 & 3 \\ -1 & 0 & 0 \\ -4 & -4 & -3 \end{bmatrix}$

21. $\begin{bmatrix} 3 & 6 & 3 \\ 6 & 4 & -2 \\ 0 & 1 & -1 \end{bmatrix}$

22. $\begin{bmatrix} 2 & 6 & 0 \\ 1 & 5 & 3 \\ 0 & 0 & 1 \end{bmatrix}$

23. $\begin{bmatrix} -1 & -1 & -1 \\ 4 & 5 & 0 \\ 0 & 1 & -3 \end{bmatrix}$

24. $\begin{bmatrix} 2 & 0 & 4 \\ 3 & 1 & 5 \\ -1 & 1 & -2 \end{bmatrix}$

25. $\begin{bmatrix} -.4 & .1 & .2 \\ 0 & .6 & .8 \\ .3 & 0 & -.2 \end{bmatrix}$

26. $\begin{bmatrix} .8 & .2 & .1 \\ -.2 & 0 & .3 \\ 0 & 0 & .5 \end{bmatrix}$

27. $\begin{bmatrix} 2 & 1 & 2 \\ 5 & 10 & 5 \\ 3 & 6 & 3 \end{bmatrix}$

28. $\begin{bmatrix} 5 & -3 & 2 \\ -5 & 3 & -2 \\ 1 & 0 & 1 \end{bmatrix}$

29. Refer to Example 4 and show that $A^{-1}A = I_3$.

30. Explain why a matrix with a row or column of all zeros must have 0 as the value of its determinant.

*R*elating Concepts

Determinants can be used to find the equation of a line passing through two given points. The next two exercises show how this is done.

31. Find the determinant below and show that the result is the equation of the line through $(2, 3)$ and $(-1, 4)$.

$$\det \begin{bmatrix} x & y & 1 \\ 2 & 3 & 1 \\ -1 & 4 & 1 \end{bmatrix} = 0$$

32. (a) Write the equation of the line through the points (x_1, y_1) and (x_2, y_2) using the point-slope formula.
(b) Expanding the determinant in the equation below, show that the equation is equivalent to the equation in part (a).

$$\det \begin{bmatrix} x & y & 1 \\ x_1 & y_1 & 1 \\ x_2 & y_2 & 1 \end{bmatrix} = 0$$

Solve each of the following equations for x.

33.
$$\det \begin{bmatrix} -2 & 0 & 1 \\ -1 & 3 & x \\ 5 & -2 & 0 \end{bmatrix} = 3$$

34.
$$\det \begin{bmatrix} 4 & 3 & 0 \\ 2 & 0 & 1 \\ -3 & x & -1 \end{bmatrix} = 5$$

35.
$$\det \begin{bmatrix} 5 & 3x & -3 \\ 0 & 2 & -1 \\ 4 & -1 & x \end{bmatrix} = -7$$

36.
$$\det \begin{bmatrix} 2x & 1 & -1 \\ 0 & 4 & x \\ 3 & 0 & 2 \end{bmatrix} = x$$

Use the inverse matrix method to solve each system.

37.
$$\begin{aligned} 2x + 4z &= 14 \\ 3x + y + 5z &= 19 \\ -x + y - 2z &= -7 \end{aligned}$$

38.
$$\begin{aligned} 3x + 6y + 3z &= 12 \\ 6x + 4y - 2z &= -4 \\ y - z &= -3 \end{aligned}$$

39.
$$\begin{aligned} x + 3y + z &= 2 \\ x - 2y + 3z &= -3 \\ 2x - 3y - z &= 34 \end{aligned}$$

40.
$$\begin{aligned} x + y - z &= 6 \\ 2x - y + z &= -9 \\ x - 2y + 3z &= 1 \end{aligned}$$

41.
$$\begin{aligned} x + 3y - 2z - w &= 9 \\ 4x + y + z + 2w &= 2 \\ -3x - y + z - w &= -5 \\ x - y - 3z - 2w &= 2 \end{aligned}$$

42.
$$\begin{aligned} 3x + 2y - w &= 0 \\ 2x + z + 2w &= 5 \\ x + 2y - z &= -2 \\ 2x - y + z + w &= 2 \end{aligned}$$

In Exercises 43 and 44, use the method of Example 6 to find the cubic polynomial $P(x)$ that defines the curve shown in the four figures.

43.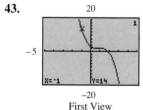

First View

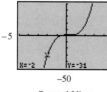

Second View

Third View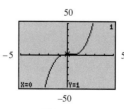

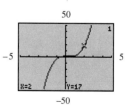

Fourth View

44.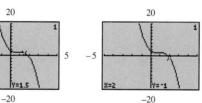

First View   Second View   Third View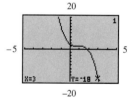

Fourth View

45. Find the fourth degree polynomial $P(x)$ satisfying the following conditions:
$P(-2) = 13$, $P(-1) = 2$, $P(0) = -1$, $P(1) = 4$, $P(2) = 41$.

46. Find the fifth degree polynomial $P(x)$ satisfying the following conditions:
$P(-2) = -8$, $P(-1) = -1$, $P(0) = -4$, $P(1) = -5$, $P(2) = 8$, $P(3) = 167$.

A formal study of determinants usually examines the following properties.

> ## PROPERTIES OF DETERMINANTS
>
> 1. If every element in a row or column of a matrix is 0, then the determinant equals 0.
> 2. If corresponding rows and columns of a matrix are interchanged, the determinant is not changed.
> 3. Interchanging two rows (or columns) of a matrix reverses the sign of the determinant.
> 4. If every element of a row (or column) of a matrix is multiplied by the real number k, then the determinant of the new matrix is k times the determinant of the original matrix.
> 5. The determinant of a matrix with two identical rows (or columns) equals 0.
> 6. The determinant of a matrix is unchanged if a multiple of a row (or column) of the matrix is added to the corresponding elements of another row (or column).

Refer to these properties in Exercises 47–53.

47. Use property 1 to evaluate $\det \begin{bmatrix} 0 & 0 & 0 \\ 2 & 1 & 4 \\ -3 & 8 & 6 \end{bmatrix}$. Then verify your result using expansion

   by minors or a calculator.

48. Repeat Exercise 47 for $\det \begin{bmatrix} .5 & 0 & \sqrt{2} \\ 6 & 0 & -3 \\ -1 & 0 & 5 \end{bmatrix}$.

49. Use property 2 to complete the following statement: The value of $\det \begin{bmatrix} 2 & 1 & 6 \\ 3 & 0 & 5 \\ -4 & 6 & 9 \end{bmatrix}$

   is 1. Therefore, the value of $\det \begin{bmatrix} 2 & 3 & -4 \\ 1 & 0 & 6 \\ 6 & 5 & 9 \end{bmatrix}$ is _____ , because

   _____ .

50. Use property 3 and the information given in Exercise 49 to complete the following

   statement: The value of $\det \begin{bmatrix} 3 & 0 & 5 \\ 2 & 1 & 6 \\ -4 & 6 & 9 \end{bmatrix}$ is _____ , because

   _____ .

51. Use property 4 to complete the given statment: The value of $\det \begin{bmatrix} 2 & -3 \\ 4 & 1 \end{bmatrix}$

   is 14. Therefore, $\det \begin{bmatrix} 2 & -3 \\ -20 & -5 \end{bmatrix} =$ _____ , because

   _____ .

52. Use property 5 to find $\det \begin{bmatrix} -4 & 2 & 3 \\ 0 & 1 & 6 \\ -4 & 2 & 3 \end{bmatrix}$.

53. Multiply each element of the first column of the matrix $A = \begin{bmatrix} -2 & 4 & 1 \\ 2 & 1 & 5 \\ 3 & 0 & 2 \end{bmatrix}$ by 3, and

add the results to the third column to get a new matrix B.
(a) What is the new matrix B?
(b) $\det A = 37$, therefore $\det B =$ _____ .

Relating Concepts

Determinants can be used to find the area of a triangle, given the coordinates of its vertices. Given a triangle PQR with vertices $(x_1\ y_1)$, (x_2, y_2), and (x_3, y_3), as in the figure, it can be shown that the area of the triangle is given by A , where

$$A = \frac{1}{2} \det \begin{bmatrix} x_1 & y_1 & 1 \\ x_2 & y_2 & 1 \\ x_3 & y_3 & 1 \end{bmatrix}.$$

The points (x_1, y_1), (x_2, y_2), (x_3, y_3) must be taken in counterclockwise order; if this is not done, then A may have the wrong sign. Alternatively, we could define A as the absolute value of $\frac{1}{2}$ the determinant shown above. Use the formula given to find the area of the triangles in Exercises 54 and 55.

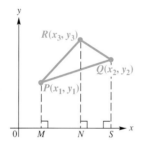

54. (a) $P(0, 0)$, $Q(0, 2)$, $R(1, 1)$
    (b) $P(0, 1)$, $Q(2, 0)$, $R(1, 3)$
    (c) $P(2, 5)$, $Q(-1, 3)$, $R(4, 0)$

55. (a) $P(2, -2)$, $Q(0, 0)$, $R(-3, -4)$
    (b) $P(4, 7)$, $Q(5, -2)$, $R(1, 1)$
    (c) $P(1, 2)$, $Q(4, 3)$, $R(3, 5)$

56. (a) Show that $\det \begin{bmatrix} 1 & 1 & 1 \\ a & b & c \\ a^2 & b^2 & c^2 \end{bmatrix} = (a - b)(b - c)(c - a)$.

   (b) Show that $\det \begin{bmatrix} 1 & 1 & 1 \\ a & b & c \\ bc & ca & ab \end{bmatrix} = (a - b)(a - c)(c - b)$.

7.5 CRAMER'S RULE

Derivation of Cramer's Rule ▌ Systems with Two Equations ▌ Systems with Three Equations

Derivation of Cramer's Rule

We have now seen several methods for solving linear systems of equations in Chapter 6 and in this chapter. Cramer's rule is a method of solving a linear system of equations using determinants. Once again, the determinant of the matrix of coefficients allows us to decide whether a single solution exists before actually solving the system.

To derive Cramer's rule, we use the elimination method to solve the general system of two equations with two variables,

$$a_1x + b_1y = c_1 \qquad \qquad \text{[1]}$$
$$a_2x + b_2y = c_2. \qquad \qquad \text{[2]}$$

To begin, eliminate y and solve for x by first multiplying both sides of equation (1) by b_2 and both sides of equation (2) by $-b_1$. Then add these results and solve for x.

$$a_1b_2x + b_1b_2y = c_1b_2$$
$$\underline{-a_2b_1x - b_1b_2y = -c_2b_1}$$
$$(a_1b_2 - a_2b_1)x = c_1b_2 - c_2b_1$$
$$x = \frac{c_1b_2 - c_2b_1}{a_1b_2 - a_2b_1}, \text{ if } a_1b_2 - a_2b_1 \neq 0$$

Solve for y by multiplying both sides of equation (1) by $-a_2$ and equation (2) by a_1 and then adding the two equations.

$$-a_1a_2x - a_2b_1y = -a_2c_1$$
$$\underline{a_1a_2x + a_1b_2y = a_1c_2}$$
$$(a_1b_2 - a_2b_1)y = a_1c_2 - a_2c_1$$
$$y = \frac{a_1c_2 - a_2c_1}{a_1b_2 - a_2b_1}, \text{ if } a_1b_2 - a_2b_1 \neq 0$$

Both numerators and the common denominator of these values for x and y can be written as determinants, since

$$c_1b_2 - c_2b_1 = \det\begin{bmatrix} c_1 & b_1 \\ c_2 & b_2 \end{bmatrix}, \qquad a_1c_2 - a_2c_1 = \det\begin{bmatrix} a_1 & c_1 \\ a_2 & c_2 \end{bmatrix},$$

$$\text{and} \quad a_1b_2 - a_2b_1 = \det\begin{bmatrix} a_1 & b_1 \\ a_2 & b_2 \end{bmatrix}.$$

Using these determinants, the solutions for x and y become

$$x = \frac{\det\begin{bmatrix} c_1 & b_1 \\ c_2 & b_2 \end{bmatrix}}{\det\begin{bmatrix} a_1 & b_1 \\ a_2 & b_2 \end{bmatrix}} \quad \text{and} \quad y = \frac{\det\begin{bmatrix} a_1 & c_1 \\ a_2 & c_2 \end{bmatrix}}{\det\begin{bmatrix} a_1 & b_1 \\ a_2 & b_2 \end{bmatrix}}, \text{ if } \det\begin{bmatrix} a_1 & b_1 \\ a_2 & b_2 \end{bmatrix} \neq 0.$$

> **CRAMER'S RULE FOR 2 × 2 SYSTEMS**
>
> For the system
>
> $$a_1x + b_1y = c_1$$
> $$a_2x + b_2y = c_2,$$
>
> $$x = \frac{D_x}{D} \quad \text{and} \quad y = \frac{D_y}{D},$$
>
> where
>
> $$D_x = \det\begin{bmatrix} c_1 & b_1 \\ c_2 & b_2 \end{bmatrix}, \quad D_y = \det\begin{bmatrix} a_1 & c_1 \\ a_2 & c_2 \end{bmatrix}, \quad \text{and} \quad D = \det\begin{bmatrix} a_1 & b_1 \\ a_2 & b_2 \end{bmatrix} \neq 0.$$

   Although this theorem is well-known as Cramer's rule, it was probably first discovered by Colin Maclaurin (1698–1746) as early as 1729 and was published under his name in 1748, two years earlier than Cramer's first publication of the rule in 1750.

   Cramer's rule is used to solve a system of linear equations by evaluating the three determinants D, D_x, and D_y and then writing the appropriate quotients for x and y.

CAUTION   As indicated above, Cramer's rule does not apply if $D = 0$. When $D = 0$, the system is inconsistent or has dependent equations. For this reason, it is a good idea to evaluate D first.

Systems with Two Equations

The first example shows how to apply Cramer's rule to a system of linear equations in two unknowns.

EXAMPLE 1

Applying Cramer's Rule to a System with Two Equations

Use Cramer's rule to solve the system

$$5x + 7y = -1$$
$$6x + 8y = 1.$$

SOLUTION   By Cramer's rule, $x = \frac{D_x}{D}$ and $y = \frac{D_y}{D}$. As mentioned above, it is a good idea to find D first, since if $D = 0$, Cramer's rule does not apply. If $D \neq 0$, then find D_x and D_y.

$$D = \det\begin{bmatrix} 5 & 7 \\ 6 & 8 \end{bmatrix} = 5(8) - 6(7) = -2$$

$$D_x = \det\begin{bmatrix} -1 & 7 \\ 1 & 8 \end{bmatrix} = (-1)(8) - (1)(7) = -15$$

$$D_y = \det\begin{bmatrix} 5 & -1 \\ 6 & 1 \end{bmatrix} = 5(1) - (6)(-1) = 11$$

From Cramer's rule,

$$x = \frac{D_x}{D} = \frac{-15}{-2} = \frac{15}{2} \quad \text{and} \quad y = \frac{D_y}{D} = \frac{11}{-2} = -\frac{11}{2}.$$

The solution set is $\{(\frac{15}{2}, -\frac{11}{2})\}$, as can be verified by substituting in the given system.   ◼

Systems with Three Equations

Cramer's rule can be generalized to systems of three equations in three variables (or n equations in n variables).

CRAMER'S RULE FOR 3 × 3 SYSTEMS

For the system

$$a_1x + b_1y + c_1z = d_1$$
$$a_2x + b_2y + c_2z = d_2$$
$$a_3x + b_3y + c_3z = d_3,$$

$$x = \frac{D_x}{D}, \quad y = \frac{D_y}{D}, \quad \text{and} \quad z = \frac{D_z}{D},$$

where

$$D_x = \det \begin{bmatrix} d_1 & b_1 & c_1 \\ d_2 & b_2 & c_2 \\ d_3 & b_3 & c_3 \end{bmatrix}, \qquad D_y = \det \begin{bmatrix} a_1 & d_1 & c_1 \\ a_2 & d_2 & c_2 \\ a_3 & d_3 & c_3 \end{bmatrix},$$

$$D_z = \det \begin{bmatrix} a_1 & b_1 & d_1 \\ a_2 & b_2 & d_2 \\ a_3 & b_3 & d_3 \end{bmatrix}, \qquad \text{and} \qquad D = \det \begin{bmatrix} a_1 & b_1 & c_1 \\ a_2 & b_2 & c_2 \\ a_3 & b_3 & c_3 \end{bmatrix} \neq 0.$$

EXAMPLE 2

Applying Cramer's Rule to a System with Three Equations

Use Cramer's rule to solve the system

$$x + y - z = -2$$
$$2x - y + z = -5$$
$$x - 2y + 3z = 4.$$

SOLUTION Verify that the required determinants are as follows:

$$D = \det \begin{bmatrix} 1 & 1 & -1 \\ 2 & -1 & 1 \\ 1 & -2 & 3 \end{bmatrix} = -3, \qquad D_x = \det \begin{bmatrix} -2 & 1 & -1 \\ -5 & -1 & 1 \\ 4 & -2 & 3 \end{bmatrix} = 7,$$

$$D_y = \det \begin{bmatrix} 1 & -2 & -1 \\ 2 & -5 & 1 \\ 1 & 4 & 3 \end{bmatrix} = -22, \qquad D_z = \det \begin{bmatrix} 1 & 1 & -2 \\ 2 & -1 & -5 \\ 1 & -2 & 4 \end{bmatrix} = -21.$$

Thus

$$x = \frac{D_x}{D} = \frac{7}{-3} = -\frac{7}{3}, \qquad y = \frac{D_y}{D} = \frac{-22}{-3} = \frac{22}{3},$$

and

$$z = \frac{D_z}{D} = \frac{-21}{-3} = 7,$$

so the solution set is $\{(-\frac{7}{3}, \frac{22}{3}, 7)\}$. ∎

EXAMPLE **3**

Verifying that
Cramer's Rule Does
Not Apply

Show why Cramer's rule does not apply to the system

$$2x - 3y + 4z = 10$$
$$6x - 9y + 12z = 24$$
$$x + 2y - 3z = 5.$$

SOLUTION   First find D by expanding about column 1.

$$D = \det\begin{bmatrix} 2 & -3 & 4 \\ 6 & -9 & 12 \\ 1 & 2 & -3 \end{bmatrix} = 2 \ \det\begin{bmatrix} -9 & 12 \\ 2 & -3 \end{bmatrix} - 6 \ \det\begin{bmatrix} -3 & 4 \\ 2 & -3 \end{bmatrix}$$

$$+ 1 \ \det\begin{bmatrix} -3 & 4 \\ -9 & 12 \end{bmatrix}$$

$$= 2(3) - 6(1) + 1(0)$$

$$= 0$$

As mentioned above, Cramer's rule does not apply if $D = 0$. When $D = 0$, the system either is inconsistent or contains dependent equations. Use the elimination method or Gaussian reduction to tell which is the case. Verify that this system is inconsistent.   ∎

7.5 EXERCISES

To the Student: Calculator Considerations

In applying Cramer's rule it is necessary to evaluate several determinants. At this point you should already know how to use your calculator to do this. As stated in this section, it is a good idea to evaluate D first, to see that $D \neq 0$.

You may also wish to explore the programming capabilities of your calculator to program Cramer's rule.

Use Cramer's rule to solve each of the following systems of equations. If $D = 0$, use another method to complete the solution.

1. $x + y = 4$
    $2x - y = 2$

2. $3x + 2y = -4$
    $2x - y = -5$

3. $4x + 3y = -7$
    $2x + 3y = -11$

4. $4x - y = 0$
    $2x + 3y = 14$

5. $5x + 4y = 10$
    $3x - 7y = 6$

6. $3x + 2y = -4$
    $5x - y = 2$

7. $2x - 3y = -5$
    $x + 5y = 17$

8. $x + 9y = -15$
    $3x + 2y = 5$

9. $3x + 2y = 4$
    $6x + 4y = 8$

10. $1.5x + 3y = 5$
    $2x + 4y = 3$

11. $12x + 8y = 3$
    $15x + 10y = 9$

12. $15x - 10y = 5$
    $9x + 6y = 3$

Use Cramer's rule to solve each of the following systems of equations. If $D = 0$, use another method to complete the solution.

13. $4x - y + 3z = -3$
    $3x + y + z = 0$
    $2x - y + 4z = 0$

14. $5x + 2y + z = 15$
    $2x - y + z = 9$
    $4x + 3y + 2z = 13$

15. $2x - y + 4z = -2$
    $3x + 2y - z = -3$
    $x + 4y + 2z = 17$

16. $x + y + z = 4$
    $2x - y + 3z = 4$
    $4x + 2y - z = -15$

17. $4x - 3y + z = -1$
    $5x + 7y + 2z = -2$
    $3x - 5y - z = 1$

18. $2x - 3y + z = 8$
    $-x - 5y + z = -4$
    $3x - 5y + 2z = 12$

19. $x + 2y + 3z = 4$
$4x + 3y + 2z = 1$
$-x - 2y - 3z = 0$

20. $2x - y + 3z = 1$
$-2x + y - 3z = 2$
$5x - y + z = 2$

21. $-2x - 2y + 3z = 4$
$5x + 7y - z = 2$
$2x + 2y - 3z = -4$

22. $-3x + 2y - 2z = 4$
$4x + y + z = 5$
$3x - 2y + 2z = 1$

23. $2x + 3y = 13$
$2y - z = 5$
$x + 2z = 4$

24. $3x - z = -10$
$y + 4z = 8$
$x + 2z = -1$

25. $5x - y = -4$
$3x + 2z = 4$
$4y + 3z = 22$

26. $3x + 5y = -7$
$2x + 7z = 2$
$4y + 3z = -8$

27. $x + 2y = 10$
$3x + 4z = 7$
$-y - z = 1$

28. $5x - 2y = 3$
$4y + z = 8$
$x + 2z = 4$

29. In your own words, explain what it means in applying Cramer's rule if $D = 0$.

30. Describe D_x, D_y, and D_z in terms of the coefficients and constants in the given system of equations.

Use Cramer's rule to solve each of the following systems.

31. $x + 3y - 2z - w = 9$
$4x + y + z + 2w = 2$
$-3x - y + z - w = -5$
$x - y - 3z - 2w = 2$

32. $3x + 2y - w = 0$
$2x + z + 2w = 5$
$x + 2y - z = -2$
$2x - y + z + w = 2$

33. $x + 2y + 4z - w = 7$
$2x + y - 3z + w = -20$
$-x + 4y - 6z + 2w = -8$
$5x - y + 2z + 3w = -36$

34. $x + 2y - z + w = 8$
$2x - y + 2w = 8$
$y + 3z = 5$
$x - z = 4$

R*elating Concepts*

A unique circle passes through the three points $(9, 0)$, $(-1, 0)$, *and* $(1, 1)$. *Recall that the equation of a circle may be written in the form*

$$x^2 + y^2 + Ax + By + C = 0.$$

Work Exercises 35–40 in order.

35. Find three equations in A, B, and C using the ordered pairs given.

36. Solve the system in Exercise 35 using Cramer's rule. What are the values of A, B, and C?

37. Use the method described in Section 6.1 to write the equation in the form $(x - h)^2 + (y - k)^2 = r^2$.

38. What are the coordinates of the center of the circle?

39. What is the radius of the circle?

40. Use the methods of Section 6.1 to graph this circle. Use a square window.

Solve each system for x and y using Cramer's rule. Assume a and b are nonzero constants.

41. $bx + y = a^2$
$ax + y = b^2$

42. $ax + by = \dfrac{b}{a}$
$x + y = \dfrac{1}{b}$

43. $b^2x + a^2y = b^2$
$ax + by = a$

44. $x + \dfrac{1}{b}y = b$
$\dfrac{1}{a}x + y = a$

45. Use Cramer's rule to solve each system in (a)–(c).

    (a) $3x + 4y = 5$ **(b)** $5x + 6y = 7$ **(c)** $13x + 14y = 15$
         $6x + 7y = 8$ $8x + 9y = 10$ $16x + 17y = 18$

    (d) Observe the coefficients and the constants in each system in (a)–(c), and make a conjecture about the solution of such a system.

    (e) Prove your conjecture using Cramer's rule.

46. Write several paragraphs describing the various methods of solving systems of linear equations you have studied. Tell which one of them is your favorite, and why. What are the advantages of each method? What are the disadvantages?

Chapter 7   SUMMARY

Matrix row transformations are used to transform an augmented matrix whose rows are the coefficients of a system of equations. The goal is to replace the matrix with one that is in triangular form, so that one variable of the corresponding system is known.

A matrix with m rows and n columns has dimension $m \times n$. Two matrices are equal if they have the same dimension and the same elements in each position. Matrices with the same dimension can be added and subtracted by adding or subtracting corresponding elements. The product of a scalar and a matrix is a matrix whose elements are found by multiplying each element in the given matrix by the scalar.

The product of two matrices is found by multiplying corresponding elements of the rows of the first matrix and the columns of the second matrix, and then adding these products. This means that the number of columns in the first matrix must be the same as the number of rows in the second matrix. It also means that matrix multiplication is not commutative.

The $n \times n$ identity matrix has ones down the main diagonal and zeros elsewhere. It has properties similar to the properties of the number 1; the product of the identity matrix and any matrix A with the same dimension is matrix A.

Every square matrix has a determinant which is a real number. The determinant is used to find the multiplicative inverse matrix of a square matrix. If its determinant is zero, a matrix has no inverse.

Linear systems can be solved using the fact that the matrix equation $AX = B$ has solution $X = A^{-1}B$. The matrix A has entries that are the coefficients of the variables in the system, X is the matrix of variables, and B is the matrix whose entries are the constants in the system.

Determinants of matrices with dimension $n \times n$ are found using the minors and cofactors of a row or column of the matrix. A cofactor of an element is the minor multiplied by either 1 or -1. Each element in the chosen row or column is multiplied by its cofactor, and the sum of these products gives the determinant. An array of signs can be used as a shortcut for finding the sign of the cofactors. It is best to use a calculator or computer to find inverses of matrices with dimension 3×3 or larger.

Cramer's rule is a method for solving linear systems that uses determinants. The values of the variables are each written as the quotient of two determinants all with the same denominator. If the denominator is zero, it indicates that the system is inconsistent or has dependent equations. In that case, one of the other methods must be used to solve the system.

Key Terms

SECTION 7.1

matrix (plural: matrices)
element (entry)
augmented matrix
triangular (echelon) form
main diagonal
Gaussian reduction method

SECTION 7.2

dimension $m \times n$
square matrix
row matrix
column matrix
zero matrix
additive inverse (negative)
scalar
transpose

SECTION 7.3

identity matrix
determinant
inverse (multiplicative)
coefficient matrix

SECTION 7.4

minor
cofactor
expanding about a row or column

SECTION 7.5

Cramer's rule

Chapter 7   REVIEW EXERCISES

Find the values of all variables in the following statements.

1. $\begin{bmatrix} 2 & z & 1 \\ m & 9 & -7 \end{bmatrix} = \begin{bmatrix} x & 5 & 1 \\ -8 & y & p \end{bmatrix}$ **2.** $\begin{bmatrix} 5 & -10 \\ a & 0 \end{bmatrix} = \begin{bmatrix} b-3 & c+2 \\ 4 & 0 \end{bmatrix}$ **3.** $\begin{bmatrix} 5 & x+2 \\ -6y & z \end{bmatrix} = \begin{bmatrix} a & 3x-1 \\ 5y & 9 \end{bmatrix}$

4. $\begin{bmatrix} 6+k & 2 & a+3 \\ -2+m & 3p & 2r \end{bmatrix} + \begin{bmatrix} 3-2k & 5 & 7 \\ 5 & 8p & 5r \end{bmatrix} = \begin{bmatrix} 5 & y & 6a \\ 2m & 11 & -35 \end{bmatrix}$

Perform each of the following operations whenever possible.

5. $\begin{bmatrix} 3 & -4 & 2 \\ 5 & -1 & 6 \end{bmatrix} + \begin{bmatrix} -3 & 2 & 5 \\ 1 & 0 & 4 \end{bmatrix}$

6. $\begin{bmatrix} 3 \\ 2 \\ 5 \end{bmatrix} - \begin{bmatrix} 8 \\ -4 \\ 6 \end{bmatrix} + \begin{bmatrix} 1 \\ 0 \\ 2 \end{bmatrix}$

7. $\begin{bmatrix} 2 & 5 & 8 \\ 1 & 9 & 2 \end{bmatrix} - \begin{bmatrix} 3 & 4 \\ 7 & 1 \end{bmatrix}$

8. $3\begin{bmatrix} 2 & 4 \\ -1 & 4 \end{bmatrix} - 2\begin{bmatrix} 5 & 8 \\ 2 & -2 \end{bmatrix}$

9. $-1\begin{bmatrix} 3 & -5 & 2 \\ 1 & 7 & -4 \end{bmatrix} + 5\begin{bmatrix} 0 & 2 \\ -1 & 3 \end{bmatrix}$

10. $10\begin{bmatrix} 2x+3y & 4x+y \\ x-5y & 6x+2y \end{bmatrix} + 2\begin{bmatrix} -3x-y & x+6y \\ 4x+2y & 5x-y \end{bmatrix}$

11. Complete the following sentence. The sum of two $m \times n$ matrices A and B is found _____ .

12. The speed limits in Italy, Britain, and the U.S. were at one time 87 mph, 70 mph, and 55 mph, respectively. The corresponding fatalities per 100 million miles driven in a recent year were 6.4, 4.0, and 3.3. Write this information as a 3 × 2 matrix and then as a 2 × 3 matrix.

13. In a recent year, Dan Marino attempted 606 passes and completed 354 for 4434 yd and 28 touchdowns. Joe Montana attempted 397 passes and completed 238 for 2981 yd and 18 touchdowns. Phil Simms attempted 479 passes and completed 263 for 3359 yd and 21 touchdowns. Write this information as a matrix in two ways.

Find each of the following matrix products, whenever possible.

14. $\begin{bmatrix} -3 & 4 \\ 2 & 8 \end{bmatrix}\begin{bmatrix} -1 & 0 \\ 2 & 5 \end{bmatrix}$

15. $\begin{bmatrix} 3 & 2 & -1 \\ 4 & 0 & 6 \end{bmatrix}\begin{bmatrix} -2 & 0 \\ 0 & 2 \\ 3 & 1 \end{bmatrix}$

16. $\begin{bmatrix} 1 & -2 & 4 & 2 \\ 0 & 1 & -1 & 8 \end{bmatrix}\begin{bmatrix} -1 \\ 2 \\ 0 \\ 1 \end{bmatrix}$

17. $\begin{bmatrix} 1 & 2 & 5 \\ -3 & 4 & 7 \\ 0 & 2 & -1 \end{bmatrix}\begin{bmatrix} 4 & 2 & 3 \\ 10 & -5 & 6 \end{bmatrix}$

18. $\begin{bmatrix} 4 & 2 & 3 \\ 10 & -5 & 6 \end{bmatrix}\begin{bmatrix} 1 & 2 & 5 \\ -3 & 4 & 7 \\ 0 & 2 & -1 \end{bmatrix}$

19. $\begin{bmatrix} 3 & -1 & 0 \end{bmatrix}\begin{bmatrix} 1 & 3 & 2 \\ 2 & -4 & 0 \\ 5 & 7 & 3 \end{bmatrix}$

20. What must be true of two matrices to find their product?

Decide whether or not each of the following pairs of matrices are multiplicative inverses.

21. $\begin{bmatrix} 2 & -3 \\ 1 & -2 \end{bmatrix}, \begin{bmatrix} 2 & -3 \\ 1 & -2 \end{bmatrix}$

22. $\begin{bmatrix} 1 & 0 \\ 2 & -3 \end{bmatrix}, \begin{bmatrix} 1 & 0 \\ \frac{2}{3} & -\frac{1}{3} \end{bmatrix}$

23. $\begin{bmatrix} 2 & 0 & 6 \\ 0 & 1 & 0 \\ 1 & 0 & 1 \end{bmatrix}, \begin{bmatrix} -1 & 0 & \frac{3}{2} \\ 0 & 1 & 0 \\ \frac{1}{4} & 0 & -1 \end{bmatrix}$

24. $\begin{bmatrix} 1 & 0 & 2 \\ 0 & 2 & 4 \\ 0 & 0 & 1 \end{bmatrix}, \begin{bmatrix} 1 & 0 & -2 \\ 0 & \frac{1}{2} & -2 \\ 0 & 0 & 1 \end{bmatrix}$

Find the inverse, if it exists, for each of the following matrices.

25. $\begin{bmatrix} 2 & 1 \\ 5 & 3 \end{bmatrix}$

26. $\begin{bmatrix} -4 & 2 \\ 0 & 3 \end{bmatrix}$

27. $\begin{bmatrix} 2 & 0 \\ -1 & 5 \end{bmatrix}$

28. $\begin{bmatrix} 2 & 0 & 4 \\ 1 & -1 & 0 \\ 0 & 1 & -2 \end{bmatrix}$

29. $\begin{bmatrix} 2 & -1 & 0 \\ 1 & 0 & 1 \\ 1 & -2 & 0 \end{bmatrix}$

30. $\begin{bmatrix} 2 & 3 & 5 \\ -2 & -3 & -5 \\ 1 & 4 & 2 \end{bmatrix}$

Use matrix inverses to solve each of the following. Identify any inconsistent systems or systems with dependent equations.

31. $\begin{aligned} x + y &= 4 \\ 2x + 3y &= 10 \end{aligned}$

32. $\begin{aligned} 5x - 3y &= -2 \\ 2x + 7y &= -9 \end{aligned}$

33. $\begin{aligned} 2x + y &= 5 \\ 3x - 2y &= 4 \end{aligned}$

34. $\begin{aligned} x - 2y &= 7 \\ 3x + y &= 7 \end{aligned}$

35. $\begin{aligned} 3x - 2y + 4z &= 1 \\ 4x + y - 5z &= 2 \\ -6x + 4y - 8z &= -2 \end{aligned}$

36. $\begin{aligned} x + 2y &= -1 \\ 3y - z &= -5 \\ x + 2y - z &= -3 \end{aligned}$

37. $\begin{aligned} x + y + z &= 1 \\ 2x - y &= -2 \\ 3y + z &= 2 \end{aligned}$

38. $\begin{aligned} x &= -3 \\ y + z &= 6 \\ 2x - 3z &= -9 \end{aligned}$

Evaluate each of the following determinants.

39. $\det \begin{bmatrix} -1 & 8 \\ 2 & 9 \end{bmatrix}$

40. $\det \begin{bmatrix} -2 & 4 \\ 0 & 3 \end{bmatrix}$

41. $\det \begin{bmatrix} -2 & 4 & 1 \\ 3 & 0 & 2 \\ -1 & 0 & 3 \end{bmatrix}$

42. $\det \begin{bmatrix} -1 & 2 & 3 \\ 4 & 0 & 3 \\ 5 & -1 & 2 \end{bmatrix}$

Solve each of the following determinant equations for x.

43. $\det \begin{bmatrix} -3 & 2 \\ 1 & x \end{bmatrix} = 5$

44. $\det \begin{bmatrix} 3x & 7 \\ -x & 4 \end{bmatrix} = 8$

45. $\det \begin{bmatrix} 2 & 5 & 0 \\ 1 & 3x & -1 \\ 0 & 2 & 0 \end{bmatrix} = 4$

46. $\det \begin{bmatrix} 6x & 2 & 0 \\ 1 & 5 & 3 \\ x & 2 & -1 \end{bmatrix} = 2x$

Exercises 47 and 48 refer to the system below.

$$3x - y = 28$$
$$2x + y = 2$$

47. Suppose you are asked to solve this system using Cramer's rule.
   (a) What is the value of D?   **(b)** What is the value of D_x?
   (c) What is the value of D_y?   **(d)** Find x and y using Cramer's rule.

48. Suppose you are asked to solve this system using the inverse matrix method. **(a)** What is A? **(b)** What is B? **(c)** Explain how you would go about solving the system using these matrices.

49. Cramer's rule has the condition that $D \neq 0$. Why is this necessary? What is true of the system when $D = 0$?

Solve each of the following systems by Cramer's rule. Identify any systems with dependent equations or any inconsistent systems.

50. $3x + \ y = -1$
   $5x + 4y = 10$

51. $3x + 7y = 2$
   $5x - \ y = -22$

52. $2x - 5y = 8$
   $3x + 4y = 10$

53. $3x + 2y + \ z = 2$
   $4x - \ y + 3z = -16$
   $x + 3y - \ z = 12$

54. $\ 5x - 2y - \ z = 8$
   $-5x + 2y + \ z = -8$
   $\ x - 4y - 2z = 0$

55. $-x + 3y - 4z = 2$
   $2x + 4y + \ z = 3$
   $3x - \ z = 9$

56. Find the equation of the quadratic polynomial $P(x)$ that defines the curve shown in the figures.

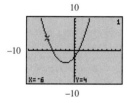

First View

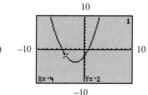

Second View

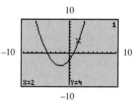

Third View

The Trigonometric Functions and Applications

8.1 ANGLES AND THEIR MEASURES

Terminology and Basic Concepts ▌ Degree Measure ▌ Angles in Standard
Position ▌ Radian Measure ▌ Linear and Angular Velocity

Terminology and Basic Concepts

A line may be drawn through the two distinct points *A* and *B*. This line is called **line
AB.** The portion of the line between *A* and *B*, including points *A* and *B* themselves,
is **segment *AB*.** The portion of line *AB* that starts at *A* and continues through *B*, and
on past *B*, is called **ray *AB*.** Point *A* is the endpoint of the ray. (See Figure 1.)

An **angle** is formed by rotating a ray around its endpoint. The ray in its initial
position is called the **initial side** of the angle, while the ray in its location after the
rotation is the **terminal side** of the angle. The endpoint of the ray is the **vertex** of
the angle. Figure 2 shows the initial and terminal sides of an angle with vertex *A*.

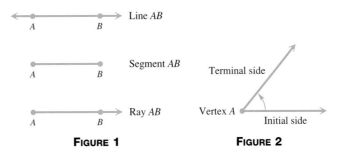

FIGURE 1	**FIGURE 2**

If the rotation of the terminal side is counterclockwise, the angle is **positive.** If the rotation is clockwise, the angle is **negative.** Figure 3 shows two angles, one positive and one negative.

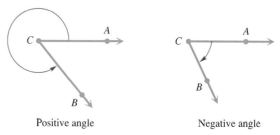

Positive angle   Negative angle

FIGURE 3

An angle can be named by using the name of its vertex. For example, either angle in Figure 3 can be called angle *C*. Alternatively, an angle can be named using three letters, with the vertex letter in the middle. For example, either angle also could be named angle *ACB* or angle *BCA*.

Degree Measure

There are two systems in common use for measuring the size of angles. The most common unit of measure is the **degree.** Degree measure was developed by the Babylonians, four thousand years ago. To use degree measure, we assign 360 degrees to a complete rotation of a ray. In Figure 4, notice that the terminal side of the angle corresponds to its initial side when it makes a complete rotation.

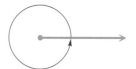

A complete rotation of a ray gives an angle whose measure is 360°.

FIGURE 4

One degree, written 1°, represents $\frac{1}{360}$ of a rotation. Therefore, 90° represents $\frac{90}{360} = \frac{1}{4}$ of a complete rotation, and 180° represents $\frac{180}{360} = \frac{1}{2}$ of a complete rotation. Angles of measure 90° and 180° are shown in Figure 5.

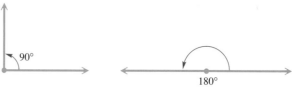

90°   180°

FIGURE 5

Angles are named as shown in the following chart.

TYPES OF ANGLES

Name	Angle Measure	Example
Acute angle	Between 0° and 90°	60° 82°
Right angle	Exactly 90°	90°
Obtuse angle	Between 90° and 180°	97° 138°
Straight angle	Exactly 180°	180°

If the sum of the measures of two angles is 90°, the angles are called **complementary angles,** or **complements.** Two angles that have a sum of 180° are called **supplementary angles,** or **supplements.**

Consider the angle shown in Figure 6. The angle is formed by the rotation of a ray about its endpoint A. The measure of the rotation is 35°. We will often use notation such as "$A = 35°$" to indicate that the measure of the angle A is 35°.

FIGURE 6

Traditionally, portions of a degree have been measured with minutes and seconds. One **minute,** written $1'$, is $\frac{1}{60}$ of a degree.

$$1' = \frac{1°}{60} \qquad \text{or} \qquad 60' = 1°$$

One **second,** $1''$, is $\frac{1}{60}$ of a minute.

$$1'' = \frac{1'}{60} = \frac{1°}{3600} \qquad \text{or} \qquad 60'' = 1' \qquad \text{or} \qquad 3600'' = 1°$$

The measure 12° 42′ 38″ represents 12 degrees, 42 minutes, 38 seconds.

The next example shows how to perform calculations with degrees, minutes, and seconds.

EXAMPLE 1

Calculating with Degrees, Minutes, and Seconds

Perform each calculation.

(a) $51° 29' + 32° 46'$

SOLUTION Add the degrees and the minutes separately.

$$51° 29' + 32° 46' = (51° + 32°) + (29' + 46') = 83° 75'$$

Since $75' = 60' + 15' = 1° 15'$, the sum is written

$$83° 75' = 83° + (1° 15') = 84° 15'.$$

(b) $90° - 73° 12'$

SOLUTION Write $90°$ as $89° 60'$. Then

$$90° - 73° 12' = 89° 60' - 73° 12' = 16° 48'.$$

TECHNOLOGICAL NOTE
If your calculator does not have the capability to convert between decimal degrees and degrees, minutes, seconds, you may wish to write a program that will perform these conversions for you.

NOTE Many modern graphics calculators have the capability to add and subtract angles given in degrees, minutes, and seconds. Likewise, they are also able to convert from degrees, minutes, seconds to *decimal* degrees (see the explanation immediately following) and vice versa. Read your owner's manual for details on these capabilities.

Because calculators are an integral part of our world today, it is now common to measure angles in **decimal degrees.** For example, $12.4238°$ represents

$$12.4238° = 12\frac{4238°}{10,000}.$$

The next example shows how to change between decimal degrees and degrees, minutes, and seconds.

EXAMPLE 2

Converting Between Decimal Degrees and Degrees, Minutes, Seconds

(a) Convert $74° 8' 14''$ to decimal degrees. Round to the nearest thousandth of a degree.

SOLUTION Since $1' = \frac{1°}{60}$ and $1'' = \frac{1°}{3600}$,

$$74° 8' 14'' = 74° + \frac{8°}{60} + \frac{14°}{3600}$$
$$= 74° + .1333° + .0039°$$
$$= 74.137° \text{ (rounded)}.$$

(b) Convert $34.817°$ to degrees, minutes, and seconds.

SOLUTION

$$34.817° = 34° + .817°$$
$$= 34° + (.817)(60') \quad \text{1 degree} = \text{60 minutes}$$
$$= 34° + 49.02'$$
$$= 34° + 49' + .02'$$
$$= 34° + 49' + (.02)(60'') \quad \text{1 minute} = \text{60 seconds}$$
$$= 34° + 49' + 1'' \text{ (rounded)}$$
$$= 34° 49' 1''$$

Angles in Standard Position

An angle is in **standard position** if its vertex is at the origin and its initial side is along the positive *x*-axis. The two angles in Figure 7 are in standard position. An angle in standard position is said to lie in the quadrant in which its terminal side lies. For example, an acute angle is in quadrant I and an obtuse angle is in quadrant II. Angles in standard position having their terminal sides along the *x*-axis or *y*-axis, such as angles with measures 90°, 180°, 270°, and so on, are called **quadrantal angles.**

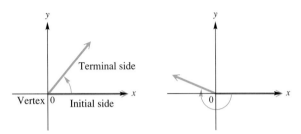

FIGURE 7

A complete rotation of a ray results in an angle of measure 360°. But there is no reason why the rotation need stop at 360°. By continuing the rotation, angles of measure larger than 360° can be produced. The angles in Figure 8(a) have measures 60° and 420°. These two angles have the same initial side and the same terminal side, but different amounts of rotation. Angles that have the same initial side and the same terminal side are called **coterminal angles.** As shown in Figure 8(b), angles with measures 110° and 830° are coterminal.

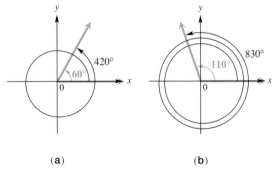

(a) (b)

FIGURE 8

EXAMPLE **3**

Finding Measures of Coterminal Angles

Find the angles of smallest possible positive measure coterminal with the following angles.

(a) 908°

SOLUTION Add or subtract 360° as many times as needed to get an angle with measure greater than 0° but less than 360°. Since 908° − 2 · 360° = 908° − 720° = 188°, an angle of 188° is coterminal with an angle of 908°. See Figure 9.

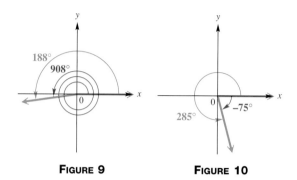

FIGURE 9   **FIGURE 10**

(b) $-75°$

SOLUTION   Use a rotation of $360° + (-75°) = 285°$. See Figure 10.   ◻

Sometimes it is necessary to find an expression that will generate all angles coterminal with a given angle. For example, suppose that we wish to do this for a $60°$ angle. Since any angle coterminal with $60°$ can be obtained by adding an appropriate integer multiple of $360°$ to $60°$, we can let n represent any integer, and the expression

$$60° + n \cdot 360°$$

will represent all such coterminal angles. The table below shows a few of these.

Value of n	Angle Coterminal with $60°$
2	$60° + 2 \cdot 360° = \mathbf{780°}$
1	$60° + 1 \cdot 360° = \mathbf{420°}$
0	$60° + 0 \cdot 360° = \mathbf{60°}$ (the angle itself)
-1	$60° + (-1) \cdot 360° = \mathbf{-300°}$

Radian Measure

Degree measure is used primarily in applications of trigonometry. For more theoretical work in mathematics, angles are measured using a more natural system of measurement: radians. To see how an angle is measured in radians, consider Figure 11. The angle θ (θ is the Greek letter *theta*) is in standard position and a circle of radius r is shown with its center at the origin. Angle θ is called a **central angle,** because its vertex is at the center of the circle. As shown in the figure, angle θ intercepts an arc of length s on the circle. If $s = r$, then θ is said to have a measure of one radian. In general, radian measure is defined as follows.

RADIAN MEASURE

Suppose that a circle has radius $r > 0$. Let θ be a central angle of the circle. If θ intercepts an arc of length s on the circle, then the radian measure of θ is given by the formula

$$\theta = \frac{s}{r}.$$

Radian measure can be thought of as a "pure number" (since the units for r and s "cancel") and consequently no symbol for radian measure is needed (as opposed to degree measure, where $°$ is needed). Therefore, if no unit of measure is indicated for an angle, radian measure is understood.

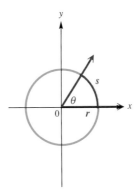

When $r = s$, θ measures
1 radian.

FIGURE 11

The circumference of a circle, the distance around the circle, is given by $C = 2\pi r$, where r is the radius of the circle. The formula $C = 2\pi r$ shows that the radius can be laid off 2π times around a circle. Therefore, an angle of 360°, which corresponds to a complete circle, intercepts an arc equal in length to 2π times the radius of the circle. Because of this, an angle of 360° has a measure of 2π radians:

$$360° = 2\pi \text{ radians.}$$

An angle of 180° is half the size of an angle of 360°, so an angle of 180° has half the radian measure of an angle of 360°.

$$180° = \frac{1}{2}(2\pi) \text{ radians}$$

DEGREE / RADIAN RELATIONSHIP

$$180° = \pi \text{ radians}$$

Dividing both sides of $180° = \pi$ radians by π leads to

$$1 \text{ radian} = \frac{180°}{\pi},$$

or approximately,

$$1 \text{ radian} \approx \frac{180°}{3.1415927} \approx 57.295779° \approx 57° \ 17' \ 45''.$$

Since $180° = \pi$ radians, dividing both sides by 180 gives

$$1° = \frac{\pi}{180} \text{ radians,}$$

or, approximately,

$$1° \approx \frac{3.1415927}{180} \text{ radians} \approx .01745329 \text{ radians.}$$

Angle measures can be converted back and forth between degrees and radians by using one of several methods described below.

CONVERTING BETWEEN DEGREES AND RADIANS

1. Proportion: $\dfrac{\text{Radian measure}}{\pi} = \dfrac{\text{Degree measure}}{180}$

2. Formulas:

From	To	Multiply by
Radians	Degrees	$\dfrac{180°}{\pi}$
Degrees	Radians	$\dfrac{\pi}{180°}$

3. If a radian measure involves a multiple of π, replace π with $180°$, and simplify in order to convert to degrees.

EXAMPLE 4

Converting Degrees to Radians

Convert each degree measure to radians.

(a) $45°$

SOLUTION By the proportion method,

$$\frac{\text{Radian measure}}{\pi} = \frac{45}{180}.$$

Multiply both sides by π.

$$\text{Radian measure} = \frac{45\pi}{180} = \frac{\pi}{4}$$

To use the formula method, multiply by $\frac{\pi}{180°}$.

$$45° = 45°\left(\frac{\pi}{180°}\right) = \frac{45\pi}{180} = \frac{\pi}{4} \text{ radians}$$

(b) $240°$

SOLUTION Using the formula, $240° = 240°\left(\dfrac{\pi}{180°}\right) = \dfrac{4\pi}{3}$ radians. ◗

EXAMPLE 5

Converting Radians to Degrees

Convert each of the following radian measures to degrees.

(a) $\dfrac{9\pi}{4}$

SOLUTION Using the proportion method.

$$\frac{\frac{9\pi}{4}}{\pi} = \frac{x}{180°}$$

$$\frac{9}{4} = \frac{x}{180°}$$

$$x = \frac{9}{4}(180°) = 405°.$$

(b) $\dfrac{11\pi}{3}$

SOLUTION   Using the formula, we multiply this radian measure by $\frac{180°}{\pi}$ to find the corresponding degree measure.

$$\frac{11\pi}{3} \cdot \frac{180°}{\pi} = \left(\frac{1980\pi}{3\pi}\right)^{\circ} = 660°$$

(c) $\dfrac{-5\pi}{6}$

SOLUTION   Fractional multiples of π often appear as radian measures. To convert to degrees, we may replace π with $180°$ and simplify.

$$\frac{-5\pi}{6} \text{ radians} = \frac{-5(180°)}{6} = -150°$$

(d)   4.2 (Write the result correct to the nearest minute.)

SOLUTION

$$4.2 \text{ radians} = 4.2\left(\frac{180°}{\pi}\right) \qquad \text{Multiply by } \tfrac{180°}{\pi}.$$

$$= \frac{4.2(180°)}{\pi}$$

$$\approx 240.642274°$$

$$= 240° + .642274°$$

$$= 240° + .642274(60')$$

$$\approx 240° + 38' + 32.186''$$

$$\approx 240° \, 39'. \quad \blacksquare$$

TECHNOLOGICAL NOTE
If your calculator does not have the capability to convert between radians and degrees, you may wish to write a program that will perform these conversions for you.

NOTE   Check your owner's manual to see if your graphics calculator has the capability of converting from radians to degrees and vice versa.

Understanding the distinction between degree measure and radian measure is essential. To see why, look at Figure 12, which shows angles measuring $30°$ and 30 radians. These angles are not at all the same.

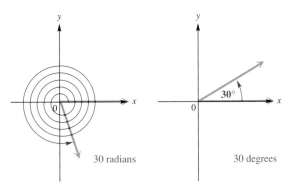

30 radians   30 degrees

FIGURE 12

The relationship $\theta = \frac{s}{r}$ can also be expressed in an alternative form which is useful in finding the length of an arc of a circle.

LENGTH OF ARC

The length s of the arc intercepted on a circle of radius r by a central angle of measure θ radians is given by the product of the radius and the radian measure of the angle, or

$$s = r\theta, \quad \theta \text{ in radians.}$$

This formula is a good example of the usefulness of radian measure. To see why, try to write the equivalent formula for an angle measured in degrees.

CAUTION When applying the formula $s = r\theta$, the value of θ *must be* expressed in *radians*.

EXAMPLE 6

Finding Arc Length Using $s = r\theta$

A circle has a radius of 18.2 centimeters. Find the length of the arc intercepted by a central angle having each of the following measures.

(a) $\frac{3\pi}{8}$ radians

SOLUTION Here $r = 18.2$ cm and $\theta = \frac{3\pi}{8}$. Since $s = r\theta$,

$$s = 18.2\left(\frac{3\pi}{8}\right) \text{ centimeters}$$

$$s = \frac{54.6\pi}{8} \text{ centimeters} \qquad \text{The exact answer}$$

or $\quad s \approx 21.4 \text{ centimeters.} \qquad \text{Calculator approximation}$

(b) $144°$

SOLUTION The formula $s = r\theta$ requires that θ be measured in radians. First, convert θ to radians.

$$144° = 144°\left(\frac{\pi}{180°}\right) \text{ radians} \qquad \text{Multiply by } \frac{\pi}{180°}.$$

$$144° = \frac{4\pi}{5} \text{ radians}$$

Now $\quad s = 18.2\left(\frac{4\pi}{5}\right) \text{ centimeters} \qquad \text{Use } s = r\theta.$

$$s = \frac{72.8\pi}{5} \text{ centimeters,}$$

or $\quad s \approx 45.7 \text{ centimeters.} \quad \blacksquare$

Reno, Nevada, is approximately due north of Los Angeles. The latitude of Reno is 40° N, while that of Los Angeles is 34° N. (The N in 34° N means *north* of the equator.) If the radius of the earth is 6400 kilometers, find the north-south distance between the two cities.

SOLUTION   Latitude gives the measure of a central angle with vertex at the earth's center whose initial side goes through the earth's equator and whose terminal side goes through the given location. As shown in Figure 13 the central angle for Reno and Los Angeles is 6°. The distance between the two cities can thus be found by the formula $s = r\theta$, after 6° is first converted to radians.

$$6° = 6°\left(\frac{\pi}{180°}\right) = \frac{\pi}{30} \text{ radians}$$

The distance between the two cities is

$$s = r\theta$$

$$s = 6400\left(\frac{\pi}{30}\right) \text{ kilometers} \qquad r = 6400, \ \theta = \frac{\pi}{30}$$

$$\approx 670 \text{ kilometers.} \qquad \blacksquare$$

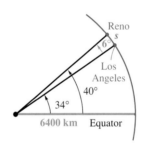

FIGURE 13

Linear and Angular Velocity

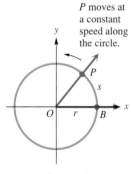

P moves at a constant speed along the circle.

FIGURE 14

Suppose that point *P* moves at a constant speed along a circle of radius *r* and center *O*. See Figure 14. The measure of how fast the position of *P* is changing is called **linear velocity.** If v represents linear velocity, then

$$v = \frac{s}{t},$$

where *s* is the length of the arc traced by point *P* at time *t*. (This formula is just a restatement of the familiar result $d = rt$ with *s* as distance, v as the rate, and *t* as time.)

Look at Figure 14 again. As point *P* moves along the circle, ray *OP* rotates around the origin. Since the ray *OP* is the terminal side of angle *POB*, the measure of the angle changes as *P* moves along the circle. The measure of how fast angle *POB* is changing is called **angular velocity.** Angular velocity, written ω, can be given as

$$\omega = \frac{\theta}{t}, \quad \theta \text{ in radians,}$$

where θ is the measure of angle *POB* at time *t*. As with the earlier formula of this type, θ must be measured in radians, with ω expressed as radians per unit of time. Angular velocity is used in physics and engineering, among other applications.

Earlier we saw that the length s of the arc intercepted on a circle of radius r by a central angle of measure θ radians was found to be $s = r\theta$. Using this formula, the formula for linear velocity, $v = \frac{s}{t}$, becomes

$$v = \frac{r\theta}{t}$$

or $v = r\omega$

This last formula relates linear and angular velocity.

EXAMPLE 8

Using the Linear and Angular Velocity Formulas

Suppose that point P is on a circle with a radius of 10 centimeters, and ray OP is rotating with angular velocity of $\frac{\pi}{18}$ radians per second.

(a) Find the angle generated by P in 6 seconds.

SOLUTION   The velocity of ray OP is $\omega = \frac{\pi}{18}$ radians per second. Since $\omega = \frac{\theta}{t}$, then in 6 seconds

$$\frac{\pi}{18} = \frac{\theta}{6},$$

or $\theta = 6\left(\frac{\pi}{18}\right) = \frac{\pi}{3}$ radians.

(b) Find the distance traveled by P along the circle in 6 seconds.

SOLUTION   In 6 seconds P generates an angle of $\frac{\pi}{3}$ radians. Since $s = r\theta$,

$$s = 10\left(\frac{\pi}{3}\right) = \frac{10\pi}{3} \text{ centimeters.}$$

(c) Find the linear velocity of P.

SOLUTION   Since $v = \frac{s}{t}$, in 6 seconds

$$v = \frac{\dfrac{10\pi}{3}}{6} = \frac{5\pi}{9} \text{ centimeters per second.}   \blacksquare$$

In practical applications, angular velocity is often given as revolutions per unit of time, which must be converted to radians per unit of time before using the formulas given in this section.

EXAMPLE 9

Using the Linear and Angular Velocity Formulas

A belt runs a pulley of radius 6 centimeters at 80 revolutions per minute.

(a) Find the angular velocity of the pulley in radians per second.

SOLUTION   In one minute, the pulley makes 80 revolutions. Each revolution is 2π radians, for a total of

$$80(2\pi) = 160\pi \text{ radians per minute.}$$

Since there are 60 seconds in a minute, ω, the angular velocity in radians per second, is found by dividing 160π by 60.

$$\omega = \frac{160\pi}{60} = \frac{8\pi}{3} \text{ radians per second}$$

(b) Find the linear velocity of the belt in centimeters per second.

SOLUTION The linear velocity of the belt will be the same as that of a point on the circumference of the pulley. Thus,

$$v = r\omega$$

$$v = 6\left(\frac{8\pi}{3}\right)$$

$$v = 16\pi \text{ centimeters per second}$$

$$v \approx 50.3 \text{ centimeters per second.} \quad \blacksquare$$

8.1 EXERCISES

To the Student: Calculator Considerations

   You should read your owner's manual to determine whether your calculator has the capability of converting from degrees, minutes, and seconds to decimal degrees, and vice versa. Also, find out whether it has a function to convert between degree measure and radian measure.

   The real number π is used extensively in trigonometry. Learn where the π key is located. Also, be aware that when exact values involving π are required, such as $\frac{\pi}{3}$ and $\frac{3\pi}{4}$, decimal approximations given by the calculator are not acceptable.

Find (a) the complement and (b) the supplement of each of the following angles.

1. 30° **2.** 60° **3.** 45° **4.** 15°

5. $\dfrac{\pi}{6}$ **6.** $\dfrac{\pi}{3}$ **7.** $\dfrac{\pi}{4}$ **8.** $\dfrac{\pi}{12}$

9. What is the degree measure of the angle that is its own complement?

10. What is the radian measure of the angle that is its own supplement?

Find the angle of smallest positive measure that is coterminal with the given angle.

11. −40° **12.** −98° **13.** 450° **14.** 539°

15. $-\dfrac{\pi}{4}$ **16.** $-\dfrac{\pi}{3}$ **17.** $-\dfrac{3\pi}{2}$ **18.** $-\pi$

Convert each of the following degree measures to radians. Leave answers as multiples of π.

19. 30° **20.** 60° **21.** 45° **22.** 90°

23. 120° **24.** 270° **25.** −315° **26.** −225°

Convert each angle measure to decimal degrees. Use a calculator, and round to the nearest thousandth of a degree.

27. 20° 54′ **28.** 38° 42′ **29.** 91° 35′ 54″

30. 34° 51′ 35″ **31.** −274° 18′ 59″ **32.** −165° 51′ 09″

Convert each angle measure to degrees, minutes, and seconds. Round seconds to whole units. Use a calculator.

33. 31.4296° **34.** 59.0854° **35.** 89.9004°

36. 102.3771° **37.** −178.5994° **38.** −122.6853°

Give an expression that generates all angles coterminal with the given angle. Let n represent any integer. Also, give the quadrant of all such angles.

39. 30° **40.** 45° **41.** 230°

42. 135° **43.** 270° **44.** −90°

Give an expression that generates all angles coterminal with the given angle. Let n represent any integer. Also, give the quadrant of all such angles. (Notice that each angle is given in radians, so you must use 2π to represent one complete revolution. For example, any angle coterminal with $\frac{\pi}{3}$ is represented as $\frac{\pi}{3} + 2n\pi$, where n is an integer.)

45. $\dfrac{\pi}{4}$ **46** $\dfrac{\pi}{6}$

47. $\dfrac{3\pi}{4}$ **48.** $-\dfrac{7\pi}{6}$

Sketch the following angles in standard position. Draw an arrow representing the correct amount of rotation. Find the measure of two other angles, one positive and one negative, that are coterminal with each angle. Give the quadrant of the angle.

49. 75° **50.** 122° **51.** −52°

52. −159° **53.** $\dfrac{5\pi}{3}$ **54.** $-\dfrac{\pi}{2}$

Find the length of the arc intercepted by a central angle θ in a circle of radius r. Give calculator approximations in your answers.

55. $r = 12.3$ cm, $\theta = \dfrac{2\pi}{3}$ **56.** $r = .892$ cm, $\theta = \dfrac{11\pi}{10}$

57. $r = 253$ m, $\theta = \dfrac{2\pi}{5}$ **58.** $r = 120$ mm, $\theta = \dfrac{\pi}{9}$

59. $r = 4.82$ m, $\theta = 60°$ **60.** $r = 71.9$ cm, $\theta = 135°$

Find the distance in kilometers between the pair of cities whose latitudes are given. Assume that the cities are on a north-south line, and that the radius of the earth is 6400 kilometers. Round answers to the nearest hundred.

61. Madison, South Dakota, 44° N, and Dallas, Texas, 33° N

62. Charleston, South Carolina, 33° N, and Toronto, Ontario, 43° N

63. Panama City, Panama, 9° N, and Pittsburgh, Pennsylvania, 40° N

64. Farmersville, California, 36° N, and Penticton, British Columbia, 49° N

65. New York City, New York, 41° N, and Lima, Peru, 12° S

66. Halifax, Nova Scotia, 45° N, and Buenos Aires, Argentina, 34° S

Solve the following problems.

67. A tire is rotating 600 times per minute. Through how many degrees does a point on the edge of the tire move in $\frac{1}{2}$ second?

68. An airplane propeller rotates 1000 times per minute. Find the number of degrees that a point on the edge of the propeller will rotate in 1 second.

69. A pulley rotates through 75° in one minute. How many rotations does the pulley make in an hour?

70. (a) How many inches will the weight in the figure rise if the pulley is rotated through an angle of 71° 50′?
(b) Through what angle, to the nearest minute, must the pulley be rotated to raise the weight 6 in.?

9.27 in

71. Find the radius of the pulley in the figure if a rotation of 51.6° raises the weight 11.4 cm.

r

72. The figure shows the chain drive of a bicycle. How far will the bicycle move if the pedals are rotated through 180°? Assume that the radius of the bicycle wheel is 13.6 in.

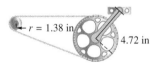

r = 1.38 in

4.72 in

73. The speedometer of a small pickup truck is designed to be accurate with tires of radius 14 in.
(a) Find the number of rotations of a tire in 1 hr if the truck is driven at 55 mph.
(b) Suppose that oversize tires of radius 16 in. are placed on the truck. If the truck is now driven for 1 hr with the speedometer reading 55 mph, how far has the truck gone? If the speed limit is 55 mph, does the driver deserve a speeding ticket?

74. A railroad track is laid along the arc of a circle of radius 1800 ft. The circular part of the track subtends a central angle of 40°. How long (in seconds) will it take a point on the front of a train traveling 30 mph to go around this portion of the track?

75. Two pulleys of diameter 4 m and 2 m, respectively, are connected by a belt. The larger pulley rotates 80 times per min. Find the speed of the belt in meters per second and the angular velocity of the smaller pulley.

76. The earth revolves on its axis once every 24 hr. Assuming that the earth's radius is 6400 km, find the following.
(a) Angular velocity of the earth in radians per day and radians per hr
(b) Linear velocity at the North Pole or South Pole
(c) Linear velocity at Quito, Equador, a city on the equator

77. The earth travels about the sun in an orbit that is almost circular. Assume that the orbit is a circle, with a radius of 93,000,000 mi. (See the figure.)
(a) Assume that a year is 365 days, and find θ, the angle formed by the earth's movement in one day.
(b) Give the angular velocity in radians per hour.
(c) Find the linear velocity of the earth in miles per hour.

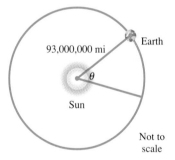

93,000,000 mi

Earth

θ

Sun

Not to scale

78. The pulley shown has a radius of 12.96 cm. Suppose that it takes 18 sec for 56 cm of belt to go around the pulley. Find the angular velocity of the pulley in radians per second.

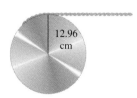

12.96 cm

79. The two pulleys in the figure have radii of 15 cm and 8 cm, respectively. The larger pulley rotates 25 times in 36 sec. Find the angular velocity of each pulley in radians per sec.

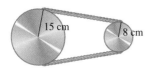

80. A gear is driven by a chain that travels 1.46 m per sec. Find the radius of the gear if it makes 46 revolutions per min.

81. A thread is being pulled off a spool at the rate of 59.4 cm per sec. Find the radius of the spool if it makes 152 revolutions per min.

A ***sector of a circle*** *is the portion of the interior of a circle intercepted by a central angle. (See the figure.) It can be shown that the area of a sector of a circle with radius r and central angle θ, where θ is in radians, is given by the formula*

$$A = \frac{1}{2} r^2 \theta.$$

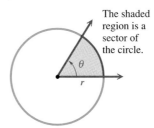

The shaded region is a sector of the circle.

Use this formula to solve the problems in Exercises 82–85.

82. Find the area of a sector of a circle having radius 29.2 m and central angle $\theta = \frac{5\pi}{6}$.

83. Find the area of a sector of a circle having radius 12.7 cm and central angle $\theta = 81°$.

84. The figure shows Medicine Wheel, an Indian structure in northern Wyoming. This circular structure is perhaps 200 years old. There are 32 spokes in the wheel, all equally spaced.

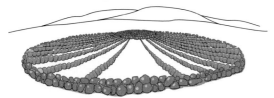

(a) Find the measure of each central angle in degrees and in radians.
(b) If the radius of the wheel is 76 ft, find the circumference.
(c) Find the length of each arc intercepted by consecutive pairs of spokes.
(d) Find the area of each sector formed by consecutive spokes.

85. The unusual corral in the figure is separated into 26 areas, many of which approximate sectors of a circle. Assume that the corral has a diameter of 50 m.

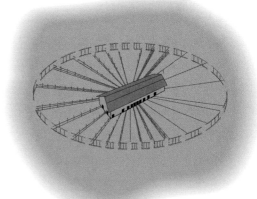

(a) Find the central angle for each region, assuming that the 26 regions are all equal sectors, with the fences meeting at the center.
(b) What is the area of each sector?

86. Eratosthenes (*ca.* 230 B.C.) made a famous measurement of the earth. He observed at Syene (the modern Aswan) at noon and at the summer solstice that a vertical stick had no shadow, while at Alexandria (on the same meridian as Syene) the sun's rays were inclined $\frac{1}{50}$ of a complete circle to the vertical. See the figure. He then calculated the circumference of the earth from the known distance of 5000 stades between Alexandria and Syene. Obtain Eratosthenes' result of 250,000 stades for the circumference of the earth. There is reason to suppose that a stade is about equal to 516.7 ft. Assuming this, use Eratosthenes' result to calculate the polar diameter of the earth in miles. (The actual polar diameter of the earth, to the nearest mile, is 7900 mi.)*

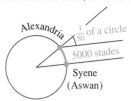

Further Explorations

You can use the TABLE feature of your graphics calculator to work problems like Exercises 61–66. In the TABLE in the figure, the *x* values are decimal approximations of increments (ΔTbl) of $\frac{\pi}{180}$.

Recall: to convert degrees to radians, multiply the degree measure by $\frac{\pi}{180}$.

The expression in Y_1 gives the degree measure of the radian measure listed in X. This means that ΔTbl is equivalent to increments of one degree. The expression in Y_2 uses the formula for distance between points on a circle ($s = r\theta$) to find the north-south distance in kilometers between the given measure of latitude and the equator. X must be in radians for the formula $s = r\theta$ to work.

X	Y₁	Y₂
0	0	0
.01745	1	111.7
.03491	2	223.4
.05236	3	335.1
.06981	4	446.8
.08727	5	558.51
.10472	6	670.21

X=0

EXAMPLE   Find the vertical distance between 2° and 5°.

SOLUTION   From the TABLE in the figure, we see that moving 5° north (or south) from the equator is equivalent to traveling 558.51 kilometers, and 2° north (or south) is equivalent to traveling 223.4 kilometers. Then the north-south distance from 2° to 5° is the difference between these two measurements.

$$558.51 \text{ km} - 223.4 \text{ km} \approx 335.1 \text{ km}$$

This can also be confirmed in the TABLE by noting that 5° − 2° = 3° and the north-south distance along the earth's surface, equivalent to 3°, is 335.1 km.

1. If your graphics calculator has a TABLE feature, find expressions for Y_1 and Y_2 that will duplicate the TABLE in the figure.

2. Support your answers to Exercises 61–66 numerically, using the TABLE feature and the expressions in Exercise 1.

---

* Mathematical exercise and figure from *A Survey of Geometry,* Vol. 1 by Howard Eves. Reprinted by permission of the author.

8.2 THE TRIGONOMETRIC FUNCTIONS AND IDENTITIES

Trigonometric Functions of an Angle ▌ The Reciprocal Identities ▌ The Pythagorean and Quotient Identities

Trigonometric Functions of an Angle

The study of trigonometry covers the six trigonometric functions defined in this section. To define these six basic functions, we start with an angle θ in standard position. Choose any point P having coordinates (x, y) on the terminal side of angle θ. (The point P must not be the vertex of the angle.) See Figure 15.

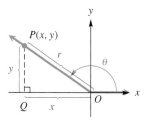

FIGURE 15

A perpendicular line from point P to the x-axis at point Q determines a triangle having vertices at O, P, and Q. The distance r from $P(x, y)$ to the origin, $(0, 0)$, can be found from the distance formula.

$$r = \sqrt{(x - 0)^2 + (y - 0)^2}$$
$$r = \sqrt{x^2 + y^2}$$

Notice that $r > 0$, since distance is never negative.

The six trigonometric functions of angle θ, called **sine, cosine, tangent, cotangent, secant,** and **cosecant,** are defined in terms of x, y, and r. We use the standard abbreviations in these definitions.

TRIGONOMETRIC FUNCTIONS

Let (x, y) be a point other than the origin on the terminal side of an angle θ in standard position. The distance from the point to the origin is $r = \sqrt{x^2 + y^2}$. The six trigonometric functions of θ are:

$$\sin \theta = \frac{y}{r} \qquad\qquad \csc \theta = \frac{r}{y} \quad (y \neq 0)$$

$$\cos \theta = \frac{x}{r} \qquad\qquad \sec \theta = \frac{r}{x} \quad (x \neq 0)$$

$$\tan \theta = \frac{y}{x} \quad (x \neq 0) \qquad \cot \theta = \frac{x}{y} \quad (y \neq 0).$$

NOTE Because of the restrictions on the denominators in the definitions of tangent, cotangent, secant, and cosecant, some angles will have undefined function values. This will be discussed in more detail later.

EXAMPLE 1

Finding the Function Values of an Angle

The terminal side of an angle α in standard position goes through the point (8, 15). Find the values of the six trigonometric functions of angle α.

FIGURE 16

SOLUTION Figure 16 shows angle α and the triangle formed by dropping a perpendicular line from the point (8, 15) to the x-axis. The point (8, 15) is 8 units to the right of the y-axis and 15 units above the x-axis, so that $x = 8$ and $y = 15$. Since $r = \sqrt{x^2 + y^2}$,

$$\begin{aligned} r &= \sqrt{8^2 + 15^2} \\ &= \sqrt{64 + 225} \\ &= \sqrt{289} \\ &= 17. \end{aligned}$$

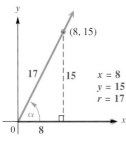

The values of the six trigonometric functions of angle α can now be found with the definitions given above.

$$\sin \alpha = \frac{y}{r} = \frac{15}{17} \qquad \csc \alpha = \frac{r}{y} = \frac{17}{15}$$

$$\cos \alpha = \frac{x}{r} = \frac{8}{17} \qquad \sec \alpha = \frac{r}{x} = \frac{17}{8}$$

$$\tan \alpha = \frac{y}{x} = \frac{15}{8} \qquad \cot \alpha = \frac{x}{y} = \frac{8}{15}$$

EXAMPLE 2

Finding the Function Values of an Angle

The terminal side of angle β in standard position goes through $(-3, -4)$. Find the values of the six trigonometric functions of β.

FIGURE 17

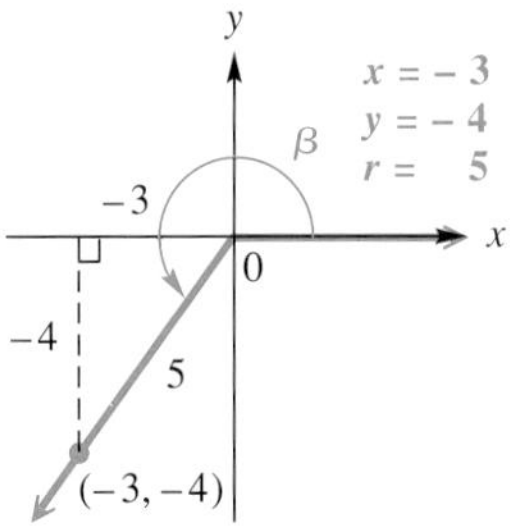

SOLUTION As shown in Figure 17, $x = -3$ and $y = -4$. The value of r is

$$\begin{aligned} r &= \sqrt{(-3)^2 + (-4)^2} \\ r &= \sqrt{25} \\ r &= 5. \end{aligned}$$

(Remember that $r > 0$.) Then by the definitions of the trigonometric functions,

$$\sin \beta = \frac{-4}{5} = -\frac{4}{5} \qquad \csc \beta = \frac{5}{-4} = -\frac{5}{4}$$

$$\cos \beta = \frac{-3}{5} = -\frac{3}{5} \qquad \sec \beta = \frac{5}{-3} = -\frac{5}{3}$$

$$\tan \beta = \frac{-4}{-3} = \frac{4}{3} \qquad \cot \beta = \frac{-3}{-4} = \frac{3}{4}.$$

Using concepts involving similar triangles, it can be shown that the trigonometric functions of an angle can be found by using *any* point on the terminal side of the angle.

If the equation of the ray forming the terminal side of an angle in standard position is known, the trigonometric function values can be found as shown in the next example.

EXAMPLE **3**
Finding the Function Values of an Angle

Find the six trigonometric function values of the angle θ in standard position, if the terminal side of θ is defined by $x + 2y = 0$, $x \geq 0$.

SOLUTION   The angle is shown in Figure 18. We can use *any* point except $(0, 0)$ on the terminal side of θ to find the trigonometric function values, so if we let $x = 2$, for example, we can find the corresponding value of y.

$$x + 2y = 0, x \geq 0$$
$$2 + 2y = 0 \qquad \text{Arbitrarily choose } x = 2.$$
$$2y = -2$$
$$y = -1$$

The point $(2, -1)$ lies on the terminal side, and the corresponding value of r is $r = \sqrt{2^2 + (-1)^2} = \sqrt{5}$. Now use the definitions of the trigonometric functions.

$$\sin \theta = \frac{y}{r} = \frac{-1}{\sqrt{5}} = -\frac{\sqrt{5}}{5} \qquad \csc \theta = \frac{r}{y} = \frac{\sqrt{5}}{-1} = -\sqrt{5}$$

$$\cos \theta = \frac{x}{r} = \frac{2}{\sqrt{5}} = \frac{2\sqrt{5}}{5} \qquad \sec \theta = \frac{r}{x} = \frac{\sqrt{5}}{2}$$

$$\tan \theta = \frac{y}{x} = \frac{-1}{2} = -\frac{1}{2} \qquad \cot \theta = \frac{x}{y} = \frac{2}{-1} = -2 \qquad \blacksquare$$

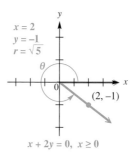

$$x + 2y = 0, \ x \geq 0$$

FIGURE 18

FOR **GROUP DISCUSSION**

Either individually or with a group of students, rework Example 3 using a different value for x. Find the corresponding y-value, and then show that the six trigonometric function values you obtain are the same as the ones shown above.

In the definition of the trigonometric functions, r is the distance from the origin to the point (x, y). Distance is never negative, so $r > 0$. If we choose a point (x, y) in quadrant I, then both x and y will be positive. Since $r > 0$, all six of the fractions used in the definitions of the trigonometric functions will be positive, so that the values of all six functions will be positive in quadrant I.

A point (x, y) in quadrant II has $x < 0$ and $y > 0$. This makes the values of sine and cosecant positive for quadrant II angles, while the other four functions take on negative values. Similar results can be obtained for the other quadrants, as summarized in the following chart.

SIGNS OF FUNCTION VALUES

θ in quadrant	$\sin \theta$	$\cos \theta$	$\tan \theta$	$\cot \theta$	$\sec \theta$	$\csc \theta$
I	+	+	+	+	+	+
II	+	−	−	−	−	+
III	−	−	+	+	−	−
IV	−	+	−	−	+	−

<table>
<tr><td>$x < 0$
$y > 0$
$r > 0$
II
Sine and cosecant
positive</td><td>$x > 0$
$y > 0$
$r > 0$
I
All functions
positive</td></tr>
<tr><td>$x < 0$
$y < 0$
$r > 0$
III
Tangent and cotangent
positive</td><td>$x > 0$
$y < 0$
$r > 0$
IV
Cosine and secant
positive</td></tr>
</table>

EXAMPLE 4

Identifying the Quadrant of an Angle

Identify the quadrant (or quadrants) for any angle θ that satisfies $\sin \theta > 0$, $\tan \theta < 0$.

SOLUTION   Since $\sin \theta > 0$ in quadrants I and II, while $\tan \theta < 0$ in quadrants II and IV, both conditions are met only in quadrant II.   ∎

Recall that a quadrantal angle has its terminal side coinciding with an axis. If the terminal side of an angle in standard position lies along the y-axis, any point on this terminal side has x-coordinate 0. Similarly, any angle with terminal side on the x-axis has y-coordinate 0 for any point on the terminal side. Since the values of x and y appear in the denominators of some of the trigonometric functions, and since a fraction is undefined if its denominator is 0, some of the trigonometric function values of quadrantal angles will be undefined.

EXAMPLE 5

Finding Trigonometric Function Values of a Quadrantal Angle

Find the values of the six trigonometric functions for an angle of 90°.

SOLUTION   First, select any point on the terminal side of a 90° angle. Let us select the point $(0, 1)$, as shown in Figure 19. Here $x = 0$ and $y = 1$. Verify that $r = 1$. Then, by the definition of the trigonometric functions,

$$\sin 90° = \frac{1}{1} = 1 \qquad\qquad \csc 90° = \frac{1}{1} = 1$$

$$\cos 90° = \frac{0}{1} = 0 \qquad\qquad \sec 90° = \frac{1}{0} \text{(undefined)}$$

$$\tan 90° = \frac{1}{0} \text{(undefined)} \qquad \cot 90° = \frac{0}{1} = 0. \quad ∎$$

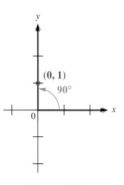

FIGURE 19

In Example 5 we chose the point on the terminal side of the angle that is 1 unit from the origin. However, we could have chosen any point on the terminal side other than the origin itself. We will see later that the idea of choosing the point 1 unit from the origin leads us to the "unit circle" concept, an important tool in the approach to *circular functions* (Chapter 9).

FOR **GROUP DISCUSSION**

Refer to Example 5 and discuss how you would go about finding the trigonometric function values of an angle of $\frac{\pi}{2}$ radians.

The conditions under which the trigonometric function values of quadrantal angles are undefined are summarized here.

If the terminal side of a quadrantal angle lies along the y-axis, the tangent and secant functions are undefined. If it lies along the x-axis, the cotangent and cosecant functions are undefined.

Since the most commonly used quadrantal angles are

$$0° = 0 \text{ radians,}$$

$$90° = \frac{\pi}{2} \text{ radians,}$$

$$180° = \pi \text{ radians,}$$

$$270° = \frac{3\pi}{2} \text{ radians,}$$

$$360° = 2\pi \text{ radians,}$$

the values of the functions of these angles are summarized in the following table. This table is for reference only; you should be able to reproduce it quickly.

QUADRANTAL ANGLE FUNCTION VALUES

θ	$\sin \theta$	$\cos \theta$	$\tan \theta$	$\cot \theta$	$\sec \theta$	$\csc \theta$
$0° = 0$	0	1	0	Undefined	1	Undefined
$90° = \frac{\pi}{2}$	1	0	Undefined	0	Undefined	1
$180° = \pi$	0	-1	0	Undefined	-1	Undefined
$270° = \frac{3\pi}{2}$	-1	0	Undefined	0	Undefined	-1
$360° = 2\pi$	0	1	0	Undefined	1	Undefined

FOR **GROUP DISCUSSION**

1. Learn how to put your calculator in the *degree mode*. Then verify the sine, cosine, and tangent values for the quadrantal angles shown above.
2. Learn how to put your calculator in the *radian mode*. Then repeat item 1.

The Reciprocal Identities

The definitions of the trigonometric functions considered earlier in this section were written so that functions directly across from one another are reciprocals. Since $\sin \theta = \frac{y}{r}$ and $\csc \theta = \frac{r}{y}$,

$$\sin \theta = \frac{1}{\csc \theta} \quad \text{and} \quad \csc \theta = \frac{1}{\sin \theta}.$$

Also, $\cos \theta$ and $\sec \theta$ are reciprocals, as are $\tan \theta$ and $\cot \theta$. In summary, we have the **reciprocal identities** that hold for any angle θ that does not lead to a zero denominator.

RECIPROCAL IDENTITIES

$$\sin \theta = \frac{1}{\csc \theta} \qquad \csc \theta = \frac{1}{\sin \theta}$$

$$\cos \theta = \frac{1}{\sec \theta} \qquad \sec \theta = \frac{1}{\cos \theta}$$

$$\tan \theta = \frac{1}{\cot \theta} \qquad \cot \theta = \frac{1}{\tan \theta}$$

NOTE   When studying identities, be aware that various forms exist. For example,

$$\sin \theta = \frac{1}{\csc \theta}$$

can also be written

$$\csc \theta = \frac{1}{\sin \theta} \quad \text{and} \quad (\sin \theta)(\csc \theta) = 1.$$

You should become familiar with all forms of these identities.

EXAMPLE 6

Using the Reciprocal Identities

Find each function value.

(a) $\cos \theta$, if $\sec \theta = \frac{5}{3}$

SOLUTION   Since $\cos \theta = \frac{1}{\sec \theta}$,

$$\cos \theta = \frac{1}{\frac{5}{3}} = \frac{3}{5}.$$

(b) $\sin \theta$, if $\csc \theta = -\frac{\sqrt{12}}{2}$

SOLUTION

$$\sin \theta = \frac{1}{-\frac{\sqrt{12}}{2}}$$

$$= \frac{-2}{\sqrt{12}}$$

$$= \frac{-2}{2\sqrt{3}} \qquad \sqrt{12} = \sqrt{4 \cdot 3} = 2\sqrt{3}$$

$$= \frac{-1}{\sqrt{3}}$$

$$= \frac{-\sqrt{3}}{3} \qquad \text{Multiply by } \frac{\sqrt{3}}{\sqrt{3}} \text{ to rationalize the denominator.} \quad \blacksquare$$

The reciprocal identities are necessary in evaluating the secant, cosecant, and cotangent function values of an angle on a calculator, because calculators do not have keys specifically marked for these functions. For example, to verify the earlier statement that $\sec 180° = -1$, we would first be sure that the calculator is in degree mode. Then we would enter $\frac{1}{\cos 180°}$ and obtain the result -1.

The Pythagorean and Quotient Identities

As shown in Figure 20, if the point (x, y) lies on the terminal side of θ in standard position, by the Pythagorean theorem,

$$x^2 + y^2 = r^2.$$

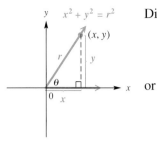

FIGURE 20

Dividing both sides of this equation by r^2 gives

$$\frac{x^2}{r^2} + \frac{y^2}{r^2} = \frac{r^2}{r^2},$$

or

$$\left(\frac{x}{r}\right)^2 + \left(\frac{y}{r}\right)^2 = 1.$$

Since $\sin \theta = \frac{y}{r}$ and $\cos \theta = \frac{x}{r}$, this result becomes

$$(\sin \theta)^2 + (\cos \theta)^2 = 1,$$

or, as it is usually written,

$$\sin^2 \theta + \cos^2 \theta = 1.$$

Starting with $x^2 + y^2 = r^2$ and dividing through by x^2 gives

$$\frac{x^2}{x^2} + \frac{y^2}{x^2} = \frac{r^2}{x^2}$$

$$1 + \left(\frac{y}{x}\right)^2 = \left(\frac{r}{x}\right)^2$$

$$1 + (\tan \theta)^2 = (\sec \theta)^2$$

or

$$\tan^2 \theta + 1 = \sec^2 \theta.$$

On the other hand, dividing through by y^2 leads to

$$1 + \cot^2 \theta = \csc^2 \theta.$$

These three identities are called the **Pythagorean identities** since the original equation that led to them, $x^2 + y^2 = r^2$, comes from the Pythagorean theorem.

PYTHAGOREAN IDENTITIES

$$\sin^2 \theta + \cos^2 \theta = 1 \qquad \tan^2 \theta + 1 = \sec^2 \theta$$
$$1 + \cot^2 \theta = \csc^2 \theta$$

As before, we have given only one form of each identity. However, algebraic transformations can be made to get equivalent identities. For example, by subtracting $\sin^2 \theta$ from both sides of $\sin^2 \theta + \cos^2 \theta = 1$ we get the equivalent identity

$$\cos^2 \theta = 1 - \sin^2 \theta.$$

You should be able to transform these identities quickly, and also recognize their equivalent forms.

Recall that $\sin \theta = \frac{y}{r}$ and $\cos \theta = \frac{x}{r}$. Consider the quotient of $\sin \theta$ and $\cos \theta$, where $\cos \theta \neq 0$.

$$\frac{\sin \theta}{\cos \theta} = \frac{\dfrac{y}{r}}{\dfrac{x}{r}} = \frac{y}{r} \div \frac{x}{r} = \frac{y}{r} \cdot \frac{r}{x} = \frac{y}{x} = \tan \theta$$

Similarly, it can be shown that $\frac{\cos \theta}{\sin \theta} = \cot \theta$, for $\sin \theta \neq 0$. Thus we have two more identities, called the **quotient identities**.

> **QUOTIENT IDENTITIES**
>
> $$\frac{\sin \theta}{\cos \theta} = \tan \theta \qquad \frac{\cos \theta}{\sin \theta} = \cot \theta$$

By using the identities discussed so far, it is possible to find all trigonometric function values of an angle in standard position if we know one function value and the quadrant of the angle. The final example of this section illustrates how this is done.

Find all other trigonometric function values of α, if $\cos \alpha = -\frac{\sqrt{3}}{4}$ and α is in quadrant II.

SOLUTION   One way to begin is to use $\sin^2 \alpha + \cos^2 \alpha = 1$, and replace $\cos \alpha$ with $-\frac{\sqrt{3}}{4}$.

$$\sin^2 \alpha + \left(-\frac{\sqrt{3}}{4}\right)^2 = 1 \qquad \text{Replace } \cos \alpha \text{ with } -\frac{\sqrt{3}}{4}.$$

$$\sin^2 \alpha + \frac{3}{16} = 1$$

$$\sin^2 \alpha = \frac{13}{16} \qquad \text{Subtract } \tfrac{3}{16}.$$

$$\sin \alpha = \pm \frac{\sqrt{13}}{4} \qquad \text{Take square roots.}$$

Since α is in quadrant II, $\sin \alpha > 0$, and

$$\sin \alpha = \frac{\sqrt{13}}{4}.$$

To find $\tan \alpha$, use the quotient identity $\tan \alpha = \dfrac{\sin \alpha}{\cos \alpha}$.

$$\tan \alpha = \frac{\sin \alpha}{\cos \alpha} = \frac{\dfrac{\sqrt{13}}{4}}{\dfrac{-\sqrt{3}}{4}} = \frac{\sqrt{13}}{4} \cdot \frac{4}{-\sqrt{3}} = \frac{\sqrt{13}}{-\sqrt{3}}$$

Rationalize the denominator as follows.

$$\frac{\sqrt{13}}{-\sqrt{3}} = \frac{\sqrt{13}}{-\sqrt{3}} \cdot \frac{\sqrt{3}}{\sqrt{3}}$$

$$= \frac{\sqrt{39}}{-3}$$

$$= -\frac{\sqrt{39}}{3}$$

Therefore, $\tan \alpha = -\dfrac{\sqrt{39}}{3}$.

The reciprocal identities can be used to find the remaining three function values:

$$\sec \alpha = \frac{1}{\cos \alpha} = \frac{1}{-\dfrac{\sqrt{3}}{4}} = -\frac{4}{\sqrt{3}} = \frac{-4\sqrt{3}}{3}$$

$$\csc \alpha = \frac{1}{\sin \alpha} = \frac{1}{\dfrac{\sqrt{13}}{4}} = \frac{4}{\sqrt{13}} = \frac{4\sqrt{13}}{13}$$

$$\cot \alpha = \frac{1}{\tan \alpha} = \frac{1}{-\dfrac{\sqrt{39}}{3}} = -\frac{3}{\sqrt{39}} = \frac{-3\sqrt{39}}{39} = \frac{-\sqrt{39}}{13}. \quad ∎$$

CAUTION When working a problem like the one in Example 7, remember that there are usually several different ways to approach it. Furthermore, one of the most common errors involves making an incorrect sign choice for a square root. Always pay close attention to the quadrant of the angle so that the correct sign choice can be made.

8.2 EXERCISES

To the Student: Calculator Considerations

 Be sure that you know how to use the reciprocal key in conjunction with the sine, cosine, and tangent keys so that cosecant, secant, and cotangent values can be found. Also be aware of how to find powers of trigonometric functions; while we usually write $\sin^2 \theta$, for example, it may be necessary to enter this as $(\sin \theta)^2$ on your calculator.

 Of utmost importance is knowing how to put the calculator in the degree or radian mode. Learn how to do this.

 To test yourself, verify these facts (to be explained in the next section): $\csc 30° = 2$; $\sin^2 30° = .25$; $\cot^2 \left(\frac{\pi}{6}\right) = 3$.

Sketch an angle θ in standard position such that θ has the smallest positive measure, and the given point is on the terminal side of θ.

1. $(-3, 4)$ **2.** $(-4, -3)$ **3.** $(5, -12)$ **4.** $(-12, -5)$

Suppose that θ is in standard position and the given point is on the terminal side of θ. Give the exact values of the six trigonometric functions of θ. If a function is not defined, say so.

5. $(-3, 4)$ **6.** $(-4, -3)$ **7.** $(5, -12)$ **8.** $(-12, -5)$

9. $(-7, 24)$ **10.** $(24, 7)$ **11.** $(0, 2)$ **12.** $(0, -9)$

13. $(-4, 0)$ **14.** $(8, 0)$ **15.** $(\sqrt{5}, -2)$ **16.** $(-\sqrt{7}, \sqrt{2})$

17. $(1, 1)$ **18.** $(-4, -4)$ **19.** $(1, \sqrt{3})$ **20.** $(-2\sqrt{3}, -2)$

21. For any nonquadrantal angle θ, $\sin \theta$ and $\csc \theta$ will have the same sign. Explain why this is so.

22. If $\cot \theta$ is undefined, what is the value of $\tan \theta$?

23. How is the value of r interpreted geometrically in the definitions of the sine, cosine, secant, and cosecant functions?

24. If the terminal side of an angle β is in quadrant III, what is the sign of each of the trigonometric function values of β?

Suppose that the point (x, y) is in the indicated quadrant. Decide whether the given ratio is positive or negative. (Hint: It may be helpful to draw a sketch.)

25. II, $\dfrac{y}{r}$ **26.** II, $\dfrac{x}{r}$ **27.** III, $\dfrac{y}{r}$ **28.** III, $\dfrac{x}{r}$

29. III, $\dfrac{y}{x}$ **30.** III, $\dfrac{x}{y}$ **31.** IV, $\dfrac{x}{r}$ **32.** IV, $\dfrac{y}{r}$

33. IV, $\dfrac{y}{x}$ **34.** IV, $\dfrac{x}{y}$ **35.** III, $\dfrac{r}{x}$ **36.** II, $\dfrac{r}{y}$

In Exercises 37–42, an equation with a restriction on x is given. This is an equation of the terminal side of an angle θ in standard position. Sketch the smallest positive such angle θ, and find the values of the six trigonometric functions of θ.

37. $y = -2x, \quad x \geq 0$ **38.** $y = -\dfrac{3}{5}x, \quad x \geq 0$ **39.** $y = \dfrac{4}{7}x, \quad x \leq 0$

40. $y = -6x, \quad x \leq 0$ **41.** $y = -\dfrac{5}{3}x, \quad x \leq 0$ **42.** $y = \dfrac{6}{5}x, \quad x \geq 0$

Use the appropriate reciprocal identity to find each function value. In Exercises 43–50, give exact values, and in Exercises 51–54, give calculator approximations.

43. $\sin \theta$, if $\csc \theta = 3$ **44.** $\cos \alpha$, if $\sec \alpha = -2.5$ **45.** $\cot \beta$, if $\tan \beta = -\dfrac{1}{5}$

46. $\sin \alpha$, if $\csc \alpha = \sqrt{15}$ **47.** $\csc \alpha$, if $\sin \alpha = \dfrac{\sqrt{2}}{4}$ **48.** $\sec \beta$, if $\cos \beta = -\dfrac{1}{\sqrt{7}}$

49. $\tan \theta$, if $\cot \theta = -\dfrac{\sqrt{5}}{3}$ **50.** $\cot \theta$, if $\tan \theta = \dfrac{\sqrt{11}}{5}$ **51.** $\sin \theta$, if $\csc \theta = 1.42716321$

52. $\cos \alpha$, if $\sec \alpha = 9.80425133$ **53.** $\tan \alpha$, if $\cot \alpha = .43900273$ **54.** $\csc \theta$, if $\sin \theta = -.37690858$

55. Can a given angle γ satisfy both $\sin \gamma > 0$ and $\csc \gamma < 0$? Explain.

56. One form of a particular reciprocal identity is

$$\tan \theta = \frac{1}{\cot \theta}.$$

Give two other equivalent forms of this identity.

57. What is wrong with the following statement? $\tan 90° = \dfrac{1}{\cot 90°}$

58. If an angle θ has undefined cotangent, which other function value of θ is undefined as well?

Identify the quadrant or quadrants for the angles satisfying the following conditions.

59. $\sin \alpha > 0$, $\cos \alpha < 0$ **60.** $\cos \beta > 0$, $\tan \beta > 0$ **61.** $\sec \theta < 0$, $\csc \theta < 0$

62. $\tan \gamma > 0$, $\cot \gamma > 0$ **63.** $\sin \beta < 0$, $\cos \beta > 0$ **64.** $\cos \beta > 0$, $\sin \beta > 0$

65. $\tan \omega < 0$, $\cot \omega < 0$ **66.** $\csc \theta < 0$, $\cos \theta < 0$ **67.** $\sin \alpha > 0$

68. $\cos \beta < 0$ **69.** $\tan \theta > 0$ **70.** $\csc \alpha < 0$

Use identities to find the indicated function value. Use a calculator in Exercises 79–82.

71. $\cos \theta$, if $\sin \theta = \dfrac{2}{3}$, with θ in quadrant II   **72.** $\tan \alpha$, if $\sec \alpha = 3$, with α in quadrant IV

73. $\csc \beta$, if $\cot \beta = -\dfrac{1}{2}$, with β in quadrant IV   **74.** $\sin \alpha$, if $\cos \alpha = -\dfrac{1}{4}$, with α in quadrant II

75. $\sec \theta$, if $\tan \theta = \dfrac{\sqrt{7}}{3}$, with θ in quadrant III

76. $\tan \theta$, if $\cos \theta = \dfrac{1}{3}$, with θ in quadrant IV

77. $\sin \theta$, if $\sec \theta = 2$, with θ in quadrant IV

78. $\cos \beta$, if $\csc \beta = -4$, with β in quadrant III

79. $\cot \alpha$, if $\csc \alpha = -3.5891420$, with α in quadrant III

80. $\sin \beta$, if $\cot \beta = 2.40129813$, with β in quadrant I

81. $\tan \beta$, if $\sin \beta = .49268329$, with β in quadrant II

82. $\csc \alpha$, if $\tan \alpha = .98244655$, with α in quadrant III

*Find all six trigonometric function values for each of the following angles. Use a calcu-
lator in Exercises 91 and 92.*

83. $\cos \alpha = -\dfrac{3}{5}$, with α in quadrant III

84. $\tan \alpha = -\dfrac{15}{8}$, with α in quadrant II

85. $\sin \beta = \dfrac{7}{25}$, with β in quadrant II

86. $\cot \gamma = \dfrac{3}{4}$, with γ in quadrant III

87. $\csc \theta = 2$, with θ in quadrant II

88. $\tan \beta = \sqrt{3}$, with β in quadrant III

89. $\cot \alpha = \dfrac{\sqrt{3}}{8}$, with $\sin \alpha > 0$

90. $\sin \beta = \dfrac{\sqrt{5}}{7}$, with $\tan \beta > 0$

91. $\sin \alpha = .164215$, with α in quadrant II

92. $\cot \theta = -1.49586$, with θ in quadrant IV

*F*urther *Explorations*

The figure shows a TABLE for $Y_1 = \cos \theta$ and $Y_2 = \sin \theta$, with $\Delta Tbl = 5°$. Note:
$\cos \theta$ and $\sin \theta$ must be entered as $\cos x$ and $\sin x$ in the $Y =$ menu of the graphics
calculator.

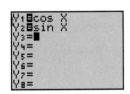
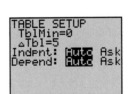

1. Use the TABLE on your graphics calculator and the up and/or down arrows to scroll
   through the TABLE to find the angle(s) between $0°$ and $360°$ in which $|\cos \theta| =
   |\sin \theta|$. (Be sure your calculator is set for DEGREE MODE.)

2. Sketch by hand the terminal sides of each of the angles in Further Explorations Exercise
   1. Find the equation for the lines containing each of these terminal sides.

3. Use the equations from Further Explorations Exercise 2 and the definitions of $\cos \theta$ and
   $\sin \theta$ ($\cos \theta = \frac{x}{r}$ and $\sin \theta = \frac{y}{r}$) to explain why $|\sin \theta| = |\cos \theta|$ for the angles found
   in Further Explorations Exercise 1.

8.3 EVALUATING TRIGONOMETRIC FUNCTIONS

Introduction to Right Triangle Trigonometry ∎ Cofunction Identities ∎
Function Values of Special Angles ∎ Calculator Approximations ∎
Reference Angles

Introduction to Right Triangle Trigonometry

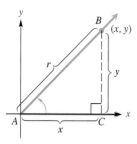

FIGURE 21

Figure 21 shows an acute angle A in standard position. The definitions of the
trigonometric function values of angle A require x, y, and r. As drawn in Figure 21,
x and y are the lengths of the two legs of right triangle ABC, and r is the length of
the hypotenuse.

The side of length y is called the **side opposite** angle A, and the side of length x
is called the **side adjacent** to angle A. The lengths of these sides can be used to
replace x and y in the definition of the trigonometric functions, with r replaced with
the length of the hypotenuse, to get the following right triangle-based definitions.

**RIGHT TRIANGLE-BASED DEFINITIONS OF
TRIGONOMETRIC FUNCTIONS**

For any acute angle A in standard position,

$$\sin A = \frac{y}{r} = \frac{\text{side opposite}}{\text{hypotenuse}} \qquad \csc A = \frac{r}{y} = \frac{\text{hypotenuse}}{\text{side opposite}}$$

$$\cos A = \frac{x}{r} = \frac{\text{side adjacent}}{\text{hypotenuse}} \qquad \sec A = \frac{r}{x} = \frac{\text{hypotenuse}}{\text{side adjacent}}$$

$$\tan A = \frac{y}{x} = \frac{\text{side opposite}}{\text{side adjacent}} \qquad \cot A = \frac{x}{y} = \frac{\text{side adjacent}}{\text{side opposite}}.$$

As seen in the first example, a coordinate system is not essential in applying
these definitions.

EXAMPLE 1

Finding
Trigonometric
Function Values of
an Acute Angle in a
Right Triangle

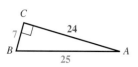

FIGURE 22

Find the values of the trigonometric functions for angles A and B in the right
triangle in Figure 22.

SOLUTION   The length of the side opposite angle A is 7. The length of the side
adjacent to angle A is 24, and the length of the hypotenuse is 25. Using the
relationships given above,

$$\sin A = \frac{\text{side opposite}}{\text{hypotenuse}} = \frac{7}{25} \qquad \csc A = \frac{\text{hypotenuse}}{\text{side opposite}} = \frac{25}{7}$$

$$\cos A = \frac{\text{side adjacent}}{\text{hypotenuse}} = \frac{24}{25} \qquad \sec A = \frac{\text{hypotenuse}}{\text{side adjacent}} = \frac{25}{24}$$

$$\tan A = \frac{\text{side opposite}}{\text{side adjacent}} = \frac{7}{24} \qquad \cot A = \frac{\text{side adjacent}}{\text{side opposite}} = \frac{24}{7}.$$

The length of the side opposite angle B is 24, while the length of the side adjacent
to B is 7, making

$$\sin B = \frac{24}{25} \qquad \tan B = \frac{24}{7} \qquad \sec B = \frac{25}{7}$$

$$\cos B = \frac{7}{25} \qquad \cot B = \frac{7}{24} \qquad \csc B = \frac{25}{24}. \qquad \blacksquare$$

Cofunction Identities

In Example 1, you may have noticed that $\sin A = \cos B$, $\cos A = \sin B$, and so on. Such relationships are always true for the two acute angles of a right triangle. Figure 23 shows a right triangle with acute angles A and B and a right angle at C. (Whenever we use A, B, and C to name the angles in a right triangle, C will be the right angle.) The length of the side opposite angle A is a, and the length of the side opposite angle B is b. The length of the hypotenuse is c.

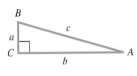

FIGURE 23

By the definitions given above, $\sin A = \frac{a}{c}$. Since $\cos B$ is also equal to $\frac{a}{c}$,

$$\sin A = \frac{a}{c} = \cos B.$$

In a similar manner,

$$\tan A = \frac{a}{b} = \cot B \quad \text{and} \quad \sec A = \frac{c}{b} = \csc B.$$

The sum of the three angles in any triangle is $180°$. Since angle C equals $90°$, angles A and B must have a sum of $180° - 90° = 90°$. As mentioned earlier, angles with a sum of $90°$ are called complementary angles. Since angles A and B are complementary and $\sin A = \cos B$, the functions sine and cosine are called **cofunctions.** Also, tangent and cotangent are cofunctions, as are secant and cosecant. Since angles A and B are complementary, $A + B = 90°$, or

$$B = 90° - A,$$

giving $\qquad \sin A = \cos B = \cos(90° - A).$

Similar results, called the cofunction identities, are true for the other trigonometric functions.

COFUNCTION IDENTITIES

If A is an acute angle measured in degrees,

$$\sin A = \cos(90° - A) \qquad \csc A = \sec(90° - A)$$
$$\cos A = \sin(90° - A) \qquad \sec A = \csc(90° - A)$$
$$\tan A = \cot(90° - A) \qquad \cot A = \tan(90° - A).$$

If A is an acute angle measured in radians,

$$\sin A = \cos\left(\frac{\pi}{2} - A\right) \qquad \csc A = \sec\left(\frac{\pi}{2} - A\right)$$

$$\cos A = \sin\left(\frac{\pi}{2} - A\right) \qquad \sec A = \csc\left(\frac{\pi}{2} - A\right)$$

$$\tan A = \cot\left(\frac{\pi}{2} - A\right) \qquad \cot A = \tan\left(\frac{\pi}{2} - A\right).$$

NOTE These identities actually apply to all angles (not just acute angles). However, for our present discussion we will only need them for acute angles.

EXAMPLE **2**
Writing Functions in Terms of Cofunctions

Write each of the following in terms of cofunctions.

(a) $\cos 52°$

SOLUTION Since $\cos A = \sin(90° - A)$,

$$\cos 52° = \sin(90° - 52°) = \sin 38°.$$

(b) $\tan \dfrac{\pi}{6}$

SOLUTION Since $\tan A = \cot(\frac{\pi}{2} - A)$,

$$\tan \frac{\pi}{6} = \cot\left(\frac{\pi}{2} - \frac{\pi}{6}\right) = \cot \frac{\pi}{3} \qquad ▯$$

FOR GROUP DISCUSSION

1. With your calculator in degree mode, find values for $\cos 52°$ and $\sin 38°$. Refer to Example 2(a), and discuss your results. (The displays are actually approximations for these functions, and we will investigate them further later in this section.)
2. With your calculator in radian mode, find values for $\tan \frac{\pi}{6}$ and $\cot \frac{\pi}{3}$. Refer to Example 2(b), and discuss your results.

Function Values of Special Angles

Certain special angles, such as

$$30° = \frac{\pi}{6}, \quad 45° = \frac{\pi}{4}, \quad \text{and} \quad 60° = \frac{\pi}{3}$$

traditionally occur quite often in the study of trigonometry. We can derive their *exact* trigonometric function values using right triangles, and in so doing, illustrate how to make the distinction between exact values and calculator approximations.

We will use degree measure in our work that follows. To find the exact trigonometric function values for 30° and 60° angles, we start with an equilateral triangle, a triangle with all sides of equal length. Each angle of such a triangle has a measure of 60°. While the results we will obtain are independent of the length, for convenience, we choose the length of each side to be 2 units. See Figure 24(a).

Bisecting one angle of this equilateral triangle leads to two right triangles, each of which has angles of 30°, 60°, and 90°, as shown in Figure 24(b). Since the hypotenuse of one of these right triangles has a length of 2, the shortest side will have a length of 1. (Why?) If x represents the length of the medium side, then, by the Pythagorean theorem,

$$2^2 = 1^2 + x^2$$
$$4 = 1 + x^2$$
$$3 = x^2$$
$$\sqrt{3} = x.$$

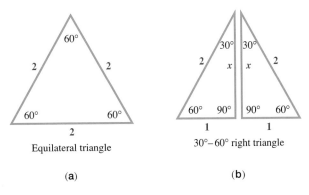

Equilateral triangle

(a)

$30°-60°$ right triangle

(b)

FIGURE 24

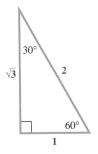

$30°-60°$ right triangle

FIGURE 25

Figure 25 summarizes our results, showing a $30°-60°$ right triangle.

As shown in the figure, the side opposite the $30°$ angle has length 1; that is, for the $30°$ angle,

$$\text{hypotenuse} = 2, \quad \text{side opposite} = 1, \quad \text{side adjacent} = \sqrt{3}.$$

Using the definitions of the trigonometric functions,

$$\sin 30° = \frac{\text{side opposite}}{\text{hypotenuse}} = \frac{1}{2} \qquad \csc 30° = \frac{2}{1} = 2$$

$$\cos 30° = \frac{\text{side adjacent}}{\text{hypotenuse}} = \frac{\sqrt{3}}{2} \qquad \sec 30° = \frac{2}{\sqrt{3}} = \frac{2\sqrt{3}}{3}$$

$$\tan 30° = \frac{\text{side opposite}}{\text{side adjacent}} = \frac{1}{\sqrt{3}} = \frac{\sqrt{3}}{3} \qquad \cot 30° = \frac{\sqrt{3}}{1} = \sqrt{3}.$$

The denominator was rationalized for $\tan 30°$ and $\sec 30°$.

In a similar manner,

$$\sin 60° = \frac{\sqrt{3}}{2} \qquad \tan 60° = \sqrt{3} \qquad \sec 60° = 2$$

$$\cos 60° = \frac{1}{2} \qquad \cot 60° = \frac{\sqrt{3}}{3} \qquad \csc 60° = \frac{2\sqrt{3}}{3}.$$

The values of the trigonometric functions for $45°$ can be found by starting with a $45°-45°$ right triangle, as shown in Figure 26. This triangle is isosceles, and, for convenience, we choose the lengths of the equal sides to be 1 unit. (As before, the results are independent of the length of the equal sides of the right triangle.) Since the shorter sides each have length 1, if r represents the length of the hypotenuse, then

$$1^2 + 1^2 = r^2$$
$$2 = r^2$$
$$\sqrt{2} = r.$$

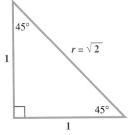

$45°-45°$ right triangle

FIGURE 26

Using the measures indicated on the 45°–45° right triangle in Figure 26, we find

$$\sin 45° = \frac{1}{\sqrt{2}} = \frac{\sqrt{2}}{2} \qquad \tan 45° = \frac{1}{1} = 1 \qquad \sec 45° = \frac{\sqrt{2}}{1} = \sqrt{2}$$

$$\cos 45° = \frac{1}{\sqrt{2}} = \frac{\sqrt{2}}{2} \qquad \cot 45° = \frac{1}{1} = 1 \qquad \csc 45° = \frac{\sqrt{2}}{1} = \sqrt{2}$$

The following chart summarizes the *exact* trigonometric function values of 30°, 45°, and 60° angles and their radian equivalents.

FUNCTION VALUES OF SPECIAL ANGLES

θ	$\sin \theta$	$\cos \theta$	$\tan \theta$	$\cot \theta$	$\sec \theta$	$\csc \theta$
$30° = \dfrac{\pi}{6}$	$\dfrac{1}{2}$	$\dfrac{\sqrt{3}}{2}$	$\dfrac{\sqrt{3}}{3}$	$\sqrt{3}$	$\dfrac{2\sqrt{3}}{3}$	2
$45° = \dfrac{\pi}{4}$	$\dfrac{\sqrt{2}}{2}$	$\dfrac{\sqrt{2}}{2}$	1	1	$\sqrt{2}$	$\sqrt{2}$
$60° = \dfrac{\pi}{3}$	$\dfrac{\sqrt{3}}{2}$	$\dfrac{1}{2}$	$\sqrt{3}$	$\dfrac{\sqrt{3}}{3}$	2	$\dfrac{2\sqrt{3}}{3}$

NOTE It is not difficult to reproduce this chart if you learn the values of sin 30°, cos 30°, and sin 45°. Then the rest of the chart can be completed by using the reciprocal identities, the cofunction identities, and the quotient identities.

Calculator Approximations

While it is interesting to see how exact trigonometric function values can be found for certain special angles, in most applications of trigonometry we rely on calculator approximations for function values.

FOR **GROUP DISCUSSION**

1. With your calculator in degree mode, find the value of sin 30°. Verify that it corresponds to the exact value of sin 30° given in the chart above.
2. With your calculator in radian mode, find the value of tan $\frac{\pi}{4}$. Verify that it corresponds to the exact value of tan $\frac{\pi}{4}$ given in the chart above.
3. With your calculator in degree mode, find an approximation for cos 30°. Then use the computational capabilities to find an approximation for the exact value given for cos 30° in the chart above. Verify that they correspond.

CAUTION One of the most common errors involving calculator use in trigonometry is working in the incorrect angle measure mode. Be sure that your calculator is in the proper mode before calculating a function value.

EXAMPLE 3

Finding Calculator
Approximations of
Trigonometric
Functions

The following approximations of trigonometric function values can be obtained with a calculator. (The number of displayed digits will vary among calculator models.)

(a) $\sin 49°12' \approx .7569950557$

SOLUTION   It may be necessary to convert to decimal degrees before calculating. Be sure that the calculator is in degree mode.

(b) $\cot 3 \approx -7.015252551$

SOLUTION   Calculators usually do not have keys for the secant, cosecant, and cotangent functions. To find this approximation, first put the calculator in radian mode, find tan 3, and then find the reciprocal of the result. Here we are using the identity $\cot \theta = \dfrac{1}{\tan \theta}$.

(c) $\sin(-147°) \approx -.544639035$

SOLUTION   Notice that this function value is negative because $-147°$ is a quadrant III angle, and the sine function is negative in quadrant III.   ◼

TECHNOLOGICAL NOTE
Students often confuse
the symbols for the in-
verse trigonometric func-
tions with the reciprocal
functions. For example,
$\sin^{-1}x$ represents an angle whose sine is x, and
not the reciprocal of sin x
(which is csc x). In order
to find reciprocal function
values, you must use the
function in conjunction
with the reciprocal func-
tion of the calculator.

Sometimes it is necessary to find an angle measure when we know a trigonometric function value of that angle, and information about the quadrant in which the angle lies. A complete discussion of inverse trigonometric functions is necessary to understand *why* a calculator will give an inverse function value in a particular quadrant. We will examine inverse trigonometric functions in more detail in the next chapter. For now, we simply present an introduction to the use of inverse trigonometric functions with calculators. (You should locate these functions on your particular model. They are usually designated $\sin^{-1}$, $\cos^{-1}$, and $\tan^{-1}$.)

EXAMPLE 4

Using Inverse
Trigonometric
Functions to Find
Angles

(a)   Use a calculator to find an angle θ in degrees that satisfies $\sin \theta \approx .9677091705$.

SOLUTION   With the calculator in degree mode, we find that an angle θ having sine value .9677091705 is 75.4°. (While there are infinitely many such angles, the calculator only gives this one.) We write this result as $\sin^{-1} .9677091705 \approx 75.4°$.

(b)   Use a calculator to find an angle θ in radians that satisfies $\tan \theta \approx .25$.

SOLUTION   With the calculator in radian mode, we find $\tan^{-1} .25 \approx .2449786631$.   ◼

Reference Angles

Associated with every non-quadrantal angle in standard position is a positive acute angle called its reference angle. A **reference angle** for an angle θ, written θ', is the positive acute angle made by the terminal side of angle θ and the x-axis. Figure 27 shows several angles θ (each less than one complete counterclockwise revolution) in quadrants II, III, and IV, respectively, with the reference angle θ' also shown. In quadrant I, θ and θ' are the same. If an angle θ is negative or has measure greater than 360°, its reference angle is found by first finding its coterminal angle that is between 0° and 360°, and then using the diagrams in Figure 27.

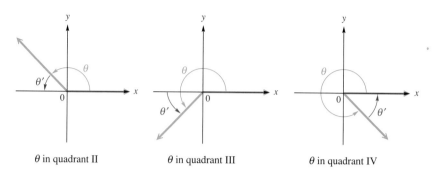

θ in quadrant II   θ in quadrant III   θ in quadrant IV

FIGURE 27

CAUTION   A very common error is to find the reference angle by using the terminal side of θ and *y*-axis. *The reference angle is always found with reference to the x-axis.*

EXAMPLE 5
Finding Reference Angles

Find the reference angles for the following three angles.

(a)   218°

SOLUTION   As shown in Figure 28, the positive acute angle made by the terminal side of this angle and the *x*-axis is 218° − 180° = 38°. For θ = 218°, the reference angle θ′ = 38°.

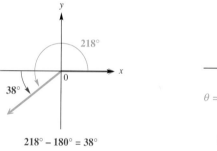

218° − 180° = 38°

FIGURE 28

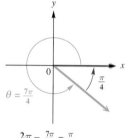

$2\pi - \frac{7\pi}{4} = \frac{\pi}{4}$

FIGURE 29

(b) $\dfrac{7\pi}{4}$

SOLUTION   As shown in Figure 29, the reference angle (the positive acute angle made by the terminal side of this angle and the *x*-axis) is $2\pi - \frac{7\pi}{4} = \frac{\pi}{4}$. Note that the angle $-\frac{\pi}{4}$, which has the same terminal ray as $\frac{7\pi}{4}$, also has a reference angle of $\frac{\pi}{4}$.

(c)   1387°

SOLUTION   First find a coterminal angle between 0° and 360°. Divide 1387° by 360° to get a quotient of about 3.9. Begin by subtracting 360° three times (because of the 3 in 3.9):

$$1387° - 3 \cdot 360° = 307°. \qquad \text{307° is in quadrant IV.}$$

The reference angle for 307° (and thus for 1387°) is 360° − 307° = 53°.   ∎

The preceding example suggests the following table for finding the reference angle θ' for any angle θ between $0°$ and $360°$ or 0 and 2π.

REFERENCE ANGLES FOR θ IN $(0°, 360°)$ OR $(0, 2\pi)$		
θ in Quadrant	**θ' Is**	**Example**
I	θ	
II	$180° - \theta$ $\pi - \theta$	
III	$\theta - 180°$ $\theta - \pi$	
IV	$360° - \theta$ $2\pi - \theta$	

If we wish to find a trigonometric function value of a non-quadrantal angle using reference angles, we use the following guidelines.

> ### FINDING TRIGONOMETRIC FUNCTION VALUES FOR ANY NON-QUADRANTAL ANGLE
>
> 1. If $\theta \geq 360°$, or if $\theta < 0°$, find a coterminal angle by adding or subtracting 360° as many times as needed to get an angle of at least 0° but less than 360°.
> 2. Find the reference angle θ'.
> 3. Find the necessary values of the trigonometric functions for the reference angle θ'.
> 4. Find the correct signs for the values found in Step 3. (Use the table of signs in Section 8.2.) This result gives the value of the trigonometric functions for angle θ.

EXAMPLE 6

Finding
Trigonometric
Function Values
Using Reference
Angles

Use reference angles to find the exact value of each of the following.

(a) $\cos(-240°)$

SOLUTION   The reference angle is 60°, as shown in Figure 30. Since the cosine is negative in quadrant II,

$$\cos(-240°) = -\cos 60° = -\frac{1}{2}.$$

(b) $\tan \frac{7\pi}{6}$

SOLUTION   The reference angle is $\frac{\pi}{6}$, as shown in Figure 31. The tangent is positive in quadrant III, so,

$$\tan \frac{7\pi}{6} = +\tan \frac{\pi}{6} = \frac{\sqrt{3}}{3}.$$

(c) $\csc 675°$

SOLUTION   Begin by subtracting 360° to get a coterminal angle between 0° and 360°.

$$675° - 360° = 315°$$

As shown in Figure 32, the reference angle is $360° - 315° = 45°$. An angle of 315° is in quadrant IV, so the cosecant is negative. Therefore,

$$\csc 675° = -\csc 45° = -\sqrt{2}. \quad \blacksquare$$

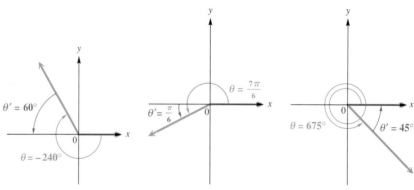

FIGURE 30   **FIGURE 31**   **FIGURE 32**

FOR **GROUP DISCUSSION**

The angles used in Example 6 were all specially chosen so that in each case the reference angle was a special angle. Therefore, we were able to find *exact* values. Calculators will, in some cases, give exact values, but in most applied work we will only need approximations.

1. Use your calculator to find cos(−240°). Compare your answer to the one in Example 6(a). Are these exact or approximate?

2. Use your calculator to find tan $\frac{7\pi}{6}$. Compare your answer to the one in Example 6(b). Are these exact or approximate? (*Hint:* One is exact and one is approximate.)

3. Use your calculator to find csc 675°. Compare your answer to the one in Example 6(c). Are these exact or approximate?

For the exact values that are irrational in these items, show that their decimal approximations as found on your calculator are the same as those obtained when you used the trigonometric functions.

The ideas discussed in this section can be reversed to find the measures of certain angles, given a trigonometric function value and an interval in which the angle must lie. We are most often interested in the interval [0°, 360°) or [0, 2π).

EXAMPLE 7

Finding Angle Measures Given an Interval and a Function Value (Degree Measure)

Find all values of θ, if θ is in the interval [0°, 360°) and cos $\theta = \frac{-\sqrt{2}}{2}$.

SOLUTION Since cosine here is negative, θ must lie in either quadrant II or III. Since the absolute value of cos θ is $\frac{\sqrt{2}}{2}$, the reference angle θ' must be 45°. The two possible angles θ are sketched in Figure 33.

The quadrant II angle θ must equal 180° − 45° = 135°, and the quadrant III angle θ must equal 180° + 45° = 225°. ∎

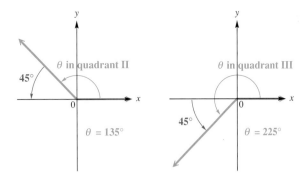

FIGURE 33

EXAMPLE 8

Finding Angle Measures Given an Interval and a Function Value (Radian Measure)

Find two angles in the interval [0, 2π) that satisfy cos $\theta \approx .3623577545$.

SOLUTION With the calculator in radian mode, we find that one such θ is

$$\cos^{-1} .3623577545 \approx 1.2.$$

Since $1.2 < \frac{\pi}{2}$, θ is in quadrant I. We must find *another* value of θ that satisfies the given condition. This other value of θ will have its reference angle θ' equal to 1.2, and must be in quadrant IV, since the angle given by the calculator is in quadrant

I and cosine is also positive in quadrant IV. The other value of θ, then, is

$$2\pi - 1.2 \approx 5.083185307.$$

Verify this result by showing that

$$\cos 5.083185307 \approx .3623577545. \quad \blacksquare$$

8.3 EXERCISES

To the Student: Calculator Considerations

Be sure that you know how to put your calculator in degree or radian mode. Also, be aware of the difference between inverse trigonometric functions (such as $\sin^{-1} \theta$) and reciprocals of trigonometric functions (such as $\csc \theta$, which is equal to $\frac{1}{\sin \theta}$).

Find the exact values of the six trigonometric functions of angle A.

1.
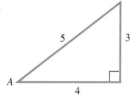

2.
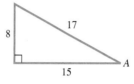

3.
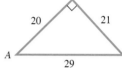

4.

5.
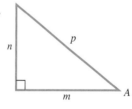

6.
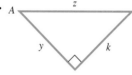

Write each of the following in terms of the cofunction.

7. $\cot 73°$

8. $\sec 39°$

9. $\sin 38° \ 29'$

10. $\tan 25° \ 43'$

11. $\cos \dfrac{\pi}{5}$

12. $\sin \dfrac{\pi}{3}$

13. $\tan .5$

14. $\csc .3$

15. A student was asked to give the exact value of sin 45°. Using this calculator, he gave the answer .7071067812. The teacher did not give him credit. What was the teacher's reason for this?

16. A student was asked to give an approximate value of sin 45. With her calculator in degree mode, she gave the value .7071067812. The teacher did not give her credit. What was her error?

*For each of the following (**a**) give the exact value, (**b**) state whether the exact value is rational or irrational, and (**c**) if the exact value is irrational, use your calculator in two ways to support your answer in part (a) by finding a decimal approximation.*

17. $\tan 30°$

18. $\cot 30°$

19. $\sin 30°$

20. $\cos 30°$

21. $\sec 30°$

22. $\csc 30°$

23. $\csc 45°$

24. $\sec 45°$

25. cos 45°   **26.** sin 45°   **27.** cot 45°   **28.** tan 45°

29. $\sin \dfrac{\pi}{3}$ **30.** $\cos \dfrac{\pi}{3}$ **31.** $\tan \dfrac{\pi}{3}$ **32.** $\cot \dfrac{\pi}{3}$

33. $\sec \dfrac{\pi}{3}$ **34.** $\csc \dfrac{\pi}{3}$

Use a calculator to find a decimal approximation for each function value. Give as many digits as your calculator displays.

35. sin 38° 40′   **36.** tan 29° 30′   **37.** cos 251° 10′

38. cot 512° 20′   **39.** sec(−108° 20′)   **40.** csc(−29° 30′)

41. tan .4538   **42.** sin .6109   **43.** csc 1.3875

44. cos(−3.0602)   **45.** sin(−17.5784)

46. (a) cos(−17.5784)
   (b) Based on your answers in Exercises 45 and 46(a), in what quadrant is an angle of −17.5784 radians?

Complete the following table with exact trigonometric function values using the methods of this section.

θ	$\sin \theta$	$\cos \theta$	$\tan \theta$	$\cot \theta$	$\sec \theta$	$\csc \theta$
47. 30°	$\frac{1}{2}$	$\frac{\sqrt{3}}{2}$	_____	_____	$\frac{2\sqrt{3}}{3}$	2
48. 45°	_____	_____	1	1	_____	_____
49. 60°	_____	$\frac{1}{2}$	$\sqrt{3}$	_____	2	_____
50. 120°	$\frac{\sqrt{3}}{2}$	_____	$-\sqrt{3}$	_____	_____	$\frac{2\sqrt{3}}{3}$
51. 135°	$\frac{\sqrt{2}}{2}$	$-\frac{\sqrt{2}}{2}$	_____	_____	$-\sqrt{2}$	$\sqrt{2}$
52. 150°	_____	$-\frac{\sqrt{3}}{2}$	$-\frac{\sqrt{3}}{3}$	_____	_____	2
53. 210°	$-\frac{1}{2}$	_____	$\frac{\sqrt{3}}{3}$	$\sqrt{3}$	_____	−2
54. 240°	$-\frac{\sqrt{3}}{2}$	$-\frac{1}{2}$	_____	_____	−2	$-\frac{2\sqrt{3}}{3}$

For each of the following (a) write the function in terms of a function of the reference angle, (b) give the exact value, and (c) use a calculator to show that the decimal value or approximation for the given function is the same as the decimal value or approximation for your answer in part (b).

55. $\sin \dfrac{7\pi}{6}$ **56.** $\cos \dfrac{5\pi}{3}$ **57.** $\tan \dfrac{3\pi}{4}$

58. $\sin \dfrac{5\pi}{3}$ **59.** $\cos \dfrac{7\pi}{6}$ **60.** $\tan \dfrac{4\pi}{3}$

Find all values of θ, if θ is in the interval [0°, 360°) and has the given function value.

61. $\sin \theta = \dfrac{1}{2}$ **62.** $\cos \theta = \dfrac{\sqrt{3}}{2}$ **63.** $\tan \theta = \sqrt{3}$ **64.** $\sec \theta = \sqrt{2}$

65. $\cos \theta = -\dfrac{1}{2}$ **66.** $\cot \theta = -\dfrac{\sqrt{3}}{3}$ **67.** $\sin \theta = -\dfrac{\sqrt{3}}{2}$ **68.** $\cos \theta = -\dfrac{\sqrt{2}}{2}$

Find all values of θ if θ is in the interval [0°, 360°) and has the given function value. Give approximations to as many decimal places as your calculator displays.

69. cos θ ≈ .68716510   **70.** cos θ ≈ .96476120   **71.** sin θ ≈ .41298643

72. sin θ ≈ .63898531   **73.** tan θ ≈ .87692035   **74.** tan θ ≈ 1.2841996

Find two angles in the interval $[0, 2\pi)$ *that satisfy the given equation. Give calculator approximations to as many digits as your calculator will display.*

75. $\tan \theta \approx .21264138$

76. $\cos \theta \approx .78269876$

77. $\sin \theta \approx .99184065$

78. $\cot \theta \approx .29949853$

79. $\csc \theta \approx 1.0219553$

80. $\cos \theta \approx .92728460$

R*elating Concepts*

In a square window of your calculator that gives a good picture of the first quadrant, graph the line $y = \sqrt{3}x$ *with* $x \geq 0$. *Then trace to any point on the line. For example, see the figure below. What we see is a simulated view of an angle in standard position, with terminal side in quadrant I. Store the values of x and y in convenient memory locations. For this group of exercises, we call them* x_1 *and* y_1. *Work the exercises in order.*

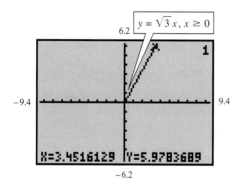

81. Find the value of $\sqrt{x_1{}^2 + y_1{}^2}$ and store it in a convenient memory location. (We will call it r.) What does this number mean geometrically?

82. With your calculator in degree mode, find $\tan^{-1}\left(\frac{y_1}{x_1}\right)$.

83. With your calculator in degree mode, find $\sin^{-1}\left(\frac{y_1}{r}\right)$.

84. With your calculator in degree mode, find $\cos^{-1}\left(\frac{x_1}{r}\right)$.

85. Your answers in Exercises 82–84 should all be the same. How does it relate to the angle formed on your screen?

86. Find the value of $\frac{y_1}{x_1}$. Now square it. What do you get? What is the exact value of $\frac{y_1}{x_1}$?

87. Look at the equation of the line you graphed, and make a conjecture: The _____ of a line passing through the origin is equal to the _____ of the angle it forms with the positive x-axis.

88. Find the value of $\left(\frac{x_1}{r}\right)^2 + \left(\frac{y_1}{r}\right)^2$. What identity does this illustrate?

89. Find $\csc 60°$. Then find the value of $\frac{r}{y_1}$. Do they agree?

90. Graph $y_2 = \sqrt{1 - x^2}$ as a second curve in the same viewing window. This is one half of a circle centered at the origin with radius 1. Now use the intersection feature of your calculator to determine the x- and y-coordinates of the points of intersection of the two graphs. What are they?

91. Find a calculator value for $\cos 60°$. How does it compare to the x-coordinate of the point you found in Exercise 90? Why is this so?

92. Find a calculator approximation for $\sin 60°$. How does it compare to the y-coordinate of the point you found in Exercise 90? Why is this so?

8.4 APPLICATIONS OF RIGHT TRIANGLE TRIGONOMETRY

Significant Digits and Accuracy ▌ Solving Right Triangles ▌ Applications

Significant Digits and Accuracy

In our discussion of applications of trigonometry in the rest of this chapter, we will be performing many calculations with trigonometric function values. Before doing so we will present a brief explanation of accuracy with such numbers.

Suppose that a wall is measured to the nearest foot and is found to be 18 feet long. Actually this means that the wall has a length between 17.5 feet and 18.5 feet. If the wall is measured more accurately and found to be 18.3 feet long, then its length is really between 18.25 feet and 18.35 feet. A measurement of 18.00 feet would indicate that the length of the wall is between 17.995 feet and 18.005 feet. The measurement 18 feet is said to have two **significant digits** of accuracy; 18.0 has three significant digits, and 18.00 has four.

A significant digit is a digit obtained by actual measurement. A number that represents the result of counting, or a number that results from theoretical work and is not the result of a measurement, is an **exact number.** For example, $\frac{\sqrt{2}}{2}$ is the *exact* value of cos 45°, while a calculator might give an *approximation* of cos 45° as .7071067812 to ten significant digits.

To perform calculations on such approximate numbers, follow the rules given below.

> ### CALCULATION WITH SIGNIFICANT DIGITS
>
> For *adding* and *subtracting*, round the answer so that the last digit you keep is in the right-most column in which all the numbers have significant digits.
>
> For *multiplying* or *dividing*, round the answer to the least number of significant digits found in any of the given numbers.

In our work to follow, we will use the following table for deciding on significant digits in angle measures.

> ### SIGNIFICANT DIGITS FOR ANGLES
>
> | Number of Significant Digits | Angle Measure to Nearest: |
> | --- | --- |
> | 2 | Degree |
> | 3 | Ten minutes, or nearest tenth of a degree |
> | 4 | Minute, or nearest hundredth of a degree |
> | 5 | Tenth of a minute, or nearest thousandth of a degree |

For example, an angle measuring 52° 30′ has three significant digits (assuming that 30′ is measured to the nearest ten minutes).

Solving Right Triangles

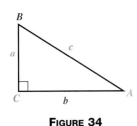

FIGURE 34

With the background that we now have, it is possible to find the measures of all angles and all sides of a right triangle if we know the measure of one acute angle and one side, or if we know the measure of two sides. Finding all measures is called **solving the right triangle**.

In using trigonometry to solve triangles, it is convenient to use a to represent the length of the side opposite angle A, b for the length of the side opposite angle B, and so on. As mentioned earlier, in a right triangle the letter c is reserved for the hypotenuse. Figure 34 shows the labeling of a typical right triangle.

EXAMPLE 1

Solving a Right Triangle Given an Angle and a Side

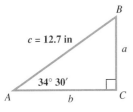

FIGURE 35

Solve right triangle ABC, with $A = 34°\ 30′$ and $c = 12.7$ in. (See Figure 35.)

SOLUTION   To solve the triangle, find the measures of the remaining sides and angles. The value of a can be found with a trigonometric function involving the known values of angle A and side c. Since the sine of angle A is given by the quotient of the side opposite A and the hypotenuse, use $\sin A$.

$$\sin A = \frac{a}{c}$$

Substituting known values gives

$$\sin 34°\ 30′ = \frac{a}{12.7},$$

or, upon multiplying both sides by 12.7,

$$a = 12.7 \sin 34°\ 30′$$
$$a \approx 12.7(.56640624) \quad \text{Use a calculator.}$$
$$a \approx 7.19 \text{ in.}$$

The value of b could be found with the Pythagorean theorem. It is better, however, to use the information given in the problem rather than a result just calculated. If a mistake were to be made in finding a, then b also would be incorrect. Also, rounding more than once may cause the result to be less accurate. Using $\cos A$ gives

$$\cos A = \frac{\text{side adjacent}}{\text{hypotenuse}} = \frac{b}{c}$$
$$\cos 34°\ 30′ = \frac{b}{12.7}$$
$$b = 12.7 \cos 34°\ 30′$$
$$b \approx 10.5 \text{ in.}$$

Once b has been found, the Pythagorean theorem could be used as a check. All that remains to solve triangle ABC is to find the measure of angle B. Since $A + B = 90°$ and $A = 34°\ 30′$,

$$A + B = 90°$$
$$B = 90° - A$$
$$B = 89°\ 60′ - 34°\ 30′$$
$$B = 55°\ 30′. \quad ∎$$

NOTE In Example 1 we could have started by finding the measure of angle B and then used the trigonometric function values of B to find the unknown sides. The process of solving a right triangle (like many problems in mathematics) can usually be done in several ways, each resulting in the correct answer. However, in order to retain as much accuracy as can be expected, always use given information as much as possible, and avoid rounding off in intermediate steps.

> **EXAMPLE 2**
>
> Solving a Right Triangle Given Two Sides

Solve right triangle ABC if $a = 29.43$ cm and $c = 53.58$ cm.

SOLUTION Draw a sketch showing the given information, as in Figure 36. One way to begin is to find angle A by using the sine.

$$\sin A = \frac{\text{side opposite}}{\text{hypotenuse}}$$

$$\sin A = \frac{29.43}{53.58}$$

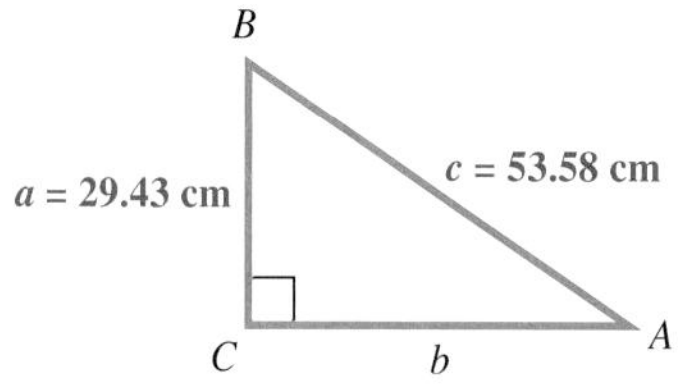

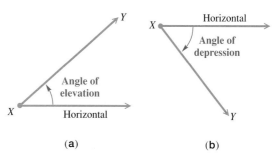

FIGURE 36

TECHNOLOGICAL NOTE
Once you have mastered the material on solving right triangles, you may wish to write a program that will accomplish this goal. You will need to consider the various cases of what is given and what must be found.

Using the inverse sine function on a calculator, we find that $A \approx 33.32°$. The measure of B is $90° - 33.32° \approx 56.68°$.

We now find b from the Pythagorean theorem, using $a^2 + b^2 = c^2$, or $b^2 = c^2 - a^2$. Since $c = 53.58$ and $a = 29.43$,

$$b^2 = 53.58^2 - 29.43^2$$

giving $\quad b \approx 44.77$ cm. ◻

Applications

The process of solving right triangles is easily adapted to solving applied problems. A crucial step in such applications involves sketching the triangle and labeling the given parts correctly. Then we can use the methods described in the earlier examples to find the unknown value or values.

Many applications of right triangles involve the angle of elevation or the angle of depression. The **angle of elevation** from point X to point Y (above X) is the angle made by ray XY and a horizontal ray with endpoint at X. The angle of elevation is always measured from the horizontal. See Figure 37(a). The **angle of depression** from point X to point Y (below X) is the angle made by ray XY and a horizontal ray with endpoint X. See Figure 37(b).

FIGURE 37

CAUTION   Errors are often made in interpreting the angle of depression. Remember that both the angle of elevation *and* the angle of depression are measured *from* the horizontal *to* the line of sight.

EXAMPLE 3

Finding a Length When the Angle of Elevation is Known

Donna Garbarino knows that when she stands 123 feet from the base of a flagpole, the angle of elevation to the top is 26° 40′. If her eyes are 5.30 feet above the ground, find the height of the flagpole.

SOLUTION   The length of the side adjacent to Donna is known and the length of the side opposite her is to be found. See Figure 38. The ratio that involves these two values is the tangent.

$$\tan A = \frac{\text{side opposite}}{\text{side adjacent}}$$

$$\tan 26° \, 40′ = \frac{a}{123}$$

$$a = 123 \tan 26° \, 40′$$

$$a \approx 61.8 \text{ feet}$$

Since Donna's eyes are 5.30 feet above the ground, the height of the flagpole is approximately

$$61.8 + 5.30 \approx 67.1 \text{ feet.} \quad \blacksquare$$

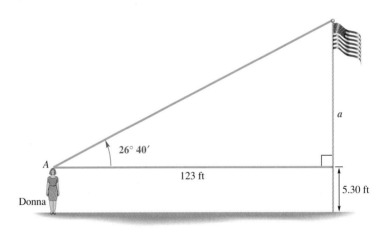

FIGURE 38

EXAMPLE 4

Finding the Angle of Elevation When Lengths Are Known

The length of the shadow of a building 34.09 meters tall is 37.62 meters. Find the angle of elevation of the sun.

SOLUTION   As shown in Figure 39, the angle of elevation of the sun is angle *B*. Since the side opposite *B* and the side adjacent to *B* are known, use the tangent ratio to find *B*.

$$\tan B = \frac{34.09}{37.62}$$

$$B \approx 42.18° \qquad \text{Use the inverse tangent function.}$$

The angle of elevation of the sun is approximately 42.18°. $\blacksquare$

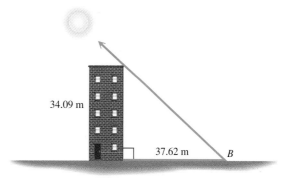

FIGURE 39

Some applications of right triangles involve **bearing,** an important idea in navigation. There are two common ways to express bearing. *When a single angle is given, such as* 164°, *it is understood that the bearing is measured in a clockwise direction from due north.* Several sample bearings using this first type of system are shown in Figure 40.

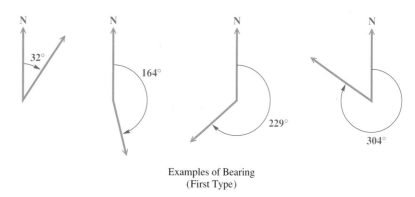

Examples of Bearing
(First Type)

FIGURE 40

<table>
<tr><td>

EXAMPLE 5

Solving a Problem
Involving Bearing
(First Type)

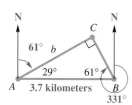

FIGURE 41

</td><td>

Radar stations A and B are on an east-west line, 3.7 kilometers apart. Station A detects a plane at C, on a bearing of 61°. Station B simultaneously detects the same plane, on a bearing of 331°. Find the distance from A to C.

SOLUTION   Draw a sketch showing the given information, as in Figure 41. Since a line drawn due north is perpendicular to an east-west line, right angles are formed at A and B, so that angles CAB and CBA can be found. Angle C is a right angle because angles CAB and CBA are complementary. (If C were not a right angle, the methods of later sections would be needed.) Find distance b by using the cosine function.

$$\cos 29° = \frac{b}{3.7}$$

$$3.7 \cos 29° = b$$

$$b \approx 3.2 \text{ kilometers}$$ Use a calculator and round to the nearest tenth.   ∎

</td></tr>
</table>

CAUTION It would be foolish to attempt to solve the problem in Example 5 without drawing a sketch. The importance of a correctly labeled sketch in applications such as this cannot be overemphasized, as some of the necessary information is often not given directly in the statement of the problem, and can only be determined from the sketch.

The second common system for expressing bearing starts with a north-south line and uses an acute angle to show the direction, either east or west, from this line. Figure 42 shows several sample bearings using this system. Either N or S always comes first, followed by an acute angle, and then E or W.

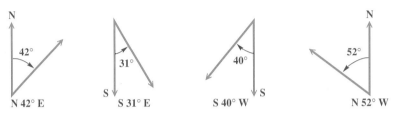

Examples of Bearing
(Second Type)

FIGURE 42

EXAMPLE 6

Solving a Problem
Involving Bearing
(Second Type)

The bearing from A to C is S 52° E. The bearing from A to B is N 84° E. The bearing from B to C is S 38° W. A plane flying at 250 miles per hour takes 2.4 hours to go from A to B. Find the distance from A to C.

SOLUTION Make a sketch of the situation. First draw the two bearings from point A. Choose a point B on the bearing N 84° E from A and draw the bearing to C. Point C will be located where the bearing lines from A and B intersect as shown in Figure 43.

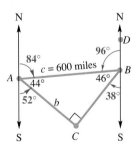

FIGURE 43

Since the bearing from A to B is N 84° E, angle ABD is 180° − 84° = 96°. Thus, angle ABC is 46°. Also, angle BAC is 180° − (84° + 52°) = 44°. Angle C is 180° − (44° + 46°) = 90°. From the statement of the problem, a plane flying at 250 miles per hour takes 2.4 hours to go from A to B. The distance from A to B is the product of rate and time, or

$$c = \text{rate} \times \text{time} = 250(2.4) = 600 \text{ miles.}$$

To find b, the distance from A to C, use the sine. (The cosine could also have been used.)

$$\sin 46° = \frac{b}{c}$$

$$\sin 46° = \frac{b}{600}$$

$$600 \sin 46° = b$$

$$b \approx 430 \text{ miles} \qquad \text{Use a calculator.} \quad \blacksquare$$

EXAMPLE 7

Solving a Problem Involving Angle of Elevation

Francisco needs to know the height of a tree. From a given point on the ground he finds that the angle of elevation to the top of the tree is $36°\,40'$. He then moves back 50 feet. From the second point, the angle of elevation to the top of the tree is $22°\,10'$. See Figure 44. Find the height of the tree.

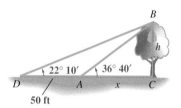

FIGURE 44

SOLUTION The figure shows two unknowns; x, the distance from the center of the trunk of the tree to the point where the first observation was made, and h, the height of the tree. Since nothing is given about the length of the hypotenuse of either triangle ABC or triangle BCD, use a ratio that does not involve the hypotenuse—the tangent.

In triangle ABC, $\tan 36°\,40' = \dfrac{h}{x}$ or $h = x \tan 36°\,40'$.

In triangle BCD, $\tan 22°\,10' = \dfrac{h}{50 + x}$ or $h = (50 + x) \tan 22°\,10'$.

Since each of these expressions equals h, these expressions must be equal. Thus,

$$x \tan 36°\,40' = (50 + x) \tan 22°\,10'.$$

Now use algebra to solve for x.

$$x \tan 36°\,40' = 50 \tan 22°\,10' + x \tan 22°\,10' \qquad \text{Distributive property}$$

$$x \tan 36°\,40' - x \tan 22°\,10' = 50 \tan 22°\,10' \qquad \text{Get } x \text{ terms on one side.}$$

$$x(\tan 36°\,40' - \tan 22°\,10') = 50 \tan 22°\,10' \qquad \text{Factor out } x \text{ on the left.}$$

$$x = \frac{50 \tan 22°\,10'}{\tan 36°\,40' - \tan 22°\,10'} \qquad \text{Divide by the coefficient of } x.$$

We saw that $h = x \tan 36° \, 40'$. Substituting for x,

$$h = \left(\frac{50 \tan 22° \, 10'}{\tan 36° \, 40' - \tan 22° \, 10'} \right) (\tan 36° \, 40').$$

From a calculator,

$$\tan 36° \, 40' \approx .74447242$$
$$\tan 22° \, 10' \approx .40741394,$$

so

$$\tan 36° \, 40' - \tan 22° \, 10' \approx .74447242 - .40741394 \approx .33705848,$$

and

$$h \approx \left(\frac{50(.40741394)}{.33705848} \right)(.74447242) \approx 45 \text{ (rounded)}.$$

The height of the tree is approximately 45 feet.   ◼

NOTE In practice we usually do not write down the intermediate calculator approximation steps. However, we have done this in Example 7 so that the reader may follow the steps more easily.

FOR **GROUP DISCUSSION**

An alternate approach to solving the problem in Example 7 uses the intersection-of-graphs capability of the graphics calculator. This approach is based on a similar solution proposed by a student, John Cree, as explained in a letter to the editor in the January 1995 issue of *Mathematics Teacher* from Cree's teacher, Robert Ruzich.*

1. Since the tangent of the angle formed by the graph of $y = mx + b$ with the x-axis is the slope of the line (m), the segment BD lies along the graph of $y_1 = (\tan 22° \, 10')x$. Enter this equation into your calculator.
2. By similar reasoning and using the fact that A lies 50 feet to the *right* of D, enter the equation $y_2 = (\tan 36° \, 40')(x - 50)$ into your calculator to represent AB.
3. Using a window of $[0, 200]$ by $[0, 100]$, find the point of intersection of the graphs of y_1 and y_2, using the intersection-of-graphs capability of your calculator.
4. Compare the y-coordinate of the point of intersection to the answer found analytically in Example 7. It should be the same.

---

*Adapted from "Letter to the Editor" by Robert Ruzich, *The Mathematics Teacher*, January 1995. Copyright © 1995 by the National Council of Teachers of Mathematics. Reprinted with the permission of The Mathematics Teacher.

8.4 EXERCISES

To the Student: Calculator Consideration

You should already be familiar with all of the necessary keystrokes needed to work the exercises in this section. If your answers seem to be "a bit off from the ones given in the back of the book," remember that we give the answers with accuracy based on the guidelines presented at the beginning of this section. In some cases, it may be necessary to convert between decimal degrees and degrees, minutes, and seconds as well.

Solve each right triangle.

1.

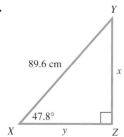

2.

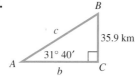

3.

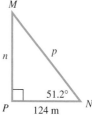

4.

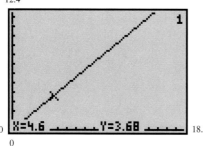

5.

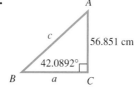

6.

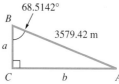

Solve the right triangle ABC, with C = 90°.

7. $A = 28° \ 00'$, $c = 17.4$ ft

8. $B = 46° \ 00'$, $c = 29.7$ m

9. $B = 73° \ 00'$, $b = 128$ in

10. $A = 61° \ 00'$, $b = 39.2$ cm

11. $a = 76.4$ yd, $b = 39.3$ yd

12. $a = 18.9$ cm, $c = 46.3$ cm

13. Can a right triangle be solved if we are given the measures of its two acute angles and no side lengths? Explain.

14. If we are given an acute angle and a side in a right triangle, what unknown part of the triangle requires the least work to find?

Find the measure of the angle formed by the line passing through the origin and the positive part of the x-axis. Use the displayed values of x and y at the bottom of the screen. (In each case, a square screen is used.)

15.

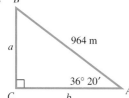

16.

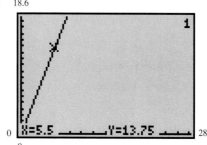

Consider the figure here in Exercises 17 and 18.

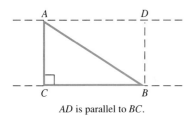

AD is parallel to BC.

17. Explain why the angle of depression *DAB* has the same measure as the angle of elevation *ABC*.

18. Why is angle *CAB not* an angle of depression?

Solve each problem.

19. A 13.5-m fire truck ladder is leaning against a wall. Find the distance the ladder goes up the wall if it makes an angle of 43° 50′ with the ground.

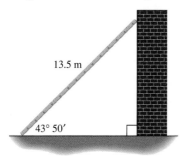

20. To measure the height of a flagpole, Kitty Pellissier finds that the angle of elevation from a point 24.73 ft from the base to the top is 38° 12′. Find the height of the flagpole.

21. A guy wire 80.1 m long is attached to the top of an antenna mast that is 71.3 m high. Find the angle that the wire makes with the ground.

22. Find the length of a guy wire that makes an angle of 45° 30′ with the ground if the wire is attached to the top of a tower 63.0 m high.

23. To find the distance *RS* across a lake, a surveyor lays off *RT* = 53.1 m, with angle *T* = 32° 10′, and angle *S* = 57° 50′. Find length *RS*.

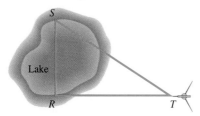

24. A surveyor must find the distance *QM* across a depressed freeway. She lays off *QN* = 769 ft along one side of the freeway, with angle *N* = 21° 50′, and with angle *M* = 68° 10′. Find *QM*.

25. The length of the base of an isosceles triangle is 42.36 in. Each base angle is 38.12°. Find the length of each of the two equal sides of the triangle. (*Hint:* Divide the triangle into two right triangles.)

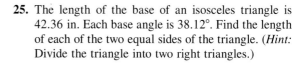

26. Find the altitude of an isosceles triangle having a base of 184.2 cm if the angle opposite the base is 68° 44′.

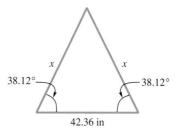

27. Suppose that the angle of elevation of the sun is 23.4°. Find the length of the shadow cast by Cindy Newman, who is 5.75 ft tall.

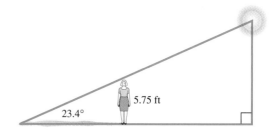

28. The shadow of a vertical tower is 40.6 m long when the angle of elevation of the sun is 34.6°. Find the height of the tower.

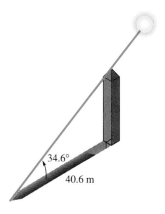

34.6°

40.6 m

29. Find the angle of elevation of the sun if a 48.6-ft flagpole casts a shadow 63.1 ft long.

30. The angle of depression from the top of a building to a point on the ground is 32° 30′. How far is the point on the ground from the top of the building if the building is 252 m high?

31. An airplane is flying 10,500 feet above the level ground. The angle of depression from the plane to the base of a tree is 13° 50′. How far horizontally must the plane fly to be directly over the tree?

10,500 ft

32. The angle of elevation from the top of a small building to the top of a nearby taller building is 46° 40′, while the angle of depression to the bottom is 14° 10′. If the smaller building is 28.0 m high, find the height of the taller building.

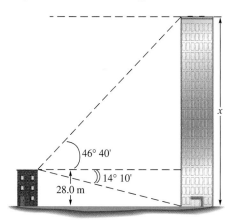

x

46° 40′

14° 10′

28.0 m

33. A television camera is to be mounted on a bank wall so as to have a good view of the head teller. (See the figure.) Find the angle of depression that the lens should make with the horizontal.

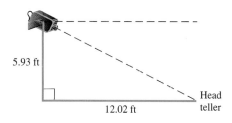

5.93 ft

12.02 ft

Head teller

34. A company safety committee has recommended that a floodlight be mounted in a parking lot so as to illuminate the employee exit. (See the figure.) Find the angle of depression of the light.

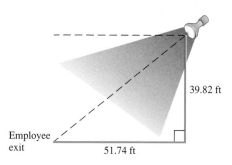

39.82 ft

Employee exit

51.74 ft

35. A tunnel is to be dug from *A* to *B*. (See the figure.) Both *A* and *B* are visible from *C*. If *AC* is 1.4923 mi and *BC* is 1.0837 mi, and if *C* is 90°, find the measures of angles *A* and *B*.

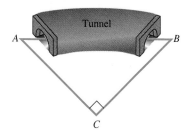

Tunnel

A

B

C

36. A piece of land has the shape shown in the figure. Find *x*.

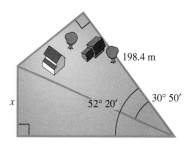

198.4 m

x

52° 20′

30° 50′

37. Find the value of x in the figure.

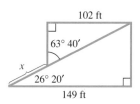

38. The leaning tower of Pisa is approximately 179 ft in height and is approximately 16.5 ft out of plumb (that is, tilted away from the vertical). Find the angle at which it deviates from the vertical.*

39. A regular pentagon (five-sided polygon with equal lengths of sides and equal angles) is inscribed in a circle of radius 7 cm. Find the length of a side of the pentagon. Give your answer to the nearest thousandth.*

40. Find h as indicated in the figure.

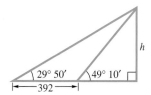

41. Find h as indicated in the figure.

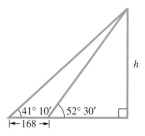

42. The angle of elevation from a point on the ground to the top of a pyramid is $35° 30'$. The angle of elevation from a point 135 ft farther back to the top of the pyramid is $21° 10'$. Find the height of the pyramid.

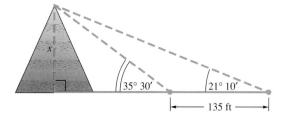

43. Debbie Glockner, a whale researcher, is watching a whale approach directly toward her as she observes from the top of a lighthouse. When she first begins watching the whale, the angle of depression of the whale is $15° 50'$. Just as the whale turns away from the lighthouse, the angle of depression is $35° 40'$. If the height of the lighthouse is 68.7 m, find the distance traveled by the whale as it approaches the lighthouse.

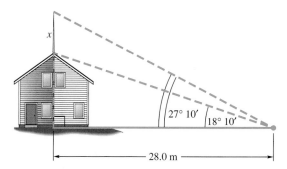

44. A scanner antenna is on top of the center of a house. The angle of elevation from a point 28.0 m from the center of the house to the top of the antenna is $27° 10'$, and the angle of elevation to the bottom of the antenna is $18° 10'$. Find the height of the antenna.

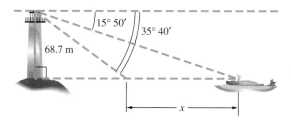

45. The angle of elevation from Lone Pine to the top of Mt. Whitney is $10° 50'$. Van Dong Le, traveling 7.00 km from Lone Pine along a straight, level road toward Mt. Whitney, finds the angle of elevation to be $22° 40'$. Find the height of the top of Mt. Whitney above the level of the road.

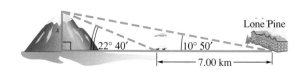

*Exercises 38 and 39 are excerpts from *Plane Trigonometry*. Revised Edition by Frank A. Rickey and J. P. Cole, copyright © 1964 by Holt, Rinehart and Winston, Inc., and renewed 1992 by Coleen C. Salley, T. E. Cole, Robert E. Cole, James P. Cole, Jr., Frank A. Rickey, Jr., Mary Ellen Rickey, Mrs. Mary E. Rickey, and W. P. Rickey, reprinted by permission of the author.

46. A plane flies 1.5 hr at 110 mph on a bearing of 40°. It then turns and flies 1.3 hr at the same speed on a bearing of 130°. How far is the plane from its starting point?

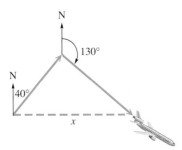

47. A ship travels 50 km on a bearing of 27°, and then travels on a bearing of 117° for 140 km. Find the distance between the starting point and the ending point.

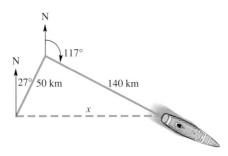

48. Two ships leave a port at the same time. The first ship sails on a bearing of 40° at 18 knots (nautical miles per hour) and the second at a bearing of 130° at 26 knots. How far apart are they after 1.5 hours?

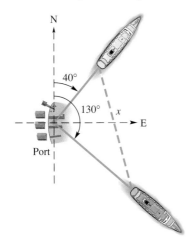

49. Two lighthouses are located on a north-south line. From lighthouse *A* the bearing of a ship 3742 m away is 129° 43′. From lighthouse *B* the bearing of the ship is 39° 43′. Find the distance between the lighthouses.

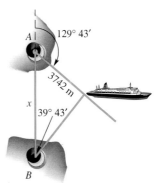

50. A ship leaves its home port and sails on a bearing of N 28° 10′ E. Another ship leaves the same port at the same time and sails on a bearing of S 61° 50′ E. If the first ship sails at 24.0 mph and the second sails at 28.0 mph, find the distance between the two ships after 4 hrs.

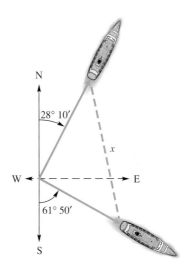

51. Radio direction finders are set up at points *A* and *B*, which are 2.50 mi apart on an east-west line. From *A* it is found that the bearing of the signal from a radio transmitter is N 36° 20′ E, while from *B* the bearing of the same signal is N 53° 40′ W. Find the distance of the transmitter from *B*.

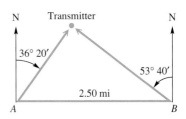

52. The bearing from Winston-Salem, North Carolina, to Danville, Virginia, is N 42° E. The bearing from Danville to Goldsboro, North Carolina, is S 48° E. A small plane piloted by Mark Ferrari, traveling at 60 mph, takes 1 hr to go from Winston-Salem to Danville and 1.8 hr to go from Danville to Goldsboro. Find the distance from Winston-Salem to Goldsboro.

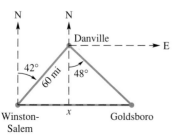

53. To determine the diameter of the sun, an astronomer might sight with a transit (a device used by surveyors for measuring angles) first to one edge of the sun and then to the other, finding that the included angle equals 1° 4′. Assuming that the distance from the earth to the sun is 92,919,800 mi, calculate the diameter of the sun. (See the figure.)

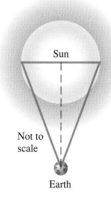

54. The figure shows a magnified view of the threads of a bolt. Find x if d is 2.894 mm.

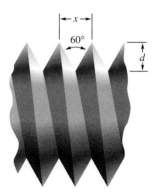

55. You have a summer job putting up outhouses in county parks. You must precut the rafter ends so that they will be vertical when in place. The front wall is 8 ft high, the back wall is $6\frac{1}{2}$ ft high and the distance between walls is 8 ft. At what angle should you cut the rafters? (Round to the nearest degree.)*

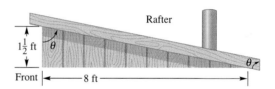

56. A degree may seem like a very small unit, but an error of one degree in measuring an angle may be very significant. For example, suppose that a laser beam is directed toward the visible center of the moon and that it misses its assigned target by 30 seconds. How far is it (in mi) from its assigned target? (Take the distance from the surface of the earth to that of the moon to be 234,000 mi.)*

8.5 THE LAW OF SINES

Introduction ▌ Congruency and Oblique Triangles ▌ Derivation of the Law of Sines ▌ Applications ▌ The Ambiguous Case

Introduction

Until now, our applied work with trigonometry has been limited to right triangles. However, the concepts developed earlier in this chapter can be extended so that our work can apply to *all* triangles. Every triangle has three sides and three angles, and we will show that if any three of the six measures of a triangle are known (provided that at least one is that of a side), then the other three measures can be found. Again, this is called *solving the triangle.*

Congruency and Oblique Triangles

Recall from geometry the following axioms that allow us to prove that two triangles are congruent (that is, their corresponding sides and angles are equal).

CONGRUENCE AXIOMS

Side-Angle-Side (SAS)

If two sides and the included angle of one triangle are equal, respectively, to two sides and the included angle of a second triangle, then the triangles are congruent.

Angle-Side-Angle (ASA)

If two angles and the included side of one triangle are equal, respectively, to two angles and the included side of a second triangle, then the triangles are congruent.

Side-Side-Side (SSS)

If three sides of one triangle are equal, respectively, to three sides of a second triangle, then the triangles are congruent.

Throughout this section and the next, keep in mind that whenever any of the groups of data described above are given, the triangle is uniquely determined; that is, all other data in the triangle are given by one and only one set of measures.

A triangle that is not a right triangle is called an **oblique triangle.** The measures of the three sides and the three angles of a triangle can be found if at least one side and any other two measures are known. There are four possible cases.

DATA REQUIRED FOR SOLVING OBLIQUE TRIANGLES

1. One side and two angles are known.
2. Two sides and one angle not included between the two sides are known. This case may lead to more than one triangle.
3. Two sides and the angle included between the two sides are known.
4. Three sides are known.

NOTE If we know three angles of a triangle, we cannot find unique side lengths, since AAA assures us only of similarity, not congruence. For example, there are infinitely many triangles ABC with $A = 35°$, $B = 65°$, and $C = 80°$.

The first two cases require the use of the *law of sines*, which is introduced in this section. The last two cases require the use of the *law of cosines*, introduced in the next section.

Derivation of the Law of Sines

To derive the law of sines, start with an oblique triangle as in Figure 45. (The discussion to follow applies to either of the triangles in Figures 45(a) or 45(b).) First, construct the perpendicular line from B to side AC or its extension. Let h be the length of this perpendicular line. Then c is the hypotenuse of right triangle ADB, and a is the hypotenuse of right triangle BDC. By results from Section 8.3,

$$\text{in triangle } ADB, \quad \sin A = \frac{h}{c} \quad \text{or} \quad h = c \sin A,$$

$$\text{in triangle } BDC, \quad \sin C = \frac{h}{a} \quad \text{or} \quad h = a \sin C.$$

Since $h = c \sin A$ and $h = a \sin C$,

$$a \sin C = c \sin A,$$

or, upon dividing both sides by $\sin A \sin C$,

$$\frac{a}{\sin A} = \frac{c}{\sin C}.$$

In a similar way, by constructing the perpendicular lines from other vertices, it can be shown that

$$\frac{a}{\sin A} = \frac{b}{\sin B} \quad \text{and} \quad \frac{b}{\sin B} = \frac{c}{\sin C}.$$

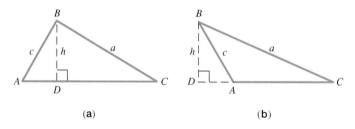

(a)   (b)

FIGURE 45

This discussion proves the following theorem.

LAW OF SINES

In any triangle ABC, with sides a, b, and c,

$$\frac{a}{\sin A} = \frac{b}{\sin B}, \quad \frac{a}{\sin A} = \frac{c}{\sin C}, \quad \text{and} \quad \frac{b}{\sin B} = \frac{c}{\sin C}.$$

This can be written in compact form as

$$\frac{a}{\sin A} = \frac{b}{\sin B} = \frac{c}{\sin C}.$$

Sometimes an alternate form of the law of sines,

$$\frac{\sin A}{a} = \frac{\sin B}{b} = \frac{\sin C}{c},$$

is convenient to use.

Applications

If two angles and the side opposite one of the angles are known, the law of sines can be used directly to solve for the side opposite the other known angle. The triangle can then be solved completely, as shown in the first example.

EXAMPLE 1

Using the Law of Sines to Solve a Triangle

Solve triangle ABC if $A = 32.0°$, $B = 81.8°$, and $a = 42.9$ centimeters. (See Figure 46.)

SOLUTION   Start by drawing a triangle, roughly to scale, and labeling the given parts as in Figure 46. Since the values of A, B, and a are known, use the part of the law of sines that involves these variables.

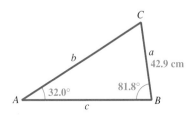

FIGURE 46

$$\frac{a}{\sin A} = \frac{b}{\sin B}$$

Substituting the known values gives

$$\frac{42.9}{\sin 32.0°} = \frac{b}{\sin 81.8°}.$$

Multiply both sides of the equation by $\sin 81.8°$.

$$b = \frac{42.9 \sin 81.8°}{\sin 32.0°}$$

$$b \approx 80.1 \text{ centimeters.} \qquad \text{Use a calculator.}$$

Find C from the fact that the sum of the angles of any triangle is $180°$.

$$A + B + C = 180°$$
$$C = 180° - A - B$$
$$C = 180° - 32.0° - 81.8° = 66.2°$$

Now use the law of sines again to find c. (Why does the Pythagorean theorem not apply?)

$$\frac{a}{\sin A} = \frac{c}{\sin C}$$

$$\frac{42.9}{\sin 32.0°} = \frac{c}{\sin 66.2°}$$

$$c = \frac{42.9 \sin 66.2°}{\sin 32.0°}$$

$$c \approx 74.1 \text{ centimeters} \quad \blacksquare$$

CAUTION   In applications of oblique triangles, such as the one that follows in Example 2, a properly labeled sketch is essential in order to set up the correct equation.

EXAMPLE 2

Using the Law of Sines in an Application

Tri Nguyen wishes to measure the distance across the Big Muddy River. (See Figure 47.) He finds that $C = 112° \, 53'$, $A = 31° \, 06'$, and $b = 347.6$ feet. Find the required distance.

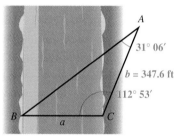

31° 06′

$b = 347.6$ ft

112° 53′

FIGURE 47

SOLUTION   To use the law of sines, one side and the angle opposite it must be known. Since the only side whose length is given is b, angle B must be found before the law of sines can be used.

$$B = 180° - A - C = 180° - 31° \, 06' - 112° \, 53' = 36° \, 01'$$

Now the required distance a can be found. Use the form of the law of sines involving A, B, and b.

$$\frac{a}{\sin A} = \frac{b}{\sin B}$$

Substitute the known values.

$$\frac{a}{\sin 31° \, 06'} = \frac{347.6}{\sin 36° \, 01'}$$

$$a = \frac{347.6 \sin 31° \, 06'}{\sin 36° \, 01'}$$

$$a \approx 305.3 \text{ feet} \qquad \text{Use a calculator.} \quad \blacksquare$$

The next example involves the use of bearing, first discussed in Section 8.4.

EXAMPLE 3

Using the Law of
Sines in an
Application

Two tracking stations are on an east-west line 110 miles apart. A forest fire is located on a bearing of N 42° E from the western station at *A* and a bearing of N 15° E from the eastern station at *B*. How far is the fire from the western station? (See Figure 48.)

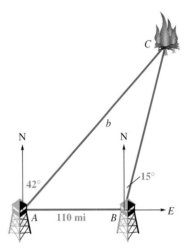

FIGURE 48

SOLUTION   Figure 48 shows the two stations at points *A* and *B* and the fire at point *C*. Angle *BAC* = 90° − 42° = 48°, the obtuse angle at *B* equals 90° + 15° = 105°, and the third angle, *C*, equals 180° − 105° − 48° = 27°. Using the law of sines to find side *b* gives

$$\frac{b}{\sin 105°} = \frac{110}{\sin 27°}$$

$$b \approx 234,$$

or approximately 230 miles (to two significant digits).   ∎

The Ambiguous Case

The law of sines can be used when given two angles and the side opposite one of these angles. Also, if two angles and the included side are known, then the third angle can be found by using the fact that the sum of the angles of a triangle is 180°, and then applying the law of sines. However, if we are given the lengths of two sides and the angle opposite one of them, it is possible that 0, 1, or 2 such triangles exist. (Recall that there is no "SSA" congruence theorem.)

To illustrate these facts, suppose that the measure of acute angle *A* of triangle *ABC*, the length of side *a*, and the length of side *b* are given. Draw angle *A* having a terminal side of length *b*. Now draw a side of length *a* opposite angle *A*. The following chart shows that there might be more than one possible outcome. This situation is called the **ambiguous case of the law of sines.**

Number of Possible Triangles	Sketch	Condition Necessary for Case to Hold
0	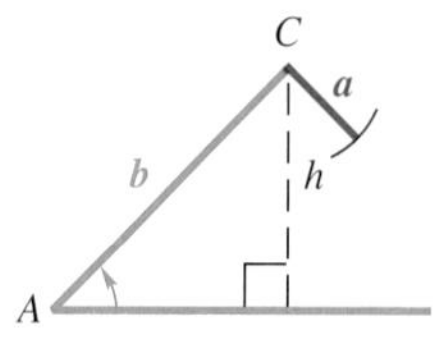	$a < h$ $(h = b \sin A)$
1	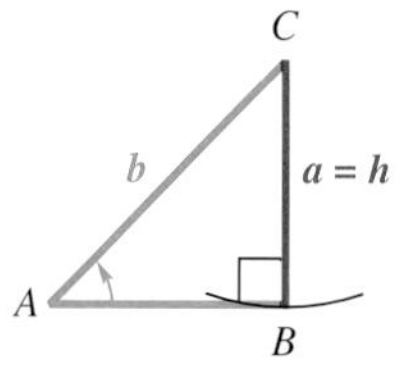	$a = h$
1	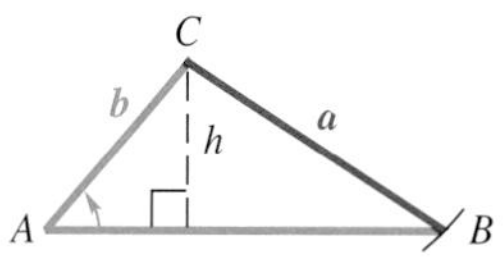	$a > b$
2	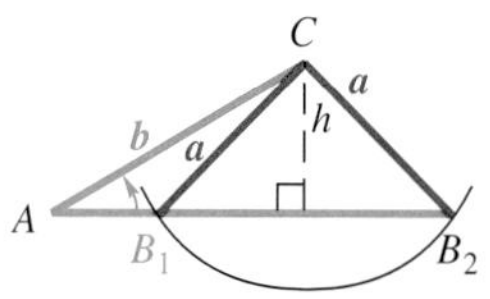	$b > a > h$

If angle A is obtuse, there are two possible outcomes, as shown in the next chart.

Number of Possible Triangles	Sketch	Condition Necessary for Case to Hold
0	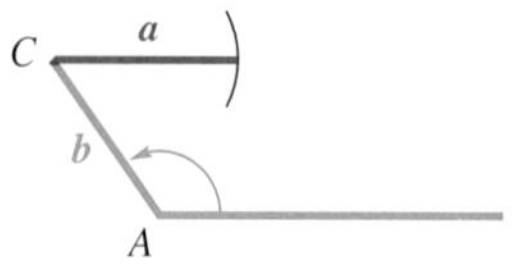	$a \leq b$
1	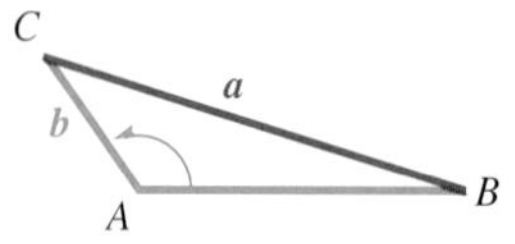	$a > b$

As the remaining examples of this section will illustrate, applying the law of sines to the values of a, b, and A and some basic properties of geometry and trigonometry will allow us to determine which of these cases applies. The following basic facts should be kept in mind.

FACTS TO REMEMBER WHEN USING THE LAW OF SINES

1. For any angle θ of a triangle, $0 < \sin \theta \leq 1$. If $\sin \theta = 1$, then $\theta = 90°$ and the triangle is a right triangle.
2. $\sin \theta = \sin(180° - \theta)$ (That is, supplementary angles have the same sine value.)
3. The smallest angle is opposite the shortest side, the largest angle is opposite the longest side, and the middle-valued angle is opposite the medium side (assuming the triangle has sides that are all of different lengths).

EXAMPLE 4

Solving a Triangle Using the Law of Sines (No Such Triangle)

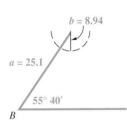

FIGURE 49

Solve the triangle ABC if $B = 55° \ 40'$, $b = 8.94$ meters, and $a = 25.1$ meters.

SOLUTION   Since we are given B, b, and a, use the law of sines to find A.

$$\frac{\sin A}{a} = \frac{\sin B}{b}$$

Substitute the given values.

$$\frac{\sin A}{25.1} = \frac{\sin 55° \ 40'}{8.94}$$

$$\sin A = \frac{25.1 \sin 55° \ 40'}{8.94}$$

$$\sin A \approx 2.3184379$$

Since $\sin A$ cannot be greater than 1, there can be no such angle A and thus no triangle with the given information. An attempt to sketch such a triangle leads to the situation seen in Figure 49.   ◼

EXAMPLE 5

Solving a Triangle Using the Law of Sines (Two Triangles)

Solve triangle ABC if $A = 55° \ 20'$, $a = 22.8$ feet, and $b = 24.9$ feet.

SOLUTION   To begin, use the law of sines to find angle B.

$$\frac{a}{\sin A} = \frac{b}{\sin B}$$

$$\frac{22.8}{\sin 55° \ 20'} = \frac{24.9}{\sin B}$$

$$\sin B = \frac{24.9 \sin 55° \ 20'}{22.8}$$

$$\sin B \approx .89822938$$

Since $\sin B \approx .89822938$, to the nearest ten minutes we have one value of B as

$$B \approx 64° \ 00'$$

using the inverse sine function of a calculator. However, since supplementary angles have the same sine value, another *possible* value of B is

$$B \approx 180° - 64° \ 00' \approx 116° \ 00'.$$

To see if $B \approx 116° \ 00'$ is a valid possibility, simply add $116° \ 00'$ to the measure of the given value of A, $55° \ 20'$. Since $116° \ 00' + 55° \ 20' = 171° \ 20'$, and this sum is less than $180°$ (the sum of the angles of a triangle), we know that it is a valid angle measure for this triangle.

To keep track of these two different values of B, let

$$B_1 \approx 116° \ 00' \quad \text{and} \quad B_2 \approx 64° \ 00'.$$

Now separately solve triangles, $AB_1 C_1$ and $AB_2 C_2$ shown in Figure 50.

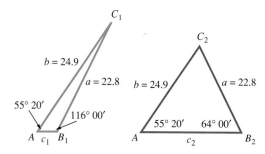

FIGURE 50

Let us begin with $AB_1 C_1$. Find C_1 first.

$$C_1 = 180° - A - B_1 \approx 8° \ 40'.$$

Now, use the law of sines to find c_1.

$$\frac{a}{\sin A} = \frac{c_1}{\sin C_1}$$

$$\frac{22.8}{\sin 55° \ 20'} = \frac{c_1}{\sin 8° \ 40'}$$

$$c_1 = \frac{22.8 \sin 8° \ 40'}{\sin 55° \ 20'}$$

$$c_1 \approx 4.18 \text{ feet}$$

To solve triangle $AB_2 C_2$, first find C_2.

$$C_2 = 180° - A - B_2 \approx 60° \ 40'$$

By the law of sines,

$$\frac{22.8}{\sin 55° \ 20'} = \frac{c_2}{\sin 60° \ 40'}$$

$$c_2 = \frac{22.8 \sin 60° \ 40'}{\sin 55° \ 20'}$$

$$c_2 \approx 24.2 \text{ feet.} \quad \blacksquare$$

CAUTION When solving a triangle using the type of data given in Example 5, do not forget to find the possible obtuse angle. The inverse sine function of the calculator will not give it directly. As we shall see in the next example, it is possible that the obtuse angle will not be a valid measure.

EXAMPLE 6

Solving a Triangle Using the Law of Sines (One Triangle)

Solve triangle ABC given $A = 43.5°$, $a = 10.7$ inches, and $b = 7.2$ inches.

SOLUTION To find angle B use the law of sines.

$$\frac{\sin B}{7.2} = \frac{\sin 43.5°}{10.7}$$

$$\sin B = \frac{7.2 \sin 43.5°}{10.7} \approx .46319186$$

The inverse sine function of the calculator gives us

$$B \approx 27.6°$$

as the acute angle. The other possible value of B is $180° - 27.6° \approx 152.4°$. However, when we add this possible obtuse angle to the given angle $A = 43.5°$, we get $152.4° + 43.5° = 195.9°$, which is greater than $180°$. So there can be only one triangle. Then angle $C \approx 180° - 27.6° - 43.5° = 108.9°$, and side c can be found with the law of sines.

$$\frac{c}{\sin 108.9°} = \frac{10.7}{\sin 43.5°}$$

$$c = \frac{10.7 \sin 108.9°}{\sin 43.5°}$$

$$c \approx 14.7 \text{ inches} \quad ▮$$

TECHNOLOGICAL NOTE
Graphics calculators can be programmed to use the law of sines to solve a triangle, given the appropriate data. You may wish to explore this further.

EXAMPLE 7

Analyzing Data Involving an Obtuse Angle

Without using the law of sines, explain why the data

$$A = 104°, \ a = 26.8 \text{ meters}, \ b = 31.3 \text{ meters}$$

cannot be valid for a triangle ABC.

SOLUTION Since A is an obtuse angle, the largest side of the triangle must be a, the side opposite A. However, we are given $b > a$, which is impossible if A is obtuse. Therefore, no such triangle ABC exists. ▮

8.5 EXERCISES

To the Student: Calculator Considerations
You should not need to use any functions on your calculator that you have not already learned. For the sake of accuracy, when making approximations, do not round off during intermediate steps—wait until the final step of your calculation to give the appropriate approximation.

Solve each of the following triangles.

1. $A = 37°$, $B = 48°$, $c = 18$ m

2. $B = 52°$, $C = 29°$, $a = 43$ cm

3. $A = 46° \ 30'$, $B = 52° \ 50'$, $b = 87.3$ mm

4. $A = 59° \ 30'$, $B = 48° \ 20'$, $b = 32.9$ m

5. $C = 74.08°$, $B = 69.38°$, $c = 45.38$ m

6. $A = 87.2°$, $b = 75.9$ yd, $C = 74.3°$

7. $B = 38° \ 40'$, $a = 19.7$ cm, $C = 91° \ 40'$

8. $B = 20° \ 50'$, $C = 103° \ 10'$, $AC = 132$ ft

9. $A = 35.3°$, $B = 52.8°$, $AC = 675$ ft

10. Explain why the law of sines cannot be used to solve a triangle if we are given the lengths of the three sides of a triangle.

11. In Example 1, we ask the question "Why does the Pythagorean theorem not apply?" Answer this question.

12. Kala Wanersdorfer, a perceptive trigonometry student, makes the statement "If we know *any* two angles and one side of a triangle, then the triangle is uniquely determined." Is this a valid statement? Explain, referring to the congruence axioms given in this section.

Solve each problem.

13. To find the distance AB across a river, a distance $BC = 354$ m is laid off on one side of the river. It is found that $B = 112°\,10'$ and $C = 15°\,20'$. Find AB.

14. To determine the distance RS across a deep canyon, Joanna lays off a distance $TR = 582$ yd. She then finds that $T = 32°\,50'$ and $R = 102°\,20'$. Find RS.

15. Radio direction finders are placed at points A and B, which are 3.46 mi apart on an east-west line, with A west of B. From A the bearing of a certain radio transmitter is 47.7°, and from B the bearing is 302.5°. Find the distance of the transmitter from A.

16. A ship is sailing due north. At a certain point the bearing of a lighthouse 12.5 km distant is N 38.8° E. Later on, the captain notices that the bearing of the lighthouse has become S 44.2° E. How far did the ship travel between the two observations of the lighthouse?

17. A folding chair is to have a seat 12.0 in deep with angles as shown in the figure. How far down from the seat should the crossing legs be joined? (Find x in the figure.)

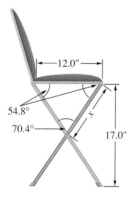

18. Mark notices that the bearing of a tree on the opposite bank of a river flowing north is 115.45°. Lisa is on the same bank as Mark, but 428.3 m away. She notices that the bearing of the tree is 45.47°. The two banks are parallel. What is the distance across the river?

19. Three gears are arranged as shown in the figure. Find angle θ.

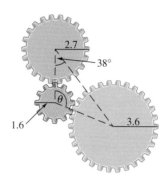

20. Three atoms with atomic radii of 2.0, 3.0, and 4.5 are arranged as in the figure. Find the distance between the centers of atoms A and C.

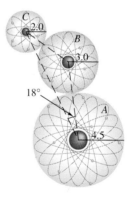

21. The bearing of a lighthouse from a ship was found to be N 37° E. After the ship sailed 2.5 miles due south, the new bearing was N 25° E. Find the distance between the ship and the lighthouse at each location.

22. A balloonist is directly above a straight road 1.5 miles long that joins two villages. She finds that the town closer to her is at an angle of depression of 35° and the farther town is at an angle of depression of 31°. How high above the ground is the balloon? (See the figure at the top of the next page.)

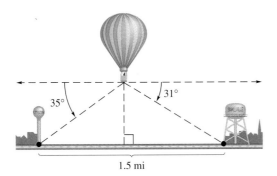

1.5 mi

23. From shore station A, a ship C is observed in the direction N 22.4° E. The same ship is observed to be in the direction N 10.6° W from shore station B, located at a distance of 25.5 km exactly southeast of A. Find the distance of the ship from station A.

24. A helicopter is sighted at the same time by two ground observers who are 3 mi apart on the same side of the helicopter. (See the figure.) They report the angles of elevation as 20.5° and 27.8°. How high is the helicopter?

20.5° 27.8°
3 mi

25. A rocket tracking station has two telescopes T_1 and T_2, placed 1.73 km apart, that lock onto the rocket and continuously transmit the angles of elevation to a computer. Find the distance to the rocket from T_1 at the moment when the angles of elevation are 28.1° and 79.5°, as shown in the figure.

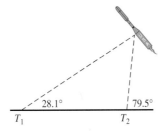

28.1° 79.5°
T_1 T_2

26. A surveyor standing 48.0 m from the base of a building measures the angle to the top of the building and finds it to be 37.4°. (See the figure.) The surveyor then measures the angle to the top of a clock tower on the building, finding that it is 45.6°. Find the height of the clock tower.

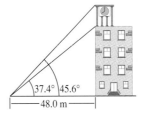

37.4° 45.6°
48.0 m

27. A slider crank mechanism is shown in the figure. Find the distance between the wrist pin W and the connecting rod center C.

Fixed Pin — 11.2 cm — P — 28.6 cm — Slider
25.5° C W Track

28. A satellite S is traveling in a circular orbit 1600 km above the earth. It will pass directly over a tracking station T at noon. The satellite takes 2 hours to make a complete orbit. Assume that the radius of the earth is 6400 km. The tracking antenna is aimed 30° above the horizon. At what time will the satellite pass through the beam of the antenna?*

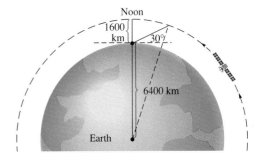

Noon
1600 km 30°
6400 km
Earth

*From *Space Mathematics* by Bernice Kastner, Ph.D. Copyright © 1972 by the National Aeronautics and Space Administration. Courtesy of NASA.

In each set of data for triangle ABC, two sides and an angle opposite one of them are given. As explained in Examples 4–6, such a set of data may lead to 0, 1, or 2 triangles. Solve for all possible triangles.

29. $A = 42.5°$, $a = 15.6$ ft, $b = 8.14$ ft

30. $C = 52.3°$, $a = 32.5$ yd, $c = 59.8$ yd

31. $B = 72.2°$, $b = 78.3$ m, $c = 145$ m

32. $C = 68.5°$, $c = 258$ cm, $b = 386$ cm

33. $A = 38° 40'$, $a = 9.72$ km, $b = 11.8$ km

34. $C = 29° 50'$, $a = 8.61$ m, $c = 5.21$ m

35. $B = 32° 50'$, $a = 7540$ cm, $b = 5180$ cm

36. $C = 22° 50'$, $b = 159$ mm, $c = 132$ mm

37. $A = 96.80°$, $b = 3.589$ ft, $a = 5.818$ ft

38. $C = 88.70°$, $b = 56.87$ yd, $c = 112.4$ yd

39. Apply the law of sines to the following: $a = \sqrt{5}$, $c = 2\sqrt{5}$, $A = 30°$. What is the value of $\sin C$? What is the measure of C? Based on its angle measures, what kind of triangle is triangle ABC?

40. In your own words, explain the condition that must exist to determine that there is no triangle satisfying the given values of a, b, and B, once the value of $\sin B$ is found.

41. Without using the law of sines, explain why no triangle ABC exists satisfying $A = 103° 20'$, $a = 14.6$ ft, $b = 20.4$ ft.

42. Apply the law of sines to the data given in Example 7. Describe in your own words what happens when you try to find the measure of angle B using a calculator.

43. A surveyor reported the following data about a piece of property: "The property is triangular in shape, with dimensions as shown in the figure." Use the law of sines to see whether such a piece of property could exist.

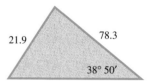

Can such a triangle exist?

44. The surveyor tries again: "A second triangular piece of property has dimensions as shown." This time it turns out that the surveyor did not consider every possible case. Use the law of sines to show why.

When a light ray travels from one medium, such as air, to another medium, such as water or glass, the speed of the light changes, and the direction that the ray is traveling changes. (This is why a fish under water is in a different position than it appears to be.) These changes are given by Snell's law

$$\frac{c_1}{c_2} = \frac{\sin \theta_1}{\sin \theta_2},$$

where c_1 is the speed of light in the first medium, c_2 is the speed of light in the second medium, and θ_1, and θ_2 are the angles shown in the figure. In the following exercises, assume that $c_1 = 3 \times 10^8$ m per sec. Find the speed of light in the second medium.

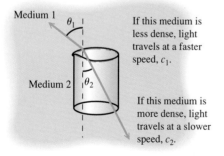

45. $\theta_1 = 46°$, $\theta_2 = 31°$

46. $\theta_1 = 39°$, $\theta_2 = 28°$

Find θ_2 for the following values of θ_1 and c_2. Round to the nearest degree.

47. $\theta_1 = 40°$, $c_2 = 1.5 \times 10^8$ m per sec

48. $\theta_1 = 62°$, $c_2 = 2.6 \times 10^8$ m per sec

49. The figure shows a fish's view of the world above the surface of the water.*

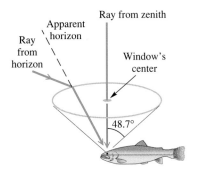

Suppose that a light ray comes from the horizon, enters the water, and strikes the fish's eye. Let us assume that this ray gives a value of 90° for angle θ_1 in the formula for Snell's law. (In a practical situation this angle would probably be a little less than 90°.) The speed of light in water is about 2.254×10^8 m per sec. Find angle θ_2.

50. What is wrong with this statement, given by a student as the law of sines?

$$\frac{a}{A} = \frac{b}{B} = \frac{c}{C}$$

8.6 THE LAW OF COSINES AND AREA FORMULAS

Introduction ▮ Derivation of the Law of Cosines ▮ Applications ▮ Heron's Formula ▮ Another Area Formula

Introduction

Recall from the previous section that if we are given two sides and the included angle or three sides of a triangle, a unique triangle is formed. These are the SAS and SSS cases, respectively. In both cases, however, we cannot begin the solution of the triangle by using the law of sines. Both of these cases require the use of the law of cosines, introduced in this section.

It will be helpful to remember the following property of triangles when applying the law of cosines.

> **RESTRICTION ON TRIANGLE SIDE LENGTHS**
>
> In any triangle, the sum of the lengths of any two sides must be greater than the length of the remaining side.

For example, it would be impossible to construct a triangle with sides of lengths 3, 4, and 10. See Figure 51.

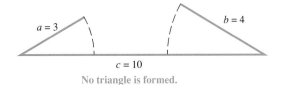

No triangle is formed.

FIGURE 51

666   CHAPTER 8   The Trigonometric Functions and Applications

Derivation of the Law of Cosines

To derive the law of cosines, let *ABC* be any oblique triangle. Choose a coordinate system so that vertex *B* is at the origin and side *BC* is along the positive *x*-axis. See Figure 52.

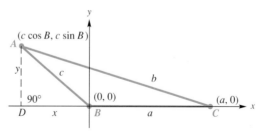

FIGURE 52

Let (x, y) be the coordinates of vertex *A* of the triangle. Verify that for angle *B*, whether obtuse or acute,

$$\sin B = \frac{y}{c} \qquad \text{and} \qquad \cos B = \frac{x}{c}.$$

(Here we assume that *x* is negative if *B* is obtuse.) From these results

$$y = c \sin B \qquad \text{and} \qquad x = c \cos B,$$

so that the coordinates of point *A* become

$$(c \cos B, c \sin B).$$

Point *C* has coordinates $(a, 0)$, and *AC* has length *b*. By the distance formula,

$$b = \sqrt{(c \cos B - a)^2 + (c \sin B)^2}.$$

Squaring both sides and simplifying gives

$$\begin{aligned}
b^2 &= (c \cos B - a)^2 + (c \sin B)^2 \\
&= c^2 \cos^2 B - 2ac \cos B + a^2 + c^2 \sin^2 B \\
&= a^2 + c^2(\cos^2 B + \sin^2 B) - 2ac \cos B \\
&= a^2 + c^2(1) - 2ac \cos B \\
&= a^2 + c^2 + 2ac \cos B.
\end{aligned}$$

This result is one form of the law of cosines. In the work above, we could just as easily have placed *A* or *C* at the origin. This would have given the same result, but with the variables rearranged. These various forms of the law of cosines are summarized in the following theorem.

LAW OF COSINES

In any triangle *ABC*, with sides *a*, *b*, and *c*,

$$a^2 = b^2 + c^2 - 2bc \cos A$$
$$b^2 = a^2 + c^2 - 2ac \cos B$$
$$c^2 = a^2 + b^2 - 2ab \cos C.$$

The law of cosines says that the square of a side of a triangle is equal to the sum of the squares of the other two sides, minus twice the product of the two sides and the cosine of the angle included between them.

NOTE If we let $C = 90°$ in the third form of the law of cosines given above, we have $\cos C = \cos 90° = 0$, and the formula becomes

$$c^2 = a^2 + b^2,$$

the familiar equation of the Pythagorean theorem. Thus, the Pythagorean theorem is a special case of the law of cosines.

Applications

We will now investigate how the law of cosines is used in applied situations.

EXAMPLE 1

Using the Law of Cosines in an Application

A surveyor wishes to find the distance between two inaccessible points A and B on opposite sides of a lake. While standing at point C, she finds that $AC = 259$ meters, $BC = 423$ meters, and angle ACB measures $132° \, 40'$. Find the distance AB. (See Figure 53.)

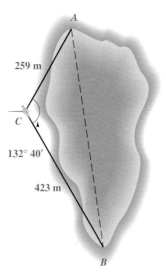

259 m

C

$132° \, 40'$

423 m

A

B

FIGURE 53

SOLUTION The law of cosines can be used here, since we know the lengths of two sides of the triangle and the measure of the included angle.

$$AB^2 = 259^2 + 423^2 - 2(259)(423) \cos 132° \, 40'$$
$$AB^2 \approx 394{,}510.6 \quad \text{Use a calculator.}$$
$$AB \approx 628 \quad \text{Take the square root and round to 3 significant digits.}$$

The distance between the points is approximately 628 meters. █

EXAMPLE 2

Using the Law of Cosines to Solve a Triangle

Solve triangle ABC if $A = 42.3°$, $b = 12.9$ meters, and $c = 15.4$ meters. (See Figure 54.)

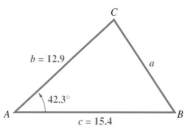

FIGURE 54

SOLUTION Start by finding a with the law of cosines.

$$a^2 = b^2 + c^2 - 2bc \cos A$$
$$a^2 = 12.9^2 + 15.4^2 - 2(12.9)(15.4) \cos 42.3°$$
$$a^2 \approx 109.7$$
$$a \approx \sqrt{109.7} \text{ meters} \qquad \text{Leave in this form for now.}$$

We now must find the measures of angles B and C. There are several approaches that can be used at this point. Let us use the law of sines to find one of these angles. Of the two remaining angles, B must be the smaller since it is opposite the shorter of the two sides b and c. Therefore, it cannot be obtuse, and we will avoid any ambiguity when we find its sine.

$$\frac{\sin 42.3°}{\sqrt{109.7}} = \frac{\sin B}{12.9}$$

$$\sin B = \frac{12.9 \sin 42.3°}{\sqrt{109.7}}$$

$$B \approx 56.0° \qquad \text{Use the inverse sine function of a calculator.}$$

The easiest way to find C is to subtract the sum of A and B from $180°$.

$$C = 180° - A - B \approx 81.7°.$$

$\sqrt{109.7} \approx 10.5$, so $a \approx 10.5$, and the triangle is solved. ◻

CAUTION Had we chosen to use the law of sines to find C rather than B in Example 2, we would not have known whether C equals $81.9°$ or its supplement, $98.1°$.

EXAMPLE 3

Using the Law of Cosines to Solve a Triangle

Solve triangle ABC if $a = 9.47$ feet, $b = 15.9$ feet, and $c = 21.1$ feet.

SOLUTION We are given the lengths of three sides of the triangle, so we may use the law of cosines to solve for any angle of the triangle. Let us solve for C, the largest angle, using the law of cosines. We will be able to tell if C is obtuse if $\cos C < 0$. Use the form of the law of cosines that involves C.

$$c^2 = a^2 + b^2 - 2ab \cos C,$$

or $\qquad \cos C = \dfrac{a^2 + b^2 - c^2}{2ab}.$

Inserting the given values leads to

$$\cos C = \frac{(9.47)^2 + (15.9)^2 - (21.1)^2}{2(9.47)(15.9)}$$

$$\cos C \approx -.34109402. \qquad \text{Use a calculator.}$$

Using the inverse cosine function of the calculator, we get the obtuse angle C. (We will see why this is so in the next chapter.)

$$C \approx 109.9°$$

We can use either the law of sines or the law of cosines to find $B \approx 45.1°$. (Verify this.) Since $A = 180° - B - C$,

$$A \approx 25.0°. \quad \blacksquare$$

As shown in this section and the previous one, four possible cases can occur when solving an oblique triangle. These cases are summarized in the chart that follows, along with a suggested procedure for solving in each case. There are other procedures that work, but we give the one that is most efficient. In all four cases, it is assumed that the given information actually produces a triangle.

Case	Suggested Procedure for Solving
One side and two angles are known. (SAA or ASA)	**1.** Find the remaining angle using the angle sum formula ($A + B + C = 180°$). **2.** Find the remaining sides using the law of sines.
Two sides and one angle (not included between the two sides) are known. (SSA)	*Be aware of the ambiguous case; there may be two triangles.* **1.** Find an angle using the law of sines. **2.** Find the remaining angle using the angle sum formula. **3.** Find the remaining side using the law of sines. *If two triangles exist, repeat Steps 1, 2, and 3.*
Two sides and the included angle are known. (SAS)	**1.** Find the third side using the law of cosines. **2.** Find the smaller of the two remaining angles using the law of sines. **3.** Find the remaining angle using the angle sum formula.
Three sides are known. (SSS)	**1.** Find the largest angle using the law of cosines. **2.** Find either remaining angle using the law of sines. **3.** Find the remaining angle using the angle sum formula.

Heron's Formula

The law of cosines can be used to derive a formula for the area of a triangle when only the lengths of the three sides are known. This formula is known as Heron's formula, named after the Greek mathematician Heron of Alexandria, who lived around A.D. 75. It is found in his work *Metrica*.

HERON'S AREA FORMULA

If a triangle has sides of lengths a, b and c, and if the **semiperimeter** is

$$s = \frac{1}{2}(a + b + c),$$

then the area A of the triangle is

$$A = \sqrt{s(s - a)(s - b)(s - c)}.$$

A proof of Heron's formula is suggested in the exercises of Section 9.4.

EXAMPLE **4**
Using Heron's Formula to Find an Area

The distance "as the crow flies" from Los Angeles to New York is 2451 miles, from New York to Montreal is 331 miles, and from Montreal to Los Angeles is 2427 miles. What is the area of the triangular region having these three cities as vertices?

SOLUTION Figure 55 shows that we can let $a = 2451$, $b = 331$, and $c = 2427$. Then the semiperimeter s is given by

$$s = \frac{1}{2}(2451 + 331 + 2427) = 2604.5.$$

Using Heron's formula, the area A is found as follows.

$$A = \sqrt{s(s - a)(s - b)(s - c)}$$
$$A = \sqrt{2604.5(2604.5 - 2451)(2604.5 - 331)(2604.5 - 2427)}$$
$$A \approx 401,700$$

The area of the triangular region is approximately 401,700 square miles. ∎

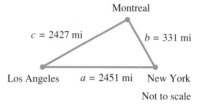

Montreal
$c = 2427$ mi $b = 331$ mi
Los Angeles $a = 2451$ mi   New York
Not to scale

FIGURE 55

FOR **GROUP DISCUSSION**

1. If your class is held in a rectangular-shaped room, measure the length and the width, and then multiply them to find the area. Now, measure a diagonal of the room, and use Heron's formula with the length, the width, and the diagonal to find the area of half the room. Double this result. Do your area calculations agree?

2. For a triangle to exist, the sum of the lengths of any two sides must exceed the length of the remaining side. Have half of the class try to calculate the area of a "triangle" with $a = 4$, $b = 8$, and $c = 12$, while the other half tries to calculate the area of the "triangle" with $a = 10$, $b = 20$, and $c = 34$. In both cases, use Heron's formula. Then, discuss the results, drawing diagrams on the chalkboard to support the results obtained.

3. A popular textbook for mathematics survey courses contains the following problem and diagram: *Find the perimeter and area of the shaded region. (See Figure 56.)*

5 cm   6 cm
2 cm
7 cm

FIGURE 56

The perimeter, obviously, is 18 cm. Now, divide the class into two groups. Have one group determine the area using area $= (\frac{1}{2})$(base)(height), and have the other group determine the area using Heron's formula. Then discuss your results. What is the problem with this problem?

Another Area Formula

If we know the measures of two sides of a triangle and the angle included between them, then we can find the area A of the triangle using the following formula.

AREA OF A TRIANGLE

In any triangle ABC, the area $\mathcal{A}$ is given by any of the following formulas:

$$\mathcal{A} = \frac{1}{2} bc \sin A, \qquad \mathcal{A} = \frac{1}{2} ab \sin C, \qquad \mathcal{A} = \frac{1}{2} ac \sin B.$$

That is, the area is given by half the product of the lengths of two sides and the sine of the angle included between them.

The derivation of this formula is outlined in Exercises 61–64.

EXAMPLE 5

Finding the Area of a Triangle Using $\mathcal{A} = \frac{1}{2} ab \sin C$

Find the area of triangle ABC if $A = 24° \, 40'$, $b = 27.3$ centimeters, and $C = 52° \, 40'$.

SOLUTION Before we can use the formula given above, we must use the law of sines to find either a or c. Since the sum of the measures of the angles of any triangle is 180°,

$$B = 180° - 24° \, 40' - 52° \, 40' = 102° \, 40'.$$

Now use the form of the law of sines that relates a, b, A, and B to find a.

$$\frac{a}{\sin A} = \frac{b}{\sin B}$$

$$\frac{a}{\sin 24° \, 40'} = \frac{27.3}{\sin 102° \, 40'}$$

Solve for a to verify that $a \approx 11.7$ centimeters. Now find the area.

$$\mathcal{A} = \frac{1}{2} ab \sin C \approx 127 \text{ (with } C = 52° \, 40')$$

The area of triangle ABC is 127 square centimeters (to three significant digits).

8.6 EXERCISES

To the Student: Calculator Considerations
The law of cosines and the two area formulas are good examples of formulas that can be programmed into modern graphics calculators. You may wish to attempt to program them. Read your owner's manual for guidelines to programming. There are also programs available from user's groups, newsletters on technology in mathematics, etc., and you may wish to investigate these sources.

Solve each of the following triangles using the laws of cosines and sines as necessary. In Exercises 7–12, give answers in degrees.

1. $C = 28.3°$, $b = 5.71$ in, $a = 4.21$ in

2. $A = 41.4°$, $b = 2.78$ yd, $c = 3.92$ yd

3. $C = 45.6°$, $b = 8.94$ m, $a = 7.23$ m

4. $A = 67.3°$, $b = 37.9$ km, $c = 40.8$ km

5. $A = 80° \, 40'$, $b = 143$ cm, $c = 89.6$ cm

6. $C = 72° \, 40'$, $a = 327$ ft, $b = 251$ ft

7. $a = 9.3$ cm, $b = 5.7$ cm, $c = 8.2$ cm

8. $a = 28$ ft, $b = 47$ ft, $c = 58$ ft

9. $a = 42.9$ m, $b = 37.6$ m, $c = 62.7$ m

10. $a = 189$ yd, $b = 214$ yd, $c = 325$ yd

11. $AB = 1240$ ft, $AC = 876$ ft, $BC = 965$ ft

12. $AB = 298$ m, $AC = 421$ m, $BC = 324$ m

13. Refer to Figure 51. If you attempt to find any angle of a triangle using the values $a = 3$, $b = 4$, and $c = 10$ with the law of cosines, what happens?

14. A familiar saying is "The shortest distance between two points is a straight line." Explain how this relates to the geometric property that states that the sum of the lengths of any two sides of a triangle must be greater than the remaining side.

Solve each problem.

15. Points A and B are on opposite sides of Lake Yankee. From a third point, C, the angle between the lines of sight to A and B is 46.3°. If AC is 350 m long and BC is 286 m long, find AB.

16. The sides of a parallelogram are 4.0 cm and 6.0 cm. One angle is 58° while another is 122°. Find the lengths of the diagonals of the parallelogram.

17. Airports A and B are 450 km apart, on an east-west line. Tom flies in a northeast direction from A to airport C. From C he flies 359 km on a bearing of 128° 40′ to B. How far is C from A?

18. Two ships leave a harbor together, traveling on courses that have an angle of 135° 40′ between them. If they each travel 402 mi, how far apart are they?

19. A ship is sailing east. At one point, the bearing of a submerged rock is 45° 20′. After sailing 15.2 mi, the bearing of the rock has become 308° 40′. Find the distance of the ship from the rock at the latter point.

20. From an airplane flying over the ocean, the angle of depression to a submarine lying just under the surface is 24° 10′. At the same moment the angle of depression from the airplane to a battleship is 17° 30′. (See the figure.) The distance from the airplane to the battleship is 5120 ft. Find the distance between the battleship and the submarine. (Assume the airplane, submarine, and battleship are in a vertical plane.)

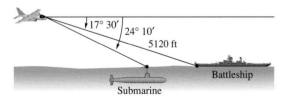

21. Two boats leave a dock together. Each travels in a straight line. The angle between their courses measures 54° 10′. One boat travels 36.2 km per hr, and the other travels 45.6 km per hr. How far apart will they be after 3 hr?

22. Find the lengths of both diagonals of a parallelogram with adjacent sides of 12 cm and 15 cm if the angle between these sides is 33°.

23. A crane with a counterweight is shown in the figure. Find the horizontal distance between points A and B.

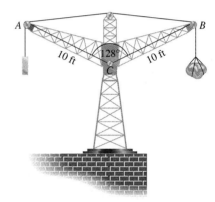

24. A weight is supported by cables attached to both ends of a balance beam, as shown in the figure. What angles are formed between the beam and the cables?

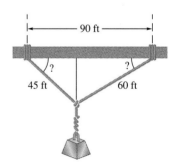

25. To measure the distance through a mountain for a proposed tunnel, a point C is chosen that can be reached from each end of the tunnel. (See the figure at the top of the next page.) If $AC = 3800$ m,

$BC = 2900$ m, and angle $C = 110°$, find the length of the tunnel.

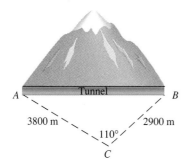

26. A baseball diamond is a square, 90 ft on a side, with home plate and the three bases as vertices. The pitcher's rubber is located 60.5 ft from home plate. Find the distance from the pitcher's rubber to each of the bases.

27. The Vietnam Veterans' Memorial in Washington, D.C., is in the shape of an unenclosed isosceles triangle (that is, V-shaped) with equal sides of length 246.75 feet and the angle between these sides measuring $125° \, 12'$. Find the distance between the ends of the two equal sides.

28. Starting at point A, a ship sails 18.5 km on a bearing of 189°, then turns and sails 47.8 km on a bearing of 317°. Find the distance of the ship from point A.

29. Two towns 21 mi apart are separated by a dense forest. (See the figure.) To travel from town A to town B, a person must go 17 mi on a bearing of 325°, then turn and continue for 9 mi to reach town B. Find the bearing of B from A.

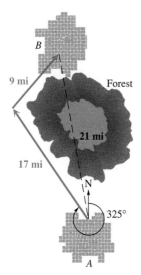

30. Two factories blow their whistles at exactly 5:00. A man hears the two blasts at 3 seconds and 6 seconds after 5:00, respectively. The angle between his lines of sight to the two factories is 42.2°. If sound travels 344 m per sec, how far apart are the factories?

31. A satellite traveling in a circular orbit 1600 km above earth is due to pass directly over a tracking station at noon.* (See the figure.) Assume that the satellite takes 2 hr to make an orbit and that the radius of the earth is 6400 km. Find the distance between the satellite and the tracking station at 12:03 P.M.

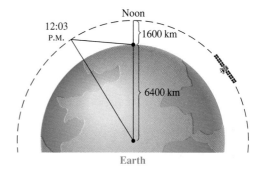

32. A ship sailing due east in the North Atlantic has been warned to change course to avoid a group of icebergs. The captain turns and sails on a bearing of 62° for a while, then changes course again to a bearing of 115° until the ship reaches its original course. (See the figure.) How much farther did the ship have to travel to avoid the icebergs?

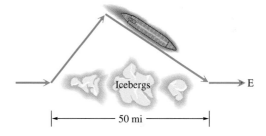

33. A parallelogram has sides of length 25.9 cm and 32.5 cm. The longer diagonal has a length of 57.8 cm. Find the angle opposite the diagonal.

34. A person in a plane flying a straight course observes a mountain at a bearing 24.1° to the right of its course. At that time the plane is 7.92 km from the mountain. A short time later, the bearing to the mountain becomes 32.7°. How far is the airplane from the mountain when the second bearing is taken?

---

*From *Space Mathematics* by Bernice Kastner, Ph.D. Copyright © 1972 by the National Aeronautics and Space Administration. Courtesy of NASA.

To help predict eruptions from the volcano Mauna Loa on the island of Hawaii, scientists keep track of the volcano's movement by using a "super triangle" with vertices on the three volcanoes shown on the map below. (For example, in a recent year, Mauna Loa moved 6 inches, a result of increasing internal pressure.) Refer to the map to work Exercises 35 and 36.

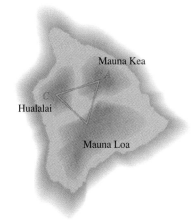

35. $AB = 22.47928$ mi, $AC = 28.14276$ mi, $A = 58.56989°$; find BC

36. $AB = 22.47928$ mi, $BC = 25.24983$ mi, $A = 58.56989°$; find B

37. The layout for a child's playhouse has the dimensions given in the figure. Find x.

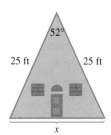

38. To find the distance between two small towns, an Electronic Distance Measuring (EDM) instrument is placed on a hill from which both towns are visible. The distance to each town from the EDM and the angle between the two lines of sight are measured. (See the figure.) Find the distance between the towns.

Find the area of each triangle by using one of the area formulas introduced in this section.

39. $A = 42.5°$, $b = 13.6$ m, $c = 10.1$ m

40. $B = 124.5°$, $a = 30.4$ cm, $c = 28.4$ cm

41. $A = 56.80°$, $b = 32.67$ in, $c = 52.89$ in

42. $A = 24° 25'$, $B = 56° 20'$, $c = 78.40$ cm

43. $a = 12$ m, $b = 16$ m, $c = 25$ m

44. $a = 154$ cm, $b = 179$ cm, $c = 183$ cm

45. $a = 76.3$ ft, $b = 109$ ft, $c = 98.8$ ft

46. $a = 22$ in, $b = 45$ in, $c = 31$ in

47. $a = 25.4$ yd, $b = 38.2$ yd, $c = 19.8$ yd

48. $a = 15.89$ in, $b = 21.74$ in, $c = 10.92$ in

Solve each problem.

49. A painter is going to apply a special coating to a triangular metal plate on a new building. Two sides measure 16.1 m and 15.2 m. She knows that the angle between these sides is 125°. What is the area of the surface she plans to cover with the coating?

50. A real estate agent wants to find the area of a triangular lot. A surveyor takes measurements and finds that two sides are 52.1 m and 21.3 m, and the angle between them is 42.2°. What is the area of the lot?

51. A painter needs to cover a triangular region 75 m by 68 m by 85 m. A can of paint covers 75 sq m of area. How many cans (to the next higher number of cans) will be needed?

52. Find the area of the Bermuda Triangle, if the sides of the triangle have the approximate lengths 850 miles, 925 miles, and 1300 miles.

Relating Concepts

In Exercises 54 and 55 of Section 7.4, we showed how determinants can be used to find the area A of a triangle in a coordinate plane, using the coordinates of the vertices of the triangle. Consider the triangle from Exercise 54(c) in that section: triangle PQR has vertices at P(2, 5), Q(−1, 3), and R(4, 0).

53. Sketch the triangle by hand in a coordinate plane.

54. Use the distance formula to find the lengths *PQ*, *QR*, and *PR*.

55. What is the semiperimeter of triangle *PQR*?

56. Use Heron's formula to find the area *A* of triangle *PQR*.

57. According to the original problem in Section 7.4, the area *A* of the triangle is given by the formula

$$A = \frac{1}{2} \det \begin{bmatrix} 2 & 5 & 1 \\ -1 & 3 & 1 \\ 4 & 0 & 1 \end{bmatrix}.$$

Evaluate the area using this formula. Does it agree with your answer in Exercise 56?

58. Find the measure of angle *PQR* by using the law of cosines and the lengths found in Exercise 54.

59. Use the second area formula given in this section, along with lengths *PQ*, *QR*, and the measure of angle *PQR* from Exercise 58 to compute the area of triangle *PQR*. Does it agree with your answers in Exercises 56 and 57?

60. Write a short paragraph explaining what you have learned by working Exercises 53–59.

Use the figures below to derive the area formula $\mathscr{A} = \frac{1}{2}bc \sin A$. Work Exercises 61–64 in order.

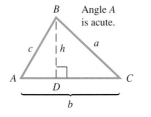
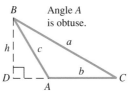

61. Using triangle *ABD*, find an expression for sin *A* in terms of *h* and *c*.

62. Solve for *h*.

63. The familiar area formula for a triangle, area = $\frac{1}{2}$(base)(height), can now be applied. Since the height to base *b* is *h*, use the expression for *h* found in Exercise 62, and write the formula in terms of *b* and that expression.

64. Explain why, if *A* = 90°, the formula $\mathscr{A} = \frac{1}{2}bc \sin A$ becomes the familiar area formula.

Chapter 8   **SUMMARY**

An angle is formed by a rotation of a ray about its endpoint. Angles are measured in degrees or radians. Rotations smaller than a degree are measured in minutes and rotations smaller than one minute are measured in seconds. Sometimes decimal degrees are used instead of minutes and seconds. Radian measure is the ratio of the arc length cut off by a central angle of a circle to the radius of the circle. We use the relationship $180° = \pi$ radians to convert between degrees and radians.

The length of an arc of a circle is given by $s = r\theta$, where θ is measured in radians and r is the radius of the circle. The linear velocity of a point moving along a circle tells how fast the point is moving. As the point moves along the circle, the positive angle in standard position also changes. The rate at which the angle changes is the angular velocity. Linear and angular velocity have many practical everyday applications.

The six trigonometric functions are defined as ratios of the quantities x, y, and r, where (x, y) is a point on the terminal side of an angle in standard position and r is the distance from the point to the origin. Since r is the square root of the sum of the squares of x and y, if the point (x, y) is known, r and the six trigonometric function values can be found. By knowing the signs of each of the trigonometric functions in each quadrant, we can find all trigonometric functions of an angle given a trigonometric function and the quadrant. The terminal side of a quadrantal angle coincides with an axis. These angles have special trigonometric function values that may be undefined.

The definitions of the six trigonometric functions indicate that there are three pairs of reciprocal functions. These reciprocal relationships are called the reciprocal identities. Three Pythagorean identities are formed from the relationship $x^2 + y^2 = r^2$. The definitions of the tangent and cotangent functions lead to two quotient identities that express these functions as quotients of two trigonometric functions. These identities are useful for finding the trigonometric functions of an angle in standard position given one function value and the quadrant of the angle.

The definitions of the trigonometric functions can be restated for any acute angle A in a right triangle in terms of the opposite side, the adjacent side, or the hypotenuse. We use these definitions to find exact trigonometric function values for 30°, 60°, and 45°. These definitions are also useful in applied problems.

Cofunctions of complementary angles have equal values. Thus, if A and B are complements, $\sin A = \cos B$, $\tan A = \cot B$, and $\sec A = \csc B$. Reference angles are used to find function values for angles in standard position by using the function values of first quadrant angles. A reference angle is the positive acute angle measured between the terminal side of the angle and the x-axis.

Solving a right triangle means finding the measures of all its angles and sides. We do this by using the trigonometric function ratios. The degree of accuracy depends on the accuracy of the given information.

Practical problems often require using or finding the angle of elevation or depression. These angles are always measured from the horizontal. Some applications involve bearing, which is measured in two ways. One way is to measure in a clockwise direction from due north. With another method we measure the angle from due north to the west or east or from due south to the west or east.

The law of sines and the law of cosines are used to solve oblique triangles. By the law of sines the ratio of any side of a triangle to the sine of the opposite angle equals the corresponding ratio of another side to the sine of its opposite angle. If we are given two sides and the angle opposite one of them, the information may lead to no triangle, one triangle, or two triangles. This situation is called the ambiguous case of the law of sines. It is useful to remember that the largest angle must be opposite the largest side, and the smallest angle must be opposite the smallest side.

The law of cosines says that the square of a side of a triangle is equal to the sum of the squares of the other two sides, minus twice the product of the two sides and the cosine of the included angle. Note the similarity to the Pythagorean theorem. When using the law of cosines, remember that the sum of the lengths of any two sides of a triangle must be greater than the length of the third side.

The laws of sines and cosines lead to two formulas for area that can be used when the height of a triangle is not known. If the lengths of the three sides of the triangle are known, Heron's formula can be used. If the measures of two sides of a triangle and the included angle are known, an alternate formula can be used. See Section 8.6 for these two formulas.

Key Terms

SECTION 8.1

line *AB*
segment *AB*
ray *AB*
angle
initial side
terminal side
vertex
positive angle
negative angle
degree
acute angle
right angle
obtuse angle
straight angle
complementary angles (complements)
supplementary angles (supplements)
minute
second
standard position
quadrantal angle
coterminal angles
central angle
radian
arc
linear velocity
angular velocity

SECTION 8.2

trigonometric functions
sine
cosine
tangent
cotangent
secant
cosecant

SECTION 8.3

side opposite an angle
side adjacent to an angle
cofunctions
reference angle

SECTION 8.4

significant digit
exact number
angle of elevation
angle of depression
bearing
congruent triangles
oblique triangle

Chapter 8   **REVIEW EXERCISES**

Let θ represent a −300° angle in Exercises 1–5.

1. Sketch θ in standard position.

2. Name the smallest positive angle coterminal with θ. Use degree measure.

3. Name a negative angle coterminal with θ. Use degree measure.

4. Give an expression that represents all angles coterminal with θ, using degree measure and letting *n* represent any integer.

5. Convert θ to radians. Leave π in your answer.

6. Find time t if central angle $\theta = \frac{5\pi}{12}$ radians and angular velocity $\omega = \frac{8\pi}{9}$ radians per second.

7. Find angular velocity ω if arc length $s = \frac{12\pi}{25}$ feet, radius $r = \frac{3}{5}$ feet, and time $t = 15$ seconds.

8. Find the linear velocity of a point on the edge of a flywheel of radius 7 meters, if the flywheel is rotating 90 times per second.

9. The radius of a circle is 15.2 centimeters. Find the length of an arc of the circle intercepted by a central angle of $\frac{3\pi}{4}$ radians.

10. What is the length of the arc intercepted by the hands of a clock at 5:00, if the radius is 12 inches?

11. Assuming that the radius of the earth is 6400 kilometers, what is the distance between cities on a north-south line that are on latitudes 28°N and 12°S, respectively?

Let θ be an angle in standard position, with the point $(-2, -7)$ on its terminal side. Find the exact value of each of the following.

12. $\sin \theta$ **13.** $\cos \theta$ **14.** $\tan \theta$

15. $\csc \theta$ **16.** $\sec \theta$ **17.** $\cot \theta$

18. For the angle θ described in the directions for Exercises 12–17 **(a)** give to the nearest hundredth the measure of θ if $0° \le \theta < 360°$, and **(b)** give the measure of the reference angle for θ to the nearest hundredth of a degree.

Consider an angle θ in standard position whose terminal side has the equation $y = -5x$, with $x \le 0$.

19. Sketch θ and use an arrow to show the rotation if $0° \le \theta < 360°$.

20. Find the exact values of $\sin \theta$ and $\cos \theta$.

21. Give the measure of θ to the nearest minute.

Find the quadrant in which the terminal side of angle θ must lie given the conditions described.

22. $\cos \theta < 0$ and $\tan \theta > 0$ **23.** $\sin \theta > 0$ and $\tan \theta < 0$

Suppose that $\sin \theta = \frac{\sqrt{3}}{5}$ and $\cos \theta < 0$. Find each of the following exact function values.

24. $\cos \theta$ **25.** $\csc \theta$ **26.** $\cot \theta$

Find the exact value for each function. If it is undefined, say so.

27. $\sin(-225°)$ **28.** $\cos(-3\pi)$ **29.** $\tan\left(-\dfrac{3\pi}{2}\right)$ **30.** $\csc\dfrac{5\pi}{3}$

31. $\sec 420°$ **32.** $\cot\left(-\dfrac{5\pi}{2}\right)$ **33.** $\sin\left(-\dfrac{15\pi}{4}\right)$ **34.** $\cos(-510°)$

35. $\tan\dfrac{13\pi}{3}$ **36.** $\csc(-13\pi)$ **37.** $\sec(180° + n \cdot 360°)$, where n is an integer   **38.** $\cot\dfrac{17\pi}{6}$

Use a calculator to find an approximation of each function value.

39. $\sin 146° \, 40'$ **40.** $\sec 5$ **41.** $\tan\left(-\dfrac{\pi}{5}\right)$ **42.** $\sin^{-1}.38$ (Give answer in degrees.)

Use the right triangle shown, and give the exact value of the specified function.

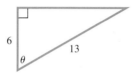

43. $\sin \theta$

44. $\sec(90° - \theta)$

45. Find the degree measure of θ correct to the nearest tenth.

Solve the right triangle with $C = 90°$.

46. $A = 39.72°$, $b = 38.97$ m

47. $a = 270.0$ m, $b = 298.6$ m

Solve the problem.

48. The angle of elevation from a point 93.2 feet from the base of a tower to the top of the tower is $38° \, 20'$. Find the height of the tower.

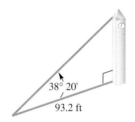

49. The angle of depression of a television tower to a point on the ground 36.0 meters from the bottom of the tower is $29.5°$. Find the height of the tower.

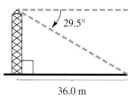

50. A rectangle has adjacent sides measuring 10.93 centimeters and 15.24 centimeters. The angle between the diagonal and the longer side is $35.65°$. Find the length of the diagonal.

51. An isosceles triangle has a base of length 49.28 meters. The angle opposite the base is $58.746°$. Find the length of each of the two equal sides.

52. The bearing of B from C is $254°$. The bearing of A from C is $344°$. The bearing of A from B is $32°$. The distance from A to C is 780 meters. Find the distance from A to B.

53. A ship leaves a pier on a bearing of S $55°$ E and travels for 80 kilometers. It then turns and continues on a bearing of N $35°$ E for 74 kilometers. How far is the ship from the pier?

54. Two cars leave an intersection at the same time. One heads due south at 55 miles per hour. The other travels due west. After two hours, the bearing of the car headed west from the car headed south is $324°$. How far apart are they at that time?

55. From the top of a building that overlooks an ocean, an observer watches a boat sailing directly toward the building. If the observer is 150 feet above sea level and if the angle of depression of the boat changes from $27°$ to $39°$ during the period of observation, approximate the distance that the boat travels.

56. Find the measure of h to the nearest unit.

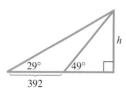

Use the law of sines and/or the law of cosines to find the indicated part of triangle ABC.

57. $C = 74° \, 10'$, $c = 96.3$ m, $B = 39° \, 30'$; find b

58. $a = 86.14$ in, $b = 253.2$ in, $c = 241.9$ in; find A

59. $A = 129° \, 40'$, $a = 127$ ft, $b = 69.8$ ft; find B

60. $B = 120.7°$, $a = 127$ ft, $c = 69.8$ ft; find b

61. $C = 51° \, 20'$, $c = 68.3$ m, $b = 58.2$ m; find B

62. $A = 51° \, 20'$, $c = 68.3$ m, $b = 58.2$ m; find a

63. $a = 165$ m, $A = 100.2°$, $B = 25.0°$; find b

64. $a = 14.8$ m, $b = 19.7$ m, $c = 31.8$ m; find B

65. $B = 39° \, 50'$, $b = 268$ m, $a = 340$ m; find A

66. $A = 46° \, 10'$, $b = 184$ cm, $c = 192$ cm; find a

67. $C = 79° \, 20'$, $c = 97.4$ mm, $a = 75.3$ mm; find A

68. $a = 7.5$ ft, $b = 12.0$ ft, $c = 6.9$ ft; find C

Use the law of sines to find all possible triangles with the given information.

69. $A = 25° \, 10'$, $a = 6.92$ yd, $b = 4.82$ yd

70. $A = 61.7°$, $a = 78.9$ m, $b = 86.4$ m

Find the area of the triangle with the given information.

71. $b = 840.6$ m, $c = 715.9$ m, $A = 149° \, 18'$

72. $a = 6.90$ ft, $b = 10.2$ ft, $C = 35° \, 10'$

73. $a = .913$ km, $b = .816$ km, $c = .582$ km

74. $a = 43$ m, $b = 32$ m, $c = 51$ m

Solve the problem.

75. The angles of elevation of a balloon from two points A and B on level ground are $24° \, 50'$ and $47° \, 20'$, respectively. As shown in the figure, points A and B are in the same vertical plane and are 8.4 miles apart. Approximate the height of the balloon above the ground to the nearest tenth of a mile.

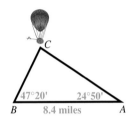

76. The pitcher's mound on a regulation softball field is 46 feet from home plate. The distance between the bases is 60 feet, as shown in the figure. How far is the pitcher's mound at point M from third base (point T)? Give your answer to the nearest foot.

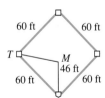

77. The course for a boat race starts at point X and goes in the direction S 48° W to point Y. It then turns and goes S 36° E to point Z, and finally returns back to point X. If point Z lies 10 kilometers directly south of point X, find the distance from Y to Z to the nearest kilometer.

78. Radio direction finders are placed at points A and B, which are 3.46 miles apart on an east-west line, with A west of B. From A the bearing of a certain illegal pirate radio transmitter is 48°, and from B the bearing is 302°. Find the distance between the transmitter and A.

79. To measure the distance AB across a canyon for a power line, a surveyor measures angles B and C and the distance BC. (See the figure.) What is the distance from A to B?

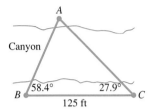

80. A banner on an 8.0-ft pole is to be mounted on a building at an angle of 115°, as shown in the figure. Find the length of the brace.

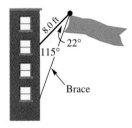

81. A hanging sculpture in an art gallery is to be hung with two wires of lengths 15.0 ft and 12.2 ft so that the angle between them is 70.3°. How far apart should the ends of the wire be placed on the ceiling?

82. A pipeline is to run between points A and B, which are separated by a protected wetlands area. To avoid the wetlands, the pipe will run from point A to C and then to B. The distances involved are $AB = 150$ km, $AC = 102$ km, and $BC = 135$ km. What angle should be used at point C?

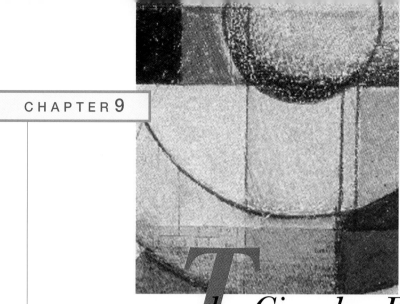

The Circular Functions and Applications

9.1 THE CIRCULAR FUNCTIONS AND IDENTITIES

The Circular Functions ▮ A Brief Discussion of Parametric Equations ▮
Fundamental Identities of Circular Functions ▮ Using the Fundamental
Identities

The Circular Functions

Our work in the previous chapter defined the six *trigonometric* functions in such a
way that the domain of each function was a set of *angles* in standard position. These
angles can be measured either in degrees or in radians. In theoretical work, it is
usually necessary to modify the trigonometric functions so that their domains
consist of sets of *real numbers* rather than angles. Because the interpretations of
these functions are based on the graph of the *unit circle* $x^2 + y^2 = 1$, we refer to
them as **circular functions.**

To define the values of the circular functions for any real number s, we use the
unit circle, shown in Figure 1(a) and (b). The calculator-generated graph in Figure 1(b) was obtained by using a square viewing window and graphing $y_1 =
\sqrt{1 - x^2}$ and $y_2 = -\sqrt{1 - x^2}$. The **unit circle** has its center at the origin and a
radius of one unit (hence the name *unit circle*). We start at the point $(1, 0)$ and
measure an arc of length s along the circle as in Figure 1(a). If $s > 0$, the arc is
measured in a counterclockwise direction, and if $s < 0$, the direction is clockwise.
(If $s = 0$, then no arc is measured.) Let the endpoint of this arc be at the point
(x, y). Then the six circular functions of s are defined as follows.

CIRCULAR FUNCTIONS

$$\sin s = y \qquad \tan s = \frac{y}{x} \ (x \neq 0) \qquad \sec s = \frac{1}{x} \ (x \neq 0)$$

$$\cos s = x \qquad \cot s = \frac{x}{y} \ (y \neq 0) \qquad \csc s = \frac{1}{y} \ (y \neq 0)$$

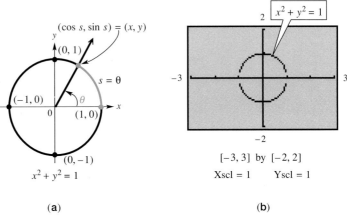

(a)

(b)

[-3, 3] by [-2, 2]
Xscl = 1   Yscl = 1

FIGURE 1

The circular functions (functions of real numbers) are closely related to the trigonometric functions of angles measured in radians. To see this, let us assume that angle θ is in standard position, superimposed on the unit circle as shown in Figure 1(a). Suppose further that θ is the *radian* measure of this angle. Using the arc length formula $s = r\theta$ with $r = 1$, we have $s = \theta$. Thus the length of the intercepted arc is the real number that corresponds to the radian measure of θ. Using the definitions of the trigonometric functions, we have

$$\sin \theta = \frac{y}{r} = \frac{y}{1} = y = \sin s,$$

$$\cos \theta = \frac{x}{r} = \frac{x}{1} = x = \cos s,$$

and so on. As shown here, the trigonometric functions and the circular functions lead to the same function values, provided we think of the angles in radian measure. This leads to the following important result concerning the evaluation of circular functions.

TECHNOLOGICAL NOTE The material of Chapter 8 involved both radian and degree measure, and as a result, you were probably often switching between degree and radian modes. The nature of the material of this chapter will necessitate using radian mode almost exclusively.

EVALUATING CIRCULAR FUNCTIONS

Circular function values of real numbers are obtained in the same manner as trigonometric function values of angles measured in radians. This applies to both methods of finding exact values (such as reference angle analysis) and calculator approximations. Calculators must be in radian mode when finding circular function values.

FOR **GROUP DISCUSSION**

With your calculator set for a square viewing window and in radian mode, graph the unit circle $x^2 + y^2 = 1$ by graphing $y_1 = \sqrt{1 - x^2}$ and $y_2 = -\sqrt{1 - x^2}$. Now graph the line $y_3 = \frac{\sqrt{3}}{3}x$. You should get graphs similar to those seen in Figure 2.

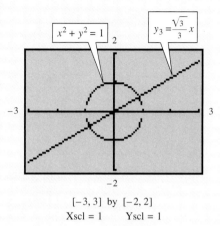

$[-3, 3]$ by $[-2, 2]$
Xscl = 1   Yscl = 1

FIGURE 2

1. Explain why the angle made by the line $y_3 = \frac{\sqrt{3}}{3}x$ makes an angle of $\frac{\pi}{6}$ radians with the positive x-axis.

2. Use the capabilities of your calculator to find the coordinates of the point of intersection of the line $y_3 = \frac{\sqrt{3}}{3}x$ and the portion of the unit circle in quadrant I.

3. What is the length of the arc from the point $(1, 0)$ to this point of intersection?

4. Find $\cos \frac{\pi}{6}$ using the cosine function of your calculator. How does this compare to the x-coordinate of the point of intersection found in item 2? How does this support the discussion earlier in this section?

5. Find $\sin \frac{\pi}{6}$ using the sine function of your calculator. How does this compare to the y-coordinate of the point of intersection found in item 2? How does this support the discussion earlier in this section?

6. If (x, y) denotes the point found in item 2, find approximations for $\frac{y}{x}$ and for $\frac{\sqrt{3}}{3}$. How do they compare? How does the tangent of the angle relate to the slope of the line?

In Figure 3, you will find the graph of the unit circle $x^2 + y^2 = 1$ with a great deal of important information. This information is based on the development of the trigonometric functions in Chapter 7, along with the discussion of this section. For many special values, degree and radian measures are given for the first counterclockwise revolution, and the coordinates of the points on the circle are also given. This figure should prove invaluable in further work.

The graph in Figure 3 can be adapted to many coterminal arcs as well. For example, the circular function values of $-\frac{3\pi}{4}$ correspond to those of $\frac{5\pi}{4}$, those of $-\frac{3\pi}{2}$ to $\frac{\pi}{2}$, and so on.

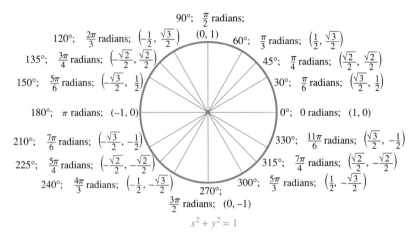

FIGURE 3

<table>
<tr><td>**EXAMPLE 1**

Finding Circular Function Values</td></tr>
</table>

(a)   Find the exact value of $\cos \frac{2\pi}{3}$ and of $\sin \frac{2\pi}{3}$.

SOLUTION   An arc of length $\frac{2\pi}{3}$ has terminal point $\left(-\frac{1}{2}, \frac{\sqrt{3}}{2}\right)$. Therefore,

$$\cos \frac{2\pi}{3} = -\frac{1}{2} \quad \text{and} \quad \sin \frac{2\pi}{3} = \frac{\sqrt{3}}{2}.$$

(b)   Find the exact value of $\tan \frac{\pi}{3}$.

SOLUTION   By the definition of the circular tangent function, $\tan s = \frac{y}{x}$. When $s = \frac{\pi}{3}$, $(x, y) = \left(\frac{1}{2}, \frac{\sqrt{3}}{2}\right)$. Therefore

$$\tan \frac{\pi}{3} = \frac{\frac{\sqrt{3}}{2}}{\frac{1}{2}} = \sqrt{3}.$$

(c)   Find a calculator approximation of $\cos 1.85$.

SOLUTION   With the calculator in *radian* mode, we find

$$\cos 1.85 \approx -.2755902468. \quad \blacksquare$$

A Brief Discussion of Parametric Equations

TECHNOLOGICAL NOTE
Refer to your owner's manual to see how to put your calculator in parametric mode.

Our treatment of graphs so far in this book has been based on functions defined in terms of the independent variable x (or in the case of the circular functions, s). In Chapter 10 we will investigate in more detail graphs defined *parametrically*. Because the circular functions can be illustrated beautifully using parametric equations, and because modern graphics calculators can graph curves parametrically, we present a brief discussion here.

PARAMETRIC EQUATIONS

Suppose that a set of points (x, y) is defined in such a way that $x = f(t)$ and $y = g(t)$. Then the set of points is said to be defined parametrically. The two equations are called **parametric equations,** and t is called the parameter. The **parameter** t is a real number in some interval I.

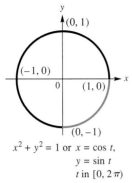

$x^2 + y^2 = 1$ or $x = \cos t,$
$y = \sin t$
t in $[0, 2\pi)$

FIGURE 4

We will investigate the parametric equations

$$x = \cos t, \qquad y = \sin t$$

where the parameter t lies in the interval $[0, 2\pi)$. Based on our earlier discussion, the graph will be that of the unit circle. See Figure 4.

Modern graphics calculators have the capability of graphing curves defined parametrically. Figure 5 shows the unit circle $x = \cos t, y = \sin t$, for t in $[0, 2\pi)$. You should read your owner's manual to see how your particular model can be used to graph parametric equations.

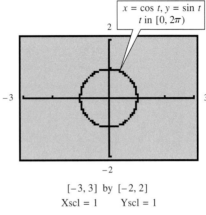

$[-3, 3]$ by $[-2, 2]$
Xscl = 1   Yscl = 1

FIGURE 5

EXAMPLE 2

Evaluating
Trigonometric
Functions Using the
Unit Circle Defined
Parametrically

(a) Find values of $\cos \frac{2\pi}{3}$ and $\sin \frac{2\pi}{3}$ by determining the coordinates of the point (x, y) on the unit circle for which $t = \frac{2\pi}{3}$.

SOLUTION   As seen in Figure 6, when $t = \frac{2\pi}{3}$ (indicated by the approximation 2.0943951) $x = -.5$ and $y = .8660254$. Therefore, $\cos \frac{2\pi}{3} = -.5$ and $\sin \frac{2\pi}{3} \approx .8660254$. These results support the exact values found in part (a) of Example 1. (The displayed value of y, which is $\sin \frac{2\pi}{3}$, is a decimal approximation for the *exact* value, $\frac{\sqrt{3}}{2}$.)

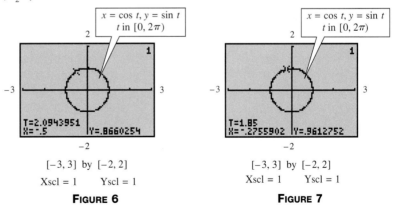

$[-3, 3]$ by $[-2, 2]$
Xscl = 1   Yscl = 1

FIGURE 6

$[-3, 3]$ by $[-2, 2]$
Xscl = 1   Yscl = 1

FIGURE 7

(b) Find an approximation of $\cos 1.85$ using the unit circle graphed parametrically.

SOLUTION   Figure 7 shows the unit circle with $t = 1.85$. Since we want to approximate $\cos 1.85$, we observe the *x-coordinate* of the terminal point of the arc with length 1.85. We find the approximation $-.2755902$, supporting our result in part (c) of Example 1.   ∎

Fundamental Identities of Circular Functions

In Chapter 7 we investigated some basic identities associated with the trigonometric functions. These identities hold for the circular functions as well. (Since the choice of the variable we use is immaterial, we will use the familiar θ (theta).) For example, using the unit circle interpretation of $\cos \theta$ and $\sin \theta$ and the fact that the equation of the unit circle is $x^2 + y^2 = 1$, we have

$$x^2 + y^2 = 1$$
$$(\cos \theta)^2 + (\sin \theta)^2 = 1$$

or, in the more familiar form,

$$\sin^2 \theta + \cos^2 \theta = 1.$$

The reciprocal identities, quotient identities, and other Pythagorean identities all follow in a similar manner. We list them here for review. In all cases, we assume that the functions are defined and that no denominators are 0.

FUNDAMENTAL IDENTITIES

Reciprocal Identities

$$\cot \theta = \frac{1}{\tan \theta} \qquad \sec \theta = \frac{1}{\cos \theta} \qquad \csc \theta = \frac{1}{\sin \theta}$$

Quotient Identities

$$\tan \theta = \frac{\sin \theta}{\cos \theta} \qquad \cot \theta = \frac{\cos \theta}{\sin \theta}$$

Pythagorean Identities

$$\sin^2 \theta + \cos^2 \theta = 1 \qquad \tan^2 \theta + 1 = \sec^2 \theta \qquad 1 + \cot^2 \theta = \csc^2 \theta$$

NOTE   The forms of the identities given above are the most commonly recognized forms. Throughout this chapter it will be necessary to recognize alternate forms of these identities as well. For example, two other forms of $\sin^2 \theta + \cos^2 \theta = 1$ are

$$\sin^2 \theta = 1 - \cos^2 \theta$$

and
$$\cos^2 \theta = 1 - \sin^2 \theta.$$

You should be able to transform the basic identities using algebraic transformations.

There is another group of identities that will prove useful throughout this chapter. They are known as the negative number (or negative angle) identities. To illustrate these identities, see the unit circle in Figure 8. As suggested in the figure, an arc of length θ having the point (x, y) as its terminal point has a corresponding arc $-\theta$ with a point $(x, -y)$ as its terminal point. From the definition of sine,

$$\sin(-\theta) = -y \qquad \text{and} \qquad \sin \theta = y$$

so that $\sin(-\theta)$ and $\sin \theta$ are negatives of each other, or

$$\sin(-\theta) = -\sin \theta.$$

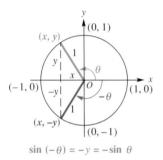

$\sin (-\theta) = -y = -\sin \theta$

FIGURE 8

The figure shows an arc θ in quadrant II, but the same result holds for θ in any quadrant. Also, by definition,

$$\cos(-\theta) = x \quad \text{and} \quad \cos\theta = x,$$

so that

$$\cos(-\theta) = \cos\theta.$$

These formulas for $\sin(-\theta)$ and $\cos(-\theta)$ can be used to find $\tan(-\theta)$ in terms of $\tan\theta$:

$$\tan(-\theta) = \frac{\sin(-\theta)}{\cos(-\theta)} = \frac{-\sin\theta}{\cos\theta} = -\frac{\sin\theta}{\cos\theta}$$

or

$$\tan(-\theta) = -\tan\theta.$$

Similar reasoning gives the remaining negative number identities:

$$\csc(-\theta) = -\csc\theta, \quad \sec(-\theta) = \sec\theta, \quad \cot(-\theta) = -\cot\theta.$$

The negative number identities are summarized below. Again, we assume that the functions are defined.

NEGATIVE NUMBER IDENTITIES

$$\sin(-\theta) = -\sin\theta \qquad \csc(-\theta) = -\csc\theta$$
$$\cos(-\theta) = \cos\theta \qquad \sec(-\theta) = \sec\theta$$
$$\tan(-\theta) = -\tan\theta \qquad \cot(-\theta) = -\cot\theta$$

FOR **GROUP DISCUSSION**

Put your calculator in radian mode. (This discussion is valid for degree-measured angles as well, but since we are studying circular functions, we will use radian measure here.) Let s be the number of letters in your last name.

1. Find $\sin s$ and then find $\sin(-s)$. How do they compare?
2. Find $\cos s$ and then find $\cos(-s)$. How do they compare?
3. Find $\tan s$ and then find $\tan(-s)$. How do they compare?

Discuss the results obtained and how they relate to the negative number identities.

Using the Fundamental Identities

Any trigonometric function of a number or angle can be expressed in terms of any other function, as shown in the following example.

EXAMPLE **3**
Expressing One Function in Terms of Another

Express $\cos x$ in terms of $\tan x$.

SOLUTION   Since $\sec x$ is related to both $\cos x$ and $\tan x$ by identities, start with $\tan^2 x + 1 = \sec^2 x$. Then take reciprocals to get

$$\frac{1}{\tan^2 x + 1} = \frac{1}{\sec^2 x}$$

or

$$\frac{1}{\tan^2 x + 1} = \cos^2 x$$

$$\pm\sqrt{\frac{1}{\tan^2 x + 1}} = \cos x \qquad \text{Take the square root of both sides.}$$

$$\cos x = \frac{\pm 1}{\sqrt{\tan^2 x + 1}}.$$

Rationalize the denominator to get

$$\cos x = \frac{\pm\sqrt{\tan^2 x + 1}}{\tan^2 x + 1}.$$

Choose the $+$ sign or the $-$ sign, depending on the quadrant of x.   ∎

Each of $\tan \theta$, $\cot \theta$, $\sec \theta$, and $\csc \theta$ can easily be expressed in terms of $\sin \theta$ and/or $\cos \theta$. For this reason, we often make such substitutions in an expression so that the expression can be simplified. The next example shows such substitutions.

EXAMPLE **4**
Rewriting an Expression in Terms of Sine and Cosine

Use the fundamental identities to write $\tan \theta + \cot \theta$ in terms of $\sin \theta$ and $\cos \theta$, and then simplify the expression.

SOLUTION   From the fundamental identities,

$$\tan \theta + \cot \theta = \frac{\sin \theta}{\cos \theta} + \frac{\cos \theta}{\sin \theta}.$$

Simplify this expression by adding the two fractions on the right side, using the common denominator $\cos \theta \sin \theta$.

$$\tan \theta + \cot \theta = \frac{\sin^2 \theta}{\cos \theta \sin \theta} + \frac{\cos^2 \theta}{\cos \theta \sin \theta}$$

$$= \frac{\sin^2 \theta + \cos^2 \theta}{\cos \theta \sin \theta}$$

Now substitute 1 for $\sin^2 \theta + \cos^2 \theta$.

$$\tan \theta + \cot \theta = \frac{1}{\cos \theta \sin \theta} \qquad ∎$$

CAUTION   When working with circular and trigonometric expressions and identities, be sure to write the argument of the function. For example, we would *not* write $\sin^2 + \cos^2 = 1$; an argument such as θ is necessary in this identity.

One of the skills required for more advanced work in mathematics (and especially in calculus) is the ability to use the identities to write expressions in alternative forms. This skill is developed by using the fundamental identities to verify that an equation is an identity (for those values of the variable for which it is defined). Here are some hints that may help you get started.

VERIFYING IDENTITIES

1. Learn the fundamental identities. Whenever you see either side of a fundamental identity, the other side should come to mind. Also, be aware of equivalent forms of the fundamental identities. For example $\sin^2 \theta = 1 - \cos^2 \theta$ is an alternate form of $\sin^2 \theta + \cos^2 \theta = 1$.

2. Try to rewrite the more complicated side of the equation so that it is identical to the simpler side.

3. It is often helpful to express all functions in the equation in terms of sine and cosine and then simplify the result.

4. Usually any factoring or indicated algebraic operations should be performed. For example, the expression $\sin^2 x + 2 \sin x + 1$ can be factored as $(\sin x + 1)^2$. The sum or difference of two expressions, such as

$$\frac{1}{\sin \theta} + \frac{1}{\cos \theta},$$

can be added or subtracted in the same way as any other rational expressions:

$$\frac{1}{\sin \theta} + \frac{1}{\cos \theta} = \frac{\cos \theta}{\sin \theta \cos \theta} + \frac{\sin \theta}{\sin \theta \cos \theta}$$

$$= \frac{\cos \theta + \sin \theta}{\sin \theta \cos \theta}.$$

5. As you select substitutions, keep in mind the side you are not changing, because it represents your goal. For example, to verify the identity

$$\tan^2 x + 1 = \frac{1}{\cos^2 x},$$

try to think of an identity that relates $\tan x$ to $\cos x$. Here, since $\sec x = \frac{1}{\cos x}$ and $\sec^2 x = \tan^2 x + 1$, the secant function is the best link between the two sides.

6. If an expression contains $1 + \sin x$, multiplying both numerator and denominator by $1 - \sin x$ would give $1 - \sin^2 x$, which could be replaced with $\cos^2 x$. Similar results for $1 - \sin x$, $1 + \cos x$, and $1 - \cos x$ may be useful.

EXAMPLE 5
Verifying an Identity (Working with One Side)

Verify that

$$\cot s + 1 = \csc s(\cos s + \sin s)$$

is an identity.

SOLUTION Use the fundamental identities to rewrite one side of the equation so that it is identical to the other side. Since the right side is more complicated, it is probably a good idea to work with it. Here we use the method of changing all the functions to sine or cosine.

Steps	Reasons

$$\csc s(\cos s + \sin s) = \frac{1}{\sin s}(\cos s + \sin s)$$

$\csc s = \frac{1}{\sin s}$

$$= \frac{\cos s}{\sin s} + \frac{\sin s}{\sin s}$$

Distributive property

$$= \cot s + 1$$

$\frac{\cos s}{\sin s} = \cot s$; $\frac{\sin s}{\sin s} = 1$

The given equation is an identity since the right side equals the left side. ∎

EXAMPLE 6

Verifying an Identity (Working with One Side)

Verify that

$$\frac{\tan t - \cot t}{\sin t \cos t} = \sec^2 t - \csc^2 t$$

is an identity.

SOLUTION   Since the left side is the more complicated one, transform the left side to equal the right side.

$$\frac{\tan t - \cot t}{\sin t \cos t}$$

$$= \frac{\tan t}{\sin t \cos t} - \frac{\cot t}{\sin t \cos t}$$

$\frac{a - b}{c} = \frac{a}{c} - \frac{b}{c}$

$$= \tan t \cdot \frac{1}{\sin t \cos t} - \cot t \cdot \frac{1}{\sin t \cos t}$$

$\frac{a}{b} = a \cdot \frac{1}{b}$

$$= \frac{\sin t}{\cos t} \cdot \frac{1}{\sin t \cos t} - \frac{\cos t}{\sin t} \cdot \frac{1}{\sin t \cos t}$$

$\tan t = \frac{\sin t}{\cos t}$; $\cot t = \frac{\cos t}{\sin t}$

$$= \frac{1}{\cos^2 t} - \frac{1}{\sin^2 t}$$

$$= \sec^2 t - \csc^2 t$$

$\frac{1}{\cos^2 t} = \sec^2 t$; $\frac{1}{\sin^2 t} = \csc^2 t$

Here, writing in terms of sine and cosine only was used in the third line. ∎

If both sides of an identity appear to be equally complex, the identity can be verified by working independently on the left side and on the right side, until each side is changed into some common third result. *Each step, on each side, must be reversible.* With all steps reversible, the procedure is as follows.

left   =   right

common third
expression

The left side leads to the third expression, which leads back to the right side. This procedure is just a shortcut for the procedure used in the first examples of this section: the left side is changed into the right side, but by going through an intermediate step.

EXAMPLE 7

Verifying an Identity
(Working with Both
Sides)

Verify that

$$\frac{\sec \alpha + \tan \alpha}{\sec \alpha - \tan \alpha} = \frac{1 + 2 \sin \alpha + \sin^2 \alpha}{\cos^2 \alpha}$$

is an identity.

SOLUTION   Both sides appear equally complex, so verify the identity by changing each side into a common third expression. Work first on the left, multiplying numerator and denominator by $\cos \alpha$.

$$\frac{\sec \alpha + \tan \alpha}{\sec \alpha - \tan \alpha} = \frac{(\sec \alpha + \tan \alpha)\cos \alpha}{(\sec \alpha - \tan \alpha)\cos \alpha} \qquad \frac{\cos \alpha}{\cos \alpha} = 1; \text{ multiplicative identity}$$

$$= \frac{\sec \alpha \cos \alpha + \tan \alpha \cos \alpha}{\sec \alpha \cos \alpha - \tan \alpha \cos \alpha} \qquad \text{Distributive property}$$

$$= \frac{1 + \tan \alpha \cos \alpha}{1 - \tan \alpha \cos \alpha} \qquad \sec \alpha \cos \alpha = 1$$

$$= \frac{1 + \dfrac{\sin \alpha}{\cos \alpha} \cdot \cos \alpha}{1 - \dfrac{\sin \alpha}{\cos \alpha} \cdot \cos \alpha} \qquad \tan \alpha = \frac{\sin \alpha}{\cos \alpha}$$

$$= \frac{1 + \sin \alpha}{1 - \sin \alpha}$$

On the right side of the original statement, begin by factoring.

$$\frac{1 + 2 \sin \alpha + \sin^2 \alpha}{\cos^2 \alpha} = \frac{(1 + \sin \alpha)^2}{\cos^2 \alpha} \qquad a^2 + 2ab + b^2 = (a + b)^2$$

$$= \frac{(1 + \sin \alpha)^2}{1 - \sin^2 \alpha} \qquad \cos^2 \alpha = 1 - \sin^2 \alpha$$

$$= \frac{(1 + \sin \alpha)^2}{(1 + \sin \alpha)(1 - \sin \alpha)} \qquad \begin{array}{l} 1 - \sin^2 \alpha = \\ (1 + \sin \alpha)(1 - \sin \alpha) \end{array}$$

$$= \frac{1 + \sin \alpha}{1 - \sin \alpha} \qquad \text{Reduce to lowest terms.}$$

We now have shown that

$$\frac{\sec \alpha + \tan \alpha}{\sec \alpha - \tan \alpha} = \frac{1 + \sin \alpha}{1 - \sin \alpha} = \frac{1 + 2 \sin \alpha + \sin^2 \alpha}{\cos^2 \alpha},$$

verifying that the original equation is an identity.   ∎

NOTE   There are usually several ways to verify a given identity. You may wish to go through the examples of this section and verify each using a method different from the one given.

In later sections, we will see how our analytic work in verifying identities can be supported graphically, using the graphs of the functions involved.

9.1 EXERCISES

Use the unit circle shown in Figure 3, along with the definitions of the circular functions, to find the exact value for each of the following. Then support your answer by finding the circular function value on your calculator. Be sure the calculator is in radian mode.

1. $\sin \dfrac{7\pi}{6}$ **2.** $\cos \dfrac{5\pi}{3}$ **3.** $\tan \dfrac{3\pi}{4}$ **4.** $\sin \dfrac{5\pi}{3}$

5. $\cos \dfrac{7\pi}{6}$ **6.** $\tan \dfrac{4\pi}{3}$ **7.** $\sec \dfrac{2\pi}{3}$ **8.** $\csc \dfrac{11\pi}{6}$

9. $\cot \dfrac{5\pi}{6}$ **10.** $\cos\left(-\dfrac{4\pi}{3}\right)$ **11.** $\sin\left(-\dfrac{5\pi}{6}\right)$ **12.** $\tan \dfrac{17\pi}{3}$

13. $\sec \dfrac{23\pi}{6}$ **14.** $\csc \dfrac{13\pi}{3}$

Find a calculator approximation of each of the following. Be sure that your calculator is in radian mode.

15. $\sin .6109$ **16.** $\sin .8203$ **17.** $\cos(-1.1519)$

18. $\cos(-5.2825)$ **19.** $\tan 4.0203$ **20.** $\tan 6.4752$

21. $\csc(-9.4946)$ **22.** $\csc 1.3875$ **23.** $\sec 2.8440$

24. $\sec(-8.3429)$ **25.** $\cot 6.0301$ **26.** $\cot 3.8426$

In Exercises 27–30, a unit circle generated by the parametric equations $x = \cos t$, $y = \sin t$ is shown, along with a display for t, x, and y. Use this information to find cos t and sin t, and then verify by using the cosine and sine functions of your calculator.

27.

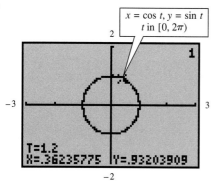

28.

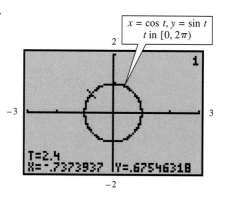

29.

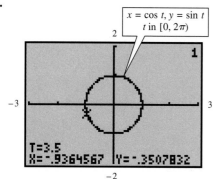

30.

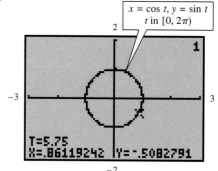

Relating Concepts

With your calculator in radian mode, respond to Exercises 31–36 in order.

31. Find a calculator approximation for cos 4.9.

32. Use a negative number identity to give an approximation for cos(−4.9), referring only to your answer in Exercise 31.

33. Find a calculator approximation for sin 4.9.

34. Use a negative number identity to give an approximation for sin(−4.9), referring only to your answer in Exercise 33.

35. Use the results of Exercises 31 and 33 to fill in the blanks: Because cos 4.9 is _____ and sin 4.9 is _____ , the arc
        (positive/negative)   (positive/negative)
on the unit circle representing the real number 4.9 must terminate in quadrant
_____ .
    (I/II/III/IV)

36. Based on your results in Exercises 32 and 34, write a statement involving −4.9 analogous to the statement in Exercise 35.

Use the negative number identities to write each of the following as a circular function of a positive number. (For example, sin (−3.4) = −sin 3.4.)

37. $\cos(-4.38)$

38. $\cos(-5.46)$

39. $\sin(-.5)$

40. $\sin(-2.5)$

41. $\tan\left(-\dfrac{\pi}{7}\right)$

42. $\tan\left(-\dfrac{4\pi}{7}\right)$

43. $\sec(-8)$

44. $\sec(-.055)$

45. $\csc\left(-\dfrac{1}{4}\right)$

46. $\csc\left(-\dfrac{5}{9}\right)$

47. $\cot(-10^5)$

48. $\cot(-5^8)$

For each expression in Column I, choose the expression from Column II that completes a fundamental identity.

    Column I   *Column II*

49. $\dfrac{\cos x}{\sin x}$ **(a)** $\sin^2 x + \cos^2 x$

50. $\tan x$ **(b)** $\cot x$

51. $\cos(-x)$ **(c)** $\sec^2 x$

52. $\tan^2 x + 1$ **(d)** $\dfrac{\sin x}{\cos x}$

53. 1 **(e)** $\cos x$

For each expression in Column I, choose the expression from Column II that completes an identity. You may have to rewrite one or both expressions, using a fundamental identity, to recognize the matches.

Column I   Column II

54. $-\tan x \cos x$ **(a)** $\dfrac{\sin^2 x}{\cos^2 x}$

55. $\sec^2 x - 1$ **(b)** $\dfrac{1}{\sec^2 x}$

56. $\dfrac{\sec x}{\csc x}$ **(c)** $\sin(-x)$

57. $1 + \sin^2 x$ **(d)** $\csc^2 x - \cot^2 x + \sin^2 x$

58. $\cos^2 x$ **(e)** $\tan x$

59. A student writes "$1 + \cot^2 = \csc^2$." Comment on this student's work.

60. Another student makes the following claim: "Since $\sin^2 \theta + \cos^2 \theta = 1$, I should be able to also say $\sin \theta + \cos \theta = 1$ if I take the square root of both sides." Comment on this student's statement.

Complete this chart, so that each function in the column at the left is expressed in terms of the functions given across the top.

	sin θ	**cos θ**	**tan θ**	**cot θ**	**sec θ**	**csc θ**
61. sin θ	$\sin \theta$	$\pm\sqrt{1 - \cos^2 \theta}$	$\dfrac{\pm\tan \theta\sqrt{1 + \tan^2 \theta}}{1 + \tan^2 \theta}$			$\dfrac{1}{\csc \theta}$
62. cos θ		$\cos \theta$	$\dfrac{\pm\sqrt{\tan^2 \theta + 1}}{\tan^2 \theta + 1}$		$\dfrac{1}{\sec \theta}$	
63. tan θ			$\tan \theta$	$\dfrac{1}{\cot \theta}$		
64. cot θ			$\dfrac{1}{\tan \theta}$	$\cot \theta$	$\dfrac{\pm\sqrt{\sec^2 \theta - 1}}{\sec^2 \theta - 1}$	
65. sec θ		$\dfrac{1}{\cos \theta}$			$\sec \theta$	
66. csc θ	$\dfrac{1}{\sin \theta}$					$\csc \theta$

Each of the following expressions simplifies to a constant, a single circular function, or a power of a circular function. Use the fundamental identities to simplify each expression.

67. $\tan \theta \cos \theta$ **68.** $\cot \alpha \sin \alpha$ **69.** $\sec r \cos r$ **70.** $\cot t \tan t$

71. $\dfrac{\sin \beta \tan \beta}{\cos \beta}$ **72.** $\dfrac{\csc \theta \sec \theta}{\cot \theta}$ **73.** $\sec^2 x - 1$ **74.** $\csc^2 t - 1$

75. $\dfrac{\sin^2 x}{\cos^2 x} + \sin x \csc x$ **76.** $\dfrac{1}{\tan^2 \alpha} + \cot \alpha \tan \alpha$ **77.** $\dfrac{1 - \cos^2 x}{\sin x}$

78. $\dfrac{1 - \sin^2 x}{\cos x}$

79. $\cos^2 x(\tan^2 x + 1)$

80. $\dfrac{\cos^2 x}{\sin x} + \sin x$

Verify each of the following identities.

81. $\dfrac{\tan^2 \gamma + 1}{\sec \gamma} = \sec \gamma$

82. $\sin^2 \beta (1 + \cot^2 \beta) = 1$

83. $\sin^2 \alpha + \tan^2 \alpha + \cos^2 \alpha = \sec^2 \alpha$

84. $\cot s + \tan s = \sec s \csc s$

85. $\dfrac{\sin^2 \gamma}{\cos \gamma} = \sec \gamma - \cos \gamma$

86. $\dfrac{\cos \alpha}{\sec \alpha} + \dfrac{\sin \alpha}{\csc \alpha} = \sec^2 \alpha - \tan^2 \alpha$

87. $\dfrac{\cos \theta}{\sin \theta \cot \theta} = 1$

88. $\sin^4 \theta - \cos^4 \theta = 2 \sin^2 \theta - 1$

89. $\tan^2 \gamma \sin^2 \gamma = \tan^2 \gamma + \cos^2 \gamma - 1$

90. $(1 - \cos^2 \alpha)(1 + \cos^2 \alpha) = 2 \sin^2 \alpha - \sin^4 \alpha$

91. $\dfrac{(\sec \theta - \tan \theta)^2 + 1}{\sec \theta \csc \theta - \tan \theta \csc \theta} = 2 \tan \theta$

92. $\dfrac{\cos \theta + 1}{\tan^2 \theta} = \dfrac{\cos \theta}{\sec \theta - 1}$

93. $\dfrac{1}{1 - \sin \theta} + \dfrac{1}{1 + \sin \theta} = 2 \sec^2 \theta$

94. $\dfrac{1 - \cos x}{1 + \cos x} = (\cot x - \csc x)^2$

95. $\dfrac{1}{\tan \alpha - \sec \alpha} + \dfrac{1}{\tan \alpha + \sec \alpha} = -2 \tan \alpha$

96. $\dfrac{\csc \theta + \cot \theta}{\tan \theta + \sin \theta} = \cot \theta \csc \theta$

97. $\dfrac{\tan s}{1 + \cos s} + \dfrac{\sin s}{1 - \cos s} = \cot s + \sec s \csc s$

98. $\dfrac{\cot \alpha + 1}{\cot \alpha - 1} = \dfrac{1 + \tan \alpha}{1 - \tan \alpha}$

99. A student claims that the equation

$$\cos \theta + \sin \theta = 1$$

is an identity, since by letting $\theta = \frac{\pi}{2}$, we get $0 + 1 = 1$, a true statement. Comment on this student's reasoning.

100. Explain why the method described in the text involving working on both sides of an identity to show that each side is equal to the same expression is a valid method of verifying an identity. When using this method, what must be true about each step taken? (*Hint:* See the discussion preceding Example 7.)

F*urther Explorations*

While radian mode is very important for connecting angle measure and arc length, the graphics calculator cannot display values like π, $\frac{\pi}{2}$, and 2π in their exact form. Instead the graphics calculator can only display and evaluate decimal approximations. In other words, it is unable to manipulate irrational numbers; it must first convert to a rational approximation. For this reason, it is sometimes preferable to use DEGREE mode when evaluating trigonometric functions. If your graphics calculator has a TABLE feature, it can be used to find patterns in trigonometric functions. The figure shows a TABLE where $Y_1 = \cos x$ and $Y_2 = \sin x$. The calculator is in DEGREE MODE and $\Delta\text{Tbl} = 15°$.

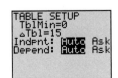

Note some patterns:

$$\cos 45° = \sin 45°$$

$\cos x$ decreases from 1 to 0 as x goes from 0° to 90°.

$\sin x$ increases from 0 to 1 as x goes from 0° to 90°.

1. Create the TABLE from the figure in your calculator. Scroll through the TABLE at least as far as $x = 360°$ in order to answer the following.
   (a) Find all values of x where $\cos x = \sin x$.
   (b) Describe all intervals where $\cos x$ is decreasing and all intervals where $\cos x$ is increasing.
   (c) Describe all intervals where $\sin x$ is decreasing and all intervals where $\sin x$ is increasing.
   (d) What are the maximum and minimum values for $\cos x$ and $\sin x$?
   (e) Find a ΔTbl and Tblmin where $\cos x$ always will equal 1. For what values of x does $\cos x = 1$?
   (f) Find a ΔTbl and Tblmin where $\cos x$ always will equal 0. For what values of x does $\cos x = 0$?
   (g) Find a ΔTbl and Tblmin where $\sin x$ always will equal 1. For what values of x does $\sin x = 1$?
   (h) Find a ΔTbl and Tblmin where $\sin x$ always will equal 0. For what values of x does $\sin x = 0$?

9.2 ANALYSIS OF THE SINE AND COSINE FUNCTIONS

Periodic Functions ∎ The Graph of the Sine Function ∎ The Graph of the Cosine Function ∎ Transformations of the Graphs of the Sine and Cosine Functions ∎ Supporting an Identity Graphically

Periodic Functions

> **FOR GROUP DISCUSSION**
>
> With your calculator in radian mode, do the following.
>
> 1. Let t represent the number of letters in your first name, and find a calculator approximation for $\cos t$. Store or write down your answer.
> 2. Let s represent the number of letters in your last name, find a calculator approximation for $\cos(t + s \cdot 2\pi)$. Compare your result to your answer in item 1.
>
> Everyone should get the same results for themselves in items 1 and 2, although the answers will vary from student to student depending upon the values of t and s. Use the unit circle interpretation from the previous section to explain why the answers in items 1 and 2 are the same. Will the same thing happen if you use the sine function rather than the cosine function?

The sine and cosine functions repeat their values over and over. They are examples of *periodic functions*.

> ## PERIODIC FUNCTION
> A **periodic function** is a function f such that
> $$f(x) = f(x + np),$$
> for every real number x in the domain of f, every integer n, and for some positive real number p. The smallest possible positive value of p is the **period** of the function.

The circumference of the unit circle is 2π, and therefore the smallest value of p for which the sine and cosine functions repeat is 2π. Therefore, the sine and cosine functions are periodic functions with period 2π.

The Graph of the Sine Function

For every real number x there is a real number y such that $y = \sin x$. This number y is a real number in the interval $[-1, 1]$, and can be determined by using a calculator in one of several ways. The value of $\sin x$ is the y-coordinate of the terminal point of an arc of length x that is represented on the unit circle; it may also be determined by using the sine function of the calculator, with the calculator in radian mode.

Because graphics calculators are capable of graphing functions by plotting points quickly, we can graph the function defined by the set of ordered pairs $(x, \sin x)$. To begin, we will graph the function $y = \sin x$ over $[0, 2\pi]$. We will set Xmin $= 0$, Xmax $= 2\pi$, Ymin $= -1.5$, Ymax $= 1.5$. See Figure 9.

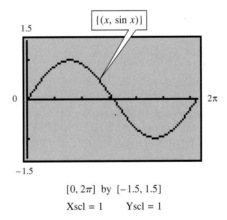

$[0, 2\pi]$ by $[-1.5, 1.5]$

Xscl $= 1$ Yscl $= 1$

FIGURE 9

TECHNOLOGICAL NOTE
While we will refer to the window $[-2\pi, 2\pi]$ by $[-4, 4]$ as the *trig viewing window* in this text, your model may use a different "standard" viewing window for the graphs of circular functions.

Since the period of the sine function is 2π, the curve shown in Figure 9 repeats over and over. This curve is called a **sine wave,** or **sinusoid.** You should learn the shape of this graph and be able to sketch it quickly by hand. Graphics calculators often have a window designated for graphing circular functions. We will refer to the window $[-2\pi, 2\pi]$ by $[-4, 4]$ with Xscl $= \frac{\pi}{2}$ and Yscl $= 1$ as the *trig viewing window*.

SINE FUNCTION

$f(x) = \sin x$ (Figure 10)

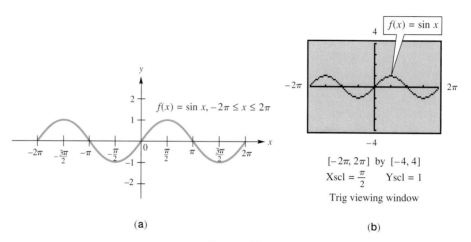

$f(x) = \sin x, -2\pi \le x \le 2\pi$

$[-2\pi, 2\pi]$ by $[-4, 4]$

$\text{Xscl} = \frac{\pi}{2}$ $\text{Yscl} = 1$

Trig viewing window

(a)   (b)

FIGURE 10

Domain: $(-\infty, \infty)$

Range: $[-1, 1]$

Over the interval $[0, 2\pi]$, the sine function exhibits the following behavior:

From 0 to $\frac{\pi}{2}$, $\sin x$ increases from 0 to 1.

From $\frac{\pi}{2}$ to π, $\sin x$ decreases from 1 to 0.

From π to $\frac{3\pi}{2}$, $\sin x$ decreases from 0 to -1.

From $\frac{3\pi}{2}$ to 2π, $\sin x$ increases from -1 to 0.

The graph is continuous over its entire domain. Its x-intercepts are of the form $n\pi$, where n is an integer. Its period is 2π. The graph is symmetric with respect to the origin.

If a periodic function has maximum and minimum values, then its **amplitude** is defined to be half the difference between the maximum and minimum function values. Thus, for the sine function, the amplitude is $\frac{1}{2}[1 - (-1)] = \frac{1}{2}(2) = 1$.

AMPLITUDE OF THE SINE FUNCTION

The amplitude of the sine function is 1.

EXAMPLE 1

Finding sin x Using Several Different Methods

Find an approximation of $\sin 1.25$ using three different methods, including the graph of the sine function.

SOLUTION We can find an approximation of $\sin 1.25$ by using the sin key of a calculator set in radian mode. Verify that $\sin 1.25 \approx .9489846194$.

As shown in the previous section, we can graph the unit circle using the parametric equations $x = \cos t$, $y = \sin t$, and letting $t = 1.25$ we find the y-coordinate is approximately .94898462. See Figure 11.

We can also find sin 1.25 by graphing $f(x) = \sin x$, and using the capabilities of the calculator to locate the point with x-coordinate 1.25. As shown in Figure 12, this point has y-coordinate approximately .94898462, supporting our other two approximations. ▯

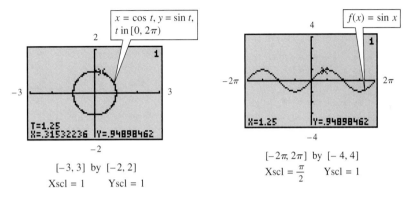

$x = \cos t$, $y = \sin t$, t in $[0, 2\pi)$

T=1.25
X=.31532236   Y=.94898462

$[-3, 3]$ by $[-2, 2]$
Xscl = 1   Yscl = 1

FIGURE 11

$f(x) = \sin x$

X=1.25   Y=.94898462

$[-2\pi, 2\pi]$ by $[-4, 4]$
Xscl = $\frac{\pi}{2}$ Yscl = 1

FIGURE 12

As we mentioned above, the graph of the sine function is symmetric with respect to the origin. Recall from Section 2.1 that the graph of a function f is symmetric with respect to the origin if $f(-x) = -f(x)$ for all x in its domain. The negative number identity $\sin(-x) = -\sin x$ is an example of this general condition. The displays in Figure 13 support this fact for $x = 4.5$.

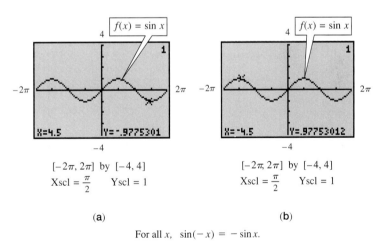

$f(x) = \sin x$

X=4.5   Y=-.9775301

$[-2\pi, 2\pi]$ by $[-4, 4]$
Xscl = $\frac{\pi}{2}$ Yscl = 1

(a)

$f(x) = \sin x$

X=-4.5   Y=.9775301 2

$[-2\pi, 2\pi]$ by $[-4, 4]$
Xscl = $\frac{\pi}{2}$ Yscl = 1

(b)

For all x, $\sin(-x) = -\sin x$.

FIGURE 13

The Graph of the Cosine Function

The same kind of analysis presented for the sine function can be applied to the cosine function, another function whose graph is a sinusoid.

COSINE FUNCTION

$f(x) = \cos x$ (Figure 14)

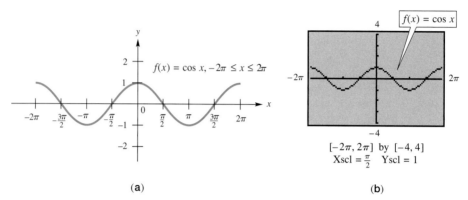

(a)

(b)

$[-2\pi, 2\pi]$ by $[-4, 4]$
Xscl $= \frac{\pi}{2}$ Yscl $= 1$

FIGURE 14

Domain: $(-\infty, \infty)$

Range: $[-1, 1]$

Over the interval $[0, 2\pi]$, the cosine function exhibits the following behavior:

From 0 to $\frac{\pi}{2}$, $\cos x$ decreases from 1 to 0.

From $\frac{\pi}{2}$ to π, $\cos x$ decreases from 0 to -1.

From π to $\frac{3\pi}{2}$, $\cos x$ increases from -1 to 0.

From $\frac{3\pi}{2}$ to 2π, $\cos x$ increases from 0 to 1.

The graph is continuous over its entire domain. Its x-intercepts are of the form $(2n + 1)\frac{\pi}{2}$, where n is an integer. Its period is 2π. The graph is symmetric with respect to the y-axis.

Verify that, like the sine function, the amplitude of the cosine function is 1.

AMPLITUDE OF THE COSINE FUNCTION

The amplitude of the cosine function is 1.

EXAMPLE 2

Interpreting the
Coordinates of a
Point on the Graph
of $f(x) = \cos x$

Use the graph of $f(x) = \cos x$ and the display in Figure 15, along with the fact that $\pi \approx 3.1415927$, to reinforce earlier concepts involving the trigonometric and circular functions.

SOLUTION   We see that $\cos \pi \approx \cos 3.1415927 = -1$. This supports the fact that the terminal point of an arc of length π on the unit circle has an x-coordinate of -1. Since the real number π corresponds to an angle of radian measure π (and thus degree measure 180°), the terminal side of such an angle in standard position will contain the point $(-r, 0)$ for any $r > 0$. The definition of the trigonometric ratio for cosine tells us that $\cos \pi = \cos 180° = \frac{x}{r} = \frac{-r}{r} = -1$, supporting the result seen in Figure 15.   ∎

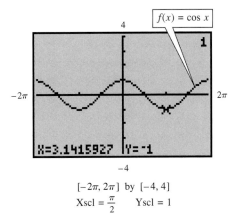

$$[-2\pi, 2\pi] \text{ by } [-4, 4]$$
$$\text{Xscl} = \frac{\pi}{2} \qquad \text{Yscl} = 1$$

FIGURE 15

FOR GROUP DISCUSSION

Graph the two functions $y_1 = \sin x$ and $y_2 = \cos(x - \frac{\pi}{2})$ in the trig viewing window of your calculator. Now answer the following items.

1. How do the two graphs compare?
2. Knowing that $\cos(-x) = \cos x$ for all x, how would the graph of $y_3 = \cos(\frac{\pi}{2} - x)$ compare to the graph of y_2? Verify this.
3. In Section 8.3, we learned that cofunctions of complementary angles are equal. How does your answer in item 2 support this for the circular function *cosine*?

While the graph of the sine function is symmetric with respect to the origin, the graph of the cosine function is symmetric with respect to the y-axis. Recall that the graph of a function f is symmetric with respect to the y-axis if $f(-x) = f(x)$ for all x in its domain. The negative number identity $\cos(-x) = \cos x$ is an example of this general condition. The displays in Figure 16 indicate this fact for $x = 4.5$.

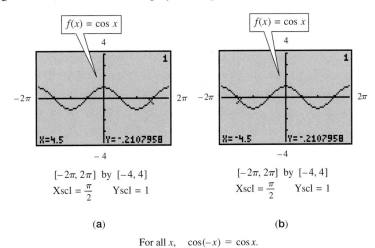

$$[-2\pi, 2\pi] \text{ by } [-4, 4]$$
$$\text{Xscl} = \frac{\pi}{2} \qquad \text{Yscl} = 1$$

(a)

$$[-2\pi, 2\pi] \text{ by } [-4, 4]$$
$$\text{Xscl} = \frac{\pi}{2} \qquad \text{Yscl} = 1$$

(b)

For all x, $\cos(-x) = \cos x$.

FIGURE 16

The most important points on the graph of the sine and cosine functions are the maximum and minimum points and the *x*-intercepts. The graph changes its concavity at these *x*-intercepts, and therefore the associated points are inflection points. We now give an interpretation of a "complete graph" of a sinusoid.

> ### COMPLETE GRAPH OF A SINUSOID
>
> A complete graph of a sinusoid will consist of at least one period of the graph. It will show the extreme points and the points of inflection in the interval.

Transformations of the Graphs of the Sine and Cosine Functions

In Chapter 2 we saw how graphs of functions may be transformed by stretching, shrinking, reflecting, and shifting. These transformations can be applied to the sine and cosine functions.

EXAMPLE 3

Analyzing the Graph of a Transformed Circular Function

The graphs of $y_1 = \sin x$ and $y_2 = 2 \sin x$ are shown in Figure 17. Notice that the graph of $y_2 = 2 \sin x$ can be obtained by stretching the graph of $y_1 = \sin x$ vertically by a factor of 2. The period of $y_2 = 2 \sin x$ is 2π, and the range is $[-2, 2]$, causing its amplitude to be 2.   ∎

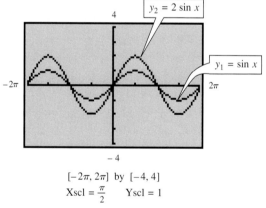

$[-2\pi, 2\pi]$ by $[-4, 4]$
$\text{Xscl} = \frac{\pi}{2}$ $\text{Yscl} = 1$

FIGURE 17

If you explore many problems similar to Example 3, you will discover the following (assume $a \neq 0$).

> ### AMPLITUDE OF SINE AND COSINE
>
> The graph of $y = a \sin x$ or $y = a \cos x$ will have the same basic shape as the graph of $y = \sin x$ or $y = \cos x$. The range of the function is $[-|a|, |a|]$ and the amplitude is $|a|$.

No matter what the value of the amplitude, the period of $y = a \sin x$ and $y = a \cos x$ is still 2π. However, the graph of a function of the form $y = \sin bx$ or $\cos bx$, for $b > 0$, $b \neq 1$, will have a period different from 2π. To see why this is so, remember that the values of sin bx or cos bx will take on all possible values as bx ranges from 0 to 2π. Therefore, to see what the period of either of these will be, we must solve the compound inequality

$$0 \leq bx \leq 2\pi.$$

Dividing by the positive number b gives

$$0 \leq x \leq \frac{2\pi}{b}.$$

Therefore, the period is $\frac{2\pi}{b}$.

PERIOD OF SINE AND COSINE

The graph of $y = \sin bx$ or $y = \cos bx$, $b > 0$, will have the same basic shape as the graph of $y = \sin x$ or $y = \cos x$. However, the period of the function is $\frac{2\pi}{b}$. The range is $[-1, 1]$ and the amplitude is 1.

By dividing the interval $[0, \frac{2\pi}{b}]$ into four equal parts, we obtain the values for which sin bx or cos bx is $-1, 0,$ or 1. These will give minimum points, x-intercepts, and maximum points on the graph. For example, consider the function $y = \sin 2x$. To determine an interval that represents one period of the graph, we solve the inequality

$$0 \leq 2x \leq 2\pi.$$

We divide by 2 to get $0 \leq x \leq \pi$. The points $(0, 0)$, $(\frac{\pi}{4}, 1)$, $(\frac{\pi}{2}, 0)$, $(\frac{3\pi}{4}, -1)$ and $(\pi, 0)$ represent the important points for this period of the graph.

The following example provides a further analysis of the graph of $y = \sin 2x$.

EXAMPLE 4

Analyzing the Graph of a Transformed Circular Function

The graphs of $y_1 = \sin x$ and $y_2 = \sin 2x$ are shown in Figure 18, graphed over the interval $[-2\pi, 2\pi]$. Notice that while the amplitude of both functions is 1, the period of y_2 is π. This period is found by dividing the period of the sine function, 2π, by the coefficient of x in y_2, namely 2. That is,

$$\text{period of } y_2 = \frac{2\pi}{2} = \pi.$$

In both cases, the range is $[-1, 1]$. ◼

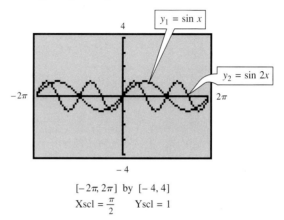

$[-2\pi, 2\pi]$ by $[-4, 4]$
$\text{Xscl} = \frac{\pi}{2}$ $\text{Yscl} = 1$

FIGURE 18

The next example shows how several transformations can be applied to the graph of $y = \cos x$.

Explain how the graph of $y_2 = -3 \cos \frac{1}{2}x$ can be obtained from the graph of $y_1 = \cos x$. Then graph both functions in the window $[-2\pi, 2\pi]$ by $[-4, 4]$.

SOLUTION The graph of $y_2 = -3 \cos \frac{1}{2}x$ can be obtained from the graph of $y_1 = \cos x$ by changing the period from 2π to $\frac{2\pi}{1/2} = 4\pi$, stretching vertically by a factor of 3, and reflecting across the x-axis (because of the negative sign). See Figure 19. ◼

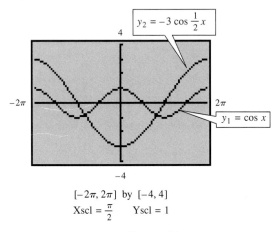

$$[-2\pi, 2\pi] \text{ by } [-4, 4]$$
$$\text{Xscl} = \frac{\pi}{2} \qquad \text{Yscl} = 1$$

FIGURE 19

The graphs of $y = \cos x$ and $y = \sin x$ can also be shifted horizontally and vertically. Recall that the graph of the function $y = f(x - d)$ is obtained by translating the graph of $y = f(x)$ d units to the right if $d > 0$ or $|d|$ units to the left if $d < 0$. In the case of a circular function, a horizontal translation is called a **phase shift**.

The graphs of $y_1 = \sin x$ and $y_2 = \sin(x - \frac{\pi}{3})$ are shown in Figure 20. The graph of y_2 can be obtained by shifting the graph of y_1 $\frac{\pi}{3}$ units to the right. The number $\frac{\pi}{3}$ is the phase shift for y_2. Notice that neither the period nor the amplitude is affected. For both functions the period is 2π and the amplitude is 1. ◼

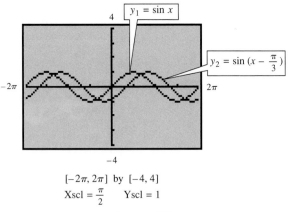

$$[-2\pi, 2\pi] \text{ by } [-4, 4]$$
$$\text{Xscl} = \frac{\pi}{2} \qquad \text{Yscl} = 1$$

FIGURE 20

EXAMPLE 7	Explain how the graph of $y_2 = -2\cos(3x + \pi)$ can be obtained from the graph of $y_1 = \cos x$. Then graph both functions in the window $[-2\pi, 2\pi]$ by $[-4, 4]$.

Analyzing the Graph of a Transformed Circular Function

SOLUTION   The graph of $y_2 = -2\cos(3x + \pi)$ can be obtained from the graph of $y_1 = \cos x$ by first noticing that the equation can be written $y_2 = -2\cos 3(x + \frac{\pi}{3})$. The graph of $y_1 = \cos x$ must be shifted $\frac{\pi}{3}$ units to the left, the period must be changed to $\frac{2\pi}{3}$, the graph will be vertically stretched by a factor of 2, and will be reflected across the x-axis because of the negative sign on the coefficient of the cosine function, -2. The amplitude of $y_2 = -2\cos(3x + \pi)$ is 2, because $\frac{1}{2}[2 - (-2)] = 2$. See Figure 21. ▯

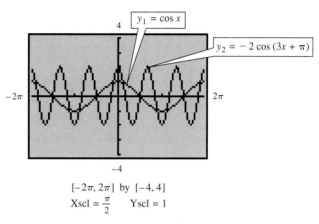

$[-2\pi, 2\pi]$ by $[-4, 4]$
$\text{Xscl} = \frac{\pi}{2}$ $\text{Yscl} = 1$

FIGURE 21

The graph of the function $y = f(x) + c$ can be obtained from the graph of $y = f(x)$ by a *vertical shift*. The shift is upward c units if $c > 0$, or downward $|c|$ units if $c < 0$.

EXAMPLE 8	The graphs of $y_1 = \cos 2x$ and $y_2 = -2 + \cos 2x$ are shown in Figure 22. The graph of y_2 can be obtained by shifting the graph of y_1 2 units down, due to the term -2 which is added to $\cos 2x$. For both functions, the period is π and the amplitude is 1. ▯

Analyzing the Graph of a Circular Function with a Vertical Shift

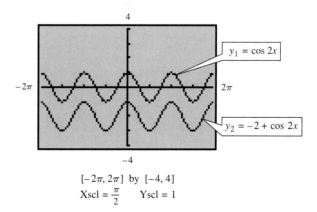

$[-2\pi, 2\pi]$ by $[-4, 4]$
$\text{Xscl} = \frac{\pi}{2}$ $\text{Yscl} = 1$

FIGURE 22

EXAMPLE 9

Analyzing the Graph of a Transformed Circular Function

Explain how the graph of $y_2 = -1 + 2 \sin 4(x + \frac{\pi}{4})$ can be obtained from the graph of $y_1 = \sin x$. Then graph both functions in the trig viewing window.

SOLUTION   The graph of $y_2 = -1 + 2 \sin 4(x + \frac{\pi}{4})$ can be obtained from the graph of $y_1 = \sin x$ by shifting $\frac{\pi}{4}$ units to the left, changing the period to $\frac{2\pi}{4} = \frac{\pi}{2}$, stretching vertically by a factor of 2, and shifting the graph down 1 unit. The amplitude of this function is 2. Notice also that its range is $[-3, 1]$. See Figure 23.   ❑

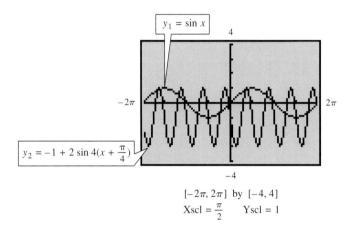

$[-2\pi, 2\pi]$ by $[-4, 4]$
$\text{Xscl} = \frac{\pi}{2}$ $\text{Yscl} = 1$

FIGURE 23

Supporting an Identity Graphically

In Section 9.1 we saw how identities involving circular functions can be verified analytically. We can support our analytic work by graphing the functions defined by the expressions on each side of an identity. If the two graphs coincide, we have provided graphical support.

CAUTION   Graphical support for an identity is not a proof, since we are only able to graph over a finite interval of the domain. In order to truly verify an identity, we must provide an analytic argument.

EXAMPLE 10

Verifying an Identity Analytically and Supporting it Graphically

Verify the identity

$$\cos^4 x - \sin^4 x = 1 - 2 \sin^2 x$$

analytically, and then support the work graphically by showing that the graphs of $y_1 = \cos^4 x - \sin^4 x$ and $y_2 = 1 - 2 \sin^2 x$ coincide in the standard trig window.

SOLUTION   We will show that the left-hand side can be transformed into the right-hand side.

$$\cos^4 x - \sin^4 x = (\cos^2 x + \sin^2 x)(\cos^2 x - \sin^2 x)$$
$$= 1(\cos^2 x - \sin^2 x)$$
$$= 1[(1 - \sin^2 x) - \sin^2 x]$$
$$= 1 - 2 \sin^2 x$$

Now if we graph $y_1 = \cos^4 x - \sin^4 x$ and $y_2 = 1 - 2 \sin^2 x$ in the standard trig window, we see that the graphs appear to coincide. This provides support for our analytic work (but it does *not* provide an actual *proof*). See Figure 24.   ❑

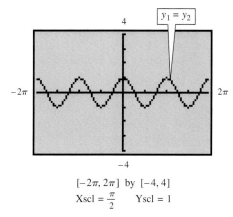

$[-2\pi, 2\pi]$ by $[-4, 4]$

$\text{Xscl} = \dfrac{\pi}{2}$ $\text{Yscl} = 1$

FIGURE 24

FOR **GROUP DISCUSSION**

Refer to Figure 24 to fill in the blanks with the correct responses.

The graph of $y = 1 - 2\sin^2 x$ can be obtained by changing the period of the graph of $y = \cos x$ to _____ . To do this, the positive coefficient b in the expression $\cos bx$ must be _____ . As a result, we have an identity

$$\cos \underline{\quad\quad} = 1 - 2\sin^2 x.$$

(This is known as a double number identity, and will be covered in detail in Section 9.4.)

9.2 EXERCISES

In Exercises 1–6, you are given the graph of either $y = \sin x$ or $y = \cos x$ in the standard trig window, along with a display. Use the graph and the display to write an equation of the form $\sin x \approx k$, $\cos x \approx k$, $\sin x = k$, or $\cos x = k$, for specific values of x and k. Then verify your result using either the unit circle graphed parametrically or the appropriate function key on your calculator. Be sure your calculator is set to radian mode.

1.

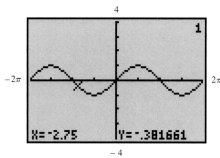

2.

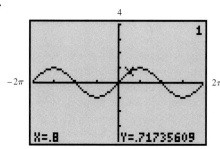

3.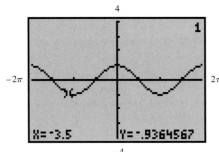

4.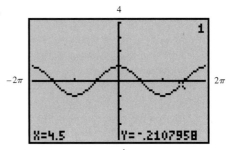

5. $\left(\dfrac{\pi}{2} \approx 1.5707963\right)$

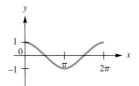

6. $\left(\dfrac{3\pi}{2} \approx 4.712389\right)$

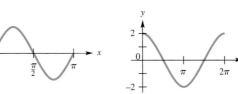

Without using a calculator to graph the functions, match each function in Exercises 7–14 with its graph.

7. $y = \sin x$

8. $y = \cos x$

9. $y = -\sin x$

10. $y = -\cos x$

11. $y = \sin 2x$

12. $y = \cos 2x$

13. $y = 2 \sin x$

14. $y = 2 \cos x$

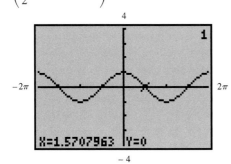

A   B   C

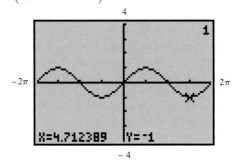

D   E   F

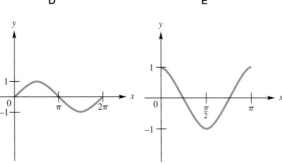

G   H

Without using a calculator to graph the functions, match each function in Exercises 15–22 with its graph.

15. $y = \sin\left(x - \frac{\pi}{4}\right)$

16. $y = \sin\left(x + \frac{\pi}{4}\right)$

17. $y = \cos\left(x - \frac{\pi}{4}\right)$

18. $y = \cos\left(x + \frac{\pi}{4}\right)$

19. $y = 1 + \sin x$

20. $y = -1 + \sin x$

21. $y = 1 + \cos x$

22. $y = -1 + \cos x$

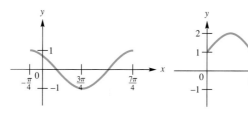

A

B

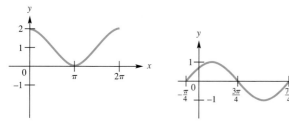

C

D

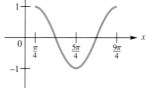

E

F

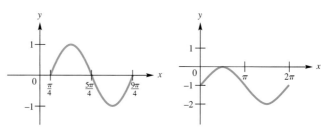

G

H

Suppose that a, b, c, and d are all positive numbers. Consider circular functions of the forms

$$y = d + a \sin b(x + c) \qquad and \qquad y = d + a \cos b(x + c)$$

and answer the questions in Exercises 23–26.

23. What is the amplitude of the function?

24. What is the period of the function?

25. What is the phase shift of the function?

26. Is there a vertical shift? If so, in which direction and by how many units?

*For each of the following circular functions, find **(a)** the amplitude **(b)** the period **(c)** the phase shift (if any) **(d)** the vertical translation (if any) and **(e)** the range of the function.*

27. $y = 2 \sin(x - \pi)$

28. $y = \frac{2}{3} \cos\left(x + \frac{\pi}{2}\right)$

29. $y = 4 \cos\left(\frac{1}{2}x + \frac{\pi}{2}\right)$

30. $y = -\cos\frac{2}{3}\left(x - \frac{\pi}{3}\right)$

31. $y = 2 - \sin\left(3x - \dfrac{\pi}{5}\right)$

32. $y = -1 + \dfrac{1}{2}\cos(2x - 3\pi)$

33. $y = 2 - 3\cos \pi x$

34. $y = 1 - \dfrac{2}{3}\sin \dfrac{3}{4}x$

Explain how the graph of the given function can be obtained from the graph of y = cos x or y = sin x by a transformation.

35. $y = -3 + 2\sin\left(x + \dfrac{\pi}{2}\right)$

36. $y = 4 - 3\cos(x - \pi)$

37. $y = -\dfrac{5}{2} + \cos 3\left(x - \dfrac{\pi}{6}\right)$

38. $y = \dfrac{1}{2} + \sin 2\left(x + \dfrac{\pi}{4}\right)$

Use a calculator to graph each function in the standard trig window.

39. $y = 3\cos 2x$

40. $y = -2\cos 4\left(x + \dfrac{\pi}{2}\right)$

41. $y = 1 - 2\cos .5x$

42. $y = -1 + 3\sin .5x$

R*elating Concepts*

Consider the function $f(x) = -5 + 3\sin 2(x - \frac{\pi}{2})$. Answer the following exercises in order.

43. (a) Because the maximum value of the sine function is ⎯⎯⎯⎯⎯⎯ , the maximum value of $\sin 2(x - \frac{\pi}{2})$ is ⎯⎯⎯⎯⎯⎯ , the maximum value of $3 \sin 2(x - \frac{\pi}{2})$ is ⎯⎯⎯⎯⎯⎯ , and thus the maximum value of $-5 + 3 \sin 2(x - \frac{\pi}{2})$ is ⎯⎯⎯⎯⎯⎯ .

(b) Because the minimum value of the sine function is ⎯⎯⎯⎯⎯⎯ , the minimum value of $\sin 2(x - \frac{\pi}{2})$ is ⎯⎯⎯⎯⎯⎯ , the minimum value of $3 \sin 2(x - \frac{\pi}{2})$ is ⎯⎯⎯⎯⎯⎯ , and thus the minimum value of $-5 + 3 \sin 2(x - \frac{\pi}{2})$ is ⎯⎯⎯⎯⎯⎯ .

44. Based on the answers in Exercise 43, what is the range of f?

45. Why will the standard trig window as defined in the text not provide a complete graph of f?

46. In order to obtain a complete graph of f, Ymax must be at least ⎯⎯⎯⎯⎯⎯ and Ymin must be at most ⎯⎯⎯⎯⎯⎯ .

47. Explain why using Xmin $= -2\pi$ and Xmax $= 2\pi$ will show exactly 4 periods of the graph of f.

48. Use your calculator to graph f in the window $[-2\pi, 2\pi]$ by $[-10, 5]$.

49. Look at the calculator-generated graph from Exercise 48. A "border" of empty space appears above and below the graph. If you did not want such a border to appear, what Ymin and Ymax values should you use so that a complete graph would fit?

50. Evaluate the approximate values of the function for $x = -2$ and for $x = -2 + \pi$. What is the value in each case? Why is this so?

*Exercises 51–56 give identities involving the sine and/or cosine functions. For each identity (**a**) verify analytically and (**b**) support graphically using the trig viewing window, letting the left side be y_1 and the right side be y_2.*

51. $(1 - \cos^2 x)(1 + \cos^2 x) = 2 \sin^2 x - \sin^4 x$

52. $2 \cos^2 x - \sin^2 x + 1 = 3 \cos^2 x$

53. $\cos^4 x + 2 \cos^2 x + 1 = (2 - \sin^2 x)^2$

54. $\sin^3 x + \cos^3 x = (\sin x + \cos x)(1 - \sin x \cos x)$

55. $(\sin x + 1)^2 - (\sin x - 1)^2 = 4 \sin x$

56. $(1 + \sin x)^2 + \cos^2 x = 2 + 2 \sin x$

A function of the form $y = a \cos bx$ or $y = a \sin bx$, where $b < 0$, can be rewritten in an equivalent form so that the coefficient of x is positive. This is done using the appropriate negative number identity. For example,

$$y = 3 \cos(-2x) \text{ is equivalent to } y = 3 \cos 2x$$

and $y = 3 \sin(-2x)$ *is equivalent to* $y = 3(-\sin 2x) = -3 \sin 2x$.

Write each of the following in an equivalent form so that the coefficient of x is positive.

57. $y = 4 \sin(-3x)$ **58.** $y = 5 \sin(-6x)$ **59.** $y = -2 \cos(-3x)$

60. $y = -\dfrac{1}{2} \cos(-\pi x)$ **61.** $y = -3 \sin(-6x)$ **62.** $y = -3 \cos(-4x)$

In Exercises 63–66, the expression on the left side of the incomplete equation is equivalent to a single circular function or a constant. Graph the expression as y_1 on your calculator and decide which function or what constant applies. Then fill in the blank on the right side, and verify the identity analytically.

63. $\dfrac{1 - \cos^2 x}{\sin x} = \underline{\hspace{2cm}}$ **64.** $\dfrac{1 - \sin^2 x}{\cos x} = \underline{\hspace{2cm}}$

65. $\cos^2 x(\tan^2 x + 1) = \underline{\hspace{2cm}}$ **66.** $\csc^2 x(\cos^2 x - 1) = \underline{\hspace{2cm}}$

*F*urther Explorations

In each of the following TABLES, $Y_1 = \sin x$ and Y_2 is some transformation of Y_1. Use what you know about transformations to find the expression for Y_2. Confirm by duplicating each TABLE on your graphics calculator in RADIAN MODE. It is possible to find more than one solution for some of these exercises. In each exercise, $\Delta \text{Tbl} = \frac{\pi}{6}$. (*Note:* The calculator displays a decimal approximation of each.) The first TABLE in each exercise shows the interval $[0, \pi]$ and the second TABLE shows the interval $[\pi, 2\pi]$.

1.

X	Y₁	Y₂
0	0	0
.5236	.5	1
1.0472	.86603	1.7321
1.5708	1	2
2.0944	.86603	1.7321
2.618	.5	1
3.1416	0	0

X=0

X	Y₁	Y₂
3.1416	0	0
3.6652	-.5	-1
4.1888	-.866	-1.732
4.7124	-1	-2
5.236	-.866	-1.732
5.7596	-.5	-1
6.2832	0	0

X=3.14159265359

2.

X	Y1	Y2
0	0	0
.5236	.5	-1.5
1.0472	.86603	-2.598
1.5708	1	-3
2.0944	.86603	-2.598
2.618	.5	-1.5
3.1416	0	0

X=0

X	Y1	Y2
3.1416	0	0
3.6652	-.5	1.5
4.1888	-.866	2.5981
4.7124	-1	3
5.236	-.866	2.5981
5.7596	-.5	1.5
6.2832	0	0

X=3.14159265359

3.

X	Y1	Y2
0	0	0
.5236	.5	.86603
1.0472	.86603	.86603
1.5708	1	0
2.0944	.86603	-.866
2.618	.5	-.866
3.1416	0	0

X=0

X	Y1	Y2
3.1416	0	0
3.6652	-.5	.86603
4.1888	-.866	.86603
4.7124	-1	0
5.236	-.866	-.866
5.7596	-.5	-.866
6.2832	0	-2E-13

X=3.14159265359

4.

X	Y1	Y2
0	0	0
.5236	.5	1
1.0472	.86603	0
1.5708	1	-1
2.0944	.86603	0
2.618	.5	1
3.1416	0	0

X=0

X	Y1	Y2
3.1416	0	0
3.6652	-.5	-1
4.1888	-.866	-2E-13
4.7124	-1	1
5.236	-.866	0
5.7596	-.5	-1
6.2832	0	2E-13

X=3.14159265359

5.

X	Y1	Y2
0	0	-.5
.5236	.5	0
1.0472	.86603	.5
1.5708	1	.86603
2.0944	.86603	1
2.618	.5	.86603
3.1416	0	.5

X=0

X	Y1	Y2
3.1416	0	.5
3.6652	-.5	0
4.1888	-.866	-.5
4.7124	-1	-.866
5.236	-.866	-1
5.7596	-.5	-.866
6.2832	0	-.5

X=3.14159265359

6.

X	Y1	Y2
0	0	0
.5236	.5	-.5
1.0472	.86603	-.866
1.5708	1	-1
2.0944	.86603	-.866
2.618	.5	-.5
3.1416	0	0

X=0

X	Y1	Y2
3.1416	0	0
3.6652	-.5	.5
4.1888	-.866	.86603
4.7124	-1	1
5.236	-.866	.86603
5.7596	-.5	.5
6.2832	0	0

X=3.14159265359

7.

X	Y1	Y2
0	0	1
.5236	.5	1.5
1.0472	.86603	1.866
1.5708	1	2
2.0944	.86603	1.866
2.618	.5	1.5
3.1416	0	1

X=0

X	Y1	Y2
3.1416	0	1
3.6652	-.5	.5
4.1888	-.866	.13397
4.7124	-1	0
5.236	-.866	.13397
5.7596	-.5	.5
6.2832	0	1

X=3.14159265359

8.

X	Y1	Y2
0	0	-2
.5236	.5	-1.5
1.0472	.86603	-1.134
1.5708	1	-1
2.0944	.86603	-1.134
2.618	.5	-1.5
3.1416	0	-2

X=0

X	Y1	Y2
3.1416	0	-2
3.6652	-.5	-2.5
4.1888	-.866	-2.866
4.7124	-1	-3
5.236	-.866	-2.866
5.7596	-.5	-2.5
6.2832	0	-2

X=3.14159265359

9.3 ANALYSIS OF OTHER CIRCULAR FUNCTIONS; EQUATIONS AND INEQUALITIES

The Graphs of the Cosecant and Secant Functions ▮ The Graphs of the Tangent and Cotangent Functions ▮ Transformation of Graphs ▮ Equations and Inequalities Involving Circular Functions

The Graphs of the Cosecant and Secant Functions

Since cosecant values are reciprocals of the corresponding sine values, the period of the function $y = \csc x$ is 2π, the same as for $y = \sin x$. When $\sin x = 1$, the value of $\csc x$ is also 1, and when $0 < \sin x < 1$, then $\csc x > 1$. Also, if $-1 < \sin x < 0$, then $\csc x < -1$. (Verify these statements with a calculator set in radian mode.) As $|x|$ approaches 0, $|\sin x|$ approaches 0, and $|\csc x|$ gets larger and larger. The graph of $\csc x$ approaches the vertical line $x = 0$ but never touches it. The line $x = 0$ is a *vertical asymptote*. In fact, the lines $x = n\pi$, where n is any integer, are all vertical asymptotes. Using this information and plotting a few points shows that the graph takes the shape of the solid curve shown in Figure 25. To show how the two graphs are related, the graph of $y = \sin x$ is also shown, as a dashed curve.

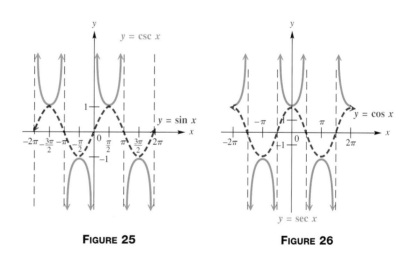

FIGURE 25   **FIGURE 26**

A similar analysis for the secant leads to the solid curve shown in Figure 26. The dashed curve, $y = \cos x$, is shown so that the relationship between these two reciprocal functions is seen.

When graphing these functions on a graphics calculator, we will use dot mode in order to get an accurate picture. (If connected mode is used, the calculator will attempt to connect points that are actually separated by vertical asymptotes.) We will enter $\csc x$ as $\frac{1}{\sin x}$, and $\sec x$ as $\frac{1}{\cos x}$.

COSECANT FUNCTION

$f(x) = \csc x$ (Figure 27)

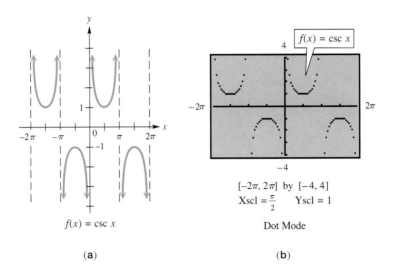

$f(x) = \csc x$

(a)

$[-2\pi, 2\pi]$ by $[-4, 4]$
$\text{Xscl} = \frac{\pi}{2}$ $\text{Yscl} = 1$

Dot Mode

(b)

FIGURE 27

Domain: $\{x \mid x \neq n\pi,$ where n is an integer$\}$

Range: $(-\infty, -1] \cup [1, \infty)$

Over the interval $(0, 2\pi)$, the cosecant function exhibits the following behavior:

From 0 to $\frac{\pi}{2}$, $\csc x$ decreases from ∞ to 1.

From $\frac{\pi}{2}$ to π, $\csc x$ increases from 1 to ∞.

From π to $\frac{3\pi}{2}$, $\csc x$ increases from $-\infty$ to -1.

From $\frac{3\pi}{2}$ to 2π, $\csc x$ decreases from -1 to $-\infty$.

The graph is discontinuous at values of x of the form $x = n\pi$, and has vertical asymptotes at these values. There are no x-intercepts. Its period is 2π. It has no amplitude, since there are no maximum and minimum values. The graph is symmetric with respect to the origin.

SECANT FUNCTION

$f(x) = \sec x$ (Figure 28)

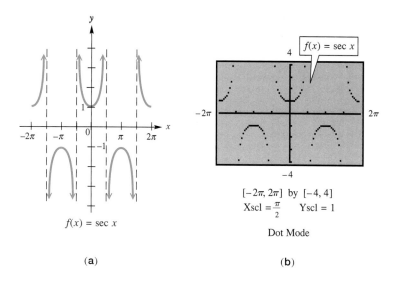

(a)

(b)

FIGURE 28

Domain: $\{x \mid x \neq (2n + 1)\frac{\pi}{2}, \text{ where } n \text{ is an integer}\}$

Range: $(-\infty, -1] \cup [1, \infty)$

Over the interval $[0, 2\pi]$, the secant function exhibits the following behavior:

From 0 to $\frac{\pi}{2}$, $\sec x$ increases from 1 to ∞.

From $\frac{\pi}{2}$ to π, $\sec x$ increases from $-\infty$ to -1.

From π to $\frac{3\pi}{2}$, $\sec x$ decreases from -1 to $-\infty$.

From $\frac{3\pi}{2}$ to 2π, $\sec x$ decreases from ∞ to 1.

The graph is discontinuous at values of x of the form $x = (2n + 1)\frac{\pi}{2}$, and has vertical asymptotes at these values. There are no x-intercepts. Its period is 2π. It has no amplitude, since there are no maximum and minimum values. The graph is symmetric with respect to the y-axis.

The Graphs of the Tangent and Cotangent Functions

Because $\tan x = \frac{\sin x}{\cos x}$, the tangent function is undefined when $\cos x = 0$ and the graph of the tangent function has x-intercepts when $\sin x = 0$. Based on our discussion of the sine and cosine functions in Section 9.2, we may conclude that the tangent is undefined when $x = (2n + 1)\frac{\pi}{2}$, where n is an integer, and has x-intercepts when $x = n\pi$. Using a calculator in dot mode, we graph $y = \tan x$. See Figure 29.

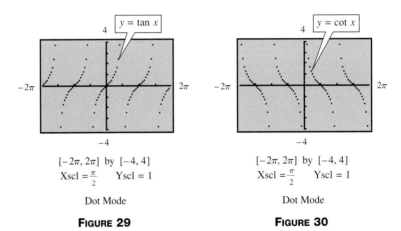

$[-2\pi, 2\pi]$ by $[-4, 4]$
Xscl $= \frac{\pi}{2}$ Yscl $= 1$

Dot Mode

FIGURE 29

$[-2\pi, 2\pi]$ by $[-4, 4]$
Xscl $= \frac{\pi}{2}$ Yscl $= 1$

Dot Mode

FIGURE 30

A similar analysis for the cotangent function leads to the graph in Figure 30. Using a graphics calculator, we may either enter $\cot x$ as $\frac{\cos x}{\sin x}$, or as $\frac{1}{\tan x}$. We see that the cotangent is undefined when $x = n\pi$, where n is an integer, and has x-intercepts when $x = (2n + 1)\frac{\pi}{2}$.

TANGENT FUNCTION

$f(x) = \tan x$ (Figure 31)

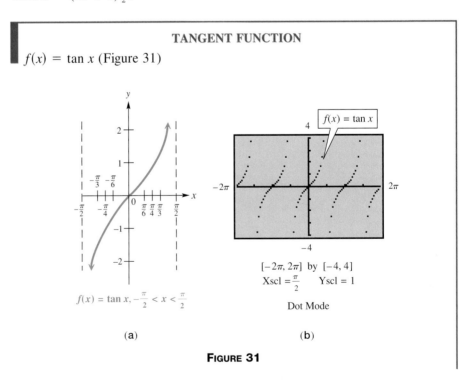

$f(x) = \tan x, -\frac{\pi}{2} < x < \frac{\pi}{2}$

(a)

$[-2\pi, 2\pi]$ by $[-4, 4]$
Xscl $= \frac{\pi}{2}$ Yscl $= 1$

Dot Mode

(b)

FIGURE 31

Domain: $\{x \mid x \neq (2n + 1)\frac{\pi}{2}, \text{ where } n \text{ is an integer}\}$

Range: $(-\infty, \infty)$

Over the interval $[0, \pi]$, the tangent function exhibits the following behavior:

From 0 to $\frac{\pi}{2}$, tan x increases from 0 to ∞.

From $\frac{\pi}{2}$ to π, tan x increases from $-\infty$ to 0.

The graph is discontinuous at values of x of the form $x = (2n + 1)\frac{\pi}{2}$, and has vertical asymptotes at these values. The x-intercepts are of the form $x = n\pi$. Its period is π. It has no amplitude, since there are no minimum and maximum values. The graph is symmetric with respect to the origin.

COTANGENT FUNCTION

$f(x) = \cot x$ (Figure 32)

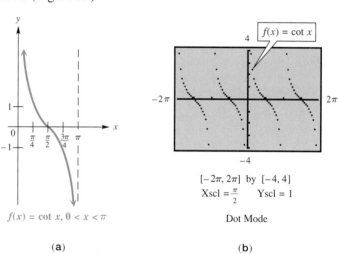

$f(x) = \cot x, 0 < x < \pi$

(a)

$[-2\pi, 2\pi]$ by $[-4, 4]$

Xscl $= \frac{\pi}{2}$ Yscl $= 1$

Dot Mode

(b)

FIGURE 32

Domain: $\{x \mid x \neq n\pi, \text{ where } n \text{ is an integer}\}$

Range: $(-\infty, \infty)$

Over the interval $(0, \pi)$, the cotangent function exhibits the following behavior:

From 0 to π, cot x decreases from ∞ to $-\infty$.

The graph is discontinuous at values of x of the form $x = n\pi$, and has vertical asymptotes at these values. The x-intercepts are of the form $x = (2n + 1)\frac{\pi}{2}$. Its period is π. It has no amplitude, since there are no minimum and maximum values. The graph is symmetric with respect to the origin.

Transformation of Graphs

In the first few examples, we show how the graphs of the secant, cosecant, tangent, and cotangent functions can be transformed.

EXAMPLE 1

Analyzing a
Transformed Secant
Graph

Figure 33 shows the graph of $y = 2 \sec \frac{1}{2}x$ in both traditional and calculator-generated forms.

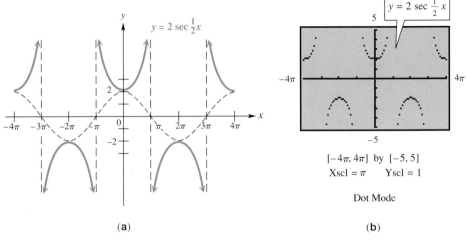

(a)   (b)

FIGURE 33

This graph is obtained by using the graph of $y = 2 \cos \frac{1}{2}x$ as a guide (shown as a dashed curve in Figure 33(a)). Its period is the same as that of this transformed cosine graph: $\frac{2\pi}{1/2} = 4\pi$. Notice that because the period is altered from the basic secant graph, the locations of the vertical asymptotes are also altered. They are of the form $x = (2n + 1)\pi$. The coefficient 2 affects the graph by stretching it vertically by a factor of 2. The y-intercept is now 2.   ∎

FOR GROUP DISCUSSION

If we know the location of the vertical asymptote $x = k$ with smallest positive value of k, along with the period of the function, we can determine the locations of all vertical asymptotes. Discuss how you might go about finding this smallest such positive value in Example 1.

EXAMPLE 2

Analyzing a
Transformed
Cosecant Graph

Figure 34 shows the graph of $y = \frac{3}{2} \csc(x - \frac{\pi}{2})$ in both traditional and calculator-generated forms.

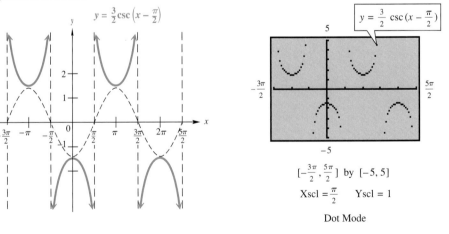

(a)   (b)

FIGURE 34

This graph is obtained by shifting the graph of $y = \csc x \frac{\pi}{2}$ units to the right (the phase shift) and stretching it vertically by a factor of $\frac{3}{2}$. The period is 2π, since this is the period of the associated reciprocal function $y = \frac{3}{2}\sin(x - \frac{\pi}{2})$. Vertical asymptotes are of the form $x = (2n + 1)\frac{\pi}{2}$. (In Figure 34(a) we show the graph of the related reciprocal function as a dashed line.)   ▐

FOR **GROUP DISCUSSION**

1. Use the graph in Figure 33 to determine the domain and the range of $y = 2 \sec \frac{1}{2}x$.
2. Use the graph in Figure 34 to determine the domain and the range of $y = \frac{3}{2}\csc(x - \frac{\pi}{2})$.
3. Use the capabilities of your calculator to support the following statement: The function $y = \frac{3}{2}\csc(x - \frac{\pi}{2})$ has a local minimum at $x = \pi$.
4. Repeat item 3 for the following statement: The point $(\frac{\pi}{2}, 2\sqrt{2})$ lies on the graph of $y = 2 \sec \frac{1}{2}x$.

EXAMPLE 3

Analyzing a
Transformed
Tangent Graph

Figure 35 shows the graph of $y = -3 \tan \frac{1}{2}x$ in both traditional and calculator-generated forms.

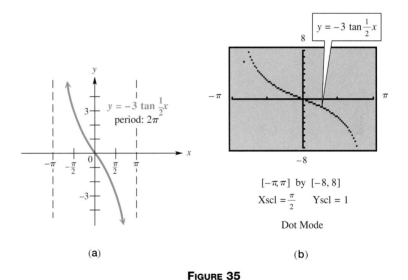

$y = -3 \tan \dfrac{1}{2}x$
period: 2π

$y = -3 \tan \dfrac{1}{2}x$

$[-\pi, \pi]$ by $[-8, 8]$

$Xscl = \dfrac{\pi}{2}$ $Yscl = 1$

Dot Mode

(a)   (b)

FIGURE 35

The graph of this function has period $\pi/(\frac{1}{2}) = 2\pi$. In addition to this, the fact that the coefficient is -3 causes a vertical stretch by a factor of 3 and a reflection across the x-axis. The vertical asymptotes are of the form $x = (2n + 1)\pi$, and the x-intercepts are of the form $2n\pi$. The domain is $\{x \mid x \neq (2n + 1)\pi\}$ and the range is $(-\infty, \infty)$.   ▐

EXAMPLE 4

Analyzing a Transformed Cotangent Graph

Figure 36 shows the graph of $y = -2 - \cot(x - \frac{\pi}{4})$ in both traditional and calculator-generated forms.

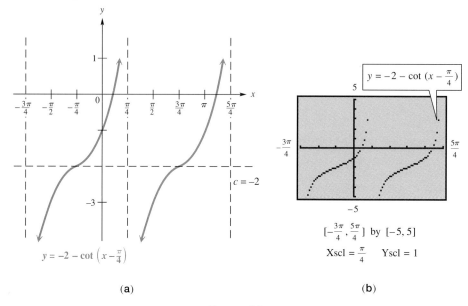

$[-\frac{3\pi}{4}, \frac{5\pi}{4}]$ by $[-5, 5]$

$\text{Xscl} = \frac{\pi}{4}$ $\text{Yscl} = 1$

(a)   (b)

FIGURE 36

This graph is obtained from the graph of $y = \cot x$ by shifting $\frac{\pi}{4}$ units to the right (the phase shift), reflecting across the x-axis, and shifting 2 units downward. The period is π, there is no amplitude, and the range is $(-\infty, \infty)$. We will investigate the location of the vertical asymptotes and the x-intercepts of the function in Exercises 31–36.

Equations and Inequalities Involving Circular Functions

We will now investigate methods of solving equations and associated inequalities involving circular functions. Because the circular functions are periodic, equations involving them will most often have infinitely many solutions unless we restrict the domain. Therefore, unless otherwise specified, we will solve equations and inequalities in this section over the interval $[0, 2\pi)$.

EXAMPLE 5

Solving an Equation Involving a Circular Function (Linear Method)

Solve $2 \sin x - 1 = 0$ over the interval $[0, 2\pi)$ and support the solution set graphically.

SOLUTION   Since this equation involves the first power of $\sin x$, it is linear in $\sin x$ and we will solve it using the usual method of solving a linear equation.

$$2 \sin x - 1 = 0$$
$$2 \sin x = 1$$
$$\sin x = \frac{1}{2}$$

The two values of x in the interval $[0, 2\pi)$ that have a sine value of $\frac{1}{2}$ are $\frac{\pi}{6}$ and $\frac{5\pi}{6}$. This can be determined by using the unit circle (see Figure 3) or reference angle analysis (see Section 8.3). Therefore, the solution set in the specified interval is $\{\frac{\pi}{6}, \frac{5\pi}{6}\}$.

To support this result graphically, we graph $y = 2 \sin x - 1$ over the interval $[0, 2\pi]$ and use the capabilities of the calculator to verify that the x-intercepts have the same decimal approximations as $\frac{\pi}{6}$ and $\frac{5\pi}{6}$. See Figure 37. ∎

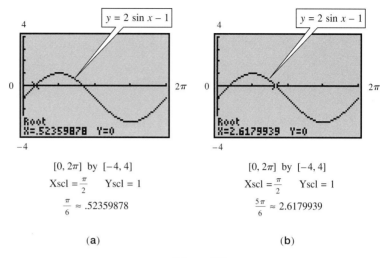

$[0, 2\pi]$ by $[-4, 4]$
$\text{Xscl} = \frac{\pi}{2}$ $\text{Yscl} = 1$
$\frac{\pi}{6} \approx .52359878$

(a)

$[0, 2\pi]$ by $[-4, 4]$
$\text{Xscl} = \frac{\pi}{2}$ $\text{Yscl} = 1$
$\frac{5\pi}{6} \approx 2.6179939$

(b)

FIGURE 37

FOR **GROUP DISCUSSION**

Refer to the equation in Example 5 as you discuss these items.

1. What is the period of the function $y = 2 \sin x - 1$?
2. Whenever an integer multiple of the period is added to a solution of the equation, another solution is found (if we allow a larger interval over which we are solving). Let k be your favorite integer between 1 and 3 inclusive. With Xmin = 0 and Xmax = 10π, graph $y_1 = 2 \sin x - 1$ and show that $\frac{\pi}{6} + 2k\pi$ and $\frac{5\pi}{6} + 2k\pi$ are also solutions of $2 \sin x - 1 = 0$.
3. Discuss why the following statement is true: The solution set of $2 \sin x - 1 = 0$ over all real numbers is $\{x \mid x = \frac{\pi}{6} + 2n\pi,\ x = \frac{5\pi}{6} + 2n\pi,\ \text{where } n \text{ is an integer}\}$.

EXAMPLE **6**

Solving Inequalities Associated with an Equation Involving a Circular Function

Use the result of Example 5 and the graph in Figure 37 to solve **(a)** $2 \sin x - 1 > 0$ and **(b)** $2 \sin x - 1 < 0$ over the interval $[0, 2\pi)$.

SOLUTION To solve the inequality in (a), we must identify the x-values in the interval $[0, 2\pi)$ for which the graph is *above* the x-axis. Similarly, to solve the inequality in (b) we must identify the x-values for which the graph is *below* the x-axis. The graph in Figure 37 indicates the following:

The solution set of $2 \sin x - 1 > 0$ is $\left(\dfrac{\pi}{6}, \dfrac{5\pi}{6} \right)$.

The solution set of $2 \sin x - 1 < 0$ is $\left[0, \dfrac{\pi}{6} \right) \cup \left(\dfrac{5\pi}{6}, 2\pi \right)$. ∎

The equation and inequalities in Examples 5 and 6 were linear in the expression $\sin x$. The next example shows how methods of solving quadratic equations and inequalities, first presented in Chapter 3, can be extended to equations and inequalities involving circular functions.

EXAMPLE 7

Solving an Equation and Associated Inequalities Involving a Circular Function (Quadratic Method)

(a) Solve $\tan^2 x + \tan x - 2 = 0$ over the interval $[0, 2\pi)$ by using a method of solving quadratic equations analytically.

SOLUTION We could use either the quadratic formula or the zero-factor property to solve this equation. We will use the latter.

$$\tan^2 x + \tan x - 2 = 0$$
$$(\tan x - 1)(\tan x + 2) = 0$$

Set each factor equal to 0.

$$\tan x - 1 = 0 \quad \text{or} \quad \tan x + 2 = 0$$
$$\tan x = 1 \quad \text{or} \quad \tan x = -2$$

The solutions for $\tan x = 1$ in the interval $[0, 2\pi)$ are $x = \frac{\pi}{4}$ or $\frac{5\pi}{4}$. To solve $\tan x = -2$ in the interval, use a calculator set in the radian mode. We find that $\tan^{-1}(-2) \approx -1.107148718$. However, due to the method in which the calculator determines this value (justified in Section 9.6), this number is not in the desired interval. Because the period of the tangent function is π, we will add π and then add 2π to $\tan^{-1}(-2)$ to obtain the solutions in the desired interval:

$$x = \tan^{-1}(-2) + \pi \approx 2.034443936$$
$$x = \tan^{-1}(-2) + 2\pi \approx 5.176036589$$

The solution set is

$$\left\{ \underbrace{\frac{\pi}{4}, \frac{5\pi}{4}}_{\substack{\text{Exact} \\ \text{values}}}, \quad \underbrace{2.03, 5.18}_{\substack{\text{Approximate} \\ \text{values to the} \\ \text{nearest} \\ \text{hundredth}}} \right\}.$$

(b) Support the solution in part (a) by graphing $y = \tan^2 x + \tan x - 2$ in the window $[0, 2\pi]$ by $[-4, 4]$, and finding the x-intercepts.

SOLUTION As seen in Figure 38, there are indeed four x-intercepts in the interval. Using the capabilities of the calculator, we can support the solutions in part (a). The figure shows the support for the solution $\frac{5\pi}{4}$, since a decimal approximation of this number is 3.9269908. The other three solutions can be supported similarly.

(c) Use the result of part (a) and Figure 38 to find the solution set of $\tan^2 x + \tan x - 2 > 0$ over the interval $[0, 2\pi)$.

SOLUTION The graph of $y = \tan^2 x + \tan x - 2$ lies *above* the x-axis for the following subset of the interval $[0, 2\pi)$:

$$\left(\frac{\pi}{4}, \frac{\pi}{2} \right) \cup \left(\frac{\pi}{2}, 2.03 \right) \cup \left(\frac{5\pi}{4}, \frac{3\pi}{2} \right) \cup \left(\frac{3\pi}{2}, 5.18 \right).$$

This is the solution set of the inequality. Notice that we used the fact that the tangent is not defined when $x = \frac{\pi}{2}$ and when $x = \frac{3\pi}{2}$. These values must be excluded from the solution set. ∎

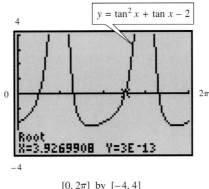

$$[0, 2\pi] \text{ by } [-4, 4]$$
$$\text{Xscl} = \frac{\pi}{2} \qquad \text{Yscl} = 1$$

FIGURE 38

FOR GROUP DISCUSSION

What is the solution set of $\tan^2 x + \tan x - 2 < 0$ over the interval $[0, 2\pi)$? (*Hint:* Refer to Figure 38.)

The final example illustrates how the quadratic formula can be used to solve an equation involving a circular function.

EXAMPLE 8

Solving an Equation Involving a Circular Function Using the Quadratic Formula

Solve the equation

$$\cot x(\cot x + 3) = 1$$

over the interval $[0, 2\pi)$. Use the quadratic formula, and support your solution graphically.

SOLUTION We begin by multiplying the factors on the left and subtracting 1 to get the equation in the standard form of a quadratic equation.

$$\cot^2 x + 3 \cot x - 1 = 0$$

Since this equation cannot be solved by factoring, use the quadratic formula, with $a = 1$, $b = 3$, $c = -1$, and $\cot x$ as the variable.

$$\cot x = \frac{-3 \pm \sqrt{9 + 4}}{2} = \frac{-3 \pm \sqrt{13}}{2}$$

Using a calculator, we find

$$\cot x \approx -3.302775638 \qquad \text{or} \qquad \cot x \approx .3027756377.$$

Since we cannot find inverse cotangent values directly on a calculator, we use the fact that $\cot x = \frac{1}{\tan x}$, and take reciprocals to get

$$\tan x \approx -.3027756377 \qquad \text{or} \qquad \tan x \approx 3.302775638$$
$$x \approx -.2940013018 \qquad\qquad\qquad x \approx 1.276795025$$

(These x-values were found by using the inverse tangent function key.)

The first of these, $-.2940013018$, is not in the desired interval. Since the period of the cotangent function is π, we add π and then add 2π to $-.2940013018$ to get 2.847591352 and 5.989184005.

The second value, 1.276795025, is in the desired interval. We add π to it to get another solution in the interval: $1.276795025 + \pi$. Rounding to the nearest hundredth, the four solutions in the interval are 1.28, 2.85, 4.42, and 5.99, and the solution set is $\{1.28, 2.85, 4.42, 5.99\}$.

The graph of $y = \cot^2 x + 3 \cot x - 1$ is shown in Figure 39. Using the capabilities of the calculator, we can support the solutions found above. The figure shows the support for the solution 1.276795025. The others can be supported similarly. ▪

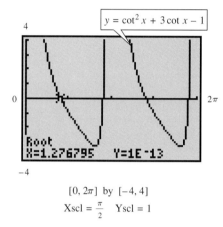

$$[0, 2\pi] \text{ by } [-4, 4]$$
$$\text{Xscl} = \frac{\pi}{2} \quad \text{Yscl} = 1$$

FIGURE 39

9.3 EXERCISES

In Exercises 1–8 you are given the graph of $y = f(x)$ in dot mode, where f is either tan, cot, sec, or csc in the standard trig window, along with a display. Use the graph and the display to write an equation of the form $f(x) \approx k$ or $f(x) = k$, for specific values of x and k. Then verify your result using the appropriate key or combination of keys on your calculator. Be sure your calculator is set to radian mode.

1.
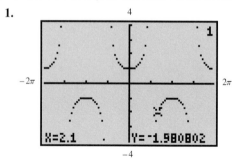

2.
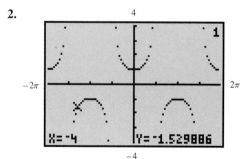

3.

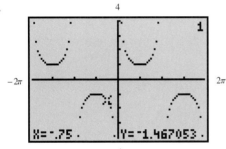

4.

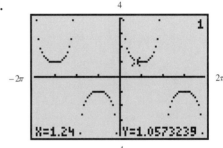

5.

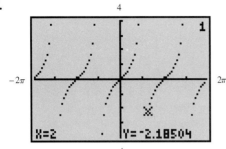

6.

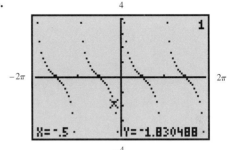

7. (*Hint:* $\dfrac{\pi}{4} \approx .78539816$)

8. (*Hint:* $-\dfrac{\pi}{4} \approx -.7853982$)

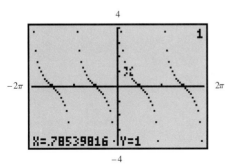

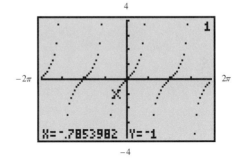

9. Between a pair of successive asymptotes, a portion of the graph of $y = \sec x$ or $y = \csc x$ resembles a parabola. However, these are not parabolas. Why not?

10. Use the graphs of $y = \sec x$ and $y = \csc x$ to respond to the following.

(a) The function $y = \sec x$ has a local maximum for any x-value of the form _____, where n is an integer.

(b) The function $y = \sec x$ has a local minimum for any x-value of the form _____, where n is an integer.

(c) The function $y = \csc x$ has a local extremum for any x-value of the form _____, where n is an integer.

Without using a calculator to graph the functions, match each function in Exercises 11–16 with its graph.

11. $y = -\csc x$

12. $y = -\sec x$

13. $y = -\tan x$

14. $y = -\cot x$

15. $y = \tan\left(x - \frac{\pi}{4}\right)$

16. $y = \cot\left(x - \frac{\pi}{4}\right)$

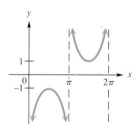

A

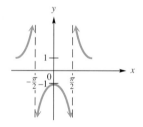

B

C

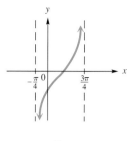

D

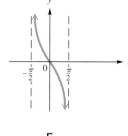

E

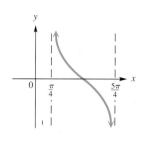

F

For each function in Exercises 17–26, do the following.
(a) *Graph the function in the trig viewing window. Use dot mode.*
(b) *Give the period of the function.*
(c) *Give the phase shift, if any.*
(d) *Give the range of the function.*
(e) *Give the smallest positive value for which the function is undefined.*

17. $y = 2 \csc \frac{1}{2}x$

18. $y = 3 \csc 2x$

19. $y = -2 \sec\left(x + \frac{\pi}{2}\right)$

20. $y = -\frac{3}{2} \sec(x - \pi)$

21. $y = \frac{5}{2} \cot \frac{1}{3}\left(x - \frac{\pi}{2}\right)$

22. $y = -3 \tan \frac{1}{2}\left(x + \frac{\pi}{4}\right)$

23. $f(x) = \frac{1}{2} \sec(2x + \pi)$

24. $f(x) = -\frac{1}{3} \csc\left(\frac{1}{2}x - \frac{\pi}{2}\right)$

25. $y = -1 - \tan\left(x + \frac{\pi}{4}\right)$

26. $y = 2 + \cot\left(2x - \frac{\pi}{3}\right)$

Explain how the graph of the given function can be obtained from the graph of one of the four circular functions analyzed in this section by a transformation.

27. $y = -2 + 3 \csc 2(x - \pi)$

28. $y = 5 + 2 \sec 3(x + \pi)$

29. $y = 4 - 2 \tan(2x - 2)$

30. $y = 5 - 3 \cot(3x + 3)$

R*elating Concepts*

Refer to Example 4 and Figure 36, which shows a portion of the graph of $y = -2 - \cot\left(x - \frac{\pi}{4}\right)$.

31. What is the smallest positive number for which the graph of $y = \cot x$ is undefined?

32. Let k represent the number you found in Exercise 31. Set $x - \frac{\pi}{4}$ equal to k and solve to find the smallest positive number for which $\cot(x - \frac{\pi}{4})$ is undefined.

33. Based on your answer in Exercise 32 and the fact that the cotangent function has period π, give the general form of the equations of the asymptotes of the graph of $y = -2 - \cot(x - \frac{\pi}{4})$. Let n represent any integer.

34. Use the capabilities of your calculator to find the smallest positive x-intercept of the graph of this function.

35. Use the fact that the period of this function is π to find the next positive x-intercept.

36. Give the solution set of the equation $-2 - \cot(x - \frac{\pi}{4}) = 0$ over all real numbers. Let n represent any integer.

For the function f in Exercises 37–46,
(a) *Solve $f(x) = 0$ analytically over the interval $[0, 2\pi)$, and give solutions in exact forms.*
(b) *Graph $y = f(x)$ in the window $[0, 2\pi]$ by $[-4, 4]$.*
(c) *Use the results of (a) and the graph in (b) to give the exact solution set of $f(x) > 0$ over $[0, 2\pi)$.*
(d) *Use the results of (a) and the graph in (b) to give the exact solution set of $f(x) < 0$ over $[0, 2\pi)$.*

37. $f(x) = -2 \cos x + 1$

38. $f(x) = 2 \sin x + 1$

39. $f(x) = \tan^2 x - 3$

40. $f(x) = \sec^2 x - 1$

41. $f(x) = 2 \cos^2 x - \sqrt{3} \cos x$

42. $f(x) = 2 \sin^2 x + 3 \sin x + 1$

43. $f(x) = \cos^2 x - \sin^2 x$
(*Hint:* Use $\cos^2 x = 1 - \sin^2 x$.)

44. $f(x) = \cos^2 x - \sin^2 x - 1$

45. $f(x) = \csc^2 x - 2 \cot x$
(*Hint:* Use $\csc^2 x = 1 + \cot^2 x$.)

46. $f(x) = \sin^2 x \cos x - \cos x$

In Exercises 47–52, repeat the directions given for Exercises 37–46, except give solutions and endpoints as approximations to the nearest hundredth when exact values cannot be determined. You will need to use the quadratic formula.

47. $f(x) = 3 \sin^2 x - \sin x - 2$

48. $f(x) = 9 \sin^2 x - 6 \sin x - 1$

49. $f(x) = \tan^2 x + 4 \tan x + 2$

50. $f(x) = 3 \cot^2 x - 3 \cot x - 1$

51. $f(x) = 2 \cos^2 x + 2 \cos x - 1$

52. $f(x) = \sin^2 x - 2 \sin x + 3$

*R*elating Concepts

Observe the basic graphs of $y = \tan x$, $y = \cot x$, $y = \sec x$, and $y = \csc x$, and recall the ideas of symmetry with respect to the y-axis and symmetry with respect to the origin to answer Exercises 53–56.

53. The graph of $y = \tan x$ is symmetric with respect to the _____ . It follows that $\tan(-x) =$ _____ . (Notice that this supports a negative number identity introduced in Section 9.1.)

54. The graph of $y = \sec x$ is symmetric with respect to the _____ . It follows that $\sec(-x) =$ _____ .

55. For all x, $\cot(-x) = -\cot x$. Based on this fact, the graph of $y = \cot x$ is symmetric with respect to the _____ . Furthermore, $\cot(-1.75) =$ _____ , and this can be supported using calculator approximations. Since $\cot 1.75 \approx -.1811469526$, $\cot(-1.75) \approx$ _____ .

56. When $\sin x = 0$, $\csc x$ is _____ . As a result, the graph of $y = \csc x$ has a(n) _____ when $\sin x = 0$. Since $\cot x = \frac{\cos x}{\sin x}$, the cotangent is undefined when the _____ is undefined, and thus also has a(n) _____ when $\sin x = 0$.

Use a strictly graphical approach to solve the equation over the interval $[0, 2\pi)$*. Express solutions to the nearest hundredth.*

57. $\cot x + 2 \csc x = 3$

58. $2 \sin x = 1 - 2 \cos x$

59. $\sin^3 x + \sin x = 1$

60. $2 \cos^3 x + \sin x = -1$

61. $e^x = \sin x + 3$

62. $\ln x = \cos x$

Further Explorations

1. Recall that if θ is an angle in standard position and the point (a, b) lies on the terminal side of θ, then $\tan \theta = \frac{b}{a}$. The table shown in the figure was generated with the calculator in RADIAN MODE, with Tblmin $= 0$ and ΔTbl $= \frac{\pi}{4}$. Use the definition above to explain why error messages appear for two values of x in the table.

X	Y₁
0	0
.7854	1
1.5708	ERROR
2.3562	-1
3.1416	0
3.927	1
4.7124	ERROR

X=0

9.4 FURTHER IDENTITIES

Sum and Difference Identities ▮ Double Number Identities ▮ Half Number Identities

Sum and Difference Identities

FOR GROUP DISCUSSION

Use different values for A and B to investigate whether the given statement is true for all values of A and B.

$$\overset{?\quad?\quad?}{\text{1. } \cos(A - B) = \cos A - \cos B} \qquad \overset{?\quad?\quad?}{\text{2. } \cos(A + B) = \cos A + \cos B}$$

The results of the preceding group discussion should convince you that for the cosine function, the cosine of the difference (or sum) of two numbers is not, in general, equal to the difference (or sum) of the cosines of the numbers.

NOTE While we will discuss the identities in this section with respect to real number domains, they apply also to degree-measured angles.

In order to derive the identity for $\cos(A - B)$ in terms of functions of A and B, consider A and B to be real numbers that correspond to radian-measured angles

A and *B*. Start by locating angles *A* and *B* in standard position on a unit circle, with *B* < *A*. Let *S* and *Q* be the points where angles *A* and *B*, respectively, intersect the circle. Locate point *R* on the unit circle so that angle *POR* equals the difference *A* − *B*. See Figure 40.

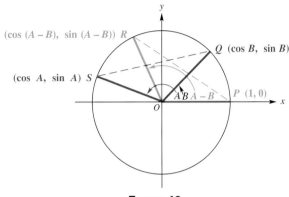

FIGURE 40

Point *Q* is on the unit circle, so by the work with circular functions, the *x*-coordinate of *Q* is given by the cosine of angle *B*, while the *y*-coordinate of *Q* is given by the sine of angle *B*:

Q has coordinates (cos *B*, sin *B*).

In the same way,

S has coordinates (cos *A*, sin *A*),

and

R has coordinates (cos(*A* − *B*), sin(*A* − *B*)).

Angle *SOQ* also equals *A* − *B*. Since the central angles *SOQ* and *POR* are equal, chords *PR* and *SQ* are equal. By the distance formula, since *PR* = *SQ*,

$$\sqrt{[\cos(A - B) - 1]^2 + [\sin(A - B) - 0]^2}$$
$$= \sqrt{(\cos A - \cos B)^2 + (\sin A - \sin B)^2}.$$

Squaring both sides and clearing parentheses gives

$$\cos^2(A - B) - 2\cos(A - B) + 1 + \sin^2(A - B)$$
$$= \cos^2 A - 2\cos A \cos B + \cos^2 B + \sin^2 A - 2\sin A \sin B + \sin^2 B.$$

Since $\sin^2 x + \cos^2 x = 1$ for any value of *x*, rewrite the equation as

$$2 - 2\cos(A - B) = 2 - 2\cos A \cos B - 2\sin A \sin B$$
$$\cos(A - B) = \cos A \cos B + \sin A \sin B.$$

This is the identity for cos(*A* − *B*). Although Figure 40 shows angles *A* and *B* in the second and first quadrants, respectively, it can be shown that this result is the same for any values of these angles.

To find a similar expression for cos(*A* + *B*), rewrite *A* + *B* as *A* − (−*B*) and use the identity for cos(*A* − *B*) found above, along with the fact that cos(−*B*) = cos *B* and sin(−*B*) = −sin *B*.

$$\cos(A + B) = \cos[A - (-B)]$$
$$= \cos A \cos(-B) + \sin A \sin(-B)$$
$$= \cos A \cos B + \sin A(-\sin B)$$
$$\cos(A + B) = \cos A \cos B - \sin A \sin B$$

The two formulas we have just derived are summarized as follows.

COSINE OF A SUM OR DIFFERENCE

$$\cos(A - B) = \cos A \cos B + \sin A \sin B$$
$$\cos(A + B) = \cos A \cos B - \sin A \sin B$$

These identities can be used to derive similar identities for the sine and tangent of the difference and the sum of two numbers. Because of the cofunction relationship, we have

$$\sin \theta = \cos\left(\frac{\pi}{2} - \theta\right).$$

Now replace θ with $A + B$.

$$\sin(A + B) = \cos\left[\frac{\pi}{2} - (A + B)\right]$$
$$= \cos\left[\left(\frac{\pi}{2} - A\right) - B\right]$$

Using the formula for $\cos(A - B)$ from the previous discussion gives

$$\sin(A + B) = \cos\left(\frac{\pi}{2} - A\right)\cos B + \sin\left(\frac{\pi}{2} - A\right)\sin B$$

or $\qquad\qquad \sin(A + B) = \sin A \cos B + \cos A \sin B.$

(The cofunction relationships were used in the last step.)

Now we write $\sin(A - B)$ as $\sin[A + (-B)]$ and use the identity for $\sin(A + B)$ to get

$$\sin(A - B) = \sin[A + (-B)]$$
$$= \sin A \cos(-B) + \cos A \sin(-B)$$
$$= \sin A \cos B - \cos A \sin B,$$

since $\cos(-B) = \cos B$ and $\sin(-B) = -\sin B$. In summary,

$$\sin(A - B) = \sin A \cos B - \cos A \sin B.$$

SINE OF A SUM OR DIFFERENCE

$$\sin(A + B) = \sin A \cos B + \cos A \sin B$$
$$\sin(A - B) = \sin A \cos B - \cos A \sin B$$

Using the identities for $\sin(A + B)$, $\cos(A + B)$, and the identity $\tan A = \frac{\sin A}{\cos A}$, we can derive the identity for $\tan(A + B)$. Start with

$$\tan(A + B) = \frac{\sin(A + B)}{\cos(A + B)}$$

$$= \frac{\sin A \cos B + \cos A \sin B}{\cos A \cos B - \sin A \sin B}.$$

To express this result in terms of the tangent function, multiply both numerator and denominator by $\frac{1}{\cos A \cos B}$.

$$\tan(A + B) = \frac{\dfrac{\sin A \cos B + \cos A \sin B}{1}}{\dfrac{\cos A \cos B - \sin A \sin B}{1}} \cdot \frac{\dfrac{1}{\cos A \cos B}}{\dfrac{1}{\cos A \cos B}}$$

$$= \frac{\dfrac{\sin A \cos B}{\cos A \cos B} + \dfrac{\cos A \sin B}{\cos A \cos B}}{\dfrac{\cos A \cos B}{\cos A \cos B} - \dfrac{\sin A \sin B}{\cos A \cos B}}$$

$$= \frac{\dfrac{\sin A}{\cos A} + \dfrac{\sin B}{\cos B}}{1 - \dfrac{\sin A}{\cos A} \cdot \dfrac{\sin B}{\cos B}}$$

Using the identity $\tan \theta = \frac{\sin \theta}{\cos \theta}$, we finally obtain

$$\tan(A + B) = \frac{\tan A + \tan B}{1 - \tan A \tan B}.$$

By replacing B with $-B$ and using the fact that $\tan(-B) = -\tan B$, we are able to find the identity for the tangent of the difference of two numbers. The two tangent identities follow.

TANGENT OF A SUM OR DIFFERENCE

$$\tan(A + B) = \frac{\tan A + \tan B}{1 - \tan A \tan B}$$

$$\tan(A - B) = \frac{\tan A - \tan B}{1 + \tan A \tan B}$$

NOTE The list that follows will help you in understanding the examples and exercises that follow in this section.

$$\frac{\pi}{3} = \frac{4\pi}{12}$$

$$\frac{\pi}{4} = \frac{3\pi}{12}$$

$$\frac{\pi}{6} = \frac{2\pi}{12}$$

Using this list, for example, we see that $\frac{\pi}{12} = \frac{\pi}{3} - \frac{\pi}{4}$ (or $\frac{\pi}{4} - \frac{\pi}{6}$).

Consider the number $\cos\frac{\pi}{12}$.

EXAMPLE 1	**(a)** Use the identity for $\cos(A - B)$ to find its exact value.

EXAMPLE 1

Finding an Exact
Function Value and
Supporting
Graphically

Consider the number $\cos\frac{\pi}{12}$.

(a) Use the identity for $\cos(A - B)$ to find its exact value.

(b) Show that a calculator approximation for $\cos\frac{\pi}{12}$ corresponds to an approximation for the exact value found in part (a).

(c) Support the result of part (a) by locating the appropriate point on the graph of $y = \cos x$.

SOLUTION

(a) Because $\frac{\pi}{12} = \frac{\pi}{3} - \frac{\pi}{4}$, we can find the *exact* value of $\cos\frac{\pi}{12}$ as follows.

$$\cos\frac{\pi}{12} = \cos\left(\frac{\pi}{3} - \frac{\pi}{4}\right)$$

$$= \cos\frac{\pi}{3}\cos\frac{\pi}{4} + \sin\frac{\pi}{3}\sin\frac{\pi}{4} \qquad \text{Use the cosine difference identity.}$$

$$= \frac{1}{2}\cdot\frac{\sqrt{2}}{2} + \frac{\sqrt{3}}{2}\cdot\frac{\sqrt{2}}{2} \qquad \text{Use exact values.}$$

$$= \frac{\sqrt{6} + \sqrt{2}}{4} \qquad \text{Exact value}$$

(b) Using a calculator in *radian* mode, we can find that an approximation for $\cos\frac{\pi}{12}$ is .9659258263. Also, if we use the square root function, along with the addition and division operations, we find that an approximation for $\frac{\sqrt{6}+\sqrt{2}}{4}$ is also .9659258263. This supports our answer in part (a).

(c) Figure 41 shows the graph of $y = \cos x$, and the display at the bottom indicates that when $x = \frac{\pi}{12}$ (as indicated by the approximation .26179939), $y \approx .96592583$, supporting our result in part (a). ∎

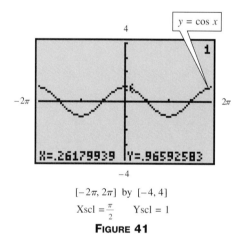

$[-2\pi, 2\pi]$ by $[-4, 4]$

$\text{Xscl} = \frac{\pi}{2} \qquad \text{Yscl} = 1$

FIGURE 41

FOR GROUP DISCUSSION

Use the graph in Figure 41 to explain why the exact value of $\cos(-\frac{\pi}{12})$ is also $\frac{\sqrt{6}+\sqrt{2}}{4}$. Support this using your own calculator.

EXAMPLE 2

Finding an Exact
Function Value and
Supporting
Graphically

Repeat Example 1 for $\tan \frac{7\pi}{12}$.

SOLUTION   To find the exact value of $\tan \frac{7\pi}{12}$, we recognize that $\frac{7\pi}{12} = \frac{\pi}{3} + \frac{\pi}{4}$. Now use the identity for $\tan(A + B)$.

$$\tan \frac{7\pi}{12} = \tan\left(\frac{\pi}{3} + \frac{\pi}{4}\right)$$

$$= \frac{\tan \frac{\pi}{3} + \tan \frac{\pi}{4}}{1 - \tan \frac{\pi}{3} \tan \frac{\pi}{4}}$$

$$= \frac{\sqrt{3} + 1}{1 - \sqrt{3} \cdot 1} \qquad \text{Use exact values.}$$

$$= \frac{\sqrt{3} + 1}{1 - \sqrt{3}} \cdot \frac{1 + \sqrt{3}}{1 + \sqrt{3}} \qquad \begin{array}{l}\text{Multiply numerator and}\\\text{denominator by the}\\\text{conjugate of the denominator.}\end{array}$$

$$= \frac{\sqrt{3} + 3 + 1 + \sqrt{3}}{1 - 3} \qquad \text{Multiply binomials.}$$

$$= \frac{4 + 2\sqrt{3}}{-2}$$

$$= -2 - \sqrt{3} \qquad \text{Reduce to lowest terms.}$$

Using a calculator in radian mode, we can show that $\tan \frac{7\pi}{12} \approx -3.732050808$, which is the same approximation found for the exact value, $-2 - \sqrt{3}$. Figure 42 shows the graph of $y = \tan x$ in the window $\left[\frac{\pi}{2}, \frac{3\pi}{2}\right]$ by $[-5, 5]$. As indicated in the display at the bottom of the screen, $\tan \frac{7\pi}{12} \approx \tan 1.8325957 \approx -3.732051$, supporting our earlier result.   ◼

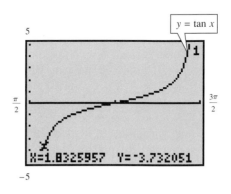

$\left[\frac{\pi}{2}, \frac{3\pi}{2}\right]$ by $[-5, 5]$

$\text{Xscl} = \frac{\pi}{2}$ $\text{Yscl} = 1$

FIGURE 42

Suppose that A and B are angles in standard position, with $\sin A = \frac{4}{5}, \frac{\pi}{2} < A < \pi$, $\cos B = -\frac{5}{13}$, and $\pi < B < \frac{3\pi}{2}$. Find **(a)** $\sin(A + B)$, **(b)** $\tan(A + B)$, and **(c)** the quadrant of $A + B$.

SOLUTION

(a) The identity for $\sin(A + B)$ requires $\sin A$, $\cos A$, $\sin B$, and $\cos B$. Two of these values are given. The two missing values, $\cos A$ and $\sin B$, must be found first. These values can be found with the identity $\sin^2 x + \cos^2 x = 1$. To find $\cos A$, use

$$\sin^2 A + \cos^2 A = 1$$

$$\frac{16}{25} + \cos^2 A = 1 \qquad \sin A = \frac{4}{5}$$

$$\cos^2 A = \frac{9}{25}$$

$$\cos A = -\frac{3}{5}. \qquad \text{Since } \frac{\pi}{2} < A < \pi, \cos A < 0.$$

In the same way, $\sin B = -\frac{12}{13}$. Now use the formula for $\sin(A + B)$.

$$\sin(A + B) = \frac{4}{5}\left(-\frac{5}{13}\right) + \left(-\frac{3}{5}\right)\left(-\frac{12}{13}\right)$$

$$= -\frac{20}{65} + \frac{36}{65} = \frac{16}{65}$$

(b) Use the values of sine and cosine from part (a) to get $\tan A = -\frac{4}{3}$ and $\tan B = \frac{12}{5}$. Then

$$\tan(A + B) = \frac{-\frac{4}{3} + \frac{12}{5}}{1 - \left(-\frac{4}{3}\right)\left(\frac{12}{5}\right)} = \frac{\frac{16}{15}}{1 + \frac{48}{15}} = \frac{\frac{16}{15}}{\frac{63}{15}} = \frac{16}{63}.$$

(c) From the results of parts (a) and (b), we find that $\sin(A + B)$ is positive and $\tan(A + B)$ is also positive. Therefore, $A + B$ must be in quadrant I, since it is the only quadrant in which both sine and tangent are positive.

Double Number Identities

Some special cases of the identities for the sum of two numbers are used often enough to be expressed as separate identities. These are the identities that result from the addition identities when $A = B$, so that $A + B = 2A$. These identities, called the **double number identities,** are now derived.

In the identity $\cos(A + B) = \cos A \cos B - \sin A \sin B$, let $B = A$ to derive an expression for $\cos 2A$.

$$\cos 2A = \cos(A + A)$$
$$= \cos A \cos A - \sin A \sin A$$
$$\cos 2A = \cos^2 A - \sin^2 A$$

Two other useful forms of this identity can be obtained by substituting either $\cos^2 A = 1 - \sin^2 A$ or $\sin^2 A = 1 - \cos^2 A$. Replace $\cos^2 A$ with $1 - \sin^2 A$ to get

$$\cos 2A = \cos^2 A - \sin^2 A$$
$$= (1 - \sin^2 A) - \sin^2 A$$
$$\cos 2A = 1 - 2 \sin^2 A,$$

and replace $\sin^2 A$ with $1 - \cos^2 A$ to get

$$\cos 2A = \cos^2 A - (1 - \cos^2 A)$$
$$= \cos^2 A - 1 + \cos^2 A$$
$$\cos 2A = 2 \cos^2 A - 1.$$

These double number identities for the cosine can be supported graphically by graphing the functions $y_1 = \cos 2x$, $y_2 = \cos^2 x - \sin^2 x$, $y_3 = 1 - 2 \sin^2 x$, and $y_4 = 2 \cos^2 x - 1$ all in the same viewing window. The graphs will all coincide, supporting our analytic work above. Figure 43 shows this single graph in the trig viewing window.

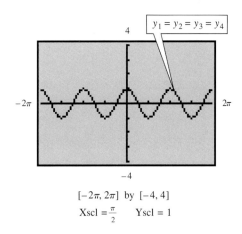

$$[-2\pi, 2\pi] \text{ by } [-4, 4]$$
$$\text{Xscl} = \frac{\pi}{2} \qquad \text{Yscl} = 1$$

FIGURE 43

We can find $\sin 2A$ with the identity $\sin(A + B) = \sin A \cos B + \cos A \sin B$, by letting $B = A$.

$$\sin 2A = \sin(A + A)$$
$$= \sin A \cos A + \cos A \sin A$$
$$\sin 2A = 2 \sin A \cos A$$

Similarly, the identity for $\tan(A + B)$ is used to find $\tan 2A$.

$$\tan 2A = \tan(A + A)$$
$$= \frac{\tan A + \tan A}{1 - \tan A \tan A}$$
$$\tan 2A = \frac{2 \tan A}{1 - \tan^2 A}$$

FOR **GROUP DISCUSSION**

---

1. Give graphical support for the identity for sin 2A by graphing both $y_1 = \sin 2x$ and $y_2 = 2 \sin x \cos x$ in the same viewing window. The graphs should overlap.

2. Repeat item 1 for tan 2A, graphing $y_1 = \tan 2x$ and $y_2 = \frac{2 \tan x}{1 - \tan^2 x}$.

A summary of the double number identities follows.

DOUBLE NUMBER IDENTITIES

$$\cos 2A = \cos^2 A - \sin^2 A \qquad \cos 2A = 1 - 2 \sin^2 A$$

$$\cos 2A = 2 \cos^2 A - 1 \qquad \sin 2A = 2 \sin A \cos A$$

$$\tan 2A = \frac{2 \tan A}{1 - \tan^2 A}$$

The next example shows how we can find function values of twice a number θ if we know the cosine of θ and the sign of sin θ.

EXAMPLE 4

Finding Function Values of 2θ Given Information About θ

Given $\cos \theta = \frac{3}{5}$ and $\sin \theta < 0$, find $\sin 2\theta$, $\cos 2\theta$, and $\tan 2\theta$.

SOLUTION In order to find sin 2θ, we must first find the value of sin θ. From the identity $\sin^2 \theta + \cos^2 \theta = 1$, we obtain

$$\sin^2 \theta + \left(\frac{3}{5}\right)^2 = 1 \qquad \cos \theta = \tfrac{3}{5}$$

$$\sin^2 \theta = \frac{16}{25}$$

$$\sin \theta = -\frac{4}{5}. \qquad \text{Choose the negative square root, since } \sin \theta < 0.$$

Using the double number identity for sine, we get

$$\sin 2\theta = 2 \sin \theta \cos \theta = 2\left(-\frac{4}{5}\right)\left(\frac{3}{5}\right) = -\frac{24}{25}.$$

Now find cos 2θ, using the first form of the identity. (Any form may be used.)

$$\cos 2\theta = \cos^2 \theta - \sin^2 \theta = \frac{9}{25} - \frac{16}{25} = -\frac{7}{25}$$

The value of tan 2θ can be found in either of two ways. We can use the double number identity, and the fact that $\tan \theta = \frac{\sin \theta}{\cos \theta} = \frac{-\frac{4}{5}}{\frac{3}{5}} = -\frac{4}{3}$.

$$\tan 2\theta = \frac{2 \tan \theta}{1 - \tan^2 \theta} = \frac{2\left(-\frac{4}{3}\right)}{1 - \frac{16}{9}} = \frac{-\frac{8}{3}}{-\frac{7}{9}} = \frac{24}{7}$$

As an alternative method, we can find $\tan 2\theta$ by finding the quotient of $\sin 2\theta$ and $\cos 2\theta$.

$$\tan 2\theta = \frac{\sin 2\theta}{\cos 2\theta} = \frac{-\dfrac{24}{5}}{-\dfrac{7}{25}} = \frac{24}{7} \quad \blacksquare$$

Other identities involving double number identities can be verified analytically and supported graphically like those first seen in Sections 9.1 and 9.2.

<table>
<tr><td>

EXAMPLE 5

Verifying an Identity Analytically and Supporting Graphically

</td><td>

Verify the identity

$$\cot x \sin 2x = 1 + \cos 2x$$

analytically, and support graphically.

</td></tr>
</table>

SOLUTION   Let us start by working on the left side.

$$\cot x \sin 2x = \frac{\cos x}{\sin x} \cdot \sin 2x \qquad \cot x = \tfrac{\cos x}{\sin x}$$

$$= \frac{\cos x}{\sin x}(2 \sin x \cos x) \qquad \sin 2x = 2 \sin x \cos x$$

$$= 2 \cos^2 x$$

$$= 1 + \cos 2x \qquad 2\cos^2 x - 1 = \cos 2x$$

TECHNOLOGICAL NOTE
The identity verified in Example 5 is not valid for values which cause the cotangent to be undefined. To illustrate this on a calculator, show that no y-value is displayed for the graph of y_1 when $x = \frac{\pi}{2}$, while for the graph of y_2, when $x = \frac{\pi}{2}$, $y = 0$.

To support our work graphically, we show that the graphs of $y_1 = \cot x \sin 2x$ and $y_2 = 1 + \cos 2x$ coincide. Notice that by observing the form of y_2, we see that the graph is obtained from that of $y = \cos x$ with period changed to $\frac{2\pi}{2} = \pi$, shifted 1 unit upward. See Figure 44. $\blacksquare$

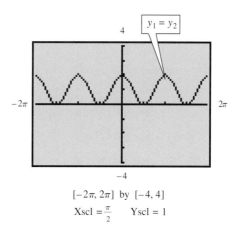

$[-2\pi, 2\pi]$ by $[-4, 4]$
$\text{Xscl} = \frac{\pi}{2} \qquad \text{Yscl} = 1$

FIGURE 44

Identities involving larger multiples of the variable can be derived by repeated use of the double number identities and the Pythagorean identities. In the following example, we find an expression for $\sin 3x$ in terms of $\sin x$.

EXAMPLE **6**
Deriving and Supporting an Identity for sin 3x

Write $\sin 3x$ in terms of $\sin x$, and then support the result graphically.

SOLUTION   Write $3x$ as $2x + x$, and then apply identities.

$$
\begin{aligned}
\sin 3x &= \sin(2x + x) \\
&= \sin 2x \cos x + \cos 2x \sin x \\
&= (2 \sin x \cos x) \cos x + (\cos^2 x - \sin^2 x) \sin x \\
&= 2 \sin x \cos^2 x + \cos^2 x \sin x - \sin^3 x \\
&= 2 \sin x(1 - \sin^2 x) + (1 - \sin^2 x) \sin x - \sin^3 x \\
&= 2 \sin x - 2 \sin^3 x + \sin x - \sin^3 x - \sin^3 x \\
&= -4 \sin^3 x + 3 \sin x
\end{aligned}
$$

By graphing $y_1 = \sin 3x$ and $y_2 = -4 \sin^3 x + 3 \sin x$ in the same viewing window, we find that the graphs coincide, supporting our analytic result. See Figure 45.   ▯

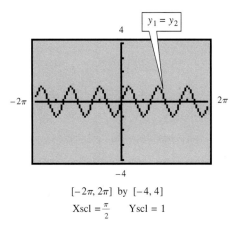

$[-2\pi, 2\pi]$ by $[-4, 4]$

$\text{Xscl} = \frac{\pi}{2}$ $\text{Yscl} = 1$

FIGURE 45

Half Number Identities

From the alternative forms of the identity for $\cos 2A$, we can derive three additional identities for $\sin \frac{A}{2}$, $\cos \frac{A}{2}$, and $\tan \frac{A}{2}$. These are known as **half number identities.**

To derive the identity for $\sin \frac{A}{2}$, start with the following double number identity for cosine.

$$\cos 2x = 1 - 2 \sin^2 x$$

Then solve for $\sin x$.

$$2 \sin^2 x = 1 - \cos 2x$$

$$\sin x = \pm \sqrt{\frac{1 - \cos 2x}{2}}$$

Now let $2x = A$, so that $x = \frac{A}{2}$, and substitute into this last expression.

$$\sin \frac{A}{2} = \pm \sqrt{\frac{1 - \cos A}{2}}$$

The $\pm$ sign in the identity above indicates that, in practice, the appropriate sign is chosen depending upon the quadrant of $\frac{A}{2}$. For example, if $\frac{A}{2}$ is a third quadrant number on the unit circle, we choose the negative sign since the sine function is negative there.

The identity for $\cos\frac{A}{2}$ is derived in a very similar way, starting with the double number identity $\cos 2x = 2\cos^2 x - 1$. Solve for $\cos x$.

$$\cos 2x + 1 = 2\cos^2 x$$

$$\cos x = \pm\sqrt{\frac{1 + \cos 2x}{2}}$$

Replacing x with $\frac{A}{2}$ gives

$$\cos\frac{A}{2} = \pm\sqrt{\frac{1 + \cos A}{2}}.$$

The $\pm$ sign is used as described earlier.

Finally, an identity for $\tan\frac{A}{2}$ comes from the half number identities for sine and cosine.

$$\tan\frac{A}{2} = \frac{\pm\sqrt{\dfrac{1 - \cos A}{2}}}{\pm\sqrt{\dfrac{1 + \cos A}{2}}}$$

or

$$\tan\frac{A}{2} = \pm\sqrt{\frac{1 - \cos A}{1 + \cos A}} \qquad \pm \text{ chosen depending upon quadrant of } \tfrac{A}{2}$$

An alternative identity for $\tan\frac{A}{2}$ can be derived using the fact that $\tan\frac{A}{2} = \frac{\sin\frac{A}{2}}{\cos\frac{A}{2}}$.

$$\tan\frac{A}{2} = \frac{\sin\dfrac{A}{2}}{\cos\dfrac{A}{2}}$$

$$= \frac{2\sin\dfrac{A}{2}\cos\dfrac{A}{2}}{2\cos^2\dfrac{A}{2}} \qquad \text{Multiply by 2 cos } \tfrac{A}{2} \text{ in numerator and denominator.}$$

$$= \frac{\sin 2\left(\dfrac{A}{2}\right)}{1 + \cos 2\left(\dfrac{A}{2}\right)} \qquad \text{Use double number identities.}$$

$$\tan\frac{A}{2} = \frac{\sin A}{1 + \cos A}$$

From this identity for $\tan\frac{A}{2}$, we can also derive

$$\tan\frac{A}{2} = \frac{1 - \cos A}{\sin A}.$$

See Exercise 61. These last two identities for $\tan\frac{A}{2}$ do not require a sign choice, as the first one does.

HALF NUMBER IDENTITIES

$$\cos \frac{A}{2} = \pm \sqrt{\frac{1 + \cos A}{2}} \qquad \sin \frac{A}{2} = \pm \sqrt{\frac{1 - \cos A}{2}}$$

$$\tan \frac{A}{2} = \pm \sqrt{\frac{1 - \cos A}{1 + \cos A}} \qquad \tan \frac{A}{2} = \frac{\sin A}{1 + \cos A}$$

$$\tan \frac{A}{2} = \frac{1 - \cos A}{\sin A}$$

In Example 1, we showed that the exact value of $\cos \frac{\pi}{12}$ is $\frac{\sqrt{6} + \sqrt{2}}{4}$. This was accomplished by using the identity for $\cos(A - B)$. Another form of the exact value of $\cos \frac{\pi}{12}$ can be found by using the identity for $\cos \frac{A}{2}$.

$$\cos \frac{\pi}{12} = \cos \frac{\frac{\pi}{6}}{2} \qquad \qquad \frac{\pi}{12} = \frac{\frac{\pi}{6}}{2}$$

$$= \sqrt{\frac{1 + \cos \frac{\pi}{6}}{2}} \qquad \qquad \text{Use the identity.}$$

$$= \sqrt{\frac{1 + \frac{\sqrt{3}}{2}}{2}} \qquad \qquad \cos \frac{\pi}{6} = \frac{\sqrt{3}}{2}$$

$$= \sqrt{\frac{\left(1 + \frac{\sqrt{3}}{2}\right) \cdot 2}{2 \cdot 2}} \qquad \text{Multiply by } \frac{2}{2} \text{ under the radical.}$$

$$= \frac{\sqrt{2 + \sqrt{3}}}{2} \qquad \qquad \sqrt{\frac{a}{b}} = \frac{\sqrt{a}}{\sqrt{b}}$$

This final expression, $\frac{\sqrt{2 + \sqrt{3}}}{2}$, has a calculator approximation of .9659258263, the same one found for $\frac{\sqrt{6} + \sqrt{2}}{4}$. Furthermore, if we graph $y = \sqrt{\frac{1 + \cos x}{2}}$ and let $x = \frac{\pi}{6} \approx .52359878$, we see that the corresponding y-value is .96592583, once again supporting our result. See Figure 46.

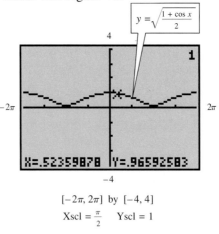

$$y = \sqrt{\frac{1 + \cos x}{2}}$$

X=.52359878   Y=.96592583

$[-2\pi, 2\pi]$ by $[-4, 4]$

$\text{Xscl} = \frac{\pi}{2}$ $\text{Yscl} = 1$

FIGURE 46

The final example shows that if the cosine of a number is known and an appropriate interval of the number is given, we can find the cosine, sine, and tangent of half the number.

<table>
<tr><td>

EXAMPLE 7

Finding Function Values of $\frac{x}{2}$ Given Information About x

</td></tr>
</table>

Given $\cos x = \frac{2}{3}$, with $\frac{3\pi}{2} < x < 2\pi$, find $\cos \frac{x}{2}$, $\sin \frac{x}{2}$, and $\tan \frac{x}{2}$.

SOLUTION Since

$$\frac{3\pi}{2} < x < 2\pi,$$

dividing through by 2 gives

$$\frac{3\pi}{4} < \frac{x}{2} < \pi,$$

showing that $\frac{x}{2}$ terminates in quadrant II on the unit circle. In this quadrant the value of $\cos \frac{x}{2}$ is negative and the value of $\sin \frac{x}{2}$ is positive. Use the appropriate half number identities to get

$$\sin \frac{x}{2} = \sqrt{\frac{1 - \frac{2}{3}}{2}} = \sqrt{\frac{1}{6}} = \frac{\sqrt{6}}{6};$$

and

$$\cos \frac{x}{2} = -\sqrt{\frac{1 + \frac{2}{3}}{2}} = -\sqrt{\frac{5}{6}} = -\frac{\sqrt{30}}{6}.$$

Also,

$$\tan \frac{x}{2} = \frac{\sin \frac{x}{2}}{\cos \frac{x}{2}} = \frac{\frac{\sqrt{6}}{6}}{-\frac{\sqrt{30}}{6}} = -\frac{\sqrt{5}}{5}.$$

Notice that it is not necessary to use a half number identity for $\tan \frac{x}{2}$ once we find $\sin \frac{x}{2}$ and $\cos \frac{x}{2}$. However, using this identity would provide an excellent check. ∎

9.4 EXERCISES

*For each of the following circular function values (**a**) find the exact value using the appropriate sum or difference identity, (**b**) find an approximation in two ways: by using the appropriate circular function of your calculator, making sure it is in radian mode, and then by finding a calculator approximation for the exact value found in part (a), and (**c**) using the appropriate circular function graph, support the fact that your answer in part (b) is the approximation for y when the domain value is entered.*

1. $\sin \frac{\pi}{12}$
2. $\tan \frac{\pi}{12}$
3. $\cos \frac{7\pi}{12}$
4. $\sin \frac{7\pi}{12}$

5. $\sin \frac{5\pi}{12}$
6. $\tan \frac{5\pi}{12}$
7. $\cos \frac{5\pi}{12}$
8. $\cos\left(\frac{-5\pi}{12}\right)$

9. $\sin\left(\frac{-5\pi}{12}\right)$
10. $\tan\left(\frac{-5\pi}{12}\right)$
11. $\sin\left(\frac{13\pi}{12}\right)$
12. $\cos\left(\frac{13\pi}{12}\right)$

*The exact value of a trigonometric function can often be found by using a sum or difference identity, along with the fact that exact function values of the special angles 30°, 45°, and 60° are known. For each of the following trigonometric function values (**a**) find the exact value using one of these identities, and (**b**) find an approximation in two ways: by using the appropriate trigonometric function of your calculator, making sure it is in degree mode, and then by finding a calculator approximation for the exact value found in part (**a**).*

13. $\sin 15°$ **14.** $\cos 15°$ **15.** $\tan 15°$ **16.** $\cos 75°$

17. $\sin 75°$ **18.** $\tan 75°$ **19.** $\sin 105°$ **20.** $\cos 105°$

21. $\tan 105°$ **22.** $\sin(-15°)$ **23.** $\cos(-15°)$ **24.** $\tan(-75°)$

Use the appropriate sum or difference identity to write the given expression as a function of x alone. (For example, using the identity for sin(A − B), it can be shown that $\sin(\frac{3\pi}{2} - x) = -\cos x$.)

25. $\cos\left(\dfrac{\pi}{2} - x\right)$ **26.** $\cos(\pi - x)$ **27.** $\cos\left(\dfrac{3\pi}{2} + x\right)$

28. $\sin(\pi - x)$ **29.** $\sin(\pi + x)$ **30.** $\tan(2\pi - x)$

31. $\tan(\pi + x)$ **32.** $\tan(\pi - x)$ **33.** $\sin\left(\dfrac{3\pi}{2} - x\right)$

34. Explain how the identities for $\sec(A + B)$, $\csc(A + B)$, and $\cot(A + B)$ can be easily found by using the sum identities given in this section.

*R*elating Concepts

35. Use a calculator in radian mode to graph the function $y_1 = \cos(x + \frac{\pi}{2})$ in the trig viewing window.

36. Explain how the graph in Exercise 35 can be obtained by a translation of the graph of $y = \cos x$.

37. Let $x = 1$ in y_1 from Exercise 35, and find an approximation for $\cos(1 + \frac{\pi}{2})$ from the graph.

38. Use the identity for $\cos(A + B)$ to show analytically that $\cos(x + \frac{\pi}{2}) = -\sin x$.

39. Graph $y_2 = -\sin x$ in the trig viewing window. How does this graph support your work in Exercise 38?

40. Let $x = 1$ in y_2 from Exercise 39, and find an approximation for $-\sin 1$. How does it compare to the result in Exercise 37?

---

*Suppose that A and B are angles in standard position. Use the given information to find (**a**) sin(A + B), (**b**) sin(A − B) (**c**) tan(A + B), (**d**) tan(A − B), (**e**) the quadrant of (A + B), and (**f**) the quadrant of (A − B).*

41. $\cos A = \dfrac{3}{5}$, $\sin B = \dfrac{5}{13}$, $0 < A < \dfrac{\pi}{2}$, $0 < B < \dfrac{\pi}{2}$

42. $\sin A = \dfrac{3}{5}$, $\sin B = -\dfrac{12}{13}$, $0 < A < \dfrac{\pi}{2}$, $\pi < B < \dfrac{3\pi}{2}$

43. $\cos A = -\dfrac{8}{17}$, $\cos B = -\dfrac{3}{5}$, $\pi < A < \dfrac{3\pi}{2}$, $\pi < B < \dfrac{3\pi}{2}$

44. $\cos A = -\dfrac{15}{17}$, $\sin B = \dfrac{4}{5}$, $\dfrac{\pi}{2} < A < \pi$, $0 < B < \dfrac{\pi}{2}$

Suppose that x is a real number. Use the given information and the double number and reciprocal identities to find all six circular function values of 2x.

45. $\cos x = -\dfrac{12}{13}$, $\sin x > 0$

46. $\tan x = 2$, $\cos x > 0$

47. $\tan x = \dfrac{5}{3}$, $\sin x < 0$

48. $\sin x = \dfrac{2}{5}$, $\cos x < 0$

For each of the following, (a) use the appropriate half number identity to find the exact value and (b) use the appropriate circular or trigonometric function key on your calculator to find an approximation. Then show that this approximation is the same as the one found for your result in part (a).

49. $\sin \dfrac{\pi}{12}$

50. $\cos \dfrac{\pi}{8}$

51. $\tan\left(-\dfrac{\pi}{8}\right)$

52. $\cos 67.5°$

53. $\sin 67.5°$

54. $\tan 195°$

Use a half number identity to find an exact value of each of the following, given the information about x.

55. $\cos \dfrac{x}{2}$, given $\cos x = \dfrac{1}{4}$, with $0 < x < \dfrac{\pi}{2}$

56. $\sin \dfrac{x}{2}$, given $\cos x = -\dfrac{5}{8}$, with $\dfrac{\pi}{2} < x < \pi$

57. $\tan \dfrac{x}{2}$, given $\sin x = \dfrac{3}{5}$, with $\dfrac{\pi}{2} < x < \pi$

58. $\cos \dfrac{x}{2}$, given $\sin x = -\dfrac{4}{5}$, with $\dfrac{3\pi}{2} < x < 2\pi$

59. $\tan \dfrac{x}{2}$, given $\tan x = \dfrac{\sqrt{7}}{3}$, with $\pi < x < \dfrac{3\pi}{2}$

60. $\tan \dfrac{x}{2}$, given $\tan x = -\dfrac{\sqrt{5}}{2}$, with $\dfrac{\pi}{2} < x < \pi$

61. Use the identity $\tan \frac{A}{2} = \frac{\sin A}{1 + \cos A}$ to derive the equivalent identity $\tan \frac{A}{2} = \frac{1 - \cos A}{\sin A}$ by multiplying both the numerator and denominator by $1 - \cos A$.

62. Consider the expression $\tan(\frac{\pi}{2} + x)$.
   (a) Why can't we use the identity for $\tan(A + B)$ to express it as a function of x alone?
   (b) Use the identity $\tan \theta = \frac{\sin \theta}{\cos \theta}$ to rewrite the expression in terms of sine and cosine.
   (c) Use the result of part (b) to show that $\tan(\frac{\pi}{2} + x) = -\cot x$.

63. The identity
$$\tan \frac{A}{2} = \pm\sqrt{\frac{1 - \cos A}{1 + \cos A}}$$
can be used to find $\tan 22.5° = \sqrt{3 - 2\sqrt{2}}$, and the identity
$$\tan \frac{A}{2} = \frac{\sin A}{1 + \cos A}$$
can be used to get $\tan 22.5° = \sqrt{2} - 1$. Show that these answers are the same, without using a calculator. (*Hint:* If $a > 0$ and $b > 0$ and $a^2 = b^2$, then $a = b$.)

64. Explain how you could use an identity of this section to find the exact value of $\sin 7.5°$.

Use the method of Example 6 to do each of the following. Then support your result graphically, using the trig viewing window of your calculator.

65. Express $\cos 3x$ in terms of $\cos x$.

66. Express $\tan 3x$ in terms of $\tan x$.

67. Express $\tan 4x$ in terms of $\tan x$.

68. Express $\cos 4x$ in terms of $\cos x$.

Verify the identity analytically.

69. $\dfrac{\cos(A - B)}{\cos A \sin B} = \tan A + \cot B$

70. $\dfrac{\sin(A + B)}{\cos A \cos B} = \tan A + \tan B$

71. $\dfrac{\sin(A - B)}{\sin(A + B)} = \dfrac{\tan A - \tan B}{\tan A + \tan B}$

72. $\dfrac{\sin(A + B)}{\cos(A - B)} = \dfrac{\cot A + \cot B}{1 + \cot A \cot B}$

73. $\dfrac{\sin(A - B)}{\sin B} + \dfrac{\cos(A - B)}{\cos B} = \dfrac{\sin A}{\sin B \cos B}$

74. $\dfrac{\tan(A + B) - \tan B}{1 + \tan(A + B) \tan B} = \tan A$

Verify the identity analytically. Then graph both sides of the equation as separate functions in the trig viewing window of your calculator, entering them as y_1, and y_2, to support your work graphically.

75. $\sin 2x = \dfrac{2 \tan x}{1 + \tan^2 x}$

76. $\cos 2x = \dfrac{2 - \sec^2 x}{\sec^2 x}$

77. $\dfrac{2 \cos 2x}{\sin 2x} = \cot x - \tan x$

78. $\cos 2x = \dfrac{1 - \tan^2 x}{1 + \tan^2 x}$

79. $\cot x = \dfrac{1 + \cos 2x}{\sin 2x}$

80. $\dfrac{2}{1 + \cos x} - \tan^2 \dfrac{x}{2} = 1$

81. $1 - \tan^2 \dfrac{x}{2} = \dfrac{2 \cos x}{1 + \cos x}$

82. $\tan \dfrac{x}{2} = \csc x - \cot x$

R*elating Concepts*

From the law of cosines, in triangle ABC, $\cos A = \frac{b^2 + c^2 - a^2}{2bc}$. Use this equation to show that each of the following is true, and from these exercises, prove Heron's Formula from Section 8.6. Work Exercises 83–88 in order.

83. $1 + \cos A = \dfrac{(b + c + a)(b + c - a)}{2bc}$

84. $1 - \cos A = \dfrac{(a - b + c)(a + b - c)}{2bc}$

85. $\cos \dfrac{A}{2} = \sqrt{\dfrac{s(s - a)}{bc}}$ $\left(Hint: \cos \dfrac{A}{2} = \sqrt{\dfrac{1 + \cos A}{2}} \right)$

86. $\sin \dfrac{A}{2} = \sqrt{\dfrac{(s - b)(s - c)}{bc}}$ $\left(Hint: \sin \dfrac{A}{2} = \sqrt{\dfrac{1 - \cos A}{2}} \right)$

87. The area of a triangle having sides b and c and angle A is given by $(\frac{1}{2})bc \sin A$. Show that this result can be written as

$$\sqrt{\dfrac{1}{2}bc(1 + \cos A) \cdot \dfrac{1}{2}bc(1 - \cos A)}.$$

88. Use the results of Exercises 83–87 to prove Heron's area formula.

Further Explorations

Identities that can be supported graphically can also be supported numerically with the TABLE. If $Y_1 = Y_2$, then the values under Y_1 correspond to the values under Y_2. However, false conclusions could occur with certain values of ΔTbl. If ΔTbl is chosen carefully so that only points of intersection for the two graphs are evaluated in the TABLE, and Tblmin is one of the points of intersection, then the TABLE could lead to the false conclusion that the expressions are equivalent. For example:

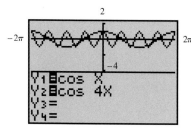

Clearly, these equations
are not equivalent.

ΔTbl $= \dfrac{2\pi}{3} \approx 2.0943951$

Tblmin $= 0$

Careful selection of ΔTbl and
Tblmin can give misleading results.

If your graphics calculator has a TABLE feature, find a ΔTbl and Tblmin for each of the following pairs of nonequivalent expressions so that the TABLE will evaluate only values for x where $Y_1 = Y_2$. (Hint: Graph each pair of expressions and find the frequency of the points where they intersect.) Use RADIAN MODE.

1. $Y_1 = \cos x$
   $Y_2 = \sin x$

2. $Y_1 = \sin(x + \pi)$
   $Y_2 = .5 \sin x$

3. $Y_1 = \cos^2 x + \sin^2 x$
   $Y_2 = \cos 2x + 1$

4. $Y_1 = \cos x$
   $Y_2 = 1 + \sin x$

5. $Y_1 = 4 \sin x \cos x$
   $Y_2 = 2 \cos 2x$

9.5 FURTHER EQUATIONS AND INEQUALITIES

Equations and Inequalities Involving Double Number Identities ∎ Equations and Inequalities Involving Half Number Identities

In this section we will study methods of solution of equations and inequalities involving circular functions of double a number (for example, $\cos 2x$) and half a number (for example, $\sin \frac{x}{2}$). As usual, we will present analytic solutions and graphical support. Because the circular functions are periodic, we will restrict our domain in most cases to the interval $[0, 2\pi)$, in order to make our work less cumbersome. If solutions over larger intervals were required, then we could use the periodic nature of these functions to write expressions for the solutions in those intervals.

Equations and Inequalities Involving Double Number Identities

EXAMPLE 1

Solving an Equation
Involving a Function
of 2x

Solve the equation $\cos 2x = \cos x$ over the interval $[0, 2\pi)$ using analytic methods. Then support the solutions with a graph.

SOLUTION   First change $\cos 2x$ to a circular function of x. Use the identity $\cos 2x = 2\cos^2 x - 1$ so that the equation involves only the cosine of x.

$$\cos 2x = \cos x$$
$$2\cos^2 x - 1 = \cos x$$
$$2\cos^2 x - \cos x - 1 = 0 \qquad \text{Standard quadratic form}$$
$$(2\cos x + 1)(\cos x - 1) = 0 \qquad \text{Factor.}$$
$$2\cos x + 1 = 0 \qquad \text{or} \qquad \cos x - 1 = 0 \qquad \text{Solve.}$$
$$\cos x = -\frac{1}{2} \qquad \text{or} \qquad \cos x = 1$$

In the required interval,

$$x = \frac{2\pi}{3} \quad \text{or} \quad \frac{4\pi}{3} \qquad \text{or} \qquad x = 0.$$

$$\underbrace{\qquad\qquad\qquad}_{\text{for } \cos x = -\frac{1}{2}} \qquad \underbrace{\qquad}_{\text{for } \cos x = 1}$$

TECHNOLOGICAL NOTE
If we let Xscl $= \frac{\pi}{3}$, we can see that the graph of $y = \cos 2x - \cos x$ intersects the x-axis at the *second* tick mark, further supporting our result that $\frac{2\pi}{3}$ is a solution in Example 1.

The solution set of exact values is $\{0, \frac{2\pi}{3}, \frac{4\pi}{3}\}$. If we graph the function $y = \cos 2x - \cos x$ (obtained by subtracting the right side of the original equation from the left side) and use the x-intercept method of solution, we can support our analytic solution. Notice that in Figure 47, there are three x-intercepts over the interval $[0, 2\pi)$. As shown on the screen, one solution is approximated by 2.0943951, which corresponds to an approximation of $\frac{2\pi}{3}$. The other solutions can be supported similarly.   ∎

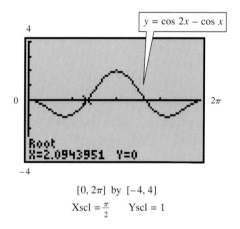

$[0, 2\pi]$ by $[-4, 4]$
Xscl $= \frac{\pi}{2}$ Yscl $= 1$

FIGURE 47

NOTE   In this section, we will round solutions of equations and endpoints of intervals of solutions of inequalities to the nearest thousandth, for convenience. Using this agreement, the solution set in Example 1 would be expressed as $\{0, 2.094, 4.189\}$.

CAUTION In Example 1 it is important to notice that cos 2x cannot be changed to cos x by dividing by 2, since 2 is not a factor of the numerator.

$$\frac{\cos 2x}{2} \neq \cos x$$

The only way to change cos 2x to a circular function of x is by using one of the identities for cos 2x.

EXAMPLE 2

Solving an
Inequality Involving
a Function of 2x

Refer to Example 1 and Figure 47 to do each of the following.

(a) Find the solution set of cos 2x < cos x over the interval $[0, 2\pi)$, giving exact values.

(b) Find the solution set of cos 2x > cos x over the interval $[0, 2\pi)$, giving approximations when appropriate.

SOLUTION

(a) The given inequality is equivalent to cos 2x − cos x < 0, and the solution set of this inequality can be determined by finding x-values for which the graph of y = cos 2x − cos x is *below* the x-axis. Referring to Figure 47 and knowing that the exact solutions of the equation cos 2x = cos x in the interval $[0, 2\pi)$ are 0, $\frac{2\pi}{3}$, and $\frac{4\pi}{3}$, we determine the solution set to be

$$\left(0, \frac{2\pi}{3}\right) \cup \left(\frac{4\pi}{3}, 2\pi\right).$$

(b) We must find the x-values for which the graph of y = cos 2x − cos x is *above* the x-axis. Using approximations as directed, we find that the solution set is the open interval (2.094, 4.189). ∎

FOR **GROUP DISCUSSION**

Refer to Example 2 and discuss how the solution sets would be affected with the following minor changes.

1. Solve cos 2x ≤ cos x.
2. Solve cos 2x ≥ cos x over the interval $[0, 4\pi)$.

Conditional equations in which a double number or half number function is involved often require an additional step to solve. This step involves adjusting the interval of solution to fit the requirements of the double number or half number. This is illustrated in the following examples.

EXAMPLE 3

Solving an Equation
Using a Double
Number Identity

Solve $4 \sin x \cos x = \sqrt{3}$ over the interval $[0, 2\pi)$.

SOLUTION The identity 2 sin x cos x = sin 2x is useful here.

$$4 \sin x \cos x = \sqrt{3}$$
$$2(2 \sin x \cos x) = \sqrt{3} \qquad \text{4 = 2 · 2}$$
$$2 \sin 2x = \sqrt{3} \qquad \text{2 sin x cos x = sin 2x}$$
$$\sin 2x = \frac{\sqrt{3}}{2} \qquad \text{Divide by 2.}$$

Because the equation now involves a function of 2x, we must be careful to observe that since $0 \le x < 2\pi$, it follows that $0 \le 2x < 4\pi$. Now we list all solutions in the interval $[0, 4\pi)$:

$$2x = \frac{\pi}{3}, \frac{2\pi}{3}, \frac{7\pi}{3}, \frac{8\pi}{3}$$

or, dividing through by 2,

$$x = \frac{\pi}{6}, \frac{\pi}{3}, \frac{7\pi}{6}, \frac{4\pi}{3},$$

giving the solution set $\{\frac{\pi}{6}, \frac{\pi}{3}, \frac{7\pi}{6}, \frac{4\pi}{3}\}$ (exact values).

In Example 1, we supported our solution by using the x-intercept method. For variety, we can support this solution by graphing $y_1 = 4 \sin x \cos x$ and $y_2 = \sqrt{3}$, and locating the points of intersection over the interval $[0, 2\pi)$. Figure 48 indicates that one of these points has x-coordinate .52359878, which is an approximation for the exact value $\frac{\pi}{6}$. The other three solutions over this interval can be supported similarly. ∎

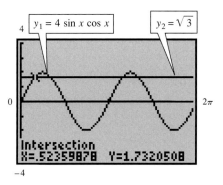

$[0, 2\pi]$ by $[-4, 4]$

$\text{Xscl} = \frac{\pi}{2}$ $\text{Yscl} = 1$

FIGURE 48

EXAMPLE 4

Solving an Equation That Involves Squaring Both Sides

Solve $\tan 3x + \sec 3x = 2$ over the interval $[0, 2\pi)$.

SOLUTION Since the tangent and secant functions are related by the identity $1 + \tan^2 \theta = \sec^2 \theta$, one way to begin is to express everything in terms of the secant. This may be done by subtracting $\sec 3x$ from both sides and then squaring.

$$\tan 3x + \sec 3x = 2$$

$$\tan 3x = 2 - \sec 3x \qquad \text{Subtract } \sec 3x.$$

$$\tan^2 3x = 4 - 4 \sec 3x + \sec^2 3x \qquad \text{Square both sides;} \\ (a - b)^2 = a^2 - 2ab + b^2$$

$$\sec^2 3x - 1 = 4 - 4 \sec 3x + \sec^2 3x \qquad \text{Replace } \tan^2 3x \text{ with } \sec^2 3x - 1.$$

$$0 = 5 - 4 \sec 3x$$

$$4 \sec 3x = 5$$

$$\sec 3x = \frac{5}{4}$$

$$\frac{1}{\cos 3x} = \frac{5}{4} \qquad \sec \theta = \frac{1}{\cos \theta}$$

$$\cos 3x = \frac{4}{5} \qquad \text{Use reciprocals.}$$

Multiply the inequality $0 \le x < 2\pi$ by 3 to find the interval for $3x$: $[0, 6\pi)$. Using a calculator and knowing that cosine is positive in quadrants I and IV, we get

$3x \approx .64350111,\ 5.6396842,\ 6.9266864,\ 11.922870,\ 13.209872,\ 18.206055.$

Dividing by 3 gives

$x \approx .21450037,\ 1.8798947,\ 2.3088955,\ 3.9742898,\ 4.4032906,\ 6.0686849.$

Recall from Section 4.4 that when both sides of an equation are squared, there is a possibility of introducing extraneous solutions. Observing the graph of $y = \tan 3x + \sec 3x - 2$ in Figure 49, we see that in the interval $[0, 2\pi)$, there are only three x-intercepts. One of these is approximately 4.4032906, which is one of the six possible solutions shown above (see the display in the figure). The other two can be verified by the calculator in a similar manner. They are .21450037 and 2.3088955. Expressing these to the nearest thousandth, the solution set in the interval $[0, 2\pi)$ is $\{.215, 2.309, 4.403\}$. ◨

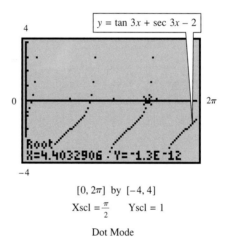

$[0, 2\pi]$ by $[-4, 4]$
$\text{Xscl} = \frac{\pi}{2}$ $\text{Yscl} = 1$

Dot Mode

FIGURE 49

Equations and Inequalities Involving Half Number Identities

EXAMPLE 5

Solving an Equation Involving a Function of $\frac{x}{2}$

Solve $2 \sin \frac{x}{2} = 1$ over the interval $[0, 2\pi)$.

SOLUTION   The interval $[0, 2\pi)$ may be written $0 \le x < 2\pi$. Dividing the three expressions by 2 gives

$$0 \le \frac{x}{2} < \pi.$$

To find all values for $\frac{x}{2}$ satisfying the equation, we begin by dividing both sides by 2.

$$\sin \frac{x}{2} = \frac{1}{2} \qquad \text{Divide by 2.}$$

In the interval $[0, \pi)$, the two numbers that have sine value of $\frac{1}{2}$ are $\frac{\pi}{6}$ and $\frac{5\pi}{6}$. To solve for the corresponding values of x, we multiply by 2.

$$\frac{x}{2} = \frac{\pi}{6} \quad \text{or} \quad \frac{x}{2} = \frac{5\pi}{6}$$

$$x = \frac{\pi}{3} \qquad\qquad x = \frac{5\pi}{3}$$

The solutions in the given interval are $\frac{\pi}{3}$ and $\frac{5\pi}{3}$.

Figure 50 shows the graph of $y = 2 \sin \frac{x}{2} - 1$ over the interval $[0, 2\pi)$. The solution $\frac{\pi}{3}$ is approximated by 1.0471976, as shown in the figure. The other solution, $\frac{5\pi}{3}$, can be supported similarly. The solution set of exact values is $\{\frac{\pi}{3}, \frac{5\pi}{3}\}$. █

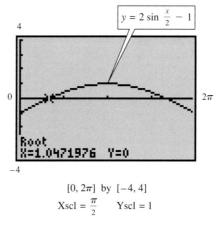

$[0, 2\pi]$ by $[-4, 4]$

$\text{Xscl} = \frac{\pi}{2} \qquad \text{Yscl} = 1$

FIGURE 50

<table>
<tr><td>**EXAMPLE 6**</td></tr>
<tr><td>Solving an
Inequality Involving
a Function of $\frac{x}{2}$</td></tr>
</table>

Refer to Example 5 and Figure 50 to do each of the following.

(a) Find the solution set of $2 \sin \frac{x}{2} > 1$ over the interval $[0, 2\pi)$, giving exact values.

(b) Find the solution set of $2 \sin \frac{x}{2} < 1$ over the interval $[0, 2\pi)$, giving approximations when appropriate.

SOLUTION

(a) Referring to Figure 50 and using the same procedure described in Example 2, we find that the graph of $y = 2 \sin \frac{x}{2} - 1$ lies *above* the x-axis between $\frac{\pi}{3}$ and $\frac{5\pi}{3}$. Therefore, the solution set of $2 \sin \frac{x}{2} > 1$ is the open interval $\left(\frac{\pi}{3}, \frac{5\pi}{3}\right)$.

(b) Using approximations as directed and approximating the other x-intercept, we determine the solution set of $2 \sin \frac{x}{2} < 1$ to be $[0, 1.047) \cup (5.236, 2\pi)$. █

If an equation or inequality involving circular functions cannot be solved analytically, a purely graphical approach can be used. The final example of this section illustrates such an equation.

<table>
<tr><td>**EXAMPLE 7**</td></tr>
<tr><td>Solving an Equation
Using a Purely
Graphical Approach</td></tr>
</table>

Solve $\sin 3x + \cos 2x = \cos \frac{x}{2}$ over the interval $[0, 2\pi)$, using a graph. Give approximate solutions when appropriate.

SOLUTION We will solve this equation using the x-intercept method. The graph of $y = \sin 3x + \cos 2x - \cos \frac{x}{2}$ is shown in Figure 51. The least positive solution, .72739787, is indicated on the screen. The solution 0 can be verified by direct substitution. The solution set for the interval $[0, 2\pi)$ is

$$\{0, .727, 2.288, 3.524, 4.189\}. \quad █$$

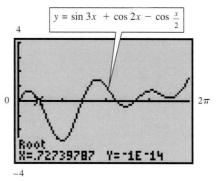

$$[0, 2\pi] \text{ by } [-4, 4]$$
$$\text{Xscl} = \frac{\pi}{2} \qquad \text{Yscl} = 1$$

FIGURE 51

FOR **GROUP DISCUSSION**

Use Figure 51 and the solution set given in Example 7 to solve the following inequalities over the interval $[0, 2\pi)$.

1. $\sin 3x + \cos 2x < \cos \dfrac{x}{2}$

2. $\sin 3x + \cos 2x > \cos \dfrac{x}{2}$

9.5 EXERCISES

*Solve the equation in part (**a**) analytically. Then use a graph to solve the inequalities in parts (**b**) and (**c**). In all cases, solve over the interval $[0, 2\pi)$, giving exact values for solutions and endpoints (that is, rational multiples of π).*

1. (a) $\cos 2x = \dfrac{\sqrt{3}}{2}$

    (b) $\cos 2x > \dfrac{\sqrt{3}}{2}$

    (c) $\cos 2x < \dfrac{\sqrt{3}}{2}$

2. (a) $\cos 2x = -\dfrac{1}{2}$

    (b) $\cos 2x > -\dfrac{1}{2}$

    (c) $\cos 2x < -\dfrac{1}{2}$

3. (a) $\sin 3x = -1$
    (b) $\sin 3x > -1$
    (c) $\sin 3x < -1$

4. (a) $\sin 3x = 0$
    (b) $\sin 3x > 0$
    (c) $\sin 3x < 0$

5. (a) $\sqrt{2} \cos 2x = -1$
    (b) $\sqrt{2} \cos 2x \geq -1$
    (c) $\sqrt{2} \cos 2x \leq -1$

6. (a) $2\sqrt{3} \sin 2x = \sqrt{3}$
    (b) $2\sqrt{3} \sin 2x \geq \sqrt{3}$
    (c) $2\sqrt{3} \sin 2x \leq \sqrt{3}$

7. (a) $\sin \dfrac{x}{2} = \sqrt{2} - \sin \dfrac{x}{2}$

    (b) $\sin \dfrac{x}{2} > \sqrt{2} - \sin \dfrac{x}{2}$

    (c) $\sin \dfrac{x}{2} < \sqrt{2} - \sin \dfrac{x}{2}$

8. (a) $\sin x = \sin 2x$

    (b) $\sin x > \sin 2x$

    (c) $\sin x < \sin 2x$

9. (a) $\cos 2x - \cos x = 0$

  (b) $\cos 2x - \cos x \leq 0$

  (c) $\cos 2x - \cos x \geq 0$

10. (a) $\sin^2 \frac{x}{2} - 1 = 0$

  (b) $\sin^2 \frac{x}{2} - 1 \leq 0$

  (c) $\sin^2 \frac{x}{2} - 1 \geq 0$

11. (a) $\sin \frac{x}{2} = \cos \frac{x}{2}$

  (b) $\sin \frac{x}{2} > \cos \frac{x}{2}$

  (c) $\sin \frac{x}{2} < \cos \frac{x}{2}$

12. (a) $\sec \frac{x}{2} = \cos \frac{x}{2}$

  (b) $\sec \frac{x}{2} > \cos \frac{x}{2}$

  (c) $\sec \frac{x}{2} < \cos \frac{x}{2}$

13. (a) $\cos 2x + \cos x = 0$

  (b) $\cos 2x + \cos x \geq 0$

  (c) $\cos 2x + \cos x \leq 0$

14. (a) $\sin x \cos x = \frac{1}{4}$

  (b) $\sin x \cos x \geq \frac{1}{4}$

  (c) $\sin x \cos x \leq \frac{1}{4}$

15. (a) $\tan 3x + \sec 3x = 1$

  (b) $\tan 3x + \sec 3x \geq 1$

  (c) $\tan 3x + \sec 3x \leq 1$

16. (a) $\csc^2 \frac{x}{2} = 2 \sec x$

  (b) $\csc^2 \frac{x}{2} > 2 \sec x$

  (c) $\csc^2 \frac{x}{2} < 2 \sec x$

17. (a) $\cos x - 1 = \cos 2x$
  (b) $\cos x - 1 \leq \cos 2x$
  (c) $\cos x - 1 \geq \cos 2x$

18. (a) $1 - \sin x = \cos 2x$
  (b) $1 - \sin x \leq \cos 2x$
  (c) $1 - \sin x \geq \cos 2x$

Use a purely graphical approach to solve each equation over the interval $[0, 2\pi)$.

19. $\sin x + \sin 3x = \cos x$

20. $\sin 3x - \sin x = 0$

21. $\cos 2x + \cos x = 0$

22. $\sin 4x + \sin 2x = 2 \cos x$

23. $\cos \frac{x}{2} = 2 \sin 2x$

24. $\sin \frac{x}{2} + \cos 3x = 0$

25. What is wrong with the following solution? Solve $\tan 2\theta = 2$ in the interval $[0, 2\pi)$.

$$\tan 2\theta = 2$$

$$\frac{\tan 2\theta}{2} = \frac{2}{2}$$

$$\tan \theta = 1$$

$$\theta = \frac{\pi}{4} \quad \text{or} \quad \theta = \frac{5\pi}{4}$$

The solutions are $\frac{\pi}{4}$ and $\frac{5\pi}{4}$.

26. The equation

$$\cot \frac{x}{2} - \csc \frac{x}{2} - 1 = 0$$

has no solution over the interval $[0, 2\pi)$. Using this information, what can be said about the graph of $y = \cot \frac{x}{2} - \csc \frac{x}{2} - 1$ over this interval? Confirm your answer by actually graphing the function over the interval.

R*elating Concepts*

It can be shown using the methods of calculus that the x-coordinates of the local extrema of the function $y_1 = \sin^2 2x + \cos \frac{1}{2}x$ are the same as the zeros of the function $y_2 = 4 \sin 2x \cos 2x - \frac{1}{2} \sin \frac{1}{2}x$. Work Exercises 27 and 28 in order.

27. Graph both y_1 and y_2 with a calculator over the interval $[0, 2\pi)$. Use the same screen for both.

28. Verify that the least positive zero of y_2 corresponds to the x-coordinate of the first local extreme point of y_1 in the interval.

9.6 THE INVERSE CIRCULAR FUNCTIONS

Preliminary Considerations ▌ The Inverse Sine Function ▌ The Inverse Cosine Function ▌ The Inverse Tangent Function ▌ Miscellaneous Problems Involving Inverse Functions

Preliminary Considerations

In our work with applications of trigonometry in Chapter 8, we learned how to use the inverse trigonometric functions to find the measure of an angle θ if we knew one of the values $\sin \theta$, $\cos \theta$, or $\tan \theta$. Now we will investigate the inverse circular functions $y = \sin^{-1} x$, $y = \cos^{-1} x$, and $y = \tan^{-1} x$. Recall from Section 4.5 that a function must be one-to-one for an inverse (function) to exist. Since the circular functions $y = \sin x$, $y = \cos x$, and $y = \tan x$ are not one-to-one if their natural domains are chosen, in order to define their inverses we must restrict their domains so that the ranges are unchanged and each y-value corresponds to one and only one x-value.

The Inverse Sine Function

From Figure 52 and the horizontal line test, it is clear that $y = \sin x$ is not one-to-one function. By suitably restricting the domain of the sine function, however, a one-to-one function can be defined. It is generally agreed upon by mathematicians that the interval $\left[-\frac{\pi}{2}, \frac{\pi}{2}\right]$ be chosen for this restriction. This gives the portion of the graph shown as a solid curve in Figure 52.

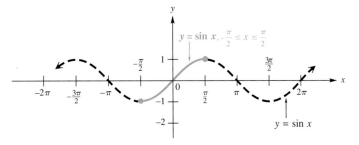

FIGURE 52

If we reflect the graph of this restricted portion of $y = \sin x$ across the line $y = x$, we obtain the graph of the inverse of the function. It is symbolized $\sin^{-1} x$. The graph of $y = \sin^{-1} x$ is shown in two forms in Figure 53; part (a) shows a traditional graph, with selected points labeled, while part (b) shows a graphics calculator-generated graph in the window $[-1, 1]$ by $[-\frac{\pi}{2}, \frac{\pi}{2}]$. (The alternate notation *arcsin x* is sometimes used to denote $\sin^{-1} x$.)

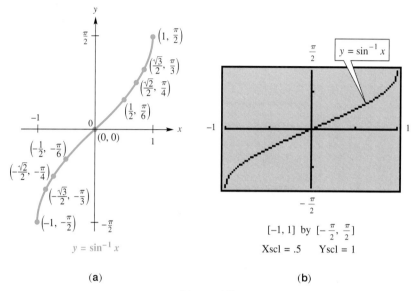

(a)

(b)

FIGURE 53

THE INVERSE SINE FUNCTION

$y = \sin^{-1} x$ or $y = \arcsin x$ means that $x = \sin y$, for $-\frac{\pi}{2} \le y \le \frac{\pi}{2}$. The domain of $y = \sin^{-1} x$ is $[-1, 1]$ and the range is $[-\frac{\pi}{2}, \frac{\pi}{2}]$.

We may think of $y = \sin^{-1} x$ or $y = \arcsin x$ as "y is the number in the interval $[-\frac{\pi}{2}, \frac{\pi}{2}]$ whose sine is x." Both notations will be used in this book.

EXAMPLE 1

Finding Inverse Sine Values

(a) Use the graph in Figure 53(a) to find $y = \sin^{-1} \frac{1}{2}$.

SOLUTION   The figure shows that the point $(\frac{1}{2}, \frac{\pi}{6})$ lies on the graph of $y = \sin^{-1} x$. Therefore, $\sin^{-1} \frac{1}{2} = \frac{\pi}{6}$.

(b) Use a graphics calculator to support the result in part (a).

SOLUTION   We use the capability of the calculator to locate the point on the graph with x-coordinate $\frac{1}{2} = .5$. We find that $y \approx .52359878$, which is an approximation for $\frac{\pi}{6}$. See Figure 54.

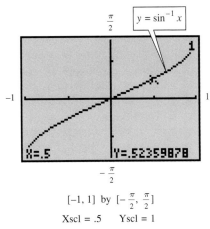

$$[-1, 1] \text{ by } [-\tfrac{\pi}{2}, \tfrac{\pi}{2}]$$
$$\text{Xscl} = .5 \qquad \text{Yscl} = 1$$

FIGURE 54

(c) Find an approximation for $\arcsin(-.36)$ in two ways, using the capabilities of a graphics calculator.

SOLUTION One way to approximate $\arcsin(-.36)$ is to put the calculator in radian mode and allow the calculator to compute it and show it in its display, as in Figure 55(a). Another way is to locate the point on the graph of $y = \sin^{-1} x$ with x-coordinate $-.36$. As seen in Figure 55(b), the y-value is approximately $-.3682679$, which corresponds to the approximation shown in part (a) of the figure.

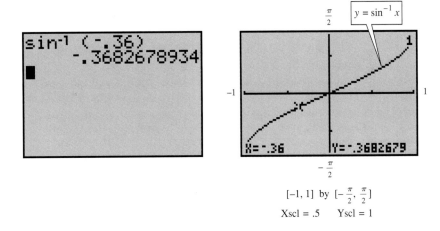

$$[-1, 1] \text{ by } [-\tfrac{\pi}{2}, \tfrac{\pi}{2}]$$
$$\text{Xscl} = .5 \qquad \text{Yscl} = 1$$

(a)   (b)

FIGURE 55

TECHNOLOGICAL NOTE
Some advanced graphics calculators *will* give a complex number display for arcsin 2. The interpretation of this result is beyond the scope of this text, and students at this level of mathematics should realize that the development of the circular and inverse circular functions as presented here is restricted to domains and ranges consisting of real numbers only.

(d) Find the exact value of $\arcsin(-1)$.

SOLUTION Because the point $(-1, -\tfrac{\pi}{2})$ lies on the graph of $y = \sin^{-1} x$, $\arcsin(-1) = -\tfrac{\pi}{2}$.

(e) Explain why arcsin 2 does not exist.

SOLUTION Because 2 is not in the domain of the inverse sine function, $\sin^{-1} 2$ does not exist. If we try to find a calculator approximation for $\sin^{-1} 2$, an error message will appear. ∎

CAUTION In Example 1(d), it is tempting to give the value of $\arcsin(-1)$ as $\frac{3\pi}{2}$, since $\sin\frac{3\pi}{2} = -1$. Notice, however, that $\frac{3\pi}{2}$ is not in the range of the inverse sine function. Be certain, in dealing with *all* inverse circular functions, that the number given for the function value is in the range of the particular inverse function being considered.

Important information about the inverse sine function is summarized in the box that follows.

INVERSE SINE FUNCTION

$y = \sin^{-1} x$ or $y = \arcsin x$
(Figure 56)
Domain: $[-1, 1]$
Range: $\left[-\frac{\pi}{2}, \frac{\pi}{2}\right]$

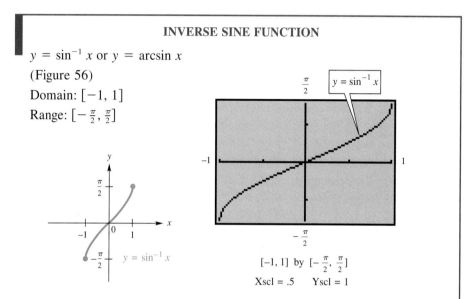

(a)   (b)

$[-1, 1]$ by $\left[-\frac{\pi}{2}, \frac{\pi}{2}\right]$
Xscl = .5   Yscl = 1

FIGURE 56

Over the interval $(-1, 1)$, the inverse sine function is increasing. Its x-intercept is 0 and its y-intercept is 0. Its graph is symmetric with respect to the origin. In general, $\sin^{-1}(-x) = -\sin^{-1} x$.

The Inverse Cosine Function

The function $y = \cos^{-1} x$ (or $y = \arccos x$) is defined by restricting the domain of the function $y = \cos x$ to the interval $[0, \pi]$, and then reversing the roles of x and y. The graph of $y = \cos^{-1} x$ is shown in Figure 57, in both traditional and calculator-generated forms. Again, some key points are shown on the traditional graph.

THE INVERSE COSINE FUNCTION

$y = \cos^{-1} x$ or $y = \arccos x$ means that $x = \cos y$, for $0 \le y \le \pi$. The domain of $y = \cos^{-1} x$ is $[-1, 1]$ and the range is $[0, \pi]$.

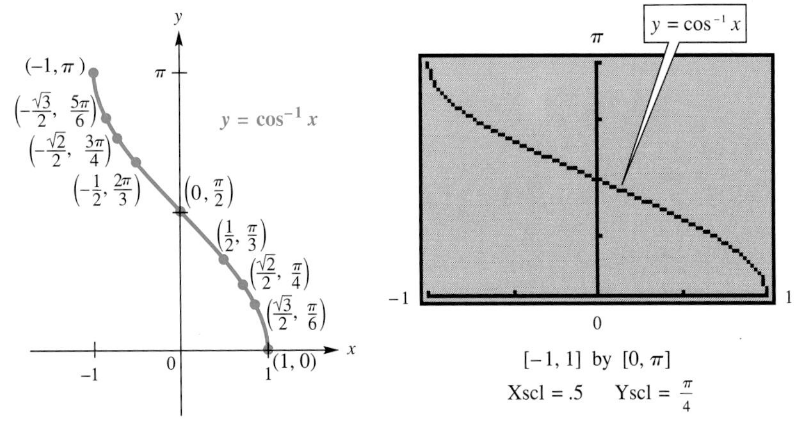

$(-1, \pi)$

$\left(-\frac{\sqrt{3}}{2}, \frac{5\pi}{6}\right)$

$\left(-\frac{\sqrt{2}}{2}, \frac{3\pi}{4}\right)$

$\left(-\frac{1}{2}, \frac{2\pi}{3}\right)$ $\left(0, \frac{\pi}{2}\right)$

$y = \cos^{-1} x$

$\left(\frac{1}{2}, \frac{\pi}{3}\right)$

$\left(\frac{\sqrt{2}}{2}, \frac{\pi}{4}\right)$

$\left(\frac{\sqrt{3}}{2}, \frac{\pi}{6}\right)$

$(1, 0)$

$y = \cos^{-1} x$

$[-1, 1]$ by $[0, \pi]$
Xscl = .5   Yscl = $\frac{\pi}{4}$

(a)

(b)

FIGURE 57

FOR GROUP DISCUSSION

Earlier in this section we stated that $y = \sin^{-1} x$ means "y is the number in the interval $\left[-\frac{\pi}{2}, \frac{\pi}{2}\right]$ whose sine is x." Make a similar statement for
$$y = \cos^{-1} x.$$

EXAMPLE 2

Finding Inverse Cosine Values

(a) Use the graph in Figure 57(a) to find y, if $y = \cos^{-1}\left(-\frac{1}{2}\right)$.

SOLUTION Since the point $\left(-\frac{1}{2}, \frac{2\pi}{3}\right)$ lies on the graph of $y = \cos^{-1} x$, $\cos^{-1}\left(-\frac{1}{2}\right) = \frac{2\pi}{3}$.

(b) Use a graphics calculator to support the result in part (a).

SOLUTION We graph $y = \cos^{-1} x$ in the window $[-1, 1]$ by $[0, \pi]$ and determine the y-value when $x = -\frac{1}{2} = -.5$. The display in Figure 58 shows that this y-value is 2.0943951, which is a decimal approximation for $\frac{2\pi}{3}$. (Verify this on your own calculator.)

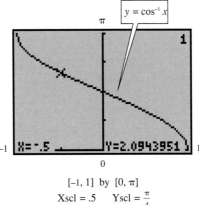

$y = \cos^{-1} x$

X=-.5   Y=2.0943951

$[-1, 1]$ by $[0, \pi]$
Xscl = .5   Yscl = $\frac{\pi}{4}$

FIGURE 58

(c) Find an approximation for $\cos^{-1}(-.75)$ in two ways, using the capabilities of a graphics calculator.

SOLUTION The display in Figure 59(a) and the graph in Figure 59(b) on the next page both indicate that $\cos^{-1}(-.75) \approx 2.4188584$. (The calculator must be in radian mode.)

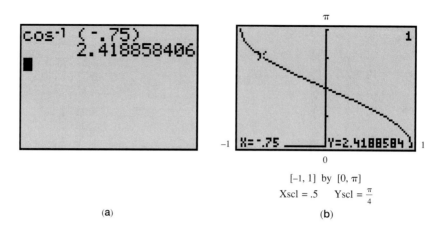

[-1, 1] by [0, π]
Xscl = .5   Yscl = $\frac{\pi}{4}$

(a)   (b)

FIGURE 59

(d)   Find the exact value of $\cos^{-1} \frac{\sqrt{2}}{2}$.

SOLUTION   The point $(\frac{\sqrt{2}}{2}, \frac{\pi}{4})$ lies on the graph of $y = \cos^{-1} x$. Therefore, $\cos^{-1} \frac{\sqrt{2}}{2} = \frac{\pi}{4}$.

(e)   Why does a calculator give an error message for $\cos^{-1} 3$?

TECHNOLOGICAL NOTE
Again some advanced
graphics calculators will
give a complex number
display for $\cos^{-1}$ 3. See
the technological note
accompanying Example 1
in this section.

SOLUTION   Because 3 is not in the domain of the inverse cosine function, the expression $\cos^{-1} 3$ is not defined. (Another way to think of this is: there is no number whose cosine is 3, because $\cos x$ must be in the interval $[-1, 1]$ for all numbers x.)   ▪

A summary of the important information about the inverse cosine function follows.

INVERSE COSINE FUNCTION

$y = \cos^{-1} x$ or $y = \arccos x$

(Figure 60)

Domain: $[-1, 1]$

Range: $[0, \pi]$

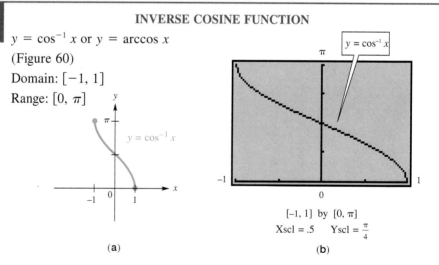

[-1, 1] by [0, π]
Xscl = .5   Yscl = $\frac{\pi}{4}$

(a)   (b)

FIGURE 60

Over the interval $(-1, 1)$, the inverse cosine function is decreasing. Its x-intercept is 1 and its y-intercept is $\frac{\pi}{2}$. The graph is neither symmetric with respect to the y-axis nor symmetric with respect to the origin.

The Inverse Tangent Function

Restricting the domain of the function $y = \tan x$ to the open interval $\left(-\frac{\pi}{2}, \frac{\pi}{2}\right)$ yields a one-to-one function. By interchanging the roles of x and y, we obtain the inverse tangent function $y = \tan^{-1} x$ or $y = \arctan x$. Both traditional and calculator-generated graphs of this function are given in Figure 61.

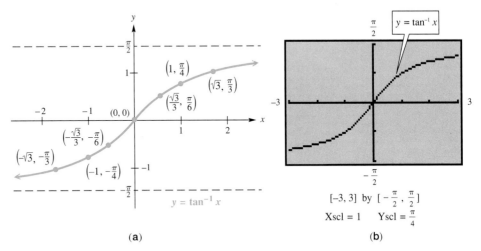

(a)   (b)

FIGURE 61

THE INVERSE TANGENT FUNCTION

$y = \tan^{-1} x$ or $y = \arctan x$ means that $x = \tan y$, for $-\frac{\pi}{2} < y < \frac{\pi}{2}$. The domain of $y = \tan^{-1} x$ is $(-\infty, \infty)$, and the range is $\left(-\frac{\pi}{2}, \frac{\pi}{2}\right)$. The lines $y = \frac{\pi}{2}$ and $y = -\frac{\pi}{2}$ are horizontal asymptotes of the graph.

FOR GROUP DISCUSSION

With calculators in hand, discuss the following.

1. Refer to Figure 61(a) to explain why $\tan^{-1} \sqrt{3} = \frac{\pi}{3}$.
2. Use your calculator to support the equality in item 1.
3. Find an approximation for $\tan^{-1} 2$ in two ways, using your calculator.
4. Find the exact value of $\tan^{-1}(-1)$.
5. Discuss the symmetry of the graph of $y = \tan^{-1} x$.
6. Discuss the following: We will never get an error message for $\tan^{-1} x$, no matter what value of x we enter into our calculators.

The following box summarizes important information about the inverse tangent function.

INVERSE TANGENT FUNCTION

$y = \tan^{-1} x$ or $y = \arctan x$

(Figure 62)

Domain: $(-\infty, \infty)$

Range: $\left(-\frac{\pi}{2}, \frac{\pi}{2}\right)$

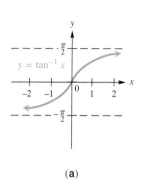

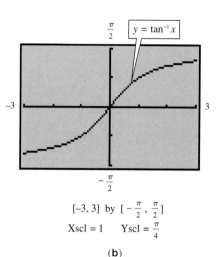

$[-3, 3]$ by $\left[-\frac{\pi}{2}, \frac{\pi}{2}\right]$

$\text{Xscl} = 1 \qquad \text{Yscl} = \frac{\pi}{4}$

(a)   (b)

FIGURE 62

Over $(-\infty, \infty)$, the inverse tangent function is increasing. Its x-intercept is 0 and its y-intercept is 0. The graph is symmetric with respect to the origin. In general, $\tan^{-1}(-x) = -\tan^{-1} x$. As $x \to \infty$, $y \to \frac{\pi}{2}$ from below, meaning that the line $y = \frac{\pi}{2}$ is a horizontal asymptote. Similarly, the line $y = -\frac{\pi}{2}$ is also a horizontal asymptote, since as $x \to -\infty$, $y \to -\frac{\pi}{2}$ from above.

Miscellaneous Problems Involving Inverse Functions

We have defined the inverse sine, cosine, and tangent functions with suitable restrictions on the domains. The other three inverse trigonometric functions are similarly defined. The six inverse trigonometric functions with their domains and ranges are given in the table.* This information, particularly the range for each function, should be learned. (The graphs of the last three inverse trigonometric functions are left for the exercises.)

Function	Domain	Range
$y = \sin^{-1} x$	$[-1, 1]$	$\left[-\frac{\pi}{2}, \frac{\pi}{2}\right]$
$y = \cos^{-1} x$	$[-1, 1]$	$[0, \pi]$
$y = \tan^{-1} x$	$(-\infty, \infty)$	$\left(-\frac{\pi}{2}, \frac{\pi}{2}\right)$
$y = \cot^{-1} x$	$(-\infty, \infty)$	$(0, \pi)$
$y = \sec^{-1} x$	$(-\infty, -1] \cup [1, \infty)$	$[0, \pi], y \neq \frac{\pi}{2}$
$y = \csc^{-1} x$	$(-\infty, -1] \cup [1, \infty)$	$\left[-\frac{\pi}{2}, \frac{\pi}{2}\right], y \neq 0$

---

* The inverse secant and inverse cosecant functions are sometimes defined differently.

Problems such as those in the remaining examples in this section are solved by using inverse function concepts. As we shall see, we can use right triangles and identities to solve them.

<table>
<tr><td>

EXAMPLE 3

Finding Function Values Using Definitions of the Trigonometric Functions

</td><td>

Evaluate each of the following without a calculator.

(a) $\sin(\tan^{-1}\frac{3}{2})$

SOLUTION   Let

</td></tr>
</table>

$$\theta = \tan^{-1}\frac{3}{2}, \text{ so that } \tan\theta = \frac{3}{2}.$$

Since $\tan^{-1}$ is defined only for angles in quadrants I and IV and since $\frac{3}{2}$ is positive, θ is in quadrant I. Sketch θ in quadrant I, and label a triangle as shown in Figure 63. The hypotenuse is $\sqrt{13}$ and the value of sine is the quotient of the side opposite and the hypotenuse, so

$$\sin\left(\tan^{-1}\frac{3}{2}\right) = \sin\theta = \frac{3}{\sqrt{13}} = \frac{3\sqrt{13}}{13}.$$

To check this result on a calculator, enter $\frac{3}{2}$ as 1.5. Then find $\tan^{-1} 1.5$, and finally find $\sin(\tan^{-1} 1.5)$. Store this result and calculate $\frac{3\sqrt{13}}{13}$, which should agree with the result for $\sin(\tan^{-1} 1.5)$. Since the values are only approximations, this check does not *prove* that the result is correct, but it is highly suggestive that it is correct.

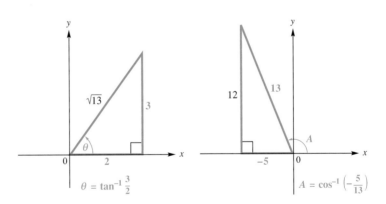

FIGURE 63   **Figure 64**

(b) $\tan(\cos^{-1}(-\frac{5}{13}))$

SOLUTION   Let $A = \cos^{-1}(-\frac{5}{13})$. Then $\cos A = -\frac{5}{13}$. Since $\cos^{-1} x$ for a negative value of x is in quadrant II, sketch A in quadrant II, as shown in Figure 64.
From the triangle in Figure 64,

$$\tan\left(\cos^{-1}\left(-\frac{5}{13}\right)\right) = \tan A = -\frac{12}{5}. \quad \blacksquare$$

> **FOR GROUP DISCUSSION**
>
> Discuss how the screen in Figure 65 supports the analytic work in Example 3(b). Does it matter whether the calculator is in radian mode or degree mode?
>
>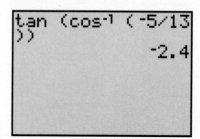
>
> **FIGURE 65**

EXAMPLE 4

Finding Function Values Using Sum and Double Number Identities

Evaluate the following without using a calculator.

(a) $\cos(\arctan \sqrt{3} + \arcsin \frac{1}{3})$

SOLUTION   Let $A = \arctan \sqrt{3}$ and $B = \arcsin \frac{1}{3}$ so that $\tan A = \sqrt{3}$ and $\sin B = \frac{1}{3}$. Sketch both A and B in quadrant I, as shown in Figure 66.

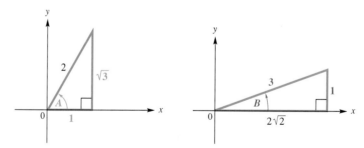

FIGURE 66

Now use the identity for $\cos(A + B)$.

$$\cos(A + B) = \cos A \cos B - \sin A \sin B$$

$$\cos\left(\arctan \sqrt{3} + \arcsin \frac{1}{3}\right) = \cos(\arctan \sqrt{3}) \cos\left(\arcsin \frac{1}{3}\right)$$

$$- \sin(\arctan \sqrt{3}) \sin\left(\arcsin \frac{1}{3}\right) \quad \text{[1]}$$

From the sketch in Figure 66,

$$\cos\left(\arctan \sqrt{3}\right) = \cos A = \frac{1}{2}, \qquad \cos\left(\arcsin \frac{1}{3}\right) = \cos B = \frac{2\sqrt{2}}{3},$$

$$\sin(\arctan \sqrt{3}) = \sin A = \frac{\sqrt{3}}{2}, \qquad \sin\left(\arcsin \frac{1}{3}\right) = \sin B = \frac{1}{3}.$$

Substitute these values into equation (1) to get

$$\cos\left(\arctan\sqrt{3} + \arcsin\frac{1}{3}\right) = \frac{1}{2} \cdot \frac{2\sqrt{2}}{3} - \frac{\sqrt{3}}{2} \cdot \frac{1}{3}$$

$$= \frac{2\sqrt{2}}{6} - \frac{\sqrt{3}}{6}$$

$$= \frac{2\sqrt{2} - \sqrt{3}}{6}.$$

(b) $\tan(2 \arcsin \frac{2}{5})$

SOLUTION Let $\arcsin \frac{2}{5} = B$. Then, from the double number tangent identity,

$$\tan\left(2 \arcsin \frac{2}{5}\right) = \tan 2B$$

$$= \frac{2 \tan B}{1 - \tan^2 B}.$$

Since $\arcsin \frac{2}{5} = B$, $\sin B = \frac{2}{5}$. Sketch a triangle in quadrant I, find the length of the third side, and then find $\tan B$. From the triangle in Figure 67 $\tan B = \frac{2}{\sqrt{21}}$, and

$$\tan\left(2 \arcsin \frac{2}{5}\right) = \frac{2\left(\dfrac{2}{\sqrt{21}}\right)}{1 - \left(\dfrac{2}{\sqrt{21}}\right)^2} = \frac{\dfrac{4}{\sqrt{21}}}{1 - \dfrac{4}{21}}$$

$$= \frac{\dfrac{4}{\sqrt{21}}}{\dfrac{17}{21}} = \frac{4\sqrt{21}}{17}. \quad \blacksquare$$

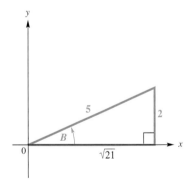

FIGURE 67

While the work shown in both parts of Example 4 does not rely on a calculator, we can support our analytic work with one. By entering $\cos(\arctan \sqrt{3} + \arcsin \frac{1}{3})$ into a calculator, we get the approximation .1827293862, which is the same approximation we get when we enter $\frac{2\sqrt{2} - \sqrt{3}}{6}$ (the exact value obtained analytically). Thus, our answer in Example 4(a) is supported. Similarly, we obtain the same approximation when we evaluate $\tan(2 \arcsin \frac{2}{5})$ and $\frac{4\sqrt{21}}{17}$, supporting our answer in Example 4(b).

EXAMPLE **5**
Writing a Function Value in Terms of *u*

Write $\sin(\tan^{-1} u)$ as an algebraic expression in *u*.

SOLUTION   Let $\theta = \tan^{-1} u$, so that $\tan \theta = u$. Here *u* may be positive or negative. Since $-\frac{\pi}{2} < \tan^{-1} u < \frac{\pi}{2}$, sketch θ in quadrants I and IV and label two triangles as shown in Figure 68. Since sine is given by the quotient of the side opposite and the hypotenuse,

$$\sin(\tan^{-1} u) = \sin \theta = \frac{u}{\sqrt{u^2 + 1}}$$

$$= \frac{u \sqrt{u^2 + 1}}{u^2 + 1}.$$

The result is positive when *u* is positive and negative when *u* is negative.   ∎

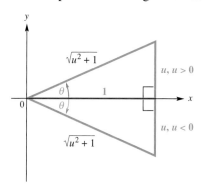

FIGURE 68

9.6   EXERCISES

Use the graph of the appropriate inverse circular function, found in Figure 53(a), 57(a), or 61(a) to find the exact value of each of the following.

1. $\arcsin\left(-\frac{1}{2}\right)$ **2.** $\arccos \frac{\sqrt{3}}{2}$ **3.** $\tan^{-1} 1$ **4.** $\sin^{-1} 0$

5. $\cos^{-1}(-1)$ **6.** $\cos^{-1} \frac{1}{2}$ **7.** $\sin^{-1}\left(-\frac{\sqrt{3}}{2}\right)$ **8.** $\cos^{-1} 0$

9. $\arctan(-1)$ **10.** $\arccos\left(-\frac{1}{2}\right)$ **11.** $\arcsin \frac{\sqrt{2}}{2}$ **12.** $\arcsin\left(-\frac{\sqrt{2}}{2}\right)$

13. $\arccos\left(-\frac{\sqrt{3}}{2}\right)$ **14.** $\arcsin\left(-\frac{\sqrt{3}}{2}\right)$ **15.** $\tan^{-1} 0$ **16.** $\tan^{-1}(-\sqrt{3})$

*For each of the following, **(a)** find an approximation of the expression using a calculator-generated graph of the appropriate inverse circular function, and **(b)** support your answer in part (a) by using the inverse function key on your calculator. Be sure the calculator is in radian mode.*

17. $\sin^{-1} .35$ **18.** $\cos^{-1} .35$ **19.** $\cos^{-1}(-.6)$ **20.** $\sin^{-1}(-.6)$

21. $\tan^{-1} 5$ **22.** $\tan^{-1}(-5)$ **23.** $\cos^{-1} 1.5$ **24.** $\sin^{-1}(-2.3)$

Draw by hand the graph of each of the following as defined in the text, and give the domain and the range.

25. $y = \cot^{-1} x$ **26.** $y = \csc^{-1} x$ **27.** $y = \sec^{-1} x$

Find the exact real number value of each of the following, using the definitions given in the text.

28. $\cot^{-1}(-1)$

29. $\sec^{-1}(-\sqrt{2})$

30. $\csc^{-1}(-2)$

31. $\cot^{-1}(-\sqrt{3})$

32. $\csc^{-1}(-1)$

33. $\sec^{-1}(-2)$

34. The following expressions were used by the mathematicians who computed the value of π to 100,000 decimal places. Use a calculator in radian mode to verify that each is (approximately) correct.

(a) $\pi = 16 \tan^{-1} \dfrac{1}{5} - 4 \tan^{-1} \dfrac{1}{239}$

(b) $\pi = 24 \tan^{-1} \dfrac{1}{8} + 8 \tan^{-1} \dfrac{1}{57} + 4 \tan^{-1} \dfrac{1}{239}$

(c) $\pi = 48 \tan^{-1} \dfrac{1}{18} + 32 \tan^{-1} \dfrac{1}{57} - 20 \tan^{-1} \dfrac{1}{239}$

35. Explain why attempting to find $\sin^{-1} 1.003$ on your calculator will result in an error message.

36. Explain why you are able to find $\tan^{-1} 1.003$ on your calculator. Why is this situation different from the one described in Exercise 35?

Decide whether the statement is true for all real numbers x in the given interval. If it is not true, say why.

37. $\sin(\sin^{-1} x) = x, \quad -1 \le x \le 1$

38. $\cos(\cos^{-1} x) = x, \quad -1 \le x \le 1$

39. $\sin^{-1}(\sin x) = x, \quad x \text{ in } (-\infty, \infty)$

40. $\cos^{-1}(\cos x) = x, \quad x \text{ in } (-\infty, \infty)$

41. $\tan(\tan^{-1} x) = x, \quad x \text{ in } (-\infty, \infty)$

42. $\tan^{-1}(\tan x) = x, \quad x \text{ in } \left(-\dfrac{\pi}{2}, \dfrac{\pi}{2}\right)$

Give the exact value of each of the following, without using a calculator. You may wish to support your answer by using your calculator as described in the discussion and in Examples 3 and 4 in the text.

43. $\tan\left(\arccos \dfrac{3}{4}\right)$

44. $\sin\left(\arccos \dfrac{1}{4}\right)$

45. $\cos(\tan^{-1}(-2))$

46. $\sec\left(\sin^{-1}\left(-\dfrac{1}{5}\right)\right)$

47. $\cot\left(\arcsin\left(-\dfrac{2}{3}\right)\right)$

48. $\cos\left(\arctan \dfrac{8}{3}\right)$

49. $\sin\left(2 \tan^{-1} \dfrac{12}{5}\right)$

50. $\cos\left(2 \sin^{-1} \dfrac{1}{4}\right)$

51. $\cos\left(2 \arctan \dfrac{4}{3}\right)$

52. $\tan\left(2 \cos^{-1} \dfrac{1}{4}\right)$

53. $\sin\left(2 \cos^{-1} \dfrac{1}{5}\right)$

54. $\cos(2 \tan^{-1}(-2))$

55. $\sec(\sec^{-1} 2)$

56. $\csc(\csc^{-1} \sqrt{2})$

57. $\cos\left(\tan^{-1} \dfrac{5}{12} - \tan^{-1} \dfrac{3}{4}\right)$

58. $\cos\left(\sin^{-1} \dfrac{3}{5} + \cos^{-1} \dfrac{5}{13}\right)$

59. $\sin\left(\sin^{-1} \dfrac{1}{2} + \tan^{-1}(-3)\right)$

60. $\tan\left(\cos^{-1} \dfrac{\sqrt{3}}{2} - \sin^{-1}\left(-\dfrac{3}{5}\right)\right)$

Write each of the following as an algebraic expression in u, u > 0.

61. $\sin(\arccos u)$

62. $\tan(\arccos u)$

63. $\cot(\arcsin u)$

64. $\cos(\arcsin u)$

65. $\sin\left(\sec^{-1} \dfrac{u}{2}\right)$

66. $\cos\left(\tan^{-1} \dfrac{3}{u}\right)$

67. $\tan\left(\sin^{-1} \dfrac{u}{\sqrt{u^2 + 2}}\right)$

68. $\sec\left(\cos^{-1} \dfrac{u}{\sqrt{u^2 + 5}}\right)$

69. $\sec\left(\operatorname{arccot} \dfrac{\sqrt{4 - u^2}}{u}\right)$

70. $\csc\left(\arctan \dfrac{\sqrt{9 - u^2}}{u}\right)$

9.7 APPLICATIONS OF THE CIRCULAR AND TRIGONOMETRIC FUNCTIONS

Solving Applications Using Equations ▮ Solving Applications Using Graphs

In Chapter 8 we saw how trigonometry can be used to solve applied problems that lead to triangles. Recall that right triangle side ratios, the law of sines, the law of cosines, and the various formulas for the area of a triangle were all used in applications. In this section we will see how other approaches involving the circular and trigonometric functions can be used to solve applications.

Solving Applications Using Equations

The first example shows how identities and equation-solving techniques can be used to solve a problem involving trajectory.

EXAMPLE 1

Solving a Problem Involving Altitude of a Projectile

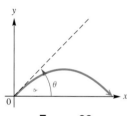

FIGURE 69

The altitude of a projectile in feet (neglecting air resistance) is given by

$$y = (\tan \theta)x - \frac{16}{v_0^2 \cos^2 \theta}x^2,$$

where x is the range (horizontal distance covered) in feet and v_0 is the initial velocity of the projectile at an angle θ from the horizontal. See Figure 69. A projectile is fired with an initial velocity of 100 feet per second. Find the firing angle of the projectile so that it strikes the ground 312.5 feet from the firing point.

SOLUTION We want to find the value of θ so that $y = 0$ when $x = 312.5$ and $v_0 = 100$. Substitute these values into the given equation.

$$y = (\tan \theta)x - \frac{16}{v_0^2 \cos^2 \theta}x^2$$

$$0 = (\tan \theta)(312.5) - \frac{16}{100^2 \cos^2 \theta}(312.5)^2$$

$$0 = \tan \theta - \frac{16}{10{,}000 \cos^2 \theta}(312.5) \qquad \text{Divide both sides by 312.5.}$$

$$0 = \tan \theta - \frac{1}{2 \cos^2 \theta} \qquad \text{Simplify.}$$

$$0 = 2 \cos^2 \theta \tan \theta - 1 \qquad \text{Multiply both sides by } 2\cos^2 \theta.$$

$$0 = 2 \cos^2 \theta \left(\frac{\sin \theta}{\cos \theta}\right) - 1 \qquad \tan \theta = \frac{\sin \theta}{\cos \theta}$$

$$0 = 2 \cos \theta \sin \theta - 1 \qquad \frac{\cos^2 \theta}{\cos \theta} = \cos \theta$$

$$0 = \sin 2\theta - 1 \qquad \begin{array}{l} 2\cos\theta\sin\theta \\ = \sin 2\theta \end{array}$$

$$\sin 2\theta = 1$$
$$2\theta = 90°$$
$$\theta = 45°$$

The projectile should be fired at an angle of 45° to meet the requirements of the problem. Note that θ must be in the interval (0°, 90°) in this situation. ▯

EXAMPLE 2

Analyzing the Concept of Mach Number of an Airplane

FIGURE 70

An airplane flying faster than sound sends out sound waves that form a cone, as shown in Figure 70. The cone intersects the ground to form a hyperbola. As this hyperbola passes over a particular point on the ground, a sonic boom is heard at that point. If α is the angle at the vertex of the cone, then

$$\sin \frac{\alpha}{2} = \frac{1}{m},$$

where m is the Mach number for the speed of the plane. (We assume $m > 1$.) The Mach number is the ratio of the speed of the plane and the speed of sound. Thus, a speed of Mach 1.4 means that the plane is flying at 1.4 times the speed of sound.

(a) Find the measure of α, if $m = 1.5$.

SOLUTION Substituting 1.5 for m in the equation and solving yields the following.

$$\sin \frac{\alpha}{2} = \frac{1}{1.5}$$

$$\sin \frac{\alpha}{2} = \frac{2}{3}$$

$$\frac{\alpha}{2} = \sin^{-1}\left(\frac{2}{3}\right)$$

$$\frac{\alpha}{2} \approx 41.8° \qquad \text{Use degree mode.}$$

$$\alpha \approx 83.6°$$

When the Mach number is 1.5, α is about 83.6°.

(b) Find the Mach number if $\alpha = 60°$.

SOLUTION Let $\alpha = 60°$ and solve for m.

$$\sin \frac{60°}{2} = \frac{1}{m}$$

$$\sin 30° = \frac{1}{m}$$

$$\frac{1}{2} = \frac{1}{m}$$

$$m = 2$$

When $\alpha = 60°$, the Mach number is 2. ∎

Solving Applications Using Graphs

The following example is an illustration of a periodic phenomenon: daylight hours at a particular location.

EXAMPLE 3

Using a Graph to Solve a Formula for Daylight Hours

The seasonal variation in the length of daylight can be modeled by a sine function. For example, the number of hours of daylight in New Orleans x days after March 21 (disregarding leap year) is approximated by the model

$$y = \frac{35}{3} + \frac{7}{3} \sin \frac{2\pi x}{365}.^*$$

*Adapted from *A Sourcebook of Applications of School Mathematics* by Donald Bushaw et al. Copyright © 1980 by The Mathematical Association of America. Reprinted by permission.

(a)   Use a graph to determine the number of daylight hours on May 1.

SOLUTION   May 1 falls exactly 41 days after March 21. We graph the model function in the window $[0, 365]$ by $[0, 24]$, and use the calculator to find y when $x = 41$. As shown in Figure 71, the number of daylight hours is approximately 13.2.

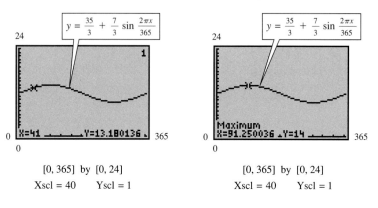

$[0, 365]$ by $[0, 24]$	$[0, 365]$ by $[0, 24]$
Xscl = 40 Yscl = 1	Xscl = 40 Yscl = 1
FIGURE 71	**FIGURE 72**

(b)   Use a graph to determine the longest day of the year.

As seen in Figure 72, the absolute maximum value of y occurs when $x \approx 91$. On that day, June 20 (91 days after March 21), the day is 14 hours in duration and is the longest of the year.   ◖◗

FOR GROUP DISCUSSION

See who in the class can be the first to determine the shortest day of the year.

9.7 EXERCISES

Use either equation-solving methods or a graph to solve each problem.

An equation for the curve describing the altitude of a projectile is

$$y = (\tan \theta)x - \frac{16}{v_0^2 \cos^2 \theta}x^2,$$

where v_0 is the initial velocity of the projectile and θ is the angle from the horizontal at which it is fired. (See Example 1.)

1. A projectile is fired with an initial velocity of 400 ft per sec at an angle of 45° with the horizontal. Find each of the following: **(a)** the range (horizontal distance covered), **(b)** the maximum altitude.

2. Repeat Exercise 1 if the projectile is fired at 800 ft per sec at an angle of 30° with the horizontal.

3. Sales of snowblowers are seasonal. Suppose the sales of snowblowers in one region of Canada are approximated by

$$S(t) = 500 + 500 \cos\left[\left(\frac{\pi}{6}\right)t\right],$$

where t is time in months, with $t = 0$ corresponding to November. For what months are sales equal to 0?

4. The equation $.342D \cos\theta + h\cos^2\theta = \frac{16D^2}{V^2}$ is used in reconstructing accidents in which a vehicle vaults into the air after hitting an obstruction. V is the velocity in feet per second of the vehicle when it hits the obstruction, D is the distance (in feet) from the obstruction to the vehicle's landing point, and h is the difference in height (in feet) between the landing point and the takeoff point. Angle θ is the takeoff angle, the angle between the horizontal and the path of the vehicle. Find θ to the nearest degree if $V = 60$, $D = 80$, and $h = 2$.

5. The temperature in Fairbanks is given by

$$T(x) = 37 \sin\left[\left(\frac{2\pi}{365}\right)(x - 101)\right] + 25,$$

where $T(x)$ is the temperature (in degrees Celsius) on day x, with $x = 1$ corresponding to January 1 and $x = 365$ corresponding to December 31.* On what days was the temperature 0°C? Below 0°C? (Round answers to the nearest whole day.)

6. The British nautical mile is defined as the length of a minute of arc of a meridian. Since the earth is flat at its poles, the nautical mile, in feet, is given by

$$L = 6,077 - 31 \cos 2\theta,$$

where θ is the latitude in degrees. (See the figure at the top of the next column.)[†]

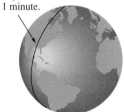

A nautical mile is the length on any of these meridians cut by a central angle of measure 1 minute.

(a) Find the latitude between 0° and 90° at which the nautical mile is 6,074 feet.

(b) At what latitude between 0° and 180° is the nautical mile 6,108 feet?

(c) In the United States the nautical mile is defined everywhere as 6,080.2 feet. At what latitude between 0° and 90° does this agree with the British nautical mile?

7. The voltage E in an electrical circuit is given by $E = 5 \cos 80\pi t$, where t is time measured in seconds.

(a) Find the amplitude and period.

(b) The reciprocal of the period, called the *frequency*, is the number of periods completed in one second. Find the frequency.

(c) Find E when $t = 0, .03, .06, .09, .12$.

8. For another electrical circuit, the current E is given by $E = 3.8 \sin 40\pi t$ where t is time measured in seconds.

(a) Find the amplitude and the period.

(b) Find the frequency. See Exercise 7(b).

(c) Find E when $t = .02, .04, .08, .12, .14$.

If an object is dropped in a vacuum, then the distance, d, the object falls in t seconds is given by

$$d = \frac{1}{2}gt^2,$$

where g is the acceleration due to gravity. At any particular point on the earth's surface, the value of g is a constant, roughly 978 cm per sec per sec. A more exact value of g at any point on the earth's surface is given by

$$g = 978.0524(1 + .005297 \sin^2\phi - .0000059 \sin^2 2\phi) - .000094h$$

in cm per second per second, where ϕ is the latitude of the point and h is the altitude of the point in feet. Find g, rounding to the nearest thousandth, given the following.

9. $\phi = 47° 12'$, $h = 387.0$ ft

10. $\phi = 68° 47'$, $h = 1145$ ft

---

*From "Is the Graph of Temperature a Sine Curve?" by Barbara Lando and Clifton Lando, *The Mathematics Teacher,* Vol. 70, September 1977, pp. 534–537. Copyright © 1977 by the National Council of Teachers of Mathematics. Reprinted with the permission of The Mathematics Teacher.

†Adapted from *A Sourcebook of Applications of School Mathematics* by Donald Bushaw et al. Copyright © 1980 by The Mathematical Association of America. Reprinted by permission.

11. In an electric circuit, let V represent the electromotive force in volts at t seconds. Assume $V = \cos 2\pi t$. Find the smallest positive value of t where $0 \leq t \leq \frac{1}{2}$ for each of the following values of V.
    (a) $V = 0$　　(b) $V = .5$　　(c) $V = .25$

12. A coil of wire rotating in a magnetic field induces a voltage given by

    $$e = 20 \sin\left(\frac{\pi t}{4} - \frac{\pi}{2}\right),$$

    where t is time in seconds. Find the smallest positive time to produce the following voltages.
    (a) 0　　(b) $10\sqrt{3}$

13. A 10-ft-high movie screen is located 2 ft above the eyes of the viewers, all of whom are sitting at the same level. A viewer seated 5 ft from the screen has the maximum viewing angle, given by x in the equation

    $$\frac{\tan x + .4}{1 - .4 \tan x} = 2.4.$$

    Find the maximum viewing angle (in degrees).

14. A certain person's blood pressure at time t (in seconds) is given by $p(t) = 105 + 15 \sin 144\pi t$ on the interval $[0, \frac{1}{72}]$. At what times is this person's blood pressure 100? 125?

15. A runner's arm swings rhythmically according to the equation

    $$y = \left(\frac{\pi}{8}\right)\cos 3\pi\left(t - \frac{1}{3}\right),$$

    where y represents the angle between the actual position of the upper arm and the downward vertical position (as shown in the figure*) and where t represents time in seconds. At what times in $[0, 2\pi)$ is the angle y equal to 0?

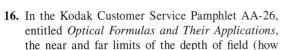

(a)　　　　　　　(b)

16. In the Kodak Customer Service Pamphlet AA-26, entitled *Optical Formulas and Their Applications*, the near and far limits of the depth of field (how close or how far away an object can be placed and still be in focus) are given by the following formulas:

    $$w_1 = \frac{u^2 (\tan \theta)}{L + u (\tan \theta)} \quad \text{and} \quad w_2 = \frac{u^2 (\tan \theta)}{L - u (\tan \theta)}.$$

    In these equations, θ represents the angle (in degrees) between the lens and the "circle of confusion," which is the circular image on the film of a point that is not exactly in focus. L is the diameter of the lens opening, and u is the distance to the object being photographed. Find the angle so that the near and far limits of the depth of field are approximately 2 m and 6 m respectively, when L is .00625 m and the object being photographed is 6 m from the camera.[†]

17. The amount of pollution in the air fluctuates with the seasons. It is lower after heavy spring rains and higher after periods of little rain. In addition to this seasonal fluctuation, the long-term trend is upward. An idealized graph of this situation is shown in the figure. Circular functions can be used to describe the fluctuating part of the pollution levels. Powers of the number e (the base of natural logarithms is e; to six decimal places, $e = 2.718282$) can be used to show the long-term growth. In fact, the pollution level in a certain area might be given by

    $$P(t) = 7(1 - \cos 2\pi t)(t + 10) + 100 e^{.2t},$$

    where t is time in years, with $t = 0$ representing January 1 of the base year. Thus, July 1 of the same year would be represented by $t = .5$, and October 1

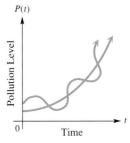

of the following year would be represented by $t = 1.75$. Find the pollution levels on the following dates:
    (a) January 1, base year
    (b) July 1, base year
    (c) January 1, following year
    (d) July 1, following year.

---

* From *Calculus for Life Sciences* by Rodolfo De Sapio. Copyright © 1978 by Rodolfo De Sapio. Reprinted by permission of the author.

† From "Optical Formulas and Their Applications," *Kodak Customer Service Pamphlet AA-26*. Reprinted by permission of Eastman Kodak Company.

18. The height in feet above ground of a weight hanging on a spring is given by

$$h = 100 + 10 \cos \left[\left(\tfrac{2\pi}{3} \right)(t - .03) \right],$$

where t is the number of seconds after the weight is released.

(a) Find h for $t = 0$, 1, and 2 sec.
(b) What are the maximum and minimum distances above ground? What is the period?

19. The graphs of equations of the form

$$y = \left(\tfrac{1}{n} \right) \cos(nt - \theta)$$

are called *harmonic waves* and are important in the study of music. Let
$y_1 = \cos(t - .5\pi)$, $y_2 = \left(\tfrac{1}{3} \right) \cos(3t - .5\pi)$,
$y_3 = \left(\tfrac{1}{5} \right) \cos(5t - .5\pi)$, $y_4 = \left(\tfrac{1}{7} \right) \cos(7t - .5\pi)$,
$y_5 = \left(\tfrac{1}{9} \right) \cos(9t - .5\pi)$. Graph each of the following harmonic waves in an appropriate window.
(a) y_1 (b) $y_1 + y_2$ (c) $y_1 + y_2 + y_3$
(d) $y_1 + y_2 + y_3 + y_4$
(e) $y_1 + y_2 + y_3 + y_4 + y_5$ (*Hint:* Just add the new term to the previous function each time.)

20. With your calculator in degree mode, follow these steps:
(a) Enter the year of your birth (all four digits).
(b) Subtract the number of years that have elapsed since 1980. For example, if it is 1996, subtract 16.
(c) Find the sine of the display.
(d) Find the inverse sine of the new display.
(e) What do you notice about the result in part (d)? (For an explanation of why this procedure works as it does, see "Sine of the Times: Your Age in a Flash" by E. John Hornsby, Jr., in the October 1985 issue of *Mathematics Teacher*.)

21. Suppose that an airplane flying faster than sound goes directly over you. Assume that the plane is flying level. At the instant that you feel the sonic

boom from the plane, the angle of elevation to the plane is given by

$$\alpha = 2 \arcsin \frac{1}{m},$$

where m is the Mach number of the plane's speed. (See Example 2.) Find α to the nearest degree for each of the following values of m.
(a) $m = 1.2$ (b) $m = 1.5$
(c) $m = 2$ (d) $m = 2.5$

22. In the study of alternating electric current, instantaneous voltage is given by

$$e = E_{\max} \sin 2\pi f t,$$

where f is the number of cycles per second, $E_{\max}$ is the maximum voltage, and t is time in seconds.
(a) Solve the equation for t.
(b) Find the smallest positive value of t if $E_{\max} = 12$, $e = 5$, and $f = 100$.

23. When a large-view camera is used to take a picture of an object that is not parallel to the film, the lens board should be tilted so that the planes containing the subject, the lens board, and the film intersect in a line. (See the figure.) This gives the best "depth of field."*

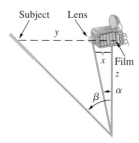

(a) Write two equations, one relating α, x, and z, and the other relating β, x, y, and z.
(b) Eliminate z from the equations in part (a) to get one equation relating α, β, x, and y.
(c) Solve the equation from part (b) for α.
(d) Solve the equation from part (b) for β.

24. One model for seasonal growth is $f(t) = 1000 e^{2 \sin t}$, where t is time in months. For what months in $[0, 12]$ is the growth equal to 4000 units? 0 units?

---

*Adapted from *A Sourcebook of Applications of School Mathematics* by Donald Bushaw et al. Copyright © 1980 by The Mathematical Association of America. Reprinted by permission.

Chapter 9   SUMMARY

The circular functions are defined in terms of real number arc lengths on a unit circle, a circle with center at the origin and radius 1. The circular functions are trigonometric functions with the argument in radians and $r = 1$ in the definitions. Thus, they have the same function values.

A set of points is defined parametrically by two equations $x = f(t)$ and $y = g(t)$, where t is the parameter. The unit circle can be graphed parametrically using $x = \cos t$ and $y = \sin t$.

The identities developed for the trigonometric functions also apply to the circular functions. The negative number identities give the function of a negative of a number in terms of the function of the number. One use of these identities, along with the reciprocal, quotient, and Pythagorean identities, is to express one circular (or trigonometric) function in terms of another. Verifying identities helps to develop this skill which is useful in more advanced work in mathematics.

The circular (trigonometric) functions are periodic—that is, the function values repeat over and over. Sine and cosine and their reciprocals have a period of 2π; tangent and cotangent have a period of π. The sine and cosine functions have domain $(-\infty, \infty)$ and range $[-1, 1]$. For both functions the amplitude is half the length of the range, 1. The sine function is symmetric with respect to the origin; the cosine function is symmetric with respect to the y-axis. Both functions have the same wavy graph, called a sinusoid. Each one is a horizontal shift of the other. The graphs of the circular functions can be transformed by stretching, shrinking, reflecting, and shifting. Horizontal shifts of the graphs of circular functions are called phase shifts.

The graphs of cosecant and secant are derived from the fact that they are the reciprocal functions of sine and cosine. Both graphs have vertical asymptotes where the graphs of sine or cosine, respectively, are zero. Both have a range of $(-\infty, -1] \cup [1, \infty)$. Cosecant is symmetric with respect to the origin; secant is symmetric with respect to the y-axis.

Because $\tan x = \frac{\sin x}{\cos x}$, its graph has x-intercepts when $\sin x = 0$ and vertical asymptotes when $\cos x = 0$. The fact that $\cot x$ is the reciprocal of $\tan x$ means that its graph has x-intercepts when $\cos x = 0$ and vertical asymptotes when $\sin x = 0$. Both functions have graphs with a range $(-\infty, \infty)$. Both have graphs that are symmetric with respect to the origin. These graphs can be transformed in the same way as the graphs of other functions.

Equations involving circular functions that are linear or quadratic in, for example, $\sin x$, can be solved with the usual linear or quadratic methods. Their corresponding inequalities can then be solved by observing where the graph is above or below the x-axis.

Identities for the sum or difference of two numbers, such as $\cos(A - B)$, twice a number, such as $\sin 2A$, or half a number, such as $\tan \frac{A}{2}$, are derived algebraically using properties from geometry and identities given earlier in this book. These identities are used to find exact function values for certain real numbers. For example, the identities are used to find exact values for $\sin(\frac{\pi}{8})$, using the fact that $\frac{\pi}{8}$ is half of $\frac{\pi}{4}$, or $\cos(\frac{7\pi}{12})$, from the fact that $\frac{7\pi}{12} = \frac{\pi}{3} + \frac{\pi}{4}$. These new identities are also used in verifying more complicated identities.

The double number and half number identities often make it possible to rewrite equations involving circular functions in a different form so that they can be solved by earlier methods. Equations or inequalities that cannot be solved analytically may be solved graphically using methods introduced in earlier chapters.

The inverse circular functions are found, like other inverse functions, by reversing the roles of x and y. For example, $y = \tan^{-1} x$ means that $x = \tan y$ for an appropriate range of y-values. The range of $\sin^{-1} x$ is $[-\frac{\pi}{2}, \frac{\pi}{2}]$, the range of $\cos^{-1} x$ is $[0, \pi]$, and the range of $\tan^{-1} x$ is $(-\frac{\pi}{2}, \frac{\pi}{2})$. The domain and range of the inverse circular functions are part of their definitions and must be memorized. The graph of an inverse circular function is found by reflecting the graph of the corresponding circular function, over a suitably restricted domain, across the line $y = x$.

Key Terms

SECTION 9.1

circular functions
unit circle
parametric equations
parameter
negative number identities

SECTION 9.2

periodic function
period
sine wave or sinusoid
amplitude
phase shift

SECTION 9.4

sum and difference identities
double number identities
half number identities

SECTION 9.6

inverse sine ($\sin^{-1}$)
arcsin x
inverse cosine ($\cos^{-1}$)
arccos x
inverse tangent ($\tan^{-1}$)
arctan x
inverse cotangent
inverse secant
inverse cosecant

C*hapter 9*   REVIEW EXERCISES

Find the exact value of each of the following.

1. $\tan \dfrac{\pi}{3}$

2. $\cos \dfrac{2\pi}{3}$

3. $\sin\left(-\dfrac{5\pi}{6}\right)$

4. $\cot \dfrac{11\pi}{6}$

5. $\tan\left(-\dfrac{7\pi}{3}\right)$

6. $\sec \dfrac{\pi}{3}$

7. $\csc\left(-\dfrac{11\pi}{6}\right)$

8. $\cot\left(-\dfrac{17\pi}{3}\right)$

Find a calculator approximation of each of the following.

9. $\sin 1.0472$

10. $\tan 7.3159$

11. $\sec .4864$

12. $\csc(-.8385)$

Use the negative number identities to write each of the following as a function of a positive number.

13. $\cos(-3)$

14. $\sin(-3)$

15. $\tan(-3)$

16. $\sec(-3)$

17. $\csc(-3)$

18. $\cot(-3)$

For each item in Column I, give the letter of the item in Column II that completes an identity.

Column I	*Column II*
19. $\sec x$	**(a)** $\dfrac{1}{\sin x}$
20. $\csc x$	**(b)** $\dfrac{1}{\cos x}$
21. $\tan x$	**(c)** $\dfrac{\sin x}{\cos x}$
22. $\cot x$	**(d)** $\dfrac{1}{\cot^2 x}$
23. $\sin^2 x$	**(e)** $\dfrac{1}{\cos^2 x}$
24. $\tan^2 x + 1$	**(f)** $\dfrac{\cos x}{\sin x}$
25. $\tan^2 x$	**(g)** $\dfrac{1}{\sin^2 x}$
	(h) $1 - \cos^2 x$

26. Graph the function $f(x) = \frac{1 - \cos 2x}{\sin 2x}$ in the trig viewing window of your calculator. It is equivalent to one of the six basic circular functions. Identify it, and complete this identity:

$$\frac{1 - \cos 2x}{\sin 2x} = \underline{\hspace{2cm}}.$$

Then verify the identity using analytic methods.

Consider the function $y = -2\sin(3x + \frac{\pi}{4}) + 5.$

27. What is the domain of this function?

28. What is its range?

29. Find the amplitude.

30. Find the period of the function.

31. What is the phase shift?

32. What is the vertical translation?

33. Use a calculator to graph the function in the window $[0, 2\pi]$ by $[0, 8]$.

Identify the one of the six circular functions that satisfies the description.

34. Period is π, x intercepts are of the form $n\pi$, where n is an integer

35. Period is 2π, passes through the origin

36. Period is 2π, passes through the point $(\frac{\pi}{2}, 0)$

37. Period is 2π, domain is $\{x \mid x \neq n\pi$, where n is an integer$\}$

38. Period is π, function is decreasing on the interval $0 < x < \pi$

39. Period is 2π, has vertical asymptotes of the form $x = (2n + 1)\frac{\pi}{2}$, where n is an integer

Consider the function $f(x) = \cos^2 x \sin x - \sin x.$

40. Solve $f(x) = 0$ analytically over the interval $[0, 2\pi)$, and give solutions in exact form.

41. Graph $y = f(x)$ in the window $[0, 2\pi]$ by $[-4, 4]$.

42. **(a)** Use the results of Exercise 40 and the graph in Exercise 41 to give the exact solution set of $f(x) > 0$ over $[0, 2\pi)$.
    (b) Repeat part (a) for $f(x) < 0$.

43. Use the quadratic formula to solve the equation $2\sin^2 x - 3\sin x - 3 = 0$ over the interval $[0, 2\pi)$. Express solutions to the nearest hundredth. Support your answer with a graph.

Suppose that $\sin A = -\frac{3}{5}$ *and* $\sin B = \frac{12}{13}$, *with* $\pi < A < \frac{3\pi}{2}$ *and* $0 < B < \frac{\pi}{2}$. *Use an appropriate identity to find the exact value of each of the following.*

44. $\sin(A + B)$ 45. $\cos(A + B)$ 46. $\tan(A + B)$

47. $\sin(A - B)$ 48. $\cos(A - B)$ 49. $\tan(A - B)$

50. $\sin 2A$ 51. $\cos 2B$ 52. $\tan 2A$

53. $\sin \dfrac{A}{2}$ 54. $\cos \dfrac{A}{2}$ 55. $\tan \dfrac{A}{2}$

Solve each equation over the interval $[0, 2\pi)$. *When a solution cannot be easily expressed as a rational multiple of* π, *give an approximation to the nearest hundredth.*

56. $\sin 2x = \cos 2x + 1$ 57. $2\sin 2x = 1$

58. $\sin 2x + \sin 4x = 0$

59. $\cos x - \cos 2x = 2 \cos x$

60. $\tan 2x = \sqrt{3}$

61. $\cos^2 \dfrac{x}{2} - 2 \cos \dfrac{x}{2} + 1 = 0$

62. Use the results of Exercise 56 and a calculator graph of the function $f(x) = \sin 2x - \cos 2x - 1$ to find the solution set of: **(a)** $\sin 2x > \cos 2x + 1$ and **(b)** $\sin 2x < \cos 2x + 1$. Give the solution set over the interval $[0, 2\pi)$.

Verify analytically each of the following identities.

63. $\sin^2 x - \sin^2 y = \cos^2 y - \cos^2 x$

64. $2 \cos^3 x - \cos x = \dfrac{\cos^2 x - \sin^2 x}{\sec x}$

65. $-\cot \dfrac{x}{2} = \dfrac{\sin 2x + \sin x}{\cos 2x - \cos x}$

66. $\dfrac{\sin^2 x}{2 - 2 \cos x} = \cos^2 \dfrac{x}{2}$

67. $\dfrac{\sin 2x}{\sin x} = \dfrac{2}{\sec x}$

68. $2 \cos A - \sec A = \cos A - \dfrac{\tan A}{\csc A}$

69. $\dfrac{2 \tan B}{\sin 2B} = \sec^2 B$

70. $\tan 4\theta = \dfrac{2 \tan 2\theta}{2 - \sec^2 2\theta}$

71. $1 + \tan^2 \alpha = 2 \tan \alpha \csc 2\alpha$

72. $\dfrac{\sin t}{1 - \cos t} = \cot \dfrac{t}{2}$

73. $\sin 2\alpha = \dfrac{2(\sin \alpha - \sin^3 \alpha)}{\cos \alpha}$

74. $\dfrac{2 \cot x}{\tan 2x} = \csc^2 x - 2$

75. $\tan \theta \cos^2 \theta = \dfrac{2 \tan \theta \cos^2 \theta - \tan \theta}{1 - \tan^2 \theta}$

76. $\tan \theta \sin 2\theta = 2 - 2 \cos^2 \theta$

77. $2 \tan x \csc 2x - \tan^2 x = 1$

78. $2 \sin^3 x - \sin x = \dfrac{\sin^2 x - \cos^2 x}{\csc x}$

Give the exact value of each of the following.

79. $\sin^{-1} \dfrac{\sqrt{2}}{2}$

80. $\arccos\left(-\dfrac{1}{2}\right)$

81. $\tan^{-1}(-\sqrt{3})$

82. $\arcsin(-1)$

83. $\cos^{-1}\left(-\dfrac{\sqrt{2}}{2}\right)$

84. $\arctan \dfrac{\sqrt{3}}{3}$

85. $\sec^{-1}(-2)$

86. $\operatorname{arccsc} \dfrac{2\sqrt{3}}{3}$

87. $\cot^{-1}(-1)$

88. Explain why $\sin^{-1}(-3)$ is not defined.

Find each of the following without the use of a calculator.

89. $\sin\left(\sin^{-1} \dfrac{1}{2}\right)$

90. $\sin\left(\cos^{-1} \dfrac{3}{4}\right)$

91. $\cos(\arctan 3)$

92. $\sec\left(2 \sin^{-1}\left(-\dfrac{1}{3}\right)\right)$

93. $\cos^{-1}\left(\cos \dfrac{3\pi}{2}\right)$

94. $\tan\left(\sin^{-1} \dfrac{3}{5} + \cos^{-1} \dfrac{5}{7}\right)$

Write each of the following as an algebraic expression in u, u > 0.

95. $\sin(\tan^{-1} u)$

96. $\cos\left(\arctan \dfrac{u}{\sqrt{1 - u^2}}\right)$

97. $\tan\left(\arccos \dfrac{u}{\sqrt{u^2 + 1}}\right)$

98. The figure shows the population of lynx and hares in Canada for the years 1847–1903. The hares are food for the lynx. An increase in hare population causes an increase in lynx population some time later. The increasing lynx population then causes a decline in hare population.
   (a) Estimate the length of one period.
   (b) Estimate maximum and minimum hare populations.

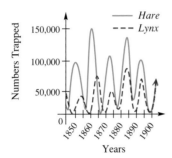

99. Many computer languages such as BASIC and FORTRAN have only the arctangent function available. To use the other inverse trigonometric functions, it is necessary to express them in terms of arctangent. This can be done as follows.
   (a) Let $u = \arcsin x$. Solve the equation for x in terms of u.
   (b) Use the result of part (a) to label the three sides of the triangle of the figure in terms of x.
   (c) Use the triangle from part (b) to write an equation for $\tan u$ in terms of x.
   (d) Solve the equation from (c) for u.

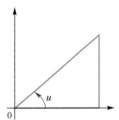

100. A riverboat's main paddle has a 30-ft radius and revolves at 20 revolutions per minute. The axis of the paddle is 10 ft above the surface of the water. (See the figure at the top of the next column.) Find a function for the height above (or below) water level of the tip of the paddle, if the paddle points straight down at $t = 0$.

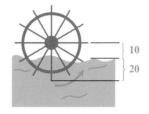

101. Suppose that the equation below describes the motion formed by a rhythmically moving arm:

$$y = \frac{1}{3} \sin \frac{4\pi t}{3}.$$

Here t is time (in seconds) and y is the angle formed.
   (a) Solve the equation for t.
   (b) At what time(s) does the arm form an angle of .3 radian?

102. Exact values of the trigonometric functions of 15° can be found by the following method, an alternative to the use of the half number formulas. Start with a right triangle ABC having a 60° angle at A and a 30° angle at B. Let the hypotenuse of this triangle have length 2. Extend side BC and draw a semicircle with diameter along BC extended, center at B, and radius AB. Draw segment AE. (See the figure.) Since any angle inscribed in a semicircle is a right angle, triangle AED is a right triangle.

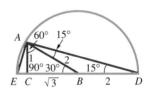

Prove each of the following statements.
   (a) Triangle ABD is isosceles.
   (b) Angle ABD is 150°.
   (c) Angle DAB and angle ADB are each 15°.
   (d) DC has length $2 + \sqrt{3}$.
   (e) Since AC has length 1, the length of AD is $AD = \sqrt{1^2 + (2 + \sqrt{3})^2}$. Reduce this to $\sqrt{8 + 4\sqrt{3}}$, and show that this result equals $\sqrt{6} + \sqrt{2}$.
   (f) Use angle ADB of triangle ADE and find $\cos 15°$.
   (g) Show that AE has length $\sqrt{6} - \sqrt{2}$. Then find $\sin 15°$.
   (h) Use triangle ACE and find $\tan 15°$.

The study of alternating electric current requires the solutions of equations of the form
$i = I_{max} \sin 2\pi ft$, *for time t in seconds, where i is instantaneous current in amperes,*
I_{max} *is maximum current in amperes, and f is the number of cycles per second. Find the*
smallest positive value of t, given the following data.

103. $i = 40, I_{max} = 100, f = 60$

104. $i = 50, I_{max} = 100, f = 120$

For each of the graphs in Exercises 105–108 give the equation of a sine function having
that graph.

105.

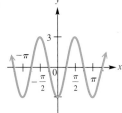

106.

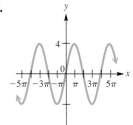

107.

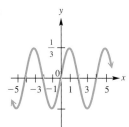

108.

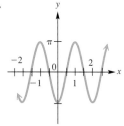

Give the equation of a cosine function having the graph of the figure.

109. Exercise 105

110. Exercise 106

111. A painting 3 ft high and 6 ft from the floor will cut off an angle θ to an observer, where

$$\theta = \tan^{-1}\left(\frac{x}{x^2 + 2}\right).$$

Assume that the observer is x ft from the wall where the painting is displayed and that the eyes of the observer are 5 ft above the ground. See the figure. Rounding to the nearest degree, find the value of θ for the following values of x:
(a) 1 **(b)** 2 **(c)** 3.

112. The slope of a line is defined as the ratio of the vertical change and the horizontal change. As shown in the figure, the tangent of the *angle of inclination* θ is given by the ratio of the side opposite and the side adjacent. This ratio is the same as that used in finding the slope, m, so that $m = \tan \theta$. In the figure on the right, let the two lines have angles of inclination α and β, and slopes m_1 and m_2, respectively. Let θ be the smallest positive angle between the lines. Show that

$$\tan \theta = \frac{m_2 - m_1}{1 + m_1 m_2}.$$

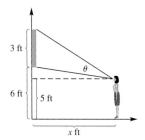

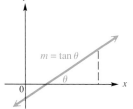

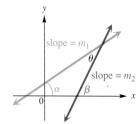

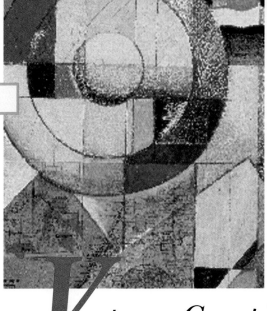

CHAPTER **10**

Vectors, Complex Numbers, and Polar and Parametric Equations

10.1 VECTORS AND APPLICATIONS

Basic Terminology ▮ Direction Angle and Components ▮ Operations with Vectors ▮ Applications of Vectors

Basic Terminology

Many quantities in mathematics involve magnitudes, such as 45 lb or 60 mph. These quantities are called **scalars.** Other quantities, called **vector quantities,** involve both magnitude and direction. Typical vector quantities are velocity, acceleration, and force.

A vector quantity is often represented with a directed line segment, called a **vector.** The length of the vector represents the magnitude of the vector quantity. The direction of the vector, indicated with an arrowhead, represents the direction of the quantity. For example, the vector in Figure 1 represents a force of 10 lb applied at an angle of 30° from the horizontal.

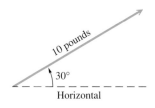

FIGURE 1

The symbol for a vector is often printed in boldface type. To write vectors by hand, it is customary to use an arrow over the letter or letters. Thus **OP** and $\overrightarrow{OP}$ both represent vector **OP**. Vectors may be named with either one lowercase or uppercase letter, or two uppercase letters. When two letters are used, the first indicates the *initial point* and the second indicates the *terminal point* of the vector. Knowing these points gives the direction of the vector. For example, vectors **OP** and **PO** in Figure 2 are not the same vectors. They have the same magnitude, but opposite directions. The magnitude of vector **OP** is written $|\mathbf{OP}|$.

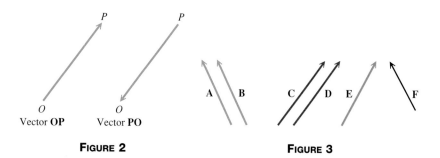

FIGURE 2	**FIGURE 3**

Two vectors are *equal* if and only if they both have the same directions and the same magnitudes. In Figure 3, vectors **A** and **B** are equal, as are vectors **C** and **D**.

To find the **sum** of two vectors **A** and **B**, we place the initial point of vector **B** at the terminal point of vector **A**, as shown in Figure 4. The vector with the same initial point as **A** and the same terminal point as **B** is the sum **A** + **B**. The sum of two vectors is also a vector.

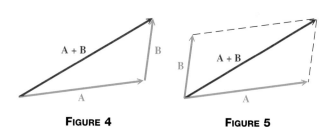

FIGURE 4	**FIGURE 5**

Another way to find the sum of two vectors is to use the **parallelogram rule.** Place vectors **A** and **B** so that their initial points coincide. Then complete a parallelogram that has **A** and **B** as two sides. The diagonal of the parallelogram with the same initial point as **A** and **B** is the same vector sum **A** + **B** found by the definition. Compare Figures 4 and 5.

Parallelograms can be used to show that vector **B** + **A** is the same as vector **A** + **B**, or that

$$\mathbf{A} + \mathbf{B} = \mathbf{B} + \mathbf{A},$$

so that vector addition is *commutative.*

The vector sum **A** + **B** is called the **resultant** of vectors **A** and **B**. Each of the vectors **A** and **B** is a **component** of vector **A** + **B**. In many practical applications, such as surveying, it is necessary to break a vector into its **vertical** and **horizontal components.** These components are two vectors, one vertical and one horizontal,

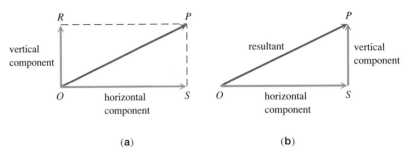

(a)   (b)

FIGURE 6

whose resultant is the original vector. As shown in Figure 6(a) vector **OR** is the vertical component and vector **OS** is the horizontal component of **OP**. Figure 6(b) shows an alternative method of indicating a resultant.

For every vector **v** there is a vector −**v** that has the same magnitude as **v** but opposite direction. Vector −**v** is called the **opposite** of **v**. (See Figure 7.) The sum of **v** and −**v** has magnitude 0 and is called the **zero vector.** As with real numbers, to *subtract* vector **B** from vector **A**, find the vector sum **A** + (−**B**). (See Figure 8.)

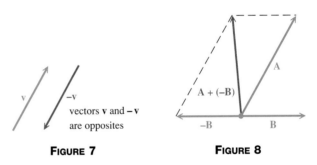

FIGURE 7   **FIGURE 8**

The **scalar product** of a real number (or scalar) k and a vector **u** is the vector $k \cdot \mathbf{u}$ which has magnitude $|k|$ times the magnitude of **u**. As suggested by Figure 9, the vector $k \cdot \mathbf{u}$ has the same direction as **u** if $k > 0$, and opposite direction if $k < 0$.

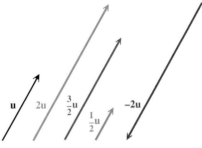

FIGURE 9

EXAMPLE 1

Finding the
Magnitude of
a Resultant

Two forces of 15 and 22 newtons (a *newton* is a unit of force used in physics) act on a point in the plane. If the angle between the forces is 100°, find the magnitude of the resultant force.

SOLUTION   As shown in Figure 10, a parallelogram that has the forces as adjacent sides can be formed. The angles of the parallelogram adjacent to angle *P* each

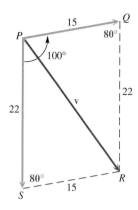

FIGURE 10

measure 80°, since adjacent angles of a parallelogram are supplementary. Opposite sides of the parallelogram are equal in length. The resultant force divides the parallelogram into two triangles. Use the law of cosines with either triangle to get

$$
\begin{aligned}
|\mathbf{v}|^2 &= 15^2 + 22^2 - 2(15)(22)\cos 80^\circ \\
&\approx 225 + 484 - 115 \\
|\mathbf{v}|^2 &\approx 594 \\
|\mathbf{v}| &\approx 24.
\end{aligned}
$$

To the nearest unit, the magnitude of the resultant force is 24 newtons. ◧

FIGURE 11

TECHNOLOGICAL NOTE
Many graphics calculators have built-in functions that will allow the user to find the horizontal and vertical components when the magnitude and direction angle are known, and also give the magnitude and direction angle when the horizontal and vertical components are known. The procedures for activating these functions vary greatly among models, and you should consult your owner's manual to see whether your model has these functions and how to use them if it does.

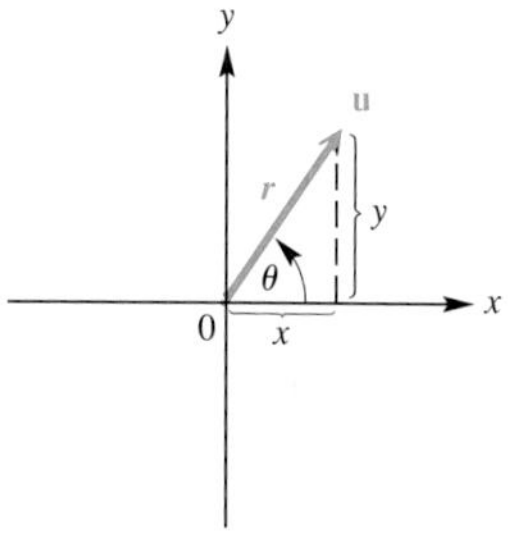

Direction Angle and Components

If a vector is placed so that its initial point coincides with the origin of a rectangular coordinate system, then the angle between the x-axis and the vector, measured in a counterclockwise direction, is called a **direction angle** for the vector. In Figure 11, **u** has direction angle θ and magnitude r. The following basic results for vectors are derived from the definition of direction angle and earlier results.

BASIC RULES FOR VECTORS

Let a vector have direction angle θ and magnitude r. Then the horizontal component of the vector has magnitude

$$x = r\cos\theta,$$

and the vertical component has magnitude

$$y = r\sin\theta.$$

Also,

$$x^2 + y^2 = r^2 \quad \text{and} \quad \tan\theta = \frac{y}{x}, \quad x \neq 0.$$

EXAMPLE 2

Finding Horizontal and Vertical Components

Vector **w** has magnitude 25.0 and direction angle 41.7°. Find the magnitudes of the horizontal and vertical components of the vector.

SOLUTION   In Figure 12 the vertical component is labeled **v** and the horizontal component is labeled **u**. Using the basic rules for vectors,

$$|\mathbf{v}| = 25.0 \sin 41.7° \approx 16.6,$$

and $$|\mathbf{u}| = 25.0 \cos 41.7° \approx 18.7.$$

To the nearest tenth, the horizontal component is 18.7 and the vertical component is 16.6.   ◼

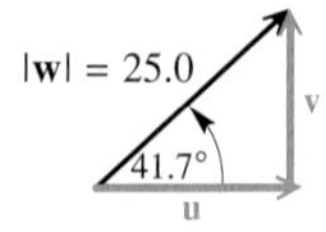

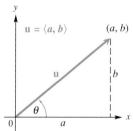

FIGURE 12

FOR GROUP DISCUSSION

Modern graphics calculators are capable of finding resultants, direction angles, and horizontal and vertical components. Refer to your instruction manual to learn how to find (a) the magnitude of the resultant and the direction angle if the horizontal and vertical components are known and (b) the horizontal and vertical components if the magnitude of the resultant and the direction angle are known. To determine whether you have learned how to do these correctly, verify the following statements.

1.   If the magnitude of the resultant is 3.5 and the direction angle is 150°, the horizontal component is −3.0380740392 and the vertical component is 1.75. (Be sure that your calculator is in degree mode.)
2.   If the horizontal component is 4.7 and the vertical component is 7.6, the magnitude of the resultant is 8.935882721 and the direction angle is 58.26648085°. (Again, be sure that your calculator is in degree mode.)

Let vector **u** be placed in a plane so that the initial point of the vector is at the origin, (0, 0), and the endpoint is at the point (a, b). A vector with initial point at the origin is called a **position vector** or (sometimes) a **radius vector.** A position vector having endpoint at the point (a, b) is called the **vector** (a, b). To avoid confusion, the vector (a, b) is written as $\langle a, b \rangle$. The numbers a and b are called the **x-component** and **y-component,** respectively. Figure 13 shows the vector **u** = $\langle a, b \rangle$.

FIGURE 13

LENGTH OR MAGNITUDE OF A VECTOR

The length or magnitude of vector **u** = $\langle a, b \rangle$ is given by

$$|\mathbf{u}| = \sqrt{a^2 + b^2}.$$

EXAMPLE **3**

Finding Magnitude
and Direction Angle

Figure 14 shows vector **u** $= \langle 3, -2 \rangle$. Find the magnitude and direction angle for **u.**

SOLUTION   The magnitude is

$$\sqrt{3^2 + (-2)^2} = \sqrt{13}.$$

To find the direction angle θ, start with

$$\tan \theta = \frac{y}{x} = \frac{-2}{3} = -\frac{2}{3}.$$

Vector **u** has positive x-component and negative y-component, placing the vector in quadrant IV. A graphics calculator gives $\tan^{-1}(-\frac{2}{3}) \approx -33.7°$. Adding 360° yields the direction angle 326.3°. This angle is shown in Figure 14. ◼

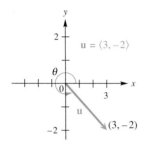

FIGURE 14

Operations with Vectors

Let vector **OM** in Figure 15 be given by $\langle a, b \rangle$, and vector **ON** be given by $\langle c, d \rangle$. Let **OP** be given by $\langle a + c, b + d \rangle$. With facts from geometry points $O, N, M,$ and P can be shown to form the vertices of a parallelogram. Since a diagonal of this parallelogram gives the resultant of **OM** and **ON**, vector **OP** is given by **OP** = **OM** + **ON,** with the resultant of $\langle a, b \rangle$ and $\langle c, d \rangle$ given by $\langle a + c, b + d \rangle$. In the same way, $k \cdot \langle a, b \rangle = \langle ka, kb \rangle$ for any real number k.

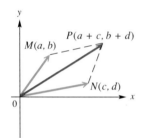

FIGURE 15

VECTOR OPERATIONS

For any real numbers $a, b, c, d,$ and k,

$$\langle a, b \rangle + \langle c, d \rangle = \langle a + c, b + d \rangle$$
$$k \cdot \langle a, b \rangle = \langle ka, kb \rangle.$$

EXAMPLE **4**

Performing Vector
Operations

Let **u** $= \langle -2, 1 \rangle$ and **v** $= \langle 4, 3 \rangle$. Find each of the following: **(a) u** + **v, (b)** -2**u,** **(c)** 4**u** + 3**v.** (See Figure 16.)

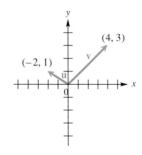

FIGURE 16

SOLUTION

(a)   **u** + **v** $= \langle -2, 1 \rangle + \langle 4, 3 \rangle = \langle -2 + 4, 1 + 3 \rangle = \langle 2, 4 \rangle$

(b) -2**u** $= -2 \cdot \langle -2, 1 \rangle = \langle -2(-2), -2(1) \rangle = \langle 4, -2 \rangle$

(c) 4**u** + 3**v** $= 4 \cdot \langle -2, 1 \rangle + 3 \cdot \langle 4, 3 \rangle = \langle -8, 4 \rangle + \langle 12, 9 \rangle$

$$= \langle -8 + 12, 4 + 9 \rangle = \langle 4, 13 \rangle \quad ◼$$

For a vector **u** with magnitude r and direction angle θ, it was shown above that the horizontal and vertical components of **u** have magnitudes

$$|\mathbf{x}| = r \cos \theta \quad \text{and} \quad |\mathbf{y}| = r \sin \theta.$$

This leads to the following result.

If a vector **u** has direction angle θ and magnitude r, then

$$\mathbf{u} = \langle r \cos \theta, r \sin \theta \rangle.$$

EXAMPLE 5

Writing Vectors in
the Form ⟨a, b⟩

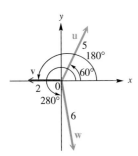

FIGURE 17

Write the vectors in Figure 17 in the form $\langle a, b \rangle$.

SOLUTION   Vector **u** in Figure 17 has a magnitude of 5 and a direction angle 60°. By the result above,

$$\mathbf{u} = \langle 5 \cos 60°, 5 \sin 60° \rangle = \left\langle 5 \cdot \frac{1}{2}, 5 \cdot \frac{\sqrt{3}}{2} \right\rangle = \left\langle \frac{5}{2}, \frac{5\sqrt{3}}{2} \right\rangle.$$

Also, $\mathbf{v} = \langle 2 \cos 180°, 2 \sin 180° \rangle = \langle 2(-1), 2(0) \rangle = \langle -2, 0 \rangle.$

Finally, $\mathbf{w} = \langle 6 \cos 280°, 6 \sin 280° \rangle$

or $\mathbf{w} \approx \langle 1.0419, -5.9088 \rangle.$ ∎

FOR GROUP DISCUSSION

Use your calculator's capability to work with vectors to support the answers found in Example 5. Note that for **u**, the calculator will give a decimal approximation for the vertical component. Verify that what the calculator gives is the same result you get if you use the calculator to approximate $\frac{5\sqrt{3}}{2}$. Be sure your calculator is in degree mode.

Applications of Vectors

The law of sines and the law of cosines, introduced in Chapter 8, are often useful in solving applied problems involving vectors. You may wish to review them at this time.

Navigation problems, such as the one in the following example, can often be solved by using vectors.

EXAMPLE 6

Applying Vectors to
a Navigation
Problem

A ship leaves port on a bearing of 28° and travels 8.2 miles. The ship then turns due east and travels 4.3 miles. How far is the ship from port? What is its bearing from port?

SOLUTION   In Figure 18, vectors **PA** and **AE** represent the ship's path. The magnitude and bearing of the resultant **PE** can be found as follows. Triangle PNA is a right triangle, so angle $NAP = 90° - 28° = 62°$. Then angle $PAE = 180° - 62° = 118°$. Use the law of cosines to find $|\mathbf{PE}|$, the magnitude of vector **PE**.

$$|\mathbf{PE}|^2 = 8.2^2 + 4.3^2 - 2(8.2)(4.3) \cos 118°$$

$$|\mathbf{PE}|^2 \approx 118.84$$

Therefore, $|\mathbf{PE}| \approx 10.9,$

or 11 miles, rounded to two significant digits.

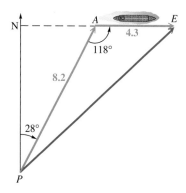

FIGURE 18

To find the bearing of the ship from port, first find angle *APE*. Use the law of sines, along with the value of $|\mathbf{PE}|$ before rounding.

$$\frac{\sin APE}{4.3} = \frac{\sin 118°}{10.9}$$

$$\sin APE = \frac{4.3 \sin 118°}{10.9}$$

$$\text{angle } APE \approx 20.4°$$

After rounding, angle *APE* is 20°, and the ship is 11 miles from port on a bearing of 28° + 20° = 48°.   ◼

In air navigation, the **airspeed** of a plane is its speed relative to the air, while the **groundspeed** is its speed relative to the ground. Because of wind, these two speeds are usually different. The groundspeed of the plane is represented by the vector sum of the airspeed and windspeed vectors. See Figure 19.

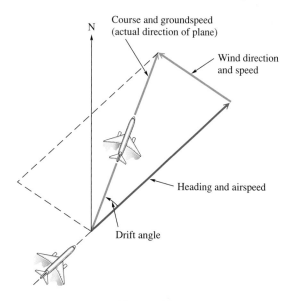

FIGURE 19

EXAMPLE **7**
Applying Vectors to a Navigation Problem

A plane with an airspeed of 192 miles per hour is headed on a bearing of 121°. A north wind is blowing (from north to south) at 15.9 miles per hour. Find the groundspeed and the actual bearing of the plane.

SOLUTION In Figure 20, the groundspeed is represented by $|\mathbf{x}|$. We must find angle α to find the bearing, which will be $121° + \alpha$. From Figure 20, angle BCO equals angle AOC, which equals 121°. Find $|\mathbf{x}|$ by the law of cosines.

$$|\mathbf{x}|^2 = 192^2 + 15.9^2 - 2(192)(15.9) \cos 121°$$
$$|\mathbf{x}|^2 \approx 40{,}261$$

Therefore, $\quad |\mathbf{x}| \approx 200.7,$

or 201 miles per hour. Now find α by using the law of sines. As before, use the value of $|\mathbf{x}|$ before rounding.

$$\frac{\sin \alpha}{15.9} = \frac{\sin 121°}{200.7}$$
$$\sin \alpha \approx .06792320$$
$$\alpha \approx 3.89°$$

After rounding, α is 3.9°. The groundspeed is about 201 miles per hour, on a bearing of approximately 125°. ◖

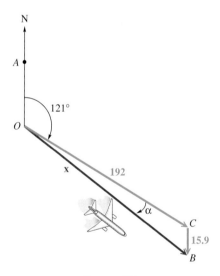

FIGURE 20

10.1 EXERCISES

Exercises 1–4 on the next page refer to the following vectors.

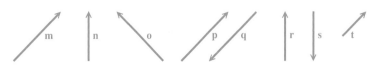

1. Name all pairs of vectors that appear to be equal.

2. Name all pairs of vectors that are opposites.

3. Name all pairs of vectors where the first is a scalar multiple of the other, with the scalar positive.

4. Name all pairs of vectors where the first is a scalar multiple of the other, with the scalar negative.

Exercises 5–22 refer to the vectors pictured. Make a copy or a careful sketch of each, and then draw a sketch to represent the following vectors. For example, find **a** + **e** *by placing* **a** *and* **e** *so that their initial points coincide. Then use the parallelogram rule to find the resultant, shown in the figure.*

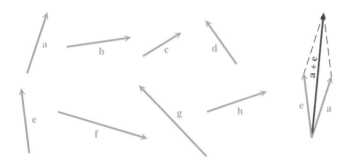

5. −**b**	**6.** −**g**	**7.** 3**a**	**8.** 2**h**	**9.** **a** + **c**
10. **a** + **b**	**11.** **h** + **g**	**12.** **e** + **f**	**13.** **a** + **h**	**14.** **b** + **d**
15. **h** + **d**	**16.** **a** + **f**	**17.** **a** − **c**	**18.** **d** − **e**	**19.** **a** + (**b** + **c**)
20. (**a** + **b**) + **c**	**21.** **c** + **d**	**22.** **d** + **c**		

23. From the results of Exercises 19 and 20, do you think vector addition is associative?

24. From the results of Exercises 21 and 22, do you think vector addition is commutative?

For each pair of vectors **u** *and* **w** *with angle θ between them, sketch the resultant.*

25. $|\mathbf{u}| = 12, |\mathbf{w}| = 20, \theta = 27°$ **26.** $|\mathbf{u}| = 8, |\mathbf{w}| = 12, \theta = 20°$ **27.** $|\mathbf{u}| = 20, |\mathbf{w}| = 30, \theta = 30°$

28. $|\mathbf{u}| = 27, |\mathbf{w}| = 50, \theta = 12°$ **29.** $|\mathbf{u}| = 50, |\mathbf{w}| = 70, \theta = 40°$

For each of the following, vector **v** *has the given magnitude and makes an angle θ with the horizontal. Find the horizontal and vertical components of* **v**, *first using the relationships* $x = r \cos θ$ *and* $y = r \sin θ$. *Then use the capabilities of your calculator to support your answers.*

30. $|\mathbf{v}| = 50, \theta = 20°$ **31.** $|\mathbf{v}| = 12, \theta = 38°$ **32.** $|\mathbf{v}| = 150, \theta = 70°$

33. $|\mathbf{v}| = 26, \theta = 50°$ **34.** $|\mathbf{v}| = 47.8, \theta = 35°50'$ **35.** $|\mathbf{v}| = 15.4, \theta = 27.5°$

36. $|\mathbf{v}| = 78.9, \theta = 59°40'$ **37.** $|\mathbf{v}| = 198, \theta = 128.5°$ **38.** $|\mathbf{v}| = 238, \theta = 146.3°$

In each of the following, two forces act at a point in the plane. The angle between the two forces is given. Find the magnitude of the resultant force.

39. Forces of 250 and 450 newtons, forming an angle of 85°

40. Forces of 19 and 32 newtons, forming an angle of 118°

41. Forces of 17.9 and 25.8 pounds, forming an angle of 105°30'

42. Forces of 75.6 and 98.2 pounds, forming an angle of 82°50'

Solve each of the following problems.

43. A force of 176 lb makes an angle of 78°50′ with a second force. The resultant of the two forces makes an angle of 41°10′ with the first force. Find the magnitude of the second force and of the resultant.

44. A force of 28.7 lb makes an angle of 42°10′ with a second force. The resultant of the two forces makes an angle of 32°40′ with the first force. Find the magnitudes of the second force and of the resultant.

45. A plane flies 650 mph on a bearing of 175.3°. A 25 mph wind, from a direction of 266.6°, blows against the plane. Find the resulting bearing of the plane.

46. A pilot wants to fly on a bearing of 74.9°. By flying due east, he finds that a 42 mph wind, blowing from the south, puts him on course. Find the airspeed and the groundspeed.

47. Starting at point *A*, a ship sails 18.5 km on a bearing of 189°, then turns and sails 47.8 km on a bearing of 317°. Find the distance of the ship from point *A*.

48. Two towns 21 mi apart are separated by a dense forest. (See the figure.) To travel from town *A* to town *B*, a person must go 17 mi on a bearing of 325°, then turn and continue for 9 mi to reach town *B*. Find the bearing of *B* from *A*.

49. An airline route from San Francisco to Honolulu is on a bearing of 233°. A jet flying at 450 mph on that bearing flies into a wind blowing at 39 mph from a direction of 114°. Find the resulting bearing and groundspeed of the plane.

50. A pilot is flying at 168 mph. She wants her flight path to be on a bearing of 57°40′. A wind is blowing from the south at 27.1 mph. Find the bearing the pilot should fly, and find the plane's groundspeed.

51. What bearing and airspeed are required for a plane to fly 400 mi due north in 2.5 hr if the wind is blowing from a direction of 328° at 11 mph?

52. A plane is headed due south with an airspeed of 192 mph. A wind from a direction of 78° is blowing at 23 mph. Find the groundspeed and resulting bearing of the plane.

53. An airplane is headed on a bearing of 174° at an airspeed of 240 km per hr. A 30 km per hr wind is blowing from a direction of 245°. Find the groundspeed and resulting bearing of the plane.

54. A ship sailing due east in the North Atlantic has been warned to change course to avoid a group of icebergs. The captain turns and sails on a bearing of 62° for a while, then changes course again to a bearing of 115° until the ship reaches its original course. (See the figure.) How much farther did the ship have to travel to avoid the icebergs?

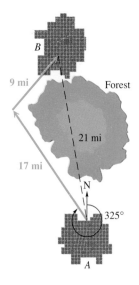

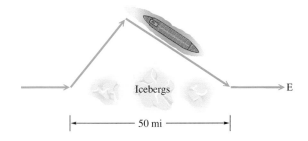

Relating Concepts

Consider the two vectors **v** *and* **u** *shown. Assume all values are exact. Work Exercises 55–60 in order.*

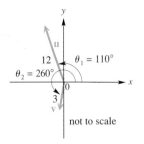

55. Use trigonometry alone (without using vector notation) to find the magnitude and direction angle of **u** + **v**. You should use the law of cosines and the law of sines in your work.

56. Find the horizontal and vertical components of **u**, using your calculator.

57. Find the horizontal and vertical components of **v**, using your calculator.

58. Find the horizontal and vertical components of **u** + **v** by adding the results you obtained in Exercises 56 and 57.

59. Use your calculator to find the magnitude and direction angle of the vector **u** + **v**.

60. Compare your answers in Exercises 55 and 59. What do you notice? Which method of solution do you prefer?

10.2 TRIGONOMETRIC OR POLAR FORM OF COMPLEX NUMBERS AND OPERATIONS

The Complex Plane and Vector Representation ▮ Trigonometric or Polar Form ▮ Products of Complex Numbers in Polar Form ▮ Quotients of Complex Numbers in Polar Form

The Complex Plane and Vector Representation

In Chapter 3 we introduced complex numbers in $a + bi$ form. We will refer to this form as rectangular form, since the real numbers a and b can be represented by a point (a, b) in the rectangular coordinate plane. We can extend the concepts of rectangular coordinates so that a complex number, such as $2 - 3i$, can be represented by a position vector $\langle 2, -3 \rangle$. In this context, the horizontal axis will be called the **real axis,** and the vertical axis the **imaginary axis.** Then complex numbers can be graphed in this **complex plane,** as shown in Figure 21 for the complex number $2 - 3i$.

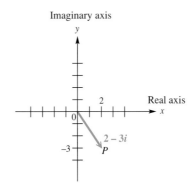

FIGURE 21

Each complex number graphed in this way determines a unique position vector. Recall from Chapter 3 that the sum of the two complex numbers $4 + i$ and $1 + 3i$ is found as follows:

$$(4 + i) + (1 + 3i) = 5 + 4i$$

Graphically, the sum of two complex numbers is represented by the vector that is the resultant of the vectors corresponding to the two numbers. The vectors representing the complex numbers $4 + i$ and $1 + 3i$ and the resultant vector that represents their sum, $5 + 4i$, are shown in Figure 22.

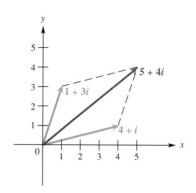

FIGURE 22

Find the sum of $6 - 2i$ and $-4 - 3i$. Graph both complex numbers and their resultant.

SOLUTION   The sum is found by adding the two numbers.

$$(6 - 2i) + (-4 - 3i) = 2 - 5i$$

The graphs are shown in Figure 23.   ◼

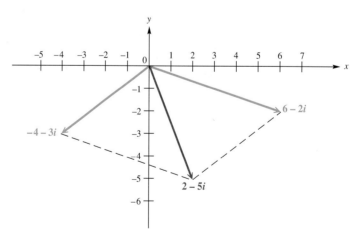

FIGURE 23

Trigonometric or Polar Form

Figure 24 shows the complex number $x + yi$ that corresponds to a vector **OP** with direction θ and magnitude r. The following relationships among r, θ, x, and y can be verified from Figure 24. Notice the similarities between these and the ones given in Section 10.1 for vectors.

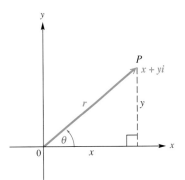

FIGURE 24

RELATIONSHIPS AMONG x, y, r, AND θ

$x = r \cos \theta \qquad r = \sqrt{x^2 + y^2}$

$y = r \sin \theta \qquad \tan \theta = \dfrac{y}{x}$, if $x \neq 0$

Substituting $x = r \cos \theta$ and $y = r \sin \theta$ from the results above into $x + yi$ gives

$$x + yi = r \cos \theta + (r \sin \theta)i$$
$$= r(\cos \theta + i \sin \theta).$$

TRIGONOMETRIC OR POLAR FORM OF A COMPLEX NUMBER

The expression

$$r(\cos \theta + i \sin \theta)$$

is called the **trigonometric form** or **polar form*** of the complex number $x + yi$. The expression $\cos \theta + i \sin \theta$ is sometimes abbreviated cis θ. Using this notation,

$$r(\cos \theta + i \sin \theta) \text{ is written as } r \text{ cis } \theta.$$

The number r is called the **modulus** or **absolute value** of $x + yi$, while θ is the **argument** of $x + yi$. In this section we will choose the value of θ in the interval $[0°, 360°)$. However, keep in mind that the angle is not unique, since any angle coterminal with it is also acceptable.

---

* The terms *trigonometric form* and *polar form* are synonymous, and will be used interchangeably in the rest of this chapter.

EXAMPLE **2**
Converting from Polar Form to Rectangular Form

Express $2(\cos 300° + i \sin 300°)$ in rectangular form.

SOLUTION   Since $\cos 300° = \frac{1}{2}$ and $\sin 300° = -\frac{\sqrt{3}}{2}$,

$$2(\cos 300° + i \sin 300°) = 2\left(\frac{1}{2} - i\frac{\sqrt{3}}{2}\right) = 1 - i\sqrt{3}. ■$$

NOTE   In most of the examples of this section, we will write arguments using degree measure. Arguments may also be written with radian measure.

TECHNOLOGICAL NOTE
The built-in functions described in the Technological Note found in Section 10.1 can be adapted to converting between polar and rectangular forms of complex numbers.

In order to convert from rectangular form to polar form, the following procedure is used.

> **STEPS FOR CONVERTING FROM RECTANGULAR TO POLAR FORM**
> 1.   Sketch a graph of the number in the complex plane.
> 2.   Find r by using the equation $r = \sqrt{x^2 + y^2}$.
> 3.   Find θ by using the equation $\tan \theta = \frac{y}{x}$, $x \neq 0$.

CAUTION   Errors often occur in Step 3 described above. Be sure that the correct quadrant of θ is chosen by referring to the graph sketched in Step 1.

EXAMPLE **3**
Converting from Rectangular Form to Polar Form

Write the following complex numbers in polar form.

(a) $-\sqrt{3} + i$

SOLUTION   Start by sketching the graph of $-\sqrt{3} + i$ in the complex plane, as shown in Figure 25.

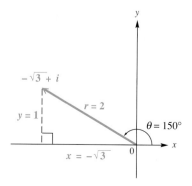

FIGURE 25

Next, find r. Since $x = -\sqrt{3}$ and $y = 1$,
$$r = \sqrt{x^2 + y^2} = \sqrt{(-\sqrt{3})^2 + 1^2} = \sqrt{3 + 1} = 2.$$
Then find θ.

$$\tan \theta = \frac{y}{x} = \frac{1}{-\sqrt{3}} = -\frac{\sqrt{3}}{3}$$

Since $\tan \theta = -\frac{\sqrt{3}}{3}$, the reference angle for θ is $30°$. From the sketch we see that θ is in quadrant II, so $\theta = 180° - 30° = 150°$. Therefore, in polar form,

$$-\sqrt{3} + i = 2(\cos 150° + i \sin 150°)$$
$$= 2 \text{ cis } 150°.$$

(b) $-3i$

SOLUTION   The sketch of $-3i$ is shown in Figure 26. (We use the fact that $-3i = 0 - 3i$.)

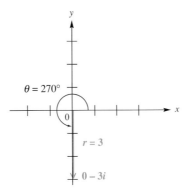

FIGURE 26

Since $-3i = 0 - 3i$, we have $x = 0$ and $y = -3$. Find r as follows.

$$r = \sqrt{0^2 + (-3)^2} = \sqrt{0 + 9} = \sqrt{9} = 3$$

We cannot find θ by using $\tan \theta = \frac{y}{x}$, since $x = 0$. In a case like this, refer to the graph and determine the argument directly from the sketch. A value for θ here is $270°$. In polar form,

$$-3i = 3(\cos 270° + i \sin 270°)$$
$$= 3 \text{ cis } 270°. \quad \blacksquare$$

NOTE   In Example 3 we gave answers in both forms: $r(\cos \theta + i \sin \theta)$ and r cis θ. These forms will be used interchangeably throughout the rest of this chapter.

FOR GROUP DISCUSSION

Use the vector conversion techniques of your calculator (described in the previous section) in degree mode to verify the following results of Examples 2 and 3.

1.   When $r = 2$ and $\theta = 300°$, we obtain $x = 1$ and $y = -\sqrt{3} \approx -1.732050808$. (from Example 2)
2.   When $x = -\sqrt{3}$ and $y = 1$, we obtain $r = 2$ and $\theta = 150°$. (from Example 3a)
3.   When $x = 0$ and $y = -3$, we obtain $r = 3$. The display for θ will depend on how your particular model is programmed. It will either equal $270°$ or an angle coterminal with $270°$, such as $-90°$. (from Example 3b).

EXAMPLE 4

Converting Between Polar and Rectangular Forms

(a)   Write the complex number $6(\cos 115° + i \sin 115°)$ in rectangular form.

SOLUTION   Since $115°$ does not have a special angle as a reference angle, we cannot find exact values for $\cos 115°$ and $\sin 115°$. Use a calculator set in the degree mode to find $\cos 115° \approx -.42261826$ and $\sin 115° \approx .90630779$. Therefore, in rectangular form,

$$6(\cos 115° + i \sin 115°) \approx 6(-.42261826 + .90630779i)$$
$$\approx -2.5357096 + 5.4378467i$$

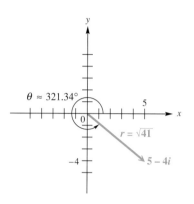

FIGURE 27

(b) Write $5 - 4i$ in polar form.

SOLUTION A sketch of $5 - 4i$ shows that θ must be in quadrant IV. See Figure 27. Here $r = \sqrt{5^2 + (-4)^2} = \sqrt{41}$ and $\tan \theta = \frac{-4}{5}$. Use a calculator to find that one measure of θ is approximately $-38.66°$. In order to express θ in the interval $[0, 360°)$, we find that $\theta \approx 360° - 38.66° \approx 321.34°$. Use these results to get

$$5 - 4i \approx \sqrt{41}(\cos 321.34° + i \sin 321.34°). \quad \blacksquare$$

FOR **GROUP DISCUSSION**

Support the results of Example 4 using the rectangular/polar conversion feature of your calculator.

Products of Complex Numbers in Polar Form

Using the method of multiplying complex numbers in rectangular form, we find the product of $1 + i\sqrt{3}$ and $-2\sqrt{3} + 2i$ as follows. (See Section 3.1.)

$$\begin{aligned}
(1 + i\sqrt{3})(-2\sqrt{3} + 2i) &= -2\sqrt{3} + 2i - 2i(3) + 2i^2\sqrt{3} \\
&= -2\sqrt{3} + 2i - 6i - 2\sqrt{3} \\
&= -4\sqrt{3} - 4i
\end{aligned}$$

This same product also can be found by first converting the complex numbers $1 + i\sqrt{3}$ and $-2\sqrt{3} + 2i$ to polar form. Using the method explained earlier in this section,

$$1 + i\sqrt{3} = 2(\cos 60° + i \sin 60°)$$

and
$$-2\sqrt{3} + 2i = 4(\cos 150° + i \sin 150°).$$

If the polar forms are now multiplied together and if the identities for the cosine and the sine of the sum of two angles are used, the result is

$$\begin{aligned}
&[2(\cos 60° + i \sin 60°)][4(\cos 150° + i \sin 150°)] \\
&= 2 \cdot 4(\cos 60° \cdot \cos 150° + i \sin 60° \cdot \cos 150° \\
&\quad + i \cos 60° \cdot \sin 150° + i^2 \sin 60° \cdot \sin 150°) \\
&= 8[(\cos 60° \cdot \cos 150° - \sin 60° \cdot \sin 150°) \\
&\quad + i(\sin 60° \cdot \cos 150° + \cos 60° \cdot \sin 150°)] \\
&= 8[\cos (60° + 150°) + i \sin(60° + 150°)] \\
&= 8(\cos 210° + i \sin 210°).
\end{aligned}$$

The modulus of the product, 8, is equal to the product of the moduli of the factors, $2 \cdot 4$, while the argument of the product, $210°$, is the sum of the arguments of the factors, $60° + 150°$.

As we would expect, the product obtained upon multiplying by the first method is the rectangular form of the product obtained upon multiplying by the second method.

$$8(\cos 210° + i \sin 210°) = 8\left(-\frac{\sqrt{3}}{2} - \frac{1}{2}i\right)$$
$$= -4\sqrt{3} - 4i$$

The work shown above is generalized in the following *product theorem.*

PRODUCT THEOREM

If $r_1(\cos \theta_1 + i \sin \theta_1)$ and $r_2(\cos \theta_2 + i \sin \theta_2)$ are any two complex numbers, then

$$[r_1(\cos \theta_1 + i \sin \theta_1)] \cdot [r_2(\cos \theta_2 + i \sin \theta_2)]$$
$$= r_1 r_2[\cos(\theta_1 + \theta_2) + i \sin(\theta_1 + \theta_2)].$$

In compact form, this is written

$$(r_1 \text{ cis } \theta_1)(r_2 \text{ cis } \theta_2) = r_1 r_2 \text{ cis}(\theta_1 + \theta_2).$$

EXAMPLE 5

Using the Product Theorem

Find the product of $3(\cos 45° + i \sin 45°)$ and $2(\cos 135° + i \sin 135°)$.

SOLUTION   Using the product theorem,

$$[3(\cos 45° + i \sin 45°)][2(\cos 135° + i \sin 135°)]$$
$$= 3 \cdot 2[\cos(45° + 135°) + i \sin(45° + 135°)]$$
$$= 6(\cos 180° + i \sin 180°),$$

which can be expressed as $6(-1 + i \cdot 0) = 6(-1) = -6$. The two complex numbers in this example are complex factors of -6.   ∎

Quotients of Complex Numbers in Polar Form

Using the method shown in Section 3.1, in rectangular form the quotient of the complex numbers $1 + i\sqrt{3}$ and $-2\sqrt{3} + 2i$ is

$$\frac{1 + i\sqrt{3}}{-2\sqrt{3} + 2i} = \frac{(1 + i\sqrt{3})(-2\sqrt{3} - 2i)}{(-2\sqrt{3} + 2i)(-2\sqrt{3} - 2i)}$$
$$= \frac{-2\sqrt{3} - 2i - 6i - 2i^2\sqrt{3}}{12 - 4i^2}$$
$$= \frac{-8i}{16} = -\frac{1}{2}i.$$

Writing $1 + i\sqrt{3}$, $-2\sqrt{3} + 2i$, and $-\frac{1}{2}i$ in polar form gives

$$1 + i\sqrt{3} = 2(\cos 60° + i \sin 60°)$$
$$-2\sqrt{3} + 2i = 4(\cos 150° + i \sin 150°)$$
$$-\frac{1}{2}i = \frac{1}{2}[\cos(-90°) + i \sin(-90°)].$$

The modulus of the quotient, $\frac{1}{2}$, is the quotient of the two moduli, 2 and 4. The argument of the quotient, $-90°$, is the difference of the two arguments, $60° - 150° = -90°$. It would be easier to find the quotient of these two complex numbers in trigonometric form than in rectangular form. Generalizing from this example leads to another theorem, the *quotient theorem*.

QUOTIENT THEOREM

If $r_1(\cos \theta_1 + i \sin \theta_1)$ and $r_2(\cos \theta_2 + i \sin \theta_2)$ are complex numbers, where $r_2(\cos \theta_2 + i \sin \theta_2) \neq 0$, then

$$\frac{r_1(\cos \theta_1 + i \sin \theta_1)}{r_2(\cos \theta_2 + i \sin \theta_2)} = \frac{r_1}{r_2}[\cos(\theta_1 - \theta_2) + i \sin(\theta_1 - \theta_2)].$$

In compact form, this is written

$$\frac{r_1 \operatorname{cis} \theta_1}{r_2 \operatorname{cis} \theta_2} = \frac{r_1}{r_2} \operatorname{cis}(\theta_1 - \theta_2).$$

EXAMPLE 6
Using the Quotient Theorem

Find the quotient

$$\frac{10 \operatorname{cis}(-60°)}{5 \operatorname{cis} 150°}.$$

Write the result in rectangular form.

SOLUTION By the quotient theorem,

$$\frac{10 \operatorname{cis}(-60°)}{5 \operatorname{cis} 150°} = \frac{10}{5} \operatorname{cis}(-60° - 150°) \qquad \text{Quotient theorem}$$

$$= 2 \operatorname{cis}(-210°) \qquad \text{Subtract.}$$

$$= 2[\cos(-210°) + i \sin(-210°)]$$

$$= 2\left[-\frac{\sqrt{3}}{2} + i\left(\frac{1}{2}\right)\right] \qquad \begin{array}{l}\cos(-210°) = -\frac{\sqrt{3}}{2}; \\ \sin(-210°) = \frac{1}{2}\end{array}$$

$$= -\sqrt{3} + i \qquad \text{Rectangular form} \qquad ∎$$

EXAMPLE 7
Using the Product and Quotient Theorems with a Calculator

Use a calculator to find the following. Write the results in rectangular form.

(a) $(9.3 \operatorname{cis} 125.2°)(2.7 \operatorname{cis} 49.8°)$

SOLUTION By the product theorem,

$$(9.3 \operatorname{cis} 125.2°)(2.7 \operatorname{cis} 49.8°) = (9.3)(2.7) \operatorname{cis}(125.2° + 49.8°)$$

$$= 25.11 \operatorname{cis} 175°$$

$$= 25.11(\cos 175° + i \sin 175°)$$

$$\approx 25.11(-.99619470 + i(.08715574))$$

$$\approx -25.014449 + 2.1884807i.$$

(b) $\dfrac{10.42\left(\cos \dfrac{3\pi}{4} + i \sin \dfrac{3\pi}{4}\right)}{5.21\left(\cos \dfrac{\pi}{5} + i \sin \dfrac{\pi}{5}\right)}$

SOLUTION   Use the quotient theorem.

$$\frac{10.42\left(\cos\dfrac{3\pi}{4} + i\sin\dfrac{3\pi}{4}\right)}{5.21\left(\cos\dfrac{\pi}{5} + i\sin\dfrac{\pi}{5}\right)} = \frac{10.42}{5.21}\left[\cos\left(\dfrac{3\pi}{4} - \dfrac{\pi}{5}\right) + i\sin\left(\dfrac{3\pi}{4} - \dfrac{\pi}{5}\right)\right]$$

$$= 2\left(\cos\dfrac{11\pi}{20} + i\sin\dfrac{11\pi}{20}\right)$$

$$\approx -.31286893 + 1.9753767i \quad \blacksquare$$

10.2 EXERCISES

Graph each of the following complex numbers as a vector in the complex plane.

1. $-2 + 3i$ **2.** $-4 + 5i$ **3.** $8 - 5i$ **4.** $6 - 5i$ **5.** $2 - 2i\sqrt{3}$

6. $4\sqrt{2} + 4i\sqrt{2}$ **7.** $-4i$ **8.** $3i$ **9.** -8 **10.** 2

11. What must be true in order for a complex number to also be a real number?

12. If a real number is graphed in the complex plane, on what axis does the vector lie?

13. A complex number of the form $a + bi$ will have its corresponding vector lying on the y-axis provided $a =$ ——————— .

14. The modulus of a complex number represents the ——————— of the vector representing it in the complex plane.

Find the resultant of each of the following pairs of complex numbers. Express in rectangular form $a + bi$.

15. $4 - 3i, -1 + 2i$ **16.** $2 + 3i, -4 - i$ **17.** $-3, 3i$

18. $6, -2i$ **19.** $2 + 6i, -2i$ **20.** $-5 - 8i, -1$

Write each of the following complex numbers in rectangular form. Give exact values for the real and imaginary parts. You may wish to use the capability of your calculator to support your answers.

21. $2(\cos 45° + i\sin 45°)$ **22.** $4(\cos 60° + i\sin 60°)$

23. 10 cis $90°$ **24.** 8 cis $270°$

25. $4(\cos 240° + i\sin 240°)$ **26.** $2(\cos 330° + i\sin 330°)$

27. $\cos\dfrac{\pi}{6} + i\sin\dfrac{\pi}{6}$ **28.** $3\left(\cos\dfrac{5\pi}{6} + i\sin\dfrac{5\pi}{6}\right)$

29. 5 cis $300°$ **30.** 6 cis $135°$

31. $\sqrt{2}$ cis $180°$ **32.** $\sqrt{3}$ cis $270°$

Write each of the following complex numbers in polar form $r(\cos\theta + i\sin\theta)$, where r is exact, and $-180° < \theta \leq 180°$. You may wish to use the capability of your calculator to support your result.

33. $3 - 3i$ **34.** $-2 + 2i\sqrt{3}$ **35.** $-3 - 3i\sqrt{3}$ **36.** $1 + i\sqrt{3}$

37. $4\sqrt{3} + 4i$ **38.** $\sqrt{3} - i$ **39.** $-\sqrt{2} + i\sqrt{2}$ **40.** $-5 - 5i$

41. $2 + 2i$ **42.** $-\sqrt{3} + i$ **43.** -4 **44.** $5i$

45. $-2i$ **46.** 7

47. Use your calculator to find the polar form of $3 + 5i$. Give as many decimal places as the calculator displays.

48. Give the smallest *positive* degree measure of θ if $r > 0$ and **(a)** r cis θ has real part equal to 0, **(b)** r cis θ has imaginary part equal to 0.

Find each of the following products, and write the products in rectangular form, using exact values.

49. $[3(\cos 60° + i \sin 60°)][2(\cos 90° + i \sin 90°)]$

50. $[4(\cos 30° + i \sin 30°)][5(\cos 120° + i \sin 120°)]$

51. $[2(\cos 45° + i \sin 45°)][2(\cos 225° + i \sin 225°)]$

52. $[8(\cos 300° + i \sin 300°)][5(\cos 120° + i \sin 120°)]$

53. $[4(\cos 60° + i \sin 60°)][6(\cos 330° + i \sin 330°)]$

54. $[8(\cos 210° + i \sin 210°)][2(\cos 330° + i \sin 330°)]$

55. $[5 \text{ cis } 90°][3 \text{ cis } 45°]$

56. $[6 \text{ cis } 120°][5 \text{ cis}(-30°)]$

57. $[\sqrt{3} \text{ cis } 45°][\sqrt{3} \text{ cis } 225°]$

58. $[\sqrt{2} \text{ cis } 300°][\sqrt{2} \text{ cis } 270°]$

Find each of the following quotients, and write the quotients in rectangular form, using exact values.

59. $\dfrac{4(\cos 120° + i \sin 120°)}{2(\cos 150° + i \sin 150°)}$

60. $\dfrac{10(\cos 225° + i \sin 225°)}{5(\cos 45° + i \sin 45°)}$

61. $\dfrac{16(\cos 300° + i \sin 300°)}{8(\cos 60° + i \sin 60°)}$

62. $\dfrac{24(\cos 150° + i \sin 150°)}{2(\cos 30° + i \sin 30°)}$

63. $\dfrac{3 \text{ cis } 305°}{9 \text{ cis } 65°}$

64. $\dfrac{12 \text{ cis } 293°}{6 \text{ cis } 23°}$

Relating Concepts

Consider the complex numbers

$$w = -1 + i \quad and \quad z = -1 - i.$$

65. Multiply w and z using their rectangular forms and the "FOIL" method from algebra. Leave the product in rectangular form.

66. Find the polar forms of w and z.

67. Multiply w and z using their polar forms and the method described in this section.

68. Use the result of Exercise 67 to find the rectangular form of wz. How does this compare to your result in Exercise 65?

69. Find the quotient $\frac{w}{z}$ using their rectangular forms and multiplying both the numerator and the denominator by the conjugate of the denominator. Leave the quotient in rectangular form.

70. Use the polar forms of w and z, found in Exercise 66, to divide w by z using the method described in this section.

71. Use the result of Exercise 70 to find the rectangular form of $\frac{w}{z}$. How does this compare to your result in Exercise 69?

72. Without actually performing the operations, state why the products $[2(\cos 45° + i \sin 45°)] \cdot [5(\cos 90° + i \sin 90°)]$ and $[2(\cos(-315°) + i \sin(-315°))][5(\cos(-270°) + i \sin(-270°))]$ are the same.

73. Notice that $(r \text{ cis } \theta)^2 = (r \text{ cis } \theta)(r \text{ cis } \theta) = r^2 \text{ cis}(\theta + \theta) = r^2 \text{ cis } 2\theta$. State in your own words how we can square a complex number in polar form. (In the next section, we will develop this idea.)

74. The alternating current in an electric inductor is

$$I = \frac{E}{Z}$$

amperes, where E is the voltage and $Z = R + X_L i$ is the impedance. If $E = 8(\cos 20° + i \sin 20°)$, $R = 6$, and $X_L = 3$, find the current. Give the answer in rectangular form.

75. The current I in a circuit with voltage E, resistance R, capacitive reactance X_c, and inductive reactance X_L is

$$I = \frac{E}{R + (X_L - X_c)i}.$$

Find I if $E = 12(\cos 25° + i \sin 25°)$, $R = 3$, $X_L = 4$, and $X_c = 6$. Give the answer in rectangular form.

76. Under what conditions is the difference between two nonreal complex numbers $a + bi$ and $c + di$ a real number?

77. Prove the product theorem for complex numbers.

78. Prove the quotient theorem for complex numbers.

10.3 POWERS AND ROOTS OF COMPLEX NUMBERS

Powers of Complex Numbers (De Moivre's Theorem) ▮ Roots of Complex Numbers

Powers of Complex Numbers (De Moivre's Theorem)

In the previous section we studied the product and quotient theorems for complex numbers in polar form. Because raising a number to a positive integer power is a repeated application of the product rule, it would seem likely that a theorem for finding powers of complex numbers exists. This is indeed the case. For example, the square of the complex number $r(\cos \theta + i \sin \theta)$ is

$$[r(\cos \theta + i \sin \theta)]^2 = [r(\cos \theta + i \sin \theta)][r(\cos \theta + i \sin \theta)]$$
$$= r \cdot r[\cos(\theta + \theta) + i \sin(\theta + \theta)]$$
$$= r^2(\cos 2\theta + i \sin 2\theta).$$

In the same way,

$$[r(\cos \theta + i \sin \theta)]^3 = r^3(\cos 3\theta + i \sin 3\theta).$$

These results suggest the plausibility of the following theorem for positive integer values of n. Although the following theorem is stated and can be proved for all n, we will use it only for positive integer values of n and their reciprocals.

DE MOIVRE'S THEOREM

If $r(\cos \theta + i \sin \theta)$ is a complex number and if n is any real number, then

$$[r(\cos \theta + i \sin \theta)]^n = r^n(\cos n\theta + i \sin n\theta).$$

In compact form, this is written

$$[r \text{ cis } \theta]^n = r^n (\text{cis } n\theta).$$

This theorem is named after the French expatriate friend of Isaac Newton, Abraham De Moivre (1667–1754), although he never explicitly stated it.

EXAMPLE 1

Finding a Power of a Complex Number

Find $(1 + i\sqrt{3})^8$ and express the result in rectangular form.

SOLUTION To use De Moivre's theorem, first convert $1 + i\sqrt{3}$ into polar form using the methods of Section 10.2.

$$1 + i\sqrt{3} = 2(\cos 60° + i \sin 60°)$$

Now apply De Moivre's theorem.

$$(1 + i\sqrt{3})^8 = [2(\cos 60° + i\sin 60°)]^8$$
$$= 2^8[\cos(8 \cdot 60°) + i\sin(8 \cdot 60°)]$$
$$= 256(\cos 480° + i\sin 480°)$$
$$= 256(\cos 120° + i\sin 120°)$$

480° and 120° are coterminal.

$$= 256\left(-\frac{1}{2} + i\frac{\sqrt{3}}{2}\right)$$

$\cos 120° = -\frac{1}{2};$
$\sin 120° = \frac{\sqrt{3}}{2}$

$$= -128 + 128i\sqrt{3}$$

Rectangular form ◼

Roots of Complex Numbers

In algebra it is shown that every nonzero complex number has exactly n distinct complex nth roots. De Moivre's theorem can be extended to find all nth roots of a complex number. An nth root of a complex number is defined as follows.

---

nTH ROOT

For a positive integer n, the complex number $a + bi$ is an **nth root** of the complex number $x + yi$ if

$$(a + bi)^n = x + yi.$$

EXAMPLE 2

Finding Complex Roots

Find the three complex cube roots of $8(\cos 135° + i\sin 135°)$.

SOLUTION To find these three complex cube roots, we must look for a complex number, say $r(\cos \alpha + i\sin \alpha)$, that will satisfy

$$[r(\cos \alpha + i\sin \alpha)]^3 = 8(\cos 135° + i\sin 135°).$$

By De Moivre's theorem, this equation becomes

$$r^3(\cos 3\alpha + i\sin 3\alpha) = 8(\cos 135° + i\sin 135°).$$

One way to satisfy this equation is to set $r^3 = 8$ and also $\cos 3\alpha + i\sin 3\alpha = \cos 135° + i\sin 135°$. The first of these conditions implies that $r = 2$, and the second implies that

$$\cos 3\alpha = \cos 135° \quad \text{and} \quad \sin 3\alpha = \sin 135°.$$

For these equations to be satisfied, 3α must represent an angle that is coterminal with 135°. Therefore, we must have

$$3\alpha = 135° + 360° \cdot k, \quad k \text{ any integer,}$$

or

$$\alpha = \frac{135° + 360° \cdot k}{3}, \quad k \text{ any integer.}$$

Now let k take on the integer values 0, 1, and 2.

$$\text{If } k = 0, \quad \alpha = \frac{135° + 0°}{3} = 45°.$$

$$\text{If } k = 1, \quad \alpha = \frac{135° + 360°}{3} = \frac{495°}{3} = 165°.$$

$$\text{If } k = 2, \quad \alpha = \frac{135° + 720°}{3} = \frac{855°}{3} = 285°.$$

In the same way, $\alpha = 405°$ when $k = 3$. But note that $\sin 405° = \sin 45°$ and $\cos 405° = \cos 45°$. If $k = 4$, $\alpha = 525°$ which has the same sine and cosine values as $165°$. To continue with larger values of k would just be repeating solutions already found. Therefore, all of the cube roots (three of them) can be found by letting $k = 0, 1,$ and 2.

When $k = 0$, the root is $2(\cos 45° + i \sin 45°)$.

When $k = 1$, the root is $2(\cos 165° + i \sin 165°)$.

When $k = 2$, the root is $2(\cos 285° + i \sin 285°)$.

In summary, $2(\cos 45° + i \sin 45°)$, $2(\cos 165° + i \sin 165°)$, and $2(\cos 285° + i \sin 285°)$ are the three cube roots of $8(\cos 135° + i \sin 135°)$.   ▯

Generalizing from Example 2, we state the following theorem that can be applied to find nth roots.

---

nTH ROOT THEOREM

If n is any positive integer and r is a positive real number, then the nonzero complex number $r(\cos \theta + i \sin \theta)$ has exactly n distinct nth roots, given by

$$\sqrt[n]{r}\,(\cos \alpha + i \sin \alpha),$$

where

$$\alpha = \frac{\theta + 360° \cdot k}{n} \quad \text{or} \quad \alpha = \frac{\theta}{n} + \frac{360° \cdot k}{n},$$

$k = 0, 1, 2, \ldots, n - 1.$

EXAMPLE 3

Finding Complex Roots

Find all fourth roots of $-8 + 8i\sqrt{3}$. Write the roots in rectangular form.

SOLUTION   First write $-8 + 8i\sqrt{3}$ in polar form as

$$-8 + 8i\sqrt{3} = 16(\cos 120° + i \sin 120°).$$

Here $r = 16$ and $\theta = 120°$. The fourth roots of this number have modulus $\sqrt[4]{16} = 2$ and arguments given as follows. Using the alternative formula for α,

$$\alpha = \frac{120°}{4} + \frac{360° \cdot k}{4} = 30° + 90° \cdot k.$$

If $k = 0$, $\alpha = 30° + 90° \cdot 0 = 30°.$

If $k = 1$, $\alpha = 30° + 90° \cdot 1 = 120°.$

If $k = 2$, $\alpha = 30° + 90° \cdot 2 = 210°.$

If $k = 3$, $\alpha = 30° + 90° \cdot 3 = 300°.$

Using these angles, the fourth roots are

$$2(\cos 30° + i \sin 30°),$$
$$2(\cos 120° + i \sin 120°),$$
$$2(\cos 210° + i \sin 210°),$$

and $\quad 2(\cos 300° + i \sin 300°).$

These four roots can be written in rectangular form as $\sqrt{3} + i$, $-1 + i\sqrt{3}$, $-\sqrt{3} - i$, and $1 - i\sqrt{3}$. ∎

An interesting geometric interpretation involving vectors can be applied to nth roots of a complex number. If we use the values of r and α found in Example 3, the four fourth roots of $-8 + 8i\sqrt{3}$ can be illustrated as position vectors whose terminal points lie on a circle of radius $r = 2$. The vectors are spaced 90° apart; the 90° comes from the fact that since $n = 4$, $\frac{360°}{n} = \frac{360°}{4} = 90°$. See Figure 28.

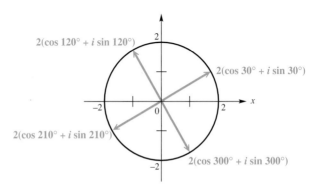

FIGURE 28

EXAMPLE 4

Solving an Equation by Finding Complex Roots

Find all complex number solutions of $x^5 - 1 = 0$. Graph them as vectors in the complex plane.

SOLUTION Write the equation as

$$x^5 - 1 = 0$$
$$x^5 = 1.$$

While there is only one real number solution, 1, there are five complex number solutions. To find these solutions, first write 1 in polar form as

$$1 = 1 + 0i = 1(\cos 0° + i \sin 0°).$$

The modulus of the fifth roots is $\sqrt[5]{1} = 1$, and the arguments are given by

$$0° + 72° \cdot k, \qquad k = 0, 1, 2, 3, \text{ or } 4.$$

By using these arguments, the fifth roots are

$$1(\cos 0° + i \sin 0°), \qquad k = 0$$
$$1(\cos 72° + i \sin 72°), \qquad k = 1$$
$$1(\cos 144° + i \sin 144°), \qquad k = 2$$
$$1(\cos 216° + i \sin 216°), \qquad k = 3$$

and $\quad 1(\cos 288° + i \sin 288°). \qquad k = 4$

The solution set of the equation may be written as {cis 0°, cis 72°, cis 144°, cis 216°, cis 288°}. The first of these roots equals 1; the others cannot easily be expressed in rectangular form. The five fifth roots all lie on a unit circle and are equally spaced around it every 72°, as shown in Figure 29. ∎

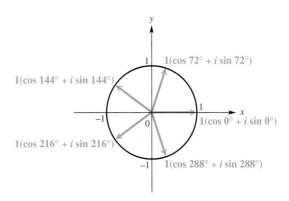

FIGURE 29

10.3 EXERCISES

Find each of the following powers. Write the answers in rectangular form.

1. $[3(\cos 30° + i \sin 30°)]^3$ **2.** $[2(\cos 135° + i \sin 135°)]^4$ **3.** $(\cos 45° + i \sin 45°)^8$

4. $[2(\cos 120° + i \sin 120°)]^3$ **5.** $[3 \text{ cis } 100°]^3$ **6.** $[3 \text{ cis } 40°]^3$

7. $(\sqrt{3} + i)^5$ **8.** $(2\sqrt{2} - 2i\sqrt{2})^6$ **9.** $(2 - 2i\sqrt{3})^4$

10. $\left(\dfrac{\sqrt{2}}{2} - \dfrac{\sqrt{2}}{2}i\right)^8$ **11.** $(-2 - 2i)^5$ **12.** $(-1 + i)^7$

Find the cube roots of each of the following complex numbers. Leave the answers in polar form. Then graph each cube root as a vector in the complex plane.

13. 1 **14.** i **15.** $8(\cos 60° + i \sin 60°)$

16. $27(\cos 300° + i \sin 300°)$ **17.** $-8i$ **18.** $27i$

19. -64 **20.** 27 **21.** $1 + i\sqrt{3}$

22. $2 - 2i\sqrt{3}$ **23.** $-2\sqrt{3} + 2i$ **24.** $\sqrt{3} - i$

25. For the real number 1, find and graph all indicated roots. Give answers in rectangular form.
   (a) fourth   **(b)** sixth   **(c)** eighth

26. For the complex number i, find and graph all indicated roots. Give answers in polar form.
   (a) square   **(b)** fourth

27. Explain why a positive real number must have a positive real nth root.

28. True or false: **(a)** Every real number must have two real square roots.
   (b) Some real numbers have three real cube roots.

Relating Concepts

We will examine how the three complex cube roots of -8 can be found in two different ways. Work Exercises 29–36 in order.

29. All complex roots of the equation $x^3 + 8 = 0$ are cube roots of -8. Factor $x^3 + 8$ as the sum of two cubes.

30. One of the factors found in Exercise 29 is linear. Set it equal to 0, solve and determine the real cube root of -8.

31. One of the factors found in Exercise 29 is quadratic. Set it equal to 0, solve, and determine the rectangular forms of the other two cube roots of -8.

32. Use the method described in this section to find the three complex cube roots of -8. Give them in polar form.

33. Convert the polar forms found in Exercise 32 to rectangular form.

34. Compare your results in Exercises 30 and 31, and Exercise 33. What do you notice?

35. Graph the function $f(x) = x^3 + 8$ in the standard window of your calculator, and find the x-intercept. How does this relate to the cube roots you found earlier?

36. The graph of $f(x) = x^6 - 1$ is shown here, along with the two x-intercepts as determined by a calculator. How do they relate to the answers in Exercise 25(b)?

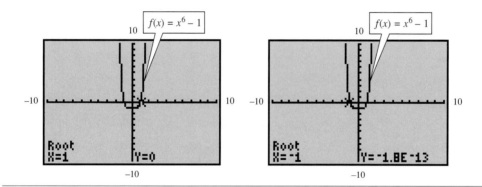

Find all complex solutions for each equation. Leave your answers in polar form.

37. $x^4 + 1 = 0$ **38.** $x^4 + 16 = 0$

39. $x^5 - i = 0$ **40.** $x^4 - i = 0$

10.4 POLAR COORDINATES AND GRAPHS OF POLAR EQUATIONS

The Polar Coordinate System ▌ Polar Equations ▌ Conversion of Equations

The Polar Coordinate System

Throughout this text we have been using the rectangular coordinate system to graph equations. Another coordinate system that is particularly useful for graphing many relations is the **polar coordinate system.** The system is based on a point, called the **pole,** and a ray, called the **polar axis.** The polar axis is usually drawn in the direction of the positive x-axis, as shown in Figure 30.

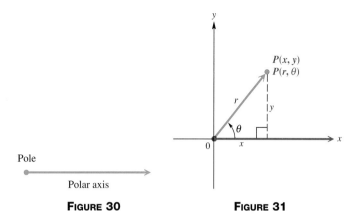

FIGURE 30   **FIGURE 31**

In Figure 31, the pole has been placed at the origin of a rectangular coordinate system, so that the polar axis coincides with the positive x-axis. Point P has coordinates (x, y) in the rectangular coordinate system. Point P can also be located by giving the directed angle θ from the positive x-axis to ray OP and the directed distance r from the pole to point P. The ordered pair (r, θ) gives the **polar coordinates** of point P.

In Sections 10.1 and 10.2 we saw how the values of x, y, r, and θ are related, as they pertained to vectors and complex numbers in polar form. These same relationships hold here.

TECHNOLOGICAL NOTE
The built-in functions described in the Technological Note found in Section 10.1 can be adapted to converting between polar and rectangular coordinates.

> **CONVERTING BETWEEN POLAR AND RECTANGULAR COORDINATES**
>
> The following relationships hold between the point (x, y) in the rectangular coordinate plane and the point (r, θ) in the polar coordinate plane.
>
> $$x = r\cos\theta \qquad r = \sqrt{x^2 + y^2}$$
>
> $$y = r\sin\theta \qquad \tan\theta = \frac{y}{x}, \quad \text{if } x \neq 0$$

EXAMPLE 1

Plotting Points with Polar Coordinates and Determining Rectangular Coordinates

For each point, plot by hand in the polar coordinate system. Then determine the rectangular coordinates of the point:

(a) $P(2, 30°)$ **(b)** $Q(-4, 120°)$ **(c)** $R(5, -45°)$.

SOLUTION

(a) In this case, $r = 2$ and $\theta = 30°$, so the point P is located 2 units from the origin in the positive direction on a ray making a $30°$ angle with the polar axis, as shown in Figure 32 on the next page.

Using the conversion equations, we find the rectangular coordinates as follows.

$$
\begin{aligned}
x &= r\cos\theta & y &= r\sin\theta \\
&= 2\cos 30° & &= 2\sin 30° \\
&= 2\left(\frac{\sqrt{3}}{2}\right) & &= 2\left(\frac{1}{2}\right) \\
&= \sqrt{3} & &= 1.
\end{aligned}
$$

The rectangular coordinates are $(\sqrt{3}, 1)$.

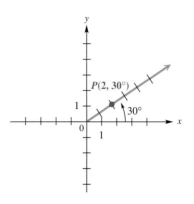

FIGURE 32

(b) Since r is negative, Q is 4 units in the negative direction from the pole on an extension of the 120° ray. See Figure 33. The rectangular coordinates are $x = -4\cos 120° = -4(-\tfrac{1}{2}) = 2$ and $y = -4\sin 120° = -4(\tfrac{\sqrt{3}}{2}) = -2\sqrt{3}$.

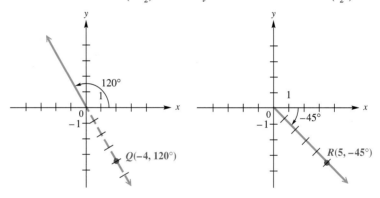

FIGURE 33   **FIGURE 34**

(c) Point R is shown in Figure 34. Since θ is negative, the angle is measured in the clockwise direction. Furthermore, we have

$$x = 5\cos(-45°)$$
$$= \frac{5\sqrt{2}}{2}$$

and

$$y = 5\sin(-45°)$$
$$= \frac{-5\sqrt{2}}{2}.$$

FOR GROUP DISCUSSION

Use the polar/rectangular conversion feature of your graphics calculator to support the results in Example 1. Be aware of decimal approximations of exact values. Also, be sure your calculator is in degree mode to correspond to the example.

One important difference between rectangular coordinates and polar coordinates is that while a given point in the plane can have only one pair of rectangular coordinates, this same point can have an infinite number of pairs of polar coordinates. For example, $(2, 30°)$ locates the same point as $(2, 390°)$ or $(2, -330°)$ or $(-2, 210°)$.

EXAMPLE **2**
Giving Alternate Forms of a Pair of Polar Coordinates

Give three other pairs of polar coordinates for the point $P(3, 140°)$.

SOLUTION   Three pairs that could be used for the point are $(3, -220°)$, $(-3, 320°)$, and $(-3, -40°)$. See Figure 35.   ▯

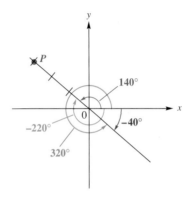

FIGURE 35

Polar Equations

So far in this text, the graphs that we have studied have been sketched or generated by a calculator using rectangular coordinates. We will now examine how equations in polar coordinates are graphed. An equation like

$$r = 3 \sin \theta, \qquad r = 2 + \cos \theta, \qquad \text{or} \qquad r = \theta,$$

where r and θ are the variables, is a *polar equation*. The simplest equation for many types of curves turns out to be a polar equation.

The traditional method of graphing polar equations is much the same as the traditional method of graphing rectangular equations. We evaluate r for various values of θ until a pattern appears, and then we join the points with a smooth curve. However, this can be a very time-consuming task, and modern graphics calculators can perform the job in a matter of seconds. In this section we will present some examples of polar curves by including a table of representative points, a traditional graph, and a calculator-generated graph.

NOTE   Refer to your owner's manual at this time to see how your particular model handles polar graphs. After reading the examples that follow, see if you can generate the curves shown. As always, pay close attention to the viewing window and the angle mode (degree or radian). Furthermore, you will need to decide on maximum and minimum values of θ. Keep in mind the periods of the functions, so that when you decide on these, the entire set of function values are generated.

Graph $r = 1 + \cos \theta$.

SOLUTION   To graph this equation in a traditional manner, we find some ordered pairs (as in the table) and then connect the points in order—from $(2, 0°)$ to $(1.9, 30°)$ to $(1.7, 45°)$ and so on.

θ	0°	30°	45°	60°	90°	120°	135°	150°	180°	270°	315°	330°	360°
$\cos \theta$	1	.9	.7	.5	0	−.5	−.7	−.9	−1	0	.7	.9	1
$r = 1 + \cos \theta$	2	1.9	1.7	1.5	1	.5	.3	.1	0	1	1.7	1.9	2

It is not necessary to choose values greater than 360° or less than 0°, since one period of the cosine function has been completely covered. Joining the points with a smooth curve yields the graph shown in Figure 36(a). The calculator-generated graph is shown in Figure 36(b). Notice that the point $(1.5, 60°)$ is indicated.

This curve is called a **cardioid** because of its heart shape.   ∎

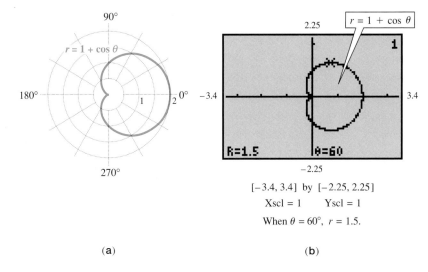

(a)   (b)

[−3.4, 3.4] by [−2.25, 2.25]
Xscl = 1   Yscl = 1
When $\theta = 60°$, $r = 1.5$.

FIGURE 36

Graph $r^2 = \cos 2\theta$.

SOLUTION   First complete a table of ordered pairs as shown. The point $(-1, 0°)$, with r negative, may be plotted as $(1, 180°)$. Also, $(-.7, 30°)$ may be plotted as $(.7, 210°)$, and so on. This curve is called a **lemniscate.**

θ	0°	30°	45°	135°	150°	180°
2θ	0°	60°	90°	270°	300°	360°
$\cos 2\theta$	1	.5	0	0	.5	1
$r = \pm\sqrt{\cos 2\theta}$	±1	±.7	0	0	±.7	±1

Values of θ for $45° < \theta < 135°$ are not included in the table because the corresponding values of cos 2θ are negative (quadrants II and III) and so do not have real square roots. Values of θ larger than 180° give 2θ larger than 360°, and would repeat the points already found.

A traditional graph of this lemniscate is shown in Figure 37(a). The calculator-generated graph shown in Figure 37(b) was obtained by letting θ take on values between 0° and 180°. Again, we show a point on the calculator graph (this time, $(0, 45°)$). It was necessary to enter two equations into the calculator: $r_1 = \sqrt{\cos 2\theta}$ and $r_2 = -\sqrt{\cos 2\theta}$.   ▣

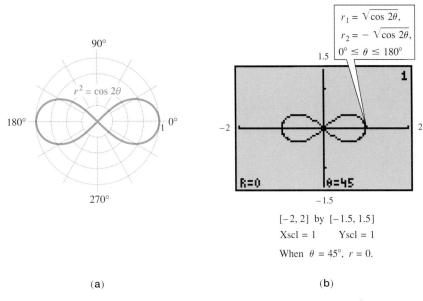

(a)   (b)

FIGURE 37

EXAMPLE 5

Graphing a Polar
Equation (Rose)

Graph $r = 3 \cos 2\theta$.

SOLUTION   Because of the 2θ, the graph requires a large number of points. A few ordered pairs are given below. You should complete the table similarly through the first 360° if you are graphing in a traditional manner.

θ	0°	15°	30°	45°	60°	75°	90°
2θ	0°	30°	60°	90°	120°	150°	180°
cos 2θ	1	.9	.5	0	$-.5$	$-.9$	-1
r	3	2.6	1.5	0	-1.5	-2.6	-3

Plotting these points in order gives the graph, called a **four-leaved rose.** Notice in Figure 38(a) how the graph is developed with a continuous curve, beginning with the upper half of the right horizontal leaf and ending with the lower half of that leaf. As the graph is traced, the curve goes through the pole four times.

This pattern is easily seen if a calculator is used to graph the rose. See Figure 38(b) for such a graph.   ▣

TECHNOLOGICAL NOTE
Figure 38(a) uses arrow-heads and circled num-bers to indicate the pat-tern in which this four-leaved rose is traced. Use your calculator to du-plicate the calculator graph seen in Figure 38(b), and watch closely to confirm this pattern. (Remember that the smaller the step, the longer it will take to trace the graph.)

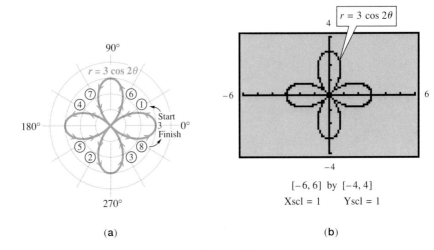

(a)   (b)

FIGURE 38

The graph in Figure 38 is one of a family of curves called **roses.** The graphs of $r = \sin n\theta$ and $r = \cos n\theta$ are roses, with n petals if n is odd, and $2n$ petals if n is even.

FOR **GROUP DISCUSSION**

Keeping in mind that the minimum and maximum values of the sine and cosine function are -1 and 1, discuss how the minimum and maximum val-ues of r in a polar equation can be found. This will help you in determining the appropriate window when using your calculator to graph a polar equation.

EXAMPLE 6

Graphing a Polar Equation (Spiral of Archimedes)

Graph $r = 2\theta$, where θ is measured in radians, $-2\pi \le \theta \le 2\pi$.
    Use a calculator to generate the graph.

SOLUTION   This graph is an example of a **spiral of Archimedes.** It is quite tedious to graph using traditional methods, so we show only a calculator graph. See Figure 39.   █

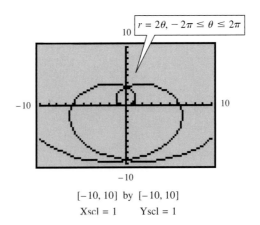

$[-10, 10]$ by $[-10, 10]$
$X\text{scl} = 1$ $Y\text{scl} = 1$

FIGURE 39

TECHNOLOGICAL NOTE
To see how the size of
the step can greatly affect
the accuracy of your
graph, try various values
such as .1, .5, 1, and 1.5
with the polar equation in
Example 6. The graph
shown in Figure 39 was
generated with
θ-step = .1.

FOR GROUP DISCUSSION

Experiment with various minimum and maximum values of θ for the spiral of Archimedes in Example 6, and discuss the various behaviors you observe.

EXAMPLE 7

Converting a Polar
Equation to a
Rectangular
Equation and
Graphing

Conversion of Equations

Using the relationships stated earlier in this section (see the box headed "Converting Between Polar and Rectangular Coordinates"), we can transform polar equations to rectangular ones and vice versa.

For the polar equation $r = \frac{4}{1 + \sin \theta}$:

(a) convert to a rectangular equation,

(b) use a graphics calculator to graph the polar equation for $0° \le \theta \le 360°$, and

(c) use a graphics calculator to graph the rectangular equation.

SOLUTION

(a) Multiply both sides of the equation by the denominator on the right, to clear the fraction.

$$r = \frac{4}{1 + \sin \theta}$$

$$r + r \sin \theta = 4$$

Now substitute $\sqrt{x^2 + y^2}$ for r and y for $r \sin \theta$.

$$\sqrt{x^2 + y^2} + y = 4$$

$$\sqrt{x^2 + y^2} = 4 - y$$

Square both sides to eliminate the radical.

$$x^2 + y^2 = (4 - y)^2$$
$$x^2 + y^2 = 16 - 8y + y^2$$
$$x^2 = -8y + 16$$
$$x^2 = -8(y - 2)$$

(b) Figure 40(a) shows a calculator-generated graph using polar coordinates.

(c) Solving $x^2 = -8(y - 2)$ for y, we obtain

$$y = 2 - \frac{1}{8}x^2.$$

The graph of this rectangular equation is shown in Figure 40(b). Notice that the two graphs are the same.   ◼

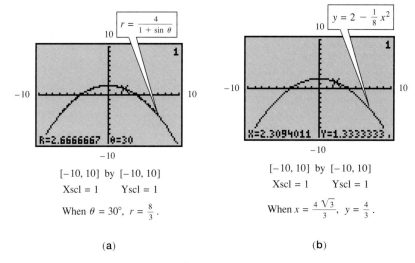

(a) $[-10, 10]$ by $[-10, 10]$
Xscl = 1   Yscl = 1
When $\theta = 30°$, $r = \frac{8}{3}$.

(b) $[-10, 10]$ by $[-10, 10]$
Xscl = 1   Yscl = 1
When $x = \frac{4\sqrt{3}}{3}$, $y = \frac{4}{3}$.

FIGURE 40

FOR GROUP DISCUSSION

The displays at the bottom of the screens in Figures 40(a) and 40(b) indicate the same point. Discuss how you would go about verifying analytically that the polar coordinates shown in (a) and the rectangular coordinates shown in (b) are equivalent.

EXAMPLE 8

Converting a Rectangular Equation to a Polar Equation and Graphing

For the rectangular equation $3x + 2y = 4$:

(a) convert to a polar equation,

(b) use a graphics calculator to graph the rectangular equation, and

(c) use a graphics calculator to graph the polar equation for $0° \le \theta \le 360°$.

SOLUTION

(a) Use $x = r \cos \theta$ and $y = r \sin \theta$ to get

$$3x + 2y = 4$$
$$3r \cos \theta + 2r \sin \theta = 4.$$

Now solve for r. First factor out r on the left.

$$r(3 \cos \theta + 2 \sin \theta) = 4$$
$$r = \frac{4}{3 \cos \theta + 2 \sin \theta}$$

The polar equation of the line $3x + 2y = 4$ is

$$r = \frac{4}{3 \cos \theta + 2 \sin \theta}.$$

(b) We solve the given rectangular equation for y to get

$$y = -\frac{3}{2}x + 2.$$

Its graph is shown in Figure 41(a).

(c) Using the polar graphing capability of the calculator and the equation found in part (a), we obtain the graph shown in Figure 41(b).   ∎

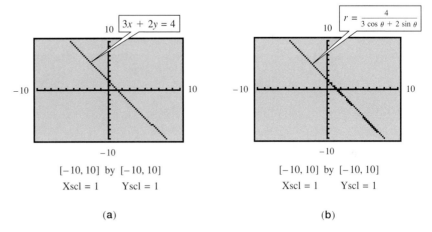

$3x + 2y = 4$

$r = \dfrac{4}{3 \cos \theta + 2 \sin \theta}$

[−10, 10] by [−10, 10]   [−10, 10] by [−10, 10]
Xscl = 1   Yscl = 1   Xscl = 1   Yscl = 1

(a)   (b)

FIGURE 41

TECHNOLOGICAL NOTE
After working Exercises 43–46, experiment with your calculator by changing the sign of a in each exercise. Notice how in one case there is a clockwise pattern, while in the other the pattern is counterclockwise.

10.4 EXERCISES

Plot by hand the graph of the point whose polar coordinates are given. Then give two other pairs of polar coordinates of the point. Finally, give the rectangular coordinates of the point.

1. $(1, 45°)$ **2.** $(3, 120°)$ **3.** $(-2, 135°)$ **4.** $(-4, 30°)$ **5.** $(5, -60°)$

6. $(2, -45°)$ **7.** $(-3, -210°)$ **8.** $(-1, -120°)$ **9.** $(3, 360°)$ **10.** $(4, 270°)$

11. If a point lies on an axis in the rectangular plane, then what kind of angle must θ be if (r, θ) represents the point in polar coordinates?

12. What will the graph of $r = k$ be, for $k > 0$?

Use a graphics calculator to graph each of the following equations for $0° \le \theta \le 360°$. Use a square viewing window.

13. $r = 2 + 2 \cos \theta$ (cardioid)   **14.** $r = 1 + 3 \cos \theta$ (limaçon with a loop)

15. $r = 8 + 6 \cos \theta$ (dimpled limaçon)   **16.** $r = 4 - 4 \sin \theta$ (cardioid)

17. $r = 4 \cos 2\theta$ (four-leaved rose)   **18.** $r = 3 \sin 4\theta$ (eight-leaved rose)

19. $r = 4 \cos 3\theta$ (three-leaved rose)   **20.** $r = 3 \sin 5\theta$ (five-leaved rose)

21. $r^2 = 4 \cos 2\theta$ (lemniscate)   **22.** $r^2 = 3 \sin 2\theta$ (lemniscate)

23. $r = 2 \sin \theta \tan \theta$ (cissoid)   **24.** $r = \dfrac{\cos 2\theta}{\cos \theta}$ (cissoid with a loop)

25. Explain the method you would use to graph (r, θ) by hand if $r < 0$.

26. Explain why, if $r > 0$, the point (r, θ) and $(-r, \theta + 180°)$ have the same graph.

For each of the following equations, find an equivalent equation in rectangular coordinates. Then use the polar graphing capability of your calculator to graph the equation in its original form. (It may be necessary to first solve for r.) Use a square window.

27. $r = 2 \sin \theta$ **28.** $r = 2 \cos \theta$ **29.** $r = \dfrac{2}{1 - \cos \theta}$ **30.** $r = \dfrac{3}{1 - \sin \theta}$

31. $r + 2 \cos \theta = -2 \sin \theta$ **32.** $r = \dfrac{3}{4 \cos \theta - \sin \theta}$ **33.** $r = 2 \sec \theta$

34. $r = -5 \csc \theta$ **35.** $r(\cos \theta + \sin \theta) = 2$ **36.** $r(2 \cos \theta + \sin \theta) = 2$

For each of the following equations, find an equivalent equation in polar coordinates.

37. $x + y = 4$ **38.** $2x - y = 5$ **39.** $x^2 + y^2 = 16$

40. $x^2 + y^2 = 9$ **41.** $y = 2$ **42.** $x = 4$

The graph of $r = a\theta$ in polar coordinates is an example of the spiral of Archimedes. With your calculator set to radian mode and polar graphing capability, use the given value of a and interval of θ to graph the spiral in the window specified.

43. $a = 1, 0 \leq \theta \leq 4\pi, [-15, 15]$ by $[-15, 15]$

44. $a = 2, -4\pi \leq \theta \leq 4\pi, [-30, 30]$ by $[-30, 30]$

45. $a = 1.5, -4\pi \leq \theta \leq 4\pi, [-20, 20]$ by $[-20, 20]$

46. $a = -1, 0 \leq \theta \leq 12\pi, [-40, 40]$ by $[-40, 40]$

47. Refer to Example 8. Would you find it easier to graph the equation using the rectangular or the polar form? Why?

48. Show that the distance between (r_1, θ_1) and (r_2, θ_2) is $\sqrt{r_1^2 + r_2^2 - 2r_1 r_2 \cos(\theta_1 - \theta_2)}$.

10.5 PARAMETRIC EQUATIONS

Basic Concepts ▪ Graphs of Parametric Equations and Their Rectangular Equivalents ▪ Alternative Forms of Parametric Equations ▪ Applications and the Cycloid

TECHNOLOGICAL NOTE
Refer to the Technological Note in Section 10.4 regarding *step* or *increment*. When graphing in parametric mode, it will be necessary to decide on what step to use for the parameter t when making the window settings. You should be aware of the *default* values built in to the calculator. The calculator-generated graphs in this section were created by using t-step $= .1$.

Basic Concepts

In Section 9.1, we saw how the graph of the unit circle can be found on a graphics calculator using parametric equations. We will now investigate parametric equations in more detail. Throughout this text, we have concentrated on graphing sets of ordered pairs of real numbers that corresponded to a function of the form $y = f(x)$ or $r = f(\theta)$. Another way to determine a set of ordered pairs involves two functions f and g defined by $x = f(t)$ and $y = g(t)$, where t is a real number in some interval I. Each value of t leads to a corresponding x-value and a corresponding y-value, and thus to an ordered pair (x, y).

PARAMETRIC EQUATIONS OF A PLANE CURVE

A **plane curve** is a set of points (x, y) such that $x = f(t)$, $y = g(t)$, and f and g are both defined on an interval I. The equations $x = f(t)$ and $y = g(t)$ are **parametric equations** with **parameter t.**

NOTE Just as graphics calculators are capable of graphing rectangular and polar equations, they are also capable of graphing plane curves defined by parametric equations. You should familiarize yourself with how your particular model handles them.

Graphs of Parametric Equations and Their Rectangular Equivalents

For the plane curve defined by the parametric equations

$$x = t^2, y = 2t + 3, \quad \text{for } t \text{ in the interval } [-3, 3],$$

graph the curve using a graphics calculator, and then find an equivalent rectangular equation.

SOLUTION Figure 42 shows the graph of the curve, which appears to be a horizontal parabola.

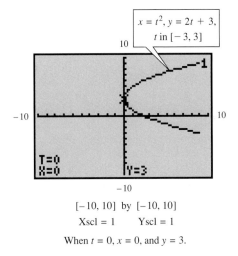

$[-10, 10]$ by $[-10, 10]$

Xscl = 1   Yscl = 1

When $t = 0, x = 0,$ and $y = 3.$

FIGURE 42

To find an equivalent rectangular equation, we analytically eliminate the parameter t. For this curve, we will solve for t in the second equation, $y = 2t + 3$, to begin.

$$y = 2t + 3$$
$$2t = y - 3$$
$$t = \frac{y - 3}{2}$$

Now substitute the result in the first equation to get

$$x = t^2 = \left(\frac{y-3}{2}\right)^2 = \frac{(y-3)^2}{4}$$

or

$$x = \frac{1}{4}(y-3)^2.$$

This is indeed an equation of a parabola. It has a horizontal axis and opens to the right. Because t is in $[-3, 3]$, x is in $[0, 9]$ and y is in $[-3, 9]$. The rectangular equation must be given with its restricted domain as

$$x = \frac{1}{4}(y-3)^2, \quad \text{for } x \text{ in } [0, 9]. \quad \blacksquare$$

FOR **GROUP DISCUSSION**

The display at the bottom of Figure 42 indicates particular values of t, x, and y. Discuss how these relate to both the parametric form and the rectangular form of the equation of this parabola.

EXAMPLE **2**

Graphing a Plane
Curve Defined
Parametrically

Repeat Example 1 for the plane curve defined by the parametric equations

$$x = 2 \sin t, \ y = 3 \cos t, \quad \text{for } t \text{ in the interval } [0, 2\pi].$$

Find a rectangular equation for the curve.

SOLUTION   Taking care to see that the calculator is in radian mode and that the minimum and maximum t values are 0 and 2π, respectively, we obtain the graph shown in Figure 43. It appears to be an ellipse. (A square window is necessary to gain the correct perspective.)

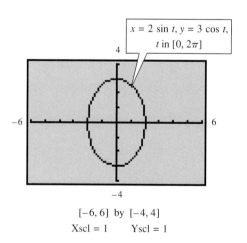

$x = 2 \sin t$, $y = 3 \cos t$,
t in $[0, 2\pi]$

$[-6, 6]$ by $[-4, 4]$
Xscl = 1   Yscl = 1

FIGURE 43

Because it is awkward to solve either equation for t, we can use another approach. Because $\sin^2 t + \cos^2 t = 1$, we can square both sides of each equation, and then solve one for $\sin^2 t$ and the other for $\cos^2 t$.

$$x = 2 \sin t \qquad\qquad y = 3 \cos t$$
$$x^2 = 4 \sin^2 t \qquad\qquad y^2 = 9 \cos^2 t$$
$$\frac{x^2}{4} = \sin^2 t \qquad\qquad \frac{y^2}{9} = \cos^2 t$$

Now add corresponding sides of the two equations to get

$$\frac{x^2}{4} + \frac{y^2}{9} = \sin^2 t + \cos^2 t$$
$$\frac{x^2}{4} + \frac{y^2}{9} = 1.$$

As we saw in Section 6.2, this is indeed an equation of an ellipse. ∎

Alternative Forms of Parametric Equations

Parametric representations of a curve are not unique. In fact, there are infinitely many parametric representations of a given curve. If the curve can be described by a rectangular equation $y = f(x)$, with domain X, then one simple parametric representation is

$$x = t, y = f(t), \quad \text{for } t \text{ in } X.$$

The next example shows how one plane curve has alternative parametric equation forms.

EXAMPLE 3

Finding Alternative Parametric Equation Forms

Give three parametric representations for the parabola

$$y = (x - 2)^2 + 1.$$

SOLUTION The simplest choice is to let

$$x = t, \quad y = (t - 2)^2 + 1, \quad \text{for } t \text{ in } (-\infty, \infty).$$

Another choice, that leads to a simpler equation for y is

$$x = t + 2, \quad y = t^2 + 1, \quad \text{for } t \text{ in } (-\infty, \infty).$$

Sometimes trigonometric functions are desirable; one choice here might be

$$x = 2 + \tan t, \quad y = \sec^2 t, \quad \text{for } t \text{ in } \left(-\frac{\pi}{2}, \frac{\pi}{2}\right). \quad ∎$$

Applications and the Cycloid

Of the many applications of parametric equations, one of the most useful allows us to determine the path of a moving object whose position is given by the function $x = f(t), y = g(t)$, where t represents time. The parametric equations give the position of the object at any time t.

EXAMPLE 4

Examining
Parametric
Equations Defining
the Position of an
Object in Motion

The motion of a projectile (neglecting air resistance) is given by

$$x = (v_0 \cos \theta)t, \quad y = (v_0 \sin \theta)t - 16t^2, \quad \text{for } t \text{ in } [0, k],$$

where t is time in seconds, v_0 is the initial speed of the projectile in the direction θ with the horizontal, x and y are in feet, and k is a positive real number. (See Figure 44.) Find the rectangular form of the equation.

SOLUTION   Solving the first equation for t and substituting the result into the second equation gives (after simplification)

$$y = (\tan \theta)x - \frac{16}{v_0{}^2 \cos^2 \theta} x^2,$$

the equation of a vertical parabola opening downward, as shown in Figure 44. ∎

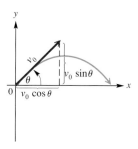

FIGURE 44

The path traced by a fixed point on the circumference of a circle rolling along a line is called a **cycloid.** See Figure 45.

The cycloid is defined by the parametric equations

$$x = at - a \sin t, \quad y = a - a \cos t, \quad \text{for } t \text{ in } (-\infty, \infty).$$

A graphics calculator provides an excellent means of graphing a cycloid.

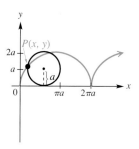

FIGURE 45

EXAMPLE 5

Graphing a Cycloid
Using a Graphics
Calculator

Graph the cycloid for $a = 3$ and t in the interval $[-2\pi, 2\pi]$.

SOLUTION   The equations

$$x = 3t - 3 \sin t, \; y = 3 - 3 \cos t$$

are used. The calculator should be set in radian mode, with minimum and maximum t values -2π and 2π, respectively. Notice that the graph never falls below the x-axis, and the maximum y-value is $2(3) = 6$. See Figure 46, where a square window is used.   ∎

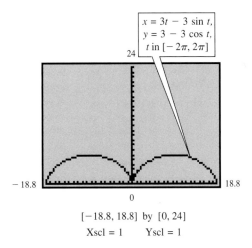

$$[-18.8, 18.8] \text{ by } [0, 24]$$
$$\text{Xscl} = 1 \qquad \text{Yscl} = 1$$

FIGURE 46

The cycloid has an interesting physical property. If a flexible cord or wire goes through points P and Q as in Figure 47 and a bead is allowed to slide without friction along this path from P to Q, the path that requires the shortest time takes the shape of the graph of an inverted cycloid. ◼

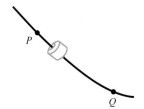

FIGURE 47

10.5 EXERCISES

For each plane curve, use a graphics calculator to generate the curve over the interval for the parameter t, using the window specified. Then find a rectangular equation for the curve.

1. $x = 2t$, $y = t + 1$,   for t in $[-2, 3]$
Window: $[-8, 8]$ by $[-8, 8]$

2. $x = t + 2$, $y = t^2$,   for t in $[-1, 1]$
Window: $[0, 4]$ by $[-2, 2]$

3. $x = \sqrt{t}$, $y = 3t - 4$,   for t in $[0, 4]$
Window: $[-6, 6]$ by $[-6, 10]$

4. $x = t^2$, $y = \sqrt{t}$,   for t in $[0, 4]$
Window: $[-2, 20]$ by $[0, 4]$

5. $x = t^3 + 1$, $y = t^3 - 1$,   for t in $[-3, 3]$
Window: $[-30, 30]$ by $[-30, 30]$

6. $x = 2t - 1$, $y = t^2 + 2$,   for t in $[-10, 10]$
Window: $[-20, 20]$ by $[0, 120]$

7. $x = 2^t$, $y = \sqrt{3t - 1}$,   for t in $\left[\dfrac{1}{3}, 4\right]$
Window: $[-2, 30]$ by $[-2, 10]$

8. $x = \ln(t - 1)$, $y = 2t - 1$,   for t in $(1, 10]$
Window: $[-5, 5]$ by $[-2, 20]$

9. $x = 2 \sin t$, $y = 2 \cos t$,   for t in $[0, 2\pi]$
Window: $[-6, 6]$ by $[-4, 4]$

10. $x = \sqrt{5} \sin t$, $y = \sqrt{3} \cos t$,   for t in $[0, 2\pi]$
Window: $[-6, 6]$ by $[-4, 4]$

11. $x = 3 \tan t$, $y = 2 \sec t$,   for t in $[0, 2\pi]$
Window: $[-6, 6]$ by $[-4, 4]$ (Use dot mode.)

12. $x = \cot t$, $y = \csc t$,   for t in $(0, 2\pi)$
Window: $[-6, 6]$ by $[-4, 4]$ (Use dot mode.)

For each plane curve, find a rectangular equation.

13. $x = \sin \theta, y = \csc \theta,$ for θ in $(0, \pi)$

14. $x = \tan \theta, y = \cot \theta,$ for θ in $\left(0, \frac{\pi}{2}\right)$

15. $x = t, y = \sqrt{t^2 + 2},$ for t in $(-\infty, \infty)$

16. $x = \sqrt{t}, y = t^2 - 1,$ for t in $[0, \infty)$

17. $x = e^t, y = e^{-t},$ for t in $(-\infty, \infty)$

18. $x = e^{2t}, y = e^t,$ for t in $(-\infty, \infty)$

19. $x = 2 + \sin \theta, y = 1 + \cos \theta,$ for θ in $[0, 2\pi]$

20. $x = 1 + 2 \sin \theta, y = 2 + 3 \cos \theta,$ for θ in $[0, 2\pi]$

21. $x = t + 2, y = \dfrac{1}{t + 2},$ $t \neq -2$

22. $x = t - 3, y = \dfrac{2}{t - 3},$ $t \neq 3$

23. $x = t^2, y = 2 \ln t,$ t in $(0, \infty)$

24. $x = \ln t, y = 3 \ln t,$ t in $(0, \infty)$

Use a graphics calculator to graph each cycloid for t in $[0, 4\pi]$. Use the window specified.

25. $x = t - \sin t, y = 1 - \cos t$
Window: $[0, 4\pi]$ by $[0, 6]$

26. $x = 2t - 2 \sin t, y = 2 - 2 \cos t$
Window: $[0, 4\pi]$ by $[0, 4]$

27. A projectile is fired with an initial velocity of 400 ft per sec. at an angle of 45° with the horizontal. Find each of the following. **(a)** the time when it strikes the ground **(b)** the range (horizontal distance covered) **(c)** the maximum altitude

28. Repeat Exercise 27 if the projectile is fired at 800 ft per sec. at an angle of 30° with the horizontal.

29. Show that the rectangular equation for the curve describing the motion of a projectile defined by
$$x = (v_0 \cos \theta)t, y = (v_0 \sin \theta)t - 16t^2,$$
for t in $[0, k]$, is
$$y = (\tan \theta)x - \frac{16}{v_0^2 \cos^2 \theta}x^2.$$

30. Find the vertex of the parabola given by the rectangular equation of Exercise 29.

31. Give two parametric representations of the line through the point (x_1, y_1) with slope m.

32. Give two parametric representations of the parabola $y = a(x - h)^2 + k$.

33. Give a parametric representation of the hyperbola
$$\frac{x^2}{a^2} - \frac{y^2}{b^2} = 1.$$

34. Give a parametric representation of the ellipse
$$\frac{x^2}{a^2} + \frac{y^2}{b^2} = 1.$$

35. The spiral of Archimedes has polar equation $r = a\theta$, where $r^2 = x^2 + y^2$. Show that a parametric representation of the spiral is
$$x = at \cos t, y = at \sin t, \text{ for } t \text{ in } (-\infty, \infty).$$
(The parameter t must be in radians.)

36. Show that the *hyperbolic spiral* $r\theta = a$, where $r^2 = x^2 + y^2$, is given parametrically by
$$x = \frac{a \cos t}{t}, y = \frac{a \sin t}{t},$$
for t in $(-\infty, 0) \cup (0, \infty)$.
(The parameter t must be in radians.)

*C*hapter 10   SUMMARY

Vectors are represented by directed line segments. Equal vectors have the same magnitude and the same direction. The parallelogram rule is used to find the sum (or resultant) of two vectors. The vector is called the resultant of its components. The horizontal and vertical components are most important in applications of vectors as forces.

A vector may also be determined by a direction angle (an angle with the horizontal) and its magnitude. The relationships of the horizontal and vertical components of a position vector to a direction angle and magnitude are basic rules for vectors. The vector operations of addition and scalar multiplication may be performed both geometrically and analytically. Vectors are particularly useful in the solution of navigation problems.

Complex numbers are graphed in a complex plane that has a real axis and an imaginary axis. Each complex number corresponds to a vector. Graphically, the sum of two complex numbers is the resultant of the corresponding vectors. Complex numbers are expressed in trigonometric or polar form by replacing the real and imaginary parts by $r \cos \theta$ and $r \sin \theta$ respectively.

Complex numbers in trigonometric (or polar) form can be multiplied and divided to give the product or quotient directly in the same form. Powers and roots of complex numbers in trigonometric (or polar) form also can be found directly from numbers expressed in that form. This is the only way to find *all* roots of many complex numbers.

A point in the plane may be graphed in the polar coordinate system by finding direction angle θ and the magnitude r of the corresponding vector. The ordered pair (r, θ) gives the polar coordinates of the point. There are many ways to express a specific point with polar coordinates. Polar equations are expressed with the variables r and θ, and may be graphed by plotting many ordered pairs (r, θ) or with a graphics calculator. Polar equations and rectangular equations may be converted from one form to the other using the relationships among x, y, r, and θ.

A third way to graph a set of ordered pairs in the plane is with parametric equations, which express x and y in terms of a third variable t. The graphs can be found by using the equations $x = f(t)$ and $y = g(t)$ to find enough ordered pairs (x, y) to determine the graph. A graphics calculator is also capable of graphing parametric equations. Alternative pairs of parametric equations can lead to the same set of ordered pairs. Parametric equations are often the simplest way to describe the position of an object in motion. A cycloid, the path traced by a fixed point on the circumference of a circle rolling along a line, is defined by parametric equations.

Key Terms

SECTION 10.1

scalar
vector
vertical component
horizontal component
zero vector
scalar product
direction angle
position vector (radius vector)
airspeed
groundspeed

SECTION 10.2

rectangular form
real axis
imaginary axis
complex plane
trigonometric (polar) form
modulus (absolute value)
argument

SECTION 10.4

rectangular coordinate system
polar coordinate system

pole
polar axis
polar coordinates
polar equation
cardioid
lemniscate
rose
spiral of Archimedes

SECTION 10.5

parametric equations
parameter
cycloid

Chapter 10   **REVIEW EXERCISES**

In Exercises 1–3, use the vectors shown here. Sketch each of the following.

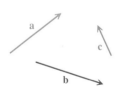

1. a + b   **2. a − b**   **3. a + 3c**

Vector **v** *has the given magnitude and makes an angle θ with the horizontal. Find the magnitudes of the horizontal and vertical components of* **v**, *first using the relationships x = r cos θ and y = r sin θ. Then use the capabilities of your calculator to support your answers.*

4. $|\mathbf{v}| = 50$, $\theta = 45°$ **5.** $|\mathbf{v}| = 69.2$, $\theta = 75°$ **6.** $|\mathbf{v}| = 964$, $\theta = 154°20'$
(give exact values)

Given two forces and the angle between them, find the magnitude of the resultant force.

7. Forces of 15 and 23 pounds, forming an angle of 87°

8. Forces of 142 and 215 newtons, forming an angle of 112°

9. Forces of 85.2 and 69.4 newtons, forming an angle of 58°20′

10. Forces of 475 and 586 pounds, forming an angle of 78°20′

Solve each of the following problems.

11. A plane has an airspeed of 520 mph. The pilot wishes to fly on a bearing of 310°. A wind of 37 mph is blowing from a bearing of 212°. What direction should the pilot fly, and what will be her actual speed?

12. A boat travels 15 km per hr in still water. The boat is traveling across a large river, on a bearing of 130°.

The current in the river, coming from the west, has a speed of 7 km per hr. Find the resulting speed of the boat and its resulting direction of travel.

13. A long-distance swimmer starts out swimming a steady 3.2 mph due north. A 5.1 mph current is flowing on a bearing of 12°. What is the swimmer's resulting bearing and speed?

Graph each complex number as a vector in the complex plane.

14. $5i$ **15.** $-4 + 2i$ **16.** $3 - 3i\sqrt{3}$

Find and graph the resultant of each pair of complex numbers.

17. $7 + 3i$ and $-2 + i$ **18.** $2 - 4i$ and $5 + i$

19. The vector representing a complex number of the form bi will lie on the _____-axis in the complex plane.

20. Explain the geometric similarity between the absolute value of a real number and the absolute value (or modulus) of a complex number.

Perform the indicated operations. Give the answer in rectangular form.

21. $[5(\cos 90° + i \sin 90°)][6(\cos 180° + i \sin 180°)]$ **22.** $[3 \text{ cis } 135°][2 \text{ cis } 105°]$

23. $\dfrac{2(\cos 60° + i \sin 60°)}{8(\cos 300° + i \sin 300°)}$ **24.** $\dfrac{4 \text{ cis } 270°}{2 \text{ cis } 90°}$ **25.** $(\sqrt{3} + i)^3$

26. $(2 - 2i)^5$ **27.** $(\cos 100° + i \sin 100°)^6$ **28.** $(\text{cis } 20°)^3$

Complete the chart in Exercises 29–36.

	Rectangular Form	Trigonometric Form
29.	$-2 + 2i$	_____
30.	_____	$3(\cos 90° + i \sin 90°)$
31.	_____	$2(\cos 225° + i \sin 225°)$
32.	$-4 + 4i\sqrt{3}$	_____
33.	$1 - i$	_____
34.	_____	$4(\cos 240° + i \sin 240°)$
35.	$-4i$	_____
36.	_____	$2 \text{ cis } 180°$

Find the indicated roots and graph as vectors in the complex plane. Leave your answers in polar form.

37. the cube roots of $-27i$ **38.** the fourth roots of $16i$ **39.** the fifth roots of 32

40. Solve the equation $x^4 + i = 0$. Leave solutions in polar form.

Convert to rectangular coordinates. Give exact values.

41. $(6, 30°)$ **42.** $(12, 225°)$ **43.** $(-8, -60°)$

Convert to polar coordinates, with $-180° < \theta \le 180°$. Give exact values.

44. $(-6, 6)$ **45.** $(0, -5)$ **46.** $(-\sqrt{5}, -\sqrt{5})$

Use a graphics calculator to graph each polar equation for $0° \le \theta \le 360°$. Use a square window.

47. $r = 4 \cos \theta$ (circle)

48. $r = -1 + \cos \theta$ (cardioid)

49. $r = 1 - 2 \sin \theta$ (limaçon with a loop)

50. $r = 2 \sin 4\theta$ (eight-leaved rose)

51. $r = 3 \cos 3\theta$ (three-leaved rose)

Find an equivalent equation in rectangular coordinates.

52. $r = \dfrac{3}{1 + \cos \theta}$ **53.** $r = \dfrac{4}{2 \sin \theta - \cos \theta}$ **54.** $r = \sin \theta + \cos \theta$ **55.** $r = 2$

Find an equivalent equation in polar coordinates.

56. $x = -3$ **57.** $y = x$ **58.** $y = x^2$ **59.** $x = y^2$

Use a graphics calculator to graph the plane curve defined by the parametric equations. Use the window specified.

60. $x = 4t - 3$, $y = t^2$,   for t in $[-3, 4]$
   Window: $[-20, 20]$ by $[-20, 20]$

61. $x = t^2$, $y = t^3$,   for t in $[-2, 2]$
   Window: $[-15, 15]$ by $[-10, 10]$

62. $x = t + \ln t$, $y = t + e^t$,   for t in $(0, 2]$
   Window: $[-10, 5]$ by $[0, 10]$

63. $x = 3t - 3 \sin t$, $y = 3 - 3 \cos t$,   for t in $[0, 2\pi]$
   Window: $[0, 6\pi]$ by $[0, 6]$

Find a rectangular equation for each plane curve with the given parametric equations.

64. $x = 3t + 2$, $y = t - 1$,   for t in $[-5, 5]$

65. $x = \sqrt{t - 1}$, $y = \sqrt{t}$,   for t in $[1, \infty)$

66. $x = 5 \tan t$, $y = 3 \sec t$,   for t in $\left(-\dfrac{\pi}{2}, \dfrac{\pi}{2}\right)$

Further Topics in Algebra

This chapter concludes our study of college algebra by discussing sequences, series, and probability. A *sequence* is a list of terms in a specific order. A *series* is the sum of the terms of a sequence. *Mathematical induction* is a powerful method used to prove that a statement is true for all positive numbers. We will use it to prove some facts about *series*. The *binomial theorem* is an example of a general statement of equality that can be proved by mathematical induction.

Counting principles and permutations and combinations are methods of finding the number of possible arrangements or groupings of the elements of a set. This information is used in *probability theory* to find the likelihood that an event will occur.

11.1 MATHEMATICAL INDUCTION

Proof by Mathematical Induction ▮ Proving Equality Statements ▮ Proving Inequality Statements

Proof by Mathematical Induction

Many results in mathematics are claimed true for any positive integer. Any of these results could be checked for $n = 1$, $n = 2$, $n = 3$, and so on, but since the set of positive integers is infinite it would be impossible to check every possible case. For example, let S_n represent the statement that the sum of the first n positive integers is $\frac{n(n+1)}{2}$,

$$S_n: 1 + 2 + 3 + \cdots + n = \frac{n(n+1)}{2}.$$

The truth of this statement can be checked quickly for the first few values of n.

If $n = 1$, S_1 is $\qquad\qquad 1 = \dfrac{1(1 + 1)}{2}$, a true statement, since $1 = 1$.

If $n = 2$, S_2 is $\qquad 1 + 2 = \dfrac{2(2 + 1)}{2}$, a true statement, since $3 = 3$.

If $n = 3$, S_3 is $\quad 1 + 2 + 3 = \dfrac{3(3 + 1)}{2}$, a true statement, since $6 = 6$.

If $n = 4$, S_4 is $1 + 2 + 3 + 4 = \dfrac{4(4 + 1)}{2}$, a true statement, since $10 = 10$.

FOR **GROUP DISCUSSION**

Have each member of the class check one of the following statements.

1. $1 = \dfrac{1(1 + 1)}{2}$

2. $1 + 2 = \dfrac{2(2 + 1)}{2}$

3. $1 + 2 + 3 = \dfrac{3(3 + 1)}{2}$

4. $1 + 2 + 3 + 4 = \dfrac{4(4 + 1)}{2}$

5. $1 + 2 + 3 + 4 + 5 = \dfrac{5(5 + 1)}{2}$

Compare the number of terms on the left in each case with the number in the numerator that is changing. What do you find? Discuss the meaning of the general statement given in the text.

Since the statement is true for $n = 1, 2, 3,$ and 4, and so on, can we conclude that the statement is true for all positive integers by observing this finite number of examples? The answer is no. However, we have an idea that it *may* be true for all positive integers.

To prove that such a statement is true for every positive integer, we use the following principle.

PRINCIPLE OF MATHEMATICAL INDUCTION

Let S_n be a statement concerning the positive integer n. Suppose that

1. S_1 is true;
2. for any positive integer k, $k \leq n$, if S_k is true, then S_{k+1} is also true.

Then S_n is true for every positive integer value of n.

A proof by mathematical induction can be explained as follows. By assumption (1) above, the statement is true when $n = 1$. By (2) above, the fact that the statement is true for $n = 1$ implies that it is true for $n = 1 + 1 = 2$. Using (2) again, the statement is thus true for $2 + 1 = 3$, for $3 + 1 = 4$, for $4 + 1 = 5$, and so on. Continuing in this way shows that the statement must be true for *every* positive integer, no matter how large.

The situation is similar to that of an infinite number of dominoes lined up as suggested in Figure 1. If the first domino is pushed over, it pushes the next, which pushes the next, and so on indefinitely.

FIGURE 1

Another example of the principle of mathematical induction might be an infinite ladder. Suppose the rungs are spaced so that, whenever you are on a rung, you know you can move to the next rung. Then *if* you can get to the first rung, you can go as high up the ladder as you wish.

Two separate steps are required for a proof by mathematical induction.

PROOF BY MATHEMATICAL INDUCTION

Step 1   Prove that the statement is true for $n = 1$.

Step 2   Show that, for any positive integer k, $k \leq n$, if S_k is true, then S_{k+1} is also true.

Proving Equality Statements

Mathematical induction is used in the next example to prove the statement discussed above: $S_n = 1 + 2 + 3 + \cdots + n = \frac{n(n + 1)}{2}$.

EXAMPLE 1

Proving an Equality Statement by Mathematical Induction

Let S_n represent the statement

$$1 + 2 + 3 + \cdots + n = \frac{n(n + 1)}{2}.$$

Prove that S_n is true for every positive integer n.

PROOF   The proof by mathematical induction is as follows.

Step 1   Show that the statement is true when $n = 1$. If $n = 1$, S_1 becomes

$$1 = \frac{1(1 + 1)}{2},$$

which is true.

Step 2   Show that S_k implies S_{k+1}, where S_k is the statement

$$1 + 2 + 3 + \cdots + k = \frac{k(k + 1)}{2},$$

and S_{k+1} is the statement

$$1 + 2 + 3 + \cdots + k + (k + 1) = \frac{(k + 1)[(k + 1) + 1]}{2}.$$

Start with S_k.

$$1 + 2 + 3 + \cdots + k = \frac{k(k + 1)}{2}$$

Add $k + 1$ to both sides of this equation.

$$1 + 2 + 3 + \cdots + k + (k + 1) = \frac{k(k + 1)}{2} + (k + 1)$$

Now factor out the common factor $k + 1$ on the right to get

$$= (k + 1)\left(\frac{k}{2} + 1\right)$$

$$= (k + 1)\left(\frac{k + 2}{2}\right)$$

$$1 + 2 + 3 + \cdots + k + (k + 1) = \frac{(k + 1)[(k + 1) + 1]}{2}.$$

This final result is the statement for $n = k + 1$; it has been shown that if S_k is true, then S_{k+1} is also true. The two steps required for a proof by mathematical induction have now been completed, so the statement S_n is true for every positive integer value of n. ∎

CAUTION Notice that the left side of the statement always includes *all* the terms up to the *n*th term, as well as the *n*th term.

<table>
<tr><td>EXAMPLE **2**</td></tr>
<tr><td>Proving an Equality
Statement by
Mathematical
Induction</td></tr>
</table>

Prove: $4 + 7 + 10 + \cdots + (3n + 1) = \dfrac{n(3n + 5)}{2}$ for all positive integers n.

PROOF

Step 1 Show that the statement is true for S_1. S_1 is

$$4 = \frac{1(3 \cdot 1 + 5)}{2}.$$

Since the right side equals 4, S_1 is a true statement.

Step 2 Show that if S_k is true, then S_{k+1} is true, where S_k is

$$4 + 7 + 10 + \cdots + (3k + 1) = \frac{k(3k + 5)}{2},$$

and S_{k+1} is

$$4 + 7 + 10 + \cdots + (3k + 1) + [3(k + 1) + 1]$$
$$= \frac{(k + 1)[3(k + 1) + 5]}{2}.$$

Start with S_k:

$$4 + 7 + 10 + \cdots + (3k + 1) = \frac{k(3k + 5)}{2}.$$

To get the left side of the equation S_k to be the left side of the equation S_{k+1}, we must add the $(k + 1)$th term. Now we try to algebraically change the right side to look like the right side of S_{k+1}. Adding $[3(k + 1) + 1]$ to both sides of S_k gives

$$4 + 7 + 10 + \cdots + (3k + 1) + [3(k + 1) + 1]$$
$$= \frac{k(3k + 5)}{2} + [3(k + 1) + 1].$$

Clear the parentheses in the new term on the right side of the equals sign and simplify.

$$= \frac{k(3k + 5)}{2} + 3k + 3 + 1$$
$$= \frac{k(3k + 5)}{2} + 3k + 4$$

Now combine the two terms on the right.

$$= \frac{k(3k + 5)}{2} + \frac{2(3k + 4)}{2}$$
$$= \frac{k(3k + 5) + 2(3k + 4)}{2}$$
$$= \frac{3k^2 + 5k + 6k + 8}{2}$$
$$= \frac{3k^2 + 11k + 8}{2}$$
$$= \frac{(k + 1)(3k + 8)}{2}$$

Since $3k + 8$ can be written as $3(k + 1) + 5$,

$$4 + 7 + 10 + \cdots + (3k + 1) + [3(k + 1) + 1]$$
$$= \frac{(k + 1)[3(k + 1) + 5]}{2}.$$

The final result is the statement S_{k+1}. Therefore, if S_k is true, then S_{k+1} is true. The two steps required for a proof by mathematical induction are completed, so the general statement S_n is true for every positive integer value of n.   ∎

Proving Inequality Statements

EXAMPLE 3

Proving an Inequality Statement by Mathematical Induction

Prove that if x is a real number between 0 and 1, then for every positive integer n,

$$0 < x^n < 1.$$

PROOF   Here S_1 is the statement

if $0 < x < 1$, then $0 < x^1 < 1$,

which is true. S_k is the statement

if $0 < x < 1$, then $0 < x^k < 1$.

To show that this implies that S_{k+1} is true, multiply all three parts of $0 < x^k < 1$ by x to get

$$x \cdot 0 < x \cdot x^k < x \cdot 1.$$

(Here the fact that $0 < x$ is used.) Simplify to get

$$0 < x^{k+1} < x.$$

Since $x < 1$,

$$x^{k+1} < x < 1$$

and

$$0 < x^{k+1} < 1.$$

By this work, if S_k is true, then S_{k+1} is true. Since both conditions for a proof have been satisfied, the given statement is true for every positive integer n. ◼

11.1 EXERCISES

1. A proof by mathematical induction allows us to prove that a statement is true for all
_____ .

2. Suppose that Step 2 in a proof by mathematical induction can be satisfied, but Step 1 cannot. May we conclude that the proof is complete? Explain.

3. What is wrong with the following proof by mathematical induction?
Prove: Any natural number equals the next natural number; that is, $n = n + 1$.
Proof. To begin, we assume the statement true for some natural number $n = k$:

$$k = k + 1.$$

We must now show that the statement is true for $n = k + 1$. If we add 1 to both sides, we have

$$k + 1 = k + 1 + 1$$
$$k + 1 = k + 2.$$

Hence, if the statement is true for $n = k$, it is also true for $n = k + 1$. Thus, the theorem is proved.

Use the method of mathematical induction to prove the following statements. Assume that n is a positive integer.

4. $2 + 4 + 6 + \cdots + 2n = n(n + 1)$

5. $1 + 3 + 5 + \cdots + (2n - 1) = n^2$

6. $3 + 6 + 9 + \cdots + 3n = \dfrac{3n(n + 1)}{2}$

7. $5 + 10 + 15 + \cdots + 5n = \dfrac{5n(n + 1)}{2}$

8. $2 + 4 + 8 + \cdots + 2^n = 2^{n+1} - 2$

9. $3 + 3^2 + 3^3 + \cdots + 3^n = \dfrac{3(3^n - 1)}{2}$

10. $1^2 + 2^2 + 3^2 + \cdots + n^2 = \dfrac{n(n + 1)(2n + 1)}{6}$

11. $1^3 + 2^3 + 3^3 + \cdots + n^3 = \dfrac{n^2(n + 1)^2}{4}$

12. $5 \cdot 6 + 5 \cdot 6^2 + 5 \cdot 6^3 + \cdots + 5 \cdot 6^n = 6(6^n - 1)$

13. $7 \cdot 8 + 7 \cdot 8^2 + 7 \cdot 8^3 + \cdots + 7 \cdot 8^n = 8(8^n - 1)$

14. $\dfrac{1}{1 \cdot 2} + \dfrac{1}{2 \cdot 3} + \dfrac{1}{3 \cdot 4} + \cdots + \dfrac{1}{n(n + 1)} = \dfrac{n}{n + 1}$

15. $\dfrac{1}{1 \cdot 4} + \dfrac{1}{4 \cdot 7} + \dfrac{1}{7 \cdot 10} + \cdots + \dfrac{1}{(3n - 2)(3n + 1)} = \dfrac{n}{3n + 1}$

16. $\dfrac{1}{2} + \dfrac{1}{2^2} + \dfrac{1}{2^3} + \cdots + \dfrac{1}{2^n} = 1 - \dfrac{1}{2^n}$

17. $\dfrac{4}{5} + \dfrac{4}{5^2} + \dfrac{4}{5^3} + \cdots + \dfrac{4}{5^n} = 1 - \dfrac{1}{5^n}$

18. $x^{2n} + x^{2n-1}y + \cdots + xy^{2n-1} + y^{2n} = \dfrac{x^{2n+1} - y^{2n+1}}{x - y}$

19. $x^{2n-1} + x^{2n-2}y + \cdots + xy^{2n-2} + y^{2n-1} = \dfrac{x^{2n} - y^{2n}}{x - y}$

20. $(a^m)^n = a^{mn}$ (Assume that a and m are constant.) **21.** $(ab)^n = a^n b^n$ (Assume that a and b are constant.)

22. If $a > 1$, then $a^n > 1$. **23.** If $a > 1$, then $a^n > a^{n-1}$.

24. If $0 < a < 1$, then $a^n < a^{n-1}$.

R*elating Concepts*

Many of the statements you are asked to prove in these exercises are generalizations of properties of algebra.

25. In the statement given in Exercise 18, replace n with 1.

26. What factorization is this a rearrangement of?

27. Write a statement similar to the one in Exercise 18 for

$$\dfrac{x^{2n+1} + y^{2n+1}}{x + y}.$$

28. The statement in Exercise 20 is the _____ rule for

_____ .

29. The statement in Exercise 21 is one of the _____ rules for

_____ .

30. Suppose that n straight lines (with $n \geq 2$) are drawn in a plane, where no two lines are parallel and no three lines pass through the same point. Show that the number of points of intersection of the lines is $\dfrac{n^2 - n}{2}$.

31. The series of sketches below starts with an equilateral triangle having sides of length 1. In the following steps, equilateral triangles are constructed on each side of the preceding figure. The lengths of the sides

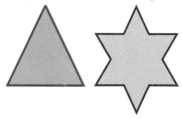

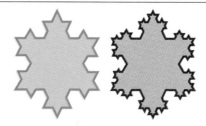

of these new triangles are $\frac{1}{3}$ the length of the sides of the preceding triangles. Develop a formula for the number of sides of the nth figure. Use mathematical induction to prove your answer.

32. Find the perimeter of the nth figure in Exercise 31.

33. Show that the area of the nth figure in Exercise 31 is

$$\sqrt{3} \left[\dfrac{2}{5} - \dfrac{3}{20} \left(\dfrac{4}{9} \right)^{n-1} \right].$$

11.2 THE BINOMIAL THEOREM

Pascal's Triangle ∎ *n*-Factorial ∎ Binomial Coefficient ∎ The Binomial Theorem ∎ *r*th Term of a Binomial Expansion

In this section we introduce a method for writing out the terms of expressions of the form $(x + y)^n$, where n is a natural number. The formula for writing out the powers of a binomial $(x + y)^n$ as a polynomial is called the *binomial theorem*. This theorem is important when working with probability and statistics. Some expansions of $(x + y)^n$, for various nonnegative integer values of n, are given below.

$$(x + y)^0 = 1$$
$$(x + y)^1 = x + y$$
$$(x + y)^2 = x^2 + 2xy + y^2$$
$$(x + y)^3 = x^3 + 3x^2y + 3xy^2 + y^3$$
$$(x + y)^4 = x^4 + 4x^3y + 6x^2y^2 + 4xy^3 + y^4$$
$$(x + y)^5 = x^5 + 5x^4y + 10x^3y^2 + 10x^2y^3 + 5xy^4 + y^5$$

Studying these results reveals a pattern that can be used to write a general expression for $(x + y)^n$. First, notice that after the special case $(x + y)^0 = 1$, each expression begins with x raised to the same power as the binomial itself. That is, the expansion of $(x + y)^1$ has a first term of x^1, $(x + y)^2$ has a first term of x^2, $(x + y)^3$ has a first term of x^3, and so on. Also, the last term in each expansion is y to the same power as the binomial. Thus the expansion of $(x + y)^n$ should begin with the term x^n and end with the term y^n.

Also, the exponents on x decrease by one in each term after the first, while the exponents on y, beginning with y in the second term, increase by one in each succeeding term. That is, the *variables* in the terms of the expansion of $(x + y)^n$ have the following pattern.

$$x^n, x^{n-1}y, x^{n-2}y^2, x^{n-3}y^3, \ldots, xy^{n-1}, y^n$$

This pattern suggests that the sum of the exponents on x and y in each term is n. For example, in the third term in the list above, the variable is $x^{n-2}y^2$, and the sum of the exponents is $n - 2 + 2 = n$.

Pascal's Triangle

Now examine the *coefficients* in the terms of the expansions shown above. Writing the coefficients alone gives the following pattern.

PASCAL'S TRIANGLE

```
              1
           1     1
        1     2     1
     1     3     3     1
  1     4     6     4     1
1     5    10    10     5     1
```

With the coefficients arranged in this way, it can be seen that each number in the triangle is the sum of the two numbers directly above it (one to the right and one to the left.) For example, if we number the rows starting with row 0, in row four, 1 is the sum of 1, the only number above it, 4 is the sum of 1 and 3, 6 is the sum of 3 and 3, and so on. This triangular array of numbers is called **Pascal's triangle,** in honor of the seventeenth-century mathematician Blaise Pascal (1623–1662), one of the first to use it extensively.

n-Factorial

To get the coefficients for $(x + y)^6$, we add row six to the array of numbers given above. Adding adjacent numbers, we find that row six is

$$1 \quad 6 \quad 15 \quad 20 \quad 15 \quad 6 \quad 1.$$

Using these coefficients, the expansion of $(x + y)^6$ is

$$(x + y)^6 = x^6 + 6x^5y + 15x^4y^2 + 20x^3y^3 + 15x^2y^4 + 6xy^5 + y^6.$$

Although it is possible to use Pascal's triangle to find the coefficients of $(x + y)^n$ for any positive integer value of n, this becomes impractical for large values of n because of the need to write out all the preceding rows. A more efficient way of finding these coefficients uses *factorial notation.* The number $n!$ (read "*n*-factorial") is defined as follows.

---

n-FACTORIAL

For any positive integer n,

$$n! = n(n - 1)(n - 2) \cdots (3)(2)(1),$$

and

$$0! = 1.$$

EXAMPLE 1

Evaluating Factorials

Evaluate each factorial: **(a)** 5!   **(b)** 7!   **(c)** 2!   **(d)** 1!

SOLUTION

(a) $5! = 5 \cdot 4 \cdot 3 \cdot 2 \cdot 1 = 120$ **(b)** $7! = 7 \cdot 6 \cdot 5 \cdot 4 \cdot 3 \cdot 2 \cdot 1 = 5040$

(c) $2! = 2 \cdot 1 = 2$ **(d)** $1! = 1$ ◘

Graphics calculators have the capability of finding $n!$. A calculator with a 10-digit display will give the exact value of $n!$ for $n \leq 13$ and approximate values of $n!$ for $14 \leq n \leq 69$. Figure 2 shows the display for 13!, 25!, and 69!.

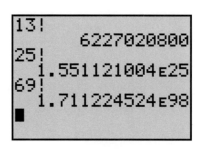

FIGURE 2

Binomial Coefficient

Now look at the coefficients of the expression

$$(x + y)^5 = x^5 + 5x^4y + 10x^3y^2 + 10x^2y^3 + 5xy^4 + y^5.$$

The coefficient of the second term, $5x^4y$, is 5, and the exponents on the variables are 4 and 1. Note that

$$5 = \frac{5!}{4!\ 1!}.$$

The coefficient of the third term is 10, with exponents of 3 and 2, and

$$10 = \frac{5!}{3!\ 2!}.$$

The last term (the sixth term) can be written as $y^5 = 1x^0y^5$, with coefficient 1, and exponents of 0 and 5. Since $0! = 1$, check that

$$1 = \frac{5!}{0!\ 5!}.$$

Generalizing from these examples, we find that the coefficient for the term of the expansion of $(x + y)^n$ in which the variable part is x^ry^{n-r} (where $r \leq n$) will be

$$\frac{n!}{r!(n - r)!}.$$

This number, called a **binomial coefficient,** is often written as $\binom{n}{r}$ (read "n choose r") or $_nC_r$. We will use both notations.

BINOMIAL COEFFICIENT

For nonnegative integers n and r, with $r \leq n$, the symbol $\binom{n}{r}$ is defined as

$$\binom{n}{r} = \frac{n!}{r!(n - r)!}.$$

These binomial coefficients are just numbers from Pascal's triangle. For example, $\binom{3}{0}$ is the first number in the row that begins 1, 3, and $\binom{7}{4}$ is the fifth number in the row that begins 1, 7. Notice that $\binom{7}{3} = \binom{7}{4}$, and in general,

$$\binom{n}{r} = \binom{n}{n - r},$$

for positive integers n and r, $n \geq r$.

EXAMPLE **2**	Evaluate each binomial coefficient.
Evaluating Binomial Coefficients	**SOLUTION**

(a) $\dfrac{8!}{6!\ 2!} = \dfrac{8 \cdot 7 \cdot 6 \cdot 5 \cdot 4 \cdot 3 \cdot 2 \cdot 1}{6 \cdot 5 \cdot 4 \cdot 3 \cdot 2 \cdot 1 \cdot 2 \cdot 1} = \dfrac{8 \cdot 7}{2 \cdot 1} = 28$

(b) $\dfrac{4!}{3!\ 1!} = \dfrac{4 \cdot 3 \cdot 2 \cdot 1}{3 \cdot 2 \cdot 1 \cdot 1} = 4$

(c) $\dbinom{6}{2} = \dfrac{6!}{2!\,4!} = \dfrac{6 \cdot 5 \cdot 4 \cdot 3 \cdot 2 \cdot 1}{2 \cdot 1 \cdot 4 \cdot 3 \cdot 2 \cdot 1} = 15$

(d) $\dbinom{7}{5} = \dfrac{7!}{5!\,2!} = \dfrac{7 \cdot 6 \cdot 5 \cdot 4 \cdot 3 \cdot 2 \cdot 1}{5 \cdot 4 \cdot 3 \cdot 2 \cdot 1 \cdot 2 \cdot 1} = 21$ ◼

The $_nC_r$ notation, usually found with the math key of a graphics calculator, gives binomial coefficients. See Figure 3 for a calculator screen of the coefficients found in Examples 2(c) and (d).

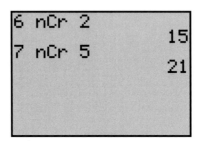

```
6 nCr 2
                    15
7 nCr 5
                    21
```

FIGURE 3

FOR GROUP DISCUSSION

1. Have members of the class compute pairs of binomial coefficients of the form $_nC_r$ and $_nC_{n-r}$, for example, $_{10}C_3$ and $_{10}C_7$.
2. Compare your results with Pascal's triangle.
3. Discuss your results and express them as a generalization.

The Binomial Theorem

Our observations about the expansion of $(x + y)^n$ are summarized as follows.

EXPANSION OF $(x + y)^n$

1. There are $(n + 1)$ terms in the expansion.
2. The first term is x^n, and the last term is y^n.
3. The exponent on x decreases by 1 and the exponent on y increases by 1 in each succeeding term.
4. The sum of the exponents on x and y in any term is n.
5. The coefficient of $x^r y^{n-r}$ is $\binom{n}{r}$.

These observations about the expansion of $(x + y)^n$ for any positive integer value of n suggest the **binomial theorem.**

BINOMIAL THEOREM

For any positive integer n:

$$(x + y)^n = x^n + \binom{n}{n-1}x^{n-1}y + \binom{n}{n-2}x^{n-2}y^2 + \binom{n}{n-3}x^{n-3}y^3$$

$$+ \cdots + \binom{n}{n-r}x^{n-r}y^r + \cdots + \binom{n}{1}xy^{n-1} + y^n.$$

The binomial theorem can be proved using mathematical induction (discussed in Section 11.1). Note that the expansion of $(x + y)^n$ has $n + 1$ terms as expected.

CAUTION Avoid the common error of expanding $(x + y)^n$ as $x^n + y^n$. Assuming that x and y are nonzero, these expressions are equivalent only if $n = 1$.

EXAMPLE 3

Expanding a Binomial

Write out the binomial expansion of $(x + y)^9$.

SOLUTION Using the binomial theorem,

$$(x + y)^9 = x^9 + \binom{9}{8} x^8 y + \binom{9}{7} x^7 y^2 + \binom{9}{6} x^6 y^3 + \binom{9}{5} x^5 y^4$$

$$+ \binom{9}{4} x^4 y^5 + \binom{9}{3} x^3 y^6 + \binom{9}{2} x^2 y^7 + \binom{9}{1} xy^8 + y^9.$$

Now evaluate each coefficient.

$$(x + y)^9 = x^9 + \frac{9!}{8!\,1!} x^8 y + \frac{9!}{7!\,2!} x^7 y^2 + \frac{9!}{6!\,3!} x^6 y^3 + \frac{9!}{5!\,4!} x^5 y^4$$

$$+ \frac{9!}{4!\,5!} x^4 y^5 + \frac{9!}{3!\,6!} x^3 y^6 + \frac{9!}{2!\,7!} x^2 y^7 + \frac{9!}{1!\,8!} xy^8 + y^9$$

$$= x^9 + 9x^8 y + 36x^7 y^2 + 84x^6 y^3 + 126x^5 y^4 + 126x^4 y^5$$

$$+ 84x^3 y^6 + 36x^2 y^7 + 9xy^8 + y^9 \quad \blacksquare$$

EXAMPLE 4

Expanding a Binomial

Expand $\left(a - \dfrac{b}{2} \right)^5$.

SOLUTION Again, use the binomial theorem.

$$\left(a - \frac{b}{2} \right)^5 = a^5 + \binom{5}{4} a^4 \left(-\frac{b}{2} \right) + \binom{5}{3} a^3 \left(-\frac{b}{2} \right)^2 + \binom{5}{2} a^2 \left(-\frac{b}{2} \right)^3$$

$$+ \binom{5}{1} a \left(-\frac{b}{2} \right)^4 + \left(-\frac{b}{2} \right)^5$$

$$= a^5 + 5a^4 \left(-\frac{b}{2} \right) + 10a^3 \left(-\frac{b}{2} \right)^2 + 10a^2 \left(-\frac{b}{2} \right)^3$$

$$+ 5a \left(-\frac{b}{2} \right)^4 + \left(-\frac{b}{2} \right)^5$$

$$= a^5 - \frac{5}{2} a^4 b + \frac{5}{2} a^3 b^2 - \frac{5}{4} a^2 b^3 + \frac{5}{16} ab^4 - \frac{1}{32} b^5 \quad \blacksquare$$

*r*th Term of a Binomial Expansion

Any single term of a binomial expansion can be determined without writing out the whole expansion. For example, the seventh term of $(x + y)^9$ has y raised to the sixth power (since y has the power 1 in the second term, the power 2 in the third term, and so on). The exponents on x and y in each term must have a sum of 9, so the exponent on x in the seventh term is $9 - 6 = 3$. Thus, writing the coefficient as given in the binomial theorem, the seventh term should be

$$\frac{9!}{3!\,6!} x^3 y^6.$$

This is in fact the seventh term of $(x + y)^9$ found in Example 3. This discussion suggests the next theorem.

---

rTH TERM OF THE BINOMIAL EXPANSION

The rth term of the binomial expansion of $(x + y)^n$, where $n \geq r - 1$, is

$$\binom{n}{n - (r - 1)} x^{n-(r-1)} y^{r-1}.$$

EXAMPLE 5

Finding a Specific Term

Find the fourth term of $(a + 2b)^{10}$.

SOLUTION In the fourth term $2b$ has an exponent of 3, while a has an exponent of $10 - 3$, or 7. Using $n = 10$, $r = 4$, $x = a$, and $y = 2b$ in the formula above, we find that the fourth term is

$$\binom{10}{7} a^7 (2b)^3 = 120a^7(8b^3) = 960a^7 b^3. \quad ∎$$

11.2 EXERCISES

Evaluate each of the following.

1. $\dfrac{6!}{3! \, 3!}$ **2.** $\dfrac{5!}{2! \, 3!}$ **3.** $\dfrac{7!}{3! \, 4!}$ **4.** $\dfrac{8!}{5! \, 3!}$ **5.** $\binom{8}{3}$

6. $\binom{7}{4}$ **7.** $\binom{10}{8}$ **8.** $\binom{9}{6}$ **9.** $\binom{n}{n-1}$ **10.** $\binom{n}{n-2}$

11. Describe in your own words how you would determine the binomial coefficient for the fifth term in the expansion of $(x + y)^8$.

12. How many terms are there in the expansion of $(x + y)^{10}$?

13. What is true of the signs (positive/negative) of the terms in the expansion of $(x - y)^n$?

14. In the expansion of $(a + 5b)^n$, what quantity replaces y in the binomial theorem, 5 or $5b$?

Write out the binomial expansion for each of the following.

15. $(x + y)^6$ **16.** $(m + n)^4$ **17.** $(p - q)^5$ **18.** $(a - b)^7$

19. $(r^2 + s)^5$ **20.** $(m + n^2)^4$ **21.** $(p + 2q)^4$ **22.** $(3r - s)^6$

23. $(7p + 2q)^4$ **24.** $(4a - 5b)^5$ **25.** $(3x - 2y)^6$ **26.** $(7k - 9j)^4$

27. $\left(\dfrac{m}{2} - 1\right)^6$ **28.** $\left(3 + \dfrac{y}{3}\right)^5$ **29.** $\left(\sqrt{2r} + \dfrac{1}{m}\right)^4$ **30.** $\left(\dfrac{1}{k} - \sqrt{3p}\right)^3$

For each of the following, write the indicated term of the binomial expansion.

31. Fifth term of $(m - 2p)^{12}$ **32.** Fourth term of $(3x + y)^6$

33. Sixth term of $(x + y)^9$ **34.** Twelfth term of $(a - b)^{15}$

35. Ninth term of $(2m + n)^{10}$ **36.** Seventh term of $(3r - 5s)^{12}$

37. Seventeenth term of $(p^2 + q)^{20}$ **38.** Tenth term of $(2x^2 + y)^{14}$

39. Eighth term of $(x^3 + 2y)^{14}$ **40.** Thirteenth term of $(a + 2b^3)^{12}$

Relating Concepts

Over the years, many interesting patterns have been discovered in Pascal's triangle. The following exercises exhibit some of these patterns.

41. *Triangular numbers* are found by counting the number of points in triangular arrangements of points. The first few triangular numbers are shown here. Count the number of points in each figure and write the results in order.

Triangular numbers

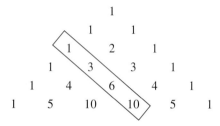

42. What special name applies to the list of numbers in the sequence from Pascal's triangle indicated below?

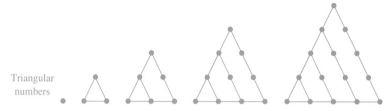

43. By counting the increase in the number of points from each triangular number to the next one, predict the next five numbers in the list in Exercise 41.

44. In the array below, Pascal's triangle is written in a different form. Find the sums along the diagonals shown. This is the *Fibonacci sequence,* an important sequence of numbers, named for Leonardo of Pisa (1170–1250). The presence of this sequence in the triangle apparently was not recognized by Pascal. Predict the next two numbers in this sequence.

$$
\begin{array}{ccccccc}
1 \\
1 & 1 \\
1 & 2 & 1 \\
1 & 3 & 3 & 1 \\
1 & 4 & 6 & 4 & 1 \\
1 & 5 & 10 & 10 & 5 & 1 \\
1 & 6 & 15 & 20 & 15 & 6 & 1
\end{array}
$$

In calculus it is shown that

$$(1 + x)^n = 1 + nx + \frac{n(n-1)}{2!}x^2 + \frac{n(n-1)(n-2)}{3!}x^3 + \cdots$$

for any real number n (not just positive integers) and any real number x, where $|x| < 1$. *This result, a generalized binomial theorem, may be used to find approximate values of powers or roots. For example,*

$$\sqrt[4]{630} = (625 + 5)^{1/4} = \left[625 \left(1 + \frac{5}{625}\right) \right]^{1/4} = 625^{1/4} \left(1 + \frac{5}{625}\right)^{1/4}.$$

45. Use the expression given above for $(1 + x)^n$ to approximate $(1 + \frac{5}{625})^{1/4}$ to the nearest thousandth. Then approximate $\sqrt[4]{630}$.

46. Approximate $\sqrt[3]{9.42}$, using this method.

47. Approximate $(1.02)^{-3}$.

48. Show that $\binom{n}{2} = \dfrac{n(n-1)}{2!}$ and $\binom{n}{3} = \dfrac{n(n-1)(n-2)}{3!}$.

Further Explorations

If your graphics calculator has a TABLE feature, it can be used to find the coefficients of the expansion of $(x + y)^n$ for any exponent n. *Note:* This will also generate a row of Pascal's triangle.

For example, to find the coefficients of the expanded form of $(x + y)^6$, we direct the calculator to construct a table for $Y_1 = 6 \,_nC_r\, X$, letting X start at 0 and increase by 1 each time. See Figures (a) and (b). In order to graph this function, we use a friendly window because only whole numbers are evaluated. See Figure (c).

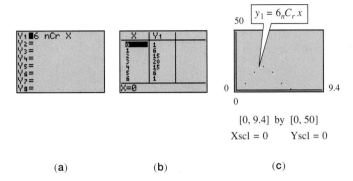

$[0, 9.4]$ by $[0, 50]$
$Xscl = 0$ $Yscl = 0$

(a)   (b)   (c)

Use the method described above to find the coefficients of the expanded form of each of the following.

1. $(x + y)^4$ **2.** $(x + y)^5$ **3.** $(x + y)^8$ **4.** $(x + y)^{16}$

11.3 SEQUENCES

Defining Sequences ▮ Recursive Sequences ▮ Arithmetic Sequences ▮ Geometric Sequences

TECHNOLOGICAL NOTE
Some modern graphics calculators have a designated SEQUENCE mode (similar to the modes FUNCTION, POLAR, and PARAMETRIC). This mode allows the user to investigate and graph sequences defined in terms of *n*, where *n* is a natural number. You should consult your owner's manual for specific instructions on how to use this feature.

Defining Sequences

If the domain of a function is the set of positive integers, the range elements can be *ordered,* as $f(1), f(2), f(3)$, and so on. This ordered list of numbers is called a **sequence.** Since the letter x has been used to suggest real numbers, the variable n is used instead with sequences to suggest the positive integer domain. Although a sequence may be defined by $f(n) = 2n + 3$, for example, it is customary to use a_n instead of $f(n)$ and write $a_n = 2n + 3$.

The elements of a sequence, called the **terms of the sequence,** are written in order as $a_1, a_2, a_3, \ldots$. The **general term,** or **nth term,** of the sequence is a_n. The general term of a sequence is used to find any term of the sequence. For example, if $a_n = n + 1$, then $a_2 = 2 + 1 = 3$.

Figure 4 shows graphs of $f(x) = 2x$ and $a_n = 2n$. Notice that f defines a "continuous" function, while a_n is "discontinuous." (See Chapter 2 for an informal discussion of continuity.)

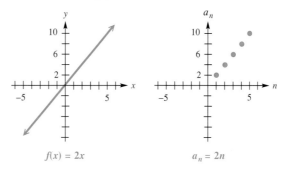

$f(x) = 2x$ $\qquad\qquad$ $a_n = 2n$

FIGURE 4

A sequence is a **finite sequence** if the domain is the set $\{1, 2, 3, 4, \ldots, n\}$, where n is a positive integer. An **infinite sequence** has the set of all positive integers as its domain. For example, the sequence of positive even integers,

$$2, 4, 6, 8, 10, 12, 14, \ldots ,$$

is infinite, but the sequence of dates in June,

$$1, 2, 3, 4, \ldots , 29, 30,$$

is finite.

EXAMPLE 1

Finding the Terms of a Sequence from the General Term

Write the first five terms for each of the following sequences.

(a) $a_n = \dfrac{n + 1}{n + 2}$

SOLUTION   Replacing n, in turn, with 1, 2, 3, 4, and 5 gives

$$\frac{2}{3}, \frac{3}{4}, \frac{4}{5}, \frac{5}{6}, \frac{6}{7}.$$

The terms of this sequence are graphed in Figure 5.

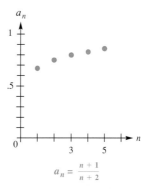

$a_n = \dfrac{n + 1}{n + 2}$

FIGURE 5

(b) $a_n = (-1)^n \cdot n$

SOLUTION   Replace n, in turn, with 1, 2, 3, 4, and 5, to get

$$a_1 = (-1)^1 \cdot \mathbf{1} = -1$$
$$a_2 = (-1)^2 \cdot \mathbf{2} = 2$$
$$a_3 = (-1)^3 \cdot \mathbf{3} = -3$$
$$a_4 = (-1)^4 \cdot \mathbf{4} = 4$$
$$a_5 = (-1)^5 \cdot \mathbf{5} = -5. \quad \blacksquare$$

TECHNOLOGICAL NOTE
Figures 6 and 7 in this
section provide examples
of how graphs of se-
quences appear on a
graphics calculator
screen.

Sequences can be graphed by using the sequence mode and dot mode of your graphics calculator. In Figure 6, we show a calculator screen that corresponds to the traditional graph in Figure 5.

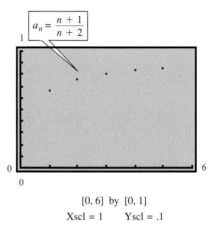

$$a_n = \frac{n + 1}{n + 2}$$

[0, 6] by [0, 1]

Xscl = 1   Yscl = .1

FIGURE 6

If the terms of an infinite sequence get closer and closer to some real number, the sequence is said to be **convergent** and to **converge** to that real number. Graphs of sequences illustrate this property. The sequence in Example 1(a) is graphed in Figures 5 and 6. What number do you think this sequence converges to? A sequence that does not converge to some number is **divergent.**

Recursive Sequences

We can define a sequence using a recursion formula. A **recursion formula** defines the nth term of a sequence in terms of the previous term. For example, if the first term of a sequence is $a_1 = 2$ and the nth term is $a_n = a_{n-1} + 3$, then $a_2 = a_1 + 3 = 2 + 3 = 5$, $a_3 = a_2 + 3 = 5 + 3 = 8$, and so on.

EXAMPLE 2

Using a Recursion
Formula

Find the first four terms for the sequences defined as follows.

(a) $a_1 = 4$; for $n \geq 2$, $a_n = 2a_{n-1} + 1$

SOLUTION   We have $a_1 = 4$, so

$$a_2 = 2 \cdot a_1 + 1 = 2 \cdot 4 + 1 = 9,$$
$$a_3 = 2 \cdot a_2 + 1 = 2 \cdot 9 + 1 = 19,$$

and
$$a_4 = 2 \cdot a_3 + 1 = 2 \cdot 19 + 1 = 39.$$

(b) $a_1 = 2$; for $n \geq 2$, $a_n = a_{n-1} + n - 1$

SOLUTION

$$a_1 = 2$$
$$a_2 = a_1 + 2 - 1 = 2 + 1 = 3$$
$$a_3 = a_2 + 3 - 1 = 3 + 2 = 5$$
$$a_4 = a_3 + 4 - 1 = 5 + 3 = 8 \quad \blacksquare$$

Arithmetic Sequences

A sequence in which each term after the first is obtained by adding a fixed number to the preceding term is called an **arithmetic sequence** (or **arithmetic progression**). The fixed number that is added is called the **common difference** and is designated d. The sequence

$$5, 9, 13, 17, 21, \ldots$$

is an arithmetic sequence since each term after the first is obtained by adding 4 to the previous term. That is,

$$9 = 5 + 4$$
$$13 = 9 + 4$$
$$17 = 13 + 4$$
$$21 = 17 + 4,$$

and so on. The common difference d is 4.

EXAMPLE 3

Finding the Terms Given a_1 and d

Write the first five terms for the arithmetic sequence with a first term of 7 and a common difference of -3.

SOLUTION   Here

$$a_1 = 7 \quad \text{and} \quad d = -3.$$
$$a_2 = 7 + (-3) = 4,$$
$$a_3 = 4 + (-3) = 1,$$
$$a_4 = 1 + (-3) = -2,$$
$$a_5 = -2 + (-3) = -5. \quad \blacksquare$$

If a_1 is the first term of an arithmetic sequence and d is the common difference, then the terms of the sequence are given by

$$a_1 = a_1$$
$$a_2 = a_1 + d$$
$$a_3 = a_2 + d = a_1 + d + d = a_1 + 2d$$
$$a_4 = a_3 + d = a_1 + 2d + d = a_1 + 3d$$
$$a_5 = a_1 + 4d$$
$$a_6 = a_1 + 5d,$$

and, by this pattern $a_n = a_1 + (n - 1)d$.

> ### nTH TERM OF AN ARITHMETIC SEQUENCE
>
> In an arithmetic sequence with first term a_1 and common difference d, the nth term, a_n, is given by
>
> $$a_n = a_1 + (n - 1)d.$$

EXAMPLE 4

Using the Formula for the nth Term

Find a_{13} and a_n for the arithmetic sequence

$$-3, 1, 5, 9, \ldots.$$

SOLUTION   Here $a_1 = -3$ and $d = 1 - (-3) = 4$. To find a_{13}, substitute 13 for n in the preceding formula.

$$a_{13} = a_1 + (13 - 1)d$$
$$a_{13} = -3 + (12)4$$
$$a_{13} = -3 + 48$$
$$a_{13} = 45$$

Find a_n by substituting values for a_1 and d in the formula for a_n.

$$a_n = -3 + (n - 1) \cdot 4$$
$$a_n = -3 + 4n - 4 \qquad \text{Distributive property}$$
$$a_n = 4n - 7 \qquad \blacksquare$$

EXAMPLE 5

Using the Formula for the nth Term

Suppose that an arithmetic sequence has $a_8 = -16$ and $a_{16} = -40$. Find a_1.

SOLUTION   We must find d first. Since $a_8 = a_1 + (8 - 1)d$, replacing a_8 with -16 gives $-16 = a_1 + 7d$ or $a_1 = -16 - 7d$. Similarly, $-40 = a_1 + 15d$ or $a_1 = -40 - 15d$. From these two equations, using the substitution method from Chapter 6,

$$-16 - 7d = -40 - 15d,$$

so $d = -3$. To find a_1, substitute -3 for d in $-16 = a_1 + 7d$:

$$-16 = a_1 + 7d$$
$$-16 = a_1 + 7(-3) \qquad \text{Let } d = -3.$$
$$a_1 = 5. \qquad \blacksquare$$

Geometric Sequences

A **geometric sequence** (or **geometric progression**) is a sequence in which each term after the first is obtained by multiplying the preceding term by a constant nonzero real number, called the **common ratio,** designated r. An example of a geometric sequence is 2, 8, 32, 128, . . . in which the first term is 2 and the common ratio r is 4. The common ratio can be found by choosing any term except the first and dividing it by the preceding term.

As mentioned above, the geometric sequence 2, 8, 32, 128, . . . has $r = 4$. Notice that

$$8 = 2 \cdot 4$$
$$32 = 8 \cdot 4 = (2 \cdot 4) \cdot 4 = 2 \cdot 4^2$$
$$128 = 32 \cdot 4 = (2 \cdot 4^2) \cdot 4 = 2 \cdot 4^3.$$

To generalize this result, assume that a geometric sequence has first term a_1 and common ratio r. The second term can be written as $a_2 = a_1 r$, the third as $a_3 = a_2 r = (a_1 r)r = a_1 r^2$, and so on. Following this pattern, the nth term is $a_n = a_1 r^{n-1}$.

---

nTH TERM OF A GEOMETRIC SEQUENCE

In the geometric sequence with first term a_1 and common ratio r, the nth term is

$$a_n = a_1 r^{n-1}.$$

EXAMPLE 6

Using the Formula for the nth Term

Find a_5 and a_n for the following geometric sequence.

$$4, 12, 36, 108, \ldots$$

SOLUTION The first term, a_1, is 4. Find r by choosing any term except the first and dividing it by the preceding term. For example,

$$r = \frac{36}{12} = 3.$$

Since $a_4 = 108$, $a_5 = 3 \cdot 108 = 324$. The fifth term also could be found using the formula for a_n, $a_n = a_1 r^{n-1}$, and replacing n with 5, r with 3, and a_1 with 4.

$$a_5 = 4 \cdot (3)^{5-1} = 4 \cdot 3^4 = 324$$

By the formula,

$$a_n = 4 \cdot 3^{n-1}. \quad \blacksquare$$

FOR GROUP DISCUSSION

Decide whether each of the following sequences are arithmetic, geometric, or neither. Explain why. Then give the general term of each sequence.

1. 2, 4, 6, 8, 10, . . .   **2.** 2, 4, 8, 16, 32, . . .   **3.** 1, 4, 9, 16, 25, . . .

EXAMPLE 7

Using the Formula for the nth Term

Find a_1 and r for the geometric sequence with $a_5 = 15$ and $a_7 = 375$.

SOLUTION First substitute $n = 5$ and then $n = 7$ into $a_n = a_1 r^{n-1}$.

$$a_5 = a_1 r^4 = 15 \quad \text{and} \quad a_7 = a_1 r^6 = 375$$

Solve the first equation for a_1 to get $a_1 = \frac{15}{r^4}$. Then substitute for a_1 in the second equation.

$$a_1 r^6 = 375$$

$$\frac{15}{r^4} \cdot r^6 = 375 \qquad \text{Let } a_1 = \tfrac{15}{r^4}.$$

$$15r^2 = 375$$

$$r^2 = 25$$

$$r = \pm 5 \qquad \text{Square root property}$$

Either 5 or -5 can be used for r. To find a_1, use

$$a_1 = \frac{15}{r^4}.$$

Replace r with ± 5.

$$a_1 = \frac{15}{(\pm 5)^4} = \frac{15}{625} = \frac{3}{125}$$

There are two sequences that satisfy the given conditions: one with $a_1 = \frac{3}{125}$ and $r = 5$ and the other with $a_1 = \frac{3}{125}$ and $r = -5$. ◻

Graphics calculators are especially useful for solving sequence problems like the following.

EXAMPLE 8

Buying a Car

Suppose you wish to buy a new car and hope to finance the price of $14,000 at 10.5% yearly interest with monthly payments of $260. Write a recursive equation that models the repayment of this loan.

SOLUTION   Each month you will pay the interest on the outstanding balance, plus $260 minus that amount (the interest) to reduce the balance. Since the interest is monthly, the interest rate will be $\frac{.105}{12}$. The first month looks like this.

Payment	Interest	Toward Balance	New Balance
1	$14{,}000\left(\frac{.105}{12}\right)$	$260 - 14{,}000\left(\frac{.105}{12}\right)$	$14{,}000 - \left(260 - 14{,}000\left(\frac{.105}{12}\right)\right)$

If we let a_n equal the balance after the nth payment, then a_{n-1} replaces 14,000 in the expressions in the chart and the new balance is

$$a_n = a_{n-1} - \left(260 - a_{n-1}\left(\frac{.105}{12}\right)\right)$$

$$= a_{n-1} - 260 + a_{n-1}\left(\frac{.105}{12}\right)$$

$$= a_{n-1}\left(1 + \frac{.105}{12}\right) - 260 \qquad \text{Factor out } a_{n-1}.$$

TECHNOLOGICAL NOTE
This section and the one that immediately follows introduce several formulas involving sequences and series. You may wish to explore the programming capability of your calculator to create short programs that utilize these formulas.

Use the capabilities of your calculator to graph the sequence. (You will need to be in sequence mode.) The graph and a portion of the table are shown in Figure 7(a) and (b). By using trace or the table feature you can answer many questions about this loan.

(a) Will the loan be paid off in 60 payments of $260 per month?
(b) If not, how many payments are needed?
(c) To the nearest dollar, what payment would pay off the loan in 60 payments?
(d) When will half the loan be paid off? ◻

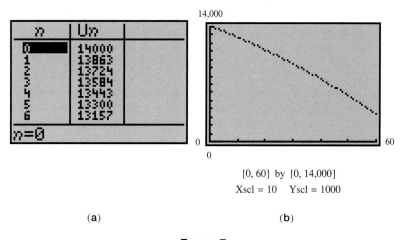

14,000

0

0   60

$[0, 60]$ by $[0, 14{,}000]$

Xscl $= 10$ Yscl $= 1000$

(a)   (b)

FIGURE 7

11.3 EXERCISES

1. Your friend does not understand what is meant by the nth term or general term of a sequence. How would you explain this idea?

2. How are sequences related to functions? Discuss some similarities and some differences.

Write the first five terms of each of the following sequences.

3. $a_n = (-1)^{n+1}$

4. $a_n = (-1)^n(n + 2)$

5. $a_n = \dfrac{2n}{n + 3}$

6. $a_n = \dfrac{-4}{n + 5}$

7. $a_n = (-2)^n(n)$

8. $a_n = \left(-\dfrac{1}{2}\right)^n (n^{-1})$

9. $a_n = x^n$

10. $a_n = n \cdot x^{-n}$

Find the first ten terms for the sequences defined as follows.

11. $a_1 = 4$;   for $n \geq 2$, $a_n = a_{n-1} + 5$

12. $a_1 = -3$;   for $n \geq 2$, $a_n = 2 \cdot a_{n-1}$

*13. $a_1 = 1$, $a_2 = 1$;   for $n \geq 3$, $a_n = a_{n-1} + a_{n-2}$

14. $a_1 = 3$, $a_2 = 2$;   for $n \geq 3$, $a_n = a_{n-1} - a_{n-2}$

For each of the following arithmetic sequences, write the indicated number of terms.

15. $a_2 = 9$, $d = -2$, $n = 4$

16. $a_3 = 7$, $d = -4$, $n = 4$

17. $a_3 = -2$, $d = -4$, $n = 4$

18. $a_2 = -12$, $d = -6$, $n = 5$

For each of the following sequences that are arithmetic, find d and a_n.

19. $8, 17, 26, 35, 44, 53, \ldots$

20. $-19, -12, -5, 2, 9, \ldots$

21. $x, x + m, x + 2m, x + 3m, x + 4m, \ldots$

22. $k + p, k + 2p, k + 3p, k + 4p, \ldots$

Find a_8 and a_n for each of the following arithmetic sequences.

23. $a_1 = 5$, $d = 2$

24. $a_1 = -3$, $d = -4$

25. $a_{10} = 6$, $a_{12} = 15$

26. $a_{15} = 8$, $a_{17} = 2$

27. $a_6 = 2m$, $a_7 = 3m$

28. $a_5 = 4p + 1$, $a_7 = 6p + 7$

29. Give an example of an arithmetic sequence with $a_4 = 12$.

30. Explain in your own words what is meant by an *arithmetic sequence*.

---

* This is the Fibonacci sequence, a well-known sequence that occurs widely in nature.

31. Which one of the following is not an arithmetic sequence?
   (a) 4, 6, 8, 10, . . .   **(b)** −2, 6, 14, 22, . . .
   (c) $\frac{1}{2}$, 1, $\frac{3}{2}$, 2, . . .   **(d)** 5, 10, 20, 40, . . .

32. Refer to the sequence in Exercise 31 that is not arithmetic. Explain in your own words how each term after the first is determined by using the previous term of the sequence.

Find a_1 for each of the following arithmetic sequences.

33. $a_9 = 47, a_{15} = 77$

34. $a_{10} = 50, a_{20} = 110$

Explain why each of the following sequences is arithmetic.

35. log 2, log 4, log 8, log 16, log 32, . . .

36. log 12, log 36, log 108, log 324, . . .

Relating Concepts

Let $f(x) = mx + b$ and $g(x) = ab^x$.

37. Find $f(1)$ and $g(1)$.   **38.** Find $f(2)$ and $g(2)$.   **39.** Find $f(3)$ and $g(3)$.

40. Consider the sequence $f(1), f(2), f(3), \ldots$. Is it an arithmetic sequence? If so, what is the common difference?

41. Consider the sequence $g(1), g(2), g(3), \ldots$. Is it a geometric sequence? If so, what is the common ratio?

For each of the following, write the first n terms of the geometric sequence that satisfies the given conditions.

42. $a_1 = -2, r = -3, n = 4$ **43.** $a_1 = -4, r = 2, n = 5$ **44.** $a_1 = 3125, r = \frac{1}{5}, n = 7$

45. $a_1 = 729, r = \frac{2}{3}, n = 5$ **46.** $a_3 = 6, a_4 = 12, n = 5$ **47.** $a_2 = 9, a_3 = 3, n = 4$

Find a_5 and a_n for each of the following geometric sequences.

48. $a_1 = 4, r = 3$ **49.** $a_1 = -5, r = 4$ **50.** $a_2 = 3, r = 2$

51. $a_3 = 6, r = 3$ **52.** $a_4 = 64, r = -4$ **53.** $a_4 = 81, r = -3$

For each of the following sequences that are geometric, find a_n.

54. 6, 12, 24, 48, . . .   **55.** 4, 16, 64, 256, . . .   **56.** $\frac{3}{4}, \frac{3}{2}, 3, 6, 12, \ldots$

57. $\frac{5}{6}, \frac{5}{3}, \frac{10}{3}, \frac{20}{3}, \frac{40}{3}, \ldots$ **58.** $-4, 2, -1, \frac{1}{2}, \ldots$ **59.** $49, -7, 1, -\frac{1}{7}, \ldots$

60. Use the sequence and/or table feature of your calculator to decide whether payments of $300 a month for 36 months are enough to pay off a $10,000 car loan at 8% annual interest. If not, what monthly payment to the nearest dollar is required? Suppose the loan is reduced to $8000. What monthly payment to the nearest dollar is needed to pay it off in 36 months? (See Example 8.)

61. John Vasquez contracts to work for thirty days, receiving $.01 on the first day of work, $.02 on the second, $.04 on the third, $.08 on the fourth, and so on, with each day's pay double that of the previous day. Write the general term of a sequence that defines John's pay on the *n*th day. Use your calculator to graph the sequence. From the table of values, what will John earn on the tenth day? the fifteenth day? the twenty-fifth day?

62. Give an example of a geometric sequence with $a_3 = 15$.

63. Explain in your own words what is meant by a *geometric sequence*.

64. Which one of the following is not a geometric sequence?
  (a) $1, -2, 4, -8, \ldots$ (c) $3, \frac{3}{2}, \frac{3}{4}, \frac{3}{8}, \ldots$
  (b) $1, 10, 100, 1000, \ldots$ (d) $1, 1, 2, 3, 5, 8, \ldots$

65. Refer to the sequence in Exercise 64 that is not geometric. Explain in your own words how each term after the first two terms is determined. (This is a famous sequence known as the Fibonacci sequence, defined recursively in Exercise 13.)

Find a_1 and r for each of the following geometric sequences.

66. $a_2 = 6, a_6 = 486$

67. $a_2 = 100, a_5 = \dfrac{1}{10}$

Explain why the following sequences are geometric.

68. $\log 6, \log 6^2, \log 6^4, \log 6^8, \ldots$

69. $\log 2, \log 4, \log 16, \log 256, \ldots$

For each sequence defined by a_n, graph the corresponding function defined by $f(x)$. Use the graph to decide whether the sequence converges, and if it does, determine the number to which it converges.

70. $a_n = \dfrac{n + 2}{2n}$

71. $a_n = 2e^n$

72. $a_n = \left(1 + \dfrac{1}{n}\right)^n$

Further Explorations

1. If your calculator has a TABLE feature, use it to find out during which year you will double your investment of $1000 at 6% annual interest compounded annually. This can be done in either the function mode or sequence mode.

 Note: If you want to know the answer to a precise fraction of a year, you must be in function mode. Function mode graphs a continuous graph whose domain in this case is all nonnegative real numbers, while sequence mode graphs a discrete graph whose domain is restricted to nonnegative integers. ΔTbl must be an integer when in sequence mode.

Use sequence mode and dot mode to analyze the sequences in Exercises 2 and 3.

2. Compare the TABLE with the graph of $u_n = \frac{1}{n}$.
  (a) What WINDOW settings give a good graph of the first 30 terms of this sequence?
  (b) Scroll through the TABLE or TRACE through the graph to find the number that this sequence approaches but never reaches.

3. Compare the TABLE with the graph of $u_n = \frac{n+1}{n}$.
  (a) What WINDOW settings give a good graph of the first 30 terms of this sequence?
  (b) Scroll through the TABLE or TRACE through the graph to find the number that this sequence approaches but never reaches.

11.4 SERIES

Defining Series ▮ Sum of the Terms of an Arithmetic Sequence ▮ Sum of the Terms of a Geometric Sequence ▮ Infinite Geometric Series ▮ Applications

Defining Series

The sum of the terms of a sequence is called a **series.** A compact shorthand notation, called **summation notation,** can be used to write a series. For example,

$$1 + 5 + 9 + 13 + 17 + 21$$

is the sum of the first six terms in the sequence with general term $a_n = 4n - 3$. This series can be written using summation notation as

$$\sum_{i=1}^{6} (4i - 3)$$

(read "the sum from $i = 1$ to 6 of $4i - 3$").

The Greek letter sigma, Σ, is used to mean "sum." To evaluate this sum, replace i in $4i - 3$ first by 1, then 2, then 3, until finally i is replaced with 6.

$$\sum_{i=1}^{6} (4i - 3) = (4 \cdot 1 - 3) + (4 \cdot 2 - 3) + (4 \cdot 3 - 3)$$
$$+ (4 \cdot 4 - 3) + (4 \cdot 5 - 3) + (4 \cdot 6 - 3)$$
$$= 1 + 5 + 9 + 13 + 17 + 21 = 66$$

In this notation, i is called the **index of summation**.

CAUTION Do not confuse this use of i with the use of i to represent an imaginary number. Other letters may be used for the index of summation.

EXAMPLE 1
Using Summation Notation

Evaluate each of the following sums: **(a)** $\sum_{i=1}^{4} i^2(i + 1)$ **(b)** $\sum_{j=3}^{6} \frac{j+1}{j-2}$.

SOLUTION

(a) $\sum_{i=1}^{4} i^2(i + 1) = 1^2(1 + 1) + 2^2(2 + 1) + 3^2(3 + 1) + 4^2(4 + 1)$
$$= 1 \cdot 2 + 4 \cdot 3 + 9 \cdot 4 + 16 \cdot 5$$
$$= 2 + 12 + 36 + 80$$
$$= 130$$

(b) $\sum_{j=3}^{6} \frac{j+1}{j-2} = \frac{3+1}{3-2} + \frac{4+1}{4-2} + \frac{5+1}{5-2} + \frac{6+1}{6-2}$
$$= \frac{4}{1} + \frac{5}{2} + \frac{6}{3} + \frac{7}{4} = \frac{41}{4} \quad \blacksquare$$

Sum of the Terms of an Arithmetic Sequence

Suppose someone borrows $3000 and agrees to pay $100 per month plus interest of 1% per month on the unpaid balance until the loan is paid off. The first month he pays $100 to reduce the loan, plus interest of $(.01)3000 = 30$ dollars. The second month he pays another $100 toward the loan and interest of $(.01)2900 = 29$ dollars. Since the loan is reduced by $100 each month, his interest payments decrease by $(.01)100 = 1$ dollar each month, forming the arithmetic sequence

$$30, 29, 28, \ldots, 3, 2, 1.$$

The total amount of interest paid is given by the sum of the terms of this sequence. A formula can be developed to find this sum without adding all thirty numbers directly.

If an arithmetic sequence has terms $a_1, a_2, a_3, a_4, \ldots, a_n$, and S_n is defined as the sum of the first n terms of the sequence, then

$$S_n = a_1 + a_2 + a_3 + \cdots + a_n.$$

A formula for S_n can be found by writing the sum of the first n terms as follows.

$$S_n = a_1 + [a_1 + d] + [a_1 + 2d] + \cdots + [a_1 + (n-1)d]$$

Next, write this same sum in reversed order.

$$S_n = [a_1 + (n-1)d] + [a_1 + (n-2)d] + \cdots + [a_1 + d] + a_1$$

Now add the corresponding sides of these two equations.

$$S_n + S_n = (a_1 + [a_1 + (n-1)d]) + ([a_1 + d] + [a_1 + (n-2)d])$$
$$+ \cdots + ([a_1 + (n-1)d] + a_1)$$

From this,

$$2S_n = [2a_1 + (n-1)d] + [2a_1 + (n-1)d] + \cdots + [2a_1 + (n-1)d].$$

Since there are n of the $[2a_1 + (n-1)d]$ terms on the right,

$$2S_n = n[2a_1 + (n-1)d]$$

$$S_n = \frac{n}{2}[2a_1 + (n-1)d].$$

Using the formula $a_n = a_1 + (n-1)d$, S_n can also be written as

$$S_n = \frac{n}{2}[a_1 + a_1 + (n-1)d],$$

or
$$S_n = \frac{n}{2}(a_1 + a_n).$$

A summary of this work with sums of arithmetic sequences follows.

SUM OF THE FIRST n TERMS OF AN ARITHMETIC SEQUENCE

If an arithmetic sequence has first term a_1 and common difference d, then the sum of the first n terms is given by

$$S_n = \frac{n}{2}(a_1 + a_n)$$

or
$$S_n = \frac{n}{2}[2a_1 + (n-1)d].$$

The first formula is used when the first and last terms are known; otherwise the second formula is used.

Either one of these formulas can be used to find the total interest on the $3000 loan discussed above. In the sequence of interest payments, $a_1 = 30$, $d = -1$, $n = 30$, and $a_n = 1$. Choosing the first formula,

$$S_n = \frac{n}{2}(a_1 + a_n),$$

gives
$$S_{30} = \frac{30}{2}(30 + 1) = 15(31) = 465,$$

so a total of $465 interest will be paid over the 30 months.

<table>
<tr><td>**EXAMPLE 2**

Using the Sum Formula (Arithmetic Sequence)</td></tr>
</table>

Find S_{12} for the arithmetic sequence

$$-9, -5, -1, 3, 7, \ldots.$$

SOLUTION We want the sum of the first twelve terms. Using $a_1 = -9$, $n = 12$, and $d = 4$ in the second formula,

$$S_n = \frac{n}{2}[2a_1 + (n-1)d],$$

gives

$$S_{12} = \frac{12}{2}[2(-9) + 11(4)] = 6(-18 + 44) = 156. \quad \blacksquare$$

<table>
<tr><td>**EXAMPLE 3**

Using the Sum Formula (Arithmetic Sequence)</td></tr>
</table>

The sum of the first 17 terms of an arithmetic sequence is 187. If $a_{17} = -13$, find a_1 and d.

SOLUTION Use the first formula for S_n, with $n = 17$, to find a_1.

$$S_{17} = \frac{17}{2}(a_1 + a_{17}) \qquad \text{Let } n = 17.$$

$$187 = \frac{17}{2}(a_1 - 13) \qquad \text{Let } S_{17} = 187, a_{17} = -13.$$

$$22 = a_1 - 13 \qquad \text{Multiply by } \tfrac{2}{17}.$$

$$a_1 = 35$$

Since $a_{17} = a_1 + (17 - 1)d$,

$$-13 = 35 + 16d \qquad \text{Let } a_{17} = -13, a_1 = 35.$$

$$-48 = 16d$$

$$d = -3. \quad \blacksquare$$

Any sum of the form

$$\sum_{i=1}^{n} (mi + p),$$

where m and p are real numbers, represents the sum of the terms of an arithmetic sequence having first term

$$a_1 = m(1) + p = m + p$$

and common difference $d = m$. These sums can be evaluated by the formulas in this section, as shown by the next example.

<table>
<tr><td>**EXAMPLE 4**

Using Summation Notation</td></tr>
</table>

Find the following sum.

$$\sum_{i=1}^{10} (4i + 8)$$

SOLUTION This sum represents the sum of the first ten terms of the arithmetic sequence having

$$a_1 = 4 \cdot 1 + 8 = 12,$$
$$n = 10,$$
and $$a_n = a_{10} = 4 \cdot 10 + 8 = 48.$$

Thus $$\sum_{i=1}^{10} (4i + 8) = S_{10} = \frac{10}{2}(12 + 48) = 5(60) = 300. \quad \blacksquare$$

Sum of the Terms of a Geometric Sequence

Just as formulas were developed to find the sum of the first n terms of an arithmetic sequence, the same can be done for a geometric sequence. We begin by writing the sum S_n as

$$S_n = a_1 + a_2 + a_3 + \cdots + a_n.$$

This can also be written as

$$S_n = a_1 + a_1 r + a_1 r^2 + \cdots + a_1 r^{n-1}. \qquad [1]$$

If $r = 1$, then $S_n = na_1$, which is a correct formula for this case. If $r \neq 1$, multiply both sides of (1) by r, obtaining

$$rS_n = a_1 r + a_1 r^2 + a_1 r^3 + \cdots + a_1 r^n. \qquad [2]$$

If equation (2) is subtracted from equation (1), the result is

$$S_n = a_1 + a_1 r + a_1 r^2 + \cdots + a_1 r^{n-1}$$
$$rS_n = \qquad a_1 r + a_1 r^2 + \cdots + a_1 r^{n-1} + a_1 r^n$$
$$\overline{S_n - rS_n = a_1 \qquad\qquad\qquad\qquad\qquad - a_1 r^n} \quad \text{Subtract.}$$

or $$S_n(1 - r) = a_1(1 - r^n), \qquad\qquad \text{Factor.}$$

which finally gives

$$S_n = \frac{a_1(1 - r^n)}{1 - r}, \quad \text{where } r \neq 1. \quad \text{Divide by } 1 - r.$$

This discussion is summarized below.

> **SUM OF THE FIRST n TERMS OF A GEOMETRIC SEQUENCE**
> If a geometric sequence has first term a_1 and common ratio r, then the sum of the first n terms is given by
> $$S_n = \frac{a_1(1 - r^n)}{1 - r}, \quad \text{where } r \neq 1.$$

EXAMPLE 5

Using the Sum Formula (Geometric Sequence)

Find S_4 for the geometric sequence

$$10, 2, \frac{2}{5}, \ldots.$$

SOLUTION Here $a_1 = 10$, $r = \frac{1}{5}$, and $n = 4$. Substitute these values into the formula for S_n.

$$S_4 = \frac{10\left[1 - \left(\frac{1}{5}\right)^4\right]}{1 - \frac{1}{5}}$$

$$= \frac{10\left(1 - \frac{1}{625}\right)}{\frac{4}{5}}$$

$$= 10\left(\frac{624}{625}\right) \cdot \frac{5}{4}$$

$$S_4 = \frac{312}{25} \qquad \blacksquare$$

A sum of the form

$$\sum_{i=1}^{n} m \cdot p^i$$

represents the sum of the terms of a geometric sequence having first term

$$a_1 = m \cdot p^1 = mp$$

and common ratio $r = p$. These sums can be found by using the formula for S_n given above.

EXAMPLE 6

Using Summation Notation

Find the sum: $\displaystyle\sum_{i=1}^{7} 2 \cdot 3^i$.

SOLUTION In this sum, $a_1 = 2 \cdot 3^1 = 6$ and $r = 3$. Thus,

$$\sum_{i=1}^{7} 2 \cdot 3^i = S_7 = \frac{6(1 - 3^7)}{1 - 3} = \frac{6(1 - 2187)}{-2} = \frac{6(-2186)}{-2} = 6558.$$

$\qquad\qquad\qquad\qquad\qquad\qquad\qquad\qquad\qquad\qquad\qquad\qquad\qquad\qquad\blacksquare$

Infinite Geometric Series

Now we consider an infinite geometric sequence such as

$$2, 1, \frac{1}{2}, \frac{1}{4}, \frac{1}{8}, \frac{1}{16}, \dots$$

with first term 2 and common ratio $\frac{1}{2}$. Using the formula for S_n gives the following sequence.

$$S_1 = 2, \quad S_2 = 3, \quad S_3 = \frac{7}{2}, \quad S_4 = \frac{15}{4}, \quad S_5 = \frac{31}{8}, \quad S_6 = \frac{63}{16}$$

As Figure 8 suggests, these sums seem to be getting closer and closer to the number 4. For no value of n is $S_n = 4$. However, if n is large enough, then S_n is as close to

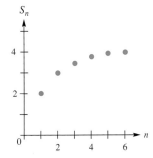

Sums for the sequence $2, 1, \frac{1}{2}, \frac{1}{4}, \frac{1}{8}, \frac{1}{16}, \cdots$

FIGURE 8

4 as desired.* As mentioned earlier, we say the sequence converges to 4. This is expressed as

$$\lim_{n \to \infty} S_n = 4.$$

(Read: "the limit of S_n as n increases without bound is 4.")

Since

$$\lim_{n \to \infty} S_n = 4,$$

the number 4 is called the *sum* of the infinite geometric sequence

$$2, 1, \frac{1}{2}, \frac{1}{4}, \cdots$$

and

$$2 + 1 + \frac{1}{2} + \frac{1}{4} + \frac{1}{8} + \cdots = 4.$$

EXAMPLE 7

Finding the Sum of an Infinite Geometric Sequence

Find $1 + \frac{1}{3} + \frac{1}{9} + \frac{1}{27} + \cdots$

SOLUTION Use the formula for the first n terms of a geometric sequence to get

$$S_1 = 1, \quad S_2 = \frac{4}{3}, \quad S_3 = \frac{13}{9}, \quad S_4 = \frac{40}{27},$$

and in general

$$S_n = \frac{1\left[1 - \left(\frac{1}{3}\right)^n\right]}{1 - \frac{1}{3}}. \qquad \text{Let } a_1 = 1, r = \tfrac{1}{3}.$$

The following chart shows the value of $\left(\frac{1}{3}\right)^n$ for larger and larger values of n.

n	1	10	100	200
$\left(\frac{1}{3}\right)^n$	$\frac{1}{3}$	1.69×10^{-5}	1.94×10^{-48}	3.76×10^{-96}

---

*These phrases "large enough" and "as close as desired" are not nearly precise enough for mathematicians; much of a standard calculus course is devoted to making them more precise.

As n gets larger and larger, $\left(\frac{1}{3}\right)^n$ gets closer and closer to 0. That is,

$$\lim_{n \to \infty} \left(\frac{1}{3}\right)^n = 0,$$

making it reasonable that

$$\lim_{n \to \infty} S_n = \lim_{n \to \infty} \frac{1\left[1 - \left(\frac{1}{3}\right)^n\right]}{1 - \frac{1}{3}} = \frac{1(1 - 0)}{1 - \frac{1}{3}} = \frac{1}{\frac{2}{3}} = \frac{3}{2}.$$

Hence, $\qquad 1 + \frac{1}{3} + \frac{1}{9} + \frac{1}{27} + \cdots = \frac{3}{2}.$ ∎

If a geometric sequence has a first term a_1 and a common ratio r, then

$$S_n = \frac{a_1(1 - r^n)}{1 - r} \quad (r \neq 1)$$

for every positive integer n. If $-1 < r < 1$, then $\lim\limits_{n \to \infty} r^n = 0$, and

$$\lim_{n \to \infty} S_n = \frac{a_1(1 - 0)}{1 - r} = \frac{a_1}{1 - r}.$$

This quotient, $\frac{a_1}{(1 - r)}$, is called the **sum of the terms of an infinite geometric sequence**. The limit $\lim\limits_{n \to \infty} S_n$ is often expressed as S_∞ or $\sum\limits_{i=1}^{\infty} a_i$. These results lead to the following definition.

SUM OF THE TERMS OF AN INFINITE GEOMETRIC SEQUENCE

The sum of the terms of an infinite geometric sequence with first term a_1 and common ratio r, where $-1 < r < 1$, is given by

$$S_\infty = \frac{a_1}{1 - r}.$$

If $|r| > 1$, the terms get larger and larger in absolute value, so there is no limit as $n \to \infty$. Hence the sequence will not have a sum.

EXAMPLE 8

Using the Sum Formula

(a) Find the sum

$$-\frac{3}{4} + \frac{3}{8} - \frac{3}{16} + \frac{3}{32} - \frac{3}{64} + \cdots.$$

SOLUTION The first term is $a_1 = -\frac{3}{4}$. To find r, divide any two adjacent terms. For example,

$$r = \frac{-\dfrac{3}{16}}{\dfrac{3}{8}} = -\frac{1}{2}.$$

Since $-1 < r < 1$, the formula above applies, and

$$S_\infty = \frac{a_1}{1-r} = \frac{-\dfrac{3}{4}}{1-\left(-\dfrac{1}{2}\right)} = -\frac{1}{2}.$$

(b) $\displaystyle\sum_{i=1}^{\infty}\left(\frac{3}{5}\right)^i = \frac{\dfrac{3}{5}}{1-\dfrac{3}{5}} = \frac{3}{2}$

Notice that the upper limit is ∞, indicating that this is an infinite geometric sequence. ◻

FOR GROUP DISCUSSION

The sum $.9 + .09 + .009 + \cdots$ may be written as $.9999\ldots$. The sum does exist, since $r = .1$. Use the formula to find this sum. What do you get? The same result can be obtained in a different way. Use the fact that $\frac{1}{3} = .3333\ldots$ to arrive at the same conclusion.

Applications

The formulas developed so far in this section and the previous one will help us to solve applied problems that can be modeled as sequences. In some cases, we might be asked to find a particular term of an arithmetic or geometric sequence that appears in an applied problem. We would then use the appropriate formula for finding a_n (developed in Section 11.3). If the problem requires finding the sum of a specified number of terms of an arithmetic or geometric sequence, we can use one of the formulas developed in this section. It is important to read carefully to determine whether we want to find a specific term or a sum of terms.

EXAMPLE 9

Solving an Applied Problem Involving a Geometric Sequence

An insect population is growing in such a way that each generation is 1.5 times as large as the previous generation. Suppose there are 100 insects in the first generation.

(a)   How many will there be in the fourth generation?

SOLUTION   The population can be written as a geometric sequence with a_1 as the first-generation population, a_2 the second-generation population, and so on. Then the fourth-generation population will be a_4. Using the formula for a_n, where $n = 4$, $r = 1.5$, and $a_1 = 100$, gives

$$a_4 = a_1 r^3 = 100(1.5)^3 = 100(3.375) = 337.5.$$

In the fourth generation, the population will number about 338 insects. By writing an expression for a_n, alternatively, we can use the table feature of the calculator to find the population in any generation. A portion of the table is shown in Figure 9. We can then also answer questions such as "In what generation will there be at least 1000 insects?" As the table shows, this will occur in the seventh generation.

TECHNOLOGICAL NOTE Figures 9 and 10 in this section provide examples of how graphics calculators can utilize sequences, tables, and lists in conjunction with each other.

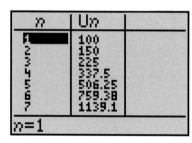

FIGURE 9

(b) What will be the total number of insects in the first four generations?

SOLUTION Since we must find the sum of the first four terms of this geometric sequence, we use the formula

$$S_n = \frac{a_1(1 - r^n)}{1 - r},$$

with $n = 4$, $a_1 = 100$, and $r = 1.5$. Use a calculator to get

$$S_4 = \frac{100(1 - 1.5^4)}{1 - 1.5} = \frac{100(1 - 5.0625)}{-.5} = 812.5.$$

The total population for the four generations will amount to about 813 insects. ❑

A sequence of equal payments made at equal periods of time is called an **annuity**. The sum of the payments and interest on the payments is the **future value** of the annuity.

EXAMPLE 10

Finding the Future Value of an Annuity

To save money for a trip to Europe, Marge deposited $1000 at the end of each year for four years in an account paying 6% interest compounded annually. What is the future value of this annuity?

SOLUTION Recall the formula for compound interest from Chapter 5, with $n = 1$: $A = P(1 + r)^t$. The first payment will earn interest for 3 years, the second payment for 2 years, and the third payment for 1 year. The last payment earns no interest. The total amount is

$$1000(1.06)^3 + 1000(1.06)^2 + 1000(1.06) + 1000.$$

TECHNOLOGICAL NOTE See the Further Explorations at the end of the exercise set in this section. They investigate how annuities can be explored via tables.

This is the sum of a geometric sequence with first term (starting at the end of the sum as written above) $a_1 = 1000$ and common ratio $r = 1.06$. Using the formula for S_4, the sum of four terms, gives

$$S_4 = \frac{1000[1 - (1.06)^4]}{1 - 1.06}$$

$$= 4374.62.$$

The future value of the annuity is $4374.62. ❑

You can use the *list* feature of your calculator to find the value of the sum of a finite series. It is convenient to save the sequence you define in a list. Then use the capability of your calculator to get the sum of the list. Typical screens for the annuity in Example 10 are shown in Figure 10.

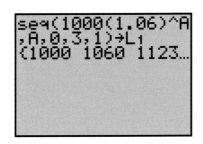 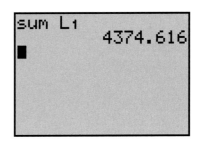

(a)   (b)

FIGURE 10

11.4 EXERCISES

Evaluate each of the following sums.

1. $\displaystyle\sum_{i=1}^{5} (2i + 1)$

2. $\displaystyle\sum_{i=1}^{6} (3i - 2)$

3. $\displaystyle\sum_{j=1}^{4} \frac{1}{j}$

4. $\displaystyle\sum_{i=1}^{5} (i + 1)^{-1}$

5. $\displaystyle\sum_{k=1}^{6} (-1)^{k} \cdot k$

6. $\displaystyle\sum_{i=1}^{7} (-1)^{i+1} \cdot i^{2}$

Find the sum of the first ten terms for each of the following arithmetic sequences.

7. $a_1 = 8, d = 3$

8. $a_1 = -9, d = 4$

9. $a_3 = 5, a_4 = 8$

10. $a_2 = 9, a_4 = 13$

11. 5, 9, 13, . . .

12. 8, 6, 4, . . .

Find a_1 and d for each of the following arithmetic sequences.

13. $S_{20} = 1090, a_{20} = 102$

14. $S_{31} = 5580, a_{31} = 360$

Evaluate each of the following series.

15. $\displaystyle\sum_{i=1}^{12} (-5 - 8i)$

16. $\displaystyle\sum_{k=1}^{19} (-3 - 4k)$

17. $\displaystyle\sum_{i=1}^{1000} i$

18. $\displaystyle\sum_{k=1}^{2000} k$

Find the sum of the first five terms for each of the following geometric sequences.

19. 3, 6, 12, 24, . . .

20. 5, 20, 80, 320, . . .

21. $12, -6, 3, -\dfrac{3}{2}, \ldots$

22. $18, -3, \dfrac{1}{2}, -\dfrac{1}{12}, \ldots$

23. $a_2 = \dfrac{1}{3}, r = 3$

24. $a_2 = -1, r = 2$

Find each of the following sums.

25. $\displaystyle\sum_{i=1}^{4} 2^{i}$

26. $\displaystyle\sum_{j=1}^{6} 3^{j}$

27. $\displaystyle\sum_{i=1}^{4} (-3)^{i}$

28. $\displaystyle\sum_{i=1}^{4} (-3^{i})$

29. $\displaystyle\sum_{i=1}^{6} 81\left(\dfrac{2}{3}\right)^{i}$

30. $\displaystyle\sum_{k=1}^{5} 9\left(\dfrac{5}{3}\right)^{k}$

Relating Concepts

The following properties of series provide useful shortcuts for evaluating series. If a_1, a_2, a_3, . . . , a_n and b_1, b_2, b_3, . . . , b_n are two sequences, and c is a constant, then for every positive integer n,

(a) $\displaystyle\sum_{i=1}^{n} c = nc$

(b) $\displaystyle\sum_{i=1}^{n} ca_i = c \sum_{i=1}^{n} a_i$

(c) $\displaystyle\sum_{i=1}^{n} (a_i + b_i) = \sum_{i=1}^{n} a_i + \sum_{i=1}^{n} b_i$

(d) $\displaystyle\sum_{i=1}^{n} (a_i - b_i) = \sum_{i=1}^{n} a_i - \sum_{i=1}^{n} b_i.$

31. Use Property (c) to write $\displaystyle\sum_{i=1}^{6} (i^2 + 3i + 5)$ as the sum of three summations.

32. Use Property (b) to rewrite the second summation from Exercise 31.

33. Use Property (a) to rewrite the third summation from Exercise 31.

34. Rewrite the sum given in Exercise 10 of 11.1 Exercises using summation notation.

35. Rewrite the sum given in Example 1 of Section 11.1 using summation notation.

36. Use the summations you wrote in Exercises 34 and 35 to evaluate the three summations from Exercises 31–33. This gives the value of $\displaystyle\sum_{i=1}^{6} (i^2 + 3i + 5)$ without writing out all six terms.

37. Use the properties given above to evaluate $\displaystyle\sum_{i=1}^{12} (i^2 - i)$.

38. Use the properties given above to evaluate $\displaystyle\sum_{i=1}^{20} (2 + i - i^2)$.

---

39. Under what conditions will the terms of an infinite geometric sequence have a sum?

40. Which one of the following does not have a sum?

(a) $\displaystyle\sum_{i=1}^{10} \left(\tfrac{1}{2}\right)^i$ (b) $\displaystyle\sum_{i=1}^{\infty} \left(\tfrac{3}{2}\right)^i$ (c) $\displaystyle\sum_{i=1}^{\infty} \left(\tfrac{1}{2}\right)^i$ (d) $\displaystyle\sum_{i=1}^{1000} 3 \cdot 4^i$

Find the sum of the terms of each geometric sequence that converges by using the formula of this section where it applies.

41. $16 + 4 + 1 + \ldots$

42. $81 + 27 + 9 + 3 + 1 + \ldots$

43. $90 + 30 + 10 + \ldots$

44. $25 + 5 + 1 + \ldots$

45. $256 - 128 + 64 - 32 + \ldots$

46. $120 - 60 + 30 - 15 + \ldots$

47. $\dfrac{3}{4} + \dfrac{3}{8} + \dfrac{3}{16} + \ldots$

48. $\dfrac{4}{5} + \dfrac{2}{5} + \dfrac{1}{5} + \ldots$

49. $3 - \dfrac{3}{2} + \dfrac{3}{4} - \ldots$

50. $1 + \dfrac{1}{1.01} + \dfrac{1}{(1.01)^2} + \ldots$

51. $\displaystyle\sum_{i=1}^{\infty} \left(\dfrac{1}{4}\right)^i$

52. $\displaystyle\sum_{i=1}^{\infty} \left(\dfrac{9}{10}\right)^i$

53. $\displaystyle\sum_{i=1}^{\infty} (1.2)^i$

54. $\displaystyle\sum_{i=1}^{\infty} (1.001)^i$

55. $\displaystyle\sum_{i=1}^{\infty} 10^{-i}$

56. $\displaystyle\sum_{i=1}^{\infty} 4^{-i}$

57. $\displaystyle\sum_{i=1}^{\infty} \left(\dfrac{1}{2}\right)^{-i}$

58. $\displaystyle\sum_{i=1}^{\infty} \left(\dfrac{3}{4}\right)^{-i}$

59. A display of stacked canned goods in a grocery store has 31 cans on the bottom, 25 in the next row, and 1 can on top. Assume the number of cans in the layers form an arithmetic sequence. How many cans are in the display?

60. A stack of telephone poles has 30 in the bottom row, 29 in the next, and so on, with one pole in the top row. How many poles are in the stack?

61. Deepwell Drilling Company charges a flat $100 set-up charge, plus $5 for the first foot, $6 for the second, $7 for the third, and so on. Find the total charge for a 70-foot well.

62. An object falling under the force of gravity falls about 16 ft the first second, 48 ft during the next second, 80 ft during the third second, and so on. How far would the object fall during the eighth second? What is the total distance the object would fall in 8 seconds?

63. The population of a city was 49,000 five years ago. Each year the zoning commission permits an increase of 580 in the population. What will the maximum population be five years from now?

64. A super slide of uniform slope is to be built on a level piece of land. There are to be twenty equally spaced supports, with the longest support 15 m long and the shortest 2 m long. Find the total length of all the supports.

65. How much material would be needed for the rungs of a ladder of 31 rungs, if the rungs taper uniformly from 18 in to 28 in?

66. Fruit and vegetable dealer Olive Greene paid 10¢ per lb for 10,000 lb of onions. Each week the price she charges increases by .1¢ per lb, while her onions lose 5% of their weight. If she sells all the onions after 6 weeks, does she make or lose money? How much?

67. The final step in processing a black and white photographic print is to immerse the print in a chemical called "fixer." The print is then washed in running water. Under certain conditions, 98% of the fixer in a print will be removed with 15 min of washing. How much of the original fixer would then be left after 1 hour?

68. A scientist has a vat containing 100 liters of a pure chemical. Twenty liters are drained and replaced with water. After complete mixing, 20 liters of the mixture are drained and replaced with water. What will be the strength of the mixture after nine such drainings?

69. Find the future value of an annuity with payments of $2430 at the end of each year for 10 years at 5.5% interest compounded annually.

70. Find the future value of 6 annual payments of $1500 at 4.2% compounded annually.

71. Mitzi drops a ball from a height of 10 m and notices that on each bounce the ball returns to about $\frac{3}{4}$ of its previous height. About how far will the ball travel before it comes to rest? (*Hint:* Consider the sum of two sequences.)

72. A sugar factory receives an order for 1000 units of sugar. The production manager thus orders production of 1000 units of sugar. He forgets, however, that the production of sugar requires some sugar (to prime the machines, for example), and so he ends up with only 900 units of sugar. He then orders an additional 100 units, and receives only 90 units. A further order for 10 units produces 9 units. Finally seeing his mistake, the manager decides to try mathematics. He views the production process as an infinite geometric progression with $a_1 = 1000$ and $r = .1$. Using this, find the number of units of sugar that he should have ordered originally.

73. After a person pedaling a bicycle removes his or her feet from the pedals, the wheel rotates 400 times the first minute. As it continues to slow down, each minute it rotates only $\frac{3}{4}$ as many times as in the previous minute. How many times will the wheel rotate before coming to a complete stop?

74. A pendulum bob swings through an arc 40 cm long on its first swing. Each swing thereafter, it swings only 80% as far as on the previous swing. How far will it swing altogether before coming to a complete stop?

75. A sequence of equilateral triangles is constructed. The first triangle has sides 2 m in length. To get the second triangle, midpoints of the sides of the original triangle are connected.
   (a) If this process could be continued indefinitely, what would be the total perimeter of all the triangles?
   (b) What would be the total area of all the triangles disregarding the overlapping?

76. Certain medical conditions are treated with a fixed dose of a drug administered at regular intervals. Suppose that a person is given 2 mg of a drug each day and that during each 24-hr period, the body utilizes 40% of the amount of drug that was present at the beginning of the period.
   (a) Show that the amount of the drug present in the body at the end of n days is

$$\sum_{i=1}^{n} 2(.6)^i.$$

   (b) What will be the approximate quantity of the drug in the body at the end of each day after the treatment has been administered for a long period of time?
   (c) What is the maximum daily dosage that will guarantee that the amount of the drug in the body never exceeds 2 mg?

Further Explorations

1. Consider the infinite sequence: $\frac{1}{2}, \frac{1}{4}, \frac{1}{8}, \frac{1}{16}, \ldots \ldots$ If your graphics calculator has sequence graphing capabilities, you can use it to calculate and graph a sequence of successive sums. To clarify:

$$\text{If } n = 1 \qquad\qquad \text{then } \frac{1}{2} = \frac{1}{2} = .5.$$

$$\text{If } n = 2 \qquad \text{then } \frac{1}{2} + \frac{1}{4} = \frac{3}{4} = .75.$$

$$\text{If } n = 3 \qquad \text{then } \frac{1}{2} + \frac{1}{4} + \frac{1}{8} = \frac{7}{8} = .875.$$

Create a sequence of the sums from $n = 1$ to $n = 20$. Graph this sequence using the SEQUENCE mode. Do the sums continue to increase? Is there a limit to the increase? Why do you think the calculator takes so long to draw this graph? Use the TABLE to numerically support what you see in the graph.

2. Sara wants to start a college account for her newborn baby. If at the beginning of each year she puts $500 into an account that pays 5% interest compounded annually, how much will she have after 1 year; 2 years; 18 years? Graph a sequence of the successive sums showing each year's growth. Use the TABLE to numerically support the graphical solution.

3. Rachel wants to start a college account for her newborn baby. If at the beginning of each year she puts $450 into an account that pays 7% interest compounded yearly, how much will she have after 1 year, 2 years, 18 years? Graph a sequence of the successive sums showing each year's growth. Use the TABLE to numerically support the graphical solution.

11.5 PERMUTATIONS AND COMBINATIONS

The Multiplication Principle of Counting ∎ Permutations ∎ Combinations ∎ Distinguishing Between Permutations and Combinations

The Multiplication Principle of Counting

If there are 3 roads from Albany to Baker and 2 roads from Baker to Creswich, in how many ways can one travel from Albany to Creswich by way of Baker? For each of the 3 roads from Albany to Baker, there are 2 different roads from Baker to Creswich, so that there are $3 \cdot 2 = 6$ different ways to make the trip, as shown in the **tree diagram** in Figure 11. This example illustrates the following property.

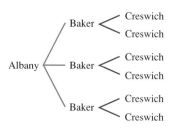

FIGURE 11

> **MULTIPLICATION PRINCIPLE OF COUNTING**
>
> If one event can occur in m ways and a second event can occur in n ways, then both events can occur in mn ways, provided the outcome of the first event does not influence the outcome of the second.

The multiplication principle of counting can be extended to any number of events, provided the outcome of no one event influences the outcome of another. Such events are called **independent events.**

EXAMPLE 1

Using the Multiplication Principle

How many telephone numbers are available with the 269 prefix?

SOLUTION We need only consider the last 4 digits. Each place can hold any digit from 0 to 9, so there are 10 choices for each, or

$$10 \cdot 10 \cdot 10 \cdot 10 = 10,000$$

possible numbers with a 269 prefix. ▪

FOR GROUP DISCUSSION

The U.S. is rapidly running out of telephone numbers. In large cities, telephone companies introduced new area codes as numbers were used up; as a result, only three unallocated area codes are left: 210, 810, and 910. At present, all area codes have a 0 or a 1 as the middle digit and the first digit cannot be 0 or 1. How many area codes are available with this scheme? How many telephone numbers does the current 7-digit sequence permit per area code? The 3-digit prefix that follows the area code cannot start with 0 or 1. Assume there are no other restrictions. The actual number of area codes is 152. Explain the discrepancy between this number and your answer. Discuss some of the alternatives that could provide additional telephone numbers.

EXAMPLE 2

Using the Multiplication Principle

Janet Branson has 5 different books that she wishes to arrange on her desk. How many different arrangements are possible?

SOLUTION Five events are involved: selecting a book for the first spot, selecting a book for the second spot, and so on. Here the outcome of an event *does* influence the outcome of the following events (since some books have already been chosen). For the first spot Branson has 5 choices, for the second spot 4 choices, for the third spot 3 choices, and so on. Now use the multiplication principle of counting to find that there are

$$5 \cdot 4 \cdot 3 \cdot 2 \cdot 1 \text{ or } 120 \text{ different arrangements.} \quad ▪$$

In using the multiplication principle of counting we often encounter such products as $5 \cdot 4 \cdot 3 \cdot 2 \cdot 1$ from Example 2. For convenience in writing these products, we use the symbol $n!$ (read "n factorial"), which was defined in Section 11.2 and is repeated here. For any positive integer n,

$$n! = n(n-1)(n-2)(n-3)\cdots(2)(1)$$

and $\qquad 0! = 1.$

By the definition, $5 \cdot 4 \cdot 3 \cdot 2 \cdot 1$ is written as $5!$. Also, $3! = 3 \cdot 2 \cdot 1 = 6$. The definition of $n!$ means that $n[(n-1)!] = n!$ for all natural numbers $n \geq 2$. It is useful to have this relation hold also for $n = 1$, so, by definition, $0! = 1$.

EXAMPLE 3

Using the
Multiplication
Principle

Suppose Branson (from Example 2) wishes to place only 3 of the 5 books on her desk. How many arrangements of 3 books are possible?

SOLUTION   She still has 5 ways to fill the first spot, 4 ways to fill the second spot, and 3 ways to fill the third. Since she wants to use 3 books, there are only 3 spots to be filled (3 events) instead of 5, so there are

$$5 \cdot 4 \cdot 3 = 60 \text{ arrangements.} \quad \blacksquare$$

Permutations

The number 60 in the example above is called the number of *permutations* of 5 things taken 3 at a time, written $P(5, 3) = 60$. Example 2 showed that the number of ways of arranging 5 elements from a set of 5 elements, written $P(5, 5)$, is 120.

A **permutation** of n elements taken r at a time is one of the ways of arranging r elements taken from a set of n elements ($r \leq n$). Generalizing from the examples above, the number of permutations of n elements taken r at a time, denoted by $P(n, r)$, is

$$P(n, r) = n(n - 1)(n - 2) \cdots (n - r + 1)$$
$$= \frac{n(n - 1)(n - 2) \cdots (n - r + 1)(n - r)(n - r - 1) \cdots (2)(1)}{(n - r)(n - r - 1) \cdots (2)(1)}$$
$$= \frac{n!}{(n - r)!}.$$

This derivation gives the following permutations formula.

> **PERMUTATIONS OF n ELEMENTS r AT A TIME**
> If $P(n, r)$ denotes the number of permutations of n elements taken r at a time, $r \leq n$, then
> $$P(n, r) = \frac{n!}{(n - r)!}.$$

Some other symbols used for permutations of n things taken r at a time are $_nP_r$ and P_r^n.

Although most graphics calculators do not have a designated key for $n!$ or permutations, they can be used to evaluate both. This function is usually found in the math or statistics menu. As mentioned earlier, for large values of n and r, the calculator display may be an approximation. A calculator screen for typical permutations calculations is shown in Figure 12.

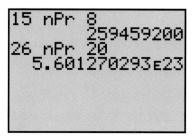

```
15 nPr 8
         259459200
26 nPr 20
    5.601270293E23
```

FIGURE 12

Telephone numbers are examples of permutations. For example, the phone number 638-8440 is not the same as the phone number 683-8440.

EXAMPLE 4

Using the Permutations Formula

Suppose 8 people enter an event in a swim meet. Assuming there are no ties, in how many ways could the gold, silver, and bronze prizes be awarded?

SOLUTION Using the multiplication principle of counting, there are 3 choices to be made giving $8 \cdot 7 \cdot 6 = 336$. However, we can also use the formula for $P(n, r)$ to get the same result.

$$P(8, 3) = \frac{8!}{(8 - 3)!}$$
$$= \frac{8!}{5!} = \frac{8 \cdot 7 \cdot 6 \cdot 5 \cdot 4 \cdot 3 \cdot 2 \cdot 1}{5 \cdot 4 \cdot 3 \cdot 2 \cdot 1}$$
$$= 8 \cdot 7 \cdot 6 = 336 \quad \blacksquare$$

EXAMPLE 5

Using the Permutations Formula and the Multiplication Principle

A televised talk show will include 4 women and 3 men as panelists.

(a) In how many ways can the panelists be seated in a row of 7 chairs?

SOLUTION Find $P(7, 7)$, the total number of ways to seat 7 panelists in 7 chairs.

$$P(7, 7) = \frac{7!}{(7 - 7)!}$$
$$= \frac{7!}{0!} = \frac{7!}{1}$$
$$= 7 \cdot 6 \cdot 5 \cdot 4 \cdot 3 \cdot 2 \cdot 1 = 5040$$

There are 5040 ways to seat the 7 panelists.

(b) In how many ways can the panelists be seated if the men and women are to be alternated?

SOLUTION Use the multiplication principle. In order to alternate men and women, a woman must be seated in the first chair (since there are 4 women and only 3 men), any of the men next, and so on. Thus, there are 4 ways to fill the first seat, 3 ways to fill the second seat, 3 ways to fill the third seat (with any of the 3 remaining women), and so on. This gives

$$4 \cdot 3 \cdot 3 \cdot 2 \cdot 2 \cdot 1 \cdot 1 = 144$$

ways to seat the panelists. $\quad \blacksquare$

EXAMPLE 6

Using the Multiplication Principle with Restrictions

In how many ways can three letters of the alphabet be arranged if a vowel cannot be used in the middle position, and repetitions of the letters are allowed?

SOLUTION We cannot use $P(26, 3)$ here, because of the restriction for the middle position. In the first and third positions, we can use any of 26 letters of the alphabet, but in the middle position, we can only use one of $26 - 5 = 21$ letters (since there are 5 vowels). Now, using the multiplication counting principle, there are $26 \cdot 21 \cdot 26 = 14{,}196$ ways to arrange the letters according to the problem. $\quad \blacksquare$

Combinations

We have discussed a method for finding the number of ways to arrange r elements taken from a set of n elements. Sometimes, however, the arrangement (or order) of the elements is not important.

For example, suppose three people (Ms. Opelka, Mr. Adams, and Ms. Jacobs) apply for 2 identical jobs. Ignoring all other factors, in how many ways can the personnel officer select 2 people from the 3 applicants? Here the arrangement or order of the people is unimportant. Selecting Ms. Opelka and Mr. Adams is the same as selecting Mr. Adams and Ms. Opelka. Therefore, there are only 3 ways to select 2 of the 3 applicants:

Ms. Opelka and Mr. Adams;

Ms. Opelka and Ms. Jacobs;

Mr. Adams and Ms. Jacobs.

These three choices are called the *combinations* of 3 elements taken 2 at a time. A **combination** of n elements taken r at a time is one of the ways in which r elements can be chosen from n elements.

In the example above, each combination of 2 applicants forms 2! permutations (Ms. Opelka and Mr. Adams and Mr. Adams and Ms. Opelka, for example). So the number of combinations in the example could be found by dividing the number of *permutations* of 3 things taken 2 at a time by 2! to get

$$\frac{P(3, 2)}{P(2, 2)} = \frac{\dfrac{3!}{(3-2)!}}{2!} = \frac{3 \cdot 2}{2 \cdot 1} = 3.$$

This agrees with the answer we found by writing out the different groups of two applicants. Similarly, the number of combinations of n elements taken r at a time is found by dividing the number of permutations, $P(n, r)$, by $r!$ to get

$$\frac{P(n, r)}{r!}$$

combinations. This expression can be rewritten as follows.

$$\frac{P(n, r)}{r!} = \frac{\dfrac{n!}{(n-r)!}}{r!} = \frac{n!}{(n-r)! \, r!}$$

We use the symbol $\binom{n}{r}$ to represent the number of combinations of n things taken r at a time. This combinations notation also represents the binomial coefficient defined in Section 11.2. That is, binomial coefficients are combinations of n elements chosen r at a time.

COMBINATIONS OF n ELEMENTS r AT A TIME

If $\binom{n}{r}$ represents the number of combinations of n elements taken r at a time, $r \le n$, then

$$\binom{n}{r} = \frac{n!}{(n-r)! \, r!}.$$

Other symbols used for $\binom{n}{r}$ are $C(n, r)$, $_nC_r$, and C_r^n. Most calculators use $_nC_r$.

EXAMPLE 7

Using the
Combinations
Formula

How many different committees of 3 people can be chosen from a group of 8 people?

SOLUTION Since the order in which the members of the committee are chosen does not affect the result, use combinations to get

$$\binom{8}{3} = \frac{8!}{5!\,3!} = \frac{8 \cdot 7 \cdot 6 \cdot 5 \cdot 4 \cdot 3 \cdot 2 \cdot 1}{5 \cdot 4 \cdot 3 \cdot 2 \cdot 1 \cdot 3 \cdot 2 \cdot 1} = 56. \quad \blacksquare$$

EXAMPLE 8

Using the
Combinations
Formula

From a group of 30 employees, 3 are to be selected to work on a special project.

(a) In how many different ways can the employees be selected?

SOLUTION The number of 3-element combinations from a set of 30 elements must be found. (Use combinations, not permutations, because order within the group of 3 does not affect the result.) Using the formula gives

$$\binom{30}{3} = \frac{30!}{27!\,3!} = 4060.$$

There are 4060 ways to select the project group.

(b) In how many different ways can the group of 3 be selected if it has already been decided that a certain employee must work on the project?

SOLUTION Since one employee has already been selected to work on the project, the problem is reduced to selecting 2 more employees from the 29 employees that are left:

$$\binom{29}{2} = \frac{29!}{27!\,2!} = 406.$$

In this case, the project group can be selected in 406 different ways.

(c) In how many ways can a (nonempty) group of *at most* 3 employees be selected from the group of 30?

SOLUTION Here, "at most 3" means "exactly 1 or exactly 2 or exactly 3." We shall find the number of ways to select employees for each case.

Case	Number of Ways
1	$\binom{30}{1} = \frac{30!}{29!\,1!} = \frac{30 \cdot 29!}{29! \cdot 1} = 30$
2	$\binom{30}{2} = \frac{30 \cdot 29 \cdot 28!}{28! \cdot 2 \cdot 1} = 435$
3	$\binom{30}{3} = \frac{30 \cdot 29 \cdot 28 \cdot 27!}{27! \cdot 3 \cdot 2 \cdot 1} = 4060$

The total number of ways to select at most 3 employees will be the sum 30 + 435 + 4060 = 4525. $\blacksquare$

Distinguishing Between Permutations and Combinations

The formulas for permutations and combinations given in this section are very useful in solving probability problems. Any difficulty in using these formulas usually comes from being unable to differentiate between them. Both permutations and combinations give the number of ways to choose r objects from a set of n objects. The differences between permutations and combinations are outlined below.

Permutations	Combinations
Different orderings or arrangements of the r objects are different permutations. $$P(n, r) = \frac{n!}{(n - r)!}$$ Clue words: Arrangement, Schedule, Order	Each choice or subset of r objects gives 1 combination. Order within the r objects does not matter. $$\binom{n}{r} = \frac{n!}{(n - r)!\, r!}$$ Clue words: Group, Committee, Sample

EXAMPLE 9

Distinguishing Between Permutations and Combinations

For each of the following problems, tell whether permutations or combinations should be used to solve the problem.

(a) How many 4-digit code numbers are possible if no digits are repeated?

SOLUTION Since changing the order of the 4 digits results in a different code, use permutations.

(b) A sample of 3 light bulbs is randomly selected from a batch of 15 items. How many different samples are possible?

SOLUTION The order in which the 3 light bulbs are selected is not important. The sample is unchanged if the items are rearranged, so combinations should be used.

(c) In a basketball tournament with 8 teams, how many games must be played so that each team plays every other team exactly once?

SOLUTION Selection of 2 teams for a game is an *unordered* subset of 2 from the set of 8 teams. Use combinations again.

(d) In how many ways can 4 patients be assigned to 6 hospital rooms so that each patient has a private room?

SOLUTION The room assignments are an *ordered* selection of 4 rooms from the 6 rooms. Exchanging the rooms of any 2 patients within a selection of 4 rooms gives a different assignment, so permutations should be used. ∎

EXAMPLE 10

Distinguishing Between Permutations and Combinations

To illustrate the differences between permutations and combinations in another way, suppose 2 cans of soup are to be selected from 4 cans on a shelf: noodle (N), bean (B), mushroom (M), and tomato (T). As shown in Figure 13(a), there are 12 ways to select 2 cans from the 4 cans if the order matters (if noodle first and bean second is considered different from bean, then noodle, for example). On the other hand, if order is unimportant, then there are 6 ways to choose 2 cans of soup from the 4, as illustrated in Figure 13(b). ∎

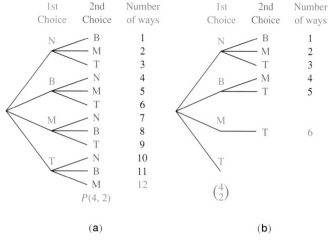

FIGURE 13

CAUTION It should be stressed that not all counting problems lend themselves to either permutations or combinations. Whenever a tree diagram or the multiplication principle can be used directly, as in Example 10, then use it.

| 11.5 EXERCISES

Evaluate each expression in Exercises 1–12.

1. $P(12, 8)$ **2.** $P(5, 5)$ **3.** $P(9, 2)$ **4.** $P(10, 9)$

5. $P(5, 1)$ **6.** $P(6, 0)$ **7.** $\binom{4}{2}$ **8.** $\binom{9}{3}$

9. $\binom{6}{0}$ **10.** $\binom{8}{1}$ **11.** $\binom{12}{4}$ **12.** $\binom{16}{3}$

Use the multiplication principle or permutations to solve the following problems.

13. In an experiment on social interaction 6 people will sit in 6 seats in a row. In how many different ways can the 6 people be seated?

14. In how many ways can 7 of 10 mice be arranged in a row for a genetics experiment?

15. For many years, the state of California used three letters followed by three digits on its automobile license plates.
  (a) How many different license plates are possible with this arrangement?
  (b) When the state ran out of new plates, the order was reversed to three digits followed by three letters. How many additional plates were then possible?
  (c) Several years ago, the plates described in (b) were also used up. The state then issued plates with one letter followed by three digits and then three letters. How many plates does this scheme provide?

16. How many 7-digit telephone numbers are possible if the first digit cannot be 0, and
  (a) only odd digits may be used?
  (b) the telephone number must be a multiple of 10 (that is, it must end in 0)?
  (c) the first three digits must be 456?

17. If your college offers 400 courses, 20 of which are in mathematics, and your counselor arranges your schedule of 4 courses by random selection, how many schedules are possible that do not include a math course?

18. In a club with 35 members, how many ways can a slate of 3 officers consisting of president, program chairman, and secretary/treasurer be chosen?

19. In how many ways can 5 players be assigned to the 5 positions on a basketball team, assuming that any player can play any position? In how many ways can 10 players be assigned to the 5 positions?

20. A softball team has 20 players. How many 9-player batting orders are possible?

Use combinations to solve the problems in Exercises 22–30.

22. A homeowners' association has 50 members. If a committee of 6 is to be selected at random, how many different committees are possible?

23. How many different samples of 4 light bulbs can be selected from a carton of 2 dozen bulbs?

24. A group of 5 students is to be selected at random from a class of 30 to participate in an experimental class. In how many ways can this be done? In how many ways can the group that will not participate be selected?

25. Harry's Hamburger Heaven sells hamburgers with cheese, relish, lettuce, tomato, onion, mustard, or ketchup. How many different hamburgers can be concocted using any 4 of the extras?

26. How many different 5-card poker hands can be dealt from a deck of 52 playing cards?

27. Seven cards are marked with the numbers 1 through 7 and are shuffled, and then 3 cards are drawn. How many different 3-card combinations are possible?

28. A bag contains 18 marbles. How many samples of 3 can be drawn from it? How many samples of 5 marbles?

Solve each of the following by using either permutations or combinations.

35. From a pool of 7 secretaries, 3 are selected to be assigned to 3 managers, with 1 secretary for each manager. In how many ways can this be done?

36. In a game of musical chairs, 12 children will sit in 11 chairs (1 will be left out). How many seatings are possible?

37. In an experiment on plant hardiness, a researcher gathers 6 wheat plants, 3 barley plants, and 2 rye plants. Four plants are to be selected at random.
(a) In how many ways can this be done?
(b) In how many ways can this be done if exactly 2 wheat plants must be included?

38. In an office with 8 men and 11 women, how many 5-member groups can be chosen that have each of the following compositions.
(a) All men   (b) All women
(c) 3 men and 2 women
(d) No more than 3 women

21. In how many ways can 6 bank tellers be assigned to 6 different windows? In how many ways can 10 tellers be assigned to the 6 windows?

29. In Exercise 28, if the bag contains 5 purple, 4 green, and 9 black marbles, how many samples of 3 can be drawn in which all the marbles are black? How many samples of 3 can be drawn in which exactly 2 marbles are black?

30. In Exercise 23, assume it is known that there are 5 defective light bulbs in the carton. How many samples of 4 can be drawn in which all are defective? How many samples of 4 can be drawn in which there are 2 good bulbs and 2 defective bulbs?

31. Explain the difference between a permutation and a combination.

32. Is choosing two kittens from a litter of six kittens an example of a permutation or a combination?

33. Padlocks with digit dials are often referred to as "combination locks." According to the mathematical definition of combination, is this an accurate description? Why or why not?

34. Determine whether each of the following is a permutation or a combination.
(a) Your five-digit postal zip code
(b) A particular five-card hand in a game of poker
(c) A committee of school board members

39. From 10 names on a ballot, 4 will be elected to a political party committee. How many different committees are possible? In how many ways can the committee of 4 be formed if each person will have a different responsibility?

40. In how many ways can 5 of 9 plants be arranged in a row on a windowsill?

41. Velma specializes in making different vegetable soups with carrots, celery, onions, beans, peas, tomatoes, and potatoes. How many different soups can she make using any 4 ingredients?

42. How many 4-letter radio-station call letters can be made if the first letter must be K or W and no letter may be repeated? How many if repeats are allowed? How many of the call letters with no repeats can end in K?

43. A group of 12 workers decide to send a delegation of 3 to their supervisor to discuss their work assignments.
   (a) How many delegations of 3 are possible?
   (b) How many are possible if one of the 12, the foreman, must be in the delegation?
   (c) If there are 5 women and 7 men in the group, how many possible delegations would include exactly one woman?

44. The Riverdale board of supervisors is composed of 2 liberals and 5 conservatives. Three members are to be selected randomly as delegates to a convention.
   (a) How many delegations are possible?
   (b) How many delegations could have all liberals?
   (c) How many delegations could have 2 conservatives and 1 liberal?
   (d) If the supervisor who serves as chairman of the board must be included, how many delegations are possible?

Prove each of the following statements for positive integers n and r, with r ≤ n.

45. $P(n, n - 1) = P(n, n)$

46. $P(n, 1) = n$

47. $P(n, 0) = 1$

48. $\binom{n}{n} = 1$

49. $\binom{n}{0} = 1$

50. $\binom{n}{n - 1} = n$

51. $\binom{n}{n - r} = \binom{n}{r}$

*F*urther *Explorations*

1. Find all values for X! as X goes from 0 to 20. You can evaluate them one at a time on the HOME screen, or if your calculator has a TABLE feature, let the TABLE generate the values (let $Y_1 = X!$). A graphical display is not appropriate for factorials, because they increase so quickly that one display window cannot include all 20 ordered pairs as distinct points. *Note:* Factorials are defined only when X is a whole number (0, 1, 2, 3, . . .); the calculator will return an error if it is commanded to find a factorial of any number that is not whole.

2. Rachel is trying to decide which of five shirts to buy. She has not decided how many of them she will get. How many combinations are possible if she were to buy 0, 1, 2, 3, 4, or all 5 shirts? Use the TABLE feature of your calculator to generate all six combinations at once.

3. You can compare combinations and permutations very easily by using the TABLE feature of your graphics calculator. Compare $_5C_x$ with $_5P_x$. Examine the resulting TABLE. Describe your observations.

11.6 PROBABILITY

The Probability of an Event ▮ The Complement of an Event ▮ Odds ▮ The Union of Two Events ▮ Binomial Probability

The Probability of an Event

The study of probability has become increasingly popular because it has a wide range of practical applications. The basic ideas of probability are introduced in this section.

Consider an experiment that has one or more possible **outcomes,** each of which is equally likely to occur. For example, the experiment of tossing a fair coin has two equally likely possible outcomes: landing heads up (*H*) or landing tails up (*T*). Also, the experiment of rolling a fair die has 6 equally likely outcomes: landing so the face that is up shows 1, 2, 3, 4, 5, or 6 points.

The set S of all possible outcomes of a given experiment is called the **sample space** for the experiment. (In this text all sample spaces are finite.) One sample space for the experiment of tossing a coin could consist of the outcomes H and T. This sample space can be written in set notation as

$$S = \{H, T\}.$$

Similarly, a sample space for the experiment of rolling a single die might be

$$S = \{1, 2, 3, 4, 5, 6\}.$$

Any subset of the sample space is called an **event.** In the experiment with the die, for example, "the number showing is a three" is an event, say E_1, such that $E_1 = \{3\}$. "The number showing is greater than three" is also an event, say E_2, such that $E_2 = \{4, 5, 6\}$. To represent the number of outcomes that belong to event E, the notation $n(E)$ is used. Then $n(E_1) = 1$ and $n(E_2) = 3$.

Another sample space might give as outcomes the number of heads in one toss of a fair coin. This sample space would be written

$$S = \{0, 1\}.$$

Are these outcomes equally likely?

PROBABILITY OF EVENT E

In a sample space with equally likely outcomes, the **probability** of an event E, written $P(E)$, is the ratio of the number of outcomes in sample space S that belong to event E, $n(E)$, to the total number of outcomes in sample space S, $n(S)$. That is,

$$P(E) = \frac{n(E)}{n(S)}.$$

This definition is used to find the probability of the event E_1 given above, by starting with the sample space for the experiment, $S = \{1, 2, 3, 4, 5, 6\}$, and the desired event, $E_1 = \{3\}$. Since $n(E_1) = 1$ and since there are 6 outcomes in the sample space,

$$P(E_1) = \frac{n(E_1)}{n(S)} = \frac{1}{6}.$$

EXAMPLE 1

Finding Probabilities of Events

A single die is rolled. Write the following events in set notation and give the probability for each event.

(a) E_3: the number showing is even

SOLUTION   Use the definition above. Since $E_3 = \{2, 4, 6\}$, $n(E_3) = 3$. As shown above, $n(S) = 6$, so

$$P(E_3) = \frac{3}{6} = \frac{1}{2}.$$

(b) E_4: the number showing is greater than 4

SOLUTION   Again $n(S) = 6$. Event $E_4 = \{5, 6\}$, with $n(E_4) = 2$. By the definition,

$$P(E_4) = \frac{2}{6} = \frac{1}{3}.$$

(c) E_5: the number showing is less than 7

SOLUTION

$$E_5 = \{1, 2, 3, 4, 5, 6\} \qquad \text{and} \qquad P(E_5) = \frac{6}{6} = 1$$

(d) E_6: the number showing is 7

SOLUTION

$$E_6 = \emptyset \qquad \text{and} \qquad P(E_6) = \frac{0}{6} = 0 \qquad \blacksquare$$

In Example 1(c), $E_5 = S$. Therefore, the event E_5 is certain to occur every time the experiment is performed. An event that is certain to occur, such as E_5, always has a probability of 1. On the other hand, $E_6 = \emptyset$ and $P(E_6)$ is 0. The probability of an impossible event, such as E_6, is always 0, since none of the outcomes in the sample space satisfy the event. For any event E, $P(E)$ is between 0 and 1 inclusive.

The Complement of an Event

The set of all outcomes in the sample space that do *not* belong to event E is called the **complement** of E, written E'. For example, in the experiment of drawing a single card from a standard deck of 52 cards, let E be the event "the card is an ace." Then E' is the event "the card is not an ace." From the definition of E', for an event E,

$$E \cup E' = S \qquad \text{and} \qquad E \cap E' = \emptyset.$$

A standard deck of 52 cards has four suits: hearts, clubs, diamonds, and spades, with thirteen cards of each suit. Each suit has an ace, king, queen, jack, and cards numbered from 2 to 10. The hearts and diamonds are red and the spades and clubs are black. We will refer to this standard deck of cards in this section.

Probability concepts can be illustrated using **Venn diagrams**, as shown in Figure 14. The rectangle in Figure 14 represents the sample space in an experiment. The area inside the circle represents event E, while the area inside the rectangle, but outside the circle, represents event E'.

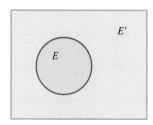

FIGURE 14

EXAMPLE 2

Using the Complement in a Probability Problem

In the experiment of drawing a card from a well-shuffled deck, find the probability of event E, the card is an ace, and event E'.

SOLUTION Since there are four aces in the deck of 52 cards, $n(E) = 4$ and $n(S) = 52$. Therefore,

$$P(E) = \frac{n(E)}{n(S)} = \frac{4}{52} = \frac{1}{13}.$$

Of the 52 cards, 48 are not aces, so

$$P(E') = \frac{n(E')}{n(S)} = \frac{48}{52} = \frac{12}{13}. \quad \blacksquare$$

In Example 2, $P(E) + P(E') = \frac{1}{13} + \frac{12}{13} = 1$. This is always true for any event E and its complement E'. That is,

$$P(E) + P(E') = 1.$$

This can be restated as

$$P(E) = 1 - P(E') \quad \text{or} \quad P(E') = 1 - P(E).$$

These two equations suggest an alternative way to compute the probability of an event. For example, if it is known that $P(E) = \frac{1}{10}$, then

$$P(E') = 1 - \frac{1}{10} = \frac{9}{10}.$$

FOR GROUP DISCUSSION

Games of chance and gambling enterprises (the earliest motivators for the study of probability) are a major force today. In July of 1992, Public Television's *MacNeil/Lehrer NewsHour* reported that 34 states now run games for public revenue. One state lottery game requires you to pick 6 different numbers from 1 to 99.

1. How many ways are there to choose 6 numbers if order is not important?
2. How many ways are there if order is important?
3. Assume order is unimportant. What is the probability of picking all 6 numbers correctly to win the big prize?
4. Discuss the probability of winning in a state lottery in your area.

Odds

Sometimes probability statements are expressed in terms of odds, a comparison of $P(E)$ with $P(E')$. The **odds** in favor of an event E are expressed as the ratio of $P(E)$ to $P(E')$ or as the fraction $\frac{P(E)}{P(E')}$. For example, if the probability of rain can be established as $\frac{1}{3}$, the odds that it will rain are

$$P(\text{rain}) \text{ to } P(\text{no rain}) = \frac{1}{3} \text{ to } \frac{2}{3} = \frac{\frac{1}{3}}{\frac{2}{3}} = \frac{1}{2} \quad \text{or} \quad 1 \text{ to } 2.$$

On the other hand, the odds that it will not rain are 2 to 1 (or $\frac{2}{3}$ to $\frac{1}{3}$.) If the odds in favor of an event are, say, 3 to 5, then the probability of the event is $\frac{3}{8}$, while the probability of the complement of the event is $\frac{5}{8}$. If the odds favoring event E are m to n, then

$$P(E) = \frac{m}{m + n} \quad \text{and} \quad P(E') = \frac{n}{m + n}.$$

EXAMPLE 3

Finding Odds in Favor of an Event

A shirt is selected at random from a dark closet containing 6 blue shirts and 4 shirts that are not blue. Find the odds in favor of a blue shirt being selected.

SOLUTION   Let E represent "a blue shirt is selected." Then $P(E) = \frac{6}{10}$ or $\frac{3}{5}$. Also, $P(E') = 1 - \frac{3}{5} = \frac{2}{5}$. Therefore, the odds in favor of a blue shirt being selected are

$$P(E) \text{ to } P(E') = \frac{3}{5} \text{ to } \frac{2}{5} = \frac{\frac{3}{5}}{\frac{2}{5}} = \frac{3}{2} \quad \text{or} \quad 3 \text{ to } 2. \quad \blacksquare$$

The Union of Two Events

We now extend the rules for probability to more complex events. Since events are sets, we can use set operations to find the union of two events. (The *union* of sets A and B includes all elements of set A in addition to all elements of set B.)

Suppose a fair die is tossed. Let H be the event "the result is a 3," and K the event "the result is an even number." From the results earlier in this section,

$$H = \{3\} \qquad\qquad P(H) = \frac{1}{6}$$

$$K = \{2, 4, 6\} \qquad\qquad P(K) = \frac{3}{6} = \frac{1}{2}$$

$$H \cup K = \{2, 3, 4, 6\} \qquad P(H \cup K) = \frac{4}{6} = \frac{2}{3}.$$

Notice that $P(H) + P(K) = P(H \cup K)$.

Before assuming that this relationship is true in general, consider another event for this experiment, "the result is a 2," event G.

$$G = \{2\} \qquad\qquad P(G) = \frac{1}{6}$$

$$K = \{2, 4, 6\} \qquad\qquad P(K) = \frac{3}{6} = \frac{1}{2}$$

$$K \cup G = \{2, 4, 6\} \qquad P(K \cup G) = \frac{3}{6} = \frac{1}{2}$$

In this case $P(K) + P(G) \neq P(K \cup G)$. See Figure 15 on the next page.

As Figure 15 suggests, the difference in the two examples above comes from the fact that events H and K cannot occur simultaneously. Such events are called **mutually exclusive events.** In fact, $H \cap K = \emptyset$, which is true for any two mutually exclusive events. Events K and G, however, can occur simultaneously. Both are satisfied if the result of the roll is a 2, the element in their intersection ($K \cap G = \{2\}$). This example suggests the property given on the next page.

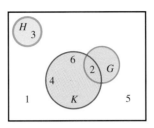

FIGURE 15

PROBABILITY OF THE UNION OF TWO EVENTS

For any events E and F:

$$P(E \text{ or } F) = P(E \cup F) = P(E) + P(F) - P(E \cap F).$$

EXAMPLE 4

Finding the Probability of a Union

One card is drawn from a well-shuffled deck of 52 cards. What is the probability of the following outcomes?

(a) The card is an ace or a spade.

SOLUTION The events "drawing an ace" and "drawing a spade" are not mutually exclusive since it is possible to draw the ace of spades, an outcome satisfying both events. The probability is

$$P(\text{ace or spade}) = P(\text{ace}) + P(\text{spade}) - P(\text{ace and spade})$$

$$= \frac{4}{52} + \frac{13}{52} - \frac{1}{52} = \frac{16}{52} = \frac{4}{13}.$$

(b) The card is a three or a king.

SOLUTION "Drawing a 3" and "drawing a king" are mutually exclusive events because it is impossible to draw one card that is both a 3 and a king. Using the rule given above,

$$P(3 \text{ or } K) = P(3) + P(K) - P(3 \text{ and } K)$$

$$= \frac{4}{52} + \frac{4}{52} - 0 = \frac{8}{52} = \frac{2}{13}. \qquad \blacksquare$$

EXAMPLE 5

Finding the Probability of a Union

Suppose two fair dice are rolled. Find each of the following probabilities.

(a) The first die shows a 2, or the sum of the points showing on the two dice is 6 or 7.

SOLUTION Think of the two dice as being distinguishable, one red and one green for example. (Actually, the sample space is the same even if they are not apparently distinguishable.) A sample space with equally likely outcomes is shown in Figure 16, where $(1, 1)$ represents the event "the first (red) die shows a 1 and the second die (green) shows a 1," $(1, 2)$ represents "the first die shows a 1 and the second die shows a 2," and so on. Let A represent the event "the first die shows a 2," and B represent the event "the sum of the results is 6 or 7." These events are indicated in Figure 16. From the diagram, event A has 6 elements, B has 11 elements, and the sample space has 36 elements. Thus,

$$P(A) = \frac{6}{36}, \quad P(B) = \frac{11}{36}, \quad \text{and} \quad P(A \cap B) = \frac{2}{36}.$$

By the union rule,

$$P(A \cup B) = P(A) + P(B) - P(A \cap B)$$
$$P(A \cup B) = \frac{6}{36} + \frac{11}{36} - \frac{2}{36} = \frac{15}{36} = \frac{5}{12}.$$

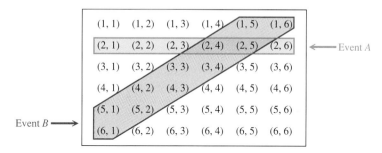

FIGURE 16

(b) The sum of the points showing is at most 4.

SOLUTION "At most 4" can be written as "2 or 3 or 4." (A sum of 1 is meaningless here.) Then

$$P(\text{at most } 4) = P(2 \text{ or } 3 \text{ or } 4)$$
$$= P(2) + P(3) + P(4),\qquad\qquad\text{[*]}$$

since the events represented by "2," "3," and "4" are mutually exclusive.

The sample space for this experiment includes the 36 possible pairs of numbers shown in Figure 16. The pair $(1, 1)$ is the only one with a sum of 2, so $P(2) = \frac{1}{36}$. Also $P(3) = \frac{2}{36}$ since both $(1, 2)$ and $(2, 1)$ give a sum of 3. The pairs $(1, 3)$, $(2, 2)$, and $(3, 1)$ have a sum of 4, so $P(4) = \frac{3}{36}$. Substituting into equation (*) above gives

$$P(\text{at most } 4) = \frac{1}{36} + \frac{2}{36} + \frac{3}{36}$$
$$= \frac{6}{36} = \frac{1}{6}. \quad \blacksquare$$

The properties of probability discussed in this section are summarized as follows.

PROPERTIES OF PROBABILITY

For any events E and F:

1. $0 \leq P(E) \leq 1$
2. $P(\text{a certain event}) = 1$
3. $P(\text{an impossible event}) = 0$
4. $P(E') = 1 - P(E)$
5. $P(E \text{ or } F) = P(E \cup F) = P(E) + P(F) - P(E \cap F)$.

CAUTION When finding the probability of a union, don't forget to subtract the probability of the intersection from the sum of the probabilities of the individual events.

Binomial Probability

If an experiment consists of repeated independent trials with only two outcomes in each trial, success or failure, it is called a **binomial experiment.** Let the probability of success in one trial be p. Then the probability of failure is $1 - p$, and the probability of r successes in n trials is given by

$$\binom{n}{r} p^r (1 - p)^{n-r}.$$

Notice that this expression is equivalent to the general term of the binomial expansion given in Section 11.2. Thus the terms of the binomial expansion give the probabilities of r successes in n trials, for $0 \le r \le n$, in a binomial experiment.

EXAMPLE 6
Finding Probabilities Using a Binomial Experiment

An experiment consists of rolling a die 10 times. Find the following probabilities.

(a) The probability that exactly 4 of the tosses result in a three.

SOLUTION The probability of a three on one roll is $p = \frac{1}{6}$. The required probability is

$$\binom{10}{4} \left(\frac{1}{6}\right)^4 \left(1 - \frac{1}{6}\right)^{10-4} = 210 \left(\frac{1}{6}\right)^4 \left(\frac{5}{6}\right)^6$$
$$\approx .054$$

TECHNOLOGICAL NOTE
Read the Further Explorations at the end of this section, and then use a table to confirm the results of Example 6.

(b) The probability that in 9 of the 10 tosses the result is not a three.

SOLUTION This probability is

$$\binom{10}{1} \left(\frac{1}{6}\right)^1 \left(\frac{5}{6}\right)^9 \approx .323. \quad \blacksquare$$

11.6 EXERCISES

Write a sample space with equally likely outcomes for each of the following experiments.

1. A two-headed coin is tossed once.

2. Two ordinary coins are tossed.

3. Three ordinary coins are tossed.

4. Five slips of paper marked with the numbers 1, 2, 3, 4, and 5 are placed in a box. After mixing well, two slips are drawn.

5. An unprepared student takes a three-question true/false quiz in which he guesses the answer to all three questions.

6. A die is rolled and then a coin is tossed.

Write the events in Exercises 7–10 in set notation and give the probability of each event.

7. In the experiment from Exercise 2:
   (a) Both coins show the same face.   **(b)** At least one coin turns up heads.

8. In Exercise 1:
   (a) The result of the toss is heads.   **(b)** The result of the toss is tails.

9. In Exercise 4:
(a) Both slips are marked with even numbers.
(b) Both slips are marked with odd numbers.
(c) Both slips are marked with the same number.
(d) One slip is marked with an odd number and the other with an even number.

10. In Exercise 5:
(a) The student gets all three answers correct.
(b) He gets all three answers wrong.
(c) He gets exactly two answers correct.
(d) He gets at least one answer correct.

11. A student gives the answer to a probability problem as $\frac{6}{5}$. Explain why this answer must be incorrect.

12. If the probability of an event is .857, what is the probability that the event will not occur?

13. A marble is drawn at random from a box containing 3 yellow, 4 white, and 8 blue marbles. Find each probability in (a)–(c).
(a) A yellow marble is drawn. (b) A blue marble is drawn.
(c) A black marble is drawn.
(d) What are the odds in favor of drawing a yellow marble?
(e) What are the odds against drawing a blue marble?

Relating Concepts

Many probability problems involve numbers that are too large to determine the number of outcomes easily, even with a tree diagram. In such cases, we can use combinations. For example, if 3 engines are tested from a shipping container packed with 12 engines, 1 of which is defective, what is $P(E)$, the probability that the defective engine will be found?

14. How many ways are there to choose the sample of 3 from the 12 engines?

15. How many ways are there to choose a sample of 3 with 1 defective and 2 good engines?

16. What is $n(E)$ in this experiment if E is the event "The defective engine is in the sample"?

17. What is $n(S)$ in this experiment?

18. Find $P(E)$.

Solve each problem.

19. A baseball player with a batting average of .300 comes to bat. What are the odds in favor of his getting a hit?

20. In Exercise 4, what are the odds that the sum of the numbers on the two slips of paper is 5?

21. If the odds that it will rain are 4 to 5, what is the probability of rain?

22. If the odds that a candidate will win an election are 3 to 2, what is the probability that the candidate will lose?

23. Ms. Bezzone invites 10 relatives to a party: her mother, two uncles, three brothers, and four cousins. If the chances of any one guest arriving first are equally likely, find the following probabilities.
(a) The first guest is an uncle or a cousin.
(b) The first guest is a brother or a cousin.
(c) The first guest is an uncle or her mother.

24. A card is drawn from a well-shuffled deck of 52 cards. Find the probability that the card is the following.
(a) A queen   (b) Red
(c) A black 3   (d) A club or red

25. In Exercise 24, find the probability of the following.
    (a) A face card (K, Q, J of any suit)
    (b) Red or a 3
    (c) Less than a four (consider aces as ones)

26. Two dice are rolled. Find the probability of the following events.
    (a) The sum of the points is at least 10.
    (b) The sum of the points is either 7 or at least 10.
    (c) The sum of the points is 3 or the dice both show the same number.

27. If a marble is drawn from a bag containing 2 yellow, 5 red, and 3 blue marbles, what are the probabilities of the following results?
    (a) The marble is yellow or blue.
    (b) The marble is yellow or red.
    (c) The marble is green.

28. The law firm of Alam, Bartolini, Chinn, Dickinson, and Ellsberg has two senior partners, Alam and Bartolini. Two of the attorneys are to be selected to attend a conference. Assuming that all are equally likely to be selected, find the following probabilities.
    (a) Chinn is selected.
    (b) Alam and Dickinson are selected.
    (c) At least one senior partner is selected.

29. The management of a firm wants to survey its workers, who are classified as follows for the purpose of an interview: 30% have worked for the company more than 5 years; 28% are female; 65% contribute to a voluntary retirement plan; half of the female workers contribute to the retirement plan. Find the following probabilities.
    (a) A male worker is selected.
    (b) A worker is selected who has been employed by the company for 5 years or less.
    (c) A worker is selected who contributes to the retirement plan or is female.

30. The table shows the probabilities of a person accumulating specific amounts of credit card charges over a 12-month period.

Charges	Probability
Under $100	.31
$100–$499	.18
$500–$999	.18
$1000–$1999	.13
$2000–$2999	.08
$3000–$4999	.05
$5000–$9999	.06
$10,000 or more	.01

Find the probabilities that a person's total charges during the period are the following.
(a) $500–$999   (b) $5000–$9999
(c) $500–$2999   (d) $3000 or more

In most animals and plants, it is very unusual for the number of main parts of the organism (arms, legs, toes, flower petals, etc.) to vary from generation to generation. Some species, however, have meristic variability, *in which the number of certain body parts varies from generation to generation. One researcher studied the front feet of certain guinea pigs and produced the following probabilities.*

$$P(\text{only four toes, all perfect}) = .77$$
$$P(\text{one imperfect toe and four good ones}) = .13$$
$$P(\text{exactly five good toes}) = .10$$

Find the probability of each of the following events.

31. No more than four good toes   32. Five toes, whether perfect or not

The probabilities for the outcomes of an experiment having sample space $S = \{s_1, s_2, s_3, s_4, s_5, s_6\}$ *are shown here.*

Outcomes	s_1	s_2	s_3	s_4	s_5	s_6
Probability	.17	.03	.09	.46	.21	.04

Let $E = \{s_1, s_2, s_5\}$, *and let* $F = \{s_4, s_5\}$. *Find each probability in Exercises 33–38.*

33. $P(E)$ 34. $P(F)$ 35. $P(E \cap F)$

36. $P(E \cup F)$ 37. $P(E' \cup F')$ 38. $P(E' \cap F)$

Suppose that a family has 5 children. Also, suppose that the probability of having a girl is $\frac{1}{2}$. Find the probabilities that the family has the following children.

39. Exactly 2 girls and 3 boys   **40.** Exactly 3 girls and 2 boys   **41.** No girls

42. No boys   **43.** At least 3 boys   **44.** No more than 4 girls

A die is rolled 12 times. Find the probabilities of rolling the following.

45. Exactly 12 ones   **46.** Exactly 6 ones   **47.** No more than 3 ones

48. No more than 1 one

*F*urther Explorations

If we let $Y_1 = 4\,_nC_r\,X\left(\dfrac{1}{2}\right)^X\left(\dfrac{1}{2}\right)^{4-X}$, we can use the TABLE feature to find the probabilities of having 0, 1, 2, 3, or 4 girls in a family of 4 children. See the figure.

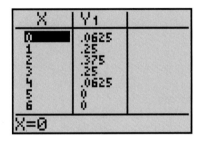

Use this approach to work each of the following.

1. Find the probabilities of having 0, 1, 2, or 3 boys in a family of 3 children.

2. Find the probabilities of having 0, 1, 2, 3, 4, 5, or 6 girls in a family of 6 children.

3. Explain why the values of Y_1 in the figure are 0 for X = 5 and for X = 6.

---

*C*hapter 11   SUMMARY

Mathematical induction is a method of proving statements that are true for all positive integers n by first showing that the statement is true when $n = 1$. We must then show that if the statement is true for $n = k$, it follows algebraically that it is also true for $n = k + 1$. If both parts of the proof are satisfied, then the statement must be true for all positive integers n.

The binomial theorem is a general statement that gives the terms in the expansion of $(x + y)^n$. Pascal's triangle is a triangular array of numbers that can be used to find the binomial coefficients, the coefficients in the expansion of the binomial theorem. The rth term of the binomial expansion can be found without writing out all the terms.

A sequence is an ordered list of numbers. The numbers in the list are the terms of the sequence. The general term gives an expression in n that determines the nth term. The terms of a convergent infinite sequence get closer and closer to some real number. A recursion formula defines each term of a sequence after the first in terms of the previous term.

In an arithmetic sequence, each term after the first is found by adding the common difference to the preceding term. In a geometric sequence, each term after the first is found by multiplying the preceding term by the common ratio.

A series is the sum of the terms of a sequence and is often described with summation notation. The sum of the first n terms of an arithmetic sequence is given in two forms:

$$S_n = \frac{n}{2}(a_1 + a_n) \quad \text{or} \quad S_n = \frac{n}{2}[2a_1 + (n-1)d].$$

The sum of the first n terms of a geometric sequence is given by

$$S_n = \frac{a_1(1 - r^n)}{1 - r}, \quad \text{where } r \neq 1.$$

The sum of an infinite geometric sequence, where $-1 < r < 1$, is found by dividing a_1 by $1 - r$. Sequences and series are quite useful in many applications, particularly those that deal with annuities. An annuity is a sequence of equal payments made at equal periods of time, such as the payments on a car loan.

The multiplication principle of counting is used to count the number of ways two or more independent events can occur. A permutation of n elements taken r at a time is one of the ways that the r elements can be *arranged*. A combination of n elements taken r at a time is one of the ways that a group of r elements can be *selected*.

The probability of an event E is the ratio of the number of outcomes in the sample space that satisfy E to the total number of outcomes in the sample space. The odds in favor of event E are the ratio of the probability of event E to the probability of the event that is the complement of E, symbolized E'. Mutually exclusive events are events that have no outcomes in common. The probability that two mutually exclusive events both occur is the sum of their individual probabilities. The probability that *any* two events E and F both occur is given by $P(E) + P(F) - P(E \text{ and } F)$.

A binomial experiment consists of repeated independent trials with only two outcomes per trial, success or failure. The probability of r successes in n trials is given by ${}_nC_r(p)^r(1 - p)^{n-r}$, where p is the probability of a success in one trial.

Key Terms

SECTION 11.1

mathematical induction

SECTION 11.2

Pascal's triangle
n-factorial
binomial coefficient
binomial expansion
binomial theorem

SECTION 11.3

sequence
term of a sequence
general (nth) term
finite sequence
infinite sequence
convergent sequence
divergent sequence
recursion formula
arithmetic sequence (progression)
common difference
geometric sequence (progression)
common ratio
annuity
future value of an annuity

SECTION 11.4

series
summation notation
index of summation

SECTION 11.5

tree diagram
independent events
multiplication principle of counting
permutation
combination

SECTION 11.6

outcome
sample space
event
probability
complement of an event
Venn diagram
odds
mutually exclusive events
binomial experiment

Chapter 11   REVIEW EXERCISES

Use mathematical induction to prove that each of the following is true for every positive integer n.

1. $2 + 6 + 10 + 14 + \cdots + (4n - 2) = 2n^2$

2. $2^2 + 4^2 + 6^2 + \cdots + (2n)^2 = \dfrac{2n(n + 1)(2n + 1)}{3}$

3. $2 + 2^2 + 2^3 + \cdots + 2^n = 2(2^n - 1)$

4. $1^3 + 3^3 + 5^3 + \cdots + (2n - 1)^3 = n^2(2n^2 - 1)$

Use the binomial theorem to expand each of the following.

5. $(x + 2y)^4$

6. $\left(\dfrac{k}{2} - g\right)^5$

Find the indicated term or terms for each of the following expansions.

7. Fifth term of $(3x - 2y)^6$

8. Eighth term of $(2m + n^2)^{12}$

9. First four terms of $(3 + x)^{16}$

10. Last three terms of $(2m - 3n)^{15}$

Write the first five terms for each of the following sequences.

11. $a_n = \dfrac{n}{n + 1}$

12. $a_n = (-2)^n$

13. $a_n = 2(n + 3)$

14. $a_n = n(n + 1)$

15. $a_1 = 5;$ for $n \geq 2$, $a_n = a_{n-1} - 3$

16. $a_1 = -2;$ for $n \geq 2$, $a_n = 3a_{n-1}$

17. $a_1 = 5, a_2 = 3;$ for $n \geq 3$, $a_n = a_{n-1} - a_{n-2}$

18. $b_1 = -2, b_2 = 2, b_3 = -4;$ for $n \geq 4$, $b_n = -2 \cdot b_{n-2}$ if n is even, and $b_n = 2 \cdot b_{n-2}$ if n is odd.

19. Arithmetic, $a_1 = 6, d = -4$

20. Arithmetic, $a_3 = 9, a_4 = 7$

21. Arithmetic, $a_1 = 3 - \sqrt{5}, a_2 = 4$

22. Arithmetic, $a_3 = \pi, a_4 = 0$

23. A certain arithmetic sequence has $a_6 = -4$ and $a_{17} = 51$. Find a_1 and a_{20}.

Find a_8 for each of the following arithmetic sequences.

24. $a_1 = 6, d = 2$

25. $a_1 = -4, d = 3$

26. $a_1 = 6x - 9, a_2 = 5x + 1$

27. $a_3 = 11m, a_5 = 7m - 4$

Write the first five terms for the geometric sequences in Exercises 28–31.

28. $a_1 = 4, r = 2$

29. $a_4 = 8, r = \dfrac{1}{2}$

30. $a_1 = -3, a_2 = 4$

31. $a_3 = 8, a_5 = 72$

32. For a given geometric sequence, $a_1 = 4$ and $a_5 = 324$. Find a_6.

Find a_5 for each of the following geometric sequences.

33. $a_1 = 3, r = 2$

34. $a_2 = 3125, r = \dfrac{1}{5}$

35. $a_1 = 5x, a_2 = x^2$

36. $a_2 = \sqrt{6}, a_4 = 6\sqrt{6}$

37. Explain the difference between an arithmetic sequence and a geometric sequence.

Find S_{12} for each of the following arithmetic sequences.

38. $a_1 = 2, d = 3$

39. $a_2 = 6, d = 10$

40. $a_1 = -4k, d = 2k$

Find S_4 for each of the following geometric sequences.

41. $a_1 = 1, r = 2$

42. $a_1 = 3, r = 3$

43. $a_1 = 2k, a_2 = -4k$

Evaluate each of the following sums.

44. $\displaystyle\sum_{i=1}^{4} \frac{2}{i}$

45. $\displaystyle\sum_{i=1}^{7} (-1)^{i+1} \cdot 6$

46. $\displaystyle\sum_{i=4}^{8} 3i(2i - 5)$

47. $\displaystyle\sum_{i=1}^{6} i(i + 2)$

48. $\displaystyle\sum_{i=1}^{6} 4 \cdot 3^i$

49. $\displaystyle\sum_{i=1}^{4} 8 \cdot 2^i$

Solve each problem.

50. A stack of canned goods in a market display requires 15 cans on the bottom, 13 in the next layer, 11 in the next layer, and so on. How many cans are needed for the display?

51. Gale Stockdale borrows $6000 at simple interest of 12% per year. He will repay the loan and interest in monthly payments of $260, $258, $256, and so on. If he makes 30 payments, what is the total amount required to pay off the loan plus the interest?

52. On a certain production line, new employees during their first week turn out $\frac{5}{4}$ as many items each day as on the previous day. If a new employee produces 48 items the first day, how many will she produce on the fifth day of work?

53. The half-life of a radioactive substance is 20 years. If 600 grams are present at the start, how much will be left after 100 years?

Evaluate the sum if it exists.

54. $18 + 9 + \dfrac{9}{2} + \dfrac{9}{4} + \cdots$

55. $20 + 15 + \dfrac{45}{4} + \dfrac{135}{16} + \cdots$

56. $-\dfrac{5}{6} + \dfrac{5}{9} - \dfrac{10}{27} + \cdots$

57. $.8 + .08 + .008 + .0008 + \cdots$

58. $\displaystyle\sum_{i=1}^{\infty} \left(\frac{5}{8}\right)^i$

59. $\displaystyle\sum_{i=1}^{\infty} -10\left(\frac{5}{2}\right)^i$

Convert each of the following repeating decimals to a quotient of integers. (See For Group Discussion *in Section 11.4.)*

60. .6666 . . .

61. .512512512 . . .

62. Explain the difference between a sequence and a series.

Evaluate each of the following.

63. $P(5, 5)$

64. $P(9, 2)$

65. $P(6, 0)$

66. $_8C_3$

67. $\dbinom{10}{5}$

68. $\dbinom{6}{0}$

Solve each problem.

69. Two people are planning their wedding. They can select from 2 different chapels, 4 soloists, 3 organists, and 2 ministers. How many different wedding arrangements are possible?

70. John Jacobs, who is furnishing his apartment, wants to buy a new couch. He can select from 5 different styles, each available in 3 different fabrics, with 6 color choices. How many different couches are available?

71. Four students are to be assigned to 4 different summer jobs. Each student is qualified for all 4 jobs. In how many ways can the jobs be assigned?

72. A student body council consists of a president, vice-president, secretary-treasurer, and 3 representatives at large. Three members are to be selected to attend a conference.
   (a) How many different such delegations are possible?
   (b) How many are possible if the president must attend?

73. Nine football teams are competing for first-, second-, and third-place titles in a statewide tournament. In how many ways can the winners be determined?

74. How many different license plates can be formed with a letter followed by 3 digits and then 3 letters? How many such license plates have no repetitions?

75. A marble is drawn at random from a box containing 4 green, 5 black, and 6 white marbles. Find the following probabilities.
  (a) A green marble is drawn.
  (b) A marble that is not black is drawn.
  (c) A blue marble is drawn.

76. Refer to Exercise 75 and answer each question.
  (a) What are the odds in favor of drawing a green marble?
  (b) What are the odds against drawing a white marble?

A card is drawn from a standard deck of 52 cards. Find the probability that each of the following is drawn.

77. A black king

78. A face card or an ace

79. An ace or a diamond

80. A card that is not a diamond

81. A card that is not a diamond or not black

A sample shipment of 5 swimming pool filters is chosen. The probability of exactly 0, 1, 2, 3, 4, or 5 filters being defective is given in the following table.

Number defective	0	1	2	3	4	5
Probability	.31	.25	.18	.12	.08	.06

Find the probability that the following numbers of filters are defective.

82. No more than 3

83. At least 2

84. More than 5

85. A die is rolled 12 times. Find the probability that exactly 2 of the rolls result in a five.

86. A coin is tossed 10 times. Find the probability that exactly 4 of the tosses result in a head.

Review of Basic Algebraic Concepts

In this appendix we present a review of topics that are usually studied in beginning and intermediate algebra courses. You may wish to refer to the various sections of this appendix from time to time if you need to refresh your memory on these basic concepts.

I. REVIEW OF RULES FOR EXPONENTS

Work with exponents can be simplified by using the rules for exponents. By definition, the notation a^m (where m is a positive integer and a is a real number) means that a appears as a factor m times. In the same way, a^n (where n is a positive integer) means that a appears as a factor n times. In the product $a^m \cdot a^n$, the number a would appear $m + n$ times.

> **PRODUCT RULE**
>
> For all positive integers m and n and every real number a:
>
> $$a^m \cdot a^n = a^{m+n}.$$

EXAMPLE 1

Using the Product Rule

Find the following products.

(a) $y^4 \cdot y^7 = y^{4+7} = y^{11}$

(b) $(6z^5)(9z^3)(2z^2)$

$$(6z^5)(9z^3)(2z^2) = (6 \cdot 9 \cdot 2) \cdot (z^5 z^3 z^2) \qquad \text{Commutative and associative properties}$$

$$= 108z^{5+3+2} \qquad \text{Product rule}$$

$$= 108z^{10}$$

(c) $(2k^m)(k^{1+m})$

$$(2k^m)(k^{1+m}) = 2k^{m+(1+m)}$$
$$= 2k^{1+2m} \quad \blacksquare$$

Product rule (if m is a positive integer)

An exponent of zero is defined as follows.

DEFINITION OF a^0

For any nonzero real number a:

$$a^0 = 1.$$

EXAMPLE 2

Using the Definition of a^0

(a) $3^0 = 1$

(b) $(-4)^0 = 1$
Replace a with -4 in the definition.

(c) $-4^0 = -1$, since -4^0 means $-(4^0) = -1$.

(d) $-(-4)^0 = -(1) = -1$

(e) $(7r)^0 = 1$, if $r \neq 0$ $\quad \blacksquare$

The expression $(2^5)^3$ can be written as

$$(2^5)^3 = 2^5 \cdot 2^5 \cdot 2^5.$$

By a generalization of the product rule for exponents, this product is

$$(2^5)^3 = 2^{5+5+5} = 2^{15}.$$

The same exponent could have been obtained by multiplying 3 and 5. This example suggests the first of the **power rules** given below. The others are found in a similar way.

POWER RULES

For all nonnegative integers m and n and all real numbers a and b:

$$(a^m)^n = a^{mn} \qquad (ab)^m = a^m b^m \qquad \left(\frac{a}{b}\right)^m = \frac{a^m}{b^m} \qquad (b \neq 0).$$

EXAMPLE 3

Using the Power Rules

(a) $(5^3)^2 = 5^{3(2)} = 5^6$

(b) $(3^4 x^2)^3 = (3^4)^3 (x^2)^3 = 3^{4(3)} x^{2(3)} = 3^{12} x^6$

(c) $\left(\dfrac{2^5}{b^4}\right)^3 = \dfrac{(2^5)^3}{(b^4)^3} = \dfrac{2^{15}}{b^{12}}$, if $b \neq 0$ $\quad \blacksquare$

The quotient rule for exponents is found in Section V of this appendix, after the discussion of negative exponents.

II. REVIEW OF OPERATIONS WITH POLYNOMIALS

Since the variables used in polynomials represent real numbers, a polynomial represents a real number. This means that all the properties of the real numbers hold for polynomials. In particular, the distributive property holds, so

$$3m^5 - 7m^5 = (3 - 7)m^5 = -4m^5.$$

Thus, polynomials are added by adding coefficients of like terms; polynomials are subtracted by subtracting coefficients of like terms.

EXAMPLE **1**
Adding and Subtracting Polynomials

Add or subtract, as indicated.

(a) $(2y^4 - 3y^2 + y) + (4y^4 + 7y^2 + 6y)$

$\quad = (2 + 4)y^4 + (-3 + 7)y^2 + (1 + 6)y$

$\quad = 6y^4 + 4y^2 + 7y$

(b) $(-3m^3 - 8m^2 + 4) - (m^3 + 7m^2 - 3)$

$\quad = (-3 - 1)m^3 + (-8 - 7)m^2 + [4 - (-3)]$

$\quad = -4m^3 - 15m^2 + 7$

(c) $8m^4p^5 - 9m^3p^5 + (11m^4p^5 + 15m^3p^5) = 19m^4p^5 + 6m^3p^5$

(d) $4(x^2 - 3x + 7) - 5(2x^2 - 8x - 4)$

$\quad = 4x^2 - 4(3x) + 4(7) - 5(2x^2) - 5(-8x) - 5(-4)$ Distributive property

$\quad = 4x^2 - 12x + 28 - 10x^2 + 40x + 20$ Associative property

$\quad = -6x^2 + 28x + 48$ Add like terms. ∎

As shown in parts (a), (b), and (d) of Example 1, polynomials in one variable are often written with their terms in *descending order*, so the term of highest degree is first, the one with the next highest degree is next, and so on.

The associative and distributive properties, together with the properties of exponents, can also be used to find the product of two polynomials. For example, to find the product of $3x - 4$ and $2x^2 - 3x + 5$, treat $3x - 4$ as a single expression and use the distributive property as follows.

$$(3x - 4)(2x^2 - 3x + 5) = (3x - 4)(2x^2) - (3x - 4)(3x) + (3x - 4)(5)$$

Now use the distributive property three separate times on the right of the equals sign to get

$$(3x - 4)(2x^2 - 3x + 5)$$
$$= (3x)(2x^2) - 4(2x^2) - (3x)(3x) - (-4)(3x) + (3x)5 - 4(5)$$
$$= 6x^3 - 8x^2 - 9x^2 + 12x + 15x - 20$$
$$= 6x^3 - 17x^2 + 27x - 20.$$

It is sometimes more convenient to write such a product vertically, as follows.

$$
\begin{array}{r}
2x^2 - 3x + 5 \\
3x - 4 \\
\hline
-8x^2 + 12x - 20 \\
6x^3 - 9x^2 + 15x \\
\hline
6x^3 - 17x^2 + 27x - 20
\end{array}
$$

Add in columns.

EXAMPLE **2**
Multiplying Polynomials

Multiply $(3p^2 - 4p + 1)(p^3 + 2p - 8)$.

SOLUTION   Multiply each term of the second polynomial by each term of the first and add these products. It is most efficient to work vertically with polynomials of more than two terms, so that like terms can be placed in columns.

$$
\begin{array}{r}
3p^2 - 4p + 1 \\
p^3 + 2p - 8 \\
\hline
-\ 24p^2 + 32p - 8 \\
6p^3 - 8p^2 + 2p \\
3p^5 - 4p^4 + p^3 \\
\hline
3p^5 - 4p^4 + 7p^3 - 32p^2 + 34p - 8
\end{array}
$$

Multiply $3p^2 - 4p + 1$ by -8.
Multiply $3p^2 - 4p + 1$ by $2p$.
Multiply $3p^2 - 4p + 1$ by p^3.
Add in columns.   ∎

The FOIL method is a convenient way to find the product of two binomials. The memory aid FOIL (for First, Outside, Inside, Last) gives the pairs of terms to be multiplied to get the product, as shown in the next examples.

EXAMPLE **3**
Using FOIL to Multiply Two Binomials

Find each product.

(a) $(6m + 1)(4m - 3) = (6m)(4m) + (6m)(-3) + 1(4m) + 1(-3)$
$$= 24m^2 - 14m - 3$$

(b) $(2x + 7)(2x - 7) = 4x^2 - 14x + 14x - 49$
$$= 4x^2 - 49$$

(c) $(2k^n - 5)(k^n + 3) = 2k^{2n} + 6k^n - 5k^n - 15$
$$= 2k^{2n} + k^n - 15$$ ∎

In parts (a) and (c) of Example 3, the product of two binomials was a trinomial, while in part (b) the product of two binomials was a binomial. The product of two binomials of the forms $x + y$ and $x - y$ is always a binomial. Check by multiplying that the following is true.

PRODUCT OF THE SUM AND DIFFERENCE OF TWO TERMS

$$(x + y)(x - y) = x^2 - y^2$$

This product is called the **difference of two squares.** Since products of this type occur frequently, it is important to *memorize this formula.*

EXAMPLE **4**
Multiplying the Sum and Difference of Two Terms

Find each product.

(a) $(3p + 11)(3p - 11)$
Using the pattern discussed above, replace x with $3p$ and y with 11.
$$(3p + 11)(3p - 11) = (3p)^2 - 11^2 = 9p^2 - 121$$

(b) $(5m^3 - 3)(5m^3 + 3) = (5m^3)^2 - 3^2 = 25m^6 - 9$

(c) $(9k - 11r^3)(9k + 11r^3) = (9k)^2 - (11r^3)^2 = 81k^2 - 121r^6$ ∎

The **squares of binomials** are also special products. The products $(x + y)^2$ and $(x - y)^2$ are shown in the following box.

> **SQUARES OF BINOMIALS**
>
> $$(x + y)^2 = x^2 + 2xy + y^2$$
> $$(x - y)^2 = x^2 - 2xy + y^2$$

This formula is also one that occurs frequently and *should be memorized.* When factoring polynomials, you will need to be able to recognize these patterns.

EXAMPLE 5

Using the Formulas for Squares of Binomials

Find each product.

(a) $(2m + 5)^2 = (2m)^2 + 2(2m)(5) + (5)^2$
$$= 4m^2 + 20m + 25$$

(b) $(3x - 7y^4)^2 = (3x)^2 - 2(3x)(7y^4) + (7y^4)^2$
$$= 9x^2 - 42xy^4 + 49y^8 \quad \blacksquare$$

III. REVIEW OF FACTORING

The process of finding polynomials whose product equals a given polynomial is called **factoring.** For example, since $4x + 12 = 4(x + 3)$, both 4 and $x + 3$ are called **factors** of $4x + 12$. Also, $4(x + 3)$ is called the **factored form** of $4x + 12$. A polynomial that cannot be written as a product of two polynomials with integer coefficients is a **prime** or **irreducible polynomial.** A polynomial is **factored completely** when it is written as a product of prime polynomials with integer coefficients.

Polynomials are factored by using the distributive property. For example, to factor $6x^2y^3 + 9xy^4 + 18y^5$, we look for a monomial that is the greatest common factor of each term. The terms of this polynomial have $3y^3$ as the greatest common factor. By the distributive property,

$$6x^2y^3 + 9xy^4 + 18y^5 = (3y^3)(2x^2) + (3y^3)(3xy) + (3y^3)(6y^2)$$
$$= 3y^3(2x^2 + 3xy + 6y^2).$$

EXAMPLE 1

Factoring Out the Greatest Common Factor

Factor out the greatest common factor from each polynomial.

(a) $9y^5 + y^2$

SOLUTION   The greatest common factor is y^2.

$$9y^5 + y^2 = y^2 \cdot 9y^3 + y^2 \cdot 1$$
$$= y^2(9y^3 + 1)$$

(b) $6x^2t + 8xt + 12t = 2t(3x^2 + 4x + 6)$

(c) $14m^4(m + 1) - 28m^3(m + 1) - 7m^2(m + 1)$

SOLUTION   The greatest common factor is $7m^2(m + 1)$. Use the distributive property as follows.

$$14m^4(m + 1) - 28m^3(m + 1) - 7m^2(m + 1)$$
$$= [7m^2(m + 1)](2m^2 - 4m - 1)$$
$$= 7m^2(m + 1)(2m^2 - 4m - 1) \quad \blacksquare$$

When a polynomial has more than three terms, it can sometimes be factored by a method called **factoring by grouping.** For example, to factor

$$ax + ay + 6x + 6y,$$

collect the terms into two groups so that each group has a common factor.

$$ax + ay + 6x + 6y = (ax + ay) + (6x + 6y)$$

Factor each group, getting

$$ax + ay + 6x + 6y = a(x + y) + 6(x + y).$$

The quantity $(x + y)$ is now a common factor, which can be factored out, producing

$$ax + ay + 6x + 6y = (x + y)(a + 6).$$

It is not always obvious which terms should be grouped. Experience and repeated trials are the most reliable tools for factoring by grouping.

EXAMPLE 2

Factoring by Grouping

Factor by grouping.

(a) $mp^2 + 7m + 3p^2 + 21$

SOLUTION   Group the terms as follows.

$$mp^2 + 7m + 3p^2 + 21 = (mp^2 + 7m) + (3p^2 + 21)$$

Factor out the greatest common factor from each group.

$$\begin{aligned}(mp^2 + 7m) + (3p^2 + 21) &= m(p^2 + 7) + 3(p^2 + 7)\\ &= (p^2 + 7)(m + 3) \qquad \text{\small p^2 + 7 is a}\\ &\qquad\qquad\qquad\qquad\qquad \text{\small common factor.}\end{aligned}$$

(b) $2y^2 - 2z - ay^2 + az$

SOLUTION   Grouping terms as above gives

$$\begin{aligned}2y^2 - 2z - ay^2 + az &= (2y^2 - 2z) + (-ay^2 + az)\\ &= 2(y^2 - z) + a(-y^2 + z).\end{aligned}$$

The expression $-y^2 + z$ is the negative of $y^2 - z$, so the terms should be grouped as follows.

$$\begin{aligned}2y^2 - 2z - ay^2 + az &= (2y^2 - 2z) - (ay^2 - az)\\ &= 2(y^2 - z) - a(y^2 - z) \qquad \text{\small Factor each group.}\\ &= (y^2 - z)(2 - a). \qquad\quad \text{\small Factor out y^2 - z.}\;\blacksquare\end{aligned}$$

Later, in Example 5(e), we show another way to factor by grouping three of the four terms.

Factoring is the opposite of multiplication. Since the product of two binomials is usually a trinomial, we can expect factorable trinomials (that have terms with no common factor) to have two binomial factors. Thus, factoring trinomials requires using FOIL backwards.

EXAMPLE 3

Factoring Trinomials

Factor each trinomial.

(a) $4y^2 - 11y + 6$

SOLUTION   To factor this polynomial, we must find integers a, b, c, and d such that

$$4y^2 - 11y + 6 = (ay + b)(cy + d).$$

By using FOIL, we see that $ac = 4$ and $bd = 6$. The positive factors of 4 are 4 and 1 or 2 and 2. Since the middle term is negative, we consider only negative factors of 6. The possibilities are -2 and -3 or -1 and -6. Now we try various arrangements of these factors until we find one that gives the correct coefficient of y.

$$(2y - 1)(2y - 6) = 4y^2 - 14y + 6 \qquad \text{Incorrect}$$
$$(2y - 2)(2y - 3) = 4y^2 - 10y + 6 \qquad \text{Incorrect}$$
$$(y - 2)(4y - 3) = 4y^2 - 11y + 6 \qquad \text{Correct}$$

The last trial gives the correct factorization.

(b) $6p^2 - 7p - 5$

SOLUTION   Again, we try various possibilities. The positive factors of 6 could be 2 and 3 or 1 and 6. As factors of -5 we have only -1 and 5 or -5 and 1. Try different combinations of these factors until the correct one is found.

$$(2p - 5)(3p + 1) = 6p^2 - 13p - 5 \qquad \text{Incorrect}$$
$$(3p - 5)(2p + 1) = 6p^2 - 7p - 5 \qquad \text{Correct}$$

Finally, $6p^2 - 7p - 5$ factors as $(3p - 5)(2p + 1)$.   ◼

Each of the special patterns of multiplication given earlier can be used in reverse to get a pattern for factoring. Perfect square trinomials can be factored as follows.

PERFECT SQUARE TRINOMIALS

$$x^2 + 2xy + y^2 = (x + y)^2$$
$$x^2 - 2xy + y^2 = (x - y)^2$$

These formulas should be memorized.

EXAMPLE 4

Factoring Perfect Square Trinomials

Factor each polynomial.

(a) $16p^2 - 40pq + 25q^2$

SOLUTION   Since $16p^2 = (4p)^2$ and $25q^2 = (5q)^2$, use the second pattern shown above with $4p$ replacing x and $5q$ replacing y to get

$$16p^2 - 40pq + 25q^2 = (4p)^2 - 2(4p)(5q) + (5q)^2$$
$$= (4p - 5q)^2.$$

Make sure that the middle term of the trinomial being factored, $-40pq$ here, is twice the product of the two terms in the binomial $4p - 5q$.

$$-40pq = 2(4p)(-5q)$$

(b) $169x^2 + 104xy^2 + 16y^4 = (13x + 4y^2)^2$, since $2(13x)(4y^2) = 104xy^2$.   ◼

The pattern for the product of the sum and difference of two terms gives the following factorization.

DIFFERENCE OF TWO SQUARES

$$x^2 - y^2 = (x + y)(x - y)$$

EXAMPLE **5**

Factoring a
Difference of
Squares

Factor each of the following polynomials.

(a) $4m^2 - 9$

SOLUTION First, recognize that $4m^2 - 9$ is the difference of two squares, since $4m^2 = (2m)^2$ and $9 = 3^2$. Use the pattern for the difference of two squares with $2m$ replacing x and 3 replacing y. Doing this gives

$$4m^2 - 9 = (2m)^2 - 3^2$$
$$= (2m + 3)(2m - 3).$$

(b) $256k^4 - 625m^4$

SOLUTION Use the difference of two squares pattern twice, as follows:

$$256k^4 - 625m^4 = (16k^2)^2 - (25m^2)^2$$
$$= (16k^2 + 25m^2)(16k^2 - 25m^2)$$
$$= (16k^2 + 25m^2)(4k + 5m)(4k - 5m).$$

(c) $(a + 2b)^2 - 4c^2 = (a + 2b)^2 - (2c)^2$
$$= [(a + 2b) + 2c][(a + 2b) - 2c]$$
$$= (a + 2b + 2c)(a + 2b - 2c)$$

(d) $y^{4q} - z^{2q} = (y^{2q} + z^q)(y^{2q} - z^q)$

(e) $x^2 - 6x + 9 - y^4$

SOLUTION Group the first three terms to get a perfect square trinomial. Then use the difference of squares pattern.

$$x^2 - 6x + 9 - y^4 = (x^2 - 6x + 9) - y^4$$
$$= (x - 3)^2 - (y^2)^2$$
$$= [(x - 3) + y^2][(x - 3) - y^2]$$
$$= (x - 3 + y^2)(x - 3 - y^2) \quad \blacksquare$$

Two other special results of factoring are listed below. Each can be verified by multiplying on the right side of the equation.

DIFFERENCE AND SUM OF TWO CUBES

Difference of two cubes $x^3 - y^3 = (x - y)(x^2 + xy + y^2)$

Sum of two cubes $x^3 + y^3 = (x + y)(x^2 - xy + y^2)$

EXAMPLE **6**

Factoring the Sum
or Difference of
Cubes

Factor each polynomial.

(a) $x^3 + 27$

SOLUTION Notice that $27 = 3^3$, so the expression is a sum of two cubes. Use the second pattern given above.

$$x^3 + 27 = x^3 + 3^3 = (x + 3)(x^2 - 3x + 9)$$

(b) $m^3 - 64n^3$

SOLUTION Since $64n^3 = (4n)^3$, the given polynomial is a difference of two cubes. To factor, use the first pattern in the box above, replacing x with m and y with $4n$.

$$m^3 - 64n^3 = m^3 - (4n)^3$$
$$= (m - 4n)[m^2 + m(4n) + (4n)^2]$$
$$= (m - 4n)(m^2 + 4mn + 16n^2)$$

(c) $8q^6 + 125p^9$

SOLUTION Write $8q^6$ as $(2q^2)^3$ and $125p^9$ as $(5p^3)^3$, so that the given polynomial is a sum of two cubes.

$$8q^6 + 125p^9 = (2q^2)^3 + (5p^3)^3$$
$$= (2q^2 + 5p^3)[(2q^2)^2 - (2q^2)(5p^3) + (5p^3)^2]$$
$$= (2q^2 + 5p^3)(4q^4 - 10q^2p^3 + 25p^6) \quad \blacksquare$$

Sometimes a polynomial can be factored by substituting one expression for another. The next example shows this **method of substitution**.

EXAMPLE 7

Factoring by Substitution

Factor each polynomial.

(a) $6z^4 - 13z^2 - 5$

SOLUTION Replace z^2 with y, so that $y^2 = (z^2)^2 = z^4$. This replacement gives

$$6z^4 - 13z^2 - 5 = 6y^2 - 13y - 5.$$

Factor $6y^2 - 13y - 5$ as

$$6y^2 - 13y - 5 = (2y - 5)(3y + 1).$$

Replacing y with z^2 gives

$$6z^4 - 13z^2 - 5 = (2z^2 - 5)(3z^2 + 1).$$

(Some students prefer to factor this type of trinomial directly using trial and error with FOIL.)

(b) $10(2a - 1)^2 - 19(2a - 1) - 15$

SOLUTION Replacing $2a - 1$ with m gives

$$10m^2 - 19m - 15 = (5m + 3)(2m - 5).$$

Now replace m with $2a - 1$ in the factored form and simplify.

$10(2a - 1)^2 - 19(2a - 1) - 15$
$= [5(2a - 1) + 3][2(2a - 1) - 5]$ Let $m = 2a - 1$.
$= (10a - 5 + 3)(4a - 2 - 5)$ Multiply.
$= (10a - 2)(4a - 7)$ Add.
$= 2(5a - 1)(4a - 7)$ Factor out the common factor.

(c) $(2a - 1)^3 + 8$

SOLUTION Let $2a - 1 = K$ to get

$$(2a - 1)^3 + 8 = K^3 + 8$$
$$= K^3 + 2^3$$
$$= (K + 2)(K^2 - 2K + 2^2).$$

Replacing K with $2a - 1$ gives

$$(2a - 1)^3 + 8 = (2a - 1 + 2)[(2a - 1)^2 - 2(2a - 1) + 4] \quad \text{Let } K = 2a - 1.$$
$$= (2a + 1)(4a^2 - 4a + 1 - 4a + 2 + 4) \quad \text{Multiply.}$$
$$= (2a + 1)(4a^2 - 8a + 7). \quad \text{Combine terms.} \quad \blacksquare$$

IV. REVIEW OF OPERATIONS WITH RATIONAL EXPRESSIONS

An expression that is the quotient of two algebraic expressions (with denominator not 0) is called a **fractional expression**. The most common fractional expressions are those that are the quotients of two polynomials; these are called **rational expressions.** Since fractional expressions involve quotients, it is important to keep track of values of the variable that satisfy the requirement that no denominator be 0. For example, $x \neq -2$ in the rational expression

$$\frac{x + 6}{x + 2}$$

because replacing x with -2 makes the denominator equal 0. Similarly, in

$$\frac{(x + 6)(x + 4)}{(x + 2)(x + 4)}$$

$x \neq -2$ and $x \neq -4$.

Just as the fraction $\frac{6}{8}$ is written in lowest terms as $\frac{3}{4}$, rational expressions may also be written in lowest terms. This is done with the fundamental principle.

FUNDAMENTAL PRINCIPLE OF FRACTIONS

$$\frac{ac}{bc} = \frac{a}{b} \quad (b \neq 0, c \neq 0)$$

EXAMPLE 1

Writing in Lowest Terms

Write each expression in lowest terms.

(a) $\dfrac{2p^2 + 7p - 4}{5p^2 + 20p}$

SOLUTION   Factor the numerator and denominator to get

$$\frac{2p^2 + 7p - 4}{5p^2 + 20p} = \frac{(2p - 1)(p + 4)}{5p(p + 4)}.$$

By the fundamental principle,

$$\frac{2p^2 + 7p - 4}{5p^2 + 20p} = \frac{2p - 1}{5p}.$$

In the original expression p cannot be 0 or -4, because $5p^2 + 20p \neq 0$, so this result is valid only for values of p other than 0 and -4. In the examples that follow, we shall always assume such restrictions when reducing rational expressions.

(b) $\dfrac{6 - 3k}{k^2 - 4}$

SOLUTION Factor to get

$$\frac{6 - 3k}{k^2 - 4} = \frac{3(2 - k)}{(k + 2)(k - 2)}.$$

The factors $2 - k$ and $k - 2$ have opposite signs. Because of this, multiply numerator and denominator by -1, as follows.

$$\frac{6 - 3k}{k^2 - 4} = \frac{3(2 - k)(-1)}{(k + 2)(k - 2)(-1)}$$

Since $(k - 2)(-1) = -k + 2$, or $2 - k$,

$$\frac{6 - 3k}{k^2 - 4} = \frac{3(2 - k)(-1)}{(k + 2)(2 - k)},$$

giving

$$\frac{6 - 3k}{k^2 - 4} = \frac{-3}{k + 2}.$$

Working in an alternative way would lead to the equivalent result

$$\frac{3}{-k - 2}. \qquad \blacksquare$$

Rational expressions are multiplied and divided using definitions from earlier experience with fractions.

MULTIPLICATION AND DIVISION

For fractions $\frac{a}{b}$ and $\frac{c}{d}$ $(b \neq 0, \, d \neq 0)$,

$$\frac{a}{b} \cdot \frac{c}{d} = \frac{ac}{bd}$$

$$\frac{a}{b} \div \frac{c}{d} = \frac{a}{b} \cdot \frac{d}{c} \quad \left(\text{if } \frac{c}{d} \neq 0\right).$$

EXAMPLE 2

Multiplying and Dividing Rational Expressions

Multiply or divide, as indicated.

(a) $\dfrac{2y^2}{9} \cdot \dfrac{27}{8y^5} = \dfrac{2y^2 \cdot 27}{9 \cdot 8y^5}$

$$= \frac{2 \cdot 9 \cdot 3 \cdot y^2}{2 \cdot 9 \cdot 4 \cdot y^2 \cdot y^3} \qquad \text{Factor.}$$

$$= \frac{3}{4y^3} \qquad \text{Fundamental principle}$$

The product was written in lowest terms in the last step.

(b) $\dfrac{3m^2 - 2m - 8}{3m^2 + 14m + 8} \cdot \dfrac{3m + 2}{3m + 4} = \dfrac{(m - 2)(3m + 4)}{(m + 4)(3m + 2)} \cdot \dfrac{3m + 2}{3m + 4}$ Factor.

$\qquad\qquad\qquad\qquad\qquad\quad = \dfrac{(m - 2)(3m + 4)(3m + 2)}{(m + 4)(3m + 2)(3m + 4)}$ Multiply fractions.

$\qquad\qquad\qquad\qquad\qquad\quad = \dfrac{m - 2}{m + 4}$ Fundamental principle

(c) $\dfrac{5}{8m + 16} \div \dfrac{7}{12m + 24} = \dfrac{5}{8(m + 2)} \div \dfrac{7}{12(m + 2)}$ Factor.

$\qquad\qquad\qquad\qquad\qquad\quad = \dfrac{5}{8(m + 2)} \cdot \dfrac{12(m + 2)}{7}$ Definition of division

$\qquad\qquad\qquad\qquad\qquad\quad = \dfrac{5 \cdot 12(m + 2)}{8 \cdot 7(m + 2)}$ Multiply.

$\qquad\qquad\qquad\qquad\qquad\quad = \dfrac{15}{14}$ Fundamental principle

(d) $\dfrac{3p^2 + 11p - 4}{24p^3 - 8p^2} \div \dfrac{9p + 36}{24p^4 - 36p^3} = \dfrac{(p + 4)(3p - 1)}{8p^2(3p - 1)} \div \dfrac{9(p + 4)}{12p^3(2p - 3)}$

$\qquad\qquad\qquad\qquad\qquad\qquad = \dfrac{(p + 4)(3p - 1)(12p^3)(2p - 3)}{8p^2(3p - 1)(9)(p + 4)}$

$\qquad\qquad\qquad\qquad\qquad\qquad = \dfrac{12p^3(2p - 3)}{9 \cdot 8p^2} = \dfrac{p(2p - 3)}{6}$ ❚

Adding and subtracting rational expressions also depends on definitions from earlier experience with fractions.

ADDITION AND SUBTRACTION

For fractions $\dfrac{a}{b}$ and $\dfrac{c}{d}$ ($b \neq 0$, $d \neq 0$),

$$\frac{a}{b} + \frac{c}{d} = \frac{ad + bc}{bd}$$

$$\frac{a}{b} - \frac{c}{d} = \frac{ad - bc}{bd}.$$

In practice, rational expressions are normally added or subtracted after rewriting all the rational expressions with a common denominator found with the steps given below.

FINDING A COMMON DENOMINATOR

1. Write each denominator as a product of prime factors.

2. Form a product of all the different prime factors. Each factor should have as exponent the *greatest* exponent that appears on that factor.

EXAMPLE **3**	Add or subtract, as indicated.

Adding or Subtracting Rational Expressions

(a) $\dfrac{5}{9x^2} + \dfrac{1}{6x}$

SOLUTION Write each denominator as a product of prime factors, as follows.

$$9x^2 = 3^2 \cdot x^2$$
$$6x = 2^1 \cdot 3^1 \cdot x^1$$

For the common denominator, form the product of all the prime factors, with each factor having the greatest exponent that appears on it. Here the greatest exponent on 2 is 1, while both 3 and x have a greatest exponent of 2. The common denominator is

$$2^1 \cdot 3^2 \cdot x^2 = 18x^2.$$

Now use the fundamental principle to write both of the given expressions with this denominator, then add.

$$\frac{5}{9x^2} + \frac{1}{6x} = \frac{5 \cdot 2}{9x^2 \cdot 2} + \frac{1 \cdot 3x}{6x \cdot 3x}$$

$$= \frac{10}{18x^2} + \frac{3x}{18x^2}$$

$$= \frac{10 + 3x}{18x^2}$$

Always check at this point to see that the answer is in lowest terms.

(b) $\dfrac{y + 2}{y^2 - y} - \dfrac{3y}{2y^2 - 4y + 2}$

SOLUTION Factor each denominator, giving

$$\frac{y + 2}{y^2 - y} - \frac{3y}{2y^2 - 4y + 2} = \frac{y + 2}{y(y - 1)} - \frac{3y}{2(y - 1)^2}.$$

The common denominator, by the method above, is $2y(y - 1)^2$. Write each rational expression with this denominator and subtract, as follows.

$$\frac{y + 2}{y(y - 1)} - \frac{3y}{2(y - 1)^2} = \frac{(y + 2) \cdot 2(y - 1)}{y(y - 1) \cdot 2(y - 1)} - \frac{3y \cdot y}{2(y - 1)^2 \cdot y}$$

$$= \frac{2(y^2 + y - 2)}{2y(y - 1)^2} - \frac{3y^2}{2y(y - 1)^2}$$

$$= \frac{2y^2 + 2y - 4 - 3y^2}{2y(y - 1)^2} \qquad \text{Subtract.}$$

$$= \frac{-y^2 + 2y - 4}{2y(y - 1)^2} \qquad \text{Combine terms.}$$

(c) $\dfrac{3}{(x - 1)(x + 2)} - \dfrac{1}{(x + 3)(x - 4)}$

SOLUTION The common denominator here is $(x - 1)(x + 2)(x + 3)(x - 4)$. Write each fraction with this common denominator, then perform the subtraction.

$$\frac{3}{(x-1)(x+2)} - \frac{1}{(x+3)(x-4)}$$

$$= \frac{3(x+3)(x-4)}{(x-1)(x+2)(x+3)(x-4)} - \frac{(x-1)(x+2)}{(x+3)(x-4)(x-1)(x+2)}$$

$$= \frac{3(x^2 - x - 12) - (x^2 + x - 2)}{(x-1)(x+2)(x+3)(x-4)}$$

$$= \frac{3x^2 - 3x - 36 - x^2 - x + 2}{(x-1)(x+2)(x+3)(x-4)}$$

$$= \frac{2x^2 - 4x - 34}{(x-1)(x+2)(x+3)(x-4)} \quad \blacksquare$$

Any quotient of two rational expressions is called a **complex fraction.** Complex fractions often can be simplified by the methods shown in the following example.

> **EXAMPLE 4**
>
> Simplifying Complex Fractions

Simplify each complex fraction.

(a) $\dfrac{6 - \dfrac{5}{k}}{1 + \dfrac{5}{k}}$

SOLUTION   Multiply both numerator and denominator by the least common denominator of all the fractions, k.

$$\frac{k\left(6 - \dfrac{5}{k}\right)}{k\left(1 + \dfrac{5}{k}\right)} = \frac{6k - k\left(\dfrac{5}{k}\right)}{k + k\left(\dfrac{5}{k}\right)} = \frac{6k - 5}{k + 5}$$

(b) $\dfrac{\dfrac{a}{a+1} + \dfrac{1}{a}}{\dfrac{1}{a} + \dfrac{1}{a+1}}$

SOLUTION   Multiply both numerator and denominator by the least common denominator of all the fractions, in this case $a(a + 1)$.

$$\frac{\dfrac{a}{a+1} + \dfrac{1}{a}}{\dfrac{1}{a} + \dfrac{1}{a+1}} = \frac{\left(\dfrac{a}{a+1} + \dfrac{1}{a}\right)a(a+1)}{\left(\dfrac{1}{a} + \dfrac{1}{a+1}\right)a(a+1)}$$

$$= \frac{\dfrac{a}{a+1}(a)(a+1) + \dfrac{1}{a}(a)(a+1)}{\dfrac{1}{a}(a)(a+1) + \dfrac{1}{a+1}(a)(a+1)} \qquad \text{Distributive property}$$

$$= \frac{a^2 + (a+1)}{(a+1) + a}$$

$$= \frac{a^2 + a + 1}{2a + 1}$$

As an alternative method of solution, first perform the indicated additions in the numerator and denominator, and then divide.

$$\frac{\dfrac{a}{a+1}+\dfrac{1}{a}}{\dfrac{1}{a}+\dfrac{1}{a+1}}=\frac{\dfrac{a^2+1(a+1)}{a(a+1)}}{\dfrac{1(a+1)+1(a)}{a(a+1)}}$$ Get a common denominator; add terms in numerator and denominator.

$$=\frac{\dfrac{a^2+a+1}{a(a+1)}}{\dfrac{2a+1}{a(a+1)}}$$ Combine terms in numerator and denominator.

$$=\frac{a^2+a+1}{a(a+1)}\cdot\frac{a(a+1)}{2a+1}$$ Definition of division

$$=\frac{a^2+a+1}{2a+1}$$ Multiply fractions and write in lowest terms. ∎

V. REVIEW OF NEGATIVE AND RATIONAL EXPONENTS

Earlier we introduced some rules for exponents: the product rule and the power rules. We now complete our review of exponential expressions, beginning with a rule for division.

In the product rule, $a^m \cdot a^n = a^{m+n}$, the exponents are *added*. If $a \neq 0$,

$$\frac{a^3}{a^7}=\frac{a\cdot a\cdot a}{a\cdot a\cdot a\cdot a\cdot a\cdot a\cdot a}=\frac{1}{a\cdot a\cdot a\cdot a}=\frac{1}{a^4}.$$

This suggests that we should *subtract* exponents when dividing. Subtracting exponents gives

$$\frac{a^3}{a^7}=a^{3-7}=a^{-4}.$$

The only way to keep these results consistent is to define a^{-4} as $\frac{1}{a^4}$. This example suggests the following definition.

NEGATIVE EXPONENTS

If a is a nonzero real number and n is any integer, then

$$a^{-n}=\frac{1}{a^n}.$$

EXAMPLE 1

Using the Definition of a Negative Exponent

Evaluate each expression in (a)–(c). In (d) and (e), write the expression without negative exponents.

SOLUTION

(a) $4^{-2}=\dfrac{1}{4^2}=\dfrac{1}{16}$

(b) $\left(\dfrac{2}{5}\right)^{-3}=\dfrac{1}{\left(\dfrac{2}{5}\right)^3}=\dfrac{1}{\dfrac{8}{125}}=\dfrac{125}{8}$

(c) $-4^{-2} = -\dfrac{1}{4^2} = -\dfrac{1}{16}$ **(d)** $x^{-4} = \dfrac{1}{x^4}$ $(x \neq 0)$

(e) $xy^{-3} = x \cdot \dfrac{1}{y^3} = \dfrac{x}{y^3}$ $(y \neq 0)$ ∎

Part (b) of Example 1 showed that

$$\left(\frac{2}{5}\right)^{-3} = \frac{125}{8} = \left(\frac{5}{2}\right)^3.$$

This result can be generalized. If $a \neq 0$ and $b \neq 0$, then

$$\left(\frac{a}{b}\right)^{-n} = \left(\frac{b}{a}\right)^n,$$

for any integer n.

The quotient rule for exponents follows from the definition of exponents, as shown above.

QUOTIENT RULE

For all integers m and n and all nonzero real numbers a,

$$\frac{a^m}{a^n} = a^{m-n}.$$

By the quotient rule, if $a \neq 0$,

$$\frac{a^m}{a^m} = a^{m-m} = a^0.$$

On the other hand, any nonzero quantity divided by itself equals 1. This is why we defined $a^0 = 1$ earlier.

EXAMPLE 2

Using the Quotient Rule

Use the quotient rule to simplify each expression. Assume that all variables represent nonzero real numbers.

SOLUTION

(a) $\dfrac{12^5}{12^2} = 12^{5-2} = 12^3$ **(b)** $\dfrac{a^5}{a^{-8}} = a^{5-(-8)} = a^{13}$

(c) $\dfrac{16m^{-9}}{12m^{11}} = \dfrac{16}{12} \cdot m^{-9-11} = \dfrac{4}{3}m^{-20} = \dfrac{4}{3} \cdot \dfrac{1}{m^{20}} = \dfrac{4}{3m^{20}}$

(d) $\dfrac{25r^7z^5}{10r^9z} = \dfrac{25}{10} \cdot \dfrac{r^7}{r^9} \cdot \dfrac{z^5}{z^1} = \dfrac{5}{2}r^{-2}z^4 = \dfrac{5z^4}{2r^2}$

(e) $\dfrac{x^{5y}}{x^{3y}} = x^{5y-3y} = x^{2y},$ if y is an integer   ∎

The rules for exponents stated earlier in this appendix also apply to negative exponents.

EXAMPLE 3

Using the Rules for Exponents

Use the rules for exponents to simplify each expression. Write answers without negative exponents. Assume that all variables represent nonzero real numbers.

SOLUTION

(a) $3x^{-2}(4^{-1}x^{-5})^2 = 3x^{-2}(4^{-2}x^{-10})$ Power rule

$= 3 \cdot 4^{-2} \cdot x^{-2+(-10)}$ Rearrange factors; use the product rule.

$= 3 \cdot 4^{-2} \cdot x^{-12}$

$= \dfrac{3}{16x^{12}}$ Write with positive exponents.

(b) $\dfrac{5m^{-3}}{10m^{-5}} = \dfrac{5}{10}m^{-3-(-5)}$ Quotient rule

$= \dfrac{1}{2}m^2$ or $\dfrac{m^2}{2}$

(c) $\dfrac{12p^3q^{-1}}{8p^{-2}q} = \dfrac{12}{8} \cdot \dfrac{p^3}{p^{-2}} \cdot \dfrac{q^{-1}}{q^1}$

$= \dfrac{3}{2} \cdot p^{3-(-2)}q^{-1-1}$ Quotient rule

$= \dfrac{3}{2}p^5q^{-2}$

$= \dfrac{3p^5}{2q^2}$ Write with positive exponents.   ◨

The definition of a^n can be extended to rational values of n by defining $a^{1/n}$ to be the nth root of a. By one of the power rules of exponents (extended to a rational exponent)

$$(a^{1/n})^n = a^{(1/n)n} = a^1 = a,$$

suggesting that $a^{1/n}$ is a number whose nth power is a.

$a^{1/n}, n$ **EVEN** $a^{1/n}, n$ **ODD**

i. If n is an *even* positive integer, and if $a > 0$, then $a^{1/n}$ is the positive real number whose nth power is a. That is, $(a^{1/n})^n = a$. In this case, $a^{1/n}$ is the principal nth root of a.

ii. If n is an *odd* positive integer, and a *is any real number*, then $a^{1/n}$ is the positive or negative real number whose nth power is a. That is, $(a^{1/n})^n = a$.

EXAMPLE 4

Using the Definition of $a^{1/n}$

Evaluate each expression.

SOLUTION

(a) $36^{1/2} = 6$ because $6^2 = 36$ **(b)** $-100^{1/2} = -10$

(c) $-(225)^{1/2} = -15$ **(d)** $625^{1/4} = 5$

(e) $(-1296)^{1/4}$ is not defined, but $-1296^{1/4} = -6$

(f) $(-27)^{1/3} = -3$

(g) $-32^{1/5} = -2$ ◨

We now discuss more general rational exponents. The notation $a^{m/n}$ should be defined so that all the past rules for exponents still hold. For the power rule to hold, $(a^{1/n})^m$ must equal $a^{m/n}$. Therefore, $a^{m/n}$ is defined as follows.

RATIONAL EXPONENTS

For all integers m, all positive integers n, and all real numbers a for which $a^{1/n}$ is defined,

$$a^{m/n} = (a^{1/n})^m.$$

EXAMPLE **5**
Using the Definition of $a^{m/n}$

Evaluate each expression.

SOLUTION

(a) $125^{2/3} = (125^{1/3})^2 = 5^2 = 25$

(b) $32^{7/5} = (32^{1/5})^7 = 2^7 = 128$

(c) $-81^{3/2} = -(81^{1/2})^3 = -9^3 = -729$

(d) $(-4)^{5/2}$ is not defined because $(-4)^{1/2}$ is not defined.

(e) $(-27)^{2/3} = [(-27)^{1/3}]^2 = (-3)^2 = 9$

(f) $16^{-3/4} = \dfrac{1}{16^{3/4}} = \dfrac{1}{(16^{1/4})^3} = \dfrac{1}{2^3} = \dfrac{1}{8}$ ◻

By starting with $(a^{1/n})^m$ and $(a^m)^{1/n}$, and raising each expression to the nth power, it can be shown that $(a^{1/n})^m$ is equal to $(a^m)^{1/n}$. This means that $a^{m/n}$ could be defined in either of the following ways.

For all real numbers a, integers m, and positive integers n for which $a^{1/n}$ is defined:

$$a^{m/n} = (a^{1/n})^m \qquad \text{or} \qquad a^{m/n} = (a^m)^{1/n}.$$

Now $a^{m/n}$ can be evaluated in either of two ways: as $(a^{1/n})^m$ or as $(a^m)^{1/n}$. It is usually easier to find $(a^{1/n})^m$. For example, $27^{4/3}$ can be evaluated in either of two ways:

$$27^{4/3} = (27^{1/3})^4 = 3^4 = 81$$
$$27^{4/3} = (27^4)^{1/3} = 531{,}441^{1/3} = 81.$$

The form $(27^{1/3})^4$ is easier to evaluate.

It can be shown that all the earlier results concerning integer exponents also apply to rational exponents. These definitions and rules are summarized here.

DEFINITIONS AND RULES FOR EXPONENTS

Let r and s be rational numbers. The results below are valid for all positive numbers a and b.

$$a^r \cdot a^s = a^{r+s} \qquad (ab)^r = a^r \cdot b^r \qquad (a^r)^s = a^{rs}$$

$$\frac{a^r}{a^s} = a^{r-s} \qquad \left(\frac{a}{b}\right)^r = \frac{a^r}{b^r} \qquad a^{-r} = \frac{1}{a^r}$$

EXAMPLE 6
Using the Definitions and Rules for Exponents

Use the definitions and rules for exponents to simplify each expression.

SOLUTION

(a) $\dfrac{27^{1/3} \cdot 27^{5/3}}{27^3} = \dfrac{27^{1/3+5/3}}{27^3}$ Product rule

$= \dfrac{27^2}{27^3} = 27^{2-3}$ Quotient rule

$= 27^{-1} = \dfrac{1}{27}$

(b) $81^{5/4} \cdot 4^{-3/2} = (81^{1/4})^5 (4^{1/2})^{-3} = 3^5 \cdot 2^{-3} = \dfrac{3^5}{2^3}$ or $\dfrac{243}{8}$.

(c) $6y^{2/3} \cdot 2y^{1/2} = 12y^{2/3+1/2} = 12y^{7/6}$, where $y \geq 0$

(d) $\left(\dfrac{3m^{5/6}}{y^{3/4}}\right)^2 \cdot \left(\dfrac{8y^3}{m^6}\right)^{2/3} = \dfrac{9m^{5/3}}{y^{3/2}} \cdot \dfrac{4y^2}{m^4} = 36m^{5/3-4}y^{2-3/2}$

$= \dfrac{36y^{1/2}}{m^{7/3}}$ $(m > 0, y > 0)$

(e) $m^{2/3}(m^{7/3} + 2m^{1/3}) = (m^{2/3+7/3} + 2m^{2/3+1/3}) = m^3 + 2m$ ∎

The next example shows how to factor when negative or rational exponents are involved.

EXAMPLE 7
Factoring an Expression with Negative or Rational Exponents

Factor out the smallest power of the variable. Assume that all variables represent positive real numbers.

(a) $9x^{-2} - 6x^{-3}$

SOLUTION   The smallest exponent here is -3. Since 3 is a common numerical factor, factor out $3x^{-3}$.

$$9x^{-2} - 6x^{-3} = 3x^{-3}(3x - 2)$$

Check by multiplying on the right. The factored form can now be written without negative exponents as $\dfrac{3(3x - 2)}{x^3}$.

(b) $4m^{1/2} + 3m^{3/2} = m^{1/2}(4 + 3m)$

SOLUTION   To check this result, multiply $m^{1/2}$ by $4 + 3m$.

(c) $y^{-1/3} + y^{2/3} = y^{-1/3}(1 + y)$

SOLUTION   The factored form can be written with only positive exponents as $\dfrac{1 + y}{y^{1/3}}$.   ∎

Negative exponents are sometimes used to write complex fractions. Recall, complex fractions are simplified either by first multiplying the numerator and denominator by the greatest common multiple of all the denominators, or by performing any indicated operations in the numerator and the denominator and then using the definition of division for fractions.

EXAMPLE 8

Simplifying a
Fraction with
Negative Exponents

Simplify $\dfrac{(x + y)^{-1}}{x^{-1} + y^{-1}}$. Write the result with only positive exponents.

SOLUTION Begin by using the definition of a negative integer exponent. Then perform the indicated operations.

$$\frac{(x + y)^{-1}}{x^{-1} + y^{-1}} = \frac{\dfrac{1}{x + y}}{\dfrac{1}{x} + \dfrac{1}{y}}$$

$$= \frac{\dfrac{1}{x + y}}{\dfrac{y + x}{xy}}$$

$$= \frac{1}{x + y} \cdot \frac{xy}{x + y}$$

$$= \frac{xy}{(x + y)^2} \quad \blacksquare$$

VI. REVIEW OF RADICALS

In part V of this appendix, the notation $a^{1/n}$ was used for the nth root of a for appropriate values of a and n. An alternative (and more familiar) notation for $a^{1/n}$ uses *radical notation*.

RADICAL NOTATION FOR $a^{1/n}$

If a is a real number, n is a positive integer, and $a^{1/n}$ is defined, then

$$\sqrt[n]{a} = a^{1/n}.$$

The symbol $\sqrt[n]{}$ is a **radical sign,** the number a is the **radicand,** and n is the **index** of the radical $\sqrt[n]{a}$. It is customary to use the familiar notation $\sqrt{a}$ instead of $\sqrt[2]{a}$ for the square root.

For even values of n (square roots, fourth roots, and so on) there are two nth roots, one positive and one negative. In such cases, the notation $\sqrt[n]{a}$ represents the positive root, the **principal nth root.** The negative root is written $-\sqrt[n]{a}$.

EXAMPLE 1

Evaluating Roots

Evaluate each root.

SOLUTION

(a) $\sqrt[4]{16} = 16^{1/4} = 2$ **(b)** $-\sqrt[4]{16} = -16^{1/4} = -2$

(c) $\sqrt[4]{-16}$ is not defined.   **(d)** $\sqrt[5]{-32} = -2$

(e) $\sqrt[3]{1000} = 10$ **(f)** $\sqrt[6]{\dfrac{64}{729}} = \dfrac{2}{3}$ $\blacksquare$

With $a^{1/n}$ written as $\sqrt[n]{a}$, $a^{m/n}$ also can be written using radicals.

> ### RADICAL NOTATION FOR $a^{m/n}$
>
> If a is a real number, m is an integer, n is a positive integer, and $\sqrt[n]{a}$ is defined, then
>
> $$a^{m/n} = (\sqrt[n]{a})^m = \sqrt[n]{a^m}.$$

EXAMPLE 2

Converting from Rational Exponents to Radicals

Write in radical form and simplify.

SOLUTION

(a) $8^{2/3} = (\sqrt[3]{8})^2 = 2^2 = 4$

(b) $(-32)^{4/5} = (\sqrt[5]{-32})^4 = (-2)^4 = 16$

(c) $-16^{3/4} = -(\sqrt[4]{16})^3 = -(2)^3 = -8$

(d) $x^{5/6} = \sqrt[6]{x^5}$ $(x \geq 0)$

(e) $3x^{2/3} = 3\sqrt[3]{x^2}$

(f) $2p^{-1/2} = \dfrac{2}{p^{1/2}} = \dfrac{2}{\sqrt{p}}$ $(p > 0)$

(g) $(3a + b)^{1/4} = \sqrt[4]{3a + b}$ $(3a + b \geq 0)$ ∎

EXAMPLE 3

Converting from Radicals to Rational Exponents

Write in exponential form.

SOLUTION

(a) $\sqrt[4]{x^5} = x^{5/4}$ $(x \geq 0)$ (b) $\sqrt{3y} = (3y)^{1/2}$ $(y \geq 0)$

(c) $10(\sqrt[5]{z})^2 = 10z^{2/5}$ (d) $5\sqrt[3]{(2x^4)^7} = 5(2x^4)^{7/3} = 5 \cdot 2^{7/3} x^{28/3}$

(e) $\sqrt{p^2 + q} = (p^2 + q)^{1/2}$ $(p^2 + q \geq 0)$ ∎

By the definition of $\sqrt[n]{a}$, for any positive integer n, if $\sqrt[n]{a}$ is defined, then

$$(\sqrt[n]{a})^n = a.$$

If a is positive, or if a is negative and n is an odd positive integer,

$$\sqrt[n]{a^n} = a.$$

Because of the conditions just given, we *cannot* simply write $\sqrt{x^2} = x$. For example, if $x = -5$,

$$\sqrt{x^2} = \sqrt{(-5)^2} = \sqrt{25} = 5 \neq x.$$

To take care of the fact that a negative value of x can produce a positive result, we use absolute value. For any real number a,

$$\sqrt{a^2} = |a|.$$

For example,

$$\sqrt{(-9)^2} = |-9| = 9, \quad \text{and} \quad \sqrt{13^2} = |13| = 13.$$

This result can be generalized to any even nth root.

If n is an even positive integer, $\sqrt[n]{a^n} = |a|$, and if n is an odd positive integer, $\sqrt[n]{a^n} = a$.

EXAMPLE **4**

Using Absolute Value to Simplify Roots

Use absolute value as applicable to simplify the following expressions.

SOLUTION

(a) $\sqrt{p^4} = |p^2| = p^2$ **(b)** $\sqrt[4]{p^4} = |p|$

(c) $\sqrt{16m^8r^6} = |4m^4r^3| = 4m^4|r^3|$ **(d)** $\sqrt[6]{(-2)^6} = |-2| = 2$

(e) $\sqrt[5]{m^5} = m$ **(f)** $\sqrt{(2k+3)^2} = |2k+3|$

(g) $\sqrt{x^2 - 4x + 4} = \sqrt{(x-2)^2} = |x-2|$ ∎

For the remainder of this review, to avoid difficulties when working with variable radicands, we usually will assume that all variables in radicands represent only nonnegative real numbers.

Three key rules for working with radicals are given below. These rules are just the power rules for exponents written in radical notation.

RULES FOR RADICALS

For all real numbers a and b, and positive integers m and n for which the indicated roots are defined,

$$\sqrt[n]{a} \cdot \sqrt[n]{b} = \sqrt[n]{ab} \qquad \sqrt[n]{\frac{a}{b}} = \frac{\sqrt[n]{a}}{\sqrt[n]{b}} \quad (b \neq 0) \qquad \sqrt[m]{\sqrt[n]{a}} = \sqrt[mn]{a}.$$

EXAMPLE **5**

Using the Rules for Radicals to Simplify Radical Expressions

Apply the rules for radicals.

SOLUTION

(a) $\sqrt{6} \cdot \sqrt{54} = \sqrt{6 \cdot 54} = \sqrt{324} = 18$

(b) $\sqrt[3]{m} \cdot \sqrt[3]{m^2} = \sqrt[3]{m^3} = m$

(c) $\sqrt{\dfrac{7}{64}} = \dfrac{\sqrt{7}}{\sqrt{64}} = \dfrac{\sqrt{7}}{8}$

(d) $\sqrt[4]{\dfrac{a}{b^4}} = \dfrac{\sqrt[4]{a}}{\sqrt[4]{b^4}} = \dfrac{\sqrt[4]{a}}{b} \quad (a \geq 0, b > 0)$

(e) $\sqrt[7]{\sqrt[3]{2}} = \sqrt[21]{2}$ Use the third rule given above.

(f) $\sqrt[4]{\sqrt{3}} = \sqrt[8]{3}$ ∎

In working with numbers, it is customary to write a number in its simplest form. For example, $\frac{10}{2}$ is written as 5, $-\frac{9}{6}$ is written as $-\frac{3}{2}$, and $\frac{4}{16}$ is written as $\frac{1}{4}$. Similarly, expressions with radicals are often written in their simplest forms.

SIMPLIFIED RADICALS

An expression with radicals is simplified when all of the following conditions are satisfied.

1. The radicand has no factor raised to a power greater than or equal to the index.
2. The radicand has no fractions.
3. No denominator contains a radical.
4. Exponents in the radicand and the index of the radical have no common factor.
5. All indicated operations have been performed (if possible).

EXAMPLE 6

Simplifying Radicals

Simplify each of the following. Assume that all variables represent nonnegative real numbers.

SOLUTION

(a) $\sqrt{175} = \sqrt{25 \cdot 7} = \sqrt{25} \cdot \sqrt{7} = 5\sqrt{7}$

(b) $-3\sqrt[5]{32} = -3\sqrt[5]{2^5} = -3 \cdot 2 = -6$

(c) $\sqrt[3]{81x^5y^7z^6} = \sqrt[3]{27 \cdot 3 \cdot x^3 \cdot x^2 \cdot y^6 \cdot y \cdot z^6}$ Factor.

           $= \sqrt[3]{(27x^3y^6z^6)(3x^2y)}$ Group all perfect cubes.

           $= 3xy^2z^2\sqrt[3]{3x^2y}$ Remove all perfect cubes from the radical.

Radicals with the same radicand and the same index, such as $3\sqrt[4]{11pq}$ and $-7\sqrt[4]{11pq}$, are called **like radicals.** Like radicals are added or subtracted by using the distributive property. Only like radicals can be combined. It is sometimes necessary to simplify radicals before adding or subtracting.

EXAMPLE 7

Adding and Subtracting Like Radicals

Add as indicated. Assume all variables are positive real numbers.

(a) $3\sqrt[4]{11pq} + (-7\sqrt[4]{11pq}) = -4\sqrt[4]{11pq}$

(b) $\sqrt{98x^3y} + 3x\sqrt{32xy}$

SOLUTION First remove all perfect square factors from under the radical. Then use the distributive property, as follows.

$$\sqrt{98x^3y} + 3x\sqrt{32xy} = \sqrt{49 \cdot 2 \cdot x^2 \cdot x \cdot y} + 3x\sqrt{16 \cdot 2 \cdot x \cdot y}$$
$$= 7x\sqrt{2xy} + (3x)(4)\sqrt{2xy}$$
$$= 7x\sqrt{2xy} + 12x\sqrt{2xy}$$
$$= 19x\sqrt{2xy} \quad \text{Distributive property}$$

If the index of the radical and an exponent in the radicand have a common factor, the radical can be simplified by writing it in exponential form, simplifying the rational exponent, then writing the result as a radical again.

EXAMPLE 8

Simplifying Radicals by Writing Them with Rational Exponents

Simplify the following radicals by first rewriting with rational exponents.

SOLUTION

(a) $\sqrt[6]{3^2} = 3^{2/6} = 3^{1/3} = \sqrt[3]{3}$

(b) $\sqrt[6]{x^{12}y^3} = (x^{12}y^3)^{1/6} = x^2y^{3/6} = x^2y^{1/2} = x^2\sqrt{y}$ $(y \geq 0)$

(c) $\sqrt[9]{\sqrt{6^3}} = \sqrt[9]{6^{3/2}} = (6^{3/2})^{1/9} = 6^{1/6} = \sqrt[6]{6}$

In Example 8(a), we simplified $\sqrt[6]{3^2}$ as $\sqrt[3]{3}$. However, to simplify $(\sqrt[6]{x})^2$, the variable x must be nonnegative. For example,

$$(-8)^{2/6} = [(-8)^{1/6}]^2.$$

This result is not a real number, since $(-8)^{1/6}$ is not a real number. On the other hand,

$$(-8)^{1/3} = -2.$$

Here, even though $2/6 = 1/3$,

$$(\sqrt[6]{x})^2 \neq \sqrt[3]{x}.$$

If a is nonnegative, then it is always true that $a^{m/n} = a^{mp/(np)}$. Reducing rational exponents on negative bases must be considered case by case.

Multiplying radical expressions is much like multiplying polynomials.

EXAMPLE 9

Multiplying Radical Expressions

Multiply.

SOLUTION

(a) $(\sqrt{2} + 3)(\sqrt{8} - 5) = \sqrt{2}(\sqrt{8}) - \sqrt{2}(5) + 3\sqrt{8} - 3(5)$ FOIL

$\qquad\qquad\qquad\qquad\qquad = \sqrt{16} - 5\sqrt{2} + 3(2\sqrt{2}) - 15$ Multiply.

$\qquad\qquad\qquad\qquad\qquad = 4 - 5\sqrt{2} + 6\sqrt{2} - 15$

$\qquad\qquad\qquad\qquad\qquad = -11 + \sqrt{2}$ Combine terms.

(b) $(\sqrt{7} - \sqrt{10})(\sqrt{7} + \sqrt{10}) = (\sqrt{7})^2 - (\sqrt{10})^2$ Product of the sum and difference of two terms

$\qquad\qquad\qquad\qquad\qquad\qquad\qquad = 7 - 10$

$\qquad\qquad\qquad\qquad\qquad\qquad\qquad = -3$ ∎

Condition 3 of the rules for simplifying radicals described above requires that no denominator contain a radical. The process of achieving this is called **rationalizing the denominator.** It is accomplished by multiplying by a form of 1, as explained in Example 10.

EXAMPLE 10

Rationalizing Denominators

Rationalize each denominator.

(a) $\dfrac{4}{\sqrt{3}}$

SOLUTION   To rationalize the denominator, multiply by $\sqrt{3}/\sqrt{3}$ (which equals 1) so that the denominator of the product is a rational number.

$$\frac{4}{\sqrt{3}} \cdot \frac{\sqrt{3}}{\sqrt{3}} = \frac{4\sqrt{3}}{3}$$

(b) $\sqrt[4]{\dfrac{3}{5}}$

SOLUTION   Start by using the fact that the radical of a quotient can be written as the quotient of radicals.

$$\sqrt[4]{\frac{3}{5}} = \frac{\sqrt[4]{3}}{\sqrt[4]{5}}$$

The denominator will be a rational number if it equals $\sqrt[4]{5^4}$. That is, four factors of 5 are needed under the radical. Since $\sqrt[4]{5}$ has just one factor of 5, three additional factors are needed, so multiply by $\sqrt[4]{5^3}/\sqrt[4]{5^3}$.

$$\frac{\sqrt[4]{3}}{\sqrt[4]{5}} = \frac{\sqrt[4]{3} \cdot \sqrt[4]{5^3}}{\sqrt[4]{5} \cdot \sqrt[4]{5^3}} = \frac{\sqrt[4]{3 \cdot 5^3}}{\sqrt[4]{5^4}} = \frac{\sqrt[4]{375}}{5} \qquad \blacksquare$$

In Example 9(b), we saw that

$$(\sqrt{7} - \sqrt{10})(\sqrt{7} + \sqrt{10}) = -3,$$

a rational number. This suggests a way to rationalize a denominator that is a binomial in which one or both terms is a radical. The expressions $a\sqrt{m} + b\sqrt{n}$ and $a\sqrt{m} - b\sqrt{n}$ are **conjugates.**

EXAMPLE 11

Rationalizing a Binomial Denominator

Rationalize the denominator of $\dfrac{1}{1 - \sqrt{2}}$.

SOLUTION   As suggested above, the best approach here is to multiply both numerator and denominator by the conjugate of the denominator, in this case $1 + \sqrt{2}$.

$$\frac{1}{1 - \sqrt{2}} = \frac{1(1 + \sqrt{2})}{(1 - \sqrt{2})(1 + \sqrt{2})} = \frac{1 + \sqrt{2}}{1 - 2} = -1 - \sqrt{2} \qquad \blacksquare$$

ANSWERS

Answers to Selected Exercises

In this section we provide the answers that we think most students will obtain when they work the exercises using the methods explained in the text. If your answer does not look exactly like the one given here, it is not necessarily wrong. In many cases there are equivalent forms of the answer that are correct. For example, if the answer section shows $\frac{3}{4}$ and your answer is .75, you have obtained the right answer but written it in a different (yet equivalent) form. Unless the directions specify otherwise, .75 is just as valid an answer as $\frac{3}{4}$. In general, if your answer does not agree with the one given, see whether it can be transformed into the other form. If it can, then it is the correct answer. If you still have doubts, talk with your instructor.

CHAPTER 1

Section 1.1 (page 8)

1. (a) 10   **(b)** 0, 10   **(c)** -6, $-\frac{12}{4}$ (or -3), 0, 10   **(d)** -6, $-\frac{12}{4}$ (or -3), $-\frac{5}{8}$, 0, .31, $.\overline{3}$, 10   **(e)** $-\sqrt{3}$, 2π, $\sqrt{17}$
(f) All are real numbers.   **3. (a)** None are natural numbers.   **(b)** None are whole numbers.   **(c)** $-\sqrt{100}$, -1
(d) $-\sqrt{100}$, $-\frac{13}{6}$, -1, 5.23, $9.\overline{14}$, 3.14, $\frac{22}{7}$ **(e)** None are irrational numbers.   **(f)** All are real numbers.
5. 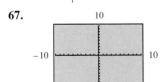   **7.** 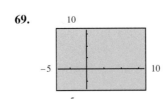   **9.**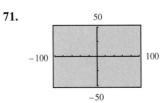

11. D   **13.** B   **15.** C   **17.** A   **19.** J   **21.** $.8\overline{3}$ **23.** $-4.\overline{3}$ **25.** $.\overline{2}$ **27.** $.0\overline{81}$
29. The rational number .87 represents $\frac{87}{100}$. The repeating decimal $.\overline{87}$ is larger than $\frac{87}{100}$, since it contains alternating 8 and 7 in all decimal places after the second. .87 contains zeros in all other decimal places.
31. 7.615773106   **33.** 3.20753433   **35.** 3.045261646   **37.** 2.236067977
39. 2.620741394   **41.** 3.66643574   **43.** $<, \leq$ **45.** $>, \geq$ **47.** $\leq, \geq$
49. I   **51.** III   **53.** no quadrant   **55.** II   **57.** no quadrant
59. I or III   **61.** II or IV   **63.** It must lie on the y-axis.
65. Answers will vary. For example, for the TI-82, use ZOOM 6.

$(-2, 4)$
$(0, 5)$
$(-1, 2)$
$(2, 3)$
$(-2, 0)$
$(3, 0)$
$(-3, -2)$
$(-2, -4)$
$(3, -3)$
$(1, -4)$

67.

```
        10



-10  +------+------+  10



       -10
```

69.

```
        10



-5   +------+------+  10



       -5
```

71.

```
        50



-100 +------+------+  100



       -50
```

73. There are no tick marks. To set a screen with no tick marks on the axes, use Xscl = 0 and Yscl = 0.

Section 1.2 (page 19)

1. $(-1, 4)$

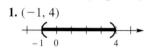

3. $(-\infty, 0)$

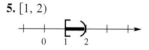

5. $[1, 2)$

7. $(-\infty, -9)$

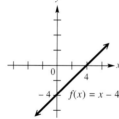

9. $\{x \mid -4 < x < 3\}$

11. $\{x \mid x \le -1\}$ **13.** $\{x \mid -2 \le x < 6\}$ **15.** $\{x \mid x \le -4\}$

17. Use a parenthesis if the symbol is $<$ or $>$, and use a square bracket if the symbol is $\le$ or $\ge$.

19. domain: $\{5, 3, 4, 7\}$; range: $\{1, 2, 9, 6\}$; function   **21.** domain: $\{2, 0\}$; range: $\{4, 2, 5\}$; not a function

23. domain: $\{-3, 4, -2\}$; range: $\{1, 7\}$; function   **25.** domain: $\{1, 4, 0, 7\}$; range: $\{3, 7, 6, 2\}$; function

27. domain: $(-\infty, \infty)$; range: $(-\infty, \infty)$; function   **29.** domain: $[3, \infty)$; range: $(-\infty, \infty)$; not a function

31. domain: $[-4, 4]$; range: $[-3, 3]$; not a function   **33.** domain: $[2, \infty)$; range: $[0, \infty)$; function

35. domain: $[-9, \infty)$; range: $(-\infty, \infty)$; not a function   **37.** domain: $\{2, 5, 11, 17, 3\}$; range: $\{1, 7, 20\}$; function

39. domain: $\{1, 2, 3, 5\}$; range: $\{10, 15, 19, 27\}$; not a function   **41.** 3   **43.** 8   **45.** 7   **47.** 0

49. -4 **51.** -3 **53.** $-4, 0, 4$ **55.** $3a - 1$ **57.** $2r^2 + r + 3$ **59.** $7p + 2s - 1$

Further Explorations

1. -15 **3.** 701

Section 1.3 (page 34)

1. yes   **3.** yes   **5.** no   **7.** no   **9.** no

11. (a) 4
  (b) -4
  (c) $(-\infty, \infty)$
  (d) $(-\infty, \infty)$
  (e) 1

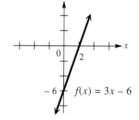

13. (a) 2
  (b) -6
  (c) $(-\infty, \infty)$
  (d) $(-\infty, \infty)$
  (e) 3

15. (a) 5
  (b) 2
  (c) $(-\infty, \infty)$
  (d) $(-\infty, \infty)$
  (e) $-\frac{2}{5}$

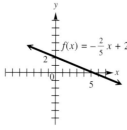

17. (a) 0
  (b) 0
  (c) $(-\infty, \infty)$
  (d) $(-\infty, \infty)$
  (e) 3

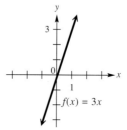

19. The point $(0, 0)$ must lie on the line.

21. (a) none
  (b) -3
  (c) $(-\infty, \infty)$
  (d) $\{-3\}$
  (e) 0

23. (a) -1.5
  (b) none
  (c) $\{-1.5\}$
  (d) $(-\infty, \infty)$
  (e) undefined

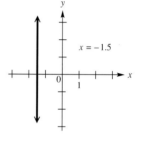

25. (a) 2
(b) none
(c) {2}
(d) $(-\infty, \infty)$
(e) undefined

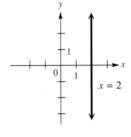

27. constant function   **29.** $y = 0$
31. Window B   **33.** Window B

35.

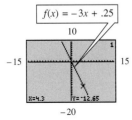

$f(4.3) = -12.65$

37.

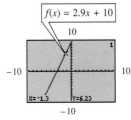

$f(-1.3) = 6.23$

39. $\frac{1}{5}$ **41.** $\frac{7}{9}$ **43.** 0   **45.** horizontal

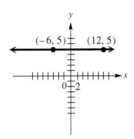

This is the graph of a function.

47. -3.5 **49. (a)** C   **(b)** B   **(c)** D   **(d)** A
51. y_1; Both slopes are positive.   **53.** y_2; Slope of y_1 is positive and slope of y_2 is negative.   **55.** y_1; Slope of y_1 is negative and slope of y_2 is positive.

57.

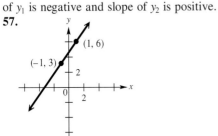

59.

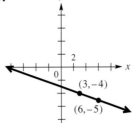

61.

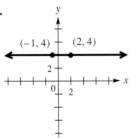

63.

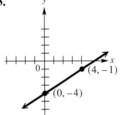

65. $y = \frac{3}{4}x - 4$ **67.** $f\left(\frac{y-b}{a}\right) = a\left(\frac{y-b}{a}\right) + b = y - b + b = y$ **69.** The line has a positive slope and a negative y-intercept.

Further Explorations

1. 3   **3.** $y = 3x - 5$

Section 1.4 (page 47)

1. $y = -2x + 5$ **3.** $y = -\frac{3}{2}x - \frac{7}{2}$ **5.** $y = -.5x - 3$ **7.** $y = \frac{1}{4}x + \frac{13}{4}$ **9.** $y = \frac{2}{3}x - 2$
11. $x = -6$ (It is not possible to write this equation in slope-intercept form.)   **13.** -2; does not; undefined; $\frac{1}{2}$; does not; 0   **15.** $x = 0$ **17.** $y = 4x - 36$ **19.** $y = -1.5x + 6.5$ **21.** See answer to Exercise 11, Section 1.3.   **23.** See answer to Exercise 13, Section 1.3.   **25.** See answer to Exercise 15, Section 1.3.
27.

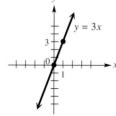

We used $(0, 0)$ and $(1, 3)$.

29.

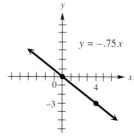

We used $(0, 0)$ and $(4, -3)$.

31. $y = -\frac{5}{3}x + 5$ **33.** $y = \frac{2}{7}x + \frac{4}{7}$ **35.** $y = -\frac{3}{4}x + \frac{25}{8}$

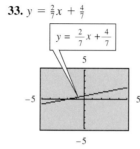

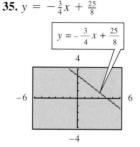

37. $c = 17$ **39.** $b = 84$ **41.** $c = \sqrt{89}$ **43.** $b = 4$ **45. (a)** $8\sqrt{2}$ **(b)** $(9, 3)$
47. (a) $\sqrt{34}$ **(b)** $(-5.5, -3.5)$ **49. (a)** $2\sqrt{185}$ **(b)** $(-1, 6)$ **51.** $(13, 10)$ **53.** $(19, 16)$
55. $y = -\frac{1}{3}x + \frac{11}{3}$ **57.** $y = \frac{5}{3}x + \frac{13}{3}$ **59.** $x = -5$ (It is not possible to write this equation in slope-intercept form.)   **61.** $y = -.2x + 7$ **63.** $y = .5x$ **65.** *Sample answer:* They are *not* parallel. Line y_1 has slope 2.3 and line y_2 has slope 2.3001. Since $2.3 \neq 2.3001$, and nonvertical parallel lines must have *equal* slopes, they can't be parallel. They only *appear* to be parallel in the figure because the slopes differ by .0001, a very small number.   **67. (a)** $a + b$; $a + b$; $a^2 + 2ab + b^2$ **(b)** $\frac{1}{2}ab$; $4(\frac{1}{2}ab) = 2ab$; c^2 **(c)** $2ab + c^2$
(d) $a^2 + 2ab + b^2$; $2ab + c^2$ **(e)** $a^2 + b^2$; c^2

Further Explorations

(a) Both have slope of 2, so they are parallel.   **(b)** Both have slope of $-.5$, so they are parallel.
(c) Y_1 has slope 3 and Y_2 has slope 4, so they are not parallel.   **(d)** Both have slope 2.6, so they are parallel.

Section 1.5 (page 64)

1. $-7x + 56$ **3.** $-.58x - 26.78$ **5.** $-\frac{7}{6}x + \frac{13}{3}$ **7.** $-12x^4y^3 + 9x^2y^2 - 19$
9. The graphs *appear* to coincide. If we graph $y_1 = 4x + 2x$ and $y_2 = 6x$, we get the figure shown here.

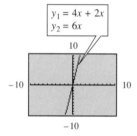

11. $\{10\}$ **13.** $\{-1.3\}$ **15.** $\{2\}$ **17.** The y-value is the function value (range value) for *both* y_1 and y_2 when the solution of the equation is substituted for x.   **19.** $\{-.8\}$ **21.** $\{-16\}$ **23.** $\emptyset$; This is called a contradiction.   **25.** $\{12\}$ **27.** $\{3\}$ **29.** $\{-2\}$ **31.** $\{7\}$ **33.** $\{75\}$ **35.** $\{-1\}$ **37.** $\{0\}$ **39.** $\{\frac{5}{4}\}$ **41.** $\{16.07\}$ **43.** $\{-1.46\}$ **45.** $\{-3.92\}$ **47.** $\{16.07\}$ **49.** $\{-1.46\}$ **51.** $\{-3.92\}$ **53.** identity   **55.** contradiction   **57.** contradiction   **59.** Graph $y_1 = .06x + .09(15 - x)$ and $y_2 = .07(15)$ to find that the coordinates of the point of intersection are $x = 10$, $y = 1.05$. (This is the intersection-of-graphs method.) The solution set contains this x-value, 10: $\{10\}$ **61.** (a); -4.05 could be a solution. In the first view, $y > 0$, and in the second, $y < 0$. The solution must therefore be between the two x-values shown. Choice (a) is the only such choice.   **63.** When subtracting the right hand side, $4x - 12$, parentheses must be inserted around it. The correct solution would be found by graphing $y_1 = 2x + 3 - (4x - 12)$.

Further Explorations

(a) Show that when $x = 12$, $Y_1 = Y_2 = 19$. **(b)** Show that when $x = 3$, $Y_1 = Y_2 = 10$. **(c)** Show that when $x = 3$, $Y_1 = Y_2 = 3.13$. **(d)** Show that when $x = -9$, $Y_1 = Y_2 = 1.74$.

Section 1.6 (page 76)

1. $\{3\}$ **3.** $(-\infty, 3)$ **5.** $(3, \infty)$ **7.** $[3, \infty)$ **9.** $\{3\}$ **11.** $\{3\}$ **13.** By subtracting $g(x)$ from both sides of $f(x) = g(x)$, we get an equivalent equation, $f(x) - g(x) = 0$ (having the same solution set).   **15. (a)** $(20, \infty)$ **(b)** $(-\infty, 20)$ **(c)** $[20, \infty)$ **(d)** $(-\infty, 20]$ **17. (a)** $[.4, \infty)$ **(b)** $(-\infty, .4)$ **(c)** $(.4, \infty)$ **(d)** $(-\infty, .4]$ **19. (a)** $\{4\}$ **(b)** $(4, \infty)$ **(c)** $(-\infty, 4)$ **21. (a)** $(-\infty, -3]$ **(b)** $(-3, \infty)$ **23. (a)** $(-\frac{3}{4}, \infty)$ **(b)** $(-\infty, -\frac{3}{4}]$ **25. (a)** $(-\infty, 15]$ **(b)** $(15, \infty)$ **27. (a)** $(-6, \infty)$ **(b)** $(-\infty, -6]$ **29. (a)** $[-8, \infty)$ **(b)** $(-\infty, -8)$ **31. (a)** $(25, \infty)$ **(b)** $(-\infty, 25]$ **33.** $(-\infty, 29.90)$ **35.** $[7.28, \infty)$ **37.** $(4.20, \infty)$ **39.** $(-\infty, 12)$ **41.** $[12, \infty)$ **43.** the open interval $(-2, 12)$ **45.** $\emptyset$ **47.** $\emptyset$ **49.** $[1, 4]$ **51.** the open interval $(-6, -4)$ **53.** $(-16, 19]$ **55.** $(-\infty, \infty)$ **57.** $\emptyset$ **59.** $\emptyset$ **61.** $(-\infty, \infty)$ **63. (a)** 6; $\{6\}$ **(b)** Every y-value is greater than 0 (i.e., positive); $(6, \infty)$ **(c)** Every y-value is less than 0 (i.e., negative); $(-\infty, 6)$

Section 1.7 (page 94)

1. $P = \frac{I}{RT}$ **3.** $W = \frac{P - 2L}{2}$ or $W = \frac{P}{2} - L$ **5.** $h = \frac{2A}{b_1 + b_2}$ **7.** $h = \frac{S - 2\pi r^2}{2\pi r}$ or $h = \frac{S}{2\pi r} - r$ **9.** $C = \frac{5}{9}(F - 32)$ **11.** 10 cm by 17 cm   **13.** length: 20 in.; width: 12 in.   **15.** $\{6\}$ **17.** 2 liters   **19.** $266\frac{2}{3}$ gal   **21.** 4 liters   **23.** $\{9\}$ **25.** 35 pounds per square foot   **27.** $51\frac{3}{7}$ feet   **29.** 12,500 fish   **31. (a)** $\{5\}$ **(b)** $(5, \infty)$ **(c)** $(-\infty, 5)$ **33. (a)** $\{5\}$ **(b)** $(-\infty, 5)$ **(c)** $(5, \infty)$ **35. (a)** $C(x) = .02x + 200$
   (b) $R(x) = .04x$
   (c) 10,000
   (d) For $x < 10,000$, a loss; for $x > 10,000$, a profit
37. (a) $C(x) = 3.00x + 2300$
   (b) $R(x) = 5.50x$
   (c) 920
   (d) For $x < 920$, a loss; for $x > 920$, a profit

39. 94 or greater   **41.** $y = \frac{3}{4}x + \frac{33}{4}$ **43. (a)** $y = 640x + 1100$ **(b)** \$17,100   **(c)** Locate on the graph of $y = 640x + 1100$ the point $(25, 17,100)$.   **45. (a)** $y = 2600x + 120,000$ **(b)** \$156,400; locate the point $(14, 156,400)$.   **(c)** It represents the annual appreciation in value of the house (\$2600 per year).   **47.** $y = 4x + 120$; 4 feet   **49.** \$160   **51.** 65 mph   **53.** 36 in.   **55.** \$7866   **57.** about 44%   **59.** about 17.678 million   **61. (a)** 500 cubic cm   **(b)** 90° C   **(c)** $-273°$ C

Further Explorations

1. See the answers to Exercises 49–52.

Chapter 1 Review Exercises (page 102)

1. $\sqrt{612}$ or $6\sqrt{17}$ **2.** $(2, 4)$ **3.** -4 **4.** $y = -4x + 12$ **5.** $-\frac{3}{4}$ **6.** 48   **7.** 36
8. Answers will vary. One possible window is $[-10, 50]$ by $[-40, 40]$.   **9.** 11   **10.** $y = \frac{1}{4}x + \frac{9}{2}$
11. C   **12.** F   **13.** A   **14.** B   **15.** E   **16.** D   **17.** I   **18.** K   **19.** B
20. A   **21.** I   **22.** M   **23.** O   **24.** K   **25.** $\{\frac{46}{7}\}$ **26.** $\{8\}$ **27.** $(-\frac{10}{3}, \frac{46}{3}]$
28. $\{-3.81\}$ **29.** $(-\infty, -3.81)$ **30.** $[-3.81, \infty)$ **31.** $(2.5, \infty)$ **32.** $\{3\}$ **33.** $(-\infty, \infty)$
34. $\{-6\}$ **35.** $(-3, \infty)$ **36.** domain: $[-6, 6]$; range: $[-6, 6]$ **37.** false   **38.** $C(x) = 30x + 150$
39. $R(x) = 37.50x$ **40.** 20   **41.** The company is losing money when $x < 20$, it is breaking even when
$x = 20$, and it is making money when $x > 20$.   **42.** $\frac{40}{7}$ liters   **43. (a)** 41°F **(b)** about 21,000 ft
(c) Graph $y = -3.52x + 58.6$. Then find the coordinates of the point where $x = 5$ to support the answer in (a),
and then find the coordinates of the point where $y = -15$ to support the answer in (b).
44. 1950   **45.** 2000   **46.** about 3.3 million

CHAPTER 2

Section 2.1 (page 119)

1. $(-\infty, \infty)$ **3.** $[0, \infty)$ **5.** $(-\infty, -3) \cup (-3, \infty)$ **7.** $(-\infty, -2) \cup (-2, \infty)$ **9.** While it seems to
be continuous at first glance, there is a discontinuity at $x = -3$, as indicated by the calculator not giving a
corresponding value of y when we trace to $x = -3$. This happens because -3 causes the denominator to equal
zero, giving an undefined expression.   **11. (a)** $(-\infty, 0)$ **(b)** $(0, \infty)$ **(c)** none   **13. (a)** $(-2, \infty)$
(b) $(-\infty, -2)$ **(c)** none   **15. (a)** $(-\infty, 1)$ **(b)** $(4, \infty)$ **(c)** $(1, 4)$ **17. (a)** none **(b)** $(-\infty, \infty)$
(c) none   **19. (a)** none **(b)** $(-\infty, -2) \cup (3, \infty)$ **(c)** $(-2, 3)$ **21.** increasing   **23.** decreasing
25. increasing   **27.** increasing   **29.** increasing   **31.** decreasing   **33. (a)** no **(b)** yes
(c) no   **35. (a)** yes **(b)** no **(c)** no   **37. (a)** yes **(b)** yes **(c)** yes   **39. (a)** yes **(b)** no **(c)** no
41. (a) no **(b)** no **(c)** yes   **43. (a)** no **(b)** no **(c)** no   **45.** $(-1.625, 2.0352051)$
47. $(.5, -.84089642)$ **49.** $(-5.092687, .9285541)$
51. symmetric with respect to the origin   **53.** symmetric with respect to the y-axis

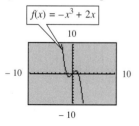

$f(x) = -x^3 + 2x$

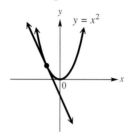

$f(x) = .5x^4 - 2x^2 + 1$

55. neither symmetric with respect
to the origin nor the y-axis

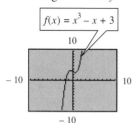

$f(x) = x^3 - x + 3$

57. The slope of the tangent
line is negative.

$y = x^2$

59. The slope of the tangent line
is 0. (Tangent line coincides
with x-axis.)

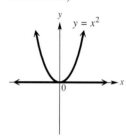

$y = x^2$

For the graphs preceding Exercises 61–69, see Figures 8, 9, 10, 13, 14, and 15 in Chapter 2.

61. true   **63.** false   **65.** true   **67.** true   **69.** true   **71. (a)** $\{0\}$ **(b)** $(-\infty, 0) \cup (0, \infty)$ **(c)** $\emptyset$
73. (a) $\{0\}$ **(b)** $(0, \infty)$ **(c)** $\emptyset$ **75. (a)** $\{0\}$ **(b)** $(-\infty, 0) \cup (0, \infty)$ **(c)** $\emptyset$ **77. (a)** $\{0, 1\}$ **(b)** the open
interval $(0, 1)$ **(c)** $(1, \infty)$ **79.** odd   **81.** even   **83.** neither

Further Explorations

1. The square root of a negative number is not real.

Section 2.2 (page 133)

1. D   **3.** C   **5.** B   **7.** A   **9.** C   **11.** B   **13.** C   **15.** A   **17.** E   **19.** B
21. D   **23.** E   **25.** A   **27. (a)** $(-\infty, \infty)$ **(b)** $[-3, \infty)$ **29. (a)** $[2, \infty)$ **(b)** $[-4, \infty)$
31. (a) $(-\infty, \infty)$ **(b)** $(-\infty, \infty)$ **33.** B   **35.** D
37.

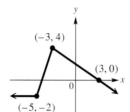

39.

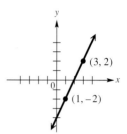

41.

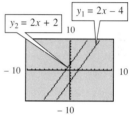

43. $y_1 = 2x - 4$ **45.** 2   **47.** The graph of y_2 can be obtained by shifting the graph of y_1 upward 6 units. The constant, 6, comes from the 6 we added to each y-value in Exercise 44.

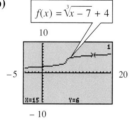

49. B   **51.** A   **53. (a)** $(-4, \infty)$ **(b)** $(-\infty, -4)$ **55. (a)** $(-\infty, \infty)$ **(b)** none
57. (a) $(-\infty, \infty)$ **(b)** none   **59.** h can be any real number, while k must equal 38.
61. (a) $\{3, 4\}$ **(b)** $(-\infty, 3) \cup (4, \infty)$ **(c)** the open interval $(3, 4)$ **63. (a)** $\{-4, 5\}$ **(b)** $(-\infty, -4] \cup [5, \infty)$
(c) $[-4, 5]$ **65.** yes   **67.** no   **69.** no
71. (a) $f(15) = \sqrt[3]{15 - 7} + 4 = 2 + 4 = 6$ **(b)**

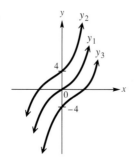

Section 2.3 (page 146)

1.

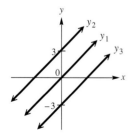

3.

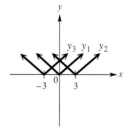

5.

7.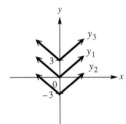

9.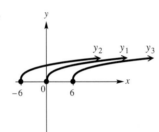

11.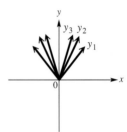

13.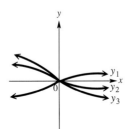

15.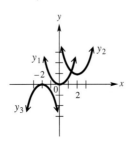

17.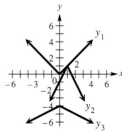

19. 4; x **21.** 2; left; $\frac{1}{4}$; x; 3; downward (or negative)   **23.** 3; right; 6   **25.** $y = \frac{1}{2}x^2 - 7$

27. $y = 4.5\sqrt{x - 3} - 6$ **29.** $g(x) = -(x - 5)^2 - 2$

31. (a) 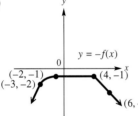   **(b)** 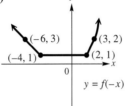   **(c)** 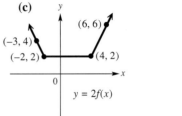   **(d)** $f(0) = 1$

33. (a) 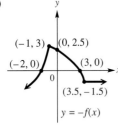   **(b)** 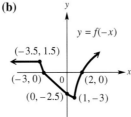   **(c)** 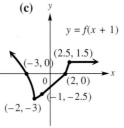   **(d)** -1 and 4

35. (a) 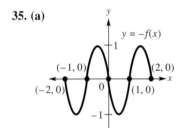   **(c)**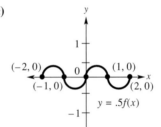

(b) The graph of $y = f(-x)$ is the same as that of $y = -f(x)$, shown in part (a).   **(d)** symmetry with respect to the origin

37. (a) r is an x-intercept.   **(b)** $-r$ is an x-intercept.   **(c)** $-r$ is an x-intercept.

39. domain: $[20, \infty)$; range: $[5, \infty)$

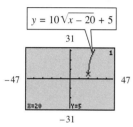

41. domain: $[-10, \infty)$; range: $(-\infty, 5]$

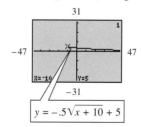

43. (a) $(-1, 2)$ **(b)** $(-\infty, -1)$ **(c)** $(2, \infty)$ **45. (a)** $(1, \infty)$ **(b)** $(-2, 1)$ **(c)** $(-\infty, -2)$
47. $y_2 = 5(2) = 10$ when $x = 8$ **49.** $y_2 = 5(-2.154435) = -10.772175$ when $x = -10$ **51.** decreases
53. increases   **55. (a)** It is symmetric with respect to the y-axis. **(b)** It is symmetric with respect to the
y-axis.   **57.** $\{1.89, 5.32\}$ **59.** $(-\infty, 1.89) \cup (5.32, \infty)$ **61.** $(-\infty, .47) \cup (.89, 2.30)$

Section 2.4 (page 159)

1.

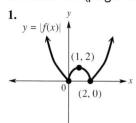

3.

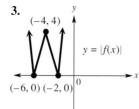

5.

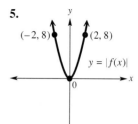

7.

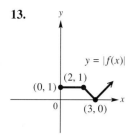

9.

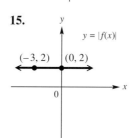

11.

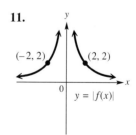

13.

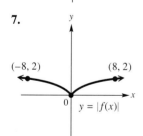

15.

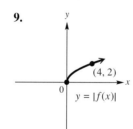

17. (a)

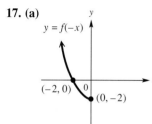

(b)

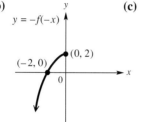

(c)

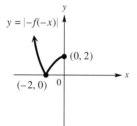

19. $[0, \infty)$ **21.** Figure (a) shows the graph of $y = f(x)$, while (b) shows the graph of $y = |f(x)|$.

23. Property 1: $|7 \cdot 14| = |7| \cdot |14|$
$|98| = 7 \cdot 14$
$98 = 98 \checkmark$

Property 2: $\left|\dfrac{7}{14}\right| = \dfrac{|7|}{|14|}$

$\left|\dfrac{1}{2}\right| = \dfrac{7}{14}$

$\dfrac{1}{2} = \dfrac{1}{2} \checkmark$

Property 3: $|7| = |-7|$
$7 = 7 \checkmark$

Property 4: $|7| + |14| \geq |7 + 14|$
$7 + 14 \geq |21|$
$21 \geq 21$
True, because $21 = 21$.

25. Property 1: $|-26 \cdot 13| = |-26| \cdot |13|$
$|-338| = 26 \cdot 13$
$338 = 338 \checkmark$

Property 2: $\left|\dfrac{-26}{13}\right| = \dfrac{|-26|}{|13|}$

$|-2| = \dfrac{26}{13}$

$2 = 2 \checkmark$

Property 3: $|-26| = |-(-26)|$
$26 = |26|$
$26 = 26 \checkmark$

Property 4: $|-26| + |13| \geq |-26 + 13|$
$26 + 13 \geq |-13|$
$39 \geq 13$
True, because $39 > 13.$

27. (a) $\{-13, 5\}$ **(b)** $(-\infty, -13) \cup (5, \infty)$ **(c)** $(-13, 5)$ **29. (a)** $\{-5, -2\}$ **(b)** $(-\infty, -5] \cup [-2, \infty)$
(c) $[-5, -2]$ **31. (a)** $\left\{-\frac{7}{6}, -\frac{5}{6}\right\}$ **(b)** $\left(-\frac{7}{6}, -\frac{5}{6}\right)$ **(c)** $\left(-\infty, -\frac{7}{6}\right) \cup \left(-\frac{5}{6}, \infty\right)$ **33. (a)** $\left\{-\frac{3}{2}, \frac{1}{2}\right\}$
(b) $\left[-\frac{3}{2}, \frac{1}{2}\right]$ **(c)** $\left(-\infty, -\frac{3}{2}\right] \cup \left[\frac{1}{2}, \infty\right)$ **35. (a)** $\left\{\frac{7}{2}\right\}$ **(b)** $\left(-\infty, \frac{7}{2}\right) \cup \left(\frac{7}{2}, \infty\right)$ **(c)** $\emptyset$
37. (a) $\emptyset$ **(b)** $\emptyset$ **(c)** $(-\infty, \infty)$ **39. (a)** $\{-1, 6\}$ **(b)** $(-1, 6)$ **(c)** $(-\infty, -1) \cup (6, \infty)$
41. (a) $\{4\}$ **(b)** $\{4\}$ **(c)** $(-\infty, \infty)$ **43. (a)** $\left\{-8, \frac{6}{5}\right\}$ **(b)** $(-\infty, -8) \cup \left(\frac{6}{5}, \infty\right)$ **(c)** $\left(-8, \frac{6}{5}\right)$
45. (a) $\left\{-3, \frac{5}{3}\right\}$ **(b)** $(-\infty, -3] \cup \left[\frac{5}{3}, \infty\right)$ **(c)** $\left[-3, \frac{5}{3}\right]$ **47. (a)** $\left\{\frac{1}{4}\right\}$ **(b)** $\left(\frac{1}{4}, \infty\right)$ **(c)** $\left(-\infty, \frac{1}{4}\right)$
49. (a) $\{-3, 8\}$ **(b)** $(-3, 8)$ **(c)** $(-\infty, -3) \cup (8, \infty)$ **51. (a)** $\{-2, 6\}$ **(b)** $[-2, 6]$
(c) $(-\infty, -2] \cup [6, \infty)$ **53.** $\{3.37, 5.41\}$ **55.** $\{-3.54, .40, .54, 2.60\}$ **57.** $\{2\}$ **59.** $(-\infty, \infty)$
61. $\emptyset$ **63.** $\{-.31, 1.44\}$ **65.** $[-140°, -28°]$ **67.** $|F - 730| \leq 50$

Section 2.5 (page 170)

1. (a) -10 **(b)** -2 **(c)** -1 **(d)** 2 **(e)** 4 **3. (a)** -10 **(b)** 2 **(c)** 5 **(d)** 3 **(e)** 5
5.

$f(x) = \begin{cases} x - 1 & \text{if } x \leq 3 \\ 2 & \text{if } x > 3 \end{cases}$

7.

$f(x) = \begin{cases} 4 - x & \text{if } x < 2 \\ 1 + 2x & \text{if } x \geq 2 \end{cases}$

9.

$f(x) = \begin{cases} 2 + x & \text{if } x < -4 \\ -x & \text{if } -4 \leq x \leq 5 \\ 3x & \text{if } x > 5 \end{cases}$

11.

$f(x) = \begin{cases} |x| & \text{if } x > -2 \\ x & \text{if } x \leq -2 \end{cases}$

13.

$(0, 2)$ $(-2, 0)$ $(2, 0)$

$f(x) = \begin{cases} -\frac{1}{2}x^2 + 2 & \text{if } x \leq 2 \\ \frac{1}{2}x & \text{if } x > 2 \end{cases}$

15.

$(2, 7)$ $(2, 5)$ $(-3, 0)$ $(0, 3)$

$f(x) = \begin{cases} x + 3 & \text{if } x \neq 2 \\ 7 & \text{if } x = 2 \end{cases}$

17.

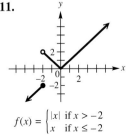

$(0, 5)$

$f(x) = \begin{cases} x^3 + 5 & \text{if } x \leq 0 \\ -x^2 & \text{if } x > 0 \end{cases}$

Although there are many possible ways to describe the functions graphed in Exercises 19 and 21, we give only the most basic rule in these answers.

19. $f(x) = \begin{cases} x \text{ if } x \leq 0 \\ 2 \text{ if } x > 0 \end{cases}$

Domain: $(-\infty, \infty)$
Range: $(-\infty, 0] \cup \{2\}$

21. $f(x) = \begin{cases} \sqrt[3]{x} \text{ if } x < 1 \\ x + 1 \text{ if } x \geq 1 \end{cases}$

Domain: $(-\infty, \infty)$
Range: $(-\infty, 1) \cup [2, \infty)$

23. Because of the choice of scale, the graph of $y = [\![x]\!]$ looks like the graph of $y = x$. If one does not know how $[\![x]\!]$ is defined, one may *incorrectly* conclude that $[\![x]\!] = x$ for *all* x, but this is true only when x is an integer.
25. The graph of $y = [\![x]\!]$ is shifted 1.5 units downward.   **27.** The graph of $y = [\![x]\!]$ is reflected across the x-axis.

29. 2 pound package: \$13; 2.5 pound package: \$16; 5.8 pound package: \$25
The range is $\{10, 13, 16, 19, 22, 25, 28\}$

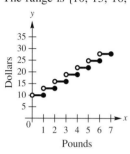

31. (a) 30¢ **(b)** 57¢ **(c)** 111¢ (or \$1.11)
(d)

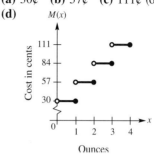

(e) domain: $(0, \infty)$; range: $\{30, 57, 84, 111, \ldots\}$

33.

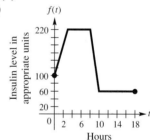

35. (a) 140 **(b)** 220 **(c)** 220 **(d)** 220
(e) 220 **(f)** 60 **(g)** 60 **(h)**

37. B   **39.** D

Further Explorations

1. \$.52   **3.** \$5.81

Section 2.6 (page 182)

1. (a) $(f + g)(x) = 10x + 2$; $(f - g)(x) = -2x - 4$; $(fg)(x) = 24x^2 + 6x - 3$ **(b)** domain is $(-\infty, \infty)$ in all cases   **(c)** $\left(\frac{f}{g}\right)(x) = \frac{4x - 1}{6x + 3}$; its domain is $\{x \mid x \neq -\frac{1}{2}\}$ **(d)** $(f \circ g)(x) = 24x + 11$; its domain is $(-\infty, \infty)$
(e) $(g \circ f)(x) = 24x - 3$; its domain is $(-\infty, \infty)$ **3. (a)** $(f + g)(x) = |x + 3| + 2x$; $(f - g)(x) = |x + 3| - 2x$; $(fg)(x) = |x + 3|(2x)$ **(b)** domain is $(-\infty, \infty)$ in all cases   **(c)** $\left(\frac{f}{g}\right)(x) = \frac{|x + 3|}{2x}$; its domain is $\{x \mid x \neq 0\}$ **(d)** $(f \circ g)(x) = |2x + 3|$; its domain is $(-\infty, \infty)$ **(e)** $(g \circ f)(x) = 2|x + 3|$; its domain is $(-\infty, \infty)$
5. (a) $(f + g)(x) = \sqrt[3]{x + 4} + x^3 + 5$; $(f - g)(x) = \sqrt[3]{x + 4} - x^3 - 5$; $(fg)(x) = (\sqrt[3]{x + 4})(x^3 + 5)$
(b) domain is $(-\infty, \infty)$ in all cases   **(c)** $\left(\frac{f}{g}\right)(x) = \frac{\sqrt[3]{x + 4}}{x^3 + 5}$; its domain is $\{x \mid x \neq \sqrt[3]{-5}\}$ **(d)** $(f \circ g)(x) = \sqrt[3]{x^3 + 9}$; its domain is $(-\infty, \infty)$ **(e)** $(g \circ f)(x) = x + 9$; its domain is $(-\infty, \infty)$
7. (a) $(f + g)(x) = \sqrt{x^2 + 3} + x + 1$; $(f - g)(x) = \sqrt{x^2 + 3} - x - 1$; $(fg)(x) = (\sqrt{x^2 + 3})(x + 1)$
(b) domain is $(-\infty, \infty)$ in all cases   **(c)** $\left(\frac{f}{g}\right)(x) = \frac{\sqrt{x^2 + 3}}{x + 1}$; its domain is $\{x \mid x \neq -1\}$
(d) $(f \circ g)(x) = \sqrt{x^2 + 2x + 4}$; its domain is $(-\infty, \infty)$ (*Note:* To see that this is the domain, graph $y = x^2 + 2x + 4$ and note that $y > 0$ for all x.) **(e)** $(g \circ f)(x) = \sqrt{x^2 + 3} + 1$; its domain is $(-\infty, \infty)$

9. (a) $(f + g)(x) = (x - 2)^2 + 2x$ or $x^2 - 2x + 4$; $(f - g)(x) = (x - 2)^2 - 2x$ or $x^2 - 6x + 4$;
$(fg)(x) = (x - 2)^2(2x)$ or $2x^3 - 8x^2 + 8x$ **(b)** domain is $(-\infty, \infty)$ in all cases
(c) $\left(\frac{f}{g}\right)(x) = \frac{(x - 2)^2}{2x}$ or $\frac{x^2 - 4x + 4}{2x}$; its domain is $\{x \mid x \neq 0\}$ **(d)** $(f \circ g)(x) = (2x - 2)^2$ or $4x^2 - 8x + 4$; its
domain is $(-\infty, \infty)$ **(e)** $(g \circ f)(x) = 2(x - 2)^2$ or $2x^2 - 8x + 8$; its domain is $(-\infty, \infty)$
11. $256x^2 + 48x + 2$ **13.** $32x^2 - 16x + 1$ **15.** 55   **17.** 1848   **19.** $-\frac{6}{7}$
21. $4m^2 - 10m - 1$ **23.** 1122   **25.** 97
33. The graph of y_2 can be obtained by *reflecting* the graph of y_1 across the line $y_3 = x$.

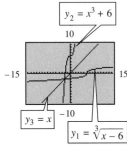

35. 1   **37.** 1   **39.** 9   **41.** 1   **43.** We are not given enough information to determine $f[g(1)]$,
which is $f(9)$.   **45.** 4   **47.** $-12x - 1 - 6h$ **49.** $3x^2 + 3xh + h^2$

We give only one of many possible correct pairs of functions f and g in Exercises 51–55.

51. $f(x) = x^2$, $g(x) = 6x - 2$ **53.** $f(x) = \sqrt{x}$, $g(x) = x^2 - 1$ **55.** $f(x) = \sqrt{x} + 12$, $g(x) = 6x$
57. (a) $C(x) = 10x + 500$ **(b)** $R(x) = 35x$ **(c)** $P(x) = 35x - (10x + 500)$ or $P(x) = 25x - 500$
(d) 21 items   **(e)** The smallest whole number value for which $P(x) > 0$ is 21. Use a window of $[0, 30]$ by
$[-1000, 500]$, for example.   **59. (a)** $C(x) = 100x + 2700$ **(b)** $R(x) = 280x$
(c) $P(x) = 280x - (100x + 2700)$ or $P(x) = 180x - 2700$ **(d)** 16 items   **(e)** The smallest whole number value
for which $P(x) > 0$ is 16. Use a window of $[0, 30]$ by $[-3000, 500]$, for example.
61. (a) $(A \circ r)(t) = 16\pi t^2$ **(b)** $(A \circ r)(t)$ is a composite function that expresses the area of the oil slick as a
function of time t (in minutes).   **(c)** 144π square feet   **(d)** Graph $y = 16\pi x^2$ and show that for $x = 3$, $y \approx 452$
(an approximation for 144π).   **63.** The right side simplifies to $16\pi x + 16\pi$, a linear expression. The slope is
16π and the y-intercept is 16π.
65. (a) $V(r) = \frac{4}{3}\pi(r + 3)^3 - \frac{4}{3}\pi r^3$ **(c)** 1168.67 cubic inches
     (b) This appears to be a portion of a parabola,   **(d)** $V(4) = \frac{4}{3}\pi(7)^3 - \frac{4}{3}\pi(4^3) \approx 1168.67$
        formed by translating the squaring function.

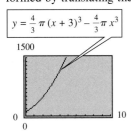

67. (a) $P = 6x$; $P(x) = 6x$; This is a linear function.
     (b) x represents the width of the   **(c)** (See graph for part (b).) (Answers   **(d)** (Answers may vary.) If the
        rectangle and y represents the   may vary.) Perimeter $= 24$. This   perimeter y of a rectangle
        perimeter.   is the y-value shown on the screen   satisfying the given conditions
                                       for the integer x-value 4.   is 36, then the width x is 6.

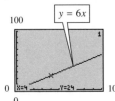      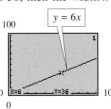

69. (a) $A(2x) = \sqrt{3}\, x^2$ **(b)** $A(16) = 64\sqrt{3}$ **(c)** On the graph of $y = \dfrac{\sqrt{3}}{4}x^2$, locate the point where $x = 16$ to find $y \approx 110.85$, an approximation for $64\sqrt{3}$.

Further Explorations

1. (a) After 1 hour (60 minutes), the radius is 120 feet and the area is 45,239 square feet. After 2 hours (120 minutes), the radius is 240 feet and the area is 180,956 square feet. **(b)** The volume is a function of the square of the time.

Chapter 2 Review Exercises (page 187)

1. F **2.** C **3.** G **4.** A **5.** B **6.** H **7.** E **8.** D **9.** $[0, \infty)$
10. $[0, \infty)$ **11.** $(-\infty, \infty)$ **12.** $(-\infty, \infty)$ **13.** $(-\infty, \infty)$ **14.** $(0, \infty)$ **15.** $[0, \infty)$ **16.** $(-\infty, \infty)$
17. $(-\infty, -2), [-2, 1], (1, \infty)$ **18.** $(-2, 1)$ **19.** $(-\infty, -2)$ **20.** $(1, \infty)$ **21.** $(-\infty, \infty)$
22. $\{-2\} \cup [-1, 1] \cup (2, \infty)$ **23.** $(-\infty, \infty)$ **24.** $(-\infty, -3)$ **25.** $(-3, \infty)$ **26.** none
27. $(-\infty, \infty)$ **28.** $(-\infty, -5]$ **29.** x-axis symmetry, y-axis symmetry, origin symmetry; not a function **30.** none of these symmetries; neither even nor odd **31.** y-axis symmetry ; even function **32.** none of these symmetries; neither even nor odd **33.** x-axis symmetry; not a function **34.** y-axis symmetry; even function **35.** Start with the graph of $y = x^2$. Shift it 4 units to the left, stretch vertically by a factor of 3, reflect across the x-axis, and shift 8 units downward. (These changes can occur in different orders.)
36. $y = -\frac{2}{3}\sqrt{-x} + 4$
37.

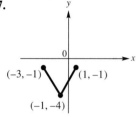

38.

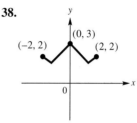

39. $\{-6, 1\}$ **40.** $[-6, 1]$ **41.** $(-\infty, -6] \cup [1, \infty)$
42. The graph of $y_1 = |2x + 5|$ intersects the graph of $y_2 = 7$ at points for which $x = -6$ and $x = 1$. Therefore, by the intersection-of-graphs method, the solution set of $|2x + 5| = 7$ is $\{-6, 1\}$. Because the graph of y_1 lies below or on the graph of y_2 for values of x between and including -6 and 1, the solution set of $|2x + 5| \le 7$ is $[-6, 1]$. Finally, because the graph of y_1 lies above or on the graph of y_2 for values of x less than or equal to -6, or greater than or equal to 1, the solution set of $|2x + 5| \ge 7$ is $(-\infty, -6] \cup [1, \infty)$.
43. $\{-3, 5\}$; when $x = -3$, we get $8 = 8$, and when $x = 5$, we also get $8 = 8$. **44. (a)** the open interval $(-3, 5)$ **(b)** $(-\infty, -3) \cup (5, \infty)$ **45.** $\{0, 2\}$ **46.** $(-\infty, 0) \cup (2, \infty)$ **47.** $[0, 2]$ **48.** $[2, \infty)$
49. $(-\infty, 0]$ **50.** $\emptyset$ **51.** -1 **52.** 3 **53.** 3
54. **55.** $(f + g)(x) = x^2 + 2x - 7$ **56.** $(f - g)(x) = -x^2 + 2x - 1$

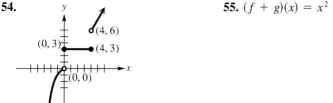

57. $(fg)(x) = 2x^3 - 4x^2 - 6x + 12$ **58.** $(-\infty, \infty)$ **59.** $\left(\frac{g}{f}\right)(x) = \frac{x^2 - 3}{2x - 4}$; domain: $(-\infty, 2) \cup (2, \infty)$
60. $(g \circ f)(x) = 4x^2 - 16x + 13$; domain: $(-\infty, \infty)$ **61.** $(f \circ g)(x) = 2x^2 - 10$; domain: $(-\infty, \infty)$
62. (a) 115 **(b)** 9 **63.** $4x + 2h$ **64.** Answers may vary. The most obvious choices for f and g are $g(x) = 5x^2 - 3$ and $f(x) = \sqrt[3]{x}$. **65.** $C(x) = 3.00x + 2300$ **66.** $R(x) = 5.50x$
67. $P(x) = 2.50x - 2300$ **68.** 921 deliveries **69.** Show that the graph of $y = P(x)$ lies above the x-axis for x-values greater than 920. **70.** It would not pass the vertical line test.

CHAPTER 3

Section 3.1 (page 196)

1. (a) 0 **(b)** -9 **(c)** imaginary **3. (a)** π **(b)** 0 **(c)** real **5. (a)** 0 **(b)** $\sqrt{6}$ **(c)** imaginary
7. (a) 2 **(b)** 5 **(c)** imaginary **9.** $10i$ **11.** $-20i$ **13.** $-i\sqrt{39}$ **15.** $5 + 2i$
17. $9 - 5i\sqrt{2}$ **19.** A real number a may be written as $a + 0i$, and thus is a complex number. However, a complex number with nonzero imaginary part, such as $2 + 3i$, is not a real number.
27. $7 - i$ **29.** 2 **31.** $1 - 10i$ **33.** $-14 + 2i$ **35.** $5 - 12i$ **37.** 13 **39.** 7
41. $25i$ **43.** $12 + 9i$ **45.** $-2 + 2i$ **47.** Verify that $(-3)^3 - (-3)^2 - 7(-3) + 15$ is equal to 0.
49. Verify that $(2 + i)^3 - (2 + i)^2 - 7(2 + i) + 15$ is equal to 0. **51.** i **53.** $\frac{7}{25} - \frac{24}{25}i$
55. $\frac{26}{29} + \frac{7}{29}i$ **57.** $-2 + i$ **59.** $-2i$ **61.** We are multiplying by 1, the identity element for multiplication. **63.** $-i$ **65.** i **67.** -1 **69.** 1

Section 3.2 (page 208)

1. (a) domain: $(-\infty, \infty)$; range: $[0, \infty)$ **(b)** $(2, 0)$ **(c)** $x = 2$ **(d)** $(2, \infty)$ **(e)** $(-\infty, 2)$ **(f)** minimum point; 0
(g) concave up **3. (a)** domain: $(-\infty, \infty)$; range: $[-4, \infty)$ **(b)** $(-3, -4)$ **(c)** $x = -3$ **(d)** $(-3, \infty)$
(e) $(-\infty, -3)$ **(f)** minimum point; -4 **(g)** concave up **5. (a)** domain: $(-\infty, \infty)$; range: $(-\infty, 2]$
(b) $(-3, 2)$ **(c)** $x = -3$ **(d)** $(-\infty, -3)$ **(e)** $(-3, \infty)$ **(f)** maximum point; 2 **(g)** concave down
7. (a) domain: $(-\infty, \infty)$; range: $(-\infty, -3]$ **(b)** $(-1, -3)$ **(c)** $x = -1$ **(d)** $(-\infty, -1)$ **(e)** $(-1, \infty)$
(f) maximum point; -3 **(g)** concave down **9. (a)** $(.5, -24.5)$ **(b)** -3 and 4 **(c)** -24 **(d)** Use $[-10, 10]$
by $[-30, 10]$ for a good view. **11. (a)** $(1, -16)$ **(b)** -3 and 5 **(c)** -15 **(d)** Use $[-10, 10]$ by $[-20, 10]$
for a good view. **13. (a)** $(1.5, 4.5)$ **(b)** 0 and 3 **(c)** 0 **(d)** Use $[-10, 10]$ by $[-10, 10]$ for a good view.
15. (a) $(2.75, -42.25)$ **(b)** $-\frac{1}{2}$ and 6 **(c)** -12 **(d)** Use $[-5, 10]$ by $[-50, 10]$ for a good view.
17. B **19.** D **21. (a)** $(2.71, 5.20)$ **(b)** -1.33 and 6.74 **23. (a)** $(1.12, .56)$ **(b)** There are no
x-intercepts. **25. (a)** $(-.52, -5.00)$ **(b)** -1.77 and .74 **27.** C and F **29.** A and D
31. $c > 0$ for A, B, and C; $c < 0$ for D, E, and F; $c = 0$ for none of these **33.** A **35.** ↖↗ **37.** ↙↘
39. ↙↘ **41.** ↖↗ **43. (a)** $\{-3, \frac{1}{2}\}$ **(b)** $(-3, \frac{1}{2})$ **(c)** $(-\infty, -3) \cup (\frac{1}{2}, \infty)$
45. $P(x) = 3x^2 + 6x - 1$ **47.** $P(x) = .5x^2 - 8x + 35$ **49.** $P(x) = -\frac{2}{3}x^2 - \frac{16}{3}x - \frac{38}{3}$ **51.** If the
vertex is in Quadrant I or Quadrant II, and the graph is concave up, there are no x-intercepts. If it is concave down, there are two x-intercepts. On the other hand, if the vertex is in Quadrant III or Quadrant IV, and the graph is concave up, there are two x-intercepts. If the graph is concave down, there are no x-intercepts.

Further Explorations

1. The other zero, to the nearest hundredth, is 3.80.
3. To the nearest hundredth, the zeros are -40.74 and .20. The vertex is approximately $(-20.27, -155.03)$.

Section 3.3 (page 226)

To support graphically the results in Exercises 1 and 3, graph the parabola defined by the left side, the line defined by the right side, and locate graphically the points of intersection. The x-values support the real solutions.

1. $\{\pm 4\}$ **3.** $\{\pm 3\}$ **5.** $\{\pm 4i\}$ **7.** $\{\pm 3i\sqrt{2}\}$

To support graphically the results in Exercises 9–21 that are real numbers, write the equation in the form $P(x) = 0$, graph $y = P(x)$, and locate the x-intercepts. Their values (or approximations) should correspond to the values (or approximations) of the solutions found by using the quadratic formula.

9. $\{3, 5\}$ **11.** $\{1 \pm \sqrt{5}\}$ **13.** $\{-\frac{1}{2} \pm \frac{1}{2}i\}$ **15.** $\{\frac{1 \pm \sqrt{5}}{2}\}$ **17.** $\{3 \pm \sqrt{2}\}$
19. $\{\frac{3}{2} \pm \frac{\sqrt{2}}{2}i\}$ **21.** $\{\frac{-1 \pm \sqrt{97}}{4}\}$ **23.** 0; one real solution (a double solution) **25.** 1; two real solutions,
both rational **27.** 84; two real solutions, both irrational **29.** -23; no real solutions
31. 2304; two real solutions, both rational **33. (b)** and **(c)** **35.** $\{2, 4\}$ **37.** $(-\infty, 2) \cup (4, \infty)$
39. $(-\infty, 3) \cup (3, \infty)$ **41.** $(-\infty, \infty)$ **43.** There are no real solutions. It has two complex solutions, both
imaginary. **45.** 3 **47.** yes; negative **49.** f

To support your answers graphically in Exercises 51–57, consider the polynomial on the left to be $P(x)$. Graph $y = P(x)$, and then determine the x-values for which the graph intersects, is above, or is below the x-axis. These values correspond to the solutions of $P(x) = 0$, $P(x) > 0$, and $P(x) < 0$, respectively.

51. (a) $(-\infty, -3] \cup [-1, \infty)$ **(b)** the interval $(-3, -1)$ **53. (a)** $(-\infty, \frac{1}{2}) \cup (4, \infty)$ **(b)** $[\frac{1}{2}, 4]$
55. (a) $[1 - \sqrt{2}, 1 + \sqrt{2}]$ **(b)** $(-\infty, 1 - \sqrt{2}) \cup (1 + \sqrt{2}, \infty)$ **57. (a)** $\emptyset$ **(b)** $(-\infty, \infty)$
59. -4 and 2   **61.** The graph of y_2 is obtained by reflecting the graph of y_1 across the x-axis.   **63.** They are
the same.   **65. (a)** $\{-1.12, .92\}$ **(b)** $(-\infty, -1.12) \cup (.92, \infty)$ **(c)** the interval $(-1.12, .92)$
67. (a) $\{.30, 2.82\}$ **(b)** the interval $(.30, 2.82)$ **(c)** $(-\infty, .30) \cup (2.82, \infty)$ **69. (a)** $\{-.79, .79\}$
(b) $(-\infty, -.79) \cup (.79, \infty)$ **(c)** the interval $(-.79, .79)$ **71.** Since $P(x) = P(-x)$, the graph is symmetric
with respect to the y-axis. Therefore, any negative x-intercept will be accompanied by a positive x-intercept of the
same absolute value. To the nearest hundredth, $-.79$ is a solution, so $.79$ is also a solution.   **73.** $\{-1, 2\}$
75. $(-\infty, -1) \cup (2, \infty)$

Section 3.4 (page 236)

1. (a) $30 - x$ **(b)** $0 < x < 30$ **(c)** $P(x) = x(30 - x)$ or $P(x) = -x^2 + 30x$ **(d)** 15 and 15; The maximum
product is 225. Locate the point $(15, 225)$ on the graph.   **3. (a)** $640 - 2x$ **(b)** $0 < x < 320$
(c) $A(x) = x(640 - 2x)$ or $A(x) = -2x^2 + 640x$ **(d)** Choose x to be between approximately 57.04 feet and 85.17
feet or between approximately 234.83 feet and 262.96 feet. **(e)** 160 feet by 320 feet; the maximum area is 51,200
square feet. Locate the point $(160, 51{,}200)$ on the graph.   **5.** 900 square feet   **7. (a)** $2x$ **(b)** length:
$2x - 4$; width: $x - 4$; $x > 4$ **(c)** $V(x) = 2(2x - 4)(x - 4)$ or $V(x) = 4x^2 - 24x + 32$
(d) 8 inches by 20 inches; Locate the point $(12, 320)$ on the graph. **(e)** 13.0 to 14.2 inches   **9. (a)** The value
of t cannot be negative since t represents time elapsed from the throw. **(b)** Since the rock was thrown from
ground level, s_0, the original height of the rock is 0. **(c)** $s(t) = -16t^2 + 90t$ **(d)** 99 feet **(e)** After 2.8125
seconds, the maximum height, 126.5625 feet is attained. Locate the vertex, $(2.8125, 126.5625)$. **(f)** 5.625 seconds;
Locate the point $(5.625, 0)$.   **11.** The ball will not reach 355 feet, because the graph of $y = -16x^2 + 150x$
does not intersect the graph of $y_2 = 355$.   **13.** 128   **15. (a)** $R(x) = x(100 - .1x)$ or $R(x) = 100x - .1x^2$
(b) 500 **(c)** \$25,000   **17. (a)** $80 - x$ **(b)** $400 + 20x$ **(c)** $R(x) = (80 - x)(400 + 20x)$
or $R(x) = 32{,}000 + 1200x - 20x^2$ **(d)** 5 or 55 **(e)** \$1000   **19. (a)** $R(x) = (75 + x)(425 - 5x)$
or $R(x) = 31{,}875 + 50x - 5x^2$ **(b)** 80 **(c)** \$32,000   **21.** about 96 feet per second   **23.** 3.5 feet
25. 1.25 feet   **27.** approximately 8723 (billion)   **29.** 1986

Further Explorations

1. When the rectangle is a square, measuring 10 inches by 10 inches, the area is greatest, 100 square inches.
3. For those values of x, the graph lies below the x-axis, indicating graphically that $y < 0$. There can be no
such rectangle.

Section 3.5 (page 253)

1. (a)   **3.** one   **5. (b)** and **(d)**   **7.** one   **9. (b)**   **11.** ⟨ ⟩ **13.** ⟨ ⟩ **15.** ⟨ ⟩ **17.** ⟨ ⟩
19. D   **21.** B   **23.** As the exponent n gets larger (but remains odd), the graph "flattens out" in the window
$[-1, 1]$ by $[-1, 1]$. The graph of $y = x^7$ will lie between the x-axis and the graph of $y = x^5$ in this window.

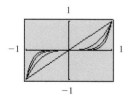

$y = x^n$ for $n = 1, 3, 5, 7$

25. They all take the shape ⟋. As n gets larger, the graphs get steeper.
27. (a) $(-\infty, \infty)$
  (b) $(2.54, -.88)$; not an absolute minimum
  (c) $(.13, 6.06)$; not an absolute maximum
  (d) $(-\infty, \infty)$
  (e) x-intercepts: $-1, 2, 3$; y-intercept: 6

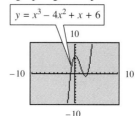

$y = x^3 - 4x^2 + x + 6$

29. (a) $(-\infty, \infty)$
  (b) $(-4.74, -27.03)$; not an absolute minimum
  (c) $(.07, 84.07)$; not an absolute maximum
  (d) $(-\infty, \infty)$
  (e) x-intercepts: $-6, -3.19, 2.19$;
    y-intercept: 84

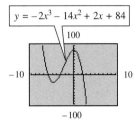

31. (a) $(-\infty, \infty)$
  (b) $(-1.73, -16.39)$; $(1.35, -3.49)$;
    neither is an absolute minimum
  (c) $(-3, 0)$; $(.17, 9.52)$; neither is an absolute
    maximum
  (d) $(-\infty, \infty)$
  (e) x-intercepts: $-3, -.62, 1, 1.62$; y-intercept: 9

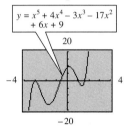

33. (a) $(-\infty, \infty)$
  (b) $(-2.63, -132.69)$ is an absolute minimum;
    $(1.68, -99.90)$
  (c) $(-.17, -71.48)$; no absolute maximum
  (d) $[-132.69, \infty)$
  (e) x-intercepts: $-4, 3$; y-intercept: -72

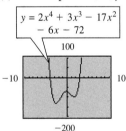

35. (a) $(-\infty, \infty)$
  (b) $(-2, 0)$; $(2, 0)$; neither is an absolute minimum
  (c) $(-3.46, 256)$, $(0, 256)$, and $(3.46, 256)$;
    all are absolute maximum points
  (d) $(-\infty, 256]$
  (e) x-intercepts: $-4, -2, 2, 4$; y-intercept: 256

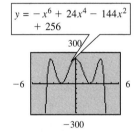

37. When n is even, $P(x) = x^n$ is always concave up.   **39.** The graph is concave up for $x > 0$, and for $x < 0$
the graph is concave down. The point $(0, 0)$ is an inflection point.

There are many possible correct windows in Exercises 41–47. We give only one in each case.

41. $[-10, 10]$ by $[-40, 10]$ **43.** $[-10, 20]$ by $[-1500, 500]$ **45.** $[-10, 10]$ by $[-20, 500]$
47. $[-10, 10]$ by $[-500, 100]$
49. Shift the graph of $y = x^4$ three units to the left,
  stretch vertically by a factor of 2, and then shift
  downward 7 units.

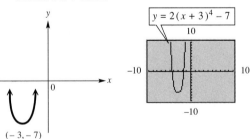

51. Shift the graph of $y = x^3$ one unit to the right, stretch
  vertically by a factor of 3, reflect across the x-axis, and
  shift upward 12 units.

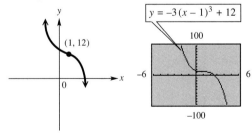

53. false   **55.** true   **57.** true   **59.** false   **61.** Possible: 0 or 2 positive real zeros; 1 negative real
zero; Actual: 0 positive; 1 negative   **63.** Possible: 1 positive real zero; 1 negative real zero;
Actual: 1 positive; 1 negative   **65.** Possible: 0 or 2 positive real zeros; 1 or 3 negative real zeros;
Actual: 0 positive; 1 negative

67. (a) Show that $P(-x) = P(x)$.
   (b) The single negative x-intercept is -3.

(c) Show that $P(-3) = 0$.
   (d) The single positive x-intercept is 3.

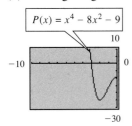

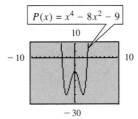

69. (a) $(-\infty, -.26); (1.93, \infty)$ **(b)** $(-.26, 1.93)$

Section 3.6 (page 272)

1. $P(1) = -5$ and $P(2) = 2$ differ in sign, so there must be a zero of P between 1 and 2. It is approximately 1.79.   **3.** $P(2) = 2$ and $P(2.5) = -.25$ differ in sign, so there must be a zero of P between 2 and 2.5. It is approximately 2.39.   **5.** $P(2) = 16$ and $P(1.5) = -.375$ differ in sign, so there must be a zero of P between 2 and 1.5. It is approximately 1.52.   **7.** There is a zero between 2 and 2.5.   **9.** $Q(x) = x^2 - 3x - 2; R = 0$
11. $Q(x) = 3x^2 + 4x; R = 3$ **13.** $Q(x) = x^3 - x^2 - 6x; R = 0$ **15.** 2   **17.** -25 **19.** $-6 - i$
21. yes   **23.** no   **25.** no   **27.** $x + 3, x - 1,$ and $x - 4$ **29.** $-3, 1,$ and 4
31. $(-3, 1) \cup (4, \infty)$ **33.** $P(x) = -\frac{1}{6}x^3 + \frac{13}{6}x + 2$ **35.** $P(x) = -\frac{1}{2}x^3 - \frac{1}{2}x^2 + x$

Answers may vary in Exercises 37–41. We give only one example.

37. $P(x) = x^2 - x - 20$ **39.** $P(x) = x^4 + x^3 - 5x^2 + x - 6$
41. $P(x) = x^4 + 2x^3 - 10x^2 - 6x + 45$ **43.** $-1 + i, -1 - i$ **45.** $-\frac{1}{2} + \frac{\sqrt{5}}{2}i, -\frac{1}{2} - \frac{\sqrt{5}}{2}i$
47. $P(x) = (x - 2)(2x - 5)(x + 3)$ **49.** $P(x) = (x + 4)(3x - 1)(2x + 1)$
51. Use synthetic division
twice, with $k = -2$.

$$
\begin{array}{r}
-2\overline{\smash{\big)}1 \quad 2 \quad -7 \quad -20 \quad -12} \\
\underline{\quad -2 \quad 0 \quad 14 \quad 12} \\
-2\overline{\smash{\big)}1 \quad 0 \quad -7 \quad -6 \quad 0} \\
\underline{\quad -2 \quad 4 \quad 6} \\
\end{array}
$$

Zeros are $-2, 3, -1$. $\quad \overline{1 \quad -2 \quad -3} \quad 0$
$x^2 - 2x - 3 = (x - 3)(x + 1)$
$P(x) = (x + 2)^2(x - 3)(x + 1)$
53. 1, 3, 5   **55.** The polynomial must also have a zero of $1 - i$, which would make it impossible for it to be of degree 3.
57.

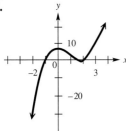

$P(x) = 2x^3 - 5x^2 - x + 6$

59.

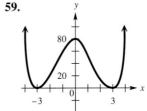

$P(x) = x^4 - 18x^2 + 81$

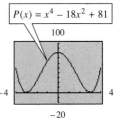

61.

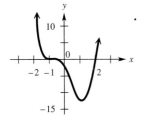

$P(x) = 2x^4 + x^3 - 6x^2 - 7x - 2$

63. ±1, ±2, ±5, ±10; The actual rational zeros are -2, -1, and 5.   **65.** ±1, ±2, ±3, ±5, ±6, ±10, ±15, ±30; The actual rational zeros are -5, -3, and 2.   **67.** ±1, ±2, ±3, ±4, ±6, ±12, $\pm\frac{1}{2}$, $\pm\frac{1}{3}$, $\pm\frac{1}{6}$, $\pm\frac{2}{3}$, $\pm\frac{3}{2}$, $\pm\frac{4}{3}$; The actual rational zeros are -4, $-\frac{1}{3}$, and $\frac{3}{2}$.   **69.** The possible rational zeros are of the form $\frac{p}{q}$, where $p \in \{\pm1, \pm2, \pm3, \pm6\}$ and $q \in \{\pm1, \pm2, \pm3, \pm4, \pm6, \pm12\}$. The actual rational zeros are $-\frac{3}{2}$, $-\frac{2}{3}$, and $\frac{1}{2}$.
71. $P(k) = (k - k) \cdot Q(k) + R$
$\qquad\ = 0 \cdot Q(k) + R$
$\qquad\ = 0 + R$
$\qquad\ = R$
Therefore, when $P(x)$ is divided by $x - k$, the remainder R is equal to $P(k)$.

Further Explorations

1. 2.34, .47   **3.** -1, 0, 5

Section 3.7 (page 287)

1. $\{-2, -1.5, 1.5, 2\}$ **3.** $\{-4, 4, -i, i\}$ **5.** $\{-8, 1, 8\}$ **7.** $\{-1.5, 0, 1\}$
9. $\{-\sqrt{7}, -1, \sqrt{7}\}$; Use $\pm\sqrt{7} \approx \pm2.65$ to support these two solutions.
11. $\{\frac{-1 - \sqrt{73}}{6}, 0, \frac{-1 + \sqrt{73}}{6}\}$; Use $\frac{-1 - \sqrt{73}}{6} \approx -1.59$ and $\frac{-1 + \sqrt{73}}{6} \approx 1.26$ to support these two solutions.
13. $\{0, -\frac{1}{2} - i\frac{\sqrt{3}}{2}, -\frac{1}{2} + i\frac{\sqrt{3}}{2}\}$ **15.** $\{-4i, -i, i, 4i\}$ **17.** $\{2, -3, -1 \pm i\sqrt{3}, \frac{3}{2} \pm \frac{3}{2}i\sqrt{3}\}$
19. $\left\{ \pm\sqrt{\dfrac{6 + \sqrt{33}}{3}}, \pm\sqrt{\dfrac{6 - \sqrt{33}}{3}} \right\}$; approximations are $\pm.29$ and ±1.98

21. It is not a complete graph, because end behavior as $x \to \infty$ is not shown, and not all x-intercepts are shown.
23. $\{-5, -\sqrt{3}, \sqrt{3}, 5\}$ **25. (a)** $\{-2, 1, 4\}$ **(b)** $(-\infty, -2) \cup (1, 4)$ **(c)** $(-2, 1) \cup (4, \infty)$
27. (a) $\{-2.5, 1, 3 \text{ (mult. 2)}\}$ **(b)** $(-2.5, 1)$ **(c)** $(-\infty, -2.5) \cup (1, 3) \cup (3, \infty)$
29. (a) $\{-3 \text{ (mult. 2)}, 0, 2\}$ **(b)** $\{-3\} \cup [0, 2]$ **(c)** $(-\infty, 0] \cup [2, \infty)$
31. The given root is approximately 2.8473221.   **33.** The given root is approximately .22102254.

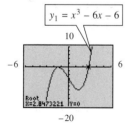

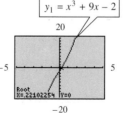

35. (a) 2 **(b)**

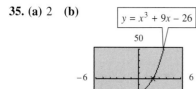

(c) $-1 \pm 2i\sqrt{3}$

37. $\{-.88, 2.12, 4.86\}$ **39.** $\{1.52\}$ **41.** $\{-.40, 2.02\}$ **43.** $\{-i, i\}$ **45.** $\{-1, \frac{1}{2} - \frac{\sqrt{3}}{2}i, \frac{1}{2} + \frac{\sqrt{3}}{2}i\}$
47. $\{3, -\frac{3}{2} - \frac{3\sqrt{3}}{2}i, -\frac{3}{2} + \frac{3\sqrt{3}}{2}i\}$ **49.** $\{-2, 2, -2i, 2i\}$ **51.** $\{-1, 1, -\frac{1}{2} \pm \frac{\sqrt{3}}{2}i, \frac{1}{2} \pm \frac{\sqrt{3}}{2}i\}$
53. (a) $0 < x < 6$ **(b)** $V(x) = x(12 - 2x)(18 - 2x)$ or $V(x) = 4x^3 - 60x^2 + 216x$ **(c)** $x \approx 2.35$; maximum volume ≈ 228.16 cubic inches **(d)** between .42 and 5

55. approximately 2.61 inches   **57. (a)** $x - 1$ **(b)** $\sqrt{x^2 - (x - 1)^2}$ **(c)** $2x^3 - 5x^2 + 4x - 28,225 = 0$
(d) hypotenuse: 25 inches; legs: 24 inches and 7 inches   **59.** approximately 2044 thousand (2,044,000)
61. after about 4.8 hours

Further Explorations

1. (a) The maximum volume is approximately 66.15 cubic inches, when $x \approx 1.59$.   **(b)** The volume is greater than 40 cubic inches when x is between .54 inches and 2.92 inches.

Chapter 3 Review Exercises (page 294)

1. $18 - 4i$ **2.** $16 + 2i$ **3.** $14 - 52i$ **4.** $288 - 34i$ **5.** $\frac{1}{10} + \frac{3}{10}i$ **6.** $2 + 5i$
7. $(-\infty, \infty)$ **8.** $(\frac{3}{2}, -\frac{25}{2})$ **9.** ↖ ↗   **10.** $-1, 4$ **11.** -8 **12.** $[-\frac{25}{2}, \infty)$
13. $(\frac{3}{2}, \infty); (-\infty, \frac{3}{2})$ **14. (a)** $\{-1, 4\}$ **(b)** $(-\infty, -1) \cup (4, \infty)$ **(c)** $[-1, 4]$ **15.** The graph intersects the x-axis at -1 and 4, supporting the answer in (a). It lies above the x-axis when $x < -1$ or $x > 4$, supporting the answer in (b). It lies on or below the x-axis when x is between -1 and 4 (inclusive), supporting the answer in (c).
16. $x = \frac{3}{2}$ **17.** The graph is concave up for all values in the domain.   **18.** Because the discriminant 67.3033 is positive, there are two x-intercepts of the graph.
19. $\{-.52, 2.59\}$ **20. (a)** the open interval $(-.52, 2.59)$ **(b)** $(-\infty, -.52) \cup (2.59, \infty)$ **21.** $(1.04, 6.37)$
22. Show that $\frac{-b}{2a} = \frac{-5.47}{2(-2.64)} \approx 1.04$, and $P(\frac{-5.47}{2(-2.64)}) \approx 6.37$.   **23. (a)** $V(x) = 4(3x - 8)(x - 8)$ **(b)** 20 inches by 60 inches   **(c)** One way is to show that the graphs of $y_1 = 2496$ and $y_2 = V(x)$ intersect at the point where $x = 20$.   **24.** It has three real solutions. The real root that is an integer is 3.
25. $(x - 3)(x^2 + x - 1)$ **26.** $\frac{-1 + \sqrt{5}}{2}, \frac{-1 - \sqrt{5}}{2}$ **27.** Show that the x-intercepts are approximations of
$\frac{-1 + \sqrt{5}}{2}$ and $\frac{-1 - \sqrt{5}}{2}$ **28. (a)** $(\frac{-1 - \sqrt{5}}{2}, \frac{-1 + \sqrt{5}}{2}) \cup (3, \infty)$ **(b)** $(-\infty, \frac{-1 - \sqrt{5}}{2}] \cup [\frac{-1 + \sqrt{5}}{2}, 3]$
29. $\{0, 1 \pm 2i\}$; The only x-intercept is 0; the solution set of $P(x) > 0$ is $(0, \infty)$ and the solution set of $P(x) < 0$ is $(-\infty, 0)$.   **30.** $(x + 2)(x - 1)(x - 3)^2$ **31.** even   **32.** odd   **33.** positive   **34.** none
35. $(-\infty, a) \cup (b, c)$ **36.** the open interval (d, h) **37.** $\{d, h\}$ **38.** $r - pi$ **39.** Its total number of zeros is three, and it has three real zeros (represented by the x-intercepts). Therefore, there can be no imaginary zeros.   **40.** 600 feet   **41.** 25 seconds   **42.** 10,600 feet   **43.** 6.3 seconds and 43.7 seconds
44. 50.7 seconds   **45.** false   **46.** true   **47.** true   **48.** true   **49.** true   **50.** false
51. two   **52.** $(2, 0)$ **53.** $Q(x) = -2x^4 + 5x^3 + 4x^2 - 12x$ **54.** $(-\infty, \infty)$ **55.** $(-.97, -54.15)$
56. $\{-\sqrt{7}, -\frac{2}{3}, \sqrt{7}\}$; Show that the x-intercepts are approximations for $-\sqrt{7}, -\frac{2}{3}$, and $\sqrt{7}$.
57. (a) $[-\sqrt{7}, -\frac{2}{3}] \cup [\sqrt{7}, \infty)$ **(b)** $(-\infty, -\sqrt{7}) \cup (-\frac{2}{3}, \sqrt{7})$ **58. (a)** $\{-1, 2, 3\}$ (*Note*: -1 is of multiplicity two.)   **(b)** the open interval $(2, 3)$ **(c)** $(-\infty, -1) \cup (-1, 2) \cup (3, \infty)$

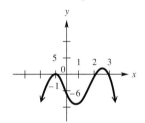

$P(x) = -x^4 + 3x^3 + 3x^2 + 17x - 6$

CHAPTER 4

Section 4.1 (page 311)

1. To obtain the graph of f, stretch the graph of $y = \frac{1}{x}$ by a factor of 2.

3. To obtain the graph of f, shift the graph of $y = \frac{1}{x}$ 2 units to the left.

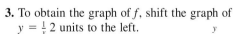
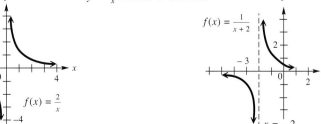

5. To obtain the graph of f, shift the graph of $y = \frac{1}{x}$ 1 unit up.

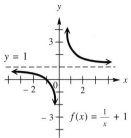

$f(x) = \frac{1}{x} + 1$

7. To obtain the graph of f reflect the graph of $y = \frac{1}{x^2}$ across the x-axis and stretch vertically by a factor of 2.

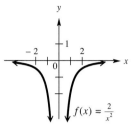

$f(x) = \frac{2}{x^2}$

11. (C)   **13.** (B)

9. To obtain the graph of f, shift the graph of $y = \frac{1}{x^2}$ 3 units to the right.

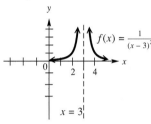

$f(x) = \frac{1}{(x-3)^2}$

$x = 3$

In Exercises 15–27, V.A. represents vertical asymptote, H.A. represents horizontal asymptote, and O.A. represents oblique asymptote.

15. V.A.: $x = 5$
   H.A.: $y = 0$

17. V.A.: $x = \frac{7}{3}$
   H.A.: $y = 0$

19. V.A.: $x = -2$
   H.A.: $y = -1$

21. V.A.: $x = -\frac{9}{2}$
   H.A.: $y = \frac{3}{2}$

23. V.A.: $x = 3, x = 1$
   H.A.: $y = 0$

25. V.A.: $x = -3$
   O.A.: $y = x - 3$

27. V.A.: $x = \frac{5}{2}, x = -2$
   H.A.: $y = \frac{1}{2}$

29. (a)

31. Set the denominator, $x^2 + x + 4$, equal to 0 and solve. Any real solutions give the x-values of the vertical asymptotes.   **33.** The complex solutions are $x = -\frac{1}{2} \pm \frac{i\sqrt{15}}{2}$. There are no real solutions. The function f has no vertical asymptotes.

35.

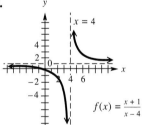

$f(x) = \frac{x+1}{x-4}$

37.

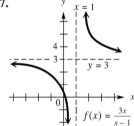

$f(x) = \frac{3x}{x-1}$

39.

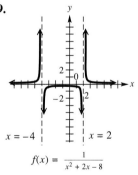

$x = -4$ $x = 2$

$f(x) = \frac{1}{x^2 + 2x - 8}$

41.

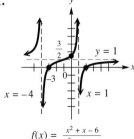

$x = -4$ $x = 1$

$f(x) = \frac{x^2 + x - 6}{x^2 + 3x - 4}$

43.

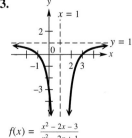

$f(x) = \frac{x^2 - 2x - 3}{x^2 - 2x + 1}$

45.

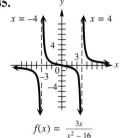

$x = -4$ $x = 4$

$f(x) = \frac{3x}{x^2 - 16}$

47.

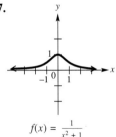

$$f(x) = \frac{1}{x^2 + 1}$$

49.

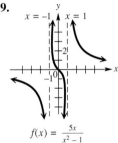

$$f(x) = \frac{5x}{x^2 - 1}$$

51. (a)

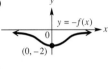

(b)

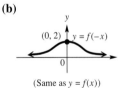

(Same as $y = f(x)$)

53. (a)

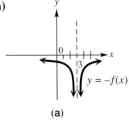

(a)

(b)

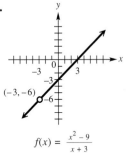

(b)

55. O.A. : $y = 2x + 8$

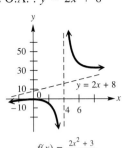

$$f(x) = \frac{2x^2 + 3}{x - 4}$$

57. O.A. : $y = x - 3$

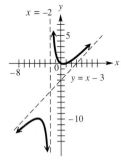

$$f(x) = \frac{x^2 - x}{x + 2}$$

59. O.A.: $y = x - 5$; The graph of f intersects it at $(\frac{1}{2}, -\frac{9}{2})$.

61.

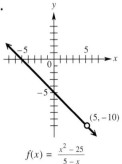

$$f(x) = \frac{x^2 - 9}{x + 3}$$

63.

$$f(x) = \frac{x^2 - 25}{5 - x}$$

65. $y = 1$ **67. (a)** $(x - 1)(x - 2)(x + 2)(x - 5)$ **(b)** $f(x) = \dfrac{(x + 4)(x + 1)(x - 3)(x - 5)}{(x - 1)(x - 2)(x + 2)(x - 5)}$

69. $-4, -1, 3$ **71.** $x = 1, x = 2, x = -2$ **73.**

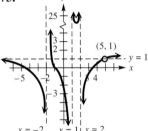

$$f(x) = \frac{x^4 - 3x^3 - 21x^2 + 43x + 60}{x^4 - 6x^3 + x^2 + 24x - 20}$$

Further Explorations

1. (a) i. $x = 1$ **ii.** The table shows ERROR for $x = 0$, because division by zero is undefined. **iii.** As $x \to \infty$, $\frac{1}{x} \to 0$ and $\frac{1}{x^2} \to 0$. **(b) i.** Both approach ∞, but Y_2 values are always larger than corresponding Y_1 values. **iii.** The values of Y_1 approach -1, while the values of Y_2 approach 1.   **3.** The horizontal asymptote is $y = -.4$.

Section 4.2 (page 324)

1. (a) $\emptyset$ **(b)** $(-\infty, -2)$ **(c)** $(-2, \infty)$ **3. (a)** $\{-1\}$ **(b)** $(-1, 0)$ **(c)** $(-\infty, -1) \cup (0, \infty)$
5. (a) $\{0\}$ **(b)** $(-2, 0) \cup (2, \infty)$ **(c)** $(-\infty, -2) \cup (0, 2)$ **7. (a)** $\{3\}$ **(b)** $(-5, 3]$
(c) $(-\infty, -5) \cup [3, \infty)$ **9. (a)** $\emptyset$ **(b)** $(-\infty, -2)$ **(c)** $(-2, \infty)$ **11. (a)** $\{\frac{9}{5}\}$ **(b)** $(-\infty, 1) \cup (\frac{9}{5}, \infty)$
(c) $(1, \frac{9}{5})$ **13. (a)** $\{-2\}$ **(b)** $(-\infty, -2] \cup (1, 2)$ **(c)** $[-2, 1) \cup (2, \infty)$ **15. (a)** $\emptyset$ **(b)** $\emptyset$
(c) $(-\infty, 2) \cup (2, \infty)$ **17. (a)** $\emptyset$ **(b)** $(-\infty, -1)$ **(c)** $(-1, \infty)$
19. If an appropriate window is not used, there may be parts of the graph that are not visible. This is particularly true for rational functions. For example, in this window, the U-shaped portion of the graph that lies in the interval $(1, 2)$ is not visible at all. We must know how to solve analytically because incorrect conclusions may be drawn from windows that do not show complete graphs.   **21.** $\{4, 9\}$ The graph of $y = 1 - \frac{1}{x} + \frac{36}{x^2}$ has these two solutions as x-intercepts.   **23.** $\left\{\dfrac{-3 - \sqrt{29}}{2}, \dfrac{-3 + \sqrt{29}}{2}\right\}$ The graph of $y = 1 + \frac{3}{x} - \frac{5}{x^2}$ has these two solutions as x-intercepts, as can be verified by the approximations -4.19 and 1.19.

In Exercises 25–33, real solutions are verified in the same manner as in Exercises 21 and 23.

25. $\left\{\dfrac{3 - \sqrt{3}}{3}, \dfrac{3 + \sqrt{3}}{3}\right\}$ **27.** $\{\pm\frac{1}{2}, \pm i\}$ **29.** $\emptyset$ **31.** $\{-10\}$ **33.** $\{\frac{27}{56}\}$ **35.** $(3, \infty)$
37. $(-1, 1]$ **39.** $\emptyset$ **41. (a)** $\{-3.54\}$ **(b)** $(-\infty, -3.54) \cup (1.20, \infty)$ **(c)** $(-3.54, 1.20)$

43. $-1.30, 1, 2.30$ **45.** The equation is $x^2 - x - 3 = 0$, and its solutions are $\dfrac{1 \pm \sqrt{13}}{2}$.

47. His solution is not correct. If $x + 2 < 0$, it is necessary to reverse the direction of the inequality symbol. He should use a sign graph or the method explained in the exposition preceding Exercise 35 in this section.
49. $\frac{220}{7}$ **51.** $\frac{32}{15}$ **53.** $\frac{18}{125}$ **55.** increases, decreases   **57.** .0444 ohm   **59.** 4.3
61. 26.67 days   **63.** 92, undernourished   **65.** .89 metric tons   **67.** 118 footcandles
69. 99.4 points   **71. (a)** 56.6 mph   **(b)** $45.25   **73.** $8.50   **75.** 5 times

Further Explorations

For the radii listed, the heights are 79.577, 19.894, 8.8419, 4.9736, 3.1831, and 2.2105 centimeters.

Section 4.3 (page 342)

1. 13   **3.** -2 **5.** 729   **7.** $\frac{1}{25}$ **9.** 100   **11.** $(-8)^{2/3} = 4$. By using $(-8)^{2/3} = [(-8)^{1/3}]^2$, we are using the rule $(a^m)^n = a^{mn}$.

The number of displayed digits may vary in Exercises 13–27.

13. 1.44224957 (rational approximation)   **15.** 2.65 (exact value)   **17.** -2.571281591 (rational approximation)   **19.** 1.464591888 (rational approximation)   **21.** 1.174618943 (rational approximation)   **23.** 1.267463962 (rational approximation)   **25.** .0322515344 (rational approximation)   **27.** 1.181352075 (rational approximation)   **29. (a)** .125 or $\frac{1}{8}$ **(b)** $(\sqrt[4]{16})^{-3} = \sqrt[4]{16^{-3}}$ (There are others.) Each is equal to .125.   **(c)** Show that $.125 = \frac{1}{8}$.

31. Because $a^{1/2} = \sqrt{a}$, $\sqrt{\sqrt{a}} = (a^{1/2})^{1/2} = a^{1/4} = \sqrt[4]{a}$, and $\sqrt{\sqrt{\sqrt{a}}} = [(a^{1/2})^{1/2}]^{1/2} = a^{1/8} = \sqrt[8]{a}$.
Therefore, $\sqrt[16]{65,536} = \sqrt{\sqrt{\sqrt{\sqrt{65,536}}}} = 65,536^{1/16} = 2$.   **33.** $2.3^{1/3} = \sqrt[3]{2.3} \approx 1.3200061$
35. $4^{-1/2} = \sqrt{4^{-1}} = .5$ **37.** $7^{3/5} = \sqrt[5]{7^3} \approx 3.2140958$ **39.** 5.35 feet   **41. (a)** 173 miles
(b) 211 miles   **43. (a)** about 133 species   **(b)** about 327 species
45. $[-\frac{5}{4}, \infty)$ **47.** $(-\infty, 6]$ **49.** $(-\infty, \infty)$ **51.** $[-7, 7]$ **53.** $[-1, 0] \cup [1, \infty)$
55. $(-\infty, -1] \cup [7, \infty)$

57. When $f(x) \geq 0$, $g(x)$ is a real number; or, the solution set of $f(x) \geq 0$ is the domain of g.

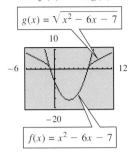

59. Because $f(-x) = f(x)$, the graph is symmetric with respect to the y-axis.
In Exercises 61–69, we do not give the graphs, only the answers to parts (a)–(d).
61. (a) $[0, \infty)$ **(b)** $\left(-\frac{5}{4}, \infty\right)$ **(c)** no interval   **(d)** $\{-1.25\}$ **63. (a)** $(-\infty, 0)$ **(b)** $(-\infty, 6)$
(c) no interval   **(d)** $\{6\}$ **65. (a)** $(-\infty, \infty)$ **(b)** $(-\infty, \infty)$ **(c)** no interval   **(d)** $\{3\}$
67. (a) $[0, 7]$ **(b)** $(-7, 0)$ **(c)** $(0, 7)$ **(d)** $\{-7, 7\}$ **69. (a)** $[0, \infty)$ **(b)** $(-1, -.58) \cup (1, \infty)$ **(c)** $(-58, 0)$
(d) $\{-1, 0, 1\}$
71. There are no x-intercepts for $f(x) = y_1 - y_2$.   **73.** The graph of $y = f(-x)$ is the reflection of the graph
of $y = f(x)$ across the y-axis.

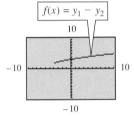

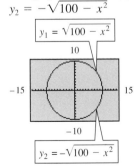

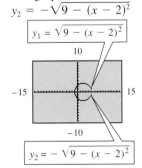

75. Shift the graph of $y = \sqrt{x}$ 3 units to the left, and stretch vertically by a factor of 3.   **77.** Shift the graph of
$y = \sqrt{x}$ 4 units to the left, stretch vertically by a factor of $\sqrt{7}$, and shift 4 units up.   **79.** Shift the graph of
$y = \sqrt[3]{x}$ 2 units to the left, stretch vertically by a factor of 3, and shift 5 units down.
81. The graph is a circle. $y_1 = \sqrt{100 - x^2}$;   **83.** The graph is a (shifted) circle. $y_1 = \sqrt{9 - (x - 2)^2}$;
$y_2 = -\sqrt{100 - x^2}$ $y_2 = -\sqrt{9 - (x - 2)^2}$

85. The graph is a horizontal parabola.
$$y_1 = -3 + \sqrt{x}; \; y_2 = -3 - \sqrt{x}$$

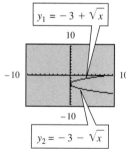

87. The graph is a horizontal parabola.
$$y_1 = -2 + \sqrt{.5x + 3.5}; \; y_2 = -2 - \sqrt{.5x + 3.5}$$

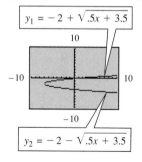

89. $y_1 = \sqrt{\dfrac{36 - 4x^2}{9}}; \; y_2 = -\sqrt{\dfrac{36 - 4x^2}{9}}$

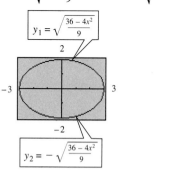

Section 4.4 (page 355)

1. (a) $\{-1\}$ **(b)** $\left[-\frac{7}{3}, -1\right)$ **(c)** $(-1, \infty)$ **3. (a)** $\{4\}$ **(b)** $(4, \infty)$ **(c)** $\left(-\frac{4}{3}, 4\right)$ **5. (a)** $\left\{\frac{5}{4}\right\}$ **(b)** $\left[1, \frac{5}{4}\right]$
(c) $\left[\frac{5}{4}, \infty\right)$ **7. (a)** $\{-2\}$ **(b)** $(-2, \infty)$ **(c)** $(-\infty, -2)$ **9. (a)** $\{4, 20\}$ **(b)** $[2, 4) \cup (20, \infty)$ **(c)** $(4, 20)$
11. (a) $\left\{\frac{1}{4}, 1\right\}$ **(b)** $\left(-\infty, \frac{1}{4}\right) \cup (1, \infty)$ **(c)** $\left(\frac{1}{4}, 1\right)$ **13. (a)** $\{31\}$ **(b)** $(31, \infty)$ **(c)** $[15, 31)$ **15. (a)** $\{-3, 1\}$
(b) $(-\infty, -3) \cup (1, \infty)$ **(c)** $[-3, -2] \cup [0, 1)$ **17. (a)** $\left\{\frac{1}{4}, 1\right\}$ **(b)** $\left(-\infty, \frac{1}{4}\right) \cup (1, \infty)$ **(c)** $\left(\frac{1}{4}, 1\right)$

19. (a) $\emptyset$ **(b)** $\emptyset$ **(c)** $\left[-\frac{2}{3}, 1\right]$ **21.** Solution set: $\left\{\dfrac{7 - \sqrt{13}}{2}\right\}$

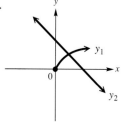

One point of intersection

23. Solution set: $\emptyset$

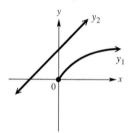

No points of intersection

25. Solution set: $\{0, 1\}$

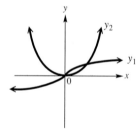

Two points of intersection

27. The graphs of $y_1 = \sqrt{x}$ and $y_2 = -x - 5$ have no points of intersection. Thus, the equation $y_1 = y_2$ (or $\sqrt{x} = -x - 5$) has no solution.

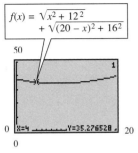

29. $(4x - 4)^{1/3} = (x + 1)^{1/2}$
31. $(4x - 4)^2 = (x + 1)^3$
33. The equation has three real roots.

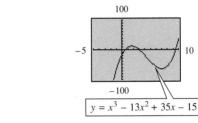

35. $P(x) = (x - 3)(x^2 - 10x + 5)$
37. $3, 5 + 2\sqrt{5}, 5 - 2\sqrt{5}$ **39.** $\{3, 5 + 2\sqrt{5}\}$; For the calculator solution, $5 + 2\sqrt{5} \approx 9.47$.
41. (a) $20 - x$ **(b)** x must be between 0 and 20.   **(c)** AP $= \sqrt{x^2 + 12^2}$; BP $= \sqrt{(20 - x)^2 + 16^2}$
(d) $f(x) = \sqrt{x^2 + 12^2} + \sqrt{(20 - x)^2 + 16^2}, 0 < x < 20$ **(e)** $f(4) \approx 35.28$. This means that when the stake is 4 feet from the base of the 12-foot pole, approximately 35.28 feet of wire will be required.

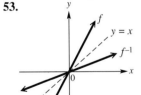

(f) When $x \approx 8.57$, $f(x)$ is a minimum (approximately 34.41 feet)   **(g)** This problem has examined how the total amount of wire used can be expressed in terms of the distance from the stake at P to the base of the 12-foot pole. We find that the amount of wire used can be minimized when the stake is approximately 8.57 feet from the 12-foot pole.
43. Since $x \approx 1.31$, the hunter must travel $8 - x \approx 8 - 1.31 = 6.69$ miles along the river.
45. At about 1:20 p.m., they will be approximately 33.28 miles apart, their minimum distance from each other.

Further Explorations

When $x < 15$, $x - 15 < 0$ and thus $f(x)$ is not real. In other words, values of x less than 15 are not in the domain of f.

Section 4.5 (page 368)

1. one-to-one   **3.** x; $(g \circ f)(x)$ **5.** (b, a) **7.** $y = x$ **9.** does not; not one-to-one
11. one-to-one   **13.** not one-to-one   **15.** one-to-one   **17.** not one-to-one   **19.** one-to-one
21. one-to-one   **23.** not one-to-one   **25.** not one-to-one   **27.** one-to-one   **29.** one-to-one
31. Because the end behavior will be either ↖↗ or ↙↘, the graph will fail the horizontal line test.
33. untying your shoelaces   **35.** leaving a room   **37.** landing in an airplane   **39.** It is not one-to-one.
For example, $f(1) = 1$ and $f(-1) = 1$, $f(3) = 81$ and $f(-3) = 81$, and so on.   **41.** yes   **43.** yes
45. no   **47.** yes   **49.** yes   **51.** yes
53.

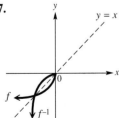

59. $f^{-1}(x) = \frac{x+4}{3}$

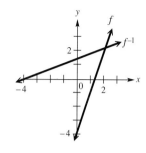

61. $f^{-1}(x) = 3x$

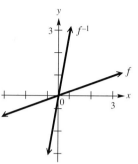

63. $f^{-1}(x) = \sqrt[3]{x-1}$

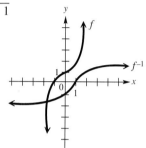

65. $f^{-1}(x) = x^2 - 6,\ x \geq 0$

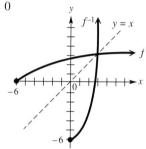

In Exercises 67–71, we give only one of several possible choices for domain restriction.

67. $[0, \infty)$　　　**69.** $[6, \infty)$　　　**71.** $[0, \infty)$

Answers to Exercises 73 and 75 are based on the restrictions chosen in Exercises 67 and 69.

73. $f^{-1}(x) = \sqrt{4-x}$　　　**75.** $f^{-1}(x) = x + 6\ (x \geq 0)$　　　**77.** The graph you see when the paper is held up to the light in this manner is the graph of the inverse of the function with which you started.

79. $f^{-1}(x) = \frac{x+2}{3}$; The message reads MIGUEL HAS ARRIVED.

Chapter 4 Review Exercises (page 373)

1.

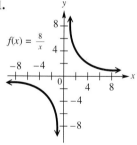

2.

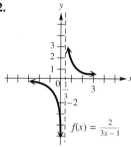

3.

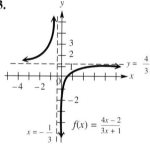

4.

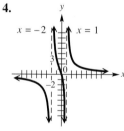

5.

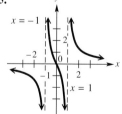

6.

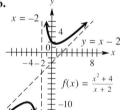

7.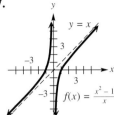

$y = x$

$f(x) = \dfrac{x^2 - 1}{x}$

8.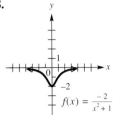

$f(x) = \dfrac{-2}{x^2 + 1}$

9.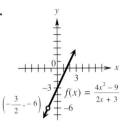

$\left(-\dfrac{3}{2}, -6\right)$ $f(x) = \dfrac{4x^2 - 9}{2x + 3}$

10. The degree of the numerator will be exactly 1 more than the degree of the denominator, and the denominator is not a factor of the numerator. **11.** $y = 2x + 3$ **12.** Shift the graph of $y = \frac{1}{x}$ three units to the left, stretch vertically by a factor of 2, and reflect across the x-axis.

13. Reflect the graph of $y = \frac{1}{x}$ across the y-axis (or equivalently here, across the x-axis), and shift 6 units upward.
14. Stretch the graph of $y = \frac{1}{x}$ 4 units vertically, and shift 3 units downward. **15.** (a) $\left\{\frac{2}{3}\right\}$ (b) the open interval $\left(-1, \frac{2}{3}\right)$ (c) $(-\infty, -1) \cup \left(\frac{2}{3}, \infty\right)$ **16.** (a) $\{-5\}$ (b) $\left(-5, -\frac{5}{2}\right) \cup (-2, \infty)$ (c) $(-\infty, -5) \cup \left(-\frac{5}{2}, -2\right)$
17. (a) $\{0\}$ (b) $(-\infty, -1) \cup [0, 2)$ (c) $(-1, 0] \cup (2, \infty)$
18. (a) $\{2, 3\}$ (b) $[2, 3]$ (c) $(-\infty, 0) \cup (0, 2] \cup [3, \infty)$
19. (a) $\{-2\}$ **20.** $(-\infty, -2) \cup (-1, \infty)$ **21.** the open interval $(-2, -1)$ **22.** $\frac{27}{2}$ **23.** 847
24. $\frac{1372}{729}$ **25.** 33,750 units **26.** 36 inches **27.** $\frac{640}{9}$ kilograms
28. 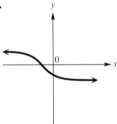 **29.** 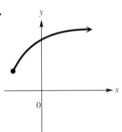 **30.**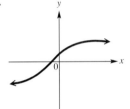

31. 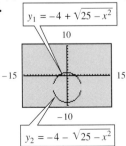 **32.**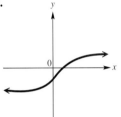

33. 2 **34.** $\frac{1}{216}$ **35.** -100 **36.** 2.429260411; (rational approximation) **37.** .5; exact
38. 2.289428485; (rational approximation) **39.** .0625; exact **40.** $[2, \infty)$ **41.** $(-\infty, 0]$
42. none **43.** $(2, \infty)$ **44.** The graph is a circle with center at $(0, -4)$ and radius 5.
45. $y_1 = -4 + \sqrt{25 - x^2}$, $y_2 = -4 - \sqrt{25 - x^2}$
46.

$y_1 = -4 + \sqrt{25 - x^2}$

$y_2 = -4 - \sqrt{25 - x^2}$

47. $\{-2\}$ **48.** $\{3, 4\}$ **49.** $\{2, 5\}$ **50.** $[-2, \infty)$ **51.** $(-\infty, 2) \cup (5, \infty)$ **52.** $(2, \infty)$

53. $(-\infty, 2) \cup (5, \infty)$ **54.** $\{3.5\}$ **55. (a)** $\{2\}$ **(b)** $[-2.5, 2)$ **(c)** $(2, \infty)$ **56. (a)** $\{0, 4\}$ **(b)** $(4, \infty)$
(c) the open interval $(0, 4)$ **57. (a)** $\{-1\}$ **(b)** $[-1, \infty)$ **(c)** $(-\infty, -1]$ **58. (a)** $\{1, 4\}$
(b) $(-\infty, 1] \cup [4, \infty)$ **(c)** $[1, 4]$ **59.** 3 inches   **60.** not one-to-one   **61.** one-to-one
62. not one-to-one   **63.** $(-\infty, \infty)$ **64.** $(-\infty, \infty)$ **65.** f is one-to-one.   **66.** $f^{-1}(x) = \frac{x^3 + 7}{2}$
67. The graphs of f and f^{-1} are reflections of each other across the line $y = x$.

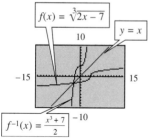

69. JUST DO IT   **70.** AWESOME DUDE

CHAPTER 5

Section 5.1 (page 387)

1. 8.952419619   **3.** .3752142272   **5.** .0868214883   **7.** 13.1207791   **9.** Show that the point
$(\sqrt{10}, 8.9524196)$ lies on the graph of $y = 2^x$.   **11.** Show that the point $(\sqrt{2}, .37521423)$ lies on the graph of
$y = (\frac{1}{2})^x$.   **13.** 2.3   **15.** .75   **17.** .31   **19.** 4   **21.** $\frac{1}{125}$ **23. (a)** $a > 1$ **(b)** domain:
$(-\infty, \infty)$; range: $(0, \infty)$
(c)   **(d)** domain: $(-\infty, \infty)$; range: $(-\infty, 0)$ **(e)**

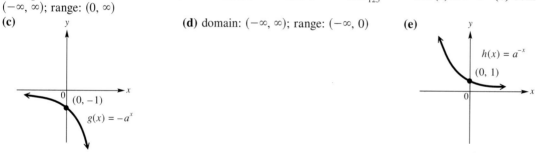

(f) domain: $(-\infty, \infty)$; range: $(0, \infty)$ **25.** Shift the graph of $y = 2^x$ five units to the left and 3 units down.
27. Reflect the graph of $y = 2^x$ across the y-axis, and shift one unit up.   **29.** Stretch the graph of $y = 2^x$ by a
factor of 3, and reflect across the x-axis.   **31. (a)** $\{\frac{1}{3}\}$ **(b)** $(\frac{1}{3}, \infty)$ **(c)** $(-\infty, \frac{1}{3})$
33. (a) $\{-2\}$ **(b)** $(-\infty, -2]$ **(c)** $[-2, \infty)$ **35. (a)** $\{0\}$ **(b)** $(-\infty, 0)$ **(c)** $(0, \infty)$ **37. (a)** $\{\frac{1}{5}\}$ **(b)** $(-\infty, \frac{1}{5})$
(c) $(\frac{1}{5}, \infty)$ **39. (a)** $\{\frac{3}{5}\}$ **(b)** $[\frac{3}{5}, \infty)$ **(c)** $(-\infty, \frac{3}{5}]$ **41. (a)** $\{-\frac{2}{3}\}$ **(b)** $(-\infty, -\frac{2}{3})$ **(c)** $(-\frac{2}{3}, \infty)$
43. 22.19795128; Show that $(3.1, 22.197951)$ lies on the graph of $y = e^x$.

(In Exercises 45 and 47, graphical support is done in a manner similar to Exercise 43.)

45. .7788007831   **47.** 4.113250379   **49. (a)** $22,510.18   **(b)** $22,529.85   **51. (a)** $33,504.35
(b) $33,504.71   **53.** Plan A is better, because it pays $102.65 more in interest.

In Exercises 55–61, we give the solution rounded to the nearest hundredth.

55. $\{1.28\}$ **57.** $\{3.58\}$ **59.** $\{2.08\}$ **61.** $\{.90\}$

Further Explorations

1. $1.16; $2.44; $7.08; $18.12; $59.34; $145.80; $318.43

Section 5.2 (page 400)

1. $\log_3 81 = 4$ **3.** $\log_{1/2} 16 = -4$ **5.** $\log_{10} .0001 = -4$ or $\log .0001 = -4$ **7.** $\log_e 1 = 0$ or
$\ln 1 = 0$ **9.** $6^2 = 36$ **11.** $\sqrt{3^8} = 81$ **13.** $10^{-3} = .001$ **15.** $10^{.5} = \sqrt{10}$
17. The expression $\log_a x$ represents the exponent to which a must be raised in order to obtain x.   **19.** 3

21. -3 **23.** 24 **25.** $-\frac{1}{6}$ **27.** $-\frac{11}{2}$ **29. (a)** 19 **(b)** 17 **(c)** $\frac{1}{3}$ **(d)** $\frac{1}{2}$
31. 1.633468456 **33.** $-.1062382379$ **35.** 4.341474094 **37.** 3.761200116
39. $-.244622583$ **41.** 9.996613531 **43. (a)** .3741982579, 1.374198258, 2.374198258, 3.374198258
(b) 2.367×10^0, 2.367×10^1, 2.367×10^2, 2.367×10^3 **(c)** In each case, the decimal digits are the same. The difference is that the whole number part corresponds to the exponent on 10 in scientific notation.
45.

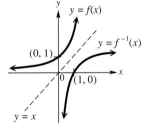

47. $f^{-1}(x) = \ln x$

49. 3.2 **51.** 8.4 **53.** 2×10^{-3} **55.** 1.6×10^{-5} **57.** 4.9 years **59.** 13.9 years
61. $\log_3 2 - \log_3 5$ **63.** $\log_2 6 + \log_2 x - \log_2 y$ **65.** $1 + \frac{1}{2}\log_5 7 - \log_5 3$ **67.** no change possible
69. $\log_k p + 2\log_k q - \log_k m$ **71.** $\frac{1}{2}(\log_m 5 + 3\log_m r - 5\log_m z)$ **73.** $\log_a \frac{xy}{m}$ **75.** $\log_m \frac{a^2}{b^6}$
77. $\log_a [(z-1)^2 (3z+2)]$ **79.** $\log_5 \frac{5^{1/3}}{m^{1/3}}$ **81.** 1.430676558 **83.** .5943161289
85. .9595390462 **87.** 1.892441722 **89.** Reflect the graph of $y = 3^x$ across the x-axis and shift 7 units upward. **91.** 1.7712437 **93.** $\frac{\log 7}{\log 3} = \frac{\ln 7}{\ln 3} \approx 1.771243749$

Section 5.3 (page 412)

1. domain: $(0, \infty)$; range: $(-\infty, \infty)$; f^{-1} increases on $(0, \infty)$; V.A.: $x = 0$ **3.** domain: $(0, \infty)$; range: $(-\infty, \infty)$; f^{-1} decreases on $(0, \infty)$; V.A.: $x = 0$ **5.** domain: $(1, \infty)$; range: $(-\infty, \infty)$; f^{-1} increases on $(1, \infty)$; V.A.: $x = 1$

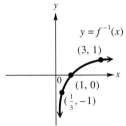

7. logarithmic **9.** $(0, \infty)$ **11.** $\left(-\frac{7}{3}, \infty\right)$ **13.** $(-\infty, \infty)$
15. $(-\infty, -3) \cup (7, \infty)$ **17.** $\left(-\infty, \frac{-1-\sqrt{5}}{2}\right) \cup \left(\frac{-1+\sqrt{5}}{2}, \infty\right)$ **19.** $(-1, 0) \cup (1, \infty)$
21. B **23.** D **25.** A **27.** C

In Exercises 29–33, other descriptions may be possible.

29. The graph of $y = \log_2 x$ is shifted 4 units to the left. **31.** The graph of $y = \log_2 x$ is stretched vertically by a factor of 3 and is shifted 1 unit up. **33.** The graph of $y = \log_2 x$ is reflected across the y-axis and is shifted 1 unit up. **35.** $(-\infty, 0) \cup \left(\frac{1}{2}, \infty\right)$ **37.** $-\frac{1}{2}$, 1 **39.** In general, to find the y-intercept for the graph of $y = f(x)$, we evaluate $f(0)$. Because 0 is not in the domain of f, $f(0)$ is not defined, and there is no y-intercept.
41. The graphs are not the same because the domain of $y = \log x^2$ is $(-\infty, 0) \cup (0, \infty)$, while the domain of $y = 2 \log x$ is $(0, \infty)$. The power rule does not apply if the argument is nonpositive.

43. (a) $\frac{3}{2}$ **(b)** $\log_9 27 = \frac{\log 27}{\log 9} = 1.5 = \frac{3}{2}$
(c)

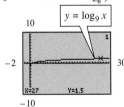

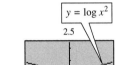

45. (a) $-\frac{3}{4}$ **(b)** $\log_{16}(\frac{1}{8}) = \dfrac{\log(\frac{1}{8})}{\log 16} = -.75 = -\frac{3}{4}$ **(c)**

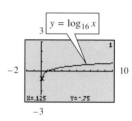

In Exercises 47–51, graph $y = \log x$ or $y = \ln x$ as appropriate. Let x equal the value given and show that y is equal to the approximation given in these answers. Check using the log or ln key.

47. .84509804   **49.** 1.9459101   **51.** 1.1673173   **53.** $\log_a x - \log_a y$

55. For any x-value, if we evaluate y_1 and y_2 and then subtract y_2 from y_1, we get approximately .6, which is an approximation for log 4.

57. The y-value is an approximation for $\log 4$, found in Exercise 56 in the form $\log 8 - \log 2$.

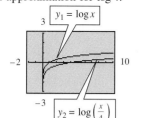

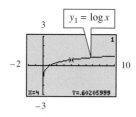

59. $f^{-1}(x) = \log_4(x + 3)$ The domain of f is equal to the range of f^{-1}, and vice-versa. The x- and y-intercepts are reversed. The horizontal asymptote for f, $y = -3$, is transformed into the vertical asymptote for f^{-1}, $x = -3$.

61. $f^{-1}(x) = \log(4 - x)$ The domain of f is equal to the range of f^{-1}, and vice-versa. The x- and y-intercepts are reversed. The horizontal asymptote for f, $y = 4$, is transformed into the vertical asymptote for f^{-1}, $x = 4$.

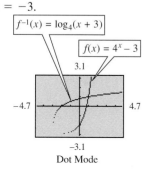

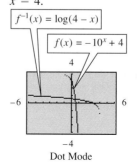

63. $\log 5.0118723 \approx .7$

Further Explorations

1. When $x \geq 4$, $\log(4 - x)$ is undefined, because the domain of the function is $(-\infty, 4)$.

Section 5.4 (page 424)

1. (a) $\left\{\dfrac{\log 10}{\log 3}\right\}$ **(b)** $\left(\dfrac{\log 10}{\log 3}, \infty\right)$ **(c)** $\left(-\infty, \dfrac{\log 10}{\log 3}\right)$ **3. (a)** $\left\{\dfrac{\log(\frac{1}{8})}{\log(\frac{2}{5})}\right\}$ **(b)** $\left(-\infty, \dfrac{\log(\frac{1}{8})}{\log(\frac{2}{5})}\right]$ **(c)** $\left[\dfrac{\log(\frac{1}{8})}{\log(\frac{2}{5})}, \infty\right)$

5. (a) $\{5\}$ **(b)** $(3, 5)$ **(c)** $(5, \infty)$ **7. (a)** $\{2.5\}$ **(b)** $(2, 2.5]$ **(c)** $[2.5, \infty)$ **9. (a)** $\{3\}$ **(b)** $[3, \infty)$

(c) $(2, 3]$ **11. (a)** $\{4\}$ **(b)** $(4, \infty)$ **(c)** $(2, 4)$ **13.** $x = \dfrac{\log 81}{\log(\frac{1}{3})} = -4$

15. (a) 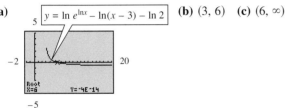 **(b)** $(3, 6)$ **(c)** $(6, \infty)$

17. (a) $\{-.535\}$ **(b)** $(-.535, \infty)$ **(c)** $(-\infty, -.535)$ **19. (a)** $\{-.123\}$ **(b)** $(-\infty, -.123)$ **(c)** $(-.123, \infty)$
21. (a) $\emptyset$ **(b)** $\emptyset$ **(c)** $(0, \infty)$ **23. (a)** $\{8\}$ **(b)** $(8, \infty)$ **(c)** $(-\infty, 8)$
25. (a) $\{1.236\}$ **(b)** $(0, 1.236]$ **(c)** $[1.236, \infty)$ **27. (a)** $\{6\}$ **(b)** $(4, 6]$ **(c)** $[6, \infty)$ **29.** $t = e^{\frac{p-r}{k}}$

The answers in Exercises 29–37 may have alternate equivalent forms.

31. $t = -\frac{1}{k} \log\left(\dfrac{T - T_0}{T_1 - T_0}\right)$ **33.** $k = \dfrac{\ln\left(\dfrac{A - T_0}{C}\right)}{-t}$ **35.** $x = \dfrac{\ln\left(\dfrac{y - A - B}{-B}\right)}{-C}$ **37.** $A = \dfrac{B}{x^C}$

39. Because $(a^m)^n = a^{mn} = a^{nm}$, $(e^x)^2 = e^{x \cdot 2} = e^{2x}$. **41.** $\{0, \ln 3\}$ **43.** $(-\infty, 0) \cup (\ln 3, \infty)$
45. $\{-.767, 2, 4\}$ **47.** $\{2.454, 5.659\}$ **49.** $\{-.443\}$ **51.** $\{17.475\}$ **53.** $\{-2, 2\}$ **55.** $\{1, 10\}$
57. $\{-3\}$ **59.** $\{\ln 2, \ln 4\}$

Section 5.5 (page 436)

Most answers given in this section are approximations.

1. (a) 440 grams **(b)** 387 grams **(c)** 264 grams **(d)** 21.66 years **(e)**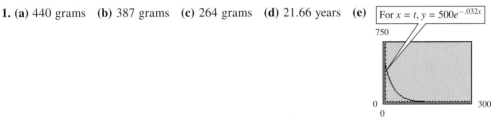

3. 1611.97 years **5.** 59,800 feet **7. (a)** 21 **(b)** 70 **(c)** 120 **9.** 89 decibels is about twice as loud as 86 decibels, for an approximate 100% increase. **11. (a)** 3.0 **(b)** 6.0 **(c)** 8.0 **13. (a)** about 200,000,000 I_0 **(b)** about 13,000,000 I_0 **(c)** The 1906 earthquake was more than 15 times as intense as the 1989 earthquake. **15.** about 9000 years ago **17.** about 16,000 years old **19.** 23 days **21.** about 1.126 billion years old **23. (a)** about 46.2 years **(b)** about 46.0 years **25.** about 1.8 years
27. about 6.14% **29.** $5583.95 **31.** 12%
33. 7% compounded quarterly; $800.31 **35. (a)** $205.52 **(b)** $1364.96 **37. (a)** $471.58
(b) $29,884.40 **39. (a)** about 2,700,000 **(b)** about 3,000,000 **(c)** about 9,500,000
41. $t = \dfrac{10^{.002F} - 3}{2}$ **(a)** about 6.4 months **(b)** about 48.5 months **43.** about 59 mg **45.** 1
47. about 2349 million, or 2.349 billion **49.** 1997 **51.** about 6.9 years **53.** about 11.6 years
55. (a) $A(20) \approx 3727$ million, which is off by about 27 million **(b)** about 5342 million, or 5.342 billion
57. (a) about 1700 years **(b)** about 3000 years **59. (a)** 11 **(b)** 12.6 **(c)** 18 **(d)**

(e) Living standards are increasing, but at a slow rate. **61.** Because $b = e^{\ln b}$, $a \cdot b^x = a \cdot (e^{\ln b})^x = a \cdot e^{(\ln b)x}$.

Further Explorations

1. 19 years, 39 days

Chapter 5 Review Exercises (page 442)

1. (c)   **2.** (a)   **3.** (d)   **4.** (b)   **5.** $0 < a < 1$ **6.** $(-\infty, \infty)$ **7.** $(0, \infty)$ **8.** 1

9.

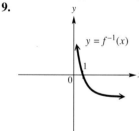

10. $f^{-1}(x) = \log_a x$ **11.**

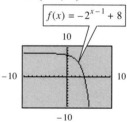

12. Shift the graph of $y = 2^x$ one unit to the right, reflect it across the x-axis, and shift 8 units upward.
13. (a) $(-\infty, \infty)$ **(b)** $(-\infty, 8)$ **14.** Yes, it has a horizontal asymptote whose equation is $y = 8$.
15. x-intercept: 4; y-intercept: 7.5. Use a calculator to find $f(0)$ to support the y-intercept, and find the x-intercept using the root-locating capability.   **16.** $(-.7666647, .58777476)$ **17.** $\{.5\}$
18. The x-coordinate is .5, supporting the solution found analytically.
19. The x-intercept is .5, supporting the solution found analytically.

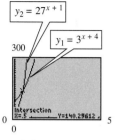

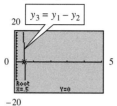

20. (a) $\{.5\}$ **(b)** $(-\infty, .5)$ **(c)** $(.5, \infty)$ **21.** The expression $\log_5 27$ is the exponent to which 5 must be raised in order to obtain 27. To find an approximation using a calculator, we use the change-of-base rule: $\log_5 27 = \frac{\log 27}{\log 5} = \frac{\ln 27}{\ln 5} \approx 2.047818583$.   **22.** 1.765668555   **23.** -2.632644079 **24.** 4.065602093   **25.** -6.061887011
26. 0   **27.** $\sqrt{6}$ **28.** 12   **29.** 13   **30.** 1.584962501   **31.** 1.464973521
32. $3 \log m + \log n - \frac{1}{2} \log y$ **33.** $x = \dfrac{\ln(3 + \sqrt{10})}{\ln 5}$ **34.** $x \approx -.485$ **35.** $x = \dfrac{1 + a}{1 - a}$
36. $c = de^{(N-a)/b}$ **37. (a)** $\left\{\dfrac{\log 18}{\log 48}\right\}$ **(b)** $\{.747\}$ **38. (a)** $\{-3 + \log 6\}$ **(b)** $\{-2.222\}$ **39.** $\{2\}$
40. -4 **41.** $y_1 = \dfrac{\log x}{\log 2} + \dfrac{\log(x + 2)}{\log 2} - 3$ **42.**

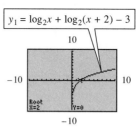

43. $(2, \infty)$ **44.** the open interval $(0, 2)$
45. (a) (B)   **(b)** (D)   **(c)** (C)   **(d)** (A)   **46. (a)** \$2322.37   **(b)** \$2323.67   **(c)** 36.6 years
47. (a) \$15 million   **(b)** \$15.6 million   **(c)** \$16.4 million   **48. (a)** .0054 grams per liter   **(b)** .00073 grams per liter   **(c)** .000013 grams per liter   **(d)** .75 miles

49. **(a)** 207 **(b)** 235 **(c)** 249 **(d)** when $t = x = 2$, $y = p(t) \approx 207$

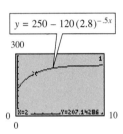

50. 6.25°C   **51.** 71.7°C   **52.** 10 seconds

CHAPTER 6

Section 6.1 (page 456)

1. C   **3.** F   **5.** G   **7.** J   **9.** B   **11.** $(x - 1)^2 + (y - 4)^2 = 9$ **13.** $x^2 + y^2 = 1$
15. $(x - \frac{2}{3})^2 + (y + \frac{4}{5})^2 = \frac{9}{49}$ **17.** $(x + 1)^2 + (y - 2)^2 = 25$ **19.** $(x + 3)^2 + (y + 2)^2 = 4$
21. $(2, -3)$ **23.** $(x - 2)^2 + (y + 3)^2 = 45$ **25.** $(x + 2)^2 + (y + 1)^2 = 41$
27. domain: $[-6, 6]$; range: $[-6, 6]$ **29.** domain: $[-6, 2]$; range: $[1, 9]$

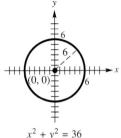

$x^2 + y^2 = 36$

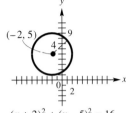

$(x + 2)^2 + (y - 5)^2 = 16$

31.

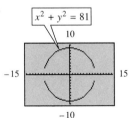

33.

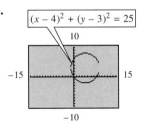

35. It is the single point $(3, 3)$.   **37.** center: $(-3, -4)$; radius: 4   **39.** center: $(6, -5)$; radius: 6
41. center: $(-4, 7)$; radius: 1   **43.** center: $(0, 1)$; radius: 7
45. The point $(-3, 4)$ lies on the circles $(x - 1)^2 + (y - 4)^2 = 16$, $(x + 6)^2 + y^2 = 25$, and $(x - 5)^2 + (y + 2)^2 = 100$.   **47.** D   **49.** C   **51.** F   **53.** E   **55.** **(a)** III **(b)** II **(c)** IV **(d)** I
57. focus: $(0, 4)$; directrix: $y = -4$; axis: y-axis   **59.** focus: $(0, -\frac{1}{8})$; directrix: $y = \frac{1}{8}$; axis: y-axis
61. focus: $(\frac{1}{64}, 0)$; directrix: $x = -\frac{1}{64}$; axis: x-axis   **63.** focus: $(-4, 0)$; directrix: $x = 4$; axis: x-axis
65. $y = -\frac{1}{8}x^2$ **67.** $x = -\frac{1}{2}y^2$ **69.** $x = \frac{1}{4}y^2$ **71.** $y = -\frac{1}{2}x^2$ **73.** $y = -x^2$

In the answers for Exercises 75–95, we provide traditional graphs.

75. vertex: $(2, 0)$; axis: $x = 2$;   **77.** vertex: $(-3, -4)$; axis: $x = -3$; domain:
    domain: $(-\infty, \infty)$; range: $[0, \infty)$ $(-\infty, \infty)$; range: $[-4, \infty)$

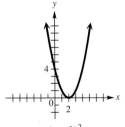

$y = (x - 2)^2$

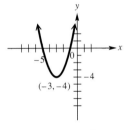

$y = (x + 3)^2 - 4$

79. vertex: $(-3, 2)$; axis: $x = -3$;
domain: $(-\infty, \infty)$; range: $(-\infty, 2]$

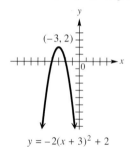

$y = -2(x + 3)^2 + 2$

81. vertex: $(-1, -3)$; axis: $x = -1$; domain: $(-\infty, \infty)$;
range: $(-\infty, -3]$

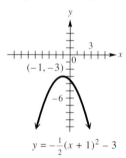

$y = -\frac{1}{2}(x + 1)^2 - 3$

83. vertex: $(1, 2)$; axis: $x = 1$; domain: $(-\infty, \infty)$;
range: $[2, \infty)$

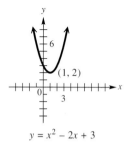

$y = x^2 - 2x + 3$

85. vertex: $(1, 3)$; axis: $x = 1$; domain: $(-\infty, \infty)$;
range: $[3, \infty)$

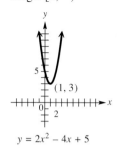

$y = 2x^2 - 4x + 5$

To obtain calculator-generated graphs in Exercises 87–95, solve for y_1 and y_2 in terms of x, and use a square window.

87. vertex: $(2, 0)$; axis: $y = 0$; domain: $[2, \infty)$;
range: $(-\infty, \infty)$

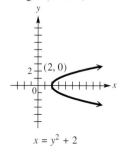

$x = y^2 + 2$

89. vertex: $(0, -1)$; axis: $y = -1$; domain: $[0, \infty)$;
range: $(-\infty, \infty)$

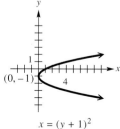

$x = (y + 1)^2$

91. vertex: $(-1, -2)$; axis: $y = -2$; domain: $[-1, \infty)$;
range: $(-\infty, \infty)$

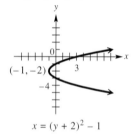

$x = (y + 2)^2 - 1$

93. vertex: $(0, -3)$; axis: $y = -3$; domain: $(-\infty, 0]$;
range: $(-\infty, \infty)$

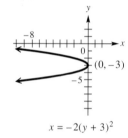

$x = -2(y + 3)^2$

95. vertex: $(-9, -1)$; axis: $y = -1$; domain: $[-9, \infty)$; range: $(-\infty, \infty)$

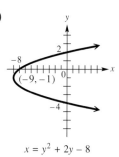

$x = y^2 + 2y - 8$

97. 60 feet   **99.** Let P be (x, y). The focus is $F(c, 0)$ and the directrix has equation $x = -c$. If $D(-c, y)$ is on the directrix, then $d(P, F) = d(P, D)$. So $\sqrt{(x - c)^2 + (y - 0)^2} = \sqrt{(x + c)^2 + (y - y)^2}$. This gives $\sqrt{(x - c)^2 + y^2} = \sqrt{(x + c)^2}$. Squaring both sides and expanding gives $x^2 - 2xc + c^2 + y^2 = x^2 + 2xc + c^2$. Subtracting x^2 and c^2 and rearranging terms gives $y^2 = 4cx$, or $x = \frac{1}{4c}y^2$.

Further Explorations

1. domain: $[-6, 6]$; points of intersection: $(-6, 0)$, $(6, 0)$
3. domain: $[1, 9]$; points of intersection: $(1, 2)$, $(9, 2)$

Section 6.2 (page 471)

1. E   **3.** H   **5.** A   **7.** I   **9.** D   **11.** A circle can be interpreted as an ellipse whose two foci actually have the same coordinates; the "coinciding foci" give the center of the circle.

In the answers for Exercises 13–23, we provide traditional graphs. To obtain calculator-generated graphs, solve for y_1 and y_2 in terms of x, and use a square window.

13. domain: $[-3, 3]$; range: $[-2, 2]$ **15.** domain: $[-\sqrt{6}, \sqrt{6}]$; range: $[-3, 3]$ **17.** domain: $[-\frac{1}{3}, \frac{1}{3}]$; range: $[-\frac{1}{4}, \frac{1}{4}]$

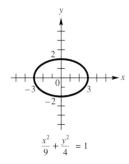

$\frac{x^2}{9} + \frac{y^2}{4} = 1$

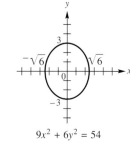

$9x^2 + 6y^2 = 54$

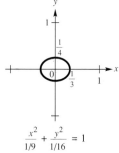

$\frac{x^2}{1/9} + \frac{y^2}{1/16} = 1$

19. domain: $[-\frac{3}{8}, \frac{3}{8}]$; range: $[-\frac{6}{5}, \frac{6}{5}]$ **21.** domain: $[-2, 4]$; range: $[-8, 2]$ **23.** domain: $[-2, 6]$; range: $[-2, 4]$

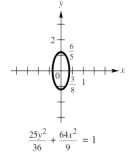

$\frac{25y^2}{36} + \frac{64x^2}{9} = 1$

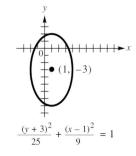

$\frac{(y + 3)^2}{25} + \frac{(x - 1)^2}{9} = 1$

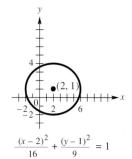

$\frac{(x - 2)^2}{16} + \frac{(y - 1)^2}{9} = 1$

25. A hyperbola centered at the origin is symmetric with respect to both axes, as well as symmetric with respect to the origin.   **27.** $16 - \frac{16(x - 2)^2}{9} \geq 0$

29.

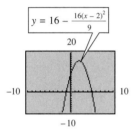

31. In Figure 18, we see that the domain is $[-1, 5]$. This corresponds to the solution set found graphically in Exercise 30.

In the answers for Exercises 33–43, we provide traditional graphs. To obtain calculator-generated graphs, solve for y_1 and y_2 in terms of x, and use a square window.

33. domain: $(-\infty, -4] \cup [4, \infty)$; range: $(-\infty, \infty)$

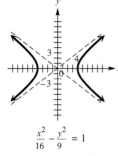

$$\frac{x^2}{16} - \frac{y^2}{9} = 1$$

35. domain: $(-\infty, \infty)$; range: $(-\infty, -6] \cup [6, \infty)$

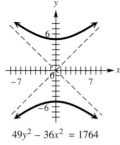

$$49y^2 - 36x^2 = 1764$$

37. domain: $\left(-\infty, -\frac{3}{2}\right] \cup \left[\frac{3}{2}, \infty\right)$; range: $(-\infty, \infty)$

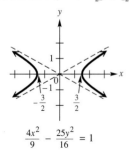

$$\frac{4x^2}{9} - \frac{25y^2}{16} = 1$$

39. domain: $(-\infty, -2] \cup [4, \infty)$; range: $(-\infty, \infty)$

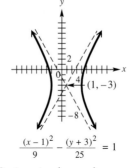

$$\frac{(x-1)^2}{9} - \frac{(y+3)^2}{25} = 1$$

41. domain: $(-\infty, -1] \cup [7, \infty)$; range: $(-\infty, \infty)$

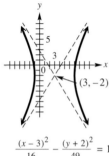

$$\frac{(x-3)^2}{16} - \frac{(y+2)^2}{49} = 1$$

43. domain: $(-\infty, \infty)$; range: $(-\infty, -6] \cup [4, \infty)$

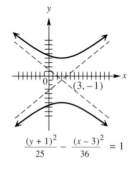

$$\frac{(y+1)^2}{25} - \frac{(x-3)^2}{36} = 1$$

45. The graph of $y = \frac{x^2}{25} - 1$ lies above or on the x-axis in $(-\infty, -5] \cup [5, \infty)$. This set is the same as the domain of the given hyperbola.

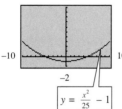

$y = \frac{x^2}{25} - 1$

47. $\frac{x^2}{16} + \frac{y^2}{12} = 1$ **49.** $\frac{x^2}{36} + \frac{y^2}{20} = 1$ **51.** $\frac{(y+2)^2}{25} + \frac{(x-3)^2}{16} = 1$ **53.** $\frac{x^2}{9} - \frac{y^2}{7} = 1$
55. $\frac{y^2}{9} - \frac{x^2}{25} = 1$ **57.** One such ellipse would have an equation of $\frac{x^2}{100} + \frac{y^2}{64} = 1$. **59.** 348.2 feet
61. It must be just under 12 feet tall.

Further Explorations

1. points of intersection: $(4, 0)$, $(-4, 0)$; domain: $[-4, 4]$
3. points of intersection: $(-8, 5)$, $(8, 5)$; domain: $(-\infty, -8] \cup [8, \infty)$
5. points of intersection: $(-6, 5)$, $(6, 5)$; domain: $[-6, 6]$
7. The domain of the ellipse is $[-2, 2]$, while the domain of the hyperbola is $(-\infty, -2] \cup [2, \infty)$.

Section 6.3 (page 483)

1. circle **3.** parabola **5.** parabola **7.** ellipse **9.** hyperbola **11.** hyperbola
13. ellipse **15.** circle **17.** line **19.** hyperbola **21.** ellipse **23.** circle **25.** parabola
27. no graph **29.** circle **31.** parabola **33.** hyperbola **35.** ellipse **37.** no graph
39. line **41.** $\frac{1}{2}$ **43.** $\sqrt{2}$ **45.** $\frac{\sqrt{21}}{7}$ **47.** $\frac{\sqrt{10}}{3}$ **49.** $\frac{\sqrt{2}}{2}$ **51.** $y = \frac{1}{32}x^2$
53. $\frac{x^2}{36} + \frac{y^2}{27} = 1$ **55.** $\frac{x^2}{36} - \frac{y^2}{108} = 1$ **57.** $y = -\frac{1}{4}x^2$ **59.** $\frac{y^2}{9} + \frac{25x^2}{81} = 1$ **61.** C, A, B, D

Section 6.4 (page 496)

1. In the first equation, $3 = 3$, and in the second equation, $1 = 1$. **3.** In both equations, $0 = 0$.
5. In the first equation, $3 = 3$, and in the second equation, $0 = 0$. **7.** single solution
9. infinitely many solutions **11.** no solution **13.** infinitely many solutions **15.** $\{(6, 15)\}$
17. $\{(2, -3)\}$ **19.** $\{(1, 2)\}$ **21.** Multiply equation (4) by 3 and equation (5) by 4.
23. $\{(-1, 4)\}$ **25.** $\emptyset$ **27.** $\{(1, 3)\}$ **29.** $\{(4, -2)\}$ **31.** $\{(\frac{y+9}{4}, y)\}$ **33.** $\{(12, 6)\}$
35. $\begin{aligned} 5t + 15u &= 16 \\ 5t + 4u &= 5 \end{aligned}$ **37.** $x = 5$, $y = 1$ **39.** $y = \frac{-4x}{5 - 5x}$ **41.** $\{(2, 2)\}$ **43.** $\{(\frac{1}{3}, \frac{1}{4})\}$
45. $\{(1.31, -6.55)\}$ **47.** $y = -3x - 5$ **49.** $y = \frac{1}{3}x + \frac{13}{3}$ **51.** $y = 4.3x - 2.1$
53. \$12,000 at 7%; \$4000 at 8% **55.** 480 kilograms of X; 320 kilograms of Y
57. $13\frac{1}{3}$ gallons of 98-octane; $26\frac{2}{3}$ gallons of 92-octane **59.** \$3 for a turkey; \$1.75 for a chicken
61. 32 days at \$88 per day **63.** \$18,000 at 4.5%; \$12,000 at 5% **65.** $x = 63$; $R = C = 346.5$
67. $x = 2100$; $R = C = 52,000$ **69.** $p = 32$, $q = 80$ **71.** $p = 9\frac{1}{3}$; $q = 28$
73. **(a)**

For $q = x$ and $p = y$,
$y = \frac{3}{2}x$

For $q = x$ and $p = y$,
$y = 81 - \frac{3}{4}x$

(b) 36 **(c)** 54

75. **(a)** Rising curve: 1983: 24%, 1993: 35%; Falling curve: 1983: 42%, 1993: 37%; $(0, 24)$, $(10, 35)$; $(0, 42)$, $(10, 37)$ **(b)** $y = 1.1x + 24$; $y = -.5x + 42$ **(c)** About one-fourth of the way through 1994, 36.4% of children will be living with each of these types of parents.

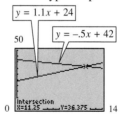

$y = 1.1x + 24$

$y = -.5x + 42$

77. Each side is equal to 10 when $x = -4$.   **79.**

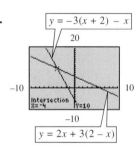

Section 6.5 (page 512)

1. In the first equation, $2 = 2$, and in the second equation, $34 = 34$.   **3.** In the first equation, $3 = 3$, and in the second equation, $25 = 25$.   **5.** In the first equation, $3 = 3$, and in the second equation, $9 = 9$.
7. In the first equation, $2 = 2$, and in the second equation, $11 = 11$.
9. $(-5, -3)$ **11.**   **13.**

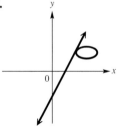

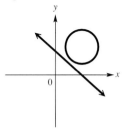

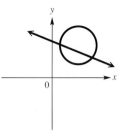

15.   **17.**   **19.**

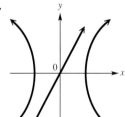

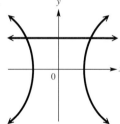

21.   **23.**   **25.**

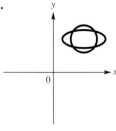

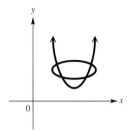

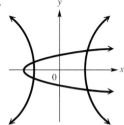

27. Substitute $x - 1$ for y in the first equation, and solve the resulting quadratic equation. When the solutions of the quadratic are found, substitute back into $y = x - 1$ to find the corresponding y-values.
29. $\{(1, 1), (-2, 4)\}$ **31.** $\{(2, 1), (\frac{1}{3}, \frac{4}{9})\}$ **33.** $\{(2, 12), (-4, 0)\}$ **35.** $\{(-\frac{3}{5}, \frac{7}{5}), (-1, 1)\}$
37. $\{(2, 2), (2, -2), (-2, 2), (-2, -2)\}$ **39.** $\{(0, 0)\}$ **41.** $\{(i, \sqrt{6}), (-i, \sqrt{6}), (i, -\sqrt{6}), (-i, -\sqrt{6})\}$
43. $\{(1, -1), (-1, 1), (1, 1), (-1, -1)\}$ **45.** $\{(x, \pm\sqrt{10 - x^2})\}$
47. $\{(\frac{i\sqrt{15}}{5}, \frac{2\sqrt{10}}{5}), (\frac{i\sqrt{15}}{5}, \frac{-2\sqrt{10}}{5}), (\frac{-i\sqrt{15}}{5}, \frac{2\sqrt{10}}{5}), (\frac{-i\sqrt{15}}{5}, \frac{-2\sqrt{10}}{5})\}$ **49.** $\{(-3, 5), (\frac{15}{4}, -4)\}$
51. $\{(3, 5), (-3, -5), (5i, -3i), (-5i, 3i)\}$ **53.** $\{(\sqrt{5}, 0), (-\sqrt{5}, 0), (\sqrt{5}, \sqrt{5}), (-\sqrt{5}, -\sqrt{5})\}$
55. $\{(3, -3), (3, 3)\}$ **57.** $\{(\frac{1 + \sqrt{13}}{2}, \frac{-1 + \sqrt{13}}{2}), (\frac{-1 - \sqrt{21}}{2}, \frac{3 + \sqrt{21}}{2})\}$ **59.** $\{(-.79, .62), (.88, .77)\}$

61. {(.06, 2.88)}   **63.** {(−1.68, −1.78), (2.12, −1.24)}   **65.** $y_1 = x^2 − 4$; parabola   **67.** 0, 1, or 2
69. {(1, −3), (−2, 0)}   **71.** The solution set of $x^2 + x − 2 = 0$ is {1, −2}.   **73.** 27 and 6 or −27 and −6
75. yes   **77.** $y = 4x − 4$ **79.** 5, −5   **81.** Find the coordinates of the two points of intersection using a system, and then use one of any of several methods discussed so far in this book to find the equation of the line joining the two points.   **83.** length: 16 inches; width: 8 inches; height: 5 inches

85.
$$\begin{array}{r} 5\overline{)10-85300} \\ \phantom{5\overline{)}}525-300 \\ \hline \phantom{5\overline{)}}15-600 \end{array}$$
← 5 is a solution.

87. Use the quadratic formula with $a = 1$, $b = 5$, and $c = −60$.   **89.** $\frac{-5 - \sqrt{265}}{2}$

Further Explorations

Stuart will overtake John after 7.5 hours at 9:00 p.m., 450 miles from Slidell.

Section 6.6 (page 525)

1.
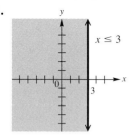
$x \le 3$

3.
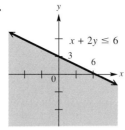
$x + 2y \le 6$

5.
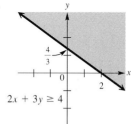
$2x + 3y \ge 4$

7.
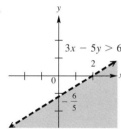
$3x − 5y > 6$

9.
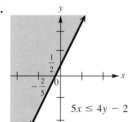
$5x \le 4y − 2$

11.
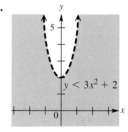
$y < 3x^2 + 2$

13.
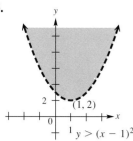
$y > (x − 1)^2 + 2$
$(1, 2)$

15.
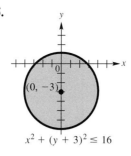
$(0, −3)$
$x^2 + (y + 3)^2 \le 16$

17.
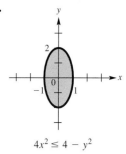
$4x^2 \le 4 − y^2$

19.
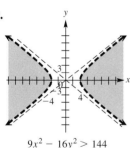
$9x^2 − 16y^2 > 144$

21. (b)   **23.** (d)   **25.** C   **27.** A

29.

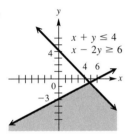

$x + y \le 4$
$x - 2y \ge 6$

31.

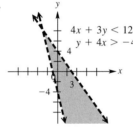

$4x + 3y < 12$
$y + 4x > -4$

33.

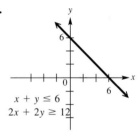

$x + y \le 6$
$2x + 2y \ge 12$

Only the points of the line are included.

35.

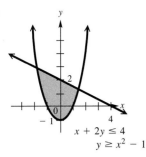

$x + 2y \le 4$
$y \ge x^2 - 1$

37.

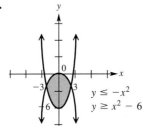

$y \le -x^2$
$y \ge x^2 - 6$

39.

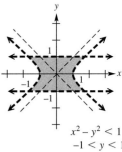

$x^2 - y^2 < 1$
$-1 < y < 1$

41.

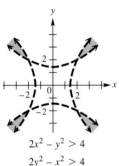

$2x^2 - y^2 > 4$
$2y^2 - x^2 > 4$

43.

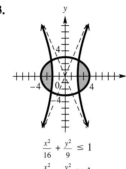

$\dfrac{x^2}{16} + \dfrac{y^2}{9} \le 1$
$\dfrac{x^2}{4} - \dfrac{y^2}{16} \ge 1$

45.

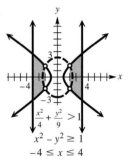

$\dfrac{x^2}{4} + \dfrac{y^2}{9} > 1$
$x^2 - y^2 \ge 1$
$-4 \le x \le 4$

47.

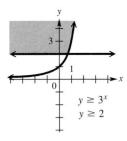

$y \ge 3^x$
$y \ge 2$

49.

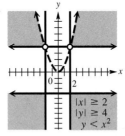

$|x| \ge 2$
$|y| \ge 4$
$y < x^2$

51.

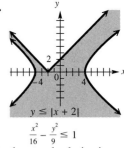

$y \le |x + 2|$
$\dfrac{x^2}{16} - \dfrac{y^2}{9} \le 1$

53. (d) **55.** A **57.** B

All graphs in Exercises 59–65 are given in the standard viewing window.

59.

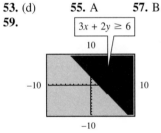

$3x + 2y \ge 6$

61.

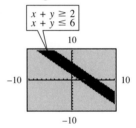

$x + y \ge 2$
$x + y \le 6$

63.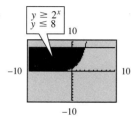

65.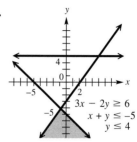

67. maximum value: 65 at $(5, 10)$; minimum value: 8 at $(1, 1)$ **69.** \$1120, with 4 pigs and 12 geese
71. 3.75 servings of A and 1.875 servings of B, for a minimum cost of \$1.69   **73.** 1600 Type 1 and 0 Type 2
for a maximum revenue of \$160   **75.** 0 medical kits and 4000 containers of water

Section 6.7 (page 534)

1. In the first equation, we get $-1 = -1$. In the second we get $-6 = -6$, and in the third we get $19 = 19$.
3. In the first equation, we get $-.4 = -.4$. In the second we get $1.8 = 1.8$, and in the third we get $-.7 = -.7$.
5. $\{(1, 2, -1)\}$ **7.** $\{(2, 0, 3)\}$ **9.** $\emptyset$ **11.** $\{(1, 2, 3)\}$ **13.** $\{(-1, 2, 1)\}$ **15.** $\{(4, 1, 2)\}$
17. $\{(\frac{1}{2}, \frac{2}{3}, -1)\}$ **19.** $\{(2, 4, 2)\}$ **21.** $\{(-1, 1, \frac{1}{3})\}$
23.   **(a)** As an example, $\begin{array}{l} x + 2y + z = 5 \\ 2x - y + 3z = 4. \end{array}$ (There are others.)   **(b)** As an example, $\begin{array}{l} x + y + z = 5 \\ 2x - y + 3z = 4. \end{array}$ (There are others.)
(c) As an example, $\begin{array}{l} 2x + 2y + 2z = 8 \\ 2x - y + 3z = 4. \end{array}$ (There are others.)
25. For example, the ceiling and two perpendicular walls intersect in a point.
27. $\{(x, -4x + 3, -3x + 4)\}$ **29.** $\{(x, -x + 15, -9x + 69)\}$ **31.** $\{(x, \frac{-15 + x}{3}, \frac{24 + x}{3})\}$
33. 12 cents; 13 nickels; 4 quarters   **35.** 30 barrels each of \$150 and \$190 glue; 210 barrels of \$120 glue
37. 28 in; 17 in; 14 in   **39.** \$50,000 at 5%; \$10,000 at 4.5%; \$40,000 at 3.75%
41. $y = .75x^2 + .25x - .5$ **43.** $y = x^2 + 2x + 1$ **45.** $y = -.5x^2 + x + .25$
47. $x^2 + y^2 + x - 7y = 0$ **49.** $a = -2, b = 20, c = 5; s(8) = 37$

Chapter 6 Review Exercises (page 538)

1. $(x + 2)^2 + (y - 3)^2 = 25$ **2.** $(x - \sqrt{5})^2 + (y + \sqrt{7})^2 = 3$ **3.** $(x + 8)^2 + (y - 1)^2 = 289$
4. $(x - 3)^2 + (y + 6)^2 = 36$ **5.** $(2, -3); r = 1$ **6.** $(3, 5); r = 2$ **7.** $(-\frac{7}{2}, -\frac{3}{2}); r = \frac{3\sqrt{6}}{2}$
8. $(-\frac{11}{2}, \frac{5}{2}); r = \frac{\sqrt{146}}{2}$ **9.** the point $(4, 5)$
10. $(-\frac{1}{6}, 0); x = \frac{1}{6}; y = 0$ **11.** $(\frac{1}{2}, 0); x = -\frac{1}{2}; y = 0$

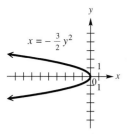

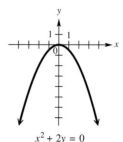

12. $(0, \frac{1}{12}); y = -\frac{1}{12}; x = 0$ **13.** $(0, -\frac{1}{2}); y = \frac{1}{2}; x = 0$

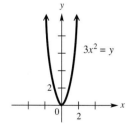

14. $x = \frac{1}{16}y^2$ **15.** $x = \frac{2}{25}y^2$ **16.** $y = -\frac{4}{9}x^2$ **17.** $x + 5 = \frac{1}{28}(y - 6)^2$ **18.** $y - 3 = \frac{1}{8}(x - 4)^2$
19. $(0, -3), (0, 3)$ **20.** $(-4, 0), (4, 0)$ **21.** $(-8, 0), (8, 0)$

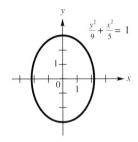

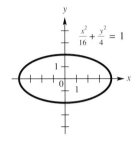

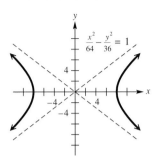

22. $(0, -5), (0, 5)$ **23.** $(1, -1), (5, -1)$ **24.** $(5, -3), (-1, -3)$

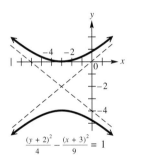

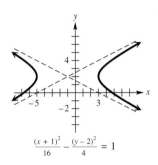

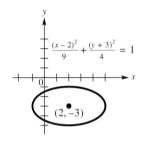

25. $(-3, 0), (-3, -4)$ **26.** $(3, 2), (-5, 2)$

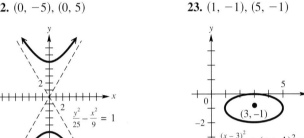

27. $\frac{y^2}{16} + \frac{x^2}{12} = 1$ **28.** $\frac{x^2}{36} + \frac{y^2}{32} = 1$ **29.** $\frac{y^2}{16} - \frac{x^2}{9} = 1$ **30.** $\frac{y^2}{4} - \frac{5x^2}{16} = 1$ **31.** $\frac{4y^2}{81} + \frac{4x^2}{45} = 1$
32. $\frac{x^2}{4} - \frac{y^2}{21} = 1$ **33. (a)** $(-1, -3)$ **(b)** 5 **(c)** $y_1 = -3 + \sqrt{24 - 2x - x^2}; \; y_2 = -3 - \sqrt{24 - 2x - x^2}$
34. F   **35.** C   **36.** A   **37.** E   **38.** B   **39.** D   **40.** $\frac{4}{5}$ **41.** $\frac{\sqrt{5}}{3}$
42. $\sqrt{10}$ **43.** $\{(5, 7)\}$ **44.** $\{(2, -4)\}$ **45.** $\{(-\frac{5}{4}, -\frac{79}{40})\}$ **46.** $\{(8, -18)\}$ **47.** $\{(\frac{2y + 11}{5}, y)\}$;
dependent   **48.** $\{(5, -3)\}$ **49.** It is not correct, because it is a system of dependent equations.
50. (a) yes **(b)** $(11.8, -1.9), (-8.6, 8.3)$ **(c)** $(\frac{8 - 8\sqrt{41}}{5}, \frac{16 + 4\sqrt{41}}{5}), (\frac{8 + 8\sqrt{41}}{5}, \frac{16 - 4\sqrt{41}}{5})$
51. $\{(-2, 3), (1, 0)\}$ **52.** $\{(1, 1), (\frac{7}{5}, -\frac{1}{5})\}$ **53.** $\{(2, 3), (-2, 3), (2, -3), (-2, -3)\}$
54. $\{(3\sqrt{3}, \sqrt{2}), (-3\sqrt{3}, \sqrt{2}), (3\sqrt{3}, -\sqrt{2}), (-3\sqrt{3}, -\sqrt{2})\}$ **55.** $\{(6, \frac{2}{3}), (-4, -1)\}$
56. $\{(-2, 0), (1, 1)\}$ **57.** It cannot, because two distinct lines will intersect in *at most* one point. If the lines
coincide, there are infinitely many solutions.   **58. (a)** $y_1 = \sqrt{2 - x^2}, \; y_2 = -\sqrt{2 - x^2}$ **(b)** $y_3 = 4 - 3x$
(c) Many answers are possible. One example is $[-3, 3]$ by $[-2, 2]$.

59.
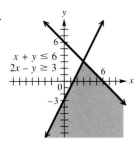
$x + y \leq 6$
$2x - y \geq 3$

60.
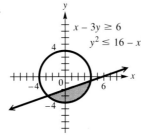
$x - 3y \geq 6$
$y^2 \leq 16 - x^2$

61.
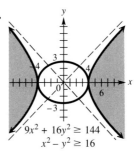
$9x^2 + 16y^2 \geq 144$
$x^2 - y^2 \geq 16$

62. maximum of 24 at $(0, 6)$ **63.** minimum of 40 at $(0, 20)$ or $(\frac{10}{3}, \frac{40}{3})$ or at any point between

64. (a)
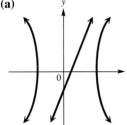

(b)
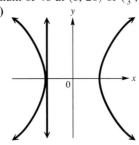

(c)
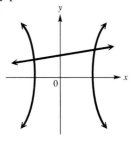

65. $\{(-1, 2, 3)\}$ **66.** $\{(6, -2, 1)\}$ **67.** $\{(1, 3, -1)\}$ **68.** $\emptyset$
69. 220 members, 80 nonmembers   **70.** $\frac{1}{3}$ cup of rice, $\frac{1}{5}$ cup of soybeans   **71.** 150 kg of the half and half
and 75 kg of the other, for a maximum revenue of \$1260   **72.** $y = \frac{3}{4}x^2 + \frac{1}{4}x - \frac{1}{2}$

CHAPTER 7

Section 7.1 (page 551)

1. $\begin{bmatrix} -4 & -8 \\ 4 & 7 \end{bmatrix}$ **3.** $\begin{bmatrix} 1 & 5 & 6 \\ -1 & 8 & 5 \\ 4 & 7 & 0 \end{bmatrix}$ **5.** $\begin{bmatrix} -3 & 1 & -4 \\ 2 & 1 & 3 \\ -17 & 0 & -13 \end{bmatrix}$

7. $\begin{bmatrix} 2 & 3 & | & 11 \\ 1 & 2 & | & 8 \end{bmatrix}$ **9.** $\begin{bmatrix} 1 & 5 & | & 6 \\ 1 & 2 & | & 8 \end{bmatrix}$ **11.** $\begin{bmatrix} 2 & 1 & 1 & | & 3 \\ 3 & -4 & 2 & | & -7 \\ 1 & 1 & 1 & | & 2 \end{bmatrix}$

13. $\begin{bmatrix} 1 & 1 & 0 & | & 2 \\ 0 & 2 & 1 & | & -4 \\ 0 & 0 & 1 & | & 2 \end{bmatrix}$ **15.** $2x + y = 1$
$3x - 2y = -9$

17. $x = 2$ **19.** $3x + 2y + z = 1$
$y = 3$ $2y + 4z = 22$
$z = -2$ $-x - 2y + 3z = 15$

21. $\{(2, 3)\}$ **23.** $\{(-3, 0)\}$ **25.** $\{(\frac{7}{2}, -1)\}$ **27.** $\emptyset$ **29.** $\{(\frac{3y+1}{6}, y)\}$ **31.** $\{(-2, 1, 3)\}$
33. $\{(-1, 23, 16)\}$ **35.** $\{(3, 2, -4)\}$ **37.** $\{(2, 1, -1)\}$ **39.** In both cases, we simply write the
coefficients and do not write the variables. This is possible because we agree beforehand on the order in which the
variables appear.
41. $\{(\frac{5z + 14}{5}, \frac{5z - 12}{5}, z)\}$ **43.** $\{(\frac{12 - z}{7}, \frac{4z - 6}{7}, z)\}$ **45.** $\emptyset$
47. wife: 40 days; husband: 32 days   **49.** model 201: 5 bicycles; model 301: 8 bicycles   **51.** \$10,000 at
8%; \$7000 at 10%; \$8000 at 9%   **53.** $\{(0, 2, -2, 1)\}$ **55.** $2 = a + b + c$ **57.** $-8 = 9a + 3b + c$
59. c must be positive because the graph intersects the y-axis above the origin. (The value of c is the y- intercept.)
61. $f(-1.5) = -8$

63. $a = 2, b = -3, c = 4, d = -2; g(4) = 94.$ The point (4, 94) lies on the graph of $g(x) = 2x^3 - 3x^2 + 4x - 2.$

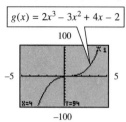

Section 7.2 (page 564)

1. Answers will vary.   **3.** 2×2; square   **5.** 3×4 **7.** 2×1; column   **9.** 3×3; square
11. $x = 2; y = 4; z = 8$ **13.** $x = -15; y = 5; k = 3$ **15.** $m = 6; n = -9$
17. $z = 18; r = 3; s = 3; p = 3; a = \frac{3}{4}$ **19.** Only matrices of the same dimensions may be added. Simply add corresponding elements.

21. $\begin{bmatrix} 8 & -43 & -18 \\ 26 & 29 & 6 \\ -2 & 10 & 43 \end{bmatrix}$ **23.** It is not possible to find the sum.

25. $\begin{bmatrix} 13 & 3 & 0 & -2 \\ 9 & -12 & 4 & 8 \\ 12 & -11 & -1 & 9 \end{bmatrix}$ **27.** $\begin{bmatrix} -12x + 8y & -x + y \\ x & 8x - y \end{bmatrix}$

29. $\begin{bmatrix} 7 & 2 \\ 9 & 0 \\ 8 & 6 \end{bmatrix}; \begin{bmatrix} 7 & 9 & 8 \\ 2 & 0 & 6 \end{bmatrix}$ **31.** $\begin{bmatrix} 5411 & 11{,}352 \\ 9371 & 15{,}956 \end{bmatrix}; \begin{bmatrix} 5411 & 9371 \\ 11{,}352 & 15{,}956 \end{bmatrix}$

33. $\begin{bmatrix} -4 & 8 \\ 0 & 6 \end{bmatrix}$ **35.** $\begin{bmatrix} 8 & -16 \\ 0 & -12 \end{bmatrix}$ **37.** $\begin{bmatrix} 2 & 6 \\ -4 & 6 \end{bmatrix}$ **39.** $\begin{bmatrix} 60 & -10 \\ -44 & 9 \end{bmatrix}$

41. $4 \times 4; 2 \times 2$ **43.** 3×2; BA is not defined.   **45.** AB is not defined; 3×2
47. Neither AB nor BA is defined.   **49.** columns; rows   **51.** (corresponding) first row; (corresponding) second column; adding

53. $\begin{bmatrix} pa + qb & pc + qd \\ ra + sb & rc + sd \end{bmatrix}$ **55.** $\begin{bmatrix} -17 \\ -1 \end{bmatrix}$ **57.** $\begin{bmatrix} 17 & -10 \\ 1 & 2 \end{bmatrix}$

59. $\begin{bmatrix} -2 & 10 \\ 0 & 8 \end{bmatrix}$ **61.** $\begin{bmatrix} -2 & 5 & 0 \\ 6 & 6 & 1 \\ 12 & 2 & -3 \end{bmatrix}$ **63.** $[2 \quad 7 \quad -4]$ **65.** $\begin{bmatrix} 20 & 0 & 8 \\ 8 & 0 & 4 \end{bmatrix}$

67. The product is not defined.   **69.** $[-8]$

71. $\begin{bmatrix} -6 & 12 & 18 \\ 4 & -8 & -12 \\ -2 & 4 & 6 \end{bmatrix}$ **73.** $\begin{bmatrix} 16 & 22 \\ 7 & 19 \end{bmatrix}$ **75.** $\begin{bmatrix} -10 & 16 & 26 \\ 5 & 7 & 22 \end{bmatrix}$

77. No; the answers are different (CA is not defined); no   **79. (a)** $\begin{bmatrix} 50 & 100 & 30 \\ 10 & 90 & 50 \\ 60 & 120 & 40 \end{bmatrix}$

(b) $\begin{bmatrix} 12 \\ 10 \\ 15 \end{bmatrix}$ (If the rows and columns are interchanged in part (a), this should be a 1×3 matrix.)

(c) $\begin{bmatrix} 2050 \\ 1770 \\ 2520 \end{bmatrix}$ (This may be a 1×3 matrix instead.) **(d)** $6340

81. always true   **83.** always true   **85.** Each side is equal to $\begin{bmatrix} a & b \\ c & d \end{bmatrix}$.

87. Each side is equal to $\begin{bmatrix} am + bp & cm + dp \\ an + bq & cn + dq \end{bmatrix}$.

Section 7.3 (page 574)

1. yes   **3.** no   **5.** yes   **7.** no   **9.** The inverse will not exist if the determinant is equal to 0.
11. -36 **13.** 7   **15.** 0   **17.** -26 **19.** -16 **21.** 0   **23.** $2x - 32$ **25.** $y^2 - 16$
27. $x = -\frac{4}{3}$ **29.** $x = 4$ or $x = -1$
31. $\begin{bmatrix} 0 & \frac{1}{2} \\ -1 & \frac{1}{2} \end{bmatrix}$ **33.** The inverse does not exist.

35. $\begin{bmatrix} 2 & 1 \\ -\frac{3}{2} & -\frac{1}{2} \end{bmatrix}$ **37.** $\begin{bmatrix} -2.5 & 5 \\ 12.5 & -15 \end{bmatrix}$ **39.** $A = \begin{bmatrix} 1 & 1 \\ 2 & -1 \end{bmatrix}$; $X = \begin{bmatrix} x \\ y \end{bmatrix}$; $B = \begin{bmatrix} 8 \\ 4 \end{bmatrix}$

41. $A = \begin{bmatrix} 4 & 5 \\ 2 & 3 \end{bmatrix}$; $X = \begin{bmatrix} x \\ y \end{bmatrix}$; $B = \begin{bmatrix} 7 \\ 5 \end{bmatrix}$ **43.** $\{(2, 3)\}$ **45.** $\{(-2, 4)\}$ **47.** $\{(4, -6)\}$

49. $\{(\frac{5}{2}, -1)\}$ **55.** $x^2 - 3x = 7$ **57.** $\{\frac{3 + \sqrt{37}}{2}, \frac{3 - \sqrt{37}}{2}\}$ **59.**

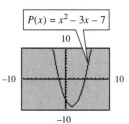

$P(x) = x^2 - 3x - 7$

Section 7.4 (page 583)

1. -10 **3.** 0   **5.** 8   **7.** 17   **9.** 166   **11.** 0   **13.** 0   **15.** -88 **17.** 1
19. $\begin{bmatrix} 1 & 0 & 0 \\ 0 & -1 & 0 \\ -1 & 0 & 1 \end{bmatrix}$ **21.** $\begin{bmatrix} -\frac{1}{24} & \frac{3}{16} & -\frac{1}{2} \\ \frac{1}{8} & -\frac{1}{16} & \frac{1}{2} \\ \frac{1}{8} & -\frac{1}{16} & -\frac{1}{2} \end{bmatrix}$ **23.** $\begin{bmatrix} 15 & 4 & -5 \\ -12 & -3 & 4 \\ -4 & -1 & 1 \end{bmatrix}$

25. $\begin{bmatrix} -\frac{10}{3} & \frac{5}{9} & -\frac{10}{9} \\ \frac{20}{3} & \frac{5}{9} & \frac{80}{9} \\ -5 & \frac{5}{6} & -\frac{20}{3} \end{bmatrix}$ **27.** The inverse does not exist.

29. $\begin{bmatrix} 0 & 0 & \frac{1}{3} \\ -\frac{1}{2} & -\frac{1}{2} & \frac{1}{2} \\ 1 & 0 & -\frac{1}{3} \end{bmatrix} \cdot \begin{bmatrix} 1 & 0 & 1 \\ 2 & -2 & -1 \\ 3 & 0 & 0 \end{bmatrix} = \begin{bmatrix} 1 & 0 & 0 \\ 0 & 1 & 0 \\ 0 & 0 & 1 \end{bmatrix}$

31. $x + 3y - 11 = 0$ is the equation of this line.   **33.** $\{-4\}$ **35.** $\{13\}$ **37.** $\{(3, 0, 2)\}$
39. $\{(12, -\frac{15}{11}, -\frac{65}{11})\}$ **41.** $\{(0, 2, -2, 1)\}$ **43.** $P(x) = -2x^3 + 5x^2 - 4x + 3$
45. $P(x) = x^4 + 2x^3 + 3x^2 - x - 1$ **47.** 0   **49.** 1; The rows of the first matrix became the columns of
the second.   **51.** -70; The second row was multiplied by -5, giving $-5(14) = -70$ as the determinant of the
new matrix.
53.   **(a)** $\begin{bmatrix} -2 & 4 & -5 \\ 2 & 1 & 11 \\ 3 & 0 & 11 \end{bmatrix}$ **(b)** 37   **55. (a)** 7   **(b)** $\frac{33}{2}$ **(c)** $\frac{7}{2}$

Section 7.5 (page 591)

1. $\{(2, 2)\}$ **3.** $\{(2, -5)\}$ **5.** $\{(2, 0)\}$ **7.** $\{(2, 3)\}$ **9.** $\{(\frac{4 - 2y}{3}, y)\}$ (Use another method.)
11. $\emptyset$ (Use another method.)   **13.** $\{(-1, 2, 1)\}$ **15.** $\{(-3, 4, 2)\}$ **17.** $\{(0, 0, -1)\}$

19. $\emptyset$ (Use another method.)   **21.** $\{(\frac{-32 + 19z}{4}, \frac{24 - 13z}{4}, z)\}$ (Use another method.)

23. $\{(2, 3, 1)\}$ **25.** $\{(0, 4, 2)\}$ **27.** $\{(6.2, 1.9, -2.9)\}$ **29.** If $D = 0$, Cramer's rule *cannot* be applied, because there is no *unique* solution.

31. $\{(0, 2, -2, 1)\}$ **33.** $\{(-6, 1, 2, -3)\}$ **35.**
$$9A + C = -81$$
$$-A + C = -1$$
$$A + B + C = -2$$

37. $(x - 4)^2 + (y + \frac{15}{2})^2 = \frac{325}{4}$ **39.** $\frac{5\sqrt{13}}{2}$ **41.** $\{(-a - b, a^2 + ab + b^2)\}$ **43.** $\{(1, 0)\}$

45. (a) $\{(-1, 2)\}$ **(b)** $\{(-1, 2)\}$ **(c)** $\{(-1, 2)\}$ **(d)** In each case, the solution set is $\{(-1, 2)\}$.

Chapter 7 Review Exercises (page 594)

1. $m = -8; p = -7; x = 2; y = 9; z = 5$ **2.** $a = 4; b = 8; c = -12$ **3.** $a = 5; x = \frac{3}{2}; y = 0; z = 9$

4. $k = 4; y = 7; a = 2; m = 3; p = 1; r = -5$

5. $\begin{bmatrix} 0 & -2 & 7 \\ 6 & -1 & 10 \end{bmatrix}$ **6.** $\begin{bmatrix} -4 \\ 6 \\ 1 \end{bmatrix}$ **7.** not possible   **8.** $\begin{bmatrix} -4 & -4 \\ -7 & 16 \end{bmatrix}$ **9.** not possible

10. $\begin{bmatrix} 14x + 28y & 42x + 22y \\ 18x - 46y & 70x + 18y \end{bmatrix}$ **11.** by adding corresponding elements

12. $\begin{bmatrix} 87 & 6.4 \\ 70 & 4.0 \\ 55 & 3.3 \end{bmatrix}$; $\begin{bmatrix} 87 & 70 & 55 \\ 6.4 & 4.0 & 3.3 \end{bmatrix}$ **13.** $\begin{bmatrix} 606 & 354 & 4434 & 28 \\ 397 & 238 & 2981 & 18 \\ 479 & 263 & 3359 & 21 \end{bmatrix}$; $\begin{bmatrix} 606 & 397 & 479 \\ 354 & 238 & 263 \\ 4434 & 2981 & 3359 \\ 28 & 18 & 21 \end{bmatrix}$ **14.** $\begin{bmatrix} 11 & 20 \\ 14 & 40 \end{bmatrix}$

15. $\begin{bmatrix} -9 & 3 \\ 10 & 6 \end{bmatrix}$ **16.** $\begin{bmatrix} -3 \\ 10 \end{bmatrix}$ **17.** not possible   **18.** $\begin{bmatrix} -2 & 22 & 31 \\ 25 & 12 & 9 \end{bmatrix}$ **19.** $[1 \quad 13 \quad 6]$

20. The number of columns of the first matrix must equal the number of rows of the second matrix.

21. yes   **22.** yes   **23.** no   **24.** yes

25. $\begin{bmatrix} 3 & -1 \\ -5 & 2 \end{bmatrix}$ **26.** $\begin{bmatrix} -\frac{1}{4} & \frac{1}{6} \\ 0 & \frac{1}{3} \end{bmatrix}$ **27.** $\begin{bmatrix} \frac{1}{2} & 0 \\ \frac{1}{10} & \frac{1}{5} \end{bmatrix}$

28. $\begin{bmatrix} \frac{1}{4} & \frac{1}{2} & \frac{1}{2} \\ \frac{1}{4} & -\frac{1}{2} & \frac{1}{2} \\ \frac{1}{8} & -\frac{1}{4} & -\frac{1}{4} \end{bmatrix}$ **29.** $\begin{bmatrix} \frac{2}{3} & 0 & -\frac{1}{3} \\ \frac{1}{3} & 0 & -\frac{2}{3} \\ -\frac{2}{3} & 1 & \frac{1}{3} \end{bmatrix}$

30. does not exist   **31.** $\{(2, 2)\}$ **32.** $\{(-1, -1)\}$ **33.** $\{(2, 1)\}$ **34.** $\{(3, -2)\}$

35. dependent equations; $\{(\frac{6z + 5}{11}, \frac{31z + 2}{11}, z)\}$ **36.** $\{(1, -1, 2)\}$ **37.** $\{(-1, 0, 2)\}$ **38.** $\{(-3, 5, 1)\}$

39. -25 **40.** -6 **41.** -44 **42.** -1 **43.** $\{-\frac{7}{3}\}$ **44.** $\{\frac{8}{19}\}$ **45.** {all real numbers}

46. $\{\frac{1}{31}\}$ **47. (a)** 5   **(b)** 30   **(c)** -50 **(d)** $x = 6, y = -10$

48. (a) $A = \begin{bmatrix} 3 & -1 \\ 2 & 1 \end{bmatrix}$ **(b)** $\begin{bmatrix} 28 \\ 2 \end{bmatrix}$ **(c)** To solve the system, multiply A^{-1} by B to get the matrix of solutions, indicating $x = 6$ and $y = -10$.   **49.** If $D = 0$, then there would be division by zero, which is undefined. When $D = 0$, the system either has no solution or infinitely many solutions.   **50.** $\{(-2, 5)\}$ **51.** $\{(-4, 2)\}$

52. $\{(\frac{82}{23}, -\frac{4}{23})\}$ **53.** $\{(-4, 6, 2)\}$ **54.** We can't use Cramer's rule, since $D = 0$. This system has dependent equations. The solution set can be found using another method. It is $\{(\frac{16}{9}, \frac{8 - 9z}{18}, z)\}$.

55. $\{(\frac{172}{67}, -\frac{14}{67}, -\frac{87}{67})\}$ **56.** $P(x) = \frac{1}{2}x^2 + 2x - 2$

CHAPTER 8

Section 8.1 (page 609)

1. (a) $60°$ **(b)** $150°$ **3. (a)** $45°$ **(b)** $135°$ **5. (a)** $\frac{\pi}{3}$ **(b)** $\frac{5\pi}{6}$ **7. (a)** $\frac{\pi}{4}$ **(b)** $\frac{3\pi}{4}$

9. $45°$ **11.** $320°$ **13.** $90°$ **15.** $\frac{7\pi}{4}$ **17.** $\frac{\pi}{2}$ **19.** $\frac{\pi}{6}$ **21.** $\frac{\pi}{4}$ **23.** $\frac{2\pi}{3}$ **25.** $-\frac{7\pi}{4}$

27. $20.900°$ **29.** $91.598°$ **31.** $-274.316°$ **33.** $31° 25' 47''$ **35.** $89° 54' 01''$

37. $-178° 35' 58''$ **39.** $30° + n \cdot 360°$; I   **41.** $230° + n \cdot 360°$; III

43. $270° + n \cdot 360°$; no quadrant   **45.** $\frac{\pi}{4} + 2n\pi$; I   **47.** $\frac{3\pi}{4} + 2n\pi$; II

Angles other than the ones listed are possible in Exercises 49–53.

49. 435°; −285°; I

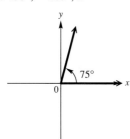

51. 308°; −412°; IV

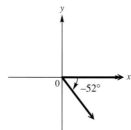

53. $\frac{11\pi}{3}$; $-\frac{\pi}{3}$; IV

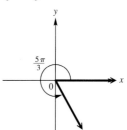

55. 25.8 cm   **57.** 318 m   **59.** 5.05 m   **61.** 1200 km   **63.** 3500 km   **65.** 5900 km
67. 1800°   **69.** 12.5 rotations per hour   **71.** 12.7 cm   **73. (a)** 39,616 rotations   **(b)** 62.9 mi; yes
75. $\frac{16\pi}{3}$ m per second; $\frac{16\pi}{3}$ radians per second   **77. (a)** $\frac{2\pi}{365}$ radians   **(b)** $\frac{\pi}{4380}$ radians per hour
(c) 66,700 miles per hour   **79.** larger pulley: $\frac{25\pi}{18}$ radians per second;   smaller pulley: $\frac{125\pi}{48}$ radians per second
81. 3.73 cm   **83.** 114 sq cm   **85. (a)** 13.85°   **(b)** 76 sq m

Further Explorations

1. $Y_1 = (180/\pi)x;\ Y_2 = 6400\ (Y_1 * (\pi/180))$

Section 8.2 (page 623)

1.

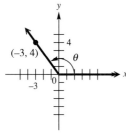

3.

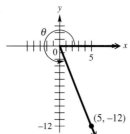

In Exercises 5–19, we give, in order: sine, cosine, tangent, cotangent, secant, and cosecant.

5. $\frac{4}{5}; -\frac{3}{5}; -\frac{4}{3}; -\frac{3}{4}, -\frac{5}{3}, \frac{5}{4}$ **7.** $-\frac{12}{13}; \frac{5}{13}; -\frac{12}{5}; -\frac{5}{12}; \frac{13}{5}; -\frac{13}{12}$ **9.** $\frac{24}{25}; -\frac{7}{25}; -\frac{24}{7}; -\frac{7}{24}; -\frac{25}{7}; \frac{25}{24}$
11. 1; 0; undefined; 0; undefined; 1   **13.** 0; −1; 0; undefined; −1; undefined
15. $-\frac{2}{3}; \frac{\sqrt{5}}{3}; -\frac{2\sqrt{5}}{5}; -\frac{\sqrt{5}}{2}; \frac{3\sqrt{5}}{5}; -\frac{3}{2}$ **17.** $\frac{\sqrt{2}}{2}; \frac{\sqrt{2}}{2}; 1; 1; \sqrt{2}; \sqrt{2}$ **19.** $\frac{\sqrt{3}}{2}; \frac{1}{2}; \sqrt{3}; \frac{\sqrt{3}}{3}; 2; \frac{2\sqrt{3}}{3}$
21. Reciprocals must always have the same sign.   **23.** It is the distance from the origin to the point (x, y) on the terminal side of the angle.   **25.** positive   **27.** negative   **29.** positive   **31.** positive
33. negative   **35.** negative

In Exercises 37–41, we give, in order: sine, cosine, tangent, cotangent, secant, and cosecant.

37. $-\frac{2\sqrt{5}}{5}; \frac{\sqrt{5}}{5}; -2; -\frac{1}{2}; \sqrt{5}; -\frac{\sqrt{5}}{2}$ **39.** $-\frac{4\sqrt{65}}{65}; -\frac{7\sqrt{65}}{65}; \frac{4}{7}; \frac{7}{4};$ $-\frac{\sqrt{65}}{7}; -\frac{\sqrt{65}}{4}$ **41.** $\frac{5\sqrt{34}}{34}; -\frac{3\sqrt{34}}{34}; -\frac{5}{3}; -\frac{3}{5}; -\frac{\sqrt{34}}{3}; \frac{\sqrt{34}}{5}$

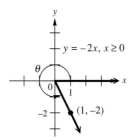

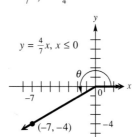

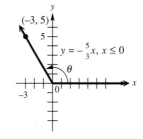

43. $\frac{1}{3}$ **45.** -5 **47.** $2\sqrt{2}$ **49.** $-\frac{3\sqrt{5}}{5}$ **51.** .70069071   **53.** 2.2778902   **55.** No, because the sine and cosecant functions are reciprocals and they must have the same sign for a particular angle. **57.** Because $\cot 90° = 0$, $\frac{1}{\cot 90°}$ is undefined. The equality symbol would indicate that the expressions represent real numbers.   **59.** II   **61.** III   **63.** IV   **65.** II or IV   **67.** I or II   **69.** I or III **71.** $-\frac{\sqrt{5}}{3}$ **73.** $-\frac{\sqrt{5}}{2}$ **75.** $-\frac{4}{3}$ **77.** $-\frac{\sqrt{3}}{2}$ **79.** 3.44701905   **81.** $-.56616682$

In Exercises 83–91 we give, in order: sine, cosine, tangent, cotangent, secant, and cosecant.

83. $-\frac{4}{5}$; $-\frac{3}{5}$; $\frac{4}{3}$; $\frac{3}{4}$; $-\frac{5}{3}$; $-\frac{5}{4}$ **85.** $\frac{7}{25}$; $-\frac{24}{25}$; $-\frac{7}{24}$; $-\frac{24}{7}$; $-\frac{25}{24}$; $\frac{25}{7}$ **87.** $\frac{1}{2}$; $-\frac{\sqrt{3}}{2}$; $-\frac{\sqrt{3}}{3}$; $-\sqrt{3}$; $-\frac{2\sqrt{3}}{3}$; 2   **89.** $\frac{8\sqrt{67}}{67}$; $\frac{\sqrt{201}}{67}$; $\frac{8\sqrt{3}}{3}$; $\frac{\sqrt{3}}{8}$; $\frac{\sqrt{201}}{3}$; $\frac{\sqrt{67}}{8}$ **91.** .164215; $-.986425$; $-.166475$; -6.00691; -1.01376; 6.08958

Further Explorations

1. 45°, 135°, 225°, 315°   **3.** Whenever x and y are equal, or whenever they are opposites ($y = -x$) of each other, $|x| = |y|$. Dividing both sides by $r = |r|$, we have $\left|\frac{x}{r}\right| = \left|\frac{y}{r}\right|$, and thus $|\cos \theta| = |\sin \theta|$.

Section 8.3 (page 636)

In Exercises 1–5 we give, in order: sine, cosine, tangent, cotangent, secant, and cosecant.

1. $\frac{3}{5}$; $\frac{4}{5}$; $\frac{3}{4}$; $\frac{4}{3}$; $\frac{5}{4}$; $\frac{5}{3}$ **3.** $\frac{21}{29}$; $\frac{20}{29}$; $\frac{21}{20}$; $\frac{20}{21}$; $\frac{29}{20}$; $\frac{29}{21}$ **5.** $\frac{n}{p}$; $\frac{m}{p}$; $\frac{n}{m}$; $\frac{m}{n}$; $\frac{p}{m}$; $\frac{p}{n}$ **7.** $\tan 17°$ **9.** $\cos 51° \ 31'$ **11.** $\sin \frac{3\pi}{10}$ **13.** $\cot(\frac{\pi}{2} - .5)$ **15.** The *exact* value of $\sin 45°$ is $\frac{\sqrt{2}}{2}$. The decimal value he gave was just an approximation.

The number of digits in part (c) of Exercises 17–33 will vary depending upon the model of calculator used.

17. (a) $\frac{\sqrt{3}}{3}$ (b) irrational (c) .5773502692   **19.** (a) $\frac{1}{2}$ (b) rational   **21.** (a) $\frac{2\sqrt{3}}{3}$ (b) irrational (c) 1.154700538   **23.** (a) $\sqrt{2}$ (b) irrational (c) 1.414213562   **25.** (a) $\frac{\sqrt{2}}{2}$ (b) irrational (c) .7071067812   **27.** (a) 1 (b) rational   **29.** (a) $\frac{\sqrt{3}}{2}$ (b) irrational (c) .8660254038   **31.** (a) $\sqrt{3}$ (b) irrational (c) 1.732050808   **33.** (a) 2 (b) rational

The number of digits displayed in Exercises 35–45 will vary.

35. .62478851   **37.** $-.32281638$ **39.** -3.1791978 **41.** .48775041   **43.** 1.0170372 **45.** .95544269   **47.** $\frac{\sqrt{3}}{3}$; $\sqrt{3}$ **49.** $\frac{\sqrt{3}}{2}$; $\frac{\sqrt{3}}{3}$; $\frac{2\sqrt{3}}{3}$ **51.** -1; -1 **53.** $-\frac{\sqrt{3}}{2}$; $-\frac{2\sqrt{3}}{3}$ **55.** (a) $-\sin \frac{\pi}{6}$ (b) $-\frac{1}{2}$ (c) $\sin \frac{7\pi}{6} = -\sin \frac{\pi}{6} = -.5$ **57.** (a) $-\tan \frac{\pi}{4}$ (b) -1 (c) $\tan \frac{3\pi}{4} = -\tan \frac{\pi}{4} = -1$ **59.** (a) $-\cos \frac{\pi}{6}$ (b) $-\frac{\sqrt{3}}{2}$ (c) $\cos \frac{7\pi}{6} = -\cos \frac{\pi}{6} \approx -.86602540$ **61.** 30°; 150°   **63.** 60°; 240° **65.** 120°; 240°   **67.** 240°; 300°

The number of displayed digits will vary in Exercises 69–73.

69. 46.593881°; 313.40612°   **71.** 24.392576°; 155.60742°   **73.** 41.248183°; 221.24818° **75.** .20952066; 3.3511133   **77.** 1.4429646; 1.6986280   **79.** 1.3631380; 1.7784546 **81.** It represents the distance from the point (x_1, y_1) to the origin.   **83.** 60°   **85.** It is a measure of the angle formed by the positive x-axis and the ray $y = \sqrt{3}x$, $x \geq 0$.   **87.** slope; tangent   **89.** They agree, as they are both approximately 1.154700538.   **91.** $\cos 60° = .5$. This is the x-coordinate of the point found in Exercise 90. Because $r = 1$, here $\cos 60° = \frac{x}{1} = x = .5$.

Section 8.4 (page 647)

1. $B = 53° \ 40'$; $a = 571$ m; $b = 777$ m   **3.** $M = 38.8°$; $n = 154$ m; $p = 198$ m **5.** $A = 47.9108°$; $c = 84.816$ cm; $a = 62.942$ cm   **7.** $B = 62° \ 00'$; $a = 8.17$ ft; $b = 15.4$ ft **9.** $A = 17° \ 00'$; $a = 39.1$ in; $c = 134$ in   **11.** $c = 85.9$ yd; $A = 62° \ 50'$; $B = 27° \ 10'$ **13.** No, because there are infinitely many such right triangles.   **15.** approximately 38.7° **17.** The transversal AB intersecting the parallel lines AD and BC forms equal alternate interior angles. **19.** 9.35 m   **21.** 62° 50'   **23.** 33.4 m   **25.** 26.92 in   **27.** 13.3 ft   **29.** 37° 40' **31.** 42,600 ft   **33.** 26° 20'   **35.** $A = 35° \ 59'$; $B = 54° \ 1'$ **37.** 52.4 ft   **39.** 8.229 cm **41.** 446   **43.** 147 m   **45.** 2.47 km   **47.** 150 km   **49.** 5856 m   **51.** 2.01 mi **53.** approximately 1,730,000 mi   **55.** 79°

Section 8.5 (page 661)

1. $C = 95°$, $b = 13$ m, $a = 11$ m **3.** $C = 80° 40'$, $a = 79.5$ mm, $c = 108$ mm
5. $A = 36.54°$, $b = 44.17$ m, $a = 28.10$ m **7.** $A = 49° 40'$, $b = 16.1$ cm, $c = 25.8$ cm
9. $C = 91.9°$, $BC = 490$ ft, $AB = 847$ ft **11.** It is not a right triangle. **13.** 118 m **15.** 1.93 mi
17. 10.4 in **19.** 111° **21.** first location: 5.1 mi; second location: 7.2 mi **23.** 26.5 km
25. 2.18 km **27.** 38.3 cm **29.** $B = 20.6°$, $C = 116.9°$, $c = 20.6$ ft **31.** There is no such triangle.
33. $B_1 = 49° 20'$, $C_1 = 92° 00'$, $c_1 = 15.5$ km; $B_2 = 130° 40'$, $C_2 = 10° 40'$, $c_2 = 2.88$ km
35. $A_1 = 52° 10'$, $C_1 = 95° 00'$, $c_1 = 9520$ cm; $A_2 = 127° 50'$, $C_2 = 19° 20'$, $c_2 = 3160$ cm
37. $B = 37.77°$, $C = 45.43°$, $c = 4.174$ ft **39.** $\sin C = 1$; $c = 90°$; ABC is a right triangle.
41. If $A = 103° 20'$, a must be the longest side of the triangle, which is not the case here, since $14.6 < 20.4$.
43. Such a piece of property cannot exist. **45.** 2×10^8 m per sec **47.** 19° **49.** 48.7°

Section 8.6 (page 671)

1. $c = 2.83$ in, $A = 44.9°$, $B = 106.8°$ **3.** $c = 6.46$ m, $A = 53.1°$, $B = 81.3°$
5. $a = 156$ cm, $B = 64° 50'$, $C = 34° 30'$ **7.** $A = 82°$, $B = 37°$, $C = 61°$
9. $A = 42° 00'$, $B = 35° 50'$, $C = 102° 10'$ **11.** $A = 50° 50'$; $B = 44° 40'$; $C = 84° 30'$
13. We get an impossible value for the cosine, leading to an error message if we use the inverse cosine function on a calculator. **15.** 257 m **17.** 281 km **19.** 10.8 mi **21.** 115 km **23.** 18 ft
25. 5500 m **27.** 438.14 ft **29.** 350° **31.** 2000 km **33.** 163.5° **35.** 25.24983 mi
37. 22 ft **39.** 46.4 m² **41.** 722.9 in² **43.** 78 m² **45.** 3650 ft² **47.** 228 yd²
49. 100 m² **51.** 33 cans
53.

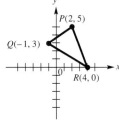

55. $\frac{\sqrt{13} + \sqrt{34} + \sqrt{29}}{2}$ **57.** 9.5(units²); yes **59.** 9.5(units²); yes **61.** $\sin A = \frac{h}{c}$
63. $\mathcal{A} = \frac{1}{2}b(c \sin A)$ or $\mathcal{A} = \frac{1}{2}bc \sin A$

Chapter 8 Review Exercises (page 677)

1.

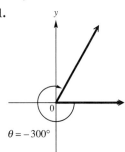

2. 60° **3.** $-660°$ (There are others.) **4.** $-300° + n \cdot 360°$ (There are others.)
5. $-\frac{5\pi}{3}$ **6.** $\frac{15}{32}$ seconds **7.** $\frac{4\pi}{75}$ radians per second **8.** 1260π meters per second **9.** 35.8 centimeters **10.** 10π inches **11.** 4500 kilometers
12. $-\frac{7\sqrt{53}}{53}$ **13.** $-\frac{2\sqrt{53}}{53}$ **14.** $\frac{7}{2}$ **15.** $-\frac{\sqrt{53}}{7}$ **16.** $-\frac{\sqrt{53}}{2}$ **17.** $\frac{2}{7}$
18. (a) 254.05° **(b)** 74.05°

19.

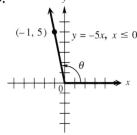

20. $\sin \theta = \frac{5\sqrt{26}}{26}$; $\cos \theta = -\frac{\sqrt{26}}{26}$ **21.** $101° \ 19'$ **22.** III **23.** II **24.** $-\frac{\sqrt{22}}{5}$ **25.** $\frac{5\sqrt{3}}{3}$
26. $-\frac{\sqrt{66}}{3}$ **27.** $\frac{\sqrt{2}}{2}$ **28.** -1 **29.** undefined **30.** $-\frac{2\sqrt{3}}{3}$ **31.** 2 **32.** 0 **33.** $\frac{\sqrt{2}}{2}$
34. $-\frac{\sqrt{3}}{2}$ **35.** $\sqrt{3}$ **36.** undefined **37.** -1 **38.** $-\sqrt{3}$

The number of digits may vary in Exercises 39–42.

39. .5495089781 **40.** 3.525320086 **41.** $-.726542528$ **42.** $22.33368266°$ **43.** $\frac{\sqrt{133}}{13}$
44. $\frac{13\sqrt{133}}{133}$ **45.** $62.5°$ **46.** $B = 50.28°$, $a = 32.38$ m, $c = 50.66$ m **47.** $A = 42° \ 07'$, $B = 47° \ 53'$,
$c = 402.6$ m **48.** 73.7 feet **49.** 20.4 meters **50.** 18.75 centimeters **51.** 50.24 meters
52. 1200 meters **53.** 110 kilometers **54.** 140 miles **55.** 110 feet **56.** 419 **57.** 63.7 m
58. $19.87°$ or $19° \ 52'$ **59.** $25° \ 00'$ **60.** 173 feet **61.** $41° \ 40'$ **62.** 55.5 m **63.** 70.9 m
64. $26.5°$ or $26° \ 30'$ **65.** $54° \ 20'$ or $125° \ 40'$ **66.** 148 cm **67.** $49° \ 30'$ **68.** $32°$
69. $B = 17° \ 10'$, $C = 137° \ 40'$, $c = 11.0$ yd
70. $B_1 = 74.6°$, $C_1 = 43.7°$, $c_1 = 61.9$ m; $B_2 = 105.4°$, $C_2 = 12.9°$, $c_2 = 20.0$ m **71.** $153,600$ m^2
72. 20.3 ft^2 **73.** $.234$ km^2 **74.** 680 m^2 **75.** 2.7 miles **76.** 43 feet **77.** 7 kilometers
78. 1.91 miles **79.** 58.6 feet **80.** 11 feet **81.** 15.8 feet **82.** $77.1°$

CHAPTER 9
Section 9.1 (page 692)

In Exercises 1–13, calculator approximations of irrational numbers are shown for the purpose of calculator support.

1. $-\frac{1}{2}$ **3.** -1 **5.** $-\frac{\sqrt{3}}{2}$ $(\approx -.8660254038)$ **7.** -2 **9.** $-\sqrt{3}$ (≈ -1.732050808) **11.** $-\frac{1}{2}$

13. $\frac{2\sqrt{3}}{3}$ (≈ 1.154700538) **15.** .5736049112 **17.** .4067524531 **19.** 1.206484913
21. 14.33376901 **23.** -1.045975716 **25.** -3.866512664 **27.** $\cos 1.2 \approx .36235775$,
$\sin 1.2 \approx .93203909$ **29.** $\cos 3.5 \approx -.9364567$, $\sin 3.5 \approx -.3507832$ **31.** .1865123694
33. $-.9824526126$ **35.** positive, negative, IV **37.** $\cos 4.38$ **39.** $-\sin .5$ **41.** $-\tan \frac{\pi}{7}$
43. $\sec 8$ **45.** $-\csc \frac{1}{4}$ **47.** $-\cot 10^5$ **49.** (b) **51.** (e) **53.** (a) **55.** (a)
57. (d) **59.** The student has neglected to write the argument of the functions (for example, θ or t).

61. $\frac{\pm\sqrt{1 + \cot^2 \theta}}{1 + \cot^2 \theta}$, $\frac{\pm\sqrt{\sec^2 \theta - 1}}{\sec \theta}$ **63.** $\frac{\pm\sin \theta \sqrt{1 - \sin^2 \theta}}{1 - \sin^2 \theta}$, $\frac{\pm\sqrt{1 - \cos^2 \theta}}{\cos \theta}$; $\pm\sqrt{\sec^2 \theta - 1}$; $\frac{\pm\sqrt{\csc^2 \theta - 1}}{\csc^2 \theta - 1}$
65. $\frac{\pm\sqrt{1 - \sin^2 \theta}}{1 - \sin^2 \theta}$; $\pm\sqrt{\tan^2 \theta + 1}$; $\frac{\pm\sqrt{1 + \cot^2 \theta}}{\cot \theta}$, $\frac{\pm\csc \theta \sqrt{\csc^2 \theta - 1}}{\csc^2 \theta - 1}$ **67.** $\sin \theta$ **69.** 1 **71.** $\tan^2 \beta$
73. $\tan^2 x$ **75.** $\sec^2 x$ **77.** $\sin x$ **79.** 1 **99.** While the equation is true for the *particular* value
$\theta = \frac{\pi}{2}$, it is not true in *general*. To be an identity, the equation must be true in *all* cases for which the functions
involved are defined.

Further Explorations

1. (a) $45°$, $225°$ **(b)** Cos x decreases from $0°$ to $180°$ and increases from $180°$ to $360°$. **(c)** Sin x decreases from
$90°$ to $270°$, and increases from $0°$ to $90°$ and from $270°$ to $360°$. **(d)** In both cases, the maximum value is 1 and
the minimum value is -1. **(e)** One example is ΔTbl $= 360$ and Tblmin $= 0$ (degrees). Cos $x = 1$ for multiples of
$360°$. **(f)** One example is ΔTbl $= 180$ and Tblmin $= 90$ (degrees). Cos $x = 0$ for odd multiples of $90°$. **(g)** One
example is ΔTbl $= 360$ and Tblmin $= 90$ (degrees). Sin $x = 1$ for $x = (4k + 1) \cdot 90°$, where k is an integer.
(h) One example is ΔTbl $= 180$ and Tblmin $= 0$ (degrees). Sin $x = 0$ for multiples of $180°$.

Section 9.2 (page 707)

In Exercises 1–5, verify the equation by using the appropriate circular function key, or the parametrically-generated
unit circle $x = \cos t$, $y = \sin t$.

1. $\sin(-2.75) \approx -.381661$ **3.** $\cos(-3.5) \approx -.9364567$ **5.** $\cos \frac{\pi}{2} = 0$ **7.** G **9.** E
11. B **13.** F **15.** D **17.** H **19.** B **21.** F **23.** a **25.** $-c$
27. (a) 2 **(b)** 2π **(c)** π **(d)** none **(e)** $[-2, 2]$ **29. (a)** 4 **(b)** 4π **(c)** $-\pi$ **(d)** none **(e)** $[-4, 4]$
31. (a) 1 **(b)** $\frac{2\pi}{3}$ **(c)** $\frac{\pi}{15}$ **(d)** up 2 units **(e)** $[1, 3]$ **33. (a)** 3 **(b)** 2 **(c)** none **(d)** up 2 units
(e) $[-1, 5]$ **35.** Shift the graph of $y = \sin x$ $\frac{\pi}{7}$ units to the left, stretch vertically by a factor of 2, and shift 3
units downward. **37.** Shift the graph of $y = \cos x$ $\frac{\pi}{6}$ units to the right, change the period to $\frac{2\pi}{3}$, and shift $\frac{5}{2}$
units downward.

39.

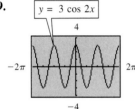

41.

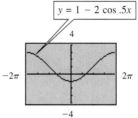

43. (a) 1; 1; 3; −2 **(b)** −1; −1; −3; −8 **45.** The standard trig window as defined in the text has Ymin = −4 and Ymax = 4. Because the range of f includes values less than −4, the local minimum points will not appear in the standard trig window. **47.** The period of f is $\frac{2\pi}{2} = \pi$, and the distance between −2π and 2π is 4π units. Since $\frac{4\pi}{\pi} = 4$, these values will show four periods of the graph. **49.** Ymin = −8; Ymax = −2

In Exercises 51–55, we provide only the graphs in parts (b). The left side of the identity is graphed as y_1, and the right side as y_2.

51. (b)

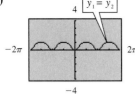

53. (b)

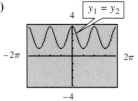

55. (b)

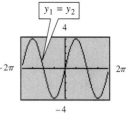

57. $y = -4 \sin 3x$ **59.** $y = -2 \cos 3x$ **61.** $y = 3 \sin 6x$ **63.** $\sin x$ **65.** 1

Further Explorations

1. $y_2 = 2 \sin x$ **3.** $y_2 = \sin 2x$ **5.** $y_2 = \sin(x - \frac{\pi}{6})$ **7.** $y_2 = 1 + \sin x$

Section 9.3 (page 724)

1. $\sec 2.1 \approx -1.980802$ **3.** $\csc(-.75) \approx -1.467053$ **5.** $\tan 2 \approx -2.18504$ **7.** $\cot \frac{\pi}{4} = 1$
9. To be a parabola, the function must be of the form $f(x) = ax^2 + bx + c$ $(a \neq 0)$. Furthermore, the graph of a vertical parabola does not lie between vertical asymptotes. **11.** B **13.** E **15.** D
17. (a)

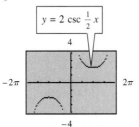

19. (a)

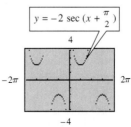

(b) 4π **(c)** none **(d)** $(-\infty, -2] \cup [2, \infty)$ **(e)** 2π **(b)** 2π **(c)** $-\frac{\pi}{2}$ **(d)** $(-\infty, -2] \cup [2, \infty)$ **(e)** π
21. (a)

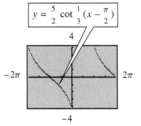

23. (a)

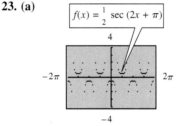

(b) 3π **(c)** $\frac{\pi}{2}$ **(d)** $(-\infty, \infty)$ **(e)** $\frac{\pi}{2}$ **(b)** π **(c)** $-\frac{\pi}{2}$ **(d)** $(-\infty, -\frac{1}{2}] \cup [\frac{1}{2}, \infty)$ **(e)** $\frac{\pi}{4}$

25. (a)

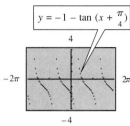

$y = -1 - \tan\left(x + \frac{\pi}{4}\right)$

(b) π **(c)** $-\frac{\pi}{4}$ **(d)** $(-\infty, \infty)$ **(e)** $\frac{\pi}{4}$

27. Start with the graph of $y = \csc x$. Shift π units to the right, change the period to π, stretch vertically by a factor of 3, and shift 2 units downward.   **29.** (The given function is equivalent to $y = 4 - 2 \tan 2(x - 1)$.) Start with the graph of $y = \tan x$. Shift 1 unit to the right, change the period to $\frac{\pi}{2}$, stretch vertically by a factor of 2, reflect across the x-axis, and shift 4 units upward.   **31.** π **33.** $\frac{5\pi}{4} + n\pi$
35. $\pi + .32175055 \approx 3.463343208$

37. (a) $\left\{\frac{\pi}{3}, \frac{5\pi}{3}\right\}$ **(b)**

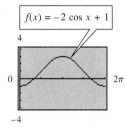

$f(x) = -2 \cos x + 1$

(c) $\left(\frac{\pi}{3}, \frac{5\pi}{3}\right)$ **(d)** $\left[0, \frac{\pi}{3}\right) \cup \left(\frac{5\pi}{3}, 2\pi\right)$

39. (a) $\left\{\frac{\pi}{3}, \frac{2\pi}{3}, \frac{4\pi}{3}, \frac{5\pi}{3}\right\}$ **(b)**

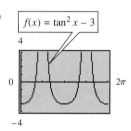

$f(x) = \tan^2 x - 3$

(c) $\left(\frac{\pi}{3}, \frac{\pi}{2}\right) \cup \left(\frac{\pi}{2}, \frac{2\pi}{3}\right) \cup \left(\frac{4\pi}{3}, \frac{3\pi}{2}\right) \cup \left(\frac{3\pi}{2}, \frac{5\pi}{3}\right)$
(d) $\left[0, \frac{\pi}{3}\right) \cup \left(\frac{2\pi}{3}, \frac{4\pi}{3}\right) \cup \left(\frac{5\pi}{3}, 2\pi\right)$

41. (a) $\left\{\frac{\pi}{6}, \frac{\pi}{2}, \frac{3\pi}{2}, \frac{11\pi}{6}\right\}$ **(b)**

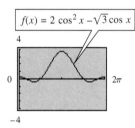

$f(x) = 2 \cos^2 x - \sqrt{3} \cos x$

(c) $\left[0, \frac{\pi}{6}\right) \cup \left(\frac{\pi}{2}, \frac{3\pi}{2}\right) \cup \left(\frac{11\pi}{6}, 2\pi\right)$ **(d)** $\left(\frac{\pi}{6}, \frac{\pi}{2}\right) \cup \left(\frac{3\pi}{2}, \frac{11\pi}{6}\right)$

43. (a) $\left\{\frac{\pi}{4}, \frac{3\pi}{4}, \frac{5\pi}{4}, \frac{7\pi}{4}\right\}$ **(b)**

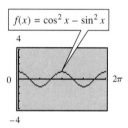

$f(x) = \cos^2 x - \sin^2 x$

(c) $\left[0, \frac{\pi}{4}\right) \cup \left(\frac{3\pi}{4}, \frac{5\pi}{4}\right) \cup \left(\frac{7\pi}{4}, 2\pi\right)$ **(d)** $\left(\frac{\pi}{4}, \frac{3\pi}{4}\right) \cup \left(\frac{5\pi}{4}, \frac{7\pi}{4}\right)$

45. (a) $\left\{\frac{\pi}{4}, \frac{5\pi}{4}\right\}$ **(b)**

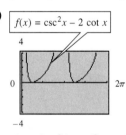

$f(x) = \csc^2 x - 2 \cot x$

(c) $\left(0, \frac{\pi}{4}\right) \cup \left(\frac{\pi}{4}, \pi\right) \cup \left(\pi, \frac{5\pi}{4}\right) \cup \left(\frac{5\pi}{4}, 2\pi\right)$ **(d)** $\emptyset$

47. (a) $\left\{\frac{\pi}{2}, 3.87, 5.55\right\}$ **(b)**

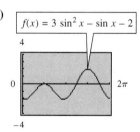

$f(x) = 3 \sin^2 x - \sin x - 2$

(c) $(3.87, 5.55)$ **(d)** $\left[0, \frac{\pi}{2}\right) \cup \left(\frac{\pi}{2}, 3.87\right) \cup (5.55, 2\pi)$

49. (a) $\{1.86, 2.61, 5.00, 5.75\}$
(b)

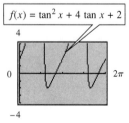

$f(x) = \tan^2 x + 4 \tan x + 2$

(c) $\left[0, \frac{\pi}{2}\right) \cup \left(\frac{\pi}{2}, 1.86\right) \cup \left(2.61, \frac{3\pi}{2}\right) \cup \left(\frac{3\pi}{2}, 5.00\right)$
$\cup (5.75, 2\pi)$ **(d)** $(1.86, 2.61) \cup (5.00, 5.75)$

51. (a) $\{1.20, 5.09\}$ **(b)**

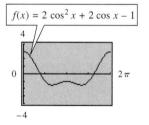

$f(x) = 2 \cos^2 x + 2 \cos x - 1$

(c) $[0, 1.20) \cup (5.09, 2\pi)$ **(d)** $(1.20, 5.09)$

53. origin; $-\tan x$ **55.** origin; $-\cot 1.75$; .1811469526   **57.** $\{1.01, 2.78\}$ **59.** $\{.75, 2.39\}$
61. $\{1.38\}$

Further Explorations

1. The two x-values that give error messages are for $x = \frac{\pi}{2}$ and $x = \frac{3\pi}{2}$. When angles of these radian measures are sketched in standard position, a point on the terminal side has $a = 0$, leading to an undefined value of $\frac{b}{a}$. Thus, an error message occurs.

Section 9.4 (page 741)

1. (a) $\frac{\sqrt{6} - \sqrt{2}}{4}$ **(b)** .2588190451   **(c)**

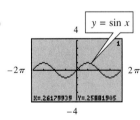

$$\frac{\pi}{12} \approx .26179939$$
$$\sin \frac{\pi}{12} \approx .25881905$$

The answer to part (c) in Exercises 3–11 is found in a manner similar to the answer to part (c) in Exercise 1.

3. (a) $\frac{-\sqrt{6} + \sqrt{2}}{4}$ **(b)** $-.2588190451$ **5. (a)** $\frac{\sqrt{6} + \sqrt{2}}{4}$ **(b)** .9659258263   **7. (a)** $\frac{\sqrt{6} - \sqrt{2}}{4}$
(b) .2588190451   **9. (a)** $\frac{-\sqrt{6} - \sqrt{2}}{4}$ **(b)** $-.9659258263$ **11. (a)** $\frac{-\sqrt{6} + \sqrt{2}}{4}$ **(b)** $-.2588190451$
13. (a) $\frac{\sqrt{6} - \sqrt{2}}{4}$ **(b)** .2588190451   **15. (a)** $2 - \sqrt{3}$ **(b)** .2679491924   **17. (a)** $\frac{\sqrt{6} + \sqrt{2}}{4}$
(b) .9659258263   **19. (a)** $\frac{\sqrt{6} + \sqrt{2}}{4}$ **(b)** .9659258263   **21. (a)** $-2 - \sqrt{3}$ **(b)** -3.732050808
23. (a) $\frac{\sqrt{6} + \sqrt{2}}{4}$ **(b)** .9659258263   **25.** $\sin x$ **27.** $\sin x$ **29.** $-\sin x$ **31.** $\tan x$ **33.** $-\cos x$
35.   **37.** $-.841471$ **39.** It is the same as the graph of $y_1 = \cos(x + \frac{\pi}{2})$.

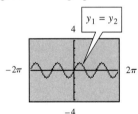

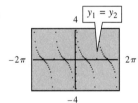

41. (a) $\frac{63}{65}$ **(b)** $\frac{33}{65}$ **(c)** $\frac{63}{16}$ **(d)** $\frac{33}{56}$
(e) I   **(f)** I   **43. (a)** $\frac{77}{85}$ **(b)** $\frac{13}{85}$ **(c)** $-\frac{77}{36}$ **(d)** $\frac{13}{84}$ **(e)** II   **(f)** I
45. $\cos 2x = \frac{119}{169}$; $\sin 2x = -\frac{120}{169}$; $\tan 2x = -\frac{120}{119}$; $\cot 2x = -\frac{119}{120}$; $\sec 2x = \frac{169}{119}$; $\csc 2x = -\frac{169}{120}$
47. $\sin 2x = \frac{15}{17}$; $\cos 2x = -\frac{8}{17}$; $\tan 2x = -\frac{15}{8}$; $\cot 2x = -\frac{8}{15}$; $\sec 2x = -\frac{17}{8}$; $\csc 2x = \frac{17}{15}$
49. (a) $\frac{\sqrt{2 - \sqrt{3}}}{2}$ **(b)** .2588190451   **51. (a)** $1 - \sqrt{2}$ (or $-\sqrt{3 - 2\sqrt{2}}$)   **(b)** $-.4142135624$
53. (a) $\frac{\sqrt{2 + \sqrt{2}}}{2}$ **(b)** .9238795325   **55.** $\frac{\sqrt{10}}{4}$ **57.** 3   **59.** $-\sqrt{7}$

Forms of answers may vary in Exercises 65 and 67.

65. $\cos 3x = 4\cos^3 x - 3\cos x$ **67.** $\tan 4x = \frac{4\tan x - 4\tan^3 x}{1 - 6\tan^2 x + \tan^4 x}$

In Exercises 75–81, we provide only the graphical support. By graphing y_1 as the left side and y_2 as the right side, we get the same graph.

75.   **77.**   **79.**

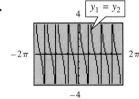

Connected Mode
(The vertical lines are
not part of the graph.)

Dot Mode

81.

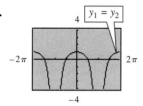

Further Explorations

Answers will vary. We give only one possibility.

1. $\Delta \text{Tbl} = \pi$, Tblmin $= \frac{\pi}{4}$ **3.** $\Delta \text{Tbl} = \frac{\pi}{2}$, Tblmin $= \frac{\pi}{4}$ **5.** $\Delta \text{Tbl} = \frac{\pi}{2}$, Tblmin $= \frac{\pi}{8}$

Section 9.5 (page 751)

1. (a) $\left\{\frac{\pi}{12}, \frac{11\pi}{12}, \frac{13\pi}{12}, \frac{23\pi}{12}\right\}$ (b) $\left[0, \frac{\pi}{12}\right) \cup \left(\frac{11\pi}{12}, \frac{13\pi}{12}\right) \cup \left(\frac{23\pi}{12}, 2\pi\right)$ (c) $\left(\frac{\pi}{12}, \frac{11\pi}{12}\right) \cup \left(\frac{13\pi}{12}, \frac{23\pi}{12}\right)$ **3.** (a) $\left\{\frac{\pi}{2}, \frac{7\pi}{6}, \frac{11\pi}{6}\right\}$
(b) $\left[0, \frac{\pi}{2}\right) \cup \left(\frac{\pi}{2}, \frac{7\pi}{6}\right) \cup \left(\frac{7\pi}{6}, \frac{11\pi}{6}\right) \cup \left(\frac{11\pi}{6}, 2\pi\right)$ (c) $\emptyset$ **5.** (a) $\left\{\frac{3\pi}{8}, \frac{5\pi}{8}, \frac{11\pi}{8}, \frac{13\pi}{8}\right\}$
(b) $\left[0, \frac{3\pi}{8}\right) \cup \left[\frac{5\pi}{8}, \frac{11\pi}{8}\right) \cup \left[\frac{13\pi}{8}, 2\pi\right)$ (c) $\left[\frac{3\pi}{8}, \frac{5\pi}{8}\right] \cup \left[\frac{11\pi}{8}, \frac{13\pi}{8}\right]$ **7.** (a) $\left\{\frac{\pi}{2}, \frac{3\pi}{2}\right\}$ (b) $\left(\frac{\pi}{2}, \frac{3\pi}{2}\right)$
(c) $\left[0, \frac{\pi}{2}\right) \cup \left(\frac{3\pi}{2}, 2\pi\right)$ **9.** (a) $\left\{0, \frac{2\pi}{3}, \frac{4\pi}{3}\right\}$ (b) $\left[0, \frac{2\pi}{3}\right) \cup \left[\frac{4\pi}{3}, 2\pi\right)$ (c) $\left[\frac{2\pi}{3}, \frac{4\pi}{3}\right]$ **11.** (a) $\left\{\frac{\pi}{2}\right\}$ (b) $\left(\frac{\pi}{2}, 2\pi\right)$
(c) $\left[0, \frac{\pi}{2}\right)$ **13.** (a) $\left\{\frac{\pi}{3}, \pi, \frac{5\pi}{3}\right\}$ (b) $\left[0, \frac{\pi}{3}\right) \cup \{\pi\} \cup \left[\frac{5\pi}{3}, 2\pi\right)$ (c) $\left[\frac{\pi}{3}, \frac{5\pi}{3}\right]$ **15.** (a) $\left\{0, \frac{2\pi}{3}, \frac{4\pi}{3}\right\}$
(b) $\left[0, \frac{\pi}{6}\right) \cup \left[\frac{2\pi}{3}, \frac{5\pi}{6}\right) \cup \left[\frac{4\pi}{3}, \frac{3\pi}{2}\right)$ (c) $\left(\frac{\pi}{6}, \frac{2\pi}{3}\right] \cup \left(\frac{5\pi}{6}, \frac{4\pi}{3}\right] \cup \left(\frac{11\pi}{6}, 2\pi\right)$, with $x \neq \frac{\pi}{2}, \frac{7\pi}{6}, \frac{11\pi}{6}$
17. (a) $\left\{\frac{\pi}{3}, \frac{\pi}{2}, \frac{3\pi}{2}, \frac{5\pi}{3}\right\}$ (b) $\left[0, \frac{\pi}{3}\right) \cup \left[\frac{\pi}{2}, \frac{3\pi}{2}\right] \cup \left[\frac{5\pi}{3}, 2\pi\right)$ (c) $\left[\frac{\pi}{3}, \frac{\pi}{2}\right] \cup \left[\frac{3\pi}{2}, \frac{5\pi}{3}\right]$ **19.** $\{.262, 1.309, 1.571, 3.403,$
$4.451, 4.712\}$ **21.** $\{1.047, 3.142, 5.236\}$ **23.** $\{.259, 1.372, 3.142, 4.911, 6.024\}$ **25.** We cannot say
that $\frac{\tan 2\theta}{2} = \tan \theta$. The 2 in the numerator on the left is not a factor of the entire numerator, so it cannot cancel
with the 2 in the denominator.
27.

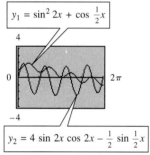

Section 9.6 (page 764)

1. $-\frac{\pi}{6}$ **3.** $\frac{\pi}{4}$ **5.** π **7.** $-\frac{\pi}{3}$ **9.** $-\frac{\pi}{4}$ **11.** $\frac{\pi}{4}$ **13.** $\frac{5\pi}{6}$ **15.** 0
17. (a)

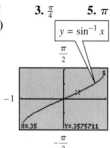

19. (a)

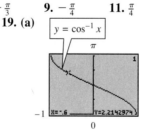

21. (a)

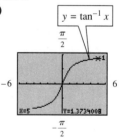

(b) .3575711036   (b) 2.214297436   (b) 1.373400767

23. The expression $\cos^{-1} 1.5$ is undefined, because 1.5   **25.** domain: $(-\infty, \infty)$; range: $(0, \pi)$
is not in the domain of the inverse cosine function.

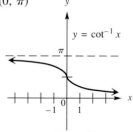

27. domain: $(-\infty, -1] \cup [1, \infty)$; range: $[0, \frac{\pi}{2}) \cup (\frac{\pi}{2}, \pi]$

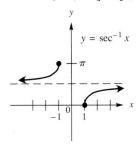

$y = \sec^{-1} x$

29. $\frac{3\pi}{4}$ **31.** $\frac{5\pi}{6}$ **33.** $\frac{2\pi}{3}$ **35.** 1.003 is not in the domain of the inverse sine function. **37.** true
39. false; The statement is true only if x is in the interval $\left[-\frac{\pi}{2}, \frac{\pi}{2}\right]$. **41.** true **43.** $\frac{\sqrt{7}}{3}$
45. $\frac{\sqrt{5}}{5}$ **47.** $-\frac{\sqrt{5}}{2}$ **49.** $\frac{120}{169}$ **51.** $-\frac{7}{25}$ **53.** $\frac{4\sqrt{6}}{25}$ **55.** 2 **57.** $\frac{63}{65}$ **59.** $\frac{\sqrt{10} - 3\sqrt{30}}{20}$
61. $\sqrt{1 - u^2}$ **63.** $\frac{\sqrt{1 - u^2}}{u}$ **65.** $\frac{\sqrt{u^2 - 4}}{u}$ **67.** $\frac{u\sqrt{2}}{2}$ **69.** $\frac{2\sqrt{4 - u^2}}{4 - u^2}$

Section 9.7 (page 768)

1. (a) 5000 feet **(b)** 1250 feet **3.** May **5.** Feb. 27 (day 58), Nov. 23 (day 327); from Nov. 24 to
Feb. 27 **7. (a)** 5, $\frac{1}{40}$ **(b)** 40 **(c)** 5, 1.55, -4.05, -4.05, 1.55 **9.** 980.799 cm per second^2
11. (a) .25 second **(b)** .17 second **(c)** .21 second **13.** about $46°$ **15.** .50 second and .83 second
17. (a) 100 **(b)** 258 **(c)** 122 **(d)** 296
19. (a)

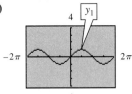

y_1

(b)

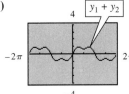

$y_1 + y_2$

(c)

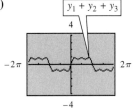

$y_1 + y_2 + y_3$

(d)

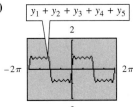

$y_1 + y_2 + y_3 + y_4$

(e)

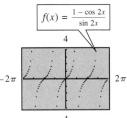

$y_1 + y_2 + y_3 + y_4 + y_5$

21. (a) $113°$ **(b)** $84°$ **(c)** $60°$ **(d)** $47°$ **23. (a)** $\tan \alpha = \frac{x}{z}$; $\tan \beta = \frac{x + y}{z}$ **(b)** $\frac{x}{\tan \alpha} = \frac{x + y}{\tan \beta}$
(c) $\alpha = \arctan\left(\frac{x \tan \beta}{x + y}\right)$ **(d)** $\beta = \arctan\left(\frac{(x + y)\tan \alpha}{x}\right)$

Chapter 9 Review Exercises (page 773)

1. $\sqrt{3}$ **2.** $-\frac{1}{2}$ **3.** $-\frac{1}{2}$ **4.** $-\sqrt{3}$ **5.** $-\sqrt{3}$ **6.** 2 **7.** 2 **8.** $\frac{\sqrt{3}}{3}$
9. .8660266282 **10.** 1.67553317 **11.** 1.131194347 **12.** -1.344734732 **13.** $\cos 3$
14. $-\sin 3$ **15.** $-\tan 3$ **16.** $\sec 3$ **17.** $-\csc 3$ **18.** $-\cot 3$ **19. (b)** **20. (a)**
21. (c) **22. (f)** **23. (h)** **24. (e)** **25. (d)** **26.** $\tan x$

$f(x) = \frac{1 - \cos 2x}{\sin 2x}$

Dot Mode

27. $(-\infty, \infty)$ **28.** $[3, 7]$ **29.** 2 **30.** $\frac{2\pi}{3}$ **31.** $-\frac{\pi}{12}$ **32.** 5 units upward

33.

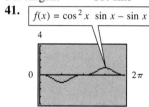

$y = -2\sin\left(3x + \frac{\pi}{4}\right) + 5$

34. tangent **35.** sine **36.** cosine **37.** cosecant **38.** cotangent **39.** secant **40.** $\{0, \pi\}$

41.

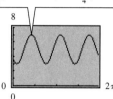

$f(x) = \cos^2 x \sin x - \sin x$

42. (a) the open interval $(\pi, 2\pi)$ **(b)** the open interval $(0, \pi)$

43. $\{3.90, 5.53\}$

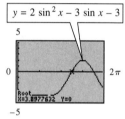 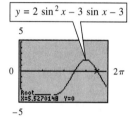

$y = 2\sin^2 x - 3\sin x - 3$

$y = 2\sin^2 x - 3\sin x - 3$

44. $-\frac{63}{65}$ **45.** $\frac{16}{65}$ **46.** $-\frac{63}{16}$ **47.** $\frac{33}{65}$ **48.** $-\frac{56}{65}$ **49.** $-\frac{33}{56}$ **50.** $\frac{24}{25}$ **51.** $-\frac{119}{169}$ **52.** $\frac{24}{7}$

53. $\frac{3\sqrt{10}}{10}$ **54.** $-\frac{\sqrt{10}}{10}$ **55.** -3 **56.** $\{\frac{\pi}{4}, \frac{\pi}{2}, \frac{5\pi}{4}, \frac{3\pi}{2}\}$ **57.** $\{\frac{\pi}{12}, \frac{5\pi}{12}, \frac{13\pi}{12}, \frac{17\pi}{12}\}$

58. $\{0, \frac{\pi}{3}, \frac{\pi}{2}, \frac{2\pi}{3}, \pi, \frac{4\pi}{3}, \frac{3\pi}{2}, \frac{5\pi}{3}\}$ **59.** $\{\frac{\pi}{3}, \pi, \frac{5\pi}{3}\}$ **60.** $\{\frac{\pi}{6}, \frac{2\pi}{3}, \frac{7\pi}{6}, \frac{5\pi}{3}\}$ **61.** $\{0\}$

62. (a) $\left(\frac{\pi}{4}, \frac{\pi}{2}\right) \cup \left(\frac{5\pi}{4}, \frac{3\pi}{2}\right)$ **(b)** $\left[0, \frac{\pi}{4}\right) \cup \left(\frac{\pi}{2}, \frac{5\pi}{4}\right) \cup \left(\frac{3\pi}{2}, 2\pi\right)$ **79.** $\frac{\pi}{4}$ **80.** $\frac{2\pi}{3}$ **81.** $-\frac{\pi}{3}$ **82.** $-\frac{\pi}{2}$

83. $\frac{3\pi}{4}$ **84.** $\frac{\pi}{6}$ **85.** $\frac{2\pi}{3}$ **86.** $\frac{\pi}{3}$ **87.** $\frac{3\pi}{4}$

88. There is no real number whose sine value is -3. **89.** $\frac{1}{2}$ **90.** $\frac{\sqrt{7}}{4}$ **91.** $\frac{\sqrt{10}}{10}$ **92.** $\frac{9}{7}$

93. $\frac{\pi}{2}$ **94.** $\frac{294 + 125\sqrt{6}}{92}$ **95.** $\frac{u\sqrt{1 + u^2}}{1 + u^2}$ **96.** $\sqrt{1 - u^2}$ **97.** $\frac{1}{u}$ **98. (a)** about 20 years **(b)** from about 5000 to about 150,000 **99. (a)** $x = \sin u$, $-\frac{\pi}{2} \le u \le \frac{\pi}{2}$ **(b)** side along x-axis: $\sqrt{1 - x^2}$; side

perpendicular to x-axis: x; hypotenuse: 1 **(c)** $\tan u = \frac{x\sqrt{1 - x^2}}{1 - x^2}$ **(d)** $u = \arctan \frac{x\sqrt{1 - x^2}}{1 - x^2}$

100. $y = 10 - 30\cos 40\pi t$ **101. (a)** $t = \frac{3}{4\pi}\arcsin 3y$ **(b)** .27 second

102. (f) $\frac{\sqrt{6} + \sqrt{2}}{4}$ **(g)** $\frac{\sqrt{6} - \sqrt{2}}{4}$ **(h)** $2 - \sqrt{3}$ **103.** .001 second **104.** .0007 second

Other answers are possible in Exercises 105–110.

105. $y = 3\sin 2\left(x - \frac{\pi}{4}\right)$ **106.** $y = 4\sin\frac{1}{2}x$ **107.** $y = \frac{1}{3}\sin\frac{\pi}{2}x$ **108.** $y = \pi\sin \pi\left(x - \frac{1}{2}\right)$

109. $y = 3\cos 2\left(x + \frac{\pi}{2}\right)$ **110.** $y = 4\cos\frac{1}{2}(x - \pi)$ **111. (a)** $18°$ **(b)** $18°$ **(c)** $15°$

CHAPTER 10

Section 10.1 (page 786)

1. m and p; n and r **3.** m and p equal $2t$, or t is $\frac{1}{2}m$ or $\frac{1}{2}p$; also $m = 1p$ and $n = 1r$

5.

7.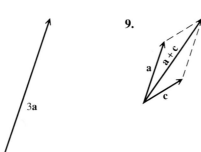

9.

11.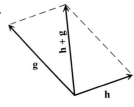

13.

15.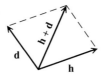

17.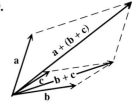

19.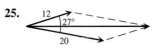

21.

23. Yes, vector addition is associative.

25.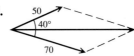

27.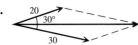

29.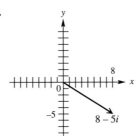

31. $\langle 9.5, 7.4 \rangle$ **33.** $\langle 17, 20 \rangle$ **35.** $\langle 13.7, 7.11 \rangle$ **37.** $\langle -123, 155 \rangle$ **39.** 530 newtons
41. 27.2 pounds   **43.** 190, 283 pounds, respectively   **45.** 173.1°   **47.** 39.2 km
49. 237°; 470 mph   **51.** 358°; 170 mph   **53.** 230 km per hr; 167°   **55.** magnitude: 9.52082827;
direction angle: 119.0646784°   **57.** $\langle -.520944533, -2.954423259 \rangle$ **59.** magnitude: 9.52082827;
direction angle: 119.0646784°

Section 10.2 (page 797)

1.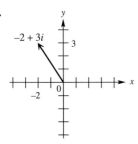

3.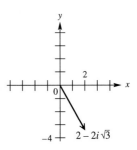

5.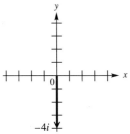

7.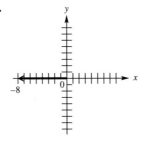

9.

11. The imaginary part must be 0.   **13.** 0   **15.** $3 - i$ **17.** $-3 + 3i$ **19.** $2 + 4i$
21. $\sqrt{2} + i\sqrt{2}$ **23.** $10i$ **25.** $-2 - 2i\sqrt{3}$ **27.** $\frac{\sqrt{3}}{2} + \frac{1}{2}i$ **29.** $\frac{5}{2} - \frac{5\sqrt{3}}{2}i$ **31.** $-\sqrt{2}$
33. $3\sqrt{2}\,(\cos(-45°) + i\,\sin(-45°))$ **35.** $6\,(\cos(-120°) + i\,\sin(-120°))$
37. $8(\cos 30° + i\,\sin 30°)$ **39.** $2(\cos 135° + i\,\sin 135°)$ **41.** $2\sqrt{2}(\cos 45° + i\,\sin 45°)$
43. $4(\cos 180° + i\,\sin 180°)$ **45.** $2(\cos(-90°) + i\,\sin(-90°))$ **47.** 5.830951895 cis 59.03624347°
49. $-3\sqrt{3} + 3i$ **51.** $-4i$ **53.** $12\sqrt{3} + 12i$ **55.** $-\frac{15\sqrt{2}}{2} + \frac{15\sqrt{2}}{2}i$ **57.** $-3i$
59. $\sqrt{3} - i$ **61.** $-1 - i\sqrt{3}$ **63.** $-\frac{1}{6} - \frac{\sqrt{3}}{6}i$ **65.** 2   **67.** 2 cis 0°   **69.** $-i$
71. $-i$; It is the same.   **73.** To square a complex number in trigonometric form, we square the absolute value r and double the argument θ.   **75.** $1.7 + 2.8i$

Section 10.3 (page 803)

1. $27i$ **3.** 1   **5.** $\frac{27}{2} - \frac{27\sqrt{3}}{2}i$ **7.** $-16\sqrt{3} + 16i$ **9.** $-128 + 128i\sqrt{3}$ **11.** $128 + 128i$
13. cis 0°, cis 120°, cis 240°   **15.** 2 cis 20°, 2 cis 140°, 2 cis 260°   **17.** 2 cis 90°, 2 cis 210°, 2 cis 330°

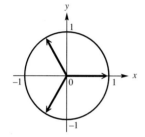
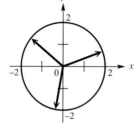
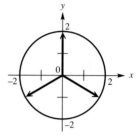

19. 4 cis 60°, 4 cis 180°, 4 cis 300°   **21.** $\sqrt[3]{2}$ cis 20°, $\sqrt[3]{2}$ cis 140°, $\sqrt[3]{2}$ cis 260°   **23.** $\sqrt[3]{4}$ cis 50°, $\sqrt[3]{4}$ cis 170°, $\sqrt[3]{4}$ cis 290°

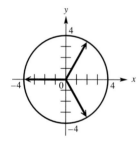
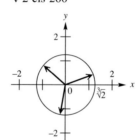
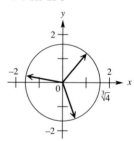

25. (a) $1, i, -1, -i$ **(b)** $1, \frac{1}{2} + \frac{\sqrt{3}}{2}i, -\frac{1}{2} + \frac{\sqrt{3}}{2}i, -1, -\frac{1}{2} - \frac{\sqrt{3}}{2}i, \frac{1}{2} - \frac{\sqrt{3}}{2}i$ **(c)** $1, \frac{\sqrt{2}}{2} + \frac{\sqrt{2}}{2}i, i, -\frac{\sqrt{2}}{2} + \frac{\sqrt{2}}{2}i, -1, -\frac{\sqrt{2}}{2} - \frac{\sqrt{2}}{2}i, -i, \frac{\sqrt{2}}{2} - \frac{\sqrt{2}}{2}i$

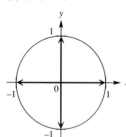
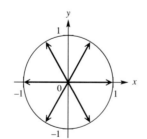
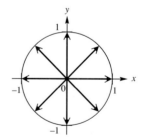

27. The trigonometric form of a positive real number will have $\theta = 0°$. When we divide $\frac{0° + 360° \cdot k}{n}$ for $k = 0$ and $n =$ the root index, we get 0°. Thus, one of the roots will also be a positive real number.

29. $(x + 2)(x^2 - 2x + 4)$ **31.** $x^2 - 2x + 4 = 0$ implies $x = 1 + i\sqrt{3}$ or $x = 1 - i\sqrt{3}$.

33. $1 + i\sqrt{3}, -2, 1 - i\sqrt{3}$ **35.** The x-intercept, -2, is the real cube root of -8.

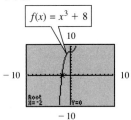

37. $\{\text{cis } 45°, \text{cis } 135°, \text{cis } 225°, \text{cis } 315°\}$ **39.** $\{\text{cis } 18°, \text{cis } 90°, \text{cis } 162°, \text{cis } 234°, \text{cis } 306°\}$

Section 10.4 (page 813)

Answers for polar coordinates may vary in Exercises 1–9.

1. $(1, 405°), (-1, 225°); \left(\frac{\sqrt{2}}{2}, \frac{\sqrt{2}}{2}\right)$ **3.** $(-2, 495°), (2, 315°); (\sqrt{2}, -\sqrt{2})$

5. $(5, 300°), (-5, 120°); \left(\frac{5}{2}, -\frac{5\sqrt{3}}{2}\right)$ **7.** $(-3, 150°), (3, -30°); \left(\frac{3\sqrt{3}}{2}, -\frac{3}{2}\right)$ **9.** $(3, 0°), (-3, 180°); (3, 0)$

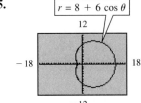

11. quadrantal

13.
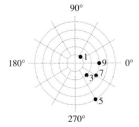
$r = 2 + 2 \cos \theta$

15.
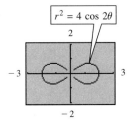
$r = 8 + 6 \cos \theta$

17.
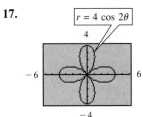
$r = 4 \cos 2\theta$

19.
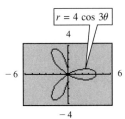
$r = 4 \cos 3\theta$

21.
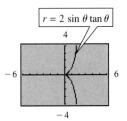
$r^2 = 4 \cos 2\theta$

23.
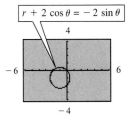
$r = 2 \sin \theta \tan \theta$

25. One way would be to change r to its opposite, add $180°$ to θ, and then move $|r|$ units along the terminal ray of angle $\theta + 180°$, when it is in standard position.

27. $x^2 + (y - 1)^2 = 1$ **29.** $y^2 = 4(x + 1)$ **31.** $(x + 1)^2 + (y + 1)^2 = 2$

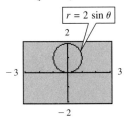
$r = 2 \sin \theta$

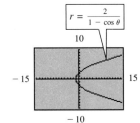
$r = \dfrac{2}{1 - \cos \theta}$

$r + 2 \cos \theta = -2 \sin \theta$

33. $x = 2$

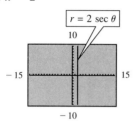

35. $x + y = 2$

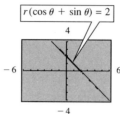

37. $r = \dfrac{4}{\cos\theta + \sin\theta}$ **39.** $r = 4$ **41.** $r = \dfrac{2}{\sin\theta}$

43.

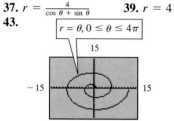

45.

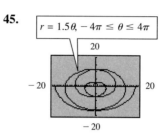

47. While a matter of preference, most people would probably prefer the rectangular form.

Section 10.5 (page 819)

1. $y = \frac{1}{2}x + 1$, for x in $[-4, 6]$

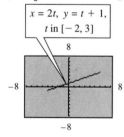

3. $y = 3x^2 - 4$, for x in $[0, 2]$

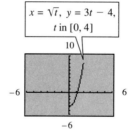

5. $y = x - 2$, for x in $[-26, 28]$

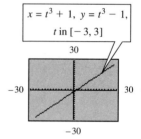

7. $x = 2^{(y^2 + 1)/3}$, for y in $[0, \sqrt{11}]$
or $y^2 = \frac{3 \ln x}{\ln 2} - 1$, for x in $[\sqrt[3]{2}, 16]$

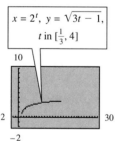

9. $x^2 + y^2 = 4$, for x in $[-2, 2]$

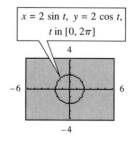

11. $\frac{y^2}{4} - \frac{x^2}{9} = 1$, for x in $(-\infty, \infty)$

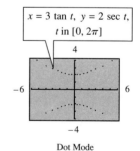

Dot Mode

13. $y = \frac{1}{x}$, for x in $(0, 1]$ **15.** $y = \sqrt{x^2 + 2}$, for x in $(-\infty, \infty)$ **17.** $y = \frac{1}{x}$, for x in $(0, \infty)$
19. $x^2 + y^2 - 4x - 2y + 4 = 0$, for x in $[1, 3]$ **21.** $y = \frac{1}{x}$, for $x \neq 0$

23. $y = \ln x$, for x in $(0, \infty)$

25.

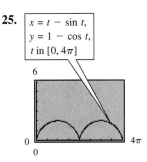

27. (a) 17.7 seconds **(b)** 5000 feet **(c)** 1250 feet

31. Many answers are possible, two of which are $x = t$, $y - y_1 = m(t - x_1)$ and $t = x - x_1$, $y = mt + y_1$.

33. Many answers are possible, two of which are $x = a \sec t$, $y = b \tan t$, and $x = t$, $y^2 = b^2(\frac{t^2}{a^2} - 1)$

Chapter 10 Review Exercises (page 821)

1.

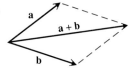

2.

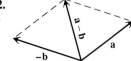

3.

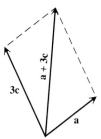

4. $\langle 25\sqrt{2}, 25\sqrt{2} \rangle$ **5.** $\langle 17.9, 66.8 \rangle$ **6.** $\langle 869, 418 \rangle$ **7.** 28 pounds **8.** 209 newtons

9. 135 newtons **10.** 826 pounds **11.** bearing: 306°; speed: 524 mph

12. speed: 21 km per hr; bearing: 118° **13.** bearing: 7° 20′; speed: 8.3 mph

14.

15.

16.

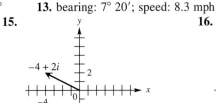

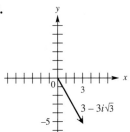

17. $5 + 4i$ **18.** $7 - 3i$

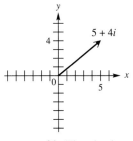

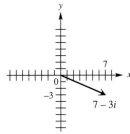

19. y **20.** The absolute value of a real number represents its distance from 0 on the number line, while the absolute value (or modulus) of a complex number represents its distance from $(0, 0)$ in the complex plane.

21. $-30i$ **22.** $-3 - 3i\sqrt{3}$ **23.** $-\frac{1}{8} + \frac{\sqrt{3}}{8}i$ **24.** -2 **25.** $8i$ **26.** $-128 + 128i$

27. $-\frac{1}{2} - \frac{\sqrt{3}}{2}i$ **28.** $\frac{1}{2} + \frac{\sqrt{3}}{2}i$ **29.** $2\sqrt{2}$ cis 135° **30.** $3i$ **31.** $-\sqrt{2} - i\sqrt{2}$

32. 8 cis 120° **33.** $\sqrt{2}$ cis$(-45°)$ **34.** $-2 - 2i\sqrt{3}$ **35.** 4 cis$(-90°)$ **36.** -2

37. 3 cis 90°, 3 cis 210°, 3 cis 330°

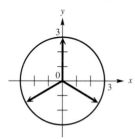

38. 2 cis 22.5°, 2 cis 112.5°, 2 cis 202.5°, 2 cis 292.5°

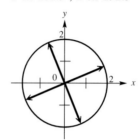

39. 2 cis 0°, 2 cis 72°, 2 cis 144°, 2 cis 216°, 2 cis 288°

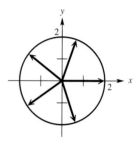

40. {cis 67.5°, cis 157.5°, cis 247.5°, cis 337.5°}
41. $(3\sqrt{3}, 3)$ **42.** $(-6\sqrt{2}, -6\sqrt{2})$ **43.** $(-4, 4\sqrt{3})$ **44.** $(6\sqrt{2}, 135°)$ **45.** $(5, -90°)$
46. $(\sqrt{10}, -135°)$

47.

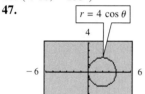

48.

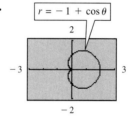

49.

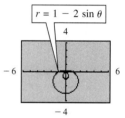

50.

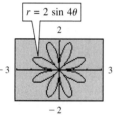

51.

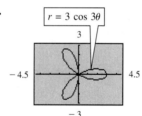

52. $y^2 + 6x - 9 = 0$ **53.** $-x + 2y = 4$ **54.** $x^2 - x + y^2 - y = 0$ **55.** $x^2 + y^2 = 4$
56. $r = \frac{-3}{\cos \theta}$ **57.** $\tan \theta = 1$ **58.** $r = \frac{\tan \theta}{\cos \theta}$ **59.** $r = \frac{\cos \theta}{\sin^2 \theta}$

60.

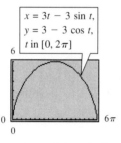

61.

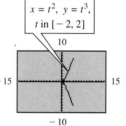

62.

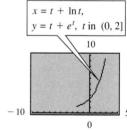

63.

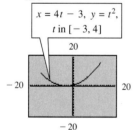

64. $x - 3y = 5$ for x in $[-13, 17]$ **65.** $y = \sqrt{x^2 + 1}$ for x in $[0, \infty)$ **66.** $\frac{y^2}{9} - \frac{x^2}{25} = 1$ for x in $(-\infty, \infty)$

CHAPTER 11

Section 11.1 (page 829)

1. positive integers   **3.** The first condition of the principle is not satisfied.

25. $x^2 + xy + y^2 = \frac{x^3 - y^3}{x - y}$ **27.** $x^{2n} - x^{2n-1}y + \cdots - xy^{2n-1} + y^{2n} = \frac{x^{2n+1} + y^{2n+1}}{x + y}$

29. power; exponents   **31.** $3 \cdot 4^{n-1}$

Section 11.2 (page 836)

1. 20   **3.** 35   **5.** 56   **7.** 45   **9.** n **11.** Use the formula preceding Example 5, with $n = 8$ and $r = 5$.   **13.** They alternate, starting with positive.

15. $x^6 + 6x^5y + 15x^4y^2 + 20x^3y^3 + 15x^2y^4 + 6xy^5 + y^6$

17. $p^5 - 5p^4q + 10p^3q^2 - 10p^2q^3 + 5pq^4 - q^5$ **19.** $r^{10} + 5r^8s + 10r^6s^2 + 10r^4s^3 + 5r^2s^4 + s^5$

21. $p^4 + 8p^3q + 24p^2q^2 + 32pq^3 + 16q^4$ **23.** $2401p^4 + 2744p^3q + 1176p^2q^2 + 224pq^3 + 16q^4$

25. $729x^6 - 2916x^5y + 4860x^4y^2 - 4320x^3y^3 + 2160x^2y^4 - 576xy^5 + 64y^6$

27. $\frac{m^6}{64} - \frac{3m^5}{16} + \frac{15m^4}{16} - \frac{5m^3}{2} + \frac{15m^2}{4} - 3m + 1$ **29.** $4r^4 + \frac{8\sqrt{2}r^3}{m} + \frac{12r^2}{m^2} + \frac{4\sqrt{2}r}{m^3} + \frac{1}{m^4}$ **31.** $7920m^8p^4$

33. $126x^4y^5$ **35.** $180m^2n^8$ **37.** $4845p^8q^{16}$ **39.** $439,296x^{21}y^7$ **41.** 1, 3, 6, 10, 15

43. 21, 28, 36, 45, 55   **45.** 1.002; 5.010   **47.** .942

Further Explorations

1. 1, 4, 6, 4, 1   **3.** 1, 8, 28, 56, 70, 56, 28, 8, 1

Section 11.3 (page 845)

1. The nth term is the term that appears in the nth position. For example, when $n = 1$, the first term is obtained. When $n = 2$, the second term is obtained, and so on.   **3.** 1, −1, 1, −1, 1   **5.** $\frac{1}{2}, \frac{4}{3}, 1, \frac{8}{7}, \frac{5}{4}$

7. −2, 8, −24, 64, −160   **9.** x, x^2, x^3, x^4, x^5 **11.** 4, 9, 14, 19, 24, 29, 34, 39, 44, 49

13. 1, 1, 2, 3, 5, 8, 13, 21, 34, 55   **15.** 11, 9, 7, 5   **17.** 6, 2, −2, −6   **19.** $d = 9, a_n = 9n - 1$

21. $d = m, a_n = x + nm - m$ **23.** $a_8 = 19; a_n = 3 + 2n$ **25.** $a_8 = -3; a_n = -39 + \frac{9n}{2}$

27. $a_8 = 4m; a_n = mn - 4m$ **29.** One example is the arithmetic sequence with $a_1 = 6$ and $d = 2$.

31. (d)   **33.** $a_1 = 7$ **35.** In each case, if we add log 2, we get the next term.   **37.** $m + b; ab$

39. $3m + b; ab^3$ **41.** yes; b **43.** −4, −8, −16, −32, −64   **45.** 729, 486, 324, 216, 144

47. 27, 9, 3, 1   **49.** $a_5 = -1280; a_n = -5 \cdot 4^{n-1}$ **51.** $a_5 = 54; a_n = (\frac{2}{3})3^{n-1}$ or $2(3)^{n-2}$

53. $a_5 = -243; a_n = (-3)^n$ **55.** $a_n = 4 \cdot 4^{n-1}$ or 4^n **57.** $a_n = (\frac{5}{6})2^{n-1}$

59. $a_n = 49(-\frac{1}{7})^{n-1}$ or $(-1)^{n-1}(7)^{-n+3}$ **61.** $a_n = .01(2)^{n-1}$; $5.12; $163.84; $167,772 (rounded)

63. A geometric sequence is a sequence obtained by multiplying each term after the first by a fixed constant.

65. The sequence in choice (d) is obtained by starting with 1 as the first term and the second term, and then adding the two previous terms to get each term after the second.

67. $a_1 = 1000; r = \frac{1}{10}$ **69.** In each case, if we multiply by 2, we get the next term.   **71.** diverges

Further Explorations

1. during the 11th year   **3. (a)** One window choice is [0, 40] by [0, 2].   **(b)** 1

Section 11.4 (page 857)

1. 35   **3.** $\frac{25}{12}$ **5.** 3   **7.** 215   **9.** 125   **11.** 230   **13.** $a_1 = 7, d = 5$ **15.** −684

17. 500,500   **19.** 93   **21.** $\frac{33}{4}$ **23.** $\frac{121}{9}$ **25.** 30   **27.** 60   **29.** $\frac{1330}{9}$

31. $\sum\limits_{i=1}^{6} i^2 + \sum\limits_{i=1}^{6} 3i + \sum\limits_{i=1}^{6} 5$ **33.** 30   **35.** $\sum\limits_{i=1}^{n} i = \frac{n(n + 1)}{2}$ **37.** 572

39. It will have a sum when $-1 < r < 1$.   **41.** $\frac{64}{3}$ **43.** 135   **45.** $\frac{512}{3}$ **47.** $\frac{3}{2}$ **49.** 2

51. $\frac{1}{3}$ **53.** The sum does not exist. (It diverges.)   **55.** $\frac{1}{9}$ **57.** The sum does not exist.

59. 96   **61.** $2865   **63.** 54,220   **65.** 713 in.   **67.** .000016%   **69.** $31,287.11   **71.** 70 m

73. 1600   **75. (a)** 12 m   **(b)** $\frac{4\sqrt{3}}{3}$ m²

Further Explorations

1. The sums continue to increase, approaching 1 as a limit. It takes the calculator a long time to graph this sequence, since each time it must calculate a sum having one more term than the previous sum.

3. After 1 year, $481.50; after 2 years, $996.71; after 18 years, $16,370.53

Section 11.5 (page 867)

1. 19,958,400   **3.** 72   **5.** 5   **7.** 6   **9.** 1   **11.** 495   **13.** 720   **15. (a)** 17,576,000
(b) 17,576,000   **(c)** 456,976,000   **17.** $2.052371412 \times 10^{10}$ **19.** 120; 30,240   **21.** 720; 151,200
23. 10,626   **25.** 35   **27.** 35   **29.** 84; 324   **31.** A permutation is an arrangement, while a
combination is a subset where order is unimportant.   **33.** It is not accurate, since order makes a difference.
35. 210   **37. (a)** 330   **(b)** 150   **39.** 210; 5040   **41.** 35   **43. (a)** 220   **(b)** 55   **(c)** 105

Further Explorations

1. We give X! for X = 1 to X = 10 here: 1! = 1, 2! = 2, 3! = 6, 4! = 24, 5! = 120, 6! = 720, 7! = 5040,
8! = 40,320, 9! = 362,880, 10! = 3,628,800.   **3.** As x goes from 0 to 5, $_5C_x$ takes on the values 1, 5, 10, 10,
5, 1. As x goes from 0 to 5, $_5P_x$ takes on the values 1, 5, 20, 60, 120, 120.

Section 11.6 (page 876)

1. Let h = heads, t = tails. $S = \{h\}$ **3.** $S = \{hhh, hht, hth, thh, htt, tht, tth, ttt\}$ **5.** Let c = correct, w =
wrong. $S = \{ccc, ccw, cwc, wcc, wwc, wcw, cww, www\}$ **7. (a)** $\{hh, tt\}, \frac{1}{2}$ **(b)** $\{hh, ht, th\}, \frac{3}{4}$
9. (a) $\{(2, 4)\}, \frac{1}{10}$ **(b)** $\{(1, 3), (1, 5), (3, 5)\}, \frac{3}{10}$ **(c)** $\emptyset, 0$ **(d)** $\{(1, 2), (1, 4), (2, 3), (2, 5), (3, 4), (4, 5)\}, \frac{3}{5}$
11. The probability of an event cannot exceed 1.   **13. (a)** $\frac{1}{5}$ **(b)** $\frac{8}{15}$ **(c)** 0   **(d)** 1 to 4   **(e)** 7 to 8
15. 55   **17.** 220   **19.** 3 to 7   **21.** $\frac{4}{9}$ **23. (a)** $\frac{3}{5}$ **(b)** $\frac{7}{10}$ **(c)** $\frac{3}{10}$ **25. (a)** $\frac{3}{13}$ **(b)** $\frac{7}{13}$ **(c)** $\frac{3}{13}$
27. (a) $\frac{1}{2}$ **(b)** $\frac{7}{10}$ **(c)** 0   **29. (a)** .72   **(b)** .70   **(c)** .79   **31.** .90   **33.** .41   **35.** .21
37. .79   **39.** .313   **41.** .031   **43.** $\frac{1}{2}$ **45.** 4.6×10^{-10} **47.** .875

Further Explorations

1. The probabilities, in order, are .125, .375, .375, .125.
3. It is impossible to have 5 or 6 girls in a family of only 4 children.

Chapter 11 Review Exercises (page 881)

5. $x^4 + 8x^3y + 24x^2y^2 + 32xy^3 + 16y^4$ **6.** $\frac{k^5}{32} - \frac{5k^4g}{16} + \frac{5k^3g^2}{4} - \frac{5k^2g^3}{2} + \frac{5kg^4}{2} - g^5$ **7.** $2160x^2y^4$
8. $25,344m^5n^{14}$ **9.** $3^{16} + 16 \cdot 3^{15}x + 120 \cdot 3^{14}x^2 + 560 \cdot 3^{13}x^3$
10. $-420 \cdot 3^{13}m^2n^{13} + 30 \cdot 3^{14}mn^{14} - 3^{15}n^{15}$ **11.** $\frac{1}{2}, \frac{2}{3}, \frac{3}{4}, \frac{4}{5}, \frac{5}{6}$ **12.** $-2, 4, -8, 16, -32$
13. 8, 10, 12, 14, 16   **14.** 2, 6, 12, 20, 30   **15.** 5, 2, -1, -4, -7 **16.** $-2, -6, -18, -54, -162$
17. 5, 3, -2, -5, -3 **18.** $-2, 2, -4, -4, -8$ **19.** 6, 2, -2, -6, -10 **20.** 13, 11, 9, 7, 5
21. $3 - \sqrt{5}, 4, 5 + \sqrt{5}, 6 + 2\sqrt{5}, 7 + 3\sqrt{5}$ **22.** $3\pi, 2\pi, \pi, 0, -\pi$ **23.** $a_1 = -29; a_{20} = 66$
24. 20   **25.** 17   **26.** $-x + 61$ **27.** $m - 10$ **28.** 4, 8, 16, 32, 64   **29.** 64, 32, 16, 8, 4
30. $-3, 4, -\frac{16}{3}, \frac{64}{9}, -\frac{256}{27}$ **31.** $\frac{8}{9}, \pm\frac{8}{3}, 8, \pm 24, 72$ **32.** 972 or -972 **33.** 48   **34.** 25
35. $\frac{x^2}{125}$ **36.** ± 36 **37.** An arithmetic sequence has a common *difference,* while a geometric sequence has
a common *ratio.*   **38.** 222   **39.** 612   **40.** 84k   **41.** 15   **42.** 120   **43.** $-10k$
44. $\frac{25}{6}$ **45.** 6   **46.** 690   **47.** 133   **48.** 4368   **49.** 240   **50.** 64 cans
51. $6930   **52.** approximately 117 items   **53.** 18.75 grams   **54.** 36   **55.** 80   **56.** $-\frac{1}{2}$
57. $\frac{8}{9}$ **58.** $\frac{5}{3}$ **59.** The sum does not exist. (It diverges.)   **60.** $\frac{2}{3}$ **61.** $\frac{512}{999}$ **62.** A sequence is a
list of terms in a specified order, while a series is a *sum of terms* of a sequence.   **63.** 120   **64.** 72
65. 1   **66.** 56   **67.** 252   **68.** 1   **69.** 48   **70.** 90   **71.** 24   **72. (a)** 20   **(b)** 10
73. 504   **74.** 456,976,000; 258,336,000   **75. (a)** $\frac{4}{15}$ **(b)** $\frac{2}{3}$ **(c)** 0   **76. (a)** 4 to 11   **(b)** 3 to 2
77. $\frac{1}{26}$ **78.** $\frac{4}{13}$ **79.** $\frac{4}{13}$ **80.** $\frac{3}{4}$ **81.** 1   **82.** .86   **83.** .44   **84.** 0
85. approximately .296   **86.** approximately .205

I N D E X